Distance Formulas

If a and b are real numbers, then $d(a, b) = |b - a| = |a - b|$.

If $P = (x_1, y_1)$ and $Q = (x_2, y_2)$ are points in the plane, then $d(P, Q) = \sqrt{(x_2 - x_1)^2 + (y_2 - y_1)^2}$.

Equations and Slope of a Line

The slope m of a line through (x_1, y_1) and (x_2, y_2) is

$$m = \frac{y_2 - y_1}{x_2 - x_1}$$

The slope-intercept form of the equation of a line with slope m and y-intercept b is

$$y = mx + b$$

The point-slope form of the equation of a line with slope m passing through the point (x_1, y_1) is

$$y - y_1 = m(x - x_1)$$

Equation of a Circle

The equation of a circle with center (a, b) and radius r is

$$(x - a)^2 + (y - b)^2 = r^2$$

Logarithms

Definition: $y = \log_a x$ if and only if $a^y = x$

(In words, $\log_a x$ is that *exponent* which, when used with the base a, gives the number x.)

Important note: $\log_a x$ is defined *only when* x is *positive.*

Properties:

$$a^{\log_a x} = x \qquad \log_a(a^x) = x$$

$$\log_a uv = \log_a u + \log_a v \qquad \log_a \frac{u}{v} = \log_a u - \log_a v$$

$$\log_a \frac{1}{v} = -\log_a v \qquad \log_a u^r = r \log_a u$$

Geometric Formulas

A = area; C = circumference; V = volume; S = surface area; r = radius; h = height; b = length of base

Circle: $A = \pi r^2$, $C = 2\pi r$

Sphere: $V = (4/3)\pi r^3$, $S = 4\pi r^2$

Right circular cylinder: $V = \pi r^2 h$, $S = 2\pi rh$

Right circular cone: $V = (1/3)\pi r^2 h$, $S = \pi r\sqrt{r^2 + h^2}$

Triangle: $A = (1/2)bh$

Parallelogram: $A = bh$

Trapezoid: $A = (1/2)(b_1 + b_2)h$

Precalculus

PRECALCULUS

Steven Roman

California State University at Fullerton

HARCOURT BRACE JOVANOVICH, PUBLISHERS
AND ITS SUBSIDIARY, ACADEMIC PRESS

San Diego New York Chicago Austin Washington, D.C.
London Sydney Tokyo Toronto

Cover: Claude Monet, *Poplars,* 1891. The National Galleries of Scotland, Edinburgh.

ISBN: 0-15-571052-4

Library of Congress Catalog Card Number: 85-82628

Printed in the United States of America

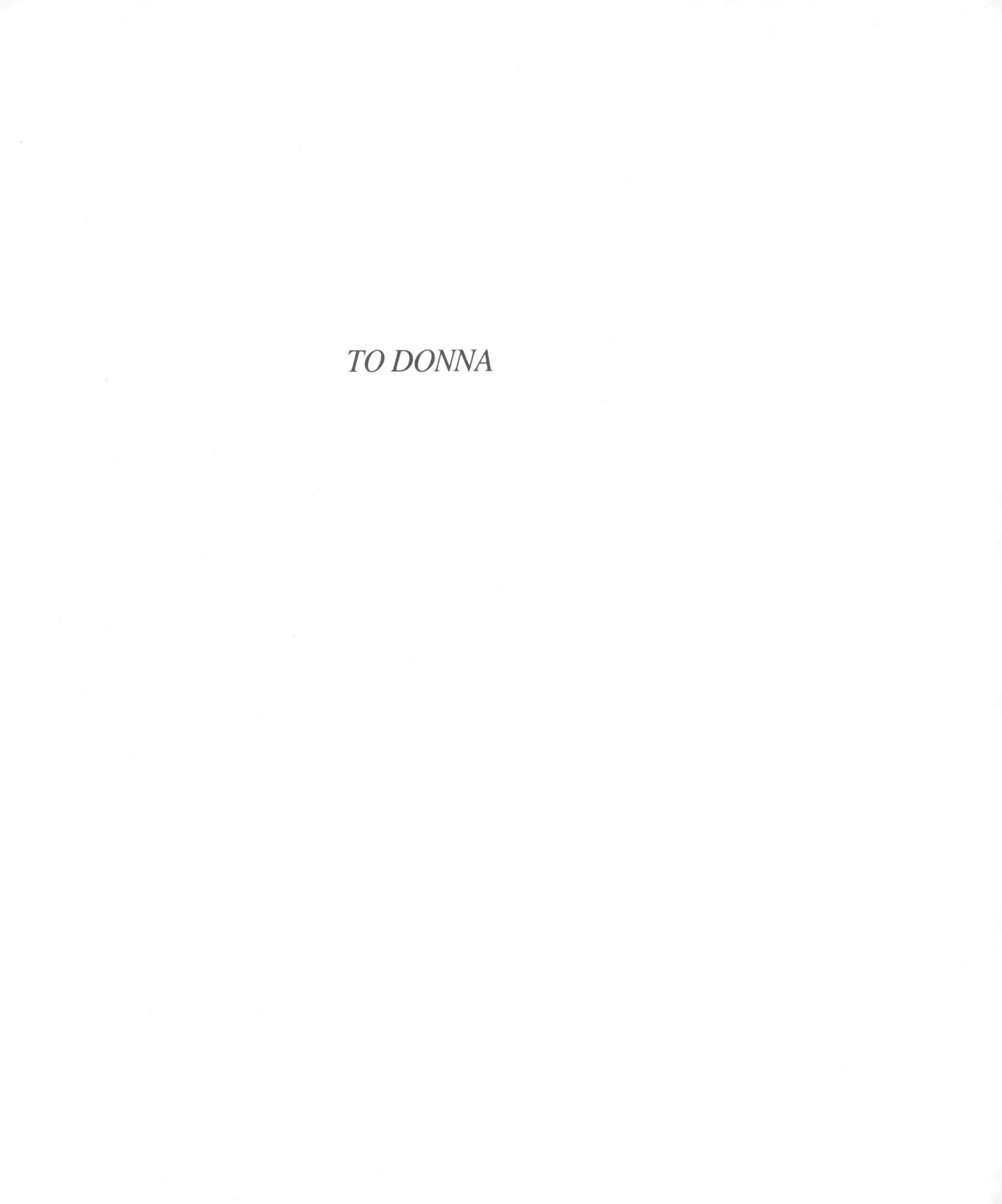

TO DONNA

THE ROMAN SERIES

College Algebra and Trigonometry
College Algebra
Precalculus

Preface

This book provides a thorough treatment of precalculus mathematics. In writing it, I have tried to maintain a "user friendly" approach to the subject by providing clear and complete explanations of the topics along with an abundance of examples and figures. At the same time, I have made every effort to maintain a level of rigor that is desirable in a mathematics textbook.

Two special features of the book are particularly noteworthy:

1. *Study Suggestions* These are exercises—about 40 per chapter—that follow most of the worked examples in the text and that can generally be solved by applying the same or similar techniques used in the corresponding examples. Thus, the Study Suggestions serve to reinforce those techniques. Answers to most of the Study Suggestions appear at the back of the book.
2. *Ideas to Remember* These are found at the end of each section. They do not represent summaries of the section, but, rather, they provide more "philosophical" remarks about the contents of the section. By remembering these ideas, the student will gain a deeper understanding of the concepts of the chapter and will be able to retain the material for a longer period of time.

Precalculus is organized as follows. Chapter 1 contains a review of elementary algebra. A section on the composition of algebraic expressions is included here for two reasons. First, it gives the student another chance to sharpen his or her skills at manipulating algebraic expressions. Second, and more important, it allows the student to concentrate more on the concepts, rather than on the computations, when studying composition of functions in Chapter 2. If desired, the first two sections in Chapter 10 on complex numbers can be covered immediately after Section 1.2, and the third section of Chapter 10 (on complex solutions to quadratic equations) can be covered after Section 1.6.

In Chapter 2 we begin the study of functions, emphasizing that the graph of a function reveals a great deal of information about the function itself.

Chapter 3 includes a brief discussion of the conic sections. The emphasis is on graphing; but we also discuss the geometric definition of each conic section as the locus of points satisfying certain properties. We have included this section to give the instructor some flexibility in covering conic sections. The instructor may choose to cover this section, and omit the more detailed discussion of conic sections in Chapter 11, or omit this section, and cover Chapter 11. *Section 3.5 and Chapter 11 are completely independent of one another.*

Chapter 4 begins with division of polynomials, followed by the Remainder and Factor Theorems and by a discussion of rational roots of a polynomial. The chapter ends with a section on partial fractions.

Chapter 5 covers the exponential and logarithmic functions, including a discussion of the logarithmic scale. The emphasis in this chapter is on applications.

Chapters 6–8 are devoted to trigonometry. We define the trigonometric functions using the unit circle, since that is the most useful for future study. Chapter 7 contains a discussion of triangle trigonometry, including the law of sines and cosines, and an introduction to polar coordinates and vectors in the plane.

Chapter 9 contains a discussion of systems of linear and nonlinear equations as well as systems of inequalities. Determinants, Cramer's Rule, and matrix algebra are also covered in this chapter.

Chapter 10 is an introduction to the topic of complex numbers. The first three sections of this chapter can be covered along with Chapter 1.

Chapter 11 is devoted to conic sections, and the final chapter contains a discussion of various topics, including a brief introduction to limits of functions.

I wish to thank the following people for reviewing the manuscript: Ann Anderson, Broward Community College; James E. Arnold, Jr., University of Wisconsin, Milwaukee; Donald W. Bellairs, Grossmont College; C. Patrick Collier, University of Wisconsin, Oshkosh; Gerry Gannon, California State University at Fullerton; Lynne B. Small, University of San Diego; Richard Semmler, North Virginia Community College; Charles N. Walter, Brigham Young University; Carroll G. Wells, Western Kentucky University; and Timothy Wilson, Honolulu Community College.

I also thank those who worked the exercises: William Hemmer, West Los Angeles College; Beverly Kawamoto, California State University at Fullerton; Gloria Langer; and David Lunsford, Grossmont College.

A Note on Calculators

To the Student

We strongly suggest that all students obtain a *scientific* calculator—that is, a calculator with keys for performing at least the following functions; y^x, $\sqrt[x]{y}$, e^x, $\log x$, $\ln x$, $\sin x$, $\cos x$, $\tan x$, $\sin^{-1} x$, $\cos^{-1} x$ and $\tan^{-1} x$. (Sometimes the last three keys are replaced by a key labeled *inv* or *arc*.) Also, we strongly suggest that you read the instruction manual that accompanies your calculator very carefully.

Let us make two important points about the use of calculators. First, since a calculator stores (and displays) only a few digits for each number entered, it generally gives only *approximations* rather than exact values. As a simple example, if you perform the operation $1 \div 3$, the display on a scientific calculator will show the number 0.333333333, which is only an approximation to 1/3.

Most scientific calculators round off answers to either eight or ten places, and while it may be true that for most applications these approximations are satisfactory, you should be aware that they are *only* approximations. To emphasize the effect of errors due to rounding, many calculators will display a small but *nonzero* number when asked to compute

$$(1 \div 3) \times 3 - 1$$

which, of course, is equal to 0.

The second, and perhaps more important, point is that a calculator is not meant to take the place of learning. As an example, in Chapter 1 we will discuss fractional exponents and what it means to raise a number to a fractional power. For instance, we will see that, by definition, $27^{2/3} = (\sqrt[3]{27})^2 = 3^2 = 9$. Now, you could determine the value of $27^{2/3}$ directly by using a calculator, which will give you the correct value. However, if you do this, you will have missed the entire point of the section on fractional exponents.

The goal of this book is to help you learn the basic concepts of algebra and how to apply them, not how to operate a calculator. Therefore, we urge you to reflect a moment before automatically reaching for your calculator.

Contents

Preface vii
A Note on Calculators ix

1 A REVIEW OF ELEMENTARY ALGEBRA 1

1.1 The Real Number System **1**
1.2 Exponents, Radicals, and Scientific Notation **10**
1.3 Algebraic Expressions, Polynomials, and Factoring **23**
1.4 Rational and Other Fractional Expressions **30**
1.5 Composition of Algebraic Expressions **37**
1.6 Solving Equations **41**
1.7 Linear and Absolute Value Inequalities: Interval Notation **49**
1.8 Quadratic and Other Inequalities **55**
Review **61**

2 FUNCTIONS AND THEIR GRAPHS 65

2.1 The Cartesian Coordinate System **65**
2.2 Graphing Sets and Equations **70**
2.3 Functions **76**
2.4 Graphing Functions **83**
2.5 Symmetry; Translating, Stretching, and Reflecting Graphs **94**
2.6 Increasing and Decreasing Functions; One-to-One Functions **102**
2.7 The Algebra of Functions; Composition of Functions **109**
2.8 The Inverse of a Function **117**
Review **125**

3 POLYNOMIAL FUNCTIONS OF DEGREE ONE AND TWO 129

3.1 The Slope of a Line **129**
3.2 Equations of a Line **137**
3.3 Quadratic Functions **142**
3.4 Applications: The Maximum and Minimum Values of a Quadratic Function **149**
3.5 Conic Sections **155**
Review **168**

4 POLYNOMIAL FUNCTIONS OF HIGHER DEGREE 171

4.1 Division of Polynomials **171**
4.2 The Remainder Theorem and the Factor Theorem **177**
4.3 Rational Roots **184**
4.4 Approximating the Roots of a Polynomial Equation **191**
4.5 Graphing Polynomial Functions **195**
4.6 Graphing Rational Functions **202**
4.7 Partial Fractions **213**
Review **218**

5 EXPONENTIAL AND LOGARITHMIC FUNCTIONS 221

5.1 The Exponential Functions **221**
5.2 The Logarithmic Functions **230**
5.3 The Natural Logarithm and the Common Logarithm **237**
5.4 The Natural Number *e* and Continuous Compounding of Interest **244**
5.5 Exponential Growth and Decay **250**
Review **257**

6 THE TRIGONOMETRIC FUNCTIONS 259

6.1 Radian Measure **259**
6.2 Unit Circles **268**
6.3 Definitions of the Trigonometric Functions **275**
6.4 Properties of the Trigonometric Functions **286**
6.5 Graphs of the Trigonometric Functions **296**
6.6 Sine Waves **304**
6.7 The Inverse Trigonometric Functions **316**
Review **325**

7 TRIANGLE TRIGONOMETRY AND ITS APPLICATIONS 329

7.1 Right Triangle Trigonometry **329**
7.2 The Law of Sines **342**
7.3 The Law of Cosines **350**
7.4 An Introduction to Polar Coordinates **356**
7.5 Vectors **364**
Review **373**

8 ANALYTIC TRIGONOMETRY 377

8.1 Trigonometric Identities **377**
8.2 Trigonometric Equations **385**
8.3 The Addition and Subtraction Formulas **394**
8.4 The Double-Angle and Half-Angle Formulas **404**
8.5 A Summary of Trigonometric Formulas **411**
Review **413**

9 SYSTEMS OF LINEAR EQUATIONS 417

9.1 Upper Triangular Systems **417**
9.2 Gaussian Elimination **422**
9.3 Matrices and Gaussian Elimination **427**
9.4 Applications Involving Systems of Linear Equations **433**
9.5 Determinants and Cramer's Rule: The 2×2 Case **437**
9.6 Determinants and Cramer's Rule: The 3×3 Case **443**
9.7 Matrix Algebra **450**
9.8 Solving Systems of Linear Equations Using Matrix Algebra **458**
9.9 Systems of Nonlinear Equations **463**
9.10 Systems of Inequalities **467**
Review **472**

10 THE COMPLEX NUMBER SYSTEM 477

10.1 The Arithmetic of Complex Numbers **477**
10.2 The Conjugate of a Complex Number and Complex Division **482**
10.3 Complex Solutions to Quadratic Equations **489**
10.4 The Trigonometric Form of a Complex Number **494**
10.5 De Moivre's Formula and the Roots of a Complex Number **500**
Review **506**

11 CONIC SECTIONS 509

11.1 Introduction **509**
11.2 The Parabola **511**
11.3 The Ellipse **518**
11.4 The Hyperbola **526**
11.5 The General Quadratic Equation and Rotation of Axes **536**
Review **547**

12 SEQUENCES, SERIES AND OTHER TOPICS 549

12.1 Mathematical Induction **549**
12.2 The Binomial Formula **556**
12.3 Infinite Sequences **561**
12.4 The Partial Sums of a Sequence **570**

12.5 Infinite Series **575**
12.6 An Introduction to Limits **581**
Review **585**

Appendix **A-1**

Answers to Study Suggestions **A-16**

Answers to Odd-Numbered Exercises **A-45**

Index **I-1**

Precalculus

1

A REVIEW OF ELEMENTARY ALGEBRA

1.1 The Real Number System

Let us begin our review of algebra by discussing the basic number systems that we will use throughout the book. With the exception of Chapter 8, we will confine our attention to the *real number system*, which consists of numbers, such as

$$0, 1, -3, \frac{5}{4}, \sqrt{2}, \pi$$

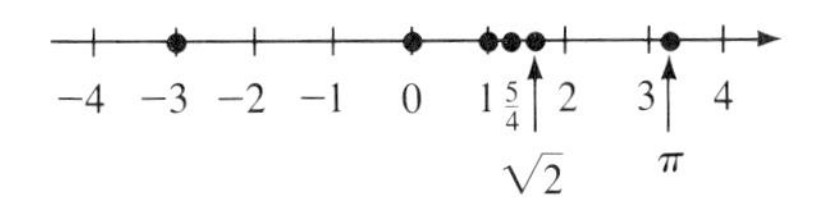

FIGURE 1.1

that can be represented as points on the *real number line*, as shown in Figure 1.1. The real number system is broken down into several subsets as follows.

The Natural Numbers: These are the numbers that we normally use for counting.

$$\mathbf{N} = \{1, 2, 3, \ldots\}$$

The Integers: These are the natural numbers, the negative natural numbers, and zero.

$$\mathbf{Z} = \{\ldots, -3, -2, -1, 0, 1, 2, 3, \ldots\}$$

The Rational Numbers: These are all numbers which can be written in the form p/q, where p and q are *integers* and $q \neq 0$. In set notation,

$$\mathbf{Q} = \left\{ \frac{p}{q} \,\middle|\, p \text{ and } q \text{ are integers and } q \neq 0 \right\}$$

The Irrational Numbers: These are the numbers that correspond to points on the real number line not associated with rational numbers. We use the symbol **I** for the irrational numbers. Some examples of irrational numbers are $\sqrt{2}$, $\sqrt{3}$, and π.

The Real Numbers: These are the rational and irrational numbers. They correspond exactly to the points on the real number line.

$$\mathbf{R} = \{\text{rational and irrational numbers}\}$$

► **Study Suggestion 1.1:** Draw a number line and plot the points 4, 2, -3, $\frac{1}{2}$, $-\frac{17}{3}$, $\sqrt{3}$, $-\sqrt{7}$, and π. (*Hint:* whenever you plot points on a number line, it is a good idea to first plot the origin, to use as a point of reference, and the point corresponding to 1, to set the scale.) ◄

Try Study Suggestion 1.1.

In order to work with the real number system, we must be familiar with its properties, the most important of which are listed below.

Properties of Addition and Multiplication

The following properties hold for all real numbers a, b, and c.

1. **The commutative properties**

$$a + b = b + a$$
$$ab = ba$$

2. **The associative properties**

$$a + (b + c) = (a + b) + c$$
$$a(bc) = (ab)c$$

3. **The identity properties.** There is a real number, namely 0, with the special property that

$$a + 0 = a \qquad \text{and} \qquad 0 + a = a$$

for all real numbers a. There is a real number, namely 1, with the special property that

$$a \cdot 1 = a \qquad \text{and} \qquad 1 \cdot a = a$$

for all real numbers a.

4. **The inverse properties.** For every real number a, there is another real number, namely $-a$, with the property that

$$a + (-a) = 0 \qquad \text{and} \qquad (-a) + a = 0$$

For every real number a, *except* 0, there is another real number, namely $1/a$, with the property that

$$a \cdot \frac{1}{a} = 1 \quad \text{and} \quad \frac{1}{a} \cdot a = 1$$

5. **The distributive properties**

$$a(b + c) = ab + ac$$
$$(a + b)c = ac + bc$$

6. **The cancellation properties**

If $a + c = b + c$ then $a = b$

If $ac = bc$ and $c \neq 0$ then $a = b$

7. **The properties of zero**

$$0 \cdot a = 0 \quad \text{and} \quad a \cdot 0 = 0$$

If $a \cdot b = 0$ then either $a = 0$ or $b = 0$ or both.

▶ **Study Suggestion 1.2:** State which property of real numbers is being illustrated in each case.

(a) $1 \cdot 2 = 2 \cdot 1$

(b) $1 \cdot 4 = 4$

(c) $(2 \cdot 3) \cdot 1 = 2 \cdot (3 \cdot 1)$

(d) $0 + 9 = 9$

(e) $(2/3)(3/2) = 1$

(f) $(3 + 4) \cdot 2 = 3 \cdot 2 + 4 \cdot 2$

(g) $4 \cdot 0 = 0$

(h) $2x = 0$ implies $x = 0$ ◀

Try Study Suggestion 1.2.

Now let us list some of the most useful properties of quotients. (All denominators are assumed to be nonzero.)

Properties of Quotients

8. $\frac{a}{b} = \frac{c}{d}$ if and only if $ad = bc$

9. $\frac{a}{b} + \frac{c}{d} = \frac{ad + bc}{bd}$

10. $\frac{a}{b} - \frac{c}{d} = \frac{ad - bc}{bd}$

11. $\frac{a}{b} \cdot \frac{c}{d} = \frac{ac}{bd}$

12. $\frac{a}{b} \div \frac{c}{d} = \frac{a}{b} \cdot \frac{d}{c} = \frac{ad}{bc}$

or in a different notation

$$\frac{\frac{a}{b}}{\frac{c}{d}} = \frac{a}{b} \cdot \frac{d}{c} = \frac{ad}{bc}$$

Property 12 can be remembered as "when dividing by a fraction, invert the fraction and multiply."

EXAMPLE 1: Perform the indicated operations and simplify.

(a) $\frac{4}{3} + \frac{5}{7}$ **(b)** $\frac{12}{5} \cdot \frac{3}{7}$ **(c)** $\frac{6}{5} \div \frac{-3}{8}$

Solutions:

(a) Using Property 9, we have

$$\frac{4}{3} + \frac{5}{7} = \frac{4 \cdot 7 + 3 \cdot 5}{3 \cdot 7} = \frac{28 + 15}{21} = \frac{43}{21}$$

(b) Using Property 11, we have

$$\frac{12}{5} \cdot \frac{3}{7} = \frac{12 \cdot 3}{5 \cdot 7} = \frac{36}{35}$$

► **Study Suggestion 1.3:** Perform the indicated operations and simplify.

(a) $\frac{1}{2} - \frac{2}{3}$ **(b)** $\frac{1}{\frac{4}{5}}$

(c) $\frac{3}{4} \div \frac{5}{8}$ ◄

(c) Using Property 12 (invert and multiply), we have

$$\frac{6}{5} \div \frac{-3}{8} = \frac{\overset{2}{\cancel{6}}}{5} \cdot \frac{8}{\underset{-1}{\cancel{-3}}} = \frac{2}{5} \cdot \frac{8}{-1} = -\frac{16}{5}$$

Try Study Suggestion 1.3. □

Now let us discuss the concept of order.

DEFINITION

A real number is said to be

1. **positive** if it lies to the right of 0 on the real number line (see Figure 1.2);
2. **negative** if it lies to the left of 0 on the real number line (see Figure 1.2);
3. **nonnegative** if it is either positive or 0
4. **nonpositive** if it is either negative or 0

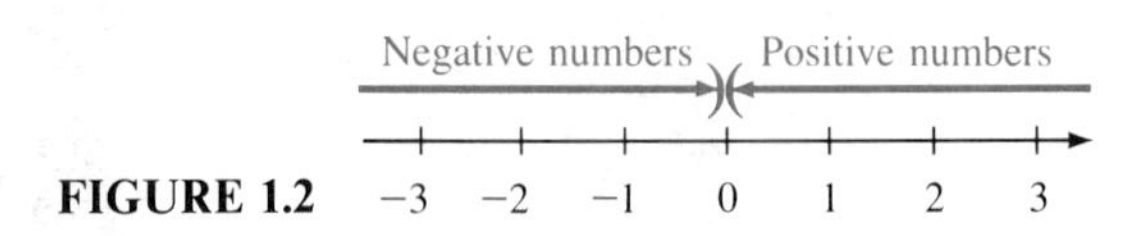

FIGURE 1.2

DEFINITION

Let a and b be real numbers. Then

1. a is **less than** b, written $a < b$, if $b - a$ is positive;
2. a is **less than or equal to** b, written $a \leq b$, if either $a < b$ or $a = b$;
3. a is **greater than** b, written $a > b$, if $b - a$ is negative;
4. a is **greater than or equal to** b, written $a \geq b$, if either $a > b$ or $a = b$.

The properties of order given in the previous definition can easily be expressed in terms of the real number line. For instance, the real number a is *less than* the real number b if and only if a lies to the *left of* b on the real number line. Similarly, a is *greater than* b if a lies to the *right of* b on the real number line.

▶ **Study Suggestion 1.4:** True or false:

(a) $-5 < -3$

(b) $-10 \leq -12$

(c) $-6 > -1$

(d) $-1/2 < -1/3$

(e) $9/10 > 13/12$ ◀

Try Study Suggestion 1.4.

We should discuss one point about positive and negative numbers that sometimes causes confusion. If a is a real number, then it is tempting to think of $-a$ as being negative, because of the presence of the negative sign. However, if a is negative, then $-a$ is positive. For example, if $a = -3$, then $-a = -(-3) = 3$ is positive. Thus, we can say that

if a is positive, then $-a$ is negative;

if a is negative, then $-a$ is positive.

The order relation possesses certain important properties with regard to addition and multiplication.

The Order Relation and Arithmetic

13. **The addition property**

$$\text{If } a \leq b \text{ and } c \leq d \text{ then } a + c \leq b + d$$

14. **Special case of Property 1**

$$\text{If } a \leq b \text{ then } a + c \leq b + c$$

15. **The multiplication property**

$$\text{If } a \leq b \text{ and if } c \text{ is } \textit{positive}, \text{ then } ac \leq bc$$
$$\text{If } a \leq b \text{ and if } c \text{ is } \textit{negative} \text{ then } ac \geq bc$$

You should pay special attention to Property 15. It says that if we multiply both sides of an inequality by a *positive* number, then we do not change the sense of the inequality. However, if we multiply both sides by a *negative* number, then we must change the sense of the inequality.

EXAMPLE 2: Since $-5 \leq -1$ and 4 is positive, Property 15 tells us that

$$(-5) \cdot 4 \leq (-1) \cdot 4$$

On the other hand, since -7 is negative, we have

$$(-5) \cdot (-7) \geq (-1) \cdot (-7)$$

Try Study Suggestion 1.5. □

► **Study Suggestion 1.5:** What do you get when you multiply both sides of the inequality $-3 < -2$ by 5?; by -5? What happens when you multiply the inequality $-3 < -2$ by 0? Is the resulting inequality still valid? ◄

The notation

$$a < x < b$$

is frequently used as a shorthand for the *two* inequalities

$$a < x \qquad \text{and} \qquad x < b$$

In other words, a real number x satisfies $a < x < b$ if and only if it satisfies *both* of the inequalities $a < x$ and $x < b$. A similar statement holds for the notation $a \leq x \leq b$.

For example, the set of all real numbers x that satisfy the inequalities

$$-2 \leq x \leq 4$$

is the set of all real numbers x that are greater than or equal to -2 *and* less than or equal to 4.

Try Study Suggestion 1.6.

► **Study Suggestion 1.6:**

(a) Describe in words which real numbers x satisfy the inequalities $-4 < x \leq 7$?

(b) Which real numbers x satisfy the inequalities $-\frac{2}{3} \leq x < -\frac{1}{2}$? ◄

Now let us turn our attention to the concept of absolute value.

DEFINITION

If a is a real number, then the **absolute value** of a is denoted by $|a|$, and defined by

$$|a| = \begin{cases} a & \text{if} \quad a \geq 0 \\ -a & \text{if} \quad a < 0 \end{cases}$$

Using the concept of absolute value, we can say that

the distance from a point a on the real number line to the origin is equal to $|a|$.

EXAMPLE 3:

(a) $|1/2| = 1/2$ since $1/2$ is nonnegative

(b) $|0| = 0$ since 0 is nonnegative

(c) $|-\sqrt{2}| = \sqrt{2}$ since $-\sqrt{2}$ is negative *Try Study Suggestion 1.7.* □

▶ **Study Suggestion 1.7:** Write each of the following without absolute value signs

(a) $|-1/2|$

(b) $|2 - 9|$

(c) $|4 + 3 - 8|$

Is it true that $|7 - 9| = |9 - 7|$? Can you generalize this? ◀

The absolute value also has some important properties.

Properties of the Absolute Value

For all real numbers a and b, we have

1. $|a| \geq 0$
2. $|-a| = |a|$
3. $|a - b| = |b - a|$
4. $|ab| = |a||b|$
5. $\left|\frac{a}{b}\right| = \frac{|a|}{|b|}$
6. $|a + b| \leq |a| + |b|$

Some care must be taken when combining the absolute value with arithmetic operations. For example,

$$\textbf{in general, } |a + b| \neq |a| + |b|.$$

(See Exercise 59.)

The absolute value can be used to find the distance between any two points on the real number line.

Distance between Two Points on the Real Number Line

The distance $d(a, b)$ between two points a and b on the real number line is given by

$$d(a, b) = |b - a|$$

Furthermore, since $|b - a| = |a - b|$, we also have

$$d(a, b) = |a - b|$$

EXAMPLE 4: Using the fact that $d(a, b) = |b - a|$, we have

(a) $d(7, 10) = |10 - 7| = |3| = 3$

(b) $d(12, 8) = |8 - 12| = |-4| = 4$

(c) $d(-22/7, -1/7) = |-(1/7) - (-22/7)|$
$= |-(1/7) + 22/7| = |21/7| = |3| = 3$

Try Study Suggestion 1.8. □

▶ **Study Suggestion 1.8:** Find
(a) $d(1, 1)$ **(b)** $d(2, -3)$
(c) $d(0, -4)$ **(d)** $d(\frac{1}{2}, -\frac{1}{3})$.
Is it true that for all real numbers a and b we have $d(a, b) = d(-a, -b)$? Justify your answer. ◀

The absolute value is very useful for describing certain subsets of real numbers. For example,

$$\{x \mid |x - 1| < 3\}$$

is the set of all real numbers x whose distance from the point 1 is less than 3. This set is pictured in Figure 1.3a. As you can see from the figure, this is the same as the set of all numbers between -2 and 4. In set notation, we have

$$\{x \mid |x - 1| < 3\} = \{x \mid -2 < x < 4\}$$

Similarly

$$\{x \mid |x - 1| \leq 3\} = \{x \mid -2 \leq x \leq 4\}$$

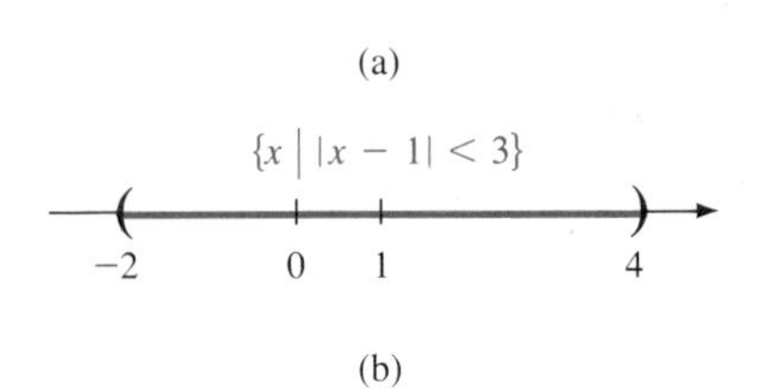

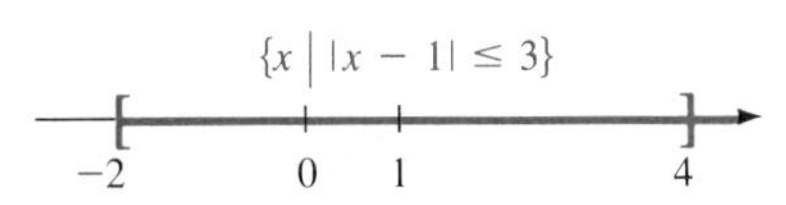

FIGURE 1.3

as shown in Figure 1.3b. Notice that we use round parentheses in Figure 1.3a to denote the fact that the endpoints $x = -2$ and $x = 4$ are *not* included in the set, and we use square brackets in Figure 1.3b to denote the fact that $x = -2$ and $x = 4$ are included in the set.

In general, if a and b are real numbers, then

$$\{x \mid |x - a| < b\}$$

is the set of all real numbers whose distance from a is less than b. It is important to notice the negative sign in the expression $|x - a|$. For example, in order to describe the set $\{x \mid |x + 1| < 2\}$, we would write $x + 1 = x - (-1)$, and so

$$\{x \mid |x + 1| < 2\} = \{x \mid |x - (-1)| < 2\}$$

is the set of all real numbers whose distance from -1 is less than 2.

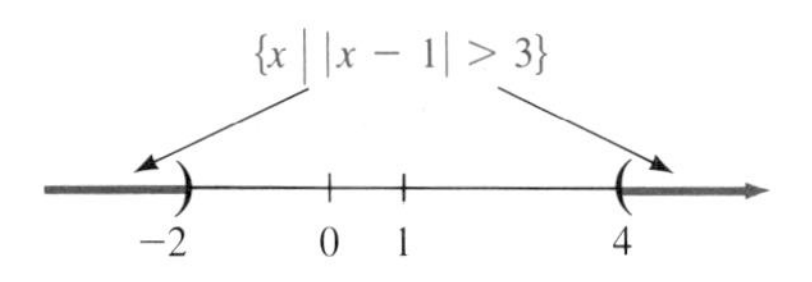

FIGURE 1.4

Try Study Suggestion 1.9.

▶ **Study Suggestion 1.9:**
(a) Show the set $\{x \mid |x - 2| < 2\}$ on a real number line.
(b) Show the set $\{x \mid |x + 3| \leq 1\}$ on a real number line. ◀

As another example of the use of absolute values,

$$\{x \mid |x - 1| > 3\}$$

is the set of all real numbers whose distance from the point 1 is greater than 3. (See Figure 1.4.) As you can see from this figure, this is the same as the set of all real numbers that are either less than -2 or greater than 4.

Try Study Suggestion 1.10.

▶ **Study Suggestion 1.10:**
(a) Show the set $\{x \mid |x - 2| \geq 2\}$ on a real number line.
(b) Show the set $\{x \mid |x + 1| > 1\}$ on a real number line. ◀

The concept of *union* can be used to describe sets such as $\{x \mid |x - 1| > 3\}$. The **union** of two sets is the set of all elements that are in either one of the two sets, or in both. Therefore, we have

$$\{x \mid |x - 1| > 3\} = \{x \mid x < -2\} \cup \{x \mid x > 4\}$$

where the symbol $\cup$ stands for union. As another example of union, the set $\{x \mid -1 < x < 1\} \cup \{x \mid 2 \le x < 3\}$ is shown in Figure 1.5.

▶ **Study Suggestion 1.11:** Show the set

$$\{x \mid -3 < x \le -2\} \cup \{x \mid x > 4\}$$

on a real number line. ◀

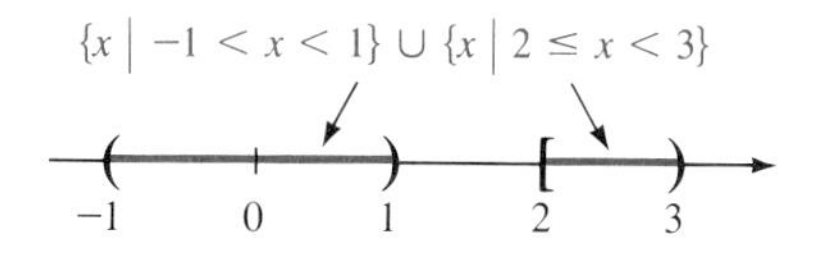

FIGURE 1.5

Try Study Suggestion 1.11.

EXERCISES

In Exercises 1–6, plot the given numbers on a number line.

1. $-1, 0, 1, \frac{1}{4}, -\frac{1}{2}$

2. $1, \frac{3}{2}, -\sqrt{2}, 0, \sqrt{3}, 2$

3. $0, 2, -2, 0.5, \pi$

4. $2, 4, -2, -4, \sqrt{5}, \frac{\pi}{2}$

5. $0, -5, -\sqrt{5}, \sqrt{21}, \pi/\sqrt{2}$

6. $\frac{3}{4}, \frac{125}{3}, -\frac{67}{4}, \frac{8}{3}, \frac{17}{5}$

In Exercises 7–22, identify which property or properties of the real number system are being illustrated.

7. $7 \cdot 3 = 3 \cdot 7$

8. $3 + 6 = 6 + 3$

9. $-7 + 7 = 0$

10. $(2 + 3) + 4 = 2 + (3 + 4)$

11. $4 + 0 = 4$

12. $(-1)5 = -5$

13. $(2 - 3) + 7 = 2 + (-3 + 7)$

14. $0 \cdot 4 = 0$

15. $1 - 1 = 0$

16. $(2 + 3) + 2 = 2 + (2 + 3)$

17. $3(2 + 5) = 3 \cdot 2 + 5 \cdot 3$

18. $2 \cdot \dfrac{1}{2} = 1$

19. $\left(-\dfrac{1}{3}\right)(-3) = 1$

20. $(12 + 1)(3 + 4) = 12 \cdot 3 + 12 \cdot 4 + 1 \cdot 3 + 1 \cdot 4$

21. $(12 + 1)(3 + 4) = 12 \cdot 3 + 12 \cdot 4 + 3 + 4$

22. $(7 + 3) \cdot 4 = 4 \cdot (7 + 3)$

In Exercises 23–36, perform the indicated operations and simplify.

23. $\dfrac{5}{2} + \dfrac{7}{3}$

24. $\dfrac{1}{2} \cdot \dfrac{3}{4}$

25. $\dfrac{1}{6} + \dfrac{1}{5} + \dfrac{1}{3}$

26. $\dfrac{2}{5} - \dfrac{3}{4} + \dfrac{8}{9}$

27. $\left(\dfrac{1}{2} + \dfrac{5}{8}\right)\left(\dfrac{1}{5} + \dfrac{2}{7}\right)$

28. $\dfrac{7}{5} - \dfrac{-9}{6}$

29. $\dfrac{\dfrac{3}{8} - \dfrac{5}{3}}{\dfrac{13}{4}}$

30. $\dfrac{7}{6} \div \dfrac{5}{6}$

31. $\dfrac{13}{12} \div 2$

32. $\dfrac{5}{2} - \dfrac{7}{5} + \dfrac{9}{8}$

33. $\dfrac{\dfrac{1}{2} - \dfrac{3}{5}}{\dfrac{2}{3} + \dfrac{5}{6}}$

34. $\dfrac{\dfrac{7}{2} \div \dfrac{-5}{9}}{\dfrac{9}{2} \div \dfrac{8}{5}}$

35. $\dfrac{\dfrac{11}{12}}{\dfrac{-3}{4}}$

36. $\left(\dfrac{13}{4} - \dfrac{5}{2}\right) \div \dfrac{6}{8}$

37. Does the operation of subtraction satisfy the commutative property? Explain your answer.

38. Does the operation of division satisfy the associative property? Explain your answer.

In Exercises 39–48, replace the question mark by one of the symbols $<$, $\le$, $>$ or $\ge$.

39. $2 ? 4$
40. $-3 ? 1$
41. $-4 ? -5$
42. $-7 ? -1$
43. $1/2 ? -\sqrt{2}$
44. $14 ? -3$
45. $-\pi ? -\sqrt{3}$
46. $-\sqrt{2} ? -\sqrt{3}$
47. $0 ? x^2$
48. $0 ? -x^2$

In Exercises 49–58, write the given expression without using absolute value signs.

49. $|-3|$
50. $|-4/3|$
51. $|\sqrt{2} - \sqrt{5}|$
52. $|4| - |5|$
53. $||3| - |-2||$
54. $||-1| - |-4||$
55. $-2/|2|$
56. $|\pi/2 - 2|$
57. $-2|-2|$
58. $(-|-2|)^2$

59. (a) Find two numbers a and b for which

$$|a + b| \neq |a| + |b|$$

(b) Find two numbers a and b for which

$$|a + b| = |a| + |b|$$

In Exercises 60–64, compute the given distances.

60. (a) $d(7, 0)$ **(b)** $d(-7, 0)$

61. (a) $d(-\pi, 0)$ **(b)** $d(-\pi, \pi)$

62. (a) $d(-2, 3)$ **(b)** $d(1, 10)$

63. (a) $d(-1/2, 1/4)$ **(b)** $d(-2/3, -4/9)$

64. (a) $d(-\sqrt{2}, \sqrt{2})$ **(b)** $d(1 + \sqrt{2}, 1 - \sqrt{2})$

In Exercises 65–72, show the given set on the real number line.

65. $\{x | -1 \le x \le 2\}$

66. $\{x | -1/2 \le x < 1/2\}$

67. $\{t | -\pi \le t \le \pi\}$

68. $\{t | -\sqrt{2} < t < \sqrt{2}\}$

69. $\{u | -4 < u < 3\} \cup \{u | -7 \le u \le -5\}$

70. $\{u | -1/2 \le u \le 1/2\} \cup \{u | \le 1\}$

71. $\{x | x \le -1\} \cup \{x | x \ge 1\}$

72. $\{x | x > 2\} \cup \{x | x > 0\}$

In Exercises 73–79, write the given set of real numbers without using absolute value signs.

73. (a) $\{x \,|\, |x| \le 5\}$ **(b)** $\{x \,|\, |x| < 5\}$

74. (a) $\{x \,|\, |x| \ge 1/2\}$ **(b)** $\{x \,|\, |x| > 1/2\}$

75. (a) $\{t \,|\, |t| \ge -1\}$ **(b)** $\{t \,|\, |t| \le -1\}$

76. (a) $\{t \,|\, |t| > 2\}$ **(b)** $\{t \,|\, |t| \ge 3\}$

77. (a) $\{x \,|\, |x - 1| < 4\}$ **(b)** $\{x \,|\, |x + 1| \ge 2\}$

78. (a) $\{t \,|\, |t - 2| > 2\}$ **(b)** $\{x \,|\, |x + 3| \ge 1\}$

79. (a) $\{x \,|\, |x - 1/2| \le 1/3\}$ **(b)** $\{y \,|\, |y + \pi| > \pi/2\}$

80. Show that if $a < b$ and if a and b have the same sign, then $\frac{1}{a} > \frac{1}{b}$. (*Hint:* start with the inequality $a < b$.)

81. Is it always true that if $a < b$ and $c < d$, then $a - c < b - d$? Justify your answer.

1.2 Exponents, Radicals, and Scientific Notation

In this section we will review the concepts of exponents, radicals, and scientific notation. Let us begin with integral exponents (that is, exponents that are integers).

DEFINITION

The following holds for all real numbers a and all positive integers n (except as noted.)

1. $a^n = \underbrace{a \cdot a \cdots a}_{na\text{'s}}$

2. $a^{-n} = \underbrace{\frac{1}{a}\frac{1}{a}\cdots\frac{1}{a}}_{n\ 1/a\text{'s}} = \frac{1}{a^n}$

3. $a^0 = 1$ provided $a \neq 0$ (0^0 is not defined)

EXAMPLE 1:

(a) $2^3 = 2 \cdot 2 \cdot 2 = 8$

(b) $2^{-4} = \frac{1}{2^4} = \frac{1}{2 \cdot 2 \cdot 2 \cdot 2} = \frac{1}{16}$

(c) $\left(-\frac{1}{4}\right)^{-3} = \frac{1}{\left(-\frac{1}{4}\right)^3} = \frac{1}{\left(-\frac{1}{4}\right)\left(-\frac{1}{4}\right)\left(-\frac{1}{4}\right)} = \frac{1}{\left(-\frac{1}{64}\right)} = -64$

Try Study Suggestion 1.12. □

▶ **Study Suggestion 1.12:** Write each of the following without exponents:
(a) $(-2)^3$ **(b)** 3^{-3}
(c) 17^0 **(d)** $(-1/2)^{-4}$ ◀

Now let us consider some of the properties of integral exponents.

Properties of Integral Exponents

The following properties hold for all nonzero real numbers a and b and all integers m and n.

1. $a^n a^m = a^{n+m}$
2. $(a^n)^m = a^{nm}$
3. $\left(\frac{a^n}{a^m}\right) = a^{n-m}$
4. $(ab)^n = a^n b^n$
5. $\left(\frac{a}{b}\right)^n = \frac{a^n}{b^n}$

Another useful property of exponents is

$$\left(\frac{a}{b}\right)^{-n} = \left(\frac{b}{a}\right)^n$$

provided, of course, that both a and b are nonzero. For instance, we have

$$\left(\frac{3}{4}\right)^{-2} = \left(\frac{4}{3}\right)^{2} = \frac{16}{9}$$

EXAMPLE 2: Simplify the following expressions.

(a) $3 \cdot 2^3 \cdot 3^{-2} \cdot 2^2$ **(b)** $\dfrac{(3^{-2}7^4)^2}{(2^{-2}3^3)^{-1}7^6}$

Solutions:

(a) $3 \cdot 2^3 \cdot 3^{-2} \cdot 2^2 = 2^3 \cdot 2^2 \cdot 3 \cdot 3^{-2} = 2^5 \cdot 3^{-1} = \dfrac{2^5}{3} = \dfrac{32}{3}$

(b)
$$\frac{(3^{-2}7^4)^2}{(2^{-2}3^3)^{-1}7^6} = \frac{(3^{-2})^2(7^4)^2}{(2^{-2})^{-1}(3^3)^{-1}7^6} = \frac{3^{-4}7^8}{2^2 3^{-3}7^6} = \frac{3^{-4-(-3)}7^{8-6}}{2^2} = \frac{3^{-1}7^2}{2^2} = \frac{49}{4 \cdot 3} = \frac{49}{12}$$

□

EXAMPLE 3: Simplify the following expressions by writing them so that no variables are repeated more than once, and so that there are no negative exponents. (Assume that all variables represent positive real numbers.)

(a) $(2x^2y^{-3})^2(4x^{-3}y^5)^{-2}$ **(b)** $\dfrac{(3p^4)^{-2}(q^2)^{-1}r}{p^{-5}(2^{-1}q^3)^2r^2}$

Solutions:

(a)
$$(2x^2y^{-3})^2(4x^{-3}y^5)^{-2} = 4x^4y^{-6}4^{-2}x^6y^{-10} = 4^{-1}x^{10}y^{-16} = \frac{x^{10}}{4y^{16}}$$

(b) $$\frac{(3p^2)^2(q^2)^{-1}r}{p^{-5}(2^{-1}q^3)^2r^2} = \frac{3^2p^4q^{-2}r}{p^{-5}2^{-2}q^6r^2}$$
$$= \frac{3^2}{2^{-2}}\frac{p^4}{p^{-5}}\frac{q^{-2}}{q^6}\frac{r}{r^2}$$
$$= 3^2 \cdot 2^2p^{4-(-5)}q^{-2-6}r^{1-2}$$
$$= 36p^9q^{-8}r^{-1}$$
$$= \frac{36p^9}{q^8r}$$

Try Study Suggestion 1.13. □

▶ **Study Suggestion 1.13:** Simplify the following expressions

(a) $(5^{-1} \cdot 3)^2(5^{-2} \cdot 3^2)^{-2}$

(b) $\dfrac{(4x^{-1})^3(3^{-1}x^2y)^2}{4x^{-2}(2y^3)^{-1}}$ ◀

Now we turn our attention to rational exponents. Our plan is to define rational exponents in such a way that the properties of integral exponents will also hold for rational exponents (at least as much as possible.)

In particular, if a is a positive real number and n is an integer, then we want $a^{1/n}$ to have the property that

$$(a^{1/n})^n = a^{(1/n) \cdot n} = a^1 = a$$

In other words, $a^{1/n}$ must be a real number whose nth power is equal to a. This leads us to the following definition.

DEFINITION

Let a be a real number.

1. If n is an odd positive integer, then

   $a^{1/n}$ = that real number whose nth power equals a

2. If n is an even positive integer, then

$$a^{1/n} = \begin{cases} \text{that } \textit{nonnegative} \text{ real number whose } n\text{th power equals } a \text{ if } a \geq 0 \\ \text{undefined if } a < 0 \end{cases}$$

Notice that, according to our definition, if n is even and $a < 0$, then the expression $a^{1/n}$ is not defined. For example, the expressions $(-1)^{1/2}$ and $(-3)^{1/4}$ are not defined.

We can also extend this definition to negative exponents by setting

$$a^{-1/n} = \frac{1}{a^{1/n}}$$

EXAMPLE 4: Evaluate the following expressions.

(a) $8^{1/3}$ **(b)** $(-27)^{1/3}$ **(c)** $9^{-1/2}$ **(d)** $(-4)^{1/2}$

Solutions:

(a) $8^{1/3}$ is that real number whose third power equals 8. Since $2^3 = 8$, we have $8^{1/3} = 2$.

(b) $(-27)^{1/3}$ is that real number whose third power equals -27. Since $(-3)^3 = -27$, we have $(-27)^{1/3} = -3$.

(c) In this case, we have

$$9^{-1/2} = \frac{1}{9^{1/2}} = \frac{1}{3}$$

(d) $(-4)^{1/2}$ is not defined, since -4 is negative and 2 is even.

Try Study Suggestion 1.14. □

▶ **Study Suggestion 1.14:** Evaluate the following expressions

(a) $(16)^{1/4}$

(b) $(16)^{-1/4}$

(c) $(-64)^{1/3}$

(d) $(-27)^{-1/3}$ ◀

When n is an integer and $n > 1$, it is common to use the notation $\sqrt[n]{a}$ for $a^{1/n}$. That is,

$$\sqrt[n]{a} = a^{1/n}$$

The symbol $\sqrt{\ }$ is called a **radical sign,** and the expression $\sqrt[n]{a}$ is read "the nth root of a." We call the integer n the **index** and the number a is called the **radicand.**

EXAMPLE 5:

(a) $\sqrt[3]{64} = 64^{1/3} = 4$ **(b)** $\sqrt[5]{32} = 32^{1/5} = 2$ □

Now let us list some of the most important properties of radicals.

Properties of Radicals

The following properties hold whenever the indicated roots exist.

1. $\sqrt[n]{a}\sqrt[n]{b} = \sqrt[n]{ab}$
2. $\sqrt[n]{\dfrac{a}{b}} = \dfrac{\sqrt[n]{a}}{\sqrt[n]{b}}$ $(b \neq 0)$
3. $\sqrt[m]{\sqrt[n]{a}} = \sqrt[nm]{a}$

There are two important points concerning radicals that we should emphasize, since they sometimes lead to confusion. First, if a is a nonzero real number, and if $\sqrt{a}$ exists, then according to the definition, $\sqrt{a}$ is *positive.*

That is,

by $\sqrt{a}$ we mean the *positive* square root of a

As to the second point, let us compute $\sqrt{(-2)^2}$

$$\sqrt{(-2)^2} = \sqrt{4} = 2$$

Notice that, since the last operation performed is a square root, the result must be positive. This is why $\sqrt{(-2)^2}$ is equal to 2, and not -2. In general, for all real numbers a, we have

$$\sqrt{a^2} = |a|$$

Thus, the square root of a^2 is equal to the *absolute value* of a.

In order to simplify an expression that involves radicals, in general we must achieve the following goals.

Procedure for Simplifying Expressions Involving Radicals

1. Remove all possible factors from under the radical sign. For example,

$$\sqrt{12} = \sqrt{4 \cdot 3} = \sqrt{4}\sqrt{3} = 2\sqrt{3}$$

2. Make sure that the index of the radical is as small as possible. For example,

$$\sqrt[4]{25} = \sqrt[4]{5^2} = (5^2)^{1/4} = 5^{2/4} = 5^{1/2} = \sqrt{5}$$

3. Remove all radicals from the *denominator* of the expression. This process is called **rationalizing the denominator.** For example, in order to rationalize the denominator of the expression $\frac{2}{\sqrt{2}}$ we multiply the numerator and denominator by $\sqrt{2}$.

$$\frac{2}{\sqrt{2}} = \frac{2\sqrt{2}}{\sqrt{2}\sqrt{2}} = \frac{2\sqrt{2}}{2} = \sqrt{2}$$

4. Perform all indicated operations. For example,

$$\begin{aligned}\sqrt{27} - \sqrt{12} &= \sqrt{9 \cdot 3} - \sqrt{4 \cdot 3} = \sqrt{9}\sqrt{3} - \sqrt{4}\sqrt{3} \\ &= 3\sqrt{3} - 2\sqrt{3} = \sqrt{3}\end{aligned}$$

EXAMPLE 6: Simplify the following expressions.

(a) $\sqrt[3]{-54}$ **(b)** $\sqrt[4]{\dfrac{3}{32}}$ **(c)** $\sqrt{8} - \sqrt{18} + \sqrt{50}$

Solutions:

(a) $\sqrt[3]{-54} = \sqrt[3]{-27 \cdot 2} = \sqrt[3]{-27}\sqrt[3]{2} = -3\sqrt[3]{2}$

(b) We begin by writing

$$\sqrt[4]{\frac{3}{32}} = \frac{\sqrt[4]{3}}{\sqrt[4]{32}} = \frac{\sqrt[4]{3}}{\sqrt[4]{16 \cdot 2}} = \frac{\sqrt[4]{3}}{\sqrt[4]{16}\sqrt[4]{2}} = \frac{\sqrt[4]{3}}{2\sqrt[4]{2}}$$

Now we must rationalize the denominator of this expression, which is done by multiplying the numerator and denominator by $\sqrt[4]{2^3}$,

$$= \frac{\sqrt[4]{3}\sqrt[4]{2^3}}{2\sqrt[4]{2}\sqrt[4]{2^3}} = \frac{\sqrt[4]{3 \cdot 2^3}}{2\sqrt[4]{2^4}} = \frac{\sqrt[4]{24}}{4}$$

(c)

$$\begin{aligned} \sqrt{8} - \sqrt{18} + \sqrt{50} &= \sqrt{4 \cdot 2} - \sqrt{9 \cdot 2} + \sqrt{25 \cdot 2} \\ &= \sqrt{4}\sqrt{2} - \sqrt{9}\sqrt{2} + \sqrt{25}\sqrt{2} \\ &= 2\sqrt{2} - 3\sqrt{2} + 5\sqrt{2} \\ &= 4\sqrt{2} \end{aligned}$$

Try Study Suggestion 1.15. □

▶ **Study Suggestion 1.15:** Simplify the following expressions

(a) $\sqrt{50}$ **(b)** $\sqrt[3]{16}$ **(c)** $\sqrt[4]{\frac{48}{5}}$

(d) $\sqrt{80} - \sqrt{125} + \sqrt{5}$ ◀

EXAMPLE 7: Simplify the expression

$$\frac{1}{\sqrt{5} - \sqrt{2}}$$

Solution: In order to simplify this expression, we must rationalize the denominator. This can be done by multiplying the numerator and denominator by $\sqrt{5} + \sqrt{2}$, since $(\sqrt{5} - \sqrt{2})(\sqrt{5} + \sqrt{2}) = (\sqrt{5})^2 - (\sqrt{2})^2 = 5 - 2 = 3$

$$\frac{1}{\sqrt{5} - \sqrt{2}} = \frac{\sqrt{5} + \sqrt{2}}{(\sqrt{5} - \sqrt{2})(\sqrt{5} + \sqrt{2})} = \frac{\sqrt{5} + \sqrt{2}}{3}$$

Try Study Suggestion 1.16. □

▶ **Study Suggestion 1.16:** Simplify the expression

$$\frac{2}{\sqrt{3} + \sqrt{2}}.$$ ◀

EXAMPLE 8: Simplify the following expressions. This includes eliminating all negative exponents. (Assume that all variables represent nonzero real numbers.)

(a) $\sqrt{2xy}\sqrt{4y^{-3}}$ **(b)** $\sqrt[3]{\frac{1}{3x^2}}$

Solutions:

(a)

$$\begin{aligned} \sqrt{2xy}\sqrt{4y^{-3}} &= \sqrt{2xy \cdot 4y^{-3}} = 2\sqrt{2xy^{-2}} = 2\sqrt{2x}\sqrt{y^{-2}} \\ &= 2\sqrt{2x}\sqrt{(y^{-1})^2} = 2|y^{-1}|\sqrt{2x} = \frac{2\sqrt{2x}}{|y|} \end{aligned}$$

(b) Since

$$\sqrt[3]{\frac{1}{3x^2}} = \frac{\sqrt[3]{1}}{\sqrt[3]{3x^2}} = \frac{1}{\sqrt[3]{3x^2}}$$

we see that in order to simplify, we must rationalize the denominator. This can be done by multiplying the numerator and denominator by $\sqrt[3]{9x}$, to give

$$= \frac{\sqrt[3]{9x}}{\sqrt[3]{3x^2}\sqrt[3]{9x}} = \frac{\sqrt[3]{9x}}{\sqrt[3]{3x^2 \cdot 9x}} = \frac{\sqrt[3]{9x}}{\sqrt[3]{27x^3}} = \frac{\sqrt[3]{9x}}{3x}$$

Try Study Suggestion 1.17. □

▶ **Study Suggestion 1.17:** Simplify the following expressions. Assume that $x \neq 0$, $y \neq 0$ and $z > 0$.

(a) $\sqrt{9x^{-2}y^8}$

(b) $\sqrt{50x^6}\sqrt{2y^2}$

(c) $\sqrt{\frac{1}{8z}}$

(d) $\sqrt[3]{\frac{x^5}{4y^4}}$ ◀

Now that we have discussed the case $a^{1/n}$, we can consider the general case $a^{m/n}$.

DEFINITION

Let a be a nonzero real number, let n be a *positive* integer, and let m/n be a rational number. Then if $a^{1/n}$ $(=\sqrt[n]{a})$ exists, we define $a^{m/n}$ by

$$a^{m/n} = (a^{1/n})^m$$

or in radical notation

$$a^{m/n} = (\sqrt[n]{a})^m$$

Before turning to examples, let us make a few remarks about this definition. The denominator n of the exponent m/n must be positive. However, since the numerator may be either positive or negative, this definition does include negative exponents. For instance, we have

$$4^{-3/2} = (4^{1/2})^{-3} = 2^{-3} = \frac{1}{8}$$

Also if $m > 0$ and if $a^{1/n}$ exists, then the properties of exponents give

$$(a^{1/n})m = \underbrace{(a^{1/n})(a^{1/n}) \cdots (a^{1/n})}_{m\ a^{1/n}\text{'s}} = \underbrace{(a \cdot a \cdots a)}_{m\ a\text{'s}}^{1/n} = (a^m)^{1/n}$$

A similar result holds when $m < 0$, and so we can write

$$a^{m/n} = (a^{1/n})^m = (a^m)^{1/n}$$

or in radical notation

$$a^{m/n} = (\sqrt[n]{a})^m = \sqrt[n]{a^m}$$

Finally, we should observe that in some cases this definition does not apply directly, but it can be made to apply with a little preliminary work.

For example, since $(-8)^{1/6}$ does not exist, we cannot apply the definition *directly* to evaluate $(-8)^{2/6}$. However, by first reducing the exponent to lowest terms, we get

$$(-8)^{2/6} = (-8)^{1/3} = -2$$

In order to avoid such problems, you should

reduce the exponent m/n to lowest terms before attempting to evaluate $a^{m/n}$

EXAMPLE 9: Evaluate the following expressions.

(a) $8^{4/6}$ **(b)** $(-27)^{2/3}$ **(c)** $(-1)^{3/4}$ **(d)** $64^{(-3/4)}$

Solutions:

(a) In order to apply the definition, we must first reduce the exponent to lowest terms

$$8^{4/6} = 8^{2/3} = (8^{1/3})^2 = 2^2 = 4$$

(b) $(-27)^{2/3} = ((-27)^{1/3})^2 = (-3)^2 = 9$

(c) $(-1)^{3/4} = ((-1)^{1/4})^3$ is not defined since $(-1)^{1/4}$ is not defined. (Notice that this exponent is already in lowest terms.)

(d) $64^{(-2/3)} = (64^{1/3})^{-2} = 4^{-2} = \dfrac{1}{4^2} = \dfrac{1}{16}$ *Try Study Suggestion 1.18.* □

▶ **Study Suggestion 1.18:** Evaluate the following expressions

(a) $(16)^{10/8}$
(b) $(16)^{-5/4}$
(c) $(-8)^{-2/3}$ ◀

EXAMPLE 10: Simplify the following expression. (Assume that all variables represent positive real numbers.)

$$\left(\frac{2x^{1/3}}{y^{1/2}}\right)^2\left(\frac{3x^{2/3}}{y^{-2}}\right)$$

Solution:

$$\left(\frac{2x^{1/3}}{y^{1/2}}\right)^2\left(\frac{3x^{2/3}}{y^{-2}}\right) = \left(\frac{4x^{2/3}}{y}\right)\left(\frac{3x^{2/3}}{y^{-2}}\right)$$
$$= \frac{12x^{4/3}}{y^{-1}}$$
$$= 12x^{4/3}y$$

Try Study Suggestion 1.19. □

▶ **Study Suggestion 1.19:** Simplify the following expressions.

(a) $(8x^3y^4z^6)^{1/2}$
(b) $\left(\dfrac{a^{1/2}}{4b^2}\right)^3\left(\dfrac{10a^{-2}}{ab^{-5}}\right)^2$ ◀

When simplifying expressions involving both radicals and rational exponents, it is usually a good idea to convert all expressions to exponent notation. Then if desired, the final expression can be rewritten using radical notation.

EXAMPLE 11: Simplify the following expression. (All variables represent positive real numbers.)

$$\frac{\sqrt{a^3b^3}}{\sqrt[3]{a^4b^{-2}}}$$

Solution:

$$\begin{aligned}\frac{\sqrt{a^3b^3}}{\sqrt[3]{a^4b^{-2}}} &= \frac{a^{3/2}b^{3/2}}{a^{4/3}b^{-2/3}}\\ &= a^{3/2-4/3}b^{3/2-(-2/3)}\\ &= a^{1/6}b^{13/6}\\ &= a^{1/6}b^2b^{1/6}\\ &= b^2\sqrt[6]{ab}\end{aligned}$$

Try Study Suggestion 1.20. □

▶ **Study Suggestion 1.20:** Simplify the following expressions.

(a) $\sqrt{ab^5}\,\sqrt[3]{a^2b}$

(b) $\dfrac{\sqrt{xy^2z^3}}{\sqrt[3]{xz^{-1}}}$ ◀

One of the most important uses for exponents is to express very large and very small numbers in a convenient way. Such numbers occur frequently in the sciences, for example. When a positive number is expressed in the form

$$a \times 10^n \tag{1.1}$$

where a is a number satisfying $1 \le a < 10$, and n is any integer, we say that the number is expressed in **scientific notation.** (In scientific notation, it is customary to use the symbol $\times$ for multiplication.)

As an example of scientific notation, light travels approximately

946,000,000,000,000,000 cm

in one year. (This is one *light year*. The abbreviation cm stands for centimeters.) In order to avoid having to write all of these zeros, we can express this number in scientific notation as

$$9.46 \times 10^{17} \text{ cm}$$

Scientific notation also makes it much easier to compare large and small numbers. Table 1.1 gives several very large and very small numbers, all expressed in scientific notation for easy comparison.

TABLE 1.1

(All sizes are measured in centimeters.)

Radius of a proton	1.00×10^{-13}
Average wavelength of visible light	1.41×10^{-8}
Size of polio virus	1.23×10^{-6}
Average size of a bacterium	1.00×10^{-4}
Diameter of the earth	1.28×10^{9}
Light year	9.46×10^{17}
Diameter of the Milky Way	1.02×10^{23}

As the next example shows, scientific-like notation (where the number a in (1.1) need not satisfy $1 \le a < 10$) can help make it easier to do arithmetic with large and small numbers.

EXAMPLE 12: Perform the indicated operations and simplify

$$\frac{(64{,}000)^{1/3}(10{,}000)^{1/2}}{(32{,}000)^{3/5}}$$

Solution: We begin by expressing all numbers in scientific-like notation. Then we can use the properties of exponents to simplify.

$$\begin{aligned}\frac{(64{,}000)^{1/3}(10{,}000)^{1/2}}{(32{,}000)^{3/5}} &= \frac{(64 \times 10^3)^{1/3}(10^4)^{1/2}}{(32 \times 10^3)^{3/5}} \\ &= \frac{(4 \times 10)(10^2)}{8 \times 10^{6/5}} \\ &= \frac{4 \times 10^3}{8 \times 10^{6/5}} \\ &= 0.5 \times 10^{3-6/5} \\ &= 0.5 \times 10^{9/5} \\ &= 5 \times 10^{4/5}\end{aligned}$$

Try Study Suggestion 1.21. □

▶ **Study Suggestion 1.21:** Simplify the expression

$$\frac{\sqrt{16{,}000}\,\sqrt[4]{8100}}{\sqrt{9000}}.$$ ◀

Let us conclude this section by remarking that most scientific calculators can display numbers (and do arithmetic) in scientific-like notation. For example, one such calculator displays the result of computing $(750{,}000)^2$ in the form

5.625E11

where the E11 means $\times 10^{11}$. (Some calculators do not display the letter "E".)

Try Study Suggestion 1.22.

Study Suggestion 1.22: If you have a calculator capable of working with scientific notation, compute the following numbers.

(a) $(450{,}000)^2$
(b) $(0.000078)^3$
(c) $(19{,}000)^3(23{,}000)^2$
(d) $(1.2 \times 10^5)(6.1 \times 10^8)$
(e) $(53 \times 10^{-7})^4$
(f) $(6.4 \times 10^6)^{-5}(1.4 \times 10^{42})$ ◀

EXERCISES

In Exercises 1–13, simplify the given expression by writing it in the form a/b where a and b are integers.

1. 3^{-3}
2. $(-2)^5$
3. $(17642)^0$
4. $(1/2)^4$
5. $(-1/3)^{-4}$
6. $\dfrac{2^{-3}}{3^{-2}}$
7. $\left(\dfrac{2}{3}\right)^{-3}\left(\dfrac{4}{3}\right)^2$
8. $2^{-3} + (-3)^2$
9. $4^0 + 0^4$
10. $12(4^2 \cdot 3^3)^{-2}$
11. $\dfrac{2^3 \cdot 3^5}{4^2 \cdot 9^2}$
12. $(3 + 2)^2 - 3^2 - 2^2$
13. $(2^{-1} - 3^{-1})(2 - 3)$

In Exercises 14–25, simplify the given expression by writing it so that no letter appears more than once, and so that it has no negative exponents.

14. $(3a)^3 - 3a^3$

15. $\dfrac{5p^2qrs^3}{10qsr^{-3}}$

16. $(a^2b^3c)^{-1}(ab^{-2}c)^2$

17. $\dfrac{(-6y^2)^3}{(9y^3)^2}$

18. $(a^{-1})^{-1}$

19. $[xy^{-1}]^{-1}x$

20. $\left(\dfrac{x^{-2}}{y^{-2}}\right)\left(\dfrac{x}{y}\right)^2$

21. $(2^{-1}x^3y^{-2}z)^{-3}$

22. $(x^{-2} - y^{-2})(x^2 + y^2)$

23. $(a^{-1}b^{-1})^{-2}$

24. $(-2xy^2z^{-1})^{-1}xy^2z$

25. $\dfrac{(a^2b^3)^{-1}}{a^{-2}b^{-3}}$

In Exercises 26–30, use a calculator to approximate the given expression.

26. $(2.34)^5$

27. $(7.86)^{-8}$

28. $(4.231)^2(6.98)^{-3}$

29. $\dfrac{(1.114)^{10} - (2.639)^9}{(4.183)^2 - (6.194)^3}$

30. $(2.413)^{-3} - (4.981)^{-4}$

31. How could you compute 3^4 using only the x^2 key (and the numeral keys) on your calculator? What other powers of 3 could you compute in this way?

In Exercises 32–61, simplify the given expression as much as possible.

32. $16^{1/4}$

33. $(-8)^{2/3}$

34. $12^{-1/2}$

35. $(-48)^{-1/3}$

36. $32^{-4/5}$

37. $\sqrt[6]{64}$

38. $\sqrt[3]{-1}$

39. $(-1)^{3/7}$

40. $\sqrt{75}$

41. $\sqrt[3]{80}$

42. $\sqrt[5]{-96}$

43. $(-21)^{3/4}$

44. $\sqrt{48} + \sqrt{12}$

45. $6\sqrt{12} - \sqrt{75} + 3\sqrt{27}$

46. $\sqrt{\sqrt{16}}$

47. $\sqrt[3]{\sqrt{128}}$

48. $\dfrac{1}{\sqrt{2}}$

49. $\dfrac{5}{\sqrt{5}}$

50. $\dfrac{6}{\sqrt{12}}$

51. $\dfrac{-3}{\sqrt[3]{9}}$

52. $\sqrt[4]{\dfrac{32}{81}}$

53. $\sqrt[5]{\dfrac{-5}{64}}$

54. $\left(\dfrac{8}{27}\right)^{-2/3}$

55. $\left(\dfrac{1}{40}\right)^{-1/3}$

56. $\dfrac{1}{1 + \sqrt{2}}$

57. $\dfrac{-3}{\sqrt{2} + \sqrt{7}}$

58. $\dfrac{\sqrt{2}}{2 + \sqrt{2}}$

59. $\dfrac{\sqrt{2} + \sqrt{3}}{\sqrt{2} - \sqrt{3}}$

60. $\dfrac{1}{\sqrt{2}} + \dfrac{2}{\sqrt{8}} - \sqrt{2}$

61. $\dfrac{2}{\sqrt[3]{3}} + \dfrac{1}{\sqrt[3]{24}} - \dfrac{1}{\sqrt[3]{81}}$

In Exercises 62–99, simplify the expression as much as possible. Assume that all variables represent positive real numbers.

62. $\sqrt{4x^{-2}y^6}$

63. $\sqrt{9a^4b^{-2}}$

64. $\sqrt[3]{8x^{-3}y^6}$

65. $\sqrt{x^6\sqrt[4]{x}}$

66. $\sqrt[3]{y\sqrt{y^3}}$

67. $\sqrt[3]{ab^2\sqrt{a^2b}}$

68. $\sqrt[5]{p^4q^3\sqrt[10]{q^{-8}}}$

69. $\sqrt[5]{16xyz^2\sqrt[5]{2x^4y^{-6}z^8}}$

70. $\sqrt{\dfrac{1}{4a^2b^{-4}}}$

71. $\sqrt{\dfrac{8x^2y}{2x^4y^{-1}}}$

72. $\sqrt[3]{\dfrac{54x^5}{y^{12}}}$

73. $\sqrt[5]{\dfrac{-64}{y^2}}$

74. $\sqrt{\left(\dfrac{2x^{-2}}{y^{-1}}\right)^{-2}}$

75. $\sqrt{\sqrt{x^2}}$

76. $\left(\dfrac{\sqrt{a}}{\sqrt[3]{b}}\right)^6$

77. $\left(\dfrac{a^{1/4}}{b^{1/6}}\right)^{-12}$

78. $(2a^{1/2}b^{-1/3})(a^{3/2}b^{-2/3})$

79. $(6^{-1}x^{-1/2}y^{1/3})^6$

80. $(4x^{4/3}y^{8/3})^{3/4}$

81. $\left(\dfrac{x^{-4/3}}{x^{-5/3}}\right)^2$

82. $(4xy^{1/2}(x^2y^{-1})^2x)^2$

83. $\dfrac{\sqrt{8x^3y^5}}{\sqrt{xy}}$

84. $\dfrac{\sqrt{a^4b^2c}}{\sqrt{\sqrt{abc}}}$

85. $\left(\dfrac{a^{-2/3}b^{1/6}}{a^{1/2}c}\right)^{-6}$

86. $\left(\dfrac{a^{-2/3}b^{1/6}c}{a^{1/4}b^{-5/4}c^{1/2}}\right)^{12}$

87. $x^{1/2}(x^{-1/2} + x^{3/2})$

88. $x^{1/2}(x^{1/2} + x^{-1/2})$

89. $(2xy^{2/3}z^{-1})^{1/2}(8x^{3/2}y^2z^{3/2})^{1/3}$

90. $\sqrt{3u^{1/2}v^{2/3}}\sqrt{3u^{1/3}v^{-1/2}}$

91. $(1 + a)^{1/2}(1 + a)^{3/4}$

92. $(x^{-3/2} + x^{3/2})(x^{-3/2} - x^{3/2})$

93. $\dfrac{c^{1/2}c^{-3/4}c^{-1/2}}{c^{-2}c^{5/4}}$

94. $\dfrac{x^{1/3}y^{2/3}z^{3/4}}{x^{4/3}y^{-1/3}z^{1/4}}$

95. $y^{1-b}y^b$

96. $\dfrac{(z^a)^a}{(z^{-a})^{-a}}$

97. $\dfrac{(p^{2b/3})(p^{-b})^{-1/3}}{(p^{5/3})^b}$

98. $\dfrac{(x^{1/n})(x^{1/m})}{x^{-m/n}}$

99. $\dfrac{((x^n)^n)^n}{x^{3n}}$

In Exercises 100–103, write the number in scientific notation.

100. **(a)** 12000 **(b)** 240300000

101. **(a)** 0.0000043 **(b)** 0.00000000001

102. **(a)** 244.00000 **(b)** 3.900000000

103. **(a)** 2.65001 **(b)** 375.00009201

104. Express the radius of a proton in decimal form. (Use Table 1.1.)

105. Express the size of the polio virus in decimal form. (Use Table 1.1.)

106. The mass of an electron is approximately 91.095×10^{-29} grams and the mass of a proton is approximately 1.673×10^{-28} kilograms. Which particle is heavier?

In Exercises 107–111, perform the indicated operations and simplify.

107. $(0.0003)^3(-12000)^2$

108. $(4{,}000{,}000{,}000)(0.0000000022)$

109. $\dfrac{\sqrt{40000}\sqrt[3]{8000000}}{\sqrt[5]{(0.00243)^3}}$

110. $\dfrac{(64{,}000{,}000{,}000)^{1/3}(16{,}000{,}000)^{1/2}}{(3{,}200{,}000)^{2/5}}$

111. $\dfrac{(0.00021)^2(2{,}000)^{-1}}{\sqrt[3]{(27{,}000)^2}}$

In Exercises 112–121, use a calculator to approximate the given value.

112. $\sqrt{1 + \sqrt{2}}$

113. $\sqrt[3]{\sqrt{2} - \sqrt{5}}$

114. $\sqrt{6 + 7\sqrt[3]{4}}$

115. $\dfrac{\sqrt{2}}{\sqrt{5} - \sqrt{12}}$

116. $\dfrac{-4\sqrt{18}}{4\sqrt{12} - 4\sqrt{25}}$

117. $(0.91)^{9/2}$

118. $(4.56)^{-10/3}$

119. $(2.776)^{-1.557}$

120. $(1.28)^{0.76} - (2.54)^{1.62}$

121. $\dfrac{(2.338)^{5.71}}{(1.998)^{6.01}}$

In Exercises 122–124, use a decimal approximation to determine which number is bigger.

122. $(1.23)^{2.45}, (1.67)^{0.99}$

123. $(3.14)^{2.78}, (2.78)^{3.14}$

124. $(0.2658)^{-0.661}, (4.993)^{0.523}$

In Exercises 125–128, perform the indicated operations, entering the numbers exactly as shown (that is, in scientific-like notation).

125. $(2.45 \times 10^5)(3.19 \times 10^4)$

126. $(256 \times 10^{-3})(486 \times 10^{-12})$

127. $(6.34 \times 10^{21})^3$

128. $\dfrac{(2.98 \times 10^{-6})^{-3}}{(4.2 \times 10^8)^2}$

1.3 Algebraic Expressions, Polynomials, and Factoring

An **algebraic expression** is any combination of variables and constants formed by using the arithmetic operations of addition, subtraction, multiplication, division, and exponentiation (this includes taking powers and roots.) For example, the following are algebraic expressions.

$$2 + x, \quad x^{-1/3} - \pi ay, \quad \frac{x^4 - y^5}{cxy}, \quad \frac{3x\sqrt{x+y}}{\sqrt{x^2 - y^2}}$$

One of the most important types of algebraic expression is the *polynomial.*

DEFINITION

Let x be a variable, let n be a nonnegative integer, and let $a_n, a_{n-1}, \ldots, a_1, a_0$ be real numbers. Then the algebraic expression

$$p(x) = a_n x^n + a_{n-1}x^{n-1} + \cdots + a_1 x + a_0$$

is called a **polynomial** in the variable x. The constants $a_n, a_{n-1}, \ldots, a_0$ are called the **coefficients** of the polynomial. (The expression a_n is read "a sub n," and similarly for the others. The expression a_0 is read "a sub 0" or "a nought.")

EXAMPLE 1: The following are polynomials.

(a) $p(x) = 5$ **(b)** $q(x) = 2x - 3$

(c) $r(x) = 4x^2 - 2x + 1$ **(d)** $s(x) = -(1/5)x^8 + \sqrt{2}x$ □

Each expression of the form $a_k x^k$ is called a **term** of the polynomial. The term a_0 is called the **constant term** and, provided that $a_n \neq 0$, the term $a_n x^n$ is called the **leading term.** In Example 1(b), for instance, the constant term is -3, and the leading term is $2x$. In Example 1(d), the constant term is 0 and the leading term is $-(1/5)x^8$.

The **degree** of a polynomial $p(x)$ is the largest exponent that appears in the polynomial *with a nonzero coefficient.* We denote the degree of $p(x)$ by $\deg(p(x))$. For example,

$$\deg(-3x^2 + 2x - 1) = 2$$
$$\deg(17x^6 - 12x^3) = 6$$
$$\deg(4x) = 1 \qquad (\text{since } 4x = 4x^1)$$
$$\deg(-3) = 0 \qquad (\text{since } -3 = -3x^0)$$

The **zero polynomial** $p(x) = 0$ is the only polynomial that does not have a degree.

▶ **Study Suggestion 1.23:** Find:

(a) $\deg(x^3 - 1)$
(b) $\deg(3x + x^2)$
(c) $\deg(-1)$
(d) $\deg(0)$ ◀

Try Study Suggestion 1.23.

A polynomial of degree 0 is called a **constant polynomial;** a polynomial of degree 1 is called a **linear polynomial;** a polynomial of degree 2 is called a **quadratic polynomial;** and a polynomial of degree 3 is called a **cubic polynomial.** These terms will be useful to us later in the book.

The following examples illustrate addition, subtraction, and multiplication of polynomials. (We will discuss division of polynomials in Chapter 4.)

EXAMPLE 2: Perform the following operations and simplify the result.

(a) $(-5x^3 - x^2 + 1) + (3x^3 + x^2 - 3x + 2)$

(b) $(2x^3 + 2x^2 - 3x) - (x^3 - 5x^2 - 3x + 4)$

Solutions:

(a) Adding like terms gives

$$\begin{aligned}(-5x^3 - x^2 + 1) &+ (3x^3 + x^2 - 3x + 2) \\ &= -5x^3 + 3x^3 - x^2 + x^2 - 3x + 1 + 2 \\ &= -2x^3 - 3x + 3\end{aligned}$$

(b) Subtracting like terms gives

$$\begin{aligned}(2x^3 + 2x^2 - 3x) &- (x^3 - 5x^2 - 3x + 4) \\ &= 2x^3 - x^3 + 2x^2 - (-5x^2) - 3x - (-3x) - 4 \\ &= 2x^3 - x^3 + 2x^2 + 5x^2 - 3x + 3x - 4 \\ &= x^3 + 7x^2 - 4\end{aligned}$$

Try Study Suggestion 1.24. □

▶ **Study Suggestion 1.24:** Perform the indicated operations and simplify.

(a) $(x^2 - 3x + 1) + (6x^2 - x + 4)$
(b) $(5x^2 - 3) - (x^2 - x - 1)$ ◀

In order to multiply polynomials, we must multiply every term of one polynomial by every term of the other.

EXAMPLE 3: Compute the product $(2x + 1)(-5x^2 + x)$. Simplify the result.

Solution:

$$\begin{aligned}(2x + 1)(-5x^2 + x) &= (2x)(-5x^2) + (2x)(x) + (1)(-5x^2) + (1)(x) \\ &= -10x^3 + 2x^2 - 5x^2 + x \\ &= -10x^3 - 3x^2 + x\end{aligned}$$

Try Study Suggestion 1.25. □

▶ **Study Suggestion 1.25:** Compute the product

$$(x^2 + 1)(3x^2 - 2x + 3)$$

and simplify the result. ◀

Some products of polynomials occur so often that it is worth making a special list.

Some Common Products of Polynomials

1. $(x + a)^2 = x^2 + 2ax + a^2$
2. $(x - a)^2 = x^2 - 2ax + a^2$
3. $(x + a)(x - a) = x^2 - a^2$
4. $(x + a)^3 = x^3 + 3ax^2 + 3a^2x + a^3$
5. $(x - a)^3 = x^3 - 3ax^2 + 3a^2x + a^3$

As the next example illustrates, we can add, subtract, and multiply polynomials in more than one variable in the same way that we did for polynomials in one variable.

EXAMPLE 4: Perform the following operations and simplify the result.

(a) $(2x^2 + xy - y^3) + (3x^2 - 4xy - 5x^2y + 6y^3)$

(b) $(x + y)(x - y + z)$

Solutions:

(a) Adding like terms gives

$$\begin{aligned}&(2x^2 + xy - y^3) + (3x^2 - 4xy - 5x^2y + 6y^3)\\ &\quad = 2x^2 + 3x^2 + xy - 4xy + -5x^2y + -y^3 + 6y^3\\ &\quad = 5x^2 - 3xy - 5x^2y + 5y^3\end{aligned}$$

(b) Using a vertical notation, and reversing the order of the factors, we get

$$\begin{array}{llll} x - y + z & & & \\ \underline{x + y} & & & \\ x^2 - xy + xz & & & \\ \underline{\quad + xy} & & \underline{- y^2 + yz} & \\ x^2 & + xz & - y^2 + yz & \end{array}$$

Try Study Suggestion 1.26. □

► **Study Suggestion 1.26:** Perform the indicated operations and simplify the result

$[(2x + 3y^2) - (4x - y^2)][4y^2 + 2x]$. ◄

We can also add, subtract, and multiply other algebraic expressions in a manner similar to that of polynomials.

EXAMPLE 5: Perform the indicated operations and simplify the result.

(a) $(2\sqrt{x} + y^{-2} + z^2) + 3(\sqrt{x} - 4y^{-2} + 2z^2 + z - 3)$

(b) $(\sqrt{x} + \sqrt{y})(\sqrt{x} - \sqrt{y})$

(c) $\left(\sqrt{u} + \frac{1}{\sqrt{u}}\right)^2$

Solutions:

(a) We distribute the coefficient 3 in the second expression and then combine like terms.

$$\begin{aligned}(2\sqrt{x} + y^{-2} + z^2) &+ 3(\sqrt{x} - 4y^{-2} + 2z^2 + z - 3)\\ &= (2\sqrt{x} + y^{-2} + z^2) + 3\sqrt{x} - 12y^{-2} + 6z^2 + 3z - 9\\ &= 5\sqrt{x} - 11y^{-2} + 7z^2 + 3z - 9\end{aligned}$$

(b) Here we have

$$\begin{aligned}(\sqrt{x} + \sqrt{y})(\sqrt{x} - \sqrt{y}) &= \sqrt{x}\sqrt{x} + \sqrt{x}\sqrt{y} - \sqrt{y}\sqrt{x} - \sqrt{y}\sqrt{y}\\ &= (\sqrt{x})^2 - (\sqrt{y})^2\\ &= x - y\end{aligned}$$

▶ **Study Suggestion 1.27:** Perform the indicated operations and simplify the result.

(a) $(x^{-2} + y^2)(x^2 + y^{-2})$

(b) $\left(\sqrt{z} - \frac{1}{\sqrt{z}}\right)^2$ ◀

(c) In this case

$$\begin{aligned}\left(\sqrt{u} + \frac{1}{\sqrt{u}}\right)^2 &= (\sqrt{u})^2 + 2\sqrt{u}\left(\frac{1}{\sqrt{u}}\right) + \left(\frac{1}{\sqrt{u}}\right)^2\\ &= u + 2 + \frac{1}{u}\end{aligned}$$

Try Study Suggestion 1.27. □

Now let us review the concept of factoring.

DEFINITION

> Let $p(x)$ be a polynomial. Suppose that $p(x)$ can be written as the product of the polynomials $p_1(x), p_2(x), \ldots, p_n(x)$, in symbols,
>
> $$p(x) = p_1(x)p_2(x) \cdots p_n(x)$$
>
> Then we say that $p(x)$ has been **factored** into the product $p_1(x)p_2(x) \cdots p_n(x)$, and that the polynomials $p_1(x), p_2(x), \ldots, p_n(x)$ are **factors** of $p(x)$.

If a polynomial cannot be factored into the product of two (or more) polynomials of *positive* degree, then we say that it is **irreducible.** For example, the polynomial $x^2 + 1$ is irreducible, whereas the polynomial $x^2 - 1$ is not irreducible, since it can be factored as follows $x^2 - 1 = (x - 1)(x + 1)$.

Some Techniques for Factoring Polynomials

Common factors. The first thing that you should look for in trying to factor a polynomial is common factors; that is, factors that are common to every term in the polynomial. For example, $2x^2$ is a common factor in the polynomial $14x^3 + 4x^2$ and so we can write

$$14x^3 + 4x^2 = 2x^2(7x + 2)$$

Some additional examples should give you the idea

$$4x^2 - 6x + 12 = 2(2x^2 - 3x + 6)$$
$$14x^5 - 12\sqrt{2}x = 2x(7x^4 - 6\sqrt{2})$$
$$3ab^2x^5 - 2a^2bx^4 + abx^3 = abx^3(3bx^2 - 2ax + 1)$$

▶ **Study Suggestion 1.28:** Factor the polynomial

$$5x^3y^4 - 15x^2y.$$ ◀

Try Study Suggestion 1.28.

Using special products. The special products given earlier in the section can be used to factor polynomials. The formula

$$x^2 - a^2 = (x + a)(x - a)$$

is sometimes called the formula for factoring a *difference of squares*, since the left side of this formula, $x^2 - a^2$ is a **difference of two squares.** Let us consider some examples of the use of this formula in factoring.

EXAMPLE 6:

(a) $x^2 - 25 = x^2 - 5^2 = (x + 5)(x - 5)$

(b) $4x^2 - 9a^2 = (2x)^2 - (3a)^2 = (2x + 3a)(2x - 3a)$

Try Study Suggestion 1.29. □

▶ **Study Suggestion 1.29:** Factor the polynomials

(a) $4x^2 - 1$

(b) $9x^2 - 16.$ ◀

Two formulas that are very useful for factoring are the formulas for factoring *the sum of two cubes*

$$x^3 + a^3 = (x + a)(x^2 - ax + a^2)$$

and *the difference of two cubes*

$$x^3 - a^3 = (x - a)(x^2 + ax + a^2)$$

For example, the first formula gives

$$x^3 + 64 = (x + 4)(x^2 - 4x + 16)$$

and the second formula gives

$$8x^3 - 27 = (2x)^3 - 3^3 = (2x - 3)(4x^2 + 6x + 9)$$

▶ **Study Suggestion 1.30:** Factor the polynomials

(a) $x^3 + 27$

(b) $125z^3 - 1.$ ◀

Try Study Suggestion 1.30.

Factoring a quadratic polynomial $ax^2 + bx + c$. If a quadratic polynomial can be factored into nonconstant factors, then each of the factors must be a linear polynomial. For example, if the polynomial $3x^2 + x - 10$ can be factored into nonconstant factors, then the factorization must have the form

$$3x^2 + x - 10 = (\_\_x + \_\_)(\_\_x + \_\_)$$

where the blanks are to be filled in by constants (real numbers). However, in taking the product on the right-hand side, we see that the product of the coefficients of x must equal 3 (which is the coefficient of x^2 on the left-hand side) and the product of the constant terms on the right must equal -10 (which is the constant term on the left)

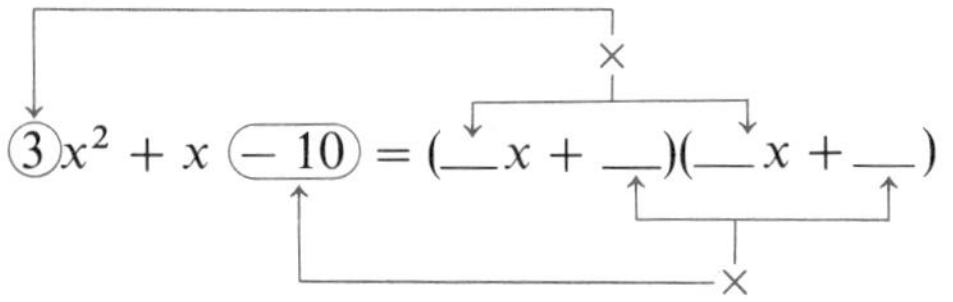

$$\boxed{3}x^2 + x \boxed{-10} = (\_\_x + \_\_)(\_\_x + \_\_)$$

This information can be used to make intelligent guesses as to what numbers to use to fill in the blanks. For example, one possible guess would be $(3x - 10)(x + 1)$, for in this case the product of the coefficients of x is $3 \cdot 1 = 3$, and the product of the constant terms is $(-10) \cdot 1 = -10$. To check this guess, we simply multiply

$$\begin{aligned}(3x - 10)(x + 1) &= 3x^2 - 10x + 3x - 10 \\ &= 3x^2 - 7x - 10\end{aligned}$$

This guess is wrong, so we try another one. This time, let us try $(3x - 5)(x + 2)$.

$$\begin{aligned}(3x - 5)(x + 2) &= 3x^2 - 5x + 6x - 10 \\ &= 3x^2 + x - 10\end{aligned}$$

This is the correct guess, and we have the factorization

$$3x^2 + x - 10 = (3x - 5)(x + 2)$$

▶ **Study Suggestion 1.31:** Factor the following polynomials

(a) $x^2 - 3x - 4$

(b) $x^2 + \frac{7}{4}x - \frac{1}{2}$

(c) $x^2 + ax - 6a^2$ ◀

Try Study Suggestion 1.31.

Miscellaneous techniques. There are a few additional techniques that can be very helpful in factoring.

1. Sometimes it is helpful to rearrange the order of the terms in a polynomial. For instance, consider the polynomial

$$ax + 2y + 2x + ay$$

We can factor this by grouping together the terms involving x, as well as those involving y

$$\begin{aligned}ax + 2y + 2x + ay &= (ax + 2x) + (ay + 2y) \\ &= (a + 2)x + (a + 2)y \\ &= (a + 2)(x + y)\end{aligned}$$

▶ **Study Suggestion 1.32:** Factor the polynomial

$$ax + 2y - x - 2ay.$$ ◀

Try Study Suggestion 1.32.

2. Sometimes it is helpful to remove a factor that is common to only some of the terms of a polynomial. As an example, we have

$$\begin{aligned} 3x^5 - 6x^4 - x + 2 &= 3x^4(x - 2) - x + 2 \\ &= 3x^4(x - 2) - (x - 2) \\ &= (3x^4 - 1)(x - 2) \end{aligned}$$

▶ **Study Suggestion 1.33:** Factor the polynomial

$$2x^4 - x^3 - 2x + 1.$$ ◀

Try Study Suggestion 1.33.

3. Consider the polynomial

$$x^2 - 4ax + 4a^2 - 4$$

The first three terms in this polynomial should catch your eye, since

$$x^2 - 4ax + 4a^2 = x^2 - 4ax + (2a)^2 = (x - 2a)^2$$

Thus, we have

$$\begin{aligned} x^2 - 4ax + 4a^2 - 4 &= (x - 2a)^2 - 4 \\ &= (x - 2a)^2 - 2^2 \\ &= (x - 2a + 2)(x - 2a - 2) \end{aligned}$$

For the last equality, we used the fact that $(x - 2a)^2 - 2^2$ is the difference of two squares.

▶ **Study Suggestion 1.34:** Factor the polynomial $x^2 - 2xy + y^2 - 1$. ◀

Try Study Suggestion 1.34.

EXERCISES

In Exercises 1–15, perform the indicated operations and simplify the result, that is, express your answer as a single polynomial. Also, state the degree of the resulting polynomial.

1. $(-6x + x^2) + (3x - 4)$

2. $(4x^3 - 3x^2 + 9x + 1) + (6x^5 - 3x^2 - 9x + 11)$

3. $(4u^3 + 5u) + (3u^2 + 2u - 1) + (5u^3 - 2u + 1)$

4. $(6y - 4) - (3y - 2)$

5. $(4x^2 - 3x + 1) - (3x^2 - 2x + 9)$

6. $(4s^5 + 3s^2 - s + 1) - (3s^5 + s^2 + 10) + (6s^5 + 3s^2 - 4s - 2)$

7. $4y(2y^2 + 1) + 3y(-4y^2 + 2y - 1)$

8. $(4x + 1)(3x - 1) + 2x(x^2 + 4)$

9. $(4x^2 - 3x + 2)(6x + 1)$

10. $(6z^2 + 1)(3z^2 - 2)$

11. $(x^2 + 2x + 1)(2x^2 - 3x + 1)$

12. $(4x^2 + 2x)(3x^2 - 4x)$

13. $(4u^9 + 1)(7u^{10} - u^2 + 1)$

14. $(2x + 1)(3x - 1)(4x + 3)$

15. $(4y^2 + 1)(3y^2 - 1)(2y^2 + 5)$

In Exercises 16–22, perform the indicated operations and simplify the result.

16. $(2x^2 + 3y - 4) + (6x - 2y + 3)$

17. $(3x^2 - 4y^2 + 2xz) - (x^2 - 4y^2 + y + 2zy - xz)$

18. $(s^2 - t^2)(2s + 3t)$

19. $(u + v)(u^2 - uv + v^2)$

20. $(3x^2 + xy + 3y^2)(x + y)$

21. $(x + 1)(y + 1)(z + 1)$

22. $\dfrac{x^2y^2 + 3x^3y^3 + x^4y^4}{xy^2}$

In Exercises 23–33, perform the indicated operations and simplify.

23. $(\sqrt{x} + 1)^2$

24. $(\sqrt{x} + \sqrt{y})^2$

25. $\left(\sqrt{y} - \dfrac{1}{\sqrt{y}}\right)^2$

26. $(3\sqrt{x} - 2\sqrt{y})(3\sqrt{x} + 2\sqrt{y})$

27. $(x^{2/3} + x^{1/3} + 1)(2x^{1/3} - 1)$

28. $(2x^{-1} + 3x)(3x^{-1} + 2x)$

29. $(x^{-2} + 2x^{-1} + 1)(3x^{-2} + 4x^{-1} - 1)$

30. $(4x - 3x^{1/2} + 1)(2x^{1/2} - 1)$

31. $(4u^{1/2} - 1)(3u^{1/4} + 1)(u^{1/4} - 2)$

32. $(2\sqrt{x} - \sqrt{y})(2\sqrt{x} + \sqrt{y})(4x + y)$

33. $\left(\sqrt{x} + \dfrac{1}{\sqrt{x}}\right)^4$

In Exercises 34–81, factor the given expression.

34. $2xy - 3y$

35. $2xy^2 + 4xy$

36. $4u^2v^3 - 2uv^2 + uv$

37. $x(x + y) + y(-x - y)$

38. $-8p^4q^5r - 10p^2qr^3s$

39. $u(v + w) + 2u^2(v + w) - 3w - 3v$

40. $x^2 + x - 6$

41. $t^2 + 3t - 10$

42. $y^2 - 13y - 30$

43. $2y^2 + 9y - 5$

44. $\frac{2}{3}x^2 - \frac{2}{3}x - \frac{1}{2}$

45. $6a^2 - 6a - 16$

46. $x^2 + (\sqrt{2} + \sqrt{3})x + \sqrt{6}$

47. $x^2 + \pi x - 2\pi^2$

48. $4x^2 - 9y^2$

49. $36a^2 - 49b^2$

50. $16s^2 - 4t^2$

51. $9p^4 - 25q^6$

52. $\frac{3}{10}x^2 + xy - \frac{4}{5}y^2$

53. $30a^2 + 2ab - 12b^2$

54. $50x^2 + 45xy - 11y^2$

55. $20x^2 + 53xy + 12y^2$

56. $25z^2 + 40z + 16$

57. $\frac{9}{4} - 3z + z^2$

58. $16z^2 + 24zw + 9w^2$

59. $36x^2 + 60xy^2 + 25y^4$

60. $8x^3 - z^3$

61. $8x^3 + 27w^6$

62. $125a^3 - 64b^3$

63. $r^3 + (r - s)^3$

64. $216 - u^6$

65. $4xy + 2xz - 6uy - 3uz$

66. $3ab + 2ac - 6b^2 - 4bc$

67. $\frac{1}{2}rst + \frac{1}{2}rsuv - rt - ruv$

68. $x^3 - x^2y + xy^2 - y^3$

69. $6x^4 + x^2 - 2$

70. $4x^{10} - 5x^5 - 6$

71. $u^6 - v^6$

72. $x^2 + 2xy + y^2 - 16$

73. $4u^2 + 12uv + 9v^2 - 9$

74. $(a + b)^3 - (a - b)^3$

75. $2x^4 + 2x^3 - x - 1$

76. $-6x^5 + 3x^4 - 2x + 1$

77. $y^7 - 4y^5 + y^2 - 4$

78. $9y^4 + 6y^2 - 8$

79. $4z^4 + 4z^2 - 35$

80. $a^4 - b^4$

81. $4s^4 - \dfrac{81}{4t^4}$

1.4 Rational and Other Fractional Expressions

A **rational expression** in the variable x is an algebraic expression of the form

$$\frac{p(x)}{q(x)}$$

where $p(x)$ and $q(x)$ are polynomials, and $q(x)$ is not the zero polynomial. Rational expressions can contain polynomials in more than one variable as well. For example, the following are rational expressions

$$\frac{2x+3}{5x^3-1}, \quad \frac{x^3+ax^2-a^2}{x^2-3a^2x+\pi}, \quad \frac{\sqrt{3}x^2y+4xy^3}{x-y}$$

In this section, we want to discuss methods for simplifying rational (and other fractional) expressions, as well as for adding, subtracting, multiplying, and dividing them.

Before beginning however, we should make one important comment. During the course of our studies, we will be interested in substituting real numbers for the variables in algebraic expressions. For example, we may want to substitute a real number for x in the rational expression

$$\frac{x^2+1}{x-1}$$

It is important to remember that we cannot substitute any value for x that would make the denominator $x - 1$ equal to 0. In this case then, we cannot make the substitution $x = 1$. As another example, we cannot make the substitution $x = 2$ in the rational expression

$$\frac{x^2-4}{x-2}$$

since that would result in the meaningless expression 0/0. Similar restrictions occur when an algebraic expression involves roots, for we cannot take the square root (or any other even root) of a negative number.

Simplifying a rational expression usually amounts to factoring both the numerator and denominator in order to find any *common factors*. By common factor, we mean a polynomial that is a factor of both the numerator and the denominator. Any common factors can be cancelled. When a rational expression has no nonconstant common factors, we say that it is in **lowest terms.**

EXAMPLE 1: Simplify the rational expression

$$\frac{x^2+x-2}{x^2-1}$$

Solution: First we factor the numerator and denominator, and then cancel any common factors

$$\frac{x^2+x-2}{x^2-1} = \frac{(x-1)(x+2)}{(x-1)(x+1)} = \frac{x+2}{x+1} \qquad (\text{for } x \neq -1, 1)$$

The last rational expression has no nonconstant common factors, and so it is in lowest terms. Thus, we say that the original rational expression has been *reduced to lowest terms.*

Notice that we have included the restrictions $x \neq -1, 1$ since these values of x will make at least one of the denominators above equal to 0. It is important to notice that even though the value $x = 1$ *can* be substituted into the last rational expression above, it cannot be substituted into the original expression, since that would result in the meaningless expression 0/0. Hence, we must also exclude this value of x. *Try Study Suggestion 1.35.* □

▶ **Study Suggestion 1.35:** Simplify

(a) $\dfrac{(x-1)(x^2-x-6)}{x^2+x-2}$

(b) $\dfrac{x^2}{x}$ ◀

Before proceeding further, let us make the following agreement.

When dealing with algebraic expressions, we will assume that all values of the variables that make any denominators equal to 0, or cause us to take any square (or other even) roots of negative numbers, have been excluded from consideration.

Multiplication and division of rational expressions is similar to multiplication and division of rational numbers, since the properties of quotients given in Section 1.1 for real numbers hold also for rational expressions.

EXAMPLE 2: Perform the indicated operation and simplify.

$$\frac{x+2}{2x-3y} \cdot \frac{4x^2-9y^2}{xy+2y}$$

Solution:

$$\begin{aligned}\frac{x+2}{2x-3y} \cdot \frac{4x^2-9y^2}{xy+2y} &= \frac{(x+2)(4x^2-9y^2)}{(2x-3y)(xy+2y)}\\ &= \frac{(x+2)(2x+3y)(2x-3y)}{y(2x-3y)(x+2)}\\ &= \frac{2x+3y}{y}\end{aligned}$$

Try Study Suggestion 1.36. □

▶ **Study Suggestion 1.36:** Perform the indicated operation and simplify

$$\frac{x^2-y^2}{x^2+4x} \cdot \frac{x+4}{x^2-xy}. \quad ◀$$

EXAMPLE 3: Perform the indicated operation and simplify.

$$\frac{x^2-3x-10}{x-3} \div \frac{x-5}{x^2-2x-3}$$

Solution: In this case we first invert and multiply

$$\begin{aligned}\frac{x^2-3x-10}{x-3} \div \frac{x-5}{x^2-2x-3} &= \frac{x^2-3x-10}{x-3} \cdot \frac{x^2-2x-3}{x-5}\\ &= \frac{(x^2-3x-10)(x^2-2x-3)}{(x-3)(x-5)}\\ &= \frac{(x-5)(x+2)(x-3)(x+1)}{(x-3)(x-5)}\\ &= (x+2)(x+1)\end{aligned}$$

Try Study Suggestion 1.37. □

▶ **Study Suggestion 1.37:** Perform the indicated operation and simplify

$$\frac{x^2-1}{y} \div \frac{x+1}{y^2}. \quad ◀$$

In order to add or subtract rational expressions, we must first arrange it so that their denominators are the same. This can be done by finding a *least common denominator*. The procedure for finding a least common denominator is as follows.

Procedure for Finding a Least Common Denominator

1. Factor all denominators completely.
2. The least common denominator is obtained by taking the product of each factor appearing in *any* of the denominators, raised to the highest power with which that factor appears.

Let us illustrate with an example.

EXAMPLE 4: Perform the indicated operation and simplify.

$$\frac{1}{2x-4} - \frac{2}{x-1} + \frac{x+1}{(x-1)^2(x-2)}$$

Solution: In this case, only the first denominator needs factoring, and we have

$$\frac{1}{2(x-2)} - \frac{2}{x-1} + \frac{x+1}{(x-1)^2(x-2)}$$

Thus we see that the factors of the denominators are 2, $x - 1$, and $x - 2$. The factors 2 and $x - 2$ appear only to the first power, but the factor $x - 1$ appears to the second power. Hence, the least common denominator is $2(x-1)^2(x-2)$. Now by multiplying the numerator and denominator of each term by the appropriate expression, we arrange it so that each term has the same (common) denominator, then take the sum and simplify

$$\begin{aligned}
\frac{1}{2(x-2)} - \frac{2}{x-1} + \frac{x+1}{(x-1)^2(x-2)} \\
&= \frac{(x-1)^2}{2(x-1)^2(x-2)} - \frac{4(x-1)(x-2)}{2(x-1)^2(x-2)} + \frac{2(x+1)}{2(x-1)^2(x-2)} \\
&= \frac{(x-1)^2 - 4(x-1)(x-2) + 2(x+1)}{2(x-1)^2(x-2)} \\
&= \frac{-3x^2 + 12x - 5}{2(x-1)^2(x-2)}
\end{aligned}$$

Try Study Suggestion 1.38. □

▶ **Study Suggestion 1.38:** Perform the indicated operation and simplify.

$$\frac{1}{3x+6} + \frac{x+1}{x^2-4x+4} - \frac{x}{x^2-4}.$$

◀

Sometimes we are called upon to simplify fractional expressions that are not rational expressions. Some of the same techniques that we use to simplify rational expressions may also work in these cases.

EXAMPLE 5: Simplify the following expression

$$\frac{\frac{1}{x} - \frac{1}{x-1}}{\frac{1}{x} + \frac{x}{x-1}}$$

Solution: First we simplify the numerator and the denominator separately

$$\frac{\frac{1}{x} - \frac{1}{x-1}}{\frac{1}{x} + \frac{x}{x-1}} = \frac{\frac{(x-1) - x}{x(x-1)}}{\frac{(x-1) + x^2}{x(x-1)}}$$

$$= \frac{\frac{-1}{x(x-1)}}{\frac{x^2 + x - 1}{x(x-1)}}$$

$$= \frac{-1}{x(x-1)} \cdot \frac{x(x-1)}{x^2 + x - 1}$$

$$= \frac{-1}{x^2 + x - 1}$$

Try Study Suggestion 1.39. □

▶ **Study Suggestion 1.39:** Simplify the expression

$$\frac{x - \frac{1}{x}}{x+1}.$$ ◀

If an algebraic expression contains radicals (or fractional exponents) in its denominator, then we may need to rationalize the denominator.

EXAMPLE 6: Simplify the expression

$$\frac{x - y}{\sqrt{x} + \sqrt{y}}.$$

Solution: We multiply the numerator and denominator by $\sqrt{x} - \sqrt{y}$. Notice that this expression resembles the denominator, except that the sign has been changed. The reason that this works is that

$$(\sqrt{x} + \sqrt{y})(\sqrt{x} - \sqrt{y}) = (\sqrt{x})^2 - (\sqrt{y})^2 = x - y$$

and so this will clear the denominator of radical signs

$$\frac{x - y}{\sqrt{x} + \sqrt{y}} = \frac{(x - y)(\sqrt{x} - \sqrt{y})}{(\sqrt{x} + \sqrt{y})(\sqrt{x} - \sqrt{y})}$$

$$= \frac{(x - y)(\sqrt{x} - \sqrt{y})}{x - y}$$

$$= \sqrt{x} - \sqrt{y}$$

Try Study Suggestion 1.40. □

▶ **Study Suggestion 1.40:** Simplify the expression

$$\frac{1}{\sqrt{x} - \sqrt{y}}.$$ ◀

EXAMPLE 7: Perform the indicated operation and simplify.

$$\frac{1}{\sqrt{x-1}} + \sqrt{x-1}$$

Solution: We begin by forming a common denominator, which in this case is $\sqrt{x-1}$,

$$\begin{aligned}\frac{1}{\sqrt{x-1}} + \sqrt{x-1} &= \frac{1}{\sqrt{x-1}} + \frac{\sqrt{x-1}\sqrt{x-1}}{\sqrt{x-1}}\\ &= \frac{1}{\sqrt{x-1}} + \frac{x-1}{\sqrt{x-1}}\\ &= \frac{1+(x-1)}{\sqrt{x-1}}\\ &= \frac{x}{\sqrt{x-1}}\end{aligned}$$

Finally, we rationalize the denominator

$$\begin{aligned}&= \frac{x\sqrt{x-1}}{\sqrt{x-1}\sqrt{x-1}}\\ &= \frac{x\sqrt{x-1}}{x-1}\end{aligned}$$

Try Study Suggestion 1.41. □

▶ **Study Suggestion 1.41:** Perform the indicated operation and simplify

$$\frac{1}{1-\sqrt{x}} - \frac{1}{1+\sqrt{x}}.$$ ◀

EXERCISES

In Exercises 1–52, simplify the given expression. (This includes performing any indicated operations.)

1. $\dfrac{x^2+x-2}{2x^2+x-3}$

2. $\dfrac{2x^2-5x-3}{6x^2-5x-4}$

3. $\dfrac{15x^2+x-2}{12x^2+2x-2}$

4. $\dfrac{2-11x+12x^2}{2x-8x^2}$

5. $\dfrac{30+7y-2y^2}{18-27y+4y^2}$

6. $\dfrac{4-12y+9y^2}{4-9y^2}$

7. $\dfrac{16z^3-z}{16z^2+8z+1}$

8. $\dfrac{6z^3+8z^2}{12z^3-8z^2-32z}$

9. $\dfrac{4x}{x+1} - \dfrac{3}{2x-1}$

10. $\dfrac{3x}{3x+1} + \dfrac{2x}{2x+1}$

11. $\dfrac{4y-5}{2y+3} - \dfrac{y+1}{2y+4}$

12. $\dfrac{a-3}{2a+1} + \dfrac{2-3a}{4+6a}$

13. $\dfrac{z}{z+1} + \dfrac{1}{z-1} - \dfrac{z+1}{z+2}$

14. $\dfrac{2b}{3b+3} - \dfrac{4}{5+2b} + \dfrac{b-1}{b+1}$

15. $\dfrac{3x+1}{2x-1}\,\dfrac{x+4}{3x-1}$

16. $\dfrac{x-1}{6x^2+x-2}\,\dfrac{2x-1}{4x^2+x-5}$

17. $\dfrac{4x^2+4x+1}{4x^2+17x-15}\,\dfrac{8x^2+22x-21}{4x^2-1}$

18. $\dfrac{x^3-1}{x^2-1} \div \dfrac{x}{x^3+1}$

19. $\dfrac{x^2+x-6}{5x^2-3x-2} \div \dfrac{x^2-9}{5x^2+2x}$

20. $\dfrac{4}{3x+1} - \dfrac{2}{(3x+1)^2}$

21. $\dfrac{12x}{(2x+5)^2} - \dfrac{6}{2x+5}$

22. $\dfrac{1}{y} - \dfrac{y+2}{y^2} + \dfrac{3}{y^3}$

23. $\dfrac{2}{3y} + \dfrac{2y+3}{y^3} - \dfrac{1-y^2}{y^4}$

24. $\dfrac{3}{3z+6} - \dfrac{4}{(z+2)^2} + \dfrac{5}{(z+2)^3}$

25. $\dfrac{1}{w} + \dfrac{1}{2w-6} + \dfrac{w}{w^2-9}$

26. $\dfrac{1}{4x^2} - \dfrac{3x}{x^2-1} + \dfrac{4-x}{x^2+x}$

27. $\dfrac{3}{2x} - \dfrac{2}{x^2} + \dfrac{1-x}{x+3} + \dfrac{x}{(x+3)^2}$

28. $\dfrac{5}{(x-1)^2} - \dfrac{4}{x-1} + \dfrac{2}{x-2} - \dfrac{3}{(x-2)^2}$

29. $\dfrac{3}{2t} - \dfrac{1+5t^2}{t^3} + \dfrac{2}{1+2t}$

30. $\dfrac{6xy+4xz-9wy-6wz}{3xy+2xz+12wy+8wz}$

31. $\left(\dfrac{1}{x+h} - \dfrac{1}{x}\right)\left(\dfrac{1}{h}\right)$

32. $\left(\dfrac{1}{2x+2h-3} - \dfrac{1}{2x-3}\right)\left(\dfrac{1}{h}\right)$

33. $\left(\dfrac{1}{4x+4h+1} - \dfrac{1}{4x+1}\right)\left(\dfrac{1}{h}\right)$

34. $\left(\dfrac{1}{(x+h)^2} - \dfrac{1}{x^2}\right)\left(\dfrac{1}{h}\right)$

35. $\dfrac{(x+h)^{-3} - x^{-3}}{h}$

36. $\dfrac{\frac{1}{x} - \frac{1}{y}}{\frac{1}{x} + \frac{1}{y}}$

37. $\dfrac{\frac{x}{y} + \frac{y}{x}}{\frac{1}{x} + \frac{1}{y}}$

38. $\dfrac{\frac{1}{2x+1} - 3}{\frac{1}{x} - x}$

39. $\dfrac{\frac{x}{y^2} - \frac{y}{x^2}}{x-y}$

40. $\dfrac{\frac{1}{a^2} - \frac{1}{b^2}}{\frac{a}{b} - \frac{b}{a}}$

41. $\dfrac{\frac{1}{x+1} - \frac{2}{3x-1}}{\frac{1}{x-1} + \frac{2}{3x+1}}$

42. $\dfrac{\frac{1}{x} + \frac{2}{x^2} - x}{\frac{1}{x} - \frac{2}{x^2} + x}$

43. $\dfrac{(s+t)^{-1} - (s-t)^{-1}}{(s+t)^{-1} + (s-t)^{-1}}$

44. $\left(\dfrac{1}{\sqrt{a+h}} - \dfrac{1}{\sqrt{a}}\right)\left(\dfrac{1}{h}\right)$

45. $\left(\dfrac{1}{\sqrt{a}} - \dfrac{1}{\sqrt{b}}\right)(\sqrt{a}+\sqrt{b})$

46. $\dfrac{1}{\sqrt{a}-\sqrt{b}}$

47. $\dfrac{\sqrt{a}-\sqrt{z}}{\sqrt{a}+\sqrt{z}}$

48. $\dfrac{u}{\sqrt{u}} - \dfrac{v}{\sqrt{v}}$

49. $\dfrac{1-x}{1-\sqrt{x}}$

50. $2\sqrt{x+2} - \dfrac{x-1}{\sqrt{x+2}}$

51. $\dfrac{x\sqrt{1-x^2} + \dfrac{x^3}{\sqrt{1-x^2}}}{1-x^2}$

52. $\dfrac{(1-3x)\sqrt{2+x} - \dfrac{6x+3x^2}{\sqrt{2+x}}}{(2+x)^{3/2}}$

1.5 Composition of Algebraic Expressions

In this chapter, we have discussed how to simplify algebraic expressions, as well as how to add, subtract, multiply, and divide them. There is one more very important operation that we can perform on algebraic expressions, namely, substituting one algebraic expression for the variable in another algebraic expression. As an example, consider the polynomial

$$p(y) = 3y + 1 \tag{1.2}$$

If we substitute the polynomial $q(x) = 5x - 2$ for the variable y in $p(y)$, the result is

$$\begin{aligned} p(q(x)) &= 3(5x - 2) + 1 \\ &= 15x - 6 + 1 \\ &= 15x - 5 \end{aligned}$$

Notice that we have replaced the variable y on *both sides* of (1.2) with the polynomial $q(x)$. On the left side, we used the symbol $q(x)$ itself, which is why we have written $p(q(x))$. (This is read "p of q of x.") On the right side, we substituted $5x - 2$ for y. Notice also that the resulting expression $p(q(x))$ is another polynomial.

Before considering other examples, we should introduce a bit of terminology that will be useful to us in Chapter 2.

DEFINITION

Let $r(x)$ and $s(y)$ be algebraic expressions in the variables x and y, respectively. (We also allow the possibility that x and y are the same variable.) Then the expression $s(r(x))$, obtained by substituting $r(x)$ for the variable y in $s(y)$, is called the **composition of $s(y)$ with $r(x)$.**

At this point, you are probably wondering why we are interested in forming the composition of two algebraic expressions. The reason is that learning how to compose algebraic expressions will make life much easier for us when we study the important concept of composition of functions in Chapter 2.

Let us consider some additional examples of composition.

EXAMPLE 1: Compute the composition $p(q(x))$, where $p(y) = y^2 - 2y + 3$ and $q(x) = 2x - 1$. Simplify your answer.

Solution: We simply replace the variable y in $p(y) = y^2 - 2y + 3$ with the expression $q(x) = 2x - 1$, and then simplify.

$$\begin{aligned} p(q(x)) &= (2x - 1)^2 - 2(2x - 1) + 3 \\ &= 4x^2 - 4x + 1 - 4x + 2 + 3 \\ &= 4x^2 - 8x + 6 \end{aligned}$$

Thus, the composition is

$$p(q(x)) = 4x^2 - 8x + 6$$

□

EXAMPLE 2: Let $p(x) = 5x + 3$ and $q(x) = 3x^2 + 2x - 4$. Compute both of the compositions $p(q(x))$ and $q(p(x))$. Are they the same?

Solution: In order to compute the composition $p(q(x))$, we replace the variable x in $p(x) = 5x + 3$ by $q(x) = 3x^2 + 2x - 4$. This gives

$$\begin{aligned} p(q(x)) &= 5(3x^2 + 2x - 4) + 3 \\ &= 15x^2 + 10x - 17 \end{aligned}$$

In order to compute the composition $q(p(x))$, we replace the variable x in $q(x) = 3x^2 + 2x - 4$ by $p(x) = 5x + 3$. This gives

$$\begin{aligned} q(p(x)) &= 3(5x + 3)^2 + 2(5x + 3) - 4 \\ &= 3(25x^2 + 30x + 9) + 10x + 6 - 4 \\ &= 75x^2 + 100x + 29 \end{aligned}$$

As you can plainly see, the two compositions are different. In fact, if $r(x)$ and $s(x)$ are algebraic expressions, then we can say that

> **in general, the composition $r(s(x))$ is *not* the same as the composition $s(r(x))$. In other words, when forming the composition of two algebraic expressions, *order is important*.**

Try Study Suggestion 1.42. □

▶ **Study Suggestion 1.42:** Let $p(x) = 3x^2 - 2x + 1$ and $q(x) = 2x + 1$.

(a) Compute $p(q(x))$

(b) Compute $q(p(x))$ ◀

EXAMPLE 3: If $p(x) = x^n$ and $q(x) = x^m$ find the composition $p(q(x))$. Compare this with the *product* $p(x)q(x)$.

Solution: In this case, we have

$$p(q(x)) = (x^m)^n = x^{mn}$$

Now, the product $p(x)q(x)$ is

$$p(x)q(x) = x^n \cdot x^m = x^{n+m}$$

and so the composition $p(q(x))$ and the product $p(x)q(x)$ are different. In fact, we can say that

> **in general, the composition $r(s(x))$ of two algebraic expressions is *not* the same as the product $r(x)s(x)$.**

□

Now let us try some examples of taking the composition of two rational expressions. Notice that, after simplifying, the result is always another rational expression.

EXAMPLE 4: Find the composition $p(r(x))$, where $p(x) = x^2 + 2x - 5$ and $r(x) = 1/x$.

Solution: In this case, we replace the variable x in $p(x) = x^2 + 2x - 5$ by the expression $r(x) = 1/x$, and then simplify

$$
\begin{aligned}
p(r(x)) &= \left(\frac{1}{x}\right)^2 + 2\left(\frac{1}{x}\right) - 5 \\
&= \frac{1}{x^2} + \frac{2}{x} - 5 \\
&= \frac{1}{x^2} + \frac{2x}{x^2} - \frac{5x^2}{x^2} \\
&= \frac{1 + 2x - 5x^2}{x^2}
\end{aligned}
$$

□

EXAMPLE 5: Find the composition $s(r(x))$, where

$$s(x) = \frac{x^2 - 2x + 1}{x + 3} \qquad \text{and} \qquad r(x) = \frac{x + 1}{x}$$

Solution: Here we replace the variable x in $s(x)$ by the expression $r(x) = (x + 1)/x$ and simplify (we leave the simplification to you)

$$
\begin{aligned}
s(r(x)) &= \frac{\left(\frac{x+1}{x}\right)^2 - 2\left(\frac{x+1}{x}\right) + 1}{\frac{x+1}{x} + 3} \\
&= \frac{1}{x(4x + 1)}
\end{aligned}
$$

□

▶ **Study Suggestion 1.43:**

(a) Compute both $r(s(x))$ and $s(r(x))$, where $r(x) = 1/x$ and $s(x) = 1/x^2$.

(b) Compute both $u(v(x))$ and $v(u(x))$, where $u(x) = \sqrt{x}$ and $v(x) = x + 1$. ◀

EXAMPLE 6: Find the composition $p(q(x))$ where $p(x) = \sqrt{x^2 + 1}$ and $q(x) = x - 1$.

Solution: In this case, we have

$$
\begin{aligned}
p(q(x)) &= \sqrt{(x - 1)^2 + 1} \\
&= \sqrt{x^2 - 2x - 1 + 1} \\
&= \sqrt{x^2 - 2x}
\end{aligned}
$$

Try Study Suggestion 1.43. □

Ideas to Remember

Forming the composition of two algebraic expressions is just as important an operation as adding, subtracting, multiplying, or dividing algebraic expressions.

EXERCISES

In Exercises 1–14, compute the compositions $p(q(x))$ and $q(p(x))$ and simplify the results. Are they the same?

1. $p(x) = 2x - 3$, $q(x) = 3x - 2$

2. $p(x) = 4x - 9$, $q(x) = (1/4)(x + 9)$

3. $p(x) = 2x + 3$, $q(x) = x$

4. $p(x) = x$, $q(x) = x^2 + 1$

5. $p(x) = x^2$, $q(x) = 1/x^2$

6. $p(x) = x^2 + 1$, $q(x) = 2x - 4$

7. $p(x) = x^2 + 1$, $q(x) = \sqrt{x + 1}$

8. $p(x) = x^3 - 1$, $q(x) = x^{1/3} + 1$

9. $p(x) = \dfrac{1}{x + 1}$, $q(x) = x^2 - 2$

10. $p(x) = \dfrac{1}{2x + 1}$, $q(x) = \dfrac{1 - x}{2x}$

11. $p(x) = \dfrac{x}{x - 1}$, $q(x) = \dfrac{x}{x - 1}$

12. $p(x) = \sqrt{x + 8}$, $q(x) = x^2 + 6x + 1$

13. $p(x) = \dfrac{x + 1}{x - 1}$, $q(x) = \dfrac{x - 1}{x + 1}$

14. $p(x) = \dfrac{x + 2}{x^2 + x}$, $q(x) = \dfrac{x - 2}{x^2 - x}$

In Exercises 15–33, find the composition $p(q(x))$. Be sure to simplify the result.

15. $p(x) = x^2 + 1$, $q(x) = 2x - 1 + 1/x$

16. $p(x) = x^4 + 4x^2 + 4$, $q(x) = x^{1/2}$

17. $p(y) = 3y - 4$, $q(x) = x^2 + 1$

18. $p(z) = 7z^2 - 4z + 1$, $q(x) = x^3$

19. $p(x) = x^4 - 3x^3 + 5x - 3$, $q(x) = x^{10}$

20. $p(t) = 4t^2 - 3\sqrt{t}$, $q(x) = 1 - \sqrt{x}$

21. $p(s) = 1 - 3s^3$, $q(x) = 1 - x^2$

22. $p(y) = \dfrac{1}{y + 1}$, $q(x) = \dfrac{1}{x + 1}$

23. $p(x) = \dfrac{x}{x^2 + x + 1}$, $q(x) = \dfrac{1}{x}$

24. $p(t) = \dfrac{1}{t^2 - 1}$, $q(x) = p(x)$

25. $p(x) = q(x)$, $q(x) = \dfrac{x}{x^2 - 1}$

26. $p(x) = \dfrac{2x + 1}{3x - 1}$, $q(x) = \dfrac{4x + 5}{3x + 1}$

27. $p(x) = \dfrac{x^2 + 3x - 1}{4x^2 - x + 1}$, $q(x) = x^2$

28. $p(x) = \sqrt{x^4 + 3x^2 - 1}$, $q(x) = \sqrt{x}$

29. $p(x) = x + \dfrac{1}{x}$, $q(x) = \dfrac{1}{x}$

30. $p(x) = x + \dfrac{1}{x} + \dfrac{1}{x^2}$, $q(x) = x + \dfrac{1}{x}$

31. $p(x) = \sqrt{x} + \dfrac{1}{\sqrt{x}}$, $q(x) = \dfrac{1}{x^2}$

32. $p(x) = (2x^2 + 3)(3x^4 - 1)$, $q(x) = x^{1/2}$

33. $p(x) = (2x + 1)^3$, $q(x) = x^{1/3}$

34. If $p(x) = 2x + 3$, find

(a) $p(p(x))$ **(b)** $p(p(p(x)))$

35. If $p(x) = x^2 + 1$, find

(a) $p(p(x))$ **(b)** $p(p(p(x)))$

36. If $p(x) = x^n$, find

(a) $p(p(x))$ **(b)** $p(p(p(x)))$ **(c)** $p(p(p(p(x))))$

37. If $p(x) = 2x + 1$, find a linear polynomial $q(x)$ with the property that $p(q(x)) = x$ and $q(p(x)) = x$.

1.6 Solving Equations

Now let us review the subject of solving **equations.** Two equations are said to be **equivalent** if they have exactly the same solutions. The usual approach to solving equations is to apply a series of algebraic operations to both sides of the given equation in order to produce equivalent equations that are easier to solve. The reason that this approach works is that applying the same algebraic operation to both sides of the equation *usually* produces an equation that is equivalent to the original one.

Let us explain why we use the term "usually." When we multiply both sides of an equation by an expression that involves the *variable*, it is possible to introduce additional solutions. By this we mean that the resulting equation may have solutions that are *not* solutions to the original equation. These numbers are called **extraneous solutions.** When extraneous solutions are introduced, the new equation is *not* equivalent to the original one.

Because extraneous solutions can be introduced, it is very important to check that each of the solutions to the final equation really is a solution to the original equation, by substituting into the original equation.

Let us consider some examples.

EXAMPLE 1: Solve the following equation.

$$\sqrt{4x + 5} = 2x + 1$$

Solution: In order to remove the square root sign, we first square both sides of the equation, which gives

$$(\sqrt{4x + 5})^2 = (2x + 1)^2$$

or

$$4x + 5 = 4x^2 + 4x + 1$$

Rearranging terms gives

$$4x^2 = 4 \qquad \text{or} \qquad x^2 = 1$$

Taking the square root of both sides of this equation gives us the two *potential* solutions $x = 1$ and $x = -1$, each of which must be checked. Substituting

$x = 1$ into the original equation gives $\sqrt{9} = 3$, which is a true statement, and so $x = 1$ is a solution. On the other hand, substituting $x = -1$ into the original equation gives $\sqrt{1} = -1$ which is *not* true. Hence, $x = -1$ is *not* a solution to the equation; that is, $x = -1$ is an extraneous solution.

Try Study Suggestion 1.44. □

▶ **Study Suggestion 1.44:** Solve the equation

$$\sqrt{2 - 6x} = 3x - 1.$$ ◀

Not only is it possible to gain extraneous solutions to an equation, but it is also possible to *lose* solutions! As a simple example, if we divide both sides of the equation

$$x^2 = x$$

by x, the result is

$$x = 1$$

But this equation has only one solution, namely $x = 1$, whereas the original equation also has the solution $x = 0$. Thus, dividing both sides by x caused us to lose the solution $x = 0$.

In order to correct for the possible loss of solutions when multiplying or dividing by an expression that involves the variable, we must check separately *in the original equation* any real numbers that make that expression equal to 0.

As you probably know, a **quadratic equation** is an equation of the form

$$ax^2 + bx + c = 0$$

where a, b, and c are constants and $a \neq 0$. There are three simple methods for solving quadratic equations.

METHOD 1—FACTORING: The method of factoring relies on the fact that if the product of two real numbers equals 0, then one or both of the numbers must equal 0. To see how this fact is used, let us consider an example.

EXAMPLE 2: Solve the equation

$$6x^2 - 3x - 3 = 0$$

by the method of factoring.

Solution: The first step is to factor the left-hand side of this equation

$$3(2x + 1)(x - 1) = 0$$

Dividing both sides by 3 gives

$$(2x + 1)(x - 1) = 0$$

Now, in order for a real number to satisfy this equation, it must make either

$2x + 1$ or $x - 1$ equal to 0; that is, it must satisfy *either of* the linear equations

$$2x + 1 = 0 \qquad \text{or} \qquad x - 1 = 0$$

But the solutions to these equations are $x = -1/2$ and $x = 1$, and so these are the solutions to the original quadratic equation as well. (You should check this by substituting into the original equation.)

Try Study Suggestion 1.45. □

▶ **Study Suggestion 1.45:** Solve the following quadratic equations by the method of factoring.

(a) $2x^2 - 4x - 6 = 0$

(b) $x^2 - 4x + 4 = 0$ ◀

METHOD 2—COMPLETING THE SQUARE: The second method for solving quadratic equations involves the concept of a perfect square. A quadratic polynomial is said to be a **perfect square** if it is the square of a linear polynomial. For example, the polynomial $x^2 + 4x + 4$ is a perfect square, since it can be written in the form

$$x^2 + 4x + 4 = (x + 2)^2$$

Now the polynomial $x^2 + kx$, where k is a constant, can always be made into a perfect square by adding the quantity $(k/2)^2$, since

$$x^2 + kx + \left(\frac{k}{2}\right)^2 = \left(x + \frac{k}{2}\right)^2$$

As an example, we can make $x^2 + 5x$ into a perfect square by adding $(5/2)^2 = 25/4$, since

$$x^2 + 5x + \frac{25}{4} = \left(x + \frac{5}{2}\right)^2$$

The process of making $x^2 + kx$ into a perfect square by adding the quantity $(k/2)^2$ and then writing it in the form $(x + k/2)^2$ is called **completing the square.** *Notice that this process applies only to quadratic polynomials of the form* $x^2 + kx$, *where the coefficient of* x^2 *is equal to* 1 *and the constant term is equal to* 0.

EXAMPLE 3: Complete the square for the following quadratic polynomials.

(a) $x^2 + x$ **(b)** $x^2 - \frac{1}{2}x$

Solutions:

(a) In this case $k = 1$ and so we add $\left(\frac{1}{2}\right)^2 = \frac{1}{4}$

$$x^2 + x + \frac{1}{4} = \left(x + \frac{1}{2}\right)^2$$

(b) Since $k = -\frac{1}{2}$, we add $\left(\frac{-1}{4}\right)^2 = \frac{1}{16}$

$$x^2 - \frac{1}{2}x + \frac{1}{16} = \left(x - \frac{1}{4}\right)^2$$

Try Study Suggestion 1.46. □

▶ **Study Suggestion 1.46:** Complete the square for the following polynomials

(a) $x^2 - 3x$

(b) $x^2 + \frac{1}{3}x$ ◀

Now let us see how we can use the process of completing the square to solve quadratic equations.

EXAMPLE 4: Solve the equation

$$2x^2 + 5x - 1 = 0$$

by completing the square.

Solution: The first step is to make the left side of this equation into an expression of the form $x^2 + kx$. This can be accomplished by first adding 1 to both sides

$$2x^2 + 5x = 1$$

and then dividing both sides by 2

$$x^2 + \frac{5}{2}x = \frac{1}{2}$$

Now we complete the square on the left-hand side. Since $k = 5/2$, this can be done by adding $(5/4)^2 = 25/16$ to the left side. *However, in order to preserve the equality, we must also add* $25/16$ *to the right side of the equation!*

$$x^2 + \frac{5}{2}x + \frac{25}{16} = \frac{1}{2} + \frac{25}{16}$$

$$\left(x + \frac{5}{4}\right)^2 = \frac{33}{16}$$

Now that the quadratic equation has this form, it is easy to solve. We begin by taking the square root of both sides, which leads to two possibilities

$$x + \frac{5}{4} = \pm\sqrt{\frac{33}{16}} = \pm\frac{\sqrt{33}}{4}$$

Subtracting 5/4 from both sides of each equation gives the solutions

$$x = \frac{-5 \pm \sqrt{33}}{4}$$

(One solution is obtained by taking the plus sign, and the other by taking the minus sign.) We will leave it to you to check that both of these are indeed solutions to the original equation. □

We can summarize the method used in the previous example as follows.

Procedure for Solving a Quadratic Equation by Completing the Square

1. Manipulate the equation so that the left-hand side has the form $x^2 + kx$, where k is a constant.
2. Complete the square on the left side. *Be sure to add the same quantity to the right side of the equation in order to preserve the equality.*
3. Solve the resulting equation. The first step will be to take the square root of both sides.

▶ **Study Suggestion 1.47:** Solve the following equations by completing the square.

(a) $x^2 - x - 12 = 0$

(b) $3x^2 + 5x + 1 = 0$ ◀

Try Study Suggestion 1.47.

METHOD 3—THE QUADRATIC FORMULA: Applying the method of completing the square to the general quadratic equation

$$ax^2 + bx + c = 0$$

gives

$$x = \frac{-b \pm \sqrt{b^2 - 4ac}}{2a} \qquad \textbf{(1.3)}$$

Formula (1.3) is known as the **quadratic formula.** It gives us the solutions to the general quadratic equation $ax^2 + bx + c = 0$ and can be used to find the solutions to any specific quadratic equation simply by substituting the appropriate values for a, b, and c. Let us consider some examples.

EXAMPLE 5: Solve the equation

$$x^2 - 3x + 1 = 0$$

by using the quadratic formula.

Solution: In this case, $a = 1$, $b = -3$, and $c = 1$. Substituting these values into the quadratic formula, we get the solutions

$$\begin{aligned} x &= \frac{-b \pm \sqrt{b^2 - 4ac}}{2a} \\ &= \frac{-(-3) \pm \sqrt{(-3)^2 - 4 \cdot 1 \cdot 1}}{2 \cdot 1} \\ &= \frac{3 \pm \sqrt{5}}{2} \end{aligned}$$

Try Study Suggestion 1.48. □

▶ **Study Suggestion 1.48:** Solve the following quadratic equations by using the quadratic formula.

(a) $2x^2 - 3x + 1 = 0$

(b) $4x^2 + 12x + 9 = 0$ ◀

EXAMPLE 6: Solve the equation

$$x^2 + 4x + 6 = 0$$

by using the quadratic formula.

Solution: Substituting the values $a = 1$, $b = 4$, and $c = 6$ into the quadratic formula gives

$$x = \frac{-b \pm \sqrt{b^2 - 4ac}}{2a} = \frac{-4 \pm \sqrt{4^2 - 4 \cdot 1 \cdot 6}}{2 \cdot 1} = \frac{-4 \pm \sqrt{-8}}{2}$$

At this point, we observe that the number under the square root sign is negative. Since we cannot take the square root of a negative number (at least not within the real number system), we conclude that this quadratic equation has no real number solutions. *Try Study Suggestion 1.49.* □

▶ **Study Suggestion 1.49:** Solve the equation $3x^2 + 3x + 1 = 0$ by using the quadratic formula. ◀

The expression $D = b^2 - 4ac$ that lies under the square root sign in the quadratic formula is the key to how many solutions a given quadratic equation has. This expression is known as the **discriminant** of the quadratic equation, and we can state the following theorem.

THEOREM

Let $D = b^2 - 4ac$ be the discriminant of the quadratic equation $ax^2 + bx + c = 0$. Then

1. If $D > 0$, then the equation has two distinct solutions.
2. If $D = 0$, then the equation has exactly one solution, called a **double root** of the equation.
3. If $D < 0$, then the equation has no (real number) solutions.

Try Study Suggestion 1.50.

▶ **Study Suggestion 1.50:** Find the discriminant of each of the following quadratic equations and state how many solutions each equation has, without computing these solutions.

(a) $9x^2 - 4x + 1 = 0$

(b) $16x^2 + 8x + 1 = 0$

(c) $-1 + x - x^2 = 0$

(d) $4x^2 - 9x + 1 = 0$ ◀

Sometimes an equation can be reduced to a quadratic equation by making a substitution. The following example illustrates this.

EXAMPLE 7: Solve the equation

$$x^4 + x^2 - 6 = 0$$

Solution: We begin by writing this equation in the form

$$(x^2)^2 + x^2 - 6 = 0$$

From this we can see that if we make the substitution $y = x^2$, we get a *quadratic* equation in the variable y,

$$y^2 + y - 6 = 0$$

The solutions to this equation are $y = 2$ and $y = -3$. Thus, converting back to the variable x, we see that the solutions to the original equation are the same as the solutions to the equations

$$x^2 = 2 \qquad \text{and} \qquad x^2 = -3$$

The first of these equations gives the solutions $x = \pm\sqrt{2}$, and the second equation has no real number solutions, since the square of a real number can never be negative. Thus, the solutions to the original equation are $x = \pm\sqrt{2}$.

Try Study Suggestion 1.51. □

▶ **Study Suggestion 1.51:**
(a) Solve the equation $2x^4 + 5x^2 - 12 = 0$.
(b) Solve the equation $x^{1/2} + 4x^{1/4} - 5 = 0$. (*Hint:* let $y = x^{1/4}$.) ◀

As the following example demonstrates, when an equation involves more than one radical sign, we may need to take powers more than once in order to eliminate them.

EXAMPLE 8: Solve the equation

$$\sqrt{2x-1} + \sqrt{x-1} = 5$$

Solution: In a situation such as this, the first step is to isolate one of the radicals. In this case, we can accomplish this by subtracting $\sqrt{x-1}$ from both sides

$$\sqrt{2x-1} = 5 - \sqrt{x-1}$$

Now we square both sides, which will eliminate the isolated radical

$$\begin{aligned}(\sqrt{2x-1})^2 &= (5 - \sqrt{x-1})^2\\ 2x - 1 &= 25 - 10\sqrt{x-1} + (x-1)\\ 2x - 1 &= 24 - 10\sqrt{x-1} + x\end{aligned}$$

Again we isolate the term involving the radical and square both sides

$$\begin{aligned}x - 25 &= -10\sqrt{x-1}\\ (x-25)^2 &= (-10\sqrt{x-1})^2\\ x^2 - 50x + 625 &= 100(x-1)\\ x^2 - 150x + 725 &= 0\\ (x-5)(x-145) &= 0\end{aligned}$$

The solutions to this equation are $x = 5$ and $x = 145$. However, after checking, we see that only $x = 5$ is a solution to the original equation.

Try Study Suggestion 1.52. □

▶ **Study Suggestion 1.52:** Solve the equation

$$\sqrt{x-2} + \sqrt{2x+4} = 6.$$ ◀

Let us conclude this section with a few examples involving absolute values.

EXAMPLE 9: Solve the equation

$$|x - 4| = 5 \tag{1.4}$$

Solution: In this case, we must ask ourselves the question, When can the absolute value of a quantity be equal to 5? Of course, the answer is that this happens in two cases—when the quantity itself equals 5 or when it equals -5. Thus, we see that a real number satisfies Equation (1.4) if and only if it satisfies *either of* the equations

$$x - 4 = 5 \qquad \text{or} \qquad x - 4 = -5$$

The solution to the first equation is $x = 9$ and the solution to the second equation is $x = -1$, and so these are the solutions to Equation (1.4), as you can easily check by substitution. *Try Study Suggestion 1.53.* □

▶ **Study Suggestion 1.53:** Solve the equation $|2x - 1| = 3$. ◀

EXAMPLE 10: Solve the equation

$$|x - 3| = |2x + 1| \tag{1.5}$$

Solution: Now we ask ourselves the question: When is the absolute value of one quantity equal to the absolute value of another quantity? Again the answer is that this happens in two cases—when the two quantities are themselves equal or else when one quantity is the negative of the other.

Thus, a real number satisfies Equation (1.5) if and only if it satisfies *either of* the equations

$$x - 3 = 2x + 1 \qquad \text{or} \qquad x - 3 = -(2x + 1)$$

The solution to the first of these equations is $x = -4$ and the solution to the second is $x = 2/3$, and so these are the solutions to Equation (1.5) as well.

Try Study Suggestion 1.54. □

▶ **Study Suggestion 1.54:** Solve the equation $|x + 9| = |2x - 3|$. ◀

EXERCISES

In Exercises 1–10, compute the discriminant of the given quadratic equation, and decide how many solutions it has (without computing the solutions).

1. $x^2 + 2x - 1 = 0$ **2.** $x^2 + 3 = 0$

3. $x^2 + 5x = 0$ **4.** $x^2 + 3x + 5 = 0$

5. $3t^2 - 7t + 4 = 0$ **6.** $3 - 3s + s^2 = 0$

7. $1 + 5u^2 = 0$ **8.** $2 - 3u + 9u^2 = 0$

9. $-2 + 4z - 2z^2 = 0$

10. $x^2 + ax + a^2 = 0$ (a constant)

In Exercises 11–18, solve the given quadratic equation by the method of factoring.

11. $x^2 - 2x - 3 = 0$ **12.** $3x - 12x^2 = 0$

13. $t^2 - 6t + 9 = 0$ **14.** $2t^2 + t = 1$

15. $1 - 3x - 10x^2 = 0$ **16.** $14 + 10x = 4x^2$

17. $x^2 - 2 = 0$ **18.** $2x^2 - 1 = 0$

In Exercises 19–28, solve the given quadratic equation by completing the square.

19. $x^2 + 6x + 5 = 0$ **20.** $x^2 + 4x + 3 = 0$

21. $y^2 + 4 = 0$ **22.** $3y - y^2 = 0$

23. $1 + t^2 = 4t$ **24.** $2t^2 = 10t + 1$

25. $4x^2 - 20x + 25 = 0$ **26.** $x^2 + x = 1$

27. $(1/2)x^2 + x - 1/5 = 0$ **28.** $x^2 - (1/3)x + 1 = 0$

In Exercises 29–36, solve the given equation by using the quadratic formula.

29. $x^2 + 7x + 10 = 0$ **30.** $x^2 - 3 = 0$

31. $t^2 + 6t = 1$ **32.** $t^2 + 3 = 2t$

33. $7x^2 + (1/2)x - 1 = 0$ **34.** $y^2 + (2/3)y = 1$

35. $9x^2 - 6\sqrt{2}x + 2 = 0$ **36.** $\sqrt{2}x^2 + 2x - 1 = 0$

In Exercises 37–48, solve the given equation by your favorite method.

37. $x^2 - x - 30 = 0$ **38.** $x^2 + 2x + 1 = 0$

39. $4x^2 + 4x - 1 = 0$ **40.** $16(x^2 - 2x + 1) = 0$

41. $x^2 + x - 1 = 0$ **42.** $x^2 + \pi x = 3$

43. $6x^2 + 5x = -1$ **44.** $x^2 = 5x - 9$

45. $172x^2 + 156x - 1 = 0$

46. $7.36x^2 - 9.47 = 0$

47. $4.93x - 126x^2 = 0$

48. $6.93x^2 + 2.29x + 1.41 = 0$

In Exercises 49–90, solve the given equation.

49. $x + 1 = 2x - 6$ **50.** $5t + 3 = 1 - t$

51. $2.2s - 3.3 = 0.3(3s + 0.1)$

52. $x(x + 1) = 2(x + 1)$

53. $t^3 - 27 = 0$ **54.** $x^2(x - 1) = 2x(x - 1)$

55. $x^4 = 16$ **56.** $x^5 - 64 = 0$

57. $x^{2/3} - 1 = 0$ **58.** $x^{5/2} - \sqrt{a} = 0$

59. $z^{-3/4} = z^{5/4}$ **60.** $z^{-1/2} = z^{1/2}$

61. $x^4 - 3x^2 - 4 = 0$ **62.** $2t^{10} - 5 = 0$

63. $3x^{2/9} = 2x^{1/9}$ **64.** $2y^{1/3} - 5y^{1/6} - 3 = 0$

65. $w + \sqrt{w} - 20 = 0$ **66.** $x^4 + 9x^2 + 9 = 0$

67. $2x^6 + 3x^3 - 1 = 0$ **68.** $-x^{-1} + x^{-2} = -6$

69. $\dfrac{2}{x + 1} = 3$ **70.** $\dfrac{4}{x^2 + 1} = 1$

71. $\dfrac{2x - 3}{4} = \dfrac{7x + 4}{9}$ **72.** $\dfrac{4y + 5}{3} = 2 - \dfrac{y}{4}$

73. $\dfrac{1}{5 - 4t} - \dfrac{1}{2(3 - t)} = 0$ **74.** $\dfrac{1}{s} + \dfrac{2}{5s} = \dfrac{5}{2s} - \dfrac{4}{3s}$

75. $\dfrac{1}{(x - 1)^2} - \dfrac{2}{x - 1} - 4 = 0$ **76.** $\sqrt{2\sqrt{x}} = \sqrt{1 + x}$

77. $\sqrt{2 + 3\sqrt{x}} = \sqrt{x + 1}$ **78.** $\sqrt{4 + 5\sqrt{x - \sqrt{x}}} = 2$

79. $\sqrt{x + 2} = \sqrt{x - 2}$ **80.** $|2x - 1| = 5$

81. $|1 - 4x| = 6$ **82.** $|2x - 3| = 0$

83. $|1 - 7x| = |2 + 3x|$ **84.** $|2 - 5x^2| = 2$

85. $|x^2 - 1| = 4$ **86.** $|5x + 3| = |x + 2|$

87. $|7x - 1| = |2x^2 + 1|$ **88.** $|x^4 + 1| = |2x^2 + 3|$

89. $\sqrt{x + 1} + \sqrt{9x - 1} = \sqrt{5}$

90. $\sqrt{2x + 3} - \sqrt{x + 1} = \sqrt{4 - x}$

1.7 Linear and Absolute Value Inequalities: Interval Notation

In the previous section, we reviewed methods for solving equations. When we replace the equality sign in an equation by one of the signs $<, \leq, >$, or $\geq$, we obtain an **inequality.** Some examples of inequalities are

$$2x - 3 \leq 0, \quad x^2 - 4x + 1 > 4, \quad \frac{x + 1}{2x - 3} < 1$$

By *solving* an inequality we mean, of course, finding *all* real numbers that can be substituted for the variable to produce a true statement.

The first inequality above is an example of a **linear inequality,** since the left-hand side is a linear polynomial. The second inequality is an example of a **quadratic inequality.** We will concentrate in this and the next section primarily on linear and quadratic inequalities, as well as inequalities that can be reduced to one of these two types. One especially important class of inequalities are those that involve the absolute value.

Solving inequalities is not really much different than solving equations. Perhaps the main difference between the two is the fact that

> **whenever we multiply or divide both sides of an inequality by a *negative* number, we *must* change the direction of the inequality.**

Another difference between equations and inequalities is that equations often have only a few solutions (if any) whereas inequalities often have an infinite number of solutions (if any). Thus, in order to describe the solutions to an inequality, we will use the concept of the **solution set,** which is simply the set of all solutions to the inequality. If an inequality has no solutions, then its solution set is the **empty set,** which is just the set with no elements in it. This set is denoted by the symbol $\varnothing$.

Sets such as

$$\{x \mid x < a\} \qquad \text{and} \qquad \{x \mid a < x < b\}$$

$\{x \mid x < a\} = (-\infty, a)$

$\{x \mid x \le a\} = (-\infty, a]$

$\{x \mid x > a\} = (a, \infty)$

$\{x \mid x \ge a\} = [a, \infty)$

$\{x \mid a < x < b\} = (a, b)$

$\{x \mid a \le x \le b\} = [a, b]$

$\{x \mid a < x \le b\} = (a, b]$

$\{x \mid a \le x < b\} = [a, b)$

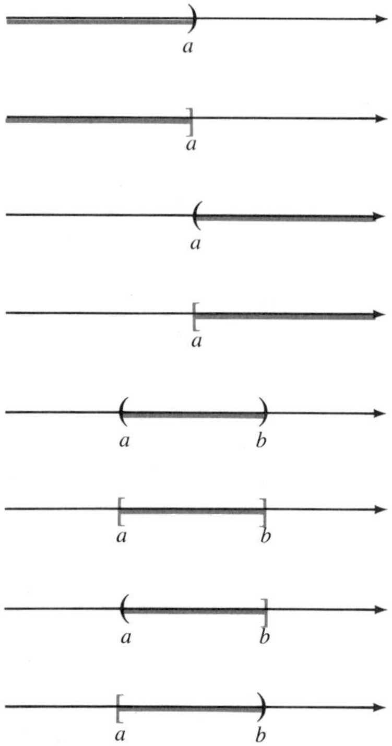

FIGURE 1.6

where a and b are real numbers, occur frequently in describing the solutions to inequalities. For this reason (and others) a special notation is used to describe these types of sets. (See Figure 1.6. We have shown each type of set on the real number line as well.) Notice that we use a square bracket to indicate that an end value is included in the set, and a round bracket to indicate that an end value is not included.

It is important to remember that these are nothing more than special notations for certain sets of real numbers. Sets of the form (a, b) are called **open intervals,** sets of the form $[a, b]$ are called **closed intervals,** and sets of the form $(a, b]$ and $[a, b)$ are called **half-open intervals.** The numbers a and b are called the **endpoints** of the interval.

Also, it is important not to confuse the concept of the *open interval* (a, b) with the concept of the *ordered pair* (a, b). Even though the notation is the same, these two concepts are totally unrelated.

In the following examples, we will usually describe the solution set of a given inequality in three ways: in set notation, in interval notation, and by showing the solution set on a real number line, since this helps to get a good "feeling" for the solutions.

EXAMPLE 1: Solve the inequality

$$-\frac{1}{3}x + 5 \leq 1$$

Solution: The first step is to subtract 5 from both sides

$$-\frac{1}{3}x \leq -4$$

Then we multiply both sides by -3, and since -3 is negative, we must remember to change the direction of the inequality. This gives

$$x \geq (-4)(-3) \qquad \text{or} \qquad x \geq 12$$

Hence, the solution set of this inequality is the set $\{x \mid x \geq 12\}$. In interval notation, this is the set $[12, \infty)$. It is pictured in Figure 1.7.

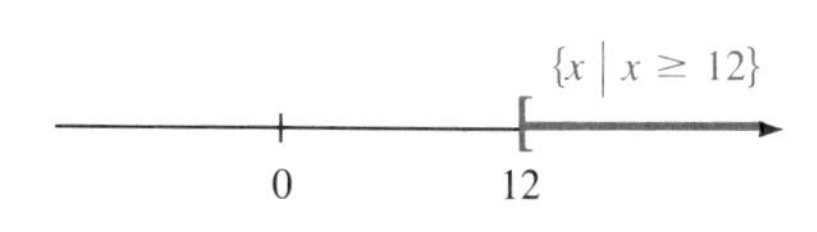

FIGURE 1.7

▶ **Study Suggestion 1.55:** Solve the following inequalities

(a) $-4x - 6 > 3$

(b) $-x + 1 \leq -3x - 5$ ◀

Try Study Suggestion 1.55. □

Sometimes we are called upon to solve *double inequalities* such as

$$2 \leq 4x - 1 \leq 6 \tag{1.6}$$

As we discussed in Section 1.1, an inequality such as this actually represents two separate inequalities,

$$2 \leq 4x - 1 \qquad \text{and} \qquad 4x - 1 \leq 6$$

In order for a real number x to be a solution to (1.6), it must be a solution to *both* of these inequalities.

The simplest way to solve inequalities such as (1.6) is to apply the same algebraic operations to all three parts of the inequality at the same time.

EXAMPLE 2: Solve the inequality

$$-3 < \frac{1-2x}{5} < 1$$

Solution: First we multiply this inequality by 5,

$$-15 < 1 - 2x < 5$$

and then subtract 1,

$$-16 < -2x < 4$$

Next, we divide by -2. But since -2 is negative, we must change the direction of *both* of the inequalities,

$$8 > x > -2 \qquad \text{or} \qquad -2 < x < 8$$

and so the solution set of this inequality is $\{x|-2 < x < 8\}$, which is the same as the open interval $(-2, 8)$. (See Figure 1.8.)

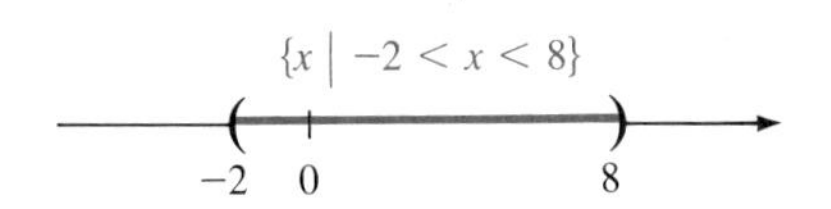

FIGURE 1.8

Try Study Suggestion 1.56. □

▶ **Study Suggestion 1.56:** Solve the double inequality

$$-5 \leq 1 - 3x \leq -3.$$ ◀

One of the most common types of inequalities is the type that involves the absolute value. The key to solving these inequalities is to remember that the absolute value of a number is the distance from that number to the origin (on the real number line.) As a simple example, a real number is a solution to the inequality

$$|x| < 7$$

if and only if its distance from the origin is less than 7, and so the solution set of this inequality is

$$\{x|-7 < x < 7\}$$

or in interval notation $(-7, 7)$. (See Figure 1.9.)

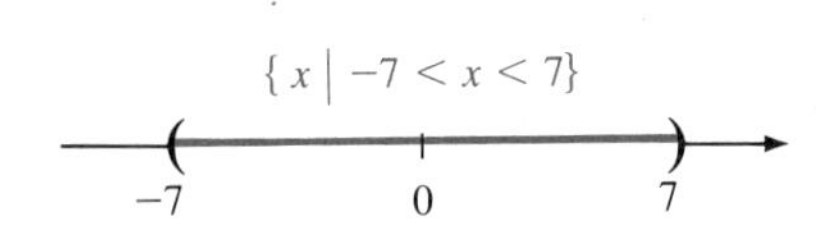

FIGURE 1.9

As another example, a real number is a solution to the inequality

$$|x| > 3$$

if and only if its distance from the origin is greater than 3. But these are the real numbers that are less than -3 or greater than 3. (See Figure 1.10.) Hence, the solution set of this inequality is the *union*

$$\{x|x < -3\} \cup \{x|x > 3\}$$

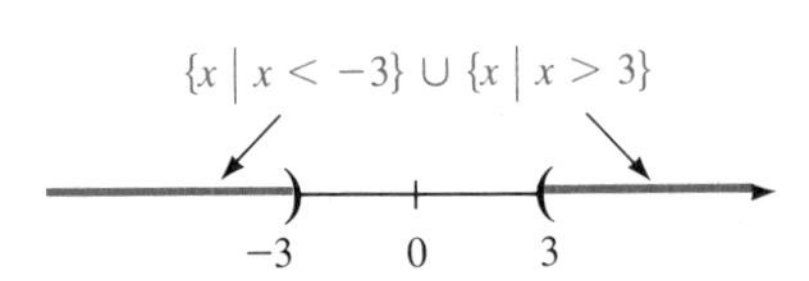

FIGURE 1.10

In interval notation, this is $(-\infty, -3) \cup (3, \infty)$.

We can generalize the results of these two examples in the following very useful theorem.

THEOREM

Let a be any nonnegative real number. Then

1. A real number satisfies the inequality

$$|x| < a$$

if and only if it satisfies the inequality

$$-a < x < a$$

A similar statement holds for $<$ replaced by $\leq$.

2. A real number satisfies the inequality

$$|x| > a$$

if and only if it satisfies *either of* the inequalities

$$x < -a \qquad \text{or} \qquad x > a$$

A similar statement holds for $>$ replaced by $\geq$ and $<$ replaced by $\leq$.

As the next examples show, we can use this theorem to solve more complicated inequalities involving the absolute value.

EXAMPLE 3: Solve the inequality

$$|x - 3| < 4 \tag{1.7}$$

Solution: Replacing x by $x - 3$ in part 1 of the theorem, we see that a real number satisfies the inequality $|x - 3| < 4$ if and only if it satisfies the inequality

$$-4 < x - 3 < 4$$

Adding 3 to all parts of this inequality gives

$$-1 < x < 7$$

Thus, the solution set of (1.7) is $\{x \mid -1 < x < 7\} = (-1, 7)$.

Try Study Suggestion 1.57. □

▶ **Study Suggestion 1.57:** Solve the inequality

$$|2x - 1| < 5$$ ◀

EXAMPLE 4: Solve the inequality

$$|2x + 1| \geq 3 \tag{1.8}$$

Solution: According to part 2 of the theorem, with x replaced $2x + 1$, a real number satisfies the inequality $|2x + 1| \geq 3$ if and only if it satisfies

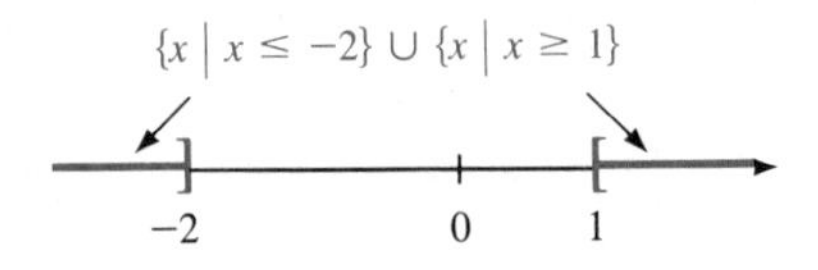

FIGURE 1.11

▶ Study Suggestion 1.58: Solve the inequality

$$|1 - 3x| \ge 5.$$ ◀

either of the inequalities

$$2x + 1 \le -3 \qquad \text{or} \qquad 2x + 1 \ge 3$$

By solving these inequalities, we find that the first inequality has solution set $\{x|x \le -2\}$ and the second has solution set $\{x|x \ge 1\}$. (You should check this yourself.) Hence, the solution set of (1.8) is the *union* of these two sets,

$$\{x|x \le -2\} \cup \{x|x \ge 1\}$$

In interval notation, this is the set $(-\infty, -2] \cup [1, \infty)$. (See Figure 1.11.)

Try Study Suggestion 1.58. □

EXERCISES

In Exercises 1–14, express the given set using interval notation.

1. $\{x|-1 \le x \le 2\}$ **2.** $\{x|-5 \le x \le -3\}$

3. $\{x|-1/2 \le y \le 0\}$ **4.** $\{y|-3 \le x \le -1/3\}$

5. $\{x|-4 < x < 7\}$ **6.** $\{x|-8 < x < -7\}$

7. $\{r|-5/3 < r < -3/5\}$ **8.** $\{r|0 < r < 1/2\}$

9. $\{x|-12 \le x < 4\}$ **10.** $\{x|-18 < x \le 0\}$

11. $\{x|x \le -5\}$ **12.** $\{x|x \ge -5\}$

13. $\{t|t < 0\}$ **14.** $\{t|-t \ge 12\}$

In Exercises 15–20, express the set using set notation.

15. $[-3, 4]$ **16.** $[-2, 7/2)$

17. $(0, \infty)$ **18.** $[0, 10)$

19. $(-1/2, 1/2]$ **20.** $(-\infty, -100)$

In Exercises 21–40, solve the given inequality. Express the answer in set notation and interval notation.

21. $4x - 1 > 0$ **22.** $-3x + 2 > 0$

23. $\frac{1}{2}x + \frac{1}{3} \ge 0$ **24.** $1 - 2x \ge 0$

25. $-7x + 9 \le 0$ **26.** $3x - \frac{1}{3} < 0$

27. $4x + 1 \ge x - 2$ **28.** $3x - 1 < 2x + 1$

29. $5x + 9 \le -5x + 19$ **30.** $4 - x > 2x - 5$

31. $-1 \le 2x + 1 \le 1$ **32.** $-5 < x + 3 < -3$

33. $2 \le -3x + 6 \le 3$ **34.** $0 \le \frac{-2x + 1}{3} \le 7$

35. $-10 < \frac{4 - x}{2} < 10$ **36.** $5 \le \frac{3x - 1}{2} \le 2$

37. $-\sqrt{2} \le 4x - 1 \le \sqrt{2}$ **38.** $-\sqrt{27} \le \sqrt{3}x \le \sqrt{12}$

39. $-1 < \frac{\sqrt{2} - x}{-\sqrt{2}} < 1$ **40.** $-\pi \le \frac{3x - 1}{\pi/2} \le \pi$

In Exercises 41–56, solve the given inequality.

41. $|x - 4| \le 4$ **42.** $|2x + 1| > 6$

43. $|1 - 2x| < 1$ **44.** $|3 - 4x| \le \frac{1}{2}$

45. $|4 - 2x| \ge 1$ **46.** $|5x - 1| > 5$

47. $\left|1 - \frac{1}{2}x\right| \le \frac{1}{4}$ **48.** $|3x + 1| \le 0$

49. $|x + 3| > -2$ **50.** $5 \le |2x - 1|$

51. $|x - 3| \le -3$ **52.** $0 < |x + 3|$

53. $|2x + 1| > 6$ **54.** $-|3x + 5| \ge -7$

55. $|x - 1| \le 2\sqrt{2}$ **56.** $\frac{|3x + 9|}{2} \le 5$

1.8 Quadratic and Other Inequalities

A **quadratic inequality** is an inequality of one of the following four types

$$ax^2 + bx + c < 0$$
$$ax^2 + bx + c \leq 0$$
$$ax^2 + bx + c > 0$$
$$ax^2 + bx + c \geq 0$$

where a, b, and c are real numbers, and $a \neq 0$. Let us begin our discussion of quadratic inequalities by making a definition. If we rcplace the inequality sign in a quadratic inequality by an equal sign, the result will be a quadratic equation. Let us call this quadratic equation the **associated quadratic equation** for the given inequality.

For instance, the associated quadratic equation for the inequality

$$x^2 + x - 2 \leq 0$$

is

$$x^2 + x - 2 = 0$$

Perhaps the best way to describe our technique for solving quadratic inequalities is to use an example.

EXAMPLE 1: Solve the inequality

$$x^2 + x - 2 \leq 0 \tag{1.9}$$

Solution: The first step is to solve the associated quadratic equation

$$x^2 + x - 2 = 0$$

We will leave it to you to show that its solutions are $x = -2$ and $x = 1$. Now we use these solutions to divide the real number line into five disjoint sets

$$(-\infty, -2), \{-2\}, (-2, 1), \{1\}, (1, \infty)$$

(**Disjoint** means that the sets have no elements in common.) Notice that three of these sets are intervals, and that the other two sets each contain one of the solutions to the associated quadratic equation. These sets are pictured in Figure 1.12.

Now, the solution set of the original inequality (1.9) is made up of one or more of these five sets. Our plan is to check each of these sets, one by one, to determine whether or not it is part of the solution set.

FIGURE 1.12

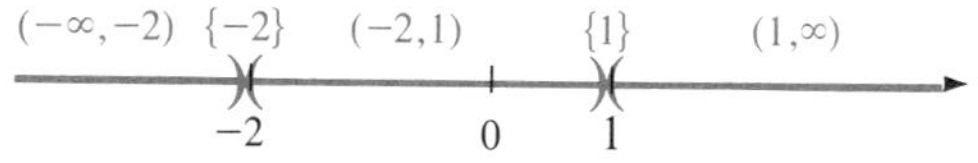

The reason that this approach works is that, in order to check whether or not one of these sets is part of the solution set of (1.9), *it is only necessary to check any one value from that set.* Thus, we may pick *any* value from the set, let us call that value the **test value** for the set. If the test value is a solution to (1.9), then so are *all* of the values in that set, and so the *entire* set is part of the solution set. On the other hand, if the test value is not a solution to (1.9), then *none* of the other values in the set are solutions, and so the set is not part of the solution set. (We will see *why* these statements are true in Chapter 3.)

For example, in order to tell whether or not the set $(-\infty, -2)$ is part of the solution set, we pick a test value, say -3. In this case, -3 is not a solution to (1.9), since if we substitute $x = -3$ into (1.9), we get the false statement $4 \leq 0$. Therefore, *none* of the values in the set $(-\infty, -2)$ are solutions.

Since we must check a total of five different sets in this procedure, it is best to organize our information in a table. Notice that for the two sets that contain only one element, the test value is that element.

Set	Test value	Result of substituting the test value into the inequality	Is the set part of the solution set?
$(-\infty, -2)$	-3	$4 \leq 0$ (false)	no
$\{-2\}$	-2	$0 \leq 0$ (true)	yes
$(-2, 1)$	0	$-2 \leq 0$ (true)	yes
$\{1\}$	1	$0 \leq 0$ (true)	yes
$(1, \infty)$	2	$4 \leq 0$ (false)	no

As you can see from this table, test values taken from the sets $\{-2\}$, $(-2, 1)$, and $\{1\}$ are solutions to (1.9), and so these sets are each part of the solution set of (1.9). However, test values taken from the sets $(-\infty, -2)$ and $(1, \infty)$ are not solutions, and so these sets are not part of the solution set. Thus, the solution set of (1.9) is

$$\{-2\} \cup (-2, 1) \cup \{1\}$$

which is the same as the closed interval $[-2, 1]$.

▶ **Study Suggestion 1.59:** Solve the inequality

$$x^2 - x - 6 \geq 0. \quad ◀$$

Try Study Suggestion 1.59. □

Now let us consider some other examples of this technique.

EXAMPLE 2: Solve the quadratic inequality

$$\frac{1}{3}x^2 > 2x - 3$$

Solution: The first step is to write this in the form $ax^2 + bx + c > 0$

$$x^2 - 6x + 9 > 0$$

Now we can solve the associated quadratic equation

$$x^2 - 6x + 9 = 0$$

In this case, the equation has only one solution $x = 3$ and so we use this solution to divide the real number line into *three* sets

$$(-\infty, 3), \{3\}, (3, \infty)$$

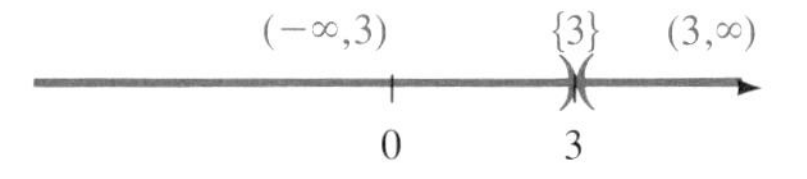

FIGURE 1.13

as pictured in Figure 1.13. Just as in Example 1, we check each of these sets to see if it is part of the solution set of the inequality. This is done in the following table.

Set	Test value	Result of substituting the test value into the inequality	Is the set part of the solution set?
$(-\infty, 3)$	0	$9 > 0$ (true)	yes
$\{3\}$	3	$0 > 0$ (false)	no
$(3, \infty)$	4	$1 > 0$ (true)	yes

According to this table, the solution set is

$$(-\infty, 3) \cup (3, \infty)$$

In words, this is the set of all real numbers *except* 3.

Try Study Suggestion 1.60. □

▶ **Study Suggestion 1.60:** Solve the inequality

$$\frac{x^2}{4} \leq -x - 1. \quad ◀$$

EXAMPLE 3: Solve the quadratic inequality

$$x(2x + 3) < -5 \qquad \textbf{(1.10)}$$

Solution: First we put this in the form $ax^2 + bx + c < 0$

$$2x^2 + 3x + 5 < 0$$

In this case, the associated quadratic equation

$$2x^2 + 3x + 5 = 0$$

has no solutions (since the discriminant is negative). As a result, there are no solutions with which to divide the real number line. Situations such as this one are actually the simplest of all. For in a sense, we have only one set to check, namely, the entire real number line. Thus, we may pick *any* real number. If it is a solution to the inequality, then *all* real numbers are solutions. On the other hand, if it is not a solution, then the inequality has no solutions.

The simplest number to pick is $x = 0$, since it is the easiest to substitute into the inequality. We see immediately that $x = 0$ is not a solution to (1.10), and so (1.10) has no solutions. In other words, the solution set is the empty set. *Try Study Suggestion 1.61.* □

▶ **Study Suggestion 1.61:** Solve the inequality

$$x^2 + 3x + 4 > 0.$$ ◀

The same technique that we have been using to solve quadratic inequalities can be used to solve certain other types of inequalities as well. Let us illustrate this with two examples.

EXAMPLE 4: Solve the inequality

$$\frac{x+6}{x-3} \leq 4$$

Solution: The first step in a case such as this is to arrange it so that one side of the inequality is 0. This can be done by subtracting 4 from each side of the inequality

$$\frac{x+6}{x-3} - 4 \leq 0$$

and then simplifying

$$\frac{x+6}{x-3} - \frac{4(x-3)}{x-3} \leq 0$$

$$\frac{x+6-4(x-3)}{x-3} \leq 0$$

$$\frac{-3x+18}{x-3} \leq 0 \qquad (1.11)$$

Now, it is important to notice that we cannot simply multiply both sides of (1.11) by $x - 3$, since if $x - 3$ is negative, then we would have to change the sense of the inequality, but if $x - 3$ is positive, then we must not change the sense of the inequality.

The next step is to find those real numbers that make either the numerator *or* the denominator equal to 0. In this case, $x = 6$ makes the numerator equal to 0 and $x = 3$ makes the denominator equal to 0.

Now we use these numbers to divide the real number line into disjoint sets, just as we did in solving quadratic inequalities. In this case, we get the sets

$$(-\infty, 3), \{3\}, (3, 6), \{6\}, (6, \infty)$$

These are pictured in Figure 1.14. At this point, we can proceed exactly as we did in the case of a quadratic inequality. The following table summarizes

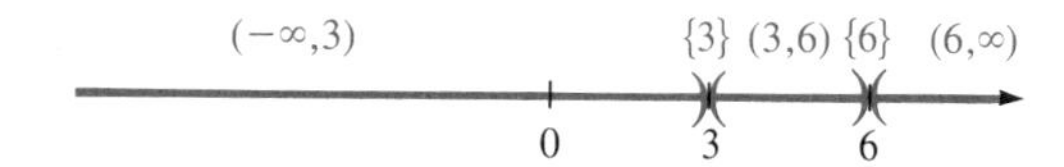

FIGURE 1.14

the needed information. In this case, we substitute our test value into (1.11), rather than into the original inequality. (You can do it either way, but it seems easier to use (1.11).) Notice also that, when we try to substitute $x = 3$ in (1.11), the result is nonsense, because it makes the denominator equal to 0. Certainly then, $x = 3$ is not a solution to the inequality.

Set	Test value	Result of substituting the test value into the inequality	Is the set part of the solution set?
$(-\infty, 3)$	0	$-6 \leq 0$ (true)	yes
$\{3\}$	3	nonsense (false)	no
$(3, 6)$	4	$6 \leq 0$ (false)	no
$\{6\}$	6	$0 \leq 0$ (true)	yes
$(6, \infty)$	7	$-3/4 \leq 0$ (true)	yes

We can see from this table that the solution set to the original inequality is

$$(-\infty, 3) \cup \{6\} \cup (6, \infty) \quad \text{or} \quad (-\infty, 3) \cup [6, \infty)$$

Try Study Suggestion 1.62. □

▶ **Study Suggestion 1.62:** Solve the following inequalities

(a) $\dfrac{x+1}{x-1} > 2$

(b) $\dfrac{x+1}{x-1} \geq 1$ ◀

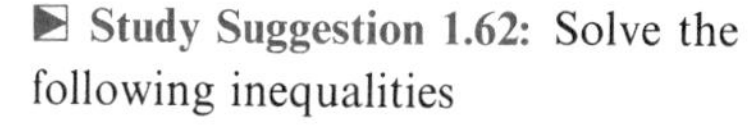

EXAMPLE 5: Solve the inequality

$$\left|\frac{x+4}{x-1}\right| < 2$$

Solution: According to the results of the previous section, this inequality is equivalent to the double inequality

$$-2 < \frac{x+4}{x-1} < 2$$

Thus, the solutions to the original inequality are those real numbers that satisfy *both* of the inequalities

$$-2 < \frac{x+4}{x-1} \quad \text{and} \quad \frac{x+4}{x-1} < 2$$

Each of these inequalities can be solved in a manner similar to the previous example. We will leave it to you to show that the solution set of the first inequality is $(-\infty, -2/3) \cup (1, \infty)$, as shown in Figure 1.15a, and the solution set of the second inequality is $(-\infty, 1) \cup (6, \infty)$, as shown in Figure 1.15b. Now, the solution set of the original inequality is the set of all real

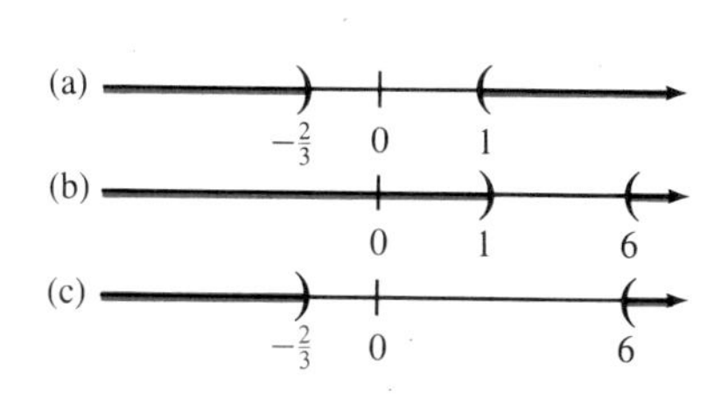

FIGURE 1.15

numbers that are in *both* of these sets. As you can see from Figure 1.15, these are the values of x in the set $(-\infty, -2/3) \cup (6, \infty)$, shown in Figure 1.15(c).

Try Study Suggestion 1.63. □

▶ **Study Suggestion 1.63:** Solve the inequality

$$\left|\frac{2x+1}{x-1}\right| \leq 2. \ ◀$$

Ideas to Remember

Quadratic inequalities can be solved in the following way.

1. Solve the associated quadratic equation.
2. Use the solutions (if any) to divide the real number line into disjoint sets.
3. Check each of these sets by means of a test value to determine which of them are part of the solution set of the inequality.

EXERCISES

In Exercises 1–24, solve the given quadratic inequality.

1. $x^2 - 3x + 2 \leq 0$
2. $x^2 - 2x < 3$
3. $2x^2 + 9x + 4 \geq 0$
4. $x^2 - 9 > 0$
5. $x^2 - 10x + 25 < 0$
6. $2x^2 - 12x + 18 \leq 0$
7. $6t^2 + t - 1 > 0$
8. $t^2 - 9t - 10 \leq 0$
9. $2x^2 - 11x + 12 < 0$
10. $20x(x - 1) < 1$
11. $u^2 - 3u + 3 > 0$
12. $4u^2 + 4u + 1 \geq 0$
13. $2y(y + 1) \leq y - 1$
14. $1 - 9y^2 \leq 0$
15. $-s^2 + 6s - 5 < 0$
16. $-2 + 2s - s^2 > 0$
17. $-x^2 + (2/3)x - (1/9) \geq 0$
18. $1 > x(2 + 3x)$
19. $-x^2 + 11x \geq 30$
20. $8 + 2x < x^2$
21. $-3x^2 + 2x + (1/2) \geq 0$
22. $1 - 2x^2 > 0$
23. $11x + 3 > 4x^2$
24. $2 + (1 - 2\pi)x - \pi x^2 < 0$

In Exercises 25–41, solve the given inequality.

25. $\dfrac{x+5}{x-2} \leq 3$
26. $\dfrac{2x+1}{2x-1} \leq 2$
27. $\dfrac{1-2x}{2-3x} > 1$
28. $\dfrac{4-x}{4+x} \geq -1$
29. $\left|\dfrac{x-1}{x+5}\right| < 1$
30. $\left|\dfrac{x+2}{x-3}\right| > 1$
31. $\left|\dfrac{x}{x+1}\right| > 1$
32. $\left|\dfrac{8x-5}{2x+1}\right| \geq 5$
33. $\left|\dfrac{3x+2}{x-3}\right| > 2$
34. $x^3 < x$
35. $\dfrac{1}{x(x-1)} \leq 0$
36. $\dfrac{x}{x^2-1} \geq 0$
37. $x^4 + 2x^2 - 5 \leq 0$
38. $(x - 1)(x + 2)(x - 3) > 0$
39. $(x + 2)(2x - 1)(2x - 3) \leq 0$
40. $(x^2 + x + 1)(2x - 3) \leq 0$
41. $(x - 1)^2(x - 2)^3(x - 3)^4 < 0$

1.9 Review

CONCEPTS FOR REVIEW

Real number line
Natural number
Integer
Rational number
Irrational number
Real number
Commutative properties
Associative properties
Identity properties
Inverse properties
Distributive properties
Cancellation properties
"Invert and multiply"
Positive
Negative
Less than
Greater than
Absolute value
Distance between two points on the number line
Union
Rationalizing the denominator
Scientific notation
Polynomial
Algebraic expression
Term
Constant term
Leading term
Degree of a polynomial
Zero polynomial
Constant polynomial
Linear polynomial
Quadratic polynomial
Cubic polynomial
Factor
Irreducible polynomial
Common factor
Difference of two squares
Sum of two cubes
Difference of two cubes
Rational expression
Composition of algebraic expressions
Equivalent equations
Extraneous solution
Quadratic equation
Method of factoring
Completing the square
The quadratic formula
The discriminant
Inequality
Solution set
Open interval
Closed interval
Half-open interval
Endpoints
Quadratic inequality
Associated quadratic equation
Disjoint
Test value

REVIEW EXERCISES

In Exercises 1 and 2, state which properties of the real number system are being used at each step.

1. $4(2x + 3) = 8$
$2x + 3 = 2$
$2x = -1$
$x = -1/2$

2. $(x - 1)(2x - 4)(3x - 9) = 0$
$x - 1 = 0$ or $2x - 4 = 0$ or $3x - 9 = 0$
$x = 1$ or $x = 2$ or $x = 3$

In Exercises 3–6, simplify the given expressions.

3. $\dfrac{7}{9} - \dfrac{9}{7}$

4. $\dfrac{\frac{2}{3} + \frac{3}{2}}{\frac{1}{4} - \frac{7}{3}}$

5. $\dfrac{x}{y^2} + \dfrac{y}{x^2}$

6. $\left(\dfrac{x^2}{y} - \dfrac{y^2}{x}\right) \div \left(\dfrac{1}{y} - \dfrac{1}{x}\right)$

7. Does the operation of subtraction satisfy the associative property? Justify your answer.

8. Does the operation of division satisfy the commutative property? Justify your answer.

9. Is it true that

$$\frac{63}{117} < \frac{62}{118}$$

Justify your answer.

10. Is it true that

$$\pi^2 \leq \frac{484}{49}$$

Justify your answer.

In Exercises 11 and 12, write the expressions without absolute value signs.

11. **(a)** $\left|\pi - \dfrac{22}{7}\right|$ **(b)** $|(\sqrt{2} - \sqrt{3})(\sqrt{2} + \sqrt{3})|$

12. **(a)** $\left|\dfrac{72}{49} - \dfrac{71}{48}\right|$ **(b)** $|x^2 + 1|$

13. Compute the following distances.

(a) $d(-5/2, 5/2)$

(b) $d(-\sqrt{2}, 1 + \sqrt{2})$

(c) $d(1/12, -11/12)$

14. Show the set $\{t \mid -2 \leq t < 3\} \cup \{t \mid -1 \leq t < 4\}$ on the real number line.

15. Write the set $\{x \mid 1 \leq |x| < 2\}$ without using absolute value signs.

In Exercises 16–31, simplify the given expression.

16. $(-2)^{-3}$

17. $(-2)^3(-4)^2(-8)^{-3}$

18. $\left(\dfrac{1}{3}\right)^{-2}\left(\dfrac{2}{3}\right)^4\left(\dfrac{3}{2}\right)^{-6}$

19. $\dfrac{a^2bc^3d^{-4}}{3^{-1}a^{-2}b^2c^{-3}}$

20. $xy^2(-3x^2y^3z^{-3})^2$

21. $(x^{-1} + y^{-1})(x + y)^{-1}$

22. $\sqrt{12}$

23. $\sqrt{288}$

24. $\sqrt{(-1)^2}$

25. $2\sqrt{20} - (8/3)\sqrt{45} + \sqrt{80}$

26. $\sqrt{21x^5y^{-3}}\sqrt{3x^{-1}y}$

27. $(\sqrt[5]{x^{1/2}}\,\sqrt[5]{x^{1/3}})^6$

28. $\left(\dfrac{1}{\sqrt{x}} + \dfrac{1}{\sqrt{y}}\right)(\sqrt{x} - \sqrt{y})$

29. $\sqrt{\left(\dfrac{-x^4}{x^6 + 1}\right)^2}$

30. $\left(\dfrac{4x^{-1/2}y^{1/3}z^{1/4}}{8x^{1/2}y^{1/3}z^{1/6}}\right)^6$

31. $(4x^2y^{1/2} + 3y^2x^{1/2})^0$

In Exercises 32 and 33, use scientific notation to help compute the indicated value.

32. $(100{,}000{,}000)^{1/2}$

33. $\dfrac{\sqrt[4]{160{,}000{,}000}}{(3{,}600{,}000{,}000{,}000)^{1/2}}$

In Exercises 34–42, perform the indicated operations and then simplify the result.

34. $(2x + 3y - 4z) + (6x - 4 + 2z) - (3x + 9 - z)$

35. $(5x^3 - 3x^2 + 2x - 1) + (6x^3 - x^2 + x + 2) - (x^5 - 3x^2 + 2x)$

36. $(3x - 1)(4x^2 + 2)$

37. $(6x^2 - 1)(3x^2 + 1) - 2x^2(4x^2 + 3x - 4)$

38. $(u^2 + v^2)(u^2 - v^2)$

39. $(a + b + c)(2a - 3b + 4c)$

40. $(3t^2 + \sqrt{t})^2$

41. $(2y - y^{-1})(2y + y^{-1})(4y^2 + y^{-2})$

42. $(\sqrt{x} + 1/\sqrt{x})^3$

In Exercises 43–54, factor the given expression.

43. $4ab^2c^3 - 6a^2c + 2abc$

44. $x(2x + y) + 3y^2(2x + y) - 2xz - yz$

45. $12x^2 + 2x - 10$

46. $12x^2 + 21x + 9$

47. $125y^3 - 1$

48. $16t^4 - 49s^6$

49. $25a^2 + 30ab + 9b^2$

50. $4xy + xz - 8wy - 2wz + 24uy + 6uz$

51. $12x^4 - 5x^2 - 2$

52. $4x^5 - x^3 - 4x^2 + 1$

53. $16y + 24\sqrt{y} + 9$

54. $9x^4 + 11x^2y^2 + 16y^4$

In Exercises 55–67, simplify the given expression.

55. $\dfrac{6x^2 + 19x - 36}{2x^2 + 7x - 9}$

56. $\dfrac{x^4 - 6x^3 + 9x^2}{x^3 - 9x}$

57. $\dfrac{2x}{x+1} - \dfrac{6}{2x+1}$

58. $\dfrac{1}{x^2+1} + \dfrac{1}{x^2-1}$

59. $\dfrac{12y^2 + 5y - 2}{6y+1} \dfrac{6y^2 - 17y - 3}{4y^2 + 3y - 1}$

60. $\dfrac{2z^3 + z^2}{3z+1} \div \dfrac{4z^3 - z}{3z^2 - 11z - 4}$

61. $\dfrac{2}{2a+1} - \dfrac{3}{(2a+1)^2} + \dfrac{4}{(2a+1)^3}$

62. $\dfrac{1}{x} + \dfrac{1-x}{x^2} + \dfrac{(1-x)^2}{x^3}$

63. $\dfrac{2uv - 3zw + 2uw - 3zv}{zv - uv - uw + zw}$

64. $\dfrac{\dfrac{1}{x^2} - \dfrac{1}{y^2}}{\dfrac{x^2}{y} - \dfrac{y^2}{x}}$

65. $(x - y)^{-2} - (x + y)^{-2}$

66. $\dfrac{\sqrt{x+1} - \sqrt{x-1}}{\sqrt{x+1} + \sqrt{x-1}}$

67. $\dfrac{(2+x)\sqrt{1+x^2} - \dfrac{x^3 - 3}{\sqrt{1+x^2}}}{x\sqrt{1+x^2}}$

In Exercises 68–74, find the composition $p(q(x))$.

68. $p(x) = 4x - 5$, $q(x) = 7x + 1$

69. $p(y) = \dfrac{y-5}{3}$, $q(x) = 3x - 5$

70. $p(x) = \dfrac{1}{2x-1}$, $q(x) = 4x + 5$

71. $p(x) = \sqrt{6x - 1}$, $q(x) = (1/6)(x^2 + 1)$

72. $p(x) = x^8 - 2x^4 - 1$, $q(x) = \sqrt{x + 2}$

73. $p(y) = \dfrac{1}{y^2 - 3}$, $q(x) = p(x)$

74. $p(x) = \dfrac{1}{x^{1/2}} + \dfrac{1}{x^{3/2}}$, $q(x) = \dfrac{1}{x^{5/2}}$

75. If $p(x) = 4x - 7$, find a linear polynomial $q(x)$ with the property that $p(q(x)) = x$ and $q(p(x)) = x$.

In Exercises 76–101, solve the given equation.

76. $5x - 9 = 12$

77. $(x + 1)^3 = 4(x + 1)^4$

78. $\dfrac{x+9}{3} = \dfrac{x-7}{5}$

79. $\dfrac{1}{2-x} = \dfrac{3}{4x+1}$

80. $\dfrac{6x-1}{6x+2} = 9 - \dfrac{1}{3x+1}$

81. $\sqrt{x} + a + 1 = (a + 1)\sqrt{ax} + 1$ (a constant)

82. $2x^2 - 5x - 3 = 0$

83. $2 - (2/3)x - (1/6)x^2 = 0$

84. $5y^2 + y + 1 = 0$

85. $t^2 + \sqrt{3}t - 1 = 0$

86. $2x^2 + x - \sqrt{2} = 0$

87. $2x^2 - 5xy + 2y^2 = 0$ (y constant)

88. $yx^2 + 2xy - 3 = 0$ (y constant)

89. $y^3x^2 - 2x\sqrt{y} = 0$ (y constant)

90. $\sqrt{x} = x^{4/5}$

91. $3x^4 + 2x^2 - 1 = 0$

92. $u - 7\sqrt{u} + 1 = 0$

93. $1 - 4t^{1/6} + 2t^{1/3} = 0$

94. $4x^6 + 20x^3 + 25 = 0$

95. $\left(\dfrac{2t}{t^2-1}\right)^2 + \dfrac{6t}{t^2-1} + 3 = 0$

96. $\sqrt{2x+1} = -\sqrt{1-3x}$

97. $|6-3x| = 6$

98. $|1+2z-z^2| = 3$

99. $|w+\sqrt{w}| = |1-\sqrt{w}|$

100. $\sqrt{4-x} - \sqrt{-x/2} = \sqrt{x+6}$

101. $|x+a| = |x-a|$ (a constant)

In Exercises 102–117, solve the given inequality.

102. $3x - 4 > 2$

103. $\frac{1}{2}x + 2 < 3x - \frac{1}{2}$

104. $-\pi \leq \frac{\pi x - 1}{2} \leq \frac{3\pi}{2}$

105. $\frac{1}{2} < |x+3| \leq 4$

106. $-\pi \leq \left|x + \frac{\pi}{2}\right| \leq \frac{5\pi}{4}$

107. $\frac{1}{2} < |x+3| - 2 \leq 4$

108. $3x^2 + x + 1 \leq 0$

109. $2x^2 + x - 1 > 0$

110. $1 - 2t - 4t^2 \geq 0$

111. $\frac{1}{2} - 2t + \frac{1}{4}t^2 \geq 1$

112. $2y^4 - 9y^2 - 5 < 0$

113. $\frac{1-3y}{2-5y} > 1$

114. $\frac{x}{2x+1} + \frac{1}{x+1} \leq 0$

115. $\frac{2x^2+1}{3x^2+1} \leq 0$

116. $\frac{x-1}{x^2+x-1} \geq 0$

117. $(x^2-2)(2x^2-1) \leq 0$

2

FUNCTIONS AND THEIR GRAPHS

2.1 The Cartesian Coordinate System

During our study of elementary algebra, we found several important uses for the real number line. For example, we used it to describe the solution set of an inequality. As we will see, the plane can be used in a similar manner, where each point represents—not just a single real number—but rather an *ordered pair* of real numbers.

We begin by setting up a **Cartesian** (or **rectangular**) **coordinate system** for the plane, which consists of two perpendicular number lines, as shown in Figure 2.1. The horizontal line is called the **horizontal axis** or the ***x*-axis** and the vertical line is called the **vertical axis** or the ***y*-axis.** The point where

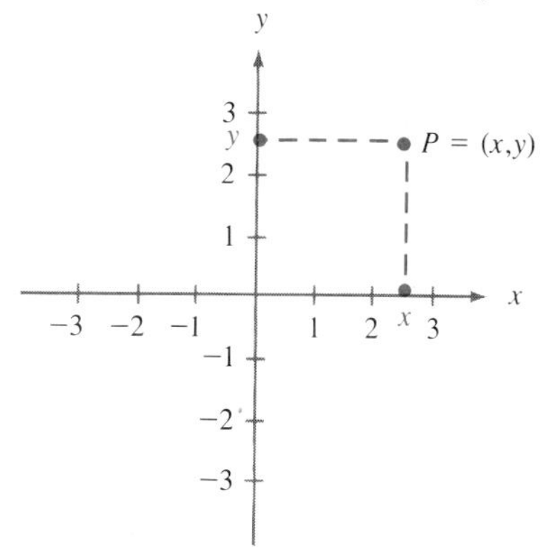

FIGURE 2.1

the two lines meet is called the **origin.** (The term *Cartesian* refers to the famous French mathematician Rene Descartes (1596–1650).)

Using a Cartesian coordinate system, we can associate each point in the plane with an ordered pair (x, y) of real numbers by "dropping perpendiculars" onto the two axes, as shown in Figure 2.1. The numbers x and y are called the **coordinates** of P, and we frequently refer to the ordered pair itself as a *point* in the plane.

When we label the point in the plane corresponding to a given ordered pair (x, y), we say that we are **plotting the point** (x, y). Several examples of points and their associated ordered pairs are given in Figure 2.2.

As you can see from Figure 2.2, the coordinate axes divide the plane into four regions, called **quadrants.** Notice that the quadrants are numbered *counterclockwise.* A point (x, y) is in the first quadrant if and only if $x > 0$ and $y > 0$. A point is in the second quadrant if and only if $x < 0$ and $y > 0$, and so on. Points that have one, or both, coordinates equal to 0 lie on one, or both, of the axes, and are not in any of the quadrants.

▶ **Study Suggestion 2.1:** Draw a Cartesian coordinate system and plot each of the following points. Also, state in which quadrant each point lies.
(a) $(0, 0)$ **(b)** $(0, -3)$ **(c)** $(1/2, 0)$
(d) $(3, -4)$ **(e)** $(-5/2, -5/2)$
(f) $(4, 1/2)$ ◀

Try Study Suggestion 2.1.

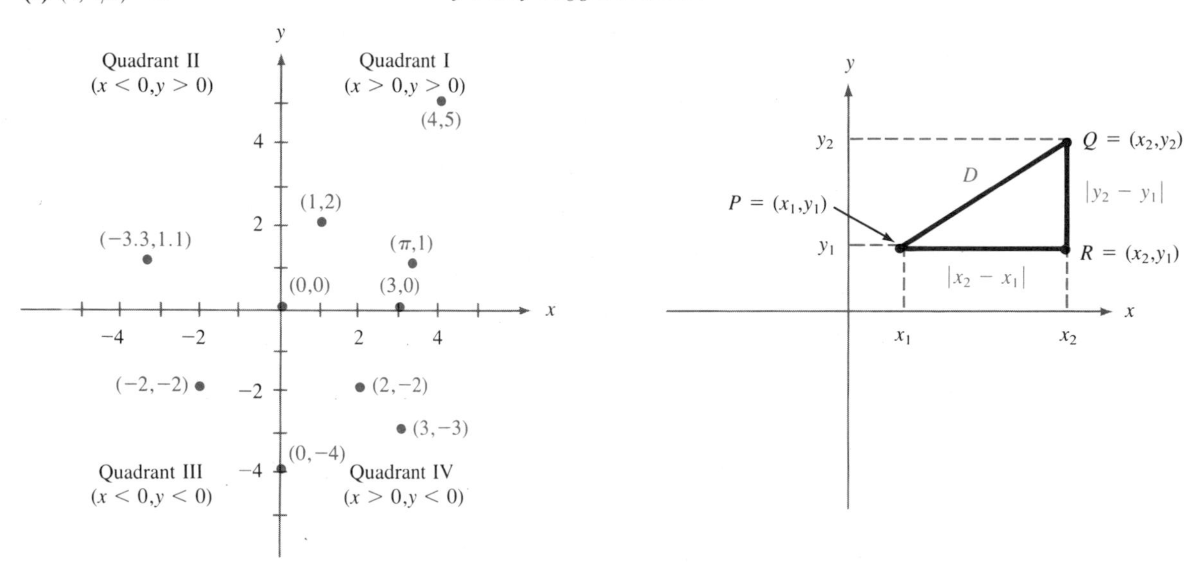

FIGURE 2.2

FIGURE 2.3

It is useful to observe that

> **points that lie on the same vertical line have the same x-coordinate, and points that lie on the same horizontal line have the same y-coordinate.**

As with points on the number line, there is a simple formula for the distance between two points $P = (x_1, y_1)$ and $Q = (x_2, y_2)$ in the plane. We can see where this formula comes from by examining Figure 2.3, where we have drawn a right triangle using P and Q as vertices, and labeled the third vertex R. Since R is on the same vertical line as Q, it has the same x-coordinate as Q, namely x_2, and since it is on the same horizontal line as P, it has the same y-coordinate as P, namely y_1. We have also labeled each of the sides of the triangle with its length. Now, according to the Pythagorean Theorem

$$D^2 = |x_2 - x_1|^2 + |y_2 - y_1|^2$$

and since $|r|^2 = r^2$ for any real number r, we can write this as

$$D^2 = (x_2 - x_1)^2 + (y_2 - y_1)^2$$

Taking the square root of both sides and using the fact that distance is always nonnegative, we get

$$D = \sqrt{(x_2 - x_1)^2 + (y_2 - y_1)^2}$$

This is our formula for the distance between two points in the plane. For reference, let us put it into a theorem.

THEOREM

The distance $d(P, Q)$ between two points $P = (x_1, y_1)$ and $Q = (x_2, y_2)$ in the plane is given by the formula

$$d(P, Q) = \sqrt{(x_2 - x_1)^2 + (y_2 - y_1)^2}$$

This formula is known as the **distance formula.**

The distance formula can be remembered by observing that $x_2 - x_1$ is the difference between the x-coordinates of the points P and Q, and $y_2 - y_1$ is the difference between the y-coordinates. Notice also that these numbers can be subtracted in either order because the result is squared.

EXAMPLE 1:

(a) If $P = (1, 2)$ and $Q = (3, -2)$ then

$$\begin{aligned} d(P, Q) &= \sqrt{(3 - 1)^2 + (-2 - 2)^2} \\ &= \sqrt{2^2 + (-4)^2} \\ &= \sqrt{20} \\ &= 2\sqrt{5} \end{aligned}$$

(b) If $P = (1/2, -4)$ and $Q = (7/2, 0)$ then

$$\begin{aligned} d(P, Q) &= \sqrt{(7/2 - 1/2)^2 + (0 - (-4))^2} \\ &= \sqrt{3^2 + 4^2} \\ &= 5 \end{aligned}$$

Try Study Suggestion 2.2. □

► **Study Suggestion 2.2:** In each case, compute the distance $d(P, Q)$.

(a) $P = (2, 2)$, $Q = (-3, -2)$

(b) $P = (0, 1)$, $Q = (1/2, 0)$

(c) $P = (a, b)$, $Q = (a, 2b)$ ◄

The distance formula can be very useful in solving certain problems in plane geometry. For example, it can tell us whether or not three points in the plane form the vertices of a right triangle. In order to see this, we use the fact that the points P, Q, and R form the vertices of a right triangle if and only if the distances $d(P, Q)$, $d(P, R)$ and $d(Q, R)$ satisfy the Pythagorean Theorem. Let us illustrate with an example.

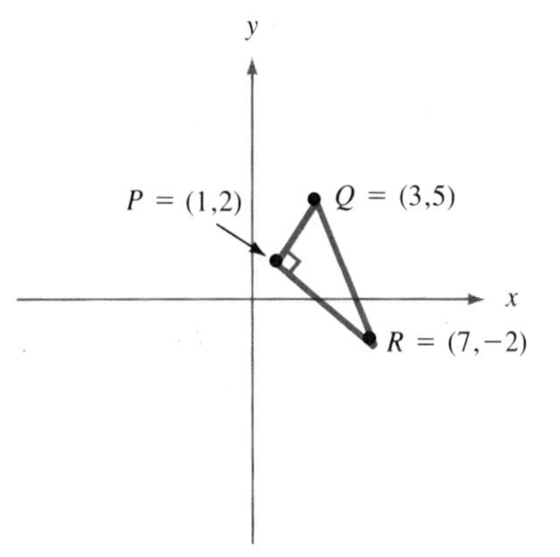

FIGURE 2.4

EXAMPLE 2: Determine whether or not the points $P = (1, 2)$, $Q = (3, 5)$, and $R = (7, -2)$ form the vertices of a right triangle.

Solution: We begin by finding the distances between each pair of points.

$$d(P, Q) = \sqrt{(3-1)^2 + (5-2)^2} = \sqrt{4+9} = \sqrt{13}$$
$$d(P, R) = \sqrt{(7-1)^2 + (-2-2)^2} = \sqrt{36+16} = \sqrt{52}$$
$$d(Q, R) = \sqrt{(7-3)^2 + (-2-5)^2} = \sqrt{16+49} = \sqrt{65}$$

Then we try to make these numbers "fit" the Pythagorean Theorem, which has the form

$$a^2 + b^2 = c^2 \tag{2.1}$$

In this case, we can fit the numbers $d(P, Q)$, $d(P, R)$, and $d(Q, R)$ into Equation (2.1) as follows

$$(\sqrt{13})^2 + (\sqrt{52})^2 = (\sqrt{65})^2 \qquad \text{(that is, } 13 + 52 = 65\text{)}$$

or

$$[d(P, Q)]^2 + [d(P, R)]^2 = [d(Q, R)]^2$$

(If the numbers are to fit at all, then the largest number must be c.) This tells us that P, Q, and R form the vertices of a right triangle, whose hypotenuse is the line segment from Q to R. This triangle is pictured in Figure 2.4. *Try Study Suggestion 2.3.* □

▶ **Study Suggestion 2.3:** In each case, determine whether or not the points P, Q, and R form the vertices of a right triangle.

(a) $P = (1, 3)$, $Q = (9, 5)$, $R = (2, -1)$
(b) $P = (4, 2)$, $Q = (7, 6)$, $R = (3, 5)$ ◀

If P and Q are points in the plane, we may wish to find the coordinates of the midpoint M of the line segment from P to Q, as shown in Figure 2.5. Let us derive a formula for the coordinates of M in terms of the coordinates of P and Q.

In Figure 2.5, we have labeled the endpoints $P = (x_1, y_1)$ and $Q = (x_2, y_2)$ and the midpoint $M = (x, y)$. According to the laws of geometry, since M divides the segment from P to Q in half, the point x also divides the segment from x_1 to x_2 on the x-axis in half. In other words, the distance from x to x_1 is the same as the distance from x to x_2. In symbols,

$$|x - x_1| = |x - x_2|$$

But since $x_1 < x < x_2$, this can be written without absolute value signs as

$$x - x_1 = x_2 - x$$

Solving this for x gives

$$x = \frac{x_1 + x_2}{2}$$

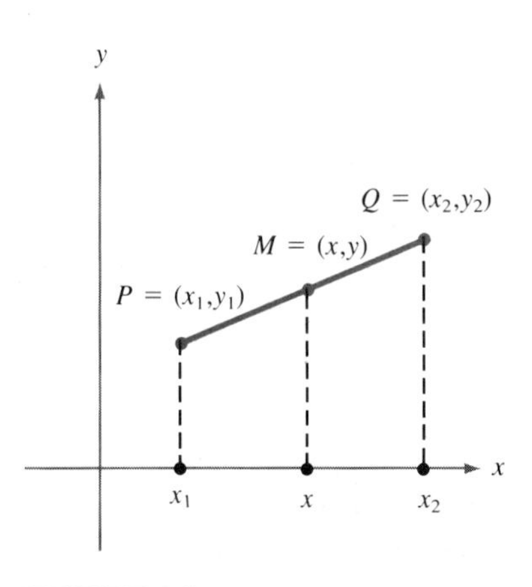

FIGURE 2.5

In a similar way, we can show that $y = (y_1 + y_2)/2$, and so we arrive at the following theorem.

THEOREM

If $P = (x_1, y_1)$ and $Q = (x_2, y_2)$ are points in the plane, then the midpoint M of the line segment from P to Q has coordinates

$$M = \left(\frac{x_1 + x_2}{2}, \frac{y_1 + y_2}{2}\right)$$

This is known as the **midpoint formula.**

Perhaps the best way to remember the midpoint formula is to remember that the x-coordinate of the midpoint is the *average* of the x-coordinates of the endpoints, and similarly for the y-coordinate.

EXAMPLE 3: Find the coordinates of the midpoint of the line segment from (2, 5) to (3, −1).

Solution: Using the midpoint formula, we have

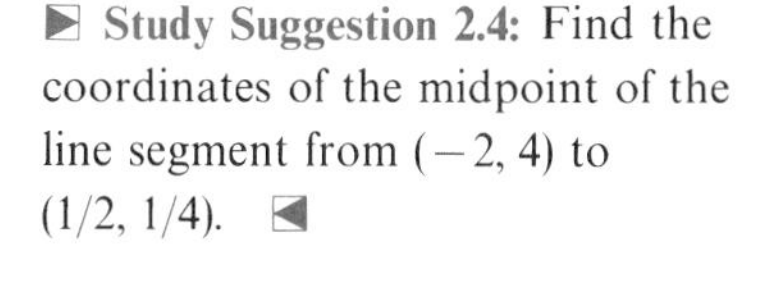

▶ **Study Suggestion 2.4:** Find the coordinates of the midpoint of the line segment from (−2, 4) to (1/2, 1/4). ◀

$$M = \left(\frac{2 + 3}{2}, \frac{5 + (-1)}{2}\right) = \left(\frac{5}{2}, 2\right)$$

Try Study Suggestion 2.4. □

Ideas to Remember

The Cartesian coordinate system allows us to represent ordered pairs of real numbers as points in the plane, just as the real number line allows us to represent real numbers as points on the line.

EXERCISES

In Exercises 1–5, plot the given points in the same plane, and identify the quadrant for each point (if it lies in a quadrant.)

1. (1, 3), (1, −4), (6, 0), (−2, 2)

2. (1/2, 1/2), (−1/2, −1/2), (1/4, 1/3), (−1/10, 1/5)

3. (12.5, −1.3), (6.12, 3.86), (−0.25, 0.55)

4. $(-2, \sqrt{2})$, $(1 + \sqrt{2}, 1 - \sqrt{2})$, $(2\sqrt{3}, 5\sqrt{3})$

5. (π, π), $(-\pi, \pi/2)$, $(\sqrt{\pi}, \pi 2)$

In Exercises 6–13, find the distance between the given pair of points.

6. (1, 2), (2, 3)

7. (2, −1), (2, 6)

8. (5, −3), (−2, −3)

9. (5, −10), (4, −8)

10. $(\sqrt{2}, 1)$, $(1, \sqrt{2})$

11. $(\sqrt{2}, \sqrt{3})$, $(1/\sqrt{2}, 1/\sqrt{3})$

12. (a, a), (b, b)

13. $(5\sqrt{3}, \sqrt{5})$, $(2\sqrt{3}, -4\sqrt{5})$

In Exercises 15–20, use the distance formula to determine whether or not the given points form the vertices of a right triangle.

15. (5, 1), (1, 3), (−1, −11)

16. (1, 1), (1/3, 3), (4, 2)

17. (6, −7), (−1, 1), (4, 2)

18. (1/3, 7/6), (−3, −1/2), (−2, −5/2)

19. $(\pi, 2\pi)$, $(3\pi, -\pi/2)$, $(5, (6/5)\pi + 4)$

20. $(\sqrt{2}, 1 + \sqrt{2})$, $(3\sqrt{2}, 2 + 4\sqrt{2})$, $(6 + 2\sqrt{2}, \sqrt{2} - 3)$

In Exercises 21–25, find the coordinates of the midpoint of the line segment from P to Q.

21. $P = (1, 2)$, $Q = (-5, 3)$

22. $P = (0, 0)$, $Q = (4, -7)$

23. $P = (1/2, 1/4)$, $Q = (1/6, 1/8)$

24. $P = (6, -10)$, $Q = (12, 17)$

25. $P = (\sqrt{2}, \sqrt{3})$, $Q = (2 + \sqrt{2}, 3 - 5\sqrt{3})$

In Exercises 26–29, find the other endpoint Q of the line segment PQ with given endpoint P and given midpoint M.

26. $P = (1, 3)$, $M = (2, 4)$

27. $P = (-2, 3)$, $M = (5, 6)$

28. $P = (-4, 5)$, $M = (-3, -10)$

29. $P = (\sqrt{2}, \pi)$, $M = (\sqrt{3}, 2\pi)$

In Exercises 30–33, determine whether or not the point A is on the perpendicular bisector of the line segment from P to Q. (*Hint*:, describe the condition that A is on the perpendicular bisector of the line segment PQ in terms of $d(A, P)$ and $d(A, Q)$.)

30. $A = (-3, 7)$, $P = (1, 1)$, $Q = (3, 3)$

31. $A = (0, -1)$, $P = (1, 3)$, $Q = (4, -2)$

32. $A = (1, 11/4)$, $P = (2, 5)$, $Q = (3, 4)$

33. $A = (4, -1)$, $P = (-1, -2)$, $Q = (3, 4)$

In Exercises 34–37, use the distance formula to determine whether or not the given points all lie on the same line. (*Hint:* three points P, Q, and R lie on the same line if and only if the sum of two of the distances $d(P, Q)$, $d(P, R)$, and $d(Q, R)$ is equal to the third.)

34. $(2, 3)$, $(1, -4)$, $(3, 10)$

35. $(1/7, 7/3)$, $(0, -2)$, $(3, 5)$

36. $(5, -1)$, $(-4, 9)$, $(3, 1)$

37. $(\pi, \sqrt{2})$, $(1 + \pi, 1 + 2\sqrt{2})$, $(\pi + \sqrt{2}, 2 + 2\sqrt{2})$

38. Find all values of x such that the distance from $(x, -3)$ to $(2, 3)$ is equal to 10.

39. Find all values of x such that the distance from $(1, 2)$ to $(x, 1 + 2x)$ is equal to 1.

40. Show that the points $(-5, -1)$, $(10, 7)$, $(18, -8)$, and $(3, -10)$ are the vertices of a square.

41. Show that the points $(-7, 0)$, $(8, 8)$, $(16, -7)$, and $(1, -9)$ are the vertices of a square.

42. Find all values of x for which the points $(-9, 1)$, $(6, 9)$, $(14, -6)$, and $(x, -8)$ form the vertices of a square.

43. Show that the diagonals of the parallelogram bisect each other. (*Hint:* position the parallelogram so that the coordinates of its vertices are $(0, 0)$, $(a, 0)$, (b, c), and $(a + b, c)$. Then find the midpoints of the diagonals.)

44. Prove that the diagonals of a rectangle have the same length.

45. Prove that the midpoint of the hypotenuse of a right triangle is equidistant from the vertices of the triangle.

2.2 Graphing Sets and Equations

The Cartesian coordinate system can be used to describe sets of ordered pairs. As a simple example, suppose that for each student in your algebra class, we form the ordered pair (a, b) whose first coordinate is the student's height, and whose second coordinate is the student's weight. Then we can obtain a "picture" of the set S of all such ordered pairs by plotting these ordered pairs in the plane, using a Cartesian coordinate system.

Suppose for the sake of simplicity that your class has five students, and that the ordered pairs are as follows:

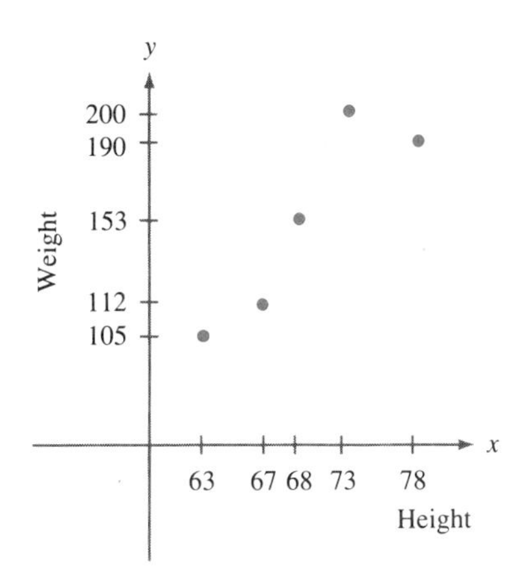

FIGURE 2.6

Joe	(68, 153)
Jane	(63, 105)
Fred	(73, 200)
Mary	(67, 112)
John	(78, 190)

(Height measurement is in inches, and weight measurement is in pounds.) In this case, the set S is

$$S = \{(68, 153), \quad (63, 105), \quad (73, 200), \quad (67, 112), \quad (78, 190)\}$$

and if we plot the ordered pairs in S, we get the picture shown in Figure 2.6. This picture of S is called the **graph** of S. The graph of a set can give us useful information about the set itself. For instance, in this case we see that the points seem to be rising as we look from left to right, and this tells us that taller people tend to weigh more than shorter people!

Let us consider some other examples of graphing sets of ordered pairs.

EXAMPLE 1: Graph the set $S = \{(x, y) | -1 \le x \le 1, 0 \le y < 2\}$.

Solution: The set S consists of all points whose first coordinate satisfies $-1 \le x \le 1$ and whose second coordinate satisfies $0 \le y < 2$. The graph of S is shown in Figure 2.7. *Try Study Suggestion 2.5.* □

► **Study Suggestion 2.5:** Graph the set $S = \{(x, y) | 1 \le x < 2, 1 \le y < 2\}$. ◄

EXAMPLE 2: Graph the set $S = \{(x, y) | y = 3\}$.

Solution: The set S consists of all points whose second coordinate is equal to 3. Hence, the graph of S is the horizontal line shown in Figure 2.8.

Try Study Suggestion 2.6. □

► **Study Suggestion 2.6:** Graph the following sets.

(a) $S = \{(x, y) | y = -4\}$

(b) $T = \{(x, y) | x = 2\}$ ◄

$S = \{(x,y) | -1 \le x \le 1, 0 \le y < 2\}$

FIGURE 2.7

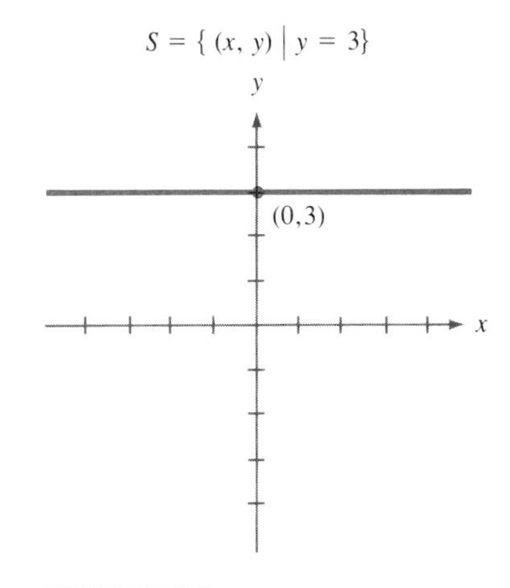

FIGURE 2.8

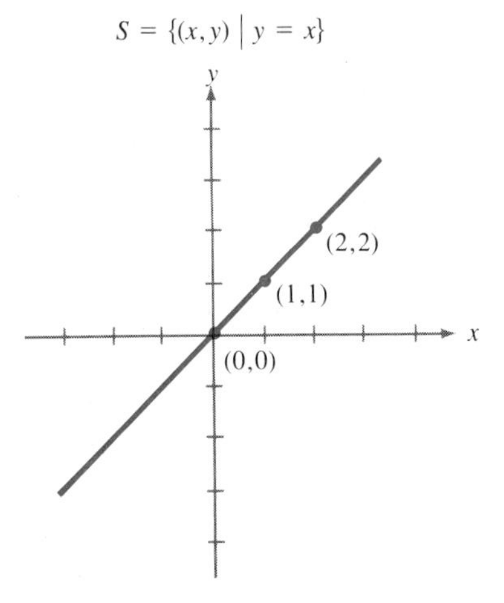

FIGURE 2.9

▶ **Study Suggestion 2.7:** Graph the set $S = \{(x, y) | y = -x\}$. ◀

EXAMPLE 3: Graph the set $S = \{(x, y) | y = x\}$.

Solution: In this case, S consists of all points whose coordinates are equal. Thus, the graph of S is the diagonal line shown in Figure 2.9.

Try Study Suggestion 2.7. □

In Example 3, we used the equation

$$y = x$$

to describe a set of ordered pairs, which we then graphed. More generally, any equation in two variables describes a set, namely the set of all ordered pairs that *satisfy* the equation. (When we say that an ordered pair (x, y) *satisfies* an equation, we mean that if we substitute the values for x and y into the equation, the result is a true statement. For example, the ordered pair $(1, 1)$ satisfies equation $y = x$.)

The graph of a set of ordered pairs that is obtained from an equation is usually called the graph of the *equation* (rather than the graph of the set). In the remainder of this section, we will consider two examples of special types of equations and their graphs.

STRAIGHT LINES

The graph of any equation of the form

$$y = ax + b \tag{2.2}$$

where a and b are constants, is a straight line. We will discuss these important graphs in detail in the next chapter. However, if we accept for now the fact that the graph of Equation (2.2) is a straight line, then it is very easy to draw this graph. Since a straight line is completely determined by giving any two of its points, we need only find two ordered pairs that satisfy Equation (2.2), plot these two points, and then draw the line through them.

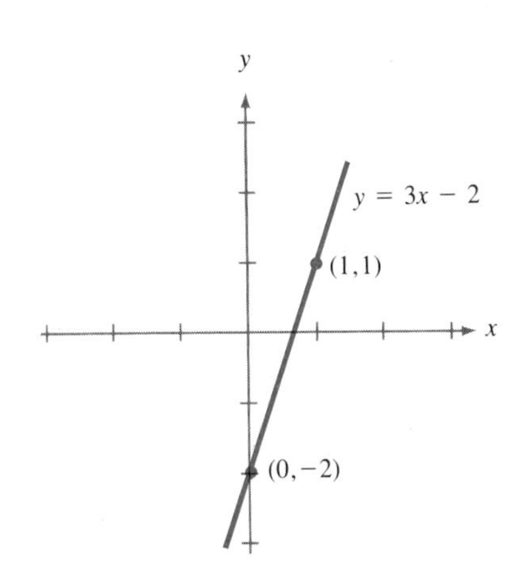

FIGURE 2.10

▶ **Study Suggestion 2.8:**

(a) Graph the equation $y = 3x + 4$.

(b) Graph the equation $4x - 2y = 1$. (*Hint:* first solve the equation for y.) ◀

EXAMPLE 4: Graph the equation

$$y = 3x - 2 \tag{2.3}$$

Solution: Substituting $x = 0$ in Equation (2.3) gives $y = -2$. Hence the ordered pair $(0, -2)$ satisfies this equation, and so it lies on the graph. Substituting $x = 1$ in Equation (2.3) gives $y = 1$, and so the point $(1, 1)$ is also on the graph. Thus, the graph of Equation (2.3) is the straight line containing the points $(0, -2)$ and $(1, 1)$, as shown in Figure 2.10. (Incidentally, we could have chosen any two distinct values of x to substitute into the equation.)

Try Study Suggestion 2.8. □

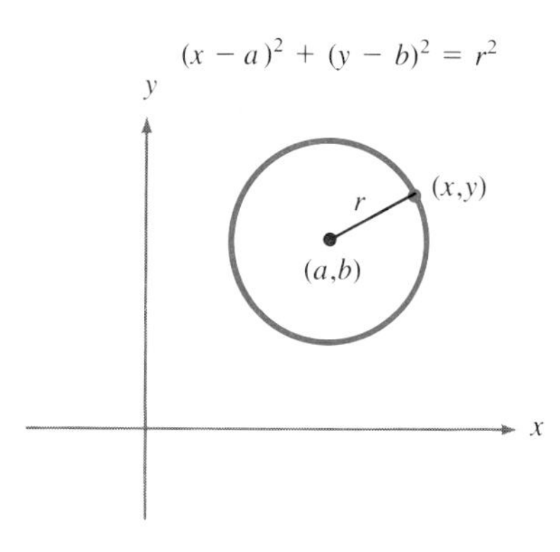

FIGURE 2.11

CIRCLES

A circle with center at the point (a, b) and radius r is the set of all points (x, y) whose distance from (a, b) is equal to r. (See Figure 2.11.) Therefore, according to the distance formula, a point (x, y) is on such a circle if and only if

$$\sqrt{(x-a)^2 + (y-b)^2} = r$$

that is, if and only if

$$(x-a)^2 + (y-b)^2 = r^2 \tag{2.4}$$

In other words, the graph of Equation (2.4) is a circle with center (a, b) and radius r. We say that Equation (2.4) is the *equation of a circle* with center (a, b) and radius r. As a special case

$$x^2 + y^2 = r^2 \tag{2.5}$$

is the equation of a circle with radius r and center at the origin. (See Figure 2.12.)

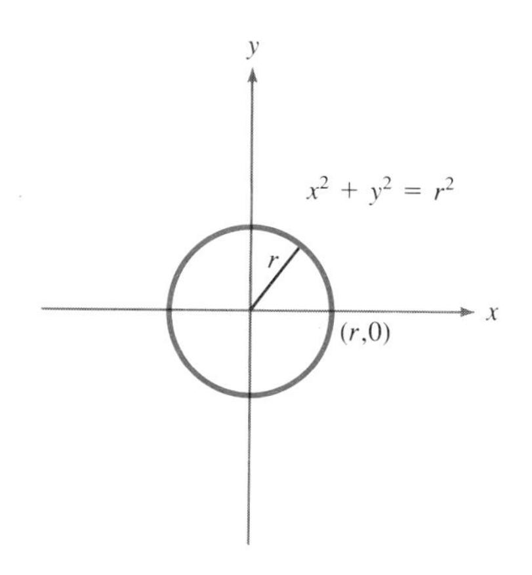

FIGURE 2.12

EXAMPLE 5: Graph the following equations

(a) $x^2 + y^2 = 9$ **(b)** $(x-2)^2 + (y+3)^2 = 5$

Solutions:

(a) Since this equation has the form of Equation (2.5) with $r = 3$, it is the equation of a circle with center at the origin and radius 3. The graph is shown in Figure 2.13.

(b) Writing this equation in the form of Equation (2.4)

$$(x-2)^2 + (y-(-3))^2 = (\sqrt{5})^2$$

shows that it is the equation of a circle with center $(2, -3)$ and radius $\sqrt{5}$. (See Figure 2.14.) *Try Study Suggestion 2.9.* □

▶ **Study Suggestion 2.9:** Graph the following equations.

(a) $x^2 + y^2 = 7$

(b) $(x+2)^2 + (y+2)^2 = 4$ ◀

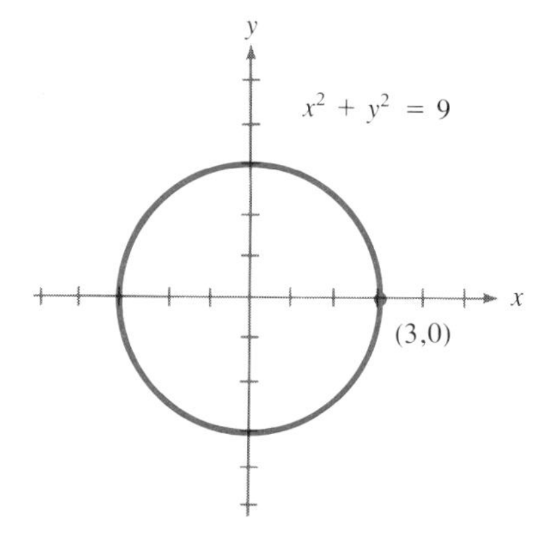

FIGURE 2.13

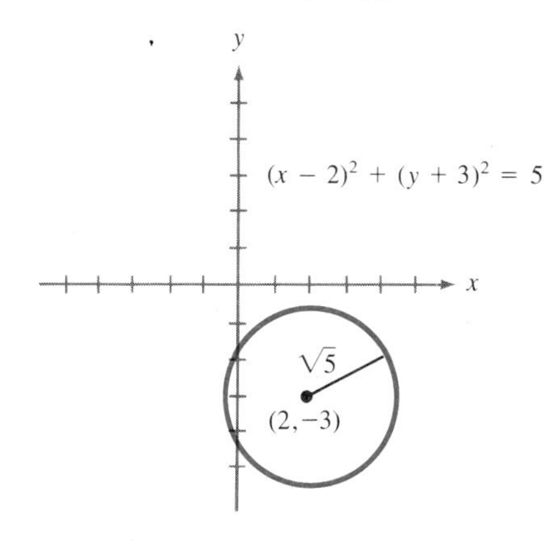

FIGURE 2.14

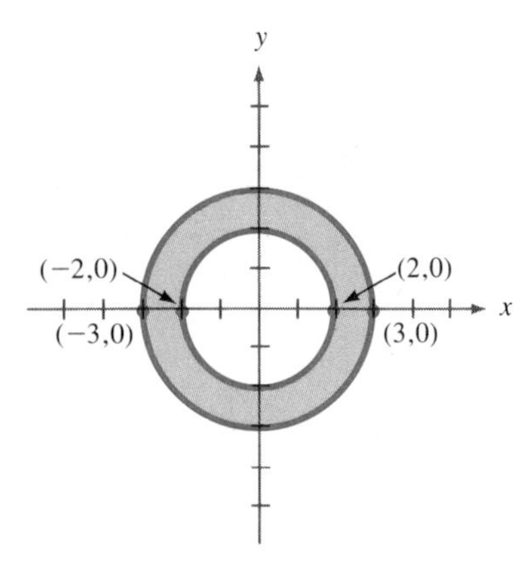

FIGURE 2.15

EXAMPLE 6: Graph the set $S = \{(x, y) | 4 \leq x^2 + y^2 \leq 9\}$.

Solution: This is the set of all points whose distance from the origin is greater than or equal to 2 but less than or equal to 3. (Why?) Hence, the graph of S has a "donut" shape, as shown in Figure 2.15. (Incidentally, this donut shape is called an *annulus*.) *Try Study Suggestion 2.10.* □

▶ **Study Suggestion 2.10:** Graph the set $S = \{(x, y) | 1 < (x - 1)^2 + (y + 2)^2 < 4\}$. ◀

EXAMPLE 7: Graph the equation

$$x^2 + y^2 - 8x + 4y + 11 = 0$$

Solution: The first step is to rewrite this equation so that it has the form of Equation (2.4), which can be done by completing the square in *both* of the variables x and y. We begin by rewriting the equation in the form

$$x^2 - 8x + y^2 + 4y = -11$$

Now the terms involving x have the form $x^2 + kx$, where $k = -8$ and so we can complete the square in x by adding $(k/2)^2 = (-8/2)^2 = 16$ to *both sides* of the equation (see Section 1.6 for a refresher on completing the square)

$$x^2 - 8x + 16 + y^2 + 4y = 5$$
$$(x - 4)^2 + y^2 + 4y = 5$$

The terms involving y also have the form $y^2 + ky$, where $k = 4$ and so we can complete the square in y by adding $(k/2)^2 = (4/2)^2 = 4$ to *both sides*

$$(x - 4)^2 + y^2 + 4y + 4 = 9$$
$$(x - 4)^2 + (y + 2)^2 = 9$$

The final step is to write this in the form of Equation (2.4)

$$(x - 4)^2 + (y - (-2))^2 = 3^2$$

This is the equation of a circle whose center is $(4, -2)$ and whose radius is equal to 3, as shown in Figure 2.16. *Try Study Suggestion 2.11.* □

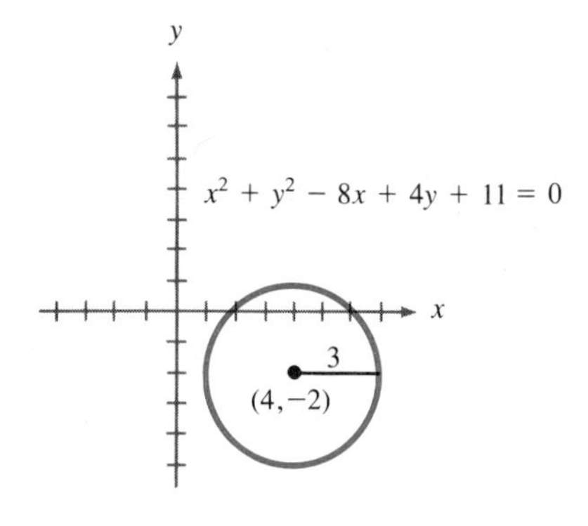

FIGURE 2.16

▶ **Study Suggestion 2.11:** Graph the equation $x^2 + y^2 - 2x + 4y + 1 = 0$. ◀

Ideas to Remember

- The Cartesian coordinate system can be used to describe sets of ordered pairs.
- The graph of an equation provides a very useful "picture" of the equation.

EXERCISES

In Exercises 1–20, graph the given set of points S.

1. $S = \{(2, 0), (-3, 4), (1/2, 1/3)\}$

2. $S = \{(\pi, -\pi)\}$

3. $S = \{(0, 0), (1, 1), (2, 2), (3, 3), (4, 4)\}$

4. $S = \{(k, k) | k = -2, -1, 0, 1, 2\}$

5. $S = \{(k, \sqrt{k}) | k = 0, 2, 4\}$

6. $S = \{(n, n^2 + n + 1) | n = -5, 2, 0, 3\}$

7. $S = \{(x, y) | -4 \le x \le 4\}$

8. $S = \{(x, y) | -2 \le x \le 2, -2 \le y < 2\}$

9. $S = \{(x, y) | x = -2\}$

10. $S = \{(x, y) | y = 0\}$

11. $S = \{(x, y) | y \le -5\}$

12. $S = \{(x, y) | xy < 0\}$

13. $S = \{(x, y) | x < y\}$

14. $S = \{(x, y) | |x| < 2, |y| < 3\}$

15. $S = \{(x, y) | |x| < 5, |y| > 1\}$

16. $S = \{(x, y) | |x| < |y|\}$

17. $S = \{(x, y) | |x - 1| < 2, |y + 1| < 2\}$

18. $S = \{(x, y) | |x + 2| > 3, |2y| < 5\}$

19. $S = \{(x, y) | x^2 + y^2 < 4\}$

20. $S = \{(x, y) | 16 \le (x - 3)^2 + (y - 3)^2 \le 25\}$

In Exercises 21–40, sketch the graph of the given equation, identifying it as a straight line or a circle.

21. $y = x$

22. $y = -x$

23. $y = x + 1$

24. $y = x - \pi$

25. $y = 5x + 3$

26. $y = -2x + 1$

27. $y = (1/2)x - 1/4$

28. $x + y = 1$

29. $x - y = 1$

30. $2x + 3y = 1$

31. $4x - 3y = 5$

32. $2x + 3y = \pi$

33. $x^2 + y^2 = 4$

34. $(x + 1)^2 + y^2 = 3$

35. $x^2 + (y - 3)^2 = 9$

36. $9x^2 + 9y^2 = 1$

37. $(x - 3)^2 + (y - 2)^2 = 7$

38. $(x - 1)^2 + (y + 2)^2 = 2$

39. $(2x + 1)^2 + 4y^2 = 16$

40. $(3x - 1)^2 + (3y + 1)^2 = 81$

In Exercises 41–51, find the equation of a circle with the given properties. (It may help to draw the circle first.)

41. center (1, 4), radius 2

42. center (2, −3), radius 10

43. center $(\sqrt{2}, -\sqrt{2})$, radius 2

44. center (1/2, 1/3), radius 1/6

45. center $(-\sqrt{3}, -\sqrt{3})$, goes through the origin

46. center at the origin, goes through (3, 5)

47. center (−4, 3), goes through (5, −2)

48. center (2, 5), tangent to the x-axis

49. center (−3, −2), tangent to the y-axis

50. endpoints of a diameter are (2, −3) and (4, 7)

51. tangent to both axes, center in the first quadrant, radius 4

In Exercises 52–61, find the center and the radius of the circle with the given equation, and draw the graph.

52. $x^2 + y^2 - 2x + 2y = 0$

53. $x^2 + y^2 - 2x + 2y = 1$

54. $x^2 + y^2 - 4x - 4y - 17 = 0$

55. $x^2 + y^2 + 6y + 5 = 0$

56. $x^2 + 14x + y^2 - 20y + 148 = 0$

57. $4x^2 + 4y^2 = 9$

58. $4x^2 + 4y^2 - 4x - 23 = 0$

59. $3x^2 + 3y^2 - x + 3y = 0$

60. $2x^2 + 2y^2 - 3x = 1$

61. $36x^2 + 36y^2 + 36x - 24y - 59 = 0$

2.3 Functions

In this section, we begin the main topic of this book—*functions.* Loosely speaking, a function is a way of associating the elements of one set A with the elements of another set B, with the restriction that an element of A is associated with one and *only* one element of B. Let us give a formal definition of the term function.

DEFINITION

A **function** consists of the following three objects:

1. A nonempty set A, called the **domain** of the function.
2. A nonempty set B. (A and B may be the same set.)
3. A rule that assigns to each element of the domain A one and *only* one element of the set B.

We will usually denote functions by letters such as f, g and h. If f is a function and A and B are the sets described in the definition, then we say that f is a *function from A to B*, and denote this by writing $f: A \to B$. In this book, we will concentrate on functions between sets of real numbers; that is, the sets A and B will be sets of real numbers.

DEFINITION

Let f be a function from A to B. If a is an element of A, then we denote the element of B that is assigned to a by $f(a)$ (read "f of a"). The element $f(a)$ is called the **image** of a under the function f. The set of all images of the elements of A is called the **range** of the function.

One way to picture a typical function $f: A \to B$ is shown in Figure 2.17. We have drawn the sets A and B as subsets of the plane, rather than the real number line, since this makes the effect of the function easier to visualize. Also, the function f is represented by a single arrow. Actually, there should be one arrow going from A to B for each element of A, but it is not in general possible to draw all of these arrows, and so this one arrow is meant to represent a "typical" arrow. The set A is the domain of the function f, and the range of f is the set consisting of all elements of B that are at the "tips" of all of the arrows of f. This points out the fact that the range of a function need not be all of B, it need only be a *subset* of B.

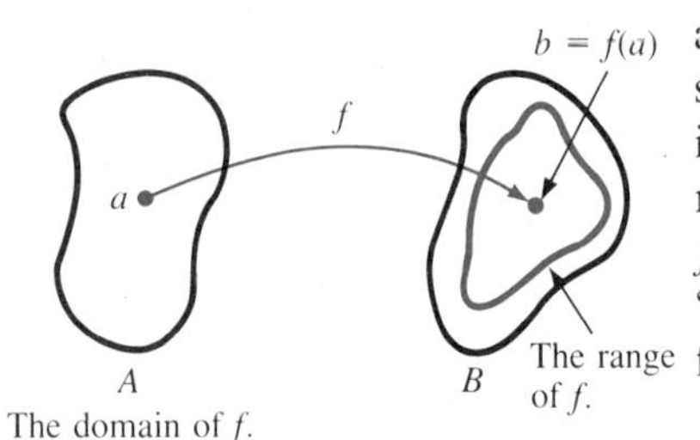

FIGURE 2.17

We should emphasize that in order to define a function, we must specify three things: the set A; the set B; and the rule for associating the elements of A and B. When A and B are sets of numbers, the *rule* is frequently given by an algebraic formula.

EXAMPLE 1: We can define a function $f:\mathbf{R} \to \mathbf{R}$ by means of the formula

$$f(x) = x^2$$

for all x in $\mathbf{R}$. Notice that by writing $f:\mathbf{R} \to \mathbf{R}$, we are in effect *defining* the sets A and B to be the set $\mathbf{R}$ of all real numbers. The function f assigns to each real number x the number x^2. In words, it assigns to each real number its square, and so we might call this the *squaring function*. For example, we have

$$f(3) = 3^2 = 9, \qquad f(-2) = (-2)^2 = 4 \qquad \text{and} \qquad f(\sqrt{2}) = (\sqrt{2})^2 = 2$$

Try Study Suggestion 2.12. □

▶ **Study Suggestion 2.12:** Define a function $g:\mathbf{R} \to \mathbf{R}$ by $g(x) = x^2 + 1$, Find **(a)** $g(2)$ **(b)** $g(-2)$ **(c)** $g(0)$. ◀

EXAMPLE 2: Let A be the set of all *nonnegative* real numbers. Then we can define a function $f:A \to \mathbf{R}$ by the formula

$$f(x) = \sqrt{x}$$

for all x in A. In this case, the image of a nonnegative number x is its *square root* $\sqrt{x}$. Since we cannot allow any negative numbers to be in the domain of this function (why?), we have chosen the domain A to be the set of all nonnegative real numbers. *Try Study Suggestion 2.13.* □

▶ **Study Suggestion 2.13:** Let A be the set of all real numbers greater than or equal to 1, and define a function $h:A \to \mathbf{R}$ by $h(x) = \sqrt{x-1}$. Find **(a)** $h(5)$ **(b)** $h(5/4)$ **(c)** $h(5/3)$. ◀

EXAMPLE 3: Let $\mathbf{R}'$ be the set of all *nonzero* real numbers, and define a function $f:\mathbf{R}' \to \mathbf{R}$ by

$$f(x) = \frac{1}{x}$$

for all x in $\mathbf{R}'$. In this case, the image of any nonzero real number x is its reciprocal $1/x$. For example, we have

$$f(2) = \frac{1}{2} \qquad \text{and} \qquad f\left(-\frac{3}{5}\right) = -\frac{5}{3}$$

Notice that we must exclude 0 from the domain of f since $1/0$ is not defined. *Try Study Suggestion 2.14.* □

▶ **Study Suggestion 2.14:** Let A be the set of all real numbers different from 1, and define $f:A \to \mathbf{R}$ by $f(x) = 1/(x-1)$. Find **(a)** $f(2)$ **(b)** $f(1/2)$ **(c)** $f(-3/4)$ **(d)** $f(0)$. Why did we exclude the number 1 from the domain of f? ◀

EXAMPLE 4: Sometimes the rule that defines a function involves more than one algebraic expression. In such a case, we must be very careful to indicate which expression is to be used to compute the value of the function for a given value of x. For instance, we can define a function $f:\mathbf{R} \to \mathbf{R}$ by the rule

$$f(x) = \begin{cases} x & \text{if} \quad x < -1 \\ x+1 & \text{if} \quad -1 \le x \le 1 \\ x+2 & \text{if} \quad x > 1 \end{cases}$$

When a function is defined in this way, it is said to be a **piecewise defined function.** Notice that in this case the rule involves three different algebraic expressions. In order to determine which expression to use for a given value of x, we must determine whether x lies in the interval $x < -1$, or the interval $-1 \leq x \leq 1$ or the interval $x > 1$. For example, because 1/2 lies in the interval $-1 \leq x \leq 1$, we see that $f(1/2) = 1/2 + 1 = 3/2$. (Notice that these intervals do not overlap, and so there is no possible ambiguity. That is, each real number lies in one and *only* one interval.) *Try Study Suggestion 2.15.* □

▶ **Study Suggestion 2.15:** Define a function $f: \mathbf{R} \to \mathbf{R}$ by

$$f(x) = \begin{cases} x & \text{if } x \leq 0 \\ x^2 & \text{if } 0 < x \leq 2 \\ x^3 & \text{if } x > 2 \end{cases}$$

Find **(a)** $f(1/2)$ **(b)** $f(0)$ **(c)** $f(-5)$ **(d)** $f(5)$ **(e)** $f(2)$ ◀

EXAMPLE 5: Express the area A of a square as a function of the length x of one of its sides. What is the domain of this function?

Solution: Since the area of a square whose sides have length x is x^2, the *area function* is given by

$$A(x) = x^2$$

Furthermore, since the length x of the side of any square is always positive, the domain of the function A is the set of all *positive* real numbers.

Try Study Suggestion 2.16. □

▶ **Study Suggestion 2.16:** Express the area A of a triangle of base 10 as a function of its height x. What is the domain of this function? ◀

It is very common to see a phrase such as "consider the function $f(x) = x + 2$." Strictly speaking, the formula $f(x) = x + 2$ *by itself* does not constitute a function. In order to define a function, we must also specify the two sets A and B. However, we do want to be able to use such phrases to define functions, and so we make the following agreement.

> **Whenever we specify a function by giving just a formula, we will assume that the domain A consists of all possible real numbers that can be substituted for the variable in the formula. We will also assume that the set B is the set R of all real numbers.**

The next example illustrates what we mean by the phrase "all possible real numbers".

EXAMPLE 6: The following functions are defined by a formula alone. Find the domain of each of these functions.

(a) $f(x) = 2x^2 + 3x - 1$ **(b)** $g(x) = \sqrt{x - 1}$ **(c)** $h(x) = \dfrac{1}{x - 2}$

(d) $k(x) = \dfrac{1}{\sqrt{x^2 - 1}}$

Solutions:

(a) Since any real number can be substituted for x in the formula $f(x) = 2x^2 + 3x - 1$, according to our agreement the domain of this function is the set **R** of all real numbers.

(b) Since we cannot take the square root of a negative number, we can only substitute values of x for which $x - 1$ is nonnegative. In other words, the domain of this function consists of all real numbers x for which $x - 1 \geq 0$, that is, all real numbers satisfying $x \geq 1$.

(c) In this case, we can substitute any real number for x except those values that make the denominator equal to 0. Since there is only one such value, namely $x = 2$, the domain of this function is all real numbers *except* $x = 2$.

(d) In this case, the quantity $x^2 - 1$ must be nonnegative in order to take its square root, and it cannot be equal to 0 since we cannot allow 0 in the denominator. Hence, we must have $x^2 - 1 > 0$. The solution set of this inequality is the set $\{x \mid x < -1\} \cup \{x \mid x > 1\}$, which is therefore the domain of the function k.

▶ **Study Suggestion 2.17:** Find the domain of each of the following functions.

(a) $f(x) = x^4 + x$

(b) $g(x) = x^2 + x^{-2}$

(c) $h(x) = \sqrt{x^2 - 1}$

(d) $k(x) = 1/\sqrt{1 - x^2}$ ◀

Try Study Suggestion 2.17. □

Another point that we should make is that it does not matter which symbol we use for the variable in a function. For example, the function

$$f(x) = x^2 + 5$$

is exactly the same as the function

$$g(y) = y^2 + 5 \tag{2.6}$$

or

$$h(z) = z^2 + 5$$

Sometimes a function such as this will be written in the form

$$y = x^2 + 5 \tag{2.7}$$

and this is where some confusion can occur. The letter y is being used in Equation (2.6) in an entirely different way than in Equation (2.7). To help avoid confusion, the letter x in Equation (2.7) is referred to as the **independent variable** of the function, whereas the letter y in Equation (2.7) is referred to as the **dependent variable** (and we say that y is a function of x). However, in Equation (2.6), the letter y is the *independent* variable, and no dependent variable is specified.

Since most of the rest of this course (as well as most of a calculus course) is concerned with manipulating functions, it is especially important that you spend some time becoming comfortable with the notation. For example, it is very common to want to replace the independent variable in a function by an algebraic expression, as the following example illustrates.

EXAMPLE 7: Let f be the function

$$f(x) = x^2 + 2 \tag{2.8}$$

(a) Compute $f(x + 1)$ and $f(x) + 1$. Are they the same?

(b) Compute $f(-x)$ and show that $f(-x) = f(x)$.

(c) Compute $f(x + h) - f(x)$, where h is a constant, and simplify the result.

Solutions:

(a) In order to compute $f(x + 1)$, we substitute $x + 1$ for x in Equation (2.8)

$$f(x + 1) = (x + 1)^2 + 2 = x^2 + 2x + 3$$

(This is very similar to taking the composition of two algebraic expressions, as we did in Section 1.5.) On the other hand,

$$f(x) + 1 = x^2 + 2 + 1 = x^2 + 3$$

As you can see, $f(x + 1) \neq f(x) + 1$.

(b) In order to compute $f(-x)$, we substitute $-x$ for x in Equation (2.8)

$$f(-x) = (-x)^2 + 2 = x^2 + 2$$

Thus we see that $f(-x) = f(x)$.

(c) In order to compute $f(x + h) - f(x)$, we proceed as follows

$$\begin{aligned} f(x + h) - f(x) &= [(x + h)^2 + 2] - [x^2 + 2] \\ &= [x^2 + 2hx + h^2 + 2] - [x^2 + 2] \\ &= 2hx + h^2 \\ &= h(2x + h) \end{aligned}$$

Try Study Suggestion 2.18. □

▶ **Study Suggestion 2.18:** Let $f: \mathbf{R} \to \mathbf{R}$ be the function defined by $f(x) = x^3$.

(a) Compute $f(2x)$ and $2f(x)$. Are they the same?

(b) Show that $f(-x)$ is *not* equal to $f(x)$.

(c) Compute $f(x + h) - f(x)$ and simplify. ◀

One of the most important concepts in calculus is that of the **difference quotient** of a function f, which is

$$\frac{f(x + h) - f(x)}{h}$$

where h is a nonzero real number.

EXAMPLE 8: Compute the difference quotient of the function $f(x) = x^2$ and simplify.

Solution: According to the definition, the difference quotient is

$$\begin{aligned} \frac{f(x + h) - f(x)}{h} &= \frac{(x + h)^2 - x^2}{h} \\ &= \frac{x^2 + 2hx + h^2 - x^2}{h} \\ &= \frac{2hx + h^2}{h} \\ &= 2x + h \end{aligned}$$

Try Study Suggestion 2.19. □

▶ **Study Suggestion 2.19:** Compute the difference quotient of the following functions and simplify the result. **(a)** $f(x) = 2x^2 + 1$ **(b)** $g(x) = x^3$ ◀

EXAMPLE 9: Compute the difference quotient of the function $f(x) = 1/x$.

Solution: In this case, the difference quotient is

$$\frac{f(x+h)-f(x)}{h} = \frac{\dfrac{1}{x+h}-\dfrac{1}{x}}{h}$$

$$= \left[\frac{1}{x+h}-\frac{1}{x}\right]\frac{1}{h}$$

$$= \left[\frac{-h}{x(x+h)}\right]\frac{1}{h}$$

$$= \frac{-1}{x(x+h)}$$

Try Study Suggestion 2.20. □

▶ **Study Suggestion 2.20:** Compute the difference quotient of the function $f(x) = 1/x^2$. ◀

Ideas to Remember

- The concept of a function is one of the most important in all of mathematics and its applications.
- By studying the behavior of functions in general, we can apply this knowledge to specific cases.

EXERCISES

1. If $f(x) = 2x - 1$, find

(a) $f(0)$ **(b)** $f(-3)$ **(c)** $f(\sqrt{2})$ **(d)** $f(x+1)$

2. If $f(x) = -x^2 + 2x - 3$, find

(a) $f(0)$ **(b)** $f(5)$ **(c)** $f(-1/2)$ **(d)** $f(2x-1)$

3. If $g(x) = \dfrac{x}{x+1}$, find

(a) $g(1)$ **(b)** $g(1/2)$ **(c)** $g(0)$ **(d)** $g(x-1)$

4. If $f(y) = \dfrac{y^2+1}{y^2-1}$, find

(a) $f(2)$ **(b)** $f(-2)$ **(c)** $f(5/2)$ **(d)** $f(2x)$

(e) Compare the values of $f(y)$ and $f(-y)$

5. If $f(x) = \sqrt{x+1}$, find

(a) $f(-1)$ **(b)** $f(0)$ **(c)** $f(24)$ **(d)** $f(4x^2-1)$

6. If $f(x) = |x|$, find

(a) $f(1)$ **(b)** $f(-3/2)$ **(c)** $f(0)$ **(d)** $f(x^2)$

7. If $F(t) = |t-1| + |t|$, find

(a) $F(0)$ **(b)** $F(1)$ **(c)** $F(1/2)$ **(d)** $F(t+1)$

8. If $f(x) = \dfrac{|x+1|}{|x|+1}$, find

(a) $f(1)$ **(b)** $f(2)$ **(c)** $f(-1)$ **(d)** $f(x^2)$

9. If $f(x) = x^x$, find

(a) $f(1)$ **(b)** $f(2)$ **(c)** $f(-3)$ **(d)** $f(1/2)$

10. Let $f(x)$ be the function defined by

$$f(x) = \begin{cases} x^2 & \text{if } x < 0 \\ 1-x & \text{if } 0 \le x < 2 \\ -3 & \text{if } x \ge 2 \end{cases}$$

Find **(a)** $f(5)$ **(b)** $f(1/2)$ **(c)** $f(-10)$

11. Let $g(z)$ be the function defined by

$$g(z) = \begin{cases} z^2/3 + 1 & \text{if } z \le -1 \\ z + 2 & \text{if } -1 < z < 3 \\ |z| & \text{if } z \ge 3 \end{cases}$$

Find **(a)** $g(3/2)$ **(b)** $g(5)$ **(c)** $g(-2)$

12. Let $f(x)$ be the function defined by

$$f(x) = \begin{cases} \dfrac{x}{2} & \text{if } x \text{ is an even integer} \\ \dfrac{x+1}{2} & \text{if } x \text{ is an odd integer} \\ 0 & \text{if } x \text{ is not an integer} \end{cases}$$

Find **(a)** $f(1/2)$ **(b)** $f(3)$ **(c)** $f(-4)$ **(d)** $f(x^2)$

In Exercises 13–17, let a and b be real numbers.

13. If $f(x) = x^2$, find

(a) $f(a)$ **(b)** $f(-a)$ **(c)** $-f(a)$ **(d)** $f(a + b)$
(e) $f(a) + f(b)$

14. If $f(x) = x^2 - 2$, find

(a) $f(a + 1)$ **(b)** $f(a) + 1$ **(c)** $f(a^2)$
(d) $[f(a)]^2$ **(e)** $f(|a|)$

15. If $h(y) = 3y^2 + 2y - 1$, find

(a) $h(a)$ **(b)** $h(-a)$ **(c)** $h(a^2)$ **(d)** $[h(a)]^2$

16. If $f(x) = \sqrt{x^2 + 1}$, find

(a) $f(a)$ **(b)** $f(-a)$ **(c)** $f(a^2)$ **(d)** $[f(a)]^2$

17. If $F(x) = \dfrac{x^2}{x + 1}$, find

(a) $F(1/a)$ **(b)** $1/F(a)$ **(c)** $F(a + 1)$ **(d)** $F(a) + 1$

18. If $f(x) = x^2$, find

(a) $f(x^3)$ **(b)** $f(x + 3)$ **(c)** $f(x^2 + 1)$ **(d)** $f(|x - 4|)$

19. If $f(x) = 2x^2 + x - 1$, find

(a) $f(1/x)$ **(b)** $f(x - 3)$ **(c)** $f(x^2) + f(x) + 1$
(d) $f(x^2 + x + 1)$

20. If $f(x) = \dfrac{1}{x^2 + 1}$, find

(a) $f(-x)$ **(b)** $f(1/x)$ **(c)** $f(1/x^2)$ **(d)** $f(x^3 + 1)$

In Exercises 21–28, compute the difference quotient of the given function, and simplify the result.

21. $f(x) = x$

22. $f(x) = x + 2$

23. $g(y) = 3y - 1$

24. $H(z) = z^2 + 1$

25. $f(x) = -2x^2 + x - 3$

26. $f(x) = \sqrt{x}$

27. $f(x) = \dfrac{2}{x + 1}$

28. $f(x) = x^4$

In Exercises 29–38, find the domain of the given function.

29. $f(x) = 3x^3 - 4x^2 + x - 1$

30. $f(y) = \dfrac{1}{y + 1}$

31. $f(x) = \sqrt{2x - 1}$

32. $f(x) = \dfrac{x}{\sqrt{2x + 1}}$

33. $f(x) = \dfrac{1}{x^2}$

34. $g(x) = \sqrt{|x + 1|}$

35. $f(x) = \dfrac{1}{x^2 + x - 6}$

36. $h(t) = \dfrac{1}{t^2 + 2t + 3}$

37. $f(x) = \dfrac{x}{2x^2 + x - 2}$

38. $f(x) = \dfrac{x - 1}{x^2 - 4x + 3}$ (be careful here)

39. Express the area of an equilateral triangle as a function of the length s of its sides. What is the domain of this function?

40. Express the area of a square as a function of its perimeter P. What is the domain of this function?

41. Express the area of a circle as a function of its circumference C. What is the domain of this function?

42. Express the temperature F in degrees Fahrenheit as a function of the temperature C in degrees Celsius. What is the domain of this function?

43. When the price of a certain record is \$5.00, the record company sells 100,000 records. But for every increase of \$1 in the price of the record, the company estimates that it loses the sale of 1000 records.

(a) Express the total number of records sold as a function of the selling price of the record. What is the domain of this function?

(b) Express the gross income to the company for the sales of this record as a function of the selling price of the record. Assume that the selling price is at least \$5.00. What is the domain of this function?

44. A company deducts 2% of an employee's salary for a pension fund, up to a maximum of \$600 per year. Express the amount withheld as a function of the employee's salary. What is the domain of this function?

2.4 Graphing Functions

If $f: A \to B$ is a function from A to B then we can get a useful "picture" of f by plotting all ordered pairs of the form $(x, f(x))$, for all real numbers x in A. This is called the **graph of the function** f, and is shown in Figure 2.18a.

Values of a for which the point $(a, 0)$ lies on the graph of f are called the **x-intercepts** of the graph. Similarly, values of b for which the point $(0, b)$ lies on the graph are called the **y-intercepts** of the graph. (We will see later that the graph of a function may have many x-intercepts, but at most one y-intercept.) These concepts are illustrated in Figure 2.18b.

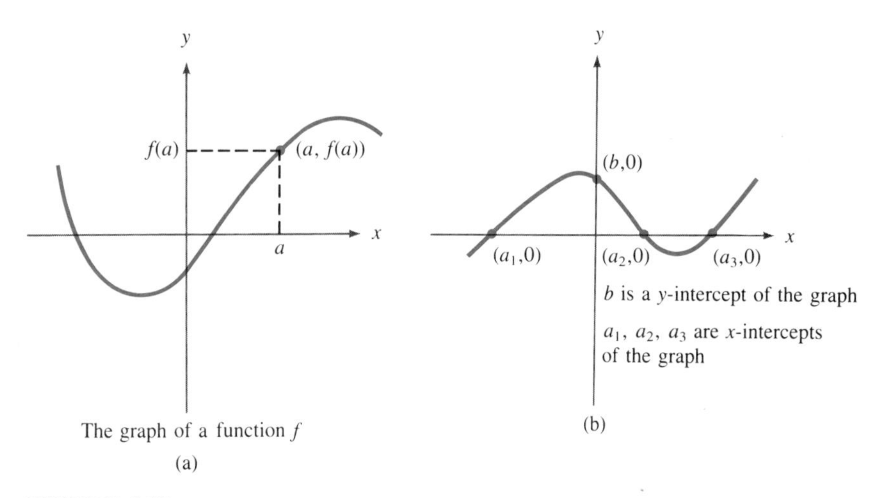

FIGURE 2.18

The simplest method for graphing a function is to plot several points of the form $(x, f(x))$, and then draw a smooth curve through them. This is referred to as the **method of plotting points.** It is the only method that we can use when we do not have any special knowledge about the shape of the graph.

Let us consider examples of some common functions and their graphs.

EXAMPLE 1: Any function of the form $f(x) = a$, where a is a constant, is called a **constant function.** For example, the function

$$f(x) = 2$$

is a constant function. The term "constant" refers to the fact that the values of the function never change. In this case, $f(x)$ is equal to 2, for all values of x. For instance,

$$f(3) = 2, \quad f(0) = 2, \quad f(1/2) = 2, \quad \text{and} \quad f(-\sqrt{3}) = 2$$

The graph of f consists of all ordered pairs of the form $(x, f(x))$; that is, all ordered pairs of the form $(x, 2)$ where x ranges over all real numbers. Thus, the graph is just a horizontal straight line, as shown in Figure 2.19.

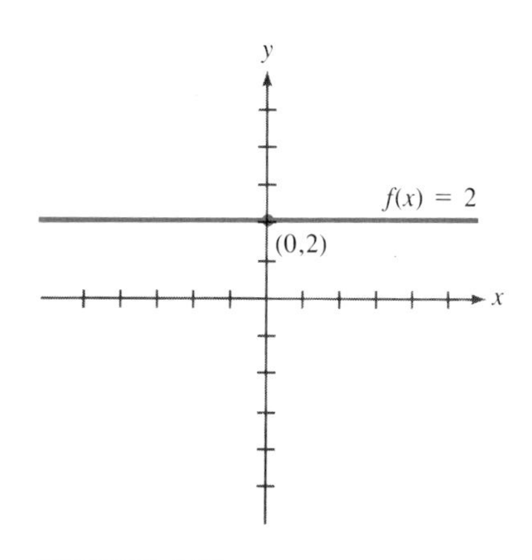

FIGURE 2.19

It is worthwhile to keep in mind that

> **the graph of any constant function is a horizontal straight line.**

▶ **Study Suggestion 2.21:** Graph the constant function $f(x) = -5$. ◀

Try Study Suggestion 2.21.

EXAMPLE 2: The function

$$f(x) = x$$

is called the **identity function.** The image of any real number under this function is just the number itself. For example,

$$f(3) = 3, \qquad f(-1/2) = -1/2, \qquad \text{and} \qquad f(\pi) = \pi$$

The graph of f consists of all ordered pairs (x, x), where x ranges over all real numbers. This graph is pictured in Figure 2.20.

▶ **Study Suggestion 2.22:** Graph the function $f(x) = -x$. ◀

Try Study Suggestion 2.22. □

EXAMPLE 3: Any function of the form

$$f(x) = ax + b$$

where a and b are constants is called a **linear function.** This name comes from the fact that the graph of a linear function is a straight line. We will see why this is true in the next chapter, but if we accept this fact for now, then we can sketch the graph of a linear function by simply finding two points on the graph and connecting them with a straight line. (In fact, the graph of the *function* $f(x) = ax + b$ is the same as the graph of the *equation* $y = ax + b$.)

For example, in order to graph the linear function

$$f(x) = 2x + 3$$

we take two values for x, and compute $f(x)$. If $x = 0$, then $f(x) = 3$ and so $(0, 3)$ is on the graph. Also, if $x = 1$, then $f(x) = 5$, and so $(1, 5)$ is on the graph. The graph of f is shown in Figure 2.21.

▶ **Study Suggestion 2.23:** Sketch the graphs of the following linear functions. **(a)** $g(x) = 3x - 4$ **(b)** $h(x) = 1 - x$ ◀

Try Study Suggestion 2.23. □

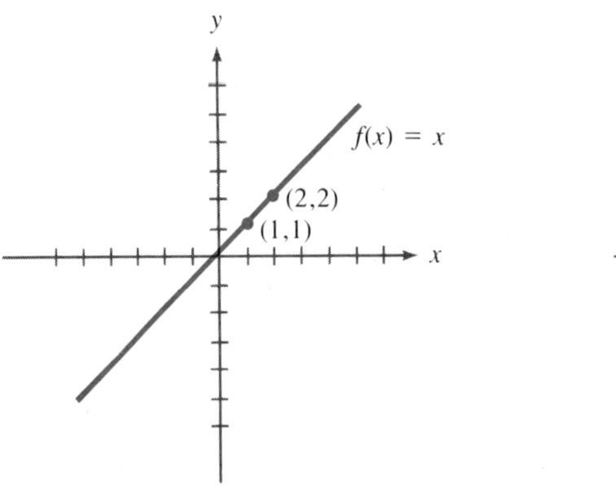

FIGURE 2.20

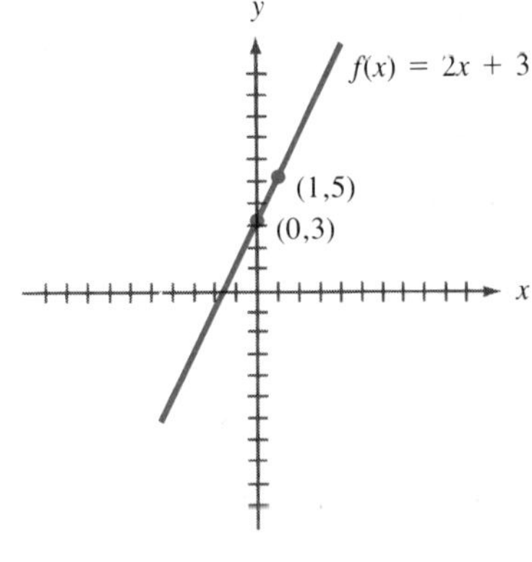

FIGURE 2.21

EXAMPLE 4: The function

$$f(x) = |x|$$

is called the **absolute value function.** In order to graph this function, we use the method of plotting points. When doing so, it is customary to first make a table of values of x and the corresponding values $f(x)$, as shown below. The points $(x, f(x))$ can then be plotted and a smooth curve drawn through them. (See Figure 2.22.) Of course, the more points that are plotted, the more accurate the graph will be. *Try Study Suggestion 2.24* □

▶ **Study Suggestion 2.24:** Graph the function $g(x) = |2x|$. ◀

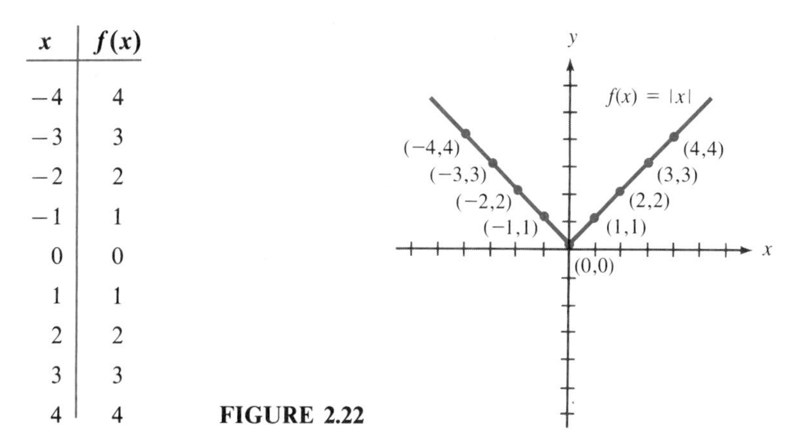

x	$f(x)$
−4	4
−3	3
−2	2
−1	1
0	0
1	1
2	2
3	3
4	4

FIGURE 2.22

EXAMPLE 5: Graph the square root function

$$f(x) = \sqrt{x}$$

Solution: First we observe that the domain of this function is the set of all *nonnegative* real numbers; that is, we can only substitute nonnegative values of x into this function. As in the previous example, we collect several points in a table, plot these points, and then draw a smooth curve through them. (See Figure 2.23.) *Try Study Suggestion 2.25.* □

▶ **Study Suggestion 2.25:** Graph the function $f(x) = \sqrt{x - 1}$. ◀

x	$f(x)$
0	0
1	1
2	$\sqrt{2} \approx 1.4$
3	$\sqrt{3} \approx 1.7$
4	2
5	$\sqrt{5} \approx 2.2$

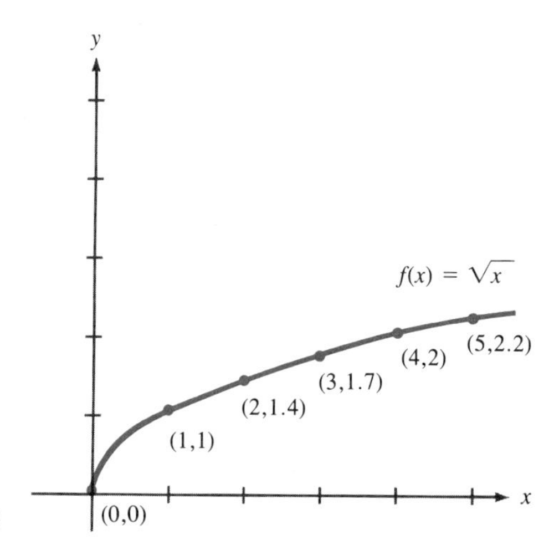

FIGURE 2.23

EXAMPLE 6: Graph the function

$$f(x) = \frac{1}{x}$$

Solution: The domain of this function is the set of all *nonzero* real numbers. Therefore, the graph of f will *not* include any points of the form $(0, y)$. In other words, the graph will not intersect the y-axis. Also, since no value of x will make $f(x)$ equal to 0, the graph will not intersect the x-axis. Using this information and the method of plotting points, we obtain the graph shown in Figure 2.24. □

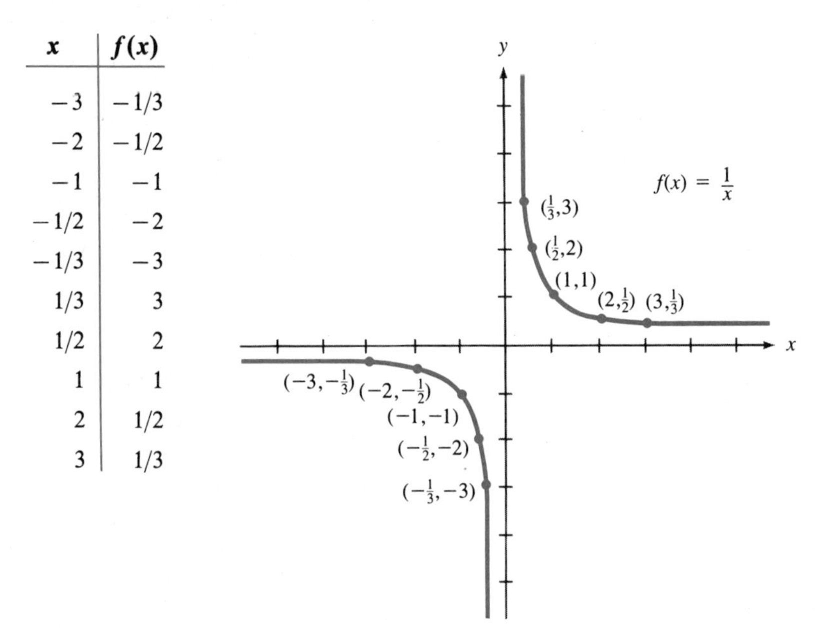

x	$f(x)$
-3	$-1/3$
-2	$-1/2$
-1	-1
$-1/2$	-2
$-1/3$	-3
$1/3$	3
$1/2$	2
1	1
2	$1/2$
3	$1/3$

FIGURE 2.24

EXAMPLE 7: Graph the function $f: A \to \mathbf{R}$ defined by

$$f(x) = x^2 + 1$$

whose domain A is the closed interval $[2, 4]$.

Solution: Since the domain of this function is the interval $[2, 4]$, we can only substitute values for x that lie in this interval. Choosing some of these values and plotting points gives the graph in Figure 2.25. Notice that the entire graph lies over the interval $[2, 4]$ on the x-axis.

Try Study Suggestion 2.26. □

▶ **Study Suggestion 2.26:** Graph the function $g: A \to \mathbf{R}$ defined by $g(x) = 1/(x + 2)$ whose domain A is the interval $(-1, 3]$. ◀

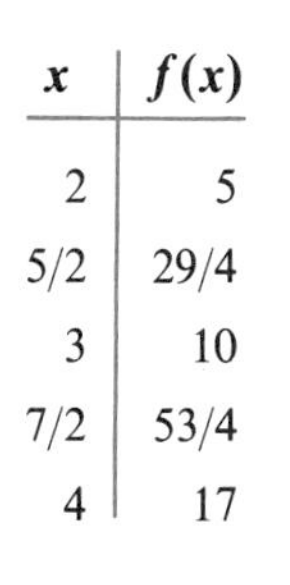

x	$f(x)$
2	5
5/2	29/4
3	10
7/2	53/4
4	17

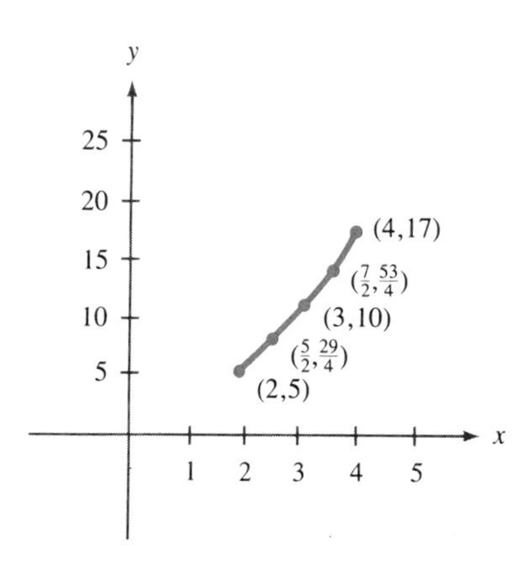

FIGURE 2.25

EXAMPLE 8: Graph the function

$$f(x) = \frac{x^2 - 1}{x - 1}$$

Solution: Since we can substitute any real number for x except $x = 1$ (which gives the meaningless expression 0/0) the domain of this function consists of all real numbers except 1. This means that the graph of f will have no point whose x-coordinate is equal to 1. In other words, the graph will not intersect the vertical line $x = 1$. Factoring the numerator, we get

$$f(x) = \frac{(x + 1)(x - 1)}{x - 1} = x + 1 \qquad \text{for } x \neq 1$$

and so, *provided that* $x \neq 1$, the graph of f is the same as the graph of the function $g(x) = x + 1$. To emphasize the difference, we first draw the graph of g, as shown in Figure 2.26(a), and then "remove" the point (1, 2) whose x-coordinate is equal to 1. This gives the graph of f, as shown in Figure 2.26(b).

Try Study Suggestion 2.27. □

▶ **Study Suggestion 2.27:** Determine the domain of the function

$$h(x) = \frac{x^3 + x^2}{x + 1}$$

and then draw its graph. ◀

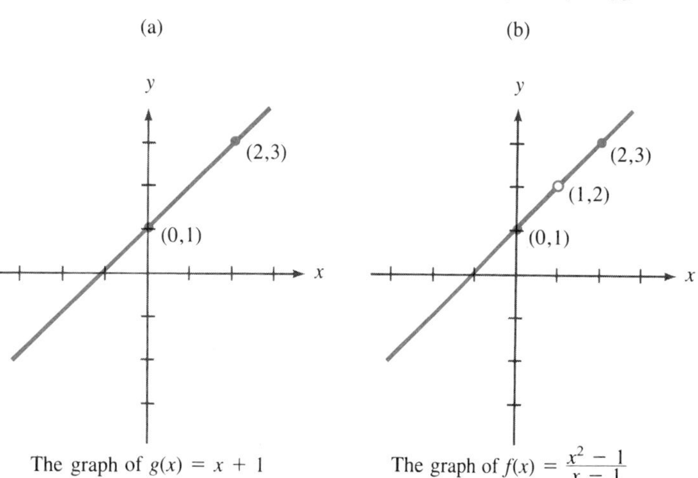

FIGURE 2.26

The graph of $g(x) = x + 1$ The graph of $f(x) = \frac{x^2 - 1}{x - 1}$

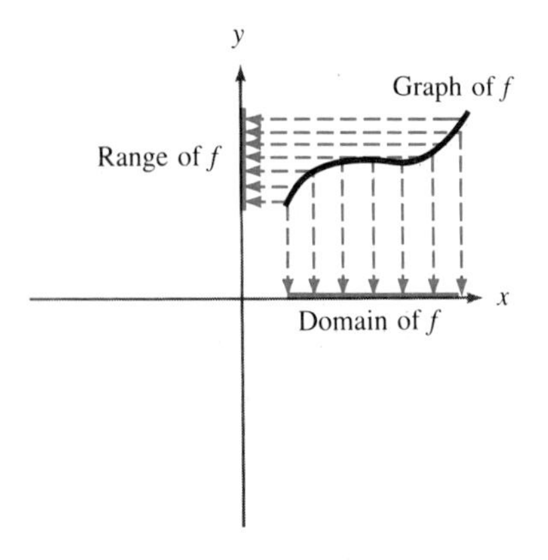

FIGURE 2.27

The graph of a function gives us not only a picture of the rule used to define the function, but also of the domain and the range of the function. This is illustrated in Figure 2.27.

EXAMPLE 9: Graph the function defined by

$$f(x) = \begin{cases} 1 & \text{if } x < 0 \\ x & \text{if } 0 \leq x \leq 2 \\ 1 - x & \text{if } x > 2 \end{cases}$$

Solution: In this case, we draw the graph in three stages, depending on the range of values of x. When $x < 0$, we have $f(x) = 1$, and this gives us that portion of the graph shown in Figure 2.28. (Notice that we use an open circle to indicate that the point (0, 1) is *not* part of the graph.) When $0 \leq x \leq 2$, we have $f(x) = x$, and so this portion of the graph can be taken from the graph of $f(x) = x$ given in Example 2. Finally, for the portion of the graph that corresponds to $x > 2$, we resort to the method of plotting points. This gives us the entire graph of f, as shown in Figure 2.29.

Try Study Suggestion 2.28. □

▶ **Study Suggestion 2.28:** Graph the function defined by

$$f(x) = \begin{cases} -x & \text{if } x \leq -1 \\ 0 & \text{if } -1 < x < 1 \\ x & \text{if } x \geq 1 \end{cases}$$ ◀

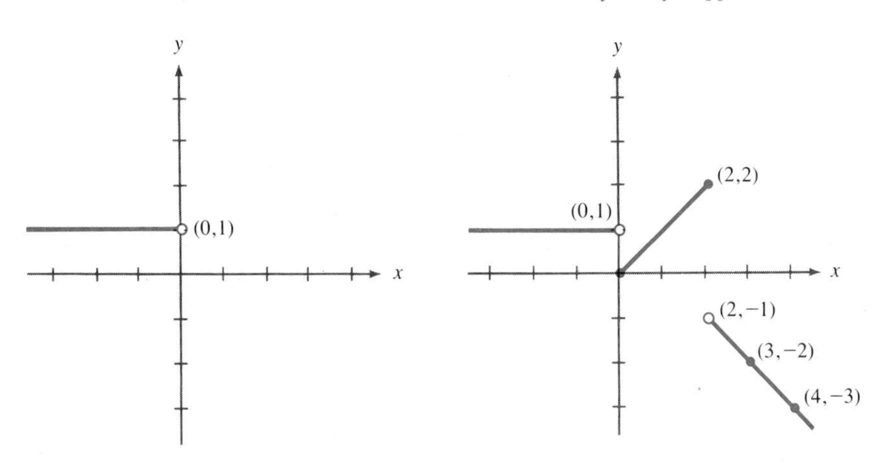

FIGURE 2.28 FIGURE 2.29

The graph of a function can give us a great deal of information about the function itself. The next example illustrates this point.

EXAMPLE 10: Suppose that the curve in Figure 2.30 is the graph of a function f. Determine for which values of x the function satisfies

(a) $f(x) = 0$ **(b)** $f(x) > 0$ **(c)** $f(x) \leq 3$

Solutions:

(a) The values of x for which $f(x) = 0$ are the x-intercepts of the graph, which in this case are $x = -9/2$, $x = 1$, and $x = 4$.

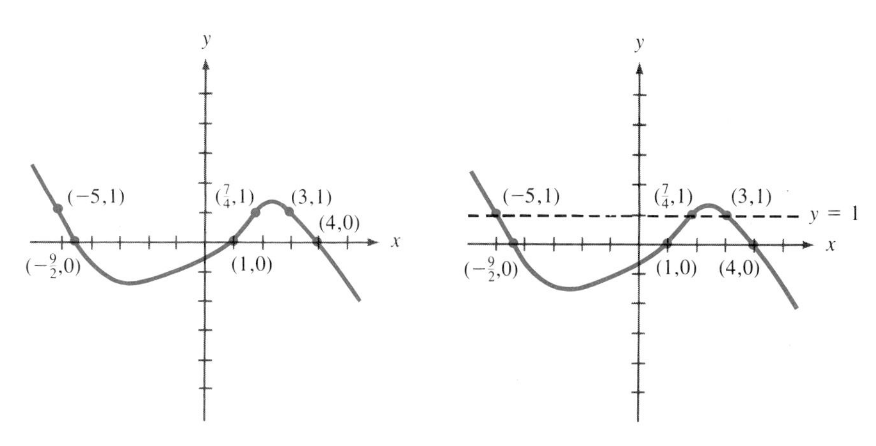

FIGURE 2.30 **FIGURE 2.31**

(b) The values of x for which $f(x) > 0$ correspond to points $(x, f(x))$ on the graph that lie *above* the x-axis. We can see from the graph that these are all x-values in the *open intervals* $(-\infty, -\frac{9}{2})$ and $(1, 4)$. Hence, the answer to this question is all values of x in the set $(-\infty, -\frac{9}{2}) \cup (1, 4)$.

(c) In order to determine the values of x for which $f(x) \leq 1$, we first draw the horizontal line $y = 1$ on the same plane as the graph of f. This is done in Figure 2.31. Now, the values of x for which $f(x) \leq 1$ correspond to points $(x, f(x))$ on the graph that lie *on or below* the line $y = 1$. We can see from Figure 2.31 that these are the x-values in the intervals $[-5, 7/4]$ and $[3, \infty)$. In other words, the answer to this question is all values of x in the set $[-5, 7/4] \cup [3, \infty)$.

Try Study Suggestion 2.29. □

▶ **Study Suggestion 2.29:** Suppose that the curve shown in Figure 2.32 is the graph of a function g. Determine for which values of x the function satisfies **(a)** $g(x) = 0$ **(b)** $g(x) \leq 0$ **(c)** $g(x) = 2$ **(d)** $g(x) > 2$ ◀

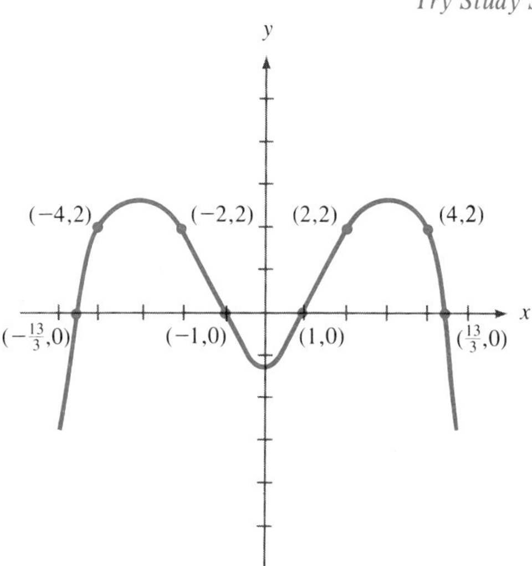

FIGURE 2.32

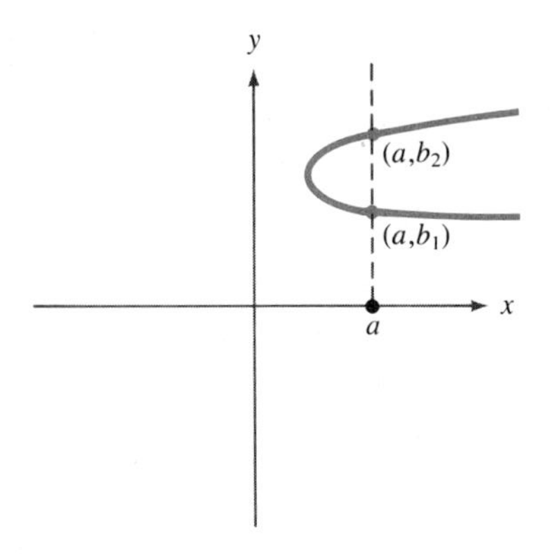

FIGURE 2.33

Let us conclude this section by pointing out that not every curve in the plane represents the graph of a function. For example, the curve shown in Figure 2.33 is not the graph of a function. We can see this by observing that there are two *different* points on this curve that have the same x-coordinate. In other words, there are points on the curve of the form (a, b_1) and (a, b_2) where $b_1 \neq b_2$. Now, if this were the graph of a function f, then we would have $f(a) = b_1$ *and* $f(a) = b_2$, which would violate the condition that a function assign one and *only* one value in B to each value in A. Hence this curve cannot be the graph of a function.

We can phrase this discussion in geometric terms by saying that the reason that the curve in Figure 2.33 is not the graph of a function is that there is at least one vertical line that intersects the curve at more than one point. (In this case at the points (a, b_1) and (a, b_2).) This leads us to the following test to determine whether or not a curve in the plane is the graph of a function.

Vertical Line Test

A curve in the plane is said to pass the **vertical line test** if no vertical line intersects the graph at more than one point. A curve in the plane is the graph of a function if and only if it passes the vertical line test.

EXAMPLE 11: Which of the following curves are the graphs of functions?

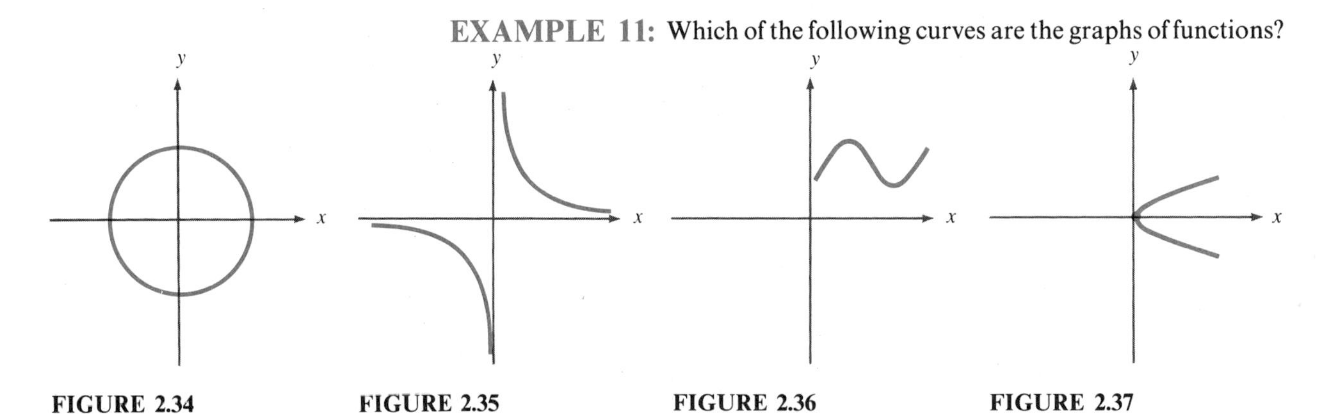

FIGURE 2.34 FIGURE 2.35 FIGURE 2.36 FIGURE 2.37

Solution: The curves in Figures 2.35 and 2.36 pass the vertical line test, and so they are the graphs of functions. However, the curves in Figures 2.34 and 2.37 do not pass the vertical line test, and so they are not the graphs of functions. □

Ideas to Remember

The graph of a function can give us a great deal of information about the function itself.

EXERCISES

In Exercises 1–8, determine whether or not the curve is the graph of a function. If it is, find the domain and range of the function.

1.

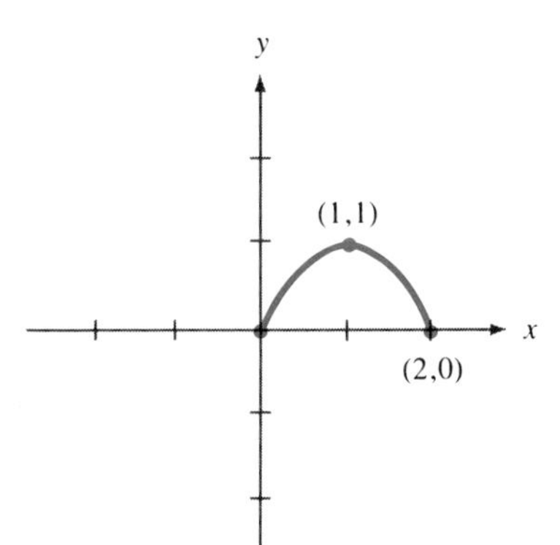

2.

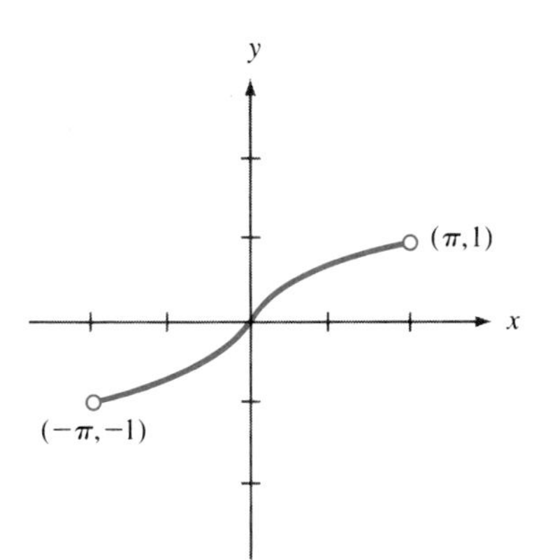

3.

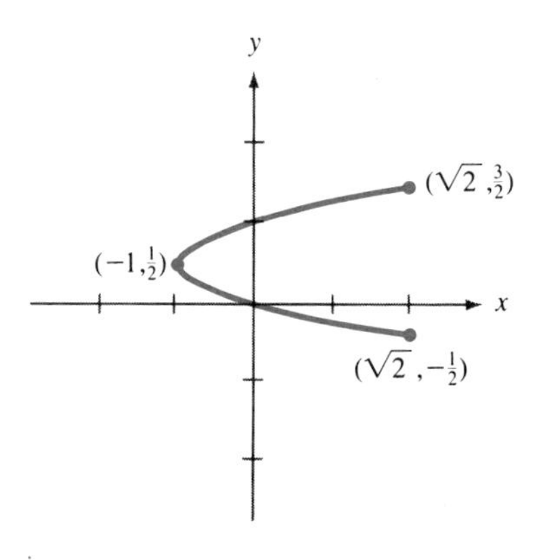

4.

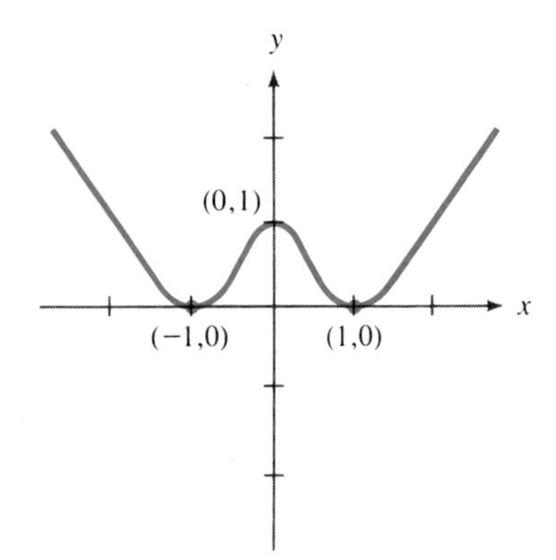

5.

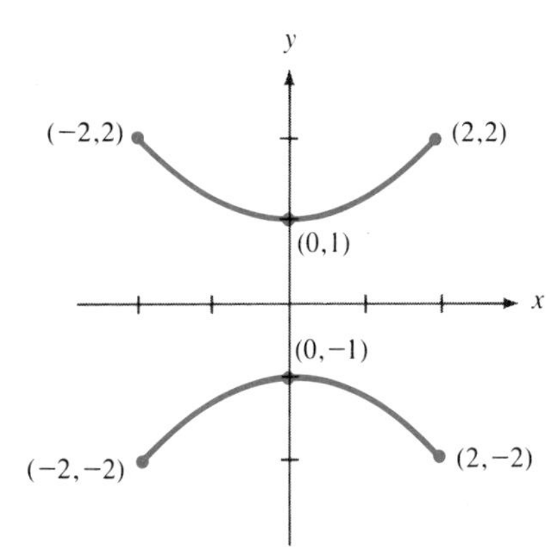

6.

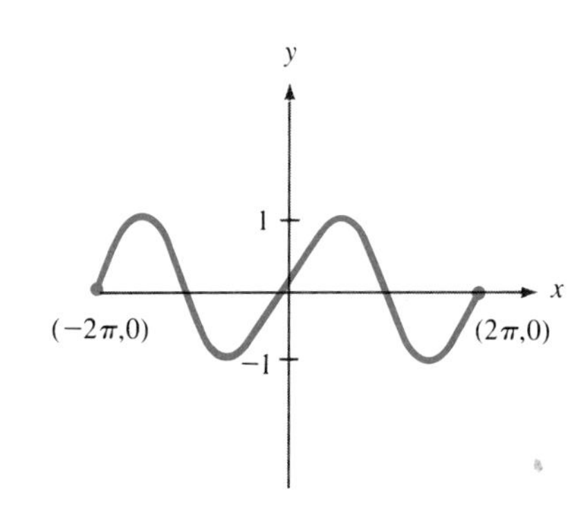

7.

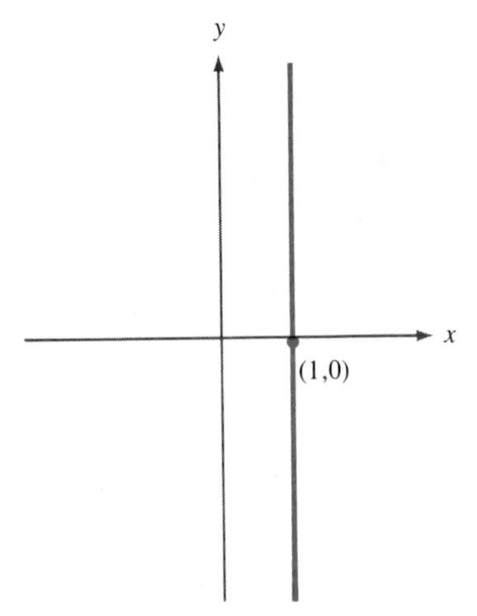

8.

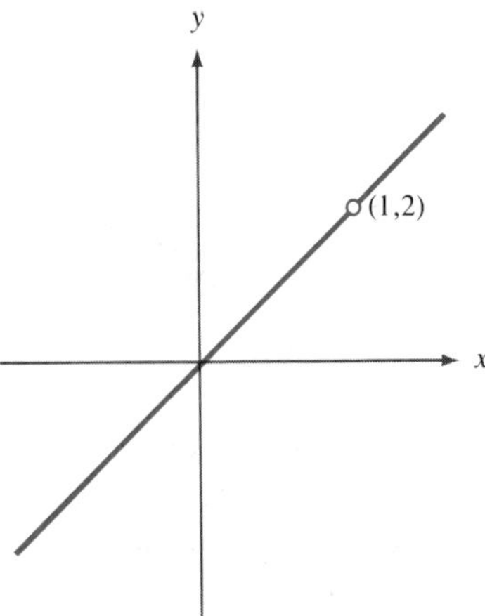

9. Suppose that the following curve is the graph of a function f. Find all values of x that satisfy the given conditions.

(a) $f(x) = 0$ **(b)** $f(x) > 1$ **(c)** $f(x) \geq 2$
(d) $1 < f(x) \leq 2$

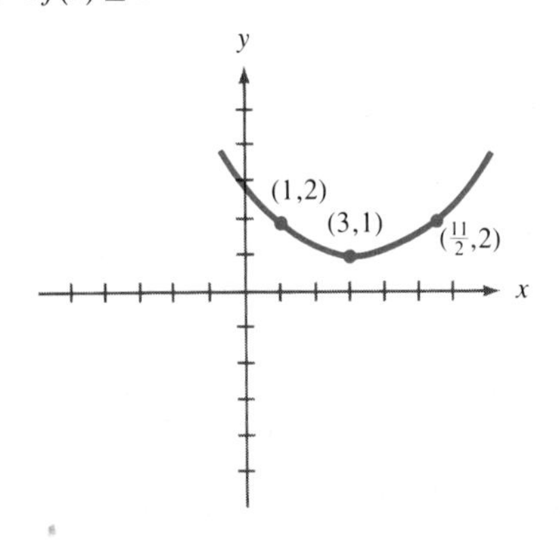

10. Suppose that the following curve is the graph of a function f. Find all values of x that satisfy the given conditions.

(a) $f(x) = 1$ **(b)** $f(x) \geq 1$ **(c)** $f(x) < 3$
(d) $1 < f(x) < 2$

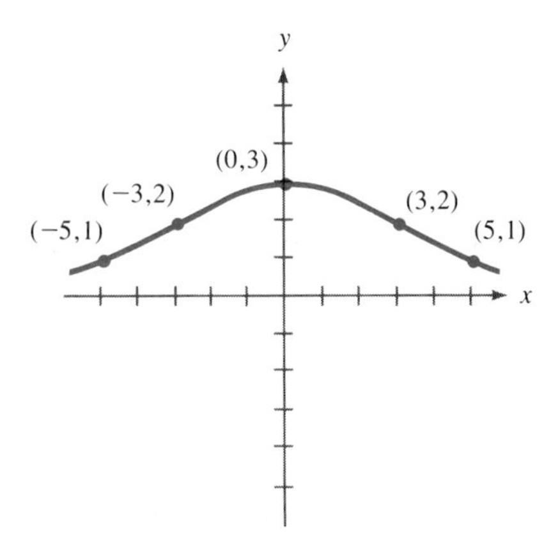

11. Suppose that the following curve is the graph of a function g. Find all values of x that satisfy the given conditions.

(a) $g(x) = 0$ **(b)** $g(x) = 1$ **(c)** $g(x) > 0$
(d) $g(x) \geq 1$

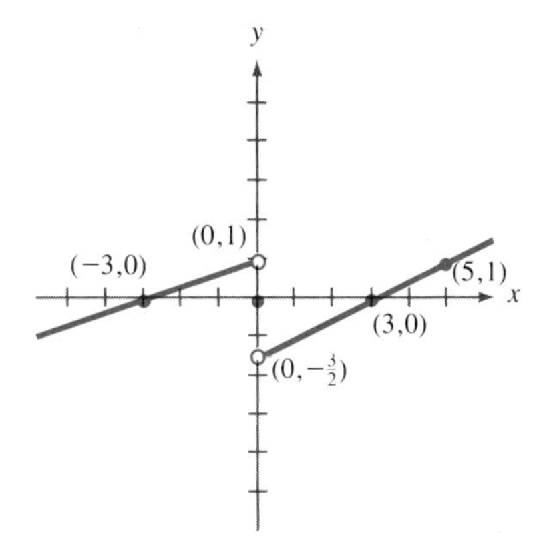

12. Suppose that the following curve is the graph of a function h. Find all values of x that satisfy the given conditions.

(a) $h(x) = 0$ **(b)** $h(x) < 1$ **(c)** $f(x) \leq 0$
(d) $0 \leq f(x) < 1$

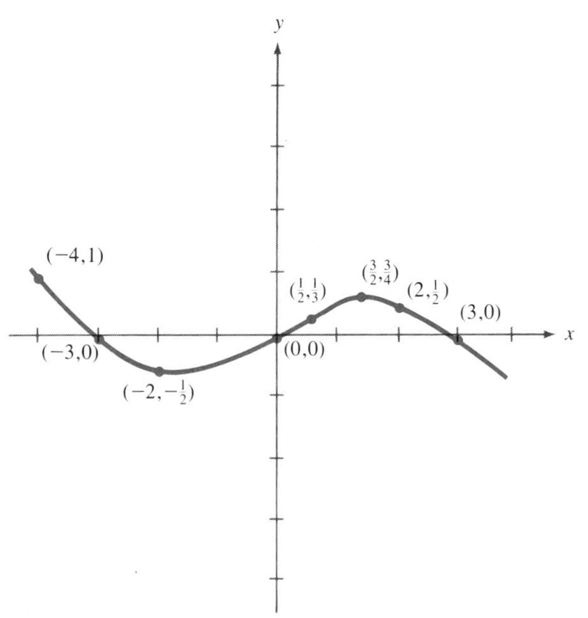

In Exercises 13–50, determine the domain of the function (if it is not given to you) and then sketch its graph.

13. $f(x) = 0$

14. $f(x) = -1/2$

15. $f(x) = \sqrt{2}$

16. $f(x) = -\pi$

17. $f(x) = 2x$

18. $f(x) = -x/3$

19. $f(x) = 3x + 1$

20. $f(x) = 1 - 4x$

21. $f(x) = 1 + x/2$

22. $f(x) = |3x|$

23. $f(x) = |-3x|$

24. $f(x) = |x + 1|$

25. $f(x) = |x| + 1$

26. $f(x) = |x|/x$

27. $f(x) = \sqrt{x - 2}$

28. $f(x) = \sqrt{2x + 3}$

29. $f(x) = \sqrt{x^2 + 1}$

30. $f(x) = \sqrt{x^2}$

31. $f(x) = \dfrac{2}{x}$

32. $f(x) = -\dfrac{1}{x}$

33. $f(x) = \dfrac{1}{x^2}$

34. $f(x) = \dfrac{1}{x + 1}$

35. $f(x) = \dfrac{x^2 + x}{x}$

36. $f(x) = \dfrac{x^2 + 4x + 4}{x + 2}$

37. $f(x) = \dfrac{x^3 + 3x}{x^2 + 3}$

38. $f(x) = \dfrac{(x + 1)^2 - 1}{x}$

39. $f(x) = 2x$; domain of f is the open interval $(-1, 1)$.

40. $f(x) = 1 - 4x$; domain of f is the open interval $(1/4, 4)$.

41. $f(x) = x^2 + 2$; domain of f is the closed interval $[0, 5]$.

42. $f(x) = 1/x$; domain of f is the half-open interval $(0, 10]$.

43. $f(x) = |x|$; domain of f is the interval $(-4, 4)$.

44. $f(x) = \begin{cases} 3 & \text{if } x \le 1 \\ x & \text{if } x > 1 \end{cases}$

45. $f(x) = \begin{cases} -x & \text{if } x < 0 \\ x & \text{if } x \ge 0 \end{cases}$

Does this look familiar?

46. $f(x) = \begin{cases} 1 & \text{if } x \text{ is an integer} \\ 0 & \text{otherwise} \end{cases}$

47. $f(x) = \begin{cases} x & \text{if } x \ne 5 \\ 0 & \text{if } x = 5 \end{cases}$

48. $f(x) = \begin{cases} 1 & \text{if } x < -1 \\ x & \text{if } -1 \le x \le 1 \\ x^2 & \text{if } x > 1 \end{cases}$

49. $f(x) = \begin{cases} 1 + x & \text{if } x < 0 \\ 1 - x/2 & \text{if } 0 \le x \le 2 \\ x - 2 & \text{if } x > 2 \end{cases}$

50. $f(x) = \begin{cases} \sqrt{-x} & \text{if } x < -1 \\ |x| & \text{if } -1 < x < 1 \\ 1/x & \text{if } x > 1 \end{cases}$

If r is a real number, then we denote by $[[r]]$ the *greatest* integer that is less than or equal to r. For example, we have $[[3/2]] = 1$ because 1 is the greatest integer that is less than or equal to $3/2$. As another example, we have $[[-3/2]] = -2$ because -2 is the greatest integer less than or equal to $-3/2$. If r is an integer, then of course $[[r]] = r$. For instance, $[[6]] = 6$ and $[[-9]] = -9$. Now we can define a function $g: \mathbf{R} \to \mathbf{Z}$ by $g(x) = [[x]]$. (Recall that $\mathbf{Z}$ is the set of all integers.) This function is called the **greatest integer function.**

51. Let g be the greatest integer function. Compute the following values.

(a) $g(1)$ **(b)** $g(-12)$ **(c)** $g(0)$

(d) $g(-3/5)$ **(e)** $g(17/3)$ **(f)** $g(-12/5)$

52. Sketch the graph of the greatest integer function.

2.5 Symmetry; Translating, Stretching and Reflecting Graphs

In this section, we will discuss some techniques that can be of great help in graphing functions.

The graph in Figure 2.38 has a very appealing property—it is **symmetric with respect to the y-axis.** By this we mean that if we were to "fold" the graph along the y-axis, the left and right halves would match exactly. Another way to say this is that the left half of the graph is the mirror image of the right half.

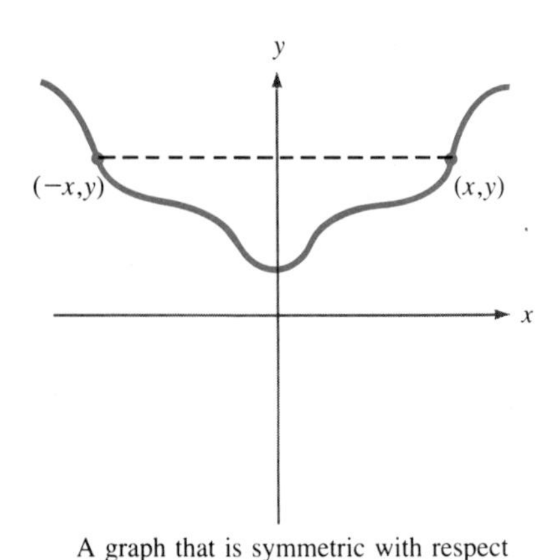

A graph that is symmetric with respect to the y-axis

FIGURE 2.38

If we know that the graph of a particular function is symmetric with respect to the y-axis, then we can sketch this graph by first sketching only that portion of the graph that lies to the right of (and on) the y-axis. Then we "reflect" this portion over the y-axis to obtain the entire graph. Thus, we would like to find a condition on the function itself that will tell us whether or not its graph is symmetric with respect to the y-axis.

In algebraic terms, a graph is symmetric with respect to the y-axis if and only if whenever the point (x, y) is on the graph, so is the point $(-x, y)$. (See Figure 2.38.) We can phrase this in terms of the function itself by saying that the graph of a function f is symmetric with respect to the y-axis if and only if $f(-x) = f(x)$ for all x in the domain of the function. This leads us to make the following definition.

DEFINITION

A function f is **even** if it has the property that

$$f(-x) = f(x)$$

for all real numbers x in its domain.

Using this definition, we can say that

the graph of a function is symmetric with respect to the y-axis if and only if the function is even.

EXAMPLE 1: Show that the function

$$f(x) = x^2$$

is even and sketch its graph.

Solution: To show that the function f is even, we compute $f(-x)$,

$$f(-x) = (-x)^2 = x^2 = f(x)$$

Thus, $f(-x) = f(x)$ for all real numbers x and so f is even. This tells us that the graph of f is symmetric with respect to the y-axis. Choosing only non-negative values of x, the method of plotting points gives the graph pictured in Figure 2.39a. Reflecting this graph over the y-axis gives the complete graph of f, as shown in Figure 2.39b. *Try Study Suggestion 2.30.* □

▶ **Study Suggestion 2.30:** Show that the function $g(x) = x^4 + 1$ is even and sketch its graph. ◀

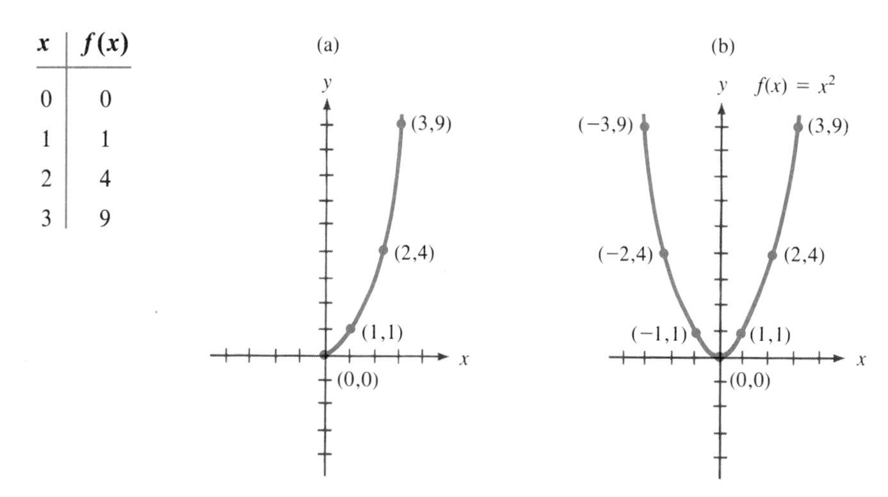

x	$f(x)$
0	0
1	1
2	4
3	9

FIGURE 2.39

There is another type of symmetry that occurs in the graphs of certain functions. The graph of a function is said to be **symmetric with respect to the origin** if whenever (x, y) is on the graph, then so is $(-x, -y)$. This is pictured in Figure 2.40. Now, a function f is symmetric with respect to the origin if and only if it satisfies the condition $f(-x) = -f(x)$ for all x in its domain. This leads us to make the following definition.

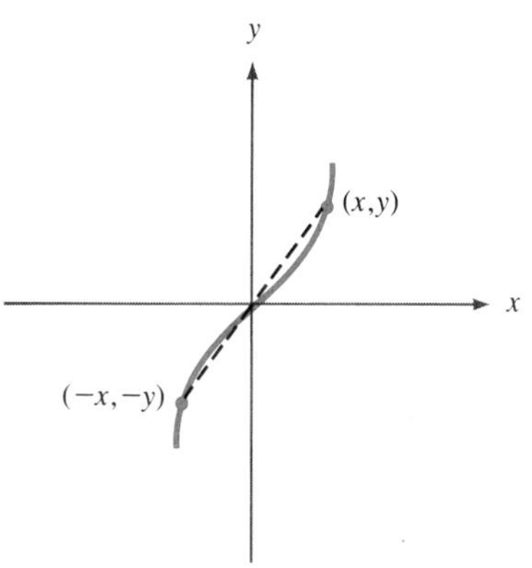

FIGURE 2.40 A graph that is symmetric with respect to the origin

DEFINITION

A function f is **odd** if it has the property that

$$f(-x) = -f(x)$$

for all real numbers x in its domain.

Using this definition, we can say that

the graph of a function is symmetric with respect to the origin if and only if the function is odd.

EXAMPLE 2: Show that the function

$$f(x) = x^3$$

is odd and sketch its graph.

Solution: In order to see that f is odd, we observe that

$$f(-x) = (-x)^3 = (-1)^3x^3 = -x^3 = -f(x)$$

and so $f(-x) = -f(x)$ for all real numbers x. This tells us that the graph of f is symmetric with respect to the origin. Therefore, we first sketch the graph for nonnegative values of x, by the method of plotting points. (See Figure 2.41(a).) Then, rather than try to reflect the entire curve in Figure 2.41(a) through the origin into the third quadrant, we reflect just the points gathered in our table, and draw a smooth curve through those points as in Figure 2.41b. (This seems a bit easier than reflecting the entire curve through a single point.) *Try Study Suggestion 2.31.* □

▶ **Study Suggestion 2.31:** Show that the function $g(x) = -x^3$ is odd, and sketch the graph. ◀

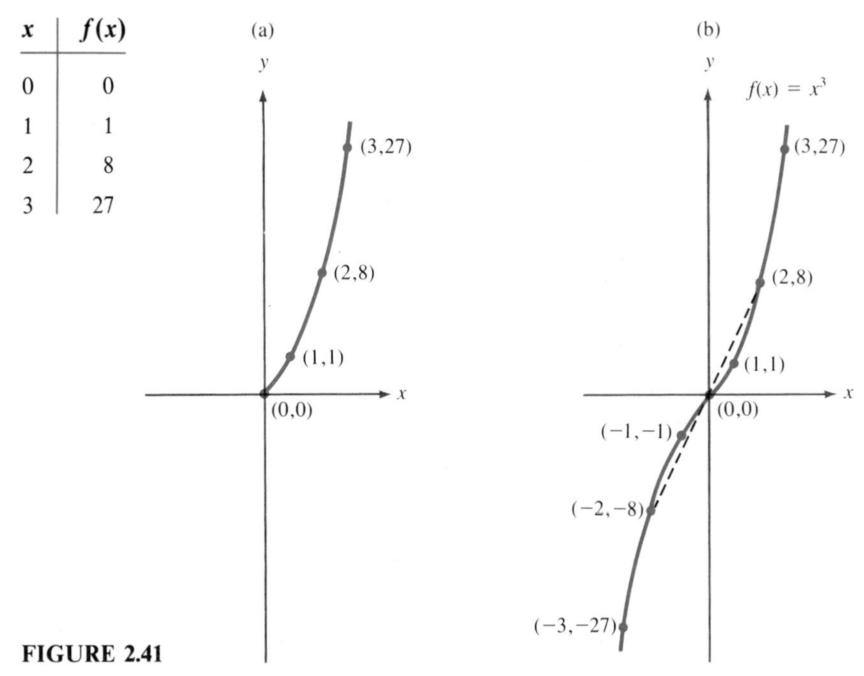

x	$f(x)$
0	0
1	1
2	8
3	27

FIGURE 2.41

We should mention that the concept of even and odd for functions is quite different than for integers. In fact, it is possible for a function to be neither even nor odd, whereas an integer must be one or the other. Also, there is actually one function that is *both* even and odd! (See Exercise 70.)

The next example illustrates the method for determining whether a function is even, odd, or neither.

EXAMPLE 3: Determine which of the following functions are even, which are odd, and which are neither.

(a) $f(x) = x^2 - 2x + 1$ **(b)** $g(x) = -4$ **(c)** $h(t) = \frac{1}{t}$

Solutions:

(a) In this case, we have

$$f(x) = x^2 - 2x + 1$$
$$-f(x) = -(x^2 - 2x + 1) = -x^2 + 2x - 1$$
$$f(-x) = (-x)^2 - 2(-x) + 1 = x^2 + 2x + 1$$

Since $f(-x)$ is not equal to $f(x)$, the function f is not even. Also, $f(-x)$ is not equal to $-f(x)$, and so f is not odd. Thus, f is neither even nor odd.

(b) Here we have

$$g(x) = -4$$
$$-g(x) = -(-4) = 4$$
$$g(-x) = -4$$

Since $g(-x) = g(x)$, the function g is even. On the other hand, since $g(-x)$ is not equal to $-g(x)$, the function g is not odd.

(c) Here we have

$$h(t) = \frac{1}{t}$$
$$-h(t) = -\frac{1}{t}$$
$$h(-t) = \frac{1}{-t} = -\frac{1}{t}$$

Since $h(t)$ is not equal to $-h(t)$, we conclude that h is not even. However, since $h(-t) = -h(t)$, the function is odd.

Try Study Suggestion 2.32. □

▶ **Study Suggestion 2.32:** Determine which of the following functions are even, which are odd, and which are neither.

(a) $f(x) = 3x^3 - x$

(b) $g(x) = x^4 + 4x$

(c) $h(y) = \frac{y^4}{y^4 + 1}$

(d) $F(x) = a$, where a is a constant ◀

Sometimes the graph of one function can be obtained directly from the graph of another function. As an example, consider the function

$$g(x) = x^3 + 2$$

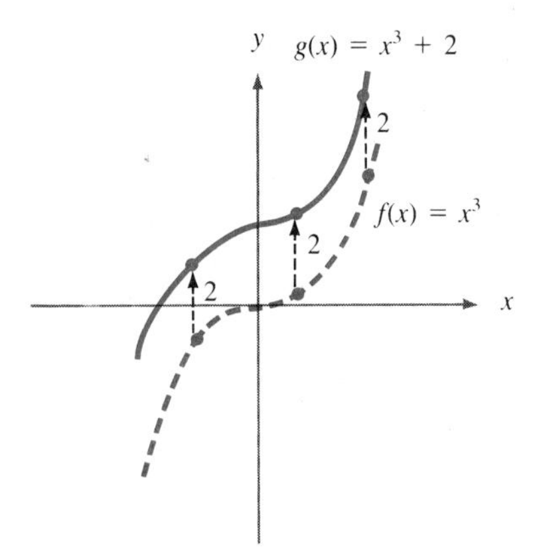

FIGURE 2.42

We can obtain the graph of this function from the graph of the function

$$f(x) = x^3$$

by observing that for each real number x, the value of $g(x)$ is 2 units *greater* than the value of $f(x)$. In terms of the graph, this means that the point $(x, g(x))$ is 2 units *higher* than the point $(x, f(x))$. Hence, the graph of $g(x)$ can be obtained by *translating* the graph of $f(x)$ up 2 units, as shown in Figure 2.42.

As another example, we can obtain the graph of the function

$$h(x) = 3x^3$$

from the graph of $f(x) = x^3$ by observing that for each real number x, the value of $h(x)$ is three times the value of $f(x)$. This means that the graph of h can be obtained by "stretching" the graph of f, as shown in Figure 2.43. Notice that the x-intercepts, being places where $f(x) = 0$, are not affected by the stretching. However, all other points on the graph of f are moved *away from* the x-axis.

As a final example, the graph of the function

$$k(x) = -x^3$$

can be obtained from the graph of $f(x) = x^3$ by observing that, for any real number x, $k(x)$ and $f(x)$ have the same *absolute value*, but are opposite in sign. Thus, the graph of $k(x) = -x^3$ can be obtained by *reflecting* the graph of $f(x) = x^3$ over the x-axis, as shown in Figure 2.44.

▶ **Study Suggestion 2.33:** Use the graph of $f(x) = x^2$ to obtain the graph of

(a) $g(x) = x^2 - 2$
(b) $h(x) = (1/2)x^2$
(c) $k(x) = -x^2$ ◀

Try Study Suggestion 2.33.

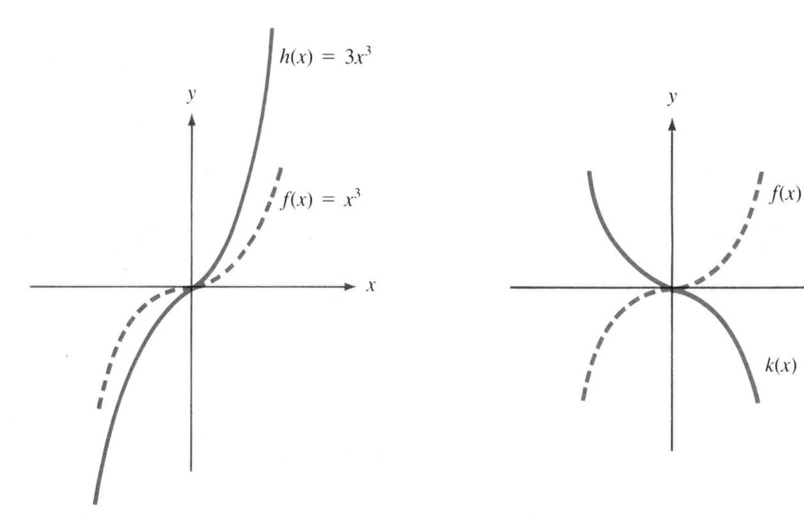

FIGURE 2.43 FIGURE 2.44

Table 2.1 gives us some general guidelines for obtaining the graph of a function g from the graph of a related function f.

TABLE 2.1

Relationship between the functions g and f		How to obtain the graph of g from the graph of f
$g(x) = f(x) + c$	$(c > 0)$	Translate up c units
$g(x) = f(x) - c$		Translate down c units
$g(x) = f(x + d)$	$(d > 0)$	Translate to the left d units
$g(x) = f(x - d)$		Translate to the right d units
$g(x) = af(x)$	$(a > 0)$	Stretch the graph away from the x-axis if $a > 1$ Shrink the graph toward the x-axis if $0 < a < 1$
$g(x) = -f(x)$		Reflect the graph over the x-axis

Let us consider some additional examples.

EXAMPLE 4: Graph the function

$$g(x) = (x + 3)^2$$

Solution: The function g has the form

$$g(x) = f(x + d)$$

where $f(x) = x^2$ and $d = 3$. Hence, according to Table 2.1, the graph of g can be obtained by translating the graph of f a distance of $d = 3$ units to the left, as shown in Figure 2.45. *Try Study Suggestion 2.34.* □

▶ **Study Suggestion 2.34:** Graph the function $g(x) = (x + 1)^3$. ◀

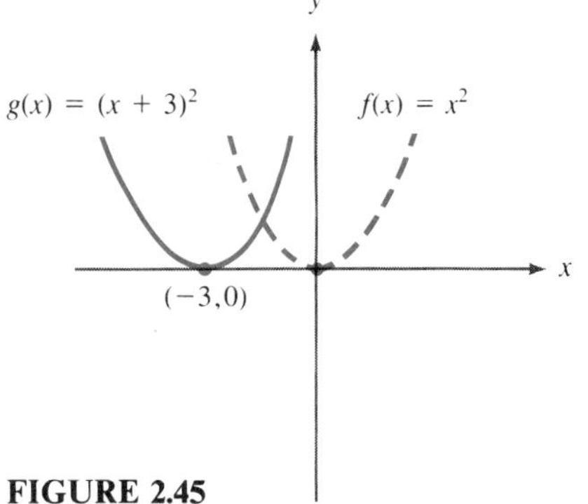

FIGURE 2.45

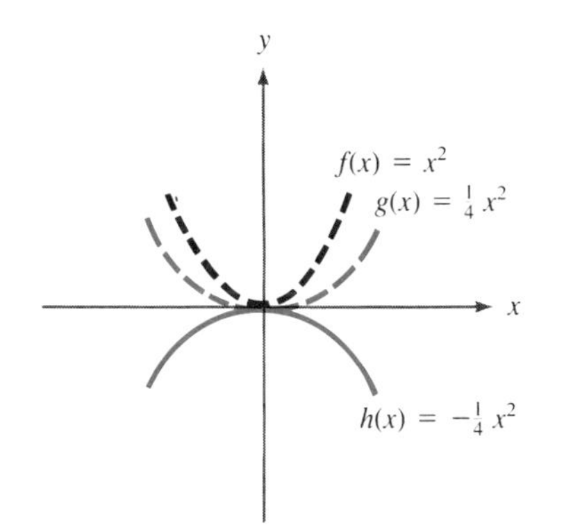

FIGURE 2.46

EXAMPLE 5: Graph the function

$$h(x) = -\frac{1}{4}x^2$$

Solution: We can obtain the graph of $h(x) = -(1/4)x^2$ from the graph of $f(x) = x^2$ in two steps. Table 2.1 tells us that the graph of $g(x) = (1/4)x^2$ can be obtained by shrinking the graph of $f(x) = x^2$. (See Figure 2.46.) Then the table tells us that the graph of $h(x) = -(1/4)x^2$ can be obtained by reflecting the graph of $g(x) = (1/4)x^2$ over the x-axis. This is also shown in Figure 2.46.

Try Study Suggestion 2.35. □

▶ **Study Suggestion 2.35:** Graph the function $g(x) = -5x^3$. ◀

EXAMPLE 6: Graph the function

$$h(x) = |x - 3| + 2$$

Solution: According to Table 2.1, we can obtain the graph of $g(x) = |x - 3|$ by translating the graph of the absolute value function $f(x) = |x|$ a distance of 3 units to the right. Then we can obtain the graph of $h(x) = |x - 3| + 2$ by translating the graph of $g(x) = |x - 3|$ a distance of 2 units up. (See Figure 2.47.)

Try Study Suggestion 2.36. □

▶ **Study Suggestion 2.36:** Graph the function $h(x) = (x + 1)^2 - 3$. ◀

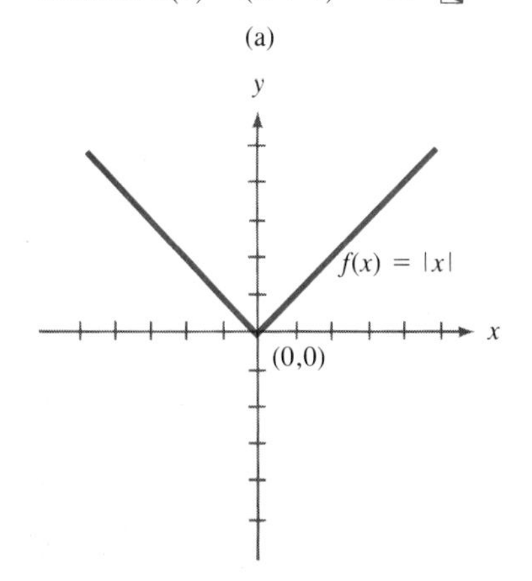

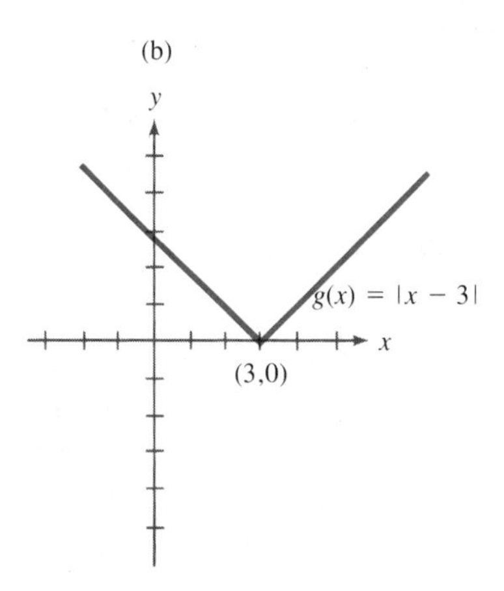

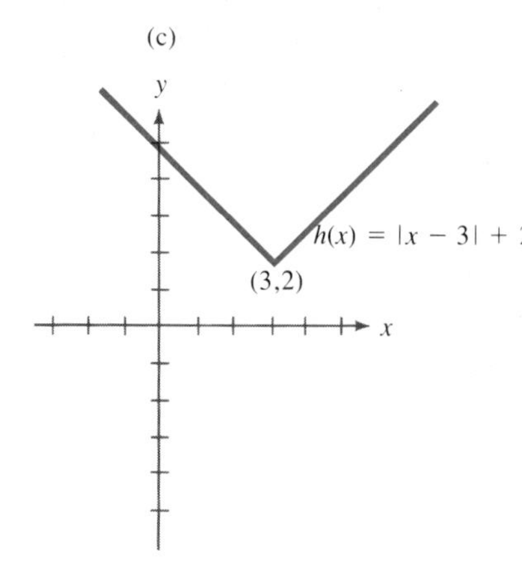

FIGURE 2.47

Ideas to Remember

- Knowing ahead of time that a graph has a certain type of symmetry can be very helpful when it comes to sketching its graph.
- Sometimes it is possible to obtain the graph of one function from the graph of another more familiar function by translating, stretching, shrinking, and/or reflecting.

EXERCISES

In Exercises 1–12, determine whether the given function is even, odd, or neither.

1. $f(x) = 3x$

2. $f(x) = 2x - 3$

3. $f(x) = x^2 + 3x - 1$

4. $f(x) = x^5 + x^2$

5. $f(x) = \dfrac{1}{x^4} + 1$

6. $f(x) = \dfrac{x^2}{x^2 + 1}$

7. $f(x) = \sqrt{x}$

8. $f(x) = 2|x| - 5$

9. $f(x) = |x + 1|$

10. $f(x) = x^3 + 3x^{1/3}$

11. $f(x) = x + x^{-1}$

12. $f(x) = \dfrac{x^2}{\sqrt{x^2 + 1}}$

In Exercises 13–22, use the graph of $f(x) = x^3$ to find the graph of the given function. (Sketch both graphs on the same plane.)

13. $g(x) = x^3 + 2$

14. $h(x) = x^3 - 4$

15. $g(x) = (x + 5)^3$

16. $k(x) = (x - 1/2)^3$

17. $G(x) = (1/2)x^3$

18. $F(x) = -x^3/3$

19. $h(x) = (x - 3)^3 + 4$

20. $g(x) = (x + 4)^3 + 2$

21. $g(x) = 1 - x^3$

22. $A(x) = -3 - x^3$

In Exercises 23–30, use the graph of $f(x) = |x|$ to find the graph of the given function. (Sketch both graphs on the same plane.)

23. $h(x) = |x + \pi|$

24. $g(x) = |x - \sqrt{2}|$

25. $F(x) = |x| + 7$

26. $p(x) = 1 - |x|$

27. $k(x) = -|x|$

28. $g(x) = 4|x| + 1$

29. $g(x) = |x - 3| + 1$

30. $H(x) = 2|x + 4| - 1$

In Exercises 31–38, use the graph of $f(x) = \sqrt{x}$ to find the graph of the given function. (Sketch both graphs on the same plane.)

31. $g(x) = \sqrt{x} + 7$

32. $g(x) = \sqrt{x} - 1/2$

33. $h(x) = \sqrt{x - 5}$

34. $g(x) = \sqrt{2x}$

35. $H(x) = -\sqrt{x}$

36. $K(x) = \sqrt{2x - \sqrt{3}}$

37. $g(x) = -\sqrt{x/2 + 1}$

38. $A(x) = \sqrt{x - 1} + 1$

In Exercises 39–44, use the graph of $f(x) = 1/x$ to find the graph of the given function. (Sketch both graphs on the same plane.)

39. $g(x) = \dfrac{1}{x + 1}$

40. $g(x) = \dfrac{1}{x - 1}$

41. $h(x) = \dfrac{2}{x}$

42. $K(x) = -\dfrac{1}{2x}$

43. $A(x) = \dfrac{1}{x} + 1$

44. $B(x) = \dfrac{3 - 2x}{x}$ (*Hint:* rewrite the function)

In Exercises 45–61, graph the given function, using the method of plotting points, aided by any of the techniques that we discussed in this section (if applicable.)

45. $f(x) = x^2 - 5$

46. $f(x) = -2x^2$

47. $f(x) = x^2 + x$

48. $f(x) = x^3 - x$

49. $f(x) = x^4$

50. $f(x) = 2(x^4 - 1)$

51. $f(x) = x^{-4}$

52. $f(x) = \dfrac{x}{x + 1}$

53. $f(x) = \dfrac{-1}{x^2 + 1}$

54. $f(x) = \sqrt{x^2 + 2}$

55. $f(x) = x^2 + |x|$

56. $f(x) = 1/\sqrt{x}$

57. $f(x) = |x^2 + 1|$

58. $f(x) = |x| + x$

59. $f(x) = \begin{cases} x^2 & \text{if } x \le 0 \\ x^4 & \text{if } x > 0 \end{cases}$

60. $f(x) = \begin{cases} x^2 & \text{if } x \text{ is an integer} \\ 0 & \text{otherwise} \end{cases}$

61. $f(x) = \begin{cases} 1 + x^2 & \text{if } -2 \le x \le -1 \text{ or } 1 \le x \le 2 \\ 1 & \text{otherwise} \end{cases}$

Figure 2.48 is the graph of a function $f(x)$ whose domain is the set $\{x | 0 \le x \le 4\}$.

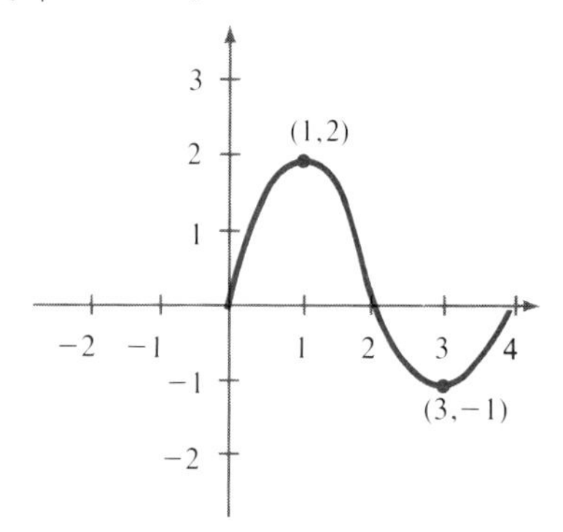

FIGURE 2.48

In Exercises 62–67, use this graph to sketch the graph of the given equation.

62. $y = -f(x)$

63. $y = f(-x)$

64. $y = f(x + 2)$

65. $y = f(x) + 2$

66. $y = |f(x)|$

67. $y = [f(x)]^2$

68. Is the greatest integer function, defined in the exercises of the previous section, even, odd, or neither? Explain your answer.

69. Under what conditions on the *integer n* is the function $f(x) = x^n$ even? Under what conditions is it odd?

70. Show that the *only* function that is both even and odd is the **zero function** $f(x) = 0$.

71. Are there any functions whose graphs are symmetric with respect to the x-axis? Justify your answer.

2.6 Increasing and Decreasing Functions; One-to-One Functions

Let us consider the graph of the function $f(x) = x^2$ shown in Figure 2.49. As we look from left to right along the x-axis, we observe that at first the graph falls. Then, once we reach the origin, the graph begins to rise. This can be translated into a statement about the values of the function. It says that as x increases, but while it is still negative, the values of the function *decrease*. However, once x becomes positive, the values of the function *increase* as x increases.

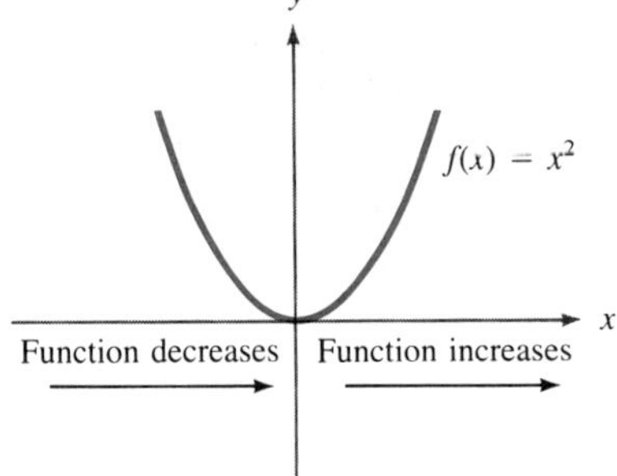

FIGURE 2.49

The concept of a function increasing and decreasing is very important, and so we want to make it more precise in a definition.

DEFINITION

1. A function $f: A \to B$ is said to be **increasing** on an interval I in its domain if whenever a and b are in I and $a < b$, then $f(a) < f(b)$. (See Figure 2.50a.)
2. A function $f: A \to B$ is said to be **decreasing** on an interval I in its domain if whenever a and b are in I and $a < b$, then $f(a) > f(b)$. (See Figure 2.50b.)
3. A function $f: A \to B$ is said to be **constant** on an interval I in its domain if $f(a) = f(b)$ for all a and b in I. (See Figure 2.50c.)

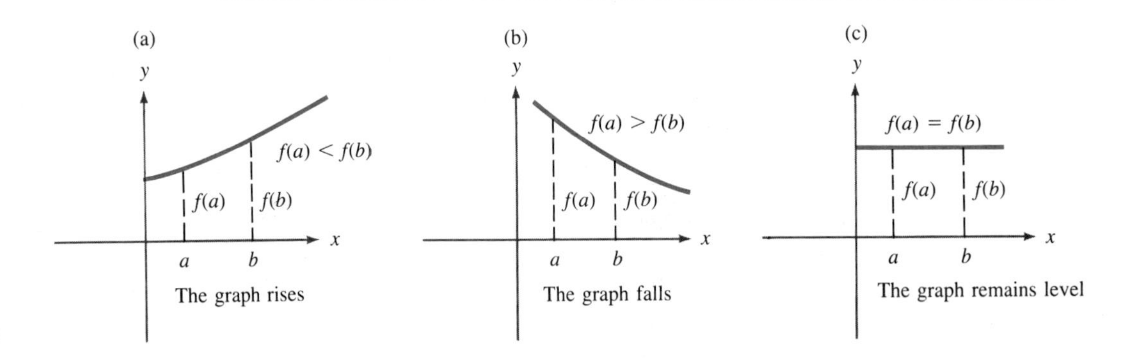

FIGURE 2.50

▶ **Study Suggestion 2.37:** Determine on what interval the function $f(x) = (x-1)^2 + 2$ is increasing and on what interval it is decreasing. (First graph the function.) ◀

EXAMPLE 1: We can see from Figure 2.49 that the function $f(x) = x^2$ is decreasing on the interval $(-\infty, 0]$ and increasing on the interval $[0, \infty)$.

Try Study Suggestion 2.37. □

▶ **Study Suggestion 2.38:** Determine on which intervals the function whose graph is given in Figure 2.52 is increasing, on which intervals it is decreasing, and on which intervals it is constant. ◀

EXAMPLE 2: Consider the graph in Figure 2.51. Let us imagine that it is the graph of a function f. Then f is increasing on the interval $(-\infty, -1]$, decreasing on the interval $[-1, 1]$, constant on the interval $[1, 2]$, and finally decreasing on the interval $[2, \infty)$. *Try Study Suggestion 2.38.* □

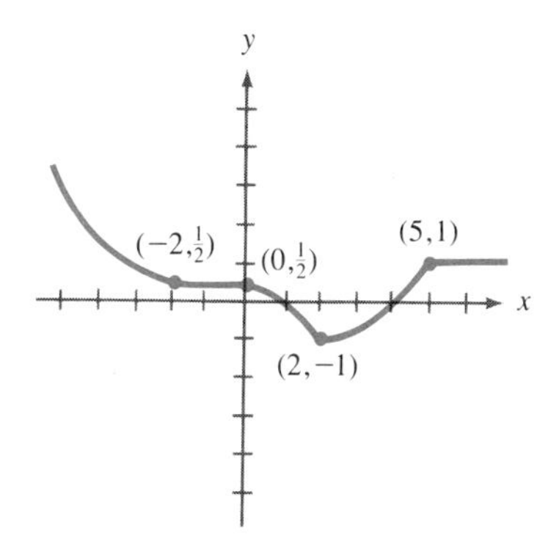

FIGURE 2.52

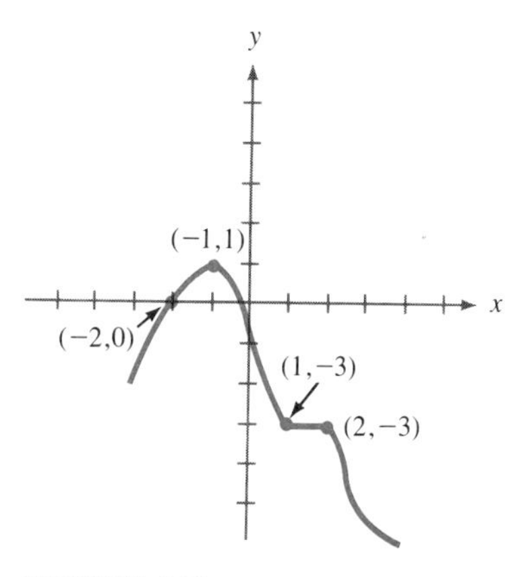

FIGURE 2.51

EXAMPLE 3: We can see from Figure 2.53 that the function $f(x) = x^3$ is increasing on the entire real line. Hence, this function is increasing *on its entire domain.* Similarly, we can see from Figure 2.54 that the function $g(x) = 1 - \sqrt{x}$ is decreasing on *its* entire domain, which is the set $\{x \mid x \geq 0\}$. □

When a function is increasing *on its entire domain,* as is the case with the function $f(x) = x^3$ in Example 3, we say that it is an **increasing function.** Similarly, when a function is decreasing *on its entire domain,* as is the case with the function $g(x) = 1 - \sqrt{x}$, we say that it is a **decreasing function.**

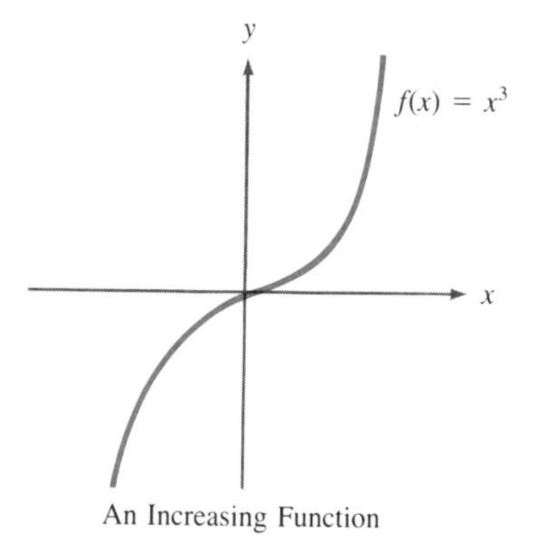

FIGURE 2.53

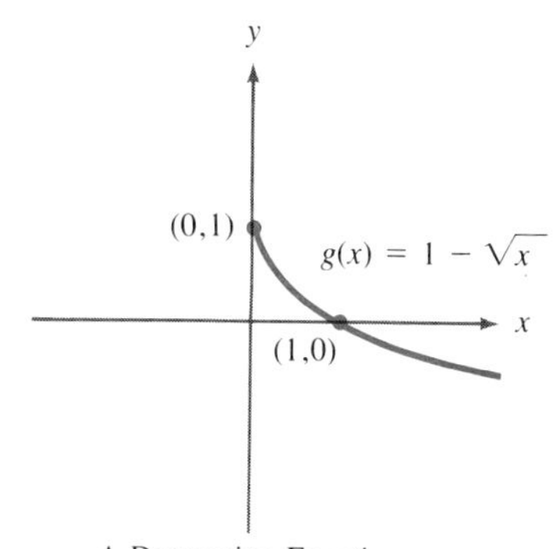

FIGURE 2.54

However, the function in Example 2 is neither an increasing nor a decreasing function, since it neither increases nor decreases *on its entire domain.*

▶ **Study Suggestion 2.39:** Determine whether the following functions are increasing, decreasing, or neither. (*Hint:* first sketch the graph.)

(a) $f(x) = 1 - x^3$

(b) $g(x) = 1 + \sqrt{x + 1}$

(c) $h(x) = 1 + 2x^4$ ◀

Try Study Suggestion 2.39.

Now let us turn to another important property that a function may possess. For purposes of illustration, let $A = \{1, 2, 3\}$ and $B = \{4, 5, 6\}$. We define a function $f: A \to B$ by setting

$$f(1) = 4, \qquad f(2) = 5, \qquad f(3) = 6$$

and a function $g: A \to B$ by setting

$$g(1) = 4, \qquad g(2) = 5, \qquad g(3) = 5$$

Now these functions may not have much practical value, but they do illustrate an important point. Namely, the function f has the property that no two distinct (that is, different) numbers in its domain have the same image. On the other hand g does not have this property, since the numbers 2 and 3 in its domain have *the same* image, namely 5. (That is, $g(2) = g(3) = 5$.)

Using these functions as a guide, we make the following definition.

DEFINITION

Let $f: A \to B$ be a function from A to B. Then f is **one-to-one** if no two *distinct* elements in its domain have the same image. Another way to say this is that f is one-to-one if distinct elements in the domain of f have distinct images. This can be phrased more algebraically in either of the following ways.

1. A function f is one-to-one if whenever a and b are in the domain of f and $a \neq b$, then $f(a) \neq f(b)$
2. A function f is one-to-one if whenever a and b are in the domain of f and $f(a) = f(b)$ then $a = b$.

Figure 2.55 illustrates the condition that is *not* allowed in order for a function to be one-to-one. As you can see, two distinct elements of A, namely a and b, have the same image.

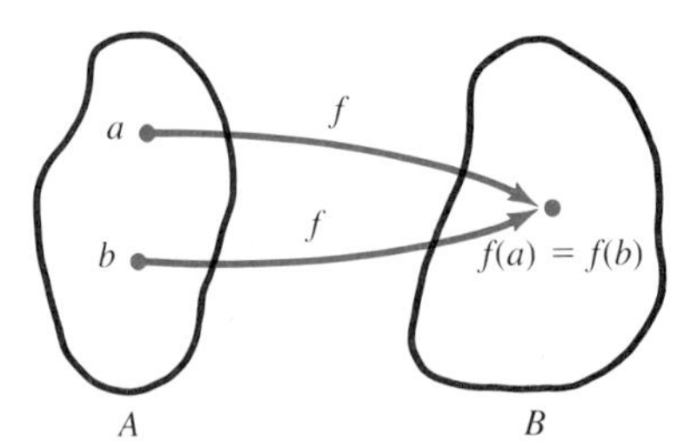

This is **not** allowed in a one-to-one function.

FIGURE 2.55

We can also describe the concept of one-to-one in terms of the graph of the function. A function f is one-to-one if no two distinct points on its graph lie on the same horizontal line. For if two such points did lie on the same horizontal line, then they would be of the form $(a, f(a))$ and $(b, f(b))$ where $a \neq b$ and $f(a) = f(b)$. But this violates the definition of being one-to-one. This gives us a simple way to test the graph of a function to see if the function is one-to-one.

The Horizontal Line Test

The graph of a function passes the **horizontal line test** if no horizontal line intersects the graph at more than one point. A function f is one-to-one if and only if its graph passes the horizontal line test.

(Don't confuse the horizontal line test with the vertical line test, discussed in Section 2.4.)

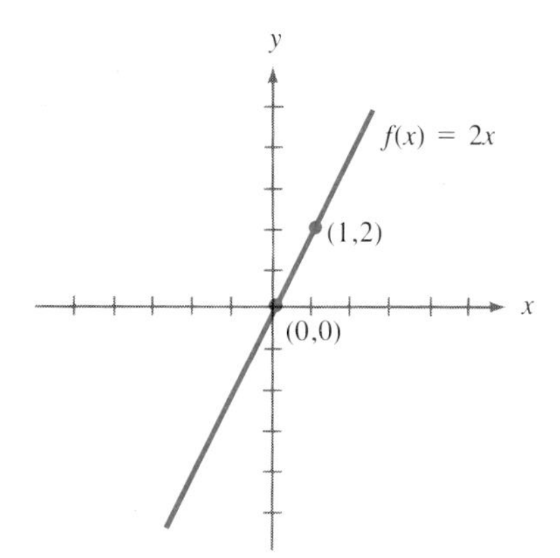

This graph passes the horizontal line test.

FIGURE 2.56

EXAMPLE 4: The function

$$f(x) = 2x$$

is one-to-one, but the function

$$g(x) = x^2$$

is not. One way to see this is to apply the horizontal line test. The graph of $f(x) = 2x$, shown in Figure 2.56, passes the horizontal line test, and so this function is one-to-one.

On the other hand, the graph of $g(x) = x^2$, shown in Figure 2.49, does not pass the horizontal line test, and so this function is not one-to-one.

Another way to see that f is one-to-one and g is not is to use the definition. In order to show that f is one-to-one, we observe that if $f(a) = f(b)$, then $2a = 2b$, and so $a = b$. Hence, according to Statement 2 of the definition, f is one-to-one. On the other hand g is not one-to-one since two distinct numbers in its domain have the same image. For example

$$g(2) = 2^2 = 4 \qquad \text{and} \qquad g(-2) = (-2)^2 = 4$$

Thus, we have $2 \neq -2$ but $g(2) = g(-2)$. *Try Study Suggestion 2.40.* □

▶ **Study Suggestion 2.40:** Use both of the methods that we used in Example 4 to show that the function $f(x) = 4x$ is one-to-one, and that the function $g(x) = x^4$ is not one-to-one. ◀

We should take the time to clear up a point that sometimes causes confusion. According to the definition, in order for a function f to be one-to-one, it must be true that if $a \neq b$ then $f(a) \neq f(b)$, for *all* numbers a and b in the domain of f. Therefore, in order to show that a function is *not* one-to-one, we only need to find *a single pair* of numbers a and b in the domain of f for which $a \neq b$ but $f(a) = f(b)$. (We did precisely this in the previous example, where we found the pair of numbers 2 and -2 for which $2 \neq -2$ but $g(2) = g(-2)$.)

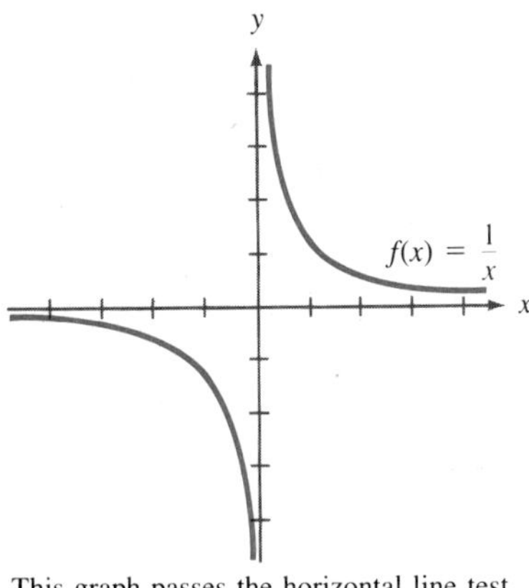

This graph passes the horizontal line test.

FIGURE 2.57

▶ **Study Suggestion 2.41:** Use the definition to show that the function $f(x) = 1/x^2$ is not one-to-one, but that the function $g(x) = 1/x^3$ is one-to-one. 

EXAMPLE 5: The function

$$f(x) = \frac{1}{x}$$

is one-to-one. For if $f(a) = f(b)$, then $1/a = 1/b$ and so $a = b$. We can also see this by observing that the graph of f, shown in Figure 2.57, passes the horizontal line test.

The function

$$g(x) = \frac{x^2}{x^2 + 1}$$

is not one-to-one, since, for example, $-1 \neq 1$ but $g(-1) = g(1)$.

Try Study Suggestion 2.41. □

EXAMPLE 6: The function

$$f(x) = \sqrt{x - 1}$$

is one-to-one, for if $f(a) = f(b)$, then $\sqrt{a - 1} = \sqrt{b - 1}$. But this implies that $a - 1 = b - 1$, and so $a = b$. We can also see that f is one-to-one by observing that its graph, shown in Figure 2.58, passes the horizontal line test. □

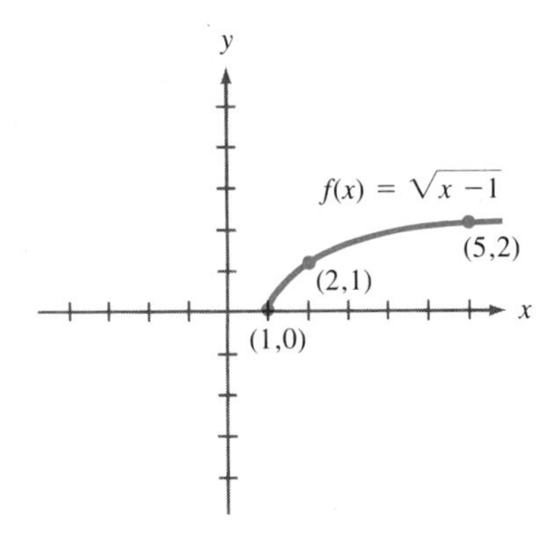

FIGURE 2.58

Finally, let us observe that if a function is an increasing function, then it certainly must pass the horizontal line test. Similarly, decreasing functions pass the horizontal line test. Therefore, we can make the following statements.

1. **Increasing functions are one-to-one.**
2. **Decreasing functions are one-to-one.**

Ideas to Remember

The graph of a function can tell us where a function is increasing and where it is decreasing. It can also tell us whether or not a function is one-to-one.

EXERCISES

In Exercises 1–8, determine on which intervals the function with the given graph is increasing, on which intervals it is decreasing, and on which intervals it is constant. Also, determine whether or not the function is one-to-one.

1.

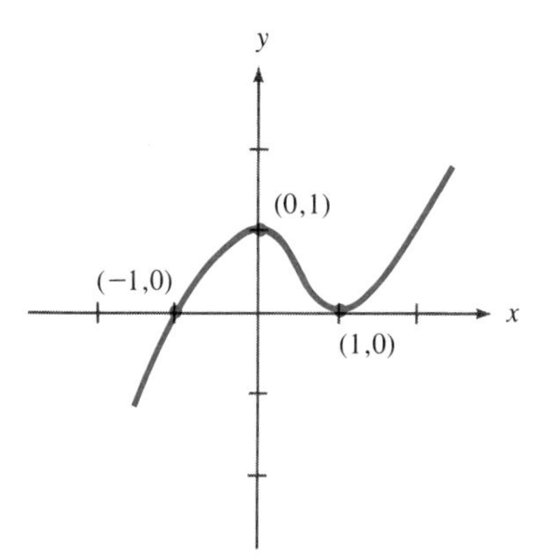

2.

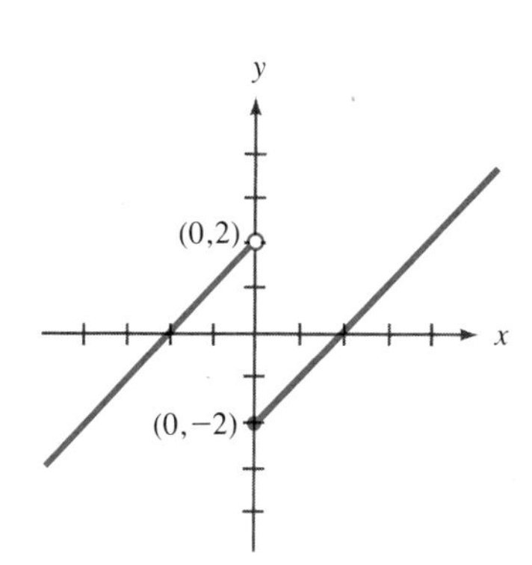

3.

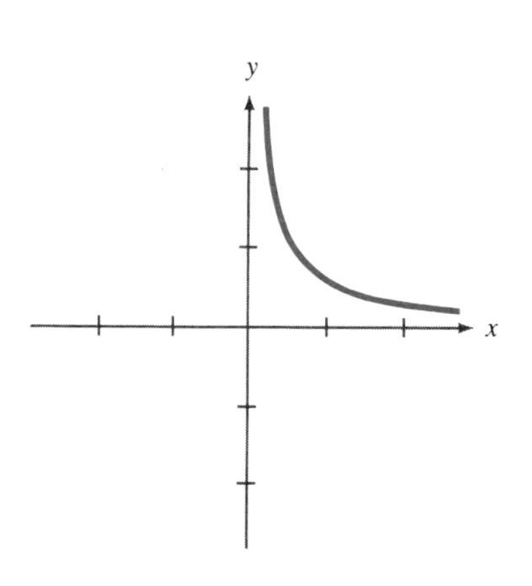

4.

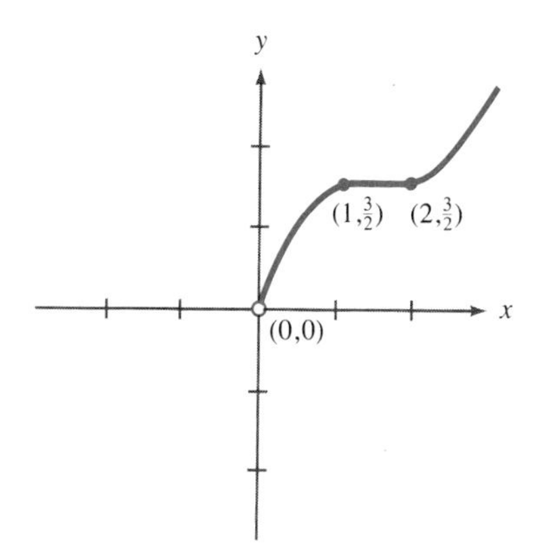

5.

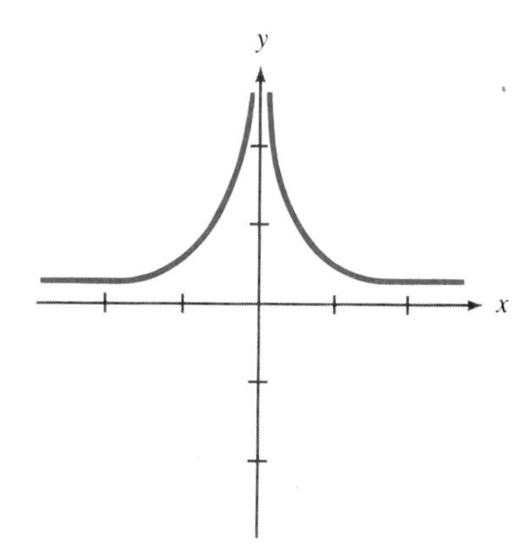

6.

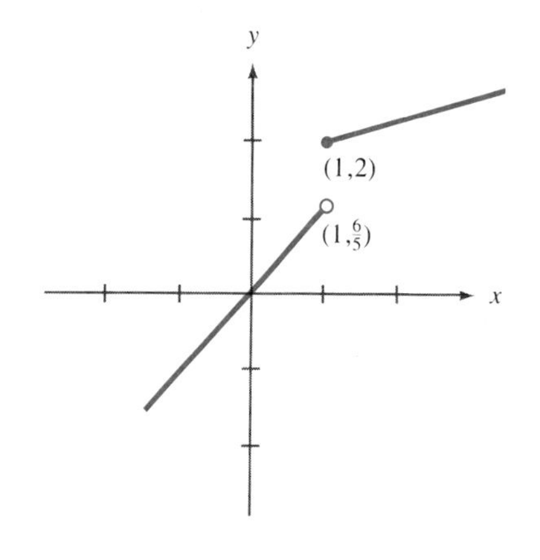

7.

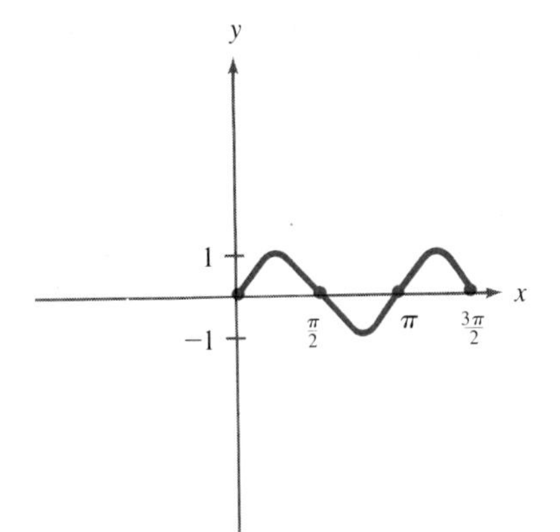

8.

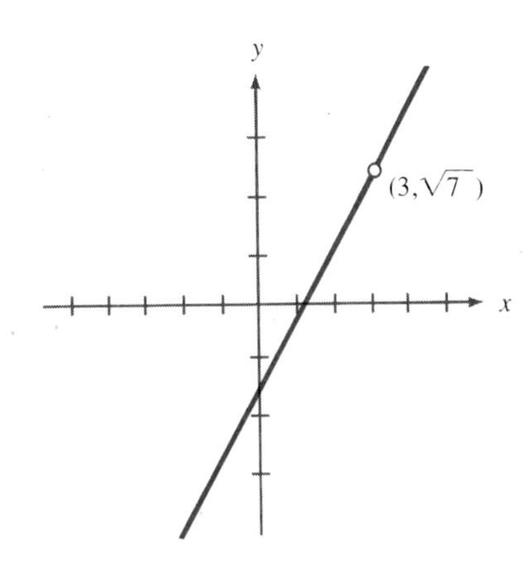

In Exercises 9–33, sketch the graph of the given function and determine on which intervals the function is increasing, on which intervals it is decreasing, and on which intervals it is constant. Also, determine whether or not the function is one-to-one by using the horizontal line test.

9. $f(x) = x/3$

10. $f(x) = -x/3$

11. $f(x) = 4x + 1$

12. $f(x) = 1 - x^2$

13. $f(x) = 2x^2 - 3$

14. $f(x) = (x + 1)^2 - 1$

15. $f(x) = (x + 1)^3 - 1$

16. $f(x) = -x^4$

17. $f(x) = \dfrac{1}{x}$

18. $f(x) = \dfrac{2}{x - 5}$

19. $f(x) = 1 - \sqrt{x}$

20. $f(x) = \sqrt{x - 2} + 2$

21. $f(x) = \sqrt{2x - 3}$

22. $f(x) = \sqrt{4 - x}$

23. $f(x) = 1 - |x|$

24. $f(x) = |x + 1|$

25. $f(x) = |x - 3| - 2$

26. $f(x) = 2 - 3|x|$

27. $f(x) = \dfrac{x}{|x|}$

28. $f(x) = x + |x|$

29. $f(x) = \begin{cases} 2x + 1 & x \geq 3 \\ -4 & 2 \leq x < 3 \\ 1 - x & x < 2 \end{cases}$

30. $f(x) = \begin{cases} x^2 & \text{for } x \geq 2 \\ x & \text{for } -2 < x < 2 \\ -x^2 & \text{for } x \leq -2 \end{cases}$

31. $f(x) = \begin{cases} x^3 & \text{if } x > 0 \\ -x^3 & \text{if } x \leq 0 \end{cases}$

32. $f(x) = \begin{cases} 1 & \text{if } x \text{ is an integer} \\ 0 & \text{if } x \text{ is not an integer} \end{cases}$

33. $f(x) = \begin{cases} 1 & \text{if } x \geq 1 \\ 0 & \text{if } 0 \leq x < 1 \\ -1 & \text{if } x < 0 \end{cases}$

In Exercises 34–39, use the definition to show that the given function is one-to-one.

34. $f(x) = x$

35. $f(x) = 2x - 3$

36. $f(x) = x^3$

37. $f(x) = \sqrt{x - 2}$

38. $f(x) = x^{3/5}$

39. $f(x) = \dfrac{x}{x - 1}$

In Exercises 40–43, use the definition to show that the given function is not one-to-one.

40. $f(x) = x^2 + 1$

41. $f(x) = -7$

42. $f(x) = |x|$

43. $f(x) = x^4 + 3x^2 - 9$

44. Show that the function $f(x) = 3x^2 + 4x - 5$ is not one-to-one. (*Hint:* solve the equation $f(x) = 0$.)

45. Show that the function

$$f(x) = \frac{x - 1}{2x^2 + 1}$$

is not one-to-one. (*Hint:* solve the equation $f(x) = -1$.)

46. Show that the function

$$f(x) = \frac{x}{\sqrt{x - 1}}$$

is not one-to-one. (*Hint:* solve the equation $f(x) = 5$.)

47. Show that the function

$$f(x) = \frac{|x-1|}{|x+1|}$$

is not one-to-one. (*Hint:* solve the equation $f(x) = 2$.)

48. Is a constant function one-to-one? Explain your answer.

49. Under what conditions on the integer n is the function $f(x) = x^n$ one-to-one?

50. Consider the function

$$f(x) = x^2$$

(a) What is the domain of f?

(b) Is f one-to-one?

(c) If we change the domain of f to be $\{x|x \geq 0\}$, is the resulting new function one-to-one? Justify your answer.

51. Consider the function

$$f(x) = x^2 - 2x$$

(a) What is the domain of f?

(b) Is f one-to-one?

(c) If we change the domain of f to be $\{x|x \geq 1\}$ is the resulting new function one-to-one? Justify your answer.

2.7 The Algebra of Functions; Composition of Functions

In Chapter 1, we discussed addition, subtraction, multiplication, and division of algebraic expressions. In this section, we will define these algebraic operations on *functions.*

In describing the domains of the functions involved, we will use the term *intersection.* The **intersection** of two sets A and B, denoted by $A \cap B$, is the set of all elements that are in *both* A and B. For example, if $A = \{1, 2, 3, 4\}$ and $B = \{2, 4, 6, 8\}$ then $A \cap B = \{2, 4\}$. Also, if $S = \{x|x \geq 0\}$ and $T = \{x|-1 \leq x \leq 1\}$ then $S \cap T = \{x|0 \leq x \leq 1\}$.

Try Study Suggestion 2.42.

▶ **Study Suggestion 2.42:** Find $A \cap B$ if

(a) $A = \{1, 5, 1/2, -3, \sqrt{2}\}$ and $B = \mathbf{Z}$ (the set of all integers)

(b) $A = \{x|-2 < x < 3\}$ and $B = \{x|-5 \leq x \leq 1\}$

(*Hint for part (b):* sketch the sets on a real number line.) ◀

Now let us define addition, subtraction, multiplication and division of functions.

DEFINITION

Let $f:A \to B$ and $g:C \to D$ be functions.

1. The **sum** of f and g, denoted by $f + g$, is the function defined by

$$(f + g)(x) = f(x) + g(x)$$

2. The **difference** of f and g, denoted by $f - g$, is the function defined by

$$(f - g)(x) = f(x) - g(x)$$

3. The **product** of f and g, denoted by fg, is the function defined by

$$(fg)(x) = f(x)g(x)$$

4. The **quotient** of f and g, denoted by f/g, is the function defined by

$$\left(\frac{f}{g}\right)(x) = \frac{f(x)}{g(x)} \qquad (g(x) \neq 0)$$

- The domain of each of the functions $f + g$, $f - g$, and fg is defined to be the intersection of the domains of f and g; that is, the set $A \cap C$.
- The domain of f/g is defined to be the set of all numbers in $A \cap C$ *for which* $g(x)$ *is not equal to* 0. (We cannot allow $g(x)$ to be equal to 0 in the case of the quotient, since we cannot divide by 0.)

The following example illustrates these definitions.

EXAMPLE 1: Let $f(x) = 2x^2 - 3x - 9$ and $g(x) = x - 3$. Find

(a) $f + g$ **(b)** $f - g$ **(c)** fg **(d)** f/g.

Also, find their domains.

Solutions:

(a) According to the definition of the sum, we have

$$\begin{aligned}(f + g)(x) &= f(x) + g(x)\\ &= (2x^2 - 3x - 9) + (x - 3)\\ &= 2x^2 - 2x - 12\end{aligned}$$

(b) According to the definition of the difference, we have

$$\begin{aligned}(f - g)(x) &= f(x) - g(x)\\ &= (2x^2 - 3x - 9) - (x - 3)\\ &= 2x^2 - 4x - 6\end{aligned}$$

(c) Using the definition of the product, we get

$$\begin{aligned}(fg)(x) &= f(x)g(x)\\ &= (2x^2 - 3x - 9)(x - 3)\\ &= 2x^3 - 9x^2 + 27\end{aligned}$$

(d) The definition of quotient gives

$$\begin{aligned}\left(\frac{f}{g}\right)(x) &= \frac{f(x)}{g(x)}\\ &= \frac{2x^2 - 3x - 9}{x - 3}\\ &= \frac{(x - 3)(2x + 3)}{x - 3}\\ &= 2x + 3\end{aligned}$$

Since the domain of f and g is the set **R** of all real numbers, the domain of $f+g$, $f-g$, and fg is also the set **R** ($\mathbf{R} \cap \mathbf{R} = \mathbf{R}$.) However, the domain of the quotient f/g consists of all real numbers *except* those for which $g(x) = 0$; that is, except 3. To emphasize this point, we have drawn the graph of the function f/g in Figure 2.59. Notice that there is no point on the graph corresponding to $x = 3$, since this value is not in the domain of f/g. (Hence, the function f/g is *not* the same as the function $h(x) = 2x + 3$, since the function h has a larger domain.) *Try Study Suggestions 2.43 and 2.44.* □

▶ **Study Suggestion 2.43:** Let $f(x) = 2x^2 - x - 1$ and $g(x) = x - 1$. Compute **(a)** $f+g$ **(b)** $f-g$ **(c)** fg **(d)** f/g. Also, find their domains, and graph the function f/g. ◀

▶ **Study Suggestion 2.44:** Let $f(x) = x^2$ and $g(x) = x$. Compute the quotient f/g, give its domain, and sketch its graph. ◀

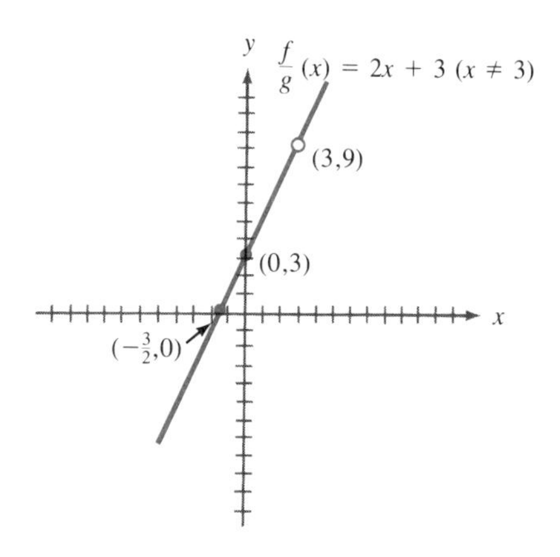

FIGURE 2.59

Another important operation that we can perform is taking the *composition* of functions. In fact, one of the reasons that we studied composition of algebraic expressions in Chapter 1 is that it will help us now with the composition of functions. Let us begin with the definition.

DEFINITION

Let $f: A \to B$ be a function from A to B, and let $g: B' \to C$ be a function from B' to C. Then the **composition** of g with f, denoted by $g \circ f$ (read "g circle f"), is the function defined by

$$(g \circ f)(x) = g(f(x))$$

for all x in A. The domain of the composition $g \circ f$ is the set of all numbers x in the domain of f for which $f(x)$ is in the domain of g.

The composition $g \circ f$ can be described in words as follows. To obtain $(g \circ f)(x)$, first apply the function f to x, to get $f(x)$. Then apply g, to obtain $g(f(x))$. In short, the composition $g \circ f$ can be described by the phrase "first apply f, then apply g."

Notice that in order for a real number x to be in the domain of $g \circ f$, it must be in the domain of f so that we can take $f(x)$. However, we must also require that $f(x)$ be in the domain of g, so that we can take $g(f(x))$.

The composition of two functions f and g is pictured in Figure 2.60. (Figure 2.60 shows the case where $B = B'$.) The function $f: A \to B$ is represented by an arrow from A to B, the function $g: B \to C$ is represented by an arrow from B to C, and the composition $g \circ f: A \to C$ is represented by an arrow from A to C.

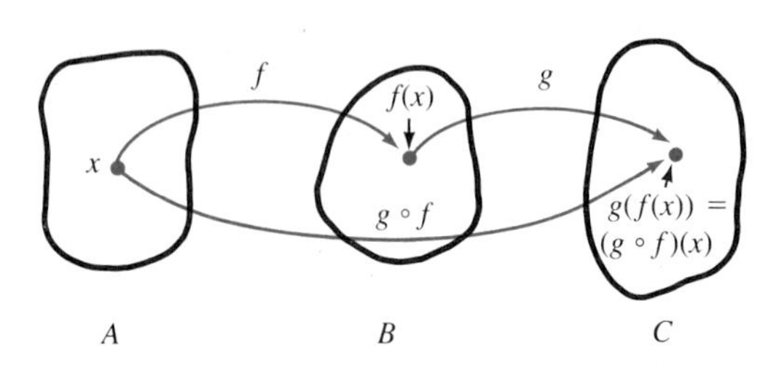

FIGURE 2.60

EXAMPLE 2: Let $f(x) = x^2 + 2x - 1$ and $g(x) = 5x - 3$. Compute the following values.

(a) $(g \circ f)(3)$ **(b)** $(f \circ g)(0)$

Solutions:

(a) Using the definition of composition, we have

$$\begin{aligned}(g \circ f)(3) &= g(f(3))\\ &= g(14)\\ &= 67\end{aligned}$$

(b) Again using the definition, we get

$$\begin{aligned}(f \circ g)(0) &= f(g(0))\\ &= f(-3)\\ &= 2\end{aligned}$$

Try Study Suggestion 2.45. □

▶ **Study Suggestion 2.45:** Let $f(x) = x^3 - 2x$ and $g(x) = 1/x$. Compute **(a)** $(f \circ g)(1/3)$ **(b)** $(g \circ f)(3)$ **(c)** $(f \circ f)(1)$ ◀

When the functions involved are given by algebraic expressions, finding the composition of the functions amounts to nothing more that finding the composition of the algebraic expressions.

EXAMPLE 3: Let $f(x) = x^2 + 2x - 1$ and $g(x) = 5x - 3$. Find the functions

(a) $g \circ f$ **(b)** $f \circ g$ **(c)** $g \circ g$

Solutions:

(a) According to the definition, we have

$$\begin{aligned}(g \circ f)(x) &= g(f(x))\\ &= g(x^2 + 2x - 1)\\ &= 5(x^2 + 2x - 1) - 3\\ &= 5x^2 + 10x - 8\end{aligned}$$

(b) In this case,

$$\begin{aligned}(f \circ g)(x) &= f(g(x))\\ &= f(5x - 3)\\ &= (5x - 3)^2 + 2(5x - 3) - 1\\ &= 25x^2 - 20x + 2\end{aligned}$$

(c) Here we have

$$\begin{aligned}(g \circ g)(x) &= g(g(x))\\ &= g(5x - 3)\\ &= 5(5x - 3) - 3\\ &= 25x - 18\end{aligned}$$

In each case, the domain of the composition is the set **R** of all real numbers.

Try Study Suggestion 2.46. □

▶ **Study Suggestion 2.46:** Let $f(x) = 2x^2 + 3x - 1$ and $g(x) = 2 - x$. Find **(a)** $f \circ g$ **(b)** $g \circ f$ **(c)** $g \circ g$ ◀

EXAMPLE 4: Let $g(x) = \sqrt{x}$ and $f(x) = 1/(1 + x)$. Find $g \circ f$. Also find its domain.

Solution: According to the definition, we have

$$\begin{aligned}(g \circ f)(x) &= g(f(x))\\ &= g\left(\frac{1}{1 + x}\right)\\ &= \sqrt{\frac{1}{1 + x}}\end{aligned}$$

The domain of f is the set of all real numbers different from -1. Hence, the domain of the composition $g \circ f$ is the set of all real numbers x different from -1 for which we can take the square root of $1/(1 + x)$. This is the set of all real numbers x for which $1/(1 + x) > 0$; that is, for which $x > -1$.

Try Study Suggestion 2.47. □

▶ **Study Suggestion 2.47:** Let $f(x) = 1/(x - 1)$ and $g(x) = \sqrt{x}$. Find the composition $f \circ g$. Also, find the domain of $f \circ g$. ◀

As you can see from Example 3, if f and g are functions, then

In general, the composition $g \circ f$ is not the same as the composition $f \circ g$. In other words, when taking the composition of two functions, order counts!

Another way to say this is to say that the operation of composition of functions does not satisfy a commutative property. (See Section 1.1.) On the other hand, composition of functions does satisfy an associative property and so we may write the expression $f \circ g \circ h$ without having to insert parentheses. In fact, we have

$$(f \circ g \circ h)(x) = f(g(h(x))) \tag{2.9}$$

EXAMPLE 5: Let $f(x) = 1 + \sqrt{x}$, $g(x) = 2x - 1$ and $h(x) = x^2$. Find

(a) $(f + g) \circ h$ **(b)** $f \circ g \circ h$

Solutions:

(a) Using the definitions of composition and sum, we have

$$\begin{aligned}[(f + g) \circ h](x) &= (f + g)(h(x)) \\ &= (f + g)(x^2) \\ &= f(x^2) + g(x^2) \\ &= (1 + \sqrt{x^2}) + (2x^2 - 1) \\ &= |x| + 2x^2 \qquad \text{(Recall that } \sqrt{x^2} = |x|)\end{aligned}$$

(b) Using Equation (2.9), we get

$$\begin{aligned}(f \circ g \circ h)(x) &= f(g(h(x))) \\ &= f(g(x^2)) \\ &= f(2x^2 - 1) \\ &= 1 + \sqrt{2x^2 - 1}\end{aligned}$$

Try Study Suggestion 2.48. □

▶ **Study Suggestion 2.48:** Let $f(x) = 2 + 1/x$, $g(x) = 1 + 3x$ and $h(x) = 1/x$. Find **(a)** $(f - g) \circ h$ **(b)** $g \circ f \circ h$ ◀

In some situations (especially in calculus) it is important to be able to "decompose" a function into simpler functions; that is, to write the function as the composition of simpler functions. For example, the function $h(x) = (x + 1)^2$ is the composition of the functions $g(x) = x^2$ and $f(x) = x + 1$, since

$$\begin{aligned}(g \circ f)(x) &= g(f(x)) \\ &= g(x + 1) \\ &= (x + 1)^2 \\ &= h(x)\end{aligned}$$

Let us try some other examples.

EXAMPLE 6: In each case, for the given function h, find two functions f and g with the property that $h = g \circ f$.

(a) $h(x) = \sqrt{2x - 5}$ **(b)** $h(x) = \dfrac{1}{(x^2 - 1)^3}$

(c) $h(x) = \dfrac{2(x - 3)^2 + 1}{(x - 3)^2 - 5}$

Solutions:

(a) In this case, we may take $g(x) = \sqrt{x}$ and $f(x) = 2x - 5$, since

$$\begin{aligned}(g \circ f)(x) &= g(f(x)) \\ &= g(2x - 5) \\ &= \sqrt{2x - 5} \\ &= h(x)\end{aligned}$$

(b) Here we recognize h as the composition of $g(x) = 1/x^3$ and $f(x) = x^2 - 1$, since

$$\begin{aligned}(g \circ f)(x) &= g(f(x)) \\ &= g(x^2 - 1) \\ &= \frac{1}{(x^2 - 1)^3}\end{aligned}$$

(c) In this case, we may take

$$g(x) = \frac{2x^2 + 1}{x^2 - 5}$$

and $f(x) = x - 3$, since

$$\begin{aligned}(g \circ f)(x) &= g(f(x)) \\ &= g(x - 3) \\ &= \frac{2(x - 3)^2 + 1}{(x - 3)^2 - 5}\end{aligned}$$

This example points out that frequently there is more than one choice for the functions g and f. For instance, if we let

$$g(x) = \frac{2x + 1}{x - 5}$$

and $f(x) = (x - 3)^2$, then the composition $g \circ f$ is also equal to h. (We will leave this for you to check.) *Try Study Suggestion 2.49.* □

▶ **Study Suggestion 2.49:** In each case, find two functions f and g with the property that $h = g \circ f$.

(a) $h(x) = \sqrt[3]{1 - 3x}$

(b) $h(x) = (x^2 - 3x + 1)^{-4}$

(c) $h(x) = \dfrac{\sqrt[4]{x - 2} + 5}{\sqrt{x - 2} - 9}$. ◀

Ideas to Remember

- Just as we did with algebraic expressions in Chapter 1, we can add, subtract, multiply, divide, and take the composition of functions.
- The action of the composition $g \circ f$ can be remembered by the phrase "first apply f, then apply g."
- Not only is it important to be able to compose functions, it is also important to be able to decompose functions.

EXERCISES

In Exercises 1–16, compute

(a) $f + g$, **(b)** $f - g$, **(c)** fg, **(d)** f/g.

Be sure to simplify the result. Also, give the domains of each function.

1. $f(x) = 1 - x$, $g(x) = 1 + x$

2. $f(x) = 1 - 2x$, $g(x) = 3x + 4$

3. $f(x) = x^2 + 2x - 3$, $g(x) = 2x^2 + 2x - 1$

4. $f(x) = x^3 + 2x^2 - 1$, $g(x) = 1 - x^3$

5. $f(x) = \dfrac{1}{x}$, $g(x) = x$

6. $f(x) = \dfrac{1}{x+1}$, $g(x) = \dfrac{1}{x-1}$

7. $f(x) = \dfrac{1}{x^2+x}$, $g(x) = \dfrac{x}{x+1}$

8. $f(x) = \dfrac{x^2+x+1}{x}$, $g(x) = \dfrac{1}{x}$

9. $f(x) = x^3$, $g(x) = x^{1/3}$

10. $f(x) = |x|$, $g(x) = |-x|$

11. $f(x) = \sqrt{2x+1}$, $g(x) = f(x)$

12. $f(x) = \sqrt{x}$, $g(x) = \sqrt{x+1}$

13. $f(x) = \sqrt{x}$, $g(x) = -\sqrt{x}$

14. $f(x) = \dfrac{1}{x^2 - 3x - 10}$, $g(x) = 3x + 6$

15. $f(x) = \dfrac{x-1}{x+1}$, $g(x) = \dfrac{x+1}{x-1}$

16. $f(x) = \dfrac{x+1}{x^2+x-12}$, $g(x) = \dfrac{x+4}{x^2-2x-3}$

Positive integral powers of functions are defined in the same way as powers of real numbers. For example, $f^2 = f \cdot f$ and $f^3 = f \cdot f \cdot f$.

17. Let $f(x) = x\sqrt{x^2+1}$. Compute f^2, f^3 and f^4.

18. If $f(x) = x + 1$, $g(x) = x^2$ and $h(x) = 1/x$, compute

(a) f^2g **(b)** fg^2h^3 **(c)** $(f+g)^2$

(d) $(f+g)h$ **(e)** $\dfrac{f}{gh}$

In Exercises 19–22, let $f(x) = 2x - 3$ and $g(x) = 4x^2 + 1$ and find the indicated value.

19. $(g \circ f)(4)$ **20.** $(g \circ g)(-2)$

21. $(f \circ g)(1/2)$ **22.** $(f \circ f \circ f)(0)$

In Exercises 23–26, let $h(x) = 5x - \sqrt{x+1}$ and $k(x) = 1/x$ and find the indicated values.

23. $(h \circ k)(-1)$ **24.** $(k \circ h)(0)$

25. $(k \circ h \circ k)(1/4)$ **26.** $(k \circ k \circ k \circ k)(100)$

In Exercises 27–41, compute **(a)** $f \circ g$ **(b)** $g \circ f$ **(c)** $f \circ f$. Be sure to simplify the result. Also, give the domain of each composition.

27. $f(x) = x + 1$, $g(x) = x - 1$

28. $f(x) = x$, $g(x) = x^2 + 2x - 1$

29. $f(x) = x^2 + 1$, $g(x) = 2x - 3$

30. $f(x) = x^2 + 2$, $g(x) = 3 - x^2$

31. $f(x) = \dfrac{1}{x}$, $g(x) = x^2 + 1$

32. $f(x) = \dfrac{1}{x+1}$, $g(x) = \dfrac{1}{x-1}$

33. $f(x) = \dfrac{x}{x-1}$, $g(x) = \dfrac{x}{x-1}$

34. $f(x) = \dfrac{x^2+1}{x}$, $g(x) = \dfrac{1}{x^2}$

35. $f(x) = \sqrt{x}$, $g(x) = x^2$

36. $f(x) = \sqrt{x-1}$, $g(x) = (x-1)^2$

37. $f(x) = |x|$, $g(x) = |x+1|$

38. $f(x) = x^{2/3}$, $g(x) = x^3$

39. $f(x) = x^{1/2}$, $g(x) = f(x)$

40. $f(x) = x^2$, $g(x) = 2$

41. $f(x) = \dfrac{1}{x}$, $g(x) = f(x)$

42. Let $f(x) = x^2 + 1$, $g(x) = 2x^2 - 3$ and $h(x) = x - 1$. Compute

(a) $(f+g) \circ h$ **(b)** $f \circ (g - h)$ **(c)** $f \circ (g \circ h)$

43. Let $f(x) = 2x + 1$, $g(x) = x^2 - 1$ and $h(x) = 1/x$. Compute
(a) $f \circ (g \circ h)$ **(b)** $(f \circ g) \circ h$
Are they the same?

In Exercises 44–51, find two functions f and g for which $h = g \circ f$. Verify your answer by taking the composition.

44. $h(x) = \sqrt{5x^2 + 3}$

45. $h(x) = |x - 1|$

46. $h(x) = (2x^2 - 3x)^{-9}$

47. $h(x) = \sqrt{1 - \sqrt{x + 5}}$

48. $h(x) = \dfrac{3|x + 2| - 4}{|x + 2|}$

49. $h(x) = \dfrac{(x - 1)^2}{(x + 1)^2}$

50. $h(x) = \dfrac{6}{(x^2 - 1)^3}$

51. $h(x) = |x + 1| - \sqrt{x + 1}$

52. Under what conditions on the constants a, b, c, and d is it true that $f \circ g = g \circ f$, where $f(x) = ax + b$ and $g(x) = cx + d$?

53. Suppose that you currently have two bank accounts earning interest. The amount of money in the first account is given by the function $A_1(t) = 1000 + 0.1t$, and the amount of money in the second account is given by the function $A_2(t) = 500 + 0.12t$, where t is the time since the opening of the accounts. What function represents the total amount of money in both accounts?

54. The gross income of a company that sells computers is given by a function $G(x) = 2500x + 100x^2$, where x is the number of computers sold. The total cost is given by the function $C(x) = 1000x + 15x^2$. Compute the function $G - C$. What does it represent?

55. An oil spill in the shape of a rectangle is growing in size. Its width is given by the function $W(t) = 10 + t + t^2$, and its length is given by the function $L(t) = 2 - t + 3t^2$, where t represents time. Find a function that gives the area of the spill.

56. A balloon made in the shape of a soft drink can is expanding. If the radius of its base is given by the function $R(t) = 1 + 2t$, and its volume is given by the function $V(t) = 2t^3 + t^2 + 6t + 3$, find a function that gives its height.

57. It is known that barracuda feed on bass, and that bass feed on shrimp. Suppose that the size of the barracuda population is given by the function $f(x) = 1000 + \sqrt{2x}$, where x is the size of the bass population. Suppose also that the size of the bass population is given by the function $g(x) = 2000 + \sqrt{x}$, where x is the size of the shrimp population.

(a) Express the size of the barracuda population in terms of the size of the shrimp population.

(b) What is the size of the barracuda population when the size of the shrimp population is 1,000,000 shrimp per cubic mile?

2.8 The Inverse of a Function

In this section, we continue our study of the composition of functions. Let us consider the functions f and g defined by

$$f(x) = x^3 \qquad \text{and} \qquad g(x) = \sqrt[3]{x}$$

Taking the composition $g \circ f$, we get

$$\begin{aligned}(g \circ f)(x) &= g(f(x))\\ &= g(x^3)\\ &= \sqrt[3]{x^3}\\ &= x\end{aligned}$$

That is

$$(g \circ f)(x) = x \tag{2.10}$$

for all real numbers x. Also,

$$\begin{aligned}(f \circ g)(x) &= f(g(x)) \\ &= f(\sqrt[3]{x}) \\ &= (\sqrt[3]{x})^3 \\ &= x\end{aligned}$$

and so

$$(f \circ g)(x) = x \tag{2.11}$$

for all real numbers x.

We can describe Equation (2.10) in words by saying that if we start with any real number x, apply the function f, and then apply g to that, we get x back again. In a sense g has the effect of "undoing" the function f. Similarly, Equation (2.11) says that f has the effect of "undoing" the function g.

When this situation occurs, we say that the functions f and g are *inverses* of each other. Let us describe this concept more formally in a definition.

DEFINITION

Let $f: A \to B$ be a function with domain A and *range* B. If $g: B \to A$ is a function with domain B and range A and if

$$(g \circ f)(x) = x \tag{2.12}$$

for all x in A, and

$$(f \circ g)(x) = x \tag{2.13}$$

for all x in B, then we say that g is the **inverse** of f (and also that f is the inverse of g.)

When a function f has an inverse (and not all functions do), we say that the function is **invertible,** and denote the inverse by f^{-1}. Thus, the function f^{-1} is characterized by the fact that

$$(f^{-1} \circ f)(x) = x$$

and

$$(f \circ f^{-1})(x) = x$$

A typical function f and its inverse function f^{-1} are pictured in Figure 2.61.

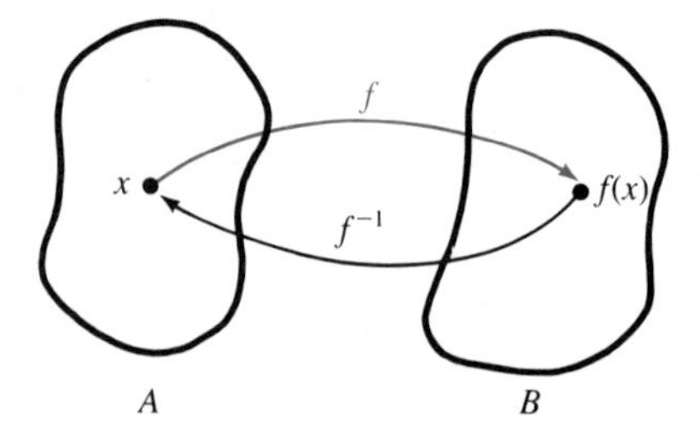

FIGURE 2.61

According to the definition, the domain and range of f^{-1} are related to the domain and range of f as follows

$$\textbf{Domain of } f^{-1} = \textbf{Range of } f$$
$$\textbf{Range of } f^{-1} = \textbf{Domain of } f$$

These relationships will prove useful to us when we study certain inverse functions in Chapter 6.

We should make a comment about the notation f^{-1}. The -1 appearing in f^{-1} is *not* an exponent. It is there simply to denote the inverse function. In order to avoid confusion, the *reciprocal* of a function f should be denoted by $1/f$. It is important to remember that for functions, the concepts of inverse and reciprocal are *totally unrelated.* (As you know, if a is a nonzero real number, then the notation a^{-1} stands for the reciprocal $1/a$. However, this is *not* the case for functions.)

EXAMPLE 1: The inverse of the function

$$f(x) = 2x - 3$$

is the function

$$g(x) = \frac{x+3}{2}$$

This follows from the fact that

$$\begin{aligned}(g \circ f)(x) &= g(f(x)) \\ &= g(2x - 3) \\ &= \frac{(2x-3)+3}{2} \\ &= \frac{2x}{2} \\ &= x\end{aligned}$$

and

$$\begin{aligned}(f \circ g)(x) &= f(g(x)) \\ &= f\left(\frac{x+3}{2}\right) \\ &= 2\left(\frac{x+3}{2}\right) - 3 \\ &= x + 3 - 3 \\ &= x\end{aligned}$$

Thus $(g \circ f)(x) = x$ and $(f \circ g)(x) = x$, and according to the definition, g is the inverse function of f, in symbols, $g = f^{-1}$. □

It is important to keep in mind that, in order for a function g to be the inverse of a function f, *both* of the conditions in Equations (2.12) and (2.13) must hold. This is why we checked both conditions in the previous example.

Study Suggestion 2.50: Show that the inverse of $f(x) = 4 - 3x$ is $g(x) = (1/3)(4 - x)$. (Only *after* this has been done are we allowed to write $f^{-1}(x) = (1/3)(4 - x)$.)

Try Study Suggestion 2.50.

EXAMPLE 2: The function $f(x) = 1/x$ is its own inverse! This follows from the fact that

$$\begin{aligned}(f \circ f)(x) &= f(f(x)) \\ &= f\left(\frac{1}{x}\right) \\ &= \frac{1}{\frac{1}{x}} \\ &= x\end{aligned}$$

Try Study Suggestion 2.51. □

Study Suggestion 2.51: Show that the function

$$f(x) = \frac{x}{x-1}$$

is its own inverse.

We mentioned earlier that not all functions have an inverse, and this brings up the question of how to tell whether or not a given function has an inverse. Fortunately, we can answer this question in terms of a concept that we studied earlier in this chapter. Let us give this answer in a theorem.

THEOREM

A function $f: A \to B$ with domain A and range B has an inverse function $f^{-1}: B \to A$ if and only if f is one-to-one.

As we know, increasing functions are one-to-one, and so according to this theorem, they have inverses. Similarly, decreasing functions are one-to-one, and so they too have inverses.

1. **Increasing functions have inverses.**
2. **Decreasing functions have inverses.**

In particular, since a nonconstant linear function must be either increasing or decreasing, we can say that *all nonconstant linear functions have inverses.*

There is a technique for computing the inverse of a function that works in simple cases (provided, of course, that the function has an inverse.) Let us illustrate this technique with an example.

EXAMPLE 3: Find the inverse of the linear function

$$f(x) = 5x + 7$$

Solution: First we write the equation $y = f(x)$, which in this case is

$$y = 5x + 7$$

Then we solve this equation for x

$$y - 7 = 5x$$

$$x = \frac{y - 7}{5}$$

The next step is to interchange x and y

$$y = \frac{x - 7}{5}$$

Finally, we replace y by $f^{-1}(x)$,

$$f^{-1}(x) = \frac{x - 7}{5}$$

We will leave it to you to check that this is the inverse of f by taking the necessary compositions. □

The procedure used in the previous example is outlined below.

Procedure for Finding the Inverse of a Function

Let f be a function defined by a single algebraic formula. If f has an inverse, then the following procedure will produce that inverse.

1. First write $y = f(x)$.
2. Solve the resulting equation for x.
3. Interchange the symbols x and y. (That is, replace all appearances of x by y and all appearances of y by x.)
4. Replace y by $f^{-1}(x)$.

EXAMPLE 4: Find the inverse of the function

$$f(x) = \frac{x}{x + 1}$$

Solution: First we write

$$y = \frac{x}{x + 1}$$

Then we solve this equation for x,

$$y(x+1) = x$$
$$yx + y = x$$
$$yx - x = -y$$
$$x(y-1) = -y$$
$$x = \frac{-y}{y-1}$$

Next we interchange x and y

$$y = \frac{-x}{x-1}$$

and finally we write

$$f^{-1}(x) = \frac{-x}{x-1}$$

We will leave it to you to check that this is the inverse function by computing the necessary compositions. *Try Study Suggestion 2.52.* □

▶ **Study Suggestion 2.52:** Find the inverses of the following functions. **(a)** $f(x) = 10x - 1$ **(b)** $f(x) = x^5 - 1$ **(c)** $g(x) = 2x/(x-1)$ ◀

EXAMPLE 5: Find the inverse of the function

$$f(x) = x^2 + 3$$

whose domain is the set $\{x \mid x \geq 0\}$.

Solution: We begin by discussing why we restricted the domain of f to the set of nonnegative real numbers. The graph of the function $g(x) = x^2 + 3$, whose domain is the set of *all* real numbers, is shown in Figure 2.62. As you can see, this graph does not pass the horizontal line test, and so the function g is not one-to-one. Hence, it does not have an inverse. On the other hand, the graph of the function f, shown in Figure 2.63, passes the horizontal line test, and so it is one-to-one. Hence, it does have an inverse, which can be computed by following the four-step procedure outlined above.

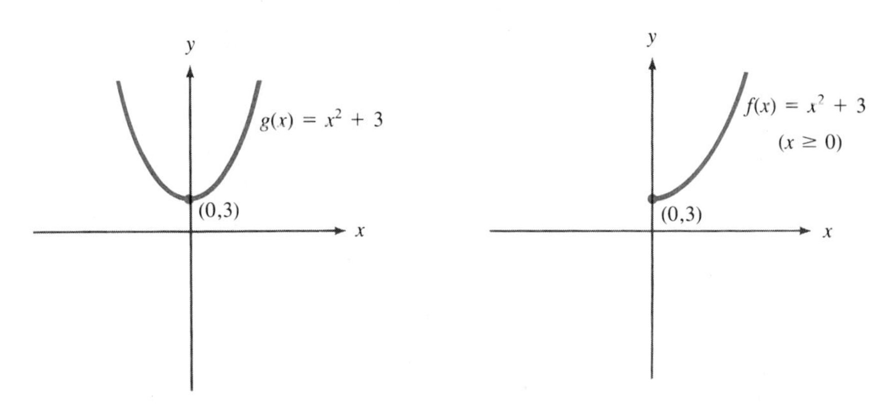

FIGURE 2.62 **FIGURE 2.63**

The first step is to write the equation $y = f(x)$,

$$y = x^2 + 3 \qquad (x \geq 0)$$

Then we solve for x,

$$y - 3 = x^2 \qquad (x \geq 0)$$

$$x = \sqrt{y - 3} \qquad (x \geq 0)$$

Notice that we take only the *positive* square root, since we are restricted by the fact that $x \geq 0$. Next we interchange x and y,

$$y = \sqrt{x - 3} \qquad (y \geq 0, \text{ that is, } x \geq 3)$$

Finally, we replace y by $f^{-1}(x)$

$$f^{-1}(x) = \sqrt{x - 3} \qquad (x \geq 3)$$

Again we suggest that you verify that this is the inverse of f.

Try Study Suggestion 2.53. □

▶ **Study Suggestion 2.53:** Let f be the function $f(x) = 2x^2 - 5$, with domain $\{x | x \geq 0\}$. Show that this function is one-to-one and then find its inverse. ◀

There is a very interesting relationship between the graph of an invertible function and the graph of its inverse function. In particular, if f has an inverse, then we may obtain the graph of f^{-1} by reflecting the graph of f over the diagonal line $y = x$. As an example, the function $f(x) = x^2$, with domain $\{x | x \geq 0\}$, has inverse $f^{-1}(x) = \sqrt{x}$. The graph of f is pictured in Figure 2.64, along with its reflection about the diagonal line $y = x$. As you can see, this reflected graph is the graph of the inverse function $f^{-1}(x) = \sqrt{x}$.

Try Study Suggestion 2.54.

▶ **Study Suggestion 2.54:** Let f be the function defined by $f(x) = 2x - 5$.

(a) Show that f has an inverse.

(b) Find f^{-1}.

(c) Use the graph of f to find the graph of f^{-1}. ◀

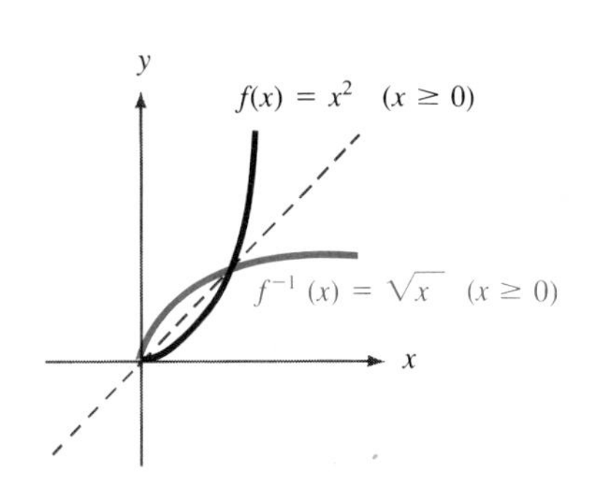

FIGURE 2.64

Ideas to Remember

- The inverse of a function has the effect of "undoing" whatever the function "does."
- Not all functions have inverses.

EXERCISES

In Exercises 1–8, verify that the function $g(x)$ is the inverse of the function $f(x)$.

1. $f(x) = 3x - 1$, $g(x) = \frac{1}{3}(x + 1)$
2. $f(x) = 8 + 4x$, $g(x) = \frac{1}{4}x - 2$
3. $f(x) = x^{1/5}$, $g(x) = x^5$
4. $f(x) = \sqrt{x}$, $g(x) = x^2 \quad (x \geq 0)$
5. $f(x) = \dfrac{1}{x^3}$, $g(x) = \dfrac{1}{x^{1/3}}$
6. $f(x) = \dfrac{1}{x + 1}$, $g(x) = \dfrac{1 - x}{x}$
7. $f(x) = x^2/2 + 4 \quad (x \geq 0)$, $g(x) = \sqrt{2x - 8}$
8. $f(x) = x^2 + 3x - 2 \quad (x \geq -3/2)$, $g(x) = (1/2)(-3 + \sqrt{17 + 4x})$

In Exercises 9–22, find the inverse of the given function and verify your result by taking the necessary compositions.

9. $f(x) = -5x + 2$
10. $f(x) = 1 - x^2 \quad (x \geq 0)$
11. $f(x) = x^3 - 3$
12. $f(x) = |x| \quad (x \geq 0)$
13. $f(x) = \dfrac{1}{7x - 2}$
14. $f(x) = \dfrac{1}{x^3 - 4}$
15. $f(x) = x^{-5} - 1$
16. $f(x) = \dfrac{x}{2x - 1}$
17. $f(x) = \sqrt{2x - 3} \quad (x \geq 3/2)$
18. $f(x) = \sqrt{1 - x^2} \quad (0 \leq x \leq 1)$
19. $f(x) = \dfrac{x + 1}{x - 1}$
20. $f(x) = \dfrac{2x^3 - 1}{x^3 + 1}$
21. $f(x) = \sqrt{x} + 1 \quad (x \geq 0)$
22. $f(x) = x^{7/13}$

In Exercises 23–32, compute f^{-1} and use the graph of f to find the graph of f^{-1}.

23. $f(x) = 4x - 5$
24. $f(x) = 3 - x$
25. $f(x) = (x - 3)^2 \quad (x \geq 3)$
26. $f(x) = x^3$
27. $f(x) = \dfrac{-1}{x}$
28. $f(x) = \dfrac{2}{x - 1}$
29. $f(x) = 2x^3 + 1$
30. $f(x) = \sqrt{x + 5}$
31. $f(x) = \sqrt[3]{x + 8}$
32. $f(x) = (x - 1)^5 + 1$
33. Show that the function $f(x) = \sqrt{9 - x^2}$, with domain $\{x | 0 \leq x \leq 3\}$ is its own inverse.
34. Under what conditions does the linear function $f(x) = ax + b$ have an inverse? Find f^{-1} when it exists.
35. Let f be the function defined by

$$f(x) = \frac{ax + b}{cx + d}$$

where $ac \neq bd$. Find the inverse of f.

36. Let f be an invertible function.
    (a) Show that if $y = f(x)$, then $x = f^{-1}(y)$.
    (b) Show that if $x = f^{-1}(y)$, then $y = f(x)$.
37. Let f be an invertible function. Show that a point (a, b) is on the graph of f if and only if the point (b, a) is on the graph of f^{-1}.
38. If f is a one-to-one function, show that f^{-1} is also one-to-one.

2.9 Review

CONCEPTS FOR REVIEW

Cartesian coordinate system
x-axis
y-axis
Origin
Coordinate
Plotting a point
Quadrant
Distance formula
Midpoint formula
Graph of a set
Graph of an equation
Function
Domain
Image
Range
Piecewise defined function
Independent variable
Dependent variable
Difference quotient
Graph of a function
x-intercepts
y-intercepts
Method of plotting points
Constant function
Identity function
Linear function
Absolute value function
Vertical line test
Symmetry with respect to the y-axis
Even function
Symmetry with respect to the origin
Odd function
Increasing
Decreasing
Constant
Increasing function
Decreasing function
One-to-one
Horizontal line test
Intersection of two sets
Sum of two functions
Difference of two functions
Product of two functions
Quotient of two functions
Composition of functions
Inverse of a function

REVIEW EXERCISES

In Exercises 1–4, find the distance between the given pairs of points.

1. $(2, -3), (10, 12)$

2. $(\pi, \sqrt{2}), (2 + \pi, 1 + \sqrt{2})$

3. $(a, a^2), (2a, 2a^2)$

4. $(2\sqrt{7}, -3\sqrt{6}), (4\sqrt{7}, -5\sqrt{6})$

In Exercises 5–8, determine the missing point, where M is the midpoint of the line segment PQ.

5. $P = (2, -7), Q = (-1, 12)$

6. $P = (\sqrt{2}, \pi + 1), Q = (1 + 2\sqrt{2}, -2\pi)$

7. $P = (5, -3), M = (7, 2)$

8. $Q = (15, 12), M = (1, 0)$

9. Show that if the diagonals of a parallelogram have equal length, then the parallelogram must be a rectangle.

In Exercises 10–13, graph the set S.

10. $S = \{(x, y) | x \leq 3, y > 4\}$

11. $S = \{(x, y) | x = 4, |y - 2| \leq 3\}$

12. $S = \{(x, y) | 2 \leq (x - 5)^2 + (y + 3)^2 \leq 25\}$

13. $S = \{(x, y) | |x - y| < 1\}$

In Exercises 14–24, sketch the graph of the given equation.

14. $x = 5y + 1$

15. $4y + 3x - 2 = 0$

16. $(x + 3)^2 + (y - 4)^2 = 1$

17. $x^2 + y^2 + 6x + 2 = 0$

18. $4x^2 + 4y^2 - 40x - 4y + 99 = 0$

19. $x = y^2$

20. $y = \sqrt{x + 2}$

21. $y = x^6$

22. $y = -x^7$

23. $y = 1/x^2$

24. $y = -\sqrt{1 - x}$

25. Find the equation of the circle that is tangent to both coordinate axes, has center in the second quadrant, and radius equal to 5.

26. If $f(x) = \dfrac{x + |x|}{\sqrt{x + 1}}$, find

(a) $f(0)$ **(b)** $f(1)$ **(c)** $f(-1/2)$ **(d)** $f(x - 1)$

27. If $f(x) = x^{-\sqrt{x}}$, find

(a) $f(1)$ **(b)** $f(4)$ **(c)** $f(2)$ **(d)** $f(x^2)$

28. Let $f(x) = ax + b$, where a and b are constants. Find all values of a and b for which

$$f(x + y) = f(x) + f(y)$$

for all real numbers x and y.

29. Let $f(x) = \dfrac{x + a}{ax + 1}$, where a is a constant. Show that

$$f\left(\frac{1}{x}\right) = \frac{1}{f(x)}$$

30. Let $f(x) = \sqrt{x}$. Show that

$$\frac{f(x + h) - f(x)}{h} = \frac{1}{\sqrt{x + h} + \sqrt{x}}$$

31. Find the domain of the function $f(x) = \sqrt{2x - x^2}$.

32. Find the domain of the function $f(x) = \sqrt{3 + 2x - x^2}$.

33. Find the range of the function $f(x) = 1 + \sqrt{x}$.

In Exercises 34–39, determine whether the given function is even, odd, both, or neither. Justify your answer.

34. $f(x) = 7x - 12$ **35.** $g(u) = 1/(6|u| + 3)$

36. $h(x) = x^5 + x^{1/5}$ **37.** $f(x) = x^4 + x^{1/4}$

38. $f(x) = \sqrt[3]{1 + \sqrt[3]{t}}$ **39.** $f(t) = \sqrt{1 + \sqrt{t}}$

40. For the function f whose graph is pictured below determine all values of x for which

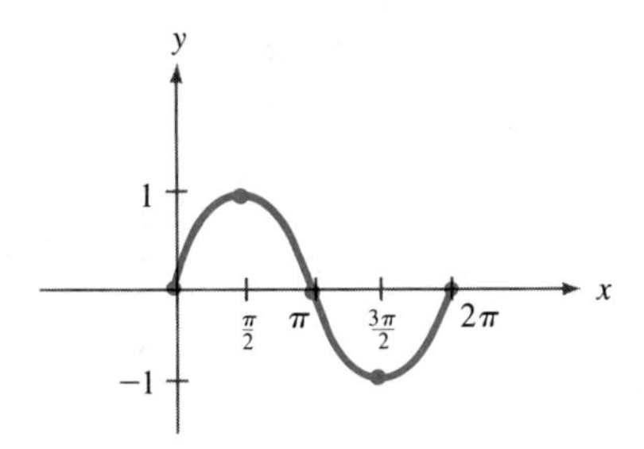

(a) $f(x) = 0$ **(b)** $f(x) > 0$ **(c)** $-1 < f(x) < 1$

41. For the function f whose graph is pictured below determine all values of x for which

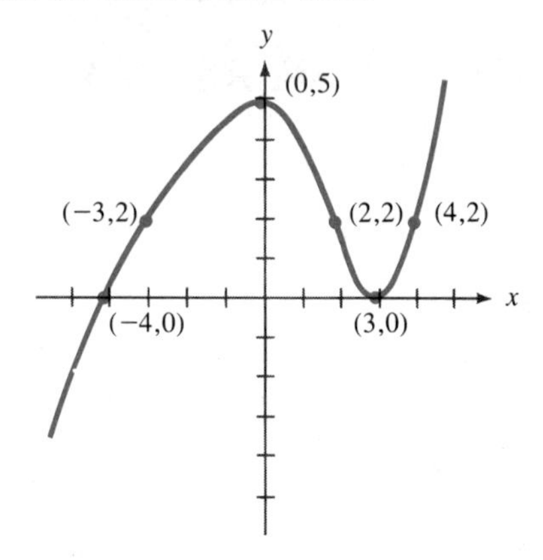

(a) $f(x) = 2$ **(b)** $f(x) \le 0$ **(c)** $2 < f(x) < 5$

In Exercises 42–47, use the graph of $f(x) = \dfrac{1}{x^2}$ to find the graph of the given function.

42. $f(x) = \dfrac{1}{(x + 1)^2}$ **43.** $f(x) = \dfrac{1}{x^2 + 4x + 4}$

44. $f(x) = \dfrac{1}{x^2} + 1$ **45.** $g(x) = \dfrac{1 - x^2}{2x^2}$

46. $h(y) = \dfrac{2y^2 - 1}{y^2}$ **47.** $F(z) = -\dfrac{4}{z^2}$

48. Graph the function $f(x) = |x + 3| - 2$ and use it to find the graph of $g(x) = 1 - |x + 3|/2$.

In Exercises 49–57, sketch the graph of the given function, and determine on which intervals it is increasing, on which intervals it is decreasing, and on which intervals it is constant.

49. $f(x) = 5x^2 + 1$ **50.** $f(x) = \dfrac{1}{2x - 1}$

51. $g(t) = \sqrt{2t + 1}$ **52.** $g(s) = |s - 1|$

53. $h(u) = 1 - u^3$ **54.** $h(y) = |1 - y^2|$

55. $f(x) = \begin{cases} |x| & \text{if } x < 0 \\ 1 - |x| & \text{if } x \ge 0 \end{cases}$

56. $f(x) = \begin{cases} 1 & \text{if } x \text{ is an even positive integer} \\ -1 & \text{if } x \text{ is an odd positive integer} \\ x^2 & \text{otherwise} \end{cases}$

57. $f(x) = \begin{cases} 1 & \text{if } x \text{ is rational} \\ 0 & \text{if } x \text{ is irrational} \end{cases}$

In Exercises 58–65, graph the given function, and use the horizontal line test to determine whether or not the function is one-to-one.

58. $f(x) = 5 - x$

59. $g(x) = 1/2 + \pi x$

60. $f(x) = \sqrt{2}$

61. $h(u) = \sqrt{2}u + 1$

62. $f(x) = \dfrac{3}{x + 1}$

63. $f(t) = \begin{cases} 1/t^2 & \text{if } t \neq 0 \\ 0 & \text{if } t = 0 \end{cases}$

64. $f(x) = \begin{cases} \sqrt{x} & \text{if } x \geq 0 \\ |x| & \text{if } x < 0 \end{cases}$

65. $f(x) = \begin{cases} \sqrt{x + 1} & \text{if } x \geq -1 \\ x^3 - 1 & \text{otherwise} \end{cases}$

In Exercises 66–71, use the definition to determine whether or not the given function is one-to-one.

66. $f(x) = 2x + 3$

67. $g(x) = -x^2 + \pi$

68. $f(t) = \dfrac{t}{t + 1}$

69. $g(y) = \sqrt{1 - y^3}$

70. $h(s) = \dfrac{2s^4 + 1}{3s^4 + 1}$

71. $f(x) = |x + 1|$

72. Show that the function

$$f(x) = \frac{x + 1}{x^2 + 1}$$

is not one-to-one. (*Hint:* solve the equation $f(x) = 1/2$.)

73. Under what conditions on the constants a and b and the *integer* n is the function

$$f(x) = \frac{ax^n + 1}{bx^n + 1}$$

one-to-one? Justify your answer.

74. Let $f: \mathbf{R} \to \mathbf{R}$ be the function defined by

$$f(x) = \begin{cases} x & \text{if } x \geq 1 \\ 2x - 3 & \text{if } x < 1 \end{cases}$$

Is f one-to-one? Justify your answer.

In Exercises 75–77, find $f + g$ and $f - g$. Also, give their domains.

75. $f(x) = x^2 - 3x + \sqrt{x}$, $g(x) = x^3 - x^2 + x + 2\sqrt{x}$

76. $f(t) = |t + 1|$, $g(t) = |t - 1|$

77. $f(x) = \dfrac{x}{x + 1}$, $g(x) = \dfrac{1}{x + 1}$

In Exercises 78–80, find fg and f/g. Also, give their domains.

78. $f(x) = x^2 + x$, $g(x) = x/(x + 1)$

79. $f(t) = |t + 1|$, $g(t) = |t^2 - 1|$

80. $f(y) = |y + 2|$, $g(y) = f(y)$

In Exercises 81–84, find the composition $f \circ g$. Also, give its domain.

81. $f(x) = x^2 + 2x - 10$, $g(x) = 3x + 10$

82. $f(x) = 1/x$, $g(x) = 1/x^4$

83. $f(x) = \sqrt{t + 1}$, $g(t) = \frac{1}{2}t - 1$

84. $f(x) = 1/\sqrt{x}$, $g(x) = f(x)$

In Exercises 85–88, find two functions f and g with the property that $h = g \circ f$.

85. $h(x) = \sqrt[3]{x^2 + x}$

86. $h(x) = (x + \sqrt{x})^{-4}$

87. $h(y) = \sqrt{y - \sqrt{y + 1}}$

88. $h(x) = \dfrac{\sqrt{2x - 9} + 5}{1 - (2x - 9)^2}$

89. Let $f(x) = x^2 + x$, $g(x) = 1/(x + 1)$ and $h(x) = x + 1/x$. Compute and compare the following functions.
(a) $(f + g) \circ h$ **(b)** $f \circ h + g \circ h$

What property of composition does this illustrate?

In Exercises 90–95, find the inverse of the given function.

90. $f(x) = 7 - 3x$

91. $f(x) = \sqrt{2x}$

92. $f(x) = \dfrac{2}{3 - 5x}$

93. $f(t) = \dfrac{4t^5 - 1}{3t^5 + 1}$

94. $f(u) = \dfrac{\sqrt[3]{u + 1}}{\sqrt[3]{u - 1}}$

95. $g(x) = \dfrac{ax^n - 1}{bx^n + 1}$ (n odd)

96. **(a)** Find a function f for which $f^2 = f \circ f$.
(b) Find a function g for which $g^2 \neq g \circ g$.

3

POLYNOMIAL FUNCTIONS OF DEGREE ONE AND TWO

3.1 The Slope of a Line

There are various ways to describe a straight line in the plane. One way is simply to give two distinct points in the plane, since this uniquely determines a straight line. Another way is to give only one point in the plane, together with some measure of the direction of the line. The question is: How do we measure the direction of a line?

We could measure the direction of a line by using the angle A that the line makes with the x-axis. (See Figure 3.1.) However, there is another way to measure direction that does not directly involve the concept of an angle.

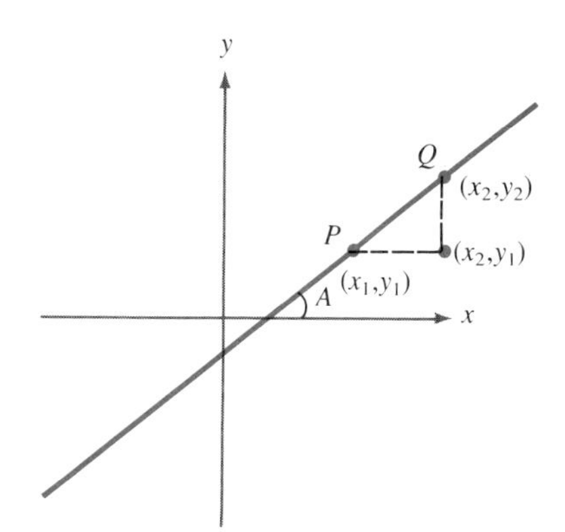

FIGURE 3.1

DEFINITION

Let L be a *nonvertical* line in the plane, and let $P = (x_1, y_1)$ and $Q = (x_2, y_2)$ be distinct points on L. (See Figure 3.1.) Then the quantity

$$m = \frac{y_2 - y_1}{x_2 - x_1}$$

which is simply the difference in the y-coordinates of the two points, divided by the difference in their x-coordinates, is called the **slope** of the line L. If L is a vertical line, then we say that its slope is *undefined*, or does not exist.

EXAMPLE 1: In each case, compute the slope of the line containing the given pair of points.

(a) $(2, 4), (5, 10)$ **(b)** $(-3, 4), (3/2, 1/2)$

Solutions:

(a) Applying the definition of slope, we get

$$m = \frac{y_2 - y_1}{x_2 - x_1} = \frac{10 - 4}{5 - 2} = 2$$

(b) Again using the definition, we get

$$m = \frac{y_2 - y_1}{x_2 - x_1} = \frac{1/2 - 4}{3/2 - (-3)} = \frac{-7/2}{9/2} = -\frac{7}{9}$$ □

There are two remarks that we should make concerning the computation of slopes. First, it does not matter in which order we take the two points used in computing the slope. For instance, we could have computed the slope in part (a) of the previous example by subtracting *both* pairs of coordinates in the opposite order,

$$m = \frac{4 - 10}{2 - 5} = 2$$

Second, it does not matter which two distinct points on a given line we use to compute the slope—we will always get the same result. This is especially important, since if it were possible to get different slopes by choosing different points, then it would not make sense to say "*the* slope of a line."

▶ **Study Suggestion 3.1:** In each case, compute the slope of the line containing the given points.

(a) $(5, 3), (2, -4)$

(b) $(1/2, 1/3), (1/4, -1/6)$

(c) $(1, \sqrt{2}), (0, \sqrt{2})$ ◀

Try Study Suggestion 3.1.

The quantity $y_2 - y_1$ is called the **change in y** from P to Q and the quantity $x_2 - x_1$ is called the **change in x** from P to Q. Of course, each of these changes can be either positive, negative, or zero. Using these terms, the slope of the line can be described by the formula

$$\text{slope} = \frac{\text{change in } y \text{ from } P \text{ to } Q}{\text{change in } x \text{ from } P \text{ to } Q}$$

provided that the change in x is not equal to 0, in which case we say that the slope is *undefined.*

If P is a point with coordinates (x, y), and m is a real number, then we can sketch the line through P with slope m as follows. Starting at the point P we move 1 unit to the right. Then we move m units up if $m \geq 0$, or $|m|$ units down if $m < 0$. This brings us to a point, call it Q, with coordinates $(x + 1, y + m)$. Now, the change in x from P to Q is $(x + 1) - x = 1$ and the change in y from P to Q is $(y + m) - y = m$. Hence, the slope of the line through P and Q is indeed equal to $m/1 = m$.

EXAMPLE 2:

(a) Sketch the line that goes through the point (3, 1) and has slope 2.

(b) Sketch the line that goes through the point (−2, 1) and has slope −3/2.

Solutions:

(a) Let P be the point whose coordinates are (3, 1). (See Figure 3.2a.) Then we start at P, move 1 unit to the right and then $m = 2$ units up, arriving at the point Q whose coordinates are $(3 + 1, 1 + 2) = (4, 3)$. As we can see from the figure, the line through P and Q has slope equal to $2/1 = 2$.

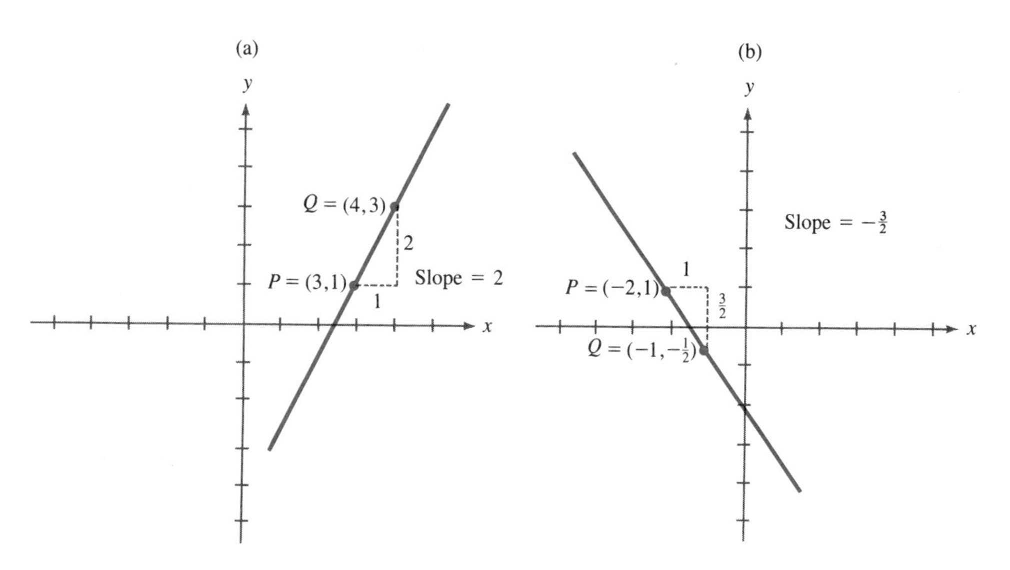

FIGURE 3.2

(b) In this case, we start at the point P whose coordinates are $(-2, 1)$, move 1 unit to the right, and then $|m| = 3/2$ units *down*, arriving at the point Q whose coordinates are $(-2 + 1, 1 - 3/2) = (-1, -1/2)$. (See Figure 3.2b.) As we can see from the figure, the line through P and Q has slope $-(3/2)/1 = -3/2$. *Try Study Suggestion 3.2.* □

Study Suggestion 3.2:

(a) Sketch the line through the point (2, 3) with slope 1/2.

(b) Sketch the line through the point (−2, −1) with slope −1.

In order to see that the slope of a line provides a good measure of the direction of the line, let us compare the slopes of several lines, as shown in Figure 3.3. We have labeled these lines L_1 through L_7, and computed each of their slopes, using the origin and the labeled point. These slopes are denoted by m_1 through m_7, respectively. For example, the slope of the line L_4 is

$$m_4 = \frac{10 - 0}{1 - 0} = 10$$

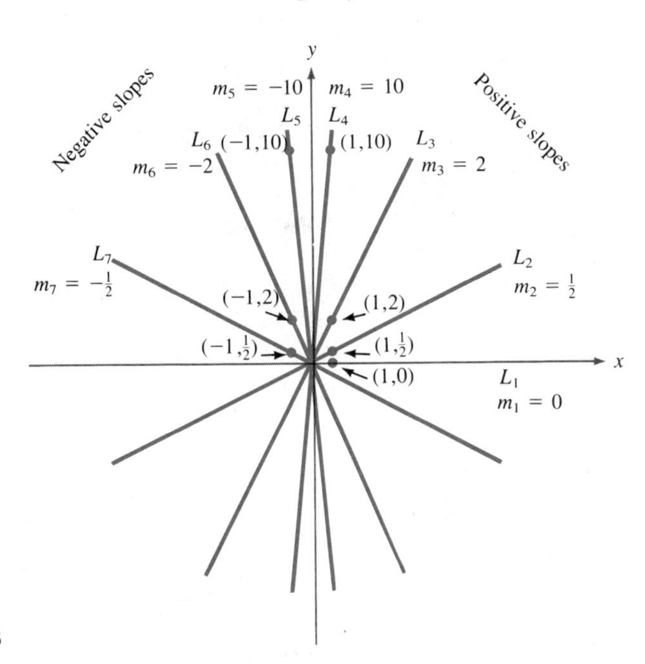

FIGURE 3.3

Using these examples as a guide, we can make the following observations about the slope of a line.

1. **Lines that slant upward as we move from left to right have positive slopes, and lines that slant downward have negative slopes. Also, horizontal lines have slope equal to 0.**

2. **The closer a line is to being vertical, the larger is the *absolute value* of its slope, and the closer the line is to being horizontal, the smaller is the *absolute value* of its slope.**

Our next result confirms the fact that the slope measures the direction of a line.

THEOREM

> Two *nonvertical* lines are parallel if and only if they have the same slope.

Proof: We will give the proof in the case where the lines have positive slope. (We will leave it to you to modify the proof for negative slope.) Suppose that L_1 and L_2 are parallel lines as shown in Figure 3.4. We want to show that they have the same slope. Notice that we have chosen the points P_1 and Q_1 in such a way that they lie on the same vertical line. Similarly, the points P_2 and Q_2 lie on the same vertical line. Now we can see that triangle $P_1P_2P_3$ is congruent to triangle $Q_1Q_2Q_3$. This follows from the fact that side a has the same length as side c, and that, since the lines L_1 and L_2 are parallel, angle A is equal to angle B. In addition, the angles at P_3 and Q_3 are equal (being right angles) and so the triangles are congruent. This implies that side b has the same length as side d, and so

$$\begin{aligned}\text{slope of } L_1 &= \frac{\text{length of side } b}{\text{length of side } a}\\ &= \frac{\text{length of side } d}{\text{length of side } c}\\ &= \text{slope of } L_2\end{aligned}$$

Hence, the lines L_1 and L_2 have the same slope. Since the same argument applies to any pair of nonvertical parallel lines, we have proven that nonvertical parallel lines have the same slope. We will leave it as an exercise for you to show that if two lines have the same slope, then they must be parallel. (See Exercise 34.) These two statements taken together prove the theorem. ■

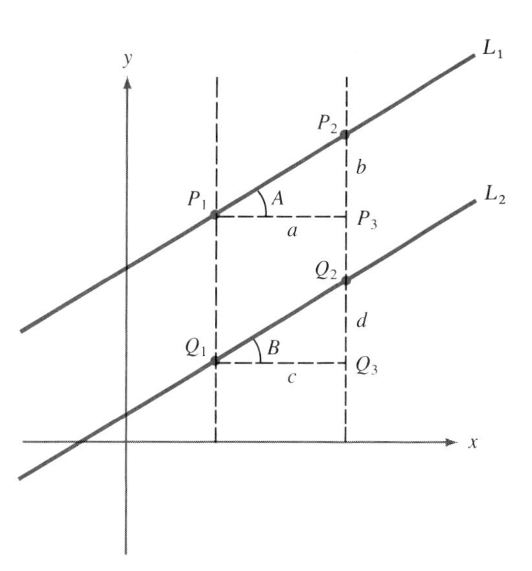

FIGURE 3.4

The previous theorem has a companion that tells us how to use the slope to determine when two lines are perpendicular. (We will not prove the following result, however.)

THEOREM

> Two nonvertical lines are perpendicular if and only if the product of their slopes is equal to -1. That is, if L_1 is a nonvertical line with slope m_1, and L_2 is a nonvertical line with slope m_2, then L_1 and L_2 are perpendicular if and only if $m_1m_2 = -1$.

Another way of saying that $m_1m_2 = -1$ is to say that $m_2 = -1/m_1$ (or $m_1 = -1/m_2$.) In words, this says that m_2 is the *negative reciprocal* of m_1 (or that m_1 is the negative reciprocal of m_2.) Now we can summarize both theorems.

Two nonvertical lines are

1. **parallel if and only if their slopes are equal;**
2. **perpendicular if and only if their slopes are negative reciprocals of each other.**

EXAMPLE 3: In each of the following parts, we give two pairs of points. Compute the slopes of the lines determined by each pair of points, and decide whether the lines are parallel, perpendicular, or neither. Finally, sketch the graphs of the two lines on the same plane.

(a) $(1, -1), (3, 4); (2, 1), (3, -1)$

(b) $(2, 4), (4, 0); (5, 6), (-3, 2)$

(c) $(2, -3), (4, 1/2); (-4, -5), (4, 9)$

Solution:

(a) The slope of the line containing the points $(1, -1)$ and $(3, 4)$ is

$$m_1 = \frac{4-(-1)}{3-1} = 5/2$$

and the slope of the line containing the points $(2, 1)$ and $(3, -1)$ is

$$m_2 = \frac{-1-1}{3-2} = -2$$

Since m_1 is not equal to m_2, these lines are not parallel. Also, since $m_1m_2 = (5/2)(-2) = -5 \neq -1$, the lines are not perpendicular. The lines are pictured in Figure 3.5.

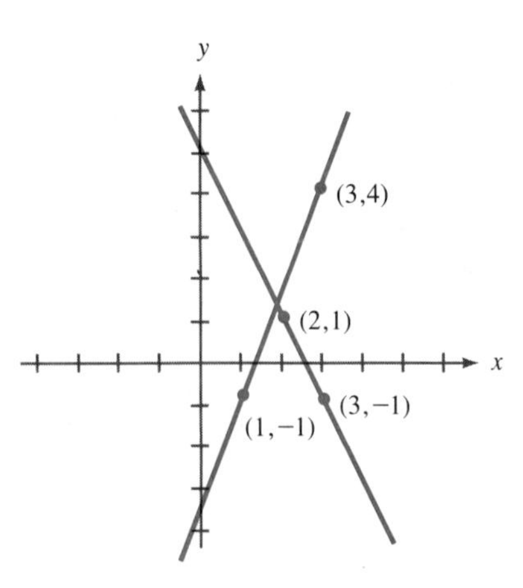

FIGURE 3.5

(b) The slope of the line containing (2, 4) and (4, 0) is

$$m_1 = \frac{0 - 4}{4 - 2} = -2$$

and the slope of the line containing (5, 6) and (−3, 2) is

$$m_2 = \frac{2 - 6}{-3 - 5} = \frac{1}{2}$$

Since $m_1 m_2 = (-2)(1/2) = -1$, these slopes are negative reciprocals of one another, and so the lines are perpendicular. (See Figure 3.6.)

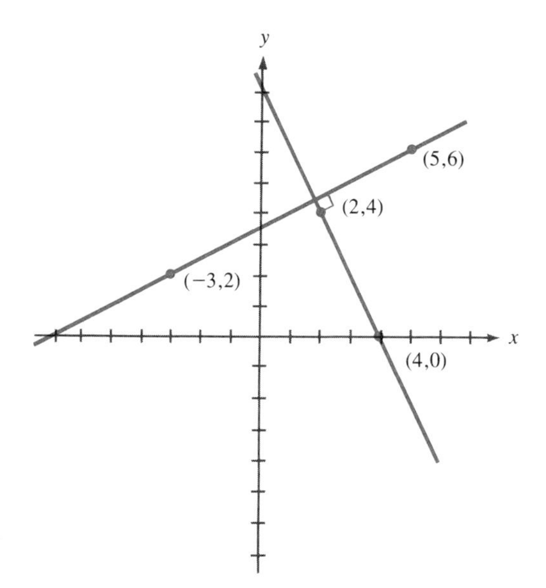

FIGURE 3.6

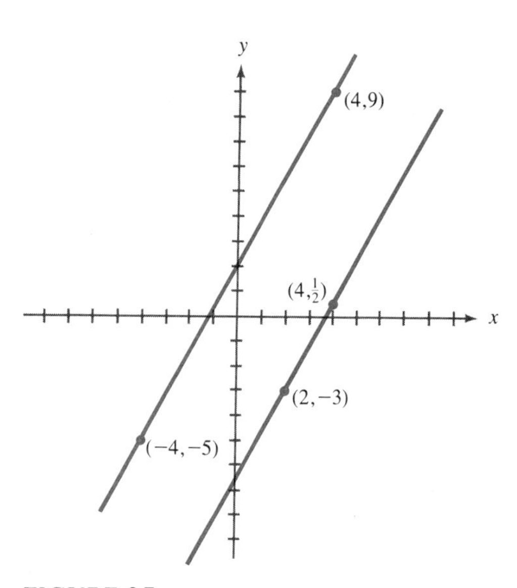

FIGURE 3.7

(c) The slope of the line containing (2, −3) and (4, 1/2) is

$$m_1 = \frac{1/2 - (-3)}{4 - 2} = \frac{7}{4}$$

and the slope of the line containing (−4, −5) and (4, 9) is

$$m_2 = \frac{9 - (-5)}{4 - (-4)} = \frac{7}{4}$$

Since $m_1 = m_2$, the lines are parallel. (See Figure 3.7.)

Try Study Suggestion 3.3. □

▶ **Study Suggestion 3.3:** Follow the instructions given in Example 3 for each pair of points.

(a) (2, 1), (3, 3); (0, −5), (1, −3)
(b) (5, 2), (−2, 3); (−2, 2), (0, 16)
(c) (0, 0), (1, −3); (4, 1/2), (1/2, 4) ◀

EXAMPLE 4: Sketch the line through the point (2, 1) that is perpendicular to the line through (2, 1/2) and (1, 4).

Solution: The line through (2, 1/2) and (1, 4) has slope equal to

$$\frac{4 - 1/2}{1 - 2} = -\frac{7}{2}$$

Since the negative reciprocal of $-7/2$ is $-1/(-7/2) = 2/7$, the previous theorem tells us that the line we are looking for has slope 2/7. Using the fact that it goes through the point (2, 1), we get the graph shown in Figure 3.8.

Try Study Suggestion 3.4. □

Study Suggestion 3.4:

(a) Sketch the line that goes through the point $(-4, 1)$ and is perpendicular to the line through the points (0, 2) and (1, 1/2).

(b) Sketch the line that goes through the origin and is parallel to the line through the points (5, 2) and (4, 4). ◀

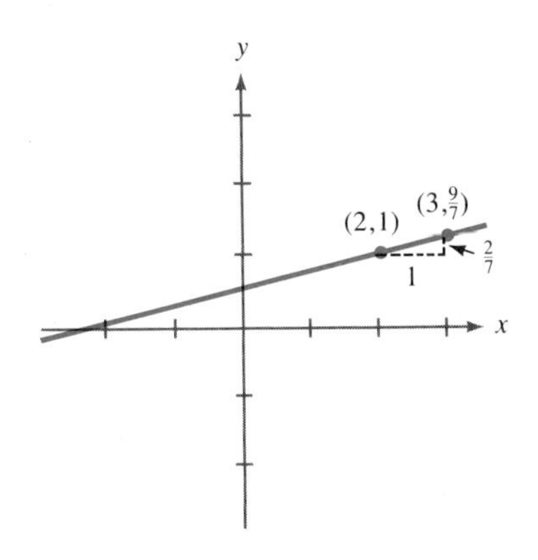

FIGURE 3.8

Ideas to Remember

- The slope of a line is a good measure of the direction of the line.
- The slope can be used to tell whether two nonvertical lines are parallel, perpendicular, or neither.

EXERCISES

In Exercises 1–10, find the slope of the line through the given points. In Exercises 1–7, sketch the line.

1. $(1, 2), (3, -1)$
2. $(0, 4), (4, 0)$
3. $(2, -2), (2, -3)$
4. $(7, 6), (-4, 2)$
5. $(1/2, 1/4), (1/3, 1/6)$
6. $(\sqrt{2}, -\sqrt{2}), (1, 1)$
7. $(1/\sqrt{2}, \sqrt{2}), (1/\sqrt{3}, \sqrt{3})$
8. $(a, a), (ma, ma)$
9. $(a, b), (ma, mb)$
10. $(a, b), (a + m, b + m)$

In Exercises 11–17, two pairs of points are given. Compute the slopes of the lines through each pair of points, decide whether the lines are parallel, perpendicular or neither, and sketch the two lines on the same plane.

11. $(1, 2), (3, 4); (-1, 2), (-3, 4)$
12. $(3, 3), (1, -2); (5, 1), (-2, 3)$
13. $(-1, 0), (2, 1); (4, 3), (6, 2)$
14. $(5, -5), (4, -4); (3, 7), (2, 8)$

15. (17, 12), (16, −10); (4, 22), (5, 0)

16. (1/2, 1/4), (3/2, 1/2); (3, 2), (6, −10)

17. $(\sqrt{2}, -\sqrt{2}), (\sqrt{3}, -\sqrt{3}); (1/\sqrt{2}, -1/\sqrt{2}), (1/\sqrt{3}, -1/\sqrt{3})$

In Exercises 18–25, sketch the graph of the line that goes through the given point P, and has the given slope m. Be sure to label at least two points on the line.

18. $P = (1, 2)$, $m = 4$

19. $P = (-1, -2)$, $m = -4$

20. $P = (0, 2)$, $m = 1/2$

21. $P = (-5, 3)$, $m = -1/2$

22. $P = (1/2, -1/2)$, $m = 50$

23. $P = (6, -2)$, $m = -50$

24. $P = (\sqrt{2}, 1/\sqrt{2})$, $m = \sqrt{2}$

25. $P = (0, \pi)$, $m = 1/\pi$

In Exercises 26–29, sketch the line through the given point P, with the given property. Label at least two points on the line.

26. $P = (1, 1)$, parallel to the line through (1, 2) and (4, 1).

27. $P = (2, 4)$, parallel to the line through (2, 5) and (−1, 0).

28. $P = (-2, 3)$, perpendicular to the line through (−2, 3) and (−5, 6).

29. $P = (5, 5)$, perpendicular to the line through (−5, 5) and (7, −3).

30. Show that the points (1, 2), (4, 17), and (16, −1) form the vertices of a right triangle. (*Hint:* plot the points to get an idea where the right angle might be, and then use slopes to check for perpendicularity.)

31. Do the points (1, 3), (6, 4), and (3, 2) form the vertices of a right triangle? Justify your answer.

32. Do the points (0, 5), (1, −2), (0, 15), and (2, 5) form the vertices of a parallelogram? Justify your answer. (*Hint:* first plot the points. Then use slopes to check for parallelism.)

33. Describe the figure formed by connecting the points (1, 1), (4, −4) (6, 4), and $(\frac{138}{17}, \frac{8}{17})$ by straight line segments. Justify your description by using slopes.

34. Show that two distinct lines with the same slope must be parallel. (*Hint:* if they were not, then they would meet at some point. Start at that point and move 1 unit to the right. How far do you have to move up (or down) from there in order to reach each line?)

35. Use slopes to prove that the diagonals of a rectangle are perpendicular if and only if the rectangle is a square. (*Hint:* assume that the coordinates of the rectangle are (0, 0), $(a, 0)$, $(0, b)$, and (a, b).

3.2 Equations of a Line

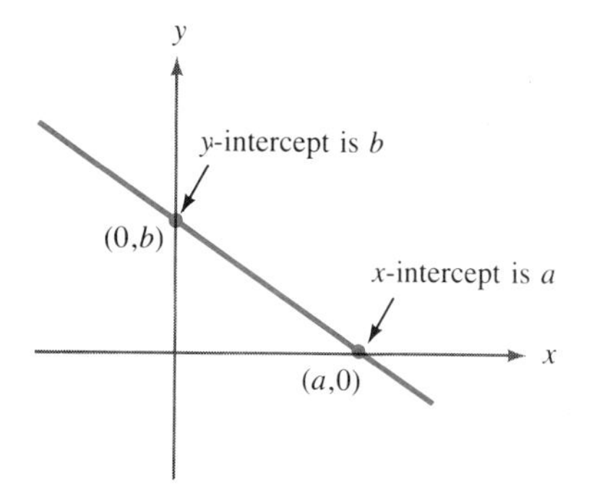

FIGURE 3.9

Given a straight line, there are several ways to find an equation whose graph is that line. Before discussing this, however, let us make two important definitions.

The **x-intercept** of a line is the x-coordinate of the point where the line intersects the x-axis. (See Figure 3.9.) Thus, a real number a is the x-intercept of a line if and only if the point $(a, 0)$ is on the line. The **y-intercept** of a line is the y-coordinate of the point where the line intersects the y-axis. (See Figure 3.9.) A real number b is the y-intercept of a given line if and only if the point $(0, b)$ is on the line. Of course, every nonhorizontal line has exactly one x-intercept, and every nonvertical line has exactly one y-intercept.

THE SLOPE-INTERCEPT FORM OF THE EQUATION OF A LINE

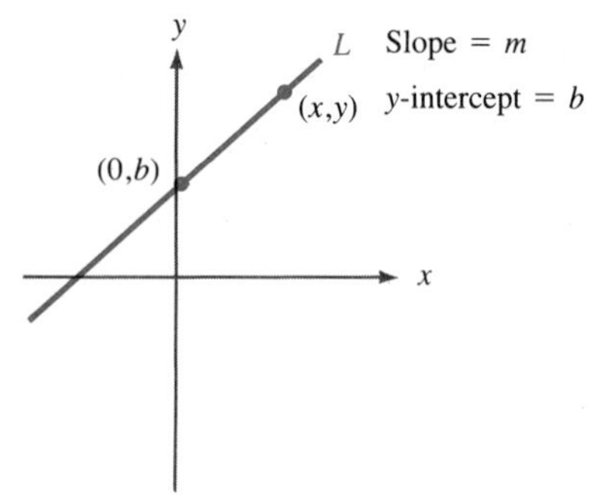

FIGURE 3.10

Suppose that we are given a line L by being given its slope m and its y-intercept b, as shown in Figure 3.10. (We have drawn the line as though it had a positive slope, but the same discussion would hold if $m \leq 0$.)

We can find the equation of the line L by a very simple and very useful idea. All we have to do is label an arbitrary point on the line with the variable coordinates (x, y), as in Figure 3.10, and then compute the slope of the line by using this variable point and the point $(0, b)$. Of course, the result of this computation must equal m, since m is the slope of the line. In other words, we must have

$$\frac{y-b}{x-0} = m$$

Solving this equation for y gives

$$y = mx + b \tag{3.1}$$

This equation is known as the **slope-intercept form of the equation of a line.** It is one of the nicest forms of the equation of a line, since much of the information about the line can be obtained simply by looking at the equation.

For instance, we can see immediately that the equation

$$y = 3x + 7$$

is the equation of a straight line whose slope is 3 and whose y-intercept is 7. From this information, we can sketch the graph without having to substitute values of x into the equation. (See Figure 3.11.)

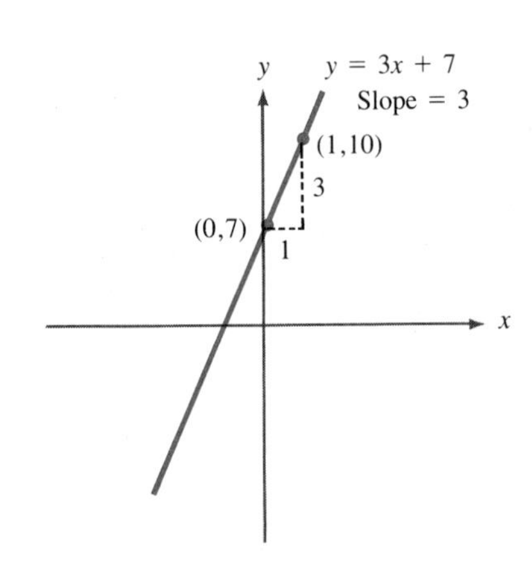

FIGURE 3.11

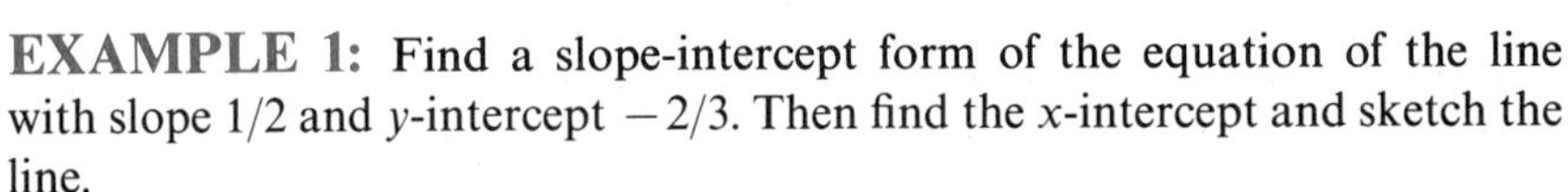

EXAMPLE 1: Find a slope-intercept form of the equation of the line with slope 1/2 and y-intercept $-2/3$. Then find the x-intercept and sketch the line.

Solution: Substituting the values $m = 1/2$ and $b = -2/3$ into Equation (3.1) gives the slope-intercept form

$$y = \frac{1}{2}x - \frac{2}{3} \tag{3.2}$$

Setting $y = 0$ in Equation (3.2), and solving for x gives

$$\frac{1}{2}x - \frac{2}{3} = 0$$

$$\frac{1}{2}x = \frac{2}{3}$$

$$x = \frac{4}{3}$$

This shows that the point (4/3, 0) is on the graph, and so 4/3 is the x-intercept. Now we can sketch the graph of the line using the points (4/3, 0) and (0, −2/3), as in Figure 3.12. *Try Study Suggestion 3.5.* □

▶ **Study Suggestion 3.5:** Find a slope-intercept form of the equation of the line with slope −2 and y-intercept 5. Then find the x-intercept and sketch the graph. ◀

The slope-intercept form of the equation of a line tells us something very important about linear functions. As you know, the graph of the linear function $f(x) = ax + b$ is the same as the graph of the equation $y = ax + b$. But this is just the slope-intercept form of the equation of a line, and so we have the following result.

THEOREM The graph of any linear function $f(x) = ax + b$ is a straight line.

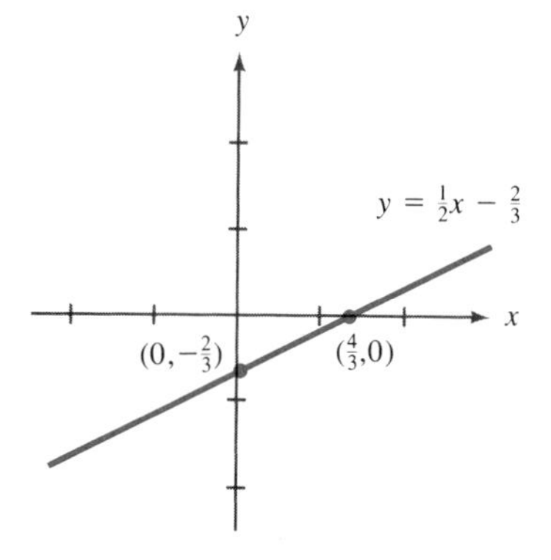

FIGURE 3.12

THE POINT-SLOPE FORM OF THE EQUATION OF A LINE

Now suppose that we are given a line L by being given its slope m and *any* point (x_1, y_1) on the line. Referring to Figure 3.13, we again label an arbitrary point on the line with the variable coordinates (x, y), and use this point and the point (x_1, y_1) to compute the slope of the line. As before, this must equal m

$$\frac{y - y_1}{x - x_1} = m$$

It is customary to multiply both sides of this by $x - x_1$, but not to solve completely for y. This gives us the equation

$$y - y_1 = m(x - x_1) \tag{3.3}$$

which is known as the **point-slope form of the equation of a line.**

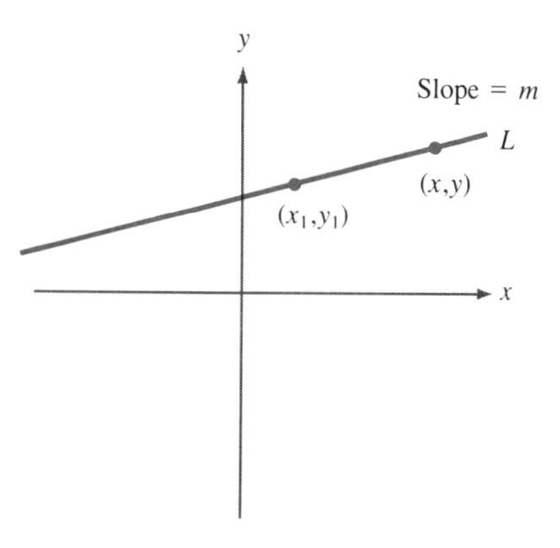

FIGURE 3.13

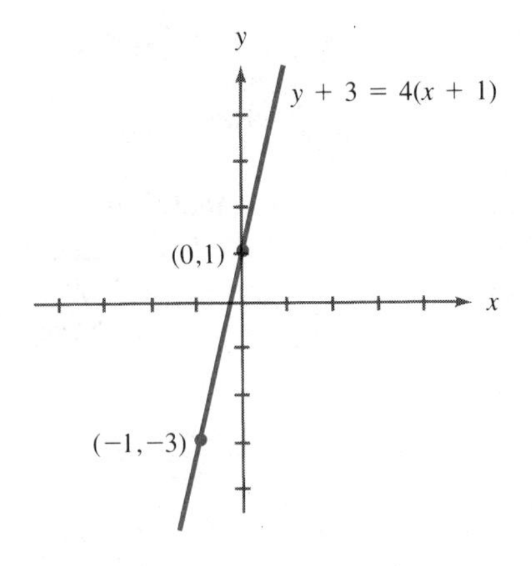

FIGURE 3.14

▶ **Study Suggestion 3.6:** Find a point-slope form of the equation of the line through the point $(-2, 1/2)$, whose slope is -2. Then find the y-intercept of the line and sketch the graph. ◀

EXAMPLE 2: Find a point-slope form of the equation of the line through the point $(-1, -3)$, whose slope is 4. Then find the y-intercept and sketch the line.

Solution: Substituting $x_1 = -1$, $y_1 = -3$ and $m = 4$ into Equation (3.3) gives

$$y - (-3) = 4(x - (-1))$$

or

$$y + 3 = 4(x + 1) \tag{3.4}$$

One way to find the y-intercept of the line is to put Equation (3.4) into the slope-intercept form, which is $y = 4x + 1$. Thus, the y-intercept is 1. Another way to find the y-intercept is to simply set $x = 0$ in Equation (3.4) and solve for y. This gives $y = 1$. In any case, the point $(0, 1)$ is on the line, and the graph is shown in Figure 3.14. *Try Study Suggestion 3.6.* □

EXAMPLE 3: Find an equation of the line through the point $(2, 4)$ that is perpendicular to the line $y = 3x - 1$.

Solution: The line $y = 3x - 1$ has slope 3, and so any line that is perpendicular to it must have slope $-1/3$ (which is the negative reciprocal of 3.) Thus, the line we are seeking has slope $-1/3$ and passes through the point $(2, 4)$. Substituting into Equation (3.3) gives

$$y - 4 = -\frac{1}{3}(x - 2)$$

or

$$y = -\frac{1}{3}x + \frac{14}{3}$$

Try Study Suggestion 3.7. □

▶ **Study Suggestion 3.7:** Find an equation of the line through the point $(-1/2, 2)$ that is perpendicular to the line with equation $3y + 2x = 4$. ◀

Suppose we are given a line L by being given two points (x_1, y_1) and (x_2, y_2) on the line. Then we can use these points to compute the slope of the line, and then use *either* one of the points (and the slope) to compute the point-slope form of the equation of the line. Let us illustrate with an example.

EXAMPLE 4: Find a point-slope form of the equation of the line through the points $(2, 3)$ and $(-1, 5)$. Put the equation into slope-intercept form and identify the slope and y-intercept.

Solution: Using both points, we compute the slope of the line to be

$$\frac{5 - 3}{-1 - 2} = -\frac{2}{3}$$

Now we choose either point, say (2, 3), to compute the point-slope form of the equation of the line

$$y - 3 = -\frac{2}{3}(x - 2)$$

The slope-intercept form is

$$y = -\frac{2}{3}x + \frac{13}{3}$$

Thus, the slope is $-2/3$ and the y-intercept is 13/3.

Try Study Suggestion 3.8. □

▶ **Study Suggestion 3.8:** Find a point-slope form of the equation of the line through the points $(-1, 3)$ and $(-3, 1)$. Put the equation into slope-intercept form and identify the slope and y-intercept. ◀

THE GENERAL FORM OF THE EQUATION OF A LINE

Neither of the forms that we have discussed so far can be used to represent the equation of a vertical line, which has the form $x = r$, where r is a constant. However, there is another form for the equation of a line that can be used to represent all lines, including vertical lines. An equation of the form

$$ax + by + c = 0$$

where a, b, and c are constants and at least one of a or b is not zero, is called the **general form of the equation of a line.** (The letter b here has no connection with the letter b as used in the slope-intercept form.)

All of the equations that we have derived so far can be put into this form with a little bit of algebraic manipulation, and the equation of a vertical line can be put into this form by choosing $b = 0$. This is one of the main advantages of the general form.

EXAMPLE 5: Show that

$$2y + 3x - 4 = 0$$

is the equation of a line, and find the slope and y-intercept of that line.

Solution: The first step is to solve this equation for y. This gives

$$y = -\frac{3}{2}x + 2$$

This is the slope-intercept form of the equation of a line, with slope $-3/2$ and y-intercept 2.

Try Study Suggestion 3.9. □

▶ **Study Suggestion 3.9:** Show that $3x + 2y = 1$ is the equation of a line, and find the slope and y-intercept of that line. ◀

Ideas to Remember

There are several ways to describe the equation of a line. The slope-intercept form is the most useful in the sense that it *directly* displays the most information about the line, namely, its slope and its y-intercept.

EXERCISES

In Exercises 1–10, show that the given equation is the equation of a line by putting it in slope-intercept form. Then identify the slope and y-intercept, and sketch the graph.

1. $y = 3x - 4$

2. $y - 4x + 1 = 8$

3. $x = 4y - 5$

4. $x + y = 1$

5. $-y/2 + x/3 = 1$

6. $2x + 3y = 4y - 3x + 1$

7. $x - y = 1$

8. $2x + 2y = 1$

9. $-x + 4y = 1$

10. $6x - 5y = 30$

In Exercises 11–14, find the slope-intercept form of the equation of the line with slope m and y-intercept b. Then find the x-intercept of the line and sketch the graph.

11. $m = 7, b = 2$

12. $m = -2, b = -1$

13. $m = \sqrt{2}, b = 1/2$

14. $m = -1/\sqrt{3}, b = \sqrt{3}$

In Exercises 15–22, find the point-slope form of the equation of the line. Then find the y-intercept of the line.

15. $P = (1, 2), m = 6$

16. $P = (1/2, 1/4), m = -2$

17. $P = (0, 3), m = 1/2$

18. $P = (-2, \pi), m = \pi$

19. $P = (-4, 0), m = -2/3$

20. $P = (\sqrt{2}, 0), m = 1/\sqrt{2}$

21. $P = (a, a), m = 1/a$

22. $P = (a, b), m = c$

In Exercises 23–30, find the slope-intercept form of the equation of the line through the given points, and give the slope and y-intercept of the line.

23. $(1, 2), (3, -1)$

24. $(2, 0), (0, 3)$

25. $(1/2, 1/3), (-2, -3)$

26. $(\sqrt{2}, 1/\sqrt{2}), (\sqrt{3}, 1/\sqrt{3})$

27. $(5, 6), (-4, 17)$

28. $(0, 0), (a, b)$

29. $(a, b), (b, a)$

30. $(a, a^2), (b, b^2)$

31. If the y-intercept of a line is b and the x-intercept is a, and if a and b are nonzero, show that the equation of the line can be written in the form

$$\frac{x}{a} + \frac{y}{b} = 1$$

This is called the **intercept form** of the equation of a line.

In Exercises 32–36, find the intercept form of the equation of the line with the given properties.

32. x-intercept $= 2$, y-intercept $= -3$

33. slope $= 2$, y-intercept $= 3$

34. slope $= -3$, x-intercept $= -2$

35. slope $= 1/2$, line goes through $(4, 5)$

36. line goes through $(4, -2)$ and $(3, 1)$

3.3 Quadratic Functions

A **quadratic function** is any function of the form

$$f(x) = ax^2 + bx + c \tag{3.5}$$

where a, b, and c are constants and $a \neq 0$. Just as the graph of a linear function $f(x) = ax + b$ is a straight line, the graph of a quadratic function has a special shape, known as a **parabola.** As you can see from Figure 3.15, the parabola has a somewhat "U" shape, but the curve extends upward and *outward* in both directions forever. Also, a parabola is symmetric about a line known as its **axis of symmetry.**

The parabolic shape has many important applications in nature. One reason is that a parabola has what is known as a *focal point*. To explain what we mean by this, imagine that we could make a one-dimensional mirror

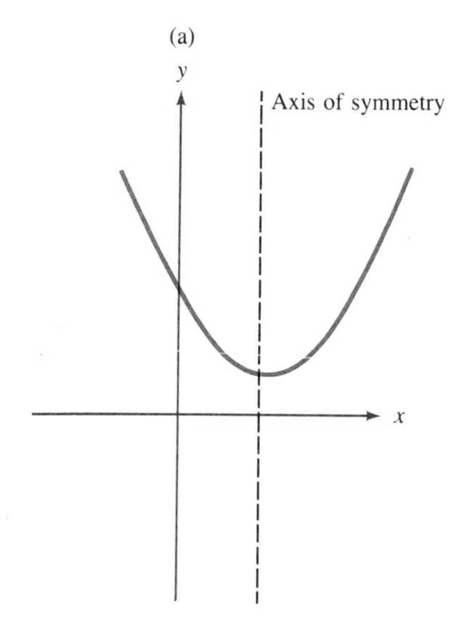

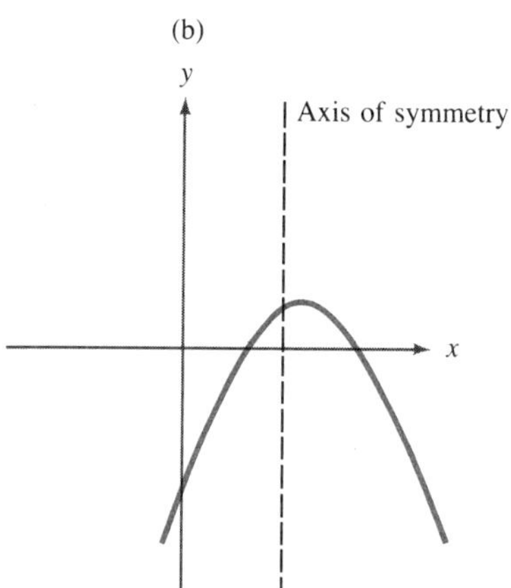

FIGURE 3.15

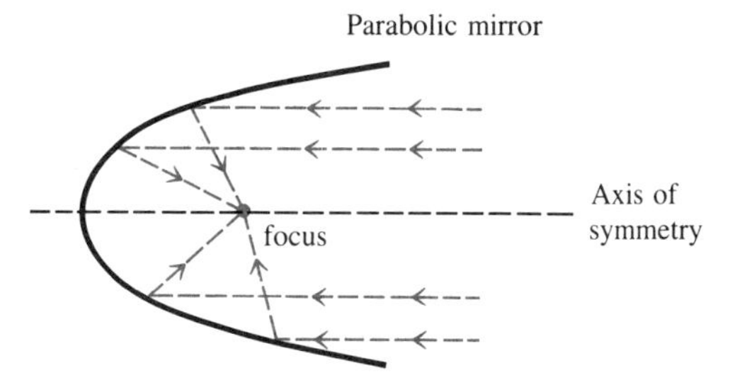

FIGURE 3.16

in the shape of a parabola, as shown in Figure 3.16. Then there would be a unique point enclosed by the mirror with the property that any beam of light traveling parallel to the axis of symmetry that reaches the mirror will be reflected to that point. This point is called the **focal point,** or **focus,** of the mirror. As a result of this unique property of parabolic mirrors, a two-dimensional version is used in the design of reflecting telescopes, where the eyepiece of the telescope is placed at the focus of the mirror. This ability to reflect light to a single focal point is not shared by mirrors of other shapes. For example, a mirror shaped like the inside of a sphere would scatter light in various directions, making it impossible to focus on any object.

The graph of the quadratic function (3.5) is the same as the graph of the quadratic equation

$$y = ax^2 + bx + c \tag{3.6}$$

In graphing such an equation, we want to distinguish between parabolas that open up, as in Figure 3.17, and those that open down, as in Figure 3.18.

When we graph a parabola that opens up, we would like to be able to identify the point which is "lowest" on the graph. More precisely, this is the

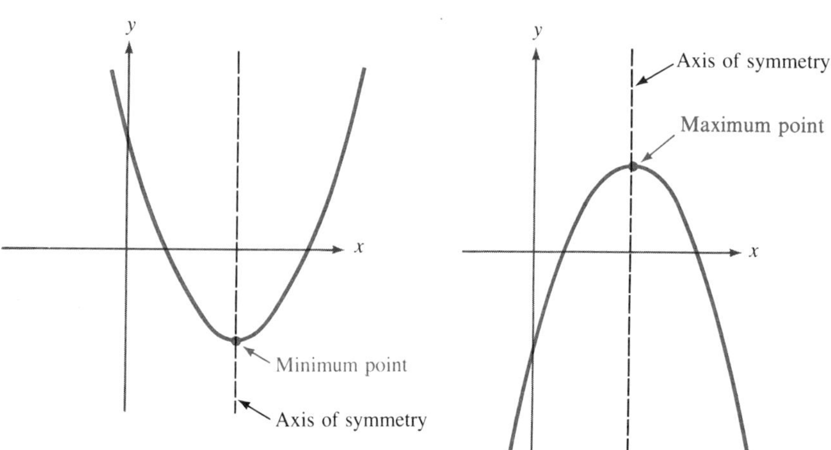

FIGURE 3.17

FIGURE 3.18

point with the smallest y-coordinate of all points on the graph. For this reason, it is called the **minimum point** of the graph. A parabola that opens up has exactly one minimum point. (See Figure 3.17.)

Similarly, when we graph a parabola that opens down, we would like to be able to identify the maximum point, which is the point that has the largest y-coordinate of all points on the graph. A parabola that opens down has exactly one maximum point. (See Figure 3.18.)

Maximum and minimum points are sometimes referred to collectively as **extreme points.** Not only is the extreme point important for its own sake, but it plays another important role as far as the graph is concerned.

The axis of symmetry of the graph of a quadratic equation goes through the extreme point.

We saw in the previous section that there are several special forms for equations of a line, and that these forms "display" certain important information about the line. Similarly, there is a special form for equations of a parabola, which also displays important information about the graph. The equation

$$y = a(x - h)^2 + k \tag{3.7}$$

where h and k are constants (note the minus sign in $x - h$) is called the **standard form of the equation of a parabola.** The standard form displays the following information about the parabola.

1. **The sign of a tells us whether the parabola opens up or down. If $a > 0$, then the parabola opens up, and so it has a minimum point. On the other hand, if $a < 0$, then the parabola opens down, and so it has a maximum point.**

2. **The coordinates of the extreme point are (h, k).**

We can see why these facts are true by looking at Equation (3.7). If a is positive, then the expression $a(x - h)^2$ is always nonnegative, regardless of the value of x. This means that y is always greater than or equal to k. Furthermore y is equal to k only when the quantity $a(x - h)^2$ is equal to 0; that is, only when $x = h$. Thus, in the case $a > 0$, the quantity y attains a *minimum* value when $x = h$, and so the parabola has a minimum point at (h, k).

On the other hand, when a is negative, the quantity $a(x - h)^2$ is always nonpositive, and so y is always less than or equal to k. Furthermore y is equal to k only when $a(x - h)^2$ is equal to 0; that is, only when $x = h$. Therefore, in this case, the quantity y attains a *maximum* value when $x = h$, and so the parabola has a maximum point at (h, k).

The procedure for putting a quadratic equation into standard form involves little more than completing the square. Let us consider some examples.

EXAMPLE 1: Put the quadratic equation

$$y = -4x^2 + 4x + 1$$

into standard form.

Solution: The first step is to manipulate the equation so that the right-hand side has the form $x^2 + kx$. This can be done by first subtracting 1 from both sides of the equation, and then dividing both sides by -4:

$$\frac{y-1}{-4} = x^2 - x$$

Now we can complete the square on the right-hand side, by adding $(-1/2)^2 = 1/4$ to both sides:

$$\frac{y-1}{-4} + \frac{1}{4} = x^2 - x + \frac{1}{4}$$

$$\frac{y-2}{-4} = \left(x - \frac{1}{2}\right)^2$$

Finally, we solve for y, to get

$$y = -4\left(x - \frac{1}{2}\right)^2 + 2$$

The equation is in standard form, where $h = 1/2$ and $k = 2$.

Try Study Suggestion 3.10. □

▶ **Study Suggestion 3.10:** Put the following quadratic equations in standard form. **(a)** $y = x^2 - 6x + 2$ **(b)** $y = 3x^2 + 2x$ ◀

In drawing the graph of a quadratic equation, we will usually label the extreme point, the axis of symmetry, and the x- and y-intercepts of the graph. Our procedure for graphing quadratic equations is as follows.

Procedure for Graphing Quadratic Equations

STEP 1. Put the equation into standard form.

STEP 2. Use the information obtained from the standard form to plot the extreme point, sketch in the axis of symmetry, and note whether the parabola opens up or down.

STEP 3. Find the intercepts of the graph and locate them on the coordinate axes.

STEP 4. Draw the graph in a smooth parabolic shape through the intercepts and the extreme point. It may be necessary to plot some additional points for this step.

The y-intercepts of the graph of Equation (3.6) can be found simply by setting $x = 0$ in Equation (3.6). Similarly, the x-intercepts can be found by setting $y = 0$ in Equation (3.6) and solving the resulting quadratic equation for x.

▶ **Study Suggestion 3.11:** How many x-intercepts can the graph of a quadratic equation have? Justify your answer. ◀

Try Study Suggestion 3.11.

EXAMPLE 2: Find the intercepts of the graph of the quadratic equation

$$y = x^2 - 2x - 5 \tag{3.8}$$

Solution: In order to find the y-intercept, we set $x = 0$ in Equation (3.8), to get $y = -5$. To find the x-intercepts of the graph we can set $y = 0$ in Equation (3.8), to obtain

$$x^2 - 2x - 5 = 0$$

Solving this quadratic equation for x gives $x = 1 \pm \sqrt{6}$.

▶ **Study Suggestion 3.12:** Find the intercepts of the graph of the quadratic equation $y = x^2 - x - 4$. ◀

Try Study Suggestion 3.12. □

Now we are ready to consider some examples of graphing quadratic equations. We will follow the four-step procedure outlined earlier.

EXAMPLE 3: Graph the quadratic equation

$$y = x^2 + 2x - 3$$

Solution:

STEP 1. The standard form of this equation is

$$y = (x + 1)^2 - 4$$

(To be perfectly correct, the standard form is $y = (x - (-1))^2 + (-4)$. However, it is customary to make the simplification that we have made here. Nevertheless, it is important to remember that $h = -1$ and $k = -4$.)

STEP 2. Since $a = 1 > 0$, the parabola opens up, and has a minimum point at $(h, k) = (-1, -4)$. We have plotted this minimum point and sketched the axis of symmetry in Figure 3.19.

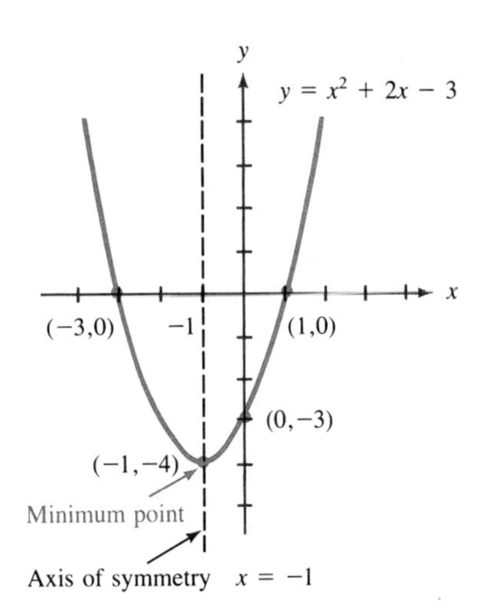

FIGURE 3.19

STEP 3. To get the y-intercept, we set $x = 0$ in the original equation, which gives $y = -3$. The x-intercepts are found by setting $y = 0$ in the standard form, and solving for x. (This is a bit easier than setting $y = 0$ in the original equation and solving for x, although both procedures are correct.) In this case, we get the equation

$$0 = (x + 1)^2 - 4$$

whose solutions are $x = -3$ and $x = 1$, and so these are the x-intercepts. (See Figure 3.19.)

STEP 4. Now we can sketch the curve, as in Figure 4.19.

Try Study Suggestion 3.13. □

▶ **Study Suggestion 3.13:** Graph the quadratic equation

$$y = -4x^2 + 8x - 3. \blacktriangleleft$$

EXAMPLE 4: Graph the quadratic equation

$$y = x^2 - 4x + 4$$

Solution:

STEP 1. The standard form in this case is

$$y = (x - 2)^2$$

STEP 2. Since $a = 1 > 0$, the parabola opens up, and has a minimum point at $(h, k) = (2, 0)$. Also, the axis of symmetry is the line $x = 2$. (See Figure 3.20.)

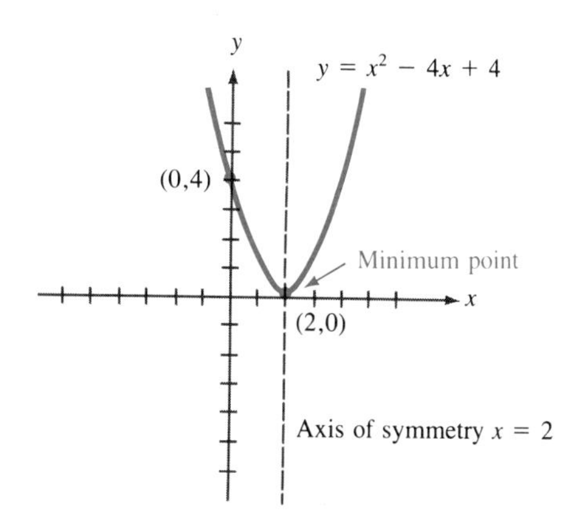

FIGURE 3.20

STEP 3. Setting $x = 0$ in the original equation gives the y-intercept 4. The x-intercepts are found by setting $y = 0$ in the standard form,

$$(x - 2)^2 = 0$$

and solving for x. In this case, the only solution is $x = 2$, and so the graph has only one x-intercept. (See Figure 3.20.)

STEP 4. Now we sketch the graph. As you can see, the fact that the graph has only one x-intercept means that it is *tangent* to the x-axis.

Try Study Suggestion 3.14. □

▶ **Study Suggestion 3.14:** Graph the quadratic equation $y = 4x^2 + 4x + 1$. ◀

EXAMPLE 5: Graph the quadratic equation

$$y = -x^2 - x - 1$$

Solution:

STEP 1. In this case, the standard form of the equation is

$$y = -\left(x + \frac{1}{2}\right)^2 - \frac{3}{4}$$

STEP 2. Since $a = -1 < 0$, the parabola opens down, and has a maximum point at $(-1/2, -3/4)$. (See Figure 3.21.)

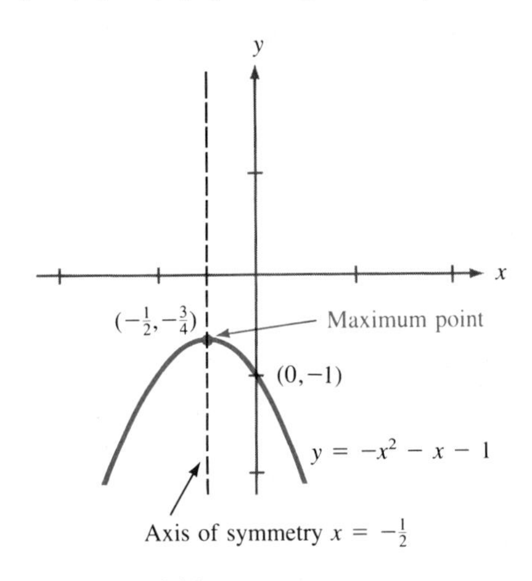

FIGURE 3.21

STEP 3. Setting $x = 0$ in the original equation gives the y-intercept -1. The x-intercepts are obtained by setting $y = 0$ in the standard form, and solving for x. However, the equation

$$-\left(x + \frac{1}{2}\right)^2 - \frac{3}{4} = 0$$

has no real solutions, and so the graph has no x-intercepts.

▶ **Study Suggestion 3.15:** Graph the quadratic equation $y = x^2 + 2x + 2$. ◀

STEP 4. The graph is sketched in Figure 3.21.

Try Study Suggestion 3.15. □

Ideas to Remember

One of the main virtues of the "special forms" of certain equations is that we can immediately read important information about the graph directly from these equations. For example, the slope-intercept form of the equation of a line enables us to read the slope and the y-intercept directly from the equation. Similarly, the standard form of the equation of a parabola enables us to read the coordinates of the extreme point, and determine whether the parabola opens up or down.

EXERCISES

In Exercises 1–20, write the given quadratic equation in standard form, and find the extreme point of its graph. Tell whether the extreme point is a maximum or minimum point, and whether the graph opens up or down.

1. $y = x^2 - 2x + 3$

2. $y = x^2 + 4$

3. $y = (x - 3)^2$

4. $y = x^2 + x + 1$

5. $y = 2x^2 - 3x + 4$

6. $y = -x^2 + 2x + 1$

7. $y = 3x^2 + x$

8. $y = -4x^2 + 5x - 2$

9. $2y = -3x^2 + 4$

10. $2y = x^2 - 3x + 1$

11. $y = \frac{1}{2}x^2 + \frac{1}{2}x - \frac{1}{2}$

12. $2y = x^2 - 3x + 1$

13. $y = \sqrt{2}x^2 - \sqrt{2}x + 1$

14. $y = 1 - 3x^2$

15. $y = 2 + x - \sqrt{2}x^2$

16. $y = \pi x^2 - \pi x + 1$

17. $\pi y = \pi + x + x^2$

18. $y = (2x + 3)^2$

19. $4y = (-7x + 1)^2$

20. $y = a^2x^2 + ax + 1$

In Exercises 21–40, graph the given quadratic function. Be sure to label all intercepts and the extreme point.

21. $y = (x + 1)^2$

22. $y = 3(x - 4)^2$

23. $y = -4(2x + 1)^2$

24. $y = x^2 + 2x - 1$

25. $y = -x^2 - 5x + 6$

26. $y = (2x + 3)(4x - 7)$

27. $y = 4x^2 + 3x + 1$

28. $y = x^2 + 2x + 3$

29. $2y = 5x^2 - x$

30. $y = 12x^2 + 17x - 1$

31. $y = 1 - x^2$

32. $y = 2 - 4x - x^2$

33. $y = 5 - 2x + x^2$

34. $y = -4 + 3x - 2x^2$

35. $y = \pi x^2 + 1$

36. $y = (\pi x - 1)(\pi x + 1)$

37. $y = x^2 + x + \pi$

38. $3y = 2x^2 + x - 1$

39. $y = 3(2x + 4)(x - 3)$

40. $y = \sqrt{2}x^2 - \sqrt{3}$

41. Find all values of k for which the graph of the equation $y = x^2 + kx + 3k + 7$ is tangent to the x-axis.

42. Find all values of k for which the graph of the equation $y = x^2 + kx + 2 - 3k$ is tangent to the x-axis.

43. Find all values of k for which the graph of the equation $y = x^2 - 3kx + 2k^2 + 1$ goes through the point $(2, 2)$.

3.4 Applications: The Maximum and Minimum Values of a Quadratic Function

In this section, we will consider two important applications of the techniques that we discussed in the previous section.

SOLVING QUADRATIC INEQUALITIES BY GRAPHING

In Section 1.8, we discussed a method for solving quadratic inequalities of the form

$$ax^2 + bx + c < 0$$

(or with $<$ replaced by $\leq$, $>$, or $\geq$) where a, b, and c are constants and $a \neq 0$. As you may recall, our method consisted of using the solutions to the *associated quadratic equation* $ax^2 + bx + c = 0$ to divide the real number line into disjoint sets. Then we chose a test value from each set and used it to determine whether or not that set was part of the solution set of the inequality.

As the next examples show, we can use graphs to save us some work in solving quadratic inequalities. (These examples should also make it clear why the test value method of Section 1.8 works.)

EXAMPLE 1: Solve the quadratic inequality

$$x^2 - x - 2 < 0$$

Solution: The first step is to graph the quadratic equation

$$y = x^2 - x - 2$$

Using the methods of the previous section, we obtain the graph in Figure 3.22. Now, the solutions to our inequality are those values of x for which the corresponding value of y is negative; that is, for which the point (x, y) lies *below* the x-axis. We can see from the graph that these are the values of x in the open interval $(-1, 2)$, and so the solution set of this inequality is the open interval $(-1, 2)$. *Try Study Suggestion 3.16.* □

▶ **Study Suggestion 3.16:** Solve the quadratic inequality $4 - 3x - x^2 \geq 0$. ◀

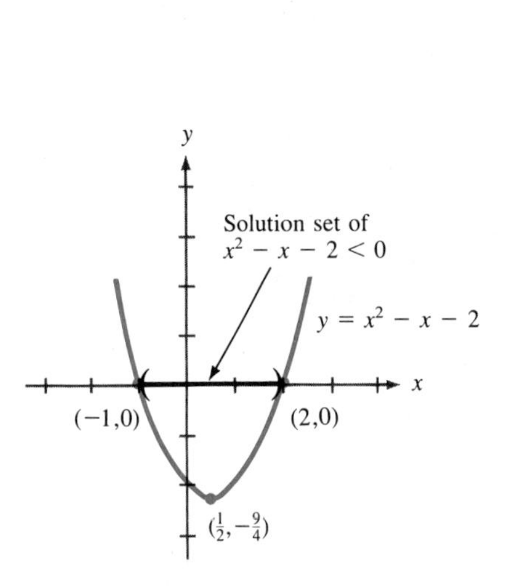

FIGURE 3.22

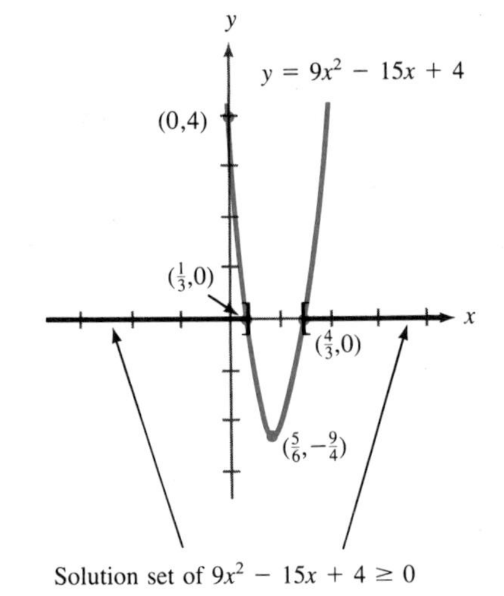

FIGURE 3.23

EXAMPLE 2: Solve the quadratic inequality

$$9x^2 - 15x + 4 \geq 0$$

Solution: The graph of the quadratic equation $y = 9x^2 - 15x + 4$ is shown in Figure 3.23. The solution set of the inequality consists of all values of x for which the corresponding value of y is nonnegative. In terms of the graph, this is the set of all values of x for which the point (x, y) lies *on or above* the x-axis. As we can see from the graph, this is the set $(-\infty, 1/3] \cup [4/3, \infty)$.

Try Study Suggestion 3.17. □

▶ **Study Suggestion 3.17:** Solve the quadratic inequality

$$-x^2 + x + 6 < 0.$$ ◀

FINDING THE EXTREME VALUES OF A QUADRATIC FUNCTION

Many problems that occur in applications involve finding the maximum or minimum values of a certain function. For instance, a company is always interested in finding the maximum value of its profit function, or the minimum value of its cost function.

Before considering specific examples, let us discuss a bit of terminology. We say that a function $f: A \to B$ has a **maximum value** at a real number a in its domain A if $f(a)$ is greater than or equal to any other value of the function. In symbols, f has a maximum value at a if

$$f(x) \le f(a)$$

for all real numbers x in A. This idea is pictured in Figure 3.24.

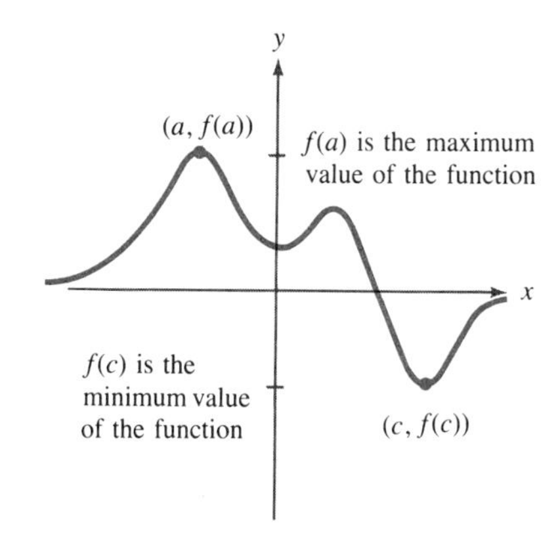

FIGURE 3.24

Similarly, a function $f: A \to B$ has a **minimum value** at a real number c in A if $f(c)$ is less than or equal to any other value of the function, in symbols,

$$f(c) \le f(x)$$

for all real numbers x in A. (See Figure 3.24.) The term **extreme value** is used to designate either a maximum value or a minimum value.

We can find the extreme value of a quadratic equation

$$f(x) = ax^2 + bx + c$$

(a quadratic function has only one extreme value) by first putting the function into standard form

$$f(x) = a(x - h)^2 + k$$

Then, according to the results of the previous section, the point (h, k) is the extreme point of the *graph* of this function. Hence $k = f(h)$ is the extreme value of the function. Let us summarize.

Procedure for Finding the Extreme Value of a Quadratic Function

STEP 1. Put the function in standard form $f(x) = a(x - h)^2 + k$.

STEP 2. Then $(h, k) = (h, f(h))$ is the extreme point on the graph and $k = f(h)$ is the extreme value of the function.

EXAMPLE 3: A farmer wishes to fence off three sides of a rectangular region whose fourth side borders a river, as shown in Figure 3.25a. If the farmer has 120 feet of fencing what dimensions should he make the region in order to maximize the total area? What is this maximum area?

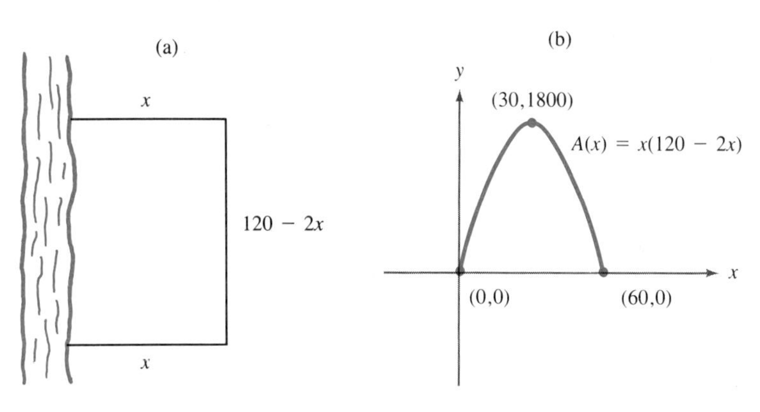

FIGURE 3.25

Solution: The key to this problem is the word "maximize." This word should tip you off to the fact that you must find a certain function, and then determine its maximum value. In this case, the function that we must maximize is the function that gives the total area enclosed by the fencing.

The first step is to label the figure, as we have done in Figure 3.25a. We have chosen x to represent the length of one of the sides of the region, and then expressed the lengths of the other two sides in terms of x. Now we can compute the area, which is a function of the variable x,

$$\begin{aligned} A(x) &= x(120 - 2x) \\ &= -2x^2 + 120x \end{aligned}$$

for $0 < x < 60$. In order to find the maximum value of this quadratic function, we put it into standard form. We will leave it to you to show that this is

$$A(x) = -2(x - 30)^2 + 1800$$

Thus the extreme point of the graph of the function A is the point $(h, k) = (30, 1800)$, and so the extreme value of the function is $A(30) = 1800$. Therefore, in order to maximize the total area enclosed by the fence, the farmer should take $x = 30$ feet, and this will give a total area of 1800 square feet. (The graph of the function A is pictured in Figure 3.25b.)

Try Study Suggestion 3.18. □

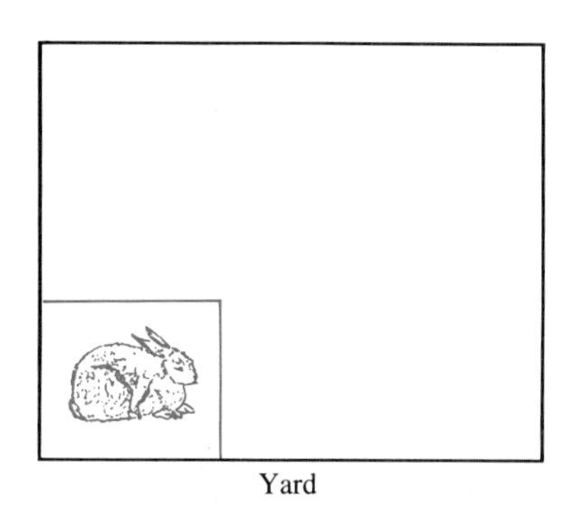

FIGURE 3.26

▶ **Study Suggestion 3.18:** Suppose that your backyard is in the shape of a rectangle, with high fencing on all sides. Suppose also that you want to fence off a corner of the yard for your pet rabbit. (See Figure 3.26.) If you have 1000 feet of fencing at your disposal, what dimensions should you use to maximize the area for your rabbit? ◀

EXAMPLE 4: A certain record store sells its records for \$4.00 each. At that price, the store sells an average of 100 records per day. Market research has shown that for each increase of 50 cents in the price per record, the store will sell an average of 10 fewer records per day. Similarly, for every decrease of 50 cents in price, the store will sell an average of 10 more records per day. At what price should the store sell its records in order to maximize revenue? What is the maximum revenue?

Solution: Let us suppose that the store sells its records at a price of $4 + x$ dollars per record, where x is the dollar amount over (or under) the present price of $4.00 per record. We want to compute the revenue R as a function of x, and find the maximum value of R. Since for every *dollar* over (or under) the $4.00 price the store will sell 20 more (or fewer) records, the total number of records sold per day will be

$$\text{Number of records sold} = 100 - 20x$$

Now we can use the fact that

$$\textbf{Total revenue} = \textbf{(number of items sold)(price per item)}$$

to get

$$R(x) = (100 - 20x)(4 + x)$$
$$= 400 + 20x - 20x^2$$

Putting the revenue function in standard form gives

$$R(x) = -20\left(x - \frac{1}{2}\right)^2 + 405$$

This tells us that the maximum value of the revenue function occurs when $x = 1/2$ dollar, and that the maximum value is $R(1/2) = 405$ dollars. Thus, the record store can increase its revenue to $405 per day (from $400 per day) by selling the records for $4.50 each.

Try Study Suggestion 3.19. □

▶ **Study Suggestion 3.19:** A candy store sells its one-pound box of assorted chocolates for $5.00 and at this price sells an average of 50 boxes per day. Research has shown that for each 25-cent increase (or decrease) in price per box of candy, the store will sell an average of 3 fewer (or more) boxes. At what price should the candy store sell each box of chocolates in order to maximize its revenue? What is the maximum revenue? ◀

EXERCISES

In Exercises 1–20, solve the given inequality.

1. $x^2 - 4 \leq 0$

2. $2x^2 + 1 > 0$

3. $(x - 7)^2 < 0$

4. $2(x + 3)^2 > 0$

5. $x^2 - x - 6 \geq 0$

6. $x^2 - 5x + 4 \leq 0$

7. $-6x^2 - x + 1 \leq 0$

8. $1 - x - 2x^2 > 0$

9. $x^2 + x + 1 > 0$

10. $2x^2 - x + 1 \leq 0$

11. $x^2 - 6x + 9 \leq 0$

12. $x^2 - \pi^2 \geq 0$

13. $12x^2 + 12x > 0$

14. $-10x^2 - 9x + 3 < 0$

15. $(2x + 3)(x - 4) \leq 0$

16. $(-x + 4)(2x + 1) > 0$

17. $3x^2 \geq 4x - 2$

18. $x \geq x^2 + 1$

19. $5x - x^2 \leq 2$

20. $4x - 3 \geq 4x^2 - 5$

21. A gardener wants to fence off three sides of a rectangular plot of land whose fourth side is enclosed by a wall. If the gardener has 10 feet of fencing, what dimensions should she make the plot in order to maximize the area of the plot?

22. Generalize the previous exercise as follows. A gardener wants to fence off three sides of a rectangular plot of land whose fourth side is enclosed by a wall. If the gardener has d feet of fencing, what dimensions should she make the plot in order to maximize the area of the plot?

23. A carpenter wishes to make a window in the shape of a rectangle with a semicircle sitting on top. If the perimeter of the entire window is to be 4 meters, what dimensions should he make the window in order to maximize the area of the *rectangular portion*?

24. A certain company sells microcomputers. Its profit is given by the function $P(t) = 100 + 540t - 3t^2$, where $0 \leq t \leq 1000$ and t is the number of computers sold. How many computers should the company sell in order to maximize its profit?

25. The cost for a certain computer company to manufacture computers is given by the function $C(x) = 2x^2 - 800x + 200{,}000$. How many computers should they manufacture in order to minimize their cost?

26. The number of bacteria, in millions, that grow in a certain culture medium in a 24-hour period is given by the function $A(x) = -300 + 40x - x^2$ for $10 \leq x \leq 30$, where x represents the temperature of the medium. What temperature will maximize the number of bacteria in the culture?

27. When an object is thrown upward with a velocity of 32 feet per second, it reaches a height given by the function $h(t) = 32t - 16t^2$, where t is the time since the object was released. Find the maximum height of the object and the time it takes for the object to return to the ground.

28. Find two numbers whose sum is 10 and whose product is as large as possible.

29. Meals at a certain restaurant cost an average of $8.00 per meal. At this price, the restaurant sells an average of 26 meals per day. Research has shown that every increase (or decrease) of 50 cents per meal will result in an average of 2 fewer (or more) customers per day. At what price should the restaurant sell its meals (on the average) in order to maximize revenue? What is the maximum revenue?

30. Tickets for a certain outdoor concert theater sell for $5.00 each. At this price, the theater sells an average of 3000 seats per concert. Research has shown that every increase (or decrease) of 25 cents per ticket will result in 200 fewer (or more) seats being sold. At what price should the theater management sell its tickets in order to maximize revenue? What is the maximum revenue?

31. A certain private airline company determines its fare per ticket as follows. For a given flight, each ticket costs $100 plus $5 for each empty seat on the flight. Suppose that their planes each have 50 seats.

(a) Find the total revenue to the company for a given flight, as a function of the number of occupied seats.

(b) Graph the function found in Part (a).

(c) How many seats should be occupied on a given flight in order to maximize the revenue to the company?

32. A certain archway has the shape of a parabola (opening down), with a base that is 10 feet across. It is desired to slide some rectangular concrete blocks through the archway. The blocks all have a height of 6 feet, but have varying widths. What is the widest block that can fit through the archway?

33. Show that the product $x(1 - x)$ is at most $1/4$, no matter what value x has. For what value of x is this product largest?

34. Find the maximum area of a rectangle with the property that the sum of any two of its adjacent sides is 2.

35. Find a formula for the maximum area possible among all rectangles whose perimeter is equal to P.

36. For a certain make of automobile, the distance that can be covered on one tank of gas is given by the function $D = 25v - (1/4)v^2$, where v is the speed of the car. Find the optimal speed for this car in order to obtain the best gas mileage.

37. A carpenter wants to construct a rectangular window with small rectangular pieces of glass, as shown in Figure 3.27. If she has exactly 20 feet of wood framing, what dimensions should she make the window in order to maximize the amount of light that the window will admit?

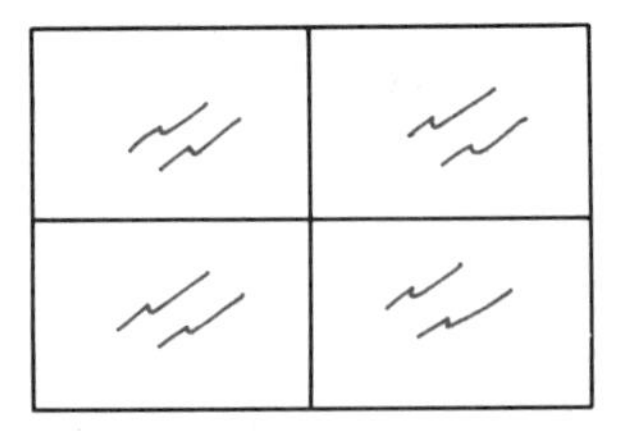

FIGURE 3.27

38. A boy starts bicycling due east on a certain road at a speed of 20 mph. A second boy starts on the same road at a point 10 miles east of the first boy, and he bicycles due south along another road at a speed of 15 mph. The boys start bicycling at exactly the same time. What is the closest distance that the boys ever come together, and when is that distance achieved? (*Hint:* minimize the square of the distance.)

39. Suppose that you have a piece of wire 8 inches long that you wish to cut into two pieces. With one piece, you want to make a circle, and with the other piece you want to make a square, as shown in Figure 3.28. How should you cut the wire so as to minimize the total area enclosed by both the circle and the square? What is this minimum area?

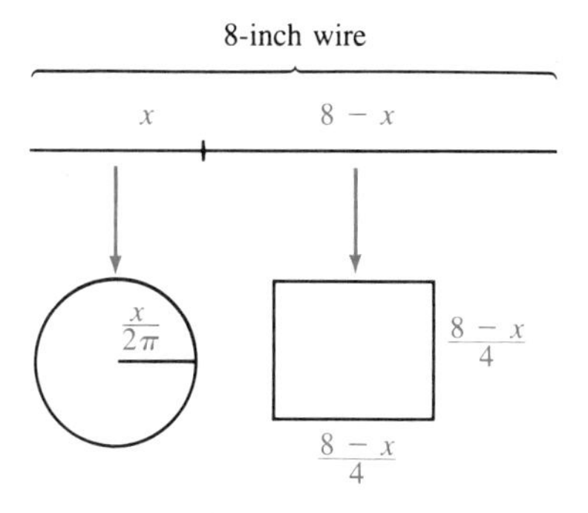

FIGURE 3.28

3.5 Conic Sections

As we have seen, a parabola is a very special type of curve. However, it is just one example of a group of curves known as *conic sections*, each of which has some very special properties.

There are four types of conic sections: the circle, the ellipse, the parabola, and the hyperbola. These curves are pictured in Figure 3.29. They are called conic sections because they arise from the intersection of a plane and a cone (with two nappes). Conic sections play an important role in the physical world. For example, the planets travel in elliptical orbits around the sun, and most comets seem to travel in parabolic orbits. (Elliptical and hyperbolic orbits do occur, but parabolic orbits are the most common for comets.)

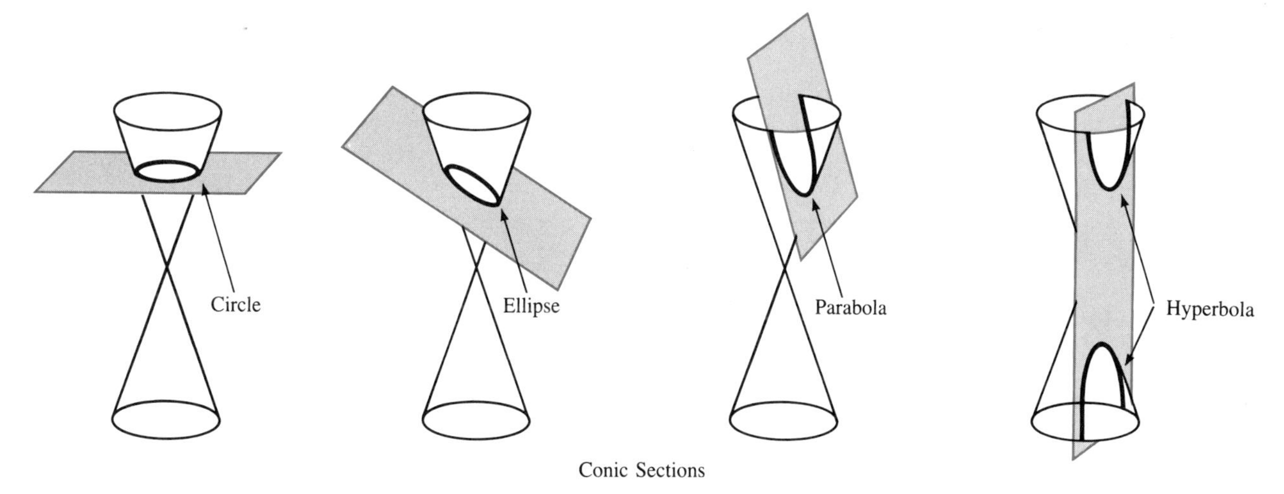

FIGURE 3.29

Each conic section can be described geometrically, as the *locus* (that is, *set*) of all points in the plane that satisfy certain conditions, as well as algebraically, as the graph of a certain equation in two variables.

For example, as we discussed in Section 2.2, a circle can be described geometrically as the locus of all points in the plane that are a fixed distance (the radius) from a given point (the center). An algebraic description of a circle can be derived from this geometric description by noting that if r is the radius and (h, k) is the center, then according to the distance formula

$$\sqrt{(x-h)^2 + (y-k)^2} = r$$

Squaring both sides of this gives

$$(x-h)^2 + (y-k)^2 = r^2$$

which is the equation of a circle of radius r and center (h, k).

In the remainder of this section we will discuss the other three conic sections. In each case, we will first give the geometric description and then the algebraic description. It is possible to derive each of the algebraic descriptions from the geometric description, but since the computations are a bit involved, we will perform them only in the case of the parabola.

Incidentally, the technique of using algebra to describe geometric concepts is known as **analytic geometry,** and was given its major development by the French mathematicians René Descartes (1596–1650), whose name is associated with the Cartesian coordinate system, and Pierre de Fermat (1601–1665). It is interesting to note however, that the study of conic sections dates back to the mathematician and astronomer Apollonius (ca. 262–190 B.C.), who wrote an eight volume treatise entitled *Conic Sections*, for which he received the title *The Great Geometer*. Apollonius also coined the terms "ellipse," "parabola," and "hyperbola" for the conic sections.

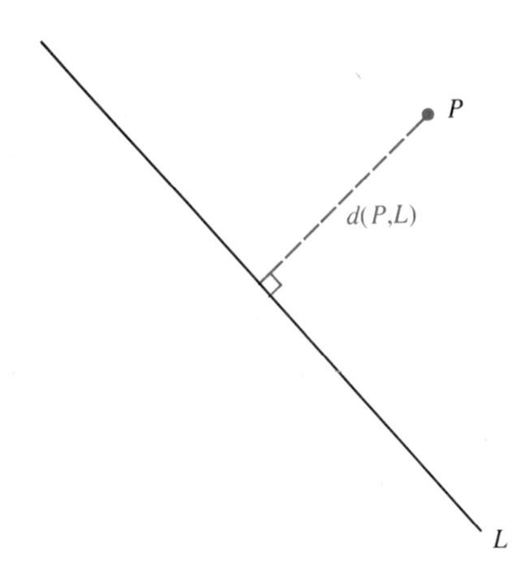

FIGURE 3.30

THE PARABOLA

If P is a point in the plane, and L is a line that does not contain P, the **distance** $d(P, L)$ from P to L is the length of the line segment from P to L that is perpendicular to L. This is pictured in Figure 3.30. Using this concept, we can give the geometric definition of a parabola.

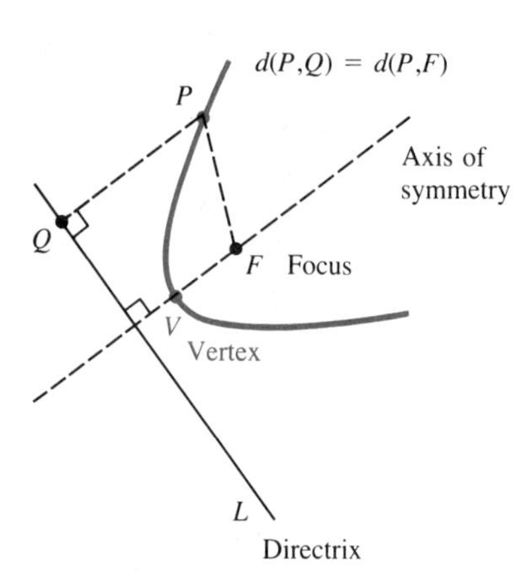

FIGURE 3.31

DEFINITION

A **parabola** is the locus of all points in the plane that are equidistant from a given point F and a given line L (that does not contain F). (See Figure 3.31.)

The point F is the **focus** of the parabola, the line L is called the **directrix,** and the line through the focus perpendicular to the directrix is the **axis of**

symmetry (or simply axis) of the parabola. Finally, the point V that lies on both the parabola and the axis of symmetry is called the **vertex** of the parabola. (See Figure 3.31.)

Now let us derive the equation of a parabola whose vertex is at the origin, as illustrated in Figures 3.32 and 3.33. If we label the focus $F = (0, p)$, then the directrix is the line $y = -p$. (Figure 3.32 shows the case $p > 0$ and Figure 3.33 shows the case $p < 0$.)

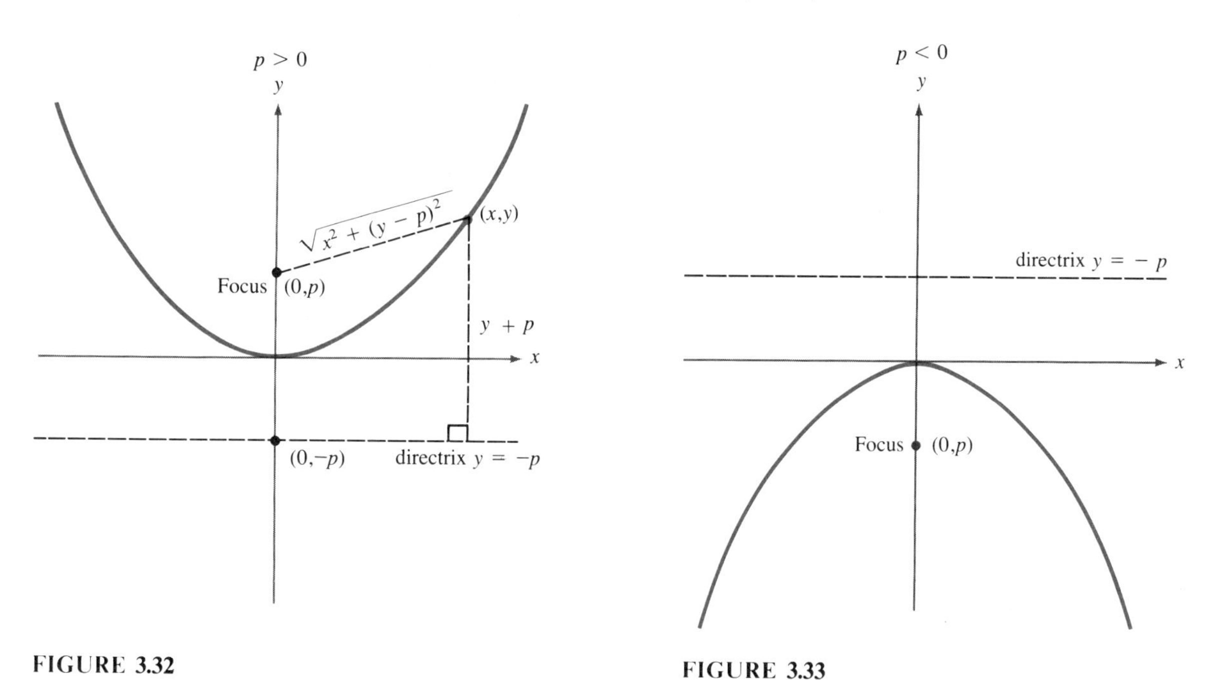

FIGURE 3.32

FIGURE 3.33

In order to derive the equation of this parabola, we pick a "variable" point $P = (x, y)$, and apply the geometric definition of a parabola. According to the distance formula, the distance from the point (x, y) to the focus $(0, p)$ is

$$\sqrt{(x-0)^2 + (y-p)^2} = \sqrt{x^2 + (y-p)^2}$$

On the other hand, as you can see from Figure 3.32, the distance from the point (x, y) to the line whose equation is $y = -p$ is $y + p$. Equating these two distances gives

$$\sqrt{x^2 + (y-p)^2} = y + p$$

Squaring both sides of this equation gives

$$x^2 + (y-p)^2 = (y+p)^2$$

or

$$x^2 + y^2 - 2py + p^2 = y^2 + 2py + p^2$$

and this simplifies further to

$$y = \frac{1}{4p}x^2$$

As we know from our discussion in the previous section, this is the standard form of the equation of a parabola that opens up (since $1/4p > 0$) and whose extreme point is at the origin.

Thus we see, at least in this case, that the geometric description of a parabola does lead us to the type of equation that we studied in Section 3.3. By placing the focus and directrix at other locations in the plane, we can obtain more general equations of a parabola.

There is one additional matter concerning the equation of a parabola that we did not discuss in Section 3.3. As you know, the equation

$$y = ax^2 + bx + c \tag{3.9}$$

is the equation of a parabola that opens up if $a > 0$ and down if $a < 0$. By reversing the roles of x and y, we get

$$x = ay^2 + by + c \tag{3.10}$$

which is the equation of a parabola that opens to the *right* if $a > 0$ and to the *left* if $a < 0$, as shown in Figure 3.34a and b. (We will see in a moment why this is the case.)

The procedure for graphing an equation of the form (3.10) is the same as for graphing an equation of the form (3.9), except that the roles of x and y are reversed. In this case, the standard form is

$$x = a(y - k)^2 + h$$

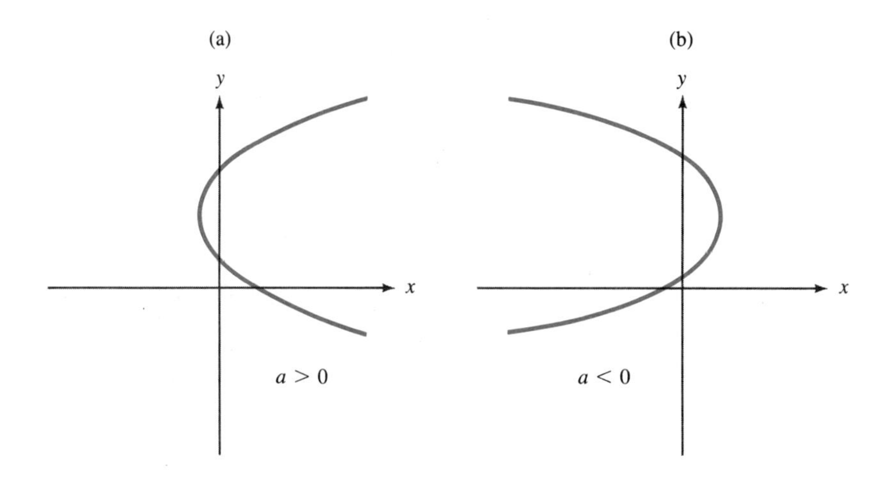

The graph of $x = ay^2 + by + c$ for $a > 0$ and $a < 0$

FIGURE 3.34

which can be obtained from Equation (3.10) by completing the square in the variable y. From this standard form, we can see that

1. **if $a > 0$ the point (h, k) is the *leftmost* point on the parabola, and so the parabola opens to the right;**
2. **if $a < 0$ the point (h, k) is the *rightmost* point on the parabola, and so the parabola opens to the left.**

Let us consider an example.

EXAMPLE 1: Sketch the graph of the equation

$$x = 2y^2 - 5y + 3$$

Solution: This equation can be put into standard form as follows

$$\frac{x-3}{2} = y^2 - \frac{5}{2}y$$

$$\frac{x-3}{2} + \frac{25}{16} = y^2 - \frac{5}{2}y + \frac{25}{16}$$

$$\frac{x-3}{2} + \frac{25}{16} = \left(y - \frac{5}{4}\right)^2$$

$$\frac{x-3}{2} = \left(y - \frac{5}{4}\right)^2 - \frac{25}{16}$$

$$x - 3 = 2\left(y - \frac{5}{4}\right)^2 - \frac{25}{8}$$

$$x = 2\left(y - \frac{5}{4}\right)^2 - \frac{1}{8}$$

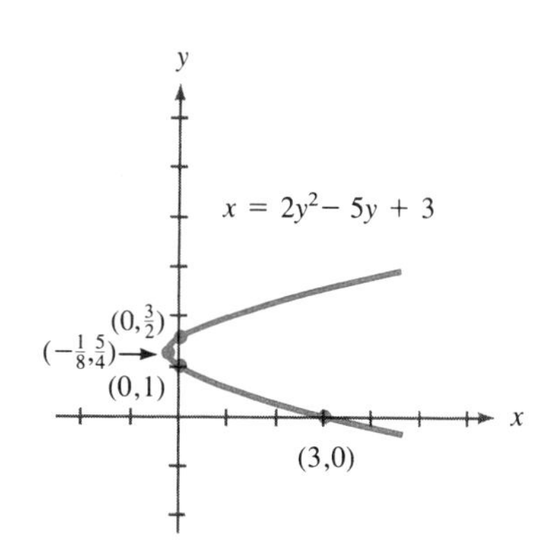

FIGURE 3.35

Since the coefficient of $(y - 5/4)^2$ is positive (that is, $a > 0$), the point $(-1/8, 5/4)$ is the leftmost point on the graph, and the parabola opens to the right.

Before drawing the graph, we should find its intercepts. The x-intercept is found by setting $y = 0$ in the original equation, and this gives $x = 3$. The y-intercepts are most easily found by setting $x = 0$ in the standard form

$$0 = 2\left(y - \frac{5}{4}\right)^2 - \frac{1}{8}$$

and solving for y. We will leave it to you to show that the solutions are $y = 1$ and $y = 3/2$. Now we are ready to draw the graph, as shown in Figure 3.35.

Try Study Suggestion 3.20. □

▶ **Study Suggestion 3.20:** Graph the equation $x = -2y^2 + 4y + 1$. ◀

THE ELLIPSE

Now let us turn to the ellipse.

DEFINITION An **ellipse** is the locus of all points in the plane the *sum* of whose distances from two fixed points F_1 and F_2 is equal to a fixed positive constant. (See Figure 3.36.)

FIGURE 3.36

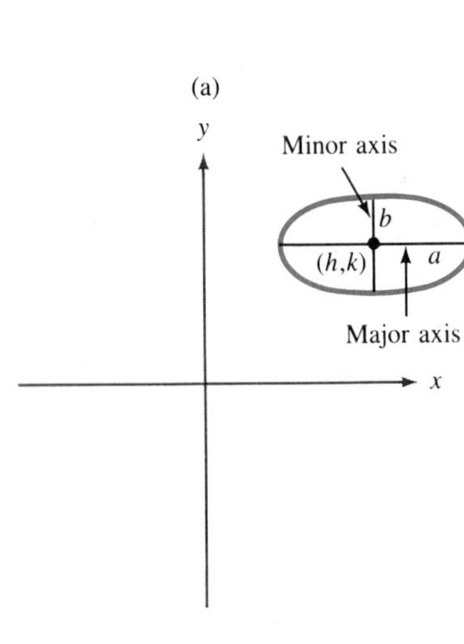

The points F_1 and F_2 are called the **foci** (plural of focus) of the ellipse. The **center** of the ellipse lies at the midpoint of the line segment connecting the two foci. The two solid lines in Figure 3.36 are called the **axes** of the ellipse. The longer of the two is the **major axis,** and the shorter of the two is the **minor axis.** (The foci of the ellipse always lie on the *major* axis.)

An ellipse can also be described algebraically by the equation

$$\frac{(x-h)^2}{a^2} + \frac{(y-k)^2}{b^2} = 1 \tag{3.11}$$

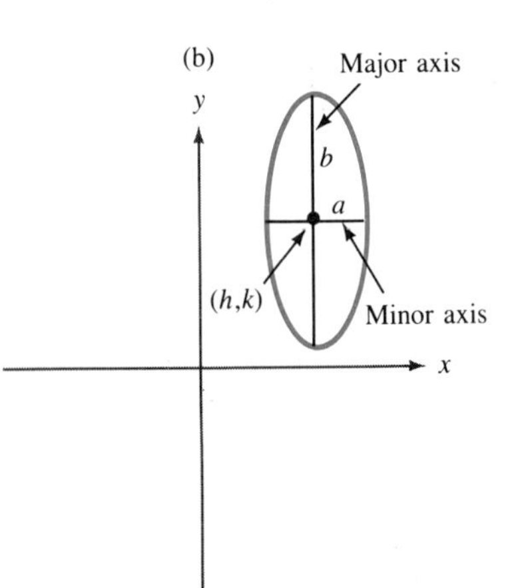

FIGURE 3.37

where a and b are positive constants and $a \neq b$. This is called the **standard form of the equation of an ellipse.** (If $a = b$, then we have a circle.) Referring to Equation (3.11), the center of the ellipse is at the point (h, k), and the role of the constants a and b is shown in Figure 3.37a and b. As you can see, the quantity a is *one-half* the length of the axis of the ellipse that is parallel to the x-axis, and the quantity b is *one-half* the length of the axis of the ellipse that is parallel to the y-axis.

EXAMPLE 2: Find the equation of the ellipse with center $(3, -4)$ whose major axis is parallel to the x-axis and has length 12, and whose minor axis has length 5.

Solution: Since $a = 6$ and $b = 5/2$ the standard form of the equation of this ellipse is

$$\frac{(x-3)^2}{6^2} + \frac{(y+4)^2}{(5/2)^2} = 1$$

or

$$\frac{(x-3)^2}{36} + \frac{(y+4)^2}{25/4} = 1$$

Try Study Suggestion 3.21. □

▶ **Study Suggestion 3.21:** Find the equation of the ellipse whose center is at the point (0, 2), whose major axis is parallel to the y-axis and has length 4, and whose minor axis has length 2. ◀

EXAMPLE 3: Graph the equation

$$9x^2 + 4y^2 - 18x - 24y + 9 = 0$$

Solution: The first step is to put the equation into standard form by completing the square in both x and y. Subtracting 9 from both sides and rearranging terms gives

$$9x^2 - 18x + 4y^2 - 24y = -9$$

Next, we factor out the coefficients of x^2 and y^2

$$9(x^2 - 2x) + 4(y^2 - 6y) = -9$$

and then complete the square inside each of the parentheses. As always, we must perform the same operations on both sides of the equation, and since we are actually adding 9 and 36 to the left side, we must also add these numbers to the right side

$$9(x^2 - 2x + 1) + 4(y^2 - 6y + 9) = -9 + 9 + 36$$

This is equivalent to the equation

$$9(x-1)^2 + 4(y-3)^3 = 36$$

Now we divide both sides by 36

$$\frac{(x-1)^2}{4} + \frac{(y-3)^2}{9} = 1$$

This is the standard form of the equation of an ellipse, with center $(h, k) = (1, 3)$ and $a = 2$, $b = 3$. (See Figure 3.38.) *Try Study Suggestion 3.22.* □

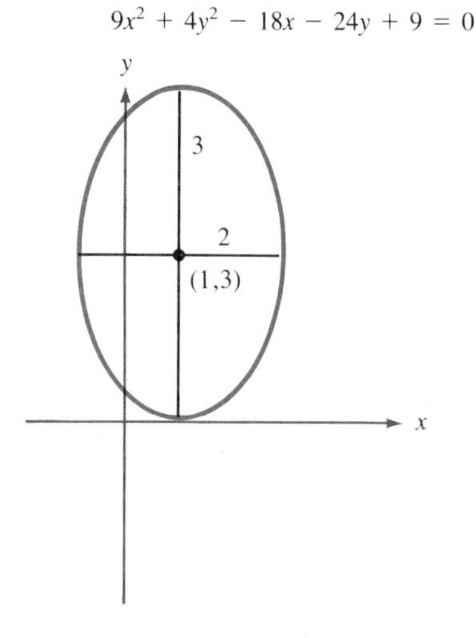

FIGURE 3.38

▶ **Study Suggestion 3.22:** Graph the equation

$x^2 + 2y^2 - 2x - 4y - 3 = 0.$ ◀

We should mention in conclusion that the equation of an ellipse whose center is at the origin has the particularly simple form

$$\frac{x^2}{a^2} + \frac{y^2}{b^2} = 1$$

Try Study Suggestion 3.23.

▶ **Study Suggestion 3.23:** Graph the equation $4x^2 + 5y^2 - 20 = 0.$ ◀

THE HYPERBOLA

Now let us discuss the last of the conic sections.

DEFINITION

A **hyperbola** is the locus of all points in the plane the *difference* of whose distances from two distinct points F_1 and F_2 is equal to a fixed positive constant. (See Figure 3.39.)

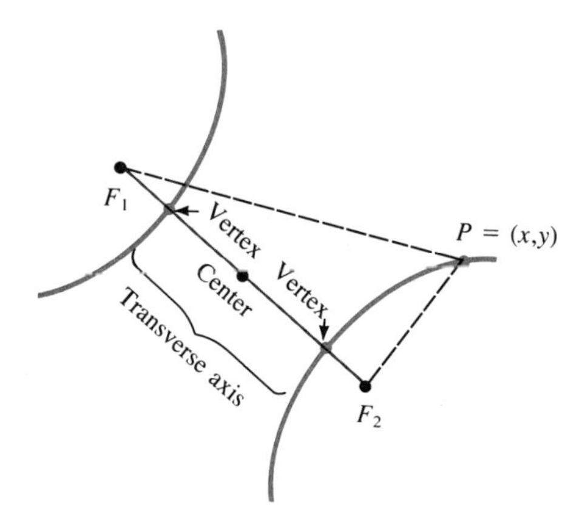

FIGURE 3.39

The points F_1 and F_2 are called the **foci** of the hyperbola. The midpoint of the line segment that connects the two foci is the **center** of the hyperbola, and the points where this line segment meet the curve itself are the **vertices** of the hyperbola. Finally, the line segment between the two vertices is called the **transverse axis** of the hyperbola.

The equation of a hyperbola is quite similar to the equation of an ellipse. However, instead of taking the sum of two terms on the left-hand side, we take the difference. This leads to two different **standard forms** for the equation of a hyperbola, namely

$$\frac{(x-h)^2}{a^2} - \frac{(y-k)^2}{b^2} = 1 \tag{3.12}$$

and

$$\frac{(y-k)^2}{b^2} - \frac{(x-h)^2}{a^2} = 1 \tag{3.13}$$

where a and b are positive constants. The difference between the Equations (3.12) and (3.13) is that in one equation, the minus sign is associated with the term involving x, whereas in the other the minus sign is associated with the term involving y.

The graph of Equation (3.12) is pictured in Figure 3.40 and the graph of Equation (3.13) is pictured in Figure 3.41. In each case, the point (h, k)

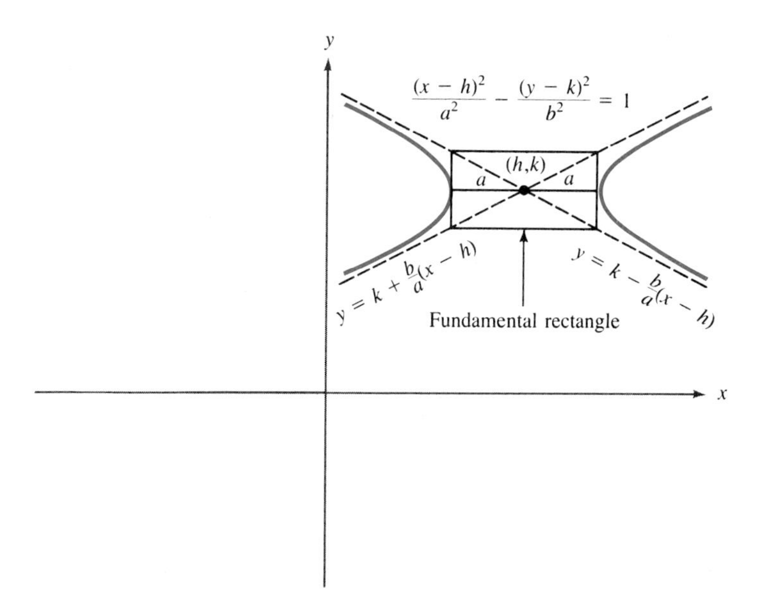

FIGURE 3.40

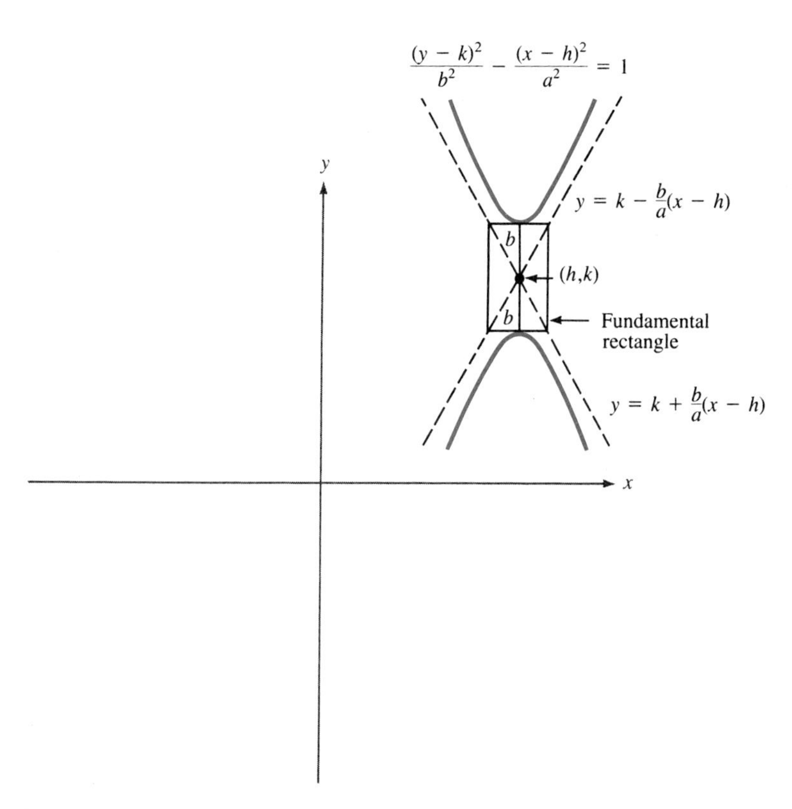

FIGURE 3.41

is the center of the hyperbola. When the minus sign is associated with the term involving y as in Equation (3.12), the hyperbola "opens" to the right and left. When the minus sign is associated with the term involving x as in Equation (3.13), the hyperbola "opens" up and down.

The broken lines in Figures 3.40 and 3.41 are called the **asymptotes** of the hyperbola. The significance of the asymptotes is that the hyperbola gets closer and closer to these lines as we move farther out along the curve. The equations of these asymptotes are

$$y = k + \frac{b}{a}(x - h) \tag{3.14}$$

and

$$y = k - \frac{b}{a}(x - h) \tag{3.15}$$

Perhaps the simplest way to remember the location of the asymptotes of a hyperbola is to observe that they go through the points $(h \pm a, k \pm b)$. In fact, the asymptotes are simply an extension of the diagonals of a rectangle whose center is at the center (h, k) of the hyperbola and whose corners are at $(h \pm a, k \pm b)$. This rectangle is called the **fundamental rectangle** of the hyperbola, as is also shown in Figures 3.40 and 3.41. The hyperbola is actually *tangent* to the fundamental rectangle at the vertices of the hyperbola.

The following steps will make it easier to sketch the graph of a hyperbola.

Procedure for Sketching the Graph of a Hyperbola

STEP 1. Put the equation into standard form. This may involve completing the square in one or both variables.

STEP 2. Locate the center of the hyperbola.

STEP 3. Sketch the fundamental rectangle, and extend the diagonals of this rectangle to get the asymptotes of the hyperbola.

STEP 4. Sketch the hyperbola, using the asymptotes and the fact that the curve is tangent to the fundamental rectangle at the vertices.

EXAMPLE 4: Graph the equation

$$\frac{(x-2)^2}{4} - \frac{(y+1)^2}{16} = 1$$

Solution:

STEP 1. The equation is already in standard form.

STEP 2. The center of the hyperbola is the point $(h, k) = (2, -1)$. (See Figure 3.42.)

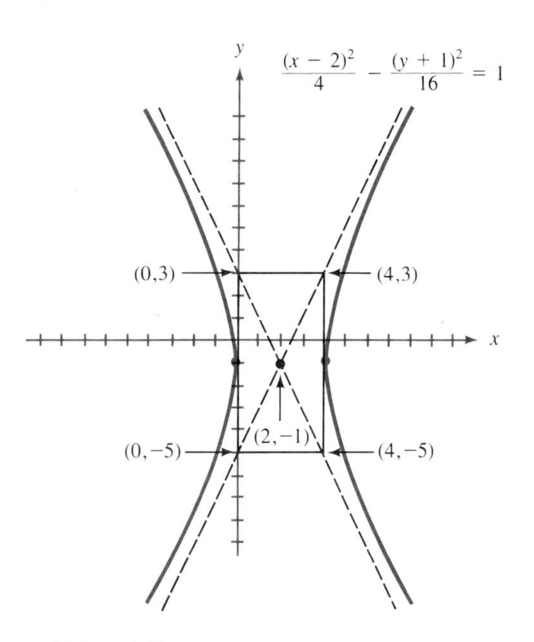

FIGURE 3.42

STEP 3. The corners of the fundamental rectangle are $(h \pm a, k \pm b) = (2 \pm 2, -1 \pm 4)$ or

$$(0, -5), (0, 3), (4, -5), \text{ and } (4, 3)$$

The fundamental rectangle and the asymptotes are shown in Figure 3.42.

STEP 4. Now we can draw the graph shown in Figure 3.42. (Since the minus sign is associated with the term involving y, the hyperbola opens to the right and left.) *Try Study Suggestion 3.24.* □

▶ **Study Suggestion 3.24:** Graph the equation

$$\frac{(y+1)^2}{4} - \frac{(x-2)^2}{9} = 1$$ ◀

EXAMPLE 5: Graph the equation

$$4y^2 - x^2 - 24y + 6x + 23 = 0$$

Solution:

STEP 1. The first step is to put this equation into standard form

$$4y^2 - x^2 - 24y + 6x + 23 = 0$$
$$4y^2 - 24y - x^2 + 6x = -23$$
$$4(y^2 - 6y) - (x^2 - 6x) = -23$$
$$4(y^2 - 6y + 9) - (x^2 - 6x + 9) = -23 + 36 - 9$$
$$4(y-3)^2 - (x-3)^2 = 4$$
$$\frac{(y-3)^2}{1} - \frac{(x-3)^2}{4} = 1$$

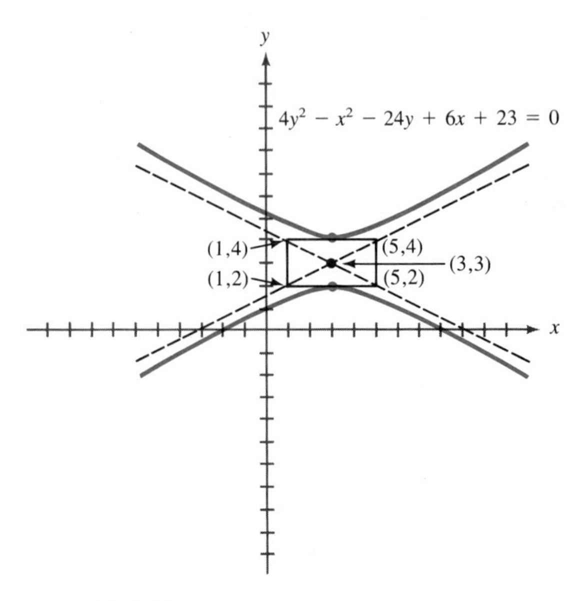

FIGURE 3.43

STEP 2. The center of the hyperbola is at the point $(h, k) = (3, 3)$. (See Figure 3.43.)

STEP 3. The corners of the fundamental rectangle are $(h \pm a, k \pm b) = (3 \pm 2, 3 \pm 1)$ or

$$(1, 2), (1, 4), (5, 2), \text{ and } (5, 4)$$

The fundamental rectangle and the asymptotes are shown in Figure 3.43.

STEP 4. Now we can draw the graph shown in Figure 3.43. (Since the minus sign is associated with the terms involving x, the hyperbola opens up and down.) *Try Study Suggestion 3.25.* □

▶ **Study Suggestion 3.25:** Graph the equation

$$x^2 - 3y^2 - 2x + 6y - 5 = 0.$$ 

Ideas to Remember

Each of the conic sections can be described geometrically, as the locus of all points in the plane satisfying certain conditions, as well as algebraically, as the graph of a certain equation in two variables.

EXERCISES

In Exercises 1–41, graph the given equation.

1. $(x-2)^2+(y+1)^2=6$
2. $x^2+y^2-2x-3=0$
3. $x^2+y^2-4x+2y=2$
4. $3x^2+3y^2-3x+y=0$
5. $2(x-1)^2+2(y+1)^2=1$
6. $\sqrt{2}x^2+\sqrt{2}y^2-x-y-1=0$
7. $5x^2+5y^2-2x-y+5=0$
8. $12x^2+12y^2-12x-16y-101=0$
9. $x^2+y^2-\sqrt{6}x-\sqrt{6}y=0$
10. $\pi x^2+\pi y^2=2x$
11. $2x=y^2$
12. $(x-2)^2=3y$
13. $x=y^2+y+1$
14. $y-1=4x^2+x$
15. $-2x+3y=y^2$
16. $2y=x^2+2x$
17. $x=y^2+10y+1$
18. $2x-2y^2=1$
19. $x-2y^2-3y+1=0$
20. $\dfrac{(x-3)^2}{4}+\dfrac{(y+1)^2}{9}=1$
21. $3(x+1)^2+4(y-2)^2=12$
22. $(x+1)^2=4-2(y-2)^2$
23. $x^2+2y^2-2x-4y=1$
24. $(2x+1)^2+(y-4)^2=2$
25. $7x^2+6y^2-42=0$
26. $-9x^2=3y^2+18y$
27. $4x^2-4x=7-8y^2$
28. $10x^2+8y^2-10x-4y-37=0$
29. $2x^2+3y^2-12x+18y+39=0$
30. $\dfrac{(x-3)^2}{9}-\dfrac{(y-1)^2}{25}=1$
31. $\dfrac{(y-1)^2}{8}-x^2=1$
32. $y^2-x^2=1$
33. $(y+2)^2=3+(x+1)^2$
34. $2(x-1)^2-3(y+2)^2=1$
35. $(6y-1)^2-2(3x+2)^2=2$
36. $2x^2-y^2=2$
37. $4y^2-3x^2+2x=0$
38. $x^2-y^2-10x-6y+15=0$
39. $x^2-3y^2+2x-y=0$
40. $x^2-y^2-2\pi x+2\pi y=1$
41. $3x^2-2y^2-6x-8y-11=0$
42. Graph the equation $x^2-y^2=0$.
43. Graph the equation $xy=1$. Do you recognize the shape of this curve?
44. Find the equation of the ellipse with center $(0, 0)$, x-intercepts ± 3, and y-intercepts ± 10.
45. Find the equation of the ellipse with center $(0, 0)$, x-intercepts ± 5, and y-intercepts ± 2.
46. Find the equation of the hyperbola having asymptotes $y=\pm 2x$ and x-intercepts ± 5.
47. Find the equation of the hyperbola having asymptotes $y=2x$ and $y=4-2x$, center $(1, 2)$, and x-intercept $1+\sqrt{2}$.
48. A square whose sides are parallel to the coordinate axes is inscribed inside the ellipse whose equation is

$$\frac{x^2}{a^2}+\frac{y^2}{b^2}=1$$

Express the area of the square in terms of a and b.

49. The graphs of the equations

$$\frac{x^2}{a^2}-\frac{y^2}{b^2}=1 \quad \text{and} \quad \frac{y^2}{b^2}-\frac{x^2}{a^2}=1$$

are called **conjugate hyperbolas.** Sketch these two hyperbolas on the same plane for $a=4$ and $b=9$. Describe the relationship between these curves.

3.6 Review

CONCEPTS FOR REVIEW

Slope of a line
Negative reciprocal
Slope-intercept form of the equation of a line
Point-slope form of the equation of a line
General form of the equation of a line
Quadratic function
Parabola
Axis of symmetry of a parabola
Extreme point of a parabola
Standard form of the equation of a parabola
Maximum value of a function
Minimum value of a function
Extreme value of a function
Ellipse
Foci of an ellipse
Major axis of an ellipse
Minor axis of an ellipse
Standard form of the equation of an ellipse
Hyperbola
Foci of a hyperbola
Center of a hyperbola
Vertices of a hyperbola
Transverse axis of a hyperbola
Standard form of the equation of a hyperbola
Asymptotes of a hyperbola

REVIEW EXERCISES

1. Find the slope of the line through the points (5, 1) and (7, −2). Sketch the line.

2. Find the slope of the line through the point (6, 2) that is parallel to the line through the points (6, 4) and (2, −3). Sketch the line.

3. Find the slope of the line through the point (−5, 4) that is perpendicular to the line through the points (2, 4) and (9, 6). Sketch the line.

4. Find all values of a for which the line through the points $(a, 1)$ and $(2, -a)$ is perpendicular to the line through $(1, 1)$ and $(-3a, 2a)$.

In Exercises 5–10, find the **(a)** slope-intercept form; **(b)** point-slope form; and **(c)** general form of the equation of the line with the given properties.

5. slope = 12, y-intercept = 7

6. slope = 4, line goes through (−3, 1/2)

7. line goes through (1, 4) and $(-1/2, \sqrt{2})$

8. line goes through (1, 7) and is parallel to the line that has equation $5x + 3y - 4 = 0$

9. line goes through (−2, 4) and is perpendicular to the line that has equation $3x - 2y + 7 = 0$

10. the line is the perpendicular bisector of the line segment from (4, 2) to (8, −9).

In Exercises 11–16, graph the given equation, labeling all intercepts and the extreme point.

11. $y = 7x^2 + 3x + 1$

12. $y = 2(x - 3)(3x - 2)$

13. $y = -x^2 - x + 1$

14. $y = \pi(x^2 + 2x + 1)$

15. $y = 12 - 12x - x^2$

16. $y = 5 - 4x^2$

In Exercises 17–20, solve the given inequality.

17. $2x^2 - 4x + 3 \geq 0$

18. $x^2 - 5x + 1 \geq 0$

19. $-1 \leq x^2 + 3x - 1 \leq 4$

20. $-\sqrt{2} < 1 - x^2 < \sqrt{2}$

21. A gardener wishes to fence off a rectangular plot of land and divide the plot into three pieces, as shown below. If he has 100 feet of fence, by what dimensions should he make the plot in order to maximize its area?

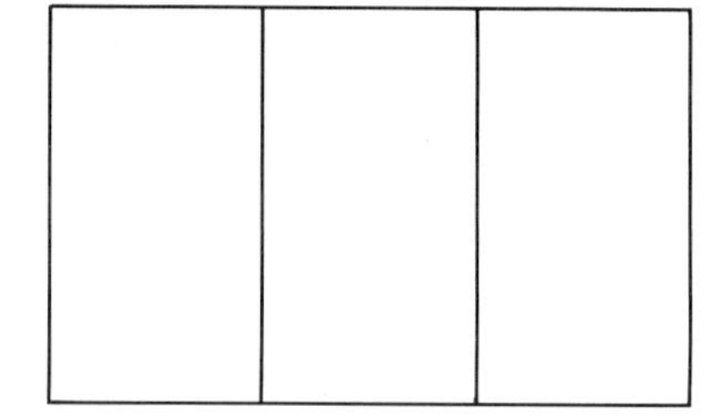

22. A restaurant charges \$10.00 for its special dinner, and at this price, the restaurant usually sells 100 dinners per night. The owner of the restaurant estimates that for each 25 cents that she raises the price of the special, she will lose 3 customers. At what price should she set the special in order to maximize her revenue? What is the maximum revenue?

23. Suppose that you have a piece of wire 6 feet long, which you want to cut into two pieces. With one piece, you want to make a square, and with the other piece you want to make a rectangle whose length is 3 times its width. How should you cut the wire in order to minimize the total area enclosed by the square and the rectangle?

In Exercises 24–34, sketch the graph of the given equation.

24. $(x-4)^2+(y+1)^2=7$

25. $3x^2+3y^2-\sqrt{2}x+\sqrt{2}y=0$

26. $x=2y^2+y-1$

27. $x+3y^2-4y=0$

28. $(x-20)^2+(3y-1)^2=1$

29. $\dfrac{(x-1)^2}{3}+y^2=1$

30. $3x^2+2y^2-4x-1=0$

31. $x^2+2y^2-3x+4y=0$

32. $x^2-y^2=5$

33. $y^2-x^2+2\sqrt{2}y+6x-13=0$

34. $y^2-2x^2+(2/3)y-4x-89/9=0$

The **eccentricity** of an ellipse whose equation is

$$\frac{x^2}{a^2}+\frac{y^2}{b^2}=1$$

is defined to be the number $e=\sqrt{|a^2-b^2|}/a$. (Notice the absolute value bars to make the quantity under the square root sign nonnegative.) As the next exercises show, the eccentricity is a measure of how "flat" the ellipse is.

35. Show that the eccentricity satisfies $0 \le e \le 1$.

36. What is the eccentricity of a circle? (Take $a=b=r$.)

37. What is the eccentricity of the "flat" ellipse

$$\frac{x^2}{100}+y^2=1$$

Sketch the graph of this ellipse.

38. Give an argument to show that the "flatter" the ellipse, the closer the eccentricity is to 1, and the "rounder" the ellipse, the closer the eccentricity is to 0.

4
POLYNOMIAL FUNCTIONS OF HIGHER DEGREE

4.1 Division of Polynomials

In this chapter, we continue our study of polynomials and polynomial functions. We will concentrate primarily on the problem of finding the (real number) solutions to **polynomial equations,** that is, to equations of the form

$$a_nx^n + a_{n-1}x^{n-1} + \cdots + a_0 = 0 \tag{4.1}$$

where $p(x) = a_nx^n + a_{n-1}x^{n-1} + \cdots + a_0$ is a polynomial. If $p(x)$ has degree n, then we will say that the *equation* $p(x) = 0$ has degree n as well.

If $p(x)$ is either linear or quadratic, then we can find the solutions to Equation (4.1) by using the methods of Chapter 1. However, if the degree of $p(x)$ is greater than or equal to 3, the problem of finding the solutions to Equation (4.1) becomes much more difficult. In the case where the degree of $p(x)$ is equal to 3 or 4, there are formulas similar to the quadratic formula, for the solutions to Equation (4.1). But these formulas are *very* complicated, and consequently not of much use.

The situation is even worse when the degree of $p(x)$ exceeds 4. Not only are there no general formulas similar to the quadratic formula, for solving polynomial equations of degree greater than 4, but it can be shown that *no such formula can ever exist*!

This is quite a remarkable statement, and to verify it takes some very advanced mathematics. Nevertheless it is true, and it leaves us to face the fact that there is no general method for finding the solutions to polynomial equations. However, there are some techniques that can be of great help in solving specific polynomial equations, and our goal in this chapter is to discuss some of these techniques.

Let us begin our discussion with a bit of terminology. A solution to the polynomial equation $p(x) = 0$ is also called a **root** of the equation, or a **zero** of the polynomial $p(x)$. Thus, for example, the number 2 is a *root* of the polynomial equation

$$x^2 - x - 2 = 0$$

and a *zero* of the polynomial $p(x) = x^2 - x - 2$.

We will devote the remainder of this section to a discussion of division of polynomials. In order to set our terminology, let us look at a problem in division of real numbers

$$\begin{array}{r@{\;}r@{\;}l}
 & 156 & \longleftarrow \text{quotient} \\
\text{divisor} \longrightarrow 15 & \overline{)\,2344} & \longleftarrow \text{dividend} \\
 & \underline{15} & \\
 & 84 & \\
 & \underline{75} & \\
 & 94 & \\
 & \underline{90} & \\
 & 4 & \longleftarrow \text{remainder}
\end{array}$$

In this example, the number 15 is called the **divisor,** 2344 is called the **dividend,** 156 is called the **quotient,** and 4 is called the **remainder.** Of course, the remainder must satisfy the inequalities

$$0 \leq \text{remainder} < \text{divisor}$$

The result of this division can be expressed in either of the forms

$$\frac{2344}{15} = 156 + \frac{4}{15}$$

or

$$2344 = 156 \cdot 15 + 4$$

The last form is particularly important, and we can generalize it by writing

$$\begin{gathered}\textbf{dividend} = \textbf{quotient} \cdot \textbf{divisor} + \textbf{remainder} \\ \textbf{where } 0 \leq \textbf{remainder} < \textbf{divisor}\end{gathered} \tag{4.2}$$

A formula similar to (4.2) also holds for polynomials, and this formula will help us a great deal when it comes to solving polynomial equations. In fact, division of polynomials follows the same general pattern as division of numbers. Let us consider some examples.

EXAMPLE 1: Divide the polynomial $2x^3 + x^2 - 3x + 4$ (the dividend) by the polynomial $x + 3$ (the divisor). Then express the result in the form (4.2).

Solution: First we must be certain that the terms of each polynomial are arranged in descending powers of x. This has already been done for us in this example.

$$x + 3 \overline{\big)\, 2x^3 + x^2 - 3x + 4}$$

Now we divide the *first term* of the divisor into the *first term* of the dividend, and put the result above the first term of the dividend.

$$\begin{array}{r l} & 2x^2 \\ x + 3 & \overline{\big)\, 2x^3 + x^2 - 3x + 4} \end{array}$$

Then we multiply the last term of the quotient obtained so far (in this case $2x^2$) by the *entire* divisor, and place the result below the dividend. Then we subtract and bring down the next term in the dividend.

$$\begin{array}{r l} & 2x^2 \\ x + 3 & \overline{\big)\, 2x^3 + x^2 - 3x + 4} \\ & \underline{2x^3 + 6x^2} \\ & - 5x^2 - 3x \end{array}$$

Repeating these steps, we have

$$\begin{array}{r l} & 2x^2 - 5x + 12 \\ x + 3 & \overline{\big)\, 2x^3 + x^2 - 3x + 4} \\ & \underline{2x^3 + 6x^2} \\ & - 5x^2 - 3x \\ & \underline{- 5x^2 - 15x} \\ & 12x + 4 \\ & \underline{12x + 36} \\ & - 32 \end{array}$$

The division process stops when the remainder is equal to 0, or has degree *less* than the degree of the divisor. Thus, in this case, the quotient is $2x^2 - 5x + 12$ and the remainder is the constant polynomial -32. This can be expressed in the form (4.2) as follows

$$\underset{\text{dividend}}{\underset{\uparrow}{2x^3 + x^2 - 3x + 4}} = \underset{\text{quotient}}{\underset{\uparrow}{(2x^2 - 5x + 12)}}\underset{\text{divisor}}{\underset{\uparrow}{(x + 3)}} - \underset{\text{remainder}}{\underset{\uparrow}{32}}$$

Try Study Suggestion 4.1. ☐

▶ **Study Suggestion 4.1:** Divide the polynomial $3x^3 - 2x^2 + x - 1$ by $x - 2$. Then express the result in the form (4.2). ◀

EXAMPLE 2: Divide $x^4 + 2x^3 + x - 2$ by $x^2 - 1$ and express the result in the form (4.2).

Solution: This division follows the same pattern as the previous one, and so we will not include detailed comments. Notice, however, that we have included a term of the form $0x$ in the divisor, and a term of the form $0x^2$

in the dividend. These terms help keep proper alignment during the division process.

$$\begin{array}{r|l} & x^2 + 2x + 1 \\ \hline x^2 + 0x - 1 & x^4 + 2x^3 + 0x^2 + x - 2 \\ & x^4 + 0x^3 - x^2 \\ \hline & \quad 2x^3 + x^2 + x \\ & \quad 2x^3 + 0x^2 - 2x \\ \hline & \qquad x^2 + 3x - 2 \\ & \qquad x^2 + 0x - 1 \\ \hline & \qquad\qquad 3x - 1 \end{array}$$

In this case, the quotient is $x^2 + 2x + 1$ and the remainder is $3x - 1$. We can express this in the form (4.2) as follows

$$x^4 + 2x^3 + x - 2 = (x^2 + 2x + 1)(x^2 - 1) + 3x - 1$$

Try Study Suggestion 4.2. □

▶ **Study Suggestion 4.2:** Divide $2x^5 - 3x^2 + x + 1$ by $2x^2 + 1$. Then express the result in the form (4.2). ◀

For future reference, let us summarize division of polynomials in a theorem.

THEOREM

Let $a(x)$ and $b(x)$ be polynomials. Then there are polynomials $q(x)$ and $r(x)$ for which

$$b(x) = q(x)a(x) + r(x)$$

where either $r(x) = 0$ or

$$0 \le \deg r(x) < \deg a(x)$$

The polynomial $a(x)$ is called the **divisor,** $b(x)$ is called the **dividend,** $q(x)$ is called the **quotient** and $r(x)$ is called the **remainder.**

We will have several occasions in this chapter to divide by a polynomial of the form $x - a$, where a is a constant. For such divisions, there is a shortcut method, which is known as **synthetic division.** Let us demonstrate synthetic division with two examples, the first of which is a repeat of Example 1.

EXAMPLE 3: Use synthetic division to divide $2x^3 + x^2 - 3x + 4$ by $x + 3$.

Solution: We begin all synthetic division problems by writing the divisor in the form $x - a$. In this case, we have $x + 3 = x - (-3)$, and so $a = -3$. (It would be *incorrect* to use $a = 3$.) Making certain that the terms in the dividend are arranged in descending powers of x, we write a in place of the divisor, and the coefficients of the dividend in place of the dividend itself.

$$-3 \,|\, 2 \quad 1 \quad -3 \quad 4$$

Now we leave a blank row under the dividend, draw a horizontal line, and bring down the first coefficient of the dividend.

$$\begin{array}{r|rrrrl} -3 & 2 & 1 & -3 & 4 & \leftarrow \text{Row 1} \\ & & & & & \leftarrow \text{Row 2} \\ \hline & 2 & & & & \leftarrow \text{Row 3} \end{array}$$

The next step is to multiply the last number in the third row (2) by a (-3), and place the result (-6) in the second row. Then we add the numbers in the second column.

$$\begin{array}{r|rrrr} -3 & 2 & 1 & -3 & 4 \\ & & -6 & & \\ \hline & 2 & -5 & & \end{array}$$

This process is repeated until the third row has the same number of elements as the first row. (The next step is to multiply -5 by -3, put the result in the third column, and then add.)

$$\begin{array}{r|rrrr} -3 & 2 & 1 & -3 & 4 \\ & & -6 & 15 & -36 \\ \hline & 2 & -5 & 12 & -32 \end{array}$$

quotient $= 2x^2 - 5x + 12$

remainder $= -32$

When the synthetic division process is complete, the third row will contain the coefficients of the quotient and the remainder. Of course, we obtain the same result here as in Example 1. *Try Study Suggestion 4.3.* □

▶ **Study Suggestion 4.3:** Repeat Study Suggestion 4.1 using synthetic division. ◀

EXAMPLE 4: Use synthetic division to divide $x^7 - 3x^5 + 2x^2 - 1$ by $x - 2$.

Solution: In this case, $x - a = x - 2$, and so $a = 2$. The procedure is the same as in the previous example. Notice however, that we have included 0's to keep the place of "missing" terms in the dividend.

$$\begin{array}{r|rrrrrrrr} 2 & 1 & 0 & -3 & 0 & 0 & 2 & 0 & -1 \\ & & 2 & 4 & 2 & 4 & 8 & 20 & 40 \\ \hline & 1 & 2 & 1 & 2 & 4 & 10 & 20 & 39 \end{array}$$

Thus, the quotient is

$$q(x) = x^6 + 2x^5 + x^4 + 2x^3 + 4x^2 + 10x + 20$$

and the remainder is $r(x) = 39$. To appreciate the time and space saved by using synthetic division, all we have to do is compare this with the ordinary method of division

$$
\begin{array}{r}
x^6 + 2x^5 + x^4 + 2x^3 + 4x^2 + 10x + 20 \\
x - 2 \overline{\big)\, x^7 + 0x^6 - 3x^5 + 0x^4 + 0x^3 + 2x^2 + 0x - 1} \\
\underline{x^7 - 2x^6} \qquad\qquad\qquad\qquad\qquad\qquad\qquad\qquad\quad \\
2x^6 - 3x^5 \qquad\qquad\qquad\qquad\qquad\qquad\quad \\
\underline{2x^6 - 4x^5} \qquad\qquad\qquad\qquad\qquad\qquad\quad \\
x^5 + 0x^4 \qquad\qquad\qquad\qquad\qquad \\
\underline{x^5 - 2x^4} \qquad\qquad\qquad\qquad\qquad \\
2x^4 + 0x^3 \qquad\qquad\qquad\quad \\
\underline{2x^4 - 4x^3} \qquad\qquad\qquad\quad \\
4x^3 + 2x^2 \qquad\qquad\quad \\
\underline{4x^3 - 8x^2} \qquad\qquad\quad \\
10x^2 + 0x \qquad\quad \\
\underline{10x^2 - 20x} \qquad\quad \\
20x - 1 \\
\underline{20x - 40} \\
39
\end{array}
$$

Try Study Suggestion 4.4. □

▶ **Study Suggestion 4.4:** Use synthetic division to divide $4x^5 - 2x^2 + x - 1$ by $x + 3$. Then perform the same division without using synthetic division. Which method do you prefer? ◀

Ideas to Remember

- Division of polynomials is very similar to division of real numbers. In fact, we use the same terms—quotient, remainder, dividend, and divisor—in both cases.
- When dividing by a polynomial of the form $x - a$, synthetic division can save both time and space.

EXERCISES

In Exercises 1–25, divide $b(x)$ by $a(x)$. Do *not* use synthetic division. Express the results in the form (4.2).

1. $b(x) = 6x^2 + 5x - 4$, $a(x) = 3x + 4$

2. $b(x) = 9x^2 + 1$, $a(x) = x^2 - 2x + 1$

3. $b(x) = x^3 - 2x^2 + x - 4$, $a(x) = x - 1$

4. $b(x) = 3x^3 - 2x^2 + 1$, $a(x) = x + 3$

5. $b(x) = x^3 - 4x^2 + 3x$, $a(x) = x - 1$

6. $b(x) = 5x^3 - 17x^2 + 31x - 10$, $a(x) = 5x - 2$

7. $b(x) = -12x^3 + 4x^2 - 1$, $a(x) = 4x^2 + 1$

8. $b(x) = x^5 - 1$, $a(x) = x^4 - 1$

9. $b(x) = 2x^3 + 11x^2 - 7x - 6$, $a(x) = x^2 + 5x - 6$

10. $b(x) = 10x^3 + 9x^2 - 3x - 2$, $a(x) = 10x^2 - x - 2$

11. $b(x) = x - 1$, $a(x) = 2x^2 + 3x - 1$

12. $b(x) = 4x^2 - 3x + 1$, $a(x) = 4x^2 + 3x + 1$

13. $b(x) = a^2x^2 - 3ax - 10$, $a(x) = ax - 5$

14. $b(x) = x^3 + a^3$, $a(x) = x - a$

15. $b(x) = x^3 - a^3$, $a(x) = x^2 + ax + a$

16. $b(x) = 4x^4 - 4x^3 + 29x^2 - 5x - 30$, $a(x) = 4x^2 + 5$

17. $b(x) = x^4 - 3x^2 + 2x - 1$, $a(x) = x^3 - 1$

18. $b(x) = -2x^4 + x^2 - 2x + 1$, $a(x) = x^4 + 1$

19. $b(x) = x^4 + 2x^3 - 3x^2 + x + 1$, $a(x) = x - 1$

20. $b(x) = x^5 - 3x^4 + 2x - 1$, $a(x) = x^2 - 1$

21. $b(x) = -x^4 - x^3 + 2x^2 + x + 3$, $a(x) = x^2 - 2x$

22. $b(x) = 2x^4 + 3x^3 - x^2 + 1$, $a(x) = 4x^3 - 2x + 1$

23. $b(x) = 2x^{12} + 3x^2 - 9$, $a(x) = x^4 - 3$

24. $b(x) = x^4 + a^3x + a^2x^2 + a^3x + a^4$, $a(x) = x - a$

25. $b(x) = x^5 - a$, $a(x) = x^2 - a$

In Exercises 26–42, use synthetic division to divide $b(x)$ by $a(x)$. Express the results in the form (4.2).

26. $b(x) = 7x^2 - 4x + 1$, $a(x) = x - 3$

27. $b(x) = x^2$, $a(x) = x - 2$

28. $b(x) = 24 - 13x - 2x^2$, $a(x) = 1 + x$

29. $b(x) = x^3 + 2x^2 - x + 1$, $a(x) = x + 2$

30. $b(x) = x^3 - 12$, $a(x) = x + 1$

31. $b(x) = 2x^3 + x^2 - x + 1$, $a(x) = x + 3$

32. $b(x) = x^4 - 1$, $a(x) = x - 1$

33. $b(x) = 32 - x^5$, $a(x) = x - 2$

34. $b(x) = x^{10} + 3x^5 - x^2 + 1$, $a(x) = x + 1$

35. $b(x) = x^4 - a^4$, $a(x) = x - a$

36. $b(x) = x^4 + 2ax + a^2$, $a(x) = x - a$

37. $b(x) = 2x^3 - x^2 + 3x - 1$, $a(x) = x - 1/2$

38. $b(x) = (1/3)x^2 - (2/3)x + 1/6$, $a(x) = x + 1/3$

39. $b(x) = 4x^3 + 2x^2 - 6x + 8$, $a(x) = 2x + 4$

40. $b(x) = 3x^3 - 6x^2 + 9$, $a(x) = 3x - 9$

41. $b(x) = 4x^2 - x - 1$, $a(x) = 2x - 1$ (*Hint:* First divide $b(x)$ and $a(x)$ by a certain constant.)

42. $b(x) = 4x^3 - 2x^2 + 1$, $a(x) = 2x - 1$

43. Use synthetic division to divide $p(x) = ax^2 + bx + c$ by $q(x) = x - r$, where a, b, c, and r are constants. How can you express the remainder? Express the result in the form (4.2).

4.2 The Remainder Theorem and the Factor Theorem

Let us begin this section by taking a close look at the theorem of the previous section in the case where the divisor has the form $x - a$. According to that theorem, if $p(x)$ is a polynomial, then there are polynomials $q(x)$ and $r(x)$ for which

$$p(x) = q(x)(x - a) + r(x) \tag{4.3}$$

where either $r(x) = 0$ or

$$0 \leq \deg r(x) < 1 \tag{4.4}$$

But condition (4.4) is the same as saying that $\deg r(x) = 0$. Thus, in either case, the remainder $r(x)$ must be a *constant*. If we denote this constant by r, then we can rewrite Equation (4.3) as follows

$$p(x) = q(x)(x - a) + r \tag{4.5}$$

where $q(x)$ is a polynomial of degree *one less* than the degree of $p(x)$.

Making the substitution $x = a$ in Equation (4.5) gives

$$p(a) = q(a)(a - a) + r$$

or

$$p(a) = r$$

This says that the remainder r, when $p(x)$ is divided by $x - a$, is equal to $p(a)$. This important fact is known as the **Remainder Theorem.**

THE REMAINDER THEOREM

If a polynomial $p(x)$ is divided by $x - a$, the remainder is equal to $p(a)$.

The Remainder Theorem has several important consequences, one of which is that it gives us another way to compute $p(a)$. Let us consider an example.

EXAMPLE 1: Suppose that we want to compute $p(5)$, where

$$p(x) = x^4 + 3x^3 - 7x^2 + 9x - 1$$

Of course, one way to do this is simply to make the substitution $x = 5$,

$$\begin{aligned} p(5) &= 5^4 + 3 \cdot 5^3 - 7 \cdot 5^2 + 9 \cdot 5 - 1 \\ &= 625 + 375 - 175 + 45 - 1 \\ &= 869 \end{aligned}$$

Another way, however, is to use the Remainder Theorem. According to this theorem, $p(5)$ is equal to the remainder obtained by dividing $p(x)$ by $x - 5$. Using synthetic division, we have

$$\begin{array}{r|rrrrr} 5 & 1 & 3 & -7 & 9 & -1 \\ & & 5 & 40 & 165 & 870 \\ \hline & 1 & 8 & 33 & 174 & 869 \end{array}$$

Thus, the remainder is 869, and so $p(5) = 869$. We will leave it to you to decide which method you think is easier. However, we think you will agree that, at least in this case, the two methods are competitive.

Try Study Suggestion 4.5. □

▶ **Study Suggestion 4.5:** Use the Remainder Theorem to compute $p(-3)$, where $p(x) = x^5 - 4x^3 + x^2 - x + 1$. ◀

Recall that our main goal in this chapter is to learn what we can about solving polynomial equations. The Remainder Theorem can help us with this goal as well. For a real number a is a solution to the polynomial equation

$$p(x) = 0 \tag{4.6}$$

if and only if $p(a) = 0$. But according to the Remainder Theorem, $p(a)$ is the remainder when $p(x)$ is divided by $x - a$. Thus $x = a$ is a solution to Equation (4.6) if and only if the remainder when $p(x)$ is divided by $x - a$ is equal to 0; that is, if and only if $x - a$ is a *factor* of $p(x)$. This statement is one of the keys to solving polynomial equations, and is known as the **Factor Theorem.**

THE FACTOR THEOREM

Let $p(x)$ be a polynomial. Then $x = a$ is a solution to the equation

$$p(x) = 0 \tag{4.7}$$

if and only if $x - a$ is a factor of $p(x)$. Put another way, a is a solution to Equation (4.7) if and only if $p(x)$ can be written in the form

$$p(x) = q(x)(x - a)$$

where $q(x)$ is a polynomial of degree *one less* than the degree of $p(x)$.

The Factor Theorem can be very useful when it comes to solving polynomial equations. For if we know one solution to the polynomial equation

$$p(x) = 0 \tag{4.8}$$

say $x = a$, then according to the Factor Theorem this equation can be written in the form

$$q(x)(x - a) = 0$$

where $q(x)$ has degree one less than $p(x)$. But the solutions to this equation are $x = a$ and the solutions to the polynomial equation

$$q(x) = 0 \tag{4.9}$$

Since Equation (4.9) has degree one less than the degree of Equation (4.8); it should (at least in principle) be easier to solve than Equation (4.8). Let us consider an example.

EXAMPLE 2: Given the fact that $x = 2$ is a solution to the equation

$$x^3 + x^2 - 11x + 10 = 0 \tag{4.10}$$

find all solutions to this equation.

Solution: Since $x = 2$ is a root of the polynomial $p(x) = x^3 + x^2 - 11x + 10$, the Factor Theorem tells us that $x - 2$ is a factor of $p(x)$. Using synthetic division, we get

$$\begin{array}{r|rrrr} 2 & 1 & 1 & -11 & 10 \\ & & 2 & 6 & -10 \\ \hline & 1 & 3 & -5 & 0 \end{array}$$

(As expected, the remainder is 0.) This gives us the factorization

$$p(x) = (x^2 + 3x - 5)(x - 2)$$

and Equation (4.10) can be written in the form

$$(x^2 + 3x - 5)(x - 2) = 0$$

Now, the solutions to this equation are $x = 2$, as well as the solutions to the polynomial equation

$$x^2 + 3x - 5 = 0$$

But this is just a quadratic equation, whose solutions can be found by the quadratic formula. We will leave it to you to show that they are

$$x = \frac{-3 - \sqrt{29}}{2} \quad \text{and} \quad x = \frac{-3 + \sqrt{29}}{2}$$

and so the solutions to Equation (4.10) are

$$x = 2, \quad x = \frac{-3 - \sqrt{29}}{2}, \quad \text{and} \quad x = \frac{-3 + \sqrt{29}}{2}$$ □

As the previous example shows, the process of "reducing" the degree of the equation pays off especially well in the case of a cubic equation (degree $= 3$). For in this case, the "reduced" Equation (4.9) is a quadratic equation, and we can always find its solutions (if any) by using the quadratic formula. Thus, *if we know one solution to a cubic equation, we can always find all of its solutions.*

▶ **Study Suggestion 4.6:** Given that $x = -1$ is a solution to the equation $x^3 + 5x^2 + 2x - 2 = 0$, find all the solutions to this equation. ◀

Try Study Suggestion 4.6.

The same principle can be applied to cases where we know several of the solutions to a polynomial equation.

EXAMPLE 3: Given the fact that both $x = -2$ and $x = 3$ are solutions to the equation

$$6x^4 - 11x^3 - 35x^2 + 34x + 24 = 0 \qquad \textbf{(4.11)}$$

find all the solutions to this equation.

Solution: Since $x = -2$ is a solution to Equation (4.11), we know from the Factor Theorem that $x - (-2) = x + 2$ is a factor of the polynomial

$$p(x) = 6x^4 - 11x^3 - 35x^2 + 34x + 24$$

Synthetic division gives

−2	6	−11	−35	34	24
		−12	46	−22	−24
	6	−23	11	12	0

and so we have the factorization

$$p(x) = (x + 2)(6x^3 - 23x^2 + 11x + 12)$$

Hence, the solutions to Equation (4.11) are $x = -2$, together with the solutions to the *cubic* equation

$$6x^3 - 23x^2 + 11x + 12 = 0$$

Now, we are given the fact that $x = 3$ is a solution to Equation (4.11), and so $x = 3$ must also be a solution to this cubic equation. (Why?) Therefore, we may again use the Factor Theorem to conclude that $x - 3$ is a factor of the cubic polynomial

$$q(x) = 6x^3 - 23x^2 + 11x + 12$$

Synthetic division gives

$$\begin{array}{r|rrrr} 3 & 6 & -23 & 11 & 12 \\ & & 18 & -15 & -12 \\ \hline & 6 & -5 & -4 & 0 \end{array}$$

and so

$$6x^3 - 23x^2 + 11x + 12 = (x - 3)(6x^2 - 5x - 4)$$

We can use this to write $p(x)$ in the form

$$p(x) = (x + 2)(x - 3)(6x^2 - 5x - 4)$$

and so the solutions to Equation (4.11) are $x = -2$, $x = 3$, and the solutions to the *quadratic* equation

$$6x^2 - 5x - 4 = 0$$

This quadratic has solutions $x = -1/2$ and $x = 4/3$, and so the solutions to Equation (4.11) are

$$x = -2, \qquad x = 3, \qquad x = -1/2, \qquad \text{and} \qquad x = 4/3$$

(Of course, these solutions should be verified by direct substitution in order to catch any potential mistakes in arithmetic.) *Try Study Suggestion 4.7.* □

▶ **Study Suggestion 4.7:** Given that $x = 1$ and $x = -4$ are roots of the polynomial $x^4 + 2x^3 - 13x^2 - 14x + 24$, find all the roots of this polynomial. ◀

This is a good time to discuss another matter relating to the solutions of a polynomial equation. Consider, for example, the polynomial equation

$$x^3 - 3x^2 + 4 = 0 \tag{4.12}$$

The polynomial on the left-hand side can be factored to give

$$(x + 1)(x - 2)^2 = 0$$

Thus the solutions to (4.12) are $x = -1$ and $x = 2$. However, because the factor $x - 2$ is raised to the second power, we say that $x = 2$ is a solution of *multiplicity* 2.

More generally, if a polynomial equation can be written in the form

$$(x - a)^n q(x) = 0$$

where $x - a$ is *not* a factor of $q(x)$; that is, where $q(a)$ is *not* 0, then we say that $x = a$ is a solution of **multiplicity** n. Loosely speaking, the multiplicity of a solution $x = a$ to the equation $p(x) = 0$ is the *largest* power of $x - a$ that can be factored out of the polynomial $p(x)$.

When a solution to a polynomial equation has multiplicity 1, it is called a **simple root.** Solutions of multiplicity greater than 1 are called **multiple roots.** Solutions of multiplicity 2 are referred to as **double roots,** and solutions of multiplicity 3 are referred to as **triple roots.**

For example, the polynomial equation

$$x^6 - 10x^5 + 36x^4 - 54x^3 + 27x^2 = 0$$

can be factored as follows

$$x^2(x - 3)^3(x - 1) = 0$$

Thus, $x = 1$ is a simple root, $x = 0$ is a double root, and $x = 3$ is a triple root.

▶ **Study Suggestion 4.8:** State the multiplicity of the root $x = 2$ in each of the following equations.

(a) $(x - 2)^2(x^2 + x + 1) = 0$

(b) $(x - 2)^3(x^2 - x - 2) = 0$ ◀

Try Study Suggestion 4.8.

The Factor Theorem conceals a very interesting fact about the number of solutions to a polynomial equation

$$p(x) = 0 \tag{4.13}$$

where $p(x)$ is a polynomial of degree n. As we now know, if $x = a$ is a solution to this equation, then $x - a$ is a factor of $p(x)$. In fact, if $x = a$ is a root of multiplicity m, then we can factor out m linear factors, each of the form $x - a$. But $p(x)$, being a polynomial of degree n, can have *at most n* linear factors. (It may have less than n linear factors. For example, the quadratic polynomial $x^2 + 1$ has *no* linear factors.)

Therefore, Equation (4.13) can have *at most n* solutions, where we are counting multiple roots as many times as their multiplicities indicate. Let us state this important fact in a theorem.

THEOREM

If $p(x)$ is a polynomial of degree n, then the polynomial equation

$$p(x) = 0$$

can have *at most n* solutions, where we count any multiple roots as many times as their multiplicities indicate.

Ideas to Remember

- If $p(x)$ is a polynomial and a is a real number, then $p(a)$ can be computed by dividing $p(x)$ by $x - a$, rather than by using direct evaluation.
- There is a very close association between the solutions of a polynomial equation and its linear factors. This association is described by the Factor Theorem.

EXERCISES

In Exercises 1–10, use the Remainder Theorem to compute $p(a)$.

1. $p(x) = 2x^2 - 3x + 1$, $a = 5$

2. $p(x) = -x^2 + 3x - 10$, $a = 20$

3. $p(x) = x^3 - 4x^2 + 5x - 1$, $a = 3$

4. $p(x) = 2x^3 - x^2 + x - 5$, $a = -1$

5. $p(x) = x^4 - 3x^3 + 2x^2 - 1$, $a = 2$

6. $p(x) = 12x^4 - 7x^3 + 5x^2 - x + 10$, $a = -2$

7. $p(x) = x^{10} - 1$, $a = 1$

8. $p(x) = x^{10} - x^5 + 1$, $a = 2$

9. $p(x) = 3x^2 + 2x - 1$, $a = 1/2$

10. $p(x) = 4x^3 - 7x^2 + 2x - 1$, $a = 1/4$

In Exercises 11–18, use the Factor Theorem to show that the given value of a is a solution to the given equation.

11. $x^2 - 3x + 2 = 0$, $a = 1$

12. $2x^2 - 3x - 14$, $a = -2$

13. $9x^2 + 18x - 7$, $a = 1/3$

14. $x^3 - 2x^2 + 3x - 2$, $a = 1$

15. $-2x^3 + x^2 - 4x - 7$, $a = -1$

16. $4x^3 - 7x^2 - 45$, $a = 3$

17. $x^4 - x^3 - 3x^2 + 3x - 2$, $a = 2$

18. $16x^4 + 16x^3 + 16x^2 + 16x + 5 = 0$, $a = -1/2$

In Exercises 19–23, use the Factor Theorem to show that $q(x)$ is a factor of $p(x)$.

19. $p(x) = 2x^2 - 3x - 2$, $q(x) = x - 2$

20. $p(x) = x^3 + 3x^2 - 2x - 4$, $q(x) = x + 1$

21. $p(x) = 8x^3 - 16x + 7$, $q(x) = x - 1/2$

22. $p(x) = 27x^3 - 81x + 26$, $q(x) = 3x - 1$

23. $p(x) = 25x^2 - 75x + 26$, $q(x) = 5x - 2$

24. Prove that $x - y$ is a factor of $x^n - y^n$.

25. Under what conditions is $x + y$ a factor of $x^n + y^n$?

26. Prove that $x^4 + 3x^2 + 10$ has no factors of the form $x - a$.

27. Find all values of a for which $6x^2 - a^2x - a$ is divisible by $x - 1$.

28. Find all values of a for which $x^2 - 3a^2x + a$ is divisible by $x - 2$.

29. Show that $x = 2$ is a triple root of the equation $x^3 - 6x^2 + 12x - 8 = 0$.

30. Show that $x = -1$ is a double root of the equation $x^4 - x^3 - 4x^2 - x + 1 = 0$.

31. Show that $x = 1/2$ is a root of multiplicity 1 of the equation $2x^5 - x^4 + 2x^3 - x^2 + 6x - 3 = 0$.

In Exercises 32–40, the given numbers are roots of the given polynomial $p(x)$. Find all of the roots of the polynomial. Give the multiplicity of each root and write the polynomial as a product of irreducible factors. (Recall from Section 1.3 that a polynomial is *irreducible* if it has no nontrivial factors.)

32. $p(x) = x^3 - 2x^2 - 5x + 6$, $r = 1$

33. $p(x) = x^3 - 12x + 16$, $r = 2$

34. $p(x) = 6x^3 + 7x^2 - 9x + 2$, $r = 1/2$

35. $p(x) = 7x^3 + 11x^2 + 18x + 8$, $r = -4/7$

36. $p(x) = 27x^3 - 108x^2 + 144x - 64$, $r = 4/3$

37. $p(x) = x^4 - 7x^3 + 11x^2 + 7x - 12$, $r_1 = 1$, $r_2 = -1$

38. $p(x) = x^4 - 3x^3 + x^2 + 3x - 2$, $r_1 = 1$, $r_2 = 2$

39. $p(x) = x^4 - 2x^3 - 3x^2 + 4x + 4$, $r_1 = -1$, $r_2 = 2$

40. $p(x) = x^4 - 1$, $r = 1$

4.3 Rational Roots

Many of the polynomial equations that occur in applications have the property that all of their coefficients are rational numbers. For a polynomial equation with rational *coefficients*, there is a systematic method for finding its rational solutions. (By rational solutions, we mean solutions that are rational numbers.)

A polynomial equation with rational coefficients may have solutions that are not rational. For example, the equation

$$x^3 + x^2 - 2x - 2 = 0 \tag{4.14}$$

has solutions $x = 1$, $x = -\sqrt{2}$ and $x = \sqrt{2}$. Hence, it has one rational solution and two irrational solutions. Nevertheless, being able to find all of the rational solutions to a polynomial equation is a big step forward. For instance, in the case of Equation (4.14), once we have found the rational solution $x = 1$, we can use the methods of the previous section to find the two irrational solutions.

Let us begin our discussion by considering polynomial equations whose coefficients are all integers. Then it will be very easy to extend our results to equations with rational coefficients.

The key to the method of finding the rational solutions of a polynomial equation with integral coefficients lies in the following theorem, known as the **Rational Root Theorem.**

THE RATIONAL ROOT THEOREM

Suppose that

$$a_n x^n + a_{n-1}x^{n-1} + \cdots + a_1 x + a_0 = 0$$

is a polynomial equation whose coefficients $a_n, a_{n-1}, \ldots, a_1, a_0$, are integers. If p/q is a rational solution to this equation, and if *p/q has been reduced to lowest terms* (that is, if p and q have no common factors other than ± 1) then p must divide the coefficient a_0 and q must divide the coefficient a_n.

We will indicate why the Rational Root Theorem is true at the end of this section. (If you wish, you can read the explanation now.) Let us turn instead to some examples of the use of this theorem.

EXAMPLE 1: Find a rational solution, if one exists, to the equation

$$3x^3 - 2x^2 - 6x + 4 = 0 \tag{4.15}$$

Solution: The first step is to observe that the polynomial

$$p(x) = 3x^3 - 2x^2 - 6x + 4$$

has integral coefficients, and so we may apply the Rational Root Theorem. According to this theorem, if p/q is a rational solution to Equation (4.15), and if p/q is in lowest terms, then p must divide $a_0 = 4$ and q must divide $a_3 = 3$. This means that p must be one of the numbers

$$\pm 1, \qquad \pm 2, \qquad \pm 4 \tag{4.16}$$

and q must be one of the numbers

$$\pm 1, \qquad \pm 3 \tag{4.17}$$

Thus, the possibilities for p/q are

$$\pm 1, \qquad \pm\frac{1}{3}, \qquad \pm 2, \qquad \pm\frac{2}{3}, \qquad \pm 4, \qquad \pm\frac{4}{3} \tag{4.18}$$

We obtained this list of possibilities by combining each of the possible numerators in (4.16) with each of the possible denominators in (4.17). This may seem like a rather long list of possibilities, but remember that there are an *infinite* number of rational numbers, and we have narrowed down the list of possible solutions to only 12 numbers.

Let us emphasize that the Rational Root Theorem says that *if* there is a rational solution to Equation (4.15), reduced to lowest terms, then it *must* be among the numbers in the list (4.18). Thus, all we need to do is check these numbers. Either we will discover that one or more of these rational numbers is a solution, or else we can conclude that the equation has no rational solutions. As you can see, the Rational Root Theorem is quite a powerful theorem!

Trying the numbers in the list (4.18) in turn, we finally hit upon $x = 2/3$ as a rational solution to Equation (4.15). (You should check this for yourself by substituting into Equation (4.15)).

At this point, we could find all of the *distinct* rational solutions to Equation (4.15) simply by checking the other numbers in the list (4.18). However, this will not give us any information about the *multiplicities* of the solutions, and knowing the multiplicity of each solution is very important. After all, in a sense, a solution with multiplicity greater than one actually represents more than one solution! We will discuss the problem of how best to find all rational solutions a bit later in this section. □

EXAMPLE 2: Find a rational solution, if one exists, to the equation

$$x^4 - 5x^2 + 6 = 0 \tag{4.19}$$

Solution: Since this equation has only integral coefficients, we can apply the Rational Root Theorem. If p/q is a rational solution to Equation (4.19), expressed in lowest terms, then p must divide $a_0 = 6$ and q must divide $a_4 = 1$. Thus, the only possibilities for p are

$$\pm 1, \quad \pm 2, \quad \pm 3, \quad \pm 6$$

and the only possibilities for q are

$$\pm 1$$

Hence, the only possibilities for p/q are

$$\pm 1, \quad \pm 2, \quad \pm 3, \quad \pm 6$$

Trying each of these numbers, we discover that none of them is a solution to Equation (4.19). (You should do this yourself.) Therefore, Equation (4.19) has no rational solutions. (The solutions to Equation (4.19) are $x = \pm\sqrt{2}$ and $x = \pm\sqrt{3}$.)

Try Study Suggestion 4.9. □

▶ **Study Suggestion 4.9:** Find a rational solution, if one exists, to each of the following equations.

(a) $4x^3 - 20x^2 - x + 5 = 0$

(b) $x^4 - 7x^2 + 10 = 0$ ◀

Now we can extend our results to polynomial equations with rational coefficients. In this case, all we have to do is multiply both sides of the equation by a number that will "clear" all of the denominators. Let us demonstrate this with an example.

EXAMPLE 3: Find a rational solution, if one exists, to the equation

$$x^3 - \frac{11}{6}x^2 - \frac{2}{3}x + \frac{2}{3} = 0 \tag{4.20}$$

Solution: The first step is to multiply both sides of this equation by 6, which will clear it of all denominators.

$$6x^3 - 11x^2 - 4x + 4 = 0$$

Now we can proceed as in the previous examples. Any rational solution p/q in lowest terms must have the property that p divides $a_0 = 4$ and q divides $a_3 = 6$. The possibilities for p are

$$\pm 1, \quad \pm 2, \quad \pm 4$$

and the possibilities for q are

$$\pm 1, \quad \pm 2, \quad \pm 3, \quad \pm 6$$

and so the possibilities for p/q are

$$\pm 1, \quad \pm\frac{1}{2}, \quad \pm\frac{1}{3}, \quad \pm\frac{1}{6}, \quad \pm 2, \quad \pm\frac{2}{3}, \quad \pm 4, \quad \pm\frac{4}{3}$$

▶ **Study Suggestion 4.10:** Find a rational solution, if one exists, to the equation $x^3 + (5/2)x^2 - 25x - 125/4 = 0$. ◀

Testing these numbers in turn, we discover that $x = 1/2$ is a solution to Equation (4.20). *Try Study Suggestion 4.10.* □

In the previous examples, we found only one rational solution to the given polynomial equation. As we mentioned at the end of Example 1, we could have found all of the rational solutions by simply checking each of the possibilities in our list. Once we exhausted the list, the Rational Root Theorem would guarantee that we had found all of the rational solutions. However, as we mentioned earlier, *this gives us no information about the multiplicities of the solutions.*

In order to detect multiple roots, we first use the method of this section to find *just one* rational solution (if there is one), and then we use this solution to factor out a linear factor, as we did in the previous section. By repeating this process, we can factor the polynomial in such a way that the rational solutions, along with their multiplicities, can be easily determined. The following example will illustrate this point.

EXAMPLE 4: Find all rational solutions (along with their multiplicities) to the equation

$$8x^4 + 4x^3 - 18x^2 + 11x - 2 = 0 \tag{4.21}$$

Solution: Let us denote the polynomial on the left side of this equation by $a(x)$. If p/q is a rational solution to Equation (4.21), expressed in lowest terms, then p must divide $a_0 = -2$ and q must divide $a_4 = 8$. Thus, p must be one of the numbers

$$\pm 1 \quad \text{and} \quad \pm 2$$

and q must be one of the numbers

$$\pm 1, \quad \pm 2, \quad \pm 4, \quad \pm 8$$

and so p/q must be one of the numbers

$$\pm 1, \quad \pm\frac{1}{2}, \quad \pm\frac{1}{4}, \quad \pm\frac{1}{8}, \quad \pm 2 \tag{4.22}$$

Checking these numbers in turn, we find that $x = 1/2$ is a solution to Equation (4.21)

Therefore, according to the Factor Theorem, $x - 1/2$ is a factor of $a(x)$. Synthetic division gives

$$\begin{array}{r|rrrrr} 1/2 & 8 & 4 & -18 & 11 & -2 \\ & & 4 & 4 & -7 & 2 \\ \hline & 8 & 8 & -14 & 4 & 0 \end{array}$$

and so we have the factorization

$$\begin{aligned} a(x) &= (8x^3 + 8x^2 - 14x + 4)(x - 1/2) \\ &= 2(4x^3 + 4x^2 - 7x + 2)(x - 1/2) \end{aligned}$$

Now we look for a rational solution to the equation

$$4x^3 + 4x^2 - 7x + 2 = 0 \tag{4.23}$$

If p/q is a rational solution to Equation (4.23), expressed in lowest terms, then p must divide $a_0 = 2$ and q must divide $a_3 = 4$. Hence, p must be one of the numbers

$$\pm 1 \quad \text{and} \quad \pm 2$$

and q must be one of the numbers

$$\pm 1, \quad \pm 2, \quad \pm 4$$

and so p/q must be one of the numbers

$$\pm 1, \quad \pm\frac{1}{2}, \quad \pm\frac{1}{4}, \quad \pm 2$$

Checking these numbers, we again find that $x = 1/2$ is a solution to Equation (4.23).

Again using the Factor Theorem and synthetic division, we obtain the factorization

$$4x^3 + 4x^2 - 7x + 2 = (4x^2 + 6x - 4)(x - 1/2)$$

and so

$$\begin{aligned} a(x) &= 2(4x^3 + 4x^2 - 7x + 2)(x - 1/2) \\ &= 2(4x^2 + 6x - 4)(x - 1/2)^2 \\ &= 4(2x^2 + 3x - 2)(x - 1/2)^2 \end{aligned}$$

Incidentally, this factorization shows that $x = 1/2$ is a solution to Equation (4.21) of multiplicity *at least* 2.

Now we look for a rational solution to the quadratic equation

$$2x^2 + 3x - 2 = 0$$

We will leave it to you to show that $x = 1/2$ and $x = -2$ are the solutions to this equation, and that

$$2x^2 + 3x - 2 = 2(x + 2)(x - 1/2)$$

Therefore, we have

$$\begin{aligned} a(x) &= 4(2x^2 + 3x - 2)(x - 1/2)^2 \\ &= 8(x + 2)(x - 1/2)^3 \end{aligned}$$

At this point, we can see that the solutions to Equation (4.21) are

$$x = 1/2 \quad \text{and} \quad x = -2$$

where $x = -2$ is a simple root and $x = 1/2$ is a *triple* root—a fact that we never would have obtained directly from the list (4.22).

Try Study Suggestion 4.11. □

▶ **Study Suggestion 4.11:** Find all rational solutions (along with their multiplicities) to the equation $9x^3 - 3x^2 - 5x - 1 = 0$. ◀

EXAMPLE 5: Find all rational solutions (along with their multiplicities) to the equation

$$2x^3 = x^2 + 4x - 2$$

Solution: The first step is to put this equation in the form

$$2x^3 - x^2 - 4x + 2 = 0 \quad (4.24)$$

We will leave it to you to show, by the method that we used in the previous examples, that $x = 1/2$ is a rational solution to Equation (4.24), and so $x - 1/2$ is a factor of the left-hand side of this equation.

Synthetic division gives the factorization

$$2x^3 - x^2 - 4x + 2 = (2x^2 - 4)(x - 1/2)$$
$$= 2(x^2 - 2)(x - 1/2)$$

and so we look for a rational solution to the equation

$$x^2 - 2 = 0$$

However, this equation has no rational solutions. We can see this either by using the method of this section to look for a rational solution, or by actually solving the equation, to get the solutions $x = \pm\sqrt{2}$, neither of which is rational. Therefore, the only rational solution to Equation (4.24) is $x = 1/2$, which has multiplicity 1. *Try Study Suggestion 4.12.* □

▶ **Study Suggestion 4.12:** Find all rational solutions (along with their multiplicities) to the equation $3x^3 = 3 + 9x - x^2$. ◀

Now let us explain why the Rational Root Theorem is true. To make the discussion easier to follow, we will consider the case of a polynomial equation of degree 4

$$a_4x^4 + a_3x^3 + a_2x^2 + a_1x + a_0 = 0 \quad (4.25)$$

where the coefficients a_4, a_3, a_2, a_1, and a_0 are integers and $a_4 \neq 0$ and $a_0 \neq 0$. Our discussion can be generalized to polynomial equations of any degree.

Suppose that p/q is a solution to Equation (4.25), expressed in lowest terms. Thus, p and q have no common factors other than ± 1. We would like to show that p divides the coefficient a_0 and q divides the coefficient a_4.

Making the substitution $x = p/q$ in Equation (4.25) gives

$$a_4\left(\frac{p}{q}\right)^4 + a_3\left(\frac{p}{q}\right)^3 + a_2\left(\frac{p}{q}\right)^2 + a_1\left(\frac{p}{q}\right) + a_0 = 0$$

Multiplying both sides of this by q^4 gives

$$a_4p^4 + a_3p^3q + a_2p^2q^2 + a_1pq^3 + a_0q^4 = 0 \quad (4.26)$$

Now let us solve this equation for a_0q^4:

$$a_0q^4 = -a_4p^4 - a_3p^3q - a_2p^2q^2 - a_1pq^3 \quad (4.27)$$

As you can see, each of the terms on the right side of this equation is divisible by p, and so p is a factor of the right side of this equation. Therefore, p must also be a factor of the left side.

Although we shall not go into the details, it can be shown that because p and q have no common factors, p and q^4 also have no common factors. Hence, the only way that p can be a factor of the left side of Equation (4.27) is if p is a factor of a_0; that is, if p divides a_0. This is one-half of the Rational Root Theorem.

If we also solve Equation (4.26) for a_4p^4, the result is

$$a_4p^4 = -a_3p^3q - a_2p^2q^2 - a_1pq^3 - a_0q^4$$

By repeating the argument that we just used we can see that, since q is a factor of the right side of this equation, it must also be a factor of the left side. That is, q must be a factor of a_4p^4. Hence, since p and q have no common factors, q must be a factor of a_4. This is the other half of the Rational Root Theorem.

Ideas to Remember

If a polynomial equation has rational *coefficients*, then we can find all of its rational *solutions.*

EXERCISES

In Exercises 1–31, use the method of this section to find a rational solution, if one exists, to the given equation.

1. $4x^2 - 4x + 1 = 0$

2. $2x^2 - 5x - 3 = 0$

3. $12x^2 + x - 1 = 0$

4. $7x^2 + (31/5)x - 18/5 = 0$

5. $x^2 + 3x + 4 = 0$

6. $x^2 = (3/4)x - 1/8$

7. $x^3 - x^2 - 2x + 2 = 0$

8. $2x^3 - 3x^2 - 6x + 9 = 0$

9. $-x^3 + 3x^2 - 4 = 0$

10. $4x^3 = 24x^2 + x - 6$

11. $2x^3 - 7x^2 + 2 = 0$

12. $-6x^3 + 5x^2 + 9x = -2$

13. $27x^3 - 54x^2 + 36x - 8 = 0$

14. $x^3 - (5/6)x^2 - (1/18)x + 1/9 = 0$

15. $3x^3 - (1/10)x^2 - (6/10)x + 1/10 = 0$

16. $12x^3 + 28x^2 - 37x + 8 = -2$

17. $-2x^4 + x = 4x^3 + 5x^2 + 3x + 2$

18. $5x^2 = x^4 + 6$

19. $2x^4 - (1/3)x^3 + (1/3)x^2 - (5/3)x + 2/3 = 0$

20. $x^4 + (1/2)x^3 - (5/2)x^2 - x + 1 = 0$

21. $6x^4 - x^3 - 19x^2 + 3x + 3 = 0$

22. $37x^4 - 11x^2 + 3 = x^4 + 2x^2 + 2$

23. $9x^5 + 21x^4 + 25x^3 + 43x^2 + 40x + 12 = 0$

24. $8x^5 - 4x^4 - 18x^3 + 9x^2 + 4x - 2 = 0$

25. $2x^5 - 4x^3 + 6 = 3x^2 - 4x^3$

26. $2x^{10} + x^8 + 3x^6 + x^2 + 12 = 0$

27. $x^7 - (2/5)x^6 + 3x^3 + (6/5)x^2 - 9x + 18/5 = 0$

28. $x^{12} - x^{10} + 12x^3 - 12 = 0$

29. $x^{12} + 3x^2 - 12 = 0$

30. $2x^{85} - 3x^{84} + 4x^{64} - 6x^{63} - 2x^3 + 3x^2 + 8x - 12 = 0$

31. $x^{100} + x^2 - 2 = 0$

In Exercises 32–51, use the method of this section to find all rational solutions (along with their multiplicities) to the given equation.

32. $9x^2 + 6x + 1 = 0$

33. $6x^2 - 13x - 5 = 0$

34. $-3x^2 + (1/5)x + 1/5 = 0$

35. $4x^3 + 9 = 8x^2 + 3x$

36. $24x^3 - 10x^2 - 3x = -1$

37. $2x^3 - (16/3)x^2 + (1/2)x + 5/6 = 0$

38. $2x^3 - 6x^2 - (9/2)x + 7/2 = 0$

39. $55x^2 - 7x^3 - 97x - 15 = 0$

40. $x^4 = 2x^2 - 1$

41. $x^4 + 6x^3 + 13x^2 + 24x + 36 = 0$

42. $x^4 - 8x^2 + 16 = 0$

43. $x^4 - 6x^2 + 10 = 1$

44. $36x^4 - 12x^3 - 11x^2 + 2x + 1 = 0$

45. $3x^5 + 5x^4 = 5x^3 + 11x^2 + 2x - 2$

46. $16x^5 - 16x^4 - (13/2)x^2 + 7x - 1/2 = 0$

47. $x^5 + x^4 - 5x^3 - x^2 + 8x - 4 = 0$

48. $256x^5 + 64x^4 - 80x^3 - 4x^2 + 8x - 1 = 0$

49. $x^5 - 3x^4 + 8x^3 - 14x^2 + 15x + 5 = 0$

50. $x^{10} + x^8 = 1$

51. $2x^{12} + 7x^2 - 3x + 1 = 0$

4.4 Approximating the Roots of a Polynomial Equation

As we have discussed, the problem of finding the roots of a polynomial equation

$$a_nx^n + a_{n-1}x^{n-1} + \cdots + a_1x + a_0 = 0$$

is in general very difficult. Of course, in special cases, such as when the degree of the equation is less than or equal to 2, or when we already know some of the roots, or when the coefficients are rational and we want only the rational roots, we do have very satisfactory methods at our disposal. Nevertheless, the general case remains very difficult.

Because of the difficulty of this problem, mathematicians and scientists have spent a great deal of time and effort developing methods for *approximating* the solutions to polynomial equations, and we want to discuss one very simple method in this section, known as the **bisection method.**

Our method rests on the following fact about polynomial functions, which we can express in terms of their graphs.

The graph of any polynomial function

$$p(x) = a_nx^n + a_{n-1}x^{n-1} + \cdots + a_1x + a_0$$

is a smooth curve, in the sense that it has no "breaks" in it.

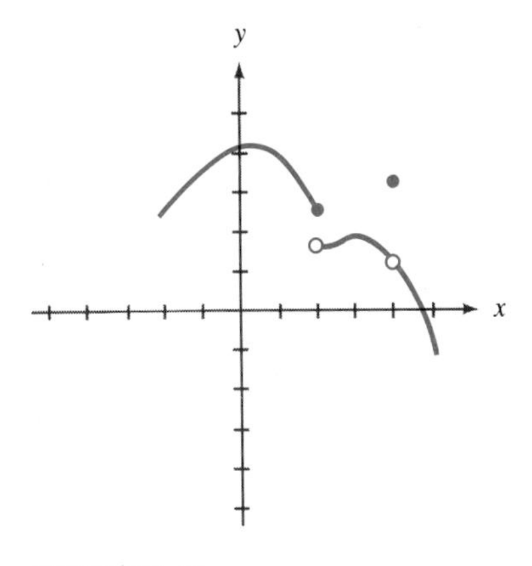

FIGURE 4.1

To give you an idea of what we mean by breaks, the graph in Figure 4.1 has a break at $x = 2$ and at $x = 4$. The graph of a polynomial function can have no such breaks.

Unfortunately, we will not be able to prove this statement, because the proof requires some knowledge of calculus. In any case, this fact does have important consequences when it comes to approximating solutions to polynomial equations. Consider for example, the equation

$$x^3 - 3x^2 + 2x - 1 = 0 \tag{4.28}$$

If we let $p(x)$ be the polynomial function

$$p(x) = x^3 - 3x^2 + 2x - 1$$

then it is easy to see that $p(2) = -1$ and $p(3) = 5$. In Figure 4.2 we have plotted the two points $(2, -1)$ and $(3, 5)$. The key point here is that, *since the graph of $p(x)$ has no breaks in it, the only way it can go through the points $(2, -1)$ and $(3, 5)$ is if it crosses the x-axis at least once somewhere between* $x = 2$ and $x = 3$. (See Figure 4.2)

Furthermore, if we denote this x-intercept by r, then we have $p(r) = 0$. Thus r is a solution to the polynomial Equation (4.28). In other words, there must be a solution to Equation (4.28) somewhere between 2 and 3; that is, somewhere in the open interval $(2, 3)$.

The principle that we have used here can apply in other situations, for it depends only on the fact that the graph does not have any breaks in it. Let us generalize this idea in the following theorem.

THEOREM

> Let $p(x)$ be a polynomial and suppose that a and b are real numbers satisfying $a < b$. If $p(a)$ and $p(b)$ have opposite signs; that is, if one is positive and the other is negative, then there must be a solution to the polynomial equation
>
> $$p(x) = 0$$
>
> somewhere in the open interval (a, b).

This theorem can be used to approximate solutions to a polynomial equation to any desired degree of accuracy. Let us look again at Equation (4.28). We know that there is a solution somewhere in the open interval $(2, 3)$. In order to narrow this interval, we evaluate the polynomial $p(x)$ at the midpoint of the interval; that is, at $x = 2.5$. Since $p(2.5) = 0.875$ is positive, we see that there must be a solution to Equation (4.28) somewhere in the interval $(2.0, 2.5)$. (See Figure 4.3.)

We can repeat this procedure by evaluating $p(x)$ at the midpoint of the interval $(2.0, 2.5)$ to determine whether the value there is positive or negative.

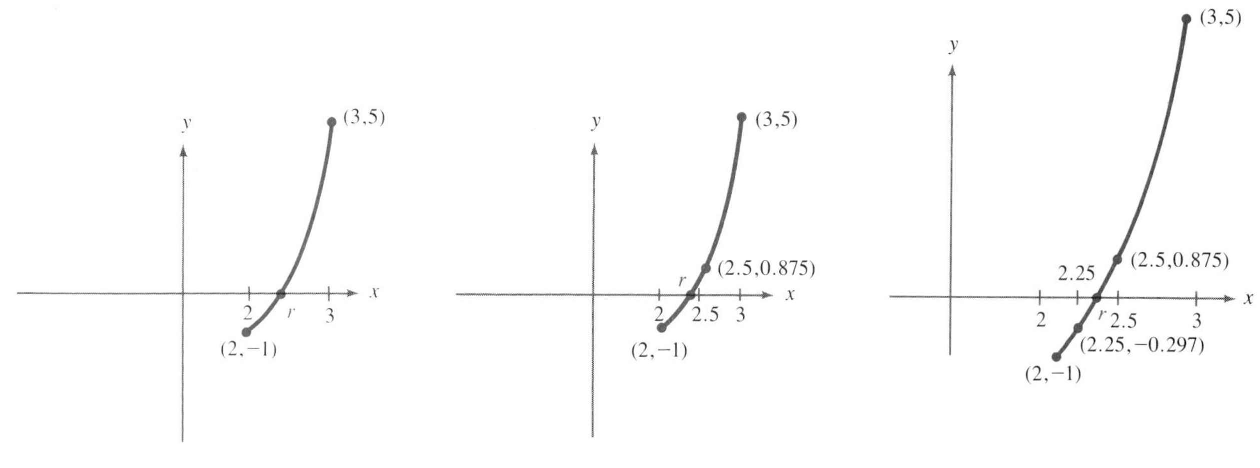

FIGURE 4.2 FIGURE 4.3 FIGURE 4.4

In this case, the midpoint is 2.25, and $p(2.25) \approx -0.297$ is negative. Hence, there must be a solution to Equation (4.28) in the interval (2.25, 2.50). (See Figure 4.4.)

Now you can see why the method is called the bisection method, for at each step we use the midpoint to *bisect* the given interval, thus obtaining a new interval containing the solution, but with half the length of the previous interval.

Continuing in this fashion, we can narrow the interval that contains a solution as much as we want, each time obtaining a more accurate idea of where the solution lies. Each time we reduce the interval by half, we say that we have performed one **iteration.** After completing the desired number of iterations, we can simply take the midpoint of the last interval as our approximation. (The number of iterations to be performed depends on the desired degree of accuracy of the approximation, a subject that we shall not discuss in this brief introduction.)

In our current example, we know that a solution to Equation (4.28) lies in the interval (2.25, 2.50) and so we can take the midpoint of this interval, namely

$$\frac{2.50 + 2.25}{2} = 2.375$$

as an approximation to the solution.

Let us consider another example of this procedure.

EXAMPLE 1: Approximate a solution to the equation

$$2x^3 + 9x^2 + 13x + 5 = 0 \qquad \textbf{(4.29)}$$

by performing three iterations of the bisection method.

Solution: We begin by computing the value of the polynomial

$$p(x) = 2x^3 + 9x^2 + 13x + 5$$

at some consecutive integral values of x. This is done in the following table.

x	$p(x)$
-3	-7
-2	-1
-1	-1 } → sign change
0	5 }
1	29
2	83
3	179

Since $p(-1) = -1$ and $p(0) = 5$ have opposite signs, we can conclude that there must be a solution to Equation (4.29) somewhere in the open interval $(-1, 0)$. Now we can start the procedure for approximating this solution.

First iteration

Current interval containing solution: $\mathbf{(-1, 0)}$

midpoint of current interval: $\mathbf{-0.5}$

values of polynomial:	$p(-1) = -1$ negative	$p(-0.5) = 0.5$ positive	$p(0) = 5$ positive

new interval containing solution: $\mathbf{(-1, -0.5)}$

Second iteration

current interval containing solution: $\mathbf{(-1, -0.5)}$

midpoint of current interval: $\mathbf{-0.75}$

values of polynomial:	$p(-1) = -1$ negative	$p(-0.75) \approx -0.53$ negative	$p(-0.5) = 0.5$ positive

new interval containing solution: $\mathbf{(-0.75, -0.5)}$

Third iteration

current interval containing solution: $\mathbf{(-0.75, -0.5)}$

midpoint of current interval: $\mathbf{-0.625}$

values of polynomial:	$p(-0.75) \approx -0.53$ negative	$p(-0.625) \approx -0.098$ negative	$p(-0.5) = 0.5$ positive

new interval containing solution: $\mathbf{(-0.625, -0.5)}$

Approximation to solution

$$\frac{-0.5 + (-0.625)}{2} = \frac{-1.125}{2} = -0.5625$$

Try Study Suggestion 4.13. □

▶ **Study Suggestion 4.13:** Approximate a solution to the equation $2x^2 - 5x + 1 = 0$ by performing three iterations of the bisection method. ◀

Ideas to Remember

The fact that the graph of a polynomial has no breaks in it can be used to approximate solutions to polynomial equations.

EXERCISES

In Exercises 1–12, approximate a solution to the given equation by performing three iterations of the bisection method. (Begin by looking for a sign change at the values $x = -3, -2, -1, 0, 1, 2, 3$.)

1. $x^2 + 5x + 3 = 0$

2. $2x^2 - 4x + 1 = 0$

3. $5x^2 + 5x - 1 = 0$

4. $x^3 - 4x^2 + 1 = 0$

5. $x^3 + 4x^2 - 12 = 0$

6. $6x^3 - x + 10 = 0$

7. $2x^4 + 3x^3 - 1 = 0$

8. $x^5 - x^4 + 1 = 0$

9. $x^3 + 2x^2 - 4x - 4 = 0$

10. $2x^3 + 4x^2 - 9x + 1 = 0$

11. $3x^3 + 2x^2 - 5x + 1 = 0$

12. $x^4 + x^3 - x^2 + x - 1 = 0$

13. Approximate a solution to the equation

$$x^3 + x^2 - 1 = 0$$

in the interval (0, 1) by performing four iterations of the bisection method.

14. Approximate a solution to the equation

$$2x^3 - 7x + 1 = 0$$

in the interval (0, 1) by performing five iterations of the bisection method.

15. Approximate a solution to the equation

$$-2x^3 + 8x^2 - 2 = 0$$

in the interval $(-1, 0)$ by performing six iterations of the bisection method.

16. Use the bisection method to approximate $\sqrt{2}$. Perform four iterations. (*Hint:* $\sqrt{2}$ is a solution to the equation $x^2 - 2 = 0$.)

17. Use the bisection method to approximate $\sqrt[3]{2}$. Perform four iterations.

18. Use the bisection method to approximate $\sqrt[5]{2}$. Perform five iterations.

4.5 Graphing Polynomial Functions

In this section, we will discuss a method for graphing polynomial functions of degree greater than 2. If you take a quick look now at the graphs in the coming examples, you can get an idea of the general shape of this type of graph. Generally speaking, the graph of a polynomial has several "humps," the number of which is always *less than* the degree of the polynomial. Also, the graph "straightens out" when $|x|$ is large. (When we say that $|x|$ is large, we mean that x is either large positive or *small* negative. Remember, for example, that -1000 is *smaller* than -1.)

Our procedure for graphing polynomial functions has four steps to it.

Procedure for Graphing Polynomial Functions

STEP 1. Plot the y-intercept and as many of the x-intercepts as possible.

STEP 2. Determine the behavior of the graph when $|x|$ is large.

STEP 3. Plot a few additional points on the graph.

STEP 4. Draw a smooth curve (that is, a curve with no breaks) through the plotted points, keeping the results of Step 2 in mind.

Let us discuss Step 2 in some detail. There are four possible types of behavior for the graph of a polynomial function when $|x|$ is large. These four possibilities are pictured in Figure 4.5a–d.

It is important to remember that these pictures describe the graph of $p(x)$ *only* when $|x|$ is large—just exactly how large will depend on the particular function. In a sense, they describe certain *trends* in the graph.

In order to determine which of the four pictures applies to a given polynomial function $p(x)$, we need only look at its *leading term*; that is, the term with the largest exponent, among those terms with nonzero coefficients. For example, if the leading term has even exponent and positive coefficient, then Figure 4.5a applies. However, if the leading term has odd exponent and negative coefficient, then Figure 4.5d applies.

The reason that the leading term of $p(x)$ determines the behavior of the graph when $|x|$ is large is that, in this case, the leading term of $p(x)$ dominates the other terms. For example, consider the polynomial

$$p(x) = 2x^3 + 2x^2 - 3x - 4$$

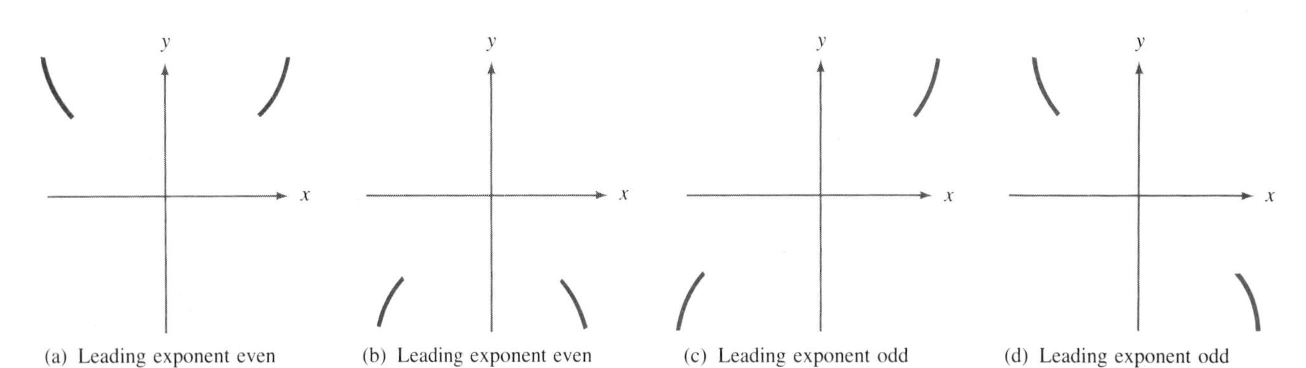

(a) Leading exponent even
Leading coefficient positive

(b) Leading exponent even
Leading coefficient negative

(c) Leading exponent odd
Leading coefficient positive

(d) Leading exponent odd
Leading coefficient negative

FIGURE 4.5

When x is very large in absolute value, the term $2x^3$ will be much more significant than any of the other terms, and so it will govern the behavior of $p(x)$. For instance, when $x = 100$, the term $2x^3$ is equal to $2 \cdot 100^3 = 2{,}000{,}000$ whereas the next largest term $2x^2$ is only equal to $2 \cdot 100^2 = 20{,}000$. Furthermore, this difference grows even more significant as $|x|$ gets larger.

Thus, when considering the behavior of $p(x)$ when $|x|$ is large, we may ignore all but the leading term. This means that $p(x)$ has the same behavior as one of the following types of polynomials

$$p_1(x) = (\text{positive})x^{\text{even}}$$
$$p_2(x) = (\text{negative})x^{\text{even}}$$
$$p_3(x) = (\text{positive})x^{\text{odd}}$$
$$p_4(x) = (\text{negative})x^{\text{odd}}$$

Now, it is not hard to see how these polynomials behave when $|x|$ is large. For convenience, let us introduce some terminology. When we say that x *increases without bound*, we mean that x gets larger and larger without bound and is positive. On the other hand, when we say that x *decreases without bound*, we mean that $|x|$ gets larger and larger without bound, but that x is negative. Thus, the numbers $-1, -10, -100, -1000, \ldots$ are *decreasing* without bound.

Now let us consider the function $p_2(x) = (\text{negative})x^{\text{even}}$. As x increases without bound, so does x^{even}, and therefore $p_2(x)$ decreases without bound, because its coefficient is negative. This accounts for the right side of Figure 4.5b. Also, as x decreases without bound, x^{even} *increases* without bound, since the exponent is even. Therefore $p_2(x)$ decreases without bound, again because its coefficient is negative. This accounts for the left side of Figure 4.5b. The other possibilities can be explained in a similar manner.

Before turning to examples, we should discuss one more point. If, as we have observed, the graph of a polynomial function has "humps," it would certainly be nice if we could identify the highest and lowest points in these humps. We are referring to points such as the ones indicated in Figure 4.6.

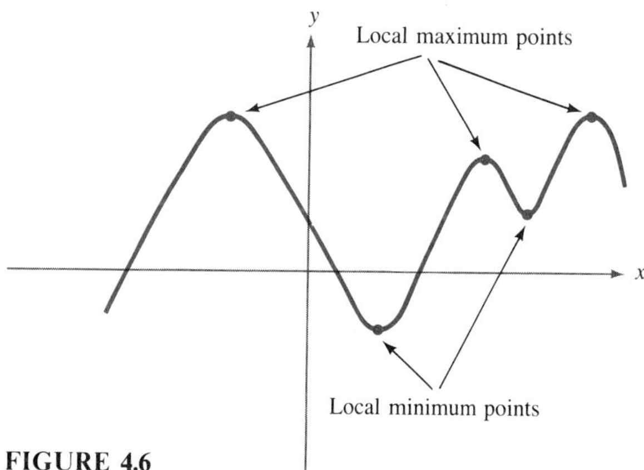

FIGURE 4.6

These points are called **local minimum points** and **local maximum points.** Together, they are referred to as **local extreme points.**

We saw in Section 3.3 that the graph of a quadratic function has only one local extreme point, and we were able to find the coordinates of this point. For polynomial equations of degree greater than 2, the situation is not quite as simple. There is a very important technique for determining the coordinates of local extreme points, and it is one of the central topics in calculus. Unfortunately we cannot take the time to discuss this technique here.

Now let us turn to some examples of our procedure for graphing polynomial functions.

EXAMPLE 1: Graph the polynomial function

$$p(x) = x^3 + x^2 - 10x + 8$$

Solution: The y-intercept is obtained by setting $x = 0$, from which we get $y = 8$. In order to obtain the x-intercepts, we set $y = 0$,

$$x^3 + x^2 - 10x + 8 = 0$$

and solve for x. We will leave it to you to show, by using the method of Section 4.3, that the solutions to this equation are $x = -4$, $x = 1$, and $x = 2$. Hence, these are the x-intercepts of the graph. (See Figure 4.7.)

Since the leading term of $p(x)$ is x^3, which has the form (positive)x^{odd}, the graph behaves as in Figure 4.5c when $|x|$ is large.

Now we plot a few additional points and then draw a smooth curve through them. (You can compute the values of $p(x)$ in the following table by using synthetic division, as we did in Example 1 of Section 4.2.)

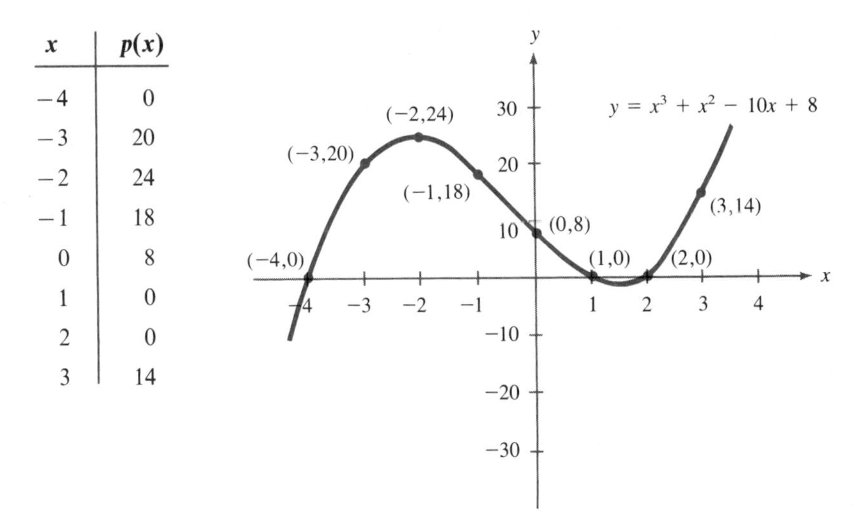

x	$p(x)$
−4	0
−3	20
−2	24
−1	18
0	8
1	0
2	0
3	14

FIGURE 4.7

Notice that, in this case, the values of the function are large in comparison with the value of x. Therefore, in order to graph the function in a limited space, we chose a different scale factor on the y-axis than on the x-axis.

▶ **Study Suggestion 4.14:** Graph the function $p(x) = x^3 - 3x^2 - x + 3$. ◀

Try Study Suggestion 4.14. □

EXAMPLE 2: Graph the polynomial function

$$p(x) = -x^4 + 4x^3 + x^2 - 6x - 18$$

Solution: The y-intercept is $p(0) = -18$. In order to find the x-intercepts, we must solve the equation

$$-x^4 + 4x^3 + x^2 - 6x - 18 = 0$$

Using the method of Section 4.3, we obtain the factorization

$$-(x-3)^2(x^2 + 2x + 2) = 0 \tag{4.30}$$

and since the quadratic $x^2 + 2x + 2$ has a negative discriminant ($b^2 - 4ac = -4$), the only solution to Equation (4.30) is $x = 3$, which is a double root. Hence, $x = 3$ is the only x-intercept. (See Figure 4.8.)

Since the leading term of $p(x)$ is $-x^4$, which is of the form (negative) x^{even}, the graph behaves as in Figure 4.5b when $|x|$ is large.

Now we plot a few additional points and sketch the graph.

x	$p(x)$
−2	−50
−1	−16
0	−18
1	−20
2	−10
3	0
4	−26

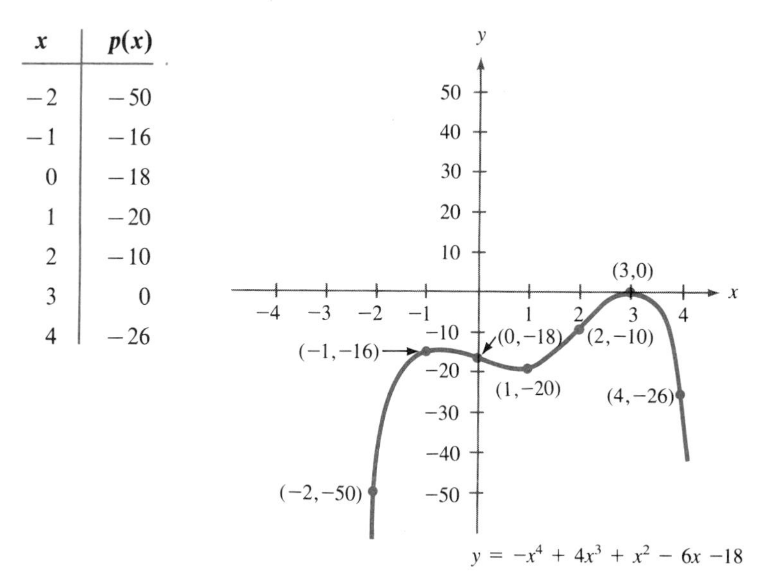

FIGURE 4.8

(Again we used a different scale factor on the y-axis than on the x-axis.) □

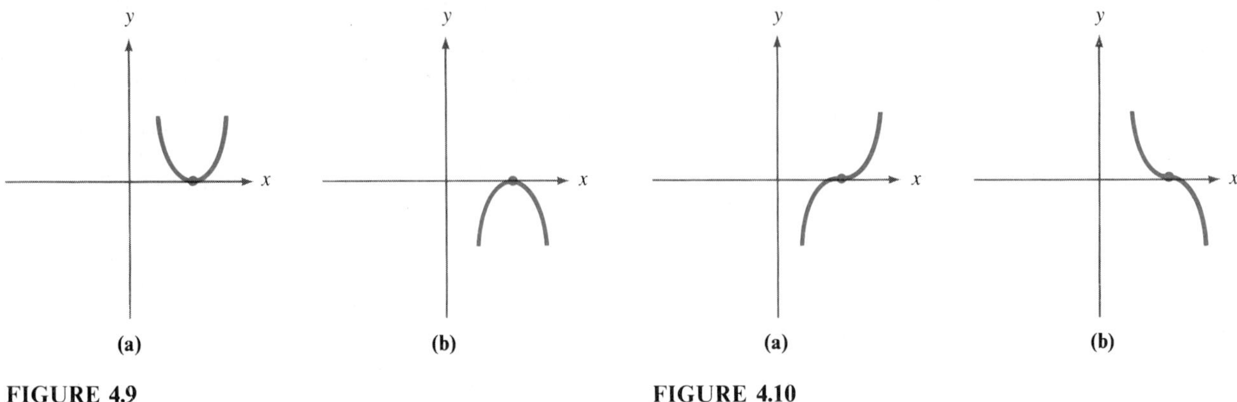

FIGURE 4.9

FIGURE 4.10

Notice that in the previous example, we have drawn the graph to be tangent to the x-axis at the point (3, 0). The reason we did this is that $x = 3$ is a double root of Equation (4.30). In general, whenever a root of a polynomial has multiplicity greater than one, its graph will be tangent at that x-intercept. (This is another point that is discussed in a course in calculus.) If the multiplicity of the root is even, then the graph has one of the forms shown in Figures 4.9a or b. However, if the multiplicity of the root is odd (and greater than 1), then the graph is still tangent to the x-axis, but crosses it as shown in Figures 4.10a or b.

▶ **Study Suggestion 4.15:** Graph the function $p(x) = -2(x-2)^2(1-x)^3$ (*Hint:* you will need to multiply out the right side—at least partially—in order to obtain the leading term.) ◀

Try Study Suggestion 4.15.

EXAMPLE 3: Graph the polynomial function

$$p(x) = 2x^4 + 4x^2 + 2$$

Solution: Before rushing ahead, we should observe that the polynomial $p(x)$ factors

$$p(x) = 2(x^2 + 1)^2$$

The y-intercept of this graph is $y = 2$, and from our factorization we see that the graph has no x-intercepts. This means that the entire graph must lie on one side of the x-axis. In order to determine which side, we simply check any one point on the graph. For example, the point (0, 2) is on the graph, and so the entire graph lies above the x-axis.

Furthermore, because $p(x)$ involves only even powers of x it is an even function and so its graph is symmetric about the y-axis. This means that if we draw the graph of $p(x)$ for nonnegative values of x, we can take the mirror image to get the graph of $p(x)$ for negative values of x. Plotting a few points and observing that $p(x)$ is increasing in the interval $[0, \infty)$, we get the graph pictured in Figure 4.11. (Incidentally, although this graph does have a shape that is similar to a parabola, it is *not* a parabola, because the function is not a quadratic.) *Try Study Suggestion 4.16.* □

▶ **Study Suggestion 4.16:** Graph the function $p(x) = -3x^4 + 6x^2 - 3$. ◀

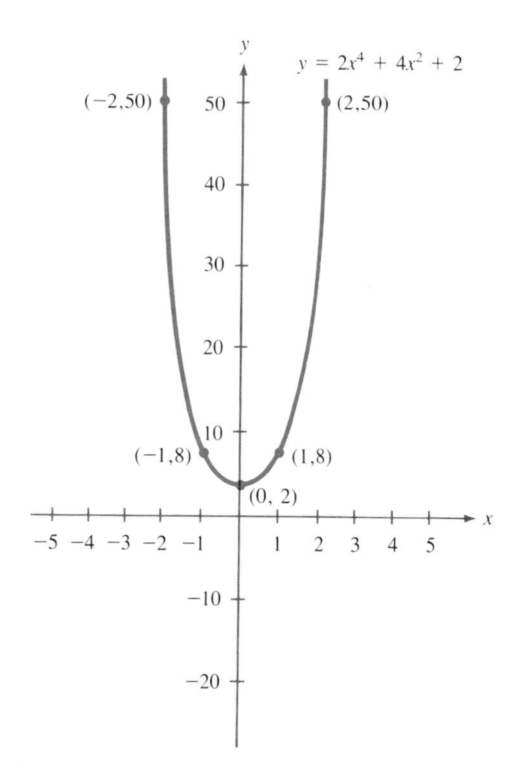

FIGURE 4.11

Ideas to Remember

When graphing a polynomial function, it is important to determine the behavior of the polynomial when $|x|$ is large.

EXERCISES

In Exercises 1–30, graph the given polynomial function.

1. $p(x) = x^3 + x$
2. $p(x) = x^3 - 2x^2 - 5x + 6$
3. $p(x) = (x - 2)^3$
4. $p(x) = 2x^3 + 3x^2 - 1$
5. $p(x) = -x^3 + 5x^2 - 2x + 10$
6. $p(x) = -6x^3 + 3x^2 - 4x + 3$
7. $p(x) = (2x - 3)(x^2 - 2x + 2)$
8. $p(x) = (x - 1)^2(x + 2)$
9. $p(x) = x^3 - x^2 - x - 2$
10. $p(x) = x^3 + x^2 + x + 1$
11. $p(x) = x^3 + 5x^2 + 16x + 20$
12. $p(x) = -2x^3 + 3x^2 + 29x - 60$
13. $p(x) = (x - 1)^3(x + 1)$

14. $p(x) = (x-1)^2(x+1)^2$

15. $p(x) = (x-1)^4$

16. $p(x) = (x-1)^5$

17. $p(x) = x^4 - 5x^2 + 4$

18. $p(x) = (x+3)(-2x^3 + 3x^2 - 2x + 3)$

19. $p(x) = -x^4 + 2x^3 + 9x^2 - 2x - 8$

20. $p(x) = -7x^5 + 2x^3$

21. $p(x) = (x-3)^2(x-1)^3$

22. $p(x) = -2x^5 + x^4 - 2x^2 - 2x + 1$

23. $p(x) = (x-1)^{10}$

24. $p(x) = x^{10} - 1$

25. $p(x) = x^3 - 4x^2 + 1$

26. $p(x) = x^3 + 2x^2 - 5x - 10$

27. $p(x) = x^4 - 5x^2 + 6$

28. $p(x) = (2x^2 + 5x - 1)(x^2 + 2x + 2)$

29. $p(x) = -2x^4 + 4x^3 + 16x^2 - 1$

30. $p(x) = x^4 + x^3 - 12x^2 - x + 1$

4.6 Graphing Rational Functions

A **rational function** is any function of the form

$$r(x) = \frac{p(x)}{q(x)}$$

where $p(x)$ and $q(x)$ are polynomials. For example, the following are rational functions.

$$r(x) = \frac{2x^2 + 3x - 1}{7x^2 - 2x + 9}, \qquad s(x) = \frac{6x^3 + 1}{x^2}$$

In this section we want to discuss a method for graphing rational functions.

One of the most important facts to keep in mind when graphing rational functions is that

the domain of a rational function

$$r(x) = \frac{p(x)}{q(x)}$$

consists of all real numbers x *except* those for which $q(x) = 0$.

In order to organize our thinking, we will divide this section into four subsections, each one devoted to a particular aspect of graphing rational functions. *Until further notice, we will assume that $p(x)$ and $q(x)$ have no nontrivial common factors; that is, we will assume that $r(x)$ is in lowest terms.*

VERTICAL ASYMPTOTES OF THE GRAPH

In the case of rational functions, not only must we determine the behavior of the function when $|x|$ is large, but also for values of x that are near the zeros of the denominator. Let us consider some examples of this behavior.

EXAMPLE 1: Describe the behavior of the rational function

$$r(x) = \frac{1}{x - 2}$$

near the zero $x = 2$ of its denominator.

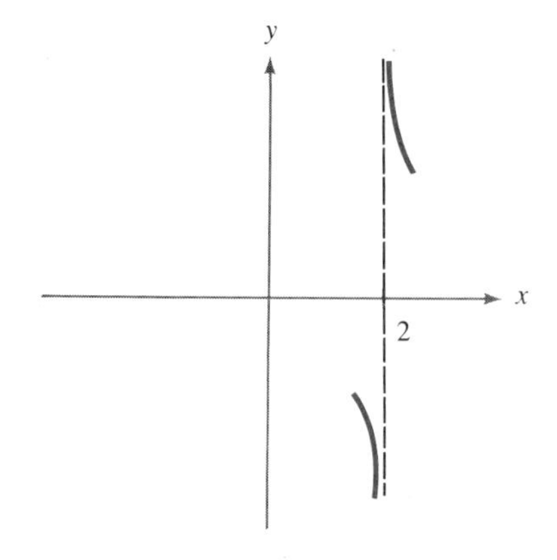

The behavior of $r(x)$ for values of x near 2

FIGURE 4.12

Solution: Figure 4.12 describes the behavior of the function for values of x that are near $x = 2$. (The function is not defined at $x = 2$ because $x = 2$ is not in its domain.) Let us see if we can justify this figure.

We begin by considering the behavior of the function for values of x that are slightly larger than 2. Let us imagine taking values of x that get closer and closer to 2, but that remain always slightly larger than 2. We say that these values *approach* 2 *from the right*. (By this we mean from the right side of 2 on the real number line.) Then the denominator $x - 2$ approaches 0 but remains always positive; that is, it approaches 0 from the right.

This implies that the quotient $r(x) = 1/(x - 2)$ increases without bound, as indicated on the right side of the vertical line $x = 2$ in Figure 4.12. We can further substantiate these statements by looking at a few numerical examples.

x	$x - 2$	$1/(x - 2)$
2.1	.1	10
2.01	.01	100
2.001	.001	1000
2.0001	.0001	10000
2.00001	.00001	100000

This table indicates that as x approaches 2 from the right, $x - 2$ approaches 0 from the right, and the quotient $1/(x - 2)$ increases without bound.

Now let us consider the behavior of $r(x)$ for values of x that are slightly smaller than 2. In this case, we imagine taking values of x that approach 2 from the left; that is, that get closer and closer to 2 but remain always slightly less than 2. Then the denominator $x - 2$ approaches 0, but always remains negative. This means that $r(x)$ decreases without bound (since $x - 2$ is negative.) This behavior is shown on the left side of the vertical line $x = 2$ in Figure 4.12. Again, we can substantiate these statements by looking at some numerical examples.

x	$x - 2$	$1/(x - 2)$
1.9	−.1	−10
1.99	−.01	−100
1.999	−.001	−1000
1.9999	−.0001	−10000
1.99999	−.00001	−100000

Thus, by examining the behavior of $r(x)$ as x approaches 2 from the right, and then from the left, we obtain the portion of the graph shown in Figure 4.12. □

The line $x = 2$ in Figure 4.12 is called a **vertical asymptote** of the graph of $r(x)$. It is not actually part of the graph, but is there only to help describe the behavior of the graph. Notice that, since $x = 2$ is not in the domain of the function $r(x)$, the graph does *not* cross this vertical asymptote.

In general, if $r(x)$ is any rational function *that is in lowest terms*, then it has a vertical asymptote at each zero of its denominator. Figure 4.13a–d describes the four possibilities for the behavior of a rational function near one of its vertical asymptotes. From these figures, we can see that

the graph of a rational function does not cross its vertical asymptotes.

(In fact, the word *asymptote* probably comes from the Latin word *asymptotus*, which means "not meeting.")

The previous example exhibits the behavior described in Figure 4.13c. Let us consider another example whose behavior near a vertical asymptote is different from this one.

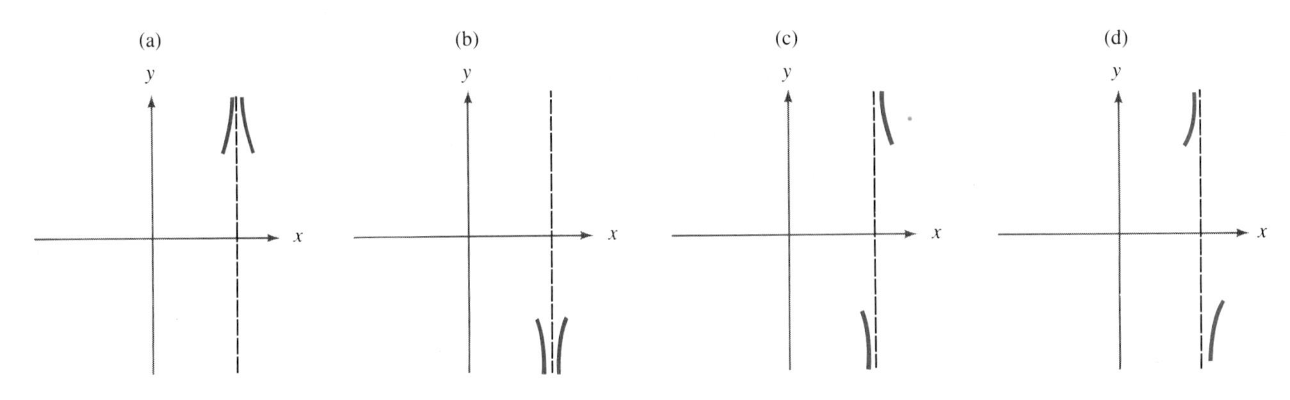

FIGURE 4.13.

EXAMPLE 2: Describe the behavior of the rational function

$$r(x) = \frac{-4}{(x+1)^2}$$

near the vertical asymptote $x = -1$.

Solution: In order to determine the behavior of $r(x)$ near this asymptote, we reason as follows. As x approaches -1 from the right, the denominator $(x + 1)^2$ approaches 0, but remains always positive. Therefore, since the numerator is negative, the quotient $r(x)$ *decreases* without bound. This behavior is pictured in Figures 4.13b and d (on the right side of the vertical asymptotes.)

As x approaches -1 from the left, $(x + 1)^2$ approaches 0—remaining always positive—because of the *even* exponent. Therefore, since the numerator is negative, the quotient $r(x)$ again decreases without bound. This behavior is shown in Figures 4.13b and c. Hence, we see that the behavior of $r(x)$ near the vertical asymptote $x = -1$ is as shown in Figure 4.13b. □

In applying the reasoning that we used in the previous examples, it may help to remember that a rational function must get large in absolute value on both sides of a vertical asymptote. We must determine whether it increases or decreases without bound and this can be done, as we did in the previous examples, by keeping track of the signs of the factors in the numerator and denominator. We will see more examples of this when we graph some rational functions a bit later in the section.

Try Study Suggestion 4.17.

▶ **Study Suggestion 4.17:**

(a) By using a reasoning similar to that in the previous examples, determine the behavior of the rational function $r(x) = -1/(x - 1)$ near the asymptote $x = 1$. (Don't forget that the numerator is negative.) Then sketch its graph near this asymptote.

(b) Repeat part *a* for the rational function $s(x) = 3x/(x - 2)^2$ near the vertical asymptote $x = 2$.

(*Hint:* for part (b), you will need to use the fact that, as x approaches 2, either from the right or from the left, the numerator $3x$ approaches $3 \cdot 2 = 6$, which is positive.) ◀

HORIZONTAL ASYMPTOTES OF THE GRAPH

Let us now turn to the problem of determining the behavior of a rational function when $|x|$ is large. The key to this behavior is in the degrees of the numerator and denominator. In order to illustrate the possibilities, consider the rational functions

$$r(x) = \frac{x + 1}{x^2 + 1} \qquad s(x) = \frac{6x + 1}{2x + 1} \qquad t(x) = \frac{x^2 + 1}{x + 1}$$

The degree of the numerator of $r(x)$ is less than the degree of the denominator. When this is the case, as $|x|$ gets large, the denominator grows large "faster" than the numerator. For example, when $x = 10$, the numerator of $r(x)$ is equal to 11, but the denominator is equal to $(10)^2 + 1 = 101$, and when $x = 1000$, the numerator is equal to 1001, but the denominator is equal to $(1000)^2 + 1 = 1{,}000{,}001$. Therefore, since the denominator grows faster than the numerator, the quotient will approach 0. The behavior of $r(x)$ when $|x|$ is large is shown in Figure 4.14. Notice that we have plotted two points in order to help determine whether the graph lies above or below the x-axis.

The degree of the numerator of $s(x)$ is equal to the degree of the denominator. In such a case, as $|x|$ gets large, the numerator and denominator "grow" at the same rate, and so the quotient $s(x)$ approaches a fixed nonzero number. This number is the quotient of the leading coefficients, which in this case is $6/2 = 3$. The behavior of $s(x)$ when $|x|$ is large is shown in Figure 4.15. In order to draw the graph more accurately, we have included the horizontal line $y = 3$. However, we should emphasize that it is *not* part of the graph of the function. Also, we have again plotted points to help determine whether the graph lies above or below this line.

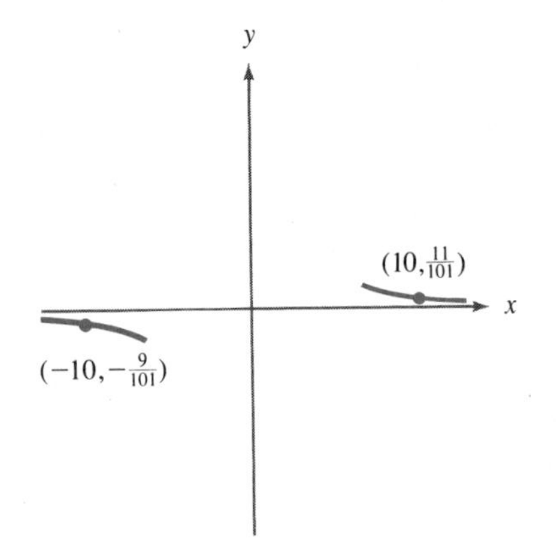

FIGURE 4.14

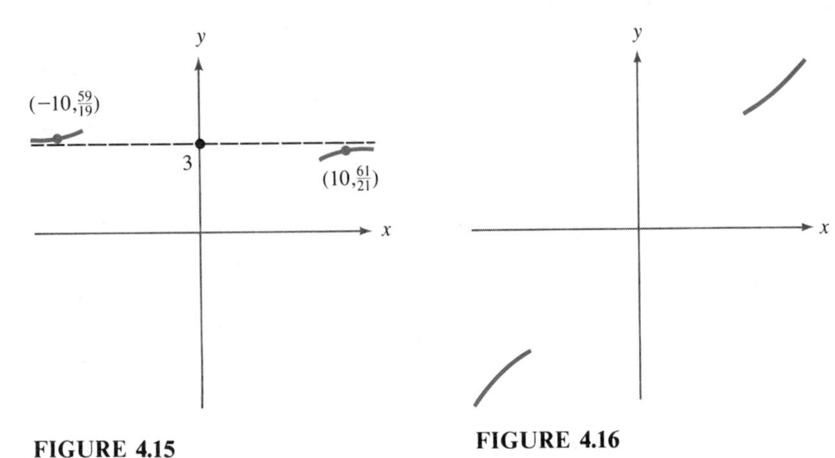

FIGURE 4.15

FIGURE 4.16

Finally, the degree of the numerator of $t(x)$ is greater than the degree of the denominator. In such a case, the numerator grows large faster than the denominator, and so the quotient $t(x)$ also grows large in absolute value. The behavior of $t(x)$ when $|x|$ is large is shown in Figure 4.16.

From this discussion, we see that when $|x|$ is large, a rational function $r(x)$ can exhibit three basic types of behavior

1. it can approach 0, as in Figure 4.14
2. it can "level off" by approaching some nonzero real number, as in Figure 4.15
3. it can get large in absolute value, as in Figure 4.16.

Furthermore, this behavior is determined by the degrees of the numerator and the denominator.

In the first two cases, the graph of the function gets closer and closer to a certain horizontal line as $|x|$ gets larger and larger. (In the first case, this line is the x-axis.) For this reason, such a line is called a **horizontal asymptote** of the graph. It is important to point out that

unlike the case of a vertical asymptote, it is possible for the graph of a rational function to cross a horizontal asymptote

Try Study Suggestion 4.18.

▶ **Study Suggestion 4.18:** Determine the behavior of the following rational functions for large values of x. Show this behavior on a graph.

(a) $r(x) = \dfrac{1}{x+1}$

(b) $s(x) = \dfrac{x^3}{x^2 - 1}$

(c) $t(x) = \dfrac{-x^2 + 1}{2x^2 - 3}$ ◀

A PROCEDURE FOR GRAPHING RATIONAL FUNCTIONS THAT ARE IN LOWEST TERMS

Now that we have discussed vertical and horizontal asymptotes, we can give a step-by-step procedure for graphing rational functions that are in lowest terms.

Procedure for Graphing Rational Functions That Are in Lowest Terms

STEP 1. Factor the numerator and denominator.

STEP 2. Locate the x- and y-intercepts of the graph. The y-intercept is found simply by setting $x = 0$. In order to find the x-intercepts, we must set $y = 0$ and solve for x. However, since a quotient is equal to 0 if and only if its numerator is equal to 0, *the x-intercepts of a rational function are just the zeros of its numerator.*

STEP 3. Locate the vertical and horizontal asymptotes (if any) on the graph.

STEP 4. Determine the behavior of the function near the vertical asymptotes.

STEP 5. Determine the behavior of the function when $|x|$ is large. This may require plotting a few points on the graph.

STEP 6. Sketch a smooth curve consistent with the information obtained in the previous steps. Remember that the *only* breaks in the graph of a rational function occur at the vertical asymptotes, and that it is possible for the graph to cross its horizontal asymptote.

EXAMPLE 3: Graph the rational function

$$r(x) = \frac{2x + 1}{x + 3}$$

Solution:

STEP 1. Nothing to do in this step.

STEP 2. The y-intercept of the graph is found by setting $x = 0$, which gives $y = 1/3$. The only zero of the numerator is $x = -1/2$, and so this is the only x-intercept. (See Figure 4.17.)

STEP 3. The graph has only one vertical asymptote, located at $x = -3$, and since the degree of the numerator is equal to the degree of the denominator, the graph has a horizontal asymptote at $y = 2/1 = 2$. (See Figure 4.17.)

STEP 4. As x approaches -3 from the right, the denominator approaches 0, but remains always positive. However, the numerator approaches $2(-3) + 1 = -5$, which is negative, and so the quotient $r(x)$ *decreases* without bound. We will leave it to you to show that,

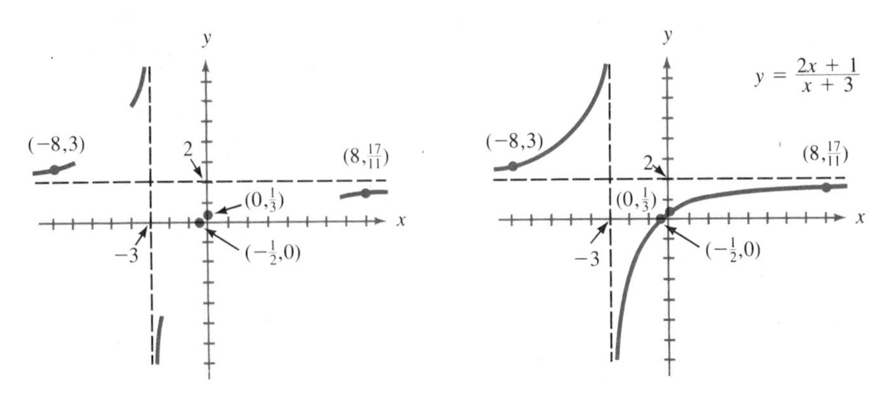

FIGURE 4.17

FIGURE 4.18

as x approaches -3 from the left, $r(x)$ increases without bound. This behavior is shown in Figure 4.17.

STEP 5. In order to determine the behavior of the function near the horizontal asymptote, we have plotted two points on the graph. (See Figure 4.17.)

STEP 6. Using Figure 4.17 as a guide, we can draw a smooth curve that fits the information that we have found in the previous steps. This is done in Figure 4.18. *Try Study Suggestion 4.19.* □

▶ **Study Suggestion 4.19:** Graph the function

$$r(x) = \frac{-x + 2}{x - 1}.$$ ◀

EXAMPLE 4: Graph the rational function

$$r(x) = \frac{x^3}{x^2 - 2x + 1}$$

Solution:

STEP 1. Factoring the denominator gives

$$r(x) = \frac{x^3}{(x - 1)^2}$$

From now on, we will use this form of the function.

STEP 2. Setting $x = 0$ gives the y-intercept 0. Similarly, the only x-intercept is at 0. Thus, the graph intersects the coordinate axes *only* at the origin. (See Figure 4.19a.)

STEP 3. The only vertical asymptote is located at $x = 1$, and since the degree of the numerator is greater than the degree of the denominator, the function itself gets large in absolute value as $|x|$ gets large. Hence, the graph has no horizontal asymptotes. (See Figure 4.19a.)

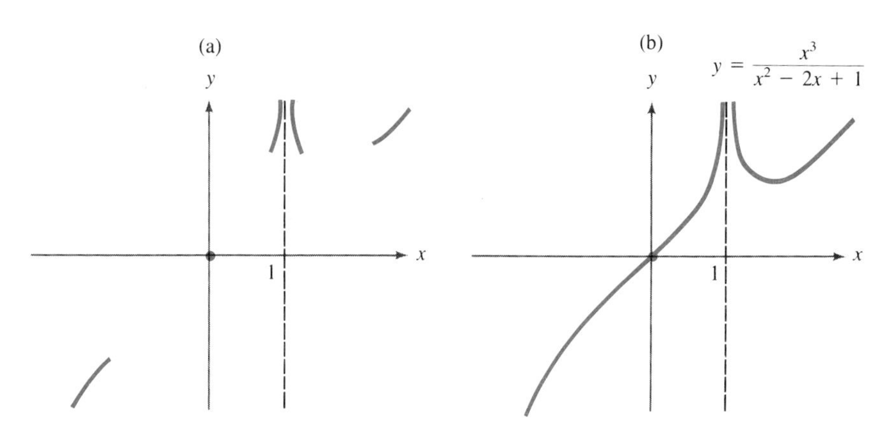

FIGURE 4.19

STEP 4. As x approaches 1—either from the right or the left—the denominator approaches 0, but remains always positive, because of the even exponent. Similarly, the numerator approaches 1, which is positive. Therefore, the quotient $r(x)$ increases without bound on *both* sides of the vertical asymptote. (See Figure 4.19a.)

STEP 5. As x increases without bound, so does $r(x)$, and as x decreases without bound, so does $r(x)$. (See Figure 4.19a.)

STEP 6. Using Figure 4.19a as a guide, we can draw a smooth curve that fits the information that we have found in the previous steps. This is done in Figure 4.19b. *Try Study Suggestion 4.20.* □

▶ **Study Suggestion 4.20:** Graph the function

$$r(x) = \frac{-x^3}{x^2 + 4x + 4}.$$ ◀

EXAMPLE 5: Graph the rational function

$$r(x) = \frac{1}{x^2 - 4}$$

Solution:

STEP 1. Factoring the denominator gives

$$r(x) = \frac{1}{(x + 2)(x - 2)}$$

STEP 2. The y-intercept is at $y = -1/4$, and since the numerator is never equal to 0, the graph has no x-intercepts. (See Figure 4.20a.)

STEP 3. The factored expression for $r(x)$ shows us that the graph has vertical asymptotes at $x = -2$ and $x = 2$. Since the degree of the numerator is less than the degree of the denominator, the function approaches 0 as $|x|$ gets large, and so the x-axis is a horizontal asymptote. (See Figure 4.20a.)

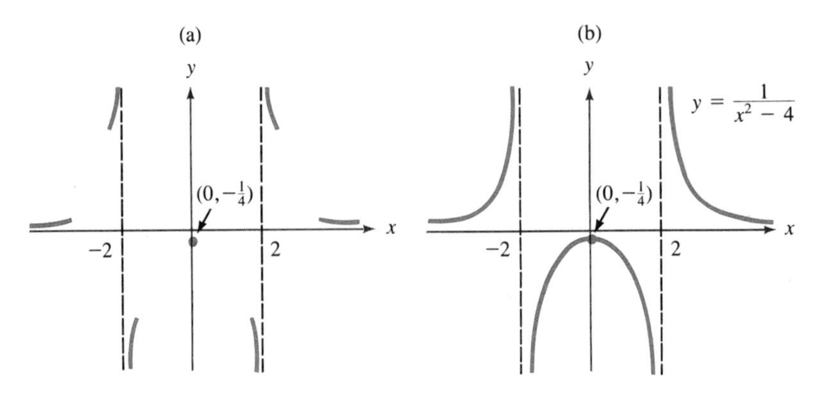

FIGURE 4.20

STEP 4. As x approaches -2 from the right, the factor $x + 2$ in the denominator approaches 0, but remains always positive. However, the factor $x - 2$ approaches -4, which is negative. Hence, the denominator approaches 0, but remains negative, and so the quotient $r(x)$ decreases without bound. This behavior is shown on the right side of the vertical asymptote $x = -2$. As x approaches -2 from the left, the factor $x + 2$ approaches 0, but remains always negative. Also, $x - 2$ approaches -4, which is negative. Hence, the denominator *increases* without bound. This is shown on the left side of the line $x = -2$. We will leave it to you to consider the behavior of the function near the asymptote $x = 2$.

STEP 5. Since the graph has no x-intercepts, it does not cross its horizontal asymptote. Hence, for large $|x|$ the graph lies above the x-axis. (See Figure 4.20a.)

STEP 6. Before drawing the graph, we should observe that the function is even, and so its graph is symmetric with respect to the y-axis. This graph is shown in Figure 4.20b. *Try Study Suggestion 4.21.* □

▶ **Study Suggestion 4.21:** Graph the rational function

$$r(x) = \frac{-2}{x^2 - 1}.$$ ◀

EXAMPLE 6: Graph the rational function

$$r(x) = \frac{1}{x^2 + 1}$$

Solution:

STEP 1. The numerator and denominator cannot be factored.

STEP 2. Setting $x = 0$, we see that the y-intercept is 1, and since the numerator is never equal to 0, there are no x-intercepts. (See Figure 4.21a.)

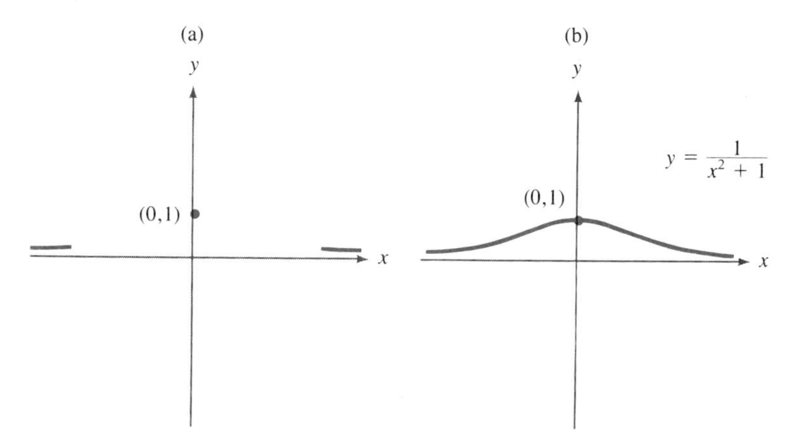

FIGURE 4.21

STEP 3. Since the denominator is never equal to 0, the graph has no vertical asymptotes. The degree of the numerator is less than the degree of the denominator, and so the x-axis is a horizontal asymptote.

STEP 4. Nothing to do here.

STEP 5. We can determine the behavior of this graph near the horizontal asymptote by observing that the function is always positive. Therefore, the graph must lie above this asymptote. This behavior is shown in Figure 4.21a.

STEP 6. Using the fact that the graph is symmetric about the y-axis, we can sketch the curve. (See Figure 4.21b.)

▶ **Study Suggestion 4.22:** Graph the rational function

$$r(x) = \frac{-2}{x^2 + 2}. \quad ◀$$

Try Study Suggestion 4.22. □

GRAPHING RATIONAL FUNCTIONS THAT ARE NOT IN LOWEST TERMS

We can sketch the graph of a rational function that is not in lowest terms by first cancelling all of the common factors of the numerator and denominator. However, *we must remember that this could have an effect on the domain of the function, and so we must determine the domain before cancelling the common factors.*

EXAMPLE 7: Graph the rational function

$$r(x) = \frac{2x^2 - x - 1}{x^2 + 2x - 3}$$

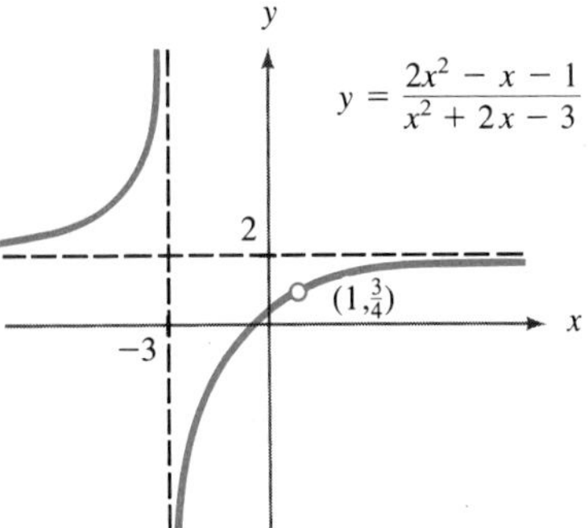
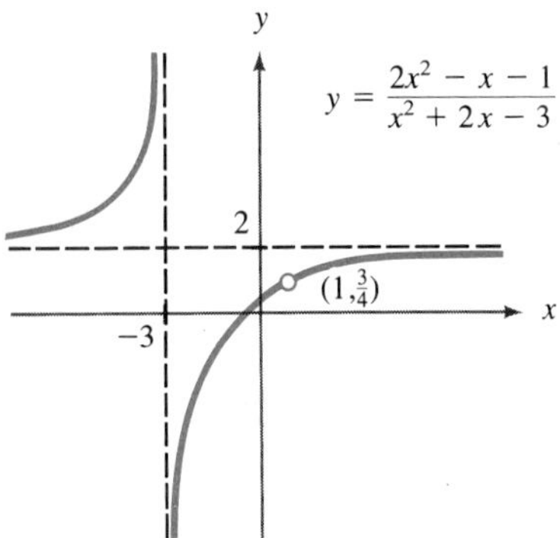

FIGURE 4.22

Solution: First we factor the numerator and denominator, to get

$$r(x) = \frac{(2x + 1)(x - 1)}{(x + 3)(x - 1)}$$

This shows that the domain of r is the set of all real numbers except -3 and 1. Once we have determined the domain, we can cancel the common factors, and write

$$r(x) = \frac{2x + 1}{x + 3} \qquad \text{for } x \neq -3, 1$$

Now we can proceed with Steps 2–6 given in the previous subsection. However, we have already done this in Example 3. Thus the graph of r is the same as the graph in Figure 4.18 *except* that since $x = 1$ is not in the domain of r, the graph of r has a hole at the point (1, 3/4). The graph of r is shown in Figure 4.22. *Try Study Suggestion 4.23.* □

▶ **Study Suggestion 4.23:** Graph the rational function

$$r(x) = \frac{x^2 - 3x}{x - 3}$$ ◀

Ideas to Remember

In order to graph a rational function, it is important to determine its behavior not only when $|x|$ is large, but also for values for x near its vertical asymptotes.

EXERCISES

In Exercises 1–8, find all vertical asymptotes of the given rational function, and sketch the graph in the neighborhood of these asymptotes.

1. $r(x) = \frac{1}{x - 4}$

2. $r(x) = \frac{-4}{2x - 3}$

3. $r(x) = \frac{x}{x^2 - 2}$

4. $r(x) = \frac{1}{(x - 3)(x + 4)}$

5. $r(x) = \frac{-2}{(x - 1)^2(x + 1)}$

6. $r(x) = \frac{x + 1}{(x - 1)(x + 2)}$

7. $r(x) = \frac{-x^2}{(x - 1)^3}$

8. $r(x) = \frac{-x}{(x + 1)^2(x + 2)}$

In Exercises 9–16, find the horizontal asymptote of the given rational function (if it has one.) Then sketch the graph for large values of x.

9. $r(x) = \frac{1}{x - 3}$

10. $r(x) = \frac{x^2}{x^4 + 1}$

11. $r(x) = \frac{x - 1}{3x - 2}$

12. $r(x) = \frac{2x^3 + 1}{x - 4}$

13. $r(x) = \frac{-3x + 2x^2}{7x - 1}$

14. $r(x) = \frac{2 - 3x}{6x}$

15. $r(x) = \frac{(3x + 1)(2x - 1)}{x + 5}$

16. $r(x) = \frac{4x - 1}{3x^2 - 7x + 1}$

In Exercises 17–36, graph the given rational function.

17. $r(x) = \frac{-1}{2x + 1}$

18. $r(x) = \frac{2}{x - 3}$

19. $r(x) = \frac{x - 2}{2x + 1}$

20. $r(x) = \frac{3 + x - 2x^2}{3 + 2x - x^2}$

21. $r(x) = \frac{2x - 3}{7x + 4}$

22. $r(x) = \frac{x^2}{x^3 + 2x}$

23. $r(x) = \frac{-2x + 8}{(x^2 - 2)(x - 4)}$

24. $r(x) = \frac{3}{x^2 - 3x - 4}$

25. $r(x) = \dfrac{x}{x^2 - 5}$

26. $r(x) = \dfrac{2x^2 - 3x + 1}{x - 2}$

27. $r(x) = \dfrac{3x^2 + x + 1}{x + 2}$

28. $r(x) = \dfrac{x - 1}{x^2}$

29. $r(x) = \dfrac{x^2 - 1}{x^2 + 3x + 2}$

30. $r(x) = \dfrac{x^2 + 4x + 4}{x^2 - 4}$

31. $r(x) = \dfrac{-x^2}{x^2 - 12x + 20}$

32. $r(x) = \dfrac{2x^2}{x^2 + 2}$

33. $r(x) = \dfrac{x^2 + 1}{x^2 + x + 1}$

34. $r(x) = \dfrac{1}{x^3 - x}$

35. $r(x) = \dfrac{x^2 + 1}{x^3 + 1}$

36. $r(x) = \dfrac{(x - 1)(x - 2)(x - 3)}{(x - 4)(x - 5)}$

4.7 Partial Fractions

In Chapter 1, we spent some time learning how to combine rational expressions. For example, by taking a common denominator, it is easy to see that

$$\frac{1}{2(x-1)} - \frac{1}{2(x+1)} = \frac{(x+1)-(x-1)}{2(x-1)(x+1)} = \frac{1}{x^2-1}$$

However, in many areas of mathematics, especially in calculus, it is important to be able to go in the other direction; that is, to "break apart" a rational expression such as $1/(x^2 - 1)$ into a sum, or difference, of rational expressions that are "simpler" in the sense that they have denominators of smaller degree. This process is called decomposing a rational expression into **partial fractions.** For instance, the equation

$$\frac{1}{x^2-1} = \frac{1}{2(x-1)} - \frac{1}{2(x+1)} \tag{4.31}$$

is the decomposition of

$$\frac{1}{x^2-1}$$

into the partial fractions

$$\frac{1}{2(x-1)} \quad \text{and} \quad \frac{1}{2(x+1)}.$$

In this section, we will discuss a method for finding the partial fractions decomposition of a given rational expression. We will not consider the most general case here, but we will be general enough so that you can get a feel for the techniques involved. We will assume here, for example, that the degree of the numerator is less than the degree of the denominator. (If you take calculus, you will study partial fractions in more detail.)

The first step in decomposing a rational expression into partial fractions is to factor the denominator as much as possible. Once factored, the denominator will be in the form of the product of powers of linear factors and powers of irreducible quadratic factors. (Recall from Section 1.3 that a polynomial is *irreducible* if it has no nontrivial factors.)

For example, the denominator might have the form $x(x - 3)^2(2x^2 + 4)$, which is the product of the first power of the linear factor x, the second power of the linear factor $x - 3$, and the first power of the irreducible quadratic factor $2x^2 + 4$.

Now, each factor of the denominator will produce one or more terms in the partial fractions decomposition. Table 4.1 gives the terms that are required in the decomposition for factors of the form $ax + b$, $(ax + b)^2$ and $ax^2 + bx + c$. (We will restrict our attention to denominators that can be written as products of just these types of factors.) In this table, the letters A and B are yet to be determined constants.

TABLE 4.1

Type of factor in the denominator	Terms in the partial fractions decomposition
first power of linear factor: $ax + b$	$\frac{A}{ax + b}$
second power of a linear factor: $(ax + b)^2$	$\frac{A}{ax + b} + \frac{B}{(ax + b)^2}$
first power of an *irreducible* quadratic factor: $ax^2 + bx + c$	$\frac{Ax + B}{ax^2 + bx + c}$

Now let us illustrate our technique for finding partial fractions decompositions with several examples.

EXAMPLE 1: Find the partial fractions decomposition of the rational expression

$$\frac{1}{x^2 - 4}$$

Solution: The first step is to factor the denominator.

$$\frac{1}{x^2 - 4} = \frac{1}{(x - 2)(x + 2)}$$

Now we can see that the denominator is the product of two factors of the form $ax + b$. Hence, according to Table 4.1, the decomposition into partial fractions has two terms of the form given in the first entry of the table. In particular, the decomposition has the form

$$\frac{1}{(x-2)(x+2)} = \frac{A}{x-2} + \frac{B}{x+2} \qquad (4.32)$$

where A and B are as yet to be determined constants.

In order to determine the values of A and B, we first clear Equation (4.32) of all denominators by multiplying both sides by $(x-2)(x+2)$. This gives

$$1 = A(x+2) + B(x-2) \qquad (4.33)$$

Now we can take advantage of the fact that this equation must hold for *all* values of x. Setting $x = 2$ makes the term $B(x-2)$ equal to 0, and gives

$$1 = A(2+2)$$

In other words, $1 = 4A$ and so $A = 1/4$. Setting $x = -2$ in Equation (4.33) makes the term $A(x+2)$ equal to 0, and gives

$$1 = B(-2-2)$$

Hence, $B = -1/4$. Putting these values of A and B into Equation (4.32) gives the decomposition

$$\frac{1}{(x-2)(x+2)} = \frac{1/4}{x-2} + \frac{-1/4}{x+2}$$

which can also be written in the form

$$\frac{1}{(x-2)(x+2)} = \frac{1}{4(x-2)} - \frac{1}{4(x+2)}$$

Try Study Suggestion 4.24. □

Study Suggestion 4.24:

(a) Use the method of Example 1 to obtain the decomposition of Equation (4.31).

(b) Find the partial fractions decomposition of the rational expression $3/(x^2-9)$.

EXAMPLE 2: Find the partial fractions decomposition of the rational expression

$$\frac{5x^2 - 3x - 6}{x(x-1)(2x-3)}$$

Solution: In this case, the denominator is already factored and we see from Table 4.1 that the partial fractions decomposition has the form

$$\frac{5x^2 - 3x - 6}{x(x-1)(2x-3)} = \frac{A}{x} + \frac{B}{x-1} + \frac{C}{2x-3} \qquad (4.34)$$

where A, B, and C are constants. In order to determine the values of A, B and C, we first clear all denominators by multiplying both sides of Equation (4.34) by $x(x-1)(2x-3)$. This gives

$$5x^2 - 3x - 6 = A(x-1)(2x-3) + Bx(2x-3) + Cx(x-1) \qquad (4.35)$$

Now we can use the same technique that we used in the previous example to make various terms on the right side of this equation equal to 0. By choosing $x = 0$, the second and third terms on the right become 0, and we are left with

$$-6 = A(-1)(-3)$$

Hence, $A = -6/3 = -2$. Choosing $x = 1$ makes the first and third terms on the right of Equation (4.35) equal to 0, and we get

$$-4 = B \cdot 1 \cdot (-1)$$

and so $B = 4$. Finally, choosing $x = 3/2$ makes the first and second terms on the right of Equation (4.34) equal to 0, and we get

$$5\left(\frac{3}{2}\right)^2 - 3\left(\frac{3}{2}\right) - 6 = C\left(\frac{3}{2}\right)\left(\frac{3}{2} - 1\right)$$

Simplifying this equation and solving for C gives $C = 1$. Now we can substitute these values of A, B, and C into Equation (4.34) to get the partial fractions decomposition

$$\frac{5x^2 - 3x - 6}{x(x-1)(2x-3)} = \frac{-2}{x} + \frac{4}{x-1} + \frac{1}{2x-3}$$

Try Study Suggestion 4.25. □

▶ **Study Suggestion 4.25:** Find the partial fractions decomposition of the rational expression

$$\frac{-11x^2 + 15x - 6}{x(x-2)(3x-1)}$$ ◀

EXAMPLE 3: Find the partial fractions decomposition of the rational expression

$$\frac{3x^2 - 10x + 4}{x(x-2)^2}$$

Solution: Table 4.1 tells us that the partial fractions decomposition has the form

$$\frac{3x^2 - 10x + 4}{x(x-2)^2} = \frac{A}{x} + \frac{B}{x-2} + \frac{C}{(x-2)^2} \tag{4.36}$$

Again we clear the denominators by multiplying both sides of Equation (4.36) by $x(x-2)^2$, to get

$$3x^2 - 10x + 4 = A(x-2)^2 + Bx(x-2) + Cx \tag{4.37}$$

Setting $x = 0$ in this equation makes the last two terms on the right side equal to 0, and gives $4 = A(-2)^2$, or $A = 1$. Setting $x = 2$ in Equation (4.37) makes the first two terms on the right side equal to 0, and gives $-4 = C(2)$, or $C = -2$.

In this case however, there is no value of x that will make *only* the first and third terms on the right side of Equation (4.37) equal to 0. Therefore, in order to determine the value of B, we substitute the values obtained for

A and C into Equation (4.37), and then solve for B

$$\begin{aligned} 3x^2 - 10x + 4 &= (x-2)^2 + Bx(x-2) - 2x \\ 3x^2 - 10x + 4 &= x^2 - 4x + 4 + B(x^2 - 2x) - 2x \\ 2x^2 - 4x &= B(x^2 - 2x) \\ 2(x^2 - 2x) &= B(x^2 - 2x) \\ 2 &= B \end{aligned}$$

Thus, $B = 2$ and the decomposition is

$$\frac{3x^2 - 10x + 4}{x(x-2)^2} = \frac{1}{x} + \frac{2}{x-2} - \frac{2}{(x-2)^2}$$

Try Study Suggestion 4.26. □

▶ **Study Suggestion 4.26:** Find the partial fractions decomposition of the rational expression

$$\frac{x^2 - 9x + 3}{x^2(2x-1)}$$ ◀

EXAMPLE 4: Find the partial fractions decomposition of the rational expression

$$\frac{5x^2 + 4x + 1}{(x+1)(x^2 + x + 1)}$$

Solution: Since the quadratic expression $x^2 + x + 1$ has a negative discriminant, it cannot be factored into linear factors. In other words, it is irreducible, and so the denominator is factored as much as possible. According to Table 4.1, the partial fractions decomposition has the form

$$\frac{5x^2 + 4x + 1}{(x+1)(x^2 + x + 1)} = \frac{A}{x+1} + \frac{Bx + C}{x^2 + x + 1} \tag{4.38}$$

Multiplying both sides of this equation by $(x + 1)(x^2 + x + 1)$ gives

$$5x^2 + 4x + 1 = A(x^2 + x + 1) + (Bx + C)(x + 1) \tag{4.39}$$

Setting $x = -1$ makes the second term on the right side of Equation (4.39) equal to 0, and gives $2 = A(1)$ and so $A = 2$. Setting $x = 0$ in Equation (4.39) gives $1 = A + C$ and since $A = 2$, we get $C = 1 - A = 1 - 2 = -1$.

Finally, we can substitute the values of A and C into Equation (4.39), and solve the resulting equation for B. We will leave it to you to show that this gives $B = 3$ and so the partial fractions decomposition in this case is

$$\frac{5x^2 + 4x + 1}{(x+1)(x^2 + x + 1)} = \frac{2}{x+1} + \frac{3x - 1}{x^2 + x + 1}$$

Try Study Suggestion 4.27. □

▶ **Study Suggestion 4.27:** Find the partial fractions decomposition of the rational expression

$$\frac{1}{(x+1)(x^2+1)}$$ ◀

Ideas to Remember

Sometimes it is important to be able to "break apart" a rational expression into a sum, or difference, of rational expressions that are "simpler" in the sense that they have denominators of smaller degree.

EXERCISES

In Exercises 1–26, find the partial fractions decomposition of the given rational expression.

1. $\dfrac{1}{(x-1)(x-3)}$

2. $\dfrac{2}{x^2-6x+8}$

3. $\dfrac{-1}{x^2-3}$

4. $\dfrac{x}{x^2+4x-5}$

5. $\dfrac{x-2}{x^2-2x-3}$

6. $\dfrac{-2}{2x^2+x}$

7. $\dfrac{2x-1}{(x+3)(2x+5)}$

8. $\dfrac{1}{x(x-1)(x+3)}$

9. $\dfrac{2}{x(x^2-1)}$

10. $\dfrac{x}{(x^2-1)(x-2)}$

11. $\dfrac{2x-1}{x(x^2-5x+6)}$

12. $\dfrac{1}{x^2(x-1)}$

13. $\dfrac{1}{x(x-1)^2}$

14. $\dfrac{-2}{x(2x-1)^2}$

15. $\dfrac{4x^2-11x+9}{(x-1)^2(x-2)}$

16. $\dfrac{5x^2+16x+4}{2x(x^2-4x+4)}$

17. $\dfrac{2x^2-x+1}{x(x^2+x+1)}$

18. $\dfrac{x^2-5x-4}{x(x^2+2x+2)}$

19. $\dfrac{4x^2-2x-1}{(x-1)(2x^2+1)}$

20. $\dfrac{1}{x^2(3x^2+1)}$

21. $\dfrac{-9x^3+7x^2-2x+1}{x^2(4x^2+1)}$

22. $\dfrac{x-1}{x^2(4x^2+1)}$

23. $\dfrac{1}{(x-1)(x^2+1)}$

24. $\dfrac{1}{(x-1)^2(x+1)^2}$

25. $\dfrac{x-2}{x(x^2+1)}$

26. $\dfrac{x}{x^4-1}$

4.8 Review

CONCEPTS FOR REVIEW

Divisor
Dividend
Quotient
Remainder
Synthetic division
The Remainder Theorem
The Factor Theorem
Multiplicity
Simple root
Multiple root
Double root
Triple root
The Rational Root Theorem
Bisection method
Iteration
Local minimum point
Local maximum point
Local extreme point
Rational function
Vertical asymptote
Horizontal asymptote
Partial fractions decomposition

REVIEW EXERCISES

In Exercises 1–3, find the quotient and remainder when $p(x)$ is divided by $q(x)$. Do *not* use synthetic division.

1. $p(x) = 2x^2 - 3x + 1$, $q(x) = x^3 - 1$

2. $p(x) = 2x^4 - 3x^2 + 1$, $q(x) = x^2 + 2x + 1$

3. $p(x) = 2x^5 - 12x^4 + 5x^3 - 3x^2 + 2x - 1$, $q(x) = x^2 - 1$

4. Repeat Exercise 1 using synthetic division.

5. Repeat Exercise 2 using synthetic division.

6. Repeat Exercise 3 using synthetic division.

7. Use the Remainder Theorem to compute $p(a)$ if $p(x) = 2x^3 - 3x^2 + 9x - 1$ and $a = -5$.

8. Use the Factor Theorem to show that $x = -4$ is a solution to the equation $x^4 + 7x^3 + 10x^2 + x + 36 = 0$.

9. Use the Factor Theorem to show that $3x + 2$ is a factor of $27x^3 + 54x^2 + 27x + 2$.

10. Find all values of a for which $2ax^2 + a^2x + 3$ is divisible by $3x - 3$.

11. The number $1/2$ is a root of the cubic polynomial $2x^3 - (1/6)x^2 - (3/4)x + 1/6$. Find the other roots of this polynomial.

In Exercises 12–14, use the method of Section 4.3 to find a rational solution to the given equation, if one exists.

12. $4x^3 - 5x^2 + 10x - 12 = 0$

13. $x^3 + (3/2)x^2 - 7x - 15/2 = 0$

14. $20x^5 - 9x^4 - 74x^3 + 30x^2 + 42x - 9 = 0$

In Exercises 15–17, find all solutions to the given equation.

15. $6x^3 + 19x^2 + x - 6 = 0$

16. $2x^4 - 3x^3 - 7x^2 + 12x - 4 = 0$

17. $x^4 + 3x^3 - 30x^2 - 6x + 56 = 0$

In Exercises 18–20, approximate a solution to the given equation by performing three iterations of the bisection method.

18. $x^3 - 4x^2 + x - 1 = 0$

19. $-2x^3 + 5x - 2 = 0$

20. $x^5 - 3x^3 + 9 = 0$

21. Use the method of Section 4.4 to approximate $\sqrt[4]{3}$. (Perform 4 iterations of the bisection method.)

In Exercises 22–28, graph the given function.

22. $p(x) = 2x^3 - 3x^2$

23. $p(x) = x^3 - x^2 - 25x + 25$

24. $p(x) = (x + 1)^3(x - 2)^4$

25. $r(x) = \dfrac{1}{x^4}$

26. $r(x) = \dfrac{2x^2 - 9x - 5}{x^2 - 9x + 20}$

27. $r(x) = \dfrac{-3x}{x^2 + 5}$

28. $r(x) = \dfrac{-x^3}{x^2 - 7x + 12}$

In Exercises 29–34, find the partial fractions decomposition of the given expression.

29. $\dfrac{1}{(x - 4)(x + 3)}$

30. $\dfrac{1}{x^2 - 5}$

31. $\dfrac{2x - 7}{(x - 1)(2x + 3)}$

32. $\dfrac{1}{x^2(x + 6)}$

33. $\dfrac{2x^2 - x - 1}{x(x^2 + 3x + 3)}$

34. $\dfrac{1}{(x - 1)(x - 2)^2}$

5 EXPONENTIAL AND LOGARITHMIC FUNCTIONS

5.1 The Exponential Functions

Up to now, we have been considering mainly polynomial and rational functions, and simple functions involving fractional powers. In this and the next three chapters, we will embark on the study of some new types of functions, which have important applications in mathematics, science and business.

Let us begin our study with the *exponential functions*. As an example, consider the function

$$f(x) = 2^x$$

Notice that the variable x appears in the *exponent*, which is why this function is called an *exponential* function. The domain of f is the set of all real numbers.

It is important not to confuse the *exponential* function $f(x) = 2^x$ with the *polynomial* function $g(x) = x^2$. As we will see, these functions are quite different.

We should say a few words about the domain of the function f. In Chapter 1 we learned how to evaluate expressions of the form $2^{p/q}$; that is, we learned how to raise a number to a rational power. In terms of the exponential function, this means that we can evaluate $f(p/q)$. But we did not discuss irrational exponents in Chapter 1, and so we do not know how to evaluate $f(r)$ when r is an irrational number. For example, how do we evaluate $f(\pi) = 2^\pi$?

In order to give a thorough discussion of irrational exponents, we need to know a bit of calculus. But the general idea is really quite simple. Since π is an irrational number, it can be represented by a nonrepeating decimal. The first few digits in this decimal are

$$\pi = 3.1415926 \cdots$$

Using this representation, we can approximate π as accurately as we wish using *rational* numbers. In fact, the rational numbers

$$3, 3.1, 3.14, 3.141, 3.1415, \ldots \qquad (5.1)$$

and so on, provide a better and better approximation to π. For this reason, we say that they *approach* π.

Since the numbers in (5.1) are rational, we know how to evaluate the expressions

$$2^3, 2^{3.1}, 2^{3.14}, 2^{3.141}, 2^{3.145}, \ldots \qquad (5.2)$$

Now it makes sense that if the numbers in (5.1) approach π, then the numbers in (5.2) should approach the number 2^π. Therefore, we can actually *define* 2^π as the number that is "approached" by the numbers in (5.2). (See Exercise 8.) This same idea can be used to define a^r where a is any positive real number and r is any irrational number.

For practical purposes however, we will need to compute, or at least approximate, numbers such as 2^π, and this is where a calculator with a y^x key is very useful. We suggest that you spend some time now familiarizing yourself with this key

Study Suggestion 5.1: Use the y^x key on your calculator to approximate the following values. **(a)** 2^π **(b)** $\sqrt{2}^{\sqrt{2}}$ **(c)** $2^{(\sqrt{2}+1)}$ **(d)** $\sqrt{3}^{-\sqrt{2}}$ ◀

Try Study Suggestion 5.1.

The number 2 in the exponential function $f(x) = 2^x$ is called the *base* of the function. There is nothing to prevent us from defining exponential functions with other bases.

DEFINITION

An **exponential function** is any function of the form

$$f(x) = a^x$$

where a is a *positive* real number and $a \neq 1$. The domain of f is the set of all real numbers. The number a is called the **base** of the function.

We exclude $a = 1$ since $f(x) = 1^x = 1$ is just a constant function.

EXAMPLE 1: Graph the exponential functions

(a) $f(x) = 2^x$ **(b)** $g(x) = (1/2)^x$

Solutions:

(a) Plotting a few points on the graph of the exponential function $f(x) = 2^x$, and drawing a smooth curve through them gives the graph shown in Figure 5.1. Notice that the function f is increasing, and that the negative portion of the x-axis is a horizontal asymptote of the graph, in the sense that as x decreases without bound the function f approaches, but never reaches, 0.

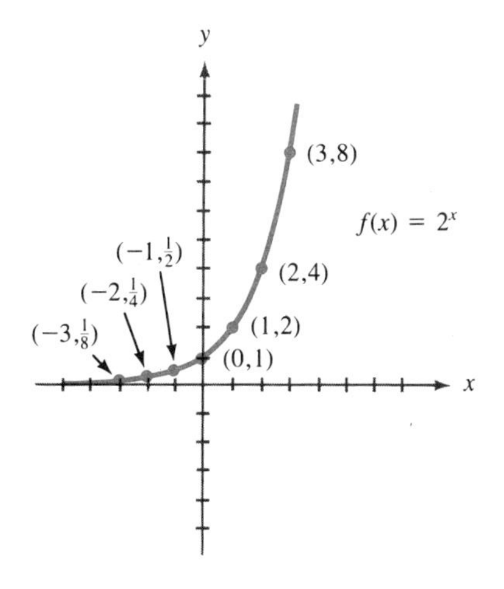

FIGURE 5.1

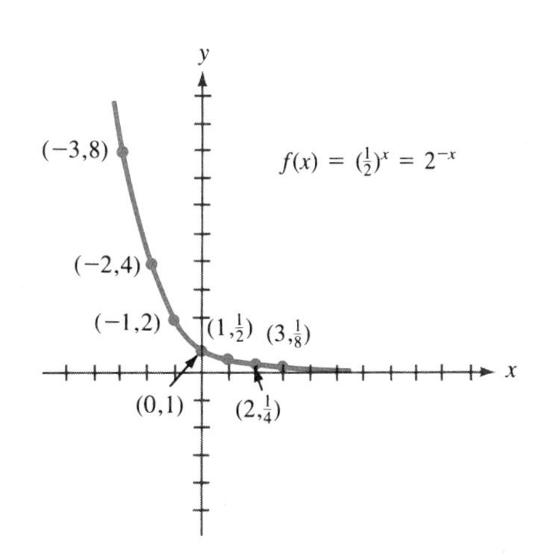

FIGURE 5.2

(b) Again by plotting points, we obtain the graph shown in Figure 5.2. We can also obtain this graph by observing that g can be written in the form

$$g(x) = 2^{-x}$$

Hence, the graph of g can be obtained by reflecting the graph of $f(x) = 2^x$ about the y-axis. The function g is decreasing, and the positive portion of the x-axis is a horizontal asymptote

Try Study Suggestion 5.2. □

▶ **Study Suggestion 5.2:** Sketch the graphs of the functions $f(x) = 3^x$ and $g(x) = (1/3)^x$. ◀

Figure 5.3a and b shows the graphs of the exponential function $f(x) = a^x$ for different values of the base a. These graphs contain a great deal of information about the exponential functions. For example, the graph in Figure 5.3a shows that for $a > 1$, the exponential function is increasing, and its range is the set of all *positive* real numbers. Thus the quantity a^x is always positive. Similarly, the graph in Figure 5.3b shows that for $0 < a < 1$, the exponential

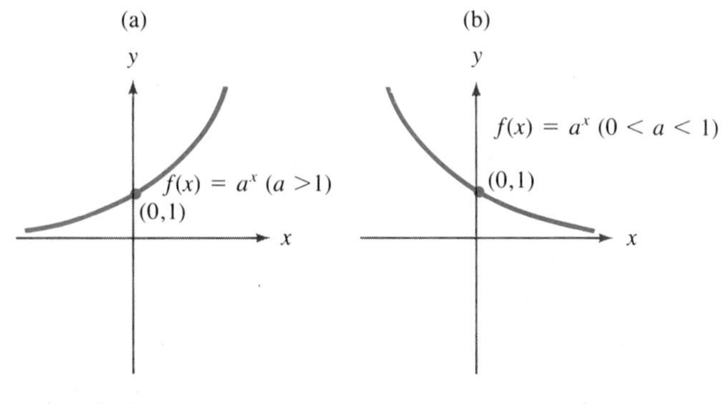

FIGURE 5.3

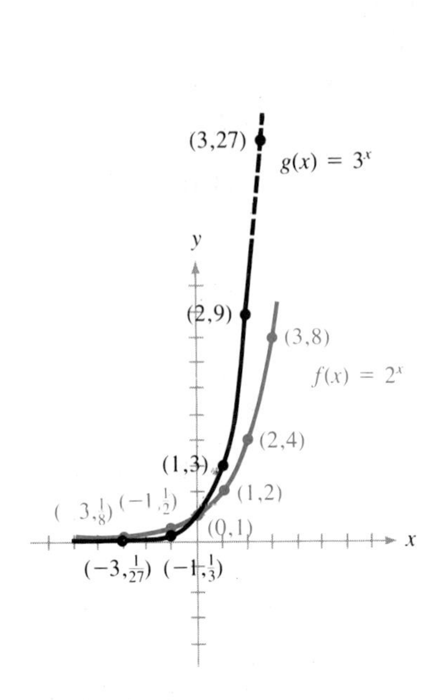

FIGURE 5.4

function is decreasing, and its range is again the set of all positive real numbers. Hence, we can say

if $a > 0$ then a^x is positive for all values of x.

EXAMPLE 2: Sketch the graphs of the exponential functions $f(x) = 2^x$ and $g(x) = 3^x$ on the same plane.

Solution: Using the method of plotting points, we obtain the graphs shown in Figure 5.4. We can see that the function $g(x) = 3^x$ grows faster than $f(x) = 2^x$ as x increases without bound, and it approaches 0 faster than f as x decreases without bound. *Try Study Suggestion 5.3.* □

▶ **Study Suggestion 5.3:** Sketch the graphs of the functions $f(x) = 2^x$ and $g(x) = 5^x$ on the same plane. (Be sure to choose an appropriate scale factor for the y-axis. The function g gets rather large even for relatively small values of x.) ◀

It is interesting to compare the graphs of $f(x) = 2^x$ and $g(x) = x^2$. This is done in Figure 5.5. Of course, they are quite different for negative values of x. But for positive values of x there is a more subtle difference, namely,

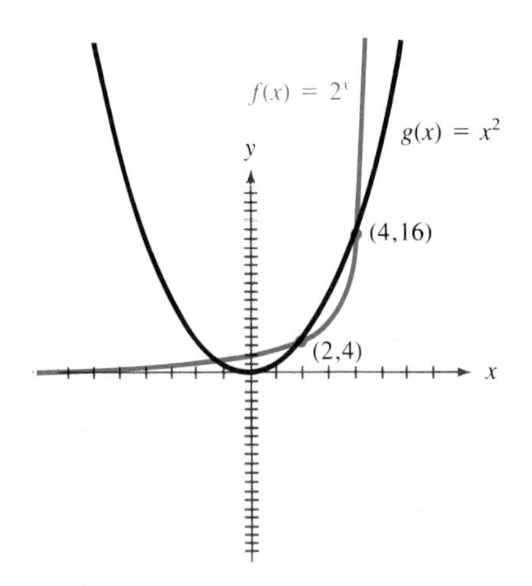

FIGURE 5.5

TABLE 5.1

x	x^2	2^x
1	1	2
3	9	8
5	25	32
10	100	1024
15	225	32,768
20	400	1,048,576
25	625	33,554,432
30	900	1,073,741,820

the exponential function $f(x) = 2^x$ grows much faster than the function $g(x) = x^2$ as x increases without bound. We can see this more clearly by comparing a few values of x^2 and 2^x, as in Table 5.1.

This table shows quite clearly that even for relatively small values of x, 2^x grows *much* faster than x^2. We will see later in this chapter that the extremely rapid growth rate of the exponential functions has very important consequences for the world we live in.

EXAMPLE 3: Graph the function

$$f(x) = 2^x + 2^{-x}$$

Solution: A convenient way to sketch this graph is to first sketch the graphs of 2^x and 2^{-x} on the same plane, and then add the y-coordinates for each value of x, as we have done in Figure 5.6. This is called the *method of adding ordinates.* (The y-coordinate of an ordered pair is sometimes called the **ordinate,** and the x-coordinate is sometimes called the **abscissa.**)

Try Study Suggestion 5.4. □

▶ **Study Suggestion 5.4:** Graph the function $g(x) = x + 2^x$. (*Hint:* use the technique of adding ordinates that we used in Example 3.) ◀

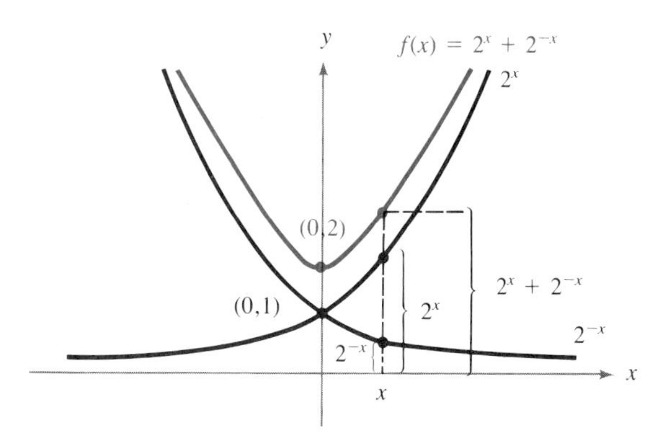

FIGURE 5.6

EXAMPLE 4: Sketch the graph of the function

$$f(x) = 2^{-x^2}$$

Solution: We begin by observing that the graph is symmetric with respect to the y-axis. (Why?) Therefore, we only need to determine the graph for nonnegative values of x, and then reflect about the y-axis. Setting $x = 0$, and observing that $2^0 = 1$, we get the y-intercept $y = 1$. Since f can be written in the form

$$f(x) = \frac{1}{2^{x^2}}$$

we see that as x increases without bound, the function approaches 0, always remaining positive. Therefore, the x-axis is a horizontal asymptote of the graph. Plotting a few points and drawing a smooth curve through them, we obtain the graph pictured in Figure 5.7. *Try Study Suggestion 5.5.* □

▶ **Study Suggestion 5.5:** Graph the function $g(x) = 2^{|x|}$. (*Hint:* check for symmetry.) ◀

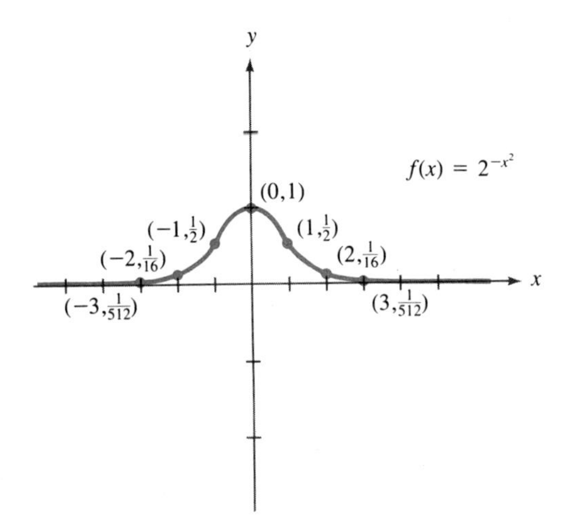

FIGURE 5.7

It is sometimes important in applications to be able to solve equations that have variables appearing in the exponents. Such equations are called **exponential equations.** Our main tool for solving such equations is the fact that, provided $a > 0$ and $a \neq 1$,

$$a^x = a^y \text{ if and only if } x = y \tag{5.3}$$

Let us consider some examples of the use of Statement (5.3).

EXAMPLE 5: Solve the following exponential equations.

(a) $2^x = 8$ **(b)** $3^{2x+1} = 27^x$ **(c)** $2^x 3^x = 36$

Solutions:

(a) Since $8 = 2^3$, this equation can be written in the form

$$2^x = 2^3$$

and Statement (5.3) then tells us that $x = 3$.

(b) Since $27 = 3^3$, this equation can be written in the form

$$3^{2x+1} = (3^3)^x$$

or

$$3^{2x+1} = 3^{3x}$$

Therefore, according to Statement (5.3), we have $2x + 1 = 3x$. Solving this for x gives $x = 1$. (You should check this by substituting into the original equation.)

(c) Since $2^x 3^x = (2 \cdot 3)^x = 6^x$, and $36 = 6^2$, we can write this equation in the form

$$6^x = 6^2$$

Hence $x = 2$. (Again you should check this solution.)

▶ **Study Suggestion 5.6:** Solve the following exponential equations.

(a) $3^x = 81$

(b) $(1/2)^{2x} = 1/4$

(c) $2^x 3^{-x} = 9/4$ ◀

Try Study Suggestion 5.6. □

Among the exponential functions, there is one that stands out as being the most important. This is the exponential function whose base is a certain irrational number, denoted by the letter e. Since e, like π, is irrational, its decimal expansion is nonrepeating. The first few digits of this expansion are

$$e = 2.71828 \cdots$$

Thus e is an irrational number between 2.5 and 3.

The exponential function $E(x)$ whose base is e is

$$E(x) = e^x$$

Many mathematicians and scientists feel that this is the most important function in all of mathematics. Most scientific calculators have a special key marked e^x, for computing values of this function. We suggest that you spend a few minutes learning how to use this key.

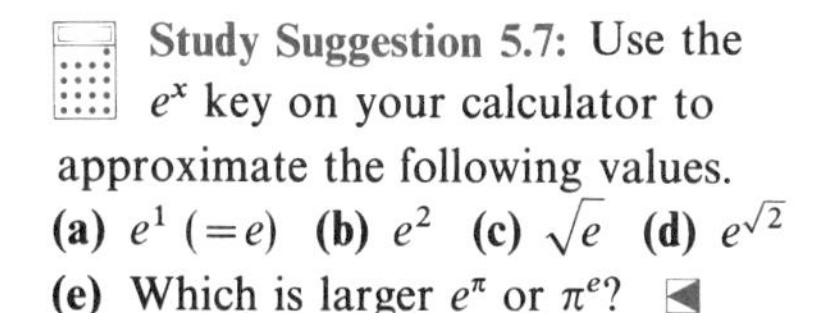

Study Suggestion 5.7: Use the e^x key on your calculator to approximate the following values.
(a) e^1 $(=e)$ **(b)** e^2 **(c)** $\sqrt{e}$ **(d)** $e^{\sqrt{2}}$
(e) Which is larger e^π or π^e? ◀

Try Study Suggestion 5.7.

The irrational number π arises in the real world in a very simple way—it is the circumference of a circle whose diameter is equal to 1. Although the number e is even more important, it is not quite as easy to describe how it occurs. We will devote the last two sections of this chapter to describing three important areas in which the number e, and the exponential function $E(x)$, occur in the real world. One of these areas is the compounding of interest, and the other two are in population growth and radioactive decay.

EXAMPLE 6: Graph the function $g(x) = e^{1/x}$.

Solution: We begin by observing that the domain of this function is the set of all *nonzero* real numbers. Therefore, since we cannot take $x = 0$, the graph has no y-intercept. Also, since $e^{1/x}$ is always positive, the graph lies above the x-axis. Now we must determine the behavior of the function for $|x|$ large, and for values of x near 0.

As x increases without bound, $1/x$ approaches 0, and so $e^{1/x}$ approaches $e^0 = 1$. In other words, the line $y = 1$ is a horizontal asymptote, at least for large *positive* values of x. This behavior is shown in Figure 5.8(a). We determined that the graph lies above this asymptote by observing that if $x > 0$ then $e^x > e^0 = 1$.

As x decreases without bound, $1/x$ still approaches 0, and so the function still approaches the horizontal asymptote $y = 1$. However, in this case, the graph lies below the line $y = 1$ since if $x < 0$ then $-x > 0$ and so $e^x = \dfrac{1}{e^{-x}} < \dfrac{1}{e^0} = 1$. (See Figure 5.8(a).)

As x approaches 0 from the right, $1/x$ increases without bound, and so $e^{1/x}$ also increases without bound, as shown in Figure 5.8(a). However, as x approaches 0 from the left, the exponent $1/x$ decreases without bound, and so the expression $e^{1/x}$ approaches 0. For example, when $x = -1/1000$, then $1/x = -1000$, and so

$$e^{1/x} = e^{-1000} = \frac{1}{e^{1000}}$$

which is very close to 0. This behavior of the graph is also shown in Figure 5.8(a). Now we can use the information in Figure 5.8(a) to complete the graph, as shown in Figure 5.8(b). *Try Study Suggestion 5.8.* □

▶ **Study Suggestion 5.8:** Graph the function $f(x) = e^{1/x^2}$. (*Hint:* first check for symmetry.) ◀

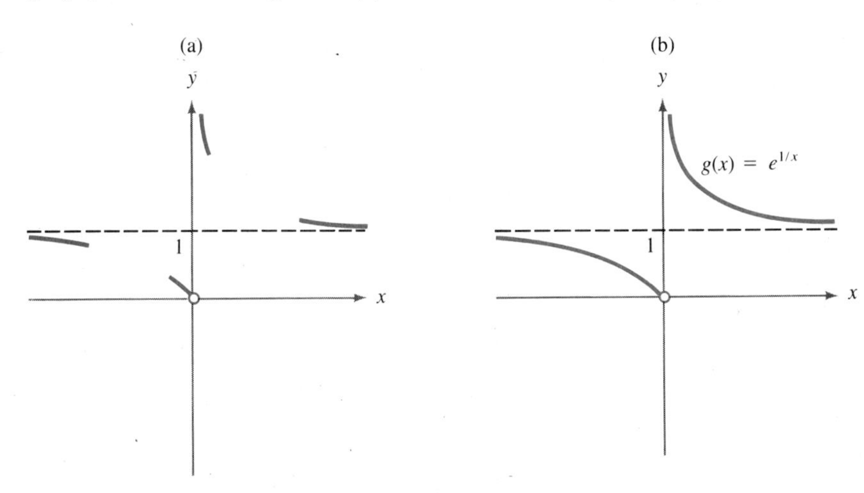

FIGURE 5.8

Ideas to Remember

There are many kinds of functions other than the polynomial and rational functions, one of which is the exponential function.

EXERCISES

In Exercises 1–6, use a calculator to approximate the given value.

1. $e^{-1/2}$

2. $e^{1/\pi}$

3. $\sqrt{\pi}^{-\sqrt{\pi}}$

4. e^{e}

5. $2^{\pi} - \pi^{2}$

6. $7.649^{9.5341}$

7. Compare the values of $(2^3)^4$ and $2^{(3^4)}$.

8. Use a calculator to approximate the values

$$2^3, 2^{3.1}, 2^{3.14}, 2^{3.141}, 2^{3.1415}, 2^{3.14159}$$

Then use the calculator to approximate the value of 2^{π}. What do you conclude from these approximations?

9. Use a calculator to approximate the values

$$2^1, 2^{1.4}, 2^{1.41}, 2^{1.414}, 2^{1.4142}, 2^{1.41421}$$

Then use the calculator to approximate the value $2^{\sqrt{2}}$. What do you conclude from these approximations?

In Exercises 10–23, graph the given function. (You may wish to use the technique of adding ordinates that we used in Example 3.)

10. $f(x) = 4^x$

11. $f(x) = 4^{-x}$

12. $f(x) = (1/5)^x$

13. $f(x) = (1/5)^{-x}$

14. $f(x) = 1 + 2^x$

15. $f(x) = 7^{-x-1}$

16. $f(x) = e^{x+1}$

17. $f(x) = e^{3x}$

18. $f(x) = e^{2x-3}$

19. $f(x) = x + e^x$

20. $f(x) = |x| + e^x$

21. $f(x) = x^2 + 2^x$

22. $f(x) = 3^x + 3^{-x}$

23. $f(x) = 2^{|x|}$

24. Graph the function $E(x) = e^x$, $f(x) = 2^x$ and $g(x) = 3^x$ on the same plane.

25. Graph the functions $f(x) = e^{-x}$, $g(x) = 2^{-x}$ and $h(x) = 3^{-x}$ on the same plane.

26. Graph the functions $f(x) = (1/2)^x$ and $g(x) = (1/3)^x$ on the same plane, and discuss the differences in these graphs.

27. Graph the functions $f(x) = x^3$ and $g(x) = 3^x$ on the same plane and discuss the differences in these graphs.

28. Graph the functions $f(x) = x^{1/2}$ and $g(x) = (1/2)^x$ on the same plane. Discuss the differences in these graphs.

29. Graph the function $f(x) = xe^x$.

30. Graph the function $f(x) = e^{-x^2}$.

31. Graph the function $f(x) = e^{-1/x}$.

32. Graph the function $f(x) = e^{-1/x^2}$.

In Exercises 33–42, solve the given equation.

33. $2^x = 32$

34. $2^{2x-1} = 16^x$

35. $3^x = 9^{x-1}$

36. $3^{x^2-1} = 27$

37. $3^x 4^x = 144$

38. $2^x 3^x = 1/36$

39. $2^x 4^{-x} = 1/8$

40. $2^x = 1$

41. $\dfrac{5^x}{25^{2x}} = 125$

42. $2^x + 3^x = -1$

43. In certain applications, scientists use the *polynomial* function

$$f(x) = 1 + x + \frac{x^2}{2} + \frac{x^3}{6}$$

to *approximate* the exponential function $E(x) = e^x$. This approximation is accurate only for values of x close to 0. Use a calculator to approximate the difference $E(x) - f(x)$ for the following values of x.
(a) $x = 1$ **(b)** $x = 0.5$ **(c)** $x = 0.1$
(d) $x = 0.01$ **(e)** $x = 0.001$ **(f)** $x = 0.0001$.

Exercises 44–48 concern the function

$$f(x) = \frac{2^x - 2^{-x}}{2^x + 2^{-x}}$$

44. Show that the function f is odd.

45. Show that f can be written in the form

$$f(x) = 1 - \frac{2}{2^{2x} + 1}$$

46. Use the results of the previous exercise to show that f is increasing. (*Hint:* if $a < b$, then $2a < 2b$ and $2^{2a} < 2^{2b}$ and $2^{2a} + 1 < 2^{2b} + 1 \ldots$. Continue in this way to arrive at $f(a) < f(b)$.

47. Use the results of Exercise 45 to find a horizontal asymptote of the graph of f.

48. Graph the function f.

5.2 The Logarithmic Functions

As you can see from Figure 5.9, the graph of the exponential function $f(x) = 2^x$ passes the horizontal line test. That is, every horizontal line intersects the graph in at most one point. Therefore, according to the results of Section 2.6, the function f is one-to-one, and so it has an inverse. This inverse is called the *logarithmic function* with base 2, and is denoted by

$$f^{-1}(x) = \log_2 x$$

($\log_2 x$ is read "log base 2 of x.")

More generally, if a is any positive real number different from 1, then we can define the logarithmic function with base a.

DEFINITION

Let a be any positive real number different from 1. Then the exponential function $f(x) = a^x$ has an inverse, which is called the **logarithmic function** with base a, and is denoted by

$$f^{-1}(x) = \log_a x$$

The domain of the logarithmic function is the set of all *positive* real numbers.

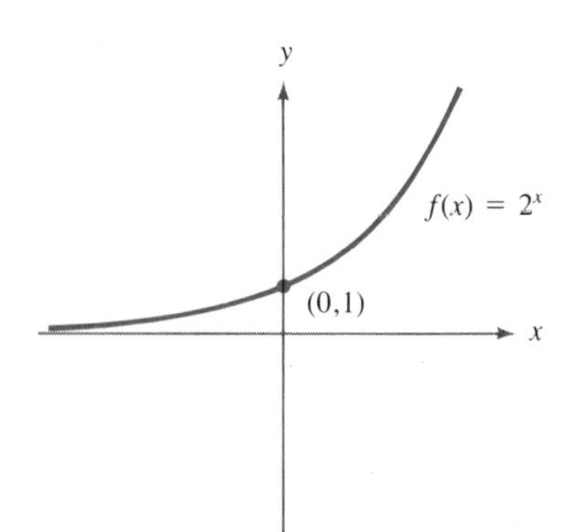

FIGURE 5.9

It is important to keep in mind that since the domain of the logarithmic function is the set of all positive real numbers,

we cannot take the logarithm of 0 or of any negative number.

According to the definition of inverse function that we gave in Section 2.8, the function f^{-1} is the inverse of f if and only if

$$f(f^{-1}(x)) = x$$

and

$$f^{-1}(f(x)) = x$$

In terms of the exponential and logarithmic functions, these equations are

$$a^{\log_a x} = x \tag{5.4}$$

valid for $x > 0$ and

$$\log_a(a^x) = x \tag{5.5}$$

valid for all real numbers x.

Equation (5.4) gives us a very useful description of the logarithm of a number. For it tells us that $\log_a x$ is that *exponent* which, when used with the base a, gives the number x. This is such a useful description that it deserves repeating.

> **$\log_a x$ is that *exponent* which, when used with the base a, gives the number x** (5.6)

This statement can be made a bit more formal as follows

$$y = \log_a x \qquad \text{if and only if} \qquad a^y = x \tag{5.7}$$

Statements (5.6) and (5.7) can be used to compute values of the logarithm function.

EXAMPLE 1: Compute the following logarithms

(a) $\log_2 8$ **(b)** $\log_{10} 10$ **(c)** $\log_5 \sqrt{5}$ **(d)** $\log_6\left(\frac{1}{36}\right)$ **(e)** $\log_e 1$

Solutions:

(a) According to Statement (5.6), $\log_2 8$ is that *exponent* which, when used with the base 2, gives 8. But $2^3 = 8$ and so $\log_2 8 = 3$.

(b) Letting $y = \log_{10} 10$. Statement (5.7) tells us that

$$10^y = 10$$

and so $y = 1$; that is, $\log_{10} 10 = 1$.

(c) According to Statement (5.6), $\log_5 \sqrt{5}$ is that *exponent* which, when used with the base 5, gives $\sqrt{5}$. But $5^{1/2} = \sqrt{5}$ and so $\log_5 \sqrt{5} = 1/2$.

(d) Letting $y = \log_6\left(\frac{1}{36}\right)$, Statement (5.7) tells us that

$$6^y = \frac{1}{36}$$

Using the fact that $1/36 = 6^{-2}$ we get

$$6^y = 6^{-2}$$

and so $y = -2$; that is, $\log_6\left(\frac{1}{36}\right) = -2$.

(e) Letting $x = \log_e 1$, Statement (5.7) tells us that

$$e^x = 1 = e^0$$

which is true if and only if $x = 0$. Hence $\log_e 1 = 0$.

Try Study Suggestion 5.9. □

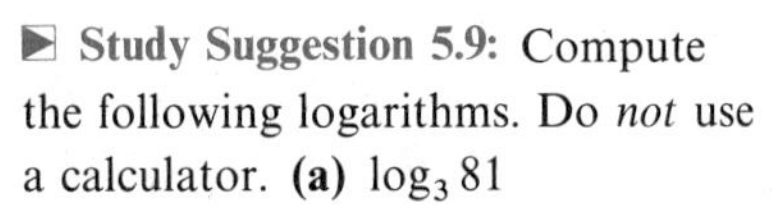

▶ **Study Suggestion 5.9:** Compute the following logarithms. Do *not* use a calculator. **(a)** $\log_3 81$ **(b)** $\log_{10}(1/1000)$ **(c)** $\log_2 1$ **(d)** $\log_e e^4$ **(e)** $\log_2 \sqrt[4]{8}$ ◀

According to Equation (5.5), for all bases a, we have $\log_a 1 = \log_a(a^0) = 0$. Thus

$$\mathbf{\log_a 1 = 0}$$

This useful fact about logarithms is worth remembering.

In Section 2.8, we pointed out that the graph of the inverse of a function can be obtained from the graph of the function by taking its mirror image with respect to the diagonal line $y = x$. This is done in Figure 5.10 for the functions $f(x) = 2^x$ and $f^{-1}(x) = \log_2 x$.

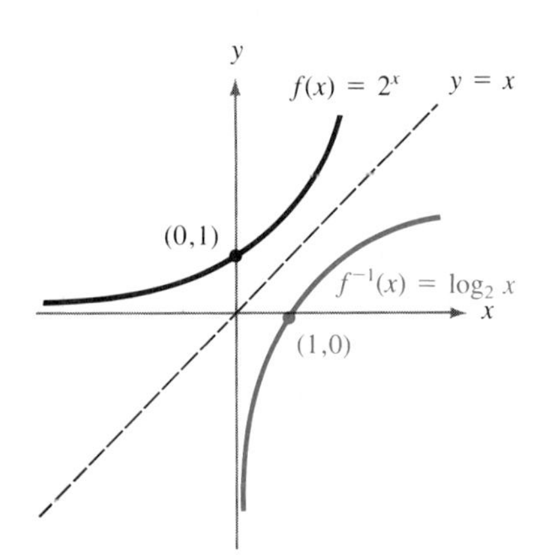

FIGURE 5.10

As you can see from this graph, the logarithmic function $f^{-1}(x) = \log_2 x$ is increasing, and its domain is the set of all *positive* real numbers. Furthermore, if $0 < x < 1$ then $\log_2 x$ is negative and if $x > 1$ then $\log_2 x$ is positive. Figure 5.11 shows the graph of the exponential function $f(x) = (1/2)^x$, and its inverse function $f^{-1}(x) = \log_{1/2} x$.

More generally, if $a > 1$ the graph of the logarithmic function $g(x) = \log_a x$ has the general shape shown in Figure 5.12(a) and if $0 < a < 1$ the graph has the general shape shown in Figure 5.12(b).

Even though we have defined the logarithmic function for any positive base other than 1, bases that are less than 1 are rarely used, and so

from now on, when we write $\log_a x$ we will assume that $a > 1$ unless otherwise noted.

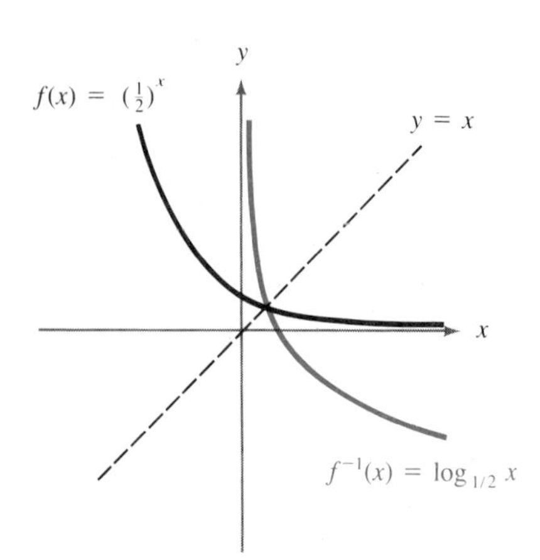

FIGURE 5.11

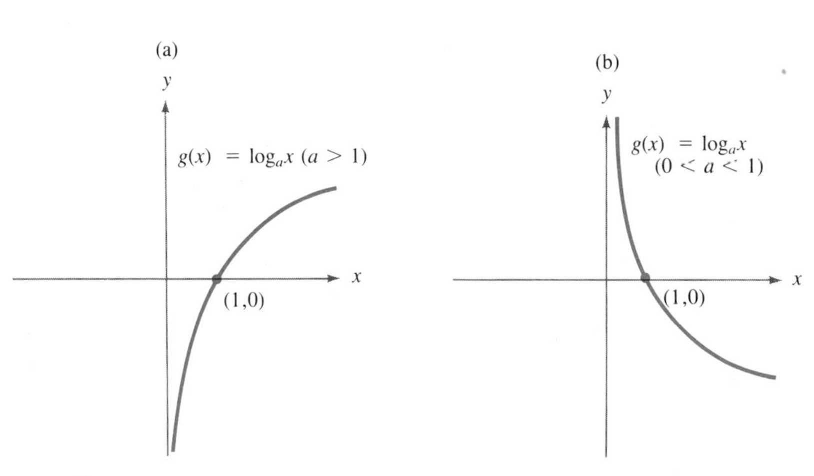

FIGURE 5.12

The logarithmic functions have some special properties, which we describe in a theorem.

THEOREM

Properties of the logarithm For any positive real numbers u and v we have

(a) $\log_a uv = \log_a u + \log_a v$

(b) $\log_a \frac{u}{v} = \log_a u - \log_a v$

(c) $\log_a \frac{1}{v} = -\log_a v$

(d) $\log_a u^r = r \log_a u$ for all real numbers r

Proof:

(a) We begin by letting $M = \log_a u$ and $N = \log_a v$. Then according to Statement (5.7),

$$a^M = u \qquad \text{and} \qquad a^N = v$$

Hence

$$uv = a^M \cdot a^N = a^{M+N}$$

and so, again using Statement (5.7), we get

$$\log_a uv = M + N$$

or

$$\log_a uv = \log_a u + \log_a v$$

which is what we wanted to prove.

(b) Property (b) has very similar proof, and we will leave it for the exercises.

(c) Property (c) follows from property (b) by setting $u = 1$. For in this case, $\log_a u = \log_a 1 = 0$, and so

$$\log_a \frac{1}{v} = \log_a 1 - \log_a v = 0 - \log_a v = -\log_a v$$

(d) Letting $M = \log_a u$, Statement (5.7) tells us that

$$a^M = u$$

Therefore, $u^r = (a^M)^r = a^{rM}$ and so, using Statement (5.5), we have

$$\log_a u^r = \log_a a^{rM} = rM = r \log_a u$$

which is what we wanted to show. ∎

It is important to notice that nowhere in our list of properties is there a formula for the logarithm of the *sum* or *difference* of two numbers. In fact,

there is no formula that expresses $\log_a(u + v)$ or $\log_a(u - v)$ in terms of $\log_a u$ and $\log_a v$

You should be especially careful here, since it is a very common mistake to think that $\log_a(u + v)$ is equal to $\log_a u + \log_a v$. This is simply *not* the case.

The properties given in the previous theorem can be used to help compute values of the logarithm.

EXAMPLE 2: In this example, let us denote $\log_{10} x$ simply by $\log x$. Using either a calculator or a table of logarithms, we find that $\log 2 \approx 0.3010$, $\log 3 \approx 0.4771$ and $\log 5 \approx 0.6990$. Use these approximations and the properties of logarithms to approximate the following values.

(a) $\log 15$ **(b)** $\log \sqrt{6}$ **(c)** $\log \frac{5}{2}$

(d) $\log \frac{1}{30}$ **(e)** $\log 125^7$

Solutions:

(a) Since $15 = 3 \cdot 5$ we can use property (a) to deduce that

$$\begin{aligned} \log 15 &= \log(3 \cdot 5) \\ &= \log 3 + \log 5 \\ &\approx 0.4771 + 0.6990 \\ &= 1.1761 \end{aligned}$$

(b) Using properties (d) and (a) we have

$$\begin{aligned} \log \sqrt{6} &= \log 6^{1/2} \\ &= (1/2) \log 6 \\ &= (1/2) \log(2 \cdot 3) \\ &= (1/2)[\log 2 + \log 3] \\ &\approx (1/2)[0.3010 + 0.4771] \\ &= 0.3891 \end{aligned}$$

(c) Using property (b), we have

$$\begin{aligned} \log \frac{5}{2} &= \log 5 - \log 2 \\ &\approx 0.6990 - 0.3010 \\ &= 0.3980 \end{aligned}$$

(d) In this case, we use property (c)

$$\begin{aligned}\log\frac{1}{30} &= -\log 30\\ &= -\log(2\cdot 3\cdot 5)\\ &= -(\log 2+\log 3+\log 5)\\ &\approx -(0.3010+0.4771+0.6990)\\ &= -1.4771\end{aligned}$$

(e) In this case, we have

$$\begin{aligned}\log 125^7 &= 7\log 125\\ &= 7\log 5^3\\ &= 21\log 5\\ &\approx 21(0.6990)\\ &= 14.679\end{aligned}$$

Try Study Suggestion 5.10. □

▶ **Study Suggestion 5.10:** Use the approximations given in Example 2, together with the properties of the logarithm, to approximate the following values. **(a)** $\log 10$ **(b)** $\log 30$ **(c)** $\log\sqrt[3]{15}$ **(d)** $\log(5/3)$ **(e)** $\log(2+3)$ **(f)** $\log 27^5$ ◀

EXAMPLE 3: If $\log_a x = 0.2$, $\log_a y = 0.4$, and $\log_a z = 0.3$, compute

$$\log_a\left(\frac{\sqrt{x}y^3}{\sqrt[3]{z}}\right)$$

Solution: Using the properties of logarithms, we have

$$\begin{aligned}\log_a\left(\frac{\sqrt{x}y^3}{\sqrt[3]{z}}\right) &= \log_a\sqrt{x}y^3-\log_a\sqrt[3]{z}\\ &= \log_a\sqrt{x}+\log_a y^3-\log_a\sqrt[3]{z}\\ &= (1/2)\log_a x+3\log_a y-(1/3)\log_a z\\ &= (1/2)(0.2)+3(0.4)-(1/3)(0.3)\\ &= 1.2\end{aligned}$$

Try Study Suggestion 5.11. □

▶ **Study Suggestion 5.11:** Using the values of the logarithm given in Example 3, compute

$$\log_a\left(\frac{\sqrt[5]{x^3}\sqrt{y^5}}{z^8}\right)$$ ◀

EXAMPLE 4: Use the properties of logarithms to write the expression

$$\log_a x^3-(1/2)\log_a y^4+\log_a 5xy$$

as a single logarithm.

Solution: Since $(1/2)\log_a y^4 = \log_a(y^4)^{1/2} = \log_a y^2$, we have

$$\begin{aligned}\log_a x^3-(1/2)\log_a y^4+\log_a 5xy &= \log_a x^3-\log_a y^2+\log_a 5xy\\ &= \log_a\left(\frac{x^3}{y^2}\cdot 5xy\right)\\ &= \log_a\frac{5x^4}{y}\end{aligned}$$

Try Study Suggestion 5.12. □

▶ **Study Suggestion 5.12:** Use the properties of logarithms to write the following expression as a single logarithm.

$$2\log_a 4xy-\log_a 6xy^3+(1/3)\log_a\frac{x^6}{y}$$ ◀

Ideas to Remember

The logarithmic function $g(x) = \log_a x$ is *defined* to be the inverse of the exponential function $f(x) = a^x$. All of the properties of the logarithm come from this definition.

EXERCISES

In Exercises 1–18, find the exact value of the given numbers. Do not use a calculator or a table of logarithms.

1. $\log_2 16$ **2.** $\log_2 \frac{1}{32}$ **3.** $\log_3 3$

4. $\log_3 1$ **5.** $\log_4 0$ **6.** $\log_5(-1)$

7. $\log_5 125$ **8.** $\log_{10} 1000$

9. $\log_{10} 1000000$ **10.** $\log_{10} 0.00001$

11. $\log_4 8$ **12.** $\log_2 \frac{\sqrt[3]{2}}{4}$

13. $\log_9 243$ **14.** $\log_5 \frac{\sqrt[4]{5}}{125}$

15. $\log_2 \frac{\sqrt[3]{16}}{8}$ **16.** $2^{\log_2 17}$

17. $9^{\log_3 5}$ **18.** $3^{\log_9 12}$

In Exercises 19–30, use the approximations $\log 2 \approx 0.3010$, $\log 5 \approx 0.6990$, and $\log 7 \approx 0.8451$ to compute the indicated logarithm.

19. $\log 10$ **20.** $\log 70$ **21.** $\log \frac{5}{7}$

22. $\log 2^5$ **23.** $\log 10^8$ **24.** $\log \sqrt{35}$

25. $\log 98$ **26.** $\log 100$ **27.** $\log(\sqrt[3]{2}\sqrt{7})$

28. $\log \frac{1}{\sqrt{10}}$ **29.** $\log \frac{\sqrt{5}}{\sqrt[3]{14}}$ **30.** $\log 5^7$

In Exercises 31–39, compute the indicated value, using the fact that $\log_e x = 0.1$, $\log_e y = 12$, and $\log_e z = -5$.

31. $\log_e xy$ **32.** $\log_e \frac{x}{z}$ **33.** $\log_e xyz$

34. $\log_e \frac{z}{xy}$ **35.** $\log_e x^5$ **36.** $\log_e x^y$

37. $\log_e \frac{\sqrt{xy^4}}{z}$ **38.** $\log_e \frac{\sqrt[3]{x}}{\sqrt{z}}$ **39.** $\log_e x^0$

In Exercises 40–45, use the properties of logarithms to write the expression as a single logarithm.

40. $\log_a 2x - \log_a x^3 + \log_a 3$

41. $\log_a \sqrt{xy} - \frac{3}{2}\log_a \frac{x}{y} - 2\log_a y + \log_a x$

42. $\log_b(x^2 - 1) + 2\log_b(x + 1) - \log_b(x - 1)$

43. $\log_a x + 4\log_a(3x - 1) - 2\log_a(9x^2 - 1)$

44. $\log_b 3 + \frac{1}{2}\log_b x + 2\log_b y + \log_b \frac{x - 1}{y} - \log_b(x^2 - 1)$

45. $5(\log_a \sqrt[3]{x} + \log_a \sqrt[5]{x}) - 2\log_a x$

In Exercises 46–53, use Statement (5.7) to solve the given equation for x.

46. $\log_2(x^2) = 4$ **47.** $\log_3(2x - 1) = 2$

48. $\log_{10}(5x + 1) = -1$ **49.** $\log_e(3x + 1) = 0$

50. $\log_2(x^2 + 2x) = 3$ **51.** $\log_5(2x^2 - 3x + 26) = 2$

52. $\log_4 \frac{x - 1}{x} = -1$ **53.** $\log_2 \frac{2x + 1}{x} = 1$

54. Use Equation (5.4) to show that if $\log_a x = \log_a y$, then $x = y$.

55. Solve the equation $\log_2(2x + 1) = \log_2 x$. (*Hint:* use the results of Exercise 54.)

56. Solve the equation $\log_e(2x^2 - 1) = \log_e(x + 1)$. (*Hint:* use the results of Exercise 54.)

57. Prove Part (b) of the theorem in this section.

5.3 The Natural Logarithm and the Common Logarithm

Among all possible bases for the logarithm, the most important by far are e and 10. Base e is important because the function $g(x) = \log_e x$ is the inverse of the exponential function $E(x) = e^x$, and base 10 is important because our number system is a base 10 number system.

Because of the importance of these bases, special names and symbols have been associated with them. In particular, $\log_e x$ is called the **natural logarithm** of x, and is denoted by $\ln x$. That is, $\ln x = \log_e x$. Also, $\log_{10} x$ is called the **common logarithm** of x, and is denoted by the symbol $\log x$. That is, $\log x = \log_{10} x$. (You should have keys marked *ln* and *log* on your scientific calculator.)

We should also mention that base 2 has become more important in recent years because of its use in certain areas of science, especially computer science. However, as yet there is no special name or symbol for $\log_2 x$.

Functions that involve the natural logarithm, such as $f(x) = \ln|x|$ and $g(x) = x + \ln x$ occur quite frequently in applications, where it is important to be able to describe their behavior.

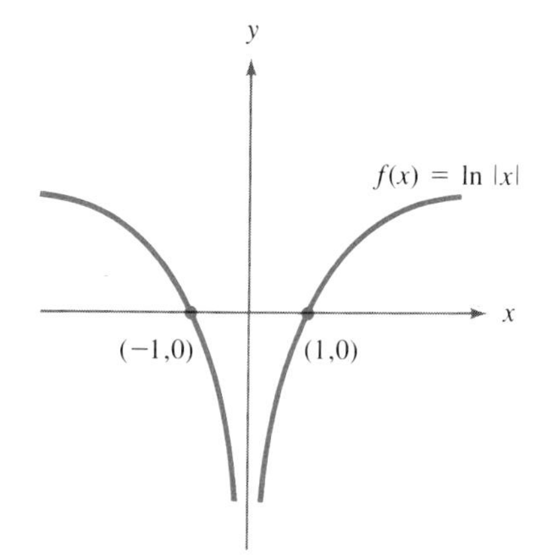

FIGURE 5.13

▶ **Study Suggestion 5.13:** Graph the function $g(x) = x + \ln x$. (*Hint:* Use the method of adding ordinates that we introduced in Example 3 of Section 5.1. First determine the domain of the function g.) ◀

EXAMPLE 1: Graph the function

$$f(x) = \ln|x|$$

Solution: First we observe that $f(-x) = \ln|-x| = \ln|x| = f(x)$, and so the function f is even. This implies that the graph is symmetric with respect to the y-axis, and so we only need to determine the graph for positive values of x. (The domain of f does not include $x = 0$.) Furthermore, when $x > 0$, we have $|x| = x$ and so $\ln|x| = \ln x$. This tells us that for $x > 0$, the graph of $f(x) = \ln|x|$ is the same as the graph of the natural logarithm function $g(x) = \ln x$. Reflecting this portion of the graph over the y-axis gives the entire graph, as shown in Figure 5.13. *Try Study Suggestion 5.13.* □

Sometimes it is useful to express the logarithm of a number with respect to one base a in terms of the logarithm with respect to a different base b. In symbols, we want to express $\log_b u$ in terms of $\log_a u$. This is the case, for example, if you need base 2 logarithms for a particular application, since most calculators do not have a key for computing these logarithms. In order to do this, we begin by letting

$$x = \log_b u$$

Then according to Statement (5.7) of the previous section, we have

$$b^x = u$$

Now we take the logarithm, to the base a, of both sides, and use the properties of logarithms to get

$$\log_a b^x = \log_a u$$
$$x \log_a b = \log_a u$$
$$x = \frac{\log_a u}{\log_a b}$$

that is,

$$\log_b u = \frac{\log_a u}{\log_a b}$$

This is the formula that we are seeking, and it is referred to as the **change of base formula.**

EXAMPLE 2: Using the common logarithm key on a calculator, and the change of base formula, compute $\log_2 12.4$.

Solution: Taking $b = 2$ and $a = 10$ in the change of base formula gives

$$\log_2 12.4 = \frac{\log 12.4}{\log 2} \approx \frac{1.0934}{0.3010} \approx 3.6326$$

Try Study Suggestion 5.14. □

Study Suggestion 5.14: Using the common logarithm key on your calculator and the change of base formula, compute $\log_2 0.235$. ◀

USING THE LOGARITHM TO SOLVE EXPONENTIAL EQUATIONS

At the end of Section 5.1, we discussed a simple method for solving exponential equations. However, the examples that we considered in that section were carefully chosen so that we could solve them using the fact that $a^x = a^y$ if and only if $x = y$ (when $a > 0$ and $a \neq 1$). Unfortunately, many exponential equations cannot be solved in this way, and we must resort to approximating their solutions using logarithms. The following examples will illustrate.

EXAMPLE 3: Solve the equation

$$3^{(x-2)} = 5^{2x}$$

Solution: We begin by taking the (common) logarithm of both sides of the equation

$$\log 3^{(x-2)} = \log 5^{2x}$$

Using the properties of logarithms, we can write this in the form

$$(x - 2)\log 3 = 2x \log 5$$

This can be solved for x as follows

$$x \log 3 - 2 \log 3 = 2x \log 5$$
$$x \log 3 - 2x \log 5 = 2 \log 3$$
$$(\log 3 - 2 \log 5)x = 2 \log 3$$
$$x = \frac{2 \log 3}{\log 3 - 2 \log 5}$$
$$\approx \frac{2 \cdot (0.4471)}{0.4771 - 2 \cdot (0.6990)}$$
$$\approx -1.0362$$

This approximation can be checked in the original equation with the aid of a calculator. *Try Study Suggestion 5.15.* □

Study Suggestion 5.15: Solve the equation $3^{(5x-1)} = 5^x$. ◀

EXAMPLE 4: Solve the equation

$$2^x 3^{(x-1)} = 5$$

Solution: Before taking logarithms, we can simplify this equation as follows

$$2^x 3^{(x-1)} = 5$$
$$2^x 3^x 3^{-1} = 5$$
$$2^x 3^x = 15$$
$$6^x = 15$$

Now we take the logarithm of both sides,

$$\log 6^x = \log 15$$
$$x \log 6 = \log 15$$
$$x = \frac{\log 15}{\log 6} \approx \frac{1.1761}{0.7782} \approx 1.5113$$

Try Study Suggestion 5.16. □

Study Suggestion 5.16: Solve the equation $2^x 3^{-x} = 2$. ◀

EXAMPLE 5: Solve the equation

$$2^{5x} = 8^{(x-1)}$$

Solution: Before taking logarithms, we should observe that $8 = 2^3$, and so the equation can be written

$$2^{5x} = 2^{3(x-1)}$$

Hence

$$5x = 3(x - 1)$$

Solving this for x gives $x = -3/2$. *Try Study Suggestion 5.17.* □

▶ **Study Suggestion 5.17:** Solve the equation $3^{2x} = 81^{(x+2)}$. ◀

THE LOGARITHMIC SCALE

The common logarithm has a special property that makes it very convenient for measuring physical quantities. As an example, let us consider the famous Richter scale for measuring the magnitude of earthquakes. (This scale was first proposed by Charles Richter, an American geologist.)

An earthquake produces several seismic waves, whose amplitudes are registered on a seismograph. For any given earthquake, we are interested in the amplitude of the largest wave. Let us simply call this the "amplitude" of the earthquake.

The Richter scale compares the amplitude of a given earthquake to some standard amplitude, which we will denote by A_0. You may think of A_0 as being the amplitude of a "standard," but rather small earthquake.

If an earthquake has amplitude A, then its **magnitude** on the Richter scale is defined to be

$$\mathbf{M = \log\left(\frac{A}{A_0}\right)}$$

where, as usual, log stands for the common logarithm.

You may be wondering why we simply don't take the ratio A/A_0 as the magnitude of the earthquake. The answer is that there is such a large variation in the amplitudes of earthquakes that such a scale would have an *extremely* large range. (We will see an example of just how large in a moment.) In a sense, the logarithm has the effect of compressing this range.

For example, suppose that we compare two earthquakes, the second of which has 10 times the amplitude of the first. If the first earthquake has amplitude A_1, its magnitude on the Richter scale is

$$M_1 = \log\left(\frac{A_1}{A_0}\right)$$

and since the amplitude of the second earthquake is $A_2 = 10A_1$, its magnitude on the Richter scale is

$$\begin{aligned} M_2 &= \log\left(\frac{A_2}{A_0}\right) \\ &= \log\left(\frac{10A_1}{A_0}\right) \\ &= \log\left(10\frac{A_1}{A_0}\right) \\ &= \log 10 + \log\left(\frac{A_1}{A_0}\right) \\ &= 1 + M_1 \end{aligned}$$

Thus we see that the magnitude of the second earthquake is only one larger than the magnitude of the first earthquake. In short, *a tenfold increase in amplitude produced only an increase of 1 on the Richter scale.*

Similarly, a hundredfold increase in amplitude would produce an increase of only 2 on the Richter scale, a thousandfold increase in amplitude would produce only an increase of 3 on the Richter scale, and so on. Thus, very large increases in amplitude produce only very small increases in magnitude on the Richter scale. This is the effect of the logarithm.

To emphasize this effect, the Richter scale is referred to as a **logarithmic scale.** Of course, there would be no need for a logarithmic scale if earthquakes did not vary much in their amplitude. The next example shows that they do in fact vary a great deal.

EXAMPLE 6: The famous San Francisco earthquake of 1906 is estimated to have measured 8.4 on the Richter scale. Compare the amplitude of this earthquake with the amplitude of an earthquake measuring 4.5 on the Richter scale, which would cause only minor damage in a small area.

Solution: Let us denote the amplitude of the moderate earthquake by A_1 and the amplitude of the 1906 earthquake by A_2. Then according to the definition of the Richter scale, we have

$$4.5 = \log\left(\frac{A_1}{A_0}\right)$$

Solving this equation for A_1 gives

$$\frac{A_1}{A_0} = 10^{4.5}$$

or

$$A_1 = A_0 10^{4.5} \tag{5.8}$$

Similarly, we have

$$8.4 = \log\left(\frac{A_2}{A_0}\right)$$

and so

$$A_2 = A_0 10^{8.4} \tag{5.9}$$

Using Equations (5.8) and (5.9), we get

$$\frac{A_2}{A_1} = \frac{A_0 10^{8.4}}{A_0 10^{4.5}} = \frac{10^{8.4}}{10^{4.5}} = 10^{(8.4-4.5)} = 10^{3.9} \approx 7943$$

Thus, the earthquake of 1906 had approximately 8000 times greater amplitude than a moderate earthquake! *Try Study Suggestion 5.18.* □

▶ **Study Suggestion 5.18:** Compare the amplitudes of an earthquake of moderate destructive potential, measuring 6 on the Richter scale, with an earthquake that is barely perceptible, measuring only 2 on the Richter scale. ◀

The measurement of the intensity of sound by decibels is another example of a logarithmic scale. (The term *decibel* is named after Alexander Graham Bell.) If a sound has intensity I, then we say that it measures

$$d = 10 \log\left(\frac{I}{I_0}\right)$$

decibels, where I_0 is the intensity of a "standard" sound that is barely audible.

EXAMPLE 7: The average voice has an intensity 10,000 times greater than the standard sound. How many decibels does the average voice measure?

Solution: Since the average voice has intensity equal to $I = 10{,}000I_0$, it measures

$$\begin{aligned} d &= 10 \log \frac{10{,}000I_0}{I_0} \\ &= 10 \log 10{,}000 \\ &= 10 \log 10^4 \\ &= 40 \log 10 \\ &= 40 \text{ decibels} \end{aligned}$$

Try Study Suggestion 5.19. □

▶ **Study Suggestion 5.19:** A person is speaking at twice the intensity of the average voice. What is the decibel measurement of his voice? ◀

EXAMPLE 8: A sound measuring approximately 120 decibels can produce pain in the ears. How much more intense is this sound than the standard sound?

Solution: If this sound has intensity I, then we have

$$120 = 10 \log\left(\frac{I}{I_0}\right)$$

We must solve this for I. Dividing by 10 gives

$$\log\left(\frac{I}{I_0}\right) = 12$$

and so

$$\frac{I}{I_0} = 10^{12}$$

Thus $I = 10^{12}I_0$ and so the intensity of this sound is 10^{12} (= 1 trillion) times greater than the standard intensity!

This example emphasizes the fact that logarithmic scales reduce the range of measurement to a level that is much easier to work with. In particular, it is much easier to work with a number such as 120 than with the number 10^{12}.

Try Study Suggestion 5.20. □

▶ **Study Suggestion 5.20:** How much more intense is a sound measuring 100 decibels than the standard sound. ◀

Ideas to Remember

- The logarithm can be used to solve, or at least approximate the solutions to exponential equations.
- The logarithmic scale is useful for measuring physical quantities that can have a large range of values, since it has the effect of greatly reducing this range.

EXERCISES

In Exercises 1–9, use a calculator to approximate the given value.

1. $\log 182$ **2.** $\ln 1.4563$

3. $\log \frac{1}{12}$ **4.** $\log 0.3349$

5. $\log(5 \times 10^{-6})$ **6.** $\log(\log 20)$

7. $\ln(\ln 1.64)$ **8.** $\ln(\ln(\ln 5.35))$

9. $\ln(\ln(\ln 2.31))$

In Exercises 10–18, use a calculator and the change of base formula to approximate the given value.

10. $\log_2 3$ **11.** $\log_2 17$ **12.** $\log_2 4.356$

13. $\log_2 \pi$ **14.** $\log_3 e$ **15.** $\log_3 2$

16. $\log_2(\log_2 6)$ **17.** $\log_5 2$ **18.** $\log_7 10$

19. Assuming that $\log_a b = 0.4$ and $\log_b x = 0.6$, find $\log_a x$.

20. Assuming that $\log_a x = 5$ and $\log_b x = 10$, find $\log_a b$.

21. If $\log_a b = 2$, find $\log_b a$.

22. If $\log_a b = -1/2$, find $\log_b a^2$.

23. Find a formula for $\log_b a$ in terms of $\log_a b$.

24. Graph the functions $f(x) = \log x$ and $g(x) = \ln x$ on the same plane and describe the differences in the two curves.

In Exercises 25–30, determine the domain of the given function and sketch its graph.

25. $f(x) = -\ln x$ **26.** $f(x) = 1 + \ln x$

27. $f(x) = 2x + \ln x$ **28.** $f(x) = x^2 + \ln x$

29. $f(x) = \ln(-x)$ **30.** $f(x) = (\ln x)^2$

In Exercises 31–42, solve or approximate the solutions to the given equation.

31. $2^{(7x-4)} = 9$ **32.** $3^{(x-1)} = -2$

33. $4^{(8x+2)} = 5^{(6x-5)}$ **34.** $2^{(-2x-6)} = 8^x$

35. $2^{2x}3^x = 7^{(x+1)}$ **36.** $4^{(2x+3)} = 5^{(x-1)}$

37. $(1/4)^x = 100$ **38.** $e^{3x} = 17$

39. $e^{(1-2x)} = 3/2$ **40.** $e^{4x} = 2^x$

41. $1.723^x = 4.987$ **42.** $4.551^{1/x} = 2.33$

43. The largest earthquake ever recorded occurred in Japan in 1933, and measured approximately 8.9 on the Richter scale. Compare the amplitude of this earthquake with the amplitude of a barely perceptible earthquake, with a magnitude of 2.0 on the Richter scale.

44. Compare the amplitude of an earthquake that measures 5.9 on the Richter scale with the amplitude of an earthquake that measures 6.0 on the Richter scale.

45. An electric typewriter produces a sound that is approximately 10^8 times the intensity of the standard sound. How many decibels does this sound measure?

46. The sound of a jet engine is approximately 10^{18} times as intense as the standard sound. How many decibels does the sound from a jet engine measure?

47. Thunder can reach a sound that measures as much as 110 decibels. How much more intense is this than the standard sound?

48. How much more intense is a sound of 80 decibels than one of 75 decibels?

49. The average ear cannot perceive a sound difference of less than 0.5 decibels. What is this difference measured in terms of intensity?

There is a logarithmic scale for measuring the acidity of chemical substances known as the pH scale. Any chemical substance contains a certain number of hydrogen ions (a hydrogen ion is a hydrogen atom whose electron has been stripped away). If we denote the hydrogen ion concentration of a substance by $[H^+]$, then the pH of that substance is given by

$$\text{pH} = -\log[H^+]$$

Water has a pH of 7, and is considered neither an acid nor a base. Any substance with a pH less than 7 is an acid, and any substance with a pH greater than 7 is a base.

50. Find the pH of each of the following substances.

(a) seawater $[H^+] = 1.1 \times 10^{-8}$

(b) milk $[H^+] = 1.77 \times 10^{-7}$

(c) lemon juice $[H^+] = 7.94 \times 10^{-3}$

Which of these substances are acids and which are bases?

51. Find the hydrogen ion concentration of each of the following substances.

(a) blood pH = 7.4 **(b)** limewater pH = 12.4

(c) beer pH = 4.1 **(d)** vinegar pH = 2.9

52. Does a higher concentration of hydrogen ions mean a higher pH, or a lower pH? Explain your answer.

53. Do bases have more hydrogen ions, or fewer hydrogen ions? Explain your answer.

54. Most substances have a pH between 1 and 14. What is the range of hydrogen ion concentration represented by this range in pH?

5.4 The Natural Number *e* and Continuous Compounding of Interest

Now we want to discuss three important areas in which the number e, and the exponential function $E(x) = e^x$, play a central role. In this section, we will discuss the compounding of interest, and this will give us a chance to discuss how the number e arises. In the next section, we will discuss the role of the exponential function in population growth and radioactive decay.

Let us suppose that we put P dollars into a savings account at an annual interest rate r. If the interest is compounded once a year, then after t years we will have

$$A(t) = P(1 + r)^t$$

dollars in our account. Notice that we have denoted the amount after t years by $A(t)$, since we wish to think of it as a function of the time t.

If the interest is compounded twice a year, at an annual interest rate equal to r, then the amount after t years is

$$A(t) = P\left(1 + \frac{r}{2}\right)^{2t}$$

We can see this as follows. Since the annual interest rate is equal to r, the rate every half year is $r/2$. Therefore, after one-half year the total amount of money has grown to $P(1 + r/2)$ dollars. After one year, the total amount of money has grown to $P(1 + r/2)(1 + r/2) = P(1 + r/2)^2$ dollars. In fact, each half-year period produces a growth factor equal to $1 + r/2$, and since there are $2t$ half-year periods in t years, after t years the total amount of money has grown to $P(1 + r/2)^{2t}$.

Using the same reasoning, we can show that if the interest is compounded k times a year, then the amount after t years is given by the formula

$$A(t) = P\left(1 + \frac{r}{k}\right)^{kt} \tag{5.10}$$

EXAMPLE 1: Suppose that you deposit \$1000 in an account that pays 10% interest (that is, $r = 0.10$), compounded quarterly (four times a year.) How much money will be in the account at the end of

(a) 1 year? **(b)** 2 years? **(c)** 6 months?

Solutions:

(a) Using Formula (5.10), with $k = 4$ and $t = 1$ gives

$$\begin{aligned} A(1) &= 1000\left(1 + \frac{0.10}{4}\right)^4 \\ &= 1000(1.025)^4 \\ &\approx 1103.81 \end{aligned}$$

Hence, there will be approximately \$1103.81 in the account at the end of one year.

(b) In this case, we use Formula (5.10) with $k = 4$ and $t = 2$,

$$\begin{aligned} A(2) &= 1000\left(1 + \frac{0.10}{4}\right)^8 \\ &= 1000(1.025)^8 \\ &\approx 1218.40 \quad \text{dollars} \end{aligned}$$

(c) Since 6 months is the same as 1/2 year, the account will contain $A(1/2)$ dollars at the end of 6 months. Using Formula (5.10) with $k = 4$ and $t = 1/2$, we get

$$\begin{aligned} A(1/2) &= 1000\left(1 + \frac{0.10}{4}\right)^2 \\ &= 1000(1.025)^2 \\ &\approx 1050.63 \quad \text{dollars} \end{aligned}$$

Try Study Suggestion 5.21. □

▶ **Study Suggestion 5.21:** Suppose that \$500 is deposited in an account that pays 12% interest compounded semiannually (that is, twice a year). How much money will be in the account at the end of **(a)** 1 year? **(b)** 3 years? **(c)** $1\frac{1}{2}$ years? ◀

It is somewhat surprising that there is not much difference in the total amount $A(t)$ when interest is compounded quarterly, monthly, daily, or even hourly. This can be seen by looking at Table 5.2, where we have computed the amount $A(1)$ of money in an account after one year, using several different methods of compounding.

TABLE 5.2 *The total amount $A(1)$ of money in an account after one year, with principal $P = 1000$ dollars and interest rate $r = 10\%$.*

method of compounding	amount $A(1)$
once a year	\$1100
four times a year	\$1103.81
once a month	\$1104.71
oncc a day	\$1105.15
once an hour	\$1105.17

As you can see from this table, the difference between compounding once a month and once an hour is only about 46 cents on \$1000.

Perhaps because of this small difference, the banks hit upon a marketing scheme to draw customers, namely, accounts that compound interest every instant! When this is the case, we say that the interest is compounded *instantaneously*, or *continuously*.

In order to see how much interest is involved in **continuous compounding of interest,** we must see what happens to Formula (5.10) as k gets larger and larger. The first step is to make the substitution $n = k/r$ in Formula (5.10). Since $r/k = 1/n$, we get

$$A(t) = P\left(1 + \frac{1}{n}\right)^{nrt}$$

Then we isolate n by writing this in the form

$$A(t) = P\left[\left(1 + \frac{1}{n}\right)^{n}\right]^{rt} \tag{5.11}$$

Now, as k gets larger and larger, so does n, and so we want to determine what happens to the expression

$$\left(1 + \frac{1}{n}\right)^{n} \tag{5.12}$$

as n gets larger and larger.

If you examine this expression closely, you can see that there are two opposing forces at work. As n gets larger, the base $1 + 1/n$ gets closer to 1, which tends to make the entire expression close to 1. (After all, 1 to any power is equal to 1.) On the other hand, as n gets larger, the exponent in expression (5.12) gets larger, which tends to make the entire expression get larger.

Now, it turns out that these two forces tend to balance each other, and that as n gets larger, expression (5.12) gets closer and closer to a certain irrational number that lies between 2.5 and 3. This can be seen by looking at Table 5.3, where we have computed the values of (5.12) for some selected values of n.

TABLE 5.3

n	approximate value of $(1 + 1/n)^n$
1	2
10	2.59374
100	2.70481
1000	2.71692
10000	2.71815
100000	2.71827
1000000	2.71828

As you can see from this table, the values of expression (5.12) seem to be approaching a number that is *approximately* equal to 2.71828. It can be shown, however, that this number is irrational. It is the number that we have been denoting by e.

We can use this information about the behavior of $(1 + 1/n)^n$ in expression (5.11). As n gets larger and larger, the expression inside the square brackets in (5.11) approaches e. Therefore, if we replace that expression by e, we get the formula

$$A(t) = Pe^{rt} \tag{5.13}$$

which we can think of as a formula for the total amount of money in an account after t years, at an interest rate equal to r, where that interest is compounded *continuously*.

EXAMPLE 2: An account pays 10% interest compounded continuously. If $1000 is deposited in this account, how much will there be in the account at the end of 1 year?

Solution: Using Formula (5.13), with the help of the e^x key on a calculator, we get

$$A(1) = 1000e^{0.10} \approx 1000(1.10517) = 1105.17 \quad \text{dollars}$$

(Actually, the difference between compounding hourly (see Table 5.2) and compounding continuously is only about 1/10 of one cent on $1000!)

Try Study Suggestion 5.22. □

Study Suggestion 5.22: An account pays 8% interest compounded continuously. If $5000 is deposited in this account, how much will there be after 2 years?

EXAMPLE 3: Suppose that you deposit $100 in an account at the beginning of each year, for 3 consecutive years. How much money will be in the account at the end of the 3-year period if the account pays 10% interest compounded

(a) annually **(b)** continuously?

Solutions:

(a) If the interest is compounded annually, then according to Formula (5.10) with $k = 1$, at the end of three years, the first deposit of $100, having been in the account for the entire period, will grow to $A(3) = 100(1 + 0.10)^3$ dollars. However, the second deposit of $100, having been in the account for only two years, will grow to $A(2) = 100(1 + 0.10)^2$ dollars, and the third deposit of \$100 will grow to only $A(1) = 100(1 + 0.10)$ dollars.

Thus, the combined deposits will grow to

$$\begin{aligned} A(3) + A(2) + A(1) &= 100(1 + 0.10)^3 + 100(1 + 0.10)^2 + 100(1 + 0.10) \\ &= 100(1.1^3 + 1.1^2 + 1.1) \\ &= 100(3.641) \\ &= 364.10 \end{aligned}$$

That is, the total amount after three years will be \$364.10.

(b) If the interest is compounded continuously, then according to Formula (5.13), after three years, the first deposit of \$100 will grow to $A(3) = 100e^{(0.10)(3)}$ dollars. Using the same formula for the other two deposits, we see that the total amount of money in the account after three years will be

$$\begin{aligned} &100e^{(0.10)(3)} + 100e^{(0.10)(2)} + 100e^{0.10} \\ &= 100(e^{(0.10)(3)} + e^{(0.10)(2)} + e^{0.10}) \\ &= 100(e^{0.3} + e^{0.2} + e^{0.1}) \\ &\approx 100(3.676) \\ &= 367.60 \quad \text{dollars} \end{aligned}$$

Try Study Suggestion 5.23. □

Study Suggestion 5.23: Suppose that you deposit \$200 in an account at the beginning of each year, for three consecutive years. How much money will be in the account at the end of three years if the account pays 12% interest compounded **(a)** annually **(b)** quarterly **(c)** continuously? ◄

EXAMPLE 4: A certain account pays interest at the rate of 10% compounded continuously. How long does it take for money to double at this rate?

Solution: If P dollars is deposited in the account, the amount of money after t years is

$$A(t) = Pe^{0.10t}$$

We want to know how long it will take for this money to double; that is, to grow to $2P$ dollars. Thus, we want to solve the equation

$$2P = Pe^{0.10t}$$

for t. Dividing both sides by P gives

$$e^{0.10t} = 2$$

Now, we can solve this equation by taking the common logarithm of both sides, and then solving for t. However, since the only base involved in this equation is the natural base e, we can save a few steps by taking the *natural* logarithm of both sides. (This only saves steps if we have a calculator with a natural logarithm key.)

$$\begin{aligned} \ln e^{0.10t} &= \ln 2 \\ 0.10t \ln e &= \ln 2 \end{aligned}$$

Now we use the fact that $\ln e = 1$

$$0.10t = \ln 2$$
$$t = 10 \ln 2 \approx 6.931$$

Thus, it takes approximately 7 years for the money in this account to double.

Try Study Suggestion 5.24. □

Study Suggestion 5.24: How long does it take money to double in an account that pays 10% interest compounded quarterly? Compare this with the doubling time when interest is compounded continuously. ◀

Ideas to Remember

The important number e plays a role in the continuous compounding of interest.

EXERCISES

1. An account pays 12% interest compounded monthly. If \$1000 is deposited in the account, how much will there be in the account after

(a) 1 month **(b)** 6 months **(c)** 1 year
(d) 5 years

2. An account pays 10% interest, compounded quarterly. How much will a deposit of \$2000 grow to in

(a) 3 months **(b)** 6 months **(c)** 1 year
(d) 5 years

3. An investor buys a \$20,000 savings certificate that yields 12% interest compounded semiannually. How much will the certificate be worth after 3 years?

4. A certain account pays 12% interest per year. How much money will be in the account after 1 year if \$1000 is deposited and the interest is compounded

(a) annually **(b)** quarterly **(c)** daily
(d) continuously

5. An account pays 9.2% interest per year. How much will \$2000 grow to in 5 years if the interest is compounded

(a) annually **(b)** quarterly **(c)** monthly
(d) daily **(e)** continuously

6. Two accounts each pay 10.25% interest, but one account compounds semiannually while the other account compounds continuously. How much more money will be in the account that compounds continuously after 1 year if \$10,000 is deposited in each account?

7. An account yields 10% interest compounded semiannually. How much should you invest in order to have \$10,000 after 1 year?

8. Suppose that you deposit \$1000 at the beginning of each year, for four consecutive years, in an account that pays 12% interest. How much money will be in the account at the end of the four-year period if the interest is compounded

(a) annually **(b)** quarterly **(c)** continuously

9. How much should you deposit in a bank account that pays 8% interest, compounded continuously, in order to have a total of \$10,000 at the end of 10 years?

10. How much should you invest in bonds that pay 9.36% interest, compounded continuously, in order to have \$1,000,000 after 30 years?

11. How long does it take money to double in an account that pays 10% interest compounded monthly?

12. How long does it take money to double at 12% interest compounded daily?

13. How long does it take money to triple at 10% compounded continuously?

14. How long does it take \$5000 to grow to \$6000 at 10% compounded continuously?

15. How long does it take \$2000 to grow to \$2500 at 12% compounded quarterly?

16. An account that contains \$1000 grows to \$1112.75 after 3 years. If the interest is compounded quarterly on the account, what is the interest rate?

17. If \$5000 grows to \$5623.12 after 1 year with interest compounded continuously, what is the interest rate? How long will it take for the original amount to grow to \$6000?

18. One thousand dollars is deposited in a certain account. At the end of 1 year, there is \$1100 in the account. If the interest is compounded continuously, find the interest rate.

19. If \$1000 is deposited in an account that pays interest quarterly, and if the money grows to \$1150 after 2 years, find the interest rate.

20. If \$850 grows to \$1125 at an interest rate of 12.6% compounded continuously, how long has the money been in the account?

The **effective interest rate** of an account is the rate of interest the bank would have to pay compounded annually in order to yield the same amount as its advertised interest rate. For example, if an account pays 10% compounded semiannually, then a \$100 deposit grows to $100(1 + .05)^2 = 110.25$ in one year. Hence, the effective rate for this account is 10.25%, since an account that pays 10.25% compounded annually will yield the same amount as this account. The effective interest rate is used to compare accounts with different methods of compounding. (This also works for loans, or anything that pays or charges interest.) Thus, for example, in order to compare a loan that pays interest compounded quarterly with one that pays interest compounded daily, we must compare their effective interest rates.

21. What is the effective interest rate of an account that pays 10% compounded quarterly? Is this better or worse than an account that pays 10.25% compounded 3 times a year? Justify your answer.

22. One bank offers to loan you money at an interest rate of 12% compounded quarterly, and another bank offers to loan you money at 11.8% compounded continuously. Which loan would you prefer, and why?

23. Find a formula that gives the effective interest rate of an account that pays interest at rate r compounded quarterly.

24. Find a formula that gives the effective interest rate of an account that pays interest at rate r compounded continuously.

5.5 Exponential Growth and Decay

In this section, we continue our discussion of exponential functions. As we will see, these functions play a role both in the growth and in the decay of natural substances.

POPULATION GROWTH

If a population of organisms, such as the bacteria in a petri dish or the human beings on a continent, is allowed to grow unchecked by either natural or man-made disasters, then it is a law of biology that the number of organisms present at any time t is given by a formula of the form

$$P(t) = Ce^{kt} \tag{5.14}$$

where C and k are positive constants. This type of growth is referred to as **exponential growth.** Equation (5.14) applies to a rather idealized situation, where there is no war, famine, or pestilence to reduce the population. However, this formula is still useful, and it tells us some rather startling things about growth rates.

The constant C has a very simple meaning in the context of exponential growth. For if we set $t = 0$, then since $e^0 = 1$, we get

$$P(0) = Ce^0 = C$$

But $P(0)$ is just the number of organisms present at the initial time $t = 0$, and so C is equal to this number as well. In short, *C is the initial population.*

Since there are two constants in Equation (5.14), we need two pieces of information in order to determine its exact form for a given population. Let us illustrate this with some examples.

EXAMPLE 1: A certain population of microorganisms is growing exponentially in a petri dish. (A petri dish is a flat dish used to grow bacterial cultures.) If there are 1000 organisms initially, and 25,000 organisms after 2 hours, determine the formula for their growth, and compute the number of microorganisms present after 10 hours.

Solution: We know that the number of microorganisms present after t hours is given by $P(t) = Ce^{kt}$ and since C is the initial population, which in this case is 1000, we have

$$P(t) = 1000e^{kt} \tag{5.15}$$

We also know that at time $t = 2$, there are 25,000 microorganisms, that is $P(2) = 25{,}000$. Thus,

$$25000 = 1000e^{2k}$$

Let us solve this for e^k. Dividing by 1000 gives

$$25 = e^{2k}$$

and since $e^{2k} = (e^k)^2$, we can write this in the form

$$25 = (e^k)^2$$

Taking the square root of both sides gives

$$e^k = 5 \tag{5.16}$$

(We have taken the positive square root on the right because e^k is positive.) Now since Equation (5.15) can be written in the form

$$P(t) = 1000(e^k)^t$$

we can use Equation (5.16) to get

$$P(t) = 1000 \cdot 5^t$$

This is the growth function for this particular population. Now we can compute the value of $P(t)$ for any time t. In particular,

$$P(10) = 1000 \cdot 5^{10} = 9{,}765{,}625{,}000$$

and so after 10 hours there will be almost 10 billion microorganisms in the petri dish! This example shows quite clearly that if exponential growth is not checked, it can easily get out of hand. *Try Study Suggestion 5.25.* □

▶ **Study Suggestion 5.25:** A certain species of fish is growing exponentially in an aquarium. In 5 days, the population of the aquarium doubles from an initial population of 25 fish. How many more days will it be until there are 800 fish in the aquarium? ◀

EXAMPLE 2: In 1950, the world needed approximately 2×10^9 acres (2 billion acres) of fertile land in order to grow enough food to feed its population. In 1980, the world needed approximately 4×10^9 acres for this purpose. Assuming that the amount of land needed grows exponentially, and that the earth has a total of 8×10^9 acres of fertile land, how long will it take to reach the point where we require all of the available fertile land?

Solution: If we let $t = 0$ correspond to the year 1950, then the initial amount of fertile land is 2×10^9 acres. Therefore, the amount of fertile land needed to feed the population is given by the function

$$A(t) = (2 \times 10^9)e^{kt}$$

Using the fact that 1980 corresponds to $t = 30$ years, we have $A(30) = 4 \times 10^9$, that is

$$4 \times 10^9 = (2 \times 10^9)e^{30k}$$

Dividing both sides of this by 2×10^9 gives

$$2 = e^{30k}$$

and so

$$e^k = 2^{1/30}$$

Substituting this into the growth function, we get

$$A(t) = (2 \times 10^9)2^{t/30}$$

In order to determine when the earth will need 8×10^9 acres of fertile land, we must solve the equation

$$8 \times 10^9 = (2 \times 10^9)2^{t/30}$$

Dividing both sides of this equation by 2×10^9 gives

$$4 = 2^{t/30}$$

or

$$2^2 = 2^{t/30}$$

and so $2 = t/30$. That is, $t = 60$, and the world will require 8×10^9 acres of fertile land $t = 60$ years after the initial time $t = 0$; that is, in the year 2010.

Try Study Suggestion 5.26. □

▶ **Study Suggestion 5.26:** The population on a certain tropical island has been growing at an exponential rate. From 1760 to 1860, the population grew to a total of 1000 people, and from 1860 to 1960, the population grew to 4000 people. Using this information, determine the growth function for this population, and determine the year in which there will be 32,000 people on the island. (*Hint:* let the initial time $t = 0$ correspond to the year 1860.) ◀

RADIOACTIVE DECAY

Radioactive substances decay by emitting three types of particles: alpha particles (the nuclei of Helium atoms), beta particles (electrons) and gamma rays (composed of photons). According to the laws of physics, the amount of a radioactive substance left after t years is given by a formula of the form

$$A(t) = Ce^{-kt} \tag{5.17}$$

where C and k are positive constants. (Notice the negative sign in the exponent, to account for *decay*, rather than growth.) This type of decay is called **exponential decay.** As is the case with exponential growth, the constant C is the *initial* amount of the radioactive substance.

A common way to measure the rate at which a particular radioactive substance decays is to give its **half-life,** which is the time that it takes for one-half of the total amount of the substance to decay. One of the special, and rather remarkable, properties of exponential decay is that the time it takes for one-half of the total amount of a radioactive substance to decay does not depend on how much of the substance is present initially. For example, it takes the same amount of time for a 1-ounce supply of Uranium 238 to decay to 1/2 ounce as it does for a 1000-pound supply of Uranium 238 to decay to 500 pounds. (This time is approximately 1 billion years!)

Half-lives range from millionths of a second to billions of years, depending on the substance. Table 5.4 contains some half-lives for various radioactive substances.

TABLE 5.4

Radioactive substance	Half-life (approximate)
Polonium 212	3×10^{-7} (=0.0000003) seconds
Iridium 198	50 seconds
Lead 212	11 minutes
Gold 196	6 days
Polonium 210	138 days
Lead 210	19 years
Radium 226	1,620 years
Plutonium 239	24,400 years
Uranium 234	248,000 years
Uranium 238	4.5×10^{9} (=4,500,000,000) years

EXAMPLE 3: Plutonium 239, with a half-life of approximately 24,400 years, is produced in the so-called *fast nuclear reactors.* If 100 pounds of Plutonium 239 is produced by a certain reactor, how long will it take for the substance to decay to the point where there is only 1 pound left?

Solution: Since the initial amount of the substance is 100 pounds, we can write

$$A(t) = 100e^{-kt} \tag{5.18}$$

Also, since the half-life of this substance is approximately 24,400 years, there will be exactly 50 pounds of the substance left after 24,400 years; that is,

$$50 = 100e^{-24400k}$$

Let us solve this for e^k. Dividing both sides by 100 gives

$$e^{-24400k} = \frac{1}{2}$$

or

$$e^{-24400k} = 2^{-1}$$

Taking the reciprocal of both sides gives

$$e^{24400k} = 2$$

and since $e^{24400k} = (e^k)^{24400}$, we have

$$e^k = 2^{1/24400}$$

Using this in Equation (5.18), we get

$$A(t) = 100 \cdot 2^{-t/24400}$$

Now, in order to determine how long it will be before the substance has decayed to the point where there is only 1 pound remaining, we must solve the exponential equation

$$1 = 100 \cdot 2^{-t/24400}$$

Dividing both sides by 100 gives

$$2^{-t/24400} = \frac{1}{100} = 100^{-1}$$

or

$$2^{t/24400} = 100$$

Now we take the logarithm of both sides, and solve for t

$$\log(2^{t/24400}) = \log 100$$

$$\frac{t}{24400} \log 2 = \log 10^2 = 2$$

$$t = \frac{(2)(24400)}{\log 2} \approx 162{,}110$$

Thus, it takes approximately 162,110 years for the substance to decay to the point where there is only 1 pound left! *Try Study Suggestion 5.27.* □

Study Suggestion 5.27: The half-life of Radium 226 is approximately 1,620 years. How long does it take for 10 pounds of this substance to decay to 1/2 pound?

Following the same reasoning as in Example 3, we can determine a formula for the amount of a radioactive substance in terms of its half-life. Suppose that a certain substance has a half-life of h years. This means that an initial amount C of the substance decays to an amount equal to $C/2$ after h years. Thus, according to Formula (5.17), we have

$$\frac{C}{2} = Ce^{-kh}$$

Let us solve this equation for e^{-k}. Dividing both sides by C gives

$$e^{-kh} = \frac{1}{2} = 2^{-1}$$

Now we write e^{-kh} in the form $(e^{-k})^h$ and take the $1/h$ power of both sides, to get

$$e^{-k} = 2^{-1/h}$$

Substituting this into Formula (5.17) gives

$$A(t) = C2^{-t/h} \tag{5.19}$$

Now if we are given the half-life of a radioactive substance, we can determine the function $A(t)$ by substituting into Formula (5.19).

EXAMPLE 4: The half-life of Lead 210 is approximately 19 years. How long does it take 12 pounds of this substance to decay to 1/4 pound?

Solution: Setting $h = 19$ in Formula (5.19) gives

$$A(t) = C2^{-t/19}$$

Hence, in order to answer this question, we must solve the exponential equation

$$\frac{1}{4} = 12 \cdot 2^{-t/19}$$

We will leave it to you to show that its solution is

$$t = \frac{19 \log 48}{\log 2} \approx 106.11$$

Hence, it takes approximately 106.11 years for 12 pounds of Lead 210 to decay to 1/4 pound. *Try Study Suggestion 5.28.* □

Study Suggestion 5.28: The half-life of Polonium 210 is approximately 138 days. How long does it take 20 pounds of Polonium 210 to decay to 1/10 pound?

Ideas to Remember

- Exponential growth occurs at an extremely fast rate, and exponential decay occurs at an extremely slow rate.
- The irrational number e is perhaps the most important number in nature, and the exponential function $E(x) = e^x$ is perhaps the most important function in nature.

EXERCISES

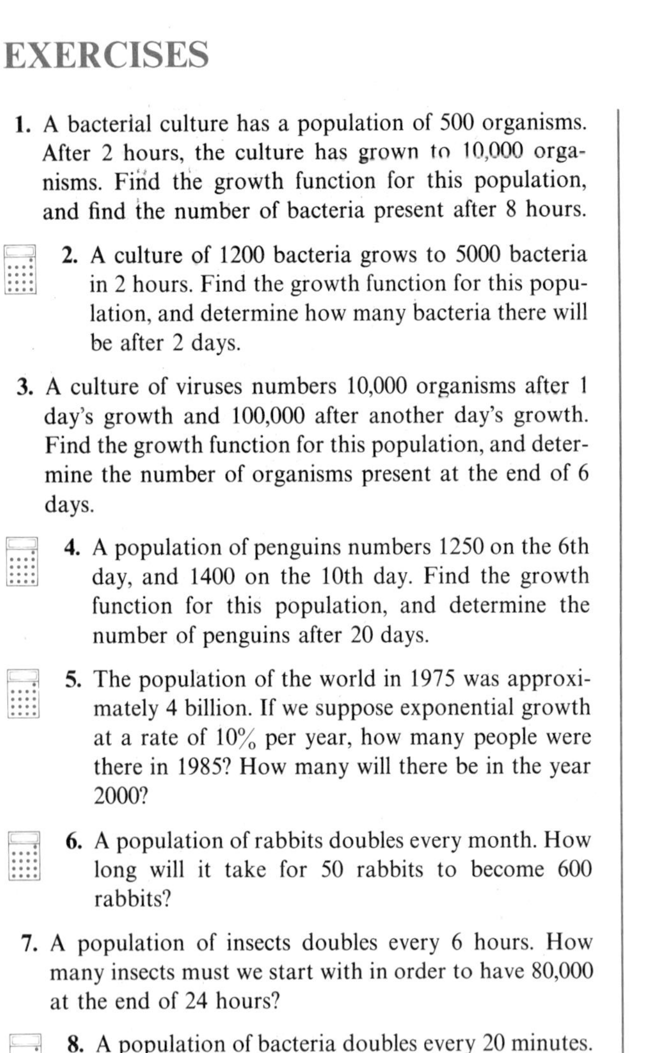

1. A bacterial culture has a population of 500 organisms. After 2 hours, the culture has grown to 10,000 organisms. Find the growth function for this population, and find the number of bacteria present after 8 hours.

2. A culture of 1200 bacteria grows to 5000 bacteria in 2 hours. Find the growth function for this population, and determine how many bacteria there will be after 2 days.

3. A culture of viruses numbers 10,000 organisms after 1 day's growth and 100,000 after another day's growth. Find the growth function for this population, and determine the number of organisms present at the end of 6 days.

4. A population of penguins numbers 1250 on the 6th day, and 1400 on the 10th day. Find the growth function for this population, and determine the number of penguins after 20 days.

5. The population of the world in 1975 was approximately 4 billion. If we suppose exponential growth at a rate of 10% per year, how many people were there in 1985? How many will there be in the year 2000?

6. A population of rabbits doubles every month. How long will it take for 50 rabbits to become 600 rabbits?

7. A population of insects doubles every 6 hours. How many insects must we start with in order to have 80,000 at the end of 24 hours?

8. A population of bacteria doubles every 20 minutes. How many times more bacteria will there be after 2 days than there was at the beginning of the 2-day period?

9. Population A doubles every 2 hours, and population B triples every 3 hours. Without doing any calculations, which population do you think grows faster? Now find the growth function for each population. Use these functions to decide which population grows faster, and support your conclusions.

10. The half-life of Lead 212 is approximately 11 minutes. How long will it take for 10 pounds of Lead 212 to decay to 5/8 of a pound?

11. The half-life of Uranium 234 is approximately 248,000 years. How long will it take for 1000 pounds of Uranium 234 to decay to 1 pound?

12. The half life of Uranium 238 is approximately 4.5 billion years. How much of a 1000-pound supply of Uranium 238 will be left after 100 years?

13. A 100-pound supply of a certain radioactive substance decays to 99.8 pounds in 5 years. How much of this substance will be left after 1000 years?

14. Radioactive Iodine 131, which has a half-life of approximately 8 days, is a product of nuclear fallout. If a cow eats hay that is contaminated with Iodine 131, the iodine makes its way into the cow's milk. Suppose that a certain batch of hay contains 8 times the "safe" levels of Iodine 131. How long is it necessary to wait before the cow's milk will be safe to drink?

15. Radioactive Strontium 90 is a particularly dangerous by-product of nuclear explosions, for it is very similar in chemical composition to calcium, and is therefore absorbed into the bones and tissues of man. Furthermore, the half-life of Strontium 90 is approximately 26 years. How long does it take a

100-pound supply of Strontium 90 to decay to 1 pound?

16. One hundred pounds of a certain radioactive substance decays to 90 pounds after 6 months. Determine how much of the substance is left after 2 years.

It is possible to estimate the age of once living organisms by measuring the amount of radioactive C^{14} (carbon-14) in the organisms. The method is based on the work of W.F. Libby, a chemist who later won the Noble Prize. In 1946, Libby discovered that the amount of radioactive C^{14} remained constant in living organisms, but that once the organism died, the C^{14} began to decay at a constant rate. Thus, organisms with less C^{14} should be older than organisms of a similar type with more C^{14}. Furthermore, there is evidence that the amount of C^{14} in the environment has not significantly changed over time. Therefore, an organism that is extremely old should have contained the same amount of C^{14} when it was alive as a similar organism does today. This fact can be used to determine the amount of decay of C^{14} in very old organisms. In the following exercises, use the fact that the half-life of C^{14} is approximately 5600 years.

17. Using the half-life, find the constant k in the decay function for C^{14}.

18. Comparing a fossil leg bone to the leg bone of a modern man, it is observed that the fossil leg bone contains about 10% as much C^{14}. How old is the leg bone?

19. A fossil tree has about 5% as much C^{14} as a similar living tree. How old is the tree?

According to Newton's Law of Cooling, the difference D in temperature between an object and its surrounding medium is given by the exponential function

$$D = D_0 e^{-kt}$$

where D_0 is the difference in temperature when $t = 0$, that is, D_0 is the initial difference in temperature, and the constant k depends on the type of object and the type of surrounding medium (air, water, dirt, etc.).

20. The coffee in a certain pot is at a temperature of 210°. The coffee is poured into a mug which is at 60°. If we assume that in this case $k = 0.16$, find the temperature of the coffee after 10 minutes.

21. On a summer day, the temperature of a glass of ice tea rises from 40° to 80° in 20 minutes. If $k = 0.05$ degrees/min, how hot is it on that day?

5.6 Review

CONCEPTS FOR REVIEW

The exponential functions
Base of an exponential function
Exponential equation
The logarithmic functions
Natural logarithm
Common logarithm
Richter scale
Logarithmic scale
Continuous compounding of interest
Exponential growth
Exponential decay

REVIEW EXERCISES

In Exercises 1–4, approximate the given value.

1. e^{-e} **2.** $e^{\pi} - \pi^{e}$

3. $\sqrt{e}^{\sqrt{e}}$ **4.** $\sqrt{e+1}^{\sqrt{e+1}}$

In Exercises 5–10, graph the given function.

5. $f(x) = (2/5)^{-x}$ **6.** $f(x) = 2^x - 2^{-x}$

7. $f(x) = e^x + e^{-x}$ **8.** $f(x) = xe^x$

9. $f(x) = e^{-(x+1)^2}$

10. $f(x) = \dfrac{2}{e^x - e^{-x}}$

In Exercises 11–14, solve the given equation without using logarithms.

11. $6^x = 36^{2x+1}$

12. $9^{2x} = 27^{x-1}$

13. $8^{3x}4^{-x+1} = 16$

14. $4^x - 6 \cdot 2^x + 8 = 0$ (*Hint:* let $y = 2^x$)

In Exercises 15–18, compute the given logarithm without using a calculator or table of logarithms.

15. $\log_9 27$

16. $\log_8 2\sqrt{2}$

17. $\log_5 25\sqrt{5}$

18. $\log(\log_2 1024))$

19. Solve the equation $\log(x + 1) = 10$.

20. Solve the equation $\log_2(2x + 4) = 2$.

21. Solve the equation $\log_x 3 = 2$.

22. Solve the equation $\log_x(x + 1) = 2$.

23. Use the fact that $\log x = 2.1$, $\log y = -1.3$ and $\log z = 0.4$ to compute

$$\log \frac{\sqrt[3]{x^2 y}}{z^5}$$

24. Graph the functions $f(x) = \ln x$ and $g(x) = (x + 1)\ln x \cdot$ on the same plane.

25. Graph the function $f(x) = 1/\ln x$

In Exercises 26–31, solve or approximate the solutions to the given equation.

26. $2^x = 7$

27. $4^{2x-3} = 6^{x+1}$

28. $e^{x+2} = 2^4$

29. $5^{2x}6^{x-3} = 2^x$

30. $(1/2)^{x-1} = 1$

31. $(1/4)^{-x} = 4^x$

The magnitude of a celestial object is measured on a logarithmic scale,

$$m = 1 - 2.5\log\left(\frac{I}{I_0}\right)$$

where I_0 is the intensity of light from the bright stars Altair and Aldebaran, which are taken as "standards" for this scale. Thus, for example, a star that is 10 times as intense as Altair has a magnitude of $m = 1 - 2.5 \log 10 = -1.5$. Notice that, due to the negative sign in the formula for m, brighter objects have *smaller* magnitudes, and dimmer objects have *larger* magnitudes.

32. Mars at its brightest has magnitude -2. How much more intense is the light from Mars than from the star Altair?

33. The full moon has intensity -12.5. How much brighter is it than Mars? (See Exercise 32.)

34. The sun has magnitude -26.5. How much more intense is the light from the sun than from the moon? (See Exercise 33.)

35. The human eye is barely capable of discerning objects of magnitude 6.5. How much less intense is the light from such an object than is the light from the moon? (See Exercise 33.)

36. The 200-inch telescope at Mt. Palomar can discern an object of magnitude 20. How much less intense is the light from such an object than is the light from the star Altair?

37. A savings account pays 12% interest compounded quarterly. What interest rate would a second account have to pay in order to yield the same amount of money if it compounds continuously?

38. An account grows from \$1000 to \$1150 in 1 year. How long does it take for the account to double?

39. Find a formula for the length of time required for an account to double if the interest rate is r and the interest is compounded n times a year.

40. A certain population grows exponentially.

(a) Find an expression for the time d that it takes the population to double.

(b) Find a formula for the growth function $A(t)$ of this population in terms of the doubling time d.

41. The so-called *fast reactors* produce Uranium 238, which has a half-life of approximately 4.5 billion ($=4.5 \times 10^9$) years. How long does it take for 1 pound of Uranium 238 to decay to 1 ounce? (1 ounce = 1/32 pound). How does this time period compare with the age of the earth, which is estimated to be only 4.6×10^9 years?

6 THE TRIGONOMETRIC FUNCTIONS

6.1 Radian Measure

In this and the next two chapters, we are going to study a class of functions known as the *trigonometric* or *circular functions.* The fact that there are two different names for these functions reflects the fact that there are two different approaches to defining them, one of which involves triangles, and the other of which involves circles.* Our plan in this chapter is to introduce the trigonometric functions using circles, and then discuss the triangle approach in the next chapter.

In order to begin our study of the trigonometric functions, we must first set some terminology regarding angles and their measurement. If A and B are distinct points in the plane, then the unbounded line segment that begins at A and extends *through* B is called the **ray** from A through B. The point A is called the **endpoint** of the ray. (See Figure 6.1(a).)

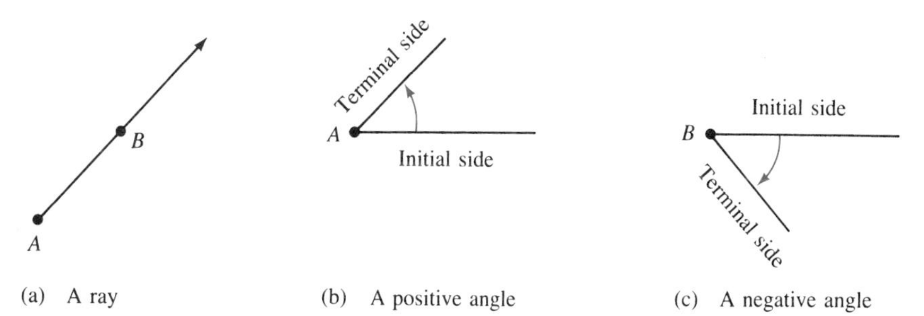

FIGURE 6.1 (a) A ray (b) A positive angle (c) A negative angle

* The term *trigonometry* comes from the Greek word *trigonon*, meaning *triangle*, and the suffix *metria*, meaning *the art or science of measuring*. Thus, *trigonometry* refers to the art or science of measuring triangles.

By rotating a ray with endpoint A, we obtain an **angle,** with **vertex** A. When there is no possibility of confusion, we will also denote the angle by the letter A. The original position of the ray is called the **initial side** of the angle, and the rotated position of the ray is called the **terminal side** of the angle. Angles formed by rotating a ray in a *counterclockwise* direction are said to be **positive** angles, and angles formed by rotating in a *clockwise* direction are said to be **negative** angles. Thus, angle A in Figure 6.1(b) is positive and angle B in Figure 6.1(c) is negative.

An angle that corresponds to one complete counterclockwise revolution is called a **full angle,** an angle that corresponds to one-half of a complete counterclockwise revolution is called a **straight angle,** and an angle that corresponds to one-fourth of a complete counterclockwise revolution is called a **right angle.** These angles are pictured in Figure 6.2.

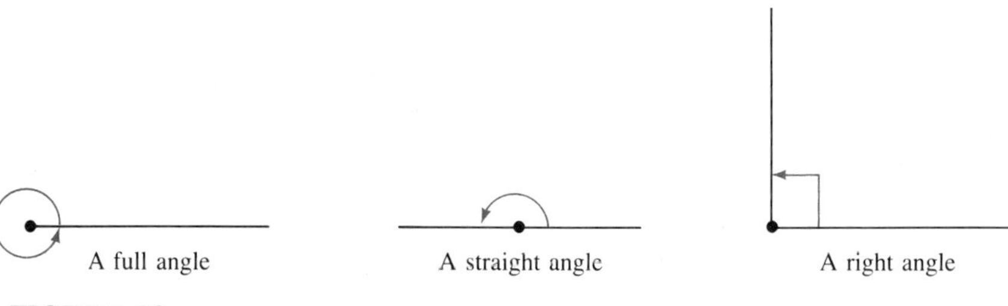

FIGURE 6.2

When an angle is placed in the plane in such a way that its vertex is at the origin and its initial side is along the positive x-axis, we say that the angle is in **standard position.** For example, the angle A in Figure 6.3 is in standard position.

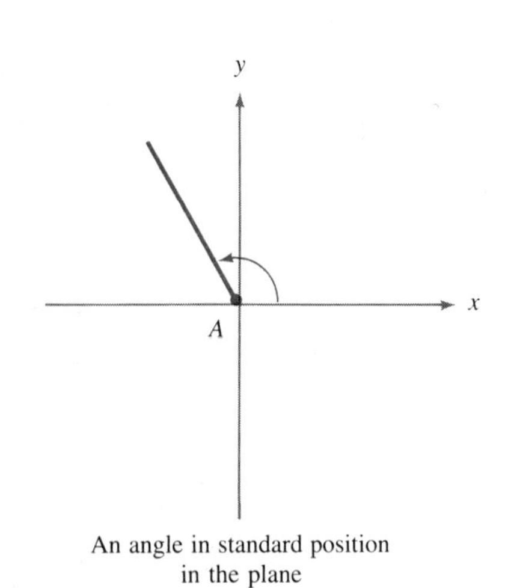

An angle in standard position in the plane

FIGURE 6.3

The angle A shown in Figure 6.4(a) is placed in such a way that its vertex is at the center of a circle. In this case, we say that A is a **central angle,** and that it **subtends** the arc PQ. The negative angle B in Figure 6.4(b) subtends the arc QP.

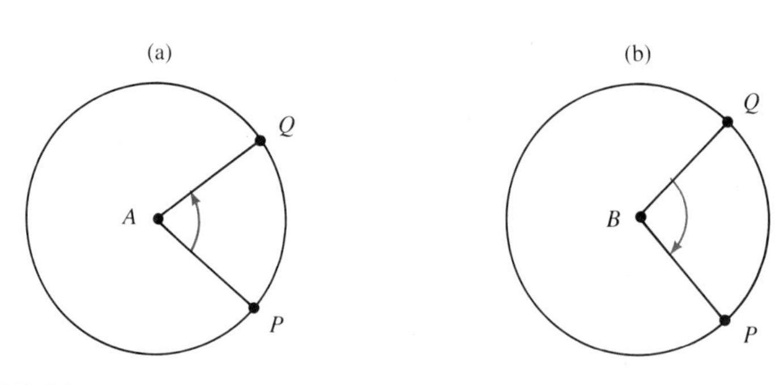

FIGURE 6.4

One of the most common units of measurement of angles is the *degree.* Since a full angle measures 360 degrees, an angle measures 1 degree if it can be obtained by rotating a ray exactly 1/360 of a complete counterclockwise revolution. (See Figure 6.5.)

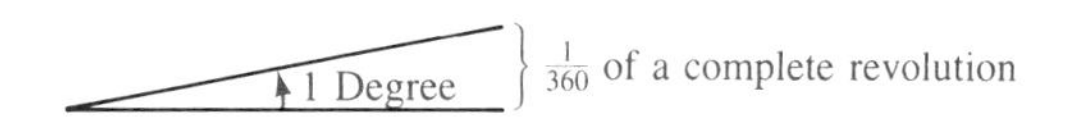

FIGURE 6.5

Degrees are not the only important unit of measurement of angles, however. For our purposes, we will require another unit, known as the *radian.* In order to motivate the definition of this new unit of measurement, let us look again at the degree.

Figure 6.6(a) shows an angle of one degree placed at the center of a circle of radius 1. (A circle of radius 1 is known as a **unit circle.**) Since a degree corresponds to 1/360 of a complete revolution around the circle, and since the circumference of a unit circle has length 2π, the arc PQ in Figure 6.6(a) has length

$$\frac{1}{360}(2\pi) = \frac{\pi}{180}$$

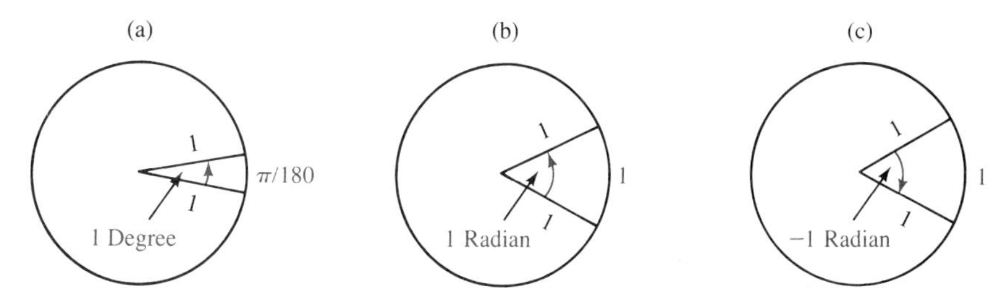

A positive angle of 1 degree subtends an arc of length $\pi/180$.
A positive angle of 1 radian subtends an arc of length 1.

FIGURE 6.6

Now, we could use this fact to *define* the term degree, by saying that a positive angle measures 1 degree if, when that angle is placed at the center of a unit circle, it subtends an arc of length $\pi/180$. Although this definition is a bit awkward, it does lead us to the idea of defining a new unit of measurement as follows.

DEFINITION

> A positive angle measures 1 **radian** if, when that angle is placed at the center of a *unit* circle, it subtends an arc of length 1.

Thus, the angle A in Figure 6.6(b) measures one radian. In keeping with our definition of positive and negative angles, we will say that a negative angle that subtends an arc of length 1, such as the one in Figure 6.6(c), measures -1 radian.

More generally, if a positive angle subtends an arc of length s, when placed at the center of a unit circle, then the angle measures s radians. (See Figure 6.7(a).) On the other hand, if a negative angle subtends an arc of length s, when placed at the center of a unit circle, then it measures $-s$ radians. (See Figure 6.7(b).)

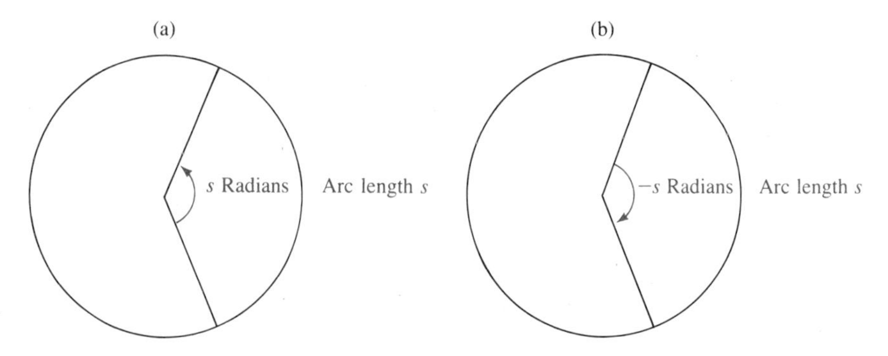

FIGURE 6.7

As you can see from this discussion, *the definition of radian measure is designed so that there is a direct equivalence between the radian measure of a (positive) central angle and the length of arc subtended by that angle on a unit circle.* For example, a central angle of 3 radians subtends an arc of length 3 inches, on a circle of radius 1 inch. (See Figure 6.8.) This is certainly not the case for degree measure. For instance, a central angle of 60 degrees does *not* subtend an arc of 60 inches, on a circle of radius 1 inch.

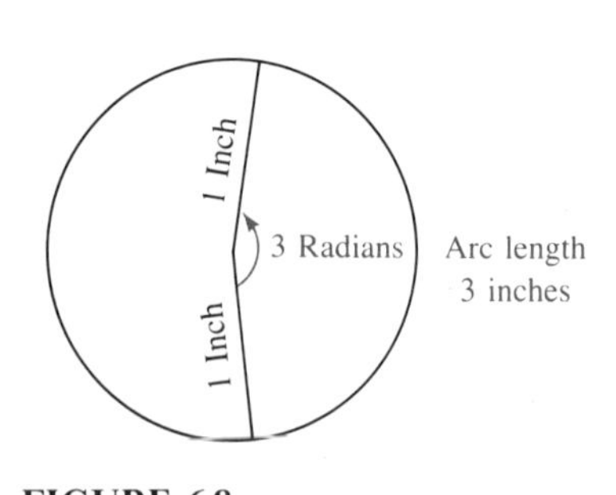

FIGURE 6.8

This equivalence between radian measure and arc length is the main reason that radian measure is so important, and it is the reason that we will use radian measure when we define the trigonometric functions in Section 6.3.

We will discuss the connection between radian measure and arc length in more detail later in the section, but first let us compare our new unit of measurement with our old unit, the degree. Since a full angle subtends the entire circumference of a unit circle (when placed at its center) and since the circumference of a unit circle has length 2π, the radian measure of a full angle is equal to 2π. Therefore, we have

$$2\pi \text{ radians} = 360 \text{ degrees}$$

This gives the conversion factors

$$1 \text{ radian} = \frac{180}{\pi} \text{ degrees} \quad \text{and} \quad 1 \text{ degree} = \frac{\pi}{180} \text{ radians}$$

Hence,

$$\textbf{an angle of } r \textbf{ radians measures } \left(\frac{180}{\pi}\right) r \textbf{ degrees;} \qquad (6.1)$$

$$\textbf{an angle of } d \textbf{ degrees measures } \left(\frac{\pi}{180}\right) d \textbf{ radians.} \qquad (6.2)$$

A calculator gives us the approximation $\pi/180 \approx 57.29$, and so we see that 1 radian is *approximately* equal to 57 degrees. However, let us emphasize that this is only an approximation. In order to obtain exact conversions, we must use Equations (6.1) and (6.2). As a result, angles with common degree measurements, such as 30° and 45°, will have radian measures that are rational multiples of π.

EXAMPLE 1: Convert the following angles from degree measure to radian measure.

(a) 30° **(b)** 45° **(c)** -150°

Solutions: In order to convert degrees to radians, we multiply by $\pi/180$. Therefore,

(a) An angle of 30° measures

$$\left(\frac{\pi}{180}\right) \cdot 30 = \frac{\pi}{6} \text{ radians}$$

(b) An angle of 45° measures

$$\left(\frac{\pi}{180}\right) \cdot 45 = \frac{\pi}{4} \text{ radians}$$

(c) An angle of -150° measures

$$\left(\frac{\pi}{180}\right)\left(-150\right) = -\frac{5\pi}{6} \text{ radians}$$

Try Study Suggestion 6.1. □

▶ **Study Suggestion 6.1:** Convert the following from degrees to radians.
(a) -60° **(b)** 270° **(c)** -40° ◀

EXAMPLE 2: Convert the following angles from radians to degrees.

(a) π **(b)** $-\dfrac{2\pi}{3}$ **(c)** 4

Solutions: In order to convert from radians to degrees, we multiply by $180/\pi$. Therefore,

(a) An angle of π radians measures

$$\left(\frac{180}{\pi}\right) \cdot \pi = 180 \text{ degrees}$$

(b) An angle of $-2\pi/3$ radians measures

$$\left(\frac{180}{\pi}\right)\left(-\frac{2\pi}{3}\right) = -120 \text{ degrees}$$

(c) An angle of 4 radians measures

$$\left(\frac{180}{\pi}\right) \cdot 4 = \frac{720}{\pi} \text{ degrees}$$

Try Study Suggestion 6.2 □

▶ **Study Suggestion 6.2:** Convert the following from radians to degrees.
(a) $\pi/3$ **(b)** 12π **(c)** 2 ◀

Since we will use radian measure when defining the trigonometric functions, we strongly suggest that you familiarize yourself with the radian measure of some common angles, as given in Table 6.1. The angles in this table are also shown in standard position in Figure 6.9. It is common practice when giving the radian measure of an angle to omit the word radian, and we have followed this practice in Figure 6.9. Therefore, *when there is no mention of the units involved, you should assume that the angle is given in radian measure.*

TABLE 6.1

degrees	radians
30	$\pi/6$
45	$\pi/4$
60	$\pi/3$
90	$\pi/2$
180	π
270	$3\pi/2$
360	2π

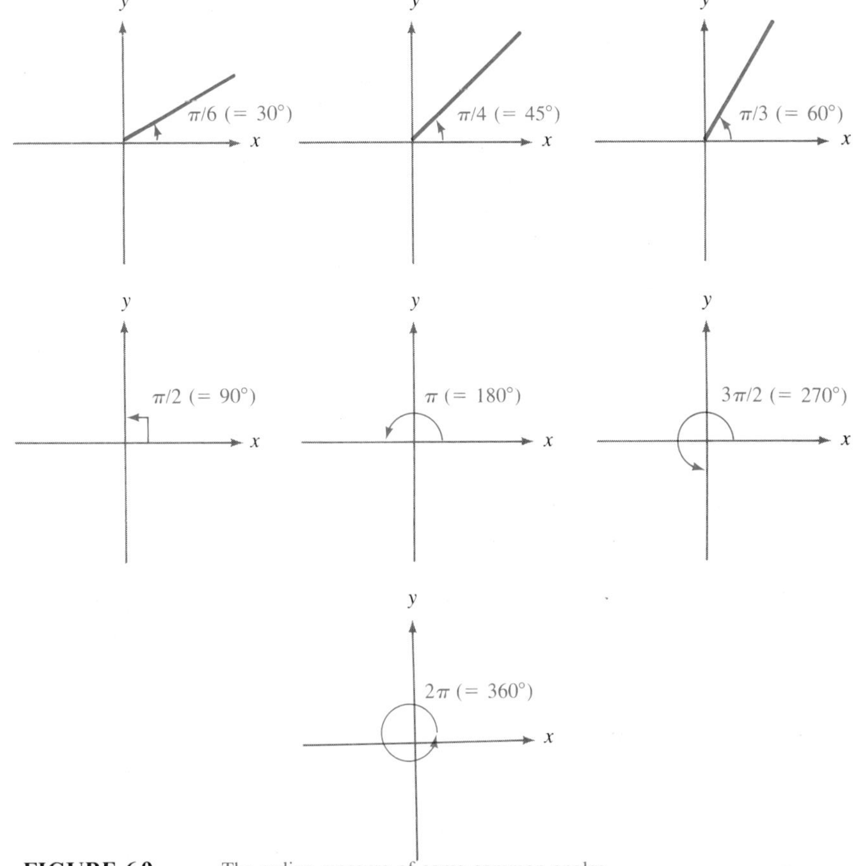

FIGURE 6.9 The radian measure of some common angles

Table 6.1 can be used to convert angles other than those given in the table. For example, using this table, we get

$$\frac{4\pi}{3} \text{ radians} = 4\left(\frac{\pi}{3}\right) \text{ radians} = 4 \cdot 60^\circ = 240^\circ$$

Now let us continue our discussion of the connection between radian measure and arc length. Figure 6.10 shows two concentric circles, of radius 1 and r, along with a central angle A of θ radians. (θ is the Greek letter *theta*. Greek letters are often used to denote the measure of angles.)

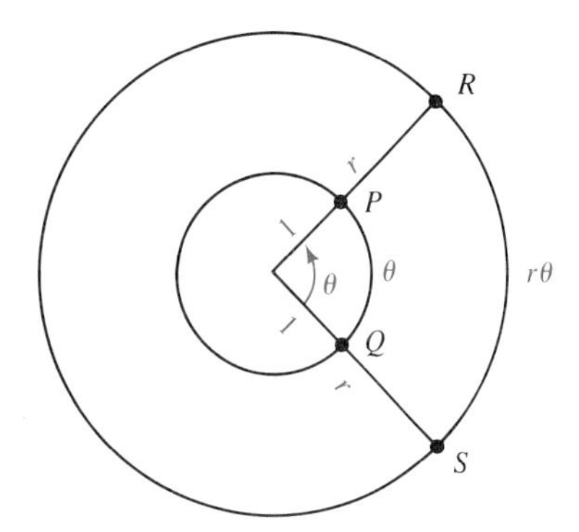

A central angle of θ radians subtends an arc of length $r\theta$ on a circle of radius r

FIGURE 6.10

We know from our earlier discussion that the arc PQ has length θ. Let us compute the length of the arc RS. According to the laws of geometry, angle A subtends the same *proportion* of arc on the unit circle as it does on the circle of radius r. Thus we have

$$\frac{\text{length of arc } RS}{2\pi r} = \frac{\text{length of arc } PQ}{\text{circumference of a unit circle}} = \frac{\theta}{2\pi}$$

and so

$$\text{length of arc } RS = \frac{\theta}{2\pi}(2\pi r) = r\theta$$

Let us summarize in a theorem.

THEOREM

A positive angle of θ radians, placed at the center of a circle of radius r, subtends an arc of length

$$s = r\theta \tag{6.3}$$

We should emphasize that *in order to use Formula (6.3), the measure of the angle must be in radians, and not degrees.* (See Exercise 37.)

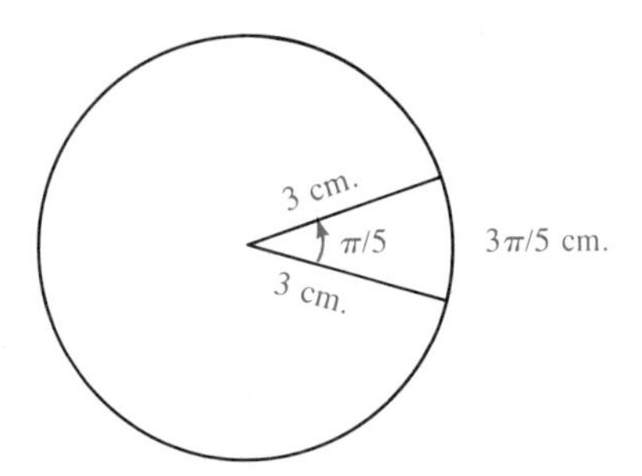

FIGURE 6.11

EXAMPLE 3: An angle measuring 36° has its vertex at the center of a circle of radius 3 centimeters. What is the length of arc subtended by this angle?

Solution: In order to use equation (6.3), we must first convert the degree measure to radian measure. An angle of 36° measures $36(\pi/180) = \pi/5$ radians, and so, according to (6.3), this angle subtends an arc of length $s = r\theta = 3(\pi/5) = 3\pi/5$ centimeters. (See Figure 6.11.) *Try Study Suggestion 6.3.* □

Study Suggestion 6.3: A roulette wheel whose radius is 10 inches has been rotated through an angle of 300°. How much of the circumference of the roulette wheel has passed a fixed point adjacent to the wheel?

EXAMPLE 4: A truck whose tires have radius 12 inches is pushed a distance of 30 inches. What is the size of the angle through which the tires have rotated?

Solution: Since the truck has moved 30 inches, the tires have rotated so that 30 inches of arc on the tire's surface has touched the ground. If we denote the radian measure of the angle of rotation by θ, then, since $s = 30$ and $r = 12$, Formula (6.3) tells us that

$$30 = 12\theta$$

and so $\theta = 30/12 = 5/2$ radians. Thus, the tires have rotated through an angle of 5/2 radians. (Using a calculator, we see that this is $(5/2)(180/\pi) \approx 143°$.) *Try Study Suggestion 6.4.* □

Study Suggestion 6.4: A cement roller whose wheel has a radius of 2.5 feet has just smoothed out 12 yards of cement. How many times did the roller rotate?

Formula (6.3) can also be used to approximate the distances between points.

EXAMPLE 5: A surveyor wants to measure the distance between two points A and B on opposite sides of a river, as shown in Figure 6.12. In order to do so, he positions himself at point C, which is 1000 yards from point A, and about the same distance from point B. Then he measures the angle subtended by the points A and B, with himself at the vertex. If this angle measures 40°, approximately how far is it from point A to point B?

Study Suggestion 6.5: A boat lies offshore equidistant from two lighthouses. If the distance from the boat to each of the lighthouses is 5000 yards, and if the angle subtended by the lighthouses, with vertex at the boat, measures 130°, approximately how far apart are the lighthouses?

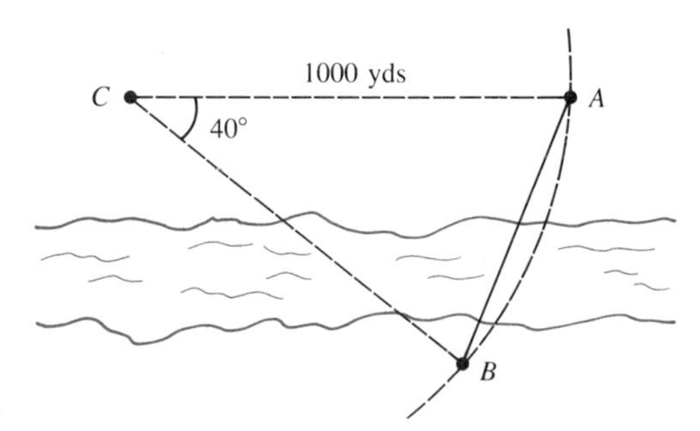

FIGURE 6.12

Solution: If we think of the point C as the center of a large circle that goes through A and B, then the length s of arc subtended by angle ACB can be computed using Formula (6.3). Since 40° is $(40)(\pi/180) \approx 0.698$ radians, Formula (6.3) tells us that $s \approx 1000(0.698) = 698$ yards. But since the circle is large, the arc length from A to B is not much greater than the length of the chord from A to B, and so the distance from A to B is *approximately* 698 yards.

Try Study Suggestion 6.5. □

Ideas to Remember

Degrees and radians are simply two different units of measurement for angles, much as inches and centimeters are two different units of measurement for lengths. However, radian measure is more directly related to arc length, and for this reason, it is much more important than degree measure for the applications that we have in mind (such as defining the trigonometric functions.)

EXERCISES

In Exercises 1–8, convert the given degree measure to radian measure. Do not use a calculator.

1. $-30°$ **2.** $120°$

3. $-80°$ **4.** $-90°$

5. $15°$ **6.** $135°$

7. $225°$ **8.** $-300°$

In Exercises 9–16, convert the given radian measure to degree measure. Do not use a calculator.

9. $\pi/6$ **10.** $5\pi/3$

11. $3\pi/4$ **12.** $-\pi/6$

13. $-\pi/3$ **14.** $7\pi/4$

15. $\pi/12$ **16.** 17π

In Exercises 17–25, use a calculator to convert degrees to radians, or radians to degrees. Express the answer as a decimal. (Of course, your answers will only be approximations.)

17. $123°$ **18.** $-17.29°$

19. $498.13°$ **20.** $0.059°$

21. $\pi/10$ radians **22.** 0.49π radians

23. 7.21 radians **24.** -0.493 radians

25. 300 radians

26. Find the length of arc subtended by an angle of $150°$ placed at the center of a circle of radius 6 feet.

27. A woman is standing at the center of a circular room of diameter 50 feet. If the woman has a field of vision measuring $145°$, how much of the wall can she see without moving her head?

28. Find the degree measure and the radian measure of an angle that subtends an arc of length 30 inches when placed at the center of a circle of radius 20 inches.

29. One hundred yards of thread is wound around a spool whose radius is 1/2 inch. How many revolutions must the spool undergo in order to completely unwind the thread?

30. A central angle of 0.3 radians subtends an arc of length 12 cm on a certain circle. What is the radius of this circle?

31. A central angle of $240°$ subtends an arc of length 18 feet on a certain circle. What is the radius of the circle?

32. How large should a roulette wheel be if a central angle of $8°$ is to subtend an arc of length 2.65 inches?

33. Two towers are each approximately 9 miles from point A. Find the approximate distance between the towers if the angle subtended by the towers, with vertex at point A, is $6°$.

34. A man sees a building approximately 5 miles away, and measures the angle of inclination of the top of the building to be $1°$. Find the approximate height of the building.

35. The angle subtended by San Francisco and Sacramento, with vertex at Los Angeles, is approximately $12.7°$. San Francisco and Sacramento are both approximately 360 miles from Los Angeles. Use this information to approximate the distance between San Francisco and Sacramento.

36. The following figure shows a belt stretched between two wheels.

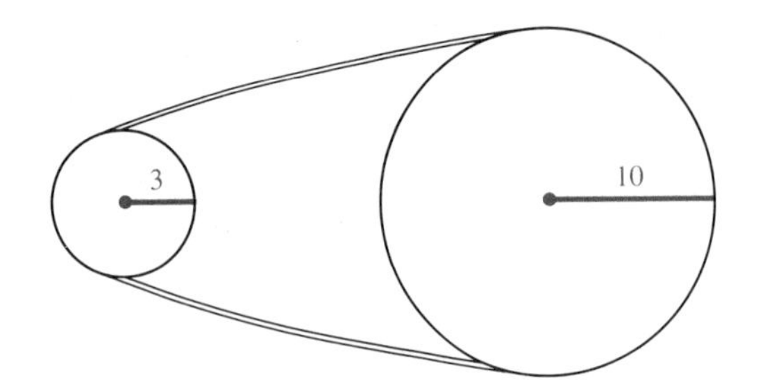

(a) If the smaller wheel is rotated through an angle of θ radians, through what angle is the larger wheel rotated?
(b) If the smaller wheel is rotated through an angle of 90°, through what angle is the larger wheel rotated?
(c) If the smaller wheel undergoes one complete revolution, through what angle does the larger wheel rotate?
(d) Through what angle must the smaller wheel be rotated in order for the larger wheel to undergo exactly one revolution?

37. Find a formula for the length s of arc subtended by an angle of d *degrees*, placed at the center of a circle of radius r. (That is, find a formula for s in terms of r and d.)

38. If a radial line is rotated through an angle of θ radians, it sweeps out an area known as a **sector.** This is pictured in the following figure. Show that the area of this sector is given by the formula $A = (1/2)r^2\theta$. (*Hint:* the area A of the sector is in the same proportion to the area of the entire circle as θ is to the radian measure of a full angle. Translate this statement into symbols and solve for A.)

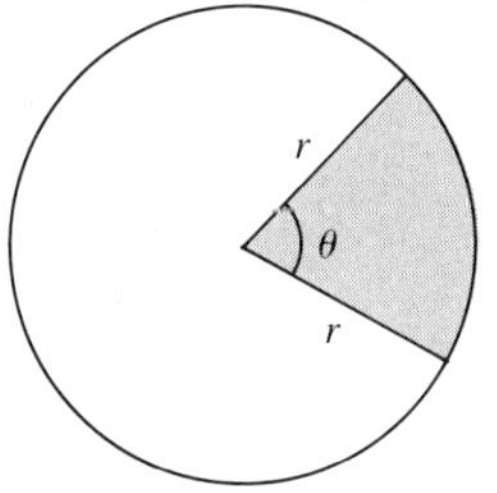

Area of shaded sector $= \frac{1}{2}r^2\theta$

39. A searchlight sweeps out an angle of 110° in heavy fog. If the light can reach a distance of 2000 yards into the fog, what is the total area that the searchlight illuminates? (See Exercise 38.)

6.2 Unit Circles

Let us consider a unit circle whose center is at the origin of a Cartesian coordinate system, as pictured in Figure 6.13(a and b). If t is a real number, then **sweeping out a central angle of t radians** means starting at the point $A = (1, 0)$ and traveling along the circle to the point, denoted by $P(t)$, with the property that the central angle $P(t)OA$ has radian measure t. This is done in Figure 6.13(a) for t positive and in Figure 6.13(b) for t negative.

EXAMPLE 1: In each case, draw a picture to show the effect of sweeping out a central angle of the given radian measure.

(a) $\frac{\pi}{2}$ **(b)** $-\frac{\pi}{6}$ **(c)** $\frac{2\pi}{3}$ **(d)** $-\frac{4\pi}{7}$

Solution: Figure 6.14 shows the effects of sweeping out these central angles. The angles corresponding to parts (a)–(c) are easily recognizable. In the case of part *d*, we observe that $-\pi < -4\pi/7 < -\pi/2$ and so $P(-4\pi/7)$ lies in the third quadrant. We can get a more accurate estimate of the location of

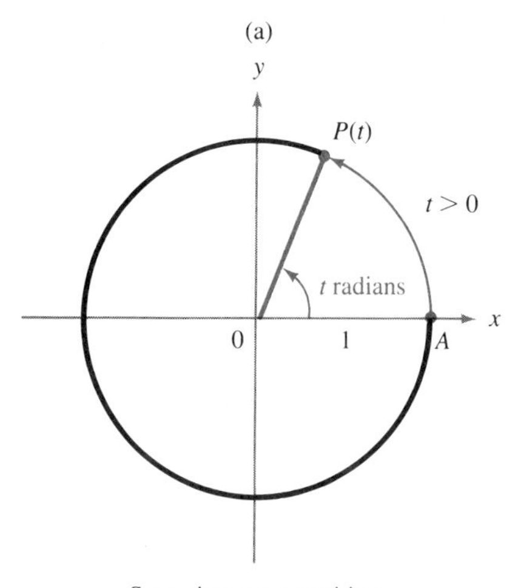

Sweeping out a positive central angle t on a unit circle

$P(-4\pi/7)$ by observing that since $-3\pi/4 < -4\pi/7 < -\pi/2$, the point $P(-4\pi/7)$ lies in that half of the third quadrant closest to the y-axis, as shown in Figure 6.14(d).

Try Study Suggestion 6.6. □

Since t is the *radian* measure of the angle $P(t)OA$, the length of arc from A to $P(t)$ is equal to $|t|$, where the absolute value sign is necessary since t may be negative, but the length of an arc is always positive. Thus,

sweeping out a central angle of radian measure t is the same as sweeping out an arc of length $|t|$ on the unit circle, where the arc is swept out in the counterclockwise direction if t is positive and in the clockwise direction if t is negative.

Our goal in this section is to determine the coordinates of certain of the points $P(t)$, obtained by sweeping out a central angle of t radians. These coordinates are the key to computing values of the trigonometric functions, which we will do in the next section.

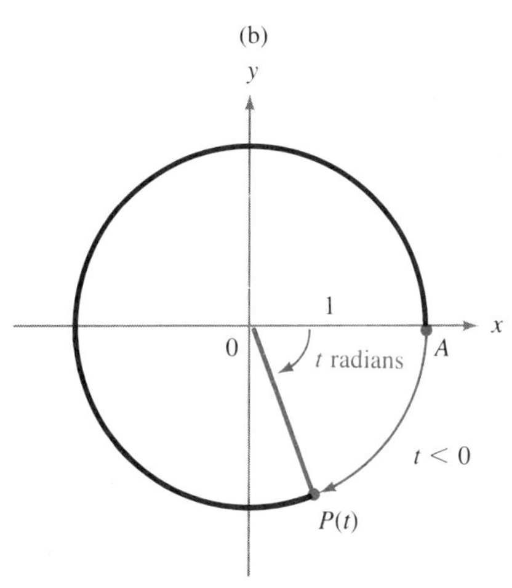

Sweeping out a negative central angle t on a unit circle

FIGURE 6.13

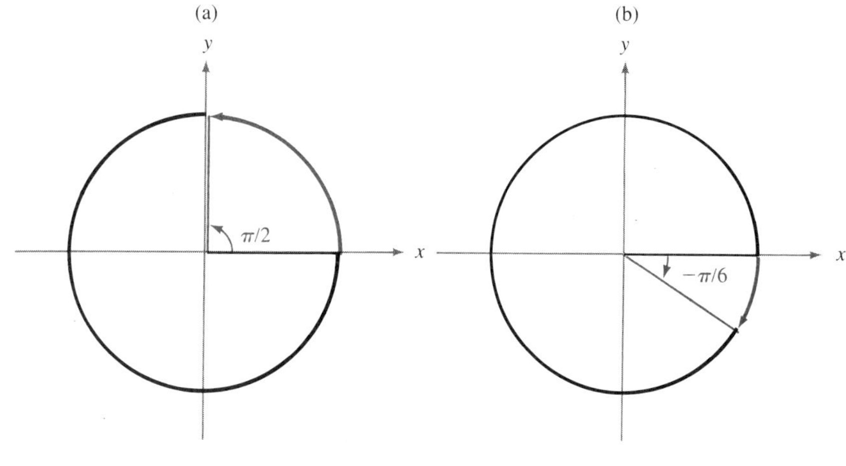

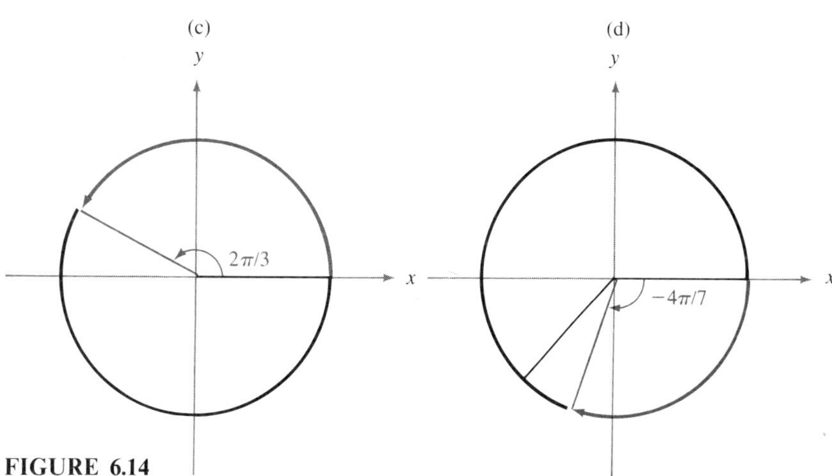

FIGURE 6.14

▶ **Study Suggestion 6.6:** Draw a picture to show the effect of sweeping out central angles of each of the following radian measures.

(a) $\pi/3$ **(b)** $-3\pi/4$ **(c)** $-7\pi/6$ **(d)** $8\pi/5$

(Assume that each quadrant is divided into two equal sections, and locate the point $P(t)$ in the proper section, as we did in Example 1.) ◀

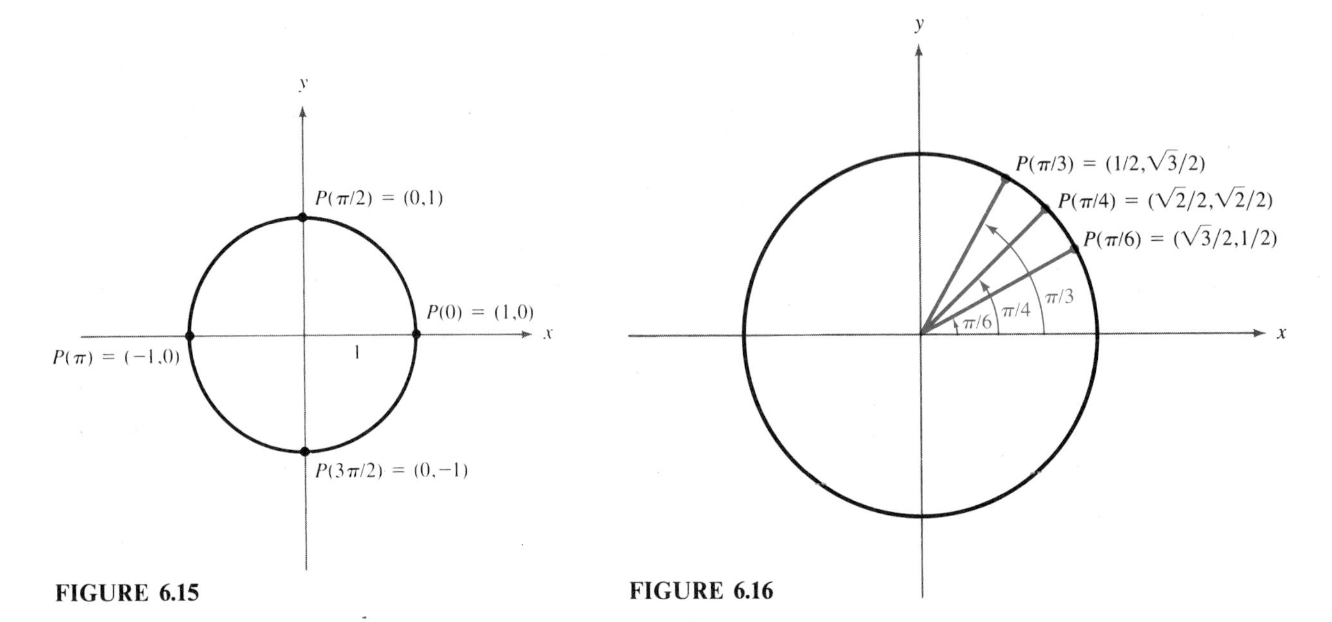

FIGURE 6.15

FIGURE 6.16

The coordinates of the points $P(0)$, $P(\pi/2)$, $P(\pi)$ and $P(3\pi/2)$, shown in Figure 6.15, can be easily determined from the fact that these points are the x and y intercepts of a circle of radius 1.

In order to determine the coordinates of other points on the unit circle, we use the fact that $P(t)$ lies on this circle, and so if $P(t) = (x, y)$ then we must have

$$x^2 + y^2 = 1 \tag{6.4}$$

(This is the equation of a circle of radius 1.) For example, Figure 6.16 shows the coordinates of the points $P(\pi/6)$, $P(\pi/4)$ and $P(\pi/3)$. These coordinates can be determined by using Equation (6.4) along with some simple geometry.

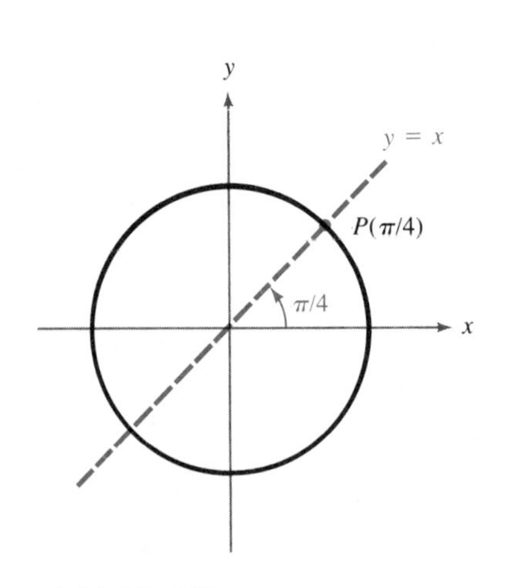

FIGURE 6.17

EXAMPLE 2: Determine the coordinates of the point $P(\pi/4)$.

Solution: Since the point $P(\pi/4) = (x, y)$ is on the unit circle, we know that Equation (6.4) holds. Also, since $\pi/4$ is halfway between 0 and $\pi/2$, the point $P(t) = (x, y)$ is halfway between the points $P(0)$ and $P(\pi/2)$ on the circle. Hence, it must lie on the line $y = x$. (See Figure 6.17.) Setting $y = x$ in Equation (6.4) gives

$$2x^2 = 1$$

and so x must be one of the numbers $\pm\sqrt{2}/2$. But since $P(t)$ lies in the first quadrant, its x-coordinate is positive, and so $x = \sqrt{2}/2$. Finally, since $y = x$, we have

$$P\left(\frac{\pi}{4}\right) = \left(\frac{\sqrt{2}}{2}, \frac{\sqrt{2}}{2}\right)$$

□

EXAMPLE 3: Determine the coordinates of the point $P\left(\frac{\pi}{6}\right)$.

Solution: Since $P(\pi/6) = (x, y)$ is on the unit circle, Equation (6.4) holds. In order to determine x and y, we need one more piece of information. Since $\pi/6 = (1/3)(\pi/2)$, the point $P(\pi/6)$ is one third of the way around the circle from $P(0)$ to $P(\pi/2)$. (See Figure 6.18.) Thus, chord AB has the same length as chord BC, and this implies that angle AOB equals angle BOC. But the length of BC is $2y$ and, according to the distance formula, the length of AB is

$$\sqrt{(x-0)^2 + (y-1)^2}$$

Equating these distances, we get

$$2y = \sqrt{(x-0)^2 + (y-1)^2}$$

Squaring both sides of this equation gives

$$4y^2 = x^2 + y^2 - 2y + 1$$

Now we can use Equation (6.4) to replace $x^2 + y^2$ by 1, giving

$$4y^2 = -2y + 2$$

or

$$4y^2 + 2y - 2 = 0$$

This equation has solutions $y = -1$ and $y = 1/2$, but since the y-coordinate of the point $P(\pi/6)$ is positive, we are interested only in the solution $y = 1/2$. Substituting this into Equation (6.4) and solving for x gives $x = \sqrt{3}/2$, and so

$$P\left(\frac{\pi}{6}\right) = \left(\frac{\sqrt{3}}{2}, \frac{1}{2}\right)$$

Try Study Suggestion 6.7. □

▶ **Study Suggestion 6.7:** Determine the coordinates of the point $P(\pi/3)$. (*Hint:* follow the same reasoning as in the previous example, using the following figure.) ◀

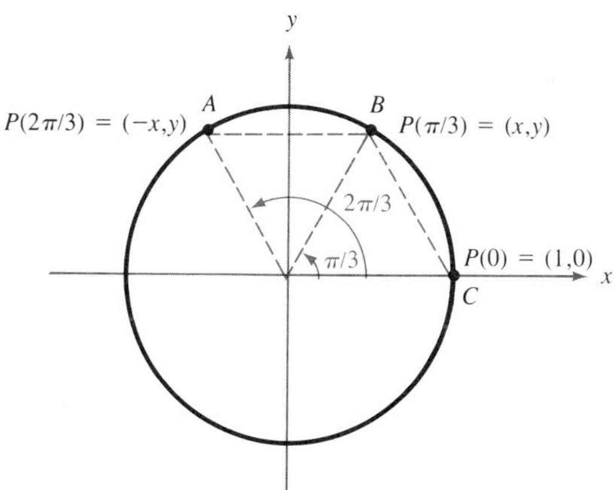

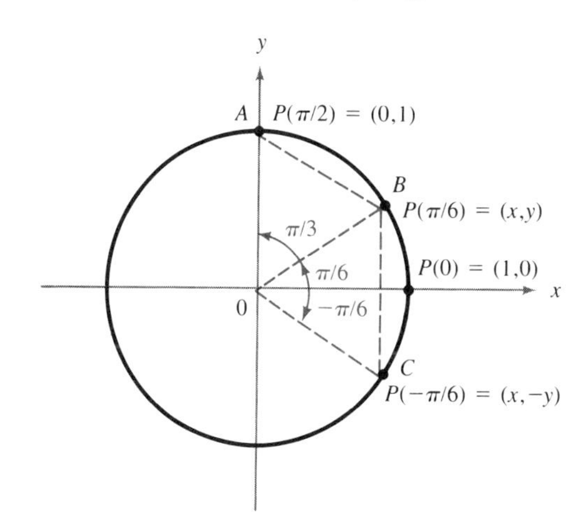

FIGURE 6.18

The coordinates of certain other points on the unit circle can be determined by using the symmetry of the circle, and the coordinates of the points in Figure 6.16.

EXAMPLE 4: Determine the coordinates of $P\left(\frac{3\pi}{4}\right)$, $P\left(\frac{5\pi}{4}\right)$, and $P\left(\frac{7\pi}{4}\right)$. (See Figure 6.19.)

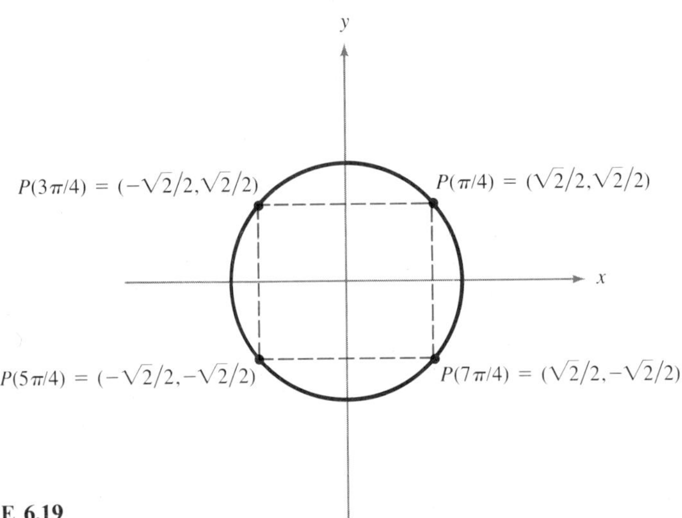

FIGURE 6.19

Solution: We can compute the coordinates of these points using the coordinates of $P(\pi/4)$. Figure 6.19 shows that the four points $P(\pi/4)$, $P(3\pi/4)$, $P(5\pi/4)$, and $P(7\pi/4)$ are located in symmetrical positions around the circle. Thus, for example, the y-coordinate of $P(3\pi/4)$ is the same as the y-coordinate of $P(\pi/4)$, but its x-coordinate is the negative of the x-coordinate of $P(\pi/4)$. Therefore, since $P(\pi/4) = (\sqrt{2}/2, \sqrt{2}/2)$, we have

$$P\left(\frac{3\pi}{4}\right) = \left(-\frac{\sqrt{2}}{2}, \frac{\sqrt{2}}{2}\right)$$

By a similar reasoning we get

$$P\left(\frac{5\pi}{4}\right) = \left(-\frac{\sqrt{2}}{2}, -\frac{\sqrt{2}}{2}\right) \quad \text{and} \quad P\left(\frac{7\pi}{4}\right) = \left(\frac{\sqrt{2}}{2}, -\frac{\sqrt{2}}{2}\right)$$

Try Study Suggestions 6.8 and 6.9. □

▶ **Study Suggestion 6.8:** Use symmetry to determine the coordinates of the points $P(5\pi/6)$, $P(7\pi/6)$, and $P(11\pi/6)$. ◀

▶ **Study Suggestion 6.9:** Use symmetry to determine the coordinates of the points $P(2\pi/3)$, $P(4\pi/3)$, and $P(5\pi/3)$. ◀

For the sake of reference, Figure 6.20 shows the coordinates of all of the points that we have obtained so far (including those obtained from the Study Suggestions.)

Up to now, we have been dealing only with values of t that satisfy $0 \leq t < 2\pi$. Let us consider the case where t does not lie in this interval.

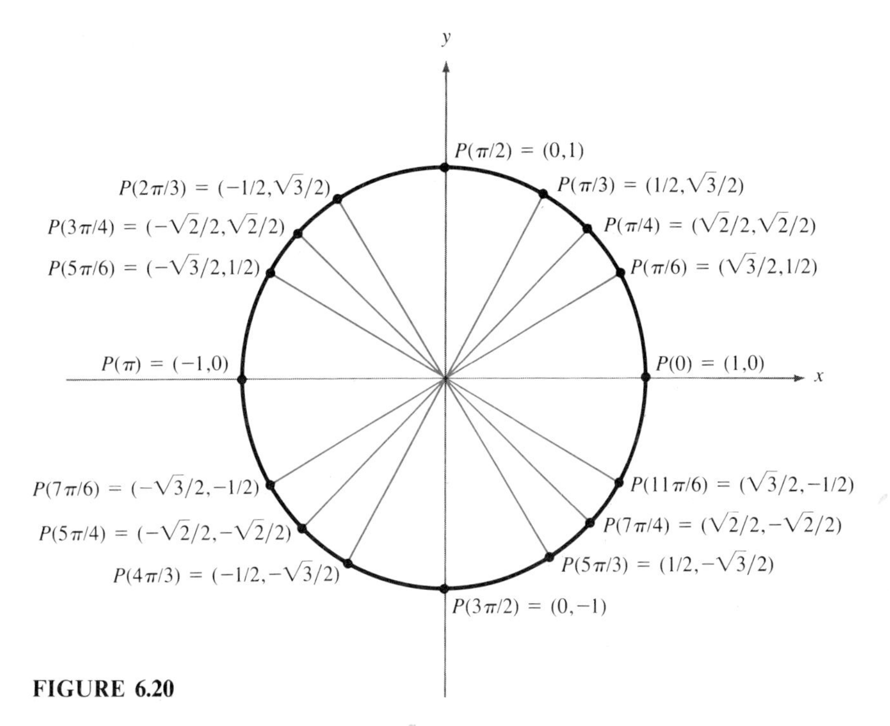

FIGURE 6.20

As we know, the point $P(t)$ is determined by sweeping out a central angle of t radians. But since a full angle has radian measure 2π, sweeping out a central angle of $t + 2\pi$ radians will take us to the same point as sweeping out a central angle of t radians. In other words, the points $P(t)$ and $P(t + 2\pi)$ are the same.

In fact, if n is any integer, then sweeping out a central angle of $t + 2\pi n$ radians will take us to the same point on the unit circle as sweeping out a central angle of t radians—we just go around the circle n additional times to get to this point. In symbols

$$P(t + 2\pi n) = P(t) \tag{6.5}$$

for all real numbers t and all integers n (both positive and negative.)

Equation (6.5) tells us that we can add (when $n > 0$), or subtract (when $n < 0$), any integral multiple of 2π to t without affecting the position of the point $P(t)$ (which means, of course, that we do not affect the value of its coordinates.) This fact can be used to compute the coordinates of $P(t)$ when t does not lie in the interval $0 \le t < 2\pi$. For we can first add (or subtract) an appropriate integral multiple of 2π to t in order to bring it into that interval. Let us illustrate this with some examples.

EXAMPLE 5: Find the coordinates of the following points.

(a) $P\left(\frac{5\pi}{2}\right)$ **(b)** $P\left(-\frac{2\pi}{3}\right)$ **(c)** $P\left(\frac{25\pi}{4}\right)$

Solutions:

(a) In order to determine the coordinates of $P(5\pi/2)$, we first subtract 2π from $5\pi/2$, since $5\pi/2 - 2\pi = \pi/2$ is in the interval $0 \leq t < 2\pi$,

$$P\left(\frac{5\pi}{2}\right) = P\left(\frac{5\pi}{2} - 2\pi\right) = P\left(\frac{\pi}{2}\right) = (0, 1)$$

(b) In this case, we first add 2π,

$$P\left(-\frac{2\pi}{3}\right) = P\left(-\frac{2\pi}{3} + 2\pi\right) = P\left(\frac{4\pi}{3}\right) = \left(-\frac{1}{2}, -\frac{\sqrt{3}}{2}\right)$$

(Here we used Figure 6.20 as a reference for the coordinates of $P(4\pi/3)$.)

(c) In this case, we first subtract $3 \cdot 2\pi = 6\pi$,

$$P\left(\frac{25\pi}{4}\right) = P\left(\frac{25\pi}{4} - 6\pi\right) = P\left(\frac{\pi}{4}\right) = \left(\frac{\sqrt{2}}{2}, \frac{\sqrt{2}}{2}\right)$$

Try Study Suggestion 6.10. □

▶ **Study Suggestion 6.10:** Find the coordinates of the following points.

(a) $P(-7\pi/2)$ **(b)** $P(29\pi/6)$ **(c)** $P(25\pi)$ ◀

EXERCISES

In Exercises 1–16, draw a picture to show the effect of sweeping out an angle of the given radian measure. (Assume that each quadrant is divided into two equal sections, and locate the point $P(t)$ in the proper section, as we did in Example 1.)

1. $\pi/6$ **2.** $5\pi/4$ **3.** π

4. $-\pi/5$ **5.** $\pi/9$ **6.** $25\pi/13$

7. $-2\pi/3$ **8.** $-\pi$ **9.** $-5\pi/3$

10. $5\pi/8$ **11.** $7\pi/3$ **12.** $-20\pi/9$

13. 2π **14.** $-8\pi/7$ **15.** $5\pi/6$

16. -3π

In Exercises 17–34, you are given $P(t)$ for some value of t. Find the coordinates of $P(t)$ by first finding a value s satisfying $0 \leq s < 2\pi$ with the property that $P(t) = P(s)$, and then finding the coordinates of $P(s)$.

17. $P(-\pi/3)$ **18.** $P(-7\pi/6)$ **19.** $P(-\pi/6)$

20. $P(-5\pi/3)$ **21.** $P(-\pi/4)$ **22.** $P(-4\pi/3)$

23. $P(-5\pi/6)$ **24.** $P(3\pi)$ **25.** $P(7\pi/2)$

26. $P(-10\pi)$ **27.** $P(17\pi)$ **28.** $P(-19\pi/3)$

29. $P(-11\pi/6)$ **30.** $P(10000\pi)$ **31.** $P(-10000\pi)$

32. $P(35\pi)$ **33.** $P(8\pi/3)$ **34.** $P(11\pi/4)$

In Exercises 35–43, approximate the position of the given point on the unit circle. (Assume that each quadrant is divided into two equal sections, and locate the point $P(t)$ in the proper section, as we did in Example 1.)

35. $P(1)$ **36.** $P(-2)$ **37.** $P(3)$

38. $P(-4)$ **39.** $P(7)$ **40.** $P(1.2)$

41. $P(-1.7)$ **42.** $P(3.3)$ **43.** $P(100)$

44. If $P(t) = (1/4, y)$, what are the possible values of y?

45. If $P(t) = (x, -1/3)$, what are the possible values of x?

46. If $P(t) = (x, 2x)$, what are the possible values of x?

47. If $P(t) = (x, x + 1)$, what are the possible values of x?

48. If $P(t) = (x, x + 2)$, what are the possible values of x?

49. **(a)** If $P(t) = (a, b)$, what are the coordinates of $P(-t)$? Justify your answer by drawing an appropriate figure.

(b) Find $P(-t)$ if $P(t) = (3/5, -4/5)$.

50. **(a)** If $P(t) = (a, b)$, what are the coordinates of $P(t + \pi)$? Justify your answer by drawing an appropriate figure.

(b) Find $P(t + \pi)$ if $P(t) = (-5/13, 12/13)$.

51. **(a)** If $P(t) = (a, b)$, what are the coordinates of $P(t + \pi/2)$? (*Hint:* consider the following figure. Show that triangle $OBP(t)$ is congruent to triangle $OAP(t + \pi/2)$.)

(b) Find $P(t + \pi/2)$ if $P(t) = (1/6, -\sqrt{35}/6)$.

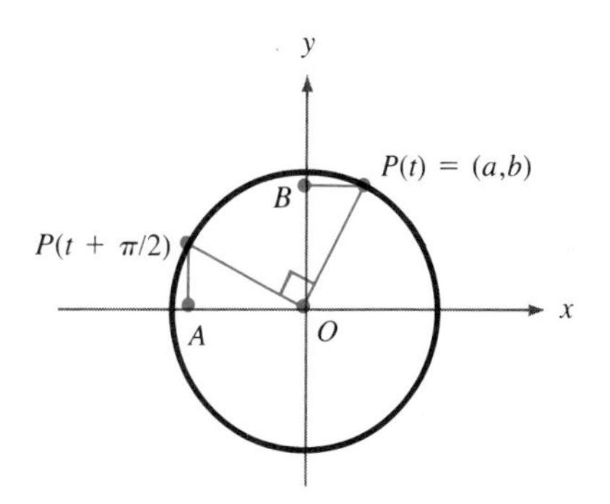

6.3 Definitions of the Trigonometric Functions

Now we are ready to define the trigonometric functions. We will begin with the definitions of the *sine*, and *cosine functions.* Then, after discussing methods for evaluating these two functions, we will define the other trigonometric functions.

DEFINITION

Let t be a real number. We define the **sine** of t, denoted by $\sin t$ and the **cosine** of t, denoted by $\cos t$, as follows. Referring to Figure 6.21(a and b), we begin at the point $(1, 0)$ on the unit circle, and sweep out a central angle of t radians, coming to rest at the point $P(t)$. The x-coordinate of $P(t)$ is defined to be $\cos t$ and the y-coordinate of $P(t)$ is defined to be $\sin t$. In short,

$$P(t) = (\cos t, \sin t)$$

The function $f(t) = \sin t$ is called the **sine function,** and the function $g(t) = \cos t$ is called the **cosine function.** Both of these functions have domain equal to the set of all real numbers.

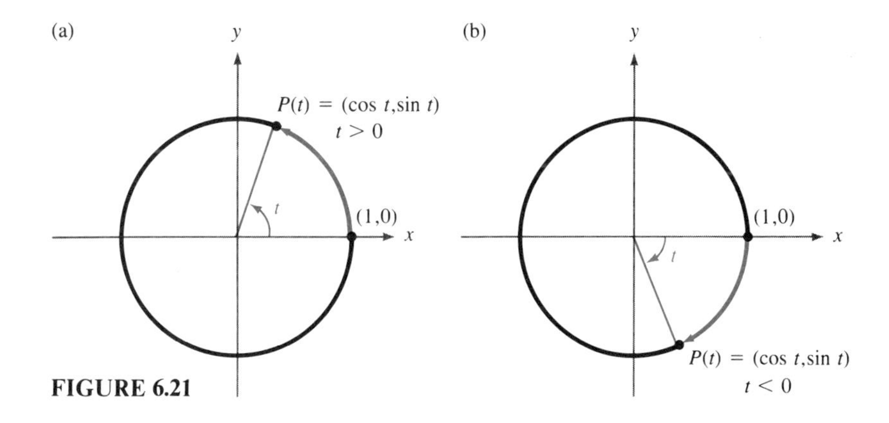

FIGURE 6.21

Now it should be clear why we devoted the previous section to finding the coordinates of points of the form $P(t)$—these coordinates are just the values of $\sin t$ and $\cos t$. For example, since

$$P\left(\frac{3\pi}{4}\right) = \left(-\frac{\sqrt{2}}{2}, \frac{\sqrt{2}}{2}\right)$$

we have

$$\cos \frac{3\pi}{4} = -\frac{\sqrt{2}}{2} \quad \text{and} \quad \sin \frac{3\pi}{4} = \frac{\sqrt{2}}{2}$$

We will determine other values of the sine and cosine functions in a moment. However, it turns out that sometimes we only need to know, for a given value of t, whether $\sin t$ or $\cos t$ is positive, negative, or zero. This is easy to do if we keep in mind that $\sin t$ is the y-coordinate and $\cos t$ is the x-coordinate of the point $P(t)$. Figure 6.22 shows the signs of $\sin t$ and $\cos t$.

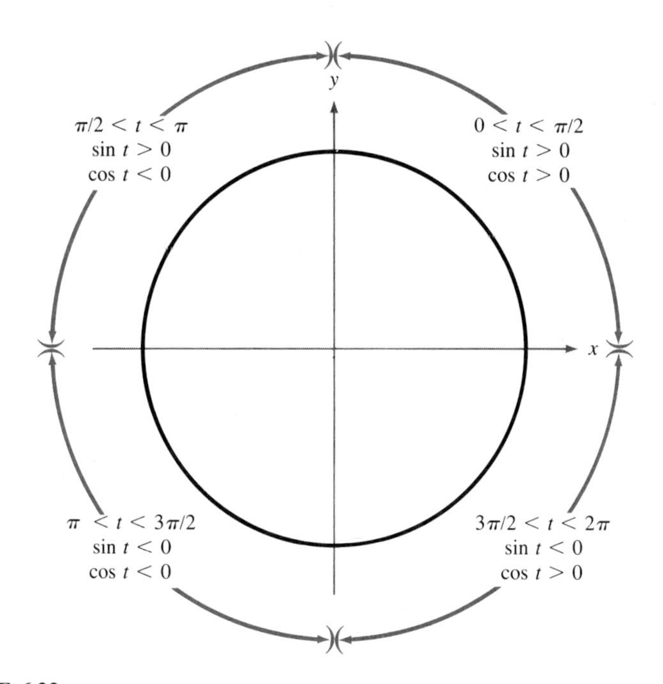

FIGURE 6.22

EXAMPLE 1: Determine whether each of the following values is positive, negative, or zero.

(a) $\sin \frac{\pi}{3}$ and $\cos \frac{\pi}{3}$ **(b)** $\sin\left(-\frac{4\pi}{7}\right)$ and $\cos\left(-\frac{4\pi}{7}\right)$

(c) $\sin \frac{3\pi}{2}$ and $\cos \frac{3\pi}{2}$

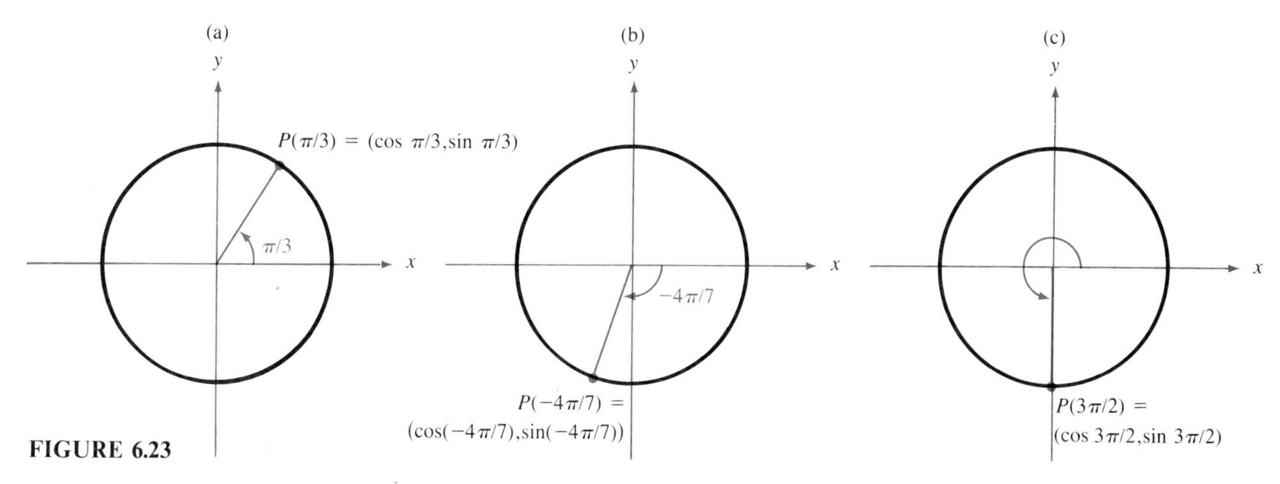

FIGURE 6.23

Solutions:

(a) Since $P(\pi/3)$ lies in the first quadrant (see Figure 6.23a), both of its coordinates are positive; that is, $\sin \pi/3 > 0$ and $\cos \pi/3 > 0$.

(b) Since $P(-4\pi/7)$ lies in the third quadrant (see Figure 6.23b), both of its coordinates are negative; that is, $\sin(-4\pi/7) < 0$ and $\cos(-4\pi/7) < 0$.

(c) Since $P(3\pi/2)$ lies on the negative half of the y-axis (see Figure 6.23c), its y-coordinate is negative and its x-coordinate is equal to 0; that is, $\sin 3\pi/2 < 0$ and $\cos 3\pi/2 = 0$. *Try Study Suggestion 6.11.* □

TABLE 6.2

t	$\sin t$	$\cos t$
0	0	1
$\frac{\pi}{6}$	$\frac{1}{2}$	$\frac{\sqrt{3}}{2}$
$\frac{\pi}{4}$	$\frac{\sqrt{2}}{2}$	$\frac{\sqrt{2}}{2}$
$\frac{\pi}{3}$	$\frac{\sqrt{3}}{2}$	$\frac{1}{2}$
$\frac{\pi}{2}$	1	0

Using Figure 6.20 of Section 6.2, we can compile the values of $\sin t$ and $\cos t$ shown in Table 6.2, for t in the range $0 \le t \le \pi/2$. Now we could use Figure 6.20 to enlarge this table by including values of $\sin t$ and $\cos t$ for t outside the interval $0 \le t \le \pi/2$. However, this is not necessary since there is a method for computing the values of $\sin t$ and $\cos t$ for *any* value of t, using the values of $\sin t$ and $\cos t$ for t in the interval $0 \le t \le \pi/2$. This method relies on the concept of a *reference number*.

DEFINITION

The **reference number** of a real number t, denoted by t_R, is the *length* of the *shortest* arc along the unit circle from the point $P(t)$ to the x-axis.

▶ **Study Suggestion 6.11:** Determine whether each of the following values is positive, negative, or zero.

(a) $\sin 2\pi/3$ and $\cos 2\pi/3$

(b) $\sin 7\pi/4$ and $\cos 7\pi/4$

(c) $\sin \pi$ and $\cos \pi$ ◀

Figure 6.24 shows the reference number t_R in four cases, depending on which quadrant contains the point $P(t)$. (If $P(t)$ lies on the x-axis, then the reference number of t is equal to 0, and if it lies on the y-axis, then the reference number of t is $t_R = \pi/2$.)

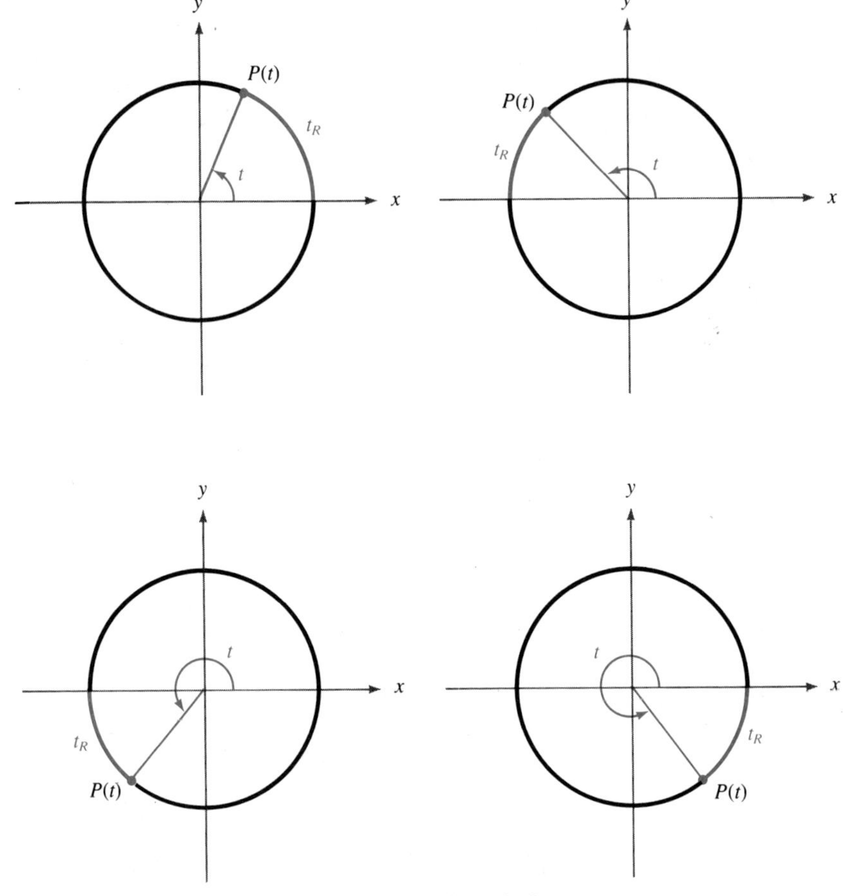

The reference number t_R is the *length* of the colored arc.

FIGURE 6.24

We should emphasize that, since a reference number is defined to be the *length* of a certain arc, it is always a nonnegative number. In fact, as you can see from Figure 6.24 *all reference numbers lie in the range* $0 \leq t_R \leq \pi/2$.

EXAMPLE 2: Find the reference number for each of the following numbers.

▶ Study Suggestion 6.12: Find the reference number for each of the following numbers.

(a) $t = 5\pi/6$ **(b)** $t = \pi/6$
(c) $t = -\pi/5$ **(d)** $t = 5\pi/3$
(e) $t = -6\pi$ ◀

(a) $t = \dfrac{\pi}{3}$ **(b)** $t = \dfrac{7\pi}{3}$ **(c)** $t = -\dfrac{\pi}{6}$ **(d)** $t = \dfrac{2\pi}{3}$ **(e)** $t = \dfrac{5\pi}{4}$

Solutions: Reference numbers are most easily found by plotting the point $P(t)$ on the unit circle. The solutions for this example are shown in Figure 6.25.

Try Study Suggestion 6.12. □

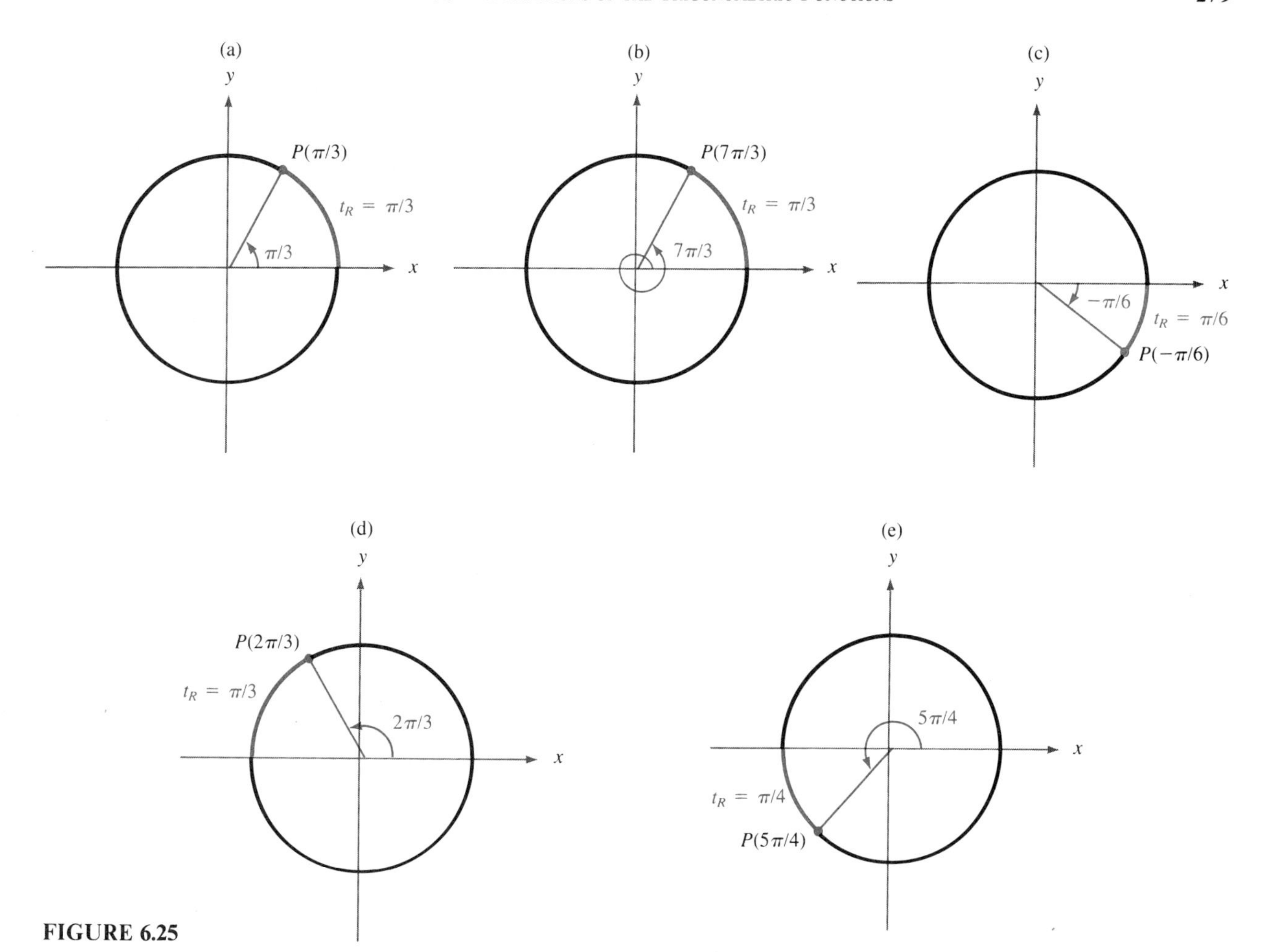

FIGURE 6.25

The following theorem shows why reference numbers are useful for computing values of the sine and cosine.

THE REFERENCE NUMBER THEOREM

Let t be a real number, and let t_R be its reference number. Then

$$|\sin t| = \sin t_R \qquad \text{and} \qquad |\cos t| = \cos t_R$$

In words, the *absolute value* of $\sin t$ is equal to $\sin t_R$, and the *absolute value* of $\cos t$ is equal to $\cos t_R$.

In view of the reference number theorem, we can compute $\sin t$ (or $\cos t$) for any value of t by first computing $\sin t_R$, which gives us the *absolute value* of $\sin t$, and then determining the sign of $\sin t$, as we did in Example 1.

EXAMPLE 3: Use the reference number theorem to compute the following values.

(a) $\sin\frac{7\pi}{6}$ **(b)** $\cos\left(-\frac{7\pi}{4}\right)$ **(c)** $\sin\frac{8\pi}{3}$

Solutions:

(a) The first step is to plot the point $P(7\pi/6)$, as in Figure 6.26(a). From this we see that the reference number of $7\pi/6$ is $\pi/6$, and so

$$\left|\sin\frac{7\pi}{6}\right| = \sin\frac{\pi}{6} = \frac{1}{2}$$

But we also see from this figure that $P(7\pi/6)$ lies in the third quadrant, and so its y-coordinate, which is $\sin 7\pi/6$, is negative. Hence, $\sin 7\pi/6 = -1/2$.

(b) Figure 6.26(b) shows the point $P(-7\pi/4)$. From this figure, we see that the reference number of $-7\pi/4$ is $\pi/4$, and so

$$\left|\cos\left(-\frac{7\pi}{4}\right)\right| = \cos\frac{\pi}{4} = \frac{\sqrt{2}}{2}$$

Furthermore, since $P(-7\pi/4)$ lies in the first quadrant, its x-coordinate $\cos(-7\pi/4)$ is positive, and so $\cos(-7\pi/4) = \sqrt{2}/2$.

(c) Figure 6.26(c) shows the point $P(8\pi/3)$. From this figure, we see that the reference number of $8\pi/3$ is $\pi/3$. Also, since $P(8\pi/3)$ lies in the second quadrant, its y-coordinate $\sin 8\pi/3$ is positive. Therefore, $\sin 8\pi/3 = \sin\pi/3 = \sqrt{3}/2$. *Try Study Suggestion 6.13.* □

▶ **Study Suggestion 6.13:** Use the reference number theorem to complete the following values.
(a) $\cos 3\pi/4$ **(b)** $\sin 5\pi/3$
(c) $\cos(-5\pi/6)$ **(d)** $\sin(-\pi/4)$ ◀

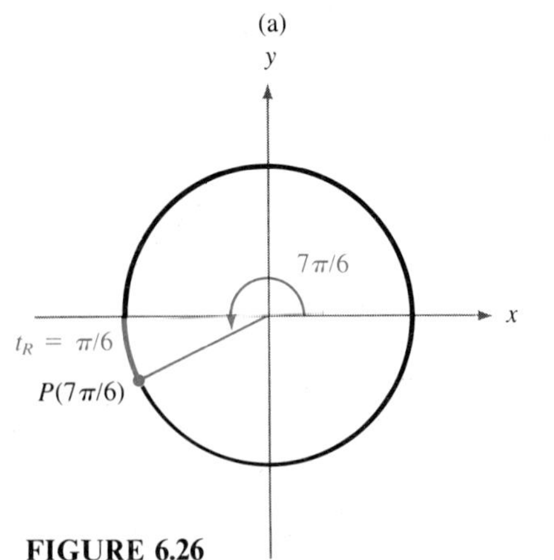

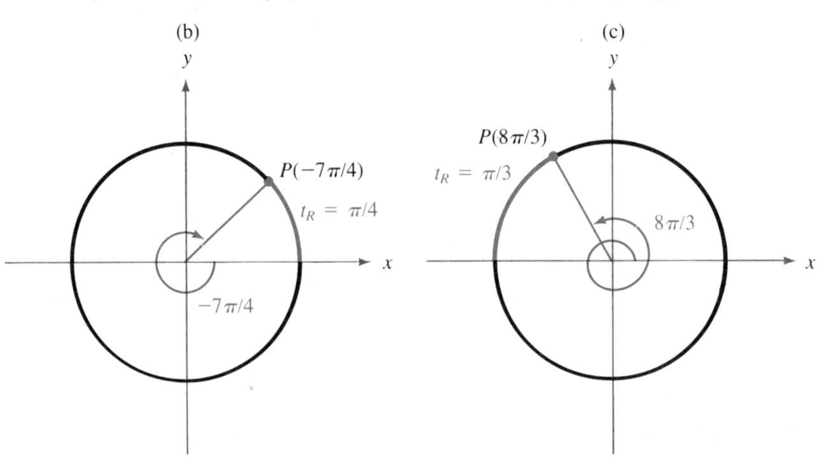

FIGURE 6.26

Let us make a few remarks about the use of calculators to compute values of $\sin t$ and $\cos t$. In general, exact values of $\sin t$ and $\cos t$ cannot be written in a simple form, such as $1/2$, $\sqrt{2}/2$ or $\sqrt{3}/2$. In fact, the values of t

that we considered in the previous section were very carefully chosen so that the coordinates of $P(t)$ (that is, $\sin t$ and $\cos t$) could be easily determined by geometric means. Therefore, if we need the value of, say, $\sin \pi/5$ or $\cos 1$, we must in general settle for an approximation, and this is where the calculator is invaluable.

For this reason, we suggest that you spend some time becoming familiar with the *sin* and *cos* keys on your calculator (as well as the keys relating to the other trigonometric functions, as we come to them in the book.) It is especially important to be aware of the fact that scientific calculators have both a *degree* mode and a *radian* mode for computing values of the trigonometric functions. In order for your calculator to give correct answers, *you must be sure that it is in the radian mode.* We will use the degree mode in the next chapter, when we learn how to take the sine and cosine of an *angle* that is measured in degrees.

Study Suggestion 6.14: Use a calculator (set in the radian mode) to approximate the following values.

(a) $\sin \pi/5$ **(b)** $\cos 1$ **(c)** $\sin(-3\pi/7)$
(d) $\sin 12.65$ **(e)** $\cos 3.8\pi$ ◀

Try Study Suggestion 6.14.

Now let us define the other trigonometric functions. As you can see, they are defined in terms of the sine and cosine.

DEFINITION

Let t be a real number. Then we make the following definitions

name	**abbreviation**	**definition**
tangent	$\tan t$	$\tan t = \dfrac{\sin t}{\cos t}$
secant	$\sec t$	$\sec t = \dfrac{1}{\cos t}$
cosecant	$\csc t$	$\csc t = \dfrac{1}{\sin t}$
cotangent	$\cot t$ or $\operatorname{ctn} t$	$\cot t = \dfrac{\cos t}{\sin t}$

These values are defined for all real numbers t for which the denominators are not equal to 0. The function $f(t) = \tan t$ is called the **tangent function,** the function $g(t) = \sec t$ is called the **secant function**, the function $h(t) = \csc t$ is called the **cosecant function,** and the function $k(t) = \cot t$ is called the **cotangent function.**

The tangent function has a very interesting interpretation with regard to the unit circle. Figure 6.27 shows a point $P(t)$ on the unit circle, along with the line L through $P(t)$ and the origin. According to the definition of

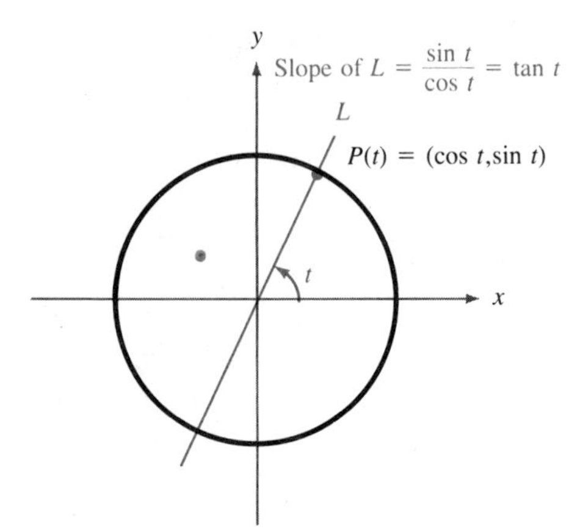

An interpretation of tan t as the slope of line L.

FIGURE 6.27

the sine and cosine, $P(t)$ has coordinates $(\cos t, \sin t)$. Using this point, along with the point $(0, 0)$, we can compute the slope of the line L

$$\text{slope of } L = \frac{\sin t - 0}{\cos t - 0} = \frac{\sin t}{\cos t} = \tan t$$

Thus we see that

> **$\tan t$ is the slope of the line through the origin and the point $P(t) = (\cos t, \sin t)$.**

We will give another interpretation of the tangent in Exercise 82.

Computing values of the tangent, secant, cosecant, and cotangent usually amount to first computing values of the sine and cosine, and then applying the definition.

EXAMPLE 4: Compute the following values.

(a) $\tan\frac{\pi}{4}$ **(b)** $\sec\left(-\frac{\pi}{3}\right)$ **(c)** $\csc\frac{4\pi}{3}$ **(d)** $\cot\left(-\frac{5\pi}{6}\right)$

Solutions: In the following solutions, we used the reference number theorem whenever necessary to obtain the values of $\sin t$ and $\cos t$.

(a) Since $\sin \pi/4 = \sqrt{2}/2$ and $\cos \pi/4 = \sqrt{2}/2$, we have

$$\tan\frac{\pi}{4} = \frac{\sin \pi/4}{\cos \pi/4} = \frac{\sqrt{2}/2}{\sqrt{2}/2} = 1$$

(b) Since $\cos(-\pi/3) = \cos(\pi/3) = 1/2$, we have

$$\sec\left(-\frac{\pi}{3}\right) = \frac{1}{\cos(-\pi/3)} = \frac{1}{1/2} = 2$$

(c) Since $\sin 4\pi/3 = -\sin \pi/3 = -\sqrt{3}/2$, we have

$$\csc\left(\frac{4\pi}{3}\right) = \frac{1}{\sin(4\pi/3)} = \frac{1}{-\sqrt{3}/2} = -\frac{2}{\sqrt{3}} = -\frac{2\sqrt{3}}{3}$$

(Notice that we rationalized the denominator in the last expression.)

(d) Since $\sin(-5\pi/6) = -\sin \pi/6 = -1/2$ and $\cos(-5\pi/6) = -\cos \pi/6 = -\sqrt{3}/2$, we have

$$\cot\left(-\frac{5\pi}{6}\right) = \frac{\cos(-5\pi/6)}{\sin(-5\pi/6)} = \frac{-\sqrt{3}/2}{-1/2} = \sqrt{3}$$

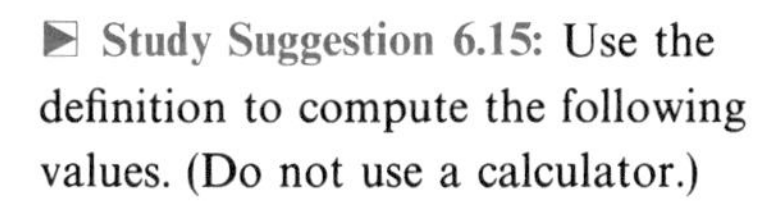

▶ **Study Suggestion 6.15:** Use the definition to compute the following values. (Do not use a calculator.)
(a) $\tan(-\pi/6)$ **(b)** $\sec 3\pi/4$
(c) $\csc(-3\pi/2)$ **(d)** $\cot 2\pi/3$ ◀

Try Study Suggestion 6.15. □

Most scientific calculators have a *tan* key for direct computation of the tangent function. However, computing values of the secant, cosecant, or cotangent usually requires extra steps. For example, in order to compute $\sec \pi/3$, we first use the calculator to compute $\cos \pi/3$, and then use the *1/x* key to obtain $\sec \pi/3 = 1/\cos(\pi/3)$.

Study Suggestion 6.16: Use a calculator to approximate the following values. **(a)** $\tan \pi/5$
(b) $\sec 1.43\pi$ **(c)** $\csc(-2.14)$
(d) $\text{ctn}(-7\pi/12)$ ◀

Try Study Suggestion 6.16.

Let us conclude this section by giving a simple application of the trigonometric functions. (We will give some additional applications as the chapter progresses, when we have developed some of the properties of these functions.)

Figure 6.28 shows two circles with centers at the origin. As we know, the point $P(t)$ on the *unit* circle has coordinates $(\cos t, \sin t)$. We will leave it as an exercise to show that the point $Q(t)$ on the circle of radius r has coordinates $(r \cos t, r \sin t)$. (See Exercise 80.) Now we are ready for our application.

EXAMPLE 5: Figure 6.29 shows a Ferris wheel whose radius is 25 feet and whose center is 30 feet above the ground. For convenience, we have

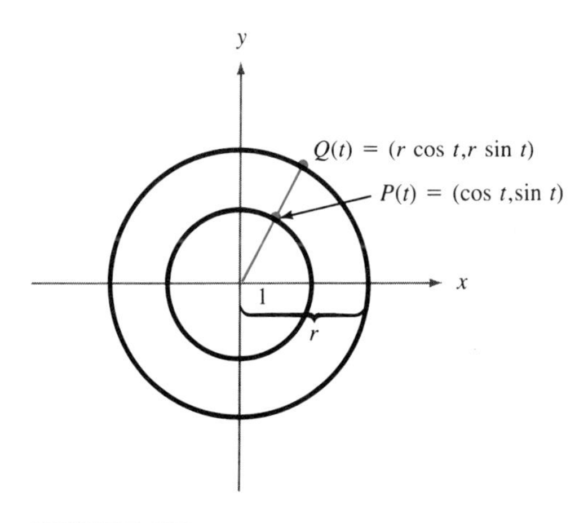

FIGURE 6.28

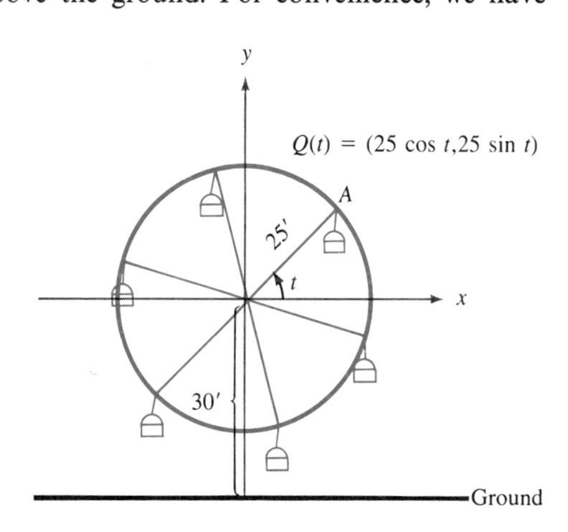

FIGURE 6.29

placed a coordinate system so that the center of the Ferris wheel is at the origin. Suppose that you were in car A. Then your position with respect to the coordinate system is $Q(t) = (25\cos t, 25\sin t)$, where t is the *radian* measure of the angle that you make with the positive x-axis, as shown in the figure.

Now, the distance $D(t)$ from you to the ground is equal to 30 feet *plus* the y-coordinate of the point $Q(t)$. In other words, this distance is given by the formula

$$D(t) = 30 + 25\sin t$$

For example, if you are inclined at an angle of 45° from the positive x-axis, then since in this case $t = \pi/4$, your distance from the ground is

$$D\left(\frac{\pi}{4}\right) = 30 + 25\sin\frac{\pi}{4}$$

$$= 30 + 25\left(\frac{\sqrt{2}}{2}\right)$$

$$\approx 47.68 \text{ feet}$$

Try Study Suggestion 6.17. □

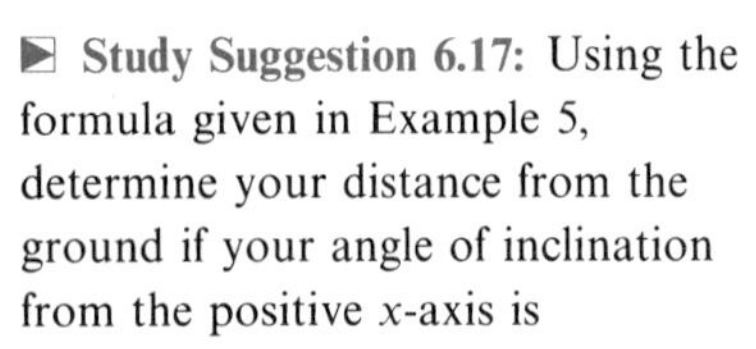

▶ **Study Suggestion 6.17:** Using the formula given in Example 5, determine your distance from the ground if your angle of inclination from the positive x-axis is

(a) 0° **(b)** 30° **(c)** 90°

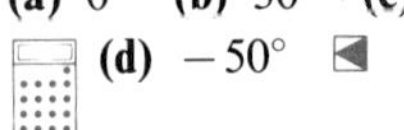

(d) −50° ◀

Ideas to Remember

- Because of the connection between the sine and cosine functions and circles (of all sizes), these functions have many applications to the real world.
- The reference number theorem enables us to determine the values of the sine and cosine functions (and hence of the other trigonometric functions), from their values on the numbers between 0 and $\pi/2$.

EXERCISES

In Exercises 1–12, plot the point $P(t)$ on the unit circle and find the reference number of t.

1. $\pi/4$ **2.** $-\pi/4$ **3.** $4\pi/3$

4. $7\pi/4$ **5.** $2\pi/5$ **6.** $-17\pi/3$

7. $9\pi/2$ **8.** $9\pi/4$ **9.** $\pi/2$

10. 0 **11.** $-\pi/2$ **12.** 13π

In Exercises 13–24, for the given value of t, determine whether each of the six trigonometric functions is positive, negative, zero, or undefined at that value of t. (Some values of t may not be in the domain of a particular trigonometric function, in which case the function is undefined at that value.)

13. $\pi/7$ **14.** $8\pi/9$ **15.** $-3\pi/2$

16. $9\pi/2$ **17.** 12π **18.** $19\pi/2$

19. $-7\pi/3$ **20.** $11\pi/5$ **21.** $-7\pi/4$

22. -13π **23.** $-\pi/12$ **24.** $-13\pi/3$

In Exercises 25–41, use reference numbers to fill in the table below. Do *not* use a calculator. If a value of t is not in the domain of a trigonometric function, put ND (for *Not Defined*) in the table corresponding to that entry. For example, since $\cos\pi/2 = 0$, we see that $t = \pi/2$ is not in the domain of the secant or the cotangent functions. Hence we would put ND in the row corresponding to $t = \pi/2$ under the headings $\sec t$ and $\cot t$.

	t	$\sin t$	$\cos t$	$\tan t$	$\sec t$	$\csc t$	$\cot t$
25.	0						
26.	$\pi/6$						
27.	$\pi/4$						
28.	$\pi/3$						
29.	$\pi/2$						
30.	$2\pi/3$						
31.	$3\pi/4$						
32.	$5\pi/6$						
33.	π						
34.	$7\pi/6$						
35.	$5\pi/4$						
36.	$4\pi/3$						
37.	$3\pi/2$						
38.	$5\pi/3$						
39.	$7\pi/4$						
40.	$11\pi/6$						
41.	2π						

In Exercises 42–53, compute the given value. Do not use a calculator.

42. $\sin(-\pi/3)$ **43.** $\sin(-\pi)$ **44.** $\cos(-4\pi/3)$

45. $\cos(-7\pi/4)$ **46.** $\tan(-\pi/3)$ **47.** $\tan(-2\pi/3)$

48. $\sec(-3\pi/4)$ **49.** $\sec(-2\pi/3)$ **50.** $\csc(-9\pi/2)$

51. $\csc(-\pi/3)$ **52.** $\cot(-4\pi/3)$ **53.** $\cot(-\pi/4)$

In Exercises 54–61, identify the quadrant that contains the point $P(t)$ if the given conditions are true.

54. $\cos t < 0$, $\sin t > 0$

55. $\cos t < 0$, $\tan t < 0$

56. $\cos t > 0$, $\csc t > 0$

57. $\csc t < 0$, $\sec t > 0$

58. $\sec t < 0$, $\tan t > 0$

59. $\csc t > 0$, $\cot t > 0$

60. $\tan t < 0$, $\tan t + \cos t > 0$

61. $\sin t > 0$, $\cos t - \sin t > 0$

In Exercises 62–79, use a calculator to approximate the indicated value.

62. $\sin 2.341$ **63.** $\sin \pi/12$

64. $\sin(-7\pi/13)$ **65.** $\cos(-0.024)$

66. $\cos(2\pi/5)$ **67.** $\cos(-12\pi/5)$

68. $\tan 17.632$ **69.** $\tan 5.3\pi$

70. $\tan(-14\pi/13)$ **71.** $\sec(-3.411)$

72. $\sec(7\pi/3)$ **73.** $\sec(-13\pi/15)$

74. $\csc 4.001$ **75.** $\csc 9.5\pi$

76. $\csc(-12.552\pi)$ **77.** $\cot 0.00509$

78. $\cot 0.002\pi$ **79.** $\cot(-1.002\pi)$

80. Show that the coordinates of the point $Q(t)$ in Figure 6.28 are $(r\cos t, r\sin t)$. (*Hint:* One method is to find the equation of the line through the origin and the point $P(t)$. Then determine where this line intersects the circle of radius r.)

81. Referring to Example 5, determine your distance from the ground when your angle of inclination from the positive x-axis is

(a) $60°$ **(b)** $120°$ **(c)** $-45°$ **(d)** $17.2°$

82. Referring to the following figure show that the line segment PQ has length $\tan t$. (*Hint:* first find the lengths of the line segments OA and AB. Then use the properties of similar triangles.)

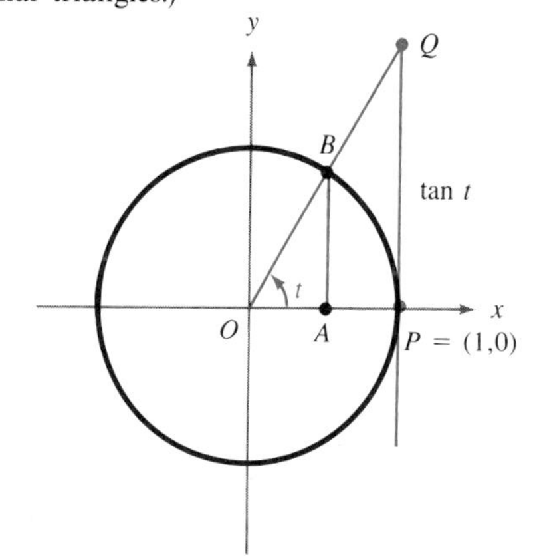

6.4 Properties of the Trigonometric Functions

Now that we have acquired some skill in computing values of the trigonometric functions, we should consider some of the more important properties of these functions. We will begin with a discussion of the sine and cosine functions, since their properties will help us obtain the properties of the other trigonometric functions.

One of the most important properties of the sine and cosine comes from the fact that

$$P(t + 2\pi n) = P(t)$$

for all real numbers t and all integers n. (See Section 6.2.) Writing this out in terms of components gives

$$(\cos(t + 2\pi n), \sin(t + 2\pi n)) = (\cos t, \sin t)$$

and so we have

$$\cos(t + 2\pi n) = \cos t \tag{6.6}$$

$$\sin(t + 2\pi n) = \sin t \tag{6.7}$$

for all real numbers t and all integers n.

Equations (6.6) and (6.7) tell us that we may add (or subtract) any integral multiple of 2π to t without affecting the value of $\sin t$ or $\cos t$. For example,

$$\sin 7\pi = \sin(7\pi - 6\pi) = \sin \pi = 0$$

and

$$\cos\left(-\frac{15\pi}{4}\right) = \cos\left(-\frac{15\pi}{4} + 4\pi\right) = \cos\frac{\pi}{4} = \frac{\sqrt{2}}{2}$$

This special property is described by saying that the sine and cosine functions are **periodic.** Loosely speaking, a function is *periodic* if it repeats itself at regular intervals. As we have just seen, the sine and cosine functions repeat themselves every 2π units. We will see a bit later in this section that the tangent function repeats itself every π units. In fact, all of the trigonometric functions are periodic, and this fact will be of great help to us when we graph these functions in the next section.

Many of the properties of the sine and cosine follow directly from the fact that the point $P(t) = (\cos t, \sin t)$ is on the unit circle. For instance, in Figure 6.30, we have plotted two points of the form $P(t)$ and $P(-t)$. As you can see from this figure, the x-coordinates of these points are equal, and the y-coordinates are negatives of each other. Thus, we have

$$\cos(-t) = \cos t \tag{6.8}$$

$$\sin(-t) = -\sin t \tag{6.9}$$

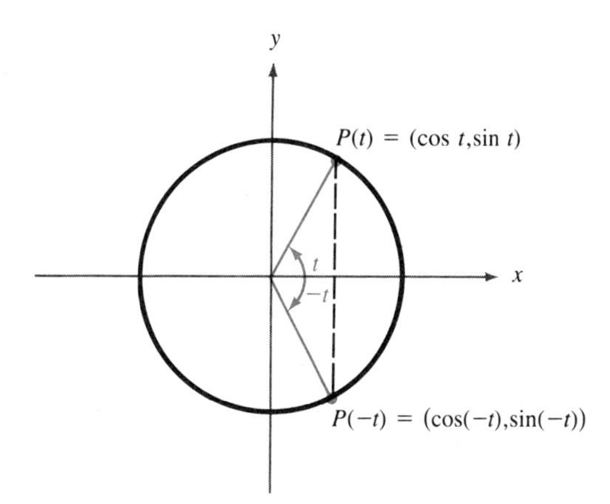

FIGURE 6.30

In other words,

> **the cosine function $f(t) = \cos t$ is even, and the sine function $g(t) = \sin t$ is odd.**

Perhaps the most efficient way to compute exact values of the sine and cosine is to use Equations (6.8) and (6.9), together with the periodicity of these functions, and the reference number theorem. For example, in order to compute $\sin(-22\pi/3)$, first we use Equation (6.9) to write

$$\sin\left(-\frac{22\pi}{3}\right) = -\sin\frac{22\pi}{3}$$

Then we use the periodicity of the sine function (Equation 6.7) to bring the value of t into the range $0 \le t < 2\pi$. In this case, since $t = 22\pi/3 = 6\pi + 4\pi/3$, subtracting $2 \cdot 3\pi = 6\pi$ will bring t into the desired range

$$\begin{aligned}\sin\left(-\frac{22\pi}{3}\right) &= -\sin\frac{22\pi}{3}\\ &= -\sin\left(\frac{22\pi}{3} - 6\pi\right)\\ &= -\sin\frac{4\pi}{3}\end{aligned}$$

Now since $P(4\pi/3)$ lies in the third quadrant, $\sin 4\pi/3$ is negative, and so the reference number theorem gives $\sin 4\pi/3 = -\sin\pi/3 = -\sqrt{3}/2$. Hence, we finally get

$$\begin{aligned}\sin\left(-\frac{22\pi}{3}\right) &= -\sin\frac{4\pi}{3}\\ &= -\left(-\frac{\sqrt{3}}{2}\right)\\ &= \frac{\sqrt{3}}{2}\end{aligned}$$

With a little practice, this type of computation will become routine.

EXAMPLE 1: Use the periodicity of the sine and cosine, together with Equations (6.8) and (6.9) and the reference number theorem, to compute the following values.

(a) $\sin \frac{5\pi}{2}$ **(b)** $\cos\left(-\frac{13\pi}{4}\right)$ **(c)** $\tan\left(-\frac{25\pi}{6}\right)$

Solutions:

(a) Subtracting 2π from $5\pi/2$ will bring it into the range $0 \leq t < \pi/2$,

$$\sin \frac{5\pi}{2} = \sin\left(\frac{5\pi}{2} - 2\pi\right) = \sin \frac{\pi}{2} = 1$$

(b) First we notice that $\cos(-13\pi/4) = \cos 13\pi/4$, and since $13\pi/4 = 2\pi + 5\pi/4$, we subtract 2π,

$$\begin{aligned} \cos\left(-\frac{13\pi}{4}\right) &= \cos \frac{13\pi}{4} \\ &= \cos\left(\frac{13\pi}{4} - 2\pi\right) \\ &= \cos \frac{5\pi}{4} \\ &= -\cos \frac{\pi}{4} \qquad \text{(by reference number theorem)} \\ &= -\frac{\sqrt{2}}{2} \end{aligned}$$

(c) We begin by expressing $\tan(-25\pi/6)$ in terms of $\sin(-25\pi/6)$ and $\cos(-25\pi/6)$

$$\begin{aligned} \tan\left(-\frac{25\pi}{6}\right) &= \frac{\sin(-25\pi/6)}{\cos(-25\pi/6)} \\ &= \frac{-\sin 25\pi/6}{\cos 25\pi/6} \end{aligned}$$

and since $25\pi/6 = 4\pi + \pi/6$, we subtract 4π,

$$\begin{aligned} &= \frac{-\sin(25\pi/6 - 4\pi)}{\cos(25\pi/6 - 4\pi)} \\ &= \frac{-\sin \pi/6}{\cos \pi/6} \\ &= \frac{-1/2}{\sqrt{3}/2} \\ &= -\frac{\sqrt{3}}{3} \end{aligned}$$

Try Study Suggestion 6.18. □

▶ **Study Suggestion 6.18:** Use the periodicity of the sine and cosine, together with Equations (6.8) and (6.9) and the reference number theorem, to compute the following values.

(a) $\sin 19\pi/2$ **(b)** $\cos(-35\pi/6)$
(c) $\cot 17\pi/3$ **(d)** $\csc(-7\pi/6)$ ◀

Since the point $P(t) = (\cos t, \sin t)$ is on the unit circle, and since the equation of the unit circle is

$$x^2 + y^2 = 1$$

we have

$$(\sin t)^2 + (\cos t)^2 = 1 \qquad \textbf{(6.10)}$$

for all real numbers t. This equation provides a very useful connection between the sine and the cosine.

Since powers of the trigonometric functions occur quite frequently, a special notation has been invented for denoting these powers. If n is different from -1, then we write $\sin^n t$ in place of $(\sin t)^n$, and similarly for the other trigonometric functions. This saves us having to write parentheses. (We cannot simply drop the parentheses, since $\sin t^2$ means $\sin(t^2)$, and not $(\sin t)^2$.) The symbols $\sin^{-1} t$, $\cos^{-1} t$ and so on, have been reserved for the inverse trigonometric functions, which we will discuss later in the chapter. Using this notation, Equation (6.10) becomes

$$\sin^2 t + \cos^2 t = 1 \qquad (6.11)$$

Equation (6.11) is known as a **Pythagorean identity.** There are two other Pythagorean identities, involving the other trigonometric functions, which we will derive later in the section, and we will explain in Chapter 8 why these identities are associated with the name of Pythagoras.

Study Suggestion 6.19: Use a calculator to approximate the values $\sin^2 2.5$ and $\sin(2.5)^2$. Are they the same? ◀

Try Study Suggestion 6.19.

The following example illustrates one use of the Pythagorean identity (6.11).

EXAMPLE 2: If $\sin t = 4/5$ and $\tan t < 0$, find the value of all of the trigonometric functions at t.

Solution: The first step is to determine the value of $\cos t$. Once this has been done, we can compute the values of the other trigonometric functions directly from the definitions. Substituting $\sin t = 4/5$ into Equation (6.11) gives

$$\left(\frac{4}{5}\right)^2 + \cos^2 t = 1$$

Solving this for $\cos^2 t$ gives

$$\cos^2 t = \frac{9}{25}$$

and so

$$|\cos t| = \frac{3}{5}$$

But since $\tan t = \sin t/\cos t$ is negative, we see that $\sin t$ and $\cos t$ must have opposite signs. Hence, $\cos t$ must be negative and so $\cos t = -3/5$. Now we can use the fact that $\sin t = 4/5$ and $\cos t = -3/5$ to determine the values of the other trigonometric functions:

$$\tan t = \frac{\sin t}{\cos t} = \frac{4/5}{-3/5} = -\frac{4}{3} \qquad \sec t = \frac{1}{\cos t} = \frac{1}{-3/5} = -\frac{5}{3}$$

$$\csc t = \frac{1}{\sin t} = \frac{1}{4/5} = \frac{5}{4} \qquad \cot t = \frac{\cos t}{\sin t} = \frac{-3/5}{4/5} = -\frac{3}{4}$$

▶ **Study Suggestion 6.20:** If $\cos t = -1/3$ and $\cot t > 0$, find the values of all of the trigonometric functions at t. ◀

Try Study Suggestion 6.20. □

Since the point $P(t) = (\cos t, \sin t)$ is on the unit circle, its coordinates must lie in the range from -1 to 1, and so we have

$$\mathbf{-1 \leq \sin t \leq 1}$$
$$\mathbf{-1 \leq \cos t \leq 1}$$

for all real numbers t. This information will also be useful when we sketch the graphs of the sine and cosine functions in the next section.

Another interesting property of the sine and cosine can be obtained from the fact that the points $P(t)$ and $P(t + \pi)$ lie on opposite ends of a diameter on the unit circle. (See Figure 6.31.) Therefore, the coordinates of $P(t)$ are equal in *absolute value* to the coordinates of $P(t + \pi)$, but have opposite signs, that is,

$$\sin(t + \pi) = -\sin t \qquad \textbf{(6.12)}$$

$$\cos(t + \pi) = -\cos t \qquad \textbf{(6.13)}$$

for all real numbers t.

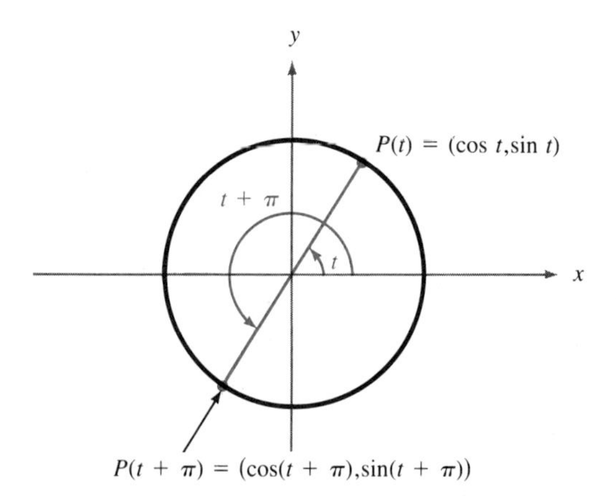

FIGURE 6.31

▶ **Study Suggestion 6.21:** Verify Formula (6.14) for the following values of t. (That is, show that both sides of this formula give the same value for the given value of t.)

(a) $t = 0$ **(b)** $t = \pi/4$
(c) $t = -\pi/2$ **(d)** $t = 2\pi$ ◀

Finally, we have the following connection between the sine and cosine functions, which we will verify in Chapter 8

$$\cos t = \sin\left(t + \frac{\pi}{2}\right) \tag{6.14}$$

Try Study Suggestion 6.21. □

Let us summarize the properties of the sine and cosine functions that we have discussed so far. We strongly suggest that you commit these important properties to memory.

THEOREM

Properties of the sine and cosine functions

1. Definitions:

$$P(t) = (\cos t, \sin t)$$

2. Range:

$$-1 \le \sin t \le 1$$
$$-1 \le \cos t \le 1$$

3. Periodicity:

$$\sin(t + 2\pi n) = \sin t$$
$$\cos(t + 2\pi n) = \cos t$$

for all integers n.

4. Pythagorean identity:

$$\sin^2 t + \cos^2 t = 1$$

5. Relationship between the sine and cosine:

$$\cos t = \sin\left(t + \frac{\pi}{2}\right)$$

6. Sine odd and cosine even:

$$\sin(-t) = -\sin t$$
$$\cos(-t) = \cos t$$

7. Miscellaneous formulas:

$$\sin(t + \pi) = -\sin t$$
$$\cos(t + \pi) = -\cos t$$

Properties of the other trigonometric functions can be obtained from the definitions and the corresponding properties of the sine and cosine. For example, since the tangent function is defined in terms of the sine and cosine, it should come as no surprise that it is also periodic. However, Equations (6.12) and (6.13) can be used to show that the tangent function actually repeats itself every π units, rather than every 2π units. Using these equations, we have

$$\begin{aligned} \tan(t+\pi) &= \frac{\sin(t+\pi)}{\cos(t+\pi)} \\ &= \frac{-\sin t}{-\cos t} \\ &= \frac{\sin t}{\cos t} \\ &= \tan t \end{aligned}$$

and so $\tan(t+\pi) = \tan t$ for all values of t in the domain of the tangent function. As with the sine and cosine functions, we may extend this and write

$$\mathbf{\tan(t+\pi n) = \tan t}$$

for all real numbers t in the domain of the tangent function, and all *integers* n.

In a similar way using the definitions, we can derive the following formulas

$$\begin{aligned} \sec(t+2\pi n) &= \sec t \\ \csc(t+2\pi n) &= \csc t \\ \cot(t+\pi n) &= \cot t \end{aligned}$$

▶ **Study Suggestion 6.22:** Show that $\sec(t+2\pi) = \sec t$ for all values of t in the domain of the cosecant function. (*Hint:* use the definition of secant to express $\sec(t+2\pi)$ in terms of $\cos(t+2\pi)$.) ◀

Try Study Suggestion 6.22.

Equation (6.11) can be used to derive a relationship between the tangent and the secant, as well as a relationship between the cotangent and the cosecant. In particular, if we divide both sides of Equation (6.11),

$$\sin^2 t + \cos^2 t = 1$$

by $\cos^2 t$ (assuming that $\cos t \neq 0$), we get

$$\frac{\sin^2 t}{\cos^2 t} + \frac{\cos^2 t}{\cos^2 t} = \frac{1}{\cos^2 t}$$

or

$$\left(\frac{\sin t}{\cos t}\right)^2 + 1 = \left(\frac{1}{\cos t}\right)^2$$

Using the definition of the tangent and secant, this can be written in the form

$$\mathbf{\tan^2 t + 1 = \sec^2 t} \qquad (6.15)$$

In a similar way (see Exercise 46) we can derive the formula

$$\mathbf{\cot^2 t + 1 = \csc^2 t} \qquad (6.16)$$

Equations (6.15) and (6.16) are the other two Pythagorean identities that we mentioned earlier. (The first Pythagorean identity is equation 6.11.)

Now let us summarize the properties of the secant, cosecant, tangent, and cotangent functions in a theorem. We should emphasize that, by knowing the definitions of these functions, and the properties of the sine and cosine, you should be able to derive any of these properties when they are needed.

THEOREM

Properties of the secant, cosecant, tangent, and cotangent

1. Definitions:

$$\sec t = \frac{1}{\cos t} \qquad \csc t = \frac{1}{\sin t}$$

$$\tan t = \frac{\sin t}{\cos t} \qquad \cot t = \frac{\cos t}{\sin t}$$

2. Periodicity:

$$\sec(t + 2\pi n) = \sec t, \qquad \csc(t + 2\pi n) = \csc t$$

$$\tan(t + \pi n) = \tan t, \qquad \cot(t + \pi n) = \cot t$$

for all integers n.

3. Pythagorean identities:

$$\tan^2 t + 1 = \sec^2 t$$

$$\cot^2 t + 1 = \csc^2 t$$

4. Evenness and oddness:

$$\sec(-t) = \sec t, \qquad \csc(-t) = -\csc t$$

$$\tan(-t) = -\tan t, \qquad \cot(-t) = -\cot t$$

Let us conclude this section with another application of the trigonometric functions.

EXAMPLE 3: Figure 6.32 shows a piston that is being driven by a rod of length 7 inches that is attached to a wheel of radius 2 inches. In the next

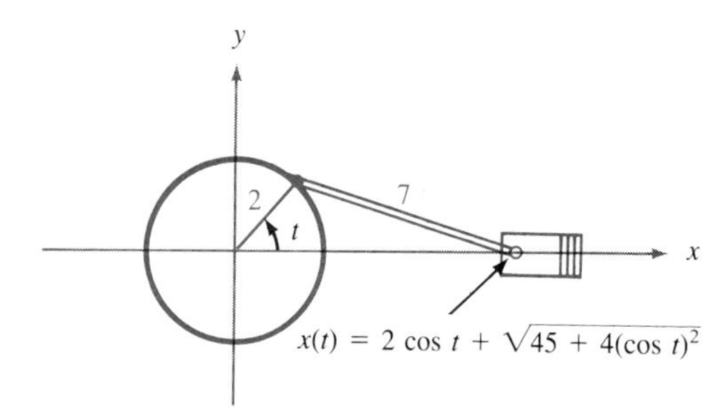

FIGURE 6.32

chapter, we will ask you to show that the position $x(t)$ of the piston along the x-axis is given by the formula

$$x(t) = 2\cos t + \sqrt{45 + 4\cos^2 t}$$

(Exercise 32 of Section 7.3.) Thus, for example, when the angle t is equal to 180° ($=\pi$ radians), the position of the piston is

$$\begin{aligned} x(\pi) &= 2\cos\pi + \sqrt{45 + 4\cos^2\pi} \\ &= -2 + \sqrt{45 + 4(-1)^2} \\ &= -2 + \sqrt{49} \\ &= 5 \text{ inches} \end{aligned}$$

just as we would expect when the point of attachment of the rod of the wheel is at the far left.

Now we can use the properties of the cosine function to determine properties of the motion of the piston. For example, we can derive a formula for the total displacement of the piston when the wheel rotates 180° ($=\pi$ radians). This displacement is

$$\begin{aligned} D(t) &= x(t + \pi) - x(t) \\ &= 2\cos(t + \pi) + \sqrt{45 + 4\cos^2(t + \pi)} - (2\cos t + \sqrt{45 + 4\cos^2 t}) \end{aligned}$$

Using the fact that $\cos(t + \pi) = -\cos t$ (Equation (6.13)), we get

$$\begin{aligned} &= -2\cos t + \sqrt{45 + 4\cos^2 t} - (2\cos t + \sqrt{45 + 4\cos^2 t}) \\ &= -4\cos t \end{aligned}$$

Hence, the displacement is

$$D(t) = -4\cos t$$

For instance, if the wheel rotates 180° from an initial position given by $t = \pi/4$ ($=45°$), then the displacement of the piston is

$$D\left(\frac{\pi}{4}\right) = -4\cos\frac{\pi}{4} = -4\left(\frac{\sqrt{2}}{2}\right) = -2\sqrt{2}$$

▶ **Study Suggestion 6.23:** Referring to Example 3, find the position of the piston when the angle t is

(a) 45° **(b)** 90° **(c)** 270° **(d)** $-60°$

Find the displacement of the piston when the wheel rotates 180° from an initial position given by $t = 30°$. ◀

The negative sign indicates that the displacement is in the direction of decreasing values of x; that is, it is in the direction from right to left along the x-axis. *Try Study Suggestion 6.23.* □

Ideas to Remember

Most of the properties of the sine and cosine functions come directly from the fact that $P(t) = (\cos t, \sin t)$ is the point on the unit circle obtained by sweeping out a central angle of t radians.

EXERCISES

In Exercises 1–15, use the periodicity of the sine and cosine, together with Equations (6.8) and (6.9) and the reference number theorem, to compute the given value.

1. $\sin(-3\pi)$ **2.** $\cos(-10\pi)$ **3.** $\sin(13\pi/6)$

4. $\cos(-17\pi/6)$ **5.** $\cos(19\pi/3)$ **6.** $\cos(17\pi/4)$

7. $\sin(-14\pi/3)$ **8.** $\sin(-11\pi/2)$ **9.** $\cos(100\pi/3)$

10. $\cos(55\pi/6)$ **11.** $\sin(-77\pi/4)$ **12.** $\sin 100000\pi$

13. $\cos(41\pi/2)$ **14.** $\sin(-589\pi/6)$ **15.** $\sin(-47\pi/4)$

16. Explain in words the difference between the following expressions.

(a) $\cos^2 t$ **(b)** $\cos t^2$ **(c)** $(\cos t)^2$ **(d)** $\cos(t^2)$

In Exercises 17–30, use the Pythagorean identities and the given information to compute the values of all of the trigonometric functions at t.

17. $\sin t = 1/2$, $P(t)$ lies in the first quadrant

18. $\cos t = -1/3$, $P(t)$ lies in the second quadrant

19. $\sin t = -0.3981$, $P(t)$ lies in the third quadrant

20. $\cos t = 0.7781$, $P(t)$ lies in the fourth quadrant

21. $\cos t = 1/4$, $\sin t < 0$

22. $\sin t = -1/3$, $\cos t > 0$

23. $\sin t = -0.4452$ and $\cos t > 0$

24. $\cos t = -0.6491$ and $\sin t < 0$

25. $\sin t = 2/3$, $\tan t > 0$

26. $\cos t = -1/5$, $\tan t < 0$

27. $\sin t = 1/2$, $\sin t + \cos t < 0$

28. $\cos t = -1/3$, $\sin t + \cos t < 0$

29. $\sec t = 2$ and $\tan t < 0$

30. $\csc t = 10$ and $\sec t > 0$

31. Find $\sec t$ if $\tan t = -1/2$ and $\sin t > 0$.

32. Find $\cos t$ if $\tan t = 1$ and $\sin t < 0$.

33. Find $\sin t$ if $\tan t = 2$ and $\cos t > 0$.

34. Find $\sin t$ if $\cot t = -5$ and $\cos t < 0$.

35. Find $\sin t$ if $\tan t = 6$ and $\sin t < 0$.

36. Is it true that

$$\cot t = \frac{1}{\tan t}$$

for all values of t in the domain of the cotangent function? Justify your answer. (Be careful!)

37. Derive the formula

$$\cot^2 t + 1 = \csc^2 t$$

38. Show that

$$\csc(t + 2\pi) = \csc t$$

for all real numbers t in the domain of the cosecant function.

39. Show that

$$\cot(t + \pi) = \cot t$$

for all real numbers t in the domain of the cotangent function.

40. Use the fact that $\cos t = \sin(t + \pi/2)$ to find an expression for $\sin t$ in terms of the cosine.

41. Use the fact that $-1 \le \sin t \le 1$ to show that $\csc t$ can never assume a value in the open interval $(-1, 1)$.

42. Use the fact that $-1 \le \cos t \le 1$ to show that $\sec t$ can never assume a value in the open interval $(-1, 1)$.

43. Referring to Example 3, find the position of the piston when the angle t is equal to

(a) 30° **(b)** 120° **(c)** -40° **(d)** -12.6°

44. Referring to Example 3, find the displacement of the piston when the wheel is rotated 180° from an initial position given by the following values of t.

(a) $t = 60^\circ$ **(b)** $t = 90^\circ$ **(c)** $t = 120^\circ$

(d) $t = -135^\circ$ **(e)** $t = 20^\circ$

45. Referring to Example 3, what is the displacement of the piston when the wheel rotates from a position given by $t = 45^\circ$ to a position given by $t = 135^\circ$? Compare this with the horizontal distance traveled by the point on the circumference of the wheel where the rod is connected.

6.5 Graphs of the Trigonometric Functions

At this point, we have accumulated enough information about the trigonometric functions to sketch their graphs.

EXAMPLE 1: Sketch the graph of the sine function

$$f(t) = \sin t$$

Solution: We begin by observing that since $\sin t$ satisfies the inequalities $-1 \leq \sin t \leq 1$, its graph must lie between (or on) the horizontal lines $y = 1$ and $y = -1$. (See Figure 6.33.)

Since the sine function repeats itself every 2π units, we need only determine the graph for t in the interval $0 \leq t \leq 2\pi$. Then we can repeat this portion of the graph indefinitely in both directions.

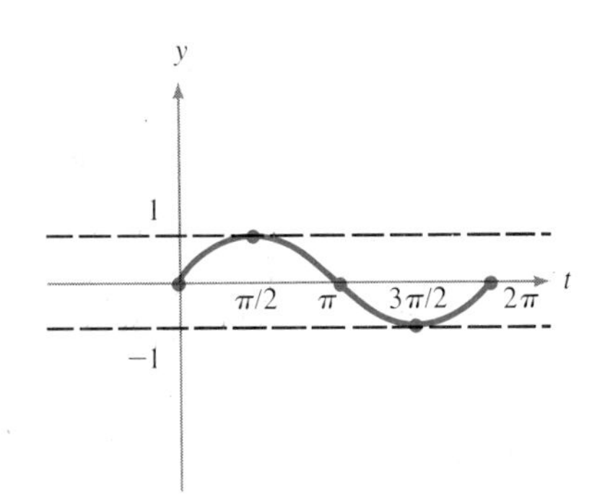

A cycle of the sine function

FIGURE 6.33

Now let us determine the t-intercepts of the graph. (We use the term *t-intercept* since the horizontal axis is the t axis.) For t in the interval $0 \leq t \leq 2\pi$, $\sin t = 0$ precisely when $t = 0$, $t = \pi$ or $t = 2\pi$. This can be verified by drawing a unit circle and observing that the y-coordinate of $P(t)$ is 0 only for these values of t. Hence, the t-intercepts of the graph are 0, π, and 2π. (See Figure 6.33.)

Next we determine the points where the graph touches the horizontal lines $y = 1$ and $y = -1$. For $0 \leq t \leq 2\pi$, $\sin t = 1$ only when $t = \pi/2$. (Again we can use the unit circle to verify this.) Hence, the graph touches the line $y = 1$ only when $t = \pi/2$. Similarly, $\sin t = -1$ only when $t = 3\pi/2$, and so the graph touches the line $y = -1$ only when $t = 3\pi/2$.

Using this information, we get that portion of the graph of the sine function shown in Figure 6.33. This portion is called a **cycle** of the graph. We can obtain the entire graph of the sine function by repeating this cycle. This is done in Figure 6.34.

It appears from Figure 6.34 that the t-intercepts of the graph of the sine function occur when t is an integral multiple of π; that is, when t has the form $t = n\pi$, for some integer n. In fact,

$\sin t = 0$ if and only if t has the form $t = n\pi$, for some integer n.

Try Study Suggestion 6.24. □

▶ **Study Suggestion 6.24:** Use the graph of the sine function to graph the function $g(t) = 1 + \sin t$. At which values of t does $g(t) = 0$? (*Hint:* look on the unit circle.) ◀

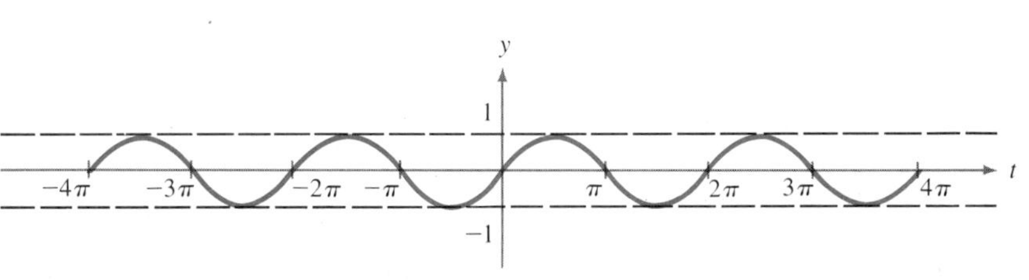

FIGURE 6.34 The graph of $f(t) = \sin t$

EXAMPLE 2: Sketch the graph of the cosine function

$$f(t) = \cos t$$

Solution: In order to sketch the graph of the cosine function, we use the fact that

$$\cos t = \sin\left(t + \frac{\pi}{2}\right)$$

for all values of t. This tells us that we can obtain the graph of the cosine function by translating the graph of the sine function to the left $\pi/2$ units. (Translating graphs is discussed in Section 2.5.) This is done in Figure 6.35.

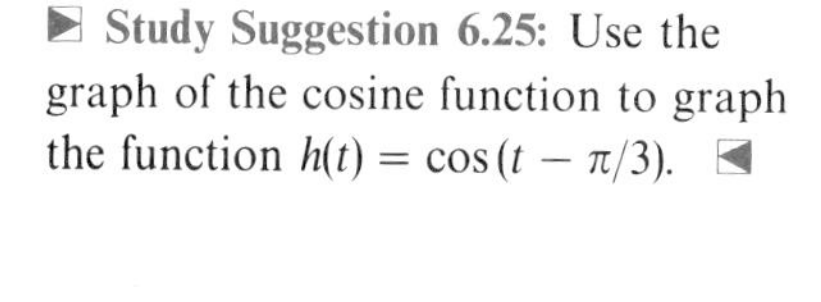

▶ **Study Suggestion 6.25:** Use the graph of the cosine function to graph the function $h(t) = \cos(t - \pi/3)$. ◀

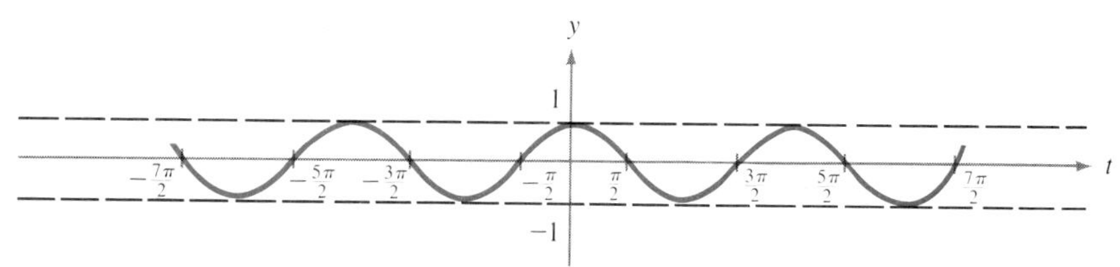

FIGURE 6.35 The graph of $f(t) = \cos t$

It appears from Figure 6.35 that the t-intercepts of the graph of the cosine occur when t is an *odd* integral multiple of $\pi/2$; that is, when t has the form $t = n\pi/2$, where n is an *odd* integer. In fact,

$\cos t = 0$ if and only if t has the form $t = \frac{n\pi}{2}$, where n is an *odd* integer.

Try Study Suggestion 6.25. □

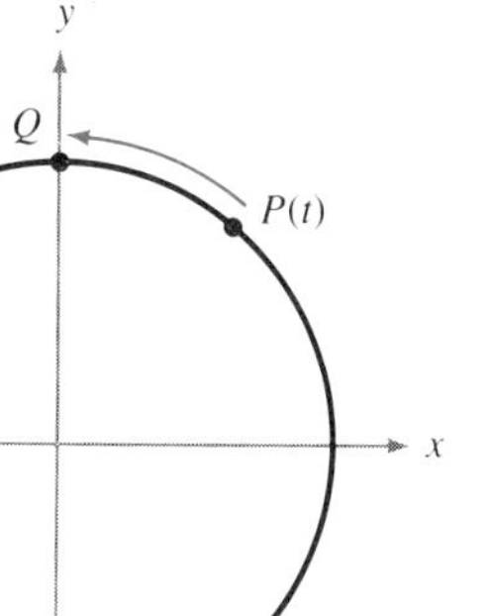

$P(t)$ approaches Q as t approaches $\pi/2$ from the left.

FIGURE 6.36

EXAMPLE 3: Sketch the graph of the tangent function

$$f(t) = \tan t$$

Solution: Since the tangent function repeats itself every π units; that is, since $\tan(t + \pi) = \tan t$, we need only sketch the graph for some interval of length π say $-\pi/2 < t < \pi/2$. Then we can repeat this portion to obtain the entire graph of the tangent function.

First we examine the behavior of $\tan t$ as t approaches $\pi/2$ from the left, since that is the side of $\pi/2$ that lies inside the interval $-\pi/2 < t < \pi/2$. Since $\tan t = \sin t/\cos t$ we must first examine the behavior of $\sin t$ and $\cos t$ as t approaches $\pi/2$ from the left.

Looking at Figure 6.36, we see that as t approaches $\pi/2$ from the left; that is, as t gets close to $\pi/2$, but remains always less than $\pi/2$, the point $P(t) = (\cos t, \sin t)$ approaches the topmost point on the circle, in a *counter-clockwise* manner. Therefore, since the topmost point on the unit circle is $(0, 1)$, $\sin t$ approaches 1 and $\cos t$ approaches 0. Moreover, $\cos t$ remains *positive* as it approaches 0, and so $\tan t = \sin t/\cos t$ gets larger and larger without bound. This means that the line $t = \pi/2$ is a vertical asymptote, as shown in Figure 6.37.

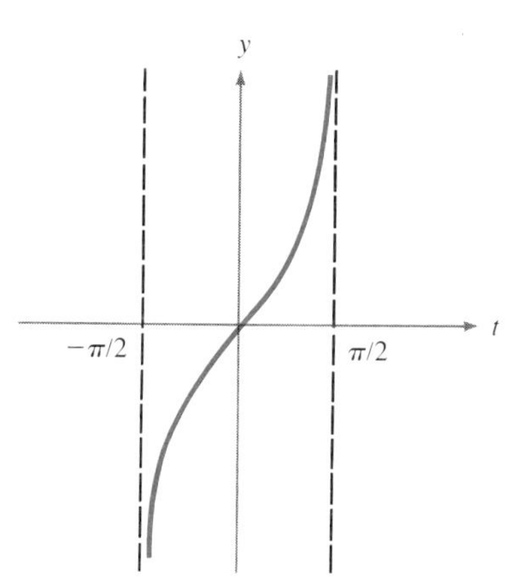

A cycle of the tangent function.

FIGURE 6.37

A similar reasoning can be used to show that as t approaches $-\pi/2$ from the right, $\tan t$ gets smaller and smaller without bound. Thus, the line $t = -\pi/2$ is also a vertical asymptote of the graph. Finally, since $\tan 0 = 0$, the graph does go through the origin. Putting these pieces of information together, we get the graph in Figure 6.37.

Now we can repeat this portion of the graph to get the graph of the tangent function. This is done in Figure 6.38.

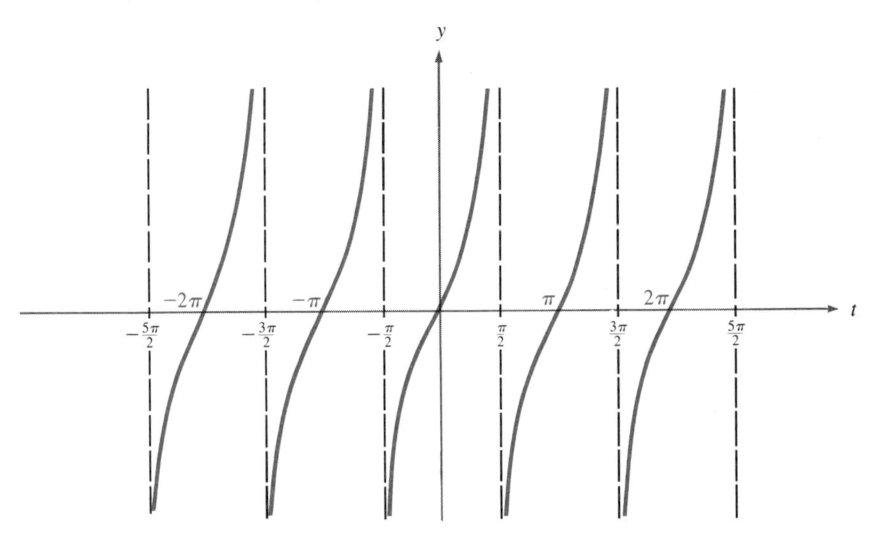

FIGURE 6.38 The graph of $f(t) = \tan t$

Notice that, unlike the graphs of the sine and cosine function, the graph of the tangent function has (infinitely) many "breaks" in it. They occur at all *odd* integral multiples of $\pi/2$; that is, at the points $t = n\pi/2$, where n is an odd integer, because this is precisely where the cosine function is equal to 0. *Try Study Suggestion 6.26.* □

▶ **Study Suggestion 6.26:** Use the graph of the tangent function to graph the function $g(t) = 2 + \tan t$. ◀

EXAMPLE 4: Sketch the graph of the secant function

$$f(t) = \sec t$$

Solution: The graph of the secant function is shown in Figure 6.39. Notice that we have included the graph of the cosine function in this picture. The graph of the secant was obtained by using the definition

$$\sec t = \frac{1}{\cos t}$$

to make the following observations about the relationship between the secant and the cosine.

1. The secant is positive whenever the cosine is positive, and the secant is negative whenever the cosine is negative.
2. When the cosine is positive, we have $0 < \cos t \leq 1$, and so $\sec t \geq 1$.

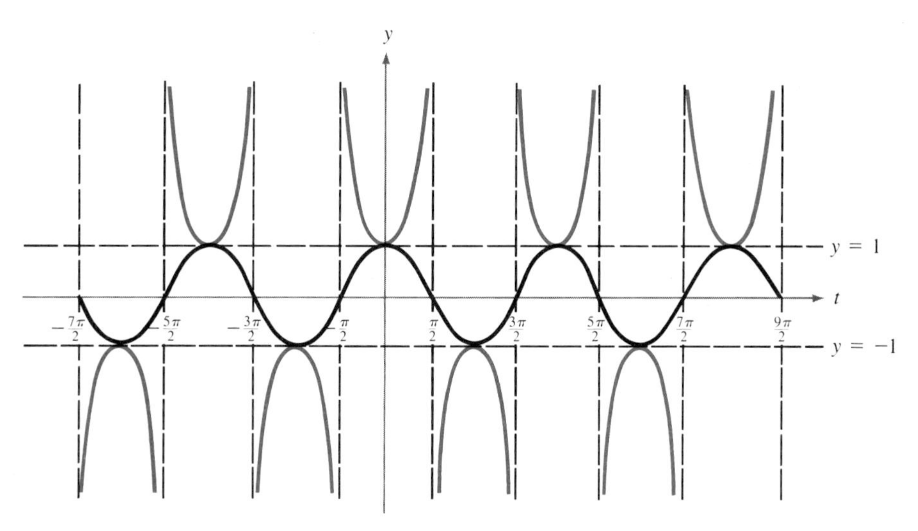

FIGURE 6.39 The graph of $f(t) = \sec t$ (blue).
The graph of $g(t) = \cos t$ (black).

Similarly, when the cosine is negative, we have $-1 \leq \cos t < 0$, and so $\sec t \leq -1$.

3. If $\cos t = 1$, then $\sec t = 1$ and if $\cos t = -1$, then $\sec t = -1$.
4. The closer $\cos t$ is to 0, the larger is $|\sec t|$. Furthermore, $\sec t$ is undefined whenever $\cos t = 0$ *Try Study Suggestion 6.27.* □

▶ **Study Suggestion 6.27:** Use the graph of the secant function to graph the function $g(t) = 2\sec(t - \pi)$. ◀

The graphs of the cosecant and cotangent functions can be obtained from the graphs of the sine and tangent functions in a similar manner. These graphs are pictured in Figures 6.40 and 6.41.

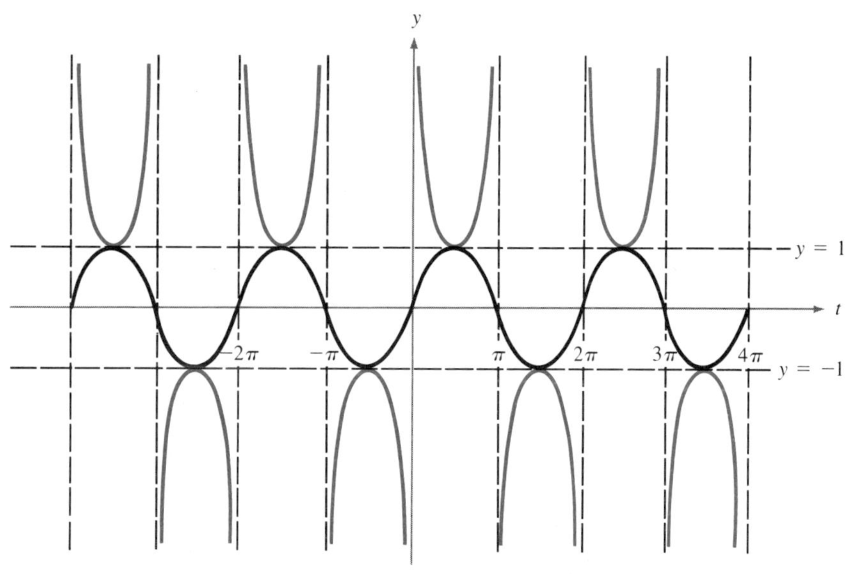

FIGURE 6.40 The graph of $f(t) = \csc t$ (blue).
The graph of $g(t) = \sin t$ (black).

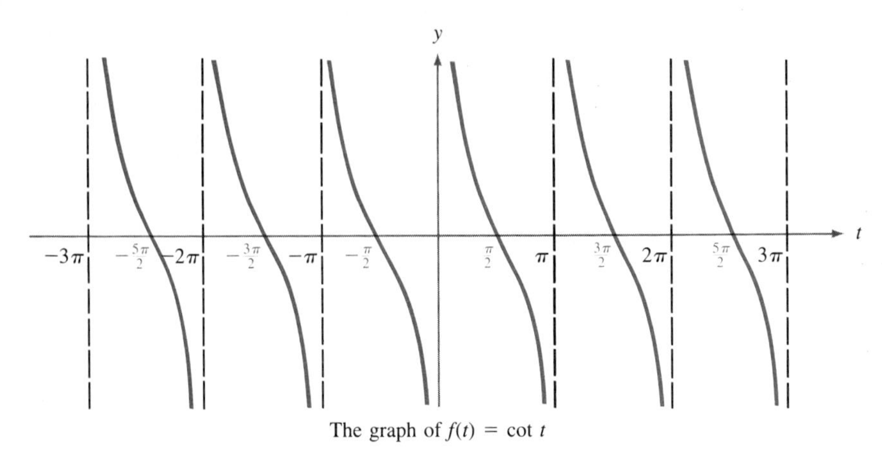

The graph of $f(t) = \cot t$

FIGURE 6.41

▶ **Study Suggestion 6.28:** Justify the graph of the cosecant function given in Figure 6.40 by making observations similar to the ones that we made for the secant function. ◀

Try Study Suggestion 6.28.

As you know if you have been working the Study Suggestions, the methods of Section 2.5 can be very useful in graphing functions that involve the trigonometric functions.

EXAMPLE 5: Graph the function

$$f(t) = -1 + 2\sec\left(t - \frac{\pi}{4}\right)$$

Solution: The graph of the function f can be obtained from the graph of the secant function in the following steps:

1. Because of the expression $t - \pi/4$, we first translate the graph of the secant function $\pi/4$ units to the right.
2. Because of the coefficient 2, we then multiply each of the y-coordinates of the graph by 2. This has the effect of "stretching" the graph in the y-direction.
3. Because of the term -1, we translate the graph down 1 unit.

In short, beginning with the graph of the secant function (Figure 6.38) we first translate $\pi/4$ units to the right, then "stretch" in the y-direction, and finally translate down 1 unit.

Perhaps the easiest way to do this is to first show the effect of these motions on the "framework" consisting of the vertical asymptotes of the secant and the horizontal lines $y = -1$ and $y = 1$. Once this framework has been repositioned, we can use it as a guide for sketching the graph of f.

The vertical asymptotes of the secant are affected by the translation $\pi/4$ units to the right, but are not affected by the stretching or the translation down. Hence, the "new" asymptotes are as shown in Figure 6.42. The lines $y = -1$ and $y = 1$ are not affected by the translation to the right, but are affected by the stretching and the translation down 1 unit. The effect of the stretching is to place the lines at the positions $y = -2$ and $y = 2$. Then the translation down 1 unit moves these lines to the positions $y = -3$ and $y = 1$. These lines are also shown in Figure 6.42.

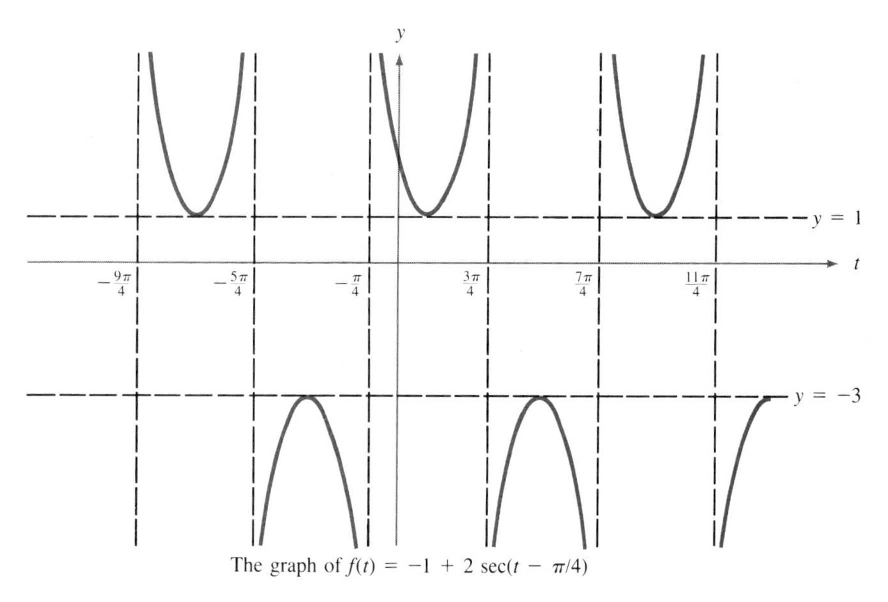

FIGURE 6.42

Now we can sketch a "stretched" version of the graph of the secant function within this new framework, as shown in Figure 6.42.

Try Study Suggestion 6.29. ☐

▶ **Study Suggestion 6.29:** Graph the function $g(t) = 1 + 2\sec(t + \pi)$ ◀

The method of "adding ordinates" that we first used in Example 3 of Section 5.1 can also be used here.

EXAMPLE 6: Graph the function

$$f(t) = \sin t + \cos t$$

in the interval $0 \leq t \leq 2\pi$.

Solution: Figure 6.43 shows the graphs of the functions $g(t) = \sin t$ and $h(t) = \cos t$ in black. By adding corresponding y-coordinates, we get the graph of $f(t) = \sin t + \cos t$, shown in color in the same figure.

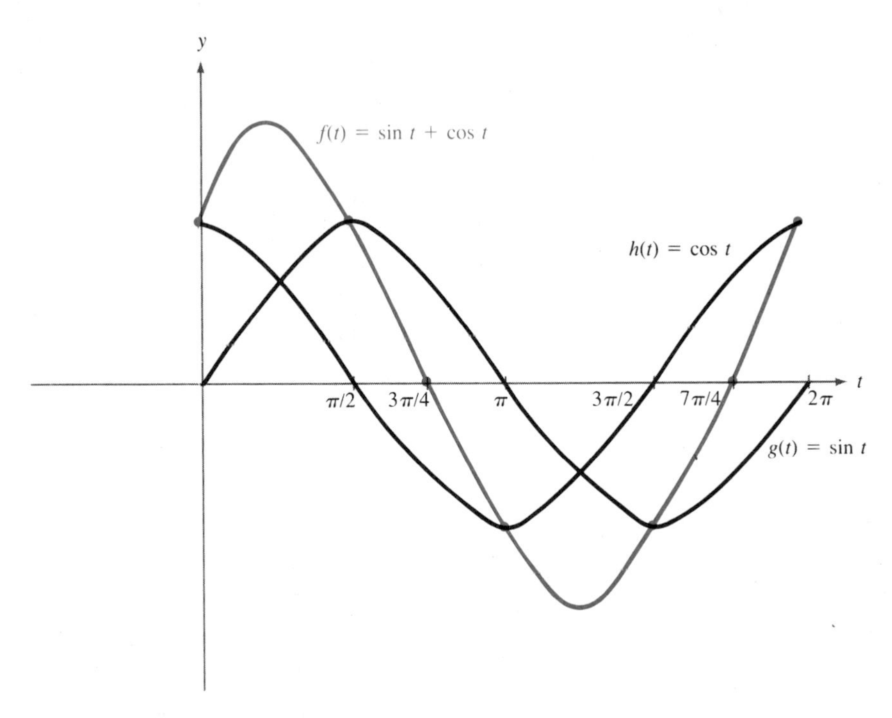

FIGURE 6.43

However, before sketching the graph of f, we should make the following observations. Since $\cos t = 0$ when $t = \pi/2$ or $t = 3\pi/2$, the graph of f intersects the graph of $g(t) = \sin t$ for these values of t. (See Figure 6.43.) Similarly, since $\sin t = 0$ when $t = 0$ or $t = \pi$, the graph of f intersects the graph of $h(t) = \cos t$ for these values of t.

Also, the graph of f intersects the t-axis when $\sin t + \cos t = 0$, that is, when $\sin t = -\cos t$, which happens when $t = 3\pi/4$ and $t = 7\pi/4$. (You can check this by referring to Figure 6.20.) These observations help us to draw a more accurate picture of f. *Try Study Suggestion 6.30.* □

▶ **Study Suggestion 6.30:** Graph the function $f(t) = \sin t - \cos t$ in the interval $0 \le t \le 2\pi$. ◀

Ideas to Remember

The graphs of the trigonometric functions give a very clear picture of the behavior of these important functions. Therefore, it pays to keep an image of these graphs in your mind.

EXERCISES

In Exercises 1–15, answer the given question. You may use the graphs of the trigonometric functions for assistance, as well as the unit circle.

1. For which values of t in the interval $-2\pi \le t \le 2\pi$ does $\sin t$ take its maximum value? What is this maximum value?

2. For which values of t in the interval $-2\pi \le t \le 2\pi$ does $\cos t$ take its maximum value? What is this maximum value?

3. For which values of t in the interval $-2\pi \le t \le 2\pi$ does $\sin t$ take its minimum value? What is this minimum value?

4. For which values of t in the interval $-2\pi \le t \le 2\pi$ does $\cos t$ take its minimum value? What is this minimum value?

5. For which values of t in the interval $-2\pi \le t \le 2\pi$ is $\sin t = \sqrt{2}/2$?

6. For which values of t in the interval $-2\pi \le t \le 2\pi$ does $\cos t = -\sqrt{3}/2$?

7. Does the secant function have either a minimum value or a maximum value?

8. For which values of t in the interval $-2\pi \le t \le 2\pi$ does $\sec t$ take on its minimum *absolute value*? What is this value?

9. For which values of t in the interval $-2\pi \le t \le 2\pi$ does $\csc t = 2$?

10. Does the tangent function have a minimum value or a maximum value?

11. For which values of t in the interval $-2\pi \le t \le 2\pi$ does $\tan t = 1$?

12. If $-1 \le a \le 1$, how many different values of t are there in the interval $-2\pi \le t \le 2\pi$ for which $\sin t = a$?

13. If a is a real number, how many different values of t are there in the interval $-2\pi \le t \le 2\pi$ for which $\tan t = a$?

14. If $a \ge 1$, how many different values of t are there in the interval $-2\pi \le t \le 2\pi$ for which $\csc t = a$?

15. For which values of t does $\cot t = 0$?

In Exercises 16–55, graph the given function.

16. $f(t) = 2 + \sin t$

17. $f(t) = -\pi + \cos t$

18. $f(t) = \dfrac{\pi}{2} - \tan t$

19. $f(t) = -\sin t$

20. $f(t) = 1 - \sec t$

21. $f(t) = -\tan t$

22. $f(t) = 4 \sin t$

23. $f(t) = -\pi \cos t$

24. $f(t) = -\left(\dfrac{1}{2}\right)\cot t$

25. $f(t) = 1 + 3 \sin t$

26. $f(t) = \pi - 2 \cos t$

27. $f(t) = 1 - 2 \cot t$

28. $f(t) = \dfrac{\pi}{4} + 2 \csc t$

29. $f(t) = \sin\left(t + \dfrac{\pi}{3}\right)$

30. $f(t) = \cos\left(t - \dfrac{\pi}{3}\right)$

31. $f(t) = 3 \tan\left(t - \dfrac{\pi}{3}\right)$

32. $f(t) = -4 \sec (t - \pi)$

33. $f(t) = -2 \cot\left(t - \dfrac{\pi}{4}\right)$

34. $f(t) = 1 + \sin\left(t - \dfrac{\pi}{4}\right)$

35. $f(t) = -1 + 2 \cos\left(t - \dfrac{\pi}{2}\right)$

36. $f(t) = \pi - \sin (t - \pi)$

37. $f(t) = \pi - 2 \sec\left(t - \dfrac{\pi}{8}\right)$

38. $f(t) = -\dfrac{\pi}{2} + \cos\left(t + \dfrac{\pi}{2}\right)$

39. $f(t) = 1 - 3 \tan\left(t + \dfrac{\pi}{4}\right)$

40. $f(t) = |\sin t|$

41. $f(t) = |\cos t|$

42. $f(t) = |\tan t|$

43. $f(t) = \sin |t|$

44. $f(t) = \cos |t|$

45. $f(t) = \tan |t|$

46. $f(t) = t + \sin t$

47. $f(t) = 2t + \cos t$

48. $f(t) = 3t - \sin t$

49. $f(t) = t + \tan t$

50. $f(t) = t + \csc t$

51. $f(t) = \tan t + \cot t$

52. $f(t) = \sin t + 2 \cos t$

53. $f(t) = 2 \sin t + 3 \cos t$

54. $f(t) = t \sin t$

55. $f(t) = t \cos t$

56. The function

$$f(t) = \frac{\sin t}{t}$$

is very important in calculus.

(a) Use a calculator to compute the value of this function at $t = 0.9$, 0.7, 0.5, 0.3, and 0.1. What do you conclude about the behavior of this function as t approaches 0?

(b) Graph this function in the interval $-\pi/2 \le t \le \pi/2$. (What is the domain of this function?)

6.6 Sine Waves

The graphs of the sine and cosine functions are examples of curves known as *sine waves*. Sine waves play a very important role in nature, and we will study them in some detail in this section.

More generally, the graph of any function of the form

$$f(t) = a \sin(bt + c) \tag{6.17}$$

$$g(t) = a \cos(bt + c) \tag{6.18}$$

where a, b, and c are real numbers, and a and b are nonzero, is called a **sine wave.** Such curves have the general appearance shown in Figure 6.44.

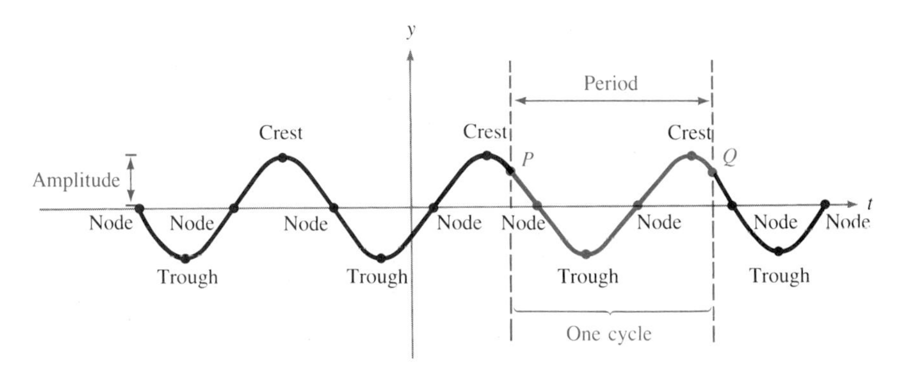

FIGURE 6.44

As you can see from the graph, the functions defined in Equations (6.17) and (6.18) are periodic; that is, they repeat themselves at regular intervals. Any portion of a sine wave that starts at some point P, and goes to a point Q where the curve *first* begins to repeat itself is called a **cycle** of the sine wave. A cycle of the graph is shown in color in Figure 6.44. The *horizontal* distance (not the distance along the curve) from the beginning of a cycle to the end of that cycle is called the **period** of the sine wave. For example, the graphs of both the sine and cosine functions have period 2π.

The **amplitude** of a sine wave is the *largest* value that the y-coordinate of the graph assumes. This quantity is also shown in Figure 6.44. Points on the wave that have the largest y-coordinate are called **crests,** and points that have the smallest y-coordinate are called **troughs.** Also, the t-intercepts of the wave are sometimes referred to as **nodes.** In order to draw sine waves accurately, it is helpful to observe that cycles that begin (and end) at either nodes, crests, or troughs have a certain symmetry. This symmetry can best be seen by dividing the cycle into four equal parts, as shown in Figure 6.45. In Figure 6.45(a), we see a cycle that goes from node to node. This cycle has another node that lies *exactly halfway* between the two outside nodes. Also,

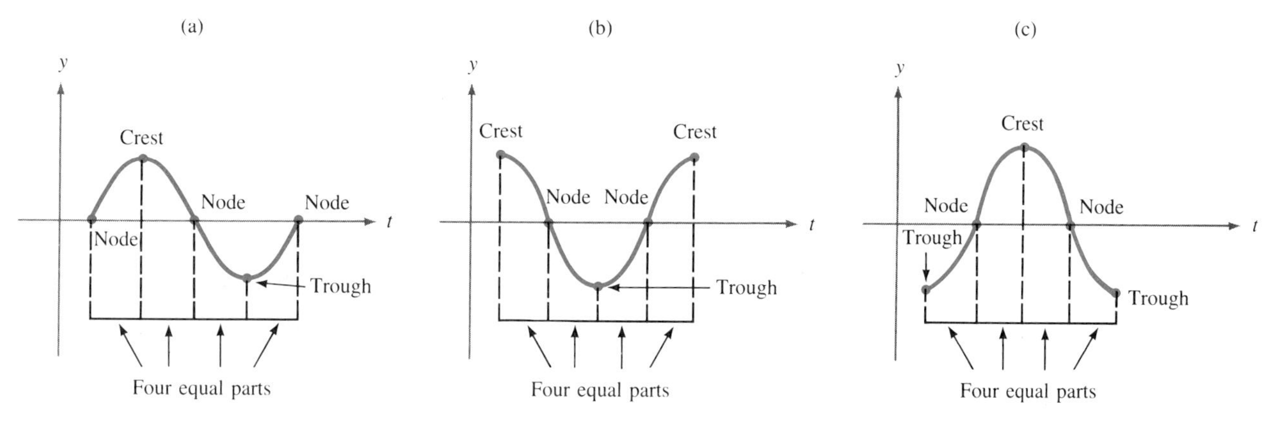

FIGURE 6.45

the crest lies *exactly halfway* between its surrounding nodes, and similarly for the trough. In fact, in all three cases shown in Figure 6.45, we can say that

the crests, nodes and troughs divide the cycle into fourths.

The cyclic behavior of sine waves makes them ideal for describing many types of cyclic phenomena that occur in nature. For example, both light and sound have a cyclic nature and can be described using sine waves. Certain types of motion, such as the motion of a pendulum, the ebb and flow of the tides, the increase and decrease in blood pressure, and the variation in light intensity from so-called *variable stars*, all behave in a cyclic manner and can be described using sine waves. Also, the vibrations of the strings of a musical instrument can be described using sine waves.

These examples make it extremely important to be able to draw accurate sine waves, and that is our goal in this section. As we will see in the coming examples, each of the numbers a, b, and c in (6.17) and (6.18) has a definite effect on the shape of the sine wave. Let us begin by discussing the effect of the coefficient a.

The graph of a function of the form

$$f(t) = a \sin t$$

where a is a nonzero real number, can be obtained from the graph of the sine function by multiplying each of the y-coordinates by the number a. If a is positive, this has the effect of "stretching" the graph (when $a > 1$), or "shrinking" the graph (when $a < 1$.) If a is negative, this will also have the effect of reflecting the graph over the t-axis. Of course, similar statements hold for the graphs of functions of the form $g(t) = a \cos t$. (Stretching, shrinking, and reflecting graphs are discussed in Section 2.5.)

EXAMPLE 1: Graph the function

$$f(t) = 2 \sin t$$

Solution: The first step is to sketch the graph of the sine function, as we have done in black in Figure 6.46. Then we multiply each of its y-coordinates by 2, which has the effect of "stretching" the graph. The graph of $f(t) = 2\sin t$, is shown in color in Figure 6.46. Notice that the amplitude of this graph is 2, which happens to be the value of the coefficient a.

Try Study Suggestion 6.31. □

▶ **Study Suggestion 6.31:** Graph the function $g(t) = (1/3)\sin t$. ◀

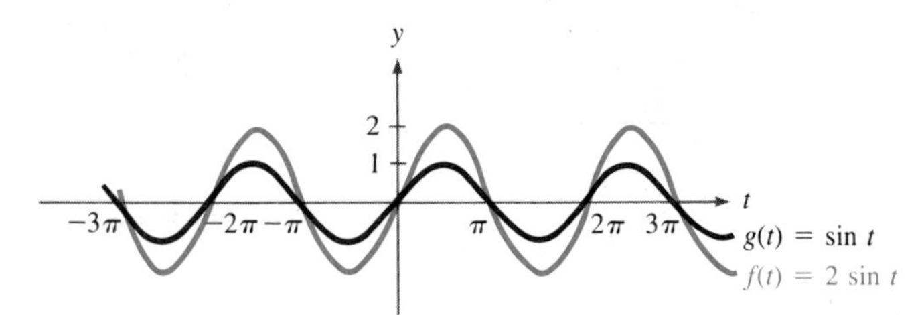

FIGURE 6.46

EXAMPLE 2: Graph the function

$$f(t) = -\frac{1}{2}\cos t$$

Solution: In this case, we multiply each y-coordinate of the graph of the cosine function by $-1/2$. This has the effect of first "shrinking" the graph of the cosine, and reflecting the resulting curve over the t-axis, as shown in Figure 6.47. Notice that the amplitude of this graph is 1/2, which happens to be the *absolute value* of the coefficient a. *Try Study Suggestion 6.32.* □

▶ **Study Suggestion 6.32:** Graph the function $h(t) = -\pi\cos t$. ◀

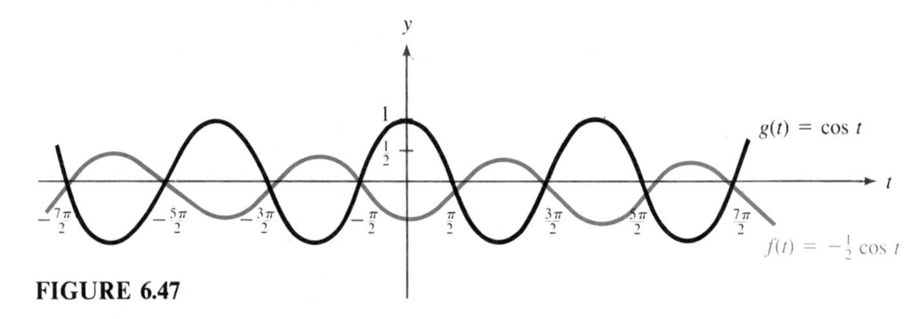

FIGURE 6.47

In each of the previous examples, the amplitude of the sine wave is equal to the absolute value of the coefficient a. This happens to be true in general, and so we can state the following theorem.

THEOREM

> The graphs of the functions $f(t) = a\sin(bt + c)$ and $g(t) = a\cos(bt + c)$ have amplitude equal to $|a|$.

In order to get a feel for the effect of the coefficient b on a sine wave, let us consider a specific example.

EXAMPLE 3: Sketch the graph of the function

$$f(t) = \cos 3t$$

Solution: As t varies from 0 to 2π, the cosine function $g(t) = \cos t$ completes one cycle, going from crest to crest. (See Figure 6.35.) Therefore, the function $f(t) = \cos 3t$ will complete one cycle, also going from crest to crest, when $3t$ ranges from 0 to 2π.

But $3t$ ranges from 0 to 2π precisely when t ranges from 0 to $2\pi/3$. Thus, the graph of $f(t) = \cos 3t$ is just a "compressed" version of the graph of the cosine function, where a complete cycle occurs in the interval $[0, 2\pi/3]$, rather than in the interval $[0, 2\pi]$. The graph of $f(t) = \cos 3t$ is pictured in Figure 6.48, along with the graph of the cosine function.

▶ **Study Suggestion 6.33:** Graph the function $g(t) = \cos 2t$ ◀

Try Study Suggestion 6.33. □

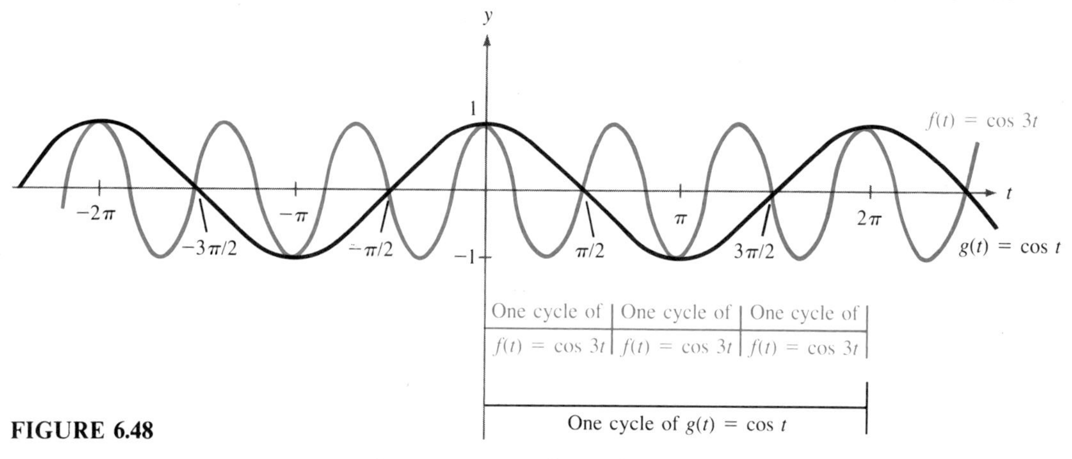

FIGURE 6.48

As you can see from the previous example, the effect of the coefficient 3 in the function $f(t) = \cos 3t$ is to change the *period* of the function from 2π, which is the period of the cosine function, to $2\pi/3$. There is nothing special about the number 3, and we can state the following general result.

THEOREM

> The graphs of the functions $f(t) = a \sin(bt + c)$ and $g(t) = a \cos(bt + c)$ have period equal to $2\pi/|b|$. (Note the absolute value sign. Periods must always be positive.)

EXAMPLE 4: Sketch the graph of the function

$$f(t) = -3 \sin \frac{t}{2}$$

Solution: First we sketch the graph of $g(t) = \sin t/2$. According to the previous theorem, the graph of this function has period equal to $2\pi/(1/2) = 4\pi$, and so it is an "expanded" version of the graph of the sine function. This graph is shown in solid black in Figure 6.49. (For the sake of comparison, we have also included the graph of the sine function.) From the graph of $g(t) = \sin t/2$ we can obtain the graph of $f(t) = -3\sin t/2$ by multiplying each y-coordinate by -3. This graph is shown in color in Figure 6.49.

▶ **Study Suggestion 6.34:** Graph the function $f(t) = -2\sin t/3$. ◀

Try Study Suggestion 6.34. □

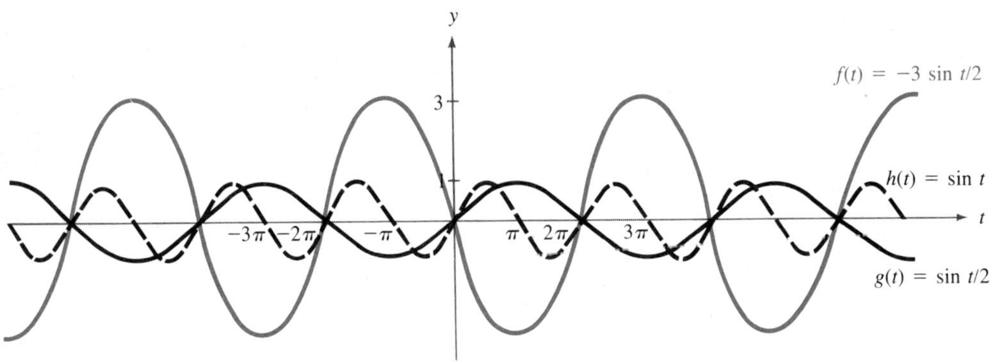

FIGURE 6.49

The effect of the number c in Equation (6.17) can be seen by writing Equation (6.17) in the form

$$f(t) = a\sin b\left[t - \left(-\frac{c}{b}\right)\right]$$

According to our discussion in Section 2.5, this shows that the graph of f can be obtained by translating the graph of

$$h(t) = a\sin bt$$

a distance of $-c/b$ units to the right if $-c/b > 0$ and $|-c/b|$ units to the left if $-c/b < 0$. Thus, we see that the effect of the number c is to cause a horizontal shift in the graph. A similar statement holds for the sine waves $g(t) = a\cos(bt + c)$. For this reason, the number c/b is sometimes referred to as the **phase shift** of the graph.

Since sine waves are periodic, we can determine the entire graph by first determining one complete cycle, and then repeating this cycle indefinitely in both directions. Fortunately, this can be done by making use of the following observations, which will save us from having to memorize exactly how far to shift the graph and in which direction.

Since the sine function $h(t) = \sin t$ completes one cycle as t ranges from 0 to 2π, the function $f(t) = a\sin(bt + c)$ completes one cycle as $bt + c$ ranges from 0 to 2π. Thus, we can determine the location of a complete cycle of the graph of f simply by solving the equations

$$bt + c = 0 \qquad \text{and} \qquad bt + c = 2\pi \tag{6.19}$$

The solutions to these equations will give us the t-coordinates of the beginning and end of a complete cycle of the wave. (Which equation gives the beginning of the cycle and which gives the end will depend on whether b is positive or negative.) A similar statement holds for the graph of the function $g(t) = a\cos(bt + c)$.

Using these observations, we can describe a simple procedure for graphing sine waves.

Procedure for Graphing Functions of the Form $f(t) = a\sin(bt + c)$ and $g(t) = a\cos(bt + c)$

STEP 1. The amplitude of the wave is $|a|$. Therefore, the graph will lie between, or on, the horizontal lines $y = -a$ and $y = a$. Sketch these lines. (The period of the wave is $2\pi/|b|$. However, knowing the period is not essential to the procedure.)

STEP 2. Solve the equations $bt + c = 0$ and $bt + c = 2\pi$. The solutions are the t-coordinates of the beginning and end of a complete cycle of the wave. Plot the points on the graph corresponding to these t coordinates.

STEP 3. Draw the cycle whose endpoints were determined in Step 2. Use the fact that the cycle has amplitude $|a|$, and has the symmetry that we discussed at the beginning of the section. (See Figure 6.45.) Label all nodes of the cycle on the t-axis.

STEP 4. The complete graph is formed by repeating this cycle indefinitely in both directions. (Of course, this is not possible to do in practice. If space permits, we will repeat the cycle at least once in each direction.) Because of the symmetry of the cycles, the positions of the crests and troughs are completely determined by the positions of the nodes. In fact, the crests and troughs lie exactly halfway between consecutive nodes.

EXAMPLE 5: Graph the function

$$f(t) = 2\sin\left(3t + \frac{\pi}{2}\right)$$

Solution:

STEP 1. This wave has amplitude $|a| = 2$ and period $2\pi/|b| = 2\pi/3$.

STEP 2. In this case, Equations (6.19) are

$$3t + \frac{\pi}{2} = 0 \qquad \text{and} \qquad 3t + \frac{\pi}{2} = 2\pi$$

The first of these equations has solution $t = -\pi/6$, and the second has solution $t = \pi/2$. Therefore, a complete cycle of the graph occurs as t ranges from $-\pi/6$ to $\pi/2$. Since $f(-\pi/6) = 0$ and $f(\pi/2) = 0$, this cycle begins at the point $(-\pi/6, 0)$ and ends at the point $(\pi/2, 0)$. (See Figure 6.50.)

▶ **Study Suggestion 6.35:** Graph the function $g(t) = 3\sin(2t + \pi/3)$. ◀

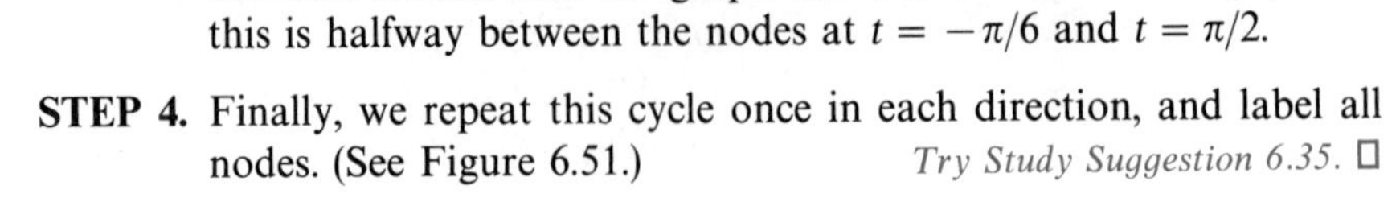

STEP 3. Using this information, we can draw the cycle, as shown in Figure 6.50. Notice that the graph also has a node at $t = \pi/6$, since this is halfway between the nodes at $t = -\pi/6$ and $t = \pi/2$.

STEP 4. Finally, we repeat this cycle once in each direction, and label all nodes. (See Figure 6.51.) *Try Study Suggestion 6.35.* □

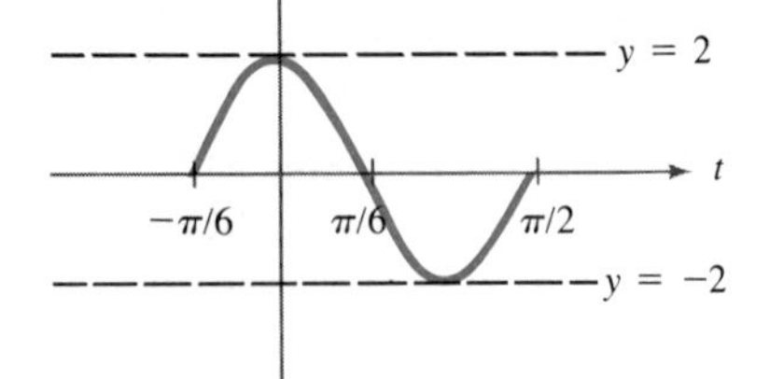

FIGURE 6.50

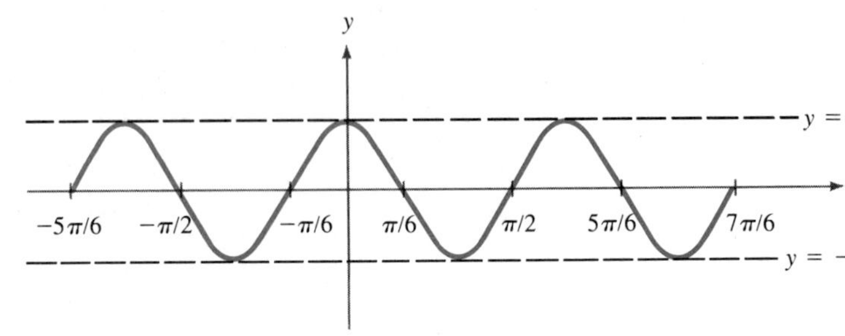

The graph of $f(t) = 2\sin(3t + \pi/2)$

FIGURE 6.51

EXAMPLE 6: Graph the function

$$f(t) = -3\cos\left(\pi - \frac{t}{2}\right)$$

Solution:

STEP 1. This wave has amplitude $|a| = 3$ and period $2\pi/|b| = 2\pi/|-1/2| = 4\pi$.

STEP 2. In this case, Equations (6.19) are

$$\pi - \frac{t}{2} = 0 \qquad \text{and} \qquad \pi - \frac{t}{2} = 2\pi$$

The first of these equations has solution $t = 2\pi$, and the second has solution $t = -2\pi$. Therefore, a complete cycle of the graph occurs as t ranges from -2π to 2π. Since $f(-2\pi) = -3$ and $f(2\pi) = -3$, this cycle begins at the point $(-2\pi, -3)$ and ends at the point $(2\pi, -3)$. (See Figure 6.52.)

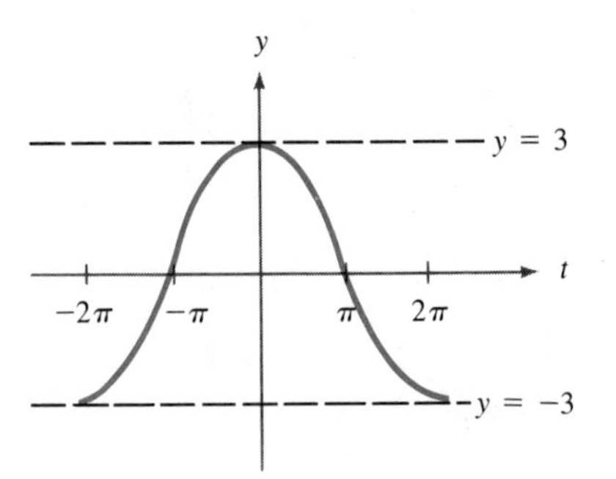

FIGURE 6.52

STEP 3. Using this information, we can draw the cycle, as shown in Figure 6.52. To determine the location of the nodes, we first notice that the graph has a crest in the middle of the cycle, at $t = 0$. Hence, its nodes lie at $t = -\pi$ and $t = \pi$ (halfway between crest and trough.)

STEP 4. Finally, we repeat this cycle once in each direction, and label all nodes. (See Figure 6.53.) *Try Study Suggestion 6.36.* □

▶ **Study Suggestion 6.36:** Graph the function $g(t) = \pi \cos(\pi/3 - t/2)$. ◀

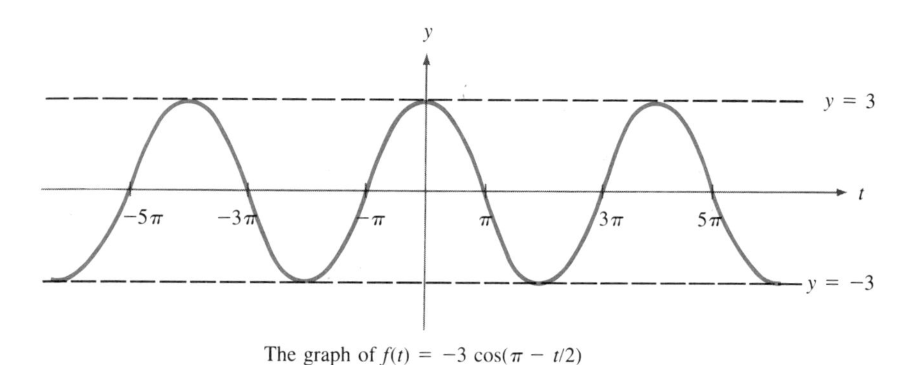

The graph of $f(t) = -3\cos(\pi - t/2)$

FIGURE 6.53

In the remainder of this section, we want to discuss one of the most important applications of sine waves. There will be no exercises on the forthcoming material, and we present it only in the hope that you will find it interesting.

For this discussion, we will think of the variable t in Equations (6.17) and (6.18) as representing time. Let us begin by establishing some terminology.

When the variable t represents time, the period $T = 2\pi/|b|$ of the wave is usually called the **wavelength.** As shown in Figure 6.54, the wavelength is

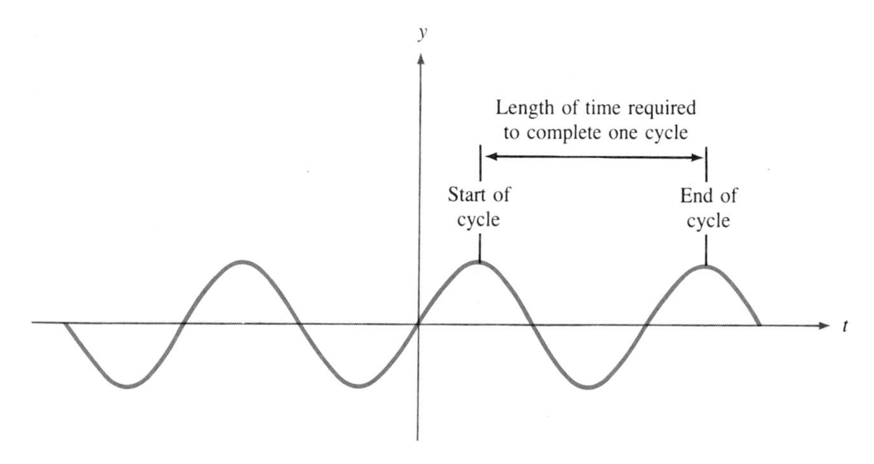

FIGURE 6.54

the length of time that it takes to complete one cycle of the wave. But, if it takes $T = 2\pi/|b|$ seconds to complete one cycle, it follows that

$$\frac{1}{T} = \frac{|b|}{2\pi}$$

cycles are completed in 1 second. For this reason, we call the quantity $f = |b|/2\pi$ the **frequency** of the wave, and say that the wave *oscillates at a frequency of f cycles per second.*

In modern times, the term *cycles per second* has been replaced by the term *hertz*, abbreviated Hz, and named after the physicist Heinrich Hertz, who discovered radio waves in 1890. Therefore, from now on, we will use this term. (You may have heard the term hertz if you have done some serious shopping for stereo equipment.)

As you probably know, waves appear in many different varieties in nature. For example, let us imagine a cork floating on the surface of a lake. If we shake the cork up and down, the effect will be to generate waves of water on the surface of the lake. Of course, these water waves travel some distance after leaving the immediate vicinity of the cork. For instance, if a second cork was floating some distance from the first, the waves generated by the first cork would eventually reach the second cork, causing it to move. Because of this effect, we say that the waves *propagate* through the water.

Just as the movement of a cork generates waves that propagate through the water, the movements inside a loudspeaker generate waves of increasing and decreasing pressure that propagate through the air. These waves are called **sound waves.** Furthermore, the sound waves coming from a speaker have the same effect on our eardrums as the water waves have on the second cork, that is, they cause our eardrums to move, and that is precisely how we hear.

Now, sound is not the only important natural phenomenon that travels in waves. For example, visible light also travels in waves. However, the medium through which light waves propagate is much more "mysterious" than the medium through which sound waves propagate. In fact, while there can be no sound in a vacuum, since there is no air to carry the waves, there can be light in a vacuum. (The light from the sun and the stars travels to us through a virtual vacuum.)

The medium through which light waves propagate is known as the **electromagnetic field.** This mysterious medium is not at all like the air through which sound waves propagate. In fact, we cannot detect the electromagnetic field directly, that is, we cannot *feel* it as we can with the air. Rather, we detect the presence of the electromagnetic field only indirectly, through the fact that light waves are able to move from one location in space to another!

The electromagnetic field carries more than just the waves of visible light. It also carries radio waves, television waves, radar waves, X-ray waves, and cosmic waves, to mention just a few. In fact, all of these waves are of the same type, the only thing that distinguishes them from one another is their *frequencies.*

Waves that propagate through the electromagnetic field are known as **electromagnetic waves,** or **electromagnetic radiation.** It is the frequency of an

TABLE 6.3

Approximate frequency of oscillation (in hertz)	Type of electromagnetic radiation
$5 \times 10^5 - 10^6$	AM radio waves
10^8	FM radio and television waves
10^{10}	Radar waves (microwaves)
$5 \times 10^{14} - 10^{15}$	Visible light waves
10^{18}	X-rays
10^{21}	Gamma rays
10^{27}	Cosmic rays

electromagnetic wave that determines its characteristics, as well as how we "preceive" it. Table 6.3 shows how we perceive electromagnetic radiation of various frequencies.

Notice that electrogmagnetic radiation at a frequency between 5×10^{14} and 10^{15} hertz appears to us as visible light. In other words, it happens that our eyes are able to react to electromagnetic radiation of these frequencies, and as a result we "see" this type of radiation.

Moreover, our eyes can distinguish between electromagnetic radiation of different frequencies within this range. Electromagnetic radiation at the lower end of this frequency range appears to us as red light, radiation in the middle frequencies appears to us as yellow or green light, and radiation at the upper end of this frequency range appears to us as blue light. In short, we distinguish light of different frequencies by seeing different colors.

On the other hand, our eyes do not respond to electromagnetic radiation in the frequency range of 5×10^5 to 10^8 hertz, and since this type of radiation seems to be harmless to us, we have chosen it to carry our radio and television signals. In other words, unlike our eyes, radios and televisions are designed to respond to electromagnetic radiation within this frequency range.

Let us discuss how radio waves are transmitted from a radio station to your home radio. This will give you a chance to see some of the concepts we have been studying, such as amplitude and frequency, in actual use.

Let us follow the progress of a pure tone, such as a note sung by a soprano, as it goes from the radio station to your ear. This pure tone is a simple sine wave of the type we have been graphing in this section. Such a wave is pictured in Figure 6.55. The amplitude of this wave determines the volume of the tone (the higher the amplitude, the louder the tone), and the frequency of the wave determines its pitch (the higher the frequency, the higher the pitch.)

Now, musical tones, such as the one sung by our soprano, have frequencies ranging between 20 and 20,000 hertz. But we have already said that radio waves have a frequency somewhere between 5×10^5 to 10^8 hertz. If this is the case, how can our musical tone by carried over the radio?

The answer is that a radio station also transmits an electromagnetic wave, in the form of a sine wave such as the one pictured in Figure 6.55,

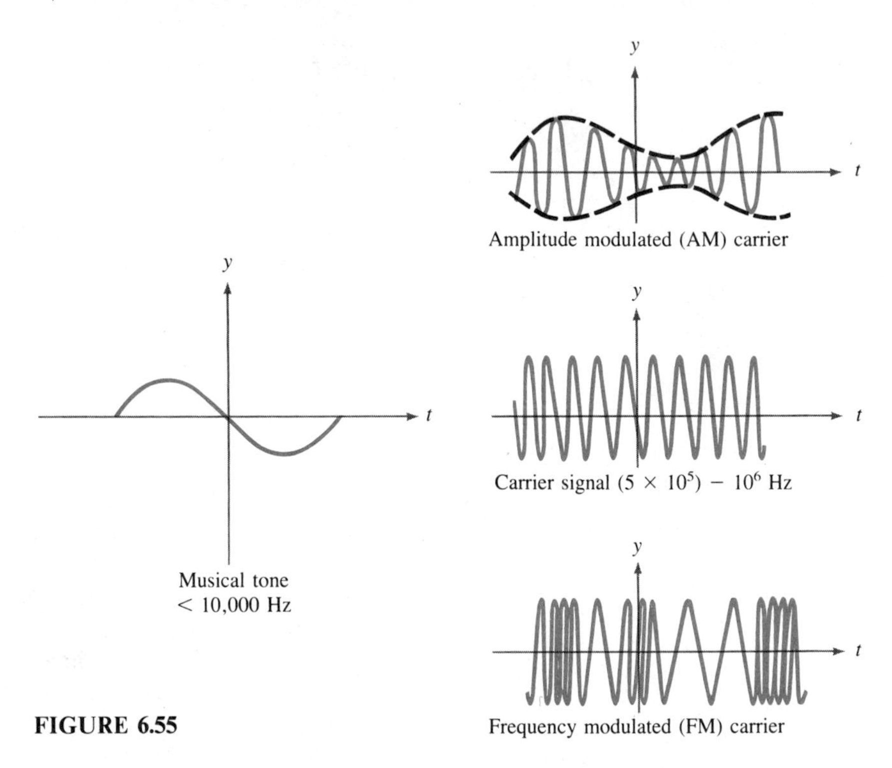

FIGURE 6.55

whose frequency is in the range of 5×10^5 to 10^8 hertz. This sine wave is called a **carrier wave.** The radio station uses the low frequency muscial tone to modify, or *modulate*, the carrier wave. When your radio receives this modulated carrier wave, it removes the carrier portion of the wave, and sends the remaining musical tone to the speakers. The speakers then turn this musical tone (which is an electromagnetic wave), into mechanical motions that create *sound waves.* As described earlier, these sound waves propagate *through the air* to your eardrums, and you hear the tone. This is the principle behind radio transmission.

The frequency of the carrier wave is what you know as the frequency of the radio station. If you look on the dial of a radio, you will see these frequencies. The AM dial has frequencies that range from about 530 to 1600 kHz (1 kHz = 1 *kilohertz* = 1000 Hz) and the FM dial has frequencies that range from about 88 to 108 MHz (1 MHz = 1 *megahertz* = 10^6 hertz.)

Actually, AM radio stations modulate (that is, modify) the carrier wave in a different manner than FM stations. AM radio stations use the musical tone (or tones) to modify the *amplitude* of the carrier wave, as shown in Figure 6.55. On the other hand, FM stations use the musical tone (or tones) to modify the *frequency* of the carrier wave, as shown in Figure 6.55. In fact, this is precisely where the radio stations get their names. The term AM is an abbreviation for *Amplitude Modulation*, and the term FM is an abbreviation for *Frequency Modulation.*

From this discussion, you can see why the concepts of amplitude and frequency (or period) are so important, and next time you hear a radio station sign off the "air," you will not only understand what they mean when they announce their *carrier frequency*, but you will also know that the announcer really means to sign "off the electromagnetic field," rather than "off the air."

Ideas to Remember

Sine waves play a very important role in nature. For example, the motion of both light and sound can be described by means of sine waves.

EXERCISES

In Exercises 1–24, graph the given function in accordance with the procedure discussed in this section. Also, identify the amplitude, period, and frequency of the wave.

1. $f(t) = \sin 3t$ **2.** $f(t) = \cos t/2$

3. $f(t) = \pi \sin \pi t$ **4.** $f(t) = \pi \cos \dfrac{2t}{\pi}$

5. $f(t) = \sin\left(-\dfrac{t}{4}\right)$ **6.** $f(t) = \cos(-\pi t)$

7. $f(t) = -2\sin(-2t)$ **8.** $f(t) = 3\cos\left(-\dfrac{t}{3}\right)$

9. $f(t) = \sin(t - \pi)$ **10.** $f(t) = \cos\left(t - \dfrac{\pi}{3}\right)$

11. $f(t) = -2\sin\left(t + \dfrac{\pi}{2}\right)$ **12.** $f(t) = -\cos\left(t + \dfrac{\pi}{4}\right)$

13. $f(t) = 4\sin\left(2t - \dfrac{\pi}{4}\right)$ **14.** $f(t) = \left(\dfrac{1}{4}\right)\cos\left(2t - \dfrac{\pi}{2}\right)$

15. $f(t) = 3\sin\left(\dfrac{\pi}{4} - t\right)$ **16.** $f(t) = -\pi\cos\left(\dfrac{\pi}{6} - t\right)$

17. $f(t) = \pi\sin\left(\dfrac{\pi}{4} - \dfrac{t}{2}\right)$ **18.** $f(t) = -5\cos(\pi - 2\pi t)$

19. $f(t) = 5\sin\left(2\pi - \dfrac{t}{4}\right)$ **20.** $f(t) = 10\cos\left(\dfrac{\pi}{6} - \dfrac{t}{7}\right)$

21. $f(t) = 1 + 2\sin t$ **22.** $f(t) = -1 + 3\cos\left(t - \dfrac{\pi}{2}\right)$

23. $f(t) = \pi - \sin(2t - \pi)$ **24.** $f(t) = -\dfrac{\pi}{2} - \pi\cos\left(\dfrac{t}{2} - \pi\right)$

25. Ordinary household current is another example of a natural phenomenon that travels in sine waves (at a frequency of 60 Hz.) Suppose that a certain electric circuit has an input voltage given by $v_i(t) = 6\sin 2t$ and an output voltage given by $v_0(t) = 3\sqrt{2}\sin(2t - \pi/4)$. Graph both of these sine waves on the same plane and justify the statement that the output voltage *lags* behind the input voltage by 22.5°.

26. The following figure shows a pendulum consisting of a mass m suspended from a fixed point by a wire of length

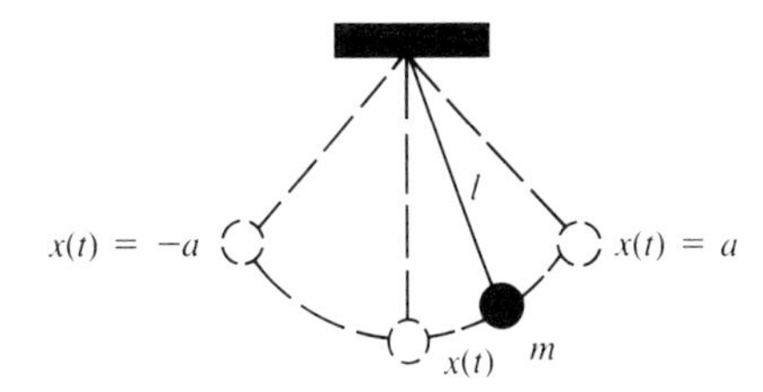

l. Suppose that the mass is pulled to one side and released at time $t = 0$. Then the displacement of the mass from its rest position, as measured along the arc of its swing, is given (approximately) by the formula $x(t) = a\cos bt$, where a is the displacement of the mass at time $t = 0$. In this situation, the period of the wave is given by

$$\text{period} = 2\pi\sqrt{\frac{l}{g}}$$

where g is the acceleration due to gravity (which at the surface of the earth is approximately 32 feet/sec^2.)

(a) Find a formula for b in terms of l and g.

(b) Sketch the graph of $x(t)$ if $a = 0.5$ feet and $l = 12$ feet.

27. The following figure shows a mass attached to the end of a spring. Suppose that the mass is at rest when its vertical position is $y = k$. If the mass is moved to the position $y = k + a$, and then released at time $t = 0$, then it can be shown that, neglecting friction and air resistance, the mass will oscillate up and down, and its position will be given by the equation $y(t) = k + a\cos bt$, where the coefficient b depends on the mass and the strength of the spring. Sketch the graph of y if $k = 3$, $a = 2$, and $b = 3/(2\pi)$.

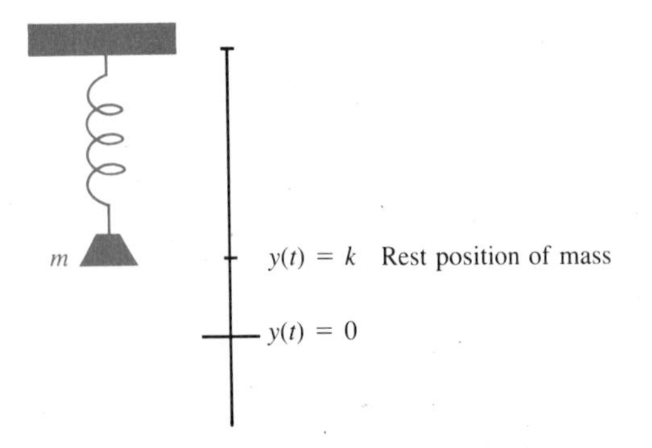

28. Graph the function $f(t) = 2^{-t}\sin t$. *Hint:* Use the fact that 2^{-t} is always positive to show that

$$-2^{-t} \leq 2^{-t}\sin t \leq 2^{-t}$$

This shows that the graph of f lies between or on the graphs of $g(t) = -2^{-t}$ and $h(t) = 2^{-t}$. First sketch these two graphs (on the same plane.) Then determine where the graph of f intersects the t-axis and the graphs of g and h. Why do you think the graph of f is called a **damped sine wave,** and 2^{-t} is called the **damping factor?**

6.7 The Inverse Trigonometric Functions

Let us begin this section by recalling a few facts from Section 2.8 concerning the inverse of a function. As you may recall from that section, the inverse of a function f, denoted by f^{-1}, is defined to be that function which satisfies

$$f^{-1}(f(x)) = x$$

for all x in the domain of f, and

$$f(f^{-1}(x)) = x$$

for all x in the domain of f^{-1}. Another way to characterize the inverse of a function is by the statement

$$y = f^{-1}(x) \qquad \textbf{if and only if} \qquad f(y) = x \tag{6.20}$$

Recall also that not all functions have inverses. In fact, a function has an inverse if and only if it is one-to-one, and a function is one-to-one if and only if its graph passes the horizontal line test (Section 2.5).

Now, it is easy to see that none of the graphs of the trigonometric functions pass the horizontal line test, and so none of the trigonometric functions are one-to-one. Therefore, none of the trigonometric functions have inverses.

Fortunately however, we can rectify this situation in a very simple way; namely, we restrict the *domain* of each of the trigonometric functions so that it becomes one-to-one. Then each of these "restricted" trigonometric functions will have an inverse.

Let us begin with the sine function, whose graph is shown in Figure 6.56. If we restrict the domain of the sine function to the closed interval $[-\pi/2, \pi/2]$, we get the graph shown in Figure 6.57. It is easy to see that

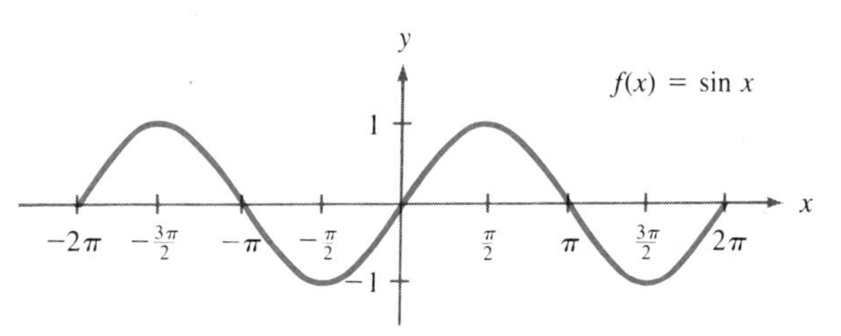

FIGURE 6.56 The sine function

this graph does pass the horizontal line test. (Other restrictions are possible, but this one is accepted by the entire mathematical and scientific community, and so we will use it here.)

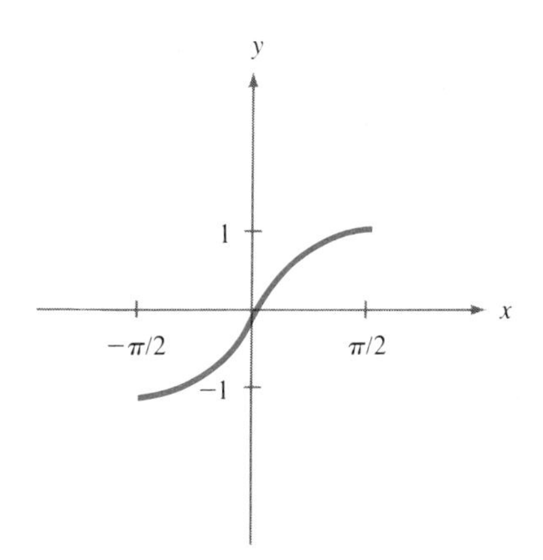

The *restricted* sine function

FIGURE 6.57

Strictly speaking, by restricting the domain of the sine function, we are actually defining a new function. For this reason, we will refer to the function $f(x) = \sin x$ pictured in Figure 6.57, whose domain is the interval $[-\pi/2, \pi/2]$, as the **restricted sine function.** (In this section, we use x to represent the independent variable, rather than t.)

Since the graph of the restricted sine function passes the horizontal line test, this function has an inverse, called the **inverse sine function** and denoted by

$$f^{-1}(x) = \sin^{-1} x$$

($\sin^{-1} x$ is read "the inverse sine of x.") The inverse sine function is also called the **arcsine function,** and denoted by

$$f^{-1}(x) = \arcsin x$$

($\arcsin x$ is read "arcsine of x.") Since both of these terms are very common, we will use them both.

The graph of the inverse sine function can be obtained by reflecting the graph of the restricted sine function over the diagonal line $y = x$. This is done in Figure 6.58. As you can see from this graph, *the domain of the inverse sine function is the interval* $[-1, 1]$ *and its range is the interval* $[-\pi/2, \pi/2]$.

The inverse sine function

FIGURE 6.58

According to the definition of inverse, we have

$$\sin^{-1}(\sin x) = x \quad \text{for} \quad -\frac{\pi}{2} \le x \le \frac{\pi}{2}$$

and

$$\sin(\sin^{-1} x) = x \quad \text{for} \quad -1 \le x \le 1$$

In arcsine notation, this is

$$\arcsin(\sin x) = x \quad \text{for} \quad -\frac{\pi}{2} \le x \le \frac{\pi}{2}$$

and

$$\sin(\arcsin x) = x \quad \text{for} \quad -1 \le x \le 1$$

Also, Equation (6.20) takes the form

$$y = \sin^{-1} x \qquad \text{if and only if} \qquad \sin y = x$$

or

$$y = \arcsin x \qquad \text{if and only if} \qquad \sin y = x$$

$$\text{for} \qquad -1 \le x \le 1 \qquad \text{and} \qquad -\frac{\pi}{2} \le y \le \frac{\pi}{2}$$

As we will see in a moment, these relationships provide the easiest method for computing values of the inverse sine function. First, let us summarize our results in a definition.

DEFINITION

The function $f(x) = \sin x$, whose domain is the interval $[-\pi/2, \pi/2]$ is called the **restricted sine function.** This function has an inverse, known as the **inverse sine function,** or the **arcsine function,** and denoted by $f^{-1}(x) = \sin^{-1} x$ or $f^{-1}(x) = \arcsin x$. The domain of the inverse sine function is the interval $[-1, 1]$ and its range is the interval $[-\pi/2, \pi/2]$. (See Figure 6.58.) The inverse sine function can be characterized by the statement

$$y = \sin^{-1} x \qquad \text{if and only if} \qquad \sin y = x$$

or

$$y = \arcsin x \qquad \text{if and only if} \qquad \sin y = x \tag{6.21}$$

$$\text{for} \qquad -1 \le x \le 1 \qquad \text{and} \qquad -\frac{\pi}{2} \le y \le \frac{\pi}{2}.$$

EXAMPLE 1: Compute the following values.

(a) $\sin^{-1} \frac{1}{2}$ **(b)** $\arcsin\left(-\frac{\sqrt{3}}{2}\right)$ **(c)** $\arcsin(-2)$

Solutions:

(a) From the relationship (6.21), we have

$$y = \sin^{-1} \frac{1}{2} \qquad \text{if and only if} \qquad \sin y = \frac{1}{2}$$

where $-\pi/2 \le y \le \pi/2$. (This condition on y is *very* important, since *y must be in the domain of the restricted sine function.*) But the only value of y in this interval for which $\sin y = 1/2$ is $y = \pi/6$. (You may wish to consult Figure 6.20.) Therefore, we have $\sin^{-1}(1/2) = \pi/6$.

(b) In this case, (6.21) gives

$$y = \arcsin\left(-\frac{\sqrt{3}}{2}\right) \quad \text{if and only if} \quad \sin y = -\frac{\sqrt{3}}{2}$$

where $-\pi/2 \le y \le \pi/2$. But the only value of y in the interval $[-\pi/2, \pi/2]$ for which $\sin y = -\sqrt{3}/2$ is $y = -\pi/3$. (Again you may wish to consult Figure 6.20.) Thus $\arcsin(\sqrt{3}/2) = -\pi/3$.

(c) Since -2 is *not* in the domain of the arcsine function, arsin(-2) is *not defined*. (There is no value of x for which $\sin x = -2$.)

▶ **Study Suggestion 6.37:** Compute the following values without using a calculator.

(a) $\sin^{-1}(-1/2)$ **(b)** $\arcsin(\sqrt{2}/2)$ ◀

Try Study Suggestion 6.37. □

As the next example illustrates, when dealing with the inverse trigonometric functions, it is very important to keep in mind the domains and the ranges of the functions involved. We have not had this problem before, since we have always taken our domains to be the largest set of real numbers for which the particular formula used to define the function made sense. However, this is not the case with the restricted trigonometric functions.

EXAMPLE 2: Compute the following values.

(a) $\sin^{-1}\left(\sin\frac{\pi}{5}\right)$ **(b)** $\sin^{-1}\left(\sin\frac{3\pi}{4}\right)$

Solutions:

(a) Since $\pi/5$ is in the domain of the restricted sine function, we have

$$\sin^{-1}\left(\sin\frac{\pi}{5}\right) = \frac{\pi}{5}.$$

(b) Since $3\pi/4$ is *not* in the domain of the restricted sine function, it is *not* true that $\sin^{-1}(\sin 3\pi/4)$ is equal to $3\pi/4$. But since $\sin 3\pi/4 = \sin \pi/4$, and $\pi/4$ is in the domain of the restricted sine function, we have

$$\sin^{-1}\left(\sin\frac{3\pi}{4}\right) = \sin^{-1}\left(\sin\frac{\pi}{4}\right) = \frac{\pi}{4}$$

▶ **Study Suggestion 6.38:** Compute the following values without using a calculator.

(a) $\sin^{-1}(\sin(-\pi/7))$

(b) $\arcsin(\sin 5\pi/4)$

(c) $\sin(\arcsin 1/4)$ ◀

Try Study Suggestion 6.38. □

Since none of the trigonometric functions are one-to-one, we must restrict the domains of each of these functions in order to be able to define the inverse functions. Unfortunately however, we cannot simply restrict the domain of each trigonometric function to the interval $[-\pi/2, \pi/2]$.

For example, when we restrict the cosine function to the interval $[-\pi/2, \pi/2]$, we get the graph shown in Figure 6.59. Clearly, this graph does not pass the horizontal line test, and so this function does not have an inverse.

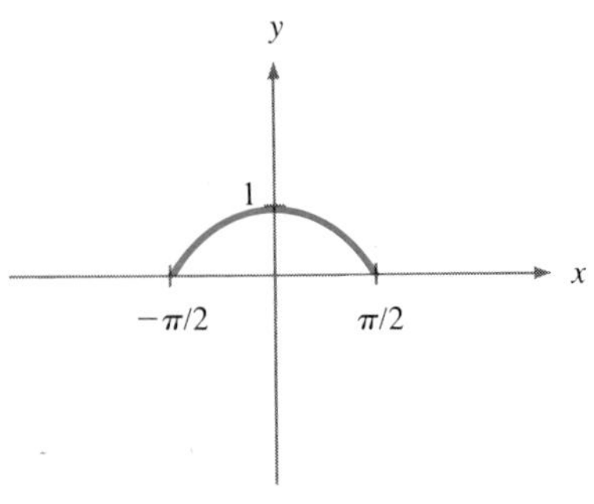

FIGURE 6.59 The cosine function restricted to the interval $[-\pi/2, \pi/2]$

In order to obtain a one-to-one function, we restrict the domain of the cosine function to the interval $[0, \pi]$, as shown in Figure 6.60. This gives us the restricted cosine function.

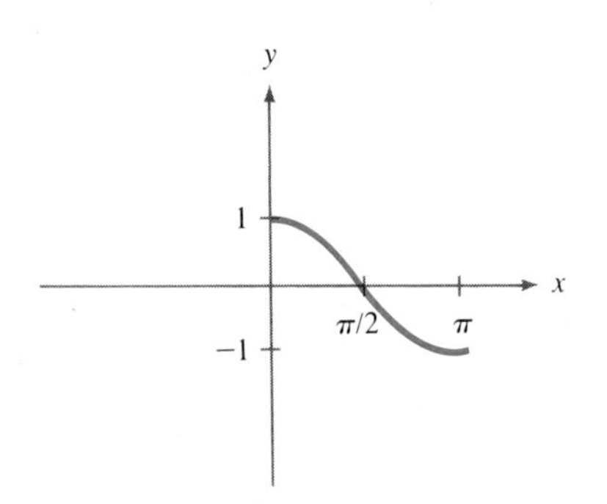

FIGURE 6.60 The *restricted* cosine function

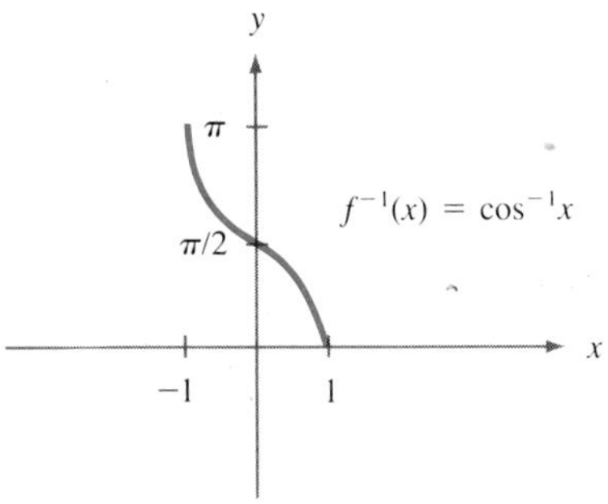

FIGURE 6.61 The inverse cosine function

DEFINITION

The function $f(x) = \cos x$, whose domain is the interval $[0, \pi]$ is called the **restricted cosine function.** This function has an inverse, known as the **inverse cosine function,** or the **arccosine function,** and denoted by $f^{-1}(x) = \cos^{-1} x$ or $f^{-1}(x) = \arccos x$. The domain of the inverse cosine function is the interval $[-1, 1]$ and its range is the interval $[0, \pi]$. The graph of the inverse cosine function is shown in Figure 6.61. The inverse cosine function can be characterized by the statement

$$y = \cos^{-1} x \quad \text{if and only if} \quad \cos y = x$$

or

$$y = \arccos x \quad \text{if and only if} \quad \cos y = x \qquad \textbf{(6.22)}$$

$$\text{for} \quad -1 \le x \le 1 \quad \text{and} \quad 0 \le y \le \pi$$

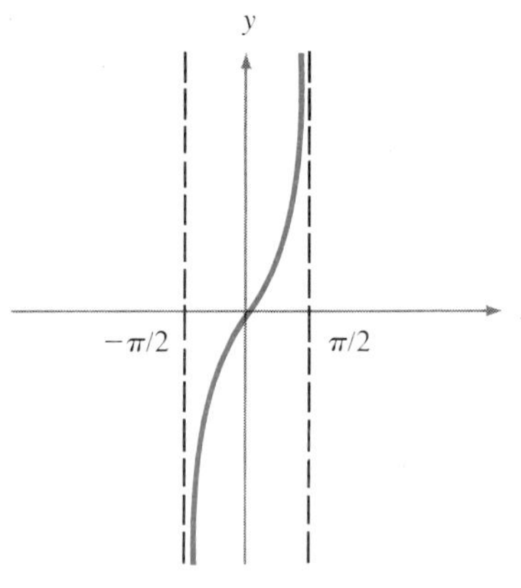

The *restricted* tangent function

FIGURE 6.62

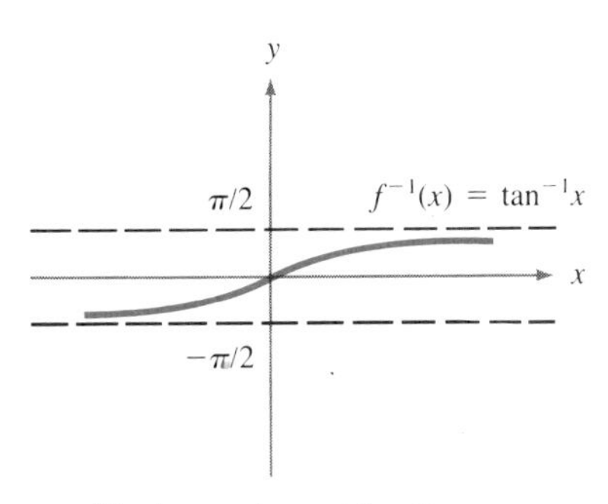

The inverse tangent function

FIGURE 6.63

EXAMPLE 3: Compute the following values.

(a) $\cos^{-1}\dfrac{\sqrt{2}}{2}$ **(b)** $\arccos\left(-\dfrac{\sqrt{3}}{2}\right)$

Solutions:

(a) According to (6.22), we have

$$y = \cos^{-1}\frac{\sqrt{2}}{2} \quad \text{if and only if} \quad \cos y = \frac{\sqrt{2}}{2}$$

where $0 \leq y \leq \pi$. But $y = \pi/4$ is the only number in the interval $[0, \pi]$ for which $\cos y = \sqrt{2}/2$. Therefore, $\cos^{-1}\sqrt{2}/2 = \pi/4$.

(b) Again using (6.22), we get

$$y = \arccos\left(-\frac{\sqrt{3}}{2}\right) \quad \text{if and only if} \quad \cos y = -\frac{\sqrt{3}}{2}$$

where $0 \leq y \leq \pi$. But $y = 5\pi/6$ is the only number in the interval $[0, \pi]$ for which $\cos y = -\sqrt{3}/2$. Hence, $\arccos(-\sqrt{3}/2) = 5\pi/6$.

Try Study Suggestion 6.39. □

Let us now consider the inverse tangent function. By restricting the domain of the tangent function to the *open* interval $(-\pi/2, \pi/2)$, we obtain the *restricted* tangent function, whose graph is shown in Figure 6.62. (Can you explain why we choose the open interval $(-\pi/2, \pi/2)$, rather than the closed interval $[-\pi/2, \pi/2]$?)

DEFINITION

The function $f(x) = \tan x$, whose domain is the *open* interval $(-\pi/2, \pi/2)$ is called the **restricted tangent function.** This function has an inverse, known as the **inverse tangent function,** or the **arctangent function,** and denoted by $f^{-1}(x) = \tan^{-1} x$ or $f^{-1}(x) = \arctan x$. The domain of the inverse tangent function is the set of *all* real numbers and its range is the interval $(-\pi/2, \pi/2)$. The graph of the inverse tangent function is shown in Figure 6.63. The inverse tangent function can be characterized by the statement

$$y = \tan^{-1} x \quad \text{if and only if} \quad \tan y = x$$

or

$$y = \arctan x \quad \text{if and only if} \quad \tan y = x \tag{6.23}$$

for $-\pi/2 < y < \pi/2$. (There are no restrictions on x.)

▶ **Study Suggestion 6.39:** Compute the following values without using a calculator. **(a)** $\cos^{-1} 0$ **(b)** $\arccos(-\sqrt{2}/2)$ ◀

EXAMPLE 4: Compute the following values.

(a) $\tan^{-1} 1$ **(b)** $\arctan\sqrt{3}$

Solutions:

(a) Using (6.23), we get

$$y = \tan^{-1} 1 \qquad \text{if and only if} \qquad \tan y = 1$$

where $-\pi/2 < y < \pi/2$. Now, $\tan y = 1$ if and only if $\sin y = \cos y$, and the only value of y in the interval $(-\pi/2, \pi/2)$ for which $\sin y = \cos y$ is $y = \pi/4$. Therefore, $\tan^{-1} 1 = \pi/4$.

(b) Again using (6.23), we get

$$y = \arctan \sqrt{3} \qquad \text{if and only if} \qquad \tan y = \sqrt{3}$$

where $-\pi/2 < y < \pi/2$. But the only value of y in the interval $(-\pi/2, \pi/2)$ for which $\tan y = \sqrt{3}$ is $y = \pi/3$. (Recall that $\sin \pi/3 = \sqrt{3}/2$ and $\cos \pi/3 = 1/2$. Hence, $\tan \pi/3 = (\sqrt{3}/2)/(1/2) = \sqrt{3}$.) Thus $\arctan \sqrt{3} = \pi/3$.

A scientific calculator can be used to approximate values of the inverse sine, cosine, and tangent functions. In fact, some scientific calculators have keys marked *sin*$^{-1}$, *cos*$^{-1}$, and *tan*$^{-1}$, whereas others require that you first push a key marked *inv*, or *arc*, and then one of the keys *sin*, *cos*, or *tan*. In any case, you should spend a few moments becoming familiar with the approach used by your calculator.

▶ **Study Suggestion 6.40:** Use a calculator to approximate the following values. **(a)** $\sin^{-1} 0.221$ **(b)** $\arccos(-1/5)$ **(c)** $\tan^{-1} 5$ ◀

Try Study Suggestion 6.40.

It often happens in applications that we need to evaluate an expression of the form $\sin(\arccos 1/4)$, or $\sec(\arctan(-5))$. The following examples will illustrate how this can be done.

EXAMPLE 5: Evaluate the expression $\sin(\arccos 1/4)$.

Solution: Since we know how to evaluate $\cos(\arccos 1/4)$, this gives us the idea of using one of the Pythagorean identities, in the form

$$\sin^2 x = 1 - \cos^2 x$$

Substituting $\arccos 1/4$ for x in this identity, and using the fact that $\cos(\arccos 1/4) = 1/4$, we get

$$\begin{aligned} \sin^2\left(\arccos \frac{1}{4}\right) &= 1 - \cos^2\left(\arccos \frac{1}{4}\right) \\ &= 1 - \left(\frac{1}{4}\right)^2 \\ &= \frac{15}{16} \end{aligned}$$

Taking square roots gives

$$\left|\sin\left(\arccos \frac{1}{4}\right)\right| = \frac{\sqrt{15}}{4}$$

In order to determine the sign of sin(arccos 1/4), we recall that the range of the arccosine function is the interval $[0, \pi]$, and so arccos 1/4 lies in this interval. But the sine of *any* number in the interval $[0, \pi]$ is positive. In particular, sin(arccos 1/4) is positive, and so sin(arccos 1/4) $= \sqrt{15}/4$.

▶ **Study Suggestion 6.41:** Evaluate the expression cos(arcsin 1/3). ◀

Try Study Suggestion 6.41. □

EXAMPLE 6: Evaluate the expression sec(arctan(−5)).

Solution: This example is similar to the previous one, but in this case we use the Pythagorean identity

$$\sec^2 x = 1 + \tan^2 x$$

Substituting $x = \arctan(-5)$, and using the fact that $\tan(\arctan(-5)) = -5$ gives

$$\begin{aligned}\sec^2(\arctan(-5)) &= 1 + \tan^2(\arctan(-5)) \\ &= 1 + (-5)^2 \\ &= 26\end{aligned}$$

and so

$$|\sec(\arctan(-5))| = \sqrt{26}$$

The sign of sec(arctan(−5)) can be determined by recalling that the range of the arctangent function is $(-\pi/2, \pi/2)$, and so arctan(−5) is in this interval. But the secant of *any* number in the interval $(-\pi/2, \pi/2)$ is positive, and so sec(arctan(−5)) $= \sqrt{26}$. *Try Study Suggestion 6.42.* □

▶ **Study Suggestion 6.42:** Evaluate the expression sec(arctan 3). ◀

EXAMPLE 7: Evaluate the expression $\csc^3(\sin^{-1} 0.1)$.

Solution: In this case, we do not need to use any special identities—only the definition of the cosecant function.

$$\begin{aligned}\csc^3(\sin^{-1} 0.1) &= \frac{1}{\sin^3(\sin^{-1} 0.1)} \\ &= \frac{1}{(0.1)^3} \\ &= \frac{1}{0.001} \\ &= 1000\end{aligned}$$

Try Study Suggestion 6.43. □

▶ **Study Suggestion 6.43:** Evaluate the expression $\sec^{1/2}(\cos^{-1} 0.04)$. ◀

The inverse secant, cosecant, and cotangent may be defined by restricting the domains of the secant, cosecant, and cotangent functions, respectively. However, since these functions are much less frequently used than the inverse sine, cosine, and tangent, we will discuss them only in the exercises. (See Exercises 59–61.)

Ideas to Remember

By suitably restricting the domains of each of the trigonometric functions, we obtain functions that are one-to-one. Hence, these *restricted* trigonometric functions have inverses. This is how the inverse trigonometric functions are obtained.

EXERCISES

In Exercises 1–9, complete the rows of the table. Do not use a calculator.

	x	$\sin^{-1} x$	$\cos^{-1} x$
1.	0		
2.	1/2		
3.	$-1/2$		
4.	$\sqrt{2}/2$		
5.	$-\sqrt{2}/2$		
6.	$\sqrt{3}/2$		
7.	$-\sqrt{3}/2$		
8.	1		
9.	-1		

In Exercises 10–16, find the exact value of the given expression. Do not use a calculator.

10. $\arcsin 1/\sqrt{2}$

11. $\arccos(-1/\sqrt{2})$

12. $\tan^{-1} 0$

13. $\tan^{-1} 1$

14. $\arctan \sqrt{3}$

15. $\arctan(-1)$

16. $\arctan(-\sqrt{3}/3)$

In Exercises 17–30, use a calculator to approximate the given value.

17. $\tan^{-1} \sqrt{3}/2$

18. $\sin^{-1} 0.3$

19. $\cos^{-1} 1.1$

20. $\tan^{-1} 12$

21. $\arctan(-\sqrt{2})$

22. $\arccos 1/5$

23. $\tan^{-1} 2.341$

24. $\sin^{-1} 0.0727$

25. $\arcsin 0.3261$

26. $\arccos(-0.5528)$

27. $\arctan 12.7321$

28. $\sin^{-1}(-0.7264)$

29. $\cos^{-1}(0.00001)$

30. $\tan^{-1}(-0.0002)$

In Exercises 31–44, compute the indicated value without using a calculator.

31. $\sin(\arcsin 0.321)$

32. $\cos(\sin^{-1} 1/3)$

33. $\sin(\arccos 1/4)$

34. $\tan^2(\tan^{-1} 5)$

35. $\sin^{-1}(\sin(-\pi/10))$

36. $\sin^{-1}(\sin 2\pi/3)$

37. $\cos^{-1}(\cos \pi/5)$

38. $\arccos(\cos(-\pi/5))$

39. $\tan(\arccos 1/2)$

40. $\tan(\sin^{-1} 1/6)$

41. $\sec(\tan^{-1} 2)$

42. $\cot(\tan^{-1} 12)$

43. $\csc(\arcsin 12)$

44. $2\sin^2(\arccos(-1/10))$

45. Is it true that

$$\arcsin x + \arcsin(-x) = 0?$$

Justify your answer.

46. Is it true that

$$\sin^{-1} x = \frac{1}{\sin x}?$$

Justify your answer.

In Exercises 47–56, graph the indicated function.

47. $f(x) = 2\sin^{-1} x$

48. $f(x) = 1 + \cos^{-1} x$

49. $f(x) = \arcsin 2x$

50. $f(x) = \pi \arccos \pi x$

51. $f(x) = 2\tan^{-1} \frac{x}{2}$

52. $f(x) = -\tan^{-1} \frac{x}{2}$

53. $f(x) = \sin^{-1}(x + 1)$

54. $f(x) = \frac{\pi}{2} + \tan^{-1}(x - 1)$

55. $f(x) = \sin(\arcsin x)$

56. $f(x) = \arcsin(\sin x)$

57. Show that $\cos(\sin^{-1} x) = \sqrt{1 - x^2}$ for $-1 \le x \le 1$.

58. Show that $\sec(\arctan x) = \sqrt{1 + x^2}$ for all real numbers x.

59. The **restricted cotangent function** is the function $f(x) = \cot x$ whose domain is the open interval $(0, \pi)$. Is this function one-to-one? Justify your answer. Use this function to define the inverse cotangent function $f^{-1}(x) = \cot^{-1} x$. Graph the function $f^{-1}(x)$.

60. **(a)** One common definition of the **restricted secant function** is the function $f(x) = \sec x$ whose domain is the set $[0, \pi/2) \cup (\pi/2, \pi]$. Is this function one-to-one? Justify your answer. Use this function to define the inverse secant function $f^{-1}(x) = \sec^{-1} x$. Graph the function $f^{-1}(x)$.

(b) Another common definition of the **restricted secant function** is the function $f(x) = \sec x$ whose domain is the set $[0, \pi/2) \cup [\pi, 3\pi/2)$. Is this function one-to-one? Justify your answer. Use this function to define the inverse secant function $f^{-1}(x) = \sec^{-1} x$. Graph the function $f^{-1}(x)$.

61. One choice for the **restricted cosecant function** is the function $f(x) = \csc x$ whose domain is the set $[-\pi/2, 0) \cup (0, \pi/2]$. Is this function one-to-one? Justify your answer. Use this function to define the inverse cosecant function $f^{-1}(x) = \csc^{-1} x$. Graph the function $f^{-1}(x)$.

62. The following figure shows a ray of light passing through a prism. If the ray of light enters the prism, making an angle θ with the perpendicular to the surface of the prism, then the ray is deflected by an angle μ from its original direction. It can be shown that the angle μ is given by the formula

$$\mu = \theta - \alpha + \arcsin\left[n \sin\left(\alpha - \arcsin\frac{\sin\theta}{n}\right)\right]$$

where n is a constant known as the **index of refraction.** This number depends on the composition of the prism and the wavelength (that is, the color) of the light. Also, all angles are measured in radians.

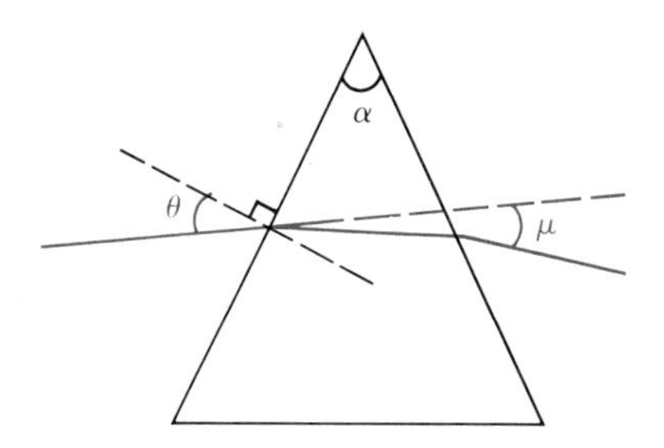

Compute the angle μ of deflection of the prism in each of the following cases.

(a) glass prism: $n = 1.75$, $\alpha = \pi/6$, $\theta = \pi/7$

(b) ice prism: $n = 1.31$, $\alpha = 45°$, $\theta = 15°$

6.8 Review

CONCEPTS FOR REVIEW

Ray
Endpoint
Angle
Vertex
Initial side
Terminal side
Positive angle
Negative angle
Full angle
Straight angle
Right angle
Standard position
Central angle
Subtend an arc
Degree
Unit circle
Radian
Sweeping out a central angle
Sine function
Cosine function
Reference number
The Reference Number Theorem
Tangent function
Secant function

Cosecant function
Cotangent function
Periodic function
Pythagorean identity
Sine wave
Cycle of a sine wave
Period
Amplitude
Crest
Trough
Node
Phase shift
Restricted sine function
Inverse sine function
Arcsine function
Restricted cosine function
Inverse cosine function
Arccosine function
Restricted tangent function
Inverse tangent function
Arctangent function

REVIEW EXERCISES

1. Find the coordinates of the point $P(-5\pi/3)$.

2. Find the coordinates of the point $P(19\pi/6)$.

3. Find the approximate position of the point $P(10)$ on the unit circle.

4. Find the approximate position of the point $P(-1)$ on the unit circle.

5. Find the coordinates of the given point if $P(t) = (1/3, 2\sqrt{2}/3)$.

(a) $P(-t)$ **(b)** $P(t + \pi)$ **(c)** $P(\pi - t)$

6. If $P(t) = (a, b)$, what are the coordinates of $P(\pi - t)$?

7. If $P(t) = (a, b)$, what are the coordinates of $P(t + \pi/2)$?

In Exercises 8–16, convert degree measure to radian measure, or vice-versa. Do *not* use a calculator.

8. $140°$ **9.** $-150°$ **10.** $570°$

11. $144°$ **12.** $-\pi/6$ **13.** $\pi/24$

14. $3\pi/8$ **15.** 1264π **16.** π^2

17. Find the degree measure and the radian measure of a central angle that subtends an arc of length 5 inches on a circle of diameter 1 foot. (Watch the units!)

18. A truck has tires of radius 24 inches. How many revolutions do the tires make if the truck moves 1 mile? (Watch the units!)

In Exercises 19–32, compute the indicated value. Do not use a calculator.

19. $\sin 17\pi/6$ **20.** $\sin(-121\pi/6)$

21. $\cos(-11\pi/4)$ **22.** $\cos(14\pi/3)$

23. $\tan 4\pi/3$ **24.** $\tan(-13\pi/2)$

25. $\sec(-5\pi)$ **26.** $\sec(-7\pi/4)$

27. $\csc \pi$ **28.** $\csc(-11\pi/4)$

29. $\cot(1728\pi + \pi/3)$ **30.** $\cot(\tan 0)$

31. $\cot(\pi \tan \pi)$ **32.** $\sec[(\pi/4)\sin \pi/2]$

In Exercises 33–40, use a calculator to approximate the given values.

33. $\sin(-0.4456)$ **34.** $\cos 12.449\pi$

35. $\tan(-17/3)$ **36.** $\cot \pi/11$

37. $\sec 17\pi/3$ **38.** $\csc(-0.0001)$

39. $\sin(\cos \pi/5)$

40. $\tan(\sin 2.3\pi) - \cot(\sec \pi/7)$

In Exercises 41–46, use the given information to compute the values of the six trigonometric functions at t.

41. $\sin t = 1/10$, $\cos t < 0$

42. $\cos t = -1/2$, $\sin^3 t > 0$

43. $\tan t = 10$, $\sec t < 0$

44. $\sec t = -5$, $\tan t > 0$

45. $\csc t = -1.55$, $P(t)$ lies in the fourth quadrant

46. $\cot t = 1/2$, $\sec t + \csc t < 0$

In Exercises 47–58, graph the given function.

47. $f(t) = \sin 3t$ **48.** $f(t) = 2\cos\left(-\frac{t}{2}\right)$

49. $f(t) = -3\sin(3t + \pi)$

50. $f(t) = \frac{1}{2}\cos\left(\frac{\pi}{3} - t\right)$

51. $f(t) = 4\tan\left(t - \frac{\pi}{3}\right)$

52. $f(t) = \cot\left(\frac{\pi}{2} - t\right)$

53. $f(t) = -2\sec(\pi - t)$

54. $f(t) = t - \sin t$

55. $f(t) = 2\sin t + \cos t$

56. $f(t) = |t| + \sin t$

57. $f(t) = |t|\sin t$

58. $f(t) = \log t + \sin t$

59. (a) Use a calculator to evaluate the function

$$f(t) = \frac{1 - \cos t}{t}$$

at $t = 0.9$, 0.7, 0.5, 0.3, and 0.1. What do you conclude about the behavior of this function as t approaches 0?

(b) Graph the function in Part (a) for $-\pi/2 \le t \le \pi/2$.

In Exercises 60–69, find the exact value of the given expression, if it exists. Do not use a calculator.

60. $\arccos\frac{1}{\sqrt{2}}$

61. $\sin^{-1}\left(-\frac{1}{2}\right)$

62. $\arctan\left(-\frac{\sqrt{3}}{3}\right)$

63. $\sin^{-1}\left(\sin\frac{7\pi}{3}\right)$

64. $\arccos\left(\cos\left(-\frac{\pi}{5}\right)\right)$

65. $\sin\left(\sin^{-1}\frac{12\pi}{5}\right)$

66. $\cos(\arccos 12)$

67. $\tan(\arctan 19)$

68. $\cos\left(\sin^{-1}\frac{1}{7}\right)$

69. $\sec(\arctan 2)$

In Exercises 70–75, use a calculator to approximate the given value, if it exists.

70. $\cos^{-1}0.9171$

71. $\sin^{-1}(-2)$

72. $\arcsin 0.9171$

73. $\arctan 13$

74. $\sin(\arccos 1)$

75. $\tan^2\left(\arccos\frac{\pi}{5}\right)$

76. Write the expression $\tan(\arccos x)$ without using trigonometric functions.

77. In certain applications, the polynomial function

$$p(t) = t - \frac{t^3}{6} + \frac{t^5}{120}$$

is used to *approximate* the sine function, for values of t close to 0. (You will learn more about this if you take calculus.) Use a calculator to compute the difference $\sin t - p(t)$ for the following values of t.

(a) $t = \pi$ **(b)** $t = 1$ **(c)** $t = 0.5$

(d) $t = 0.1$ **(e)** $t = 0.05$

78. In certain applications, the polynomial function

$$q(t) = 1 - \frac{t^2}{2} + \frac{t^4}{24}$$

is used to *approximate* the cosine function, for values of t close to 0. (You will learn more about this if you take calculus.) Use a calculator to compute the difference $\cos t - q(t)$ for the following values of t.

(a) $t = \pi$ **(b)** $t = 1$ **(c)** $t = 0.5$

(d) $t = 0.1$ **(e)** $t = 0.05$

79. In certain applications, the polynomial function

$$r(t) = t + \frac{t^3}{3} + \frac{2t^5}{15} + \frac{17t^7}{315}$$

is used to *approximate* the tangent function, for values of t close to 0. (You will learn more about this if you take calculus.) Use a calculator to compute the difference $\tan t - r(t)$ for the following values of t.

(a) $t = 1$ **(b)** $t = 0.1$ **(c)** $t = 0.01$

7

TRIANGLE TRIGONOMETRY AND ITS APPLICATIONS

7.1 Right Triangle Trigonometry

In this chapter, we will discuss the trigonometric functions from the point of view of triangles, rather than circles. This approach has a great many useful applications.

As you know, the domain of the sine and cosine functions is the set of all real numbers. According to the definition, the first step in defining $\sin\theta$ or $\cos\theta$ (where θ is a real number) is to sweep out an angle of θ radians on a unit circle. But this leads us to the idea of defining the sine (or cosine) of the *angle* A to be the sine (or cosine) of the radian measure of that angle. In symbols, if A is an angle of θ radians, then we let

$$\sin A = \sin\theta \qquad \text{and} \qquad \cos A = \cos\theta$$

This applies to *any* angle, regardless of whether or not it is drawn as the central angle of a unit circle.

As an example, consider the angle A shown in Figure 7.1. This angle measures $\pi/6$ radians, and so according to our definition, we have

$$\sin A = \sin\frac{\pi}{6} = \frac{1}{2}$$

and

$$\cos A = \cos\frac{\pi}{6} = \frac{\sqrt{3}}{2}$$

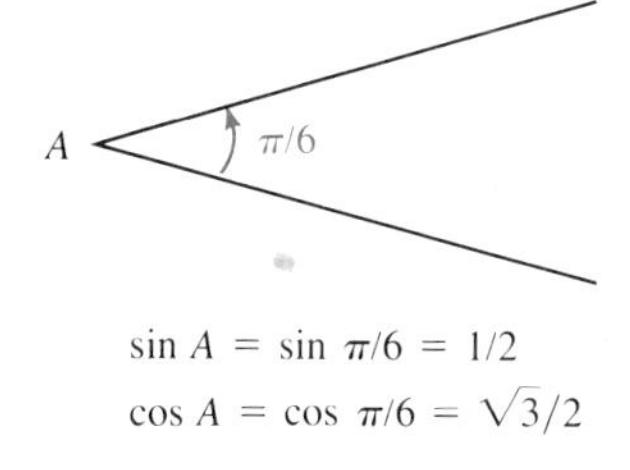

FIGURE 7.1

While we are "extending" the sine and cosine functions to angles, we may as well make another extension and define the sine and cosine of the *degree* measure of an angle. For example, if A is an angle of 30°, then we write

$$\sin A = \sin 30^\circ = \sin \frac{\pi}{6} = \frac{1}{2}$$

and

$$\cos A = \cos 30^\circ = \cos \frac{\pi}{6} = \frac{\sqrt{3}}{2}$$

Furthermore, there is no reason why we should restrict our attention to just the sine and cosine, and so we make the following definition.

DEFINITION

Let A be an angle of θ radians, and suppose that θ radians $= d$ degrees. Then

$$\sin A = \sin d^\circ = \sin \theta, \qquad \cos A = \cos d^\circ = \cos \theta$$
$$\tan A = \tan d^\circ = \tan \theta, \qquad \sec A = \sec d^\circ = \sec \theta$$
$$\csc A = \csc d^\circ = \csc \theta, \qquad \cot A = \cot d^\circ = \cot \theta$$

Notice that, according to this definition, in order to determine the sine of an angle that is given in degrees, we must first convert to radian measure.

EXAMPLE 1: Compute the following values.

(a) $\sin 60^\circ$ **(b)** $\cos(-90^\circ)$ **(c)** $\tan 120^\circ$ **(d)** $\sec 585^\circ$

Solutions:

(a) Since $60^\circ = \frac{\pi}{3}$ radians, we have

$$\sin 60^\circ = \sin \frac{\pi}{3} = \frac{\sqrt{3}}{2}$$

(b) Since $-90^\circ = \frac{-\pi}{2}$ radians, we have

$$\cos(-90^\circ) = \cos\left(-\frac{\pi}{2}\right) = 0$$

(c) Since $120^\circ = \frac{2\pi}{3}$ radians, we have

$$\tan 120^\circ = \tan \frac{2\pi}{3} = -\tan \frac{\pi}{3} = -\sqrt{3}$$

▶ **Study Suggestion 7.1:** Compute the following values. Do not use a calculator. **(a)** $\sin 45^\circ$ **(b)** $\cos(-30^\circ)$ **(c)** $\tan 135^\circ$ **(d)** $\sec(-45^\circ)$ **(e)** $\csc 300^\circ$ **(f)** $\cot(-90^\circ)$ ◀

(d) Since $585^\circ = \dfrac{13\pi}{4}$, we have

$$\sec 585^\circ = \sec \frac{13\pi}{4} = \sec \frac{5\pi}{4} = -\sec \frac{\pi}{4} = -\sqrt{2}$$

Try Study Suggestion 7.1. □

Most scientific calculators have a *degree mode* for computing (or approximating) the trigonometric functions of an angle that is measured in degrees. For example, when sin 30 is keyed into a calculator that is set in degree mode, the result will be (an approximation for) $\sin 30^\circ$. On the other hand, if sin 30 is keyed in while the calculator is in *radian* mode, the result will be (an approximation for) the sine of 30 radians!

Study Suggestion 7.2:
(a) Set your calculator to degree mode and key in sin 45. Now set your calculator to radian mode and key in $\sin \pi/4$. Do you get the same result?
(b) Set your calculator to degree mode and key in $\sin \pi/4$. What does the result represent?
(c) Set your calculator to radian mode and key in sin 45. What does the result represent? ◀

Try Study Suggestions 7.2 and 7.3.

Study Suggestion 7.3: Use a calculator to approximate the following values. **(a)** $\sin 30^\circ$ **(b)** $\cos(-20^\circ)$ **(c)** $\tan 175^\circ$ **(d)** $\sec 15^\circ$ **(e)** $\csc(-12.5^\circ)$ ◀

Now let us see how we can define the trigonometric functions using right triangles rather than circles. Consider the right triangle ABC shown in Figure 7.2. The radian measure of the angle A is denoted by θ, and the lengths of the sides are denoted by a, b, and c. Our plan is to define $\sin\theta$, $\cos\theta$, $\tan\theta$, and so on, in terms of the numbers a, b, and c.

We begin by placing the triangle in such a way that angle A is in standard position in the plane. This is done in Figure 7.3 for $b > 1$, and in Figure 7.4 for $b < 1$. We have also included a unit circle in these figures.

Notice that the hypotenuse of the triangle (or its extension in the case $b < 1$) intersects the unit circle at the point $P(\theta) = (\cos\theta, \sin\theta)$. We have also drawn a perpendicular from the point $P(\theta)$ to the x-axis, and labeled the point where it meets the x-axis with the letter D. Thus $ADP(\theta)$ is also a right triangle.

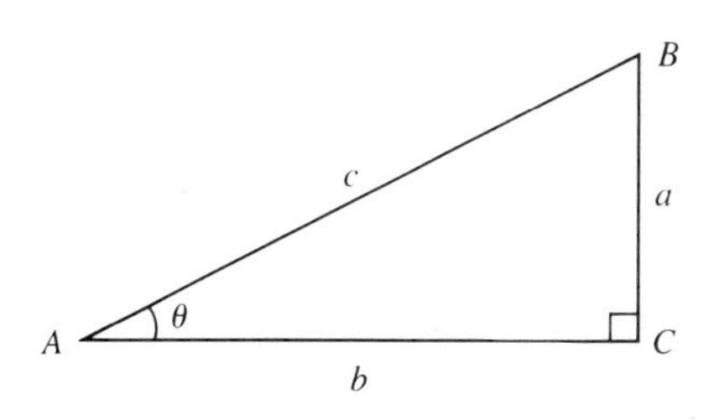

FIGURE 7.2

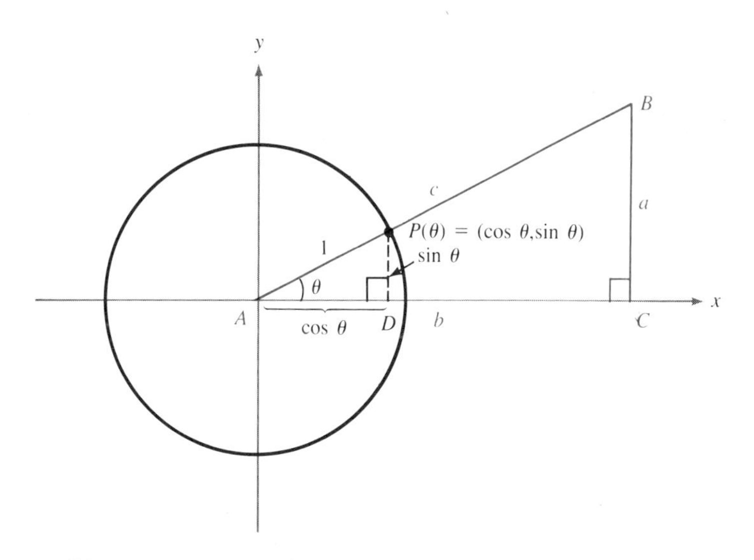

FIGURE 7.3

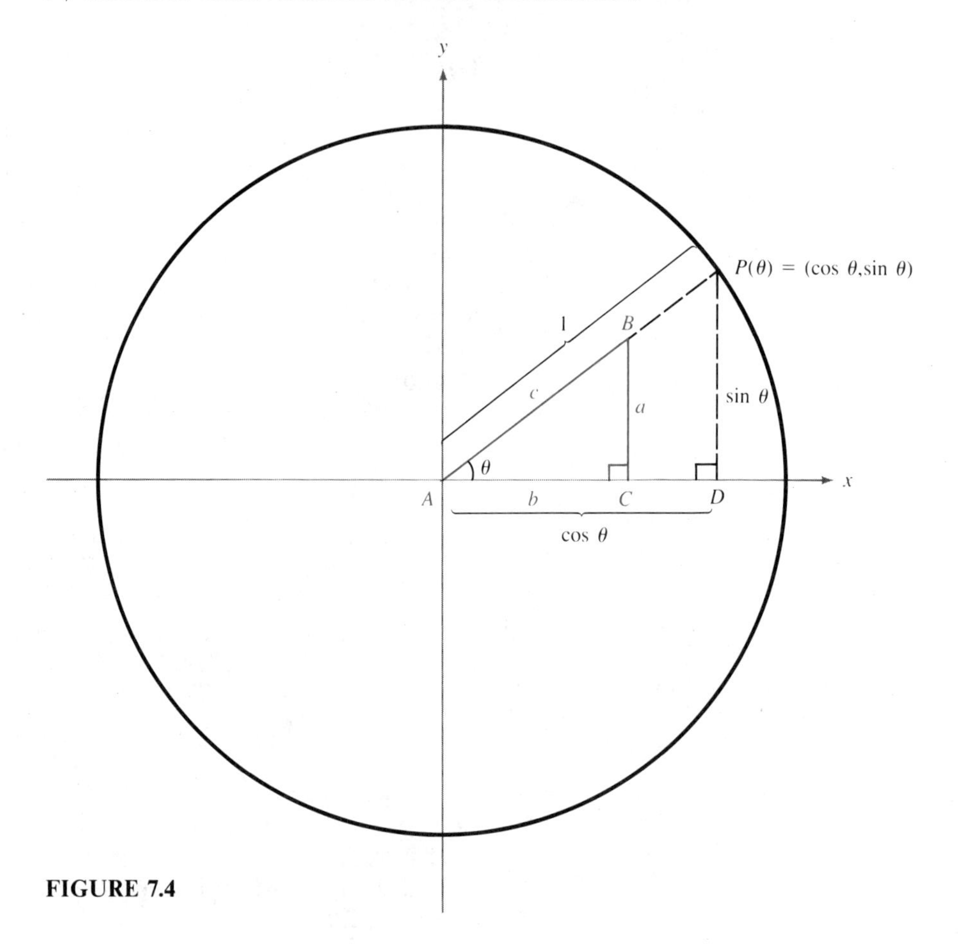

FIGURE 7.4

Now we make two simple observations about the triangles in this figure.

1. Since $P(\theta)$ has coordinates $\cos\theta$ and $\sin\theta$, the length of side AD is $\cos\theta$, and the length of side $DP(\theta)$ is $\sin\theta$. Also, since the circle has radius 1, the length of the hypotenuse $AP(\theta)$ is 1.
2. Triangles ABC and $AP(\theta)D$ are similar.

As you probably know, if two triangles are similar, the ratios of corresponding sides are equal. Thus, we have

$$\frac{\sin\theta}{a} = \frac{1}{c} \qquad \text{that is,} \qquad \sin\theta = \frac{a}{c}$$

and

$$\frac{\cos\theta}{b} = \frac{1}{c} \qquad \text{that is,} \qquad \cos\theta = \frac{b}{c}$$

From this, we get

$$\tan\theta = \frac{\sin\theta}{\cos\theta} = \frac{a/c}{b/c} = \frac{a}{b}$$

$$\cot\theta = \frac{\cos\theta}{\sin\theta} = \frac{b/c}{a/c} = \frac{b}{a}$$

$$\sec\theta = \frac{1}{\cos\theta} = \frac{1}{b/c} = \frac{c}{b}$$

and

$$\csc\theta = \frac{1}{\sin\theta} = \frac{1}{a/c} = \frac{c}{a}$$

These equations define $\sin\theta$, $\cos\theta$, $\tan\theta$, and so on, in terms of the numbers a, b, and c.

In order to help remember these equations, it is customary to refer to the *length* of side BC, which is the side opposite angle A, as the "side opposite," and the *length* of side AC, which is the side adjacent to angle A, as the "side adjacent." Using these terms, we can state the following theorem.

THEOREM

For the right triangle

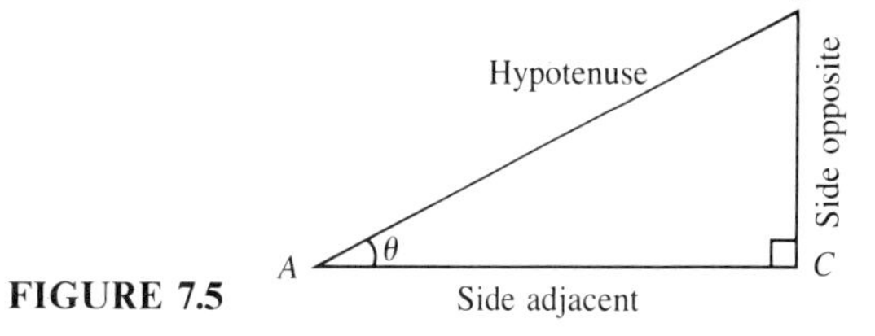

FIGURE 7.5

we have

$$\sin\theta = \frac{\text{side opposite}}{\text{hypotenuse}} \qquad \cos\theta = \frac{\text{side adjacent}}{\text{hypotenuse}}$$

$$\tan\theta = \frac{\text{side opposite}}{\text{side adjacent}} \qquad \cot\theta = \frac{\text{side adjacent}}{\text{side opposite}}$$

$$\sec\theta = \frac{\text{hypotenuse}}{\text{side adjacent}} \qquad \csc\theta = \frac{\text{hypotenuse}}{\text{side opposite}}$$

Let us see how we can use this theorem to compute values of the trigonometric functions.

EXAMPLE 2: Use the previous theorem to compute the following values.

(a) $\sin 45°$ and $\tan 45°$ **(b)** $\cos 60°$ and $\csc 60°$

Solutions:

(a) We begin by drawing the right triangle

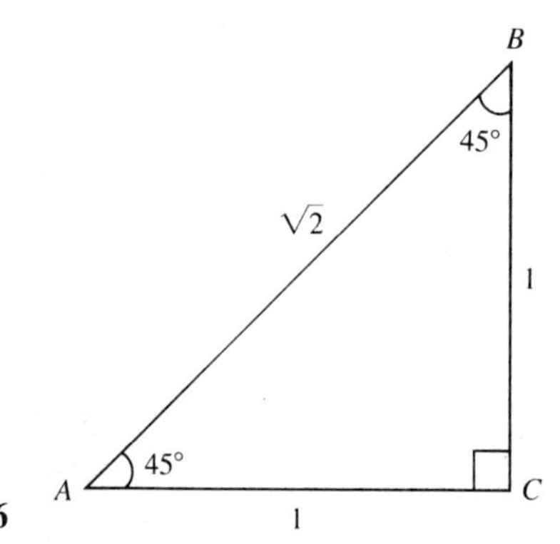

FIGURE 7.6

where the length of the hypotenuse was obtained from the lengths of the sides by using the Pythagorean Theorem. (Actually, any triangle similar to this one will do just as nicely.) Then, according to our theorem,

$$\sin 45° = \frac{\text{side opposite}}{\text{hypotenuse}} = \frac{1}{\sqrt{2}} = \frac{\sqrt{2}}{2}$$

and

$$\tan 45° = \frac{\text{side opposite}}{\text{side adjacent}} = \frac{1}{1} = 1$$

(b) In this case, we consider the equilateral triangle whose sides have length 2.

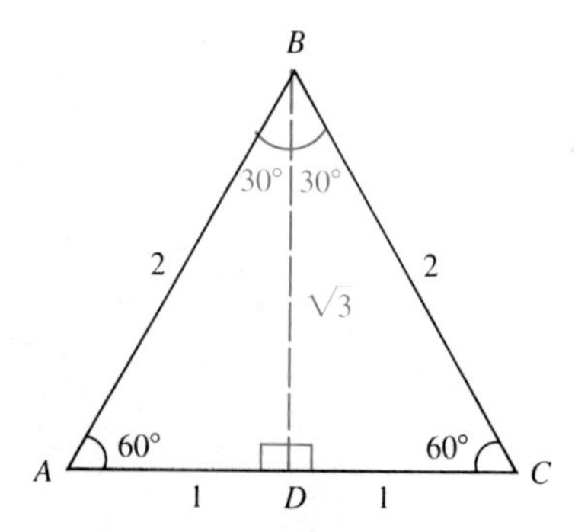

FIGURE 7.7

According to the Pythagorean Theorem, the perpendicular from vertex B to side AC has length $\sqrt{3}$. Now we can use the right triangle ABD to compute the values of the trigonometric functions at 60°. In particular, we have

$$\cos 60^\circ = \frac{\text{side adjacent}}{\text{hypotenuse}} = \frac{1}{2}$$

and

$$\csc 60^\circ = \frac{\text{hypotenuse}}{\text{side opposite}} = \frac{2}{\sqrt{3}} = \frac{2\sqrt{3}}{3}$$

We will leave it to you to obtain the values of the other trigonometric functions at 60°, using this triangle.

Try Study Suggestions 7.4 and 7.5. □

▶ **Study Suggestion 7.4:** Complete Part (b) of Example 2. ◀

▶ **Study Suggestion 7.5:** Use the triangle from Part (b) of Example 2 to compute the values of the trigonometric functions at 30°. ◀

One of the main uses of the previous theorem is in *solving a triangle;* that is, in determining the lengths of all three sides and the measures of all three angles of the triangle. (In practice however, we will settle for approximating these values.) As the next two examples show, if we know that a certain triangle is a right triangle, and if we know either

1. the length of one of its sides and the measure of one of its non-right angles, or
2. the length of two of its sides

then we can solve the triangle.

Try Study Suggestion 7.6.

▶ **Study Suggestion 7.6:** Explain why the measures of all three angles of a triangle is not enough information to solve the triangle. (*Hint:* think about similar triangles.) ◀

EXAMPLE 3: Solve the right triangle

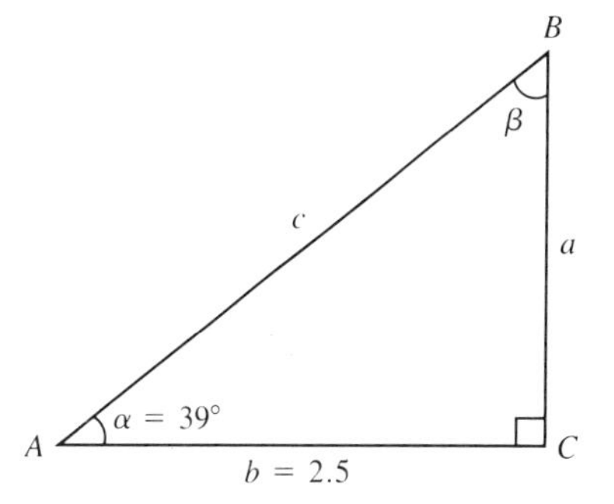

FIGURE 7.8

Solution: The most convenient trigonometric function to use in order to determine the length a is the tangent, since we have

$$\tan 39^\circ = \frac{\text{side opposite}}{\text{side adjacent}} = \frac{a}{2.5}$$

and so

$$a = 2.5 \tan 39^\circ \approx (2.5)(0.8098) = 2.025$$

We can compute the length c in two ways. One way is to use the Pythagorean Theorem, and the other is to use the cosine. Choosing the latter, we get

$$\cos 39^\circ = \frac{\text{side adjacent}}{\text{hypotenuse}} = \frac{2.5}{c}$$

and so

$$c = \frac{2.5}{\cos 39^\circ} \approx \frac{2.5}{0.7771} = 3.217$$

▶ **Study Suggestion 7.7:** Solve the following right triangles ◀

(a)

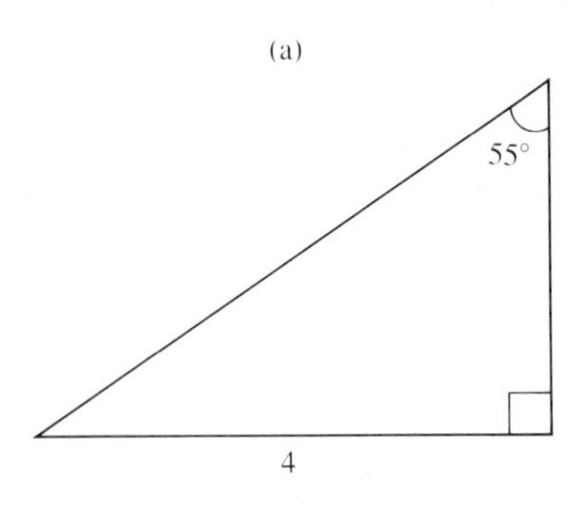

(b)

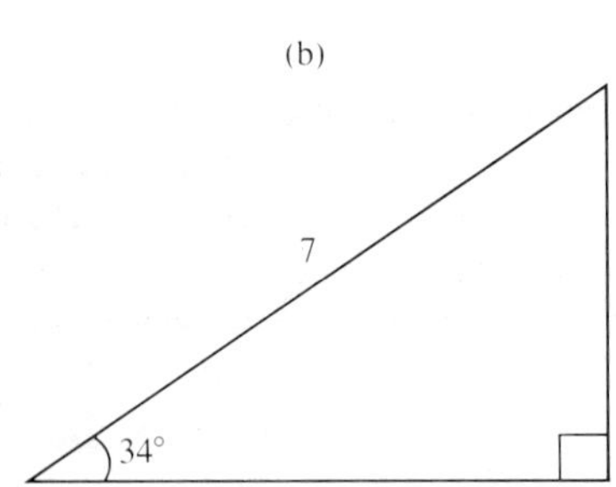

FIGURE 7.9

(Notice that we could have used the secant instead of the cosine, but we chose to use the cosine since calculators rarely have secant keys.) Finally, the measure of angle B can be determined from the fact that the sum of the measures of all three angles of any triangle is 180°. Therefore, we have

$$\beta = 180^\circ - 90^\circ - 39^\circ = 51^\circ$$

Try Study Suggestion 7.7. □

EXAMPLE 4: Solve the right triangle

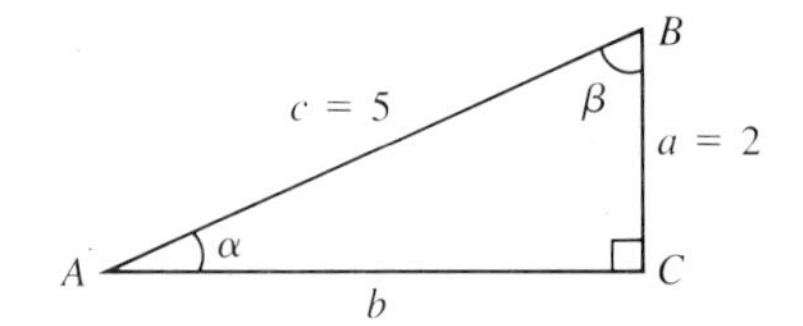

FIGURE 7.10

Solution: We can compute the length of the third side of this triangle either by using the Pythagorean Theorem or one of the trigonometric functions. For variety, let us use the Pythagorean Theorem

$$2^2 + b^2 = 5^2$$

Solving this for b gives $b = \sqrt{21}$. Now we can use our theorem to get

$$\sin \alpha = \frac{2}{5}$$

and so

$$\alpha = \sin^{-1} \frac{2}{5} \approx 23.6^\circ$$

▶ **Study Suggestion 7.8:** Solve the right triangle ◀

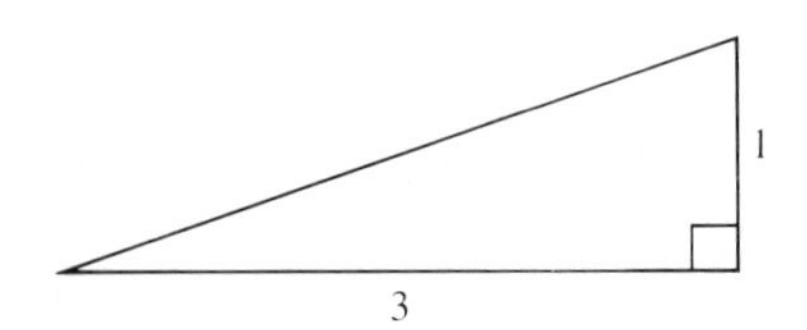

FIGURE 7.11

Finally, we have

$$\beta \approx 180^\circ - 90^\circ - 23.6^\circ = 66.4^\circ$$

Try Study Suggestion 7.8. □

Now let us consider some applications of "triangle" trigonometry.

EXAMPLE 5: A surveyor wishes to know the height of a certain building. At a distance of 100 feet from the building (on level ground), he measures the angle of inclination to the top of the building to be 72.5°. (See Figure 7.12.) How tall is the building?

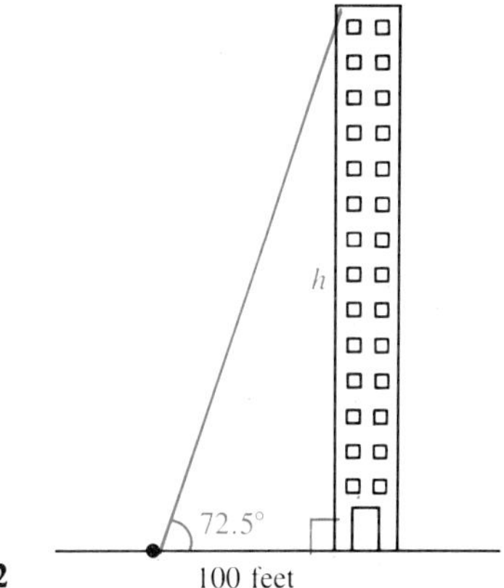

FIGURE 7.12

Solution: According to our theorem, we have

$$\tan 72.5^\circ = \frac{\text{side opposite}}{\text{side adjacent}} = \frac{h}{100}$$

where h is the height of the building. Hence,

$$h = 100 \tan 72.5^\circ \approx 100(3.1716) = 317.16$$

Thus, the building is approximately 317.16 feet high.

Try Study Suggestion 7.9. □

▶ **Study Suggestion 7.9:** Using the information given in Figure 7.13, determine the depth of the swimming pool at its deep end. ◀

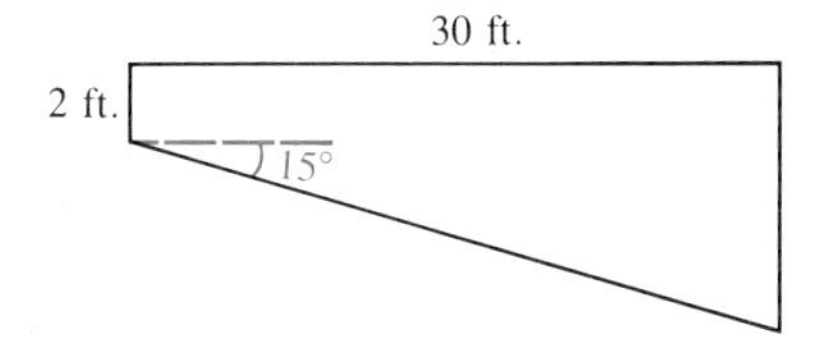

FIGURE 7.13

EXAMPLE 6: A person is standing at the top of a cliff 1000 feet high at the edge of the ocean. (See Figure 7.14.) At a certain time, she sights a boat in the water at an angle of declination of 23° (below the horizontal). A moment later, she sights the same boat at an angle of declination of 31°. How far has the boat traveled between the two sightings?

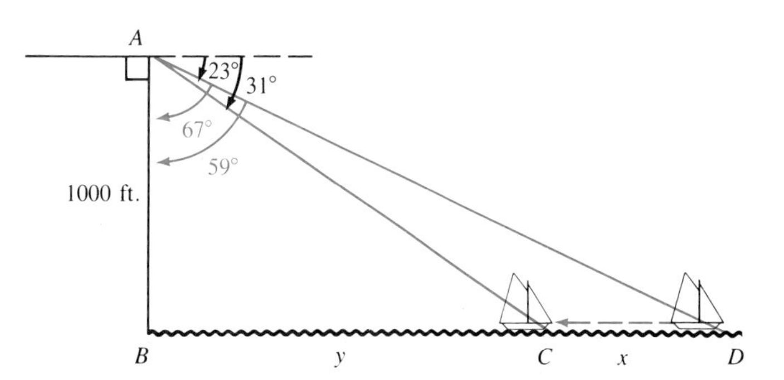

FIGURE 7.14

Solution: In Figure 7.14, we have labeled the position of the boat at the two sightings, as well as the angles BAC and BAD. Using the right triangle BAC, we get

$$\tan 59° = \frac{\text{side opposite}}{\text{side adjacent}} = \frac{y}{1000}$$

and so

$$y = 1000 \tan 59° \tag{7.1}$$

Using the right triangle BAD, we get

$$\tan 67° = \frac{\text{side opposite}}{\text{side adjacent}} = \frac{x + y}{1000}$$

and so

$$x + y = 1000 \tan 67° \tag{7.2}$$

Now we can solve for x using Equations (7.1) and (7.2)

$$\begin{aligned} x &= 1000 \tan 67° - y \\ &= 1000 \tan 67° - 1000 \tan 59° \\ &= 1000(\tan 67° - \tan 59°) \\ &\approx 1000(2.356 - 1.664) \\ &= 1000(0.692) \\ &= 692 \end{aligned}$$

Thus, the boat traveled a distance of approximately 692 feet between the two sightings. *Try Study Suggestion 7.10.* □

▶ **Study Suggestion 7.10:** Determine the height of the tree shown in Figure 7.15. ◀

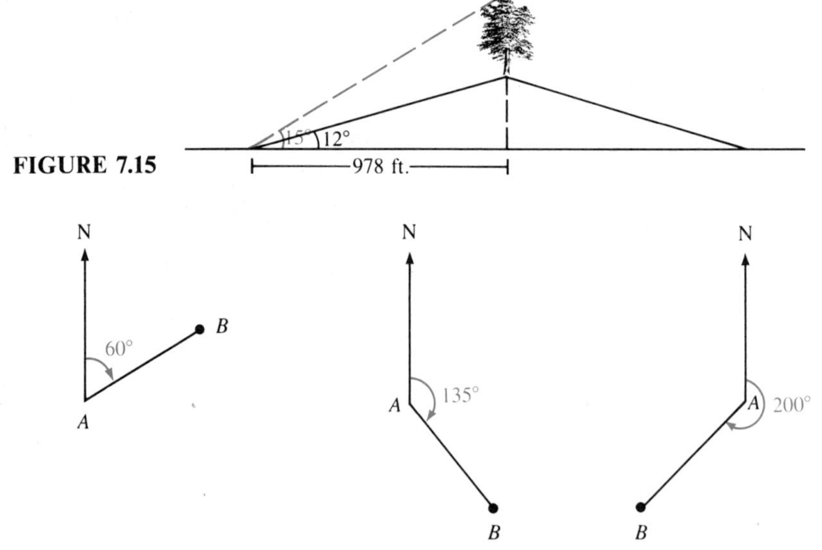

FIGURE 7.15

FIGURE 7.16 Object B has bearing 60° from point A | Object B has bearing 135° from point A | Object B has bearing 200° from point A

The next example uses the concept of the *bearing* of an object, which is its compass direction, measured in a clockwise direction from due north. Figure 7.16 shows some objects and their bearings.

EXAMPLE 7: A man is searching for buried treasure. According to his map, in order to reach the treasure starting at a certain tree, he must walk 500 yards due east then 200 yards due north and finally 300 yards due east again. How far is the treasure from the tree, and at what bearing?

Solution: Figure 7.17 shows the path that the man follows in order to reach the treasure. We must determine the distance from O to B and the angle θ. Since the line segment OB is the hypotenuse of the right triangle OAB, the Pythagorean Theorem (and a calculator) gives

$$d(O, B) = \sqrt{800^2 + 200^2} \approx 824.6 \text{ yards}$$

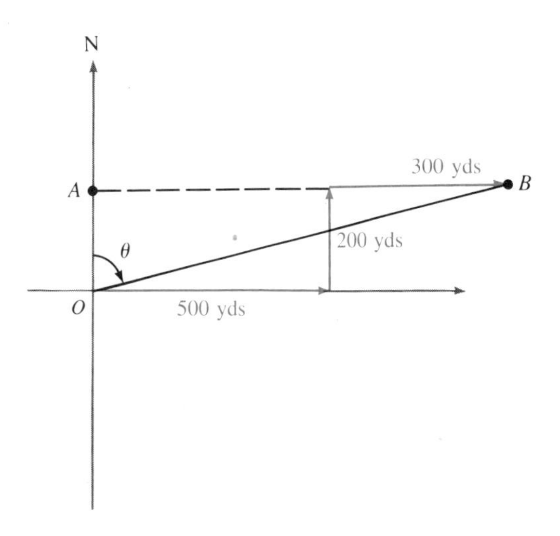

FIGURE 7.17

In order to find the bearing θ of the treasure, we observe that

$$\tan\theta = \frac{800}{200} = 4$$

and so

$$\theta = \arctan 4 \approx 75.96^\circ$$

Thus, the treasure is approximately 824.62 yards from the tree at a bearing of approximately 75.96°. *Try Study Suggestion 7.11.* □

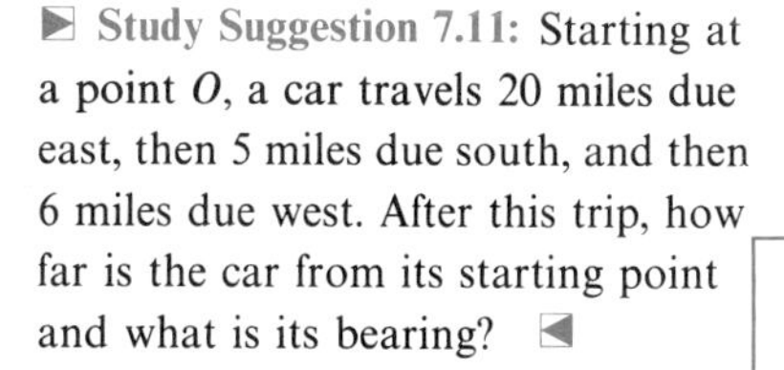

▶ **Study Suggestion 7.11:** Starting at a point O, a car travels 20 miles due east, then 5 miles due south, and then 6 miles due west. After this trip, how far is the car from its starting point and what is its bearing? ◀

Ideas to Remember

The trigonometric functions can be defined completely in terms of right triangles, and this has important practical applications.

EXERCISES

In Exercises 1–16, complete the appropriate row in the table below. Do not use a calculator.

	θ	$\sin\theta$	$\cos\theta$	$\tan\theta$	$\sec\theta$	$\csc\theta$	$\cot\theta$
1.	0°						
2.	30°						
3.	45°						
4.	60°						
5.	90°						
6.	120°						
7.	135°						
8.	150°						
9.	180°						
10.	210°						
11.	225°						
12.	240°						
13.	270°						
14.	300°						
15.	315°						
16.	330°						

In Exercises 17–31, compute the indicated value without using a calculator.

17. $\sin(-30^\circ)$ **18.** $\cos(-60^\circ)$ **19.** $\tan(-120^\circ)$

20. $\sin(-150^\circ)$ **21.** $\cos(-270^\circ)$ **22.** $\tan(-45^\circ)$

23. $\sec(-210^\circ)$ **24.** $\csc(-135^\circ)$ **25.** $\cot(-180^\circ)$

26. $\sec 390^\circ$ **27.** $\csc(-390^\circ)$ **28.** $\cot 810^\circ$

29. $\sin 420^\circ$ **30.** $\cos 495^\circ$ **31.** $\tan 945^\circ$

In Exercises 32–37, use a calculator to approximate the indicated values.

32. $\sin 13.72^\circ$ **33.** $\cos(-12.982^\circ)$

34. $\tan 85.21^\circ$ **35.** $\sec(-0.3392^\circ)$

36. $\csc 128.493^\circ$ **37.** $\cot(-199^\circ)$

38. **(a)** Is it true that $\sin d^\circ = -\sin(-d^\circ)$? Justify your answer.

(b) Is it true that $\sin(d^\circ + 2\pi) = \sin d^\circ$? Justify your answer.

39. **(a)** Is it true that $\cos d^\circ = \cos(-d^\circ)$? Justify your answer.

(b) Is it true that $\cos d^\circ = \sin(d^\circ + \pi/2)$? Justify your answer.

40. Is it true that $\sin^2 d^\circ + \cos^2 d^\circ = 1$? Justify your answer.

In Exercises 41–49, use the given right triangles to compute the value of all six trigonometric functions at θ.

41.

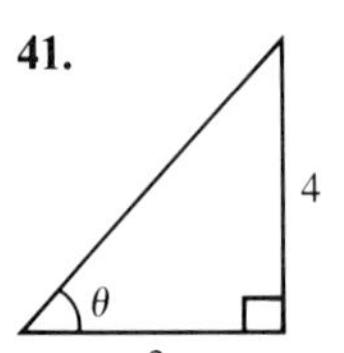

42.

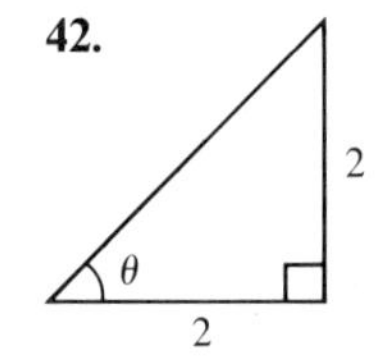

43.

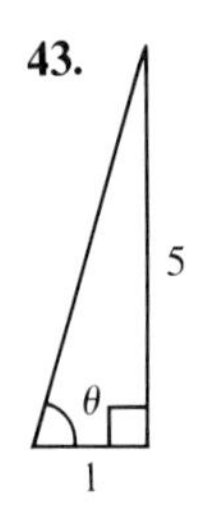

44.

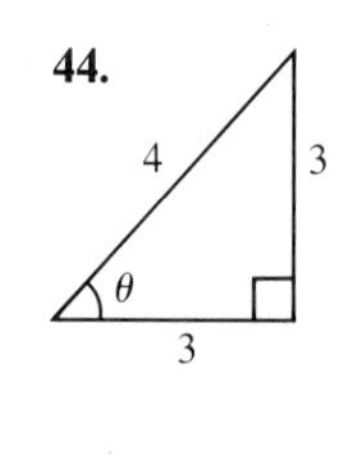

45.

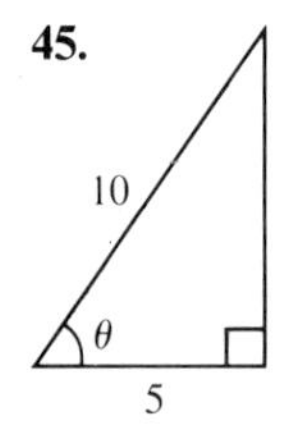

46.

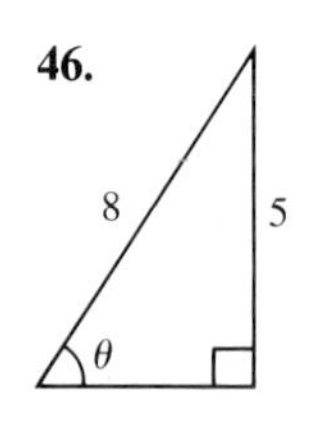

47.

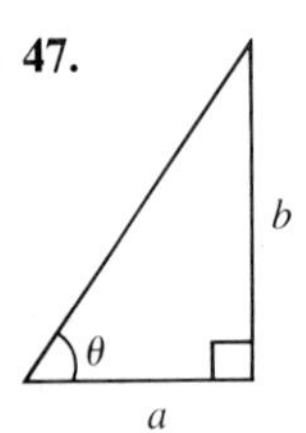

48.

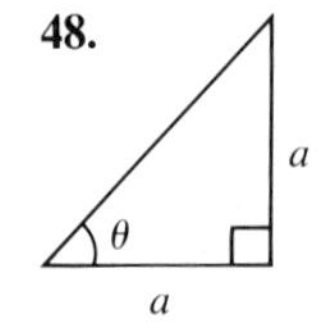

49.

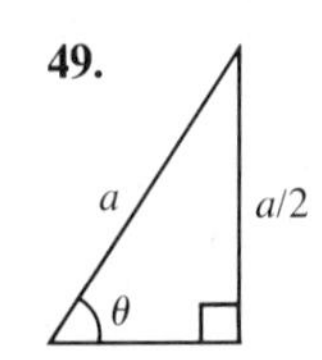

In Exercises 50–57, solve the given right triangle. Do not use a calculator.

50.

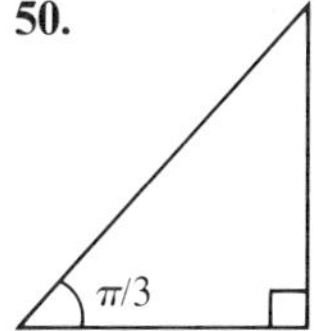

51.

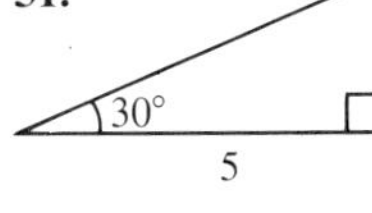

52.

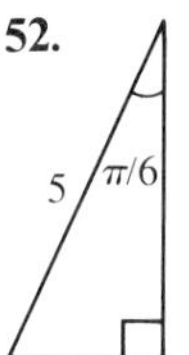

53.

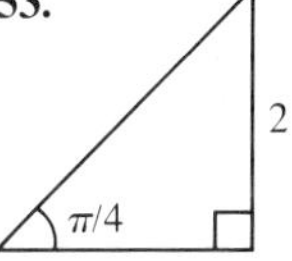

54.

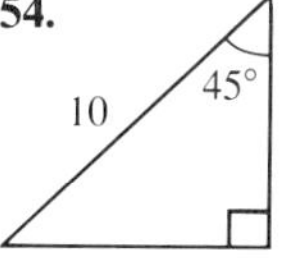

55.

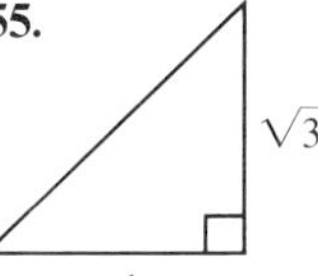

56.

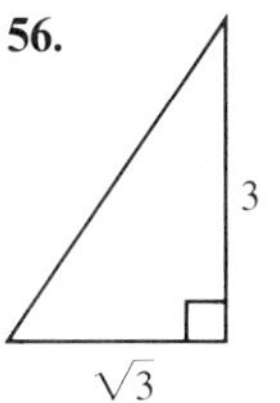

57.

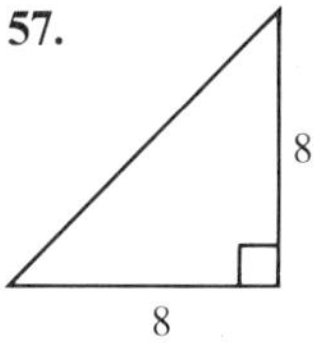

In Exercises 58–65, solve the given right triangle.

58.

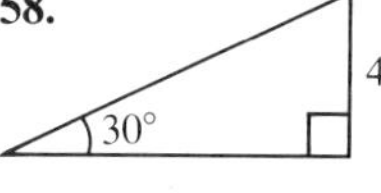

59.

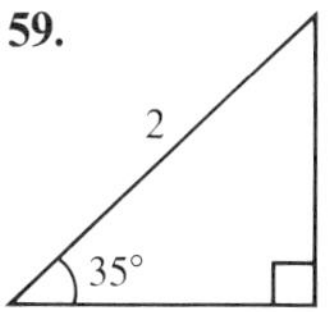

60.

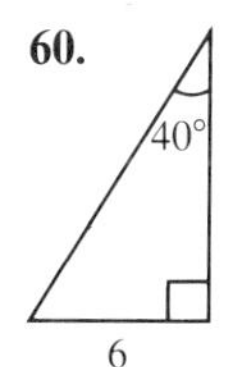

61.

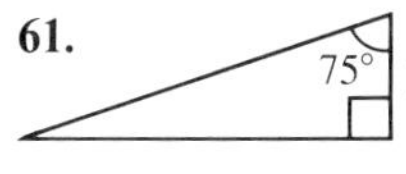

62.

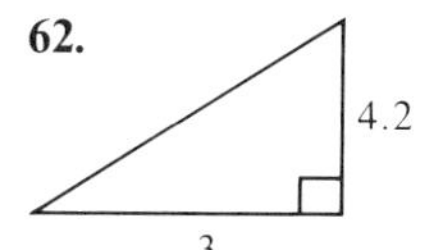

63. 2, 6

64.

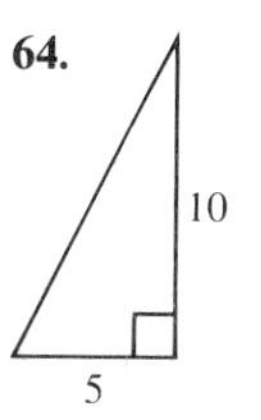

65. $\sqrt{5}$, 1

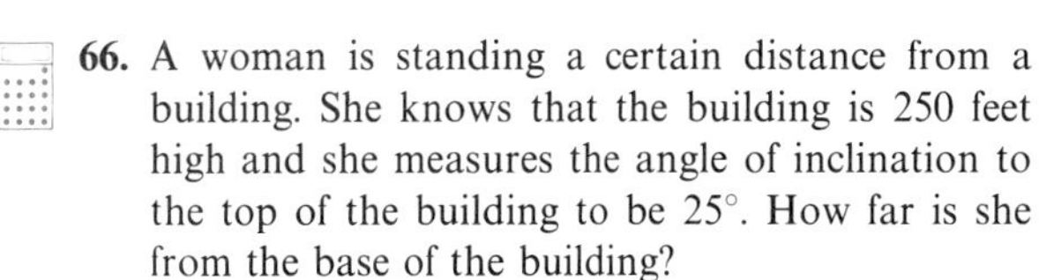

66. A woman is standing a certain distance from a building. She knows that the building is 250 feet high and she measures the angle of inclination to the top of the building to be 25°. How far is she from the base of the building?

67. A man walks 100 yards due south, 400 yards due west, and then 200 yards due south again. How far is he from his starting point and at what bearing?

68. A swimming pool has the dimensions shown below.

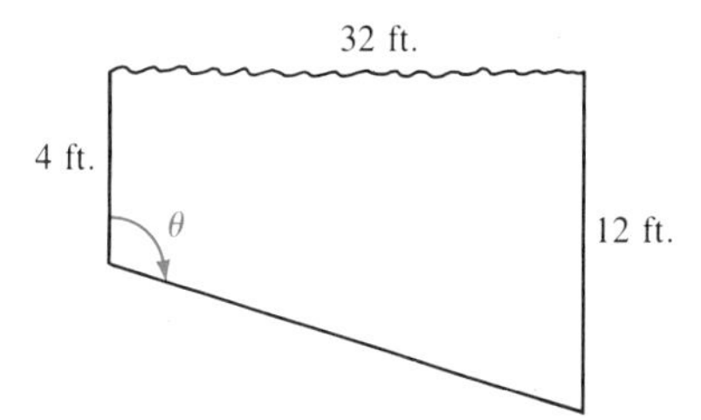

Determine the distance along the bottom of the pool and the angle θ.

69. Suppose that you are looking out over the ocean at a row of sailboats, and that the row of boats is perpendicular to your line of sight—the closest boat being 1/2 mile away. In order to see the entire row of boats, you must turn your head 30° to the right and 25° to the left. How long is the row of boats?

70. The following figure shows a swinging pendulum.

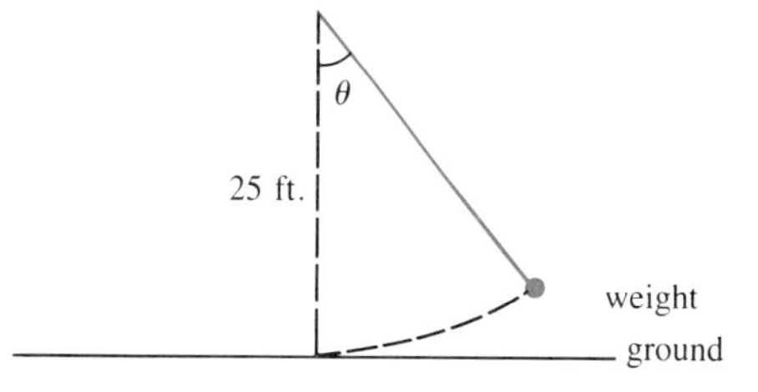

(a) Determine the angle θ when the weight is 5 feet above the ground.

(b) Determine the height of the weight above the ground when $\theta = 10°$.

71. Unfortunately, one of the tests that you must pass in your physical education class is the rope climb. In order to avoid having to climb the rope, you tell the instructor that you are afraid of heights. She says that if you can tell her the height of the rope to the nearest inch, you do not have to climb the rope. Describe in detail a method for solving this problem.

72. Suppose that you wish to reach a certain window by leaning a ladder against the side of the building. For safety reasons, you feel that the ladder should make an angle of between 65° and 75° with the ground. How long should the ladder be if the window is 12 feet from the ground?

73. A man 5′6″ tall is looking at a painting in a certain museum. If he is 10 feet from the wall on which the painting is hanging, and if he must look down at an angle of 20° to see the center of the painting, how far is the center of the painting from the floor?

74. A forest ranger is in a tower 75 feet above level ground. If she sees a fire at an angle of declination of 5°20′, how far away is the fire? ($1' = 1/60$ of a degree.)

75. In order to measure the distance from one side of a lake to the other, a marker is placed on each side of the lake directly in line with a tower that is 300 feet high. From the top of the tower, the closer marker has an angle of declination of 31° and the farther marker has an angle of declination of 5°. How far is it across the lake?

76. A man is 6 feet tall. What is the length of his shadow when the sun is at angle of 27° above the horizon?

77. A woman is searching for buried treasure. According to her map, in order to reach the treasure she must start at a certain tree, walk 50 feet due north, then walk 200 feet at a bearing of 25°, and finally walk 100 feet due east. How far is the treasure from the tree and at what bearing?

78. A dirt path travels straight up a mountainside, at an angle of inclination of 12°. Some hikers have set up camp along the path, at an elevation of 2000 feet above sea level. If the top of the mountain is 10,000 feet above sea level, how many miles is the hike from the camp to the top of the mountain?

7.2 The Law of Sines

In the previous section, we showed how the trigonometric functions can be defined using right triangles. In fact, all of the triangles that we considered in the previous section were right triangles. However, there are two important trigonometric formulas that apply to *any* type of triangle, not just right triangles. (A triangle that does not contain a right angle is called an **oblique triangle.**) These formulas are known as the *law of sines*, and the *law of cosines*. We will discuss the law of sines in this section, and the law of cosines in the next section.

Let us begin by stating the law of sines.

THEOREM

The law of sines For the triangle

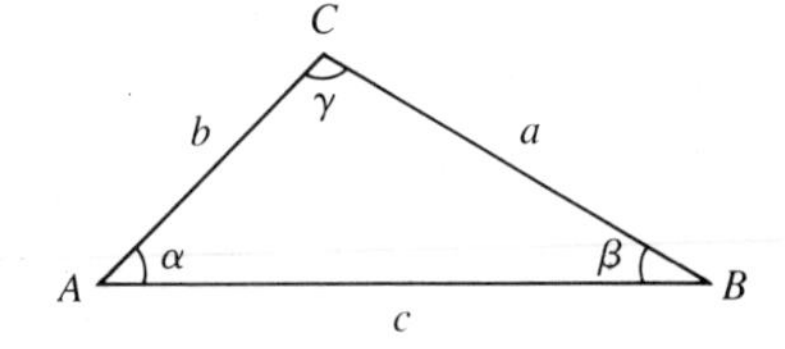

FIGURE 7.18

(which may be either a right triangle or an oblique triangle) we have

$$\frac{\sin\alpha}{a} = \frac{\sin\beta}{b} = \frac{\sin\gamma}{c} \tag{7.3}$$

Proof: The first step in the proof is to drop a perpendicular from angle C to the opposite side of the triangle, as shown in Figure 7.19. We have denoted the length of this perpendicular by x. Since triangles ACD and BCD are both *right* triangles, we have

$$\sin\alpha = \frac{x}{b} \qquad \text{and} \qquad \sin\beta = \frac{x}{a}$$

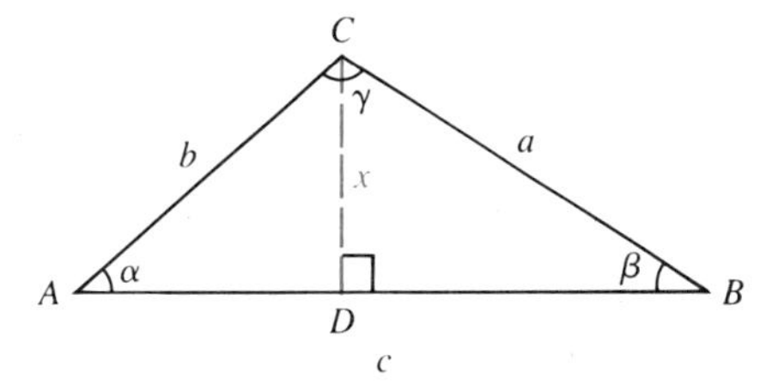

FIGURE 7.19

Solving these equations for x gives

$$x = b\sin\alpha \qquad \text{and} \qquad x = a\sin\beta$$

Equating these two values of x gives

$$b\sin\alpha = a\sin\beta$$

Now we can divide both sides of this equation by ab, to get

$$\frac{\sin\alpha}{a} = \frac{\sin\beta}{b} \tag{7.4}$$

This is one "half" of the law of sines.

The next step is to drop a perpendicular from angle A to the extension of side BC, as shown in Figure 7.20. We have denoted the length of this perpendicular by y. Notice that angle ACE has radian measure $\pi - \gamma$.

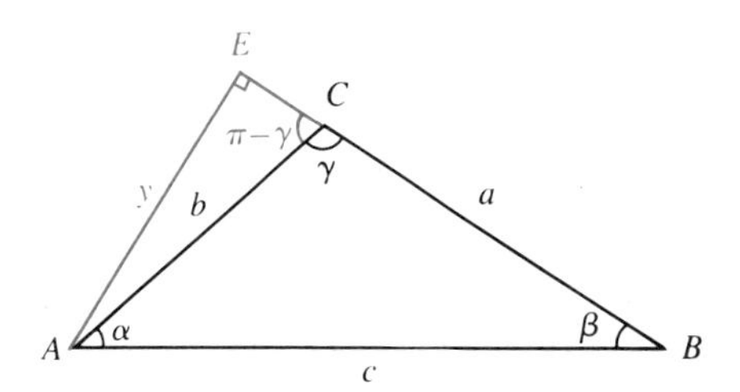

FIGURE 7.20

Now we want to compute $\sin\gamma$. Unfortunately, angle ACB is not part of a right triangle. However, angle ACE is, and using the properties of the sine function discussed in Section 6.4, we get

$$\begin{aligned}\sin(\pi-\gamma) &= \sin(-\gamma+\pi)\\ &= -\sin(-\gamma)\\ &= \sin\gamma\end{aligned}$$

Thus,

$$\sin\gamma = \sin(\pi-\gamma) = \frac{y}{b}$$

But we also have

$$\sin\beta = \frac{y}{c}$$

Solving these equations for y gives

$$y = b\sin\gamma \qquad \text{and} \qquad y = c\sin\beta$$

Now we equate these two values of y

$$b\sin\gamma = c\sin\beta$$

and divide both sides by bc to get

$$\frac{\sin\beta}{b} = \frac{\sin\gamma}{c} \tag{7.5}$$

This is the other "half" of the law of sines. Putting Equations (7.4) and (7.5) together gives Equation (7.1), and the proof is complete. ∎

Notice that the law of sines can also be written in the form

$$\frac{a}{\sin\alpha} = \frac{b}{\sin\beta} = \frac{c}{\sin\gamma}$$

Sometimes this form is a bit more convenient than the one given in the theorem. (We shall use them both.)

One of the main uses of the law of sines is to solve oblique triangles. For example, the law of sines can be used to solve a triangle, given the measures of two of its angles and the length of one of its sides.

EXAMPLE 1: Solve the triangle

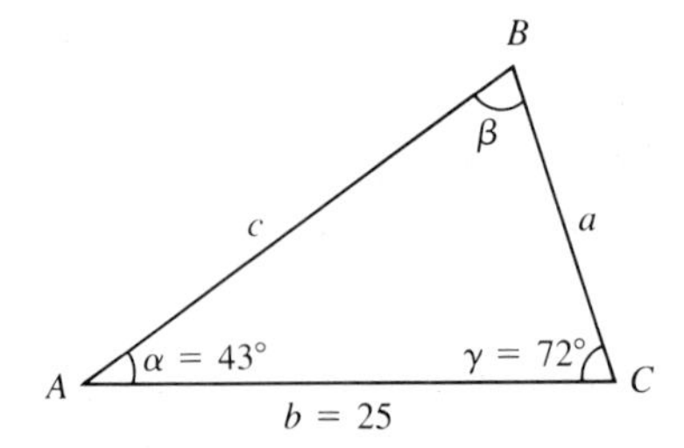

FIGURE 7.21

Solution: Since the sum of the measures of all three angles of any triangle equals 180°, we have

$$\beta = 180^\circ - \alpha - \gamma = 180^\circ - 43^\circ - 72^\circ = 65^\circ$$

Now in order to find the missing lengths, we use the law of sines. For example, since

$$\frac{a}{\sin \alpha} = \frac{b}{\sin \beta}$$

we have

$$a = \frac{b \sin \alpha}{\sin \beta} = \frac{25 \sin 43^\circ}{\sin 65^\circ} \approx \frac{(25)(0.6820)}{0.9063} \approx 18.81$$

Also, since

$$\frac{b}{\sin \beta} = \frac{c}{\sin \gamma}$$

we have

$$c = \frac{b \sin \gamma}{\sin \beta} = \frac{25 \sin 72^\circ}{\sin 65^\circ} \approx \frac{(25)(0.9511)}{0.9063} \approx 26.24$$

Hence, the triangle is solved. *Try Study Suggestion 7.12.* □

▶ **Study Suggestion 7.12:** Solve the following triangle ◀

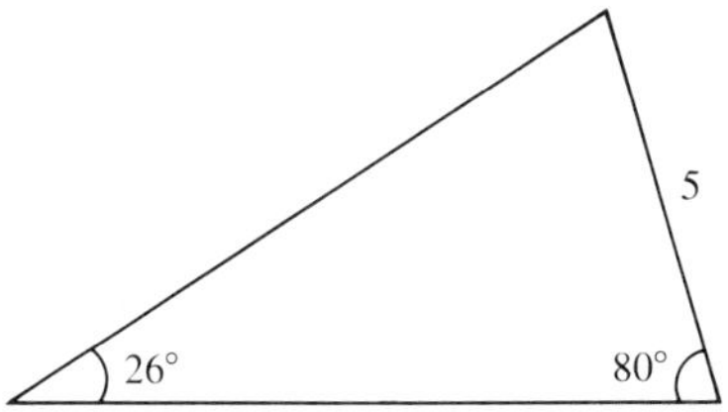

26°
80°
5

FIGURE 7.22

There are situations in which the information that we are given does not uniquely determine a triangle. In some of these situations, the law of sines can help us determine all possible triangles that fit the given information.

In particular, suppose that we are given the lengths of two sides of a triangle, along with the measure of the angle opposite one of these sides. As you can see from Figures 7.23–7.26, this information is consistent with four possibilities. Let us consider some examples of these possibilities.

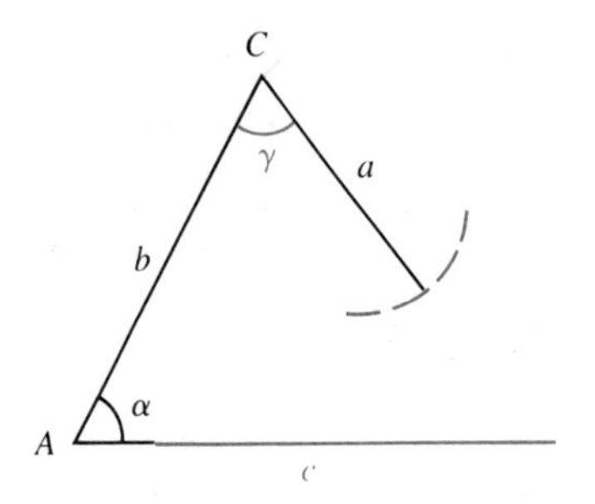

No triangle—side a is too short

FIGURE 7.23

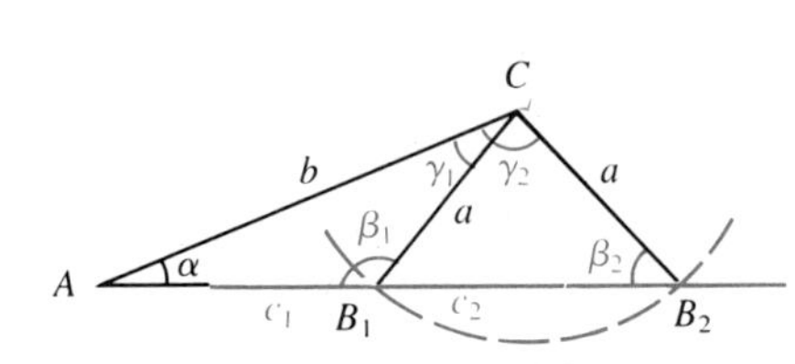

Two possible triangles

FIGURE 7.24

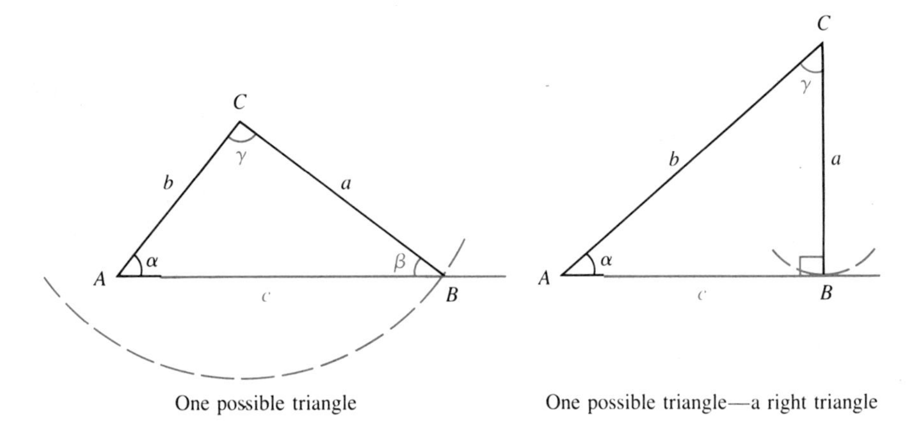

One possible triangle

FIGURE 7.25

One possible triangle—a right triangle

FIGURE 7.26

EXAMPLE 2: Find all triangles for which $a = 2$, $b = 3$, and $\alpha = 35°$.

Solution: According to the law of sines, we have

$$\frac{\sin \beta}{b} = \frac{\sin \alpha}{a}$$

and so

$$\sin \beta = \frac{b \sin \alpha}{a} = \frac{3 \sin 35°}{2} \approx 0.8604$$

As you see in Figure 7.27, there are *two* angles between 0° and 180° with this value of the sine. One angle B_1 has size $\beta_1 \approx \sin^{-1} 0.8604 \approx 59.4°$, and the other angle B_2 has size $\beta_2 \approx 180° - 59.4° = 120.6°$.

Using angle B_1, we can compute the size of the third angle C_1 as follows

$$\gamma_1 \approx 180° - 35° - 59.4° = 85.6°$$

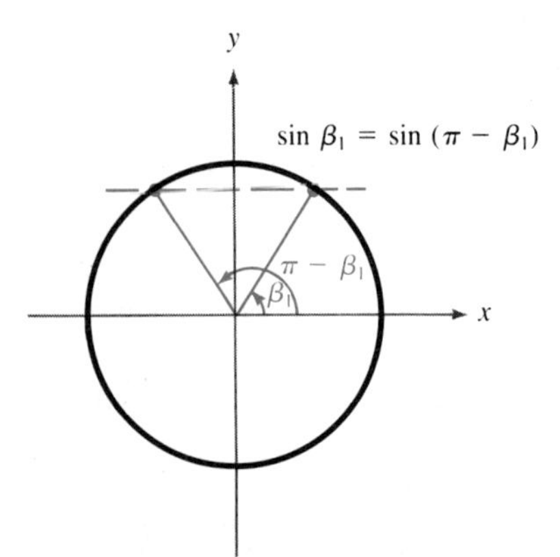

FIGURE 7.27

This gives us the triangle pictured in Figure 7.28(a). The length of the side opposite angle C_1 can now be easily computed from the law of sines,

$$c_1 = \frac{a \sin \gamma_1}{\sin \alpha} \approx \frac{2 \sin 85.6^\circ}{\sin 35^\circ} \approx 3.477$$

On the other hand, if we use angle B_2, the third angle C_2 has size

$$\gamma_2 \approx 180^\circ - 35^\circ - 120.6^\circ = 24.4^\circ$$

This triangle is pictured in Figure 7.28(b). As before, the law of sines gives us the value of c_2,

$$c_2 = \frac{a \sin \gamma_2}{\sin \alpha} \approx \frac{2 \sin 24.4^\circ}{\sin 35^\circ} \approx 1.440$$

Try Study Suggestion 7.13. □

Study Suggestion 7.13: Find all triangles for which $a = 1$, $b = 2$, and $\alpha = 25^\circ$. ◀

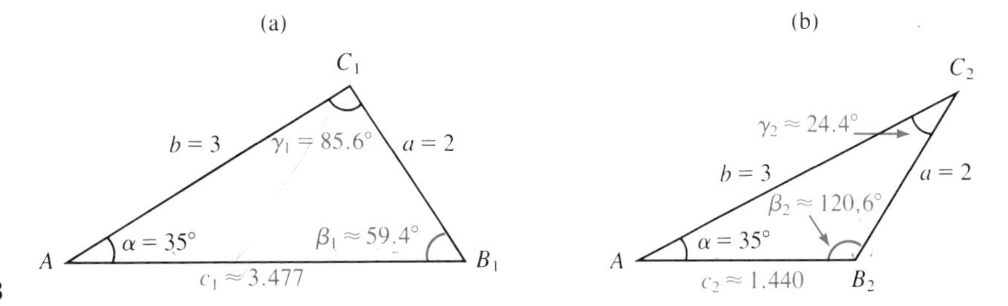

FIGURE 7.28

EXAMPLE 3: Find all triangles for which $a = 10$, $b = 5$, and $\alpha = 25^\circ$.

Solution: In this case, the law of sines gives

$$\sin \beta = \frac{b \sin \alpha}{a} = \frac{5 \sin 25^\circ}{10} \approx 0.2113$$

Again, there are two possibilities for angle B. One possibility B_1 has size $\beta_1 = \sin^{-1} 0.2113 \approx 12.2^\circ$ and the other possibility B_2 has size $\beta_2 \approx 180^\circ - 12.2^\circ = 167.8^\circ$.

Using angle B_1, we can compute the size of the third angle C_1 just as in the previous example,

$$\gamma_1 \approx 180^\circ - 25^\circ - 12.2^\circ = 142.8^\circ$$

This triangle is pictured in Figure 7.29. The law of sines also gives us the value of c_1,

$$c_1 = \frac{a \sin \gamma_1}{\sin \alpha} \approx \frac{10 \sin 142.8^\circ}{\sin 25^\circ} \approx 14.31$$

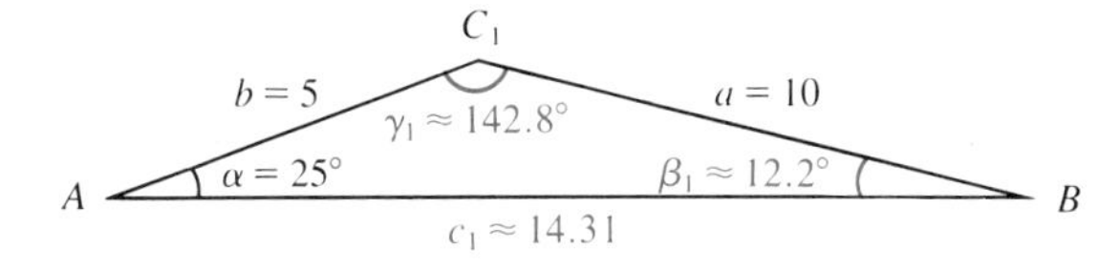

FIGURE 7.29

However, when we use angle B_2, we discover a problem in trying to compute the size of the third angle C_2. The problem is that $\alpha + \beta_2 = 25° + 167.7° = 192.7° > 180°$. In other words, angle B_2 is too large to "fit" into a triangle one of whose other angles has size 25°. Thus, in this case, there is only one triangle consistent with the given measurements.

Try Study Suggestion 7.14. □

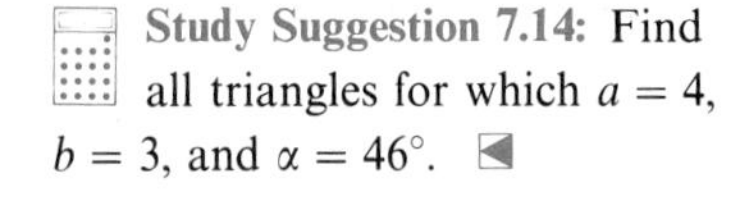

Study Suggestion 7.14: Find all triangles for which $a = 4$, $b = 3$, and $\alpha = 46°$. ◀

EXAMPLE 4: Find all triangles for which $a = 1$, $b = 2$, and $\alpha = 40°$.

Solution: In this case, the law of sines gives

$$\sin\beta = \frac{b\sin\alpha}{a} = 2\sin 40° \approx 1.286$$

But it is not possible for the sine of an angle to be greater than 1, and so in this case there are no triangles that are consistent with the given measurements. *Try Study Suggestion 7.15.* □

Study Suggestion 7.15: Find all triangles for which $a = 4$, $b = 4.2$, and $\alpha = 77°$. ◀

Ideas to Remember

The law of sines can be used to solve oblique triangles.

EXERCISES

In Exercises 1–20, find all triangles with the given measurements.

1. $\alpha = 40°$, $\beta = 60°$, $a = 10$
2. $\alpha = 25°$, $\beta = 105°$, $c = 5$
3. $\alpha = 95°$, $\gamma = 12°$, $b = 1$
4. $\beta = 65°$, $\gamma = 65°$, $a = 4$
5. $\alpha = \pi/7$, $\gamma = 4\pi/7$, $a = 2$
6. $\beta = \pi/7$, $\gamma = 5\pi/11$, $c = 3$
7. $a = 4.5$, $b = 5$, $\alpha = 60°$
8. $a = 6$, $b = 7$, $\alpha = \pi/5$
9. $a = 25$, $c = 30$, $\gamma = 10°$
10. $b = 10$, $c = 1$, $\beta = 0.2\pi$
11. $b = 45$, $c = 9$, $\gamma = 12°$
12. $b = 10$, $c = 12$, $\beta = 68.5°$
13. $a = 10$, $b = 9.9$, $\beta = 80°$
14. $b = 100$, $c = 5$, $\gamma = \pi/100$
15. $a = 5.4$, $b = 5.3$, $\alpha = 56.2°$
16. $a = 4$, $b = 5$, $\beta = 4\pi/9$
17. $a = 5.8$, $c = 6$, $\gamma = 25°18'$
18. $b = 12$, $c = 10$, $\beta = 85°10'$
19. $b = 12$, $c = 10$, $\beta = 0.47\pi$
20. $b = 10$, $c = 7$, $\gamma = 87°$
21. A man is standing at the top of a vertical cliff overlooking the ocean. From that point, he can see two boats, as shown in the following figure. The two boats are 1000 feet apart and separated by an angle of 28°. The distance between the man and Boat 1

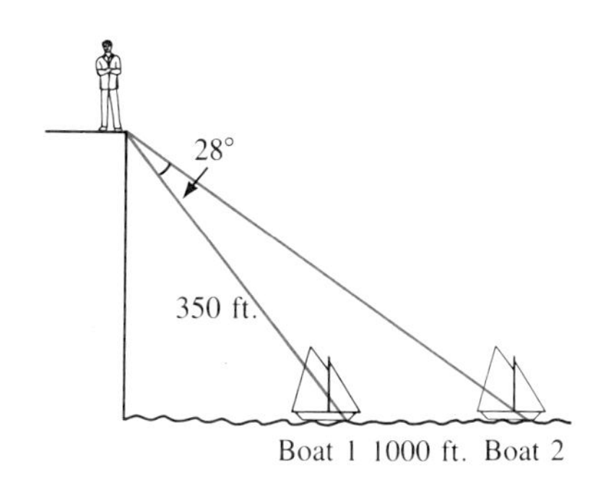

is 350 feet. Find the distance between the man and Boat 2.

22. Three towers are located at the vertices of a triangle. The distance between Tower 1 and Tower 2 is 500 feet, and the distance between Tower 2 and Tower 3 is 350 feet. The angle whose vertex is at Tower 1 measures 40.3°. Is it possible to determine the distance between Tower 1 and Tower 3 with this information? If so, find this distance, and if not, determine what is possible.

23. A woman wants to measure the length of a telephone pole that is leaning at an angle of 63° from the horizontal. In order to do this, she attaches one end of a rope of length 120 feet to the top of the pole, and determines that the other end of the rope touches the ground exactly 130 feet from the base of the pole. How long is the pole?

24. Two houses are located on opposite sides of a freeway, as shown in the following figure. In order to measure the distance between the two houses, a third point A is chosen on one side of the freeway—a distance of 1000 feet from one of the houses, and

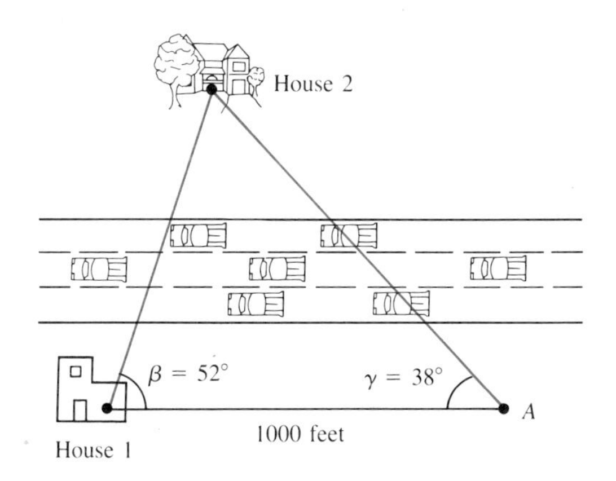

the angles α and β are measured and found to be $\alpha = 38°$ and $\beta = 52°$. Use this information to determine the distance between the two houses.

25. Camps are located at points A and B that are 200 meters apart in the woods, camp A being due west of camp B. A hiker starts at camp B, walks due east 100 meters and then due north 100 meters, ending at a point C. Find the angle ACB, as well as the distance from the hiker to each camp.

26. Two camps are located in the woods at points A and B, camp A being due west of camp B. A hiker starts at camp B, walks due east 100 meters and then due north 200 meters, ending at a point C. The hiker then measures the angle ACB to be 25°. Find the distance between the two camps, as well as the distance from the hiker to each camp.

27. A sail has the shape shown in the following figure.

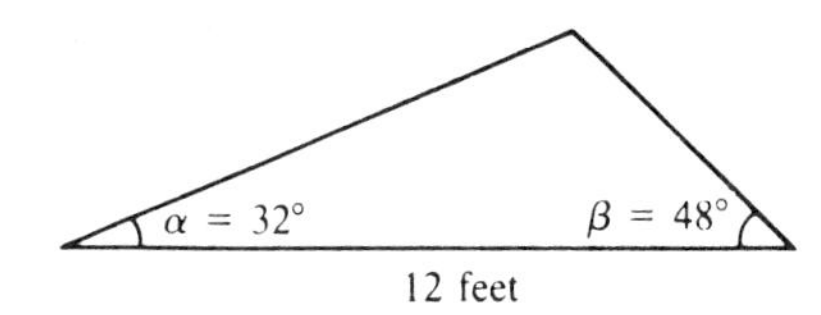

The base of the sail has length 12 feet, and the angles α and β have the measure given in the figure. Find the height of the sail and the lengths of the other two sides. Also, find the area of the sail.

28. A hot-air balloon is attached to the ground by two ropes as in the following figure. If the distance between the two anchor points is 260 feet, and the

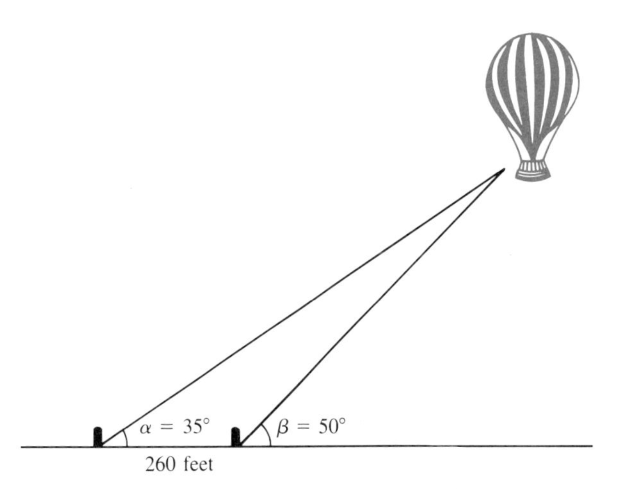

angles of inclination of the two ropes with the ground are $\alpha = 35°$ and $\beta = 50°$, find the distance from the balloon to the ground.

29. The navigator of a boat measures the distance to a lighthouse top to be 10 kilometers, and the angle of inclination of the lighthouse top above the horizon to be $38°10'$. The boat travels for two hours directly toward the lighthouse, and then the navigator measures the angle of inclination of the lighthouse top to be $78°20'$. How fast is the boat traveling, and how long will it take the boat to reach the lighthouse? ($1' = 1/60$ of a degree.)

30. Three disk shaped gears are situated as in the following figure. Using the information given in the figure, find the distance between the centers of each pair of gears.

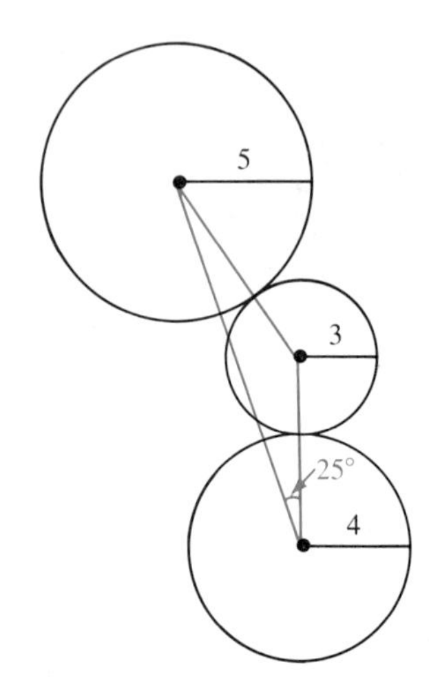

31. As we have seen, given the values of a, b, and α, there may be none, one, or two triangles that "fit" these values. Find conditions on a, b, and α that can be used to determine which of these possibilities holds.

7.3 The Law of Cosines

If we are given the information shown in Figures 7.30 or 7.31, then we should be able to solve the triangle. Unfortunately, however, the law of sines will not help us in either of these situations. For in order to apply the law of sines, we must know the measure of an angle, as well as the length of the side *opposite* that angle, so that we can determine one of the ratios in the law of sines. Hence, we must look for another formula that will help us in these situations. This formula is the *law of cosines.*

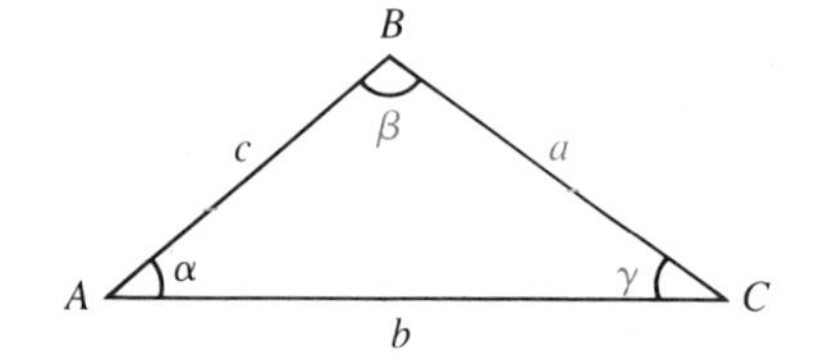

α, b, and c are known; β, γ, and a are unknown

FIGURE 7.30

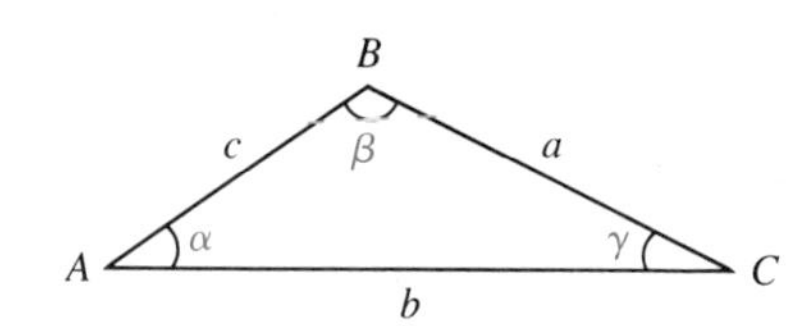

a, b, and c are known; α, β, and γ are unknown

FIGURE 7.31

Rather than simply state the law of cosines, let us see if we can't discover this law for ourselves. Suppose that we are given α, b, and c as shown in Figure 7.30. We wish to find a formula for a. Once this is done, we can then apply the law of sines to solve the triangle.

The first step is to draw a perpendicular from angle B to side AC, and label the resulting picture as in Figure 7.32. (Recall that we did the same

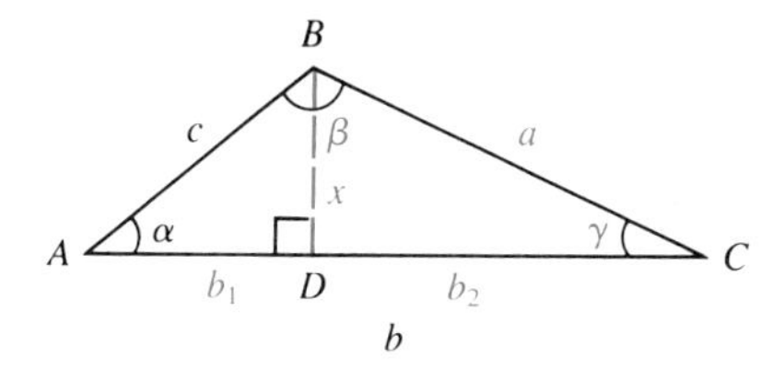

FIGURE 7.32

thing when we derived the law of sines.) Applying the Pythagorean Theorem to the right triangle BCD gives

$$a^2 = x^2 + b_2{}^2 \tag{7.6}$$

Of course, our goal is to express a strictly in terms of the information that we were given, namely, in terms of α, b, and c. To this end, we apply the Pythagorean Theorem to the right triangle ABD, to get

$$c^2 = x^2 + b_1{}^2$$

Solving this equation for x^2 and substituting into Equation (7.6) gives

$$a^2 = c^2 - b_1{}^2 + b_2{}^2 \tag{7.7}$$

Now we must find a way to express b_1 and b_2 in terms of the given quantities α, b, and c. But since ADB is a right triangle, we have

$$\cos\alpha = \frac{b_1}{c}$$

and so

$$b_1 = c\cos\alpha$$

Also, since $b_1 + b_2 = b$, we have

$$b_2 = b - b_1 = b - c\cos\alpha$$

Substituting these expressions for b_1 and b_2 into Equation (7.7) and simplifying gives

$$\begin{aligned} a^2 &= c^2 - c^2\cos^2\alpha + (b - c\cos\alpha)^2 \qquad (7.8)\\ &= c^2 - c^2\cos^2\alpha + b^2 - 2bc\cos\alpha + c^2\cos^2\alpha \\ &= b^2 + c^2 - 2bc\cos\alpha \end{aligned}$$

Now we have a formula for a (or rather a^2, which is almost as good) involving *only* the given quantities α, b, and c. This formula is one form of the law of cosines.

THEOREM

The law of cosines For the triangle

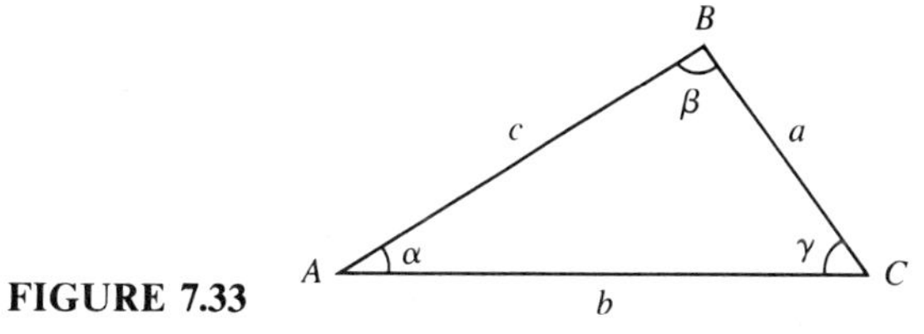

FIGURE 7.33

we have

$$\begin{aligned} a^2 &= b^2 + c^2 - 2bc\cos\alpha \qquad (7.9)\\ b^2 &= a^2 + c^2 - 2ac\cos\beta \\ c^2 &= a^2 + b^2 - 2ab\cos\gamma \end{aligned}$$

The first of these formulas is just Equation 7.8, and the other two can be derived in a similar way by drawing perpendiculars from the other two angles. (This may require extending the sides of the triangle.) Let us consider some examples of the law of cosines.

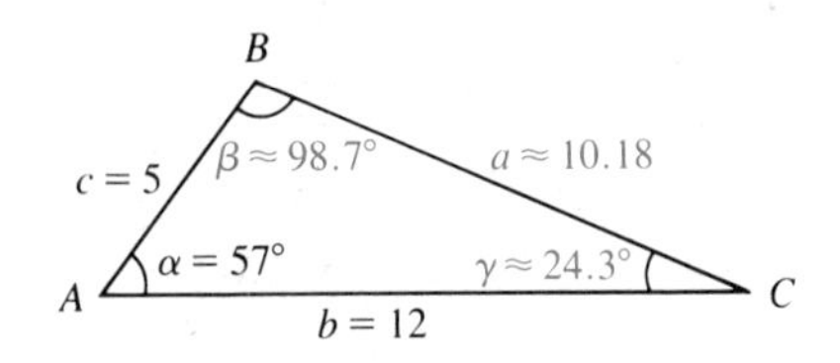

FIGURE 7.34

EXAMPLE 1: Solve the triangle for which $\alpha = 57°$, $b = 12$, and $c = 5$. (See Figure 7.34.)

Solution: According to the first form of the law of cosines,

$$\begin{aligned} a^2 &= b^2 + c^2 - 2bc\cos\alpha \\ &= (12)^2 + 5^2 - 2(12)(5)\cos 57° \\ &= 169 - 120\cos 57° \end{aligned}$$

and so

$$\begin{aligned} a &= \sqrt{169 - 120\cos 57°} \\ &\approx 10.18 \end{aligned}$$

Now we can use the law of sines to determine β and γ. As for β, we have

$$\sin\beta \approx \frac{12\sin 57°}{10.18} \approx 0.9886$$

and so β must be one of the numbers

$$\beta_1 = \sin^{-1} 0.9886 \approx 81.3°$$

or

$$\beta_2 = 180° - \beta_1 \approx 180° - 81.3° = 98.7°$$

As for γ, we have

$$\sin\gamma \approx \frac{5\sin 57°}{10.18} \approx 0.4119$$

and so γ must be one of the numbers.

$$\gamma_1 = \sin^{-1} 0.4119 \approx 24.3°$$

or

$$\gamma_2 = 180° - \gamma_1 \approx 180° - 24.3° = 155.7°$$

Now, since $\alpha = 57°$, we cannot have $\gamma = \gamma_2 \approx 155.7°$, and so $\gamma = \gamma_1 \approx 24.3°$. Then since $180° - \alpha - \gamma \approx 180° - 57° - 24.3° = 98.7°$, we also have $\beta = \beta_2 \approx 98.7°$. The triangle with these measurements is pictured in Figure 7.34.

Try Study Suggestion 7.16. □

Study Suggestion 7.16: Solve the triangle for which $a = 10$, $b = 15$, and $\gamma = 35°$. ◄

In the previous example, we were given the length of two sides and the measure of one angle. Let us consider an example where we are given the measure of all three sides.

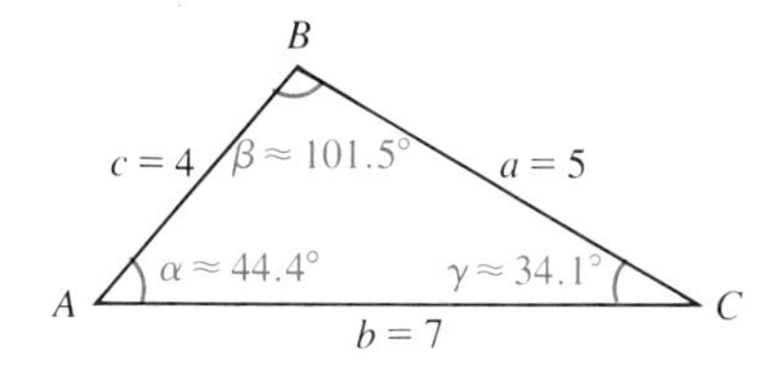

FIGURE 7.35

EXAMPLE 2: Solve the triangle for which $a = 5$, $b = 7$, and $c = 4$. (See Figure 7.35.)

Solution: Solving Equation (7.9) for $\cos\alpha$ gives

$$\cos\alpha = \left(\frac{1}{2bc}\right)(b^2 + c^2 - a^2)$$

$$= \left(\frac{1}{2(7)(4)}\right)(7^2 + 4^2 - 5^2)$$

$$= \frac{5}{7}$$

and so

$$\alpha = \cos^{-1}\frac{5}{7} \approx 44.4^\circ$$

(There is only one angle between 0° and 180° for which $\cos\alpha = 5/7$.) In a similar way, using the second form of the law of cosines, we get

$$\cos\beta = \left(\frac{1}{2ac}\right)(a^2 + c^2 - b^2) = -\frac{1}{5}$$

and so

$$\beta = \cos^{-1}\left(-\frac{1}{5}\right) \approx 101.5^\circ$$

Finally we have

$$\gamma = 180^\circ - \alpha - \beta \approx 180^\circ - 44.4^\circ - 101.5^\circ = 34.1^\circ$$

The triangle with these measurements is pictured in Figure 7.35.

Try Study Suggestion 7.17. □

Study Suggestion 7.17: Solve the triangle for which $a = 6$, $b = 9$, and $c = 5$. ◀

EXAMPLE 3: Solve the triangle for which $a = 1$, $b = 1$, and $c = 3$.

Solution: According to the law of cosines, we have

$$\cos\alpha = \left(\frac{1}{2bc}\right)(b^2 + c^2 - a^2)$$

$$= \left(\frac{1}{2(1)(3)}\right)(1^2 + 3^2 - 1^2)$$

$$= \frac{9}{6}$$

But it is not possible for the cosine of an angle to be greater than 1, and so something is wrong.

The problem is that there is no triangle whose sides have length $a = 1$, $b = 1$, and $c = 3$. After all, the sum of the lengths any two sides of a triangle must be greater than the length of the third side, and that is certainly not the case here.

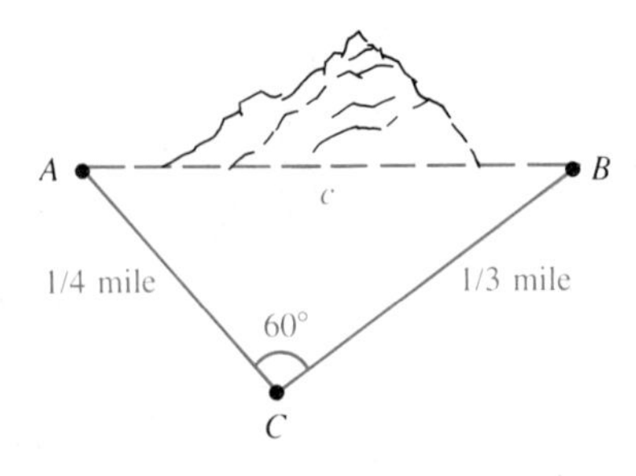

FIGURE 7.36

Study Suggestion 7.18: Two points A and B are on opposite shores of a large lake. A surveyor is located at a point C that is 2.1 miles from point A and 3.5 miles from point B. She measures the angle ACB to be 70°. How far apart are the points A and B? ◀

EXAMPLE 4: A surveyor wishes to measure the distance between two points A and B that are separated by a mountain. (See Figure 7.36.) Since he cannot measure this distance directly, he chooses a point C that is a distance of 1/4 mile from point A and 1/3 mile from point B, and measures angle ACB to be 60°. What is the distance between points A and B?

Solution: Applying the law of cosines to the triangle ABC in Figure 7.36 gives

$$\begin{aligned} c^2 &= \left(\frac{1}{4}\right)^2 + \left(\frac{1}{3}\right)^2 - 2\left(\frac{1}{4}\right)\left(\frac{1}{3}\right)\cos 60^\circ \\ &= \frac{1}{16} + \frac{1}{9} - \frac{1}{12} \\ &= \frac{13}{144} \end{aligned}$$

Hence, the distance between A and B is $c = \sqrt{13/144} \approx 0.3005$ miles.

Try Study Suggestion 7.18. □

Ideas to Remember

Just like the law of sines, the law of cosines can also be very useful for solving triangles.

EXERCISES

In Exercises 1–17, use the given information to solve the triangle (or show that no such triangle can exist).

1. $\alpha = 10^\circ,\ b = 2,\ c = 5$
2. $\alpha = 25^\circ,\ b = 6,\ c = 3$
3. $\beta = \pi/5,\ a = 7,\ c = 10$
4. $\beta = 0.4\pi,\ a = 12,\ c = 12$
5. $\gamma = 30^\circ,\ a = 100,\ b = 200$
6. $\gamma = 75^\circ,\ a = 10,\ b = 8$
7. $\alpha = \pi/4,\ b = 5.1,\ c = 3.2$
8. $\alpha = 32.4^\circ,\ b = 67.3,\ c = 12.9$
9. $a = 3,\ b = 2,\ c = 4$
10. $a = 15,\ b = 29,\ c = 13$
11. $a = 10,\ b = 17,\ c = 12$
12. $a = 100,\ b = 92,\ c = 17$
13. $a = 18,\ b = 13,\ c = 32$

14. $a = 52$, $b = 52$, $c = 52$

15. $a = 100$, $b = 100$, $c = 10$

16. $a = 73.5$, $b = 46.8$, $c = 59.3$

17. $a = 26.92$, $b = 13.87$, $c = 17.99$

18. Find conditions on the numbers a, b, and c that will guarantee that there must exist a triangle whose sides have length a, b, and c.

19. A surveyor wishes to measure the distance between two points A and B, from a third point C. In order to do this, she measures the distance from C to A to be 153 feet, and the distance from C to B to be 196 feet. Also, she measures the angle at C, subtended by the points A and B, to be 63°. Help the surveyor by using this information to compute the distance between the points A and B.

20. A ship leaves the dock, travels due north 86.7 miles and then due east 50 miles. At this point, the engine dies, and the ship is stranded. The radio operator of the ship radios the coast guard station, which happens to be 230 miles due west of the dock. How far is the ship from the coast guard station, and at what bearing?

21. A parallelogram has adjacent sides of length 5 cm and 6 cm. If one diagonal is 8 cm, how long is the other diagonal?

22. A parallelogram $ABCD$ has the property that side AB has length 20 cm, and side BC has length 12 cm. If angle ADB measures 35°, how long are the diagonals of the parallelogram?

23. A car is traveling on a straight road at a speed of 50 mph. At a certain point, someone riding in the car releases a helium filled balloon, which rises in a straight line directly over the road, at a speed of 30 mph and a bearing of 60°. How far apart are the car and the balloon after 30 minutes? How high is the balloon at that time?

24. A 100-foot telephone pole is to be anchored in a vertical position on a hill by attaching a wire to the top of the pole and anchoring the wire to the hill 50 feet above the base of the pole. Suppose that the hill has angle of inclination equal to 17°. How long must the wire be?

25. Two men start at the same point. One man walks due north at a speed of 5 mph, and the other man runs north by northwest (that is, at a bearing of 22.5°) at a speed of 9 mph. How far apart are the men after 10 hours?

26. An airplane flies 120 miles due south, makes an 80° turn to the west, and flies another 100 miles. How far is the airplane from its starting point?

27. A woman wants to measure the height of a certain building. In order to do this, she finds two points A and B on the ground, one on each side of the building, situated so that the two points and the base of the building all lie in the same line. Then she measures the distance between the two points to be 1200 feet, the distance from point A to the top of the building to be 760 feet, and the distance from point B to the top of the building to be 980 feet. Using this information, can she determine the height of the building? If so, do the computations for her, and if not, determine what further information she might need. (Ignore the width of the building.)

28. A regular pentagon (5 sides) is inscribed in a circle of radius 2. Use the law of cosines to find the perimeter of the pentagon. Is there an easier way to compute the perimeter? If so, use it as a check on your results.

29. A 100-pound weight hangs from two anchor points A and B by means of three ropes, as shown in the following figure.

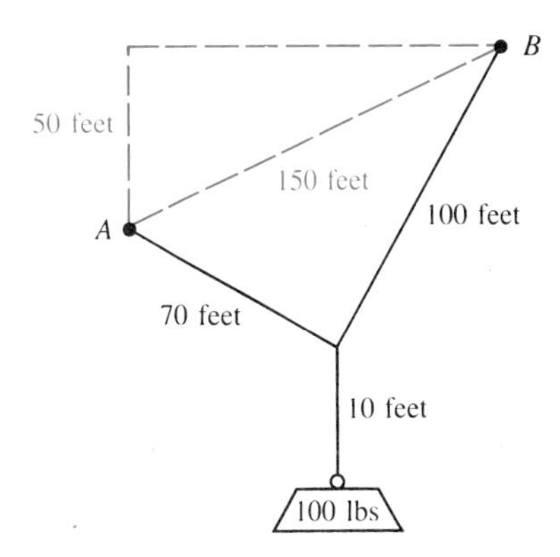

Notice that anchor point B is 50 feet higher than anchor point A. Use the information in this figure to determine the distance from the weight to each of the anchor points.

30. A 200-foot tower sits on the top of a mountain, as shown in the following figure.

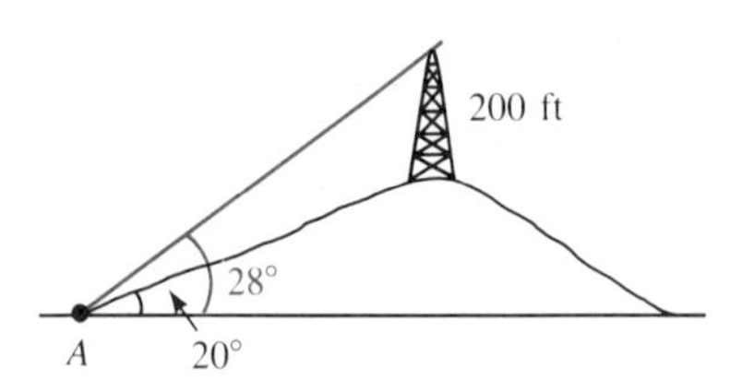

Use the information in this figure to determine the the height of the mountain at its highest point. Also, find the distance from the top of the tower to the bottom of the mountain at point A.

31. Use the law of cosines to show that

$$\frac{\cos\alpha}{a} + \frac{\cos\beta}{b} + \frac{\cos\gamma}{c} = \frac{a^2 + b^2 + c^2}{2abc}$$

32. Referring to Example 3 of Section 6.4, show that the position of the piston is given by the formula

$$x(t) = 2\cos t + \sqrt{45 + 4\cos^2 t}$$

(*Hint:* there is a triangle in Figure 6.32. Apply the law of cosines to this triangle.)

7.4 An Introduction to Polar Coordinates

In Chapter 2, we introduced a coordinate system for the plane, which we called the *Cartesian*, or *rectangular* coordinate system. Since then we have made considerable use of this coordinate system, for example, in graphing equations and functions, and in defining the trigonometric functions. However, the rectangle coordinate system is not the only important coordinate system for the plane. In this section, we will discuss another useful coordinate system.

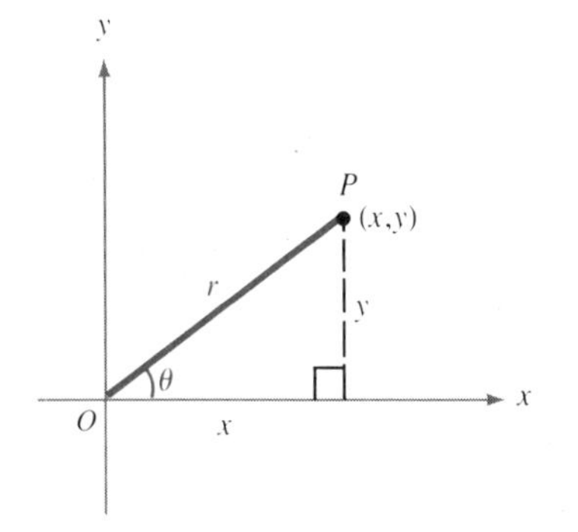

FIGURE 7.37

Figure 7.37 shows the rectangular coordinates (x, y) of a point P in the plane. This point is *uniquely determined* by the ordered pair (x, y), in the sense that there is *one and only one* point with these rectangular coordinates. However, there is another way to determine the point P uniquely, and that is by giving the distance r from the origin O to the point P, as well as the measure θ of the angle that the line segment OP makes with the positive x-axis. (θ can be either the degree measure or the radian measure.)

Thus, not only does the ordered pair (x, y) identify the point P, but so does the ordered pair (r, θ). The numbers r and θ are called the **polar coordinates** of the point P, and we write $P = (r, \theta)$, just as in the case of rectangular coordinates.

The **polar coordinate system** is defined by first choosing a point O in the plane, and then drawing a ray $\mathscr{R}$ with endpoint O, as shown in Figure 7.38. The point O is called the origin of the coordinate system, and the ray

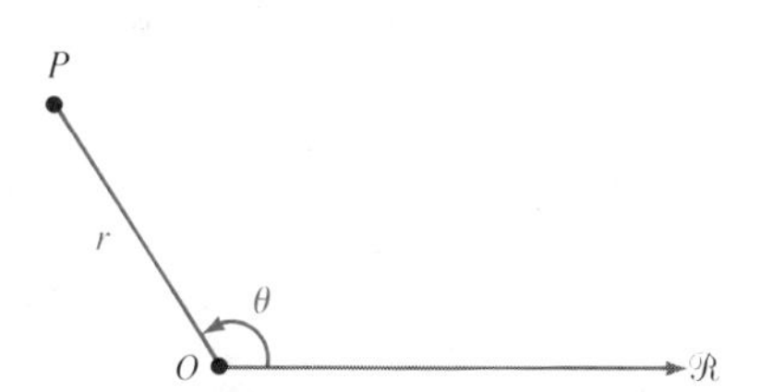

FIGURE 7.38 The polar coordinates of the point P are (r, θ).

$\mathscr{R}$ is called the **polar ray.** (The polar ray may be taken in any direction, but we will always take it to extend horizontally to the right.) Using this coordinate system, any point in the plane can be described by giving its polar coordinates, as shown in Figure 7.38. (Incidentally, polar coordinate graph paper can usually be purchased at a university book store.)

Any ordered pair of the form $(0, \theta)$, for any value of θ, describes the origin in the polar coordinate system. Also, any point on the polar ray has θ-coordinate equal to 0. Let us consider some additional examples.

EXAMPLE 1: Plot the points whose polar coordinates are given below.

(a) $\left(2, \frac{\pi}{2}\right)$ **(b)** $(3, 30°)$ **(c)** $\left(\frac{1}{2}, 0\right)$

(d) $(\sqrt{2}, -45°)$ **(e)** $\left(\frac{3}{2}, \frac{9\pi}{8}\right)$

Solution: These points are shown in Figure 7.39.

Try Study Suggestion 7.19. □

▶ **Study Suggestion 7.19:** Plot the points whose polar coordinates are **(a)** $(1/2, \pi/3)$ **(b)** $(2, -50°)$ **(c)** $(\sqrt{3}, -3\pi/4)$ **(d)** $(6, 75°)$ ◀

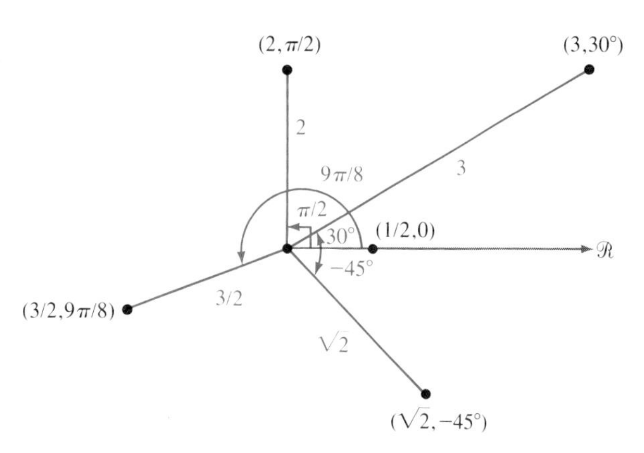

FIGURE 7.39

The polar coordinates of a point P can be described in words as follows. If P has polar coordinates (r, θ), then in order to locate the point P, we begin at the origin, facing in the direction of the ray $\mathscr{R}$. Then we turn through an angle θ, and walk a distance r. This will bring us to the point P.

Since the first coordinate r is the distance from the origin to the point $P = (r, \theta)$, we see that r must always be nonnegative. However, it is convenient to extend the definition of polar coordinates to allow r to be negative. If $r < 0$, then we can describe the location of the point $P = (r, \theta)$ in words as follows. As before, we start at the origin, facing in the direction of the ray $\mathscr{R}$. Then we turn through an angle θ, and walk *backwards* a distance $|r| = -r$.

This will bring us to the point P. Of course, this is the same as turning an additional π radians (or 180°) and then walking *forward* a distance $|r|$. Thus, we can make the following statements.

1. **If $r < 0$ and θ is measured in radians, the point (r, θ) is the same as the point $(|r|, \theta + \pi)$.**
2. **If $r < 0$ and θ is measured in degrees, the point (r, θ) is the same as the point $(|r|, \theta + 180^\circ)$.**

EXAMPLE 2: Plot the points whose polar coordinates are

(a) $\left(-1, \dfrac{\pi}{4}\right)$ **(b)** $(-5, 0)$ **(c)** $(-\sqrt{3}, 120^\circ)$ **(d)** $(-2, -105^\circ)$

Solution: These points are shown in Figure 7.40.

▶ **Study Suggestion 7.20:** Plot the points whose polar coordinates are **(a)** $(-1/2, \pi)$ **(b)** $(-2, -360^\circ)$ **(c)** $(-\sqrt{2}, \pi/4)$ ◀

Try Study Suggestion 7.20. □

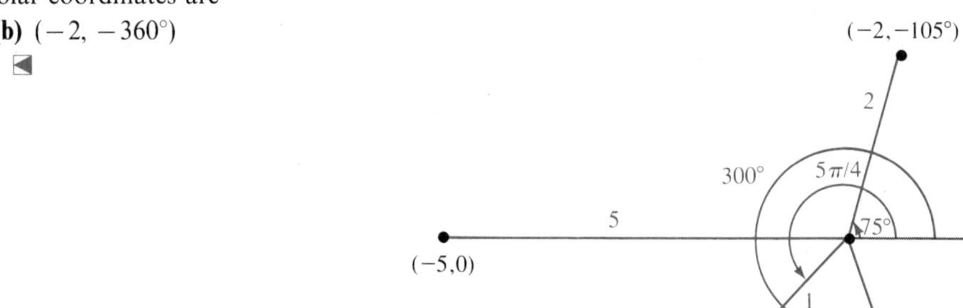

FIGURE 7.40

The polar coordinate system has one disadvantage that the rectangular coordinate system does not have. While every point in the plane has exactly one pair of rectangular coordinates, every point in the plane has infinitely many pairs of polar coordinates!

In particular, if the point P has polar coordinates (r, θ), then it also has polar coordinates $(r, \theta + 2\pi n)$ for any integer n. Furthermore, since we are allowing r to be negative, P also has polar coordinates $(-r, \theta + \pi + 2\pi n)$ for any integer n. In summary, we can say that

the polar coordinates $(r, \theta + 2\pi n)$ and $(-r, \theta + \pi + 2\pi n)$ describe the same point in the plane for all integers n. **(7.10)**

Of course, Statement (7.10) has a "degree version" obtained by replacing 2π by 360° and measuring all angles in degrees.

EXAMPLE 3: Describe all of the polar coordinates of the following points.

(a) $P = \left(3, -\frac{\pi}{5}\right)$ **(b)** $Q = (-5, 75°)$

Solutions:

(a) Statement (7.10) with $r = 3$ and $\theta = -\pi/5$ tells us that all of the polar coordinates of P are given by

$$(3, \frac{-\pi}{5} + 2\pi n) \qquad \text{and} \qquad (-3, \frac{4\pi}{5} + 2\pi n)$$

as n varies over all integers.

(b) The degree version of Statement (7.10), with $r = -5$ and $\theta = 75°$ tells us that all of the polar coordinates of Q are given by

$$(-5, 75° + 360° \cdot n) \qquad \text{and} \qquad (5, 255° + 360° \cdot n)$$

as n varies over all integers. *Try Study Suggestions 7.21.* □

▶ **Study Suggestion 7.21:** Describe all of the polar coordinates of the following points. **(a)** $P = (4, \pi/3)$ **(b)** $Q = (-2, 110°)$ ◀

In many applications, it is necessary to convert from polar coordinates to rectangular coordinates and vice-versa. In order to see how this should be done, we look again at Figure 7.37, and imagine that a polar coordinate system has been set up in such a way that the origin O coincides with the origin of the rectangular coordinate system, and the ray $\mathcal{R}$ coincides with the positive x-axis. Then we have

$$\cos\theta = \frac{x}{r} \qquad \text{and} \qquad \sin\theta = \frac{y}{r}$$

and so

$$x = r\cos\theta \qquad \text{and} \qquad y = r\sin\theta \tag{7.11}$$

It can be shown that these equations hold for all values of θ and r (both positive and negative), and they can be used to convert polar coordinates to rectangular coordinates.

EXAMPLE 4: Convert the following from polar coordinates to rectangular coordinates.

(a) $\left(2, \frac{\pi}{4}\right)$ **(b)** $(-\sqrt{2}, 170°)$

Solutions:

(a) Substituting $r = 2$ and $\theta = \pi/4$ into Equations (7.11) gives

$$x = 2\cos\frac{\pi}{4} = 2\frac{\sqrt{2}}{2} = \sqrt{2}$$

and

$$y = 2\sin\frac{\pi}{4} = 2\frac{\sqrt{2}}{2} = \sqrt{2}$$

Thus, the point with polar coordinates $(2, \pi/4)$ has rectangular coordinates $(\sqrt{2}, \sqrt{2})$.

(b) Substituting $r = -\sqrt{2}$ and $\theta = 170°$ into Equations (7.11) gives

$$x = -\sqrt{2}\cos 170° \approx -\sqrt{2}(-0.9848) \approx 1.3927$$

and

$$y = -\sqrt{2}\sin 170° \approx -\sqrt{2}(0.1736) \approx -0.2455$$

Hence, the point whose polar coordinates are $(\sqrt{2}, 50°)$ has (approximate) rectangular coordinates $(1.3927, -0.2455)$.

Try Study Suggestion 7.22. □

▶ **Study Suggestion 7.22:** Convert the following from polar coordinates to rectangular coordinates.

(a) $(3, 5\pi/6)$

(b) $(-1/2, -100°)$ ◀

Formulas for converting rectangular coordinates to polar coordinates can be obtained by looking again at Figure 7.37. The Pythagorean theorem and the definition of the tangent gives

$$r^2 = x^2 + y^2 \tag{7.12}$$

$$\tan\theta = \frac{y}{x} \tag{7.13}$$

These equations do not uniquely determine the polar coordinates r and θ from the rectangular coordinates x and y. However, by determining separately the quadrant that contains the point in question, these equations can be used to find its polar coordinates. Let us consider some examples.

EXAMPLE 5: Convert the following from rectangular coordinates to polar coordinates satisfying $r \geq 0$ and $0 \leq \theta < 2\pi$.

(a) $(2, 2)$ **(b)** $(3, -5)$

Solutions:

(a) Substituting the values $x = 2$ and $y = 2$ into Equation (7.12) gives

$$r^2 = 2^2 + 2^2 = 8$$

and so $r = \sqrt{8} = 2\sqrt{2}$. Substituting these same values into Equation (7.13) gives

$$\tan\theta = \frac{2}{2} = 1$$

Now we look for all values of θ *in the interval* $0 \leq \theta < 2\pi$ that satisfy this equation. Referring to a unit circle, we see that θ can be either of the values $\pi/4$ or $5\pi/4$. Hence, we get the *possibilities* $(2\sqrt{2}, \pi/4)$ and $(2\sqrt{2}, 5\pi/4)$. However, the point with rectangular coordinates (2, 2) lies in the first quadrant, and this eliminates the second possibility. (The point with polar coordinates $(2\sqrt{2}, 5\pi/4)$ lies in the third quadrant.) Thus, the point with rectangular coordinates (2, 2) has polar coordinates $(2\sqrt{2}, \pi/4)$.

(b) Substituting $x = 3$ and $y = -5$ into Equation (7.12) gives

$$r^2 = 3^2 + (-5)^2 = 34$$

and so $r = \sqrt{34}$. Equation (7.13) gives

$$\tan\theta = -\frac{5}{3}$$

and this leads again to two possibilities for θ. In order to obtain these possibilities, we first observe that

$$\tan^{-1}\left(-\frac{5}{3}\right) \approx -59.04^\circ$$

Now -59.04° is not in the interval $0 \le \theta < 2\pi$, but since the tangent repeats itself every 180° we can add 180° to get $\theta \approx -59.04^\circ + 180^\circ = 120.96^\circ$. Adding another 180° gives the other possibility $\theta \approx 120.96^\circ + 180^\circ = 300.96^\circ$. Thus, we have the two possibilities $(\sqrt{34}, 120.96^\circ)$ and $(\sqrt{34}, 300.96^\circ)$. But the point whose rectangular coordinates are $(3, -5)$ lies in the fourth quadrant, and so its (approximate) polar coordinates are $(\sqrt{34}, 300.96^\circ)$.

Try Study Suggestion 7.23. □

▶ **Study Suggestion 7.23:** Convert the following from rectangular coordinates to polar coordinates satisfying $r \ge 0$ and $0 \le \theta < 2\pi$.
(a) $(-\sqrt{3}, 1)$ **(b)** $(-2, 4)$ ◀

Since the polar coordinate system has the disadvantage of assigning an infinite number of different ordered pairs to the same point, you might be wondering why we bother with this coordinate system in the first place. Although you will come to understand the importance of the polar coordinate system if you continue to study mathematics and the sciences, we can give you one reason now. *The polar coordinate system can be used to describe certain types of curves in the plane much more easily than the rectangular coordinate system.*

As a simple example, the circle with center at the origin and radius 2 has the rectangular equation

$$x^2 + y^2 = 4$$

where x and y are *rectangular* coordinates. However, we can also describe this circle by using polar coordinates. Since a point P is on this circle if and only if its polar coordinates have the form $(2, \theta)$, where θ is *any* value, the circle has the polar equation

$$r = 2$$

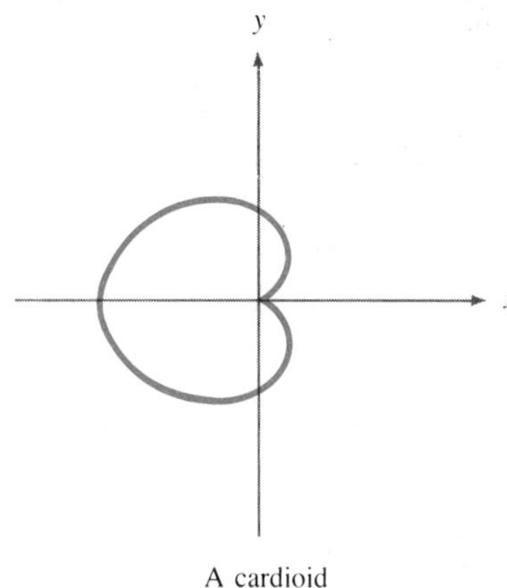

A cardioid
Polar equation: $r = 1 - \cos\theta$

FIGURE 7.41

As you can see, the polar equation of this circle is simpler than the rectangular equation. (In fact, it only involves one of the two polar coordinates.) Of course, in this simple example the rectangular equation is not really very complicated. However, let us consider the curve in Figure 7.41, known as a cardioid because of its heart-like shape. The polar equation of this cardioid is

$$r = 1 - \cos\theta$$

and its rectangular equation is

$$x^4 + 2x^2y^2 + y^4 + 2x^3 + 2xy^2 - y^2 = 0$$

Thus, in this case, the polar equation is considerably simpler than the rectangular equation.

Some other intertesting curves, along with their polar equations, are shown in Figures 7.42, 7.43, and 7.44. If you take calculus, you will get a chance to study these curves more closely.

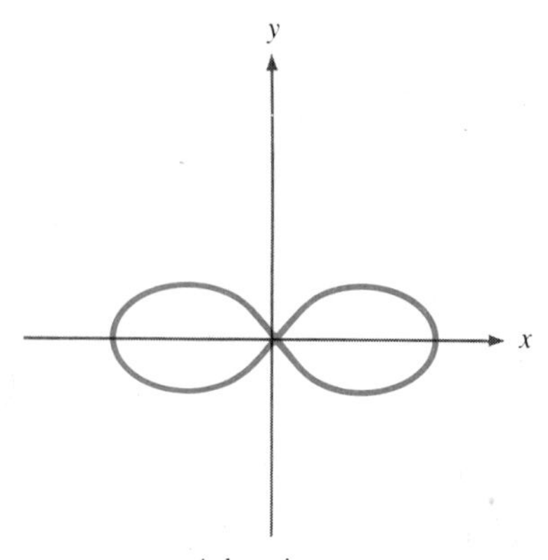

A lemniscate
Polar equation: $r^2 = 2\cos 2\theta$

FIGURE 7.42

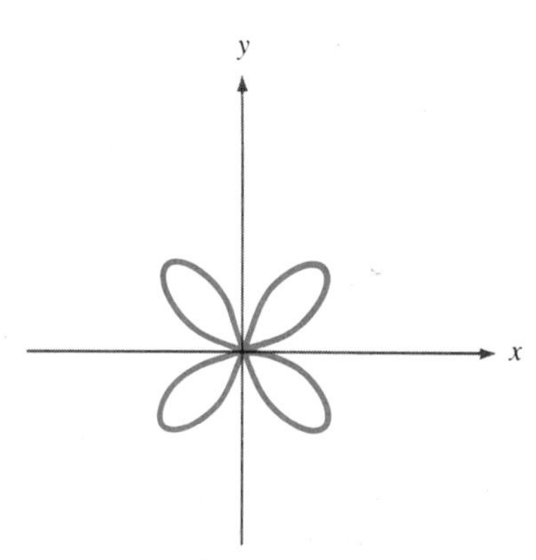

A 4-petal rose
Polar equation: $r = \sin 2\theta$

FIGURE 7.43

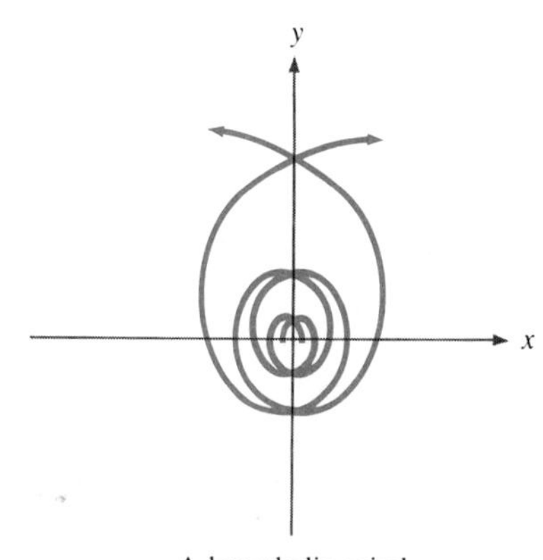

A hyperbolic spiral
Polar equation: $r = \frac{1}{\theta}$

FIGURE 7.44

Ideas to Remember

- The rectangular coordinate system is not the only useful coordinate system for the plane.
- The polar coordinate system can be used to describe certain types of curves in the plane more easily than the rectangular coordinate system.

EXERCISES

In Exercises 1–4, the polar coordinates of four points are given. Plot these points in the same plane.

1. $\left(6, -\frac{\pi}{4}\right), \left(-3, \frac{\pi}{6}\right), \left(-1, \frac{3\pi}{4}\right), \left(4, -\frac{2\pi}{3}\right)$

2. $(10, 30^\circ), (18, -40^\circ), (-20, -15^\circ), (14, 200^\circ)$

3. $\left(\frac{1}{2}, \frac{\pi}{10}\right), \left(\sqrt{2}, \frac{5\pi}{2}\right), \left(0, -\frac{\pi}{2}\right), (0, \pi)$

4. $(\pi, \pi), (0, \pi), (\pi, 0), (-\pi, -\pi)$

In Exercises 5–16, describe all of the polar coordinates of the given point.

5. $\left(2, \frac{\pi}{4}\right)$ **6.** $\left(-3, \frac{\pi}{2}\right)$ **7.** $\left(\frac{8}{3}, -\frac{\pi}{3}\right)$

8. $(1, 1)$ **9.** $\left(-\frac{1}{2}, \frac{3\pi}{4}\right)$ **10.** $(0, 0)$

11. $(5, 0)$ **12.** $(10, 10^\circ)$ **13.** $(-\sqrt{2}, 35^\circ)$

14. $(16, 135^\circ)$ **15.** $(-7, -125^\circ)$ **16.** $(-2.5, 60^\circ)$

In Exercises 17–28, convert the polar coordinates to rectangular coordinates.

17. $\left(1, \frac{\pi}{3}\right)$ **18.** $\left(2, \frac{\pi}{6}\right)$ **19.** $\left(-4, \frac{\pi}{2}\right)$

20. $\left(\frac{3}{2}, \frac{3\pi}{4}\right)$ **21.** $\left(-\sqrt{2}, -\frac{5\pi}{6}\right)$

22. $(10, 120^\circ)$ **23.** $(-5, -45^\circ)$

24. $(3, 225^\circ)$ **25.** $(2.1, -150^\circ)$

26. $(0, 12^\circ)$ **27.** $\left(5, -\frac{7\pi}{4}\right)$

28. $(1, 0^\circ)$

In Exercises 29–37, use a calculator to convert from polar coordinates to rectangular coordinates. (Some scientific calculators have special keys for this purpose. Do *not* use them for these exercises.)

29. $\left(\sqrt{5}, \frac{\pi}{5}\right)$ **30.** $\left(-2\sqrt{3}, \frac{\pi}{11}\right)$

31. $\left(12.2, -\frac{5\pi}{8}\right)$ **32.** $\left(-1, -\frac{6\pi}{7}\right)$

33. $(2.15, 0.3\pi)$ **34.** $(-2, 105^\circ)$

35. $(6.9, -12.8^\circ)$ **36.** $(7.1, 136.5^\circ)$

37. $(-4, -263^\circ)$

In Exercises 38–49, convert from rectangular coordinates to polar coordinates satisfying $r \geq 0$ and $0 \leq \theta < 2\pi$.

38. $(1, 1)$ **39.** $(-2, 2)$

40. $\left(-\frac{1}{\sqrt{5}}, -\frac{1}{\sqrt{5}}\right)$ **41.** $(\pi, -\pi)$

42. $(1, \sqrt{3})$ **43.** $(2\sqrt{3}, 2)$

44. $(-4, -4\sqrt{3})$ **45.** $\left(0, -\frac{1}{3}\right)$

46. $(14, 0)$ **47.** $\left(0, -\frac{3}{2}\right)$

48. $\left(\frac{\sqrt{3}}{3}, -1\right)$ **49.** $(-7\sqrt{3}, 7)$

In Exercises 50–58, use a calculator to convert from rectangular coordinates to polar coordinates satisfying $r \geq 0$ and $0 \leq \theta < 2\pi$. (Some scientific calculators have special keys for this purpose. Do *not* use them for these exercises.)

50. $(1, 1.1)$ **51.** $(5, 6)$ **52.** $(-8, 3)$

53. $\left(-\frac{1}{2}, -\frac{7}{3}\right)$ **54.** $(6.88, 14.22)$ **55.** $(125, 200)$

56. $(5, -19)$ **57.** $(-6.1, -3.2)$ **58.** $(1, 1000)$

59. Graph the polar equation $r = 5$.

60. Graph the polar equation $r = \sqrt{3}$.

61. Graph the polar equation $\theta = \frac{\pi}{4}$.

62. Graph the polar equation $\theta = -\frac{\pi}{3}$.

7.5 Vectors

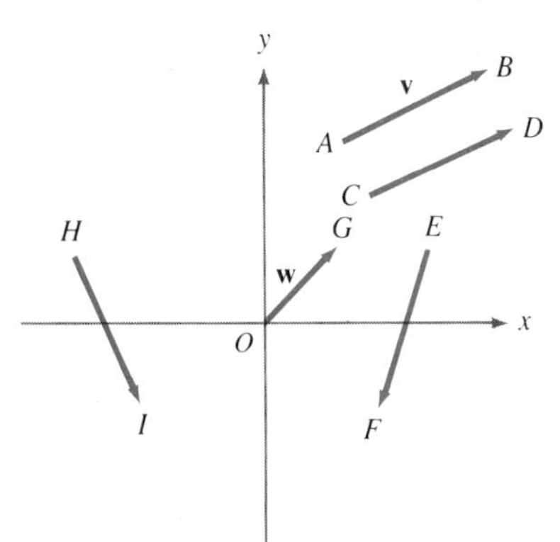

FIGURE 7.45

Many quantities in nature have both a size, or magnitude, and a *direction.* For example, if you push against an object, such as a book, then you are exerting a *force* against the object and this force has both a magnitude (how hard you push) and a direction.

Another example of a physical quantity that has both a magnitude and a direction is *velocity.* For instance, a rocket traveling in space has both a speed (measured perhaps in miles per hour) and a direction.

Physical quantities that have both a magnitude and a direction are called **vectors.** This is to distinguish them from quantities, called **scalars,** that have only a magnitude. For example, speed, length, volume, and mass are scalars, whereas velocity and force are vectors.

The Cartesian coordinate system provides us with a very convenient way to describe vectors. If we restrict our attention to directions in two dimensions only, a vector can be represented as an *arrow* in the plane. The length of the arrow gives the magnitude of the vector, and the direction of the arrow gives the direction of the vector. In fact, this representation is so common that we usually refer to the arrow itself as a vector.

Figure 7.45 gives several examples of vectors in the plane. There are two common notations for vectors. One is to use a lower case boldface letter such as **v** or **w** and the other is to give the **initial point** and the **terminal point** of the vector. For example, the initial point of the vector **v** in Figure 7.45 is the point A and the terminal point of **v** is the point B. Thus, we can denote the vector **v** by $\overrightarrow{AB}$. (Note that unlike line segments, the *vector* $\overrightarrow{AB}$ is different from the *vector* $\overrightarrow{BA}$, since they have opposite directions.)

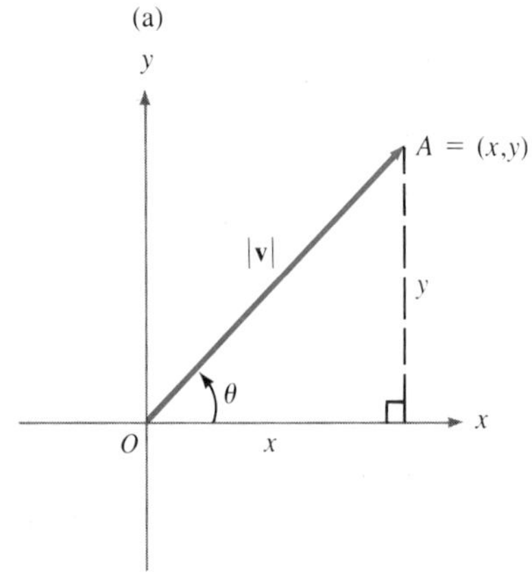

Since the arrows $\overrightarrow{AB}$ and $\overrightarrow{CD}$ in Figure 7.45 have the same magnitude and the same direction, they actually represent the same vector; that is, $\overrightarrow{AB} = \overrightarrow{CD}$. In fact, we can say that

> **two arrows represent the same vector if and only if they have the same magnitude (length) and the same direction; that is, if and only if one can be obtained from the other by a parallel translation.**

Notice that the vector $\mathbf{w} = \overrightarrow{OG}$ in Figure 7.45 has its initial point at the origin. If the point G has coordinates (x, y), then we denote the *vector* $\overrightarrow{OG}$ by $\langle x, y \rangle$, to distinguish it from the *point* $G = (x, y)$.

The **magnitude of a vector v** is denoted by $|\mathbf{v}|$. (The magnitude of a vector is similar to the absolute value of a scalar, which is why the same notation is used.)

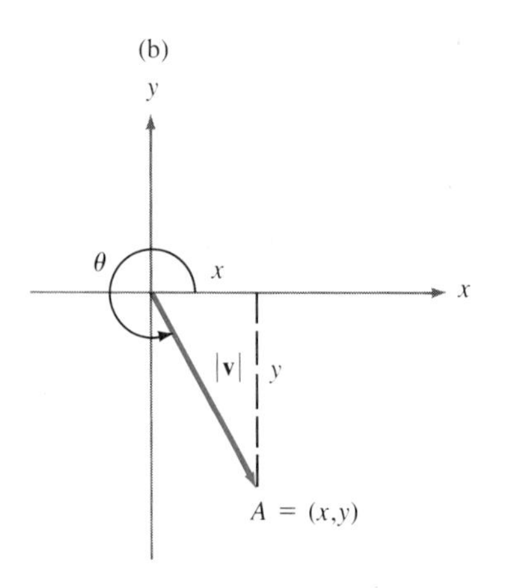

FIGURE 7.46

Figure 7.46(a) shows a vector $\mathbf{v} = \overrightarrow{OA} = \langle x, y \rangle$. The angle θ that the vector makes with the positive x-axis is called a **direction angle** of **v**. The direction angle of a vector **v** that lies in the fourth quadrant is shown in Figure 7.46(b). Of course, if θ is a direction angle for a vector **v**, then so is any angle of the form $\theta + n \cdot 360°$, where n is any integer (positive or negative.)

The vector $\mathbf{0} = \langle 0, 0 \rangle$ each of whose coordinates is equal to 0, is called the **zero vector.** The magnitude of the zero vector is 0 and it is the *only* vector that does not have a direction (and hence no direction angle.)

There are two simple ways to describe a nonzero vector $\mathbf{v} = \overrightarrow{OA} = \langle x, y \rangle$ whose initial point is at the origin. One way is to give the coordinates x and y of the point A, and the other way is to give both the magnitude $|\mathbf{v}|$ of the vector and a direction angle θ. As you can see from Figure 7.46, the quantities x, y, $|\mathbf{v}|$ and θ are related by the following equations:

$$x = |\mathbf{v}| \cos \theta \quad \text{and} \quad y = |\mathbf{v}| \sin \theta \tag{7.14}$$

$$|\mathbf{v}| = \sqrt{x^2 + y^2} \quad \text{and} \quad \tan \theta = \frac{y}{x} \quad \text{for } x \neq 0 \tag{7.15}$$

Let us consider some examples of the use of these equations.

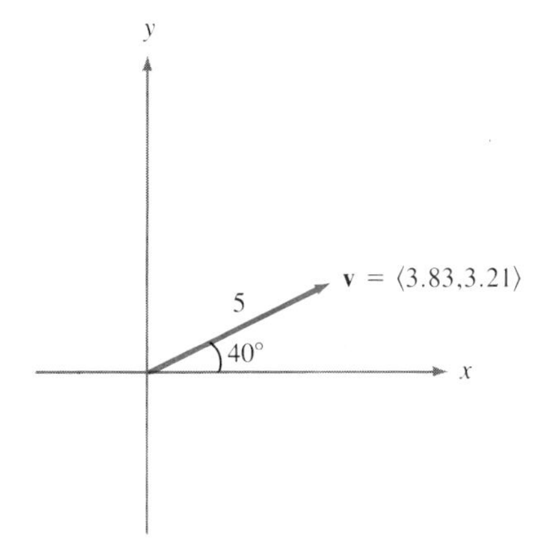

FIGURE 7.47

EXAMPLE 1: Find the coordinates of the vector $\mathbf{v}$ with magnitude 5 and direction angle 40°. (See Figure 7.47.)

Solution: Using Equations (7.14), we get

$$x = |\mathbf{v}| \cos \theta = 5 \cos 40° \approx 3.83$$

and

$$y = |\mathbf{v}| \sin \theta = 5 \sin 40° \approx 3.21$$

Hence $\mathbf{v} = \langle 3.83, 3.21 \rangle$. (These coordinates are only approximations.)

Try Study Suggestion 7.24. □

Study Suggestion 7.24: Find the coordinates of the vector $\mathbf{v}$ with magnitude 7 and direction angle 150°. Sketch the vector in the plane. ◀

EXAMPLE 2: Find the magnitude and direction angle of the vector $\mathbf{v} = \langle -2, 3 \rangle$. (See Figure 7.48.)

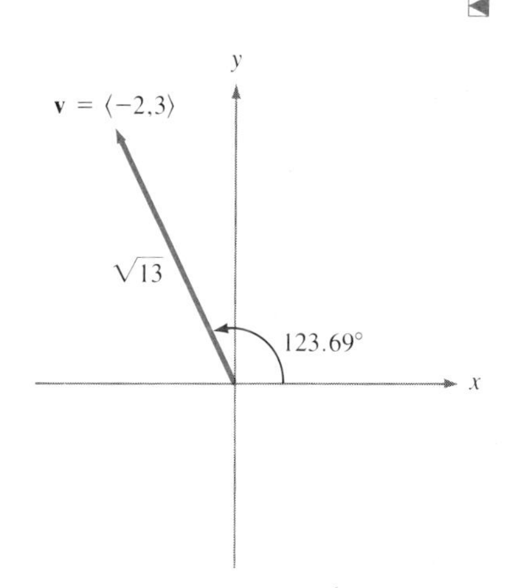

FIGURE 7.48

Solution: The magnitude of $\mathbf{v}$ can be obtained from the first equation in (7.15),

$$|\mathbf{v}| = \sqrt{x^2 + y^2} = \sqrt{4 + 9} = \sqrt{13}$$

In order to find a direction angle, we use the second equation in (7.15), which tells us that

$$\tan \theta = \frac{y}{x} = \frac{3}{-2} = -\frac{3}{2}$$

Now, since $\tan^{-1}(-3/2) \approx -56.31°$, one possibility for the direction angle of $\mathbf{v}$ is $-56.31°$, and another possibility is $-56.31° + 180° = 123.69°$. (Both of these angles have tangent equal to $-3/2$.) In order to determine which of these two possibilities is correct, we observe that $\mathbf{v} = \langle -2, 3 \rangle$ lies in the second quadrant, and so its direction angle must be 123.69°. (Any angle of the form $123.69° + n \cdot 360°$ is also a direction angle for **v**.)

Try Study Suggestion 7.25. □

Study Suggestion 7.25: Find the magnitude and direction angle of the vector $\mathbf{v} = \langle -3, -4 \rangle$. ◀

As we have said, forces can be represented by vectors in the plane. Let us imagine that we have two forces, represented by the vectors $\mathbf{v} = \langle a, b \rangle$ and $\mathbf{w} = \langle c, d \rangle$, both of which are acting at the same point on an object, such as a book. This situation is illustrated in Figure 7.49. Then it is a law of physics that these two forces combine into one force that is represented by the vector

$$\mathbf{u} = \langle a + c, b + d \rangle$$

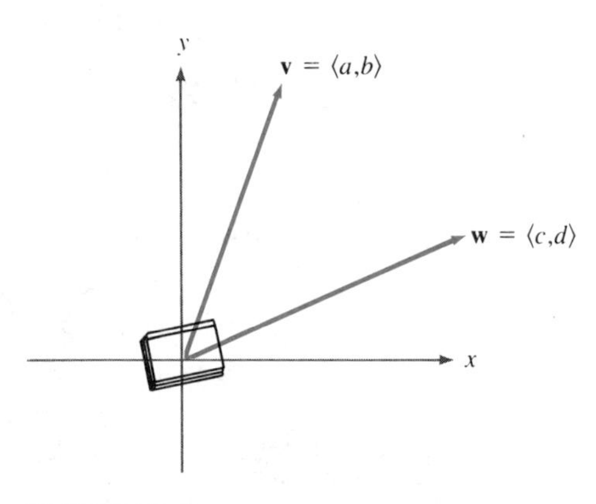

FIGURE 7.49

In other words, according to the laws of physics, if two forces are acting together, the result is equivalent to one force whose coordinates can be obtained by *adding* the corresponding coordinates of each force. For this reason, we are motivated to make the following definition.

DEFINITION

Let $\mathbf{v} = \langle a, b \rangle$ and $\mathbf{w} = \langle c, d \rangle$ be vectors. Then we define the **sum** $\mathbf{v} + \mathbf{w}$ to be the vector

$$\mathbf{v} + \mathbf{w} = \langle a + c, b + d \rangle$$

Addition of vectors can be pictured geometrically, as in Figure 7.50, where the sum $\mathbf{v} + \mathbf{w}$ is the diagonal of the parallelogram formed by the vectors **v** and **w**. For this reason, the rule for adding vectors is sometimes referred to as the **parallelogram law of addition.**

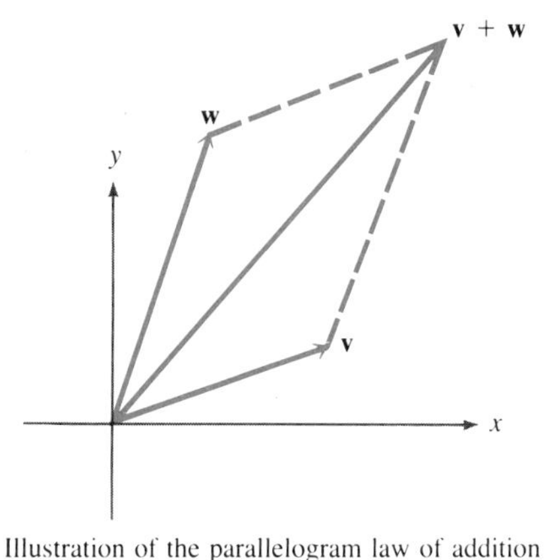

Illustration of the parallelogram law of addition of vectors

FIGURE 7.50

FIGURE 7.51

EXAMPLE 3: Let $\mathbf{v} = \langle 2, 3\rangle$ and $\mathbf{w} = \langle -1, 5\rangle$. Find the sum $\mathbf{v} + \mathbf{w}$. Sketch all three vectors in the plane.

Solution: According to the definition of addition of vectors, we have

$$\mathbf{v} + \mathbf{w} = \langle 2, 3\rangle + \langle -1, 5\rangle = \langle 2 + (-1), 3 + 5\rangle = \langle 1, 8\rangle$$

These vectors are shown in Figure 7.51.

▶ **Study Suggestion 7.26:** Let $\mathbf{v} = \langle -4, -3\rangle$ and $\mathbf{w} = \langle 2, -4\rangle$. Find the sum $\mathbf{v} + \mathbf{w}$. Sketch all three vectors in the plane. ◀

Try Study Suggestion 7.26. □

EXAMPLE 4: Two men are attempting to drag a tree stump along the ground, as shown in Figure 7.52(a). One man is exerting a force $\mathbf{v}$ of magnitude 100 newtons (a newton is a measure of force, named after Sir Issac Newton), and the other man is exerting a force $\mathbf{w}$ of magnitude 85 newtons. What is the total effect of the two forces on the tree stump? That is, what is the magnitude and direction of the total force acting on the stump?

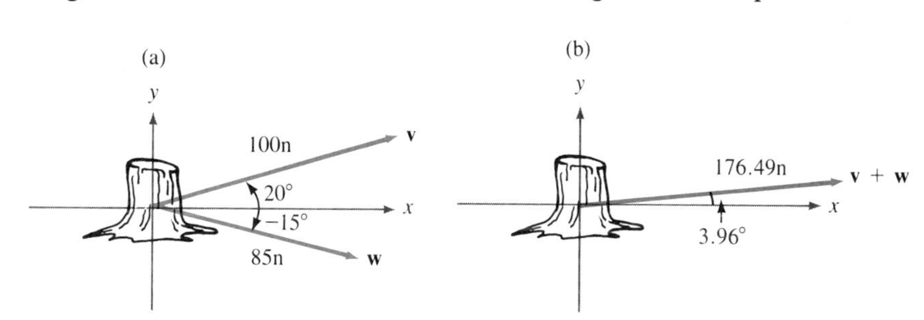

FIGURE 7.52

Solution: The force $\mathbf{v}$ has magnitude $|\mathbf{v}| = 100$ and direction angle $\theta = 20°$. Therefore, the coordinates of $\mathbf{v}$ are

$$x = 100 \cos 20° \approx 93.97$$

and

$$y = 100 \sin 20° \approx 34.20$$

and so $\mathbf{v} \approx \langle 93.97, 34.20 \rangle$. Similarly, the magnitude of the vector $\mathbf{w}$ is $|\mathbf{w}| = 85$ and its direction angle is $-15°$. Hence, the coordinates of $\mathbf{w}$ are

$$x = 85 \cos(-15°) \approx 82.10$$

and

$$y = 85 \sin(-15°) \approx -22.00$$

and so $\mathbf{w} \approx \langle 82.10, -22.00 \rangle$.

Now the total force acting on the tree stump is equal to the *sum* $\mathbf{v} + \mathbf{w}$, which is

$$\begin{aligned} \mathbf{v} + \mathbf{w} &\approx \langle 93.97, 34.20 \rangle + \langle 82.10, -22.00 \rangle \\ &= \langle 93.97 + 82.10, 34.20 - 22.00 \rangle \\ &= \langle 176.07, 12.20 \rangle \end{aligned}$$

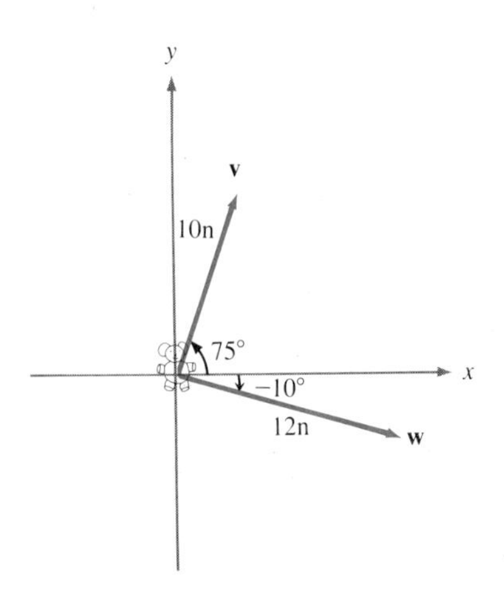

FIGURE 7.53

The vector $\mathbf{v} + \mathbf{w}$ has magnitude

$$|\mathbf{v} + \mathbf{w}| \approx \sqrt{(176.07)^2 + (12.20)^2} \approx 176.49$$

and direction angle

$$\theta \approx \tan^{-1} \frac{12.20}{176.07} \approx \tan^{-1}(0.0693) \approx 3.96°$$

(This is the correct direction angle, and not $3.96° + 180° = 183.96°$. Why?) Thus, the tree stump is acted upon by a total force of approximately 176.49 newtons, in the direction 3.96° above the positive x-axis. This force is shown in Figure 7.52(b). *Try Study Suggestion 7.27.* □

▶ **Study Suggestion 7.27:** Two children are pulling in different directions on their favorite toy. One child is exerting a force $\mathbf{v}$ on the toy, and the other child is exerting a force $\mathbf{w}$. The magnitudes and directions of these forces are given in Figure 7.53. Find the total effect of the two forces on the toy. ◀

DEFINITION

Let $\mathbf{v} = \langle a, b \rangle$ and $\mathbf{w} = \langle c, d \rangle$ be vectors. Then we define the **difference** $\mathbf{v} - \mathbf{w}$ to be the vector

$$\mathbf{v} - \mathbf{w} = \langle a - c, b - d \rangle$$

EXAMPLE 5: Let $\mathbf{v} = \langle 2, -3 \rangle$ and $\mathbf{w} = \left\langle 6, -\frac{5}{2} \right\rangle$. Find $\mathbf{v} - \mathbf{w}$ and $\mathbf{w} - \mathbf{v}$.

Solution: According to the definition, we have

$$\mathbf{v} - \mathbf{w} = \left\langle 2 - 6, (-3) - \left(-\frac{5}{2}\right) \right\rangle = \left\langle -4, -\frac{1}{2} \right\rangle$$

and

$$\mathbf{w} - \mathbf{v} = \left\langle 6 - 2, -\frac{5}{2} - (-3) \right\rangle = \left\langle 4, \frac{1}{2} \right\rangle$$

Try Study Suggestion 7.28. □

▶ **Study Suggestion 7.28:** Find $\mathbf{v} - \mathbf{w}$ and $\mathbf{w} - \mathbf{v}$ if $\mathbf{v} = \langle 12, 1/2 \rangle$ and $\mathbf{w} = \langle -10, 3/2 \rangle$. ◀

If $\mathbf{v} = \langle a, b \rangle$ is a vector and r is a real number, then we define the product $r\mathbf{v}$ to be the vector

$$r\mathbf{v} = r\langle a, b \rangle = \langle ra, rb \rangle$$

In words, the product $r\mathbf{v}$ is obtained by multiplying each coordinate of the vector $\mathbf{v}$ by the real number r. Since real numbers are sometimes referred to as scalars, this type of multiplication (of a vector by a real number) is called **scalar multiplication.**

Computing the magnitude of the scalar product $r\mathbf{v}$ gives

$$\begin{aligned} |r\mathbf{v}| &= \sqrt{(ra)^2 + (rb)^2} \\ &= \sqrt{r^2a^2 + r^2b^2} \\ &= \sqrt{r^2(a^2 + b^2)} \\ &= \sqrt{r^2}\sqrt{a^2 + b^2} \\ &= |r|\,|\mathbf{v}| \end{aligned}$$

and so

$$|r\mathbf{v}| = |r|\,|\mathbf{v}|$$

where $|r|$ is the *absolute value* of the real number r and $|\mathbf{v}|$ is the *magnitude* of the vector $\mathbf{v}$. Thus *the magnitude of* $r\mathbf{v}$ *is equal to the absolute value of* r *times the magnitude of* $\mathbf{v}$.

The effect of multiplying a vector $\mathbf{v}$ by a real number r is shown in Figure 7.54. As you can see, if r is positive the vector $r\mathbf{v}$ has the same

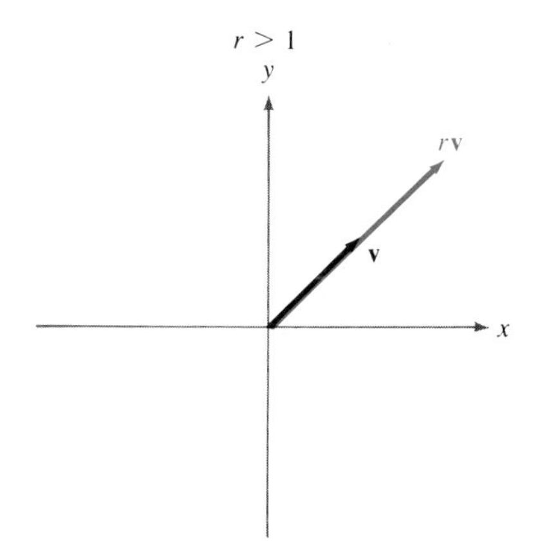

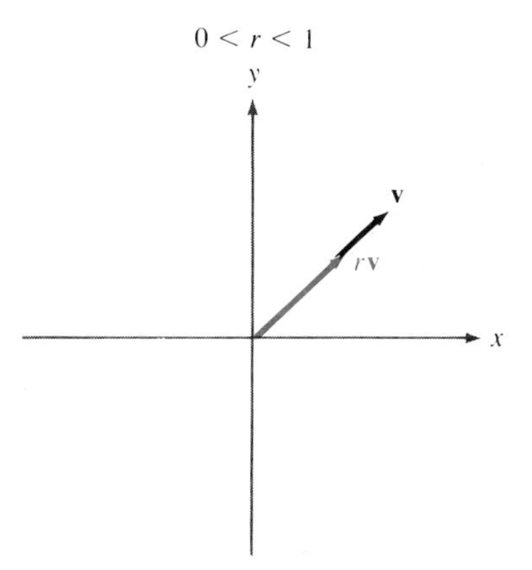

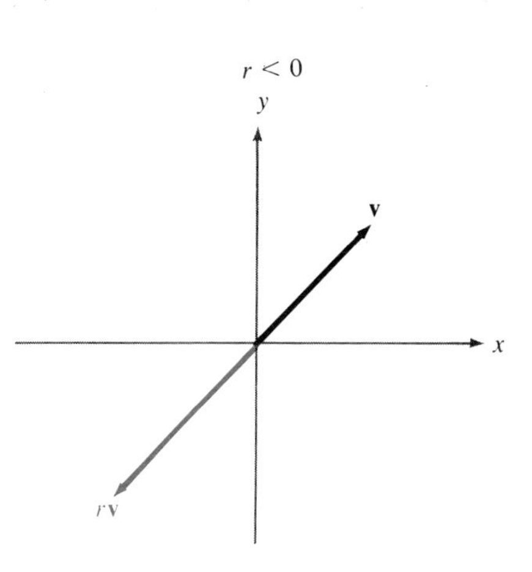

FIGURE 7.54

direction as $\mathbf{v}$, but its length is multiplied by r. However, if r is negative the vector $r\mathbf{v}$ points in the opposite direction as $\mathbf{v}$ and its length is multiplied by $|r|$.

If $r = -1$, then $r\mathbf{v} = (-1)\mathbf{v}$ is usually written $-\mathbf{v}$, and called the **negative** of the vector $\mathbf{v}$. In terms of coordinates, if $\mathbf{v} = \langle a, b\rangle$ then $-\mathbf{v} = \langle -a, -b\rangle$.

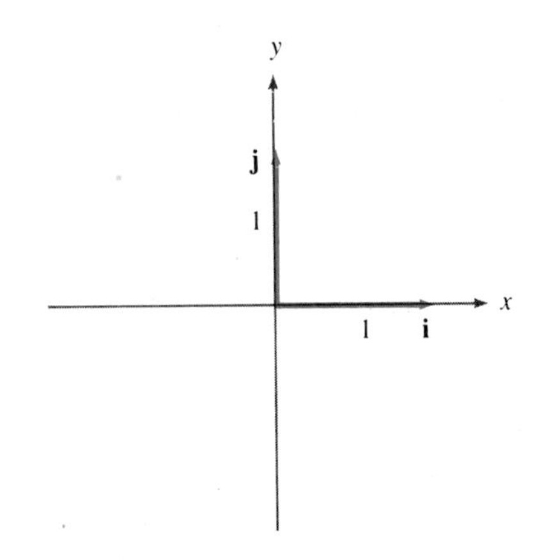

FIGURE 7.55

EXAMPLE 6: Let $\mathbf{v} = \langle 2, 5\rangle$ and $\mathbf{w} = \langle -3, 4\rangle$. Compute the following vectors.

(a) $3\mathbf{v}$ **(b)** $-2\mathbf{w}$ **(c)** $4\mathbf{v} - 6\mathbf{w}$

Solutions:

(a) $3\mathbf{v} = 3\langle 2, 5\rangle = \langle 3\cdot 2, 3\cdot 5\rangle = \langle 6, 15\rangle$

(b) $-2\mathbf{w} = -2\langle -3, 4\rangle = \langle -2\cdot(-3), -2\cdot 4\rangle = \langle 6, -8\rangle$

(c) $4\mathbf{v} - 6\mathbf{w} = 4\langle 2, 5\rangle - 6\langle -3, 4\rangle = \langle 8, 20\rangle - \langle -18, 24\rangle = \langle 26, -4\rangle$

Try Study Suggestion 7.29. □

▶ **Study Suggestion 7.29:** Let $\mathbf{u} = \langle -1, -4\rangle$ and $\mathbf{v} = \langle 1/2, -3/4\rangle$. Compute the vectors **(a)** $-\mathbf{u}$ **(b)** $-2\mathbf{v}$ **(c)** $-4\mathbf{u} - 8\mathbf{v}$ ◀

The vectors $\mathbf{i} = \langle 1, 0\rangle$ and $\mathbf{j} = \langle 0, 1\rangle$ are shown in Figure 7.55. These vectors play a very special role, and are known as the **standard basis vectors.** (This terminology comes from a more detailed study of vectors, which we will not undertake here.) The reason that the standard basis vectors are important is that any vector $\mathbf{v} = \langle a, b\rangle$ can easily be expressed in terms of these vectors. In fact, we have

$$\mathbf{v} = \langle a, b\rangle = \langle a, 0\rangle + \langle 0, b\rangle = a\langle 1, 0\rangle + b\langle 0, 1\rangle = a\mathbf{i} + b\mathbf{j}$$

Thus, for any vector $\mathbf{v} = \langle a, b\rangle$

$$\mathbf{v} = \langle a, b\rangle = a\mathbf{i} + b\mathbf{j}$$

The next theorem shows how to express addition, subtraction and scalar multiplication of vectors in terms of the standard basis vectors.

THEOREM

Let $\mathbf{v} = a\mathbf{i} + b\mathbf{j}$ and $\mathbf{w} = c\mathbf{i} + d\mathbf{j}$ be vectors. Then

1. $\mathbf{v} + \mathbf{w} = (a + c)\mathbf{i} + (b + d)\mathbf{j}$
2. $\mathbf{v} - \mathbf{w} = (a - c)\mathbf{i} + (b - d)\mathbf{j}$
3. $r\mathbf{v} = (ra)\mathbf{i} + (rb)\mathbf{j}$ for any real number r

EXAMPLE 7: Let $\mathbf{v} = 3\mathbf{i} + 2\mathbf{j}$ and $\mathbf{w} = 6\mathbf{i} - \mathbf{j}$. Compute the vectors

(a) $4\mathbf{w}$ **(b)** $2\mathbf{v} + 5\mathbf{w}$ **(c)** $\mathbf{v} - \mathbf{w}$ **(d)** $(1/2)\mathbf{v}$

Solutions:

(a) $4\mathbf{w} = 4(6\mathbf{i} - \mathbf{j}) = 24\mathbf{i} - 4\mathbf{j}$

(b) $2\mathbf{v} + 5\mathbf{w} = 2(3\mathbf{i} + 2\mathbf{j}) + 5(6\mathbf{i} - \mathbf{j}) = (6\mathbf{i} + 4\mathbf{j}) + (30\mathbf{i} - 5\mathbf{j}) = 36\mathbf{i} - \mathbf{j}$

(c) $\mathbf{v} - \mathbf{w} = (3\mathbf{i} + 2\mathbf{j}) - (6\mathbf{i} - \mathbf{j}) = -3\mathbf{i} + 3\mathbf{j}$

(d) $(1/2)\mathbf{v} = (1/2)(3\mathbf{i} + 2\mathbf{j}) = (3/2)\mathbf{i} + \mathbf{j}$

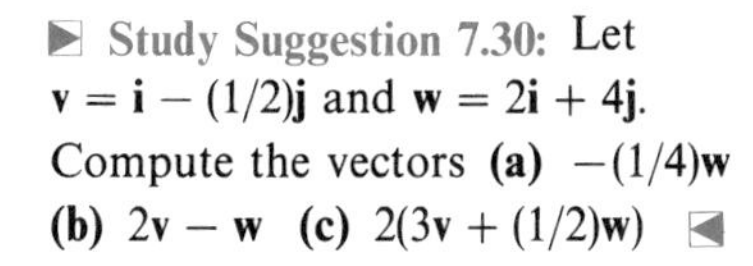

Study Suggestion 7.30: Let $\mathbf{v} = \mathbf{i} - (1/2)\mathbf{j}$ and $\mathbf{w} = 2\mathbf{i} + 4\mathbf{j}$. Compute the vectors **(a)** $-(1/4)\mathbf{w}$ **(b)** $2\mathbf{v} - \mathbf{w}$ **(c)** $2(3\mathbf{v} + (1/2)\mathbf{w})$

Try Study Suggestion 7.30. □

Ideas to Remember

Many quantities in nature have both a size, or magnitude, and a *direction*. Such quantities are known as *vectors*, and can be represented by arrows (in the plane).

EXERCISES

In Exercises 1–10, find the coordinates of the vector $\mathbf{v}$ with given magnitude and direction angle.

1. $|\mathbf{v}| = 2,\ \theta = 30°$

2. $|\mathbf{v}| = 1/2,\ \theta = 60°$

3. $|\mathbf{v}| = 1,\ \theta = 135°$

4. $|\mathbf{v}| = 5,\ \theta = -150°$

5. $|\mathbf{v}| = 10,\ \theta = 3\pi/2$

6. $|\mathbf{v}| = 5,\ \theta = \pi$

7. $|\mathbf{v}| = 9,\ \theta = -35°$

8. $|\mathbf{v}| = \sqrt{2},\ \theta = 50°$

9. $|\mathbf{v}| = 6,\ \theta = 28.5°$

10. $|\mathbf{v}| = \pi,\ \theta = 4\pi/5$

In Exercises 11–24, find the magnitude and smallest nonnegative direction angle of the vector $\mathbf{v}$.

11. $\mathbf{v} = \langle 0, 1\rangle$

12. $\mathbf{v} = \langle \sqrt{3}, 1\rangle$

13. $\mathbf{v} = 2\mathbf{i} + 2\mathbf{j}$

14. $\mathbf{v} = -\sqrt{2}\mathbf{i} + \sqrt{2}\mathbf{j}$

15. $\mathbf{v} = \langle -3, -\sqrt{3}\rangle$

16. $\mathbf{v} = \langle 15, 5\sqrt{3}\rangle$

17. $\mathbf{v} = -\mathbf{i} + \sqrt{3}\mathbf{j}$

18. $\mathbf{v} = 2\mathbf{i} - 5\mathbf{j}$

19. $\mathbf{v} = \langle -\sqrt{2}, \sqrt{3}\rangle$

20. $\mathbf{v} = \langle 10, 5\rangle$

21. $\mathbf{v} = -10\mathbf{i} - 5\mathbf{j}$

22. $\mathbf{v} = \langle 1 + \sqrt{2}, 1 - \sqrt{2}\rangle$

23. $\mathbf{v} = \langle \pi, 5\pi\rangle$

24. $\mathbf{v} = \langle 1 + \pi, \pi\rangle$

Let $\mathbf{u} = \langle 2, -3\rangle$, $\mathbf{v} = \langle 7, 1\rangle$ and $\mathbf{w} = \langle -6, 0\rangle$. Compute the following vectors.

25. $2\mathbf{u}$

26. $-3\mathbf{v}$

27. $\mathbf{u} - 4\mathbf{w}$

28. $5\mathbf{u} + 7\mathbf{w}$

29. $\mathbf{u} + \mathbf{v} + \mathbf{w}$

30. $2\mathbf{u} - 3\mathbf{v} + 4\mathbf{w}$

Let $\mathbf{u} = 5\mathbf{i} - \mathbf{j}$, $\mathbf{v} = \mathbf{i} + 2\mathbf{j}$ and $\mathbf{w} = -3\mathbf{j}$. Compute the following vectors.

31. $\mathbf{u} - 3\mathbf{v}$

32. $-4\mathbf{u} + \mathbf{w}$

33. $\mathbf{u} + \mathbf{v} + \mathbf{w}$

34. $5(\mathbf{u} + \mathbf{v}) - 3(\mathbf{u} + \mathbf{w})$

35. $(\mathbf{u} - \mathbf{v}) + (\mathbf{v} - \mathbf{w}) + (\mathbf{w} - \mathbf{u})$

36. Let $\mathbf{u} = \langle a, b\rangle$ and $\mathbf{v} = \langle c, d\rangle$. Show that $\mathbf{u} + \mathbf{v} = \mathbf{v} + \mathbf{u}$. What does this tell you about addition of vectors?

37. Let $\mathbf{u} = \langle a, b \rangle$, $\mathbf{v} = \langle c, d \rangle$ and $\mathbf{w} = \langle e, f \rangle$. Show that $(\mathbf{u} + \mathbf{v}) + \mathbf{w} = \mathbf{u} + (\mathbf{v} + \mathbf{w})$. What does this tell you about addition of vectors?

38. Show that if $\mathbf{v}$ is any vector, then $\mathbf{v} + \mathbf{0} = \mathbf{0} + \mathbf{v} = \mathbf{v}$. Describe this property of the zero vector in words.

39. Two dogs are dragging their favorite toy in two different directions, as shown in the following figure. Find the magnitude and direction of the total force acting on the toy.

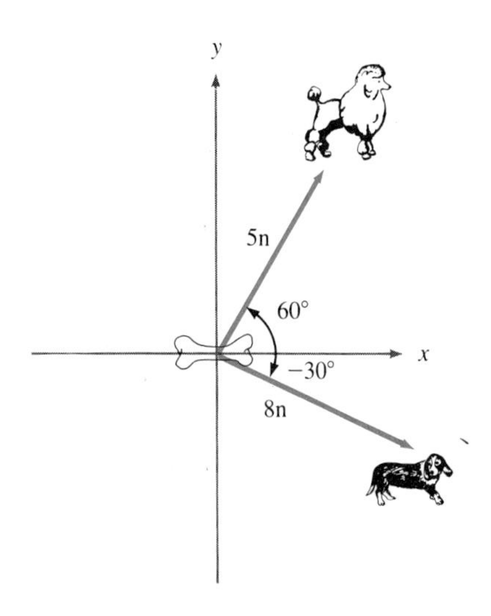

40. Three puppies are pulling on a pillow in three different directions, as shown in the following figure. In which direction will the pillow actually go?

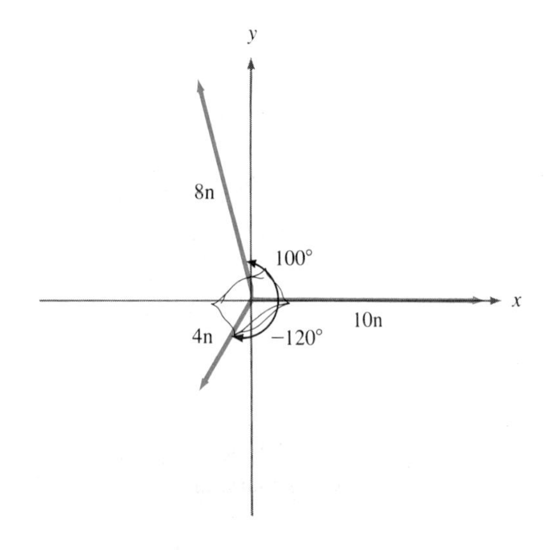

41. Two forces are acting on a certain falling object, as shown in the following figure.

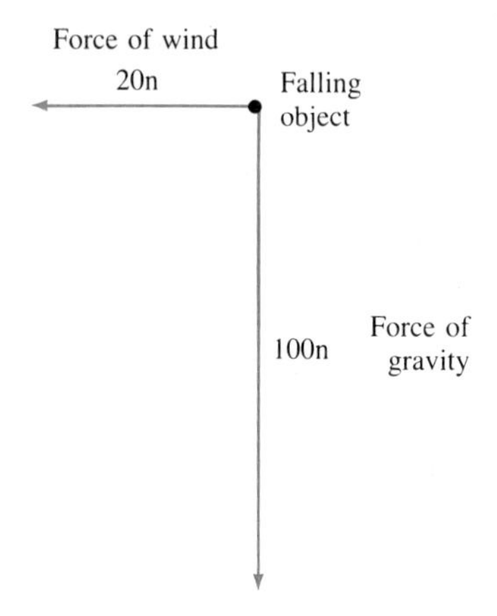

The force of gravity is acting to pull the object straight down, and the force of the wind is acting to pull the object due west. (The magnitudes of the forces are given in the figure.) What is the effect of both of these forces on the falling object? That is, in which direction does the object move, and what is the magnitude of the total force on the object?

42. A sailboat is being affected by three forces—wind currents, water currents, and the power of an outboard motor. (See the following figure.)

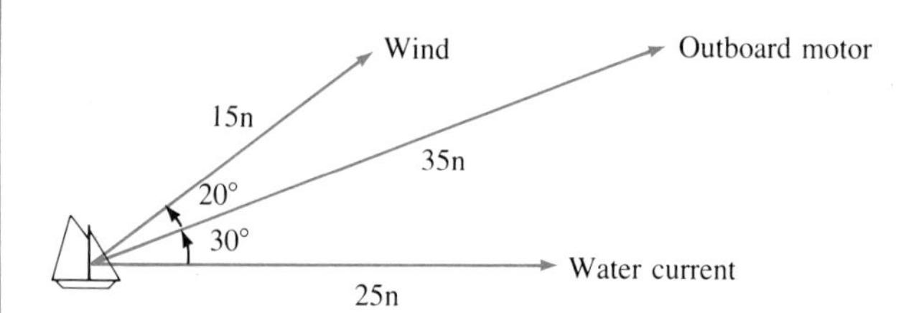

What direction is the boat actually moving and what is the magnitude of the total force on the boat?

7.6 Review

CONCEPTS FOR REVIEW

Oblique triangle
The law of sines
The law of cosines
The polar coordinates of a point
The polar coordinate system
Polar ray

Vector
Scalar
Initial point
Terminal point
Magnitude of a vector
Direction angle

Zero vector
Addition of vectors
Parallelogram law of addition
Subtraction of vectors
Scalar multiplication of vectors
Standard basis vectors

REVIEW EXERCISES

In Exercises 1–6, solve the given right triangle.

1.

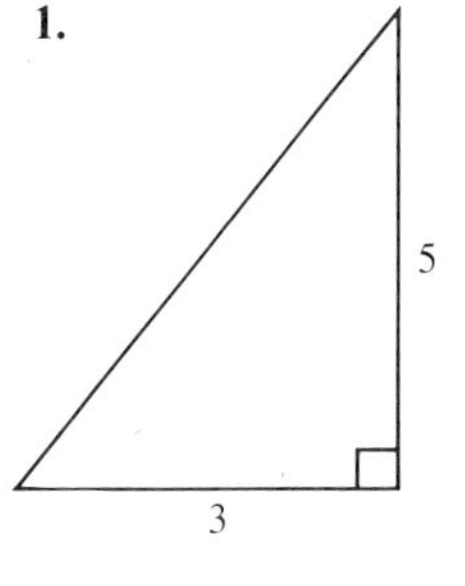

2.

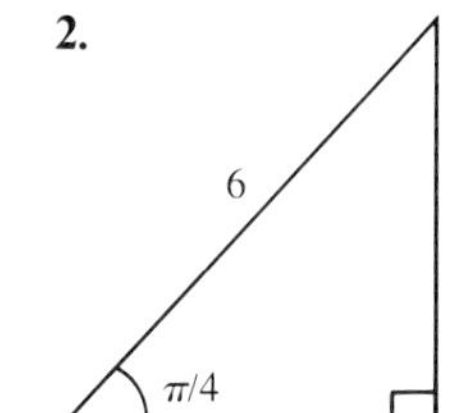

3.

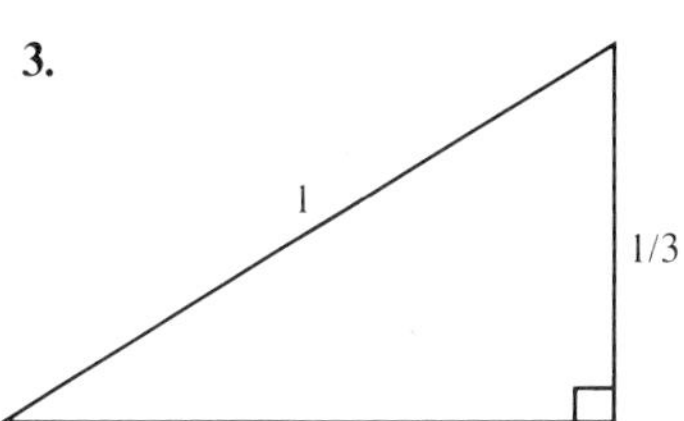

4.

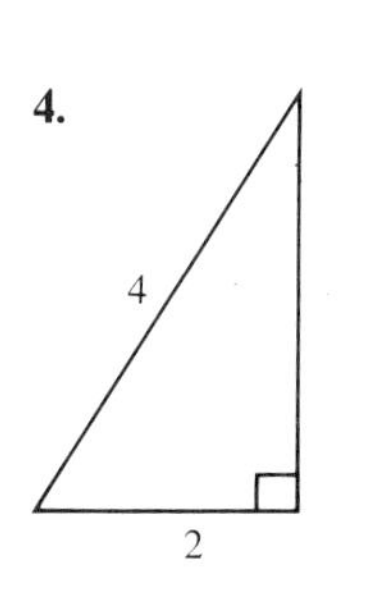

5.

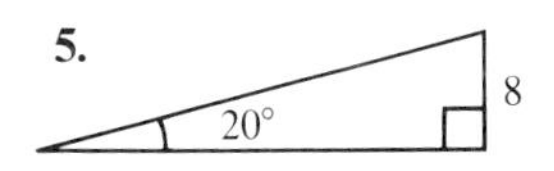

6.

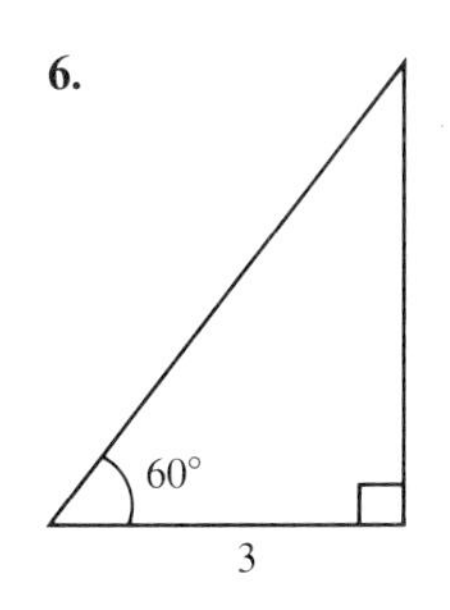

7. Is it possible to measure the height of one building from another building if you know the height of the building that you are measuring from and if you can measure angles? If so, describe how it can be done and if not, explain why not.

In the following exercises, the numbers a, b, c, and α, β and γ refer to the following figure.

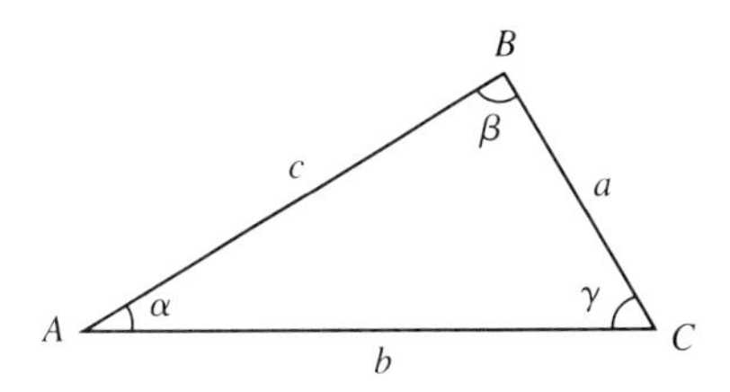

In Exercises 8–12, determine all triangles that have the given measurements.

8. $a = 12$, $b = 17$, $\alpha = 52°$

9. $a = 132$, $b = 128$, $c = 156$

10. $a = 16$, $b = 17$, $\alpha = 49°$

11. $a = 12.54$, $b = 35.61$, $\gamma = 62.39°$

12. $a = 210$, $b = 20$, $\alpha = 86°$

13. Use the law of sines to show that the area of a triangle is given by the formula

$$\text{Area} = \frac{a^2 \sin\beta \sin\gamma}{2 \sin\alpha}$$

14. What is another name for the law of cosines as applied to the right angle of a right triangle?

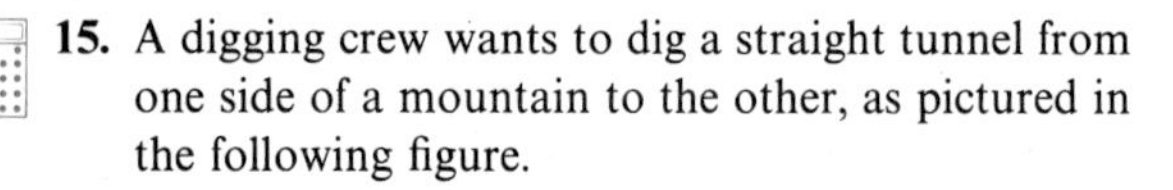

15. A digging crew wants to dig a straight tunnel from one side of a mountain to the other, as pictured in the following figure.

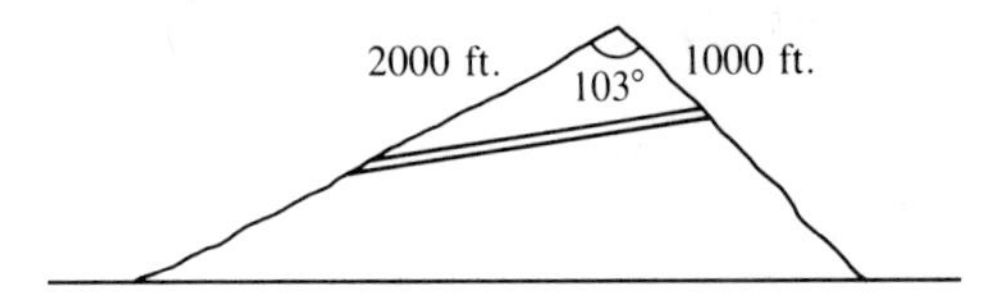

As you can see from the figure, the tunnel is 2000 feet from the top of the mountain at one entrance, and only 1000 feet from the top of the mountain at the other entrance. If the vertex at the top of the mountain measures 103°, how long will the tunnel be?

16. A man starts at a point A, walks 100 yards due east, then walks 50 yards at a bearing of 30°, and, finally, walks 100 yards at a bearing of 150°. After the trip, how far is he from the starting point and at what bearing?

17. A source of electrical interference is located at a certain point, and two federal communications commission agents are trying to pinpoint its location. Agent 1 measures the angle subtended by Agent 2 and the source to be 47° and Agent 2 measures the angle subtended by Agent 1 and the source to be 32°. Find the distance from the source to each agent if the distance between the agents is 150 feet.

18. For a triangle whose sides and angles are measured in the usual way, use the law of sines and the fact that

$$\sin \gamma = \sin [\pi - (\alpha + \beta)]$$

and

$$\sin (\alpha + \beta) = \sin \alpha \cos \beta + \cos \alpha \sin \beta$$

to show that

$$c = b \cos \alpha + a \cos \beta$$

19. The lunch whistles of two nearby factories always blow at exactly the same time. A woman standing near the two factories hears the whistle of one factory two seconds after she hears the whistle of the other factory. If the woman measures the angle subtended by the two factories to be 32°, can she determine the distance between the two factories? If so, perform the calculations for her and, if not, determine what other information she might need, and then solve the problem by assuming that this information is known.

20. Plot the following polar coordinates in the same plane.

(a) $(-2, \pi/6)$ **(b)** $(5/2, -3\pi/5)$

(c) $(\pi/3, \pi)$ **(d)** $(1/\pi, -17\pi/6)$

In Exercises 21–26, describe all polar coordinates of the given point.

21. $(1, \pi/2)$ **22.** $(-3, \pi/9)$ **23.** $(2, -4\pi/3)$

24. $(10, 165°)$ **25.** $(-20, -20°)$ **26.** $(0, 0)$

In Exercises 27–32, convert from polar coordinates to rectangular coordinates.

27. $(-\sqrt{2}, \pi/2)$ **28.** $(3, \pi/6)$ **29.** $(0, 0)$

30. $(6, -11\pi/4)$ **31.** $(2.5, -180°)$ **32.** $(1, 300°)$

In Exercises 33–36, use a calculator to convert from polar coordinates to rectangular coordinates.

33. $(12, \pi/10)$ **34.** $(-4.3, 0.8\pi)$

35. $(160, 15°)$ **36.** $(-9, -128°)$

In Exercises 37–40, convert from rectangular coordinates to polar coordinates in the range $r \geq 0$ and $0 \leq \theta < 2\pi$.

37. $(-1, \sqrt{3}/3)$ **38.** $(5, 5)$

39. $(-1/3, -1/3)$ **40.** $(-10, 10\sqrt{3}/3)$

In Exercises 41–44, use a calculator to convert from rectangular coordinates to polar coordinates satisfying $r \geq 0$ and $0 \leq \theta < 2\pi$.

41. $(4.6, -2.3)$ **42.** $(8, 9)$

43. $(7, -1/7)$ **44.** $(-5/2, -3)$

In Exercises 45–47, find the coordinates of the vector **v** with given magnitude and direction vector.

45. $|\mathbf{v}| = 1/2, \theta = 45°$

46. $|\mathbf{v}| = 10, \theta = -20°$

47. $|\mathbf{v}| = 0.33, \theta = 15.67°$

In Exercises 48–50, find the magnitude and smallest nonnegative direction angle of the vector **v**.

48. $\mathbf{v} = \langle 0, -2 \rangle$

49. $\mathbf{v} = \langle 5\sqrt{3}, -5 \rangle$

50. $\mathbf{v} = -4\mathbf{i} - 2\mathbf{j}$

Let $\mathbf{u} = \langle -6, 4 \rangle$, $\mathbf{v} = \langle 0, 3 \rangle$, and $\mathbf{w} = \langle -3, 8 \rangle$. Compute the following vectors.

51. $-2\mathbf{u}$

52. $-\mathbf{u} + \mathbf{v} - \mathbf{w}$

53. $(1/2)\mathbf{u} + (1/3)\mathbf{v} - \mathbf{w}$

54. Water currents are pulling a sailboat due north under a force of 50 newtons. Also, a wind with a force of 30 newtons is coming from the west, and a power motor is pulling the boat south by southwest with a force of 40 newtons. What is the actual direction of the boat, and what is the total force acting upon it?

8

ANALYTIC TRIGONOMETRY

8.1 Trigonometric Identities

It is often necessary to simplify complicated expressions involving various trigonometric functions. In many cases, this can be done simply by expressing the functions involved in terms of the sine and cosine alone. Let us illustrate with an example.

EXAMPLE 1: Simplify the expression

$$\csc t - \cos t \cot t$$

Solution: We begin by expressing all trigonometric functions in terms of the sine and cosine. This should give us a clue as to what to do next.

$$\begin{aligned} \csc t - \cos t \cot t &= \frac{1}{\sin t} - \cos t \frac{\cos t}{\sin t} \\ &= \frac{1 - \cos^2 t}{\sin t} \\ &= \frac{\sin^2 t}{\sin t} \\ &= \sin t \end{aligned}$$

Thus, $\csc t - \cos t \cot t = \sin t$. *Try Study Suggestion 8.1.* □

▶ **Study Suggestion 8.1:** Simplify the expression $\sec t - \sin t \tan t$. ◀

In the previous example, we used the Pythagorean identity $\sin^2 t + \cos^2 t = 1$, written in the form

$$1 - \cos^2 t = \sin^2 t$$

Since the Pythagorean identities will be very useful to us as we proceed, let us restate them here.

Pythagorean Identities

$$\sin^2 t + \cos^2 t = 1 \qquad 1 + \tan^2 t = \sec^2 t$$
$$1 + \cot^2 t = \csc^2 t$$

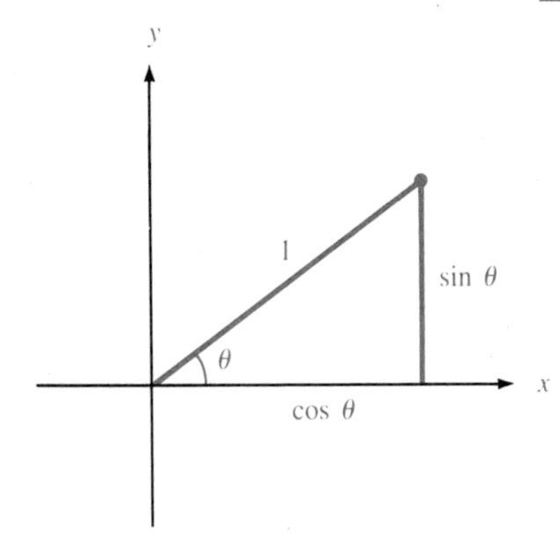

FIGURE 8.1

The first identity comes from the fact that the point $P(t) = (\cos t, \sin t)$ is on the unit circle, and as we discussed in Section 6.4, the second and third identities follow from the first one. These identities are called *Pythagorean* identities because the first one is just the Pythagorean Theorem, as applied to the right triangle in Figure 8.1.

The Pythagorean identities have many useful forms. For example, the first identity can be written in any of the forms

$$\sin^2 t = 1 - \cos^2 t, \qquad \cos^2 t = 1 - \sin^2 t$$
$$\sin t = \pm\sqrt{1 - \cos^2 t}, \qquad \cos t = \pm\sqrt{1 - \sin^2 t}$$

In the last two forms, more information is needed before we can determine whether to use the plus sign or the minus sign in any particular case. As the next example shows, these identities can be used to express one trigonometric function in terms of another (possibly with an ambiguity of sign).

EXAMPLE 2:

(a) Express $\csc t$ in terms of $\sec t$.

(b) If $\sec t = 5$ and $\csc t > 0$, find $\csc t$.

Solution:

(a) Using the fact that $\sin^2 t = 1 - \cos^2 t$, we have

$$\begin{aligned} \csc^2 t &= \frac{1}{\sin^2 t} \\ &= \frac{1}{1 - \cos^2 t} \\ &= \frac{1}{1 - \dfrac{1}{\sec^2 t}} \\ &= \frac{1}{\dfrac{\sec^2 t - 1}{\sec^2 t}} \\ &= \frac{\sec^2 t}{\sec^2 t - 1} \end{aligned}$$

and so

$$\csc t = \pm\sqrt{\frac{\sec^2 t}{\sec^2 t - 1}}$$

(b) Substituting $\sec t = 5$ into the expression for $\csc t$ that we obtained in part *a*, and using the fact that $\csc t > 0$, we get

$$\csc t = \sqrt{\frac{\sec^2 t}{\sec^2 t - 1}} = \sqrt{\frac{25}{24}} \approx 1.0206$$

► **Study Suggestion 8.2:** **(a)** Express $\sin t$ in terms of $\sec t$ **(b)** If $\sec t = -3$ and $\tan t < 0$, find $\sin t$. ◄

Try Study Suggestion 8.2. □

EXAMPLE 3: Simplify the expression

$$\frac{\sec t - \cos t}{\tan t \sin t}$$

Solution: As in Example 1, we first express all of the trigonometric functions in terms of the sine and cosine.

$$\begin{aligned}\frac{\sec t - \cos t}{\tan t \sin t} &= \frac{\dfrac{1}{\cos t} - \cos t}{\left(\dfrac{\sin t}{\cos t}\right)\sin t} \\ &= \frac{\dfrac{1 - \cos^2 t}{\cos t}}{\dfrac{\sin^2 t}{\cos t}} \\ &= \frac{1 - \cos^2 t}{\sin^2 t} \\ &= \frac{\sin^2 t}{\sin^2 t} \\ &= 1\end{aligned}$$

Thus,

$$\frac{\sec t - \cos t}{\tan t \sin t} = 1$$

for all values of t for which $\tan t \sin t \neq 0$ (and for which all functions are defined).

Try Study Suggestion 8.3. □

► **Study Suggestion 8.3:** Simplify the expression

$$\frac{\sec u}{\tan u + \cot u}$$ ◄

EXAMPLE 4: Show that

$$\frac{1}{\sec t - \tan t} = \sec t + \tan t \qquad \textbf{(8.1)}$$

Solution: Rather than express $\sec t$ and $\tan t$ in terms of $\sin t$ and $\cos t$, we instead multiply the numerator and denominator of the first expression by $\sec t + \tan t$.

$$\frac{1}{\sec t - \tan t} = \frac{1}{(\sec t - \tan t)} \frac{(\sec t + \tan t)}{(\sec t + \tan t)}$$

$$= \frac{\sec t + \tan t}{\sec^2 t - \tan^2 t}$$

$$= \sec t + \tan t$$

The last equation follows from the second Pythagorean identity, written in the form $\sec^2 t - \tan^2 t = 1$. *Try Study Suggestion 8.4.* □

▶ **Study Suggestion 8.4:** Show that

$$\frac{1}{\csc t + \cot t} = \csc t - \cot t$$ ◀

In the previous example, we showed that Equation (8.1) is valid for *all* values of the variable t for which the equation makes sense; that is, for which all functions are defined and for which the denominator $\sec t - \tan t$ is not equal to 0. More generally, equations that involve the trigonometric functions and that are valid for all possible values of the independent variable are known as **trigonometric identities.** Showing that a trigonometric equation is an identity is referred to as *verifying the identity.*

Before considering other examples of verifying trigonometric identities, let us state a few general guidelines that can be of help verifying such identities.

Guidelines for Verifying Trigonometric Identities

1. If one side of an identity is significantly more complicated than the other side, it is generally a good idea to start with the more complicated side, and try to reduce it to the simpler one. During this process, you should look frequently at the simpler side of the equation, to remind yourself what it is that you are trying to obtain.
2. If the two sides of an identity are about the same in complexity, it may be a good idea to try to reduce *both* sides to the samc (third) expression. *As long as the steps in this reduction are reversible,* this will verify the identity.
3. As with simplifying a trigonometric expression, it may be helpful to express all functions in terms of the sine and cosine.

With these principles in mind, let us consider some additional examples.

EXAMPLE 5: Verify the identity

$$\frac{\sec t + \tan t}{\cos t - \sec t} = -\csc t(\csc t + 1)$$

Solution: Since the left-hand side appears to be more complicated than the right, let us try to reduce it to the right side. The first step is to express all functions in terms of the sine and cosine.

$$\frac{\sec t + \tan t}{\cos t - \sec t} = \frac{\dfrac{1}{\cos t} + \dfrac{\sin t}{\cos t}}{\cos t - \dfrac{1}{\cos t}}$$

$$= \frac{\dfrac{1 + \sin t}{\cos t}}{\dfrac{\cos^2 t - 1}{\cos t}}$$

$$= \frac{1 + \sin t}{\cos^2 t - 1}$$

$$= \frac{1 + \sin t}{-\sin^2 t}$$

(At this point, a look at the right side of the original equation gives us an idea of what to do next.)

$$= -\frac{1}{\sin t}\,\frac{1 + \sin t}{\sin t}$$

$$= -\frac{1}{\sin t}\left(\frac{1}{\sin t} + 1\right)$$

$$= -\csc t(\csc t + 1)$$

Since this is the right-hand side, the identity is verified.

▶ **Study Suggestion 8.5:** Verify the identity

$$\frac{\tan^2 u}{1 - \cot^2 u \sin^2 u} = \sec^2 u$$ ◀

Try Study Suggestion 8.5. □

EXAMPLE 6: Verify the identity

$$\frac{\tan\theta}{\sec\theta - 1} - \frac{\sec\theta - 1}{\tan\theta} = 2\cot\theta$$

Solution: In this example, we should first combine the two fractions on the left-hand side, using the common denominator $\tan\theta(\sec\theta - 1)$.

$$\frac{\tan\theta}{\sec\theta - 1} - \frac{\sec\theta - 1}{\tan\theta} = \frac{\tan^2\theta}{\tan\theta(\sec\theta - 1)} - \frac{(\sec\theta - 1)^2}{\tan\theta(\sec\theta - 1)}$$

$$= \frac{\tan^2\theta - (\sec\theta - 1)^2}{\tan\theta(\sec\theta - 1)}$$

$$= \frac{\tan^2\theta - \sec^2\theta + 2\sec\theta - 1}{\tan\theta(\sec\theta - 1)}$$

At this point, we can use the second Pythagorean identity, in the form $\tan^2\theta - \sec^2\theta = -1$)

$$= \frac{-2 + 2\sec\theta}{\tan\theta(\sec\theta - 1)}$$

$$= \frac{2(\sec\theta - 1)}{\tan\theta(\sec\theta - 1)}$$

$$= \frac{2}{\tan\theta}$$

$$= 2\cot\theta$$

Try Study Suggestion 8.6. □

▶ **Study Suggestion 8.6:** Verify the identity

$$\frac{\sin\theta}{\csc\theta} + \frac{\cos\theta}{\sec\theta} = 1$$ ◀

EXAMPLE 7: Verify the identity

$$\frac{1 + \tan t}{\sec t} = \frac{1 + \cot t}{\csc t}$$

Solution: Both sides of the equation appear to have equal complexity, and so we will try to reduce both sides to the same expression, by using reversible steps. To this end, let us express both sides in terms of the sine and cosine functions.

LEFT-HAND SIDE	RIGHT-HAND SIDE
$\frac{1 + \tan t}{\sec t} = \frac{1 + \frac{\sin t}{\cos t}}{\frac{1}{\cos t}}$	$\frac{1 + \cot t}{\csc t} = \frac{1 + \frac{\cos t}{\sin t}}{\frac{1}{\sin t}}$
$= \frac{\frac{\cos t + \sin t}{\cos t}}{\frac{1}{\cos t}}$	$= \frac{\frac{\sin t + \cos t}{\sin t}}{\frac{1}{\sin t}}$
$= \cos t + \sin t$	$= \sin t + \cos t$
	$= \cos t + \sin t$

Since both sides of this equation have been reduced—by reversible steps—to the same expression, they must be equal. Hence, the identity is verified.

Try Study Suggestion 8.7. □

▶ **Study Suggestion 8.7:** Verify the identity

$$\frac{1 - \tan t}{1 - \cot t} = \frac{-\tan^3 t}{1 - \sec^2 t}$$ ◀

Let us conclude this section with an identity involving the inverse sine function.

EXAMPLE 8: Verify the identity

$$\sin^{-1}(-x) = -\sin^{-1} x$$

Solution: If we write $a = \sin^{-1}(-x)$ and $b = -\sin^{-1} x$ then our goal is to show that $a = b$. Let us begin by computing $\sin a$ and $\sin b$, where sin is the *restricted* sine function. We have

$$\sin a = \sin(\sin^{-1}(-x)) = -x$$

and, since the (restricted) sine function is odd,

$$\sin b = \sin(-\sin^{-1} x) = -\sin(\sin^{-1} x) = -x$$

Thus, we see that $\sin a = \sin b$.

The reason that we have gone to the trouble of showing that $\sin a = \sin b$ is that the *restricted* sine function is one-to-one. According to the definition of a one-to-one function given in Section 2.6, a function f is one-to-one if and only if whenever $f(a) = f(b)$, then $a = b$. Therefore, since $\sin a = \sin b$, we can conclude that $a = b$, which is precisely what we wanted to show, and so the identity is verified. (This example is a very good illustration of the use of the concept of one-to-oneness.) *Try Study Suggestion 8.8.* □

▶ **Study Suggestion 8.8:** Verify the identity

$$\arctan(-x) = -\arctan x$$ ◀

Ideas to Remember

- Frequently the key to simplifying trigonometric expressions is to express all of the functions involved in terms of the sine and cosine.
- Being able to verify trigonometric identities is a skill worth cultivating, since it can be useful in many different situations. Developing this skill, however, takes practice.

EXERCISES

1. **(a)** Express $\cos t$ in terms of $\csc t$.
   (b) If $\csc t = 3$ and $\cot t > 0$, find $\cos t$.
2. **(a)** Express $\sin t$ in terms of $\tan t$.
   (b) If $\tan t = -2$ and $\sin t < 0$, find $\sin t$.
3. **(a)** Express $\cos t$ in terms of $\cot t$.
   (b) If $\cot t = 5/2$ and $\sin t < 0$, find $\cos t$.
4. Express all of the trigonometric functions in terms of the sine function.
5. Express all of the trigonometric functions in terms of the cosine function.
6. Express all of the trigonometric functions in terms of the tangent function.

In Exercises 7–22, simplify the given expression.

7. $\cot t \sin t$
8. $\tan t \cot t$
9. $\dfrac{\sec t}{\tan t}$
10. $\dfrac{\cos t}{\sec t}$
11. $\dfrac{\sec \theta}{\csc \theta}$
12. $(1 + \tan^2 \theta)\cos^2 \theta$
13. $\sec^2 \theta \cot^2 \theta$
14. $(1 - \cos^2 \theta)\csc^2 \theta$

15. $\cos x\left(\dfrac{1+\sec x}{1+\sin x}\right)$

16. $\dfrac{\cos^2 x}{1-\cot^2 x\sin^2 x}$

17. $\dfrac{\cot x\sin x}{1-\tan^2 x\cos^2 x}$

18. $\dfrac{\tan x+\sin x}{1+\sec x}$

19. $\dfrac{\tan u+\sin u}{1+\cot^2 u}$

20. $\dfrac{\csc u-\cos u}{\sec u-\sin u}$

21. $\dfrac{1+\sec u}{\sin u+\tan u}$

22. $\dfrac{(1-\sin^4 u)\sec^2 u}{2-\cos^2 u}$

In Exercises 23–63, verify the given identity.

23. $\csc t\tan t=\sec t$

24. $\dfrac{\sec t}{\cot t}=\dfrac{\sin t}{1-\sin^2 t}$

25. $\sin t\sec t=\tan t$

26. $\dfrac{\tan t}{\csc t}=\sec t-\cos t$

27. $\csc x-\sin x=\cot x\cos x$

28. $1-2\sin^2 x=2\cos^2 x-1$

29. $\cos^2 x-\sin^2 x=1-2\sin^2 x$

30. $(\cos x-\sin x)^2=1-2\cos x\sin x$

31. $(1+\tan\theta)^2=2\tan\theta+\sec^2\theta$

32. $(\tan\theta+\cot\theta)^2=\sec^2\theta+\csc^2\theta$

33. $(1-\tan\theta)(1+\cot\theta)=\cot\theta-\tan\theta$

34. $(1+\tan^2\theta)\cot^2\theta=\csc^2\theta$

35. $\sec u-\cos u=\sin u\tan u$

36. $\tan u+\cot u=\sec u\csc u$

37. $(1+\sin u)(1-\sin u)=\cos^2 u$

38. $(1-\cos u)(1+\cos u)=\sin^2 u$

39. $1+\cot^2 t=\dfrac{1}{1-\cos^2 t}$

40. $\dfrac{\csc t}{\sin t}+\dfrac{\sec t}{\cos t}=\sec^2 t\csc^2 t$

41. $\sin y(\tan y+\cot y)=\sec y$

42. $(1-\sec^2 y)(1-\cot^2 y\sin^2 y)=-\sec^2 y\sin^4 y$

43. $(1-\sin^2\alpha)(1+\cot^2\alpha)=\csc^2\alpha-1$

44. $(\tan\alpha+\sec\alpha)(\cot\alpha-\cos\alpha)=\csc\alpha-\sin\alpha$

45. $(\tan^2\theta-\cot^2\theta)\cos^2\theta\sin^2\theta=\sin^2\theta-\cos^2\theta$

46. $\dfrac{\sin\theta}{1+\cos\theta}+\dfrac{1+\cos\theta}{\sin\theta}=2\csc\theta$

47. $\dfrac{\tan x+\cot x}{\sec x}-\dfrac{\sec x}{\tan x+\cot x}=\cos x\cot x$

48. $\dfrac{1+\tan^2 x}{\tan^2 x}=\csc^2 x$

49. $\dfrac{1+\tan t}{1-\tan t}+\dfrac{1-\tan t}{1+\tan t}=\dfrac{2}{2\cos^2 t-1}$

50. $\dfrac{1+\tan t}{1-\tan t}-\dfrac{1-\tan t}{1+\tan t}=\dfrac{4}{\cot t-\tan t}$

51. $\dfrac{1+\cos t}{1-\cos t}+\dfrac{1-\cos t}{1+\cos t}=2(\csc^2 t+\cot^2 t)$

52. $\dfrac{1+\cos t}{1-\cos t}-\dfrac{1-\cos t}{1+\cos t}=4\cot t\csc t$

53. $\dfrac{\sec x}{\sin x}-\cot x=\tan x$

54. $\dfrac{\cot x-\tan x}{1-\cot x}+\tan x=-1$

55. $\dfrac{\sec^4 x-\tan^4 x}{\tan^2 x+\sec^2 x}=1$

56. $\sin^4 x+\cos^2 x=\cos^4 x+\sin^2 x$

57. $\sec^4 x-\tan^4 x=\sec^2 x+\tan^2 x$

58. $\dfrac{1}{(\sin^2\theta+\cos^2\theta)^5}=\sin^2\theta+\cos^2\theta$

59. $\sqrt{\dfrac{1-\cos\theta}{1+\cos\theta}}=\dfrac{|\sin\theta|}{1+\cos\theta}$

60. $\sqrt{\dfrac{1-\sin\theta}{1+\sin\theta}}=\dfrac{|\cos\theta|}{1+\sin\theta}$

61. $\cos^{-1}(-x)=\pi-\cos^{-1}x$
(*Hint:* $\cos(\pi-t)=-\cos t$)

62. $\arcsin x=\arctan\dfrac{x}{\sqrt{1-x^2}}$

63. $\sin^{-1}x=\tan^{-1}\dfrac{x}{\sqrt{1-x^2}}$

In Exercises 64–71, show that the given equation is *not* an identity by finding a value of θ for which the equation is not valid.

64. $\sin(-\theta) = \sin\theta$

65. $(\sin\theta + \cos\theta)^2 = \sin^2\theta + \cos^2\theta$

66. $\arcsin\theta = \dfrac{1}{\sin\theta}$

67. $\arctan\theta = \dfrac{1}{\tan\theta}$

68. $(\sin^{-1}\theta)^2 + (\cos^{-1}\theta)^2 = 1$

69. $\sqrt{\sin^2\theta} = \sin\theta$

70. $\cos\theta = \sqrt{1 - \sin^2\theta}$

71. $\arctan\theta = \dfrac{\arcsin\theta}{\arccos\theta}$

8.2 Trigonometric Equations

In this section, we will consider some examples of how to solve trigonometric equations. By *solving* a trigonometric equation, we mean finding all values of the variable for which the equation is true. For example, as we saw in Section 6.5, the solutions to the equation

$$\sin t = 0 \tag{8.2}$$

are all real numbers t of the form $t = n\pi$, where n is any integer. Notice that, unlike the case of a polynomial equation, there are an infinite number of solutions to Equation (8.2).

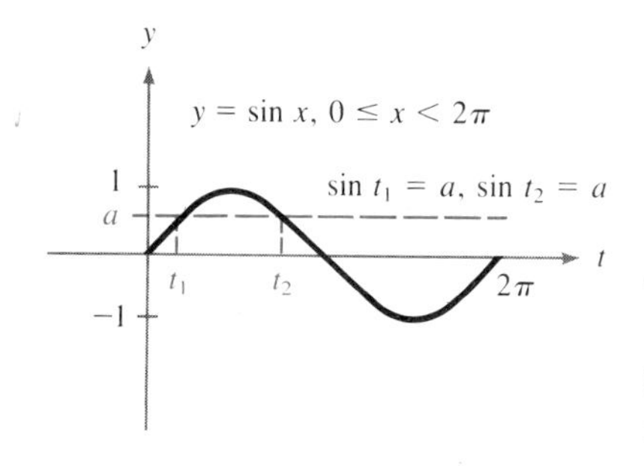

FIGURE 8.2

Frequently however, there is a restriction, or *side condition*, placed on the variable. For example, the solutions to Equation 8.2, with side condition $0 \leq t < 2\pi$, are $t = 0$ and $t = \pi$.

More generally, by looking at the graph of the sine function in Figure 8.2, we can see that if a is a real number satisfying $-1 < a < 1$ (note the *strict* inequality), then the equation

$$\sin t = a$$

has *exactly* two solutions in the interval $0 \leq t < 2\pi$. Similarly, by looking at the graphs of the other trigonometric functions in the interval $0 \leq t < 2\pi$, we get the following theorem.

THEOREM

1. If $-1 < a < 1$, then the equations

$$\sin t = a \qquad \text{and} \qquad \cos t = a$$

each have *exactly* two solutions in the interval $0 \leq t < 2\pi$.

2. For any real number a, the equations

$$\tan t = a \qquad \text{and} \qquad \cot t = a$$

each have *exactly* two solutions in the interval $0 \leq t < 2\pi$.

3. If $a > 1$ or $a < -1$, then the equations

$$\sec t = a \qquad \text{and} \qquad \csc t = a$$

each have *exactly* two solutions in the interval $0 \leq t < 2\pi$.

Furthermore, the same statements are true if we restrict t to *any* interval of the form $b \leq t < b + 2\pi$, where b is any real number.

▶ **Study Suggestion 8.9:**
(a) Graph the cosine function for $0 \leq t < 2\pi$ to convince yourself of the truth of the previous theorem in this case.
(b) Do the same for the tangent function. ◀

Try Study Suggestion 8.9.

EXAMPLE 1: Solve the equation

$$\sin t = \frac{1}{2}, \qquad 0 \leq t < 2\pi$$

Solution: Since $\sin \pi/6 = 1/2$, we see that $t = \pi/6$ is one solution to the equation. To find another solution, we plot the point $P(\pi/6)$ on the unit circle, and look for another point on the circle whose y-coordinate is equal to $1/2$. As you can see from Figure 8.3, such a point is $P(5\pi/6)$. Therefore, $t = 5\pi/6$ is another solution to our equation. The previous theorem tells us that these are the only solutions in the interval $0 \leq t < 2\pi$.

▶ **Study Suggestion 8.10:** Solve the equation $\cos t = -\sqrt{3}/2$, with side condition $0 \leq t < 2\pi$. ◀

Try Study Suggestion 8.10. □

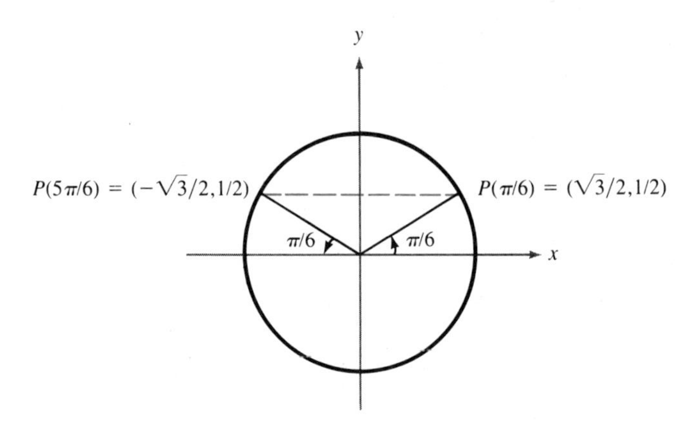

FIGURE 8.3

EXAMPLE 2: Solve the equation

$$\tan(3t - \pi) = -\sqrt{3}, \quad 0 \leq t < 2\pi \tag{8.3}$$

Solution: We begin by letting $s = 3t - \pi$. Then the side condition $0 \leq t < 2\pi$ is the same as the condition $0 \leq 3t < 6\pi$, or $-\pi \leq 3t - \pi < 5\pi$, or finally $-\pi \leq s < 5\pi$. Thus, we want to solve the equation

$$\tan s = -\sqrt{3}, \quad -\pi \leq s < 5\pi \tag{8.4}$$

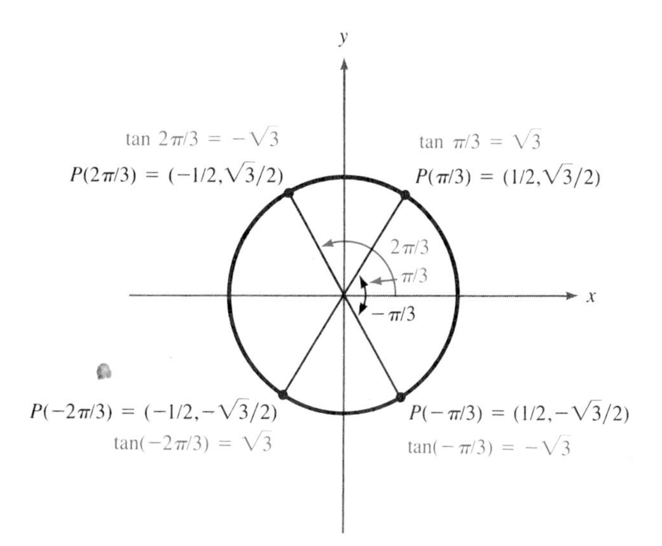

FIGURE 8.4

According to our theorem, we can expect exactly two solutions to this equation in the interval $-\pi \le s < \pi$. (Take $b = -\pi$ in the last statement of the theorem.) Also, we can expect another two solutions in the interval $\pi \le s < 3\pi$, and still another two solutions in the interval $3\pi \le s < 5\pi$, for a total of six solutions.

We begin by looking for the two solutions in the interval $-\pi \le s < \pi$. Since $\tan \pi/3 = \sqrt{3}$, we can easily locate all points $P(s)$ on the unit circle for which $\tan s = \pm\sqrt{3}$. This is done in Figure 8.4. From this figure, we see that

$$s = -\frac{\pi}{3} \quad \text{and} \quad s = \frac{2\pi}{3}$$

are the two solutions to Equation (8.4) in the interval $-\pi \le s < \pi$. By first adding 2π, and then 4π, to each of these solutions, we get all six solutions in the interval $-\pi \le s < 5\pi$,

$$s = -\frac{\pi}{3}, \frac{2\pi}{3}, \frac{5\pi}{3}, \frac{8\pi}{3}, \frac{11\pi}{3}, \frac{14\pi}{3}$$

In terms of t, this is

$$3t - \pi = -\frac{\pi}{3}, \frac{2\pi}{3}, \frac{5\pi}{3}, \frac{8\pi}{3}, \frac{11\pi}{3}, \frac{14\pi}{3}$$

First adding π, and then dividing by 3, we see that all of the solutions to equation (8.3) in the interval $-\pi \le s < 5\pi$, or $0 \le t < 2\pi$, are given by

$$t = \frac{2\pi}{9}, \frac{5\pi}{9}, \frac{8\pi}{9}, \frac{11\pi}{9}, \frac{14\pi}{9}, \frac{17\pi}{9}$$

▶ **Study Suggestion 8.11:** Solve the equation $\cot(3t + \pi) = 1$, with side condition $0 \le t < 2\pi$. ◀

Try Study Suggestion 8.11. □

EXAMPLE 3: Solve the equation

$$\sec(2t + \pi) = -\sqrt{2}$$

Solution: The first step is to write $s = 2t + \pi$, so that the equation becomes

$$\sec s = -\sqrt{2} \tag{8.5}$$

Even though there are no side conditions in this example, we next look for the two values of s in the interval $0 \leq s < 2\pi$ that satisfy Equation (8.5). (How do we know that there are exactly two such values?)

Using the fact that $\sec \pi/4 = \sqrt{2}$, we can locate all points $P(s)$ on the unit circle for which $\sec s = \pm\sqrt{2}$. This is done in Figure 8.5. From this we see that

$$s = \frac{3\pi}{4} \quad \text{and} \quad s = \frac{5\pi}{4}$$

are solutions to Equation (8.5). Hence, these are the *only* solutions in the interval $0 \leq s < 2\pi$. Now we can use the fact that the secant function repeats itself every 2π units to conclude that *all* of the solutions to Equation (8.5) are given by

$$s = \frac{3\pi}{4} + 2\pi n \quad \text{and} \quad s = \frac{5\pi}{4} + 2\pi n$$

where n is any integer. In terms of t, these solutions are

$$2t + \pi = \frac{3\pi}{4} + 2\pi n \quad \text{and} \quad 2t + \pi = \frac{5\pi}{4} + 2\pi n$$

Subtracting π from both sides of these equations gives

$$2t = -\frac{\pi}{4} + 2\pi n \quad \text{and} \quad 2t = \frac{\pi}{4} + 2\pi n$$

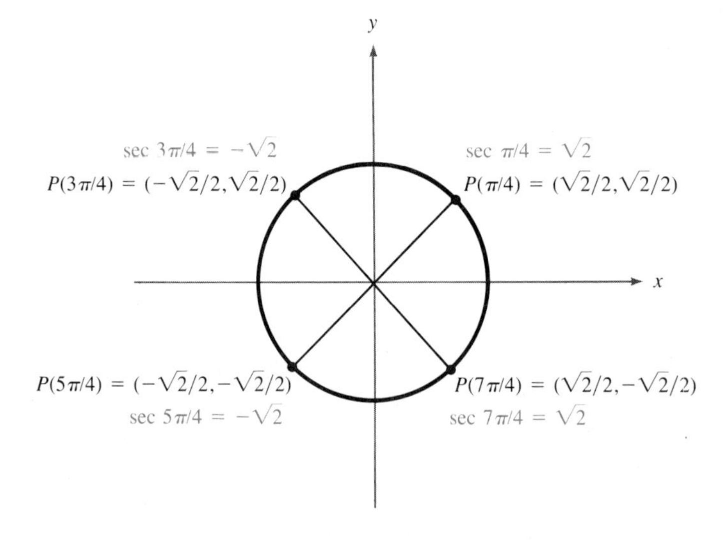

FIGURE 8.5

and dividing both sides by 2 gives the solutions

$$t = -\frac{\pi}{8} + \pi n \qquad \text{and} \qquad t = \frac{\pi}{8} + \pi n$$

where n is any integer. *Try Study Suggestion 8.12.* □

▶ **Study Suggestion 8.12:** Solve the equation

$$\csc\left(3t + \frac{\pi}{2}\right) = \sqrt{2}.$$ ◀

EXAMPLE 4: Solve the equation

$$\cot\theta = -1.688, \; 0 \le \theta < 2\pi$$

Solution: In this case, the best we can hope for is an approximation to the solution. The first step is to take the reciprocal of both sides, to get

$$\tan\theta = \frac{1}{-1.688} \tag{8.6}$$

A calculator (set in radian mode) then gives

$$\theta = \tan^{-1}\left(\frac{1}{-1.688}\right) \approx -0.5348$$

However, this value of θ is *not* a solution to our problem, since it does not lie in the interval $0 \le \theta < 2\pi$. But we know that the tangent repeats itself every π units, and so we can add any integral multiple of π to this value of θ and still have a solution to Equation (8.6). In fact, by adding both π and 2π, we get the two (approximate) solutions

$$\theta \approx -0.5348 + \pi \approx 2.6068$$

and

$$\theta \approx -0.5348 + 2\pi \approx 5.7484$$

According to our theorem, these are the *only* solutions to the equation in the interval $0 \le \theta < 2\pi$. *Try Study Suggestion 8.13.* □

Study Suggestion 8.13: Solve the equation $\tan 2\theta = 5.54$, with side condition $0 \le \theta < 2\pi$. ◀

Many trigonometric equations can be solved by using the same methods that we used for solving algebraic equations. As the next few examples show, the method of factoring is especially useful.

EXAMPLE 5: Solve the equation

$$\sin t \cos t = 0, \quad 0 \le t < 2\pi \tag{8.7}$$

Solution: The solutions to this equation are the same as the solutions to either of the equations

$$\sin t = 0 \qquad \text{and} \qquad \cos t = 0$$

For $0 \le t < 2\pi$, the first equation has solutions $t = 0$ and $t = \pi$, and the second equation has solutions $t = \pi/2$ and $t = 3\pi/2$. Hence, the solutions to

Equation (8.7) in the interval $0 \le t < 2\pi$ are

$$t = 0, \frac{\pi}{2}, \pi, \frac{3\pi}{2}$$

▶ **Study Suggestion 8.14:** Solve the equation $\tan t \sec t = 0$, with side condition $0 \le t < 2\pi$. ◀

Try Study Suggestion 8.14. □

EXAMPLE 6: Solve the equation

$$\sin x \tan x = \sin x, \quad 0 \le x < 4\pi \tag{8.8}$$

Solution: First we write this equation in the form

$$\sin x \tan x - \sin x = 0$$

and this can be factored to give

$$\sin x (\tan x - 1) = 0$$

The solutions to this equation are the solutions to either of the equations

$$\sin x = 0 \qquad \text{or} \qquad \tan x = 1$$

Now, in the interval $0 \le x < 2\pi$, the first equation has solutions $x = 0$ and $x = \pi$ and the second equation has solutions $x = \pi/4$ and $x = 5\pi/4$. Hence, all of the solutions to Equation (8.8) in the interval $0 \le x < 2\pi$ are

$$x = 0, \frac{\pi}{4}, \pi, \frac{5\pi}{4}$$

Adding 2π to these values gives us all of the solutions that lie in the interval $2\pi \le x < 4\pi$. Finally, then, all of the solutions to Equation (8.8) in the interval $0 \le x < 4\pi$ are

$$x = 0, \frac{\pi}{4}, \pi, \frac{5\pi}{4}, 2\pi, \frac{9\pi}{4}, 3\pi, \frac{13\pi}{4}$$

It is important to notice that we cannot simply divide both sides of the original equation by $\sin x$ to get the equation $\tan x = 1$. By doing so, we would lose some of the solutions to the original equation. (Dividing by $\sin x$ amounts to dividing by 0 when $x = 0$ or $x = \pi$.)

▶ **Study Suggestion 8.15:** Solve the equation $\cos x \cot x = \cos x$, with side condition $0 \le x < 2\pi$. ◀

Try Study Suggestion 8.15. □

EXAMPLE 7: Solve the equation

$$\sin^2 t + 2\cos t = -2, \quad 0 \le t < 2\pi \tag{8.9}$$

Solution: The first step is to express all functions in terms of the cosine. Using the fact that $\sin^2 t = 1 - \cos^2 t$, we get

$$1 - \cos^2 t + 2\cos t = -2$$

or after rearranging

$$\cos^2 t - 2\cos t - 3 = 0$$

Now, if we set $x = \cos t$, this equation becomes

$$x^2 - 2x - 3 = 0$$

which is a quadratic equation in the variable x. This can be factored

$$(x - 3)(x + 1) = 0$$

to get the solutions $x = 3$ and $x = -1$. Replacing x by $\cos t$, we see that the solutions to the original equation are the solutions to the equations

$$\cos t = 3 \qquad \text{and} \qquad \cos t = -1$$

The first of these equations has no solutions (why?), and the second equation has solution $t = \pi$. (This is the only solution in the interval $0 \leq t < 2\pi$.) Hence, the only solution to Equation (8.9) is $t = \pi$.

Try Study Suggestion 8.16. □

▶ **Study Suggestion 8.16:** Solve the equation $2\cos^2 t + 3\sin t = 0$, with side condition $0 \leq t < 2\pi$. ◀

EXAMPLE 8: Solve the equation

$$2\cos^2 t + 3\cos t - 1 = 0 \tag{8.10}$$

Solution: We begin by making the substitution $x = \cos t$, to get the quadratic equation

$$2x^2 + 3x - 1 = 0$$

This equation can be solved using the quadratic formula (with the aid of a calculator) to get

$$x = \frac{-3 \pm \sqrt{17}}{4} \approx 0.2808, -1.7808$$

Thus, we must solve the equations

$$\cos t = 0.2808 \qquad \text{and} \qquad \cos t = -1.7808$$

The second of these equations has no solutions. As for the first equation, we look for all solutions in the interval $0 \leq t < 2\pi$. Our theorem tells us that there are exactly two such solutions. A calculator (set in radian mode!) gives

$$\cos^{-1}(0.2808) \approx 1.2862$$

and so $t \approx 1.2862$ is one solution. Using the fact that $\cos\theta = \cos(-\theta)$, we see that $t \approx -1.2862$ is also a solution to the equation. However, it is not in the desired interval, and so we add 2π to bring it into this interval. Thus, the two solutions to the first equation in the interval $0 \leq t < 2\pi$ are $t = 1.2862$ and $t = -1.2862 + 2\pi$.

Finally, *all* of the solutions to Equation (8.10) are given by

$$t \approx 1.2862 + 2\pi n, -1.2862 + 2\pi n$$

where n varies over all integers. *Try Study Suggestion 8.17.* □

Study Suggestion 8.17: Solve the equation $\sin^2 t - 5\sin t + 3 = 0$. ◀

Let us conclude this section with an application for solving trigonometric equations. When a ray of light passes from one medium, such as the air, to

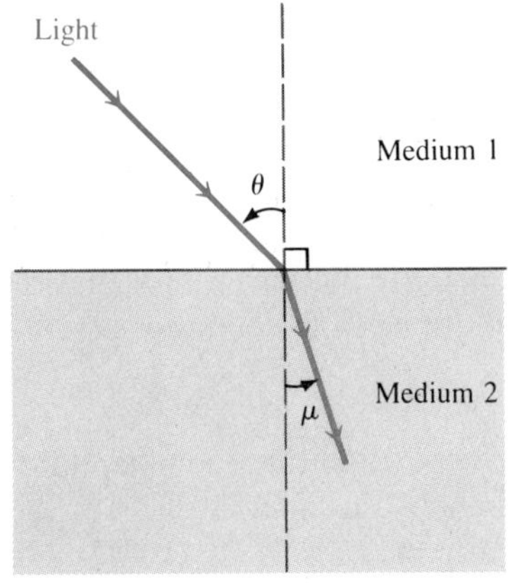

FIGURE 8.6

another medium, such as water, the ray is deflected by a certain amount, as pictured in Figure 8.6. The angle θ in this figure is called the **angle of incidence** and the angle μ (the Greek letter mu) is called the **angle of refraction.** (Of course, θ and μ satisfy the conditions $0 \leq \theta \leq \pi/2$ and $0 \leq \mu \leq \pi/2$.)

Now, according to **Snell's law of refraction,** the angles θ and μ are related by the formula

$$\frac{\sin \theta}{\sin \mu} = n$$

where n is a constant known as the **index of refraction** of the second medium with respect to the first medium. The index of refraction of water with respect to air is approximately $n = 1.333$.

EXAMPLE 9: A ray of light enters the water (from the air) at an angle of incidence equal to $\theta = 35°$. Determine the angle of refraction of this ray of light.

Solution: According to Snell's law, using the index of refraction $n = 1.333$, we have

$$\frac{\sin 35°}{\sin \mu} = 1.333$$

Hence,

$$\sin \mu = \frac{\sin 35°}{1.333} \approx 0.4303$$

and so

$$\mu \approx \sin^{-1} 0.4303 \approx 25.49°$$

(There are other angles whose sine is equal to 0.4303, but $\mu = 25.49°$ is the only angle in the interval $0 \leq \mu \leq \pi/2$.) *Try Study Suggestion 8.18.* □

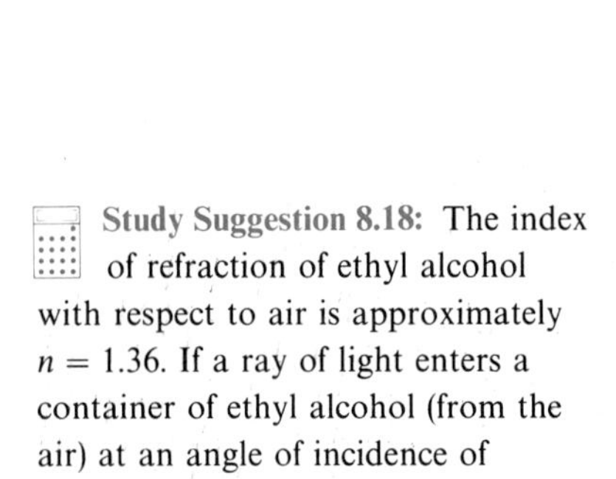

Study Suggestion 8.18: The index of refraction of ethyl alcohol with respect to air is approximately $n = 1.36$. If a ray of light enters a container of ethyl alcohol (from the air) at an angle of incidence of $\theta = 18°$, what is its angle of refraction? 

Ideas to Remember

Many of the techniques used in solving algebraic equations can also be used to solve trigonometric equations.

EXERCISES

In Exercises 1–12, solve the given equation with side condition $0 \leq t < 2\pi$.

1. $\sin t = -1/2$

2. $\sin 2t = \sqrt{2}/2$

3. $\cos t = -1$

4. $\cos t/2 = \sqrt{3}/2$

5. $\tan t = \sqrt{3}$

6. $\tan 4t = 1$

7. $\cot t = -\sqrt{3}/3$

8. $\sec t = \sqrt{2}$

9. $\sec 3t = 2$

10. $\sec t = -2\sqrt{3}/3$

11. $\csc(-t) = -2$

12. $\csc t = 1/2$

In Exercises 13–18, solve the given equation.

13. $\sin t = 1$

14. $\cos(t + \pi/3) = -\sqrt{2}/2$

15. $\tan t = \sqrt{3}/3$

16. $\cot(\pi - t) = 1/\sqrt{3}$

17. $\sec t = 2\sqrt{3}/3$

18. $\sec(2t - \pi) = -1$

In Exercises 19–30, use a calculator to approximate the solutions to the given equation with side condition $0 \le t < 2\pi$.

19. $\sin t = -0.45$

20. $\sin 2t = 1/3$

21. $\cos t = \pi/4$

22. $\cos(t - \pi/2) = -0.33$

23. $\tan t = 10$

24. $\tan t/4 = -2.344$

25. $\cot t = -0.937$

26. $\cot(-2t) = -5$

27. $\sec t = 7.101$

28. $\sec(2t - \pi) = -\sqrt{5}$

29. $\csc t = 2.998$

30. $\csc(\pi - t) = -\sqrt{2}/3$

In Exercises 31–56, find all solutions to the given equation.

31. $\sin t \tan t = 0, \quad 0 \le t < 2\pi$

32. $\cos t \cot t = 0, \quad 0 \le t < 4\pi$

33. $\tan\theta \cot\theta = \tan\theta$

34. $\sec x \sin x = \sec x, \quad -\pi \le x < \pi$

35. $(4\sin^2\theta - 3)\cos\theta = 0, \quad -\pi \le \theta < \pi$

36. $2\sec t \cos t - 3\cos t = 0, \quad 0 \le t < 2\pi$

37. $\sec x \sin x \cos x = 1, \quad 0 \le x < 2\pi$

38. $\csc t \sin t \cos t = 1, \quad -2\pi \le t < 2\pi$

39. $\cot^2\theta - \cot^2\theta\cos^2\theta = 1/2, \quad 0 \le \theta < 2\pi$

40. $4\tan\theta - 3\cot\theta = 4\sin^2\theta\tan\theta, \quad 0 \le \theta < 2\pi$
(*Hint:* use one of the Pythagorean identities.)

41. $\cot^2 t = \dfrac{1 - \sin t}{\sin t}, \quad 0 \le t < 2\pi$

(*Hint:* first express $\cot t$ in terms of $\sin t$ and $\cos t$.)

42. $2\sin^2 t - 3\sin t - 2 = 0, \quad 0 \le t < 2\pi$

43. $2\cos^2 x + (\sqrt{3} - 2)\cos x - \sqrt{3} = 0$

44. $\cos^2 x + \cos x - 1 = 0, \quad 0 \le x < 4\pi$

45. $\sin^2 x - 3\sin x - 3 = 0, \quad -2\pi \le x < 2\pi$

46. $3\sin^2 x - 2 = 0$

47. $\sec^2 t - \sec t - 12 = 0, \quad -2\pi \le t < 2\pi$

48. $\csc^4 2t = 4, \quad 0 \le t < 2\pi$

49. $\sec^4 2t - 3\sec^2 2t + 2 = 0, \quad 0 \le t < 2\pi$

50. $\tan^2\theta = 2\tan\theta - 1$

51. $9\sin^2\theta - 6\sin\theta + 1 = 0, \quad 0 \le t < 2\pi$

52. $9\cos^2 y + 12\sin y = 13$

53. $\sin^2\theta = 3\sin^3\theta$

54. $3\sin^3\theta + \cos^2\theta - 1 = 0, \quad 0 \le \theta < 2\pi$

55. $\log\tan\theta = 0$

56. $\tan\log\theta = 0$

57. The index of refraction of polyethylene with respect to air is approximately $n = 1.52$. If a ray of light enters a sheet of polyethylene (from the air) with an angle of incidence equal to $\theta = 12°$, what is its angle of refraction?

58. The index of refraction of diamond with respect to air is approximately $n = 2.419$. If a ray of light enters a diamond (from the air) with an angle of incidence equal to $\theta = 25°$, what is its angle of refraction?

59. If the angle of refraction of a ray of light is equal to $\mu = 90°$, then the ray does not pass into the second medium, but rather slides along the interface between the two media (as shown in part (a) of the following figure).

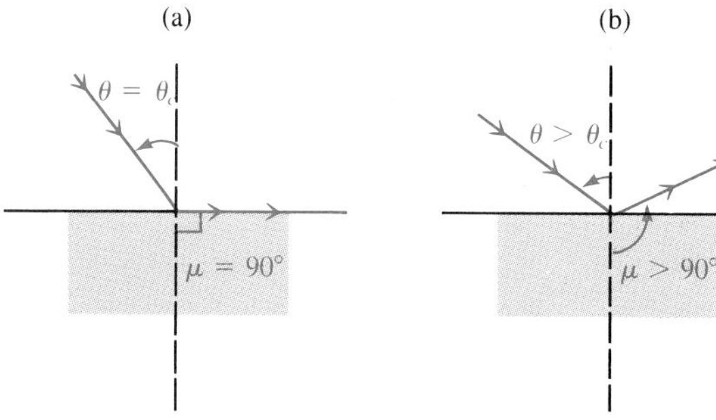

(a) When $\theta = \theta_c$, the light slides along the interface.

(b) When $\theta > \theta_c$, the light is reflected back into the first medium.

Furthermore, if the angle of refraction is greater than 90°, the light is reflected back into the first medium, as shown in part (b). For these reasons, the angle of incidence θ_c for which the angle of refraction μ is equal to 90° is called the **critical angle.** When $\theta = \theta_c$, the light slides along the interface, and when $\theta > \theta_c$, the light is reflected back into the first medium.

(a) Express the critical angle θ_c in terms of the index of refraction n.

(b) Under what conditions on the index of refraction will there be a critical angle?

(c) If the index of refraction of medium 1 with respect to medium 2 is $n = 0.233$, find the critical angle.

60. A certain generator produces voltage in accordance with the equation

$$v(t) = 6\cos\left(\frac{\pi}{2}t - \frac{\pi}{4}\right)$$

where $v(t)$ is the number of volts produced at time t. At what times does the generator produce a voltage of $3\sqrt{3}$ volts?

8.3 The Addition and Subtraction Formulas

Up to this point, we have only been able to compute the *exact* values of $\sin t$ and $\cos t$ for a relatively small number of values of t, in particular, for those values of t given in Figure 6.20. For instance, we cannot compute the exact value of $\sin 5\pi/12$ or $\cos 5\pi/12$. In this section we will discuss some very useful formulas that will enable us to compute values such as these.

It is important to realize that just because $5\pi/12 = \pi/6 + \pi/4$, does not mean that $\sin 5\pi/12$ is equal to $\sin \pi/6 + \sin \pi/4$. In fact, since $\sin 5\pi/12 \leq 1$ and

$$\sin\frac{\pi}{6} + \sin\frac{\pi}{4} = \frac{1}{2} + \frac{\sqrt{2}}{2} = \frac{1+\sqrt{2}}{2} > 1$$

these two quantities cannot possibly be equal. In fact, we can say that

in general
$$\sin(s+t) \neq \sin s + \sin t$$
$$\cos(s+t) \neq \cos s + \cos t$$
$$\tan(s+t) \neq \tan s + \tan t$$

and similarly for the other trigonometric functions. (We say "in general" because there are some values of s and t for which $\sin(s+t)$ does equal $\sin s + \sin t$.)

Fortunately, however, there are formulas for the sine, cosine, and tangent of the *sum* of two numbers. Let us begin with the cosine.

THEOREM

For any real numbers s and t, we have

$$\cos(s+t) = \cos s\cos t - \sin s\sin t \qquad (8.11)$$

and

$$\cos(s-t) = \cos s\cos t + \sin s\sin t \qquad (8.12)$$

Formula (8.11) is known as the **addition formula** for the cosine function, since it expresses the cosine of the *sum* of two numbers in terms of the sine and cosine of the numbers themselves. Similarly, Formula (8.12) is known as the **subtraction formula** for the cosine function.

Before proving these formulas, let us see how they can be used to compute some exact values of the cosine function.

EXAMPLE 1: Find the exact value of

(a) $\cos\frac{5\pi}{12}$ **(b)** $\cos 15°$

Solutions:

(a) Since $5\pi/12 = \pi/6 + \pi/4$, the addition formula for the cosine function gives

$$\begin{aligned}\cos\frac{5\pi}{12} &= \cos\left(\frac{\pi}{6}+\frac{\pi}{4}\right)\\ &= \cos\frac{\pi}{6}\cos\frac{\pi}{4} - \sin\frac{\pi}{6}\sin\frac{\pi}{4}\\ &= \left(\frac{\sqrt{3}}{2}\right)\left(\frac{\sqrt{2}}{2}\right) - \left(\frac{1}{2}\right)\left(\frac{\sqrt{2}}{2}\right)\\ &= \frac{\sqrt{6}-\sqrt{2}}{4}\end{aligned}$$

(b) Since $15° = 45° - 30°$, the subtraction formula for the cosine function gives

$$\begin{aligned}\cos 15° &= \cos(45° - 30°)\\ &= \cos 45°\cos 30° + \sin 45°\sin 30°\\ &= \left(\frac{\sqrt{2}}{2}\right)\left(\frac{\sqrt{3}}{2}\right) + \left(\frac{\sqrt{2}}{2}\right)\left(\frac{1}{2}\right)\\ &= \frac{\sqrt{2}+\sqrt{6}}{4}\end{aligned}$$

Try Study Suggestions 8.19. □

▶ **Study Suggestion 8.19:** Compute the exact value of **(a)** $\cos 7\pi/12$ **(b)** $\cos 75°$ (*Hint:* $75 = 135 - 60$.) ◀

Now let us prove the addition and subtraction formulas for the cosine.

Proof. First we prove the addition formula. Figure 8.7(a) shows the point $P(s + t)$ on the unit circle. Rotating the triangle AOB through an angle of measure $-t$ gives the picture shown in Figure 8.7(b). (For the sake of drawing this figure, we have made some assumptions about s and t. For example, we have assumed that $P(s + t)$ lies in the second quadrant, and that $t > 0$. However, the proof does not depend on these assumptions.) Since this rotation

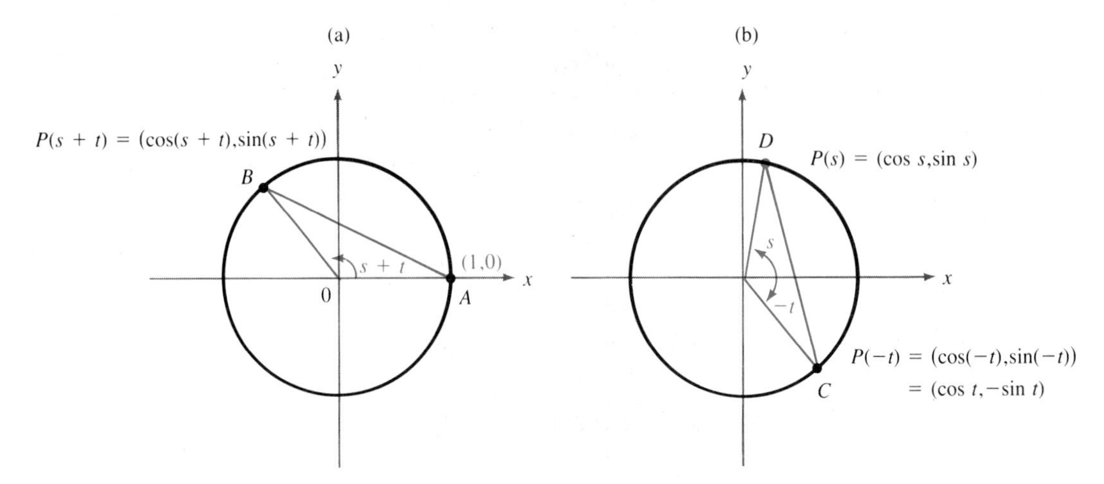

FIGURE 8.7

does not effect length, the distance $d(A, B)$ from A to B is the same as the distance $d(C, D)$ from C to D. Hence, it is also true that

$$[d(A, B)]^2 = [d(C, D)]^2 \tag{8.13}$$

(By working with the *square* of the distance, we can avoid the square root signs that come from the distance formula.)

Using the distance formula, and one of the Pythagorean identities, we get

$$\begin{aligned} [d(A, B)]^2 &= [\cos(s + t) - 1]^2 + [\sin(s + t) - 0]^2 \\ &= \cos^2(s + t) - 2\cos(s + t) + 1 + \sin^2(s + t) \\ &= [\sin^2(s + t) + \cos^2(s + t)] - 2\cos(s + t) + 1 \\ &= 1 - 2\cos(s + t) + 1 \\ &= 2 - 2\cos(s + t) \end{aligned}$$

Also, since $P(-t) = (\cos(-t), \sin(-t)) = (\cos t, -\sin t)$, the distance formula gives

$$\begin{aligned} [d(C, D)]^2 &= (\cos s - \cos t)^2 + (\sin s + \sin t)^2 \\ &= \cos^2 s - 2\cos s\cos t + \cos^2 t \\ &\quad + \sin^2 s + 2\sin s\sin t + \sin^2 t \\ &= (\sin^2 s + \cos^2 s) + (\sin^2 t + \cos^2 t) \\ &\quad - 2\cos s\cos t + 2\sin s\sin t \\ &= 2 - 2\cos s\cos t + 2\sin s\sin t \end{aligned}$$

Using these two expressions in Equation (8.13), we get

$$2 - 2\cos(s + t) = 2 - 2\cos s\cos t + 2\sin s\sin t$$

Solving for $\cos(s + t)$ gives

$$\cos(s + t) = \cos s\cos t - \sin s\sin t$$

This is the addition formula for the cosine function, and so the first part of the theorem is proved.

The subtraction formula can be proved by using the addition formula, and the fact that $\cos(-t) = \cos t$ and $\sin(-t) = -\sin t$. In particular,

$$\begin{aligned}\cos(s - t) &= \cos(s + (-t)) \\ &= \cos s \cos(-t) - \sin s \sin(-t) \\ &= \cos s \cos t + \sin s \sin t\end{aligned}$$

This is the subtraction formula, and so our proof is complete. ∎

In the next example, we use the notation $0 \leq s, t \leq \pi/2$. This stands for the *two* inequalities $0 \leq s \leq \pi/2$ and $0 \leq t \leq \pi/2$.

EXAMPLE 2: Suppose that $0 \leq s, t \leq \pi/2$. Find $\cos(s + t)$ if $\cos s = 1/5$ and $\sin t = 2/3$. Which quadrant contains the point $P(s + t)$?

Solution: The addition formula for the cosine is

$$\cos(s + t) = \cos s \cos t - \sin s \sin t \tag{8.14}$$

In order to use this formula, we must first compute $\cos t$ and $\sin s$. (The values of $\cos s$ and $\sin t$ are given.) Since $0 \leq s, t \leq \pi/2$ we know that these values are positive. Hence, the first Pythagorean identity gives

$$\cos t = \sqrt{1 - \sin^2 t} = \sqrt{1 - \left(\frac{2}{3}\right)^2} = \frac{\sqrt{5}}{3}$$

and

$$\sin s = \sqrt{1 - \cos^2 s} = \sqrt{1 - \left(\frac{1}{5}\right)^2} = \frac{2\sqrt{6}}{3}$$

Now we can use Equation (8.14).

$$\begin{aligned}\cos(s + t) &= \left(\frac{1}{5}\right)\left(\frac{\sqrt{5}}{3}\right) - \left(\frac{2\sqrt{6}}{5}\right)\left(\frac{2}{3}\right) \\ &= \frac{\sqrt{5} - 4\sqrt{6}}{15}\end{aligned}$$

Finally, since $\cos(s + t)$ is negative and $0 \leq s + t \leq \pi$, the point $P(s + t)$ must lie in the second quadrant. *Try Study Suggestion 8.20.* □

► **Study Suggestion 8.20:** Compute $\cos(s - t)$ if $0 \leq s, t \leq \pi/2$, $\cos t = 1/6$ and $\sin s = 1/5$. Which quadrant contains the point $P(s - t)$? (Explain your answer.) ◄

It is customary to refer to the sine and *co*sine functions as **cofunctions.** (The cosine is the cofunction of the sine, *and* the sine is the cofunction of the cosine.) Similarly, the tangent and cotangent are cofunctions, as are the secant and cosecant. The addition and subtraction formulas for the cosine can be used to obtain certain important relationships between cofunctions.

COFUNCTION THEOREM

For any real number t, we have

1. $\cos\left(\frac{\pi}{2} - t\right) = \sin t$
2. $\sin\left(\frac{\pi}{2} - t\right) = \cos t$
3. $\cot\left(\frac{\pi}{2} - t\right) = \tan t$
4. $\tan\left(\frac{\pi}{2} - t\right) = \cot t$
5. $\csc\left(\frac{\pi}{2} - t\right) = \sec t$
6. $\sec\left(\frac{\pi}{2} - t\right) = \csc t$

Notice that all of these formulas have the same form, namely

$$\textit{cofunction}\left(\frac{\pi}{2} - t\right) = \textit{function}\,(t)$$

Proof. We will prove formulas 1–3, and leave the others for the exercises. Formula 1 comes from a direct application of the subtraction formula for the cosine

$$\begin{aligned}\cos\left(\frac{\pi}{2} - t\right) &= \cos\frac{\pi}{2}\cos t + \sin\frac{\pi}{2}\sin t \\ &= 0 \cdot \cos t + 1 \cdot \sin t \\ &= \sin t\end{aligned}$$

Replacing t with $\frac{\pi}{2} - t$ in Formula 1 gives

$$\begin{aligned}\sin\left(\frac{\pi}{2} - t\right) &= \cos\left(\frac{\pi}{2} - \left[\frac{\pi}{2} - t\right]\right) \\ &= \cos t\end{aligned}$$

which is Formula 2. Finally, using the definition of the cotangent, along with Formulas 1 and 2, we get

$$\begin{aligned}\cot\left(\frac{\pi}{2} - t\right) &= \frac{\cos\left(\frac{\pi}{2} - t\right)}{\sin\left(\frac{\pi}{2} - t\right)} \\ &= \frac{\sin t}{\cos t} \\ &= \tan t\end{aligned}$$

and this proves Formula 3. ∎

The cofunction theorem can be used to derive the addition and subtraction formulas for the sine.

THEOREM

For all real numbers s and t, we have

$$\sin(s + t) = \sin s \cos t + \cos s \sin t$$

and

$$\sin(s - t) = \sin s \cos t - \cos s \sin t$$

Proof. The addition formula is proved by using Formulas 1 and 2 of the cofunction theorem, along with the subtraction formula for the cosine,

$$\begin{aligned}
\sin(s + t) &= \cos\left[\frac{\pi}{2} - (s + t)\right] \\
&= \cos\left[\left(\frac{\pi}{2} - s\right) - t\right] \\
&= \cos\left(\frac{\pi}{2} - s\right)\cos t + \sin\left(\frac{\pi}{2} - s\right)\sin t \\
&= \sin s \cos t + \cos s \sin t
\end{aligned}$$

The subtraction formula for the sine can be proved directly from the addition formula for the sine, and we will leave this as an exercise. ∎

The addition and subtraction formulas for the sine and cosine are among the most useful of trigonometric formulas, and if you continue to study mathematics or science, you will no doubt use them often. Therefore, *you should commit them to memory.*

EXAMPLE 3: Express $\sin(t + \pi/4)$ in terms of $\sin t$ and $\cos t$.

Solution: The addition formula for the sine function gives

$$\begin{aligned}
\sin\left(t + \frac{\pi}{4}\right) &= \sin t \cos\frac{\pi}{4} + \cos t \sin\frac{\pi}{4} \\
&= (\sin t)\left(\frac{\sqrt{2}}{2}\right) + (\cos t)\left(\frac{\sqrt{2}}{2}\right) \\
&= \left(\frac{\sqrt{2}}{2}\right)(\sin t + \cos t)
\end{aligned}$$

Try Study Suggestion 8.21. □

▶ **Study Suggestion 8.21:** Express $\cos(t + \pi/3)$ in terms of $\sin t$ and $\cos t$. ◀

EXAMPLE 4: Verify the identity

$$\sin(s+t)\sin(s-t) = \sin^2 s - \sin^2 t \qquad (8.15)$$

Solution: Using both the addition and subtraction formulas for the sine, we get

$$\begin{aligned}\sin(s+t)\sin(s-t) &= (\sin s\cos t + \cos s\sin t)(\sin s\cos t - \cos s\sin t)\\ &= (\sin s\cos t)^2 - (\cos s\sin t)^2\\ &= \sin^2 s\cos^2 t - \cos^2 s\sin^2 t\end{aligned}$$

Now we can use the first Pythagorean identity to express everything in terms of the sine,

$$\begin{aligned}&= (\sin^2 s)(1-\sin^2 t) - (1-\sin^2 s)\sin^2 t\\ &= \sin^2 s - \sin^2 s\sin^2 t - \sin^2 t + \sin^2 s\sin^2 t\\ &= \sin^2 s - \sin^2 t\end{aligned}$$

Since this is the right-hand side of Equation (8.15), the identity is proved.

▶ **Study Suggestion 8.22:** Verify the identity

$$\cos(s+t)\cos(s-t) = \cos^2 s - \sin^2 t$$ ◀

Try Study Suggestion 8.22. □

The addition and subtraction formulas for the sine and cosine can also be used to derive the addition and subtraction formulas for the tangent.

THEOREM

The following formulas hold for all real numbers s and t for which s, t, and $s+t$ are in the domain of the tangent function.

$$\tan(s+t) = \frac{\tan s + \tan t}{1 - \tan s\tan t}$$

and

$$\tan(s-t) = \frac{\tan s - \tan t}{1 + \tan s\tan t}$$

Proof. We will verify the addition formula, and leave the subtraction formula for the exercises. The definition of the tangent and the addition formulas for the sine and cosine give

$$\begin{aligned}\tan(s+t) &= \frac{\sin(s+t)}{\cos(s+t)}\\ &= \frac{\sin s\cos t + \cos s\sin t}{\cos s\cos t - \sin s\sin t}\end{aligned}$$

We would like to express this completely in terms of the tangent. To this end, we divide the numerator and denominator by the product $\cos s\cos t$. (In order to do this, we must have $\cos s \neq 0$ and $\cos t \neq 0$. But this is true since we are assuming that s and t are in the domain of the tangent function.)

Performing the division gives

$$= \frac{\left(\frac{\sin s}{\cos s}\right)\left(\frac{\cos t}{\cos t}\right) + \left(\frac{\cos s}{\cos s}\right)\left(\frac{\sin t}{\cos t}\right)}{\left(\frac{\cos s}{\cos s}\right)\left(\frac{\cos t}{\cos t}\right) - \left(\frac{\sin s}{\cos s}\right)\left(\frac{\sin t}{\cos t}\right)}$$

$$= \frac{\tan s + \tan t}{1 - \tan s \tan t}$$

and this establishes the addition formula for the tangent function. ∎

EXAMPLE 5: Evaluate the expression $\tan(\arctan 4 + \arcsin 1/3)$

Solution: Using the addition formula for the tangent function, we get

$$\tan(\arctan 4 + \arcsin 1/3) = \frac{\tan(\arctan 4) + \tan(\arcsin 1/3)}{1 - \tan(\arctan 4)\tan(\arcsin 1/3)}$$

Since $\tan(\arctan 4) = 4$, this is equal to

$$= \frac{4 + \tan(\arcsin 1/3)}{1 - 4\tan(\arcsin 1/3)}$$

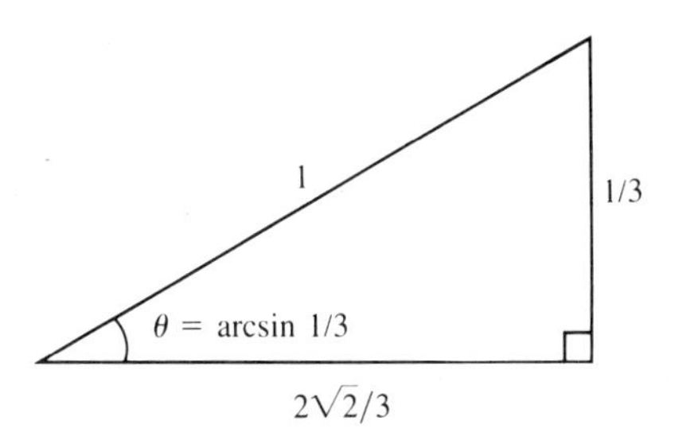

FIGURE 8.8

Now we must compute $\tan(\arcsin 1/3)$. This can be done by looking at the right triangle in Figure 8.8, where we have set $\theta = \arcsin 1/3$. This triangle shows that $\tan(\arcsin 1/3) = \tan\theta = 1/2\sqrt{2} = \sqrt{2}/4$, and so

$$\tan(\arctan 4 + \arcsin 1/3) = \frac{4 + \sqrt{2}/4}{1 - 4(\sqrt{2}/4)} = \frac{16 + \sqrt{2}}{4(1 - \sqrt{2})} = -\frac{18 + 17\sqrt{2}}{4}$$

Incidentally, another method for computing $\tan(\arcsin 1/3)$ is to use the definition of the tangent and a Pythagorean identity as follows

$$\tan(\arcsin 1/3) = \frac{\sin(\arcsin 1/3)}{\cos(\arcsin 1/3)}$$

$$= \frac{1/3}{\sqrt{1 - \sin^2(\arcsin 1/3)}}$$

$$= \frac{1/3}{\sqrt{1 - (1/3)^2}}$$

$$= \frac{1/3}{\sqrt{8/9}}$$

$$= \frac{1}{\sqrt{8}}$$

$$= \frac{\sqrt{2}}{4}$$

▶ **Study Suggestion 8.23:** Evaluate the expression

$\tan(\arctan(-5) + \arccos 1/5)$ ◀

Try Study Suggestion 8.23. □

Ideas to Remember

The addition and subtraction formulas are among the most useful trigonometric formulas. One of their many uses is to compute some exact values of the trigonometric functions that we could not compute before.

EXERCISES

1. Find $\sin(s + t)$ and $\cos(s + t)$ if $0 \leq s,\ t \leq \pi/2$, $\cos s = 1/5$, and $\sin t = 2/5$. Which quadrant contains the point $P(s + t)$?

2. Find $\sin(s + t)$ and $\cos(s + t)$ if $0 \leq s,\ t \leq \pi/2$, $\cos s = 2/5$, and $\sin t = 1/5$. Which quadrant contains the point $P(s + t)$?

3. Find $\cos(s - t)$ and $\sin(s - t)$ if $\pi/2 \leq s,\ t \leq \pi$, $\cos s = -1/4$, and $\sin t = 1/3$. Which quadrant contains the point $P(s - t)$?

4. Find $\sin s$, $\cos s$, and $\cos t$ if $\pi/2 \leq s,\ t \leq \pi$, $\tan s = 2$, $\sin t = 1/3$, and $\sin(s + t) = 2/3$.

In Exercises 5–20, use the addition and subtraction formulas to compute the exact value of the given expression.

5. $\sin \dfrac{\pi}{12}$

6. $\cos \dfrac{\pi}{12}$

7. $\sin 75°$

8. $\tan 75°$

9. $\cos \dfrac{7\pi}{12}$

10. $\csc \dfrac{7\pi}{12}$

11. $\cos \dfrac{11\pi}{12}$

$\left(\textit{Hint: } \dfrac{11\pi}{12} = \dfrac{7\pi}{6} - \dfrac{\pi}{4}\right)$

12. $\cot \dfrac{11\pi}{12}$

13. $\sin 195°$
(*Hint:* $195° = 150° + 45°$)

14. $\cos 195°$

15. $\tan \dfrac{15\pi}{12}$

16. $\tan \dfrac{15\pi}{12}$

17. $\cos \dfrac{17\pi}{12}$

$\left(\textit{Hint: } \dfrac{17\pi}{12} = \dfrac{5\pi}{4} + \dfrac{\pi}{6}\right)$

18. $\csc \dfrac{17\pi}{12}$

19. $\sin \dfrac{19\pi}{12}$

20. $\tan \dfrac{19\pi}{12}$

In Exercises 21–31, simplify the given expression.

21. $\sin\left(t + \dfrac{\pi}{2}\right)$

22. $\cos\left(t + \dfrac{\pi}{2}\right)$

23. $\tan\left(\theta + \dfrac{\pi}{2}\right)$

24. $\sec\left(\theta + \dfrac{\pi}{2}\right)$

25. $\sin(\pi - x)$

26. $\cos(\pi - x)$

27. $\tan(\pi - t)$

28. $\cot(\pi - t)$

29. $\cos\left(\dfrac{3\pi}{2} + s\right)$

30. $\sin\left(s - \dfrac{3\pi}{2}\right)$

31. $\tan\left(t + \dfrac{3\pi}{2}\right)$

In Exercises 32–37, express the given expression in terms of $\sin t$ and $\cos t$.

32. $\sin\left(t + \frac{\pi}{6}\right)$

33. $\cos\left(t - \frac{\pi}{6}\right)$

34. $\sin\left(t - \frac{3\pi}{4}\right)$

35. $\cos\left(t - \frac{3\pi}{4}\right)$

36. $\sin\left(t + \frac{5\pi}{3}\right)$

37. $\cos\left(t + \frac{5\pi}{3}\right)$

In Exercises 38–45, verify the given identity.

38. $\tan(t + \pi/4) = \dfrac{1 + \tan t}{1 - \tan t}$

39. $\tan(t - \pi/4) = \dfrac{\tan t - 1}{\tan t + 1}$

40. $\sin(s + t) + \sin(s - t) = 2 \sin s \cos t$

41. $\cos(s + t) + \cos(s - t) = 2 \cos s \cos t$

42. $\tan s + \tan t = \dfrac{\sin(s + t)}{\cos s \cos t}$

43. $\dfrac{\sin(s + t)}{\sin(s - t)} = \dfrac{\tan s + \tan t}{\tan s - \tan t}$

44. $\dfrac{\sin(s + t)}{\cos(s - t)} = \dfrac{\cot t + \cot s}{\cot t \cot s + 1}$

45. $\dfrac{\sin(s - t)}{\cos(s + t)} = \dfrac{\tan s - \tan t}{1 - \tan s \tan t}$

In Exercises 46–56, evaluate the given expression.

46. $\sin\left(\arcsin \frac{1}{3} + \arcsin\left(-\frac{2}{3}\right)\right)$

47. $\cos\left(\arctan 10 - \arcsin \frac{1}{5}\right)$

48. $\sin\left(\arccos\left(-\frac{1}{4}\right) - \arcsin \frac{3}{5}\right)$

49. $\cos\left(\frac{\pi}{3} + \arccos \frac{\pi}{4}\right)$

50. $\sin\left(\frac{\pi}{6} - \arccos \frac{\pi}{6}\right)$

51. $\tan\left(\arctan(-2) + \arcsin \frac{1}{4}\right)$

52. $\tan\left(\arcsin \frac{1}{3} + \arccos \frac{1}{6}\right)$

53. $\tan\left(\arcsin \frac{1}{3} + \arcsin\left(-\frac{1}{3}\right)\right)$

54. $\sin(\arcsin x + \arccos x)$

55. $\cos(\arctan x - \arcsin x)$

56. $\tan(\arctan x + \arcsin x)$

57. Show that $\quad \sin^{-1} x + \cos^{-1} x = \dfrac{\pi}{2}$

58. Show that $\quad \tan^{-1} x + \tan^{-1} \dfrac{1}{x} = \dfrac{\pi}{2} \quad$ for $x > 0$

59. Verify the identity $\quad \tan t = \cot\left(\frac{\pi}{2} - t\right)$

60. Verify the identity $\quad \csc t = \sec\left(\frac{\pi}{2} - t\right)$

61. Verify the identity $\quad \sec t = \csc\left(\frac{\pi}{2} - t\right)$

62. Verify the subtraction formula for the sine function.

63. Verify the subtraction formula for the tangent function.

64. If $s + t = \pi/4$, show that

$$\tan s = \frac{1 - \tan t}{1 + \tan t}$$

65. If $f(x) = \sin x$ show that

$$\frac{f(x + h) - f(x)}{h} = \sin x\left(\frac{\cos h - 1}{h}\right) + \cos x\left(\frac{\sin h}{h}\right)$$

(Recall from Section 2.3 that the expression on the left-hand side of this equation is the *difference quotient* of the function $f(x) = \sin x$. You will use this formula for the difference quotient if you take a course in calculus.)

66. Derive a formula for the difference quotient of the function $f(x) = \cos x$ that is similar to the one given in Exercise 65.

In Exercises 67–72, use a calculator to "verify" the given identities for the given values of s and t. (In general, a calculator cannot be used to verify that an identity holds for given values since calculators give only approximations. Therefore, by "verify" we mean show that the calculator gives the same approximations when each side of the identity is evaluated. However, you may ignore differences in the last digit of your display, since these differences may be due to calculator error.)

67. $\sin(s + t) = \sin s \cot t + \cos s \sin t$; $s = 3\pi/5$, $t = \pi/5$.

68. $\cos(s + t) = \cos s \cos t - \sin s \sin t$; $s = 2.41$, $t = -3.98$.

69. $\tan(s + t) = \dfrac{\tan s + \tan t}{1 - \tan s \tan t}$; $s = 17.6°$, $t = 29.8°$

70. $\sin(s - t) = \sin s \cos t - \cos s \sin t$; $s = 12°$, $t = 39°$

71. $\cos(s - t) = \cos s \cos t + \sin s \sin t$; $s = 1.98\pi$, $t = -0.364\pi$

72. $\tan(s - t) = \dfrac{\tan s - \tan t}{1 + \tan s \tan t}$; $s = 17\pi/13$, $t = -2\pi/11$

8.4 The Double-Angle and Half-Angle Formulas

The addition and subtraction formulas have several important consequences. For example, if we set $s = t$ in the addition formula for the sine function, the result is

$$\sin(t + t) = \sin t \cos t + \cos t \sin t$$

or, after a bit of simplification,

$$\mathbf{\sin 2t = 2 \sin t \cos t}$$

Thus we get a formula expressing $\sin 2t$ in terms of $\sin t$ and $\cos t$, called the **double-angle formula** for the sine function.

We can also set $s = t$ in the addition formula for the cosine function, to get

$$\cos(t + t) = \cos t \cos t - \sin t \sin t$$

or

$$\mathbf{\cos 2t = \cos^2 t - \sin^2 t} \qquad \textbf{(8.16)}$$

This is a **double-angle formula** for the cosine function. (Notice the minus sign on the right-hand side. What would we get if we accidentally replaced it with a plus sign?)

The double-angle formula for the cosine function has two other forms, one of which involves only the cosine, and the other of which involves only the sine. Using the Pythagorean identity $\sin^2 t + \cos^2 t = 1$, we may replace $\sin^2 t$ by $1 - \cos^2 t$ in Equation (8.16), to get

$$\mathbf{\cos 2t = 2\cos^2 t - 1} \qquad \textbf{(8.17)}$$

On the other hand, replacing $\cos^2 t$ by $1 - \sin^2 t$ gives

$$\mathbf{\cos 2t = 1 - 2\sin^2 t} \qquad \textbf{(8.18)}$$

The double-angle formula for the tangent can be obtained in the same way as the other double-angle formulas, and we have

$$\tan(t + t) = \frac{\tan t + \tan t}{1 - \tan t \tan t}$$

or

$$\tan 2t = \frac{2 \tan t}{1 - \tan^2 t}$$

Let us collect these formulas together for reference.

The Double-Angle Formulas

$$\sin 2t = 2 \sin t \cos t$$

$$\cos 2t = \cos^2 t - \sin^2 t = 2\cos^2 t - 1 = 1 - 2\sin^2 t$$

$$\tan 2t = \frac{2 \tan t}{1 - \tan^2 t}$$

EXAMPLE 1: Find the values of $\sin 2t$, $\cos 2t$, and $\tan 2t$ if $\sin t = -1/5$ and $P(t)$ is in the third quadrant.

Solution: First we need to compute $\cos t$. Since $P(t)$ is in third quadrant, $\cos t$ is negative, and so the Pythagorean identity gives

$$\cos t = -\sqrt{1 - \sin^2 t} = -\sqrt{1 - \left(-\frac{1}{5}\right)^2} = -\frac{2\sqrt{6}}{5}$$

With this information, we can also compute $\tan t$,

$$\tan t = \frac{\sin t}{\cos t} = \frac{-1/5}{-2\sqrt{6}/5} = \frac{1}{2\sqrt{6}} = \frac{\sqrt{6}}{12}$$

The double-angle formulas then give

$$\sin 2t = 2 \sin t \cos t = 2\left(\frac{-1}{5}\right)\left(\frac{-2\sqrt{6}}{5}\right) = \frac{4\sqrt{6}}{25}$$

$$\cos 2t = \cos^2 t - \sin^2 t = \frac{24}{25} - \frac{1}{25} = \frac{23}{25}$$

$$\tan 2t = \frac{2 \tan t}{1 - \tan^2 t} = \frac{\sqrt{6}/6}{1 - 1/24} = \frac{4\sqrt{6}}{23}$$

Try Study Suggestion 8.24. □

▶ **Study Suggestion 8.24:** Compute $\sin 2t$, $\cos 2t$, and $\tan 2t$ if $\sin t = -1/3$ and $\tan t < 0$. ◀

As the following example demonstrates, the double-angle formulas can be used to derive formulas for the sine, cosine, and tangent of other integral multiples of t as well.

EXAMPLE 2: Express $\sin 3t$ in terms of $\sin t$.

Solution: First we use the addition formula for the sine, and then the double-angle formulas for both the sine and cosine

$$\begin{aligned}\sin 3t &= \sin(2t + t)\\ &= \sin 2t \cos t + \cos 2t \sin t\\ &= (2 \sin t \cos t) \cos t + (1 - 2\sin^2 t) \sin t\\ &= 2 \sin t \cos^2 t + \sin t - 2 \sin^3 t\\ &= 2 \sin t(1 - \sin^2 t) + \sin t - 2\sin^3 t\\ &= 3 \sin t - 4 \sin^3 t\end{aligned}$$

Try Study Suggestion 8.25. □

► **Study Suggestion 8.25:** Express $\cos 3t$ in terms of $\cos t$. ◄

Solving Equations (8.17) and (8.18) for $\cos^2 t$ and $\sin^2 t$, respectively, gives us the interesting formulas

$$\sin^2 t = \frac{1 - \cos 2t}{2} \tag{8.19}$$

and

$$\cos^2 t = \frac{1 + \cos 2t}{2} \tag{8.20}$$

You will find these formulas to be very useful if you take calculus, but we will have an important use for them here as well.

EXAMPLE 3: Express $\cos^4 t$ in terms of the first power of the cosine.

Solution: In this case, we use Formula (8.20) twice,

$$\begin{aligned}\cos^4 t &= (\cos^2 t)^2\\ &= \left(\frac{1 + \cos 2t}{2}\right)^2\\ &= \frac{1}{4}(1 + 2\cos 2t + \cos^2 2t)\end{aligned}$$

Now we use Equation (8.20) with t replaced by $2t$ to get

$$= \frac{1}{4}\left(1 + 2\cos 2t + \frac{1 + \cos 4t}{2}\right)$$

$$= \frac{1}{8}(3 + 4\cos 2t + \cos 4t)$$

▶ **Study Suggestion 8.26:** Express $\sin^4 t$ in terms of the first power of the cosine. ◀

Try Study Suggestion 8.26. □

For our purposes, the most important use of Formulas (8.19) and (8.20) is to derive formulas for $\sin t/2$ and $\cos t/2$. Replacing t by $t/2$ in Formulas (8.19) and (8.20) gives

$$\sin^2 \frac{t}{2} = \frac{1 - \cos t}{2} \quad \text{and} \quad \cos^2 \frac{t}{2} = \frac{1 + \cos t}{2}$$

Taking the square root of both sides of these equations, we get the **half-angle formulas** for the sine and cosine.

The Half-Angle Formulas for the Sine and Cosine

$$\sin \frac{t}{2} = \pm\sqrt{\frac{1 - \cos t}{2}} \qquad \cos \frac{t}{2} = \pm\sqrt{\frac{1 + \cos t}{2}}$$

Notice that the half-angle formulas do not completely determine $\sin t/2$ and $\cos t/2$, for they do not determine their sign. As we will see in a moment, this can be done with additional information.

There is also a half-angle formula for the tangent, which does not have an ambiguity of sign. In order to derive this formula, we begin by writing

$$\tan t = \frac{\sin t}{\cos t} = \frac{(\sin t)(2\cos t)}{(\cos t)(2\cos t)} = \frac{2\sin t\cos t}{2\cos^2 t}$$

Using the double-angle formula for the sine in the numerator of the last expression, and the double-angle formula for the cosine (in the form $2\cos^2 t = 1 + \cos 2t$) in the denominator, we get

$$\tan t = \frac{\sin 2t}{1 + \cos 2t}$$

Finally, we replace t by $t/2$, to get the first of the following formulas. (We ask you to verify the second formula in the exercises.)

The Half-Angle Formulas for the Tangent

$$\tan\frac{t}{2} = \frac{\sin t}{1+\cos t} \qquad \tan\frac{t}{2} = \frac{1-\cos t}{\sin t}$$

EXAMPLE 4: Compute the exact value of

(a) $\sin\frac{\pi}{8}$ **(b)** $\cos 7.5°$ **(c)** $\tan\frac{\pi}{8}$

Solutions:

(a) Since $\pi/8 = (1/2)(\pi/4)$, and since $\sin \pi/8$ is positive, the half-angle formula for the sine gives

$$\sin\frac{\pi}{8} = \sqrt{\frac{1-\cos\pi/4}{2}} = \sqrt{\frac{1-\sqrt{2}/2}{2}} = \frac{\sqrt{2-\sqrt{2}}}{2}$$

(b) Since $7.5 = 15/2$, and since we computed $\cos 15°$ in the previous section (using the subtraction formula), we can use the half-angle formula to compute $\cos 7.5°$. Because $\cos 7.5°$ is positive, we have

$$\cos 7.5° = \sqrt{\frac{1+\cos 15°}{2}} = \sqrt{\frac{1+\frac{\sqrt{2}+\sqrt{6}}{4}}{2}} = \sqrt{\frac{4+\sqrt{2}+\sqrt{6}}{8}}$$

(c) Since $\pi/8 = (1/2)(\pi/4)$, the first half-angle formula for the tangent gives

$$\tan\frac{\pi}{8} = \frac{\sin\pi/4}{1+\cos\pi/4} = \frac{\sqrt{2}/2}{1+\sqrt{2}/2} = \frac{\sqrt{2}}{2+\sqrt{2}} = \sqrt{2}-1$$

Try Study Suggestion 8.27. □

▶ **Study Suggestion 8.27:** Compute the exact value of **(a)** $\sin 3\pi/8$ **(b)** $\cos 3\pi/8$ **(c)** $\tan 3\pi/8$ ◀

EXAMPLE 5: Verify the identity

$$2\cot 2x = \cot x - \tan x$$

Solution: From the definition of the cotangent, and the double-angle formulas for the sine and cosine, we get

$$\begin{aligned} 2\cot 2x &= 2\frac{\cos 2x}{\sin 2x} \\ &= 2\frac{\cos^2 x - \sin^2 x}{2\sin x \cos x} \\ &= \frac{\cos^2 x}{\sin x \cos x} - \frac{\sin^2 x}{\sin x \cos x} \\ &= \frac{\cos x}{\sin x} - \frac{\sin x}{\cos x} \\ &= \cot x - \tan x \end{aligned}$$

Try Study Suggestion 8.28. □

▶ **Study Suggestion 8.28:** Verify the identity $2\csc 2x = \sec x \csc x$. ◀

EXAMPLE 6: Solve the equation

$$\sin 2t + \cos t = 0, \quad 0 \le t < 2\pi$$

Solution: The double-angle formula for the sine function allows us to write this equation in the form

$$2\sin t \cos t + \cos t = 0$$

This can be factored to give

$$(\cos t)(2\sin t + 1) = 0$$

The solutions to this equation are the solutions to either of the equations

$$\cos t = 0 \quad \text{and} \quad \sin t = -1/2$$

The solutions to the first of these equations in the interval $0 \le t < 2\pi$ are $t = \pi/2$ and $t = 3\pi/2$, and the solutions to the second equation are $t = 7\pi/6$ and $t = 11\pi/6$. Hence, the solutions to the original equation are

$$t = \frac{\pi}{2}, \frac{7\pi}{6}, \frac{3\pi}{2}, \frac{11\pi}{6}$$

Try Study Suggestion 8.29. □

▶ **Study Suggestion 8.29:** Solve the equation $\cos 2t + \cos t = 0$, with side condition $0 \le t < 4\pi$. ◀

Ideas to Remember

The double- and half-angle formulas are consequences of the addition and subtraction formulas.

EXERCISES

In Exercises 1–8, use the given information to find $\sin 2t$, $\cos 2t$, and $\tan 2t$.

1. $\sin t = 1/3$, $P(t)$ is in the first quadrant

2. $\sin t = 1/4$, $P(t)$ is in the second quadrant

3. $\cos t = -1/5$, $\tan t < 0$

4. $\cot t = 1/3$, $P(t)$ is in the fourth quadrant

5. $\tan t = 5$, $\sin t < 0$

6. $\cot t = -1/2$, $\sec t > 0$

7. $\sin t = -0.231$, $\cos t > 0$

8. $\tan t = 312.64$, $\sin t < 0$

In Exercises 9–23, determine the exact value of the given expression.

9. $\sin 7.5°$ **10.** $\cos \pi/8$ **11.** $\tan 3\pi/8$

12. $\sin 5\pi/8$ **13.** $\cos(-5\pi/8)$ **14.** $\tan 5\pi/8$

15. $\sec 7\pi/8$ **16.** $\csc(-7\pi/8)$ **17.** $\cot 7\pi/8$

18. $\sin(-5\pi/12)$ **19.** $\tan(-7\pi/24)$ **20.** $\sin 157(1/2)°$

21. $\cos 165°$ **22.** $\sin \pi/12$ **23.** $\cot \pi/12$

In Exercises 24–29, use the given information to find $\sin t/2$, $\cos t/2$, and $\tan t/2$.

24. $\sin t = 2/3$, $P(t)$ lies in the second quadrant

25. $\cos t = -4/5$, $\sin t > 0$

26. $\tan t = 2$, $\cos t < 0$

27. $\cot t = -1/2$, $P(t)$ lies in the fourth quadrant

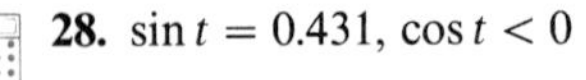

28. $\sin t = 0.431$, $\cos t < 0$

29. $\tan t = 17.693$, $P(t)$ lies in the third quadrant

30. **(a)** Express $\sec 2t$ in terms of $\sec t$.
(b) Find $\sec 2t$ if $\sec t = 7$.
(c) Find $\sec 2t$ if $\tan t = 2$ and $\sec t < 0$.

31. **(a)** Express $\csc 2t$ in terms of $\csc t$.
(b) Find $\csc 2t$ if $\csc t = -5$.
(c) Find $\csc 2t$ if $\cos t = 1/3$ and $\csc t > 0$.

32. **(a)** Express $\cot 2t$ in terms of $\cot t$.
(b) Find $\cot 2t$ if $\cot t = -1/2$.
(c) Find $\cot 2t$ if $\sin t = -1/4$ and $\sin 2t < 0$.

In Exercises 33–38, use the given information and the results of Example 2, to find $\sin 3t$.

33. $\sin t = 1/3$ **34.** $\cos t = -1/3$, $\sin 2t < 0$

35. $\tan t = 3$, $\cos t < 0$ **36.** $\cot t = -3$, $\csc t > 0$

37. $\sin t = 0.649$

38. $\tan t = 2.114$, $\cos t > 0$

39. Derive a formula for $\cos 3t$ in terms of $\cos t$.

In Exercises 40–45, use the given information, and the formula found in Exercise 39, to find $\cos 3t$.

40. $\cos t = 2/3$ **41.** $\sin t = -1/5$, $\sin 2t > 0$

42. $\tan t = 10$, $\sin t < 0$ **43.** $\cot t = -4$, $\sec t > 0$

44. $\cos t = -0.5525$

45. $\sin t = -0.216$, $\sin 2t > 0$

46. Is it possible for $\sin 2t$ to be equal to 1/3 and $\tan t$ to be negative? Justify your answer.

47. Use the formula found in Study Suggestion 8.26 to find $\sin^4 t$ if $\cos t = 1/3$.

48. Use the formula found in Study Suggestion 8.26 to find $\sin^4 t$ if $\sin t = -1/4$ and $\cos t > 0$.

49. Verify the second half-angle formula for the tangent function.

In Exercises 50–61, verify the given identity.

50. $\sin 2t = \dfrac{2 \sin t}{\sec t}$

51. $2 \sin^2 \dfrac{t}{2} = 1 - \cos t$

52. $\cos 2t = \cos^4 t - \sin^4 t$

53. $\cos 2x = \sin^2 x (\cot^2 x - 1)$

54. $\tan^2 \frac{\theta}{2} = (\csc\theta - \cot\theta)^2$

55. $\frac{\sin 2\theta}{\sin\theta} - \frac{\cos 2\theta}{\cos\theta} = \sec\theta$

56. $(\sin t + \cos t)^2 = 1 + \sin 2t$

57. $\cos 6t = 32\cos^6 t - 48\cos^4 t + 18\cos^2 t - 1$

58. $2\sin^2 2t + \cos 4t = 1$

59. $2\sin^{-1} x = \sin^{-1} 2x\sqrt{1-x^2}$ for $|x| \leq 1/\sqrt{2}$ (*Hint:* look at Example 8 of Section 8.1.)

60. $2\cos^{-1} x = \cos^{-1}(2x^2 - 1)$ for $0 \leq x \leq 1$

61. $2\tan^{-1} x = \sin^{-1} \frac{2x}{1+x^2}$

In Exercises 62–67, solve the given equation, with side condition $0 \leq t < 2\pi$.

62. $\sin 2t = \cot t$

63. $\sin 2t - 3\sin t = 0$

64. $\cos 2t - 5\cos t - 2 = 0$

65. $\tan^2 t + 3\tan 2t = 0$

66. $\tan 2t = \tan t$

67. $\sin\frac{t}{2} + \cos t = 1$

In Exercises 68–70, "verify" the identity

$$2\tan^{-1} x = \sin^{-1} \frac{2x}{1+x^2}$$

for the given value of x. (See the remark in the instructions to Exercises 67–72 of the previous section.)

68. $x = 0.249$

69. $x = -0.891$

70. $x = \sin \log 6.735$

In Exercises 71–73, "verify" the identity

$$2\sin^{-1} x = \sin^{-1} 2x\sqrt{1-x^2}$$

for the given value of x. (See the remark in the instructions to Exercises 67–72 of the previous section.)

71. $x = \cos\frac{2\pi}{5}$

72. $x = \frac{1}{\sqrt{2}}$

73. $x = -\frac{1}{\sqrt{2}}$

74. Show that the previous equation does *not* hold for $x = 0.71$.

8.5 A Summary of Trigonometric Formulas

For the sake of reference, let us list the formulas and identities that we have discussed in this chapter. The *sum* and *difference formulas* and *product formulas* listed below are new.

The Definitions

$$P(t) = (\cos t, \sin t)$$

$$\sec t = \frac{1}{\cos t} \qquad \csc t = \frac{1}{\sin t}$$

$$\tan t = \frac{\sin t}{\cos t} \qquad \cot t = \frac{\cos t}{\sin t}$$

The Pythagorean Identities

$$\sin^2 t + \cos^2 t = 1, \qquad 1 + \tan^2 t = \sec^2 t$$

$$1 + \cot^2 t = \csc^2 t$$

The Addition and Subtraction Formulas

$$\sin(s \pm t) = \sin s \cos t \pm \cos s \sin t$$
$$\cos(s \pm t) = \cos s \cos t \mp \sin s \sin t$$
$$\tan(s \pm t) = \frac{\tan s \pm \tan t}{1 \mp \tan s \tan t}$$

The Cofunction Identities

1. $\cos\left(\frac{\pi}{2} - t\right) = \sin t$
2. $\sin\left(\frac{\pi}{2} - t\right) = \cos t$
3. $\cot\left(\frac{\pi}{2} - t\right) = \tan t$
4. $\tan\left(\frac{\pi}{2} - t\right) = \cot t$
5. $\csc\left(\frac{\pi}{2} - t\right) = \sec t$
6. $\sec\left(\frac{\pi}{2} - t\right) = \csc t$

The Double-Angle Formulas

$$\sin 2t = 2 \sin t \cos t$$
$$\cos 2t = \cos^2 t - \sin^2 t = 2\cos^2 t - 1 = 1 - 2\sin^2 t$$
$$\tan 2t = \frac{2 \tan t}{1 - \tan^2 t}$$

The Half-Angle Formulas

$$\sin \frac{t}{2} = \pm\sqrt{\frac{1 - \cos t}{2}} \qquad \cos \frac{t}{2} = \pm\sqrt{\frac{1 + \cos t}{2}}$$
$$\tan \frac{t}{2} = \frac{\sin t}{1 + \cos t} = \frac{1 - \cos t}{\sin t}$$

Formulas for $\sin^2 t$ and $\cos^2 t$ in Terms of $\cos 2t$

$$\sin^2 t = \frac{1 - \cos 2t}{2}, \qquad \cos^2 t = \frac{1 + \cos 2t}{2}$$

Sum and Difference Formulas

$$\sin s + \sin t = 2\sin\frac{s+t}{2}\cos\frac{s-t}{2}$$

$$\sin s - \sin t = 2\cos\frac{s+t}{2}\sin\frac{s-t}{2}$$

$$\cos s + \cos t = 2\cos\frac{s+t}{2}\cos\frac{s-t}{2}$$

$$\cos s - \cos t = -2\sin\frac{s+t}{2}\sin\frac{s-t}{2}$$

Product Formulas

$$\sin s\cos t = \frac{1}{2}[\sin(s+t) + \sin(s-t)]$$

$$\sin s\sin t = \frac{1}{2}[\cos(s-t) - \cos(s+t)]$$

$$\cos s\cos t = \frac{1}{2}[\cos(s+t) + \cos(s-t)]$$

EXERCISES

1. Verify the first sum formula.
2. Verify the second sum formula.
3. Verify the third sum formula.
4. Verify the fourth sum formula.
5. Verify the first product formula.
6. Verify the second product formula.
7. Verify the third product formula.
8. Write down from memory as many of the formulas from this section as you can.

8.6 Review

CONCEPTS FOR REVIEW

Pythagorean Identities
Trigonometric identity
Verifying an identity
The addition formulas
The subtraction formulas
The cofunction theorem
The double-angle formulas
The half-angle formulas

REVIEW EXERCISES

In Exercises 1–3, use the given information to determine the value of each of the trigonometric functions at θ.

1. $\sin\theta = -\sqrt{3}/2$, $\tan\theta = -\sqrt{3}$

2. $\cot\theta = 12$, $\sin\theta + \sec\theta < 0$

3. $\sec\theta = -6/5$, $P(\theta)$ lies in the third quadrant

In Exercises 4–8, simplify the given expression.

4. $\dfrac{\cos^2 t}{\sin t} + \dfrac{1}{\csc t}$

5. $\dfrac{\cot t + \csc t}{\sin t + \cot(-t) + \csc(-t)}$

6. $\dfrac{1 - \tan(-\alpha)}{1 - \cot(-\alpha)}$

7. $\dfrac{\sec\beta}{(\csc\beta)(1 + \sec\beta)} + \dfrac{1 + \cos\beta}{\sin\beta}$

8. $\dfrac{1}{\cot t + \csc t} - \dfrac{1}{\cot t - \csc t}$

In Exercises 9–16, verify the given identity.

9. $\dfrac{\sin^2 x}{\cos^2 x} = \sec^2 x - 1$

10. $\cos^2 t + \cot^2 t + \sin^2 t = \csc^2 t$

11. $\dfrac{\sin^3 t - \cos^3 t}{\sin t - \cos t} = 1 + \sin t\cos t$

12. $\ln|\sec x - \tan x| = -\ln|\sec x + \tan x|$ (ln is the natural logarithm.)

13. $\dfrac{1}{\tan t - \sec t} + \dfrac{1}{\tan t + \sec t} = -2\tan t$

14. $\sec\theta + \csc\theta - \sin\theta - \cos\theta = \sin\theta\tan\theta + \cos\theta\cot\theta$

15. $(a\cos\theta - b\sin\theta)^2 + (a\sin\theta + b\cos\theta)^2 = a^2 + b^2$

16. $(1 + \sin at + \cos at)^2 = 2(1 + \sin at)(1 + \cos at)$

17. Show that

$$\cos^{-1} x = \tan^{-1}\frac{\sqrt{1 - x^2}}{x}$$

for $0 \le x \le 1$. Why is this restriction necessary?

18. Show that the equation

$$\sqrt{1 + \sin^2 x} = 1 + \sin x$$

is *not* an identity.

In Exercises 19–23, solve the given equation.

19. $\sin\theta\cos\theta\cot\theta = 1 - \sin^2\theta$

20. $\sqrt{1 + \tan^2\theta} = 1$

21. $4\cos^2 x - 20\cos x + 9 = 0$

22. $4\csc\theta\cot\theta - 9\cot\theta - 8\csc\theta + 18 = 0$

23. $(\cos\theta - 2)^2(2\sin\theta - 1)^3 = 100$ (*Hint:* how large can the left-hand side be?)

24. Find the values of $\sin(\theta + \mu)$, $\cos(\theta + \mu)$, and $\tan(\theta + \mu)$ if $\pi/2 \le \theta, \mu \le \pi$, $\cos\theta = -1/3$, and $\sin\mu = 1/3$.

25. Let $\tan\theta = x^2$, $\sec\mu = x$, $\csc\mu > 0$ and $\sin(\theta + \mu) = x$, where $x > 0$. Express $\sin\theta$, $\cos\theta$, $\sin\mu$, and $\cos\mu$ in terms of x.

26. Express $\sin(a - b + c)$ in terms of the sine and cosine of a, b, and c.

27. Express $\cot(t - \pi/3)$ in terms of $\tan t$.

28. Evaluate $\sin(\arccos 1/5 + \arccos 1/3)$.

29. Evaluate $\tan(\arctan 10 - \arcsin 1/2)$.

30. Simplify the expression $\sin(t - \pi/3) + \cos(t - \pi/6)$.

31. Express $\cot(x + y)$ in terms of $\cot x$ and $\cot y$.

In Exercises 32–34, verify the given identity.

32. $\cos(x + y)\cos(x - y) = 1 - \sin^2 x - \sin^2 y$.

33. $\dfrac{\cos(s + t)}{\sin(s - t)} = \dfrac{\cot s - \tan t}{1 - \cot s\cot t}$

34. $\dfrac{\sin\mu}{\sin\theta} + \dfrac{\cos\mu}{\cos\theta} = \csc\theta\sec\theta\sin(\theta + \mu)$

35. Find $\sin 2t$ if $\sin t = 1/4$ and $\cos t < 0$.

36. Find $\cos 2t$ if $\sin t = 1/3$ and $\sin 2t > 0$.

37. Find $\tan 2t$ if $\cot t = 18$ and $\sin t < 0$.

38. Find $\sin t/2$ if $\sec t = 5$ and $\csc t < 0$.

39. Find $\cos t/2$ if $\sin t = \pi/6$ and $\cos t > 0$.

40. Find $\tan t/2$ if $\cot t/2 = 12$.

41. Find the exact value of $\sin \dfrac{\pi}{16}$. (Do not use a calculator.)

42. Find the exact value of $\tan \dfrac{\pi}{16}$. (Do not use a calculator.)

43. Compute the following values
(a) $\cos 5\pi/24$ **(b)** $\sin(-7\pi/4)$ **(c)** $\tan(37.5°)$.

44. Verify the identity

$$\cos^{-1} x = 2\sin^{-1}\sqrt{\frac{1-x}{2}}$$

by using a half-angle formula.

45. Verify the identity

$$\tan 2t - \sec 2t = \frac{\sin t - \cos t}{\sin t + \cos t}$$

46. Solve the equation $\tan t/2 = \cot t$.

47. Find a formula for $\tan 3t$ in terms of the powers of $\tan t$.

In Exercises 48–53, use the given information and the formula found in Exercise 47 to find $\tan 3t$.

48. $\tan t = 7$

49. $\sin t = 1/4$, $\tan t < 0$

50. $\cos t = -1$

51. $\sin t = -1/3$, $\sin 2t > 0$

52. $\cos t = -0.5557$, $\tan t > 0$

53. $\sin t = 0.997$, $\tan t > 0$

Exercises 54–58 are devoted to deriving a formula for the area of a triangle in terms of its perimeter. We label the sides and angles of the triangle as shown in the following figure and let

$$s = \frac{1}{2}(a + b + c)$$

Of course, s is just one-half of the perimeter of the triangle, and for this reason it is sometimes called the **semiperimeter** of the triangle.

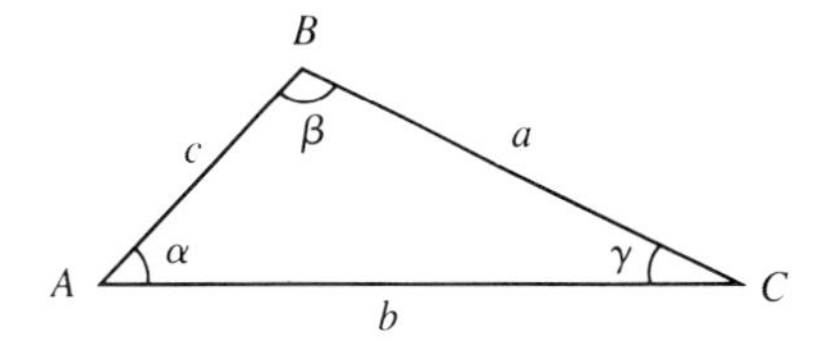

54. Use the law of cosines and the half-angle formula to show that

$$\cos\frac{\alpha}{2} = \sqrt{\frac{s(s-a)}{bc}}$$

55. Show also that

$$\sin\frac{\alpha}{2} = \sqrt{\frac{(s-b)(s-c)}{bc}}$$

56. Use the results of the previous two exercises to show that

$$\sin\alpha = \frac{2\sqrt{s(s-a)(s-b)(s-c)}}{bc}$$

57. Show that the area of a triangle is given by $(1/2)bc\sin\alpha$.

58. Use the results of the previous two exercises to show that the area of a triangle is given by

$$\text{Area} = \sqrt{s(s-a)(s-b)(s-c)}$$

This is the formula that we wanted to derive, and it is known as **Heron's formula,** named after Heron of Alexandria, who lived in the early part of the first century A.D.

59. Use Heron's formula to compute the area of a right triangle whose base has length 3 and whose height has length 4. Check the result by using the well-known formula area = (1/2) base × height.

60. Show that Heron's formula is correct for any right triangle by comparing it with the formula area = (1/2) base × height.

61. Use Heron's formula to compute the area of a triangle whose sides have length 2, 3, and 4.

62. Use Heron's formula to compute the area of a triangle whose sides have length 250, 375, and 175.

9 SYSTEMS OF LINEAR EQUATIONS

9.1 Upper Triangular Systems

In this chapter, we will study systems of linear equations and how to solve them. Let us begin with a few definitions.

A **linear equation** in two variables x and y is an equation of the form

$$ax + by = c \tag{9.1}$$

where a, b, and c are real numbers. Similarly, the equations

$$ax + by + cz = d \tag{9.2}$$

and

$$ax + by + cz + dw = e$$

where a, b, c, d, and e are real numbers, are linear equations in three and four variables, respectively.

The numbers a and b in Equation (9.1) are called the **coefficients** of the equation, and the number c is called the **constant term.** Of course, it is possible for some of the coefficients or the constant term of a linear equation to be equal to zero.

A **solution** to Equation (9.1) is a *pair* of real numbers $x = s$, $y = t$ that satisfies the equation. For example, the linear equation $2x + 3y = 8$ has solution $x = 1$, $y = 2$ (among others). Similarly, a **solution** to Equation (9.2) is a *sequence* of three numbers $x = s$, $y = t$, $z = r$ that satisfies the equation. For example, the equation $x - 2y + z = -5$ has solution $x = 2$, $y = 3$, $z = -1$ (among others).

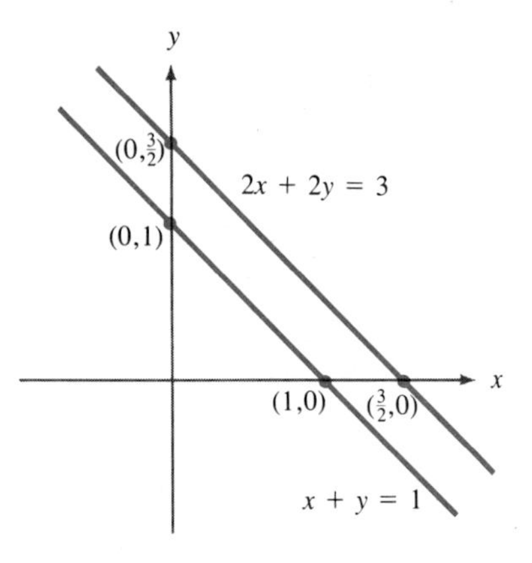

FIGURE 9.1

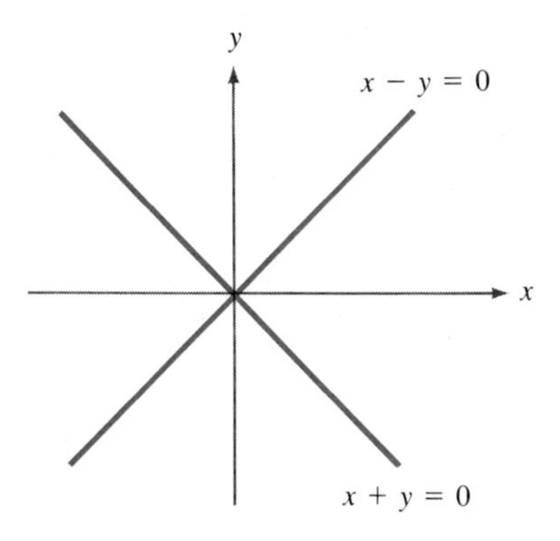

FIGURE 9.2

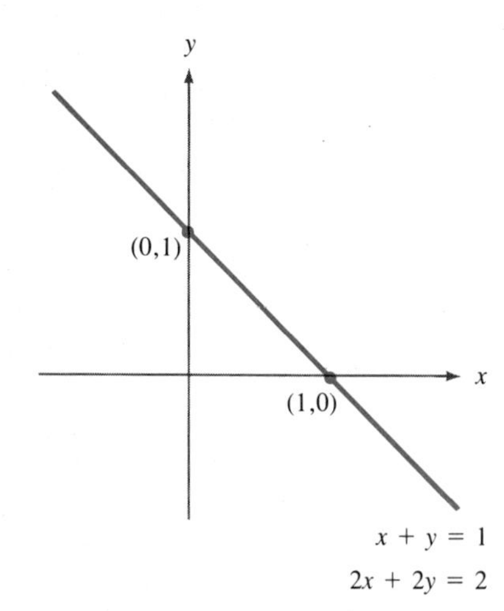

FIGURE 9.3

The following are examples of systems of linear equations

$$\begin{array}{rl} 3x + 4y = 0 \\ -x + 3y = 5 \end{array} \qquad \begin{array}{rl} x - 2y + z = -1 \\ 2x + \frac{1}{2}y - 3z = 55 \end{array} \qquad \begin{array}{rl} 2x + 3y + z = 4 \\ x - 2z = 3 \\ y + 3z = -2 \end{array} \tag{9.3}$$

Notice that the last example is a system of equations in the three variables x, y, and z, even though not all of the equations involve all three variables.

In this chapter, we will concentrate primarily on systems involving at most three equations and three variables. However, the methods that we will discuss will work on larger systems as well.

A **solution** to a *system* of equations is a sequence of numbers that satisfies *each* of the equations in the system. For example, $x = 1$, $y = 1$, $z = -1$ is a solution to the last system in (9.3), as you can check by substituting these values for the variables in this system.

For systems of two equations in two variables, we can draw some very helpful pictures. As you know, the graph of a linear equation in two variables

$$ax + by = c$$

is a straight line in the plane. (This is the general form of the equation of a line.) Therefore, the **graph** of a system of two equations in two variables consists of two straight lines in the plane. Furthermore, a point (s, t) is on one of the lines if and only if $x = s$, $y = t$ is a solution to the corresponding linear equation, and so a solution to the entire system is given by the coordinates (s, t) of any points of *intersection* of the two lines.

This leads us to three different possibilities, depending on whether the lines are distinct and parallel, nonparallel, or identical. These possibilities are illustrated in the next three examples.

EXAMPLE 1: The graph of the system

$$\begin{aligned} x + y &= 1 \\ 2x + 2y &= 3 \end{aligned}$$

is shown in Figure 9.1. Since these lines have the same slope, they are parallel. Therefore, the system has no solutions. □

EXAMPLE 2: The graph of the system

$$\begin{aligned} x - y &= 0 \\ x + y &= 0 \end{aligned}$$

is shown in Figure 9.2. In this case, the lines intersect at the origin (0, 0), and so the system has exactly one solution, namely, $x = 0$, $y = 0$. □

EXAMPLE 3: The graph of the system

$$\begin{aligned} x + y &= 1 \\ 2x + 2y &= 2 \end{aligned}$$

is pictured in Figure 9.3. In this case, both equations have the same line as their graph, and so any point on this line gives a solution to this system. Thus, there are an infinite number of solutions to this system.

Try Study Suggestion 9.1. □

▶ **Study Suggestion 9.1:** Graph each of the following systems and determine how many solutions it has.

(a) $$\begin{aligned} 2x - 3y &= 5 \\ -4x + 6y &= -10 \end{aligned}$$

(b) $$\begin{aligned} 4x - y &= 0 \\ -2x + \tfrac{1}{2}y &= 3 \end{aligned}$$

(c) $$\begin{aligned} x - 2y &= 2 \\ x - 2y &= 5 \end{aligned}$$ ◀

For a system of two linear equations in two variables, Examples 1–3 describe *all* of the possibilities. That is, such a system must have either no solutions, *exactly* one solution, or else infinitely many solutions. It may surprise you to learn that this is the case for *any* system of linear equations, of any size. That is,

given any system of linear equations, it must have either no solutions, *exactly* one solution, or else infinitely many solutions.

A system of equations with no solutions is said to be **inconsistent,** and a system of equations with *at least* one solution is said to be **consistent.**

In the next section, we will discuss a technique that will enable us to solve any system of equations (or to decide that the system has no solutions). Before doing that, however, we want to discuss how to solve a very special type of system.

DEFINITION

A system of two equations in two variables is said to be in **upper triangular form** (or simply **upper triangular**) if it has the form

$$\begin{aligned} a_1x + a_2y &= d_1 \\ b_2y &= d_2 \end{aligned}$$

Similarly, the systems

$$\begin{aligned} a_1x + a_2y + a_3z &= d_1 \\ b_2y + b_3z &= d_2 \end{aligned} \qquad \text{and} \qquad \begin{aligned} a_1x + a_2y + a_3z &= d_1 \\ b_2y + b_3z &= d_2 \\ c_3z &= d_3 \end{aligned}$$

are in **upper triangular form.** In this definition, $a_1, a_2, a_3, b_2, b_3, c_3, d_1, d_2$, and d_3 are constants (that is, real numbers), some of which may be equal to 0.

The use of the term upper triangular comes from the fact that the left-hand side of these systems has a triangular shape. Using this description, you should have no trouble extending this definition to larger systems.

There are two reasons why we want to pay special attention to upper triangular systems. First, they are very easy to solve, and second, the technique that we will discuss in the next section will allow us to replace any system of linear equations by an *equivalent* upper triangular system. (By *equivalent*, we mean having the same solutions.) Therefore, this technique will enable us to reduce the problem of solving a given system to that of solving an upper triangular system.

Let us consider some examples of how to solve upper triangular systems.

EXAMPLE 4: Solve the upper triangular system

$$\begin{aligned} 5x + y - 3z &= 0 \\ 3y + z &= 5 \\ -z &= 4 \end{aligned}$$

Solution: First we solve the *third* equation for z to get $z = -4$. Then we substitute this value of z into the *second* equation, and solve for y. This gives $y = 3$. Finally, we substitute these values for y and z into the *first* equation, and solve for x, to get $x = -3$. Hence, the solution to this system is $x = -3$, $y = 3$, $z = -4$. *Try Study Suggestion 9.2.* □

► **Study Suggestion 9.2:** Solve the following upper triangular systems

(a) $\begin{aligned} 2x - 3y &= 5 \\ 2y &= 1 \end{aligned}$

(b) $\begin{aligned} x - y - 2z &= 0 \\ y + 3z &= 4 \\ 2z &= 6 \end{aligned}$ ◄

As you can see from the previous example, upper triangular systems can be solved by starting with the last equation and working up. This approach is known as **back substitution.**

EXAMPLE 5: Solve the upper triangular system

$$\begin{aligned} x - 3y + 4z &= 2 \\ y + 2z &= 1 \end{aligned}$$

Solution: Here we have a slight complication, since we cannot solve the last equation in this system. In a case such as this, the system has infinitely many solutions. For we can choose *any* value for z, solve the second equation for y, and then solve the first equation for x. For example, if we choose $z = 1$, then the second equation gives

$$y = 1 - 2z = 1 - 2(1) = -1$$

and the first equation gives

$$x = 2 + 3y - 4z = 2 + 3(-1) - 4(1) = -5$$

and so $x = -5$, $y = -1$, $z = 1$ is one solution to the system. Clearly, we will get other solutions by choosing other values for z.

In order to express situations such as this more clearly, it is customary to pick a *new* variable, say u, which is different from any of the variables in the system, and set $z = u$. Then we can use the second equation to solve for y in terms of u,

$$y = 1 - 2u$$

and the first equation to solve for x in terms of u,

$$x = 2 + 3y - 4z = 2 + 3(1 - 2u) - 4u = 5 - 10u$$

Now we can describe *all* of the solutions to this system by writing

$$x = 5 - 10u, \qquad y = 1 - 2u, \qquad z = u$$

where u ranges over *all* real numbers. The new variable u is called a **parameter.**

Try Study Suggestion 9.3. □

▶ **Study Suggestion 9.3:** Solve the upper triangular system

$$\begin{aligned} 2x - y + z &= -4 \\ 3y - 6z &= 1 \end{aligned}$$ ◀

Ideas to Remember

- A system of linear equations has either no solutions, *exactly* one solution, or else infinitely many solutions.
- Upper triangular systems can be solved easily by the technique of back substitution.

EXERCISES

In Exercises 1–9, determine whether or not the given equation is linear. All letters represent variables unless otherwise noted.

1. $x = 7$

2. $2x - 3y = 1$

3. $2x + \sin y = 0$

4. $x^{-1} - 3y = 2$

5. $x + y + xy = 0$

6. $x + 2y = z$

7. $x - 2x = 4 + y$

8. $x + y = \cos k$ (k constant)

9. $(\log k)x + y = k$ (k constant)

In Exercises 10–13, verify that the given values of x and y form a solution to the given system.

10. $\begin{aligned} x + y &= 0 \\ 3x - 4y &= 7 \end{aligned}$ $x = 1, y = -1$

11. $\begin{aligned} 3x + 4y &= 25/6 \\ 2x - 3y &= -1 \end{aligned}$ $x = 1/2, y = 2/3$

12. $\begin{aligned} 2x + 2y &= 4a \\ 5x - 3y &= 2a + 8b \end{aligned}$ $x = a + b, y = a - b$ (a and b are constants)

13. $\begin{aligned} -2x + ky &= -23k/4 \\ (1/k)x + 4y &= 4 \end{aligned}$ $x = 3k, y = 1/4$ (k nonzero constant)

In Exercises 14–19, graph the given system, and determine how many solutions it has.

14. $\begin{aligned} x + y &= 2 \\ 2x + 3y &= 0 \end{aligned}$

15. $\begin{aligned} 3x - y &= 1 \\ -6x + 2y &= 2 \end{aligned}$

16. $\begin{aligned} 4x + 3y &= 7 \\ -4x - 3y &= 7 \end{aligned}$

17. $\begin{aligned} -x + 3y &= 2 \\ 2x - 6y &= -4 \end{aligned}$

18. $\begin{aligned} x - 3y &= -1/2 \\ 2x - 6y &= -1 \end{aligned}$

19. $\begin{aligned} x + y &= 5 \\ x + y &= -5 \end{aligned}$

In Exercises 20–31, solve the given upper triangular system. You may introduce parameters if necessary.

20. $\begin{aligned} x + y &= 1 \\ 3y &= 12 \end{aligned}$

21. $\begin{aligned} 3x - 4y &= 7 \\ 7y &= 5 \end{aligned}$

22. $\begin{aligned} x + y + z &= 1 \\ y + z &= 2 \\ z &= 3 \end{aligned}$

23. $\begin{aligned} 2x - y + 2z &= 0 \\ 3y - z &= 0 \\ 4z &= 0 \end{aligned}$

24. $\begin{aligned} 2x \quad + 6z &= 5 \\ 4z &= 2 \\ 3z &= 1 \end{aligned}$

25. $\begin{aligned} x - 3y \quad &= 6 \\ 5y \quad &= 10 \\ 2z &= 12 \end{aligned}$

26. $\begin{aligned} x \quad + z &= 4 \\ y + z &= 3 \end{aligned}$

27. $\begin{aligned} x - 3y \quad &= 1 \\ y - z &= 0 \end{aligned}$

28. $x + 2z = 4$
$z = 2$

29. $x + y + z = 1$
$y + z = 2$

30. $2x - 3y + 4z = 7$
$4y + z = -2$

31. $5x + 3y - 2z = 4$
$4y = 2$

32. For which values of the constant k does the system

$$\begin{aligned} x + y &= 1 \\ 2x + 2y &= k \end{aligned}$$

have no solutions, exactly one solution, infinitely many solutions?

33. For a system of three linear equations in two variables, draw pictures to indicate the possible relative positions of the three lines. For each picture, describe the number of solutions.

34. A system of equations all of whose constant terms are equal to zero is called a **homogeneous** system. Does a homogeneous system always have at least one solution? What is that solution?

9.2 Gaussian Elimination

In this section, we discuss a technique for solving systems of equations known as **Gaussian elimination.** This technique is named after Carl Friedrich Gauss (1777–1855), who is generally considered to be one of the most brilliant mathematicians of all time. Even though this technique is very old and very simple, it is still the best method that we know for solving systems of equations.

The idea behind Gaussian elimination is to apply certain simple operations to the equations of the system in order to reduce it to an *equivalent* upper triangular system. (*Equivalent* means having the same solutions.) Then we can solve the upper triangular system using back substitution, as in the previous section.

The operations that we will use to reduce a given system to upper triangular form are called **elementary operations,** and they consist of three types.

Type 1. Multiply both sides of an equation by a nonzero real number.
Type 2. Interchange any two equations.
Type 3. Add a multiple of one equation to another equation.

Let us give an example of each of these elementary operations, as applied to the following system of equations

$$\begin{aligned} x + 2y + 3z &= 1 \\ 2x - y + 4z &= 0 \\ 2x + 6y + 8z &= 2 \end{aligned} \qquad (9.4)$$

An example of a Type 1 operation applied to this system is to multiply both sides of the third equation by 1/2, giving the new (and equivalent) system

$$\begin{aligned} x + 2y + 3z &= 1 \\ 2x - y + 4z &= 0 \\ x + 3y + 4z &= 1 \end{aligned}$$

An example of a Type 2 operation applied to the system (9.4) is to interchange the second and third equations, giving

$$\begin{aligned} x + 2y + 3z &= 1 \\ 2x + 6y + 8z &= 2 \\ 2x - y + 4z &= 0 \end{aligned}$$

An example of a Type 3 operation applied to the system (9.4) is to add -2 times the first equation to the third equation, to get

$$\begin{aligned} x + 2y + 3z &= 1 \\ 2x - y + 4z &= 0 \\ 2y + 2z &= 0 \end{aligned}$$

It is important to observe that we do *not* change the first equation in this operation, but only the third equation.

Now let us do some examples of Gaussian elimination. Remember that the idea is to reduce the system to one that is in upper triangular form. We will do this by using elementary operations to eliminate certain terms from the equations of the system.

EXAMPLE 1: Solve the system

$$\begin{aligned} x + 2y &= 3 \\ 2x - 5y &= 2 \end{aligned}$$

Solution: In order to get an upper triangular system, we would like to eliminate the term $2x$ from the second equation. Type 3 elementary operations are the ones to use for eliminating terms. In fact, adding -2 times the first equation to the second equation gives the equivalent system

$$\begin{aligned} x + 2y &= 3 \\ -9y &= -4 \end{aligned}$$

Now we have an upper triangular system, which can be solved by back substitution, to get

$$x = \frac{19}{9}, \qquad y = \frac{4}{9}$$

You should check that this is indeed a solution to the *original* system of equations by substituting into that system. *Try Study Suggestion 9.4.* □

▶ **Study Suggestion 9.4:** Solve the system

$$\begin{aligned} 2x + y &= 6 \\ -x + 2y &= 7 \end{aligned}$$ ◀

EXAMPLE 2: Solve the system

$$\begin{aligned} 2x + 2y - 4z &= 0 \\ 2x + 2y - z &= 1 \\ 3x + 2y - 3z &= 3 \end{aligned}$$

Solution: The arithmetic involved in performing Gaussian elimination can usually be made simpler by arranging it so that the first equation of the system has coefficient of x equal to 1. To accomplish this, we multiply the first equation by 1/2.

$$\begin{aligned} x + y - 2z &= 0 \\ 2x + 2y - z &= 1 \\ 3x + 2y - 3z &= 3 \end{aligned}$$

Now we want to eliminate the terms involving x from both the second and third equations. To do this, we add -2 times the first equation to the second equation.

$$\begin{aligned} x + y - 2z &= 0 \\ 3z &= 1 \\ 3x + 2y - 3z &= 3 \end{aligned}$$

Then we add -3 times the first equation to the third equation.

$$\begin{aligned} x + y - 2z &= 0 \\ 3z &= 1 \\ -y + 3z &= 3 \end{aligned}$$

Now we have the first "column" of the system in the proper form. Let us proceed to the second column. All we have to do here is interchange the second and third equations, to get

$$\begin{aligned} x + y - 2z &= 0 \\ -y + 3z &= 3 \\ 3z &= 1 \end{aligned}$$

This system is in upper triangular form, and so it can be solved by using back substitution. We will leave it for you to show that the solution to this system, and hence also to the original system, is

$$x = \frac{8}{3}, \qquad y = -2, \qquad z = \frac{1}{3}$$

Try Study Suggestion 9.5. □

▶ **Study Suggestion 9.5:** Solve the system

$$\begin{aligned} 2x + 4y - 6z &= 12 \\ 3x - y + z &= 0 \\ -x + 2y - 3z &= 5 \end{aligned}$$ ◀

EXAMPLE 3: Solve the system

$$\begin{aligned} -3x + 4y - 10z &= -19 \\ x + 2y \qquad &= 3 \\ -x \qquad - 2z &= -5 \end{aligned}$$

Solution: In order to make the coefficient of x in the first equation equal to 1, we could multiply the first equation by $-1/3$. However, this will make the other coefficients in the first equation difficult to handle. A better way is simply to interchange the first and second equations.

$$\begin{aligned} x + 2y \quad\quad &= 3 \\ -3x + 4y - 10z &= -19 \\ -x \quad\quad - 2z &= -5 \end{aligned}$$

Next, we add 3 times the first equation to the second equation.

$$\begin{aligned} x + 2y \quad\quad &= 3 \\ 10y - 10z &= -10 \\ -x \quad\quad - 2z &= -5 \end{aligned}$$

Then we add the first equation to the third equation.

$$\begin{aligned} x + 2y \quad\quad &= 3 \\ 10y - 10z &= -10 \\ 2y - 2z &= -2 \end{aligned}$$

To simplify, we multiply the second equation by 1/10 and the third equation by 1/2.

$$\begin{aligned} x + 2y \quad\quad &= 3 \\ y - z &= -1 \\ y - z &= -1 \end{aligned}$$

At this point, we notice that the second and third equations are identical, and so we can drop the third equation, since it contributes no additional information to the system.

$$\begin{aligned} x + 2y \quad\quad &= 3 \\ y - z &= -1 \end{aligned}$$

This system is upper triangular, and its solution calls for the introduction of a parameter. So we let $z = u$, solve the second equation for y in terms of u,

$$y = -1 + z = -1 + u$$

and the first equation for x in terms of u,

$$x = 3 - 2y = 3 - 2(-1 + u) = 5 - 2u$$

Thus, the solutions to the original system are given by

$$x = 5 - 2u, \qquad y = -1 + u, \qquad z = u$$

where u ranges over all real numbers. In this case, the system has infinitely many solutions. For example, one solution is found by choosing $u = 1$, to get $x = 3$, $y = 0$, $z = 1$, and another solution is found by setting $u = 0$, to get $x = 5$, $y = -1$, $z = 0$. *Try Study Suggestion 9.6.* □

▶ **Study Suggestion 9.6:** Solve the system

$$\begin{aligned} 2x + y + 2z &= 4 \\ 4x - y - 5z &= 2 \\ 2x + 3y + 8z &= 8 \end{aligned}$$ ◀

EXAMPLE 4: Solve the system

$$\begin{aligned} x + y + 2z &= 1 \\ 3x \quad\quad + 3z &= 5 \\ 5y + 5z &= 0 \end{aligned}$$

Solution: In order to eliminate the term involving x in the second equation, we add -3 times the first equation to the second equation.

$$\begin{aligned} x + y + 2z &= 1 \\ -3y - 3z &= 2 \\ 5y + 5z &= 0 \end{aligned}$$

In order to eliminate the term involving y in the third equation, we add 5/3 times the second equation to the third equation, giving

$$\begin{aligned} x + y + 2z &= 1 \\ -3y - 3z &= 2 \\ 0 &= 10/3 \end{aligned}$$

But the third equation in this system can never be true. Hence, this system has no solutions, and since it is equivalent to the original system, it too has no solutions. This is a typical example of what happens when there are no solutions to the original system. *Try Study Suggestion 9.7.* □

▶ **Study Suggestion 9.7:** Solve the system

$$\begin{aligned} x + y - 2z &= -2 \\ x + 4y - 3z &= 0 \\ -x + 2y + z &= 3 \end{aligned}$$ ◀

Examples 2–4 cover the three possibilities for the number of solutions to a system, as we discussed in the previous section. Any system of three linear equations in three variables that you encounter will be similar to one of these examples.

Ideas to Remember

Gaussian elimination can be used to reduce a system of linear equations to an equivalent system that is in upper triangular form, which can be solved easily by back substitution.

EXERCISES

Solve the given system of equations by using Gaussian elimination. The letter k is meant to denote a constant. All other letters are variables.

1. $3x - 2y = 20$, $2x - 3y = 20$

2. $3x + 4y = 0$, $-3x + 7y = 0$

3. $2x - y = 7$, $3x + y = 6$

4. $\frac{1}{2}x - \frac{1}{3}y = \frac{1}{5}$, $x + y = 0$

5. $3x - y = 2$, $-6x + 2y = -4$

6. $u - 4v = 2$, $2u - 8v = 6$

7. $2s + 3t = 1$, $-4s + 6t = 0$

8. $2x - 3y = -5k - 3$, $x + y = 1$

9. $kx + k^2y = 5$, $2k^2x - 5k^2y = 4k - 15$

10. $x + 2y = \sin k$, $2x + y = \cos k$

11. $x + y - 2z = 0$, $2x + y - 3z = 0$

12. $3x + 3y - 4z = 2$, $x + y - z = 5$

13. $x - y + 3z = 4$, $2x - 2y + 6z = 5$

14. $-x + 7y = 13$, $x + 4z = 13$

15. $2u - 3v + 4w = k$, $u + 2v - w = 2k$

16. $-3x + 6y + kz = 1$, $4x - 3y - kz = 2$

17. $3x - 2y = 0$, $2x - 3y + z = 0$, $3x - 7y + 3z = 0$

18. $4x + y - z = 6$, $3x - y + 2z = 7$, $5x + 3y - 4z = 2$

19. $$\begin{aligned} 6x + 4y + 4z &= 2 \\ 7x + 5y + z &= 14 \\ 5x + 4y + 3z &= 4 \end{aligned}$$

20. $$\begin{aligned} 2x - 3y + 4z &= 3 \\ x - y + 10z &= 10 \\ 3x + z &= 4 \end{aligned}$$

21. $$\begin{aligned} 5x + y - z &= 4 \\ 2x + y - z &= 4 \\ 2x + 2y - 6z &= 14 \end{aligned}$$

22. $$\begin{aligned} x - 2y - 3z &= -1 \\ 2x - y - 2z &= 2 \\ 3x - y - 3z &= 3 \end{aligned}$$

23. $$\begin{aligned} y + 3z &= 2 \\ 3x + y - z &= 1 \\ 2x - y - 5z &= -3 \end{aligned}$$

24. $$\begin{aligned} 2x + y - 3z &= 1 \\ x - y + 2z &= -2 \\ 4x - y - z &= -3 \end{aligned}$$

25. $$\begin{aligned} 3x + y + 6z &= 6 \\ x + y + 2z &= 2 \\ 2x + y + 4z &= 3 \end{aligned}$$

26. $$\begin{aligned} x - 2y + z &= 0 \\ 2x - 4y + 2z &= 0 \\ x - 2y + z &= 0 \end{aligned}$$

27. $$\begin{aligned} 2x + y - 2z &= 1 \\ 5x + 5y + 5z &= 11 \\ 5x + 4y + z &= 6 \end{aligned}$$

28. $$\begin{aligned} x - y &= 1 \\ y - z &= 1 \\ x - z &= 2 \end{aligned}$$

29. $$\begin{aligned} x + y &= 2 \\ y + z &= 3 \\ x - z &= 1 \end{aligned}$$

30. $$\begin{aligned} x - y + 2z &= 5 \\ 4x - y + 9z &= 24 \\ 3x + 3y + 8z &= 23 \end{aligned}$$

31. $$\begin{aligned} 2x - 3y + 4z &= 0 \\ 3x - 3y + 9z &= 0 \\ 3x - 5y + 5z &= 0 \end{aligned}$$

32. $$\begin{aligned} 6x - 4y - 2z &= 4 \\ 8x + y - 5z &= 8 \\ 7x - 11y &= 5 \end{aligned}$$

33. $$\begin{aligned} x + y + z &= 6 \\ 2x + 3y - z &= 5 \\ 3x + 2y + z &= 10 \end{aligned}$$

34. $$\begin{aligned} 5x + 3y + z &= 7 \\ 6x + 7y - z &= 15 \\ -2x + 4y - 5z &= 11 \end{aligned}$$

35. $$\begin{aligned} 5x + 2y + 3z &= 9k \\ 7x - 3y + 7z &= 24k \\ 6x + 7y - 2z &= -5k \end{aligned}$$

36. $$\begin{aligned} 2u - 8v - 5w &= -2k \\ 3u - 6v - 4w &= 0 \\ 3u + 2v + 2w &= 4k \end{aligned}$$

37. $$\begin{aligned} 3x + 2y - z &= 6 \\ x - y + z &= -1 \\ 2x + 3y - 4z &= 9 \end{aligned}$$

38. $$\begin{aligned} 2x + 3y - 4z + w &= 0 \\ 2x + 3y - 4z + 2w &= 8 \\ 2x + 2y - 4z + 3w &= 3 \end{aligned}$$

39. $$\begin{aligned} 2x + 3y + 2z + w &= 0 \\ 3x - y - 2z - w &= 0 \\ x + 3y + 3z + w &= 0 \end{aligned}$$

40. $$\begin{aligned} 2x + y - 2z + 3w &= 1 \\ 5x + 5y + 5z - 5w &= 11 \\ 5x + 4y + z &= 6 \end{aligned}$$

41. $$\begin{aligned} u + v + w - x &= 1 \\ 2u - 3v + w - 2x &= -4 \\ 3u + v - 2w + x &= 4 \end{aligned}$$

42. $$\begin{aligned} 5x - 2y + 3z + w &= 8 \\ x + y - z - w &= 3 \end{aligned}$$

43. $$\begin{aligned} x + y &= 2k \\ y + z &= 3k \\ z + w &= 4k \end{aligned}$$

44. $$\begin{aligned} x + y + z + u &= k + 1 \\ y + z + u &= k + 2 \\ z + u &= k + 3 \end{aligned}$$

45. $$\begin{aligned} x + y + z + w &= 6 \\ 2x - y + 2z + 3w &= 13 \\ x + 3y - 2z - w &= 3 \\ 3x - y - z + 4w &= 22 \end{aligned}$$

46. $$\begin{aligned} x + y - z - w &= -1 \\ 3x + y + z - 2w &= -9 \\ 3x + 3y - 4z &= 10 \\ x - 3y - 2z + 3w &= 4 \end{aligned}$$

47. $$\begin{aligned} x + y + z + w + u &= 3 \\ y + z + u &= 2 \end{aligned}$$

48. $x + y + z + u + v + w = 1$

49. $$\begin{aligned} x + y &= -1 \\ 2x - y &= 7 \\ 3x + 2y &= 0 \\ 7x + 5y &= -1 \\ x - y &= 5 \end{aligned}$$

(*Hint:* solve the system consisting of the first two equations, and then check to see if the solution satisfies the other equations.)

50. $$\begin{aligned} 2x + y &= 3 \\ x - 2y &= -1 \\ 7x + 2y &= 9 \\ \tfrac{3}{2}x + \tfrac{1}{2}y &= 2 \\ 6x - 5y &= 2 \end{aligned}$$

9.3 Matrices and Gaussian Elimination

In this section we will discuss a "shorthand" notation that can be useful when using Gaussian elimination to solve a system of equations. As an example, consider the system

$$\begin{aligned} 3x + 4y - 2z &= 3 \\ x - y + 5z &= -2 \\ -x + 2y + z &= 7 \end{aligned} \tag{9.5}$$

Since the first column of this system contains the terms involving x, the second column contains the terms involving y, the third column contains the terms involving z, and the last column contains the constant terms, we may as well just write down the coefficients and constant terms of the system in a rectangular array as follows

$$\begin{pmatrix} 3 & 4 & -2 & 3 \\ 1 & -1 & 5 & -2 \\ -1 & 2 & 1 & 7 \end{pmatrix} \tag{9.6}$$

As long as we are consistent about the positioning of the numbers in the array, we can always reconstruct the system. Hence, this array is a kind of "shorthand" notation for the system.

Since arrays of numbers such as these are used quite frequently in mathematics and the sciences, they have been given a special name.

DEFINITION

A **matrix** is a rectangular array of real numbers. Each of the numbers in the matrix is called an **entry.** If an entry is in the ith row and the jth column of the matrix, we call it the (i, j)th entry of the matrix. If a matrix has n rows and m columns, then we say that it has **size** $n \times m$, or that it is an $n \times m$ matrix.

As an example, the matrix

$$A = \begin{pmatrix} 2 & 6 & -1 & 4 \\ 7 & 3 & -2 & 0 \\ 9 & 0 & 3 & 1 \end{pmatrix}$$

has size 3×4. The $(3, 2)$th entry of A is 0 and the $(2, 3)$th entry is -2.

DEFINITION

Given a system of equations, the matrix whose ith row consists of the coefficients and constant term of the ith equation is called the **augmented matrix** of the system.

As an example, the matrix A above is the augmented matrix of the system

$$\begin{aligned} 2x + 6y - z &= 4 \\ 7x + 3y - 2z &= 0 \\ 9x \phantom{{}+ 3y} + 3z &= 1 \end{aligned}$$

To remind ourselves that we are dealing with an augmented matrix, it is customary to draw a vertical line between the last two columns of the matrix. (Not all books do this, however.) For instance, the augmented matrix of the system

$$\begin{aligned} 2x + 3y - 4z &= 8 \\ 2y + 4z &= -3 \\ 5x \qquad - 2z &= 4 \end{aligned} \tag{9.7}$$

is

$$\left(\begin{array}{ccc|c} 2 & 3 & -4 & 8 \\ 0 & 2 & 4 & -3 \\ 5 & 0 & -2 & 4 \end{array}\right)$$

Notice that we have put zeros in this matrix for the "missing" terms in the system.

Using augmented matrices, our plan for solving systems of linear equations will be as follows.

Gaussian Elimination

1. Find the augmented matrix of the system.
2. Perform the elementary operations on the *rows* of this matrix, until it has the form of the augmented matrix of an upper triangular system.
3. Convert back to "system" notation, and solve the system by the method of back substitution.

Let us consider some examples.

EXAMPLE 1: Solve the system

$$\begin{aligned} x + 3y - 2z &= 4 \\ 2x + 7y + z &= 6 \\ 3x + 10y \qquad &= 0 \end{aligned}$$

Solution: For this example (only), we will perform Gaussian elimination on the augmented matrix, and at the same time indicate the effect on the system of equations. We begin by writing the system to the right of its augmented matrix

$$\left(\begin{array}{ccc|c} 1 & 3 & -2 & 4 \\ 2 & 7 & 1 & 6 \\ 3 & 10 & 0 & 0 \end{array}\right) \qquad \begin{aligned} x + 3y - 2z &= 4 \\ 2x + 7y + z &= 6 \\ 3x + 10y \qquad &= 0 \end{aligned}$$

The first step is to add -2 times the first row (of the augmented matrix) to the second row. This has the same effect as adding -2 times the first equation of the system to the second equation.

$$\left(\begin{array}{ccc|c} 1 & 3 & -2 & 4 \\ 0 & 1 & 5 & -2 \\ 3 & 10 & 0 & 0 \end{array}\right) \qquad \begin{aligned} x + 3y - 2z &= 4 \\ y + 5z &= -2 \\ 3x + 10y \qquad &= 0 \end{aligned}$$

Then we add -3 times the first row (equation) to the third row (equation)

$$\left(\begin{array}{ccc|c} 1 & 3 & -2 & 4 \\ 0 & 1 & 5 & -2 \\ 0 & 1 & 6 & -12 \end{array}\right) \qquad \begin{aligned} x + 3y - 2z &= 4 \\ y + 5z &= -2 \\ y + 6z &= -12 \end{aligned}$$

Next we add -1 times the second row (equation) to the third row (equation)

$$\left(\begin{array}{ccc|c} 1 & 3 & -2 & 4 \\ 0 & 1 & 5 & -2 \\ 0 & 0 & 1 & -10 \end{array}\right) \qquad \begin{aligned} x + 3y - 2z &= 4 \\ y + 5z &= -2 \\ z &= -10 \end{aligned}$$

Now, this is the matrix of a system that is in upper triangular form, and so the elimination procedure is complete. We will leave it to you to solve the resulting system by the method of back substitution, to obtain the solution $x = -160$, $y = 48$, $z = -10$. *Try Study Suggestion 9.8.* □

▶ **Study Suggestion 9.8:** Solve the system

$$\begin{aligned} x + y + z &= 0 \\ 3x - y + 2z &= -5 \\ 2x - y + 3z &= -9 \end{aligned}$$

using augmented matrices. ◀

EXAMPLE 2: Solve the system

$$\begin{aligned} x + 2y - z + 3w &= -4 \\ 2x + 4y + 3z - w &= 11 \\ 3x - 2y - 4z - w &= -9 \\ -5x + 6y + 2z + 8w &= -1 \end{aligned}$$

Solution: The augmented matrix of this system is

$$\left(\begin{array}{cccc|c} 1 & 2 & -1 & 3 & -4 \\ 2 & 4 & 3 & -1 & 11 \\ 3 & -2 & -4 & -1 & -9 \\ -5 & 6 & 2 & 8 & -1 \end{array}\right)$$

As always, we will work on one column at a time. We begin by adding -2 times the first row to the second row

$$\left(\begin{array}{cccc|c} 1 & 2 & -1 & 3 & -4 \\ 0 & 0 & 5 & -7 & 19 \\ 3 & -2 & -4 & -1 & -9 \\ -5 & 6 & 2 & 8 & -1 \end{array}\right)$$

Then we add -3 times the first row to the third row

$$\left(\begin{array}{cccc|c} 1 & 2 & -1 & 3 & -4 \\ 0 & 0 & 5 & -7 & 19 \\ 0 & -8 & -1 & -10 & 3 \\ -5 & 6 & 2 & 8 & -1 \end{array}\right)$$

Next we add 5 times the first row to the fourth row

$$\left(\begin{array}{cccc|c} 1 & 2 & -1 & 3 & -4 \\ 0 & 0 & 5 & -7 & 19 \\ 0 & -8 & -1 & -10 & 3 \\ 0 & 16 & -3 & 23 & -21 \end{array}\right)$$

Now that the first column has the proper form, we proceed to the second column. Interchanging the second and third rows gives

$$\left(\begin{array}{cccc|c} 1 & 2 & -1 & 3 & -4 \\ 0 & -8 & -1 & -10 & 3 \\ 0 & 0 & 5 & -7 & 19 \\ 0 & 16 & -3 & 23 & -21 \end{array}\right)$$

and adding 2 times the second row to the fourth row gives

$$\left(\begin{array}{cccc|c} 1 & 2 & -1 & 3 & -4 \\ 0 & -8 & -1 & -10 & 3 \\ 0 & 0 & 5 & -7 & 19 \\ 0 & 0 & -5 & 3 & -15 \end{array}\right)$$

Finally, we add the third row to the fourth row

$$\left(\begin{array}{cccc|c} 1 & 2 & -1 & 3 & -4 \\ 0 & -8 & -1 & -10 & 3 \\ 0 & 0 & 5 & -7 & 19 \\ 0 & 0 & 0 & -4 & 4 \end{array}\right)$$

This is the augmented matrix of the upper triangular system

$$\begin{aligned} x + 2y - z + 3w &= -4 \\ -8y - z - 10w &= 3 \\ 5z - 7w &= 19 \\ -4w &= 4 \end{aligned}$$

Using back substitution, we obtain the solution

$$x = \frac{1}{4}, \qquad y = \frac{23}{40}, \qquad z = \frac{12}{5}, \qquad w = -1$$

Try Study Suggestion 9.9. □

▶ **Study Suggestion 9.9:** Solve the system

$$\begin{aligned} x + y + 2z + w &= 8 \\ x + z + w &= 6 \\ y + w &= 4 \\ x - y - z - 2w &= 6 \end{aligned}$$

using augmented matrices. ◀

Performing Gaussian elimination on the augmented matrix of a system can save a great deal of writing. In fact, it is so common to perform the elimination on the augmented matrix that the elementary operations are usually called **elementary row operations,** to reflect the fact that they are performed on the *rows* of this matrix.

Ideas to Remember

The augmented matrix is a very convenient shorthand notation for a system of linear equations. In fact, Gaussian elimination is usually performed on this matrix, rather than on the original system.

EXERCISES

In Exercises 1–5, find the augmented matrix of the system.

1. $\begin{aligned} 2x - 4y &= 16 \\ 2x + \tfrac{1}{3}y &= -4 \end{aligned}$

2. $\begin{aligned} 2x - 3y + 4z &= 0 \\ -\ y + 7z &= 1 \\ 8x + 2y - 6z &= 4 \end{aligned}$

3. $\begin{aligned} x \quad + 2z &= 0 \\ y + 3z &= 5 \\ z &= 6 \end{aligned}$

4. $\begin{aligned} x \qquad\qquad &= 0 \\ y \qquad &= 0 \\ z \quad &= 0 \\ w &= 0 \end{aligned}$

5. $\begin{aligned} 2x + 3y - 2z \qquad\quad &= k \\ 4y \qquad - \tfrac{1}{2}w &= \tfrac{2}{3} \\ -\tfrac{1}{3}z + \tfrac{1}{12}w &= 13k \\ x \qquad - kz \qquad &= -1 \end{aligned}$ (k constant)

In Exercises 6–13, find the system corresponding to each augmented matrix.

6. $\left(\begin{array}{cc|c} 2 & 3 & 4 \\ 9 & -3 & 5 \end{array}\right)$

7. $\left(\begin{array}{ccc|c} 3 & 6 & 7 & 5 \\ 4 & 0 & 6 & -2 \end{array}\right)$

8. $\left(\begin{array}{ccc|c} 1 & 3 & 1 & 1 \\ 0 & 5 & 0 & 1 \\ 0 & 0 & 3 & 3 \end{array}\right)$

9. $\left(\begin{array}{c|c} 1 & 1 \\ 2 & 3 \\ 4 & -8 \end{array}\right)$

10. $\left(\begin{array}{ccc|c} 1 & 0 & 0 & 0 \\ 0 & 2 & 0 & 0 \\ 0 & 0 & 3 & 0 \end{array}\right)$

11. $\left(\begin{array}{cccc|c} 1 & 0 & 0 & 0 & 0 \end{array}\right)$

12. $\left(\begin{array}{ccc|c} -3 & \pi & \sqrt{2} & -6 \\ 2 & 4 & -\sqrt{3} & 7 \\ 5 & 2 & \tfrac{1}{2} & \tfrac{1}{4} \\ 4 & -2 & 7 & \pi^2 \end{array}\right)$

13. $\left(\begin{array}{ccc|c} k & k+1 & k+2 & 0 \\ 2k & 3k & 4k & 0 \\ 5 & 7 & 4 & 0 \\ k & k^2 & k^3 & k^4 \end{array}\right)$

k constant

14. Can any matrix be an augmented matrix? Explain your answer.

In Exercises 15–19, write out the matrix with 3 rows and 3 columns that satisfies the given properties.

15. The (i, j)th entry is equal to $i + j$.

16. The (i, j)th entry is equal to i.

17. The (i, j)th entry is equal to 1 if $i = j$ and 0 if $i \neq j$.

18. The (i, j)th entry is equal to $(-1)^{i+j}$.

19. The (i, j)th entry is equal to $\sin(i + j)\pi$.

20. How many entries does a matrix with m rows and n columns have?

Repeat Exercises 1–50 of Section 9.2 by using the augmented matrix of the given system.

9.4 Applications Involving Systems of Linear Equations

In this section, we will consider some examples of how systems of equations arise in various contexts.

As you know, two distinct points uniquely determine a straight line. In a similar way, three points that do not lie on the same line will uniquely determine a parabola. Our first example shows how to find the quadratic function whose graph goes through three such points.

EXAMPLE 1: Find the quadratic function $f(x) = ax^2 + bx + c$ whose graph goes through the points $(2, 1)$, $(1, -1)$, and $(-1, 4)$.

Solution: Since the point $(2, 1)$ is on the graph, we must have $f(2) = 1$; that is,

$$4a + 2b + c = 1$$

Similarly, since $(1, -1)$ is on the graph, we have $f(1) = -1$, that is

$$a + b + c = -1$$

Finally since $(-1, 4)$ is on the graph, we have

$$a - b + c = 4$$

Thus, in order for the graph of f to go through all three of these points, the coefficients a, b, and c must satisfy the *system* of equations

$$\begin{aligned} 4a + 2b + c &= 1 \\ a + b + c &= -1 \\ a - b + c &= 4 \end{aligned}$$

in the variables a, b, and c. We will leave it to you to show that the solution to this system is

$$a = \frac{3}{2}, \qquad b = -\frac{5}{2}, \qquad c = 0$$

Therefore, the desired function is $f(x) = (3/2)x^2 - (5/2)x$. We will leave it to you to check the result by computing $f(2)$, $f(1)$, and $f(-1)$.

Try Study Suggestion 9.10. □

▶ **Study Suggestion 9.10:** Find a quadratic function whose graph goes through the points $(0, 1)$, $(2, 4)$, and $(3, 3)$. ◀

EXAMPLE 2: Two numbers have the property that their sum is equal to 8 and their quotient is equal to 12. Find the numbers.

Solution: If we denote the numbers by x and y, then we have

$$x + y = 8$$

and

$$\frac{x}{y} = 12$$

Now, the second equation is not linear. However, since $y \neq 0$ this equation can be written in the form $x = 12y$, or $x - 12y = 0$, which is linear, and so we get the system

$$\begin{aligned} x + y &= 8 \\ x - 12y &= 0 \end{aligned}$$

with the side condition $y \neq 0$. We will leave it to you to show that the solution to this system is $x = 96/13$, $y = 8/13$. (The side condition is clearly satisfied by this solution. Don't forget to check that $x + y = 8$ and $x/y = 12$.)

Study Suggestion 9.11: Two numbers have the property that their sum is equal to 7/6 and one of their quotients is equal to 3/4. Find the numbers.

Try Study Suggestion 9.11. □

EXAMPLE 3: A tobacco merchant sells three types of tobacco—Virginia dark, which sells for \$3.00 per pound, Indiana one-sucker, which sells for \$4.00 per pound, and North Carolina bright, which sells for \$5.00 per pound. A customer wants a blend of the three types of tobacco, for a total of 10 pounds costing \$36.00. How much of each type of tobacco should the merchant use?

Solution: We begin by letting

x = number of pounds of Virginia dark
y = number of pounds of Indiana one-sucker
z = number of pounds of North Carolina bright

Then since the customer wants a total of 10 pounds of tobacco, we have

$$x + y + z = 10$$

and since the total cost is \$36.00, we have

$$3x + 4y + 5z = 36$$

Thus, the tobacco merchant must solve the system

$$\begin{aligned} x + y + z &= 10 \\ 3x + 4y + 5z &= 36 \end{aligned}$$

We will leave it to you to show that this system is equivalent to the upper triangular system

$$\begin{aligned} x + y + z &= 10 \\ y + 2z &= 6 \end{aligned}$$

In order to solve this system, we introduce a parameter u. Setting $z = u$, we get $y = 6 - 2z = 6 - 2u$ and $x = 10 - y - z = 10 - (6 - 2u) - u = 4 + u$. Hence, the solutions to the original system are given by

$$x = 4 + u, \qquad y = 6 - 2u, \qquad z = u$$

where u ranges over all nonnegative real numbers. Thus, we see that there are infinitely many possibilities for the tobacco merchant to choose from. For example, one possibility is $x = 5$, $y = 4$, $z = 1$; that is,

5 pounds of Virginia dark
4 pounds of Indiana one-sucker
1 pound of North Carolina bright

Try Study Suggestion 9.12. □

Study Suggestion 9.12:

(a) Referring to Example 3, what mixture of tobaccos acceptable to the customer contains the most North Carolina bright?

(b) What mixture contains the most Virginia dark?

(c) What mixture contains the most Indiana one-sucker?

EXAMPLE 4: A nutritionist is planning a meal and wants to make certain that it contains an adequate amount of protein. The ingredients of the meal that contain protein are milk, cheese, and eggs. Assuming that the following information on protein content, calorie content, and cost is accurate, how much of each food should the meal contain if it is to have 66 grams of protein, 1350 calories, and cost $1.80?

100 gm units of	calories	gms of protein	cost in dollars
milk	50	4	.20
cheese	600	24	.60
eggs	200	28	.80

Solution: We begin by letting

x = amount of milk in the meal (in 100 gm units)
y = amount of cheese in the meal (in 100 gm units)
z = amount of eggs in the meal (in 100 gm units)

(For example, $x = 5$ means that the meal contains 500 grams of milk.) Then the total number of calories in the meal is $50x + 600y + 200z$, and since this must equal 1350, we get the equation

$$50x + 600y + 200z = 1350$$

In a similar way, we can get two other equations by computing the total amount of protein and the total cost, and this leads to the system

$$\begin{aligned} 50x + 600y + 200z &= 1350 \\ 4x + 24y + 28z &= 66 \\ .20x + .60y + .80z &= 1.80 \end{aligned}$$

whose solution is $x = 1$, $y = 2$, $z = 1/2$. Hence, the meal should consist of 100 grams of milk, 200 grams of cheese, and 50 grams of eggs.

Try Study Suggestion 9.13. □

▶ **Study Suggestion 9.13:** Is it possible for the meal referred to in Example 4 to contain 115 grams of protein, 1350 calories, and cost \$1.80? Justify your answer. ◀

Ideas to Remember

Applied problems often lead to systems of linear equations, which can be solved by Gaussian elimination.

EXERCISES

In Exercises 1–4, find a linear function f with the given properties.

1. The graph of f goes through the points $(1, 0)$ and $(2, 0)$.

2. $f(0) = -1$, $f(3) = 4$

3. $f(3) = -4$, $f(7) = 7$

4. $f(2) = -5f(1) + 2$, $f(-1) = 3f(-2) + 1$

In Exercises 5–8, find a quadratic function f with the given properties, if one exists. If not, explain why not.

5. $f(1) = 1$, $f(2) = 2$, $f(3) = -1$

6. $f(1) = 1$, $f(2) = 2$, $f(3) = 3$

7. The graph of f goes through the points $(-1, 4)$, $(2, -1)$, and $(3, 5)$.

8. $f(1) = f(2) + 1$, $f(3) = 1$, $f(-2) = 2f(4)$

In Exercises 9–11, find a function of the form $f(x) = a\sin x + b\cos x$ with the given properties, if one exists.

9. $f(\pi/2) = 3$ and $f(\pi/3) = 2$

10. $f(\arcsin(\sqrt{2}/2)) = 0$, $f(\arccos(-\sqrt{3}/2)) = 0$

11. $f(\arctan(-\sqrt{2}/2)) = 0$, $f(\pi/6) = 1/2$

12. Two numbers have the property that their sum is equal to 10 and their product is equal to 2. Find the numbers.

13. Three numbers have the property that their average is equal to 7 and the first number is equal to the sum of twice the second number and three times the third number. Find the numbers.

14. A two-digit number has the property that the sum of its digits is equal to 5. If the digits in the number are reversed, the resulting number is 9 larger than the original number. Find the original number. (*Hint:* if x is the tens digit and y is the ones digit of the number, then the number is $10x + y$.)

15. A three-digit number has the property that the sum of its digits is equal to 11. If the tens and hundreds digits of the number are reversed, the resulting number is larger than the original number by 90. If the ones and hundreds digits are reversed, the resulting number is larger than the original number by 396. Find the original number. (See the hint for Exercise 14.)

16. A tea merchant wishes to mix three types of tea. Flowering pekoe, which comes from the leaf buds, costs \$12.00 per pound. Orange pekoe, which comes from young leaves, costs \$5.00 per pound, and souchong first, which comes from more mature leaves, costs only \$2.00 per pound. Is it possible for the merchant to mix these teas together to make a total of 5 pounds of mixture, costing \$6.00 per pound, containing twice as much souchong first as the other two types combined? Justify your answer.

17. A theater sells 2 types of tickets. Tickets for adults cost \$5.00 and tickets for children cost \$2.00. On a certain night, the theater sells 200 tickets and takes in a total of \$769.00. How many tickets of each type did the theater sell that night?

18. A theater sells 3 types of tickets. Tickets for adults cost \$5.00, tickets for children cost \$2.00, and tickets for senior citizens cost \$3.00. On a certain night, the theater sells 195 tickets and takes in $715.00. Can you determine exactly how many tickets of each type were sold? Explain your answer.

19. A pharmacist must mix 2 types of drugs together for a particular prescription. The first drug is in a 10% solution, and the second drug is in a 25% solution. The prescription calls for 100 ml. of a 15% solution. How much of each solution should the pharmacist use?

20. A chemist has 3 acid solutions. The first solution is 10% acid, the second is 15% acid and the third is 40% acid. How much of each solution should the chemist use if he wants 10 liters of a 15% acid solution and if he must use three times as much of the 10% solution as the other two solutions combined?

21. Suppose that you have invested all of your money in two bank accounts, one of which pays 8% interest and the other of which pays 10% interest. If you invested twice as much in the account with the higher interest rate, and if you earned $1000 in interest last year, how much did you invest in each account?

22. A certain candy store sells both dark chocolate and milk chocolate. A 3-pound box that contains 2 pounds of dark chocolate and 1 pound of milk chocolate sells for $32.00 and a 3-pound box that contains 1 pound of dark chocolate and 2 pounds of milk chocolate sells for $28.00. What is the price per pound of each type of chocolate?

23. Suppose that you want to invest in two types of bonds. The first type is the AAA bond, which is very safe, but only returns 10% per year on your investment. The second type is the B bond, which is not as safe, but returns 18% per year on your investment. If you want to invest $100,000 and need a return of $15,000 per year, how much should you invest in each type of bond?

24. You have decided to invest $1,200,000 in three types of investments—AAA bonds, A bonds, and uranium mine stock. The AAA bonds are very secure, but return only 10% per year. The A bonds are fairly secure, and return 14% per year. The uranium mine stock is very risky, but returns 30% per year. To be on the safe side, you have decided to invest as much in AAA bonds as in the other two investments combined. How much money should you put in each type of investment in order to make a return of $160,000?

25. An airplane can travel from Los Angeles to New York in 5 hours with a certain tail wind. However, when it flies from New York to Los Angeles, against this same wind, it takes 6 hours to make the trip. Find the speed of the plane and the speed of the wind. (Assume that the distance from Los Angeles to New York is 3000 miles and that the plane travels at a constant speed.)

26. A fast-food company sells two types of franchises—one for $25,000 and one for $35,000. In one month the company reported income from the sale of 11 franchises to be $355,000. How many franchises of each type did the company sell that month?

27. A woman wants to invest in two types of stocks. At the present time, stock A pays a dividend of $5 per share per year, and stock B pays a dividend of $12 per share per year. The price of stock A is $50 per share and the price of stock B is $75 per share. How many shares of each type must the woman buy in order to get a return of 14.5% on a $10,000 investment?

28. Suppose that you have a total of 30 pennies, nickels, and dimes, and that the number of dimes is one less than three times the number of pennies. If the coins are worth $2.00, how many of each type of coin do you have?

9.5 Determinants and Cramer's Rule: The 2 × 2 Case

Our goal in this section is to derive a formula for the solution to certain systems of two linear equations in two variables. You may recall that the quadratic formula can be derived by solving the "general" quadratic equation $ax^2 + bx + c = 0$. Let us follow the same approach here and see if we can solve the "general" system of two equations in two variables

$$\begin{aligned} a_1x + b_1y &= c_1 \\ a_2x + b_2y &= c_2 \end{aligned} \tag{9.8}$$

In order to solve the system (9.8), we perform Gaussian elimination on its augmented matrix

$$\left(\begin{array}{cc|c} a_1 & b_1 & c_1 \\ a_2 & b_2 & c_2 \end{array}\right)$$

Assuming that a_1 is not equal to 0, we can add $-a_2/a_1$ times the first row to the second row, to get

$$\left(\begin{array}{cc|c} a_1 & b_1 & c_1 \\ 0 & b_2 - \dfrac{a_2b_1}{a_1} & c_2 - \dfrac{a_2c_1}{a_1} \end{array}\right)$$

The second row of this matrix can be simplified by multiplying it by a_1

$$\left(\begin{array}{cc|c} a_1 & b_1 & c_1 \\ 0 & a_1b_2 - a_2b_1 & a_1c_2 - a_2c_1 \end{array}\right)$$

This is the matrix of the upper triangular system

$$\begin{aligned} a_1x + \qquad\qquad b_1y &= c_1 \\ (a_1b_2 - a_2b_1)y &= a_1c_2 - a_2c_1 \end{aligned}$$

Assuming that $a_1b_2 - a_2b_1 \neq 0$, we can solve the second equation in this system, to get

$$y = \frac{a_1c_2 - a_2c_1}{a_1b_2 - a_2b_1}$$

Substituting this value of y into the first equation in the system and solving for x gives the solution

$$x = \frac{b_2c_1 - b_1c_2}{a_1b_2 - a_2b_1}, \qquad y = \frac{a_1c_2 - a_2c_1}{a_1b_2 - a_2b_1} \tag{9.9}$$

provided that $a_1 \neq 0$ and $a_1b_2 - a_2b_1 \neq 0$. This solution can be checked, by substituting into (9.8), and doing so will show that (9.9) is the solution to the system (9.8) even when $a_1 = 0$. (However, we still must have $a_1b_2 - a_2b_1 \neq 0$.)

At first glance, it may seem that formula (9.9) is too complicated to be of much use. However, by taking advantage of the pattern in the numerators and denominators (9.9), we can write it in a form that will make it much easier to remember and to use. Let us begin with a definition.

DEFINITION

Referring to the system (9.8), the matrix

$$\begin{pmatrix} a_1 & b_1 \\ a_2 & b_2 \end{pmatrix}$$

is called the **matrix of coefficients** of the system. The matrix

$$\begin{pmatrix} x \\ y \end{pmatrix}$$

is called the **matrix of variables** of the system and the matrix

$$\begin{pmatrix} c_1 \\ c_2 \end{pmatrix}$$

is called the **matrix of constant terms** of the system. (These terms are also used for larger systems of equations.)

Now, the denominator $a_1b_2 - a_2b_1$ in formula (9.9) can be obtained from the *matrix of coefficients* of the system (9.8) by the scheme shown in Figure 9.4.

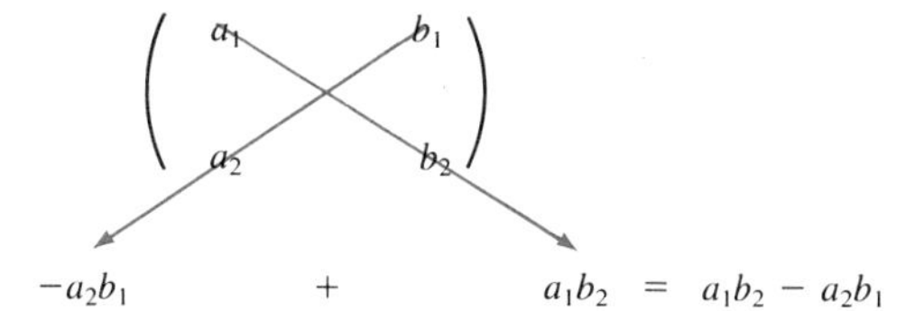

FIGURE 9.4

The idea is to draw two arrows, and take the product of the entries along each arrow. Then we put a minus sign in front of the product associated with the arrow that goes from right to left, and add the two products.

This device for remembering the denominators in (9.9) leads us to make the following definition.

DEFINITION

Let

$$A = \begin{pmatrix} a & b \\ c & d \end{pmatrix}$$

be any 2 × 2 matrix. Then the **determinant** of A, denoted by $|A|$, is defined by

$$|A| = \begin{vmatrix} a & b \\ c & d \end{vmatrix} = ad - bc$$

Notice that the determinant of a matrix is a *number*, not another matrix.

EXAMPLE 1: Evaluate the following determinants.

(a) $\begin{vmatrix} 1 & 4 \\ 5 & -2 \end{vmatrix}$ **(b)** $\begin{vmatrix} -3 & 6 \\ -4 & 8 \end{vmatrix}$

Solution: Using the pattern in Figure 9.4, we get

(a) $\begin{vmatrix} 1 & 4 \\ 5 & -2 \end{vmatrix} = 1(-2) - 4(5) = -22$

(b) $\begin{vmatrix} -3 & 6 \\ -4 & 8 \end{vmatrix} = -3(8) - 6(-4) = 0$

Try Study Suggestion 9.14. □

▶ **Study Suggestion 9.14:** Evaluate the following determinants.

(a) $\begin{vmatrix} 5 & 9 \\ 3 & 0 \end{vmatrix}$ **(b)** $\begin{vmatrix} 1 & -1 \\ 2 & -4 \end{vmatrix}$ **(c)** $\begin{vmatrix} a & b \\ -b & a \end{vmatrix}$ ◀

Now, all of the numerators and denominators in formula (9.9) can be expressed in terms of determinants. In fact, we have the following theorem.

THEOREM

Let

$$a_1x + b_1y = c_1$$
$$a_2x + b_2y = c_2$$

be a system of two equations in two variables. Then, provided that the determinant of the matrix of coefficients of this system is not zero; that is, provided that

$$\begin{vmatrix} a_1 & b_1 \\ a_2 & b_2 \end{vmatrix} \neq 0$$

the system has exactly one solution, given by

$$x = \frac{\begin{vmatrix} c_1 & b_1 \\ c_2 & b_2 \end{vmatrix}}{\begin{vmatrix} a_1 & b_1 \\ a_2 & b_2 \end{vmatrix}}, \quad y = \frac{\begin{vmatrix} a_1 & c_1 \\ a_2 & c_2 \end{vmatrix}}{\begin{vmatrix} a_1 & b_1 \\ a_2 & b_2 \end{vmatrix}} \qquad (9.10)$$

Formula (9.10) is known as **Cramer's rule,** and can be remembered by using the following procedure.

Procedure for Implementing Cramer's Rule

STEP 1. Begin by writing

$$\frac{\begin{vmatrix} a_1 & b_1 \\ a_2 & b_2 \end{vmatrix}}{\begin{vmatrix} a_1 & b_1 \\ a_2 & b_2 \end{vmatrix}} \tag{9.11}$$

(This is the fraction whose numerator and denominator are both the determinant of the matrix of coefficients of the system.)

STEP 2. To obtain the formula for x, replace the *first* column in the *numerator* of (9.11) by the column of constant terms

$$\begin{pmatrix} c_1 \\ c_2 \end{pmatrix}$$

To obtain the formula for y, replace the *second* column in the *numerator* of (9.11) by the column of constant terms

$$\begin{pmatrix} c_1 \\ c_2 \end{pmatrix}$$

EXAMPLE 2: Use Cramer's rule to solve the system

$$\begin{aligned} 2x + 3y &= 1 \\ 4x - y &= 9 \end{aligned}$$

Solution: According to Cramer's rule, we have

$$x = \frac{\begin{vmatrix} c_1 & b_1 \\ c_2 & b_2 \end{vmatrix}}{\begin{vmatrix} a_1 & b_1 \\ a_2 & b_2 \end{vmatrix}} = \frac{\begin{vmatrix} 1 & 3 \\ 9 & -1 \end{vmatrix}}{\begin{vmatrix} 2 & 3 \\ 4 & -1 \end{vmatrix}} = \frac{1(-1) - 3(9)}{2(-1) - 3(4)} = \frac{-28}{-14} = 2$$

and

$$y = \frac{\begin{vmatrix} a_1 & c_1 \\ a_2 & c_2 \end{vmatrix}}{\begin{vmatrix} a_1 & b_1 \\ a_2 & b_2 \end{vmatrix}} = \frac{\begin{vmatrix} 2 & 1 \\ 4 & 9 \end{vmatrix}}{\begin{vmatrix} 2 & 3 \\ 4 & -1 \end{vmatrix}} = \frac{2(9) - 1(4)}{2(-1) - 3(4)} = \frac{14}{-14} = -1$$

Thus, the solution to this system is $x = 2$, $y = -1$. (You should check this by substitution.) *Try Study Suggestion 9.15.* □

▶ **Study Suggestion 9.15:** Use Cramer's rule to solve the system

$$\begin{aligned} 2x + y &= -1 \\ 3x - 4y &= 15 \end{aligned}$$ ◀

We should emphasize that Cramer's rule does not apply when the determinant of the matrix of coefficients of the system is equal to 0. In this case, the expressions for x and y in (9.10) are meaningless, since their denominators would be equal to zero. In fact, we have the following theorem (which we will not prove.)

THEOREM

Let the determinant of the matrix of coefficients of the system (9.8) be denoted by D. Then

1. If $D \neq 0$ the system has a unique solution, given by Cramer's rule.
2. If $D = 0$ the system has either no solutions or infinitely many solutions.

Ideas to Remember

The determinant makes it much easier to remember, and therefore to use, the formula for the solution to a system of two linear equations in two variables.

EXERCISES

In Exercises 1–9, compute the determinant.

1. $\begin{vmatrix} 2 & 3 \\ -1 & 4 \end{vmatrix}$ **2.** $\begin{vmatrix} 1 & 2 \\ 3 & 4 \end{vmatrix}$ **3.** $\begin{vmatrix} 5 & 2 \\ 5 & 2 \end{vmatrix}$

4. $\begin{vmatrix} 2 & -1 \\ -6 & 3 \end{vmatrix}$ **5.** $\begin{vmatrix} 1 & 4 \\ 1 & 5 \end{vmatrix}$ **6.** $\begin{vmatrix} a & a \\ a & a \end{vmatrix}$

7. $\begin{vmatrix} 1 & x \\ 1 & y \end{vmatrix}$ **8.** $\begin{vmatrix} u & v \\ v & u \end{vmatrix}$ **9.** $\begin{vmatrix} x & y \\ 0 & 0 \end{vmatrix}$

In Exercises 10–19, solve the given system of equations by using Cramer's rule, if possible.

10. $\begin{aligned} 2x - 3y &= 1 \\ x + y &= 2 \end{aligned}$

11. $\begin{aligned} x - 3y &= 4 \\ -x + 2y &= 0 \end{aligned}$

12. $\begin{aligned} 3x - 2y &= 1 \\ -9x + 6y &= 4 \end{aligned}$

13. $\begin{aligned} 4x - y &= a \\ 2x + y &= b \end{aligned}$

14. $\begin{aligned} 5x + 6y &= 1 \\ -10x - 12y &= -2 \end{aligned}$

15. $\begin{aligned} 3x &= 4 \\ 5x - 6y &= 4 \end{aligned}$

16. $\begin{aligned} 4x + 3y &= k \\ x - 2y &= k + 1 \end{aligned}$

17. $\begin{aligned} -2kx &= 6 \\ 3k^2x + 2y &= 5 \end{aligned}$

18. $\begin{aligned} x + y &= a \\ x - y &= a^2 \end{aligned}$

19. $\begin{aligned} 5x - 23y &= 2 \\ 3x - 12y &= -2 \end{aligned}$

20. Show that

$$\begin{vmatrix} a & b \\ c & d \end{vmatrix} = -\begin{vmatrix} c & d \\ a & b \end{vmatrix}$$

Express this property of the determinant in words.

21. Show that

$$\begin{vmatrix} ra & rb \\ c & d \end{vmatrix} = r\begin{vmatrix} a & b \\ c & d \end{vmatrix}$$

Express this property of the determinant in words.

22. Suppose that the determinant of the matrix of coefficients of a certain system of equations is equal to 1. Suppose also that the coefficients and the constant terms of the system are integers. Use Cramer's rule to show that the solutions to the system must also be integers.

9.6 Determinants and Cramer's Rule: The 3 × 3 Case

In this section, we discuss the analog of Cramer's rule for systems of three equations in three variables. The first step is to define the determinant of a matrix of size 3 × 3, and to do this we must make some other definitions.

DEFINITION

Let A be a matrix of size 3 × 3. Then the (i, j)th **minor** of A, denoted by $M_{i,j}$, is the *determinant* of the 2 × 2 matrix obtained by deleting the ith row and the jth column from A. The (i, j)th **cofactor** of A, denoted by $C_{i,j}$, is defined by

$$C_{i,j} = (-1)^{i+j} M_{i,j}$$

EXAMPLE 1: Compute the numbers $M_{2,1}$, $C_{2,1}$, $M_{2,2}$, and $C_{2,2}$ for the 3 × 3 matrix

$$A = \begin{pmatrix} 1 & 2 & 0 \\ -1 & 6 & 5 \\ 2 & 4 & 9 \end{pmatrix}$$

Solution: In order to obtain the minor $M_{2,1}$, we delete the second row and the first column of A,

$$\begin{pmatrix} 1 & 2 & 0 \\ -1 & 6 & 5 \\ 2 & 4 & 9 \end{pmatrix}$$

and take the determinant of the remaining 2 × 2 matrix,

$$M_{2,1} = \begin{vmatrix} 2 & 0 \\ 4 & 9 \end{vmatrix} = 18 - 0 = 18$$

The cofactor $C_{2,1}$ is then

$$C_{2,1} = (-1)^{2+1} M_{2,1} = (-1)^3 18 = -18$$

The minor $M_{2,2}$ is obtained by deleting the second row and the second column of A

$$\begin{pmatrix} 1 & 2 & 0 \\ -1 & 6 & 5 \\ 2 & 4 & 9 \end{pmatrix}$$

and taking the determinant of the remaining 2×2 matrix

$$M_{2,2} = \begin{vmatrix} 1 & 0 \\ 2 & 9 \end{vmatrix} = 9 - 0 = 9$$

Also,

$$C_{2,2} = (-1)^{2+2} M_{2,2} = (-1)^4 (9) = 9$$

Try Study Suggestion 9.16. □

▶ **Study Suggestion 9.16:** Compute the numbers $M_{1,1}$, $C_{1,1}$, $M_{2,3}$, $C_{2,3}$, $M_{3,2}$, and $C_{3,2}$ for the matrix

$$\begin{pmatrix} 4 & 0 & 1 \\ -2 & 1 & -1 \\ 0 & 5 & 3 \end{pmatrix}$$ ◀

Now we are ready to define the determinant of a 3×3 matrix.

DEFINITION

The **determinant** of the 3×3 matrix

$$A = \begin{pmatrix} a & b & c \\ d & e & f \\ g & h & i \end{pmatrix}$$

is defined by

$$|A| = \begin{vmatrix} a & b & c \\ d & e & f \\ g & h & i \end{vmatrix} = aC_{1,1} + bC_{1,2} + cC_{1,3} \qquad \textbf{(9.12)}$$

where $C_{1,1}$, $C_{1,2}$, and $C_{1,3}$ are cofactors of A.

Equation (9.12) is called the **expansion of $|A|$ along the first row of A.** This is because the coefficients on the far right-hand side of Equation (9.12) are just the entries of the first row of A, and each entry is multiplied by the corresponding cofactor.

The determinant $|A|$ can also be expressed in terms of minors. In particular, since

$$C_{1,1} = (-1)^2 M_{1,1} = M_{1,1}$$
$$C_{1,2} = (-1)^3 M_{1,2} = -M_{1,2}$$

and

$$C_{1,3} = (-1)^4 M_{1,3} = M_{1,3}$$

we can write Equation (9.12) in the form

$$|A| = \begin{vmatrix} a & b & c \\ d & e & f \\ g & h & i \end{vmatrix} = aM_{1,1} - bM_{1,2} + cM_{1,3} \qquad \textbf{(9.13)}$$

Notice that the signs alternate (starting with a plus sign) in the expansion on the right side of (9.13).

EXAMPLE 2: Evaluate the following determinant

$$\begin{vmatrix} 6 & 2 & -3 \\ 4 & 5 & 0 \\ 3 & 0 & 7 \end{vmatrix}$$

Solution: According to Equation (9.13), we have

$$\begin{aligned}\begin{vmatrix} 6 & 2 & -3 \\ 4 & 5 & 0 \\ 3 & 0 & 7 \end{vmatrix} &= 6M_{1,1} - 2M_{1,2} + (-3)M_{1,3} \\ &= 6\begin{vmatrix} 5 & 0 \\ 0 & 7 \end{vmatrix} - 2\begin{vmatrix} 4 & 0 \\ 3 & 7 \end{vmatrix} - 3\begin{vmatrix} 4 & 5 \\ 3 & 0 \end{vmatrix} \\ &= 6(35 - 0) - 2(28 - 0) - 3(0 - 15) \\ &= 6(35) - 2(28) - 3(-15) \\ &= 199\end{aligned}$$

Try Study Suggestion 9.17. □

▶ **Study Suggestion 9.17:** Evaluate the following determinant

$$\begin{vmatrix} 1 & -2 & 3 \\ -1 & 0 & 4 \\ 5 & -3 & 2 \end{vmatrix}$$ ◀

We mentioned that Equation (9.12) is called the expansion of $|A|$ along the *first row* of A. Actually, we can use either of the other two rows, or for that matter any of the columns, to evaluate $|A|$. For example, the formula

$$|A| = \begin{vmatrix} a & b & c \\ d & e & f \\ g & h & i \end{vmatrix} = dC_{2,1} + eC_{2,2} + fC_{2,3}$$

is called the **expansion of $|A|$ along the second row of A.** Notice that it has exactly the same pattern as Equation (9.12). The coefficients on the right side are the entries of the second row, and each entry is multiplied by the corresponding cofactor. In terms of minors, this expansion is

$$|A| = -dM_{2,1} + eM_{2,2} - fM_{2,3} \tag{9.14}$$

In this case, the signs alternate as in (9.13), but they start with a minus sign (since $C_{2,1} = -M_{2,1}$).

As another example, the formula

$$|A| = \begin{vmatrix} a & b & c \\ d & e & f \\ g & h & i \end{vmatrix} = aC_{1,1} + dC_{2,1} + gC_{3,1}$$

is called the **expansion of $|A|$ along the first column of A.** In this case, the coefficients on the right side are the entries of the first *column* of A, and each is multiplied by the corresponding cofactor. In terms of minors, this expansion is

$$|A| = aM_{1,1} - dM_{2,1} + gM_{3,1} \tag{9.15}$$

EXAMPLE 3: Evaluate the determinant in Example 2 by expanding along the second row, and then by expanding along the first column.

Solution: Expanding along the second row using Equation (9.14), we get

$$\begin{aligned}\begin{vmatrix} 6 & 2 & -3 \\ 4 & 5 & 0 \\ 3 & 0 & 7 \end{vmatrix} &= -4M_{2,1} + 5M_{2,2} - 0M_{2,3} \\ &= -4\begin{vmatrix} 2 & -3 \\ 0 & 7 \end{vmatrix} + 5\begin{vmatrix} 6 & -3 \\ 3 & 7 \end{vmatrix} \\ &= -4(14 - 0) + 5(42 + 9) \\ &= -4(14) + 5(51) \\ &= 199\end{aligned}$$

Expanding along the first column using Equation (9.15), we get

$$\begin{aligned}\begin{vmatrix} 6 & 2 & -3 \\ 4 & 5 & 0 \\ 3 & 0 & 7 \end{vmatrix} &= 6M_{1,1} - 4M_{2,1} + 3M_{3,1} \\ &= 6\begin{vmatrix} 5 & 0 \\ 0 & 7 \end{vmatrix} - 4\begin{vmatrix} 2 & -3 \\ 0 & 7 \end{vmatrix} + 3\begin{vmatrix} 2 & -3 \\ 5 & 0 \end{vmatrix} \\ &= 6(35 - 0) - 4(14 - 0) + 3(0 + 15) \\ &= 6(35) - 4(14) + 3(15) \\ &= 199\end{aligned}$$

Try Study Suggestion 9.18. □

▶ **Study Suggestion 9.18:** Evaluate the determinant in Study Suggestion 9.17 by expanding along the third row, and then by expanding along the second column. ◀

The previous example points out one of the advantages of being able to expand along any row or column. In this example, by expanding along the second row, instead of the first row, we only need to evaluate two 2×2 determinants, rather than three. The reason for this is that one of the entries in the second row is 0, and so we don't need to evaluate the corresponding minor. In general, when taking the determinant of a 3×3 matrix, we can save time by expanding along a row or column with the largest number of zeros.

Now that we have defined the determinant of a 3×3 matrix, we can state Cramer's rule for systems of three equations in three variables.

THEOREM

Cramer's rule Let

$$\begin{aligned} a_1x + b_1y + c_1z &= d_1 \\ a_2x + b_2y + c_2z &= d_2 \\ a_3x + b_3y + c_3z &= d_3 \end{aligned} \tag{9.16}$$

be a system of three equations in three variables. Then provided that the determinant of the matrix of coefficients of this system is not zero, the system has exactly one solution, given by

$$x = \frac{\begin{vmatrix} d_1 & b_1 & c_1 \\ d_2 & b_2 & c_2 \\ d_3 & b_3 & c_3 \end{vmatrix}}{\begin{vmatrix} a_1 & b_1 & c_1 \\ a_2 & b_2 & c_2 \\ a_3 & b_3 & c_3 \end{vmatrix}}, \qquad y = \frac{\begin{vmatrix} a_1 & d_1 & c_1 \\ a_2 & d_2 & c_2 \\ a_3 & d_3 & c_3 \end{vmatrix}}{\begin{vmatrix} a_1 & b_1 & c_1 \\ a_2 & b_2 & c_2 \\ a_3 & b_3 & c_3 \end{vmatrix}}, \qquad z = \frac{\begin{vmatrix} a_1 & b_1 & d_1 \\ a_2 & b_2 & d_2 \\ a_3 & b_3 & d_3 \end{vmatrix}}{\begin{vmatrix} a_1 & b_1 & c_1 \\ a_2 & b_2 & c_2 \\ a_3 & b_3 & c_3 \end{vmatrix}}$$

Notice that this has exactly the same form as Cramer's rule for systems of two equations in two variables. In particular, each of the denominators is the determinant of the matrix of coefficients of the system, and the determinants in the numerators are obtained by replacing the appropriate column of the matrix of coefficients by the column of constant terms. We have highlighted this column in color to make it easier to see.

It is customary to denote the determinant of the matrix of coefficients of a system by D. Also, the determinant in the numerator in the expression for x is denoted by D_x; the determinant in the numerator in the expression for y is denoted by D_y, and similarly for D_z. Then Cramer's rule can be written

$$x = \frac{D_x}{D}, \qquad y = \frac{D_y}{D}, \qquad z = \frac{D_z}{D}$$

provided that $D \neq 0$.

EXAMPLE 4: Use Cramer's rule to solve the system

$$\begin{aligned} x + 2y - z &= 9 \\ 3x - y \phantom{{}+3z} &= 3 \\ -x + 2y + 3z &= 1 \end{aligned}$$

Solution: The first step is to compute the necessary determinants, beginning with the determinant of the matrix of coefficients. (If the determinant of the matrix of coefficients is zero, then Cramer's rule does not apply, and so we do not need to compute the other determinants. We will expand the first three determinants along the second row, since that row contains only two nonzero entries.)

$$D = \begin{vmatrix} 1 & 2 & -1 \\ 3 & -1 & 0 \\ -1 & 2 & 3 \end{vmatrix} = -3\begin{vmatrix} 2 & -1 \\ 2 & 3 \end{vmatrix} + (-1)\begin{vmatrix} 1 & -1 \\ -1 & 3 \end{vmatrix}$$
$$= -3 \cdot 8 + (-1) \cdot 2$$
$$= -26$$

$$D_x = \begin{vmatrix} 9 & 2 & -1 \\ 3 & -1 & 0 \\ 1 & 2 & 3 \end{vmatrix} = -3\begin{vmatrix} 2 & -1 \\ 2 & 3 \end{vmatrix} + (-1)\begin{vmatrix} 9 & -1 \\ 1 & 3 \end{vmatrix}$$
$$= -3 \cdot 8 + (-1) \cdot 28$$
$$= -52$$

$$D_y = \begin{vmatrix} 1 & 9 & -1 \\ 3 & 3 & 0 \\ -1 & 1 & 3 \end{vmatrix} = -3\begin{vmatrix} 9 & -1 \\ 1 & 3 \end{vmatrix} + 3\begin{vmatrix} 1 & -1 \\ -1 & 3 \end{vmatrix}$$
$$= -3 \cdot 28 + 3 \cdot 2$$
$$= -78$$

$$D_z = \begin{vmatrix} 1 & 2 & 9 \\ 3 & -1 & 3 \\ -1 & 2 & 1 \end{vmatrix} = \begin{vmatrix} -1 & 3 \\ 2 & 1 \end{vmatrix} - 2\begin{vmatrix} 3 & 3 \\ -1 & 1 \end{vmatrix} + 9\begin{vmatrix} 3 & -1 \\ -1 & 2 \end{vmatrix}$$
$$= -7 - 2(6) + 9(5)$$
$$= 26$$

Thus, according to Cramer's rule, the solution to this system is

$$x = \frac{D_x}{D} = \frac{-52}{-26} = 2, \qquad y = \frac{D_y}{D} = \frac{-78}{-26} = 3, \qquad z = \frac{D_z}{D} = \frac{26}{-26} = -1$$

(You should check this solution by substitution into the original system.)

Try Study Suggestion 9.19. □

▶ **Study Suggestion 9.19:** Use Cramer's rule to solve the system

$$\begin{aligned} x - 3y + z &= 0 \\ 2x \qquad - 2z &= 1 \\ x + y - z &= 3 \end{aligned}$$ ◀

The last theorem in the previous section has an analog for systems of three equations and three variables.

THEOREM

Let the determinant of the matrix of coefficients of the system (9.16) be denoted by D. Then

1. If $D \neq 0$ the system (9.16) has unique solution, given by Cramer's rule.
2. If $D = 0$ the system (9.16) has either no solutions or else infinitely many solutions.

Let us conclude this section by mentioning that the concept of a determinant can be extended to larger matrices. (However, the determinant is defined only for *square* matrices; that is, only for matrices with the same number of rows as columns.) Also, Cramer's rule can be extended to larger systems of equations. However, as you can see from the previous example, the computations involved in using Cramer's rule become rather lengthy, and so the value of this particular approach diminishes as the size of the system increases.

Ideas to Remember

- The determinant of a 3 × 3 matrix is defined in terms of the determinants of 2 × 2 matrices by expanding along a row or column of the matrix.
- Cramer's rule has the same general form for systems of three equations in three variables as it does for systems of two equations in two variables.

EXERCISES

1. Compute the numbers $M_{2,3}$, $C_{2,3}$, $M_{1,3}$, and $C_{1,3}$ for the matrix

$$\begin{pmatrix} 1 & 2 & 3 \\ 4 & 5 & 6 \\ 7 & 8 & 9 \end{pmatrix}$$

2. Compute the numbers $M_{1,1}$, $C_{1,1}$, $M_{2,2}$, $C_{2,2}$, $M_{3,3}$, and $C_{3,3}$ for the matrix

$$\begin{pmatrix} -2 & 3 & -4 \\ 1 & -2 & 0 \\ 6 & 5 & 8 \end{pmatrix}$$

3. Evaluate

$$\begin{vmatrix} -1 & 3 & -3 \\ 4 & -1 & 6 \\ 3 & 7 & 9 \end{vmatrix}$$

(a) by expanding along the second row.

(b) by expanding along the third column.

4. Evaluate

$$\begin{vmatrix} 2 & -1 & 5 \\ 5 & 2 & 13 \\ 1 & 4 & 3 \end{vmatrix}$$

(a) by expanding along the third row.

(b) by expanding along the first column.

5. Evaluate

$$\begin{vmatrix} 2 & 3 & 8 \\ 3 & 13 & 22 \\ 1 & 5 & 5 \end{vmatrix}$$

(a) by expanding along the first row.

(b) by expanding along the second row.

In Exercises 6–20, evaluate the determinant.

6. $\begin{vmatrix} 2 & 1 & 2 \\ 3 & 1 & -3 \\ 4 & 2 & 1 \end{vmatrix}$

7. $\begin{vmatrix} a & 0 & 0 \\ 0 & b & 0 \\ 0 & 0 & c \end{vmatrix}$

8. $\begin{vmatrix} 2 & -4 & 5 \\ 1 & -1 & 4 \\ 1 & -5 & -2 \end{vmatrix}$

9. $\begin{vmatrix} 5 & 4 & 0 \\ \frac{1}{2} & 2 & 4 \\ 0 & 0 & 0 \end{vmatrix}$

10. $\begin{vmatrix} 2 & 0 & 10 \\ 8 & 1 & 9 \\ 5 & 0 & 4 \end{vmatrix}$

11. $\begin{vmatrix} 3 & 9 & 12 \\ 0 & 1 & 0 \\ 12 & 8 & 9 \end{vmatrix}$

12. $\begin{vmatrix} 1 & -1 & 2 \\ 2 & 5 & -3 \\ 3 & 4 & 6 \end{vmatrix}$

13. $\begin{vmatrix} 3 & 2 & 4 \\ 1 & 2 & 3 \\ 5 & 6 & 13 \end{vmatrix}$

14. $\begin{vmatrix} 1 & 2 & 3 \\ -1 & -2 & -3 \\ 4 & 2 & 9 \end{vmatrix}$

15. $\begin{vmatrix} \frac{1}{2} & \frac{1}{4} & \frac{1}{6} \\ \frac{1}{3} & \frac{1}{2} & 1 \\ \frac{1}{2} & \frac{1}{5} & 1 \end{vmatrix}$

16. $\begin{vmatrix} -2 & -2 & -2 \\ -3 & -3 & -3 \\ -4 & -4 & -4 \end{vmatrix}$

17. $\begin{vmatrix} 1 & 1 & 1 \\ x & y & z \\ x^2 & y^2 & z^2 \end{vmatrix}$

18. $\begin{vmatrix} 2 & 3 & 4 \\ 4 & -1 & 3 \\ 2 & a & 0 \end{vmatrix}$

19. $\begin{vmatrix} a & b & c \\ 0 & d & e \\ 0 & 0 & f \end{vmatrix}$

20. $\begin{vmatrix} x-1 & 3 & 4 \\ 0 & x-2 & 0 \\ 0 & 0 & x+9 \end{vmatrix}$

21. Evaluate the determinant of the 3×3 matrix whose (i, j)th entry is

$$a_{i,j} = \begin{cases} (-1)^{i+j} & \text{if} \quad i \leq j \\ 0 & \text{if} \quad i > j \end{cases}$$

22. Evaluate the determinant of the 3×3 matrix whose (i, j)th entry is

$$a_{i,j} = \begin{cases} 2 & \text{if} \quad i = j \\ 0 & \text{if} \quad i \neq j \end{cases}$$

In Exercises 23–32, solve the given system of equations by using Cramer's rule, if applicable.

23. $\begin{aligned} x - 8y + z &= 12 \\ 3x + 17y - z &= 4 \end{aligned}$

24. $\begin{aligned} 2x + 4y - 7z &= 1 \\ x - y + z &= 2 \\ y + 3z &= 4 \end{aligned}$

25. $\begin{aligned} 4x + 3y - z &= 2 \\ x + y + z &= 3 \\ 4x + 5y + 2z &= 6 \end{aligned}$

26. $\begin{aligned} 2x + 3y + 4z &= 0 \\ x - y + z &= 0 \\ 7x + y + z &= 0 \end{aligned}$

27. $\begin{aligned} x + y + z &= 3 \\ 2x - 3y + z &= 4 \\ -4x + 11y - z &= 5 \end{aligned}$

28. $\begin{aligned} 5x + 3y - 2z &= 12 \\ 9x - y + 3z &= 13 \\ z &= 1 \end{aligned}$

29. $\begin{aligned} 3x + 4y - 2z &= 7 \\ x - 2y + 4z &= 1 \\ 4x + y - 3z &= 3 \end{aligned}$

30. $\begin{aligned} x - 3y + 2z &= 0 \\ 2x + y - z &= 1 \\ 3x + 2y - 2z &= 0 \end{aligned}$

31. $\begin{aligned} x + 2y - 3z &= 5 \\ 4x + y + 2z &= 5 \\ 5x + 3y + 5z &= 5 \end{aligned}$

32. $\begin{aligned} x + 7y \quad &= 2 \\ 2x \quad - z &= 1 \\ 3y \quad &= 6 \end{aligned}$

33. Solve the system in Exercise 24 by Gaussian elimination. Which method do you think is easier?

34. Solve the system in Exercise 25 by Gaussian elimination. Which method do you think is easier?

9.7 Matrix Algebra

In Section 9.3 we used matrices as a convenient tool to help simplify the process of Gaussian elimination. However, matrices play a much more important role in the study of systems of equations than just being a convenient notational device. In order to understand this role, we must define addition and multiplication of matrices, and that is the goal of this section.

Recall that if a matrix has m rows and n columns, then we say that it has **size $m \times n$,** or that it is an **$m \times n$ matrix.** If a matrix has the same number of rows as columns, it is said to be **square.** Also, the entry in the ith row and jth column of a matrix is called the (i, j)th **entry** of the matrix.

Two matrices are said to be **equal** if they have the same size and if corresponding entries in the two matrices are the same. Now let us define addition and subtraction of matrices.

DEFINITION

Let A and B be matrices of the *same size* $m \times n$. Then the **sum** $A + B$ is defined to be the $m \times n$ matrix whose (i, j)th entry is the sum of the (i, j)th entries of A and B. The **difference** $A - B$ is defined to be the $m \times n$ matrix whose (i, j)th entry is the difference of the (i, j)th entries of A and B.

EXAMPLE 1: Compute the sum $A + B$ and the difference $A - B$ of the 2×2 matrices

$$A = \begin{pmatrix} 2 & -3 \\ 6 & 4 \end{pmatrix} \quad \text{and} \quad B = \begin{pmatrix} 7 & 1 \\ 0 & 9 \end{pmatrix}$$

Solution: According to the definitions, we have

$$A + B = \begin{pmatrix} 2 & -3 \\ 6 & 4 \end{pmatrix} + \begin{pmatrix} 7 & 1 \\ 0 & 9 \end{pmatrix} = \begin{pmatrix} 2+7 & -3+1 \\ 6+0 & 4+9 \end{pmatrix} = \begin{pmatrix} 9 & -2 \\ 6 & 13 \end{pmatrix}$$

and

$$A - B = \begin{pmatrix} 2 & -3 \\ 6 & 4 \end{pmatrix} - \begin{pmatrix} 7 & 1 \\ 0 & 9 \end{pmatrix} = \begin{pmatrix} 2-7 & -3-1 \\ 6-0 & 4-9 \end{pmatrix} = \begin{pmatrix} -5 & -4 \\ 6 & -5 \end{pmatrix}$$

Try Study Suggestion 9.20. □

▶ **Study Suggestion 9.20:** Compute $A + B$ and $A - B$ for the matrices

$$A = \begin{pmatrix} 2 & -4 \\ 5 & \frac{1}{2} \end{pmatrix}$$

and

$$B = \begin{pmatrix} 6 & 0 \\ 3 & -\frac{3}{4} \end{pmatrix}$$ ◀

If r is a real number, and A is a matrix, then we define rA to be the matrix obtained by multiplying *each* entry of A by r. The matrix $(-1)A$ is usually denoted by $-A$, and called the **negative** of A. For example, if A is the matrix in Example 1, then

$$2A = 2\begin{pmatrix} 2 & -3 \\ 6 & 4 \end{pmatrix} = \begin{pmatrix} 2 \cdot 2 & 2 \cdot (-3) \\ 2 \cdot 6 & 2 \cdot 4 \end{pmatrix} = \begin{pmatrix} 4 & -6 \\ 12 & 8 \end{pmatrix}$$

and

$$-A = (-1)\begin{pmatrix} 2 & -3 \\ 6 & 4 \end{pmatrix} = \begin{pmatrix} -2 & 3 \\ -6 & -4 \end{pmatrix}$$

Try Study Suggestion 9.21. □

▶ **Study Suggestion 9.21:** Compute the matrices $-4B$ and $-B$, where B is the matrix in Example 1. ◀

Addition of matrices satisfies many of the same properties as addition of real numbers. For example, addition of matrices satisfies the **associative property**

$$A + (B + C) = (A + B) + C$$

and the **commutative property**

$$A + B = B + A$$

provided, of course, that the matrices involved all have the same size, so that the sums are defined.

A matrix all of whose entries are equal to 0 is called a **zero matrix.** For example, the following matrices are zero matrices

$$\begin{pmatrix} 0 & 0 \\ 0 & 0 \end{pmatrix} \quad \begin{pmatrix} 0 & 0 & 0 \\ 0 & 0 & 0 \end{pmatrix} \quad \begin{pmatrix} 0 & 0 & 0 \\ 0 & 0 & 0 \\ 0 & 0 & 0 \end{pmatrix}$$

Zero matrices play a similar role for matrix addition as the number 0 plays for addition of real numbers. In particular, if $\mathbf{0}_{m,n}$ is the $m \times n$ zero matrix, then

$$A + \mathbf{0}_{m,n} = A \quad \text{and} \quad \mathbf{0}_{m,n} + A = A$$

for all $m \times n$ matrices A. (A must have the same size as $\mathbf{0}_{m,n}$ in order for the sum $A + \mathbf{0}_{m,n}$ to be defined.)

At this point, we could define the product of two matrices to be the matrix obtained by multiplying corresponding entries in each matrix. However, there is another way to define the product of two matrices, which is much more useful when it comes to solving systems of linear equations. (We will see why in the next section.)

In order to help give a clear definition of the product of two matrices, we begin by defining the product of a row of a matrix and a column of a matrix. If

$$R = a_1 \quad a_2 \quad a_3 \quad \cdots \quad a_n$$

is a row of a matrix A and

$$C = \begin{matrix} b_1 \\ b_2 \\ b_3 \\ \vdots \\ b_n \end{matrix}$$

is a column of a matrix B, then the *product* RC is the *real number*

$$a_1 \cdot b_1 + a_2 \cdot b_2 + a_3 \cdot b_3 + \cdots + a_n \cdot b_n$$

Notice that this product is defined only if the row and column have the same number of elements.

▶ **Study Suggestion 9.22:** Compute the product RC where

$$R = 4 \quad 0 \quad -5 \quad 7$$

and

$$C = \begin{matrix} 10 \\ -1 \\ -2 \\ 0 \end{matrix}$$ ◀

EXAMPLE 2: Find the product RC if

$$R = 2 \quad -3 \quad 7 \quad \text{and} \quad C = \begin{matrix} 3 \\ 4 \\ 1 \end{matrix}$$

Solution: According to the definition, we have

$$RC = 2 \cdot 3 + (-3) \cdot 4 + 7 \cdot 1 = 1$$

Try Study Suggestion 9.22. □

Now we are ready for the definition of the product of two matrices.

DEFINITION

Let A be a matrix of size $m \times k$ and let B be a matrix of size $k \times n$. Then the **product** AB is the matrix of size $m \times n$ whose (i, j)th entry is the product of the ith *row* of A and the jth *column* of B.

Notice that in order for the product AB *to be defined*, the number of columns of A must equal the number of rows of B. Some examples of matrix multiplication should help make the definition clear.

EXAMPLE 3: Compute the product AB where

$$A = \begin{pmatrix} 2 & 3 & 4 \\ 5 & 6 & 7 \end{pmatrix} \quad \text{and} \quad B = \begin{pmatrix} 3 & 1 \\ 4 & 2 \\ 1 & 9 \end{pmatrix}$$

Solution: According to the definition, we have

$$2 \cdot 3 + 3 \cdot 4 + 4 \cdot 1 = 22$$
$$2 \cdot 1 + 3 \cdot 2 + 4 \cdot 9 = 44$$

$$\begin{pmatrix} 2 & 3 & 4 \\ 5 & 6 & 7 \end{pmatrix}\begin{pmatrix} 3 & 1 \\ 4 & 2 \\ 1 & 9 \end{pmatrix} = \begin{pmatrix} 22 & 44 \\ 46 & 80 \end{pmatrix}$$

$$5 \cdot 1 + 6 \cdot 2 + 7 \cdot 9 = 80$$
$$5 \cdot 3 + 6 \cdot 4 + 7 \cdot 1 = 46$$

□

EXAMPLE 4:

$$\begin{pmatrix} 6 & 2 \\ 4 & 3 \end{pmatrix}\begin{pmatrix} 2 & 1 \\ 3 & 2 \end{pmatrix} = \begin{pmatrix} 6 \cdot 2 + 2 \cdot 3 & 6 \cdot 1 + 2 \cdot 2 \\ 4 \cdot 2 + 3 \cdot 3 & 4 \cdot 1 + 3 \cdot 2 \end{pmatrix} = \begin{pmatrix} 18 & 10 \\ 17 & 10 \end{pmatrix}$$

□

EXAMPLE 5:

$$\begin{pmatrix} 1 & 2 & 3 \\ 0 & 4 & 2 \\ 9 & 5 & 1 \end{pmatrix}\begin{pmatrix} 5 & 4 & -3 \\ 3 & 1 & 0 \\ -1 & 2 & 1 \end{pmatrix}$$

$$= \begin{pmatrix} 1 \cdot 5 + 2 \cdot 3 + 3 \cdot (-1) & 1 \cdot 4 + 2 \cdot 1 + 3 \cdot 2 & 1 \cdot (-3) + 2 \cdot 0 + 3 \cdot 1 \\ 0 \cdot 5 + 4 \cdot 3 + 2 \cdot (-1) & 0 \cdot 4 + 4 \cdot 1 + 2 \cdot 2 & 0 \cdot (-3) + 4 \cdot 0 + 2 \cdot 1 \\ 9 \cdot 5 + 5 \cdot 3 + 1 \cdot (-1) & 9 \cdot 4 + 5 \cdot 1 + 1 \cdot 2 & 9 \cdot (-3) + 5 \cdot 0 + 1 \cdot 1 \end{pmatrix}$$

$$= \begin{pmatrix} 8 & 12 & 0 \\ 10 & 8 & 2 \\ 59 & 43 & -26 \end{pmatrix}$$

Try Study Suggestion 9.23. □

▶ **Study Suggestion 9.23:** Compute the following products

(a) $\begin{pmatrix} 1 & -1 \\ 3 & 4 \end{pmatrix}\begin{pmatrix} 2 & -5 \\ 2 & 6 \end{pmatrix}$

(b) $\begin{pmatrix} 1 & 1 & 1 \\ -1 & 2 & 0 \\ 3 & 0 & 3 \end{pmatrix}\begin{pmatrix} 2 & 2 & 2 \\ 5 & -1 & 2 \\ 6 & 1 & 1 \end{pmatrix}$ ◀

Multiplication of matrices also satisfies the **associative property**

$$A(BC) = (AB)C$$

provided that all of the products are defined. However, as the next example shows, it does *not* satisfy the commutative property; that is,

$$\text{in general, } AB \neq BA$$

EXAMPLE 6: Compare the product AB with the product BA, where

$$A = \begin{pmatrix} 1 & 2 \\ 3 & 4 \end{pmatrix} \quad \text{and} \quad B = \begin{pmatrix} 3 & -1 \\ 2 & 1 \end{pmatrix}$$

Solution: According to the definition of matrix multiplication, we have

$$AB = \begin{pmatrix} 1 & 2 \\ 3 & 4 \end{pmatrix}\begin{pmatrix} 3 & -1 \\ 2 & 1 \end{pmatrix} = \begin{pmatrix} 1 \cdot 3 + 2 \cdot 2 & 1 \cdot (-1) + 2 \cdot 1 \\ 3 \cdot 3 + 4 \cdot 2 & 3 \cdot (-1) + 4 \cdot 1 \end{pmatrix} = \begin{pmatrix} 7 & 1 \\ 17 & 1 \end{pmatrix}$$

but

$$BA = \begin{pmatrix} 3 & -1 \\ 2 & 1 \end{pmatrix}\begin{pmatrix} 1 & 2 \\ 3 & 4 \end{pmatrix} = \begin{pmatrix} 3 \cdot 1 + (-1) \cdot 3 & 3 \cdot 2 + (-1) \cdot 4 \\ 2 \cdot 1 + 1 \cdot 3 & 2 \cdot 2 + 1 \cdot 4 \end{pmatrix} = \begin{pmatrix} 0 & 2 \\ 5 & 8 \end{pmatrix}$$

Try Study Suggestion 9.24. □

▶ **Study Suggestion 9.24:** Compare the product AB with the product BA, where

$$A = \begin{pmatrix} 0 & 0 \\ 1 & 0 \end{pmatrix}$$

and

$$B = \begin{pmatrix} 0 & 0 \\ 0 & 1 \end{pmatrix}$$ ◀

A *square* matrix, all of whose entries on the diagonal running from upper left to lower right are equal to 1, and all of whose other entries are equal to 0, is called an **identity matrix.** For example, the matrices

$$\mathbf{I}_2 = \begin{pmatrix} 1 & 0 \\ 0 & 1 \end{pmatrix}, \quad \mathbf{I}_3 = \begin{pmatrix} 1 & 0 & 0 \\ 0 & 1 & 0 \\ 0 & 0 & 1 \end{pmatrix}, \quad \mathbf{I}_4 = \begin{pmatrix} 1 & 0 & 0 & 0 \\ 0 & 1 & 0 & 0 \\ 0 & 0 & 1 & 0 \\ 0 & 0 & 0 & 1 \end{pmatrix}$$

are identity matrices. We will denote the identity matrix of size $n \times n$ by $\mathbf{I}_n$. It is important to notice that, while zero matrices are defined for all sizes, identity matrices must be square.

The identity matrices play a similar role for matrix multiplication that the number 1 plays for multiplication of real numbers. In particular, we have

$$\mathbf{I}_n A = A \quad \text{and} \quad A\mathbf{I}_n = A$$

for all square $n \times n$ matrices A.

Now that we have defined the identity matrices, we can say what it means for a matrix to have an inverse.

DEFINITION

A square $n \times n$ matrix A is said to be **invertible** if there is another matrix B, also of size $n \times n$, with the property that

$$AB = \mathbf{I}_n \quad \text{and} \quad BA = \mathbf{I}_n$$

When this is the case, we say that B is the **inverse** of the matrix A, and write $B = A^{-1}$. (Another term for invertible is **nonsingular.**)

EXAMPLE 7: Verify that the inverse of the matrix

$$A = \begin{pmatrix} 1 & 2 \\ 1 & 3 \end{pmatrix}$$

is the matrix

$$B = \begin{pmatrix} 3 & -2 \\ -1 & 1 \end{pmatrix}$$

Solution: We must show that $AB = \mathbf{I}_2$ and $BA = \mathbf{I}_2$. But

$$AB = \begin{pmatrix} 1 & 2 \\ 1 & 3 \end{pmatrix}\begin{pmatrix} 3 & -2 \\ -1 & 1 \end{pmatrix} = \begin{pmatrix} 1 \cdot 3 + 2(-1) & 1(-2) + 2 \cdot 1 \\ 1 \cdot 3 + 3(-1) & 1(-2) + 3 \cdot 1 \end{pmatrix}$$
$$= \begin{pmatrix} 1 & 0 \\ 0 & 1 \end{pmatrix} = \mathbf{I}_2$$

We will leave it to you to show that $BA = \mathbf{I}_2$. □

There is a simple technique for finding the inverse of an invertible matrix that uses elementary row operations. Let us illustrate this technique with an example.

EXAMPLE 8: Find the inverse of the invertible matrix

$$A = \begin{pmatrix} -1 & 2 & -3 \\ 2 & 1 & 0 \\ 4 & -2 & 5 \end{pmatrix}$$

Solution: We begin by forming a matrix of size 3×6 by placing the entries of A in the first three columns and the entries of the identity matrix $\mathbf{I}_3$ in the last three columns, as shown below.

$$\left(\begin{array}{ccc|ccc} -1 & 2 & -3 & 1 & 0 & 0 \\ 2 & 1 & 0 & 0 & 1 & 0 \\ 4 & -2 & 5 & 0 & 0 & 1 \end{array}\right)$$

Then we apply elementary row operations to this matrix, in an attempt to change the *left* half into the identity $\mathbf{I}_3$. The steps are as follows.

We begin by multiplying the first row by -1

$$\left(\begin{array}{ccc|ccc} 1 & -2 & 3 & -1 & 0 & 0 \\ 2 & 1 & 0 & 0 & 1 & 0 \\ 4 & -2 & 5 & 0 & 0 & 1 \end{array}\right)$$

Then we add -2 times the first row to the second row, and -4 times the first row to the third row

$$\left(\begin{array}{ccc|ccc} 1 & -2 & 3 & -1 & 0 & 0 \\ 0 & 5 & -6 & 2 & 1 & 0 \\ 0 & 6 & -7 & 4 & 0 & 1 \end{array}\right)$$

(The first column is in the proper form, so we turn our attention to the second column.) Next, we multiply the second row by 1/5

$$\left(\begin{array}{ccc|ccc} 1 & -2 & 3 & -1 & 0 & 0 \\ 0 & 1 & -\frac{6}{5} & \frac{2}{5} & \frac{1}{5} & 0 \\ 0 & 6 & -7 & 4 & 0 & 1 \end{array}\right)$$

Then we add 2 times the second row to the first row, and -6 times the second row to the third row

$$\left(\begin{array}{ccc|ccc} 1 & 0 & \frac{3}{5} & -\frac{1}{5} & \frac{2}{5} & 0 \\ 0 & 1 & -\frac{6}{5} & \frac{2}{5} & \frac{1}{5} & 0 \\ 0 & 0 & \frac{1}{5} & \frac{8}{5} & -\frac{6}{5} & 1 \end{array}\right)$$

(The second column is in the proper form, and so we proceed to the third column.) Now we multiply the third row by 5

$$\left(\begin{array}{ccc|ccc} 1 & 0 & \frac{3}{5} & -\frac{1}{5} & \frac{2}{5} & 0 \\ 0 & 1 & -\frac{6}{5} & \frac{2}{5} & \frac{1}{5} & 0 \\ 0 & 0 & 1 & 8 & -6 & 5 \end{array}\right)$$

Finally, we add $-3/5$ times the third row to the first row, and 6/5 times the third row to the second row

$$\left(\begin{array}{ccc|ccc} 1 & 0 & 0 & -5 & 4 & -3 \\ 0 & 1 & 0 & 10 & -7 & 6 \\ 0 & 0 & 1 & 8 & -6 & 5 \end{array}\right)$$

Now that the left half of this matrix is $\mathbf{I}_3$, the right half is A^{-1}, that is

$$A^{-1} = \begin{pmatrix} -5 & 4 & -3 \\ 10 & -7 & 6 \\ 8 & -6 & 5 \end{pmatrix}$$

(You might wish to verify this by taking the necessary products.)

Try Study Suggestion 9.25. □

► **Study Suggestion 9.25:** Use the technique in the previous example to find the inverse of the matrix

$$\begin{pmatrix} 1 & 2 & 1 \\ 3 & 0 & -1 \\ 1 & 1 & 2 \end{pmatrix}$$ ◄

Ideas to Remember

- Matrix multiplication may seem a bit strange at first, but as we will see in the next section, it can be very useful for solving systems of linear equations.
- Matrix addition and multiplication satisfy many, but not all, of the same properties that addition and multiplication of real numbers satisfy.

EXERCISES

In Exercises 1–6, let A and B be the matrices

$$A = \begin{pmatrix} 1 & 3 \\ 4 & 0 \end{pmatrix} \quad \text{and} \quad B = \begin{pmatrix} 2 & -1 \\ 1 & 7 \end{pmatrix}$$

Compute the following matrices.

1. $A + B$ **2.** $A - B$ **3.** $3A$

4. $-B$ **5.** $2A + B$ **6.** $3A - 2B$

In Exercises 7–12, let A and B be the matrices

$$A = \begin{pmatrix} 2 & 1 \\ 8 & -6 \\ 4 & 0 \end{pmatrix} \quad \text{and} \quad B = \begin{pmatrix} -3 & 9 \\ 0 & 2 \\ 5 & -5 \end{pmatrix}$$

Compute the following matrices.

7. $A + B$ **8.** $A - B$ **9.** $B - A$

10. $7A$ **11.** $4B - 3A$ **12.** $(1/2)A - B$

In Exercises 13–18, let A and B be the matrices

$$A = \begin{pmatrix} -2 & 1 & 3 \\ 5 & 0 & -1 \\ 1 & -1 & 0 \end{pmatrix} \quad \text{and} \quad B = \begin{pmatrix} 1 & -1 & 1 \\ 0 & 3 & 0 \\ 1 & 3 & 4 \end{pmatrix}$$

Compute the following matrices.

13. $A + B$ **14.** $B - A$ **15.** πA

16. $-2A$ **17.** $3A - 4B$ **18.** $(1/2)A - B$

In Exercises 19–23, let A, B, and C be the matrices

$$A = \begin{pmatrix} 2 & 3 \\ -1 & 4 \end{pmatrix} \quad B = \begin{pmatrix} 1 & 3 \\ -5 & 2 \end{pmatrix} \quad C = \begin{pmatrix} 3 & 6 \\ 1 & 1 \end{pmatrix}$$

In each case, verify the given formula and state the property (associative, commutative, distributive, etc.) that it illustrates.

19. Verify that $(A + B) + C = A + (B + C)$

20. Verify that $(AB)C = A(BC)$

21. Verify that $A - B = A + (-B)$

22. Verify that $A(B + C) = AB + AC$

23. Show that $AB \neq BA$

24. For the matrix

$$A = \begin{pmatrix} 2 & 3 & 5 \\ 5 & -6 & 7 \\ 8 & 10 & 0 \end{pmatrix}$$

show that

$$\mathbf{I}_3 A = A\mathbf{I}_3 = A$$

In Exercises 25–37, find the indicated products.

25. $\begin{pmatrix} 2 & 1 \\ 3 & 9 \end{pmatrix}\begin{pmatrix} 6 & 4 \\ 2 & -8 \end{pmatrix}$

26. $\begin{pmatrix} 2 & 0 \\ 0 & 2 \end{pmatrix}\begin{pmatrix} 1 & -1 \\ 9 & 9 \end{pmatrix}$

27. $\begin{pmatrix} 2 & 3 \\ 4 & 6 \end{pmatrix}\begin{pmatrix} 3 & -6 \\ -2 & 4 \end{pmatrix}$

28. $\begin{pmatrix} a & 0 \\ 0 & b \end{pmatrix}\begin{pmatrix} x & 0 \\ 0 & y \end{pmatrix}$

29. $\begin{pmatrix} 4 & 1 \\ -2 & 3 \end{pmatrix}\begin{pmatrix} 2 & -3 & 1 \\ 1 & 0 & 2 \end{pmatrix}$

30. $\begin{pmatrix} x & x \\ y & y \end{pmatrix}\begin{pmatrix} 1 & 1 & 1 \\ 1 & 1 & 1 \end{pmatrix}$

31. $\begin{pmatrix} 3 & -1 & 0 \\ 2 & 1 & 4 \\ 1 & 2 & -1 \end{pmatrix}\begin{pmatrix} 3 & -1 \\ 2 & 1 \\ 4 & 5 \end{pmatrix}$

32. $\begin{pmatrix} 4 & 2 & 1 \\ 1 & -1 & 10 \\ 6 & -6 & 0 \end{pmatrix}\begin{pmatrix} 6 & 0 & 1 \\ 5 & 2 & -4 \\ 2 & 1 & -8 \end{pmatrix}$

33. $\begin{pmatrix} 3 & 1 & 2 \\ 6 & 2 & 4 \\ 9 & 3 & 6 \end{pmatrix}\begin{pmatrix} 2 & 1 & 1 \\ 2 & 3 & -1 \\ -4 & -3 & -1 \end{pmatrix}$

34. $\begin{pmatrix} 1 & 2 & 3 \\ 4 & 5 & 6 \\ 7 & 8 & 9 \end{pmatrix}\begin{pmatrix} -1 & -2 & -3 \\ -4 & -5 & -6 \\ -7 & -8 & -9 \end{pmatrix}$

35. $\begin{pmatrix} k & 0 & 0 \\ 0 & k & 0 \\ 0 & 0 & k \end{pmatrix}\begin{pmatrix} 1 & 1 & 1 \\ 1 & 1 & 1 \\ 1 & 1 & 1 \end{pmatrix}$

36. $\begin{pmatrix} r & 0 & 0 \\ 0 & 0 & 0 \\ 0 & 0 & 0 \end{pmatrix}\begin{pmatrix} a & b & c \\ d & e & f \\ g & h & i \end{pmatrix}$

37. $\begin{pmatrix} a & 0 & 0 \\ 0 & b & 0 \\ 0 & 0 & c \end{pmatrix}\begin{pmatrix} 1 & 2 & 3 \\ 1 & 2 & 3 \\ 1 & 2 & 3 \end{pmatrix}$

38. Suppose that A and B are matrices, A has size $m \times n$ and both products AB and BA are defined. What size is B?

39. Verify that the matrix

$$A = \begin{pmatrix} -1 & 2 & -3 \\ 2 & 1 & 0 \\ 4 & -2 & 5 \end{pmatrix}$$

has inverse

$$A^{-1} = \begin{pmatrix} -5 & 4 & -3 \\ 10 & -7 & 6 \\ 8 & -6 & 5 \end{pmatrix}$$

40. Let

$$A = \begin{pmatrix} 1 & 2 \\ 2 & 4 \end{pmatrix} \qquad B = \begin{pmatrix} -1 & -6 \\ \frac{1}{2} & 3 \end{pmatrix}$$

show that AB is a zero matrix. Compute BA.

41. Let

$$A = \begin{pmatrix} 1 & 2 \\ 2 & 4 \end{pmatrix} \qquad B = \begin{pmatrix} 1 & 1 \\ 2 & 3 \end{pmatrix} \qquad C = \begin{pmatrix} 3 & 3 \\ 1 & 2 \end{pmatrix}$$

Show that $AB = AC$, but that $B \neq C$. This shows that cancellation of matrices is not in general permitted!

In Exercises 42–55, find the inverse of the given matrix. Then verify your answer by taking the necessary matrix products.

42. $\begin{pmatrix} 1 & 2 \\ 2 & 1 \end{pmatrix}$

43. $\begin{pmatrix} 3 & 1 \\ 0 & -1 \end{pmatrix}$

44. $\begin{pmatrix} 1 & 3 \\ 2 & 2 \end{pmatrix}$

45. $\begin{pmatrix} 2 & 0 \\ 3 & 5 \end{pmatrix}$

46. $\begin{pmatrix} 5 & -6 \\ 1 & 4 \end{pmatrix}$

47. $\begin{pmatrix} 2 & 5 \\ 1 & 10 \end{pmatrix}$

48. $\begin{pmatrix} 1 & 2 & 1 \\ 2 & 0 & 1 \\ 1 & 1 & -1 \end{pmatrix}$

49. $\begin{pmatrix} 1 & 2 & 3 \\ 2 & 1 & 3 \\ 1 & 2 & 4 \end{pmatrix}$

50. $\begin{pmatrix} 1 & 2 & 3 \\ 2 & 1 & 2 \\ -4 & 1 & -2 \end{pmatrix}$

51. $\begin{pmatrix} -2 & 0 & -1 \\ \frac{1}{2} & 1 & -1 \\ 4 & 0 & 1 \end{pmatrix}$

52. $\begin{pmatrix} 1 & 1 & 1 \\ 1 & 0 & 2 \\ 1 & 1 & 3 \end{pmatrix}$

53. $\begin{pmatrix} 1 & -3 & 2 \\ 4 & 1 & 1 \\ 0 & 2 & -1 \end{pmatrix}$

54. $\begin{pmatrix} 1 & 0 & 2 \\ 2 & -1 & 3 \\ 4 & 1 & 8 \end{pmatrix}$

55. $\begin{pmatrix} 1 & 3 & 4 \\ 1 & 3 & 0 \\ -1 & 5 & 1 \end{pmatrix}$

56. Let A be the matrix

$$A = \begin{pmatrix} a & b \\ c & d \end{pmatrix}$$

Find a "formula" for A^{-1} by using the method of Example 8 on the matrix

$$\left(\begin{array}{cc|cc} a & b & 1 & 0 \\ c & d & 0 & 1 \end{array}\right)$$

Use your results to explain why the following statement is true: If the determinant of a 2×2 matrix is nonzero, then the matrix is invertible.

9.8 Solving Systems of Linear Equations Using Matrix Algebra

Now that we have defined the product of matrices, we can make the connection between matrix multiplication and systems of linear equations. We begin by observing that, according to the definition of matrix multiplication,

$$\begin{pmatrix} a_1 & b_1 \\ a_2 & b_2 \end{pmatrix}\begin{pmatrix} x \\ y \end{pmatrix} = \begin{pmatrix} a_1x + b_1y \\ a_2x + b_2y \end{pmatrix}$$

and so the system

$$\begin{aligned} a_1x + b_1y &= k_1 \\ a_2x + b_2y &= k_2 \end{aligned}$$

of two equations in two variables can be written in the *matrix form*

$$\begin{pmatrix} a_1 & b_1 \\ a_2 & b_2 \end{pmatrix}\begin{pmatrix} x \\ y \end{pmatrix} = \begin{pmatrix} k_1 \\ k_2 \end{pmatrix}$$

Also, since

$$\begin{pmatrix} a_1 & b_1 & c_1 \\ a_2 & b_2 & c_2 \\ a_3 & b_3 & c_3 \end{pmatrix}\begin{pmatrix} x \\ y \\ z \end{pmatrix} = \begin{pmatrix} a_1x + b_1y + c_1z \\ a_2x + b_2y + c_2z \\ a_3x + b_3y + c_3z \end{pmatrix}$$

the system

$$\begin{aligned} a_1x + b_1y + c_1z &= k_1 \\ a_2x + b_2y + c_2z &= k_2 \\ a_3x + b_3y + c_3z &= k_3 \end{aligned}$$

of three equations in three variables can be written in the *matrix form*

$$\begin{pmatrix} a_1 & b_1 & c_1 \\ a_2 & b_2 & c_2 \\ a_3 & b_3 & c_3 \end{pmatrix}\begin{pmatrix} x \\ y \\ z \end{pmatrix} = \begin{pmatrix} k_1 \\ k_2 \\ k_3 \end{pmatrix}$$

Put more simply, *both* of these systems can be written in the matrix form

$$AX = K \tag{9.17}$$

where A is the matrix of the coefficients of the system, X is the matrix of variables, and K is the matrix of constant terms. In fact, *any* system, regardless of its size, can be put in the matrix form (9.17).

EXAMPLE 1: Write the following systems in matrix form

(a)
$$\begin{aligned} 2x - 3y &= 5 \\ -4x + 9y &= 0 \end{aligned}$$

(b)
$$\begin{aligned} 2x - y + 3z &= 0 \\ 5y - 9z &= 2 \\ -4x + 3y - z &= -1 \end{aligned}$$

Solutions:

(a) The matrix form of this system is

$$\begin{pmatrix} 2 & -3 \\ -4 & 9 \end{pmatrix}\begin{pmatrix} x \\ y \end{pmatrix} = \begin{pmatrix} 5 \\ 0 \end{pmatrix}$$

(b) The matrix form of this system is

$$\begin{pmatrix} 2 & -1 & 3 \\ 0 & 5 & -9 \\ -4 & 3 & -1 \end{pmatrix} \begin{pmatrix} x \\ y \\ z \end{pmatrix} = \begin{pmatrix} 0 \\ 2 \\ -1 \end{pmatrix}$$

Try Study Suggestion 9.26. □

▶ **Study Suggestion 9.26:** Write the following systems in matrix form.

(a) $3x - 2y = 0$
$4x + y = 3$

(b) $5x - 4y + z = 5$
$2x - 3z = 8$

(c) $x - 2y + 3z = 0$
$x + 2y - 3z = 0$
$-x + z = 0$ ◀

Now let us consider a system of two equations in two variables

$$a_1x + b_1y = k_1$$
$$a_2x + b_2y = k_2$$

This system can be written in the matrix form

$$AX = K \tag{9.18}$$

where

$$A = \begin{pmatrix} a_1 & b_1 \\ a_2 & b_2 \end{pmatrix} \qquad x = \begin{pmatrix} x \\ y \end{pmatrix} \qquad \text{and} \qquad K = \begin{pmatrix} k_1 \\ k_2 \end{pmatrix}$$

If the matrix A is invertible, then we can solve Equation (9.18) in the following way. First, we multiply both sides of this equation on the left by A^{-1}

$$A^{-1}(AX) = A^{-1}K$$

Then we use the associative property of matrix multiplication to write this as

$$(A^{-1}A)X = A^{-1}K$$

But $A^{-1}A = \mathbf{I}_2$, and so we have

$$\mathbf{I}_2X = A^{-1}K$$

Finally, since $\mathbf{I}_2$ is an identity matrix, we have $\mathbf{I}_2X = X$, and so

$$X = A^{-1}K \tag{9.19}$$

Thus, the matrix $A^{-1}K$ gives the solution to the system (9.18).

Since the same argument works for any system of n equations in n variables (with $\mathbf{I}_2$ replaced by $\mathbf{I}_n$), we have the following theorem.

THEOREM

If A is an *invertible* $n \times n$ matrix, then the system

$$AX = K$$

of n equations in n variables has exactly one solution, namely,

$$X = A^{-1}K$$

EXAMPLE 2: Use matrix inversion to solve the system

$$\begin{aligned} x + 2y &= 4 \\ x + 3y &= -5 \end{aligned}$$

Solution: This system has the matrix form

$$\begin{pmatrix} 1 & 2 \\ 1 & 3 \end{pmatrix}\begin{pmatrix} x \\ y \end{pmatrix} = \begin{pmatrix} 4 \\ -5 \end{pmatrix}$$

and so

$$A = \begin{pmatrix} 1 & 2 \\ 1 & 3 \end{pmatrix}, \qquad X = \begin{pmatrix} x \\ y \end{pmatrix}, \qquad \text{and} \qquad K = \begin{pmatrix} 4 \\ -5 \end{pmatrix}$$

In Example 7 of the previous section, we showed that

$$A^{-1} = \begin{pmatrix} 3 & -2 \\ -1 & 1 \end{pmatrix}$$

Hence, according to Equation (9.19), $X = A^{-1}K$; that is,

$$\begin{pmatrix} x \\ y \end{pmatrix} = \begin{pmatrix} 3 & -2 \\ -1 & 1 \end{pmatrix}\begin{pmatrix} 4 \\ -5 \end{pmatrix} = \begin{pmatrix} 3 \cdot 4 + (-2)(-5) \\ (-1) \cdot 4 + 1(-5) \end{pmatrix} = \begin{pmatrix} 22 \\ -9 \end{pmatrix}$$

and so the solution to the system is $x = 22$, $y = -9$. (We suggest that you check this by substitution.) *Try Study Suggestion 9.27.* □

▶ **Study Suggestion 9.27:** Use matrix inversion to solve the system

$$\begin{aligned} x + 2y &= -1 \\ x + 3y &= 8 \end{aligned}$$ ◀

EXAMPLE 3: Use matrix inversion to solve the system

$$\begin{aligned} -x + 2y - 3z &= 4 \\ 2x + y &= 3 \\ 4x - 2y + 5z &= 1 \end{aligned}$$

Solution: The matrix form of this system of equations is

$$\begin{pmatrix} -1 & 2 & -3 \\ 2 & 1 & 0 \\ 4 & -2 & 5 \end{pmatrix}\begin{pmatrix} x \\ y \\ z \end{pmatrix} = \begin{pmatrix} 4 \\ 3 \\ 1 \end{pmatrix}$$

and so the coefficient matrix is

$$A = \begin{pmatrix} -1 & 2 & -3 \\ 2 & 1 & 0 \\ 4 & -2 & 5 \end{pmatrix}$$

We showed in Example 8 of the previous section that

$$A^{-1} = \begin{pmatrix} -5 & 4 & -3 \\ 10 & -7 & 6 \\ 8 & -6 & 5 \end{pmatrix}$$

Therefore, according to our theorem, the solution to the system is given in matrix form by

$$\begin{pmatrix} x \\ y \\ z \end{pmatrix} = \begin{pmatrix} -5 & 4 & -3 \\ 10 & -7 & 6 \\ 8 & -6 & 5 \end{pmatrix} \begin{pmatrix} 4 \\ 3 \\ 1 \end{pmatrix} = \begin{pmatrix} -11 \\ 25 \\ 19 \end{pmatrix}$$

That is, the solution is $x = -11$, $y = 25$, $z = 19$, as you can check by substitution. *Try Study Suggestion 9.28.* □

▶ **Study Suggestion 9.28:** Use matrix inversion to solve the system

$$\begin{aligned} x + 2y + 3z &= 2 \\ 2x + 3y + 4z &= 5 \\ 3x + 4y + 6z &= 4 \end{aligned}$$ ◀

It may seem from these examples that the present method of solving systems of equations is much simpler than the method of Gaussian elimination. However, this apparent simplicity is a little deceptive. For if we wish to solve a given system of equations by the present method, then we must first find the matrix A^{-1} and this is usually as time consuming as performing Gaussian elimination. Furthermore, the matrix method only applies when the matrix A is invertible, whereas Gaussian elimination can be used on any system of equations.

Ideas to Remember

Matrix algebra can be used to solve certain systems of linear equations. Nevertheless, the most *practical* method is Gaussian elimination.

EXERCISES

In Exercises 1–14, write the given system of equations in matrix form.

1. $\begin{aligned} 2x + 3y &= 1 \\ x - 2y &= 0 \end{aligned}$

2. $\begin{aligned} -3x + y &= \tfrac{1}{2} \\ 2x - 4y &= 5 \end{aligned}$

3. $\begin{aligned} 10x - 11y &= 12 \\ 17x + 12y &= \sqrt{2} \end{aligned}$

4. $\begin{aligned} 2y &= 5 \\ 2x \quad &= 5 \end{aligned}$

5. $\begin{aligned} 2x - 3y &= 7 \\ x + 4y &= 2 \end{aligned}$

6. $\begin{aligned} -x + 3y &= 0 \\ y &= 0 \end{aligned}$

7. $\begin{aligned} 3x + 2y &= r \\ 5x \quad &= r^2 \end{aligned}$

8. $\begin{aligned} ax - by &= c \\ dx + ey &= f \end{aligned}$

9. $\begin{aligned} \pi x + \sqrt{2}y - z &= 3 \\ 4x - \sqrt{3}y + z &= \sqrt{5} \\ x - \quad y + z &= \pi \end{aligned}$

10. $\begin{aligned} \tfrac{1}{2}x - 4y + 2z &= 0 \\ y - \quad z &= 0 \\ z &= 4 \end{aligned}$

11. $\begin{aligned} 2x + 3y - 5z &= 9 \\ 3x - 2y + 9z &= \tfrac{1}{2} \\ -8x \quad - \quad z &= -\tfrac{1}{2} \end{aligned}$

12. $\begin{aligned} x + y + z &= 1 \\ x + y + z &= 1 \\ x + y + z &= 1 \end{aligned}$

13. $\begin{aligned} x \quad &= 9 \\ y \quad &= 9 \\ z &= 9 \end{aligned}$

14. $\begin{aligned} 5x + 3y - 2z + 4w &= 7 \\ 2x + 5y - \quad z - \quad w &= 0 \\ 7x - 2y + 3z - 8w &= 8 \\ x + \quad y - \quad z + \quad w &= 0 \end{aligned}$

The inverse of the matrix

$$\begin{pmatrix} 1 & 2 \\ 1 & 1 \end{pmatrix}$$

is the matrix

$$\begin{pmatrix} -1 & 2 \\ 1 & -1 \end{pmatrix}$$

Use this information to solve the systems of equations in Exercises 15–18.

15. $\begin{aligned} x + 2y &= 0 \\ x + \quad y &= 0 \end{aligned}$

16. $\begin{aligned} x + 2y &= 1 \\ x + \quad y &= -2 \end{aligned}$

17. $\begin{aligned} x + 2y &= -4 \\ x + \quad y &= 7 \end{aligned}$

18. $\begin{aligned} x + 2y &= a \\ x + \quad y &= b \end{aligned}$

Find the inverse of the matrix

$$\begin{pmatrix} 2 & 4 \\ -1 & 3 \end{pmatrix}$$

and use it to solve the systems of equations in Exercises 19–22.

19. $\begin{aligned} 2x + 4y &= 7 \\ -x + 3y &= 3 \end{aligned}$

20. $\begin{aligned} 2x + 4y &= 3 \\ -x + 3y &= 4 \end{aligned}$

21. $\begin{aligned} 2x + 4y &= 0 \\ -x + 3y &= 0 \end{aligned}$

22. $\begin{aligned} x + 2y &= -1 \\ -x + 3y &= 5 \end{aligned}$

The inverse of the matrix

$$\begin{pmatrix} 1 & 2 & 3 \\ 2 & 3 & 4 \\ 3 & 4 & 6 \end{pmatrix}$$

is the matrix

$$\begin{pmatrix} -2 & 0 & 1 \\ 0 & 3 & -1 \\ 1 & -2 & 1 \end{pmatrix}$$

Use this information to solve the systems of equations in Exercises 23–26.

23. $\begin{aligned} x + 2y + 3z &= 0 \\ 2x + 3y + 4z &= 0 \\ 3x + 4y + 6z &= 0 \end{aligned}$

24. $\begin{aligned} x + 2y + 3z &= 4 \\ 2x + 3y + 4z &= -3 \\ 3x + 4y + 6z &= 2 \end{aligned}$

25. $\begin{aligned} x + 2y + 3z &= 0 \\ 2x + 3y + 4z &= 2 \\ 3x + 4y + 6z &= 3 \end{aligned}$

26. $\begin{aligned} x + 2y + 3z &= a \\ 2x + 3y + 4z &= b \\ 3x + 4y + 6z &= c \end{aligned}$

Find the inverse of the matrix

$$\begin{pmatrix} 1 & 0 & 1 \\ 1 & 2 & -1 \\ 0 & 0 & -1 \end{pmatrix}$$

and use it to solve the systems of equations in Exercises 27–30.

27. $\begin{aligned} x \qquad + z &= 0 \\ x + 2y - z &= 0 \\ -z &= 0 \end{aligned}$

28. $\begin{aligned} x \qquad + z &= 1 \\ x + 2y - z &= 1 \\ -z &= 1 \end{aligned}$

29. $\begin{aligned} x \qquad + z &= 5 \\ x + 2y - z &= -2 \\ -z &= 8 \end{aligned}$

30. $\begin{aligned} x \qquad + z &= a \\ x + 2y - z &= b \\ -z &= c \end{aligned}$

Find the inverse of the matrix

$$\begin{pmatrix} 1 & -2 & 2 \\ 3 & -1 & 2 \\ -1 & 0 & 1 \end{pmatrix}$$

and use it to solve the systems of equations in Exercises 31–34.

31. $\begin{aligned} x - 2y + 2z &= 1 \\ 3x - y + 2z &= 1 \\ -x \qquad + z &= 1 \end{aligned}$

32. $\begin{aligned} x - 2y + 2z &= 3 \\ 3x - y + 2z &= -4 \\ -x \qquad + z &= 6 \end{aligned}$

33. $\begin{aligned} x - 2y + 2z &= 0 \\ 3x - y + 2z &= 0 \\ -x \qquad + z &= 0 \end{aligned}$

34. $\begin{aligned} 3x - y + 2z &= 2 \\ -x \qquad + z &= 3 \\ x - 2y + 2z &= -6 \end{aligned}$

Find the inverse of the matrix

$$\begin{pmatrix} -5 & 4 & -3 \\ 10 & -7 & 6 \\ 8 & -6 & 5 \end{pmatrix}$$

and use it to solve the systems of equations in Exercises 35–38.

35. $\begin{aligned} -5x + 4y - 3z &= 1 \\ 10x - 7y + 6z &= 1 \\ 8x - 6y + 5z &= 1 \end{aligned}$

36. $\begin{aligned} -5x + 4y - 3z &= 0 \\ 10x - 7y + 6z &= 1 \\ 8x - 6y + 5z &= 2 \end{aligned}$

37. $\begin{aligned} -5x + 4y - 3z &= 1 \\ 10x - 7y + 6z &= a \\ 8x - 6y + 5z &= a^2 \end{aligned}$

38. $\begin{aligned} -10x + 8y - 6z &= 0 \\ 8x - 6y + 5z &= 1 \\ 10x - 7y + 6z &= 2 \end{aligned}$

9.9 Systems of Nonlinear Equations

The methods that we have been discussing in this chapter apply to systems of *linear* equations. In general we cannot use these methods to solve systems such as

$$\begin{aligned} x^2 + y^2 &= 1 \\ x + 2y &= -1 \end{aligned} \tag{9.20}$$

where one or both of the equations is not linear. However, quite often we can solve systems such as these by solving one of the equations for one of the variables, and then substituting into the other equation. Let us consider some examples, beginning with the system (9.20).

EXAMPLE 1: Solve the system

$$\begin{aligned} x^2 + y^2 &= 1 \\ x + 2y &= -1 \end{aligned}$$

Solution: Solving the second equation for x gives

$$x = -1 - 2y \tag{9.21}$$

Substituting this into the first equation and simplifying, we get

$$\begin{aligned} (-1 - 2y)^2 + y^2 &= 1 \\ 1 + 4y + 4y^2 + y^2 &= 1 \\ 5y^2 + 4y &= 0 \end{aligned}$$

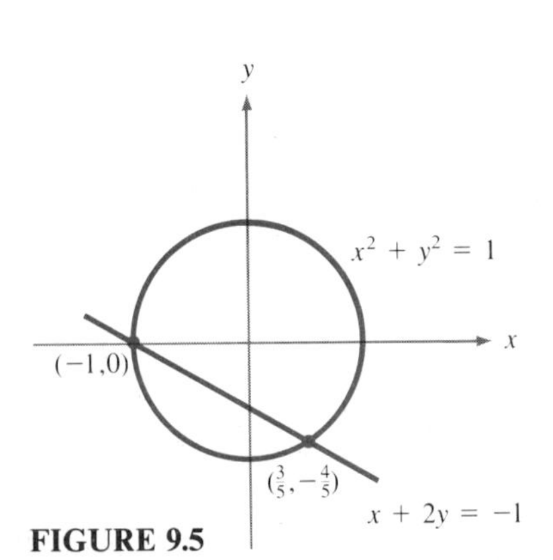

FIGURE 9.5

Now this is a quadratic equation in y, whose solutions are $y = 0$ and $y = -4/5$. Substituting these values for y into Equation (9.21), we get the *potential* solutions

$$x = -1,\ y = 0 \qquad \text{and} \qquad x = \frac{3}{5},\ y = \frac{-4}{5}$$

These solutions must be checked in the *original* system, which will show that they are both solutions. Figure 9.5 shows the graphs of the equations of the system. The solutions to the system are the coordinates of the points of intersection of the two graphs. *Try Study Suggestion 9.29.* □

▶ **Study Suggestion 9.29:** Solve the system

$$\begin{aligned} x^2 - 2y^2 &= 3 \\ x + 3y &= 0 \end{aligned}$$

Then sketch the graphs of these two equations on the same plane, and label the points of intersection. ◀

EXAMPLE 2: Solve the system

$$\begin{aligned} x^2 - y^2 &= 2 \\ 2x^2 + y^2 &= 10 \end{aligned}$$

Solution: In this case, we solve the first equation for x^2, to get

$$x^2 = 2 + y^2 \tag{9.22}$$

and substitute this into the second equation

$$2(2 + y^2) + y^2 = 10$$

Simplifying and solving for y gives $y = \pm\sqrt{2}$. Substituting each of these values into Equation (9.22) give $x^2 = 4$ or $x = \pm 2$. Hence, there are four potential solutions to the system,

$$\begin{aligned} x = 2,\ y = \sqrt{2}; &\quad x = 2,\ y = -\sqrt{2} \\ x = -2,\ y = \sqrt{2}; &\quad x = -2,\ y = -\sqrt{2} \end{aligned}$$

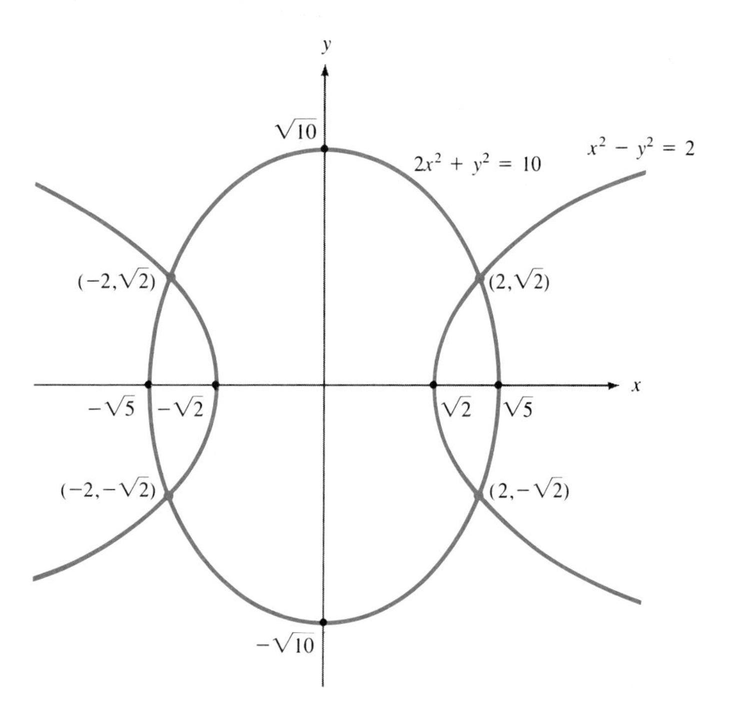

FIGURE 9.6

After checking, we find that each of these is a solution. Figure 9.6 shows the graphs of the two equations as well as the points of intersection.

Try Study Suggestion 9.30. □

► **Study Suggestion 9.30:** Solve the system

$$x^2 + 3y^2 = 1$$
$$3x^2 + y^2 = 1$$

Then sketch the graphs of each equation on the same plane, and label the points of intersection. ◄

EXAMPLE 3: Solve the system

$$x - \log(5y + 5) = 3$$
$$x + \log y = 4$$

where log is the common logarithm (base 10).

Solution: First we solve the second equation for x

$$x = 4 - \log y \qquad (9.23)$$

Then we substitute this into the first equation and simplify, using the properties of the logarithm

$$(4 - \log y) - \log(5y + 5) = 3$$
$$\log y + \log(5y + 5) = 1$$
$$\log[y(5y + 5)] = 1$$
$$y(5y + 5) = 10^1 = 10$$
$$y^2 + y - 2 = 0$$
$$(y - 1)(y + 2) = 0$$
$$y = 1, y = -2$$

Now, we cannot substitute $y = -2$ into equation (9.23), since this would require taking the logarithm of a negative number. Therefore, we do not get a solution when $y = -2$. However, substituting $y = 1$ into Equation (9.23) gives $x = 4$, and so the only solution to this system is $x = 4$, $y = 1$. (As usual, this solution must be checked in the original system.)

Try Study Suggestion 9.31. □

▶ **Study Suggestion 9.31:** Solve the system

$$2y^2 + \sin^2 x = 1$$
$$y - \cos^2 x = 0$$ ◀

We should mention in conclusion that the systems we have discussed in these examples were carefully chosen. Unfortunately, there is no general method (such as Gaussian elimination) for solving systems that involve nonlinear equations.

Ideas to Remember

Sometimes a system of nonlinear equations in two variables can be solved by solving one of the equations for one of the variables and then substituting into the other equation.

EXERCISES

In Exercises 1–14, solve the system of equations. Also, graph the two equations and label the points of intersection. (You may wish to graph the equations before solving the system.)

1. $x^2 + y^2 = 4$, $2x - 3y = 4$

2. $x^2 + y^2 = 2$, $x - y = 5$

3. $x^2 - 2y = 1$, $x^2 + y^2 = 2$

4. $y = x^2 - 4$, $x^2 = 9 - y^2$

5. $x^2 - y^2 = 3$, $2x^2 + y^2 = 6$

6. $2y^2 = 1 + x^2$, $2x - 3y = 1$

7. $x^2 + 2y^2 = 9$, $7x + 3y^2 = 19$

8. $x - 2y^2 = -3$, $x - y^2 = -2$

9. $2x^2 - 8x - y = -4$, $2x + 2 = x^2 + y$

10. $(x-1)^2 + (y-2)^2 = 4$, $x^2 - 2x + 3 = y$

11. $x^2 + y^2 = 9$, $y + 2x = 8$

12. $x - 3y = -3$, $x^2 + y^2 - 2y = 3$

13. $2x^2 + 3y^2 = 4$, $3x^2 + 2y^2 = 4$

14. $x^2 + \sqrt{2}y^2 = 2$, $2x^2 - y - 1 = 3$

In Exercises 15–28, solve the system of equations.

15. $xy = 4$, $x + y = 6$

16. $10y - 1/x^2 = 0$, $3y + 1/2x^2 = 2$

17. $\sqrt{3}y^2 - x^2 = 2$, $x^4 - 3y^4 = 1$

18. $xy = 1$, $x^2 - 2x + 2y = 1$

19. $1/x + 2/y = 3$, $xy = -1$

20. $x - \log(3y + 2) = 4$, $x + \log y = 3$

21. $x - \log(1 + y) = 3$, $x - \log(10 - y) = 4$

22. $3x - \log(1 + y^3) = -6$, $x - \log(1 + y) = -2$

23. $y + 3^x = 7$, $2y - 3^{2x} = 4$

24. $2y + 3^x = 13$, $y - 3^{2x+1} = -22$

25. $2x + y - z^2 = 3$, $x^2 - xz = 0$, $x - 3y + z^2 = 1$
(*Hint:* factor the second equation. What does this tell you about x? Use this information in the other two equations of the system.)

26. $x^2 + y^2 + 4x - 4y = -2$
$2x^2 + 2y^2 - x + y = 5$
(*Hint:* you may need to introduce a parameter here.)

27. $x = \sin^2 y + z^2$
$z = \cos y$
$\pi y - x = 0$

28. $y = x + b$ (b constant)
$x^2 + y^2 = 1$

29. Is it possible to find two real numbers whose difference is 1 and whose product is 1? If so, find all such numbers.

30. Is it possible to find two real numbers whose difference is 1 and whose product is -1? If so, find all such numbers.

31. The sum of two numbers is 1 and the difference of their squares is equal to -5. Find the numbers.

32. The product of two numbers is 3/2 and the sum of their reciprocals is 7/3. Find the numbers.

33. A rectangle has area equal to 1 square foot and perimeter equal to 5 feet. Find the dimensions of the rectangle.

34. A box with a square base has volume equal to 20 cubic centimeters. The sum of the lengths of all 12 edges is equal to 36 centimeters. Find the dimensions of the box.

35. A right circular cylinder has volume equal to 400π cubic meters. If the radius of its base is 2 less than 3 times its height, find the height of the cylinder and the radius of its base.

36. Find the equation of the line passing through the points of intersection of the graphs of $x^2 + y^2 = 5$ and $y - 2x^2 - 1 = 0$.

37. Find the equation of the circle going through the three points (2, 0), (0, 1), and (1, 1).

38. Find the equations of all lines passing through the points of intersections of the graphs of the equations $x^2 - y^2 = 1$ and $x^2 + 2y^2 = 10$.

39. A rectangle is inscribed in a circle of radius $\sqrt{5}$. If the area of the rectangle is 8 square centimeters, find the dimensions of the rectangle.

40. A right circular cylinder is inscribed inside a sphere of radius $\sqrt{13}$ meters. If the volume of the cylinder is 36π cubic meters, find the height of the cylinder and the radius of its base.

9.10 Systems of Inequalities

In many cases, a practical problem leads not to a system of equations, but rather to a system of *inequalities*. The following case is an example.

Suppose that you and a friend decide to take a vacation. Naturally, you are concerned about not spending too much money. Let's say that you do not want to spend more than \$250 for five nights in a hotel and three meals a day. Your friend, however, does not want to spend more than \$180, and since he (or she) wants to stay in the same hotel and spend the same amount on meals, he decides to stay only four nights and skip breakfasts. How much can you afford to pay for the hotel and for meals?

In order to answer this question, we let x stand for the room rate (per person, per night) and let y stand for the *average* cost per meal. Then, since you will spend $5x$ dollars on the hotel room and $15y$ dollars on meals, the variables x and y must satisfy the *inequality*

$$5x + 15y \leq 250$$

On the other hand, your friend will spend $4x$ dollars on a hotel room and $8y$ dollars on meals, and so x and y must also satisfy the inequality

$$4x + 8y \leq 180$$

Therefore, in order to determine what hotel rate you can afford, and what class of restaurant you can patronize, you must solve the *system of inequalities*

$$\begin{aligned} 5x + 15y &\leq 250 \\ 4x + 8y &\leq 180 \end{aligned}$$

A system of inequalities such as this has many solutions, in fact, infinitely many. Because of this, we will write the solutions as ordered pairs of real numbers, and think of them as points in the plane. This gives a nice graphical representation of the solution set. (All our systems of inequalities will involve only two variables.) Let us consider some examples, beginning with single inequalities.

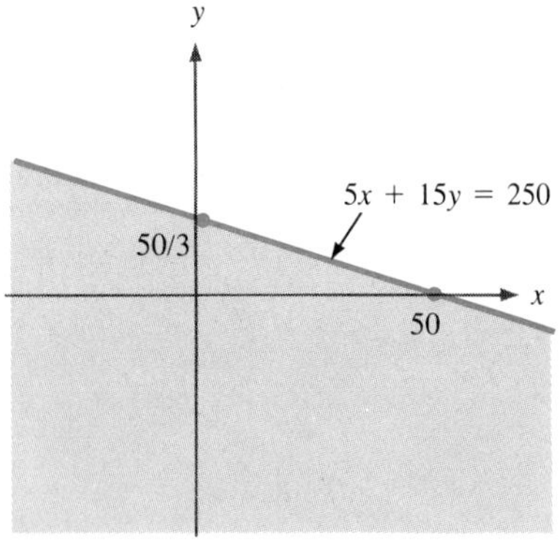

EXAMPLE 1: Solve the inequality

$$5x + 15y \leq 250$$

Solution: We begin by graphing the **associated equation**

$$5x + 15y = 250$$

FIGURE 9.7

This is the straight line shown in Figure 9.7. Then the solutions to the *inequality* consist of all points on this line (since the inequality is not strict), as well as all points on *one side* of the line. In order to determine which side, we simply check any point that is not on the line. For example, it is easy to see that the point (0, 0) is a solution to the inequality (that is, $x = 0$, $y = 0$ is a solution), and so all points on the same side of the line as (0, 0) are also solutions. These points are shown in color in Figure 9.7.

Try Study Suggestion 9.32. □

▶ **Study Suggestion 9.32:** Solve the inequality $3x - 2y \geq 1$. ◀

EXAMPLE 2: Solve the inequality

$$x^2 + y^2 > 3$$

Solution: In this case, the associated equation is $x^2 + y^2 = 3$, which is the equation of a circle of radius $\sqrt{3}$. (See Figure 9.8.) Hence, since the inequality is strict, the solutions to the inequality consist of one of the two parts of the plane determined by this circle—either the points interior to the circle or the points exterior to the circle. In order to determine which part, we check the origin. Since the point (0, 0) is not a solution, the solution set consists of that portion of the plane which lies exterior to the circle, as shown in color in Figure 9.8. *Try Study Suggestion 9.33.* □

▶ **Study Suggestion 9.33:** Solve the inequality $2x^2 + y^2 \geq 1$. ◀

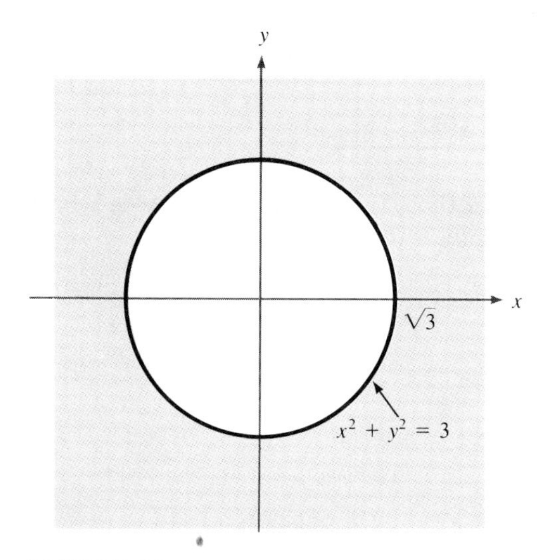

FIGURE 9.8

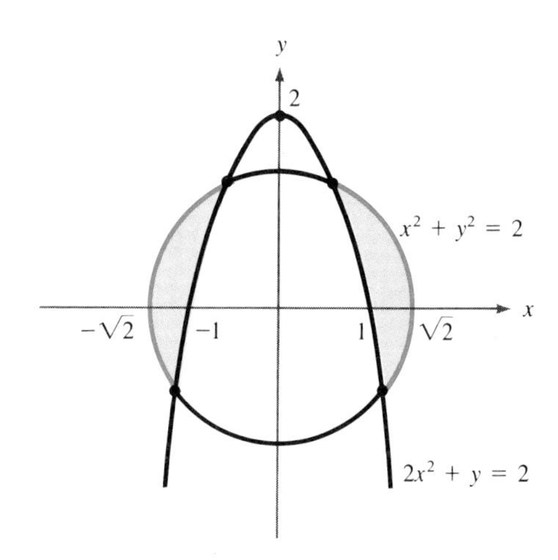

FIGURE 9.9

In order to solve a system of inequalities, we graph the solution sets for each of the inequalities of the system, and then take those points in the plane that are common to all of these solution sets.

EXAMPLE 3: Solve the system

$$x^2 + y^2 \leq 2$$
$$2x^2 + y > 2$$

Solution: The solutions to the first inequality consist of all points interior to or on the circle $x^2 + y^2 = 2$, and since the second inequality can be written in the form

$$y > -2x^2 + 2$$

its solutions consist of all points *above* the parabola

$$y = -2x^2 + 2$$

Therefore, the solutions to the system consist of all points that satisfy *both* of these conditions. This set is shown in color in Figure 9.9.

Try Study Suggestion 9.34. □

▶ **Study Suggestion 9.34:** Solve the system

$$x^2 + y^2 > 4$$
$$y^2 < 1$$ ◀

EXAMPLE 4: Solve the system

$$\sin x - y > 0$$
$$\cos x - y < 0$$

Solution: The first of these inequalities can be written in the form

$$y < \sin x$$

and so its solutions consist of all points *below* the curve $y = \sin x$. Similarly, the second inequality can be written

$$y > \cos x$$

and so its solutions consist of all points *above* the curve $y = \cos x$. Hence, the solutions to the system consist of those points that lie "between" these two curves, as shown in color in Figure 9.10.

Try Study Suggestion 9.35. □

▶ **Study Suggestion 9.35:** Solve the system

$$\tan x - y > 0$$
$$x^2 + y^2 \leq \frac{\pi^2}{4}$$ ◀

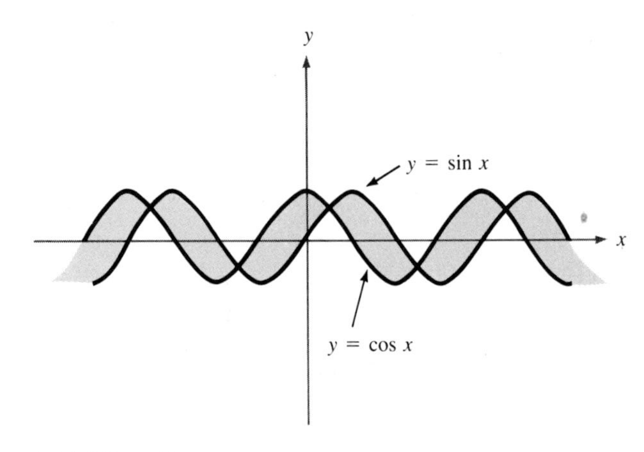

FIGURE 9.10

Now let us consider the problem that we discussed at the beginning of this section.

EXAMPLE 5: Solve the system

$$5x + 15y \leq 250$$
$$4x + 8y \leq 180$$

Solution: Figure 9.11 shows the solutions for this system of inequalities. Notice that the point of intersection of the two lines in Figure 9.11 is just the solution to the *associated system of equations*

$$5x + 15y = 250$$
$$4x + 8y = 180$$

and this solution is (35, 5). (That is, $x = 35$, $y = 5$ is a solution to this system.)

In terms of the problem stated at the beginning of this section, if you and your friend each decide to spend your maximum alloted amount on the vacation, you could each spend \$35 per night on the hotel and an average of \$5 per meal. Also, if you decide to spend less money, then any values for x and y for which the point (x, y) lies in the shaded region in Figure 9.11 will allow you to remain within your budgets. □

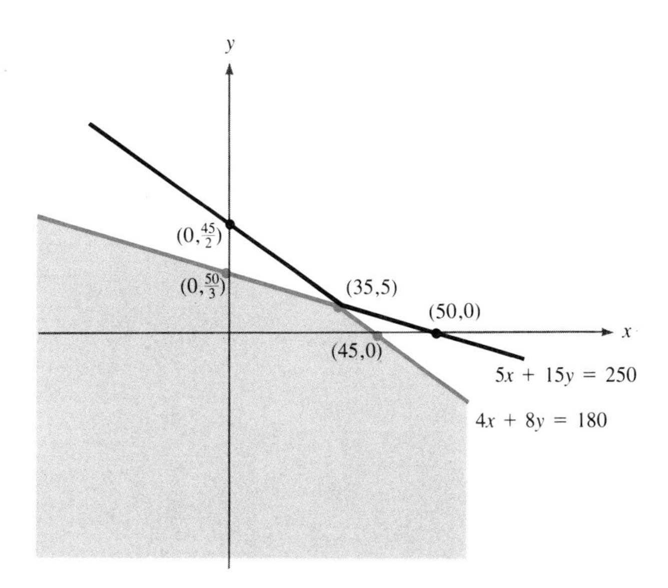

FIGURE 9.11

Ideas to Remember

Systems of inequalities frequently have infinitely many solutions. For this reason, we write the solutions as ordered pairs of real numbers, and think of them as points in the plane. In this way, we can get a nice picture of the solution set of the system.

EXERCISES

In Exercises 1–26, solve the given inequality by sketching its solution set in the plane.

1. $x + 2y \leq 3$

2. $-3x + y > 4$

3. $x - y < 6$

4. $2x - 4y \geq 1$

5. $x^2 + y^2 \leq 4$

6. $x^2 + y^2 > 3$

7. $(x - 4)^2 + (y - 2)^2 \leq 3$

8. $x^2 + y^2 - 2x < 0$

9. $x^2 + y^2 - 4x - 4y > -7$

10. $2y \leq x^2 + 1$

11. $y - x^2 \leq 2x + 3$

12. $x + 1 \geq 2y^2 + y$

13. $x^2 \leq x + y$

14. $y^2 - y > x$

15. $2x^2 + 3y^2 \leq 1$

16. $x^2/2 + y^2/4 > 2$

17. $x^2 - y^2 \leq 1$

18. $2y^2 - x^2 \geq 2$

19. $xy \leq 1$

20. $3xy > -4$

21. $y \geq \sec x$

22. $y \leq \tan x$

23. $y - 2^x \leq 0$

24. $2^x + y \leq 2$

25. $y \leq x + \log x \quad (x > 0)$

26. $2y - \log x^3 > 0 \quad (x > 0)$

In Exercises 27–47, solve the given system of inequalities by sketching its solution set in the plane.

27. $x \leq -y$
$x \leq y$

28. $y \leq 3x + 4$
$y > 3x - 8$

29. $-3x + 1 - y \geq 0$
$y \geq 4x + 5$

30. $2y + 5x - 1 < 0$
$6y > -15x + 2$

31. $x^2 + y^2 \leq 4$
$x^2 + y^2 > 2$

32. $x^2 - x + 1 \leq y$
$4(2x - 1)^2 + (4y - 3)^2 \leq 100$

33. $y \geq (x - 1)^2$
$y \leq 2x + 1$

34. $y \geq 2x^2 - 4x + 1$
$3x^2 + 2y^2 \geq 6$

35. $y^2 + 2y - 3 \leq x$
$x - 1 \leq 0$

36. $y + x^2 \leq 3x - 4$
$y - x \geq 0$

37. $(x - 1)^2 + (2y - 4)^2 \leq 3$
$(x - 1)^2 + (2y + 3)^2 \leq 3$

38. $x^2 - 4x + y^2 - 6y \leq 1$
$x^2 + 4x + y^2 + 6y \leq 1$

39. $y \leq \sin x$
$y \leq \cos x$

40. $y - \tan x \geq 2$
$x^2 + y^2 = \pi^2/4$

41. $y + \log x \leq 1$
$x > 0$
$y - \log x \leq 0$

42. $y - 2^x \leq 0$
$y + 2^x \geq 0$

43. $y - e^{-x} \leq 0$
$y + e^{-x} \geq 0$

44. $x^2 + y^2 \leq 4$
$y \geq x^2 - 2$
$y + x^2 \leq 0$

45. $x^2 + y^2 \leq 5$
$y \leq 3x + 2$
$y \geq 3x - 2$

46. $x^2 + y^2 \leq 16$
$y \geq 4x + 3$
$y \leq 4x - 3$

47. $(x - 1)^2 + (y - 1)^2 \leq 4$
$4y \leq 2x + 5$
$4y \geq 2x + 3$

48. Two types of seats are being sold for a certain outdoor concert. The inexpensive seats are \$5.00 per ticket, and the expensive seats are \$20.00 per ticket. In order for the concert to be held, at least 10,000 tickets must be sold, including at least 6,000 inexpensive tickets and 2,000 expensive tickets. Also, the concert must gross at least \$90,000. Show the possibilities for the number of tickets of each kind sold by setting up an appropriate system of inequalities and graphing the solution set.

49. A woman wants to deposit at least \$10,000 in two bank accounts. One of the accounts returns 8% interest per year and the other returns 10% per year. Furthermore, the woman wants to deposit at least twice as much money in the account that returns 10% interest. Show the various possibilities for the amount of money she can deposit in the two accounts by setting up a system of inequalities and graphing the solution set.

50. Is it possible for a point in the first quadrant of the plane to be a distance less than 2 from the point (2, 1) and a distance less than $\sqrt{5}$ from the point (3, −1)? Justify your answer.

51. Is it possible to have a square and a circle with the property that the area of the square is greater than the area of the circle, but the perimeter of the square is less than the circumference of the circle? Justify your answer.

52. Is it possible to have a square and a circle with the property that the area of the square is less than the area of the circle, but the perimeter of the square is greater than the circumference of the circle? Justify your answer. Compare this with the previous exercise.

9.11 Review

CONCEPTS FOR REVIEW

System of linear equations
Linear equation
Coefficient
Constant term
Solution to a system of linear equations
Inconsistent
Consistent
Upper Triangular Form
Back substitution
Parameter
Gaussian elimination

Elementary operations
Matrix
Entry
Augmented matrix
Elementary row operations
Matrix of coefficients
Matrix of variables
Matrix of constant terms
Determinant
Cramer's rule
Minor
Cofactor
Expansion of the determinant
Size of a matrix
Square matrix
Addition of matrices
Subtraction of matrices
Zero matrix
Product
Identity matrix
Invertible matrix

REVIEW EXERCISES

In Exercises 1–3 decide whether or not the equation is linear.

1. $x + xy = 3^2$

2. $\sqrt{x^2} = 4$

3. $2kx + 3k^2y = 1$ (k constant)

In Exercises 4–12, solve the system of equations.

4. $\begin{aligned} x - 3y &= 4 \\ 2x + y &= 6 \end{aligned}$

5. $\begin{aligned} 3x - 4y &= 7 \\ -6x + 8y &= -14 \end{aligned}$

6. $\begin{aligned} 4x - y &= 5 \\ x - \tfrac{1}{2}y &= 1 \end{aligned}$

7. $\begin{aligned} x - 4y + 6z &= -14 \\ 2x + 3y - 2z &= 12 \\ -3x + 4z &= -5 \end{aligned}$

8. $\begin{aligned} x + 2y - 3z &= 0 \\ 2x - 3y + z &= 0 \\ 8x - 5y - 3z &= 0 \end{aligned}$

9. $\begin{aligned} -2x + 3y - z &= 1 \\ 2x + y - 4z &= 2 \\ 6x + 7y - 16z &= 10 \end{aligned}$

10. $\begin{aligned} 2x - 3y &= 3 \\ x + y &= 4 \\ 7x - 8y &= 13 \\ 2x + 4y &= 9 \end{aligned}$

11. $\begin{aligned} x + 2y - 3z + w &= -6 \\ 2x + y + 2z - w &= 7 \\ 3x - 4y + z + w &= 4 \\ -x + y - 4z + 2w &= -11 \end{aligned}$

12. $\begin{aligned} x + y - 2z + w - u &= -6 \\ 2y + z - 2w + u &= 11 \\ 2x - y + 4z - 2u &= 8 \\ -x + y - 3z + 6w &= -14 \\ x + y + z + w + 2u &= 9 \end{aligned}$

In Exercises 13–16, determine for which values of k the given system has no solutions, exactly one solution, and infinitely many solutions. When the system does have solutions, find them.

13. $\begin{aligned} 3x - 2y &= 4 \\ 6x - 4y &= k + 1 \end{aligned}$

14. $\begin{aligned} x - 3y &= 2 \\ 4x + y &= 3k - 1 \end{aligned}$

15. $\begin{aligned} x - y + z &= 4 \\ 3x + 2y - z &= 6 \\ 6x - y + 2z &= 7k + 4 \end{aligned}$

16. $\begin{aligned} x - 2y + 3z &= k \\ -2x + y + z &= 2k^2 \\ 7x - 8y + 7z &= k - 1 \end{aligned}$

17. Determine all values of a and b for which the system

$$\begin{aligned} x + 3y &= -4 \\ 3ax + y &= 2 \\ x + y &= ab - 4/3 \\ 6x + y &= 10 \\ 2x - 3y &= b^2 - 3b + 12 \end{aligned}$$

has a solution, and find that solution.

18. Write out the matrix with the following properties.

1. 3 rows, 3 columns
2. the (i, j)th entry is equal to the (j, i)th entry
3. the (i, j)th entry is equal to 2^{i-j}, for $i \geq j$

19. Find all functions of the form $f(x) = ax + b\log x$ for which $f(10) = -17$ and $f(\frac{1}{10}) = -\frac{16}{5}$.

20. Find all polynomial functions of degree 3 for which $f(0) = 1$, $f(1) = 3$, $f(-1) = 1$, and $f(2) = 13$.

21. A three-digit number has the following properties: The average of its digits is equal to 3; the product of its digits is equal to 24; and the hundreds digit is twice the ones digit. Find the number.

22. A candy store sells milk chocolate, dark chocolate, and white chocolate. A 4-pound box of chocolate containing 2 pounds of milk chocolate and 1 pound each of dark and white chocolate costs \$41.00. A 4-pound box containing 2 pounds of dark chocolate and 1 pound each of milk and white chocolate costs \$44.00. A 4-pound box containing 2 pounds of white chocolate and 1 pound each of milk and dark chocolate costs \$47.00. How much does each type of chocolate cost?

In Exercises 23–28, evaluate the given determinant.

23. $\begin{vmatrix} 2 & 3 \\ 5 & -5 \end{vmatrix}$ **24.** $\begin{vmatrix} a & a-1 \\ a^2 & (a-1)^2 \end{vmatrix}$

25. $\begin{vmatrix} \sin x & \cos x \\ -\cos x & \sin x \end{vmatrix}$ **26.** $\begin{vmatrix} 3 & -2 & 9 \\ 5 & 9 & 8 \\ 2 & -1 & 4 \end{vmatrix}$

27. $\begin{vmatrix} a & a^2 & a^3 \\ 1 & 2 & 3 \\ 1 & 4 & 9 \end{vmatrix}$ **28.** $\begin{vmatrix} \sin x & \cos x & 0 \\ -\cos x & \sin x & 0 \\ 0 & 0 & x \end{vmatrix}$

29. Evaluate the determinant of the matrix with 3 rows and 3 columns whose (i, j)th entry $a_{i,j}$ is

$$a_{i,j} = \begin{cases} (-1)^i & \text{if} \quad i = j \\ 0 & \text{if} \quad i \neq j \end{cases}$$

30. Evaluate the determinant of the matrix with 3 rows and 3 columns whose (i, j)th entry is equal to $i + j$.

31. Evaluate the determinant of the matrix with 3 rows and 3 columns whose (i, j)th entry is equal to ij.

In Exercises 32–35, solve the given system of equations by using Cramer's rule, if possible.

32. $\begin{aligned} 2x - 3y &= 7 \\ 4x + y &= 6 \end{aligned}$ **33.** $\begin{aligned} 2x + 8y &= 1 \\ x - 3y &= 1/2 \end{aligned}$

34. $\begin{aligned} 2x + y - z &= 1 \\ 4x + 2y - z &= 2 \\ -x + 3y - 2z &= 6 \end{aligned}$ **35.** $\begin{aligned} x - 3y + 4z &= 1 \\ 2x + y - 6z &= -2 \\ 4x - 5y + 2z &= 0 \end{aligned}$

36. Show that

$$\begin{vmatrix} a & b & c \\ d & e & f \\ g & h & i \end{vmatrix} = -\begin{vmatrix} d & e & f \\ a & b & c \\ g & h & i \end{vmatrix}$$

(*Hint:* make a wise choice of row to expand the two determinants, not necessarily the same row in each case.)

In Exercise 37–42, let A and B be the matrices

$$A = \begin{pmatrix} 1 & 2 & 3 \\ 2 & 0 & 0 \\ 4 & -1 & 2 \end{pmatrix} \quad \text{and} \quad B = \begin{pmatrix} -3 & -5 & -\sqrt{2} \\ 0 & \pi & 1 \\ 2 & 2 & -9 \end{pmatrix}$$

Perform the indicated operations and simplify the result.

37. $A + B$ **38.** $A - B$ **39.** $2A$

40. $-\pi B$ **41.** $2B - 3A$ **42.** $aA + bB$

We define the **average** of the n matrices $A_1, A_2, \ldots, A_n$ to be the matrix

$$\frac{1}{n}(A_1 + A_2 + \cdots + A_n)$$

43. Compute the average of the matrices

$$A_1 = \begin{pmatrix} 2 & 5 \\ 6 & 0 \end{pmatrix} \qquad A_2 = \begin{pmatrix} -5 & 9 \\ 1 & -4 \end{pmatrix}$$
$$A_3 = \begin{pmatrix} 7 & -7 \\ -2 & 2 \end{pmatrix} \qquad A_4 = \begin{pmatrix} 3 & -5 \\ 5 & 1 \end{pmatrix}$$

44. Compute the average of the matrices

$$A = \begin{pmatrix} 1 & 3 & 5 \\ 3 & 4 & 0 \\ 6 & -3 & -5 \end{pmatrix} \qquad B = \begin{pmatrix} 3 & -1 & 0 \\ 5 & -6 & 8 \\ 3 & 2 & 1 \end{pmatrix}$$
$$C = \begin{pmatrix} 2 & 0 & -11 \\ 1 & 2 & 1 \\ 1 & -2 & 0 \end{pmatrix}$$

In Exercises 45–50, let A and B be the 2×2 matrices

$$A = \begin{pmatrix} \begin{pmatrix} 0 & 2 \\ 4 & 1 \end{pmatrix} & \begin{pmatrix} 0 & 2 \\ 1 & -2 \end{pmatrix} \\ \begin{pmatrix} 2 & 4 \\ 1 & 2 \end{pmatrix} & \begin{pmatrix} 3 & 0 \\ 0 & 3 \end{pmatrix} \end{pmatrix}$$

$$B = \begin{pmatrix} \begin{pmatrix} 2 & 5 \\ 1 & 1 \end{pmatrix} & \begin{pmatrix} 3 & 6 \\ 4 & 2 \end{pmatrix} \\ \begin{pmatrix} 1 & 2 \\ 1 & 2 \end{pmatrix} & \begin{pmatrix} -3 & 1 \\ 1 & -1 \end{pmatrix} \end{pmatrix}$$

whose entries are themselves 2×2 matrices. Perform the indicated operations and simplify the result.

45. $A + B$ **46.** $A - B$ **47.** $3A$

48. rA **49.** $2A - B$ **50.** $rA + sB$

51. Find a 2×2 matrix A, which is not the zero matrix $\mathbf{0}_{2,2}$ but for which $A^2 = \mathbf{0}_{2,2}$. ($A^2 = AA$.) *Hint:* solve the matrix equation

$$\begin{pmatrix} a & b \\ c & d \end{pmatrix}\begin{pmatrix} a & b \\ c & d \end{pmatrix} = \begin{pmatrix} 0 \\ 0 \end{pmatrix}$$

In Exercises 52–55, find the indicated product.

52. $\begin{pmatrix} 2 & -3 \\ 4 & 1 \end{pmatrix}\begin{pmatrix} 5 \\ 6 \end{pmatrix}$ **53.** $\begin{pmatrix} 1 & 2 \\ 3 & 6 \end{pmatrix}\begin{pmatrix} 4 \\ -2 \end{pmatrix}$

54. $\begin{pmatrix} 2 & 3 & 1 \\ 8 & 7 & 2 \\ 5 & 6 & 1 \end{pmatrix} \begin{pmatrix} 4 & 0 & 9 \\ -1 & 1 & 1 \\ 0 & 0 & 0 \end{pmatrix}$

55. $\begin{pmatrix} 1 & 1 & 2 & 1 \\ 8 & 0 & 0 & 0 \\ 7 & -3 & -2 & -1 \\ 0 & 9 & 10 & 1 \end{pmatrix} \begin{pmatrix} -3 & 2 & -3 & 1 \\ 1 & 1 & 1 & 1 \\ 3 & 0 & 3 & 0 \\ 6 & -2 & 1 & -2 \end{pmatrix}$

In Exercises 56–59, write the given system of equations in matrix form.

56. $\begin{aligned} 2x + 3y &= 4 \\ x - 5y &= 5 \end{aligned}$

57. $\begin{aligned} 3x + 2y &= 3 \\ x + 4y &= 0 \end{aligned}$

58. $\begin{aligned} x + 3y - 4z &= 0 \\ 7x - 5y + z &= -2 \\ y - z &= 1 \end{aligned}$

59. $\begin{aligned} 9x + 2y - 3z + 9w &= 12 \\ x - 8y + z &= 10 \\ x + y - 2z - w &= 3 \\ 2x + 9y - 7z &= 0 \end{aligned}$

In Exercises 60–63, find the inverse of the given matrix.

60. $\begin{pmatrix} 2 & 4 \\ 9 & 1 \end{pmatrix}$

61. $\begin{pmatrix} 1 & -10 \\ 2 & -10 \end{pmatrix}$

62. $\begin{pmatrix} 2 & 1 & -1 \\ 0 & 2 & 1 \\ 5 & 2 & -3 \end{pmatrix}$

63. $\begin{pmatrix} -1 & 2 & -3 \\ 6 & -1 & 5 \\ 4 & -2 & 5 \end{pmatrix}$

Find the inverse of the matrix

$$\begin{pmatrix} 2 & 3 & 1 \\ 4 & 6 & 0 \\ -1 & 1 & 4 \end{pmatrix}$$

and use it to solve the systems of equations in Exercises 64 and 65.

64. $\begin{aligned} 2x + 3y + z &= 3 \\ 4x + 6y &= -4 \\ -x + y + 4z &= 3 \end{aligned}$

65. $\begin{aligned} 2x + 3y + z &= 5 \\ 4x + 6y &= 9 \\ -x + y + 4z &= 10 \end{aligned}$

Find the inverse of the matrix

$$\begin{pmatrix} 1 & 1 & 1 & 1 \\ 0 & 1 & 1 & 1 \\ 0 & 0 & 1 & 1 \\ 0 & 0 & 0 & 1 \end{pmatrix}$$

and use it to solve the systems of equations in Exercises 66–69.

66. $\begin{aligned} x + y + z + w &= 1 \\ y + z + w &= 1 \\ z + w &= 1 \\ w &= 1 \end{aligned}$

67. $\begin{aligned} x + y + z + w &= -3 \\ y + z + w &= 5 \\ z + w &= 21 \\ w &= 54 \end{aligned}$

68. $\begin{aligned} x + y + z + w &= a \\ y + z + w &= b \\ z + w &= c \\ w &= d \end{aligned}$

69. $\begin{aligned} 2x + 2y + 2z + 2w &= 4 \\ 3y + 3z + 3w &= 5 \\ z + w &= 7 \\ w &= 8 \end{aligned}$

70. Solve the system

$$\frac{1}{x} + \frac{1}{y} = 1$$
$$\frac{1}{x} + \frac{1}{z} = 1$$
$$\frac{1}{y} + \frac{1}{z} = 1$$

71. Solve the system

$$x \log y = 2$$
$$y10^x = 1000$$

72. Solve the system

$$\frac{1}{x} + x = \frac{1}{y} + y$$
$$xy = 1$$

73. An equilateral triangle is inscribed in a circle. If the area of the triangle is $\sqrt{3}/2$, find the radius of the circle.

74. Solve the following system of inequalities

$$x^2 + x \leq y$$
$$x^2 - x \leq y$$

75. Solve the following system of inequalities

$$(x - 1)^2 + 4y^2 \leq 1$$
$$(x + 1)^2 + 4y^2 \leq 1$$

76. Solve the following system of inequalities

$$x^2 + y^2 \leq 4$$
$$|xy| \geq 1$$

10 THE COMPLEX NUMBER SYSTEM

10.1 The Arithmetic of Complex Numbers

As you know, not all quadratic equations have solutions within the real number system. For example, the quadratic equation

$$x^2 + 1 = 0 \tag{10.1}$$

which is equivalent to the equation $x^2 = -1$, has no real solutions, since no real number has the property that its square is negative.

This situation has led mathematicians to extend the real number system to a larger system, known as the *complex number system*, which is so "complete" that *all* quadratic equations have solutions within this system. Our goal in this chapter is to discuss some of the elementary properties of the complex number system.

Let us begin by asking ourselves the question "How can we arrange it so that the quadratic Equation (10.1) has a solution?" Since no real number can be a solution to Equation (10.1), the answer is to invent a new "number" for this purpose. We will denote this new number by the letter i, a notation first used by the famous Swiss mathematician Leonhard Euler in 1748. Thus, i is a number (but not a *real* number) with the property that

$$i^2 = -1$$

or, to put it another way, $i = \sqrt{-1}$.

Now, in order to be able to do arithmetic with the number i, we must also include "numbers" such as

$$2i, \quad -3i, \quad \pi i, \quad 6 + i, \quad 7 - \frac{1}{2}i$$

in our number system. In general, we must include all "numbers" of the form

$$a + bi \tag{10.2}$$

where a and b are real numbers. This leads us to make the following definition.

DEFINITION

Any number of the form $a + bi$, where a and b are real numbers, and $i = \sqrt{-1}$ is called a **complex number.** The set of all complex numbers is denoted by C, and called the **complex number system.** In symbols,

$$C = \{a + bi \mid a \text{ and } b \text{ are real numbers and } i = \sqrt{-1}\}$$

Complex numbers are usually denoted by letters from the end of the alphabet. If $z = a + bi$ is a complex number, then a is called the **real part** of z and b is called the **imaginary part** of z. We denote the real part of a complex number z by $Re(z)$ and the imaginary part by $Im(z)$.

EXAMPLE 1: Find $Re(z)$ and $Im(z)$ for the following complex numbers.

(a) $z = 2 + \frac{1}{3}i$ **(b)** $z = -3 - i$ **(c)** $z = 4i$

Solutions:

(a) $Re(2 + (1/3)i) = 2$ and $Im(2 + (1/3)i) = 1/3$

(b) Since $-3 - i = -3 + (-1)i$, we have $Re(-3 - i) = -3$ and $Im(-3 - i) = -1$

(c) Since $4i = 0 + 4i$, we have $Re(4i) = 0$ and $Im(4i) = 4$

Try Study Suggestion 10.1. □

▶ **Study Suggestion 10.1:** Find $Re(z)$ and $Im(z)$ for the following complex numbers.

(a) $z = 2 - 3i$

(b) $z = 1/2 + \sqrt{2}i$

(c) $z = -i/\sqrt{2}$ ◀

Since any real number a can be written in the form $a + 0i$, we see that *a real number is also a complex number*. Thus, the complex number system is truly an enlargement of the real number system. Any number of the form bi (where $b \neq 0$) is said to be a **purely imaginary number.**

We should also remark that two complex numbers $w = a + bi$ and $z = c + di$ are *equal* if and only if their real parts are equal and their imaginary parts are equal. In symbols,

$$a + bi = c + di \quad \text{if and only if} \quad a = c \text{ and } b = d$$

With the complex numbers at our disposal, we can now take the square root of *any* number, positive or negative. For example, we have

$$\sqrt{-4} = \sqrt{4 \cdot (-1)} = \sqrt{4}\sqrt{-1} = 2i$$

$$\sqrt{-12} = \sqrt{12 \cdot (-1)} = \sqrt{12}\sqrt{-1} = 2\sqrt{3}i$$

and

$$\sqrt{-\frac{1}{2}} = \sqrt{\frac{1}{2} \cdot (-1)} = \sqrt{\frac{1}{2}}i = \frac{1}{\sqrt{2}}i = \frac{\sqrt{2}}{2}i$$

Try Study Suggestion 10.2.

▶ **Study Suggestion 10.2:** Take the following square roots.

(a) $\sqrt{-16}$ **(b)** $\sqrt{-2}$

(c) $\sqrt{-75}$ **(d)** $\sqrt{-2/3}$ ◀

If $w = a + bi$ and $z = c + di$ are complex numbers, then the **sum** $w + z$ is defined to be the complex number

$$w + z = (a + bi) + (c + di) = (a + c) + (b + d)i$$

Thus, in order to add two complex numbers, we simply add the corresponding real and imaginary parts.

Subtraction of complex numbers is defined in a similar manner, except that we subtract the corresponding real and imaginary parts,

$$w - z = (a + bi) - (c + di) = (a - c) + (b - d)i$$

EXAMPLE 2: Perform the following operations

(a) $(2 + 3i) + (6 - i)$ **(b)** $(-3 - 4i) - \left(9 - \frac{1}{2}i\right)$

Solutions:

(a) Adding real and imaginary parts gives

$$(2 + 3i) + (6 - i) = (2 + 6) + (3 - 1)i = 8 + 2i$$

(b) Subtracting real and imaginary parts gives

$$\begin{aligned}(-3 - 4i) - \left(9 - \frac{1}{2}i\right) &= (-3 - 9) + \left(-4 - \left(-\frac{1}{2}\right)\right)i \\ &= -12 - \frac{7}{2}i\end{aligned}$$

▶ **Study Suggestion 10.3:** Compute $z + w$ and $z - w$ for the following values of z and w.

(a) $z = 1 - i$, $w = 2 + (1/2)i$

(b) $z = 6i$, $w = 4$ ◀

Try Study Suggestion 10.3. □

Complex numbers are multiplied by using the same rules that are used for real numbers, together with the fact that $i^2 = -1$. Thus if $w = a + bi$ and $z = c + di$, we have

$$\begin{aligned} wz &= (a + bi)(c + di) \\ &= ac + adi + bci + bdi^2 \\ &= ac + adi + bci - bd \\ &= (ac - bd) + (ad + bc)i \end{aligned}$$

and so

$$wz = (a + bi)(c + di) = (ac - bd) + (ad + bc)i$$

As a special case, if a is a real number and $z = c + di$ is a complex number, then

$$az = a(b + ci) = ab + aci$$

There is really no need to memorize the definition of multiplication if you remember that all of the ordinary properties of multiplication hold for complex numbers, and that $i^2 = -1$. We should also note that

$$(-i)^2 = (-i)(-i) = i^2 = -1$$

EXAMPLE 3: Perform the following operations.

(a) $(3 + 7i)(6 - 2i)$ **(b)** $7(2 - \pi i)$ **(c)** $(1 + i)^2$ **(d)** $2i(3 - 4i)$

Solutions:

(a)
$$\begin{aligned}(3 + 7i)(6 - 2i) &= 3 \cdot 6 + 6 \cdot 7i - 3 \cdot 2i - 2 \cdot 7i^2 \\ &= 18 + 42i - 6i + 14 \qquad (\text{using } i^2 = -1) \\ &= 32 + 36i\end{aligned}$$

(b) $7(2 - \pi i) = 7 \cdot 2 - 7 \cdot \pi i = 14 - 7\pi i$

(c) Exponents are used in the same way for complex numbers as for real numbers, and so we have

$$\begin{aligned}(1 + i)^2 &= (1 + i)(1 + i) \\ &= 1 + i + i + i^2 \\ &= 1 + i + i - 1 \\ &= 2i\end{aligned}$$

(d)
$$\begin{aligned}2i(3 - 4i) &= (2i) \cdot 3 - (2i)(4i) \\ &= 6i - 8i^2 \\ &= 6i + 8 \\ &= 8 + 6i\end{aligned}$$

Try Study Suggestion 10.4. □

▶ **Study Suggestion 10.4:** Compute the product zw for the following values of z and w.

(a) $z = 2 - i$, $w = 3 + 4i$
(b) $z = 1/2 - i$, $w = 5$
(c) $z = 2i$, $w = 1 - i$ ◀

EXAMPLE 4: Simplify each of the following expressions.

(a) $7(-2 + i) + 3(i - 4)^2$

(b) $1 + i + i^2 + i^3 + i^4 + i^5 + i^6 + i^7$

(c) i^{75}

Solutions:

(a) (Notice that we have written $i - 4$ instead of $-4 + i$. This is perfectly acceptable.)

$$\begin{aligned}7(-2 + i) + 3(i - 4)^2 &= -14 + 7i + 3(i^2 - 8i + 16) \\ &= -14 + 7i - 3 - 24i + 48 \\ &= 31 - 17i\end{aligned}$$

(b) First we observe that

$$\begin{aligned}&i^2 = -1, && i^3 = i^2 i = (-1)i = -i \\ &i^4 = (i^2)^2 = (-1)^2 = 1, && i^5 = (i^4)i = (1)i = i \\ &i^6 = (i^4)i^2 = (1)(-1) = -1, && i^7 = (i^4)i^3 = (1)(-i) = -i\end{aligned}$$

Thus,

$$\begin{aligned} 1 + i + i^2 + i^3 + i^4 + i^5 + i^6 + i^7 &= 1 + i - 1 - i + 1 + i - 1 - i \\ &= 0 \end{aligned}$$

(c) Taking advantage of the fact that $i^4 = 1$, we have

$$i^{75} = (i^4)^{18} i^3 = 1 \cdot i^3 = -i$$

□

All of the properties of the arithmetic of real numbers that we discussed in Chapter 1 hold also for the complex numbers. For example, complex addition and multiplication satisfy the **commutative properties**

$$z + w = w + z$$
$$zw = wz$$

and the **associative properties**

$$z + (w + u) = (z + w) + u$$
$$z(wu) = (zw)u$$

Also, if the product of two complex numbers is equal to 0, then at least one of the numbers must equal 0, in symbols,

if $zw = 0$ then either $z = 0$ or $w = 0$ or both

As a result, we can work with complex numbers in much the same way as we work with real numbers. (However, see Exercise 54.)

Ideas to Remember

- The complex number system is so "complete" that all quadratic equations have solutions within this number system. In fact, all polynomial equations have solutions within this system.
- The properties of the arithmetic of real numbers that we discussed in Chapter 1 also hold in the complex number system.

EXERCISES

In Exercises 1–12, simplify the expression as much as possible.

1. $Re(2 + i)$ **2.** $Re(\pi - 2i)$ **3.** $Re(-i)$

4. $Re(i - 1)$ **5.** $Im(2 + i)$ **6.** $Im(2 - i)$

7. $Im(\sqrt{2}i + 3)$ **8.** $Im(-i)$ **9.** i^9

10. $(-i)^{23}$ **11.** i^{100} **12.** $(-i)^{139}$

In Exercises 13–45, perform the indicated operations and then write the result in the form $a + bi$.

13. $(6 + i) + (2 - 3i)$ **14.** $3(7 - i) + (3 - 6i)$

15. $i + (3 - i)$ **16.** $(7 - i) + (3 + \sqrt{2}i)$

17. $(\pi - i) - 2(\sqrt{2} - 2i)$ **18.** $(3i - 4) - i + (2i - 1)$

19. $(2 + 3i)(4 - i)$ **20.** $(i - 1)(1 + i)$

21. $(\sqrt{2} + 3i)(\sqrt{2} - 3i)$

22. $(-2 - \pi i)\left(\frac{1}{2} + \frac{i}{\pi}\right)$

23. $3i(6 - 2i) + 4i(2 - i)$

24. $(4 + i) + (3 - 2i)(5 - i)$

25. $(2i + 1)(3 - i)(1 - i)$

26. $(1 - i)^2$

27. $(2 - i)^3$

28. $(4 - 4i)^2$

29. $(2 - i)(1 + i) + (2 - i)(1 - i) - 2(2 - i)$

30. $1 + i + i^2$

31. $1 + i + i^2 + i^3$

32. $1 + i + i^2 + i^3 + i^4 + i^5$

33. $-1 + i - i^3 + i^8$

34. $1 - i^{10} + i^{100} - i^{1000}$

35. $(1 - i)(1 - i^2)(1 - i^3)(1 - i^4)(1 - i^5)(1 - i^6)$

36. $\sqrt{-250}$

37. $\sqrt{-3}\sqrt{-4}$

38. $(1 + \sqrt{-2})(1 - \sqrt{-2})$

39. $\sqrt{-3}(1 - i)$

40. $\sqrt{-9}i$

41. $(4 + \sqrt{-1}) + (3 - \sqrt{-2})$

42. $\sqrt{2}(2 + \sqrt{-2}) - \sqrt{3}i(\sqrt{3} - \sqrt{-3})$

43. $\sqrt{-2} + 2i$

44. $(2 + \sqrt{-4})(1 + \sqrt{-5})$

45. $(3 - \sqrt{-3})^3$

In Exercises 46–53, find all *real* numbers x and y that satisfy the given equation. (Remember that $z = w$ if and only if $Re(z) = Re(w)$ and $Im(z) = Im(w)$.)

46. $2x + 1 - 3yi = 7x + 2i$

47. $(x + i)^2 = 1$

48. $(x + i)^2 = -1$

49. $x^2 + y^2 i = 4 + i$

50. $(6x + 3xi - 3i)^2 = -9 + 40i$

51. $(x + 3yi)^2 = 8 + 6i$

52. $x^2 - 2y^2 i = \sqrt{2} - 3i$

53. $x^2 - i = -1$

54. **(a)** Compute $\sqrt{-2}\sqrt{-8}$

(b) Compute $\sqrt{(-2)(-8)}$

(c) Are the results of parts *a* and *b* the same. What do you conclude?

In Exercises 55–66, find all *complex* numbers z that satisfy the given equation. (Remember that the properties of complex arithmetic are the same as the properties of real arithmetic.)

55. $z^2 = -3$

56. $z^2 = -4$

57. $z^2 = -z$

58. $(z + i)^2 = 1$

59. $(z + i)^2 = -1$

60. $\dfrac{z + 1}{z - 1} = 1$

61. $\dfrac{(z + i)^3}{2iz - 3} = 0$

62. $\dfrac{z + 2i}{iz - 3} = -2i$

63. $z^2 + 2z + 1 = 1$

64. $(z - i)(z + i) = z^2$

65. $(z - i)^2 = z^2$

66. $(z - 3)(z + i) = 0$

10.2 The Conjugate of a Complex Number and Complex Division

Let us continue our discussion of the arithmetic of complex numbers.

DEFINITION

If $z = a + bi$ is a complex number, then the **conjugate** of z is the complex number $\bar{z} = a - bi$. In words, the conjugate of a complex number is obtained by changing the sign of the imaginary part of the number.

For example,

if $z = 2 + 3i$	then	$\bar{z} = 2 - 3i$
if $z = 4 - 5i$	then	$\bar{z} = 4 + 5i$
if $z = \pi i$	then	$\bar{z} = -\pi i$
if $z = 7$	then	$\bar{z} = 7$

As you can see from the definition, and the last example, the conjugate of a *real* number z is z itself.

▶ **Study Suggestion 10.5:** Compute the conjugate of the following complex numbers.
(a) $2 - i$ **(b)** $\sqrt{2} + \sqrt{3}i$ **(c)** $-i$ **(d)** π ◀

Try Study Suggestion 10.5.

In order to appreciate the importance of the conjugate of a complex number $z = a + bi$, let us take the product $z\bar{z}$,

$$z\bar{z} = (a + bi)(a - bi) = (a^2 + b^2) + (ab - ab)i = a^2 + b^2$$

Thus

$$\textbf{if } z = a + bi \textbf{ then } z\bar{z} = a^2 + b^2 \qquad \textbf{(10.3)}$$

This shows that the product $z\bar{z}$ is always a *real* number, regardless of the value of z. In fact, the product $z\bar{z}$ is always *nonnegative*, and equals 0 if and only if both a and b equal 0; that is, if and only if z equals 0.

EXAMPLE 1: Compute $z\bar{z}$ for the following values of z.

(a) $z = 2 + 3i$ **(b)** $z = 1 - i$

Solutions:

(a) In this case $a = 2$ and $b = 3$, and so

$$z\bar{z} = (2 + 3i)(2 - 3i) = 2^2 + 3^2 = 13$$

(b) In this case, $a = 1$ and $b = -1$, and so

$$z\bar{z} = (1 - i)(1 + i) = 1^2 + (-1)^2 = 2$$

▶ **Study Suggestion 10.6:** Compute $z\bar{z}$ for
(a) $z = 2 + i$ **(b)** $z = 3 - 5i$ ◀

Try Study Suggestion 10.6. □

Statement (10.3) can be used to help divide complex numbers. For example, suppose that we want to divide $2 + i$ by $3 - 2i$; that is, suppose that we want to write the expression

$$\frac{2 + i}{3 - 2i}$$

in the form $a + bi$. Then we first multiply the numerator and denominator by the *conjugate* of the *denominator*. According to (10.3), we have

$(3-2i)(3+2i) = 3^2 + (-2)^2$, and so

$$\begin{aligned}\frac{2+i}{3-2i} &= \frac{(2+i)(3+2i)}{(3-2i)(3+2i)} \\ &= \frac{2\cdot 3 + 2\cdot 2i + 3i + 2i^2}{3^2+(-2)^2} \\ &= \frac{4+7i}{13} \\ &= \frac{4}{13} + \frac{7}{13}i\end{aligned}$$

We can summarize this approach to division of complex numbers as follows.

In order to simplify the expression, w/z, where w and z are complex numbers and $z \neq 0$, first multiply the numerator and denominator by the conjugate $\bar{z}$ of the denominator.

EXAMPLE 2: Simplify the following expressions. (That is, put them in the form $a + bi$.)

(a) $\dfrac{1-i}{1+i}$ **(b)** $\dfrac{1}{2-5i}$ **(c)** $\dfrac{1}{i}$

Solutions:

(a) The conjugate of the denominator is $1 - i$, and so we have

$$\frac{1-i}{1+i} = \frac{(1-i)(1-i)}{(1+i)(1-i)} = \frac{1-i-i+i^2}{1^2+1^2} = \frac{-2i}{2} = -i$$

(b) The conjugate of the denominator is $2 + 5i$, and so we have

$$\frac{1}{2-5i} = \frac{2+5i}{(2-5i)(2+5i)} = \frac{2+5i}{2^2+(-5)^2} = \frac{2+5i}{29} = \frac{2}{29} + \frac{5}{29}i$$

(c) In this case, the conjugate of the denominator is $-i$, and so

$$\frac{1}{i} = \frac{-i}{i\cdot(-i)} = \frac{-i}{1} = -i$$

Thus, $1/i = -i$. (This is worth remembering.)

Try Study Suggestion 10.7. □

▶ **Study Suggestion 10.7:** Simplify the following expressions.

(a) $\dfrac{1}{2+3i}$ **(b)** $\dfrac{4+i}{4-i}$ **(c)** $\dfrac{1}{-i}$ ◀

The next theorem gives some of the more useful properties of the conjugate.

THEOREM

For all complex numbers w and z, we have

1. $\overline{\overline{z}} = z$
2. $\overline{w + z} = \overline{w} + \overline{z}$ and $\overline{w - z} = \overline{w} - \overline{z}$
3. $\overline{wz} = \overline{w}\,\overline{z}$
4. $\overline{(z^n)} = (\overline{z})^n$ for any nonnegative integer n
5. z is real if and only if $z = \overline{z}$

Proof: We begin by writing $w = a + bi$ and $z = c + di$. In order to prove property 1, we observe that

$$\begin{aligned}\overline{\overline{z}} &= \overline{\overline{c + di}}\\ &= \overline{c - di}\\ &= c + di\\ &= z\end{aligned}$$

Property 2 is proved by observing that

$$\begin{aligned}\overline{w + z} &= \overline{(a + bi) + (c + di)}\\ &= \overline{(a + c) + (b + d)i}\\ &= (a + c) - (b + d)i\\ &= (a - bi) + (c - di)\\ &= \overline{w} + \overline{z}\end{aligned}$$

We will leave the second part of property 2 as an exercise.
Property 3 is proved as follows,

$$\begin{aligned}\overline{wz} &= \overline{(a + bi)(c + di)}\\ &= \overline{(ac - bd) + (ad + bc)i}\\ &= (ac - bd) - (ad + bc)i\\ &= (a - bi)(c - di)\\ &= \overline{w}\,\overline{z}\end{aligned}$$

Property 4 can be proved by a technique known as *mathematical induction*, which we will discuss in Chapter 12.

In order to prove property 5, we must actually prove two things. First, we must prove that if z is a real number, then $z = \overline{z}$. But if z is a real number, then it can be written in the form $z = a = a + 0i$ where a is real and so

$$\overline{z} = \overline{a + 0i} = a - 0i = a = z$$

Second, we must prove that if z is a complex number with the property that $z = \overline{z}$, then z must actually be real; that is, the imaginary part of z must be 0.

But if $z = c + di$, and $z = \bar{z}$, then

$$c + di = c - di$$

and this can be true if and only if $d = -d$; that is, if and only if $d = 0$. Hence, the imaginary part of z is equal to 0, and so z is real. This completes the proof of property 5. ▮

As you know, real numbers can be represented as points on the real number line. Fortunately, we can get a similar representation of complex numbers by using points in the plane. The idea is simply to associate with each *complex* number $z = a + bi$ the *ordered pair* of real numbers (a, b),

$$a + bi \longleftrightarrow (a, b)$$

By making this association, every complex number corresponds to exactly one point *in the plane*, and vice versa. Figure 10.1 shows a few points in the plane, labeled with their associated complex numbers.

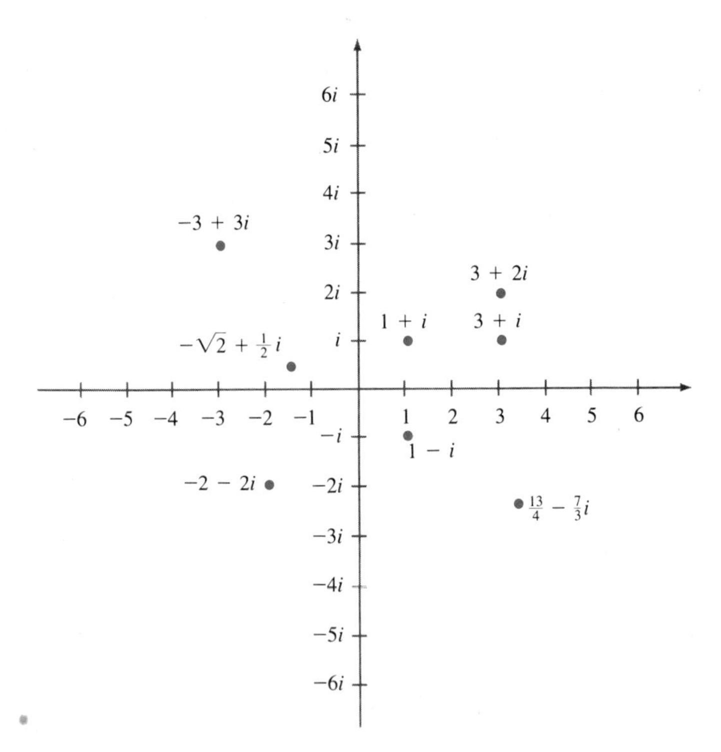

FIGURE 10.1

When we use the plane in this way to describe the set of complex numbers, we refer to it as the **complex plane,** by analogy with the *real number line* (which, incidentally, is also called the *real line*). In this case, the horizontal axis is called the **real axis,** since any point on this axis corresponds to a real

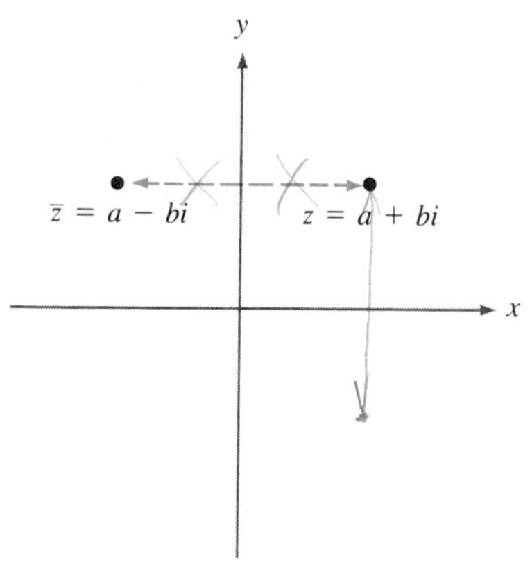

A complex number z and its conjugate $\bar{z}$ are mirror images of each other across the y-axis.

FIGURE 10.2

number, and the vertical axis is called the **imaginary axis,** since any point on this axis (except the origin) corresponds to a purely imaginary number.

Figure 10.2 shows a point $z = a + bi$ in the complex plane, along with its conjugate $\bar{z} = a - bi$. As you can see from this figure, the points z and $\bar{z}$ are reflections of each other across the y-axis.

The **absolute value,** or **modulus,** of a complex number $z = a + bi$ is defined to be the number

$$|z| = \sqrt{a^2 + b^2} \tag{10.4}$$

Notice that the absolute value of a complex number is always a nonnegative real number. In fact,

$|z|$ is the distance from the point z in the complex plane to the origin.

The next theorem (which we will not prove) describes some of the properties of the absolute value.

THEOREM

For all complex numbers w and z, we have

1. $|z| = 0$ if and only if $z = 0$
2. $|wz| = |w|\,|z|$
3. $\left|\dfrac{z}{w}\right| = \dfrac{|z|}{|w|}$ for $w \neq 0$
4. $|z^n| = |z|^n$ for all integers n
5. $|\bar{z}| = |z|$

EXAMPLE 3: Compute the following numbers.

(a) $|2 - 9i|$ **(b)** $|i|$ **(c)** $|(2 + i)(1 - i)|$ **(d)** $\left|\left(\dfrac{1}{\sqrt{2}} + \dfrac{1}{\sqrt{2}}i\right)^{50}\right|$

Solutions:

(a) According to the definition, we have

$$|2 - 9i| = \sqrt{2^2 + (-9)^2} = \sqrt{4 + 81} = \sqrt{85}$$

(b) In this case,

$$|i| = |0 + 1i| = \sqrt{0^2 + 1^2} = \sqrt{1} = 1$$

(c) Using part 2 of the previous theorem, we get

$$\begin{aligned}|(2+i)(1-i)| &= |2+i|\,|1-i| \\ &= \sqrt{2^2+1^2}\sqrt{1^2+(-1)^2} \\ &= \sqrt{5}\sqrt{2} \\ &= \sqrt{10}\end{aligned}$$

(d) According to part 4 of the previous theorem, we have

$$\left|\left(\frac{1}{\sqrt{2}}+\frac{1}{\sqrt{2}}i\right)^{50}\right| = \left|\frac{1}{\sqrt{2}}+\frac{1}{\sqrt{2}}i\right|^{50}$$

But

$$\left|\frac{1}{\sqrt{2}}+\frac{1}{\sqrt{2}}i\right| = \sqrt{\frac{1}{2}+\frac{1}{2}} = \sqrt{1} = 1$$

and so

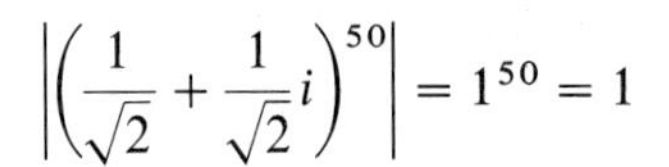

$$\left|\left(\frac{1}{\sqrt{2}}+\frac{1}{\sqrt{2}}i\right)^{50}\right| = 1^{50} = 1$$

Try Study Suggestion 10.8. □

▶ **Study Suggestion 10.8:** Compute the following absolute values.
(a) $|2-4i|$ **(b)** $|-i|$ **(c)** $|(1+i)^{20}|$ ◀

Ideas to Remember

Just as the real numbers can be represented by the points on the (real) number line, the complex numbers can be represented by the points in the (complex) plane.

EXERCISES

In Exercise 1–4, find $\bar{z}$.

1. $z = 2 + 3i$

2. $z = -1 - i$

3. $z = i^3$

4. $z = 2z + 3$

In Exercises 5–8, find z.

5. $\bar{z} = -i$

6. $\bar{z} = 2 + 3i$

7. $\bar{z} = -i/2 - 7$

8. $\bar{z} = 3\bar{z} - 2$

In Exercises 9–19, find the absolute value of the given complex number.

9. $1 - i$

10. i

11. $3 - \sqrt{2}i$

12. i^{10}

13. $(1 - 3i)^{20}$

14. $(2 + i)(1 - i)^3$

15. $\dfrac{(2+i)^2}{3-i}$

16. $1 - i^{10}$

17. i^{-100}

18. $\dfrac{1}{2+i} - \dfrac{1}{1-i}$

19. $\dfrac{1}{1-i} + \dfrac{1}{2-i}$

In Exercises 20–23, find the distance from the given point in the complex plane to the origin.

20. $3 + 2i$

21. $-i - 1$

22. $-\dfrac{1}{2} - \dfrac{\sqrt{3}}{2}i$

23. -6

In Exercises 24–39, write the expression in the form $a + bi$.

24. $\dfrac{1}{2i}$

25. $\dfrac{1}{5 - 6i}$

26. $\dfrac{1}{3 + 2i}$

27. $\dfrac{7}{3 - 4i}$

28. $\dfrac{2 + i}{5 - i}$

29. $\dfrac{3 - 2i}{3 + 2i}$

30. $\dfrac{1 - \pi i}{i}$

31. $\dfrac{i}{3 + \sqrt{2}i}$

32. $\dfrac{1}{-i - 1} + \dfrac{2}{3i}$

33. $\dfrac{2 + 3i}{1 - 4i} + \dfrac{1 - i}{2 - i}$

34. $\dfrac{1 + i}{2 - i} - \dfrac{1 - i}{4 - 2i}$

35. $1 + \dfrac{1}{i} + \dfrac{1}{i^2}$

36. $1 - \dfrac{1}{i} + \dfrac{1}{i^2} - \dfrac{1}{i^3}$

37. $\dfrac{(2 - i)^2}{2 + i}$

38. $\dfrac{1 + i^2}{(1 + i)^2}$

39. $\dfrac{i}{(1 + 2i)^2}$

In Exercises 40–47, solve for z. (*Hint:* write $z = a + bi$.)

40. $z = \bar{z}$

41. $z = i\bar{z}$

42. $z = \bar{z} + 1$

43. $z\bar{z} = 0$

44. $z\bar{z} = -1$

45. $z = |z|$

46. $2z + i = 5iz + 3$

47. $3iz + 2 = -i + z$

48. **(a)** Show that $|\bar{z}| = |z|$, for all complex numbers z.

(b) Show that $|-z| = |z|$ for all complex numbers z.

(c) Show that $|z| = 0$ if and only if $z = 0$.

(*Hint:* write $z = a + bi$.)

49. Show that $|zw| = |z|\,|w|$ for all complex numbers z and w. (*Hint:* write $z = a + bi$ and $w = c + di$.)

50. **(a)** Show that $|1/z| = 1/|z|$ for all complex numbers $z \neq 0$.

(b) Show that $|z/w| = |z|/|w|$ for all complex numbers z and w with $w \neq 0$.

(*Hint*: for part (a), use the fact that $|z \cdot (1/z)| = |z|\,|1/z|$.)

51. **(a)** Show that $\overline{z - w} = \bar{z} - \bar{w}$ for all complex numbers z and w.

(b) Is it true that $|z - w| = |z| - |w|$ for all complex numbers z and w? Justify your answer.

10.3 Complex Solutions to Quadratic Equations

Now let us turn our attention to solving quadratic equations. Up to now, in using the quadratic formula

$$x = \frac{-b \pm \sqrt{b^2 - 4ac}}{2a}$$

whenever we encountered the situation of having to take the square root of a negative number, we have simply said that there were no (real number) solutions to the equation.

But now that we have the complex number system at our disposal, we can take the square root of a negative number. Hence, there is nothing standing in the way of our using the quadratic formula to find the solutions to *any* quadratic equation. (We can even use the quadratic formula when the coefficients are complex numbers. However, our main interest is in solving quadratic equations with real coefficients.)

EXAMPLE 1: Solve the quadratic equation

$$x^2 - 2x + 10 = 0$$

Solution: According to the quadratic formula, we have

$$\begin{aligned} x &= \frac{-b \pm \sqrt{b^2 - 4ac}}{2a} \\ &= \frac{2 \pm \sqrt{4 - 40}}{2} \\ &= \frac{2 \pm \sqrt{-36}}{2} \\ &= \frac{2 \pm 6i}{2} \\ &= 1 \pm 3i \end{aligned}$$

Thus, the solutions to this quadratic equation are $x = 1 + 3i$ and $x = 1 - 3i$. We suggest that you verify this by substitution.

▶ **Study Suggestion 10.9:** Solve the quadratic equation $x^2 + 2x + 26 = 0$ ◀

Try Study Suggestion 10.9. □

EXAMPLE 2: Solve the quadratic equation

$$x^2 + x + 2 = 0$$

Solution: Again we use the quadratic formula

$$\begin{aligned} x &= \frac{-b \pm \sqrt{b^2 - 4ac}}{2a} \\ &= \frac{-1 \pm \sqrt{1 - 8}}{2} \\ &= \frac{-1 \pm \sqrt{-7}}{2} \\ &= \frac{-1 \pm \sqrt{7}i}{2} \\ &= -\frac{1}{2} \pm \frac{\sqrt{7}}{2}i \end{aligned}$$

Thus, the solutions to this equation are $x = -(1/2) + (\sqrt{7}/2)i$ and $x = -(1/2) - (\sqrt{7}/2)i$.

▶ **Study Suggestion 10.10:** Solve the quadratic equation $x^2 + x + 1 = 0$ ◀

Try Study Suggestion 10.10. □

You may have noticed that, in both of the preceding examples, the solutions came in conjugate pairs. In particular, $1 + 3i$ and $1 - 3i$ are conjugates, and so are $-(1/2) + (\sqrt{7}/2)i$ and $-(1/2) - (\sqrt{7}/2)i$.

As we will see in a moment, these examples are typical. That is, if a quadratic equation with *real* coefficients has one complex solution with nonzero imaginary part, call it z, then the other solution is the conjugate $\bar{z}$. The following theorem describes the situation more fully.

THEOREM

The solutions to the quadratic equation

$$ax^2 + bx + c = 0 \tag{10.5}$$

where a, b, and c are real numbers and $a \neq 0$ are given by the *quadratic formula*

$$x = \frac{-b \pm \sqrt{b^2 - 4ac}}{2a}.$$

Furthermore, the *discriminant* $b^2 - 4ac$ gives us the following information about these solutions.

1. If $b^2 - 4ac > 0$, then Equation (10.5) has two distinct solutions, both of which are real numbers.
2. If $b^2 - 4ac = 0$, then Equation (10.5) has exactly one solution (a double root), which is a real number.
3. If $b^2 - 4ac < 0$, then Equation (10.5) has two distinct solutions, each of which is a complex number with a nonzero imaginary part. Furthermore, the solutions are conjugates of one another.

Proof: We have known for some time now that the first two parts of this theorem are true, so let us prove the third part. According to the quadratic formula, the solutions to the quadratic equation are

$$x = \frac{-b \pm \sqrt{b^2 - 4ac}}{2a} \tag{10.6}$$

Now, since we are assuming that the discriminant $b^2 - 4ac$ is negative, we can write

$$b^2 - 4ac = -(4ac - b^2)$$

where $4ac - b^2$ is *positive*. Substituting this into Equation 10.6 gives

$$\begin{aligned} x &= \frac{-b \pm \sqrt{b^2 - 4ac}}{2a} \\ &= \frac{-b \pm \sqrt{-(4ac - b^2)}}{2a} \\ &= \frac{-b \pm \sqrt{4ac - b^2}i}{2a} \\ &= -\frac{b}{2a} \pm \frac{\sqrt{4ac - b^2}}{2a}i \end{aligned}$$

But since $4ac - b^2$ is positive, its square root is a positive real number. Hence, these two solutions are complex numbers with nonzero imaginary parts, and they are also conjugates of one another. This proves the third part of the theorem. ■

We should point out that this theorem is true *only* if the coefficients of the quadratic equation are *real* numbers, even though the quadratic formula holds for quadratic equations whose coefficients are complex numbers.

Also, we should mention that whenever *any* polynomial equation with real coefficients, regardless of its degree, has a solution z with nonzero imaginary part, then the conjugate $\bar{z}$ is also a solution. For example, consider the cubic equation

$$5x^3 + 2x^2 - 3x + 1 = 0 \tag{10.7}$$

Then if z is a complex solution to this equation, we have

$$5z^3 + 2z^2 - 3z + 1 = 0$$

Now we can take the conjugate of both sides of this equation, and use the properties of conjugates to get

$$\overline{5z^3 + 2z^2 - 3z + 1} = \bar{0}$$

or

$$\bar{5}(\bar{z})^3 + \bar{2}(\bar{z})^2 - \bar{3}\bar{z} + \bar{1} = \bar{0}$$

But the conjugate of a real number is just the number itself, and so we have

$$5(\bar{z})^3 + 2(\bar{z})^2 - 3\bar{z} + 1 = 0$$

This shows that the conjugate $\bar{z}$ is also a solution to Equation (10.7). Furthermore, a similar argument will work for any polynomial equation, provided that it has *real* coefficients.

EXAMPLE 3: Solve the equation

$$x^4 + 3x^2 - 4 = 0 \tag{10.8}$$

Solution: We begin by making the substitution $y = x^2$, to get

$$y^2 + 3y - 4 = 0$$

or

$$(y - 1)(y + 4) = 0$$

The solutions to this equation are $y = 1$ and $y = -4$, and so the solutions to Equation (10.8) are the solutions to

$$x^2 = 1 \qquad \text{and} \qquad x^2 = -4$$

which are $x = \pm\sqrt{1} = \pm 1$ and $x = \pm\sqrt{-4} = \pm 2i$.

▶ **Study Suggestion 10.11:** Solve the equation $x^4 + 3x^2 + 2 = 0$ ◀

Try Study Suggestion 10.11. □

Given a complex number z and its conjugate $\bar{z}$, it is a simple matter to find a quadratic equation with *real* coefficients whose solutions are these two complex numbers. The quadratic equation

$$(x - z)(x - \bar{z}) = 0$$

has solutions $x = z$ and $x = \bar{z}$, and if we take the product on the left side, we get

$$x^2 - (z + \bar{z})x + z\bar{z} = 0$$

Now we already know from Section 10.2 that $z\bar{z}$ is a real number, and we will leave it to you to show that $z + \bar{z}$ is also real, even if z is not (Exercise 40). Hence this quadratic equation actually does have real coefficients.

EXAMPLE 4: Find a quadratic equation with real coefficients whose solutions are $z = 3 + 4i$ and $\bar{z} = 3 - 4i$.

Solution: According to our discussion, the quadratic equation

$$[x - (3 + 4i)][x - (3 - 4i)] = 0 \tag{10.9}$$

is such an equation. Taking the product on the left-hand side and simplifying gives

$$x^2 - 6x + 25 = 0$$

Try Study Suggestion 10.12. □

Study Suggestion 10.12: Find a quadratic equation with real coefficients one of whose solutions is $2 + 3i$. How many such quadratic equations are there?

Ideas to Remember

The quadratic formula gives all solutions—both real and complex—to any quadratic equation, and the discriminant tells us the number and type of the solutions.

EXERCISES

In Exercises 1–30, solve the given equation.

1. $4x^2 + 1 = 0$

2. $x^2 + x - 2 = 0$

3. $x^2 + 4 = 0$

4. $4x^2 - 1 = 0$

5. $x^2 - 2x + 2 = 0$

6. $x^2 + 4x + 13 = 0$

7. $2x^2 - 7x + 3 = 0$

8. $3x^2 + 4x - 7 = 0$

9. $36x^2 - 36x + 13 = 0$

10. $x^2 + 10 = 0$

11. $x^2 + 1/3 = 0$

12. $x^2 + \pi x - 2\pi^2 = 0$

13. $x^2 + \pi^2 = 0$

14. $9x^2 + 6x + 10 = 0$

15. $(x - 2)^2 = -5$

16. $x^2 + \sqrt{5} = 0$

17. $x^2 - 2x + 6 = 0$

18. $4x^2 + 8x + 7 = 0$

19. $2\sqrt{2}x^2 - 4x + 9\sqrt{2} = 0$

20. $144x^2 - 96x + 25 = 0$

21. $9x^2 - 4x + 1 = 0$

22. $4x^2 + 4\sqrt{3}x + 7 = 0$

23. $(x^2 - 3)^2 = 1$

24. $x^4 + x^2 - 12 = 0$

25. $24x^4 + 31x^2 + 10 = 0$

26. $4x^4 + 25x^2 + 36 = 0$

27. $5x^4 - 11x^2 - 2 = 0$

28. $x^3 + 9x = 0$

29. $x^3 - 3x^2 + 7x - 5 = 0$

30. $x^3 + x^2 + x = 0$

In Exercises 31–39, find a quadratic equation with *real* coefficients that has the given complex number as a solution.

31. i

32. $2 - 3i$

33. $3 - \frac{1}{2}i$

34. $\pi + \sqrt{2}i$

35. $\frac{1}{i}$

36. $-\sqrt{2}i(1 + i)$

37. $1 - \frac{1}{i}$

38. $\frac{1 + i}{2 - i}$

39. 2

40. Show that for all complex numbers z, the quantity $z + \bar{z}$ is real. What can you say about $z - \bar{z}$?

10.4 The Trigonometric Form of a Complex Number

In Section 10.2 we saw how complex numbers can be represented as points in the plane. In this section, we will use this representation to express complex numbers in terms of the trigonometric functions.

Consider the complex number $z = a + bi$, pictured in Figure 10.3. We have drawn a right triangle using the point z as a vertex, and labeled it in the usual way (except that the length of the hypotenuse is labeled r rather than c).

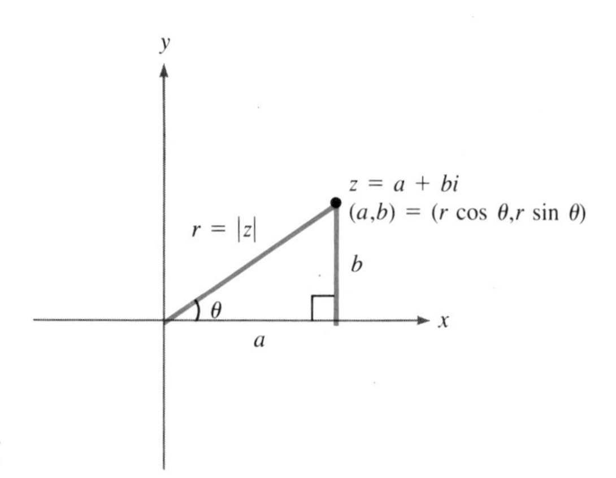

FIGURE 10.3

As you can see from this figure, the point (a, b), which corresponds to the complex number z, also has coordinates $(r\cos\theta, r\sin\theta)$. In other words, $a = r\cos\theta$ and $b = r\sin\theta$, and so

$$z = a + bi = (r\cos\theta) + (r\sin\theta)i = r(\cos\theta + i\sin\theta)$$

(In order to avoid any possible confusion, it is customary to place the i in front of $\sin\theta$.) Thus, the complex number z can be written in the form

$$z = r(\cos\theta + i\sin\theta) \tag{10.10}$$

where $r = |z|$ is the modulus of z and θ is the angle that the line segment from the origin to the point z makes with the positive real axis. The same formula holds if z lies anywhere in the complex plane and not just in the first quadrant.

Equation (10.10) is called a **trigonometric form,** or **polar form,** of the complex number z. The angle θ is called an **argument** of z, and the number r is the modulus of z. (In this context, the term modulus is much more common than the term absolute value, although both terms are correct.)

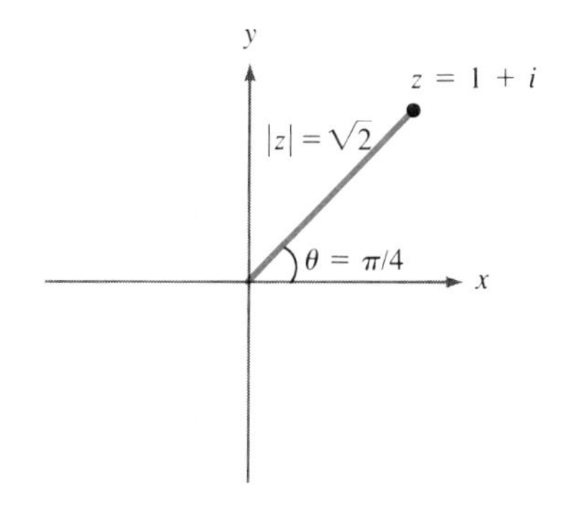

FIGURE 10.4

EXAMPLE 1: Find a trigonometric form of the complex number $z = 1 + i$.

Solution: The number $z = 1 + i$ is shown in Figure 10.4. The modulus of z is easily computed to be

$$|z| = |1 + i| = \sqrt{1^2 + 1^2} = \sqrt{2}$$

and, as you can see from the figure, an argument of z is $\pi/4$. Therefore, a trigonometric form of z is

$$z = \sqrt{2}\left(\cos\frac{\pi}{4} + i\sin\frac{\pi}{4}\right)$$

This can be easily checked by computing $\cos\pi/4$ and $\sin\pi/4$ and then substituting into the right-hand side of this expression.

▶ **Study Suggestion 10.13:** Find a trigonometric form of the complex number $z = -1 + i$. ◀

Try Study Suggestion 10.13. □

A complex number has many different arguments, since if θ is one argument of z, then so is $\theta + 2\pi n$, where n is any integer. As a result, a complex number can be written in trigonometric form in many different ways.

For example, since $\pi/4$ is one argument of the complex number $z = 1 + i$, all angles of the form

$$\frac{\pi}{4} + 2\pi n$$

where n is any integer, are also arguments of z. Thus, we can express z in the trigonometric form

$$z = \sqrt{2}\left[\cos\left(\frac{\pi}{4} + 2\pi n\right) + i\sin\left(\frac{\pi}{4} + 2\pi n\right)\right]$$

where n is any integer.

However, if z is any complex number, then it has *only one* argument θ satisfying the inequalities.

$$-\pi < \theta \leq \pi$$

This argument is called the **principal value** of the argument of z, and we will use it whenever we express a complex number in trigonometric form. In fact,

▶ **Study Suggestion 10.14:** Find all of the arguments of the complex number $z = \sqrt{3}/2 + (1/2)i$. What is the principal value of the argument? ◀

from now on, *whenever we use the term argument, we will be referring to the principal value of the argument.*

Try Study Suggestion 10.14.

Looking at Figure 10.3, we can see that if θ is *any* argument of $z = a + bi$, and if $a \neq 0$, then

$$\tan\theta = \frac{b}{a}$$

This fact can be of great help in determining the argument of a complex number (although it does not determine it completely).

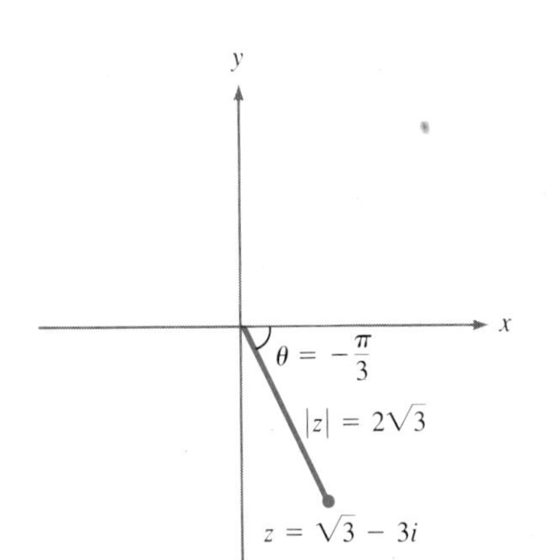

FIGURE 10.5

EXAMPLE 2: Express the following complex numbers in trigonometric form.

(a) $\sqrt{3} - 3i$ **(b)** $1 + 5i$

Solutions:

(a) The number $z = \sqrt{3} - 3i$ is pictured in Figure 10.5. The modulus of z is easily computed as follows

$$|z| = |\sqrt{3} - 3i| = \sqrt{(\sqrt{3})^2 + 3^2} = \sqrt{12} = 2\sqrt{3}$$

In order to compute the argument of z, we use the fact that

$$\tan\theta = \frac{b}{a} = \frac{-3}{\sqrt{3}} = -\sqrt{3}$$

Therefore, since θ is clearly between 0 and $-\pi/2$, we have $\theta = -\pi/3$. (This is the only angle in the fourth quadrant whose tangent is equal to $-\sqrt{3}$.) Thus, the trigonometric form of z is

$$z = 2\sqrt{3}\left(\cos\left(-\frac{\pi}{3}\right) + i\sin\left(-\frac{\pi}{3}\right)\right)$$

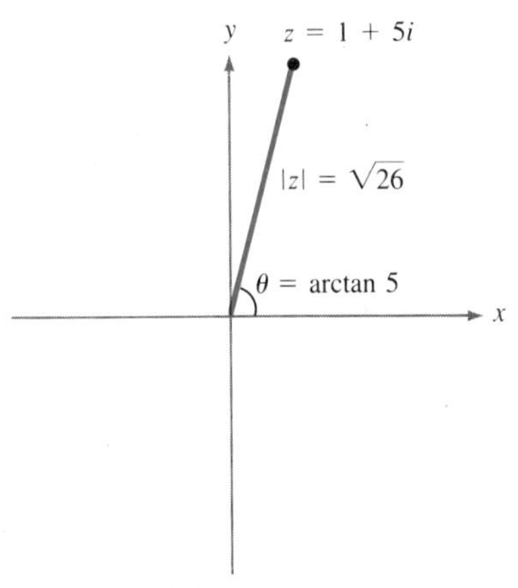

FIGURE 10.6

(b) In this case, we have $|z| = |1 + 5i| = \sqrt{26}$ and $\theta = \arctan 5$ (see Figure 10.6) and so

$$1 + 5i = \sqrt{26}\,[\cos(\arctan 5) + i\sin(\arctan 5)]$$

Try Study Suggestion 10.15. □

▶ **Study Suggestion 10.15:** Find a trigonometric form of the following complex numbers.

(a) $z = -2 + 2\sqrt{3}i$

(b) $z = -1 + 2i$ ◀

As you can see from Part (b) of the previous example, the trigonometric form of a complex number is not always simple. Why, then, are we interested in this form?

The answer is that multiplication and division of complex numbers can sometimes be easier to perform when all numbers are expressed in trigonometric form. Let us consider multiplication first. Suppose that we want to take the product zw, where

$$z = r(\cos\theta + i\sin\theta)$$

and

$$w = s(\cos\mu + i\sin\mu)$$

are complex numbers, expressed in trigonometric form (μ is the Greek letter mu). Then we may use the addition formulas for the sine and cosine function to obtain

$$\begin{aligned} zw &= r(\cos\theta + i\sin\theta)s(\cos\mu + i\sin\mu) \\ &= rs(\cos\theta\cos\mu + i\cos\theta\sin\mu + i\sin\theta\cos\mu - \sin\theta\sin\mu) \\ &= rs[(\cos\theta\cos\mu - \sin\theta\sin\mu) + i(\cos\theta\sin\mu + \sin\theta\cos\mu)] \\ &= rs[\cos(\theta+\mu) + i\sin(\theta+\mu)] \end{aligned}$$

Thus, the product zw is given by

$$zw = rs[\cos(\theta+\mu) + i\sin(\theta+\mu)] \tag{10.11}$$

This formula tells us that

> **in order to multiply two complex numbers that are expressed in trigonometric form, we multiply their moduli and *add* their arguments.**

EXAMPLE 3: Use trigonometric forms to compute the product of the complex numbers $z = 1 + i$ and $w = \sqrt{3} - 3i$.

Solution: We have already shown that the trigonometric forms of these numbers are

$$1 + i = \sqrt{2}\left(\cos\frac{\pi}{4} + i\sin\frac{\pi}{4}\right)$$

and

$$\sqrt{3} - 3i = 2\sqrt{3}\left(\cos\left(-\frac{\pi}{3}\right) + i\sin\left(-\frac{\pi}{3}\right)\right)$$

Thus, since $\sqrt{2}\cdot 2\sqrt{3} = 2\sqrt{6}$ and $\pi/4 + (-\pi/3) = -\pi/12$, Equation (10.11) tells us that

$$(1+i)(\sqrt{3} - 3i) = 2\sqrt{6}\left[\cos\left(-\frac{\pi}{12}\right) + i\sin\left(-\frac{\pi}{12}\right)\right]$$

Try Study Suggestion 10.16. □

► **Study Suggestion 10.16:** Use trigonometric forms to compute the product $(1 - i)(-1 + \sqrt{3}i)$. ◄

Of course, there is a formula analogous to (10.11) for division of complex numbers. Let us state both formulas in a theorem.

THEOREM

For any two complex numbers

$$z = r(\cos\theta + i\sin\theta) \qquad \text{and} \qquad w = s(\cos\mu + i\sin\mu)$$

that are expressed in trigonometric form, we have

1. $zw = rs[\cos(\theta + \mu) + i\sin(\theta + \mu)]$
2. $\dfrac{z}{w} = \dfrac{r}{s}[\cos(\theta - \mu) + i\sin(\theta - \mu)]$ (for $w \neq 0$)

Part 2 of this theorem says that

in order to divide one complex number by another, we divide their moduli and *subtract* their arguments.

This formula can be proved using the subtraction formulas for the sine and cosine.

EXAMPLE 4: Compute the product zw and the quotient z/w if

$$z = 8\left(\cos\frac{2\pi}{3} + i\sin\frac{2\pi}{3}\right)$$

and

$$w = 2\left(\cos\frac{3\pi}{4} + i\sin\frac{3\pi}{4}\right)$$

Solution: According to part 1 of our theorem

$$\begin{aligned} zw &= 8\cdot 2\left[\cos\left(\frac{2\pi}{3} + \frac{3\pi}{4}\right) + i\sin\left(\frac{2\pi}{3} + \frac{3\pi}{4}\right)\right] \\ &= 16\left(\cos\frac{17\pi}{12} + i\sin\frac{17\pi}{12}\right) \\ &= 16\left(\cos\left(-\frac{7\pi}{12}\right) + i\sin\left(-\frac{7\pi}{12}\right)\right) \end{aligned}$$

(Remember that the principal value of the argument lies between $-\pi$ and π, and this is why we have replaced $17\pi/12$ with $17\pi/12 - 2\pi = -7\pi/12$, which is in the proper range.) Part 2 of our theorem gives

$$\begin{aligned} \frac{z}{w} &= \frac{8}{2}\left[\cos\left(\frac{2\pi}{3} - \frac{3\pi}{4}\right) + i\sin\left(\frac{2\pi}{3} - \frac{3\pi}{4}\right)\right] \\ &= 4\left(\cos\left(-\frac{\pi}{12}\right) + i\sin\left(-\frac{\pi}{12}\right)\right) \end{aligned}$$

Try Study Suggestion 10.17. □

▶ **Study Suggestion 10.17:** Use the previous theorem to compute zw and z/w for the complex numbers $z = 1 + i$ and $w = 1 - i$. (You must first find the trigonometric form of these numbers.) ◀

Ideas to Remember

Multiplying and dividing complex numbers can sometimes be made easier when the numbers are expressed in trigonometric form.

EXERCISES

In Exercises 1–8, change the given complex number from trigonometric form to the form $a + bi$.

1. $3\left(\cos\frac{\pi}{2} + i\sin\frac{\pi}{2}\right)$

2. $10\left(\cos\frac{\pi}{3} + i\sin\frac{\pi}{3}\right)$

3. $5(\cos 45^\circ + i\sin 45^\circ)$

4. $\cos\left(-\frac{\pi}{4}\right) + i\sin\left(-\frac{\pi}{4}\right)$

5. $2(\cos\pi + i\sin\pi)$

6. $7(\cos 570^\circ + i\sin 570^\circ)$

7. $4\left(\cos\frac{5\pi}{6} + i\sin\frac{5\pi}{6}\right)$

8. $\cos(-135^\circ) + i\sin(-135^\circ)$

In Exercises 9–29, find the trigonometric form for the given complex number.

9. $-1 - i$ **10.** i **11.** $-i$

12. $2 + 2i$ **13.** $1 + \sqrt{3}i$ **14.** $3 + 4i$

15. $\frac{3}{2} + i$ **16.** $\frac{\sqrt{3}}{2} + \frac{i}{2}$ **17.** 8

18. $\pi + \sqrt{2}$ **19.** $-9 + 9i$ **20.** $-3\sqrt{3} + 3i$

21. 0 **22.** -1 **23.** πi

24. $i(1 - \sqrt{3}i)$ **25.** $-83(3 + 4i)$ **26.** 5260

27. $2 + i$ **28.** $\frac{\sqrt{3}}{2}$ **29.** $\frac{i}{324}$

In Exercises 30–33, find the product zw and the quotient z/w. Leave the results in polar form.

30. $z = 3\left(\cos\frac{\pi}{4} + i\sin\frac{\pi}{4}\right)$, $w = \cos\frac{\pi}{3} + i\sin\frac{\pi}{3}$

31. $z = \cos 15^\circ + i\sin 15^\circ$, $w = 4(\cos 2^\circ + i\sin 2^\circ)$

32. $z = \cos\left(-\frac{\pi}{4}\right) + i\sin\left(-\frac{\pi}{4}\right)$, $w = \cos\left(-\frac{\pi}{2}\right) + i\sin\left(-\frac{\pi}{2}\right)$

33. $z = 9(\cos\pi + i\sin\pi)$, $w = \sqrt{2}\left(\cos\frac{\pi}{2} + i\sin\frac{\pi}{2}\right)$

In Exercises 34–41, use trigonometric forms to compute the product zw and the quotient z/w. Express the results in the form $a + bi$.

34. $z = 4$, $w = -3$

35. $z = 2i$, $w = 1 + i$

36. $z = -i$, $w = \sqrt{2}i$

37. $z = 1 - i$, $w = 1 + i$

38. $z = -1 - \sqrt{3}i$, $w = -5i$

39. $z = \sqrt{3} - i$, $w = -\sqrt{3} - i$

40. $z = -4 - 4i$, $w = -5 - 5\sqrt{3}i$

41. $z = i$, $w = i^2$

42. **(a)** Extend Part 1 of the theorem of this section to the case of three complex numbers.

(b) Use the formula that you obtained in Part (a) to compute the product $(1 - i)\left(\frac{1}{2} + \frac{\sqrt{3}}{2}i\right)(\sqrt{3} - i)$. Leave the answer in trigonometric form.

43. If $z = r(\cos\theta + i\sin\theta)$, find a formula for z^2 and z^3. Can you generalize this to a formula for z^n, where n is any positive integer?

44. Prove Part 2 of the theorem of this section.

10.5 De Moivre's Formula and the Roots of a Complex Number

The theorem of the previous section can be used to obtain a very useful formula for the powers of a complex number. For example, if $z = r(\cos\theta + i\sin\theta)$, then according to that theorem we have

$$\begin{aligned} z^2 = z \cdot z &= [r(\cos\theta + i\sin\theta)][r(\cos\theta + i\sin\theta)] \\ &= r^2[\cos(\theta + \theta) + i\sin(\theta + \theta)] \\ &= r^2(\cos 2\theta + i\sin 2\theta) \end{aligned}$$

and

$$\begin{aligned} z^3 = z^2 \cdot z &= [r^2(\cos 2\theta + i\sin 2\theta)][r(\cos\theta + i\sin\theta)] \\ &= r^3(\cos 3\theta + i\sin 3\theta) \end{aligned}$$

Even now a pattern is beginning to emerge, and we are lead to the following theorem.

THEOREM

Let $z = r(\cos\theta + i\sin\theta)$ be a complex number, expressed in trigonometric form. Then for all positive integers n, we have

$$z^n = r^n(\cos n\theta + i\sin n\theta)$$

We can also write this in the form

$$[r(\cos\theta + i\sin\theta)]^n = r^n[\cos n\theta + i\sin n\theta] \qquad \textbf{(10.12)}$$

Equation (10.12) is known as **De Moivre's formula,** named after the French mathematician Abraham De Moivre, who discovered it in the early 1700s. In words, De Moivre's formula says that

in order to take the nth power of a complex number that is expressed in trigonometric form, take the nth power of its modulus, and multiply its argument by n.

EXAMPLE 1: Express the following complex numbers in the form $a + bi$.

(a) $(1 + i)^{10}$ **(b)** $(\sqrt{3} - 3i)^5$

Solutions:

(a) In the previous section, we determined that

$$1 + i = \sqrt{2}\left(\cos\frac{\pi}{4} + i\sin\frac{\pi}{4}\right)$$

and so, according to De Moivre's formula,

$$\begin{aligned}(1+i)^{10} &= (\sqrt{2})^{10}\left(\cos\frac{10\pi}{4} + i\sin\frac{10\pi}{4}\right)\\ &= 2^5\left(\cos\frac{5\pi}{2} + i\sin\frac{5\pi}{2}\right)\\ &= 2^5\left(\cos\frac{\pi}{2} + i\sin\frac{\pi}{2}\right)\\ &= 32i\end{aligned}$$

(b) We also determined in the previous section that

$$\sqrt{3} - 3i = 2\sqrt{3}\left[\cos\left(-\frac{\pi}{3}\right) + i\sin\left(-\frac{\pi}{3}\right)\right]$$

and so

$$\begin{aligned}(\sqrt{3} - 3i)^5 &= (2\sqrt{3})^5\left[\cos\left(-\frac{5\pi}{3}\right) + i\sin\left(-\frac{5\pi}{3}\right)\right]\\ &= 32 \cdot 3^{5/2}\left[\cos\frac{\pi}{3} + i\sin\frac{\pi}{3}\right]\\ &= 32 \cdot 3^{5/2}\left[\frac{\sqrt{3}}{2} + \frac{1}{2}i\right]\\ &= 16 \cdot 3^{5/2}(\sqrt{3} + i)\\ &= 432 + 144\sqrt{3}i\end{aligned}$$

Try Study Suggestion 10.18. □

▶ **Study Suggestion 10.18:** Use De Moivre's formula to compute $(2 - 2\sqrt{3}i)^6$. ◀

De Moivre's formula can be used to find the roots of a complex number, as well as its powers. If z is a complex number, and if n is a positive integer, then we say that a complex number w is an ***n*th root** of z if $w^n = z$. Of course, the number 0 has only one nth root, namely, 0 itself. However, it is possible to show that every nonzero complex number has exactly n different nth roots. In fact, we can even derive a formula for these n roots. Let us do that now.

Suppose that we are given a nonzero complex number $z = r(\cos\theta + i\sin\theta)$, and that we want to find its nth roots, that is, we want to find all complex numbers $w = s(\cos\mu + i\sin\mu)$ for which

$$w^n = z$$

Expressing this equation in trigonometric form, we get

$$[s(\cos\mu + i\sin\mu)]^n = r(\cos\theta + i\sin\theta)$$

Applying De Moivre's formula to the left-hand side gives

$$s^n(\cos n\mu + i\sin n\mu) = r(\cos\theta + i\sin\theta) \tag{10.13}$$

But in order for two complex numbers to be equal, their moduli must be equal, and their arguments can differ only by an integral multiple of 2π. Therefore, in order for Equation (10.13) to hold, we must have

$$s^n = r \qquad \text{and} \qquad n\mu - \theta = 2\pi k$$

for some integer k. These equations are equivalent to

$$s = \sqrt[n]{r} \qquad \text{and} \qquad \mu = \frac{\theta + 2\pi k}{n}$$

Thus, the nth roots of $z = r(\cos\theta + i\sin\theta)$ are given by

$$w = \sqrt[n]{r}\left[\cos\left(\frac{\theta + 2\pi k}{n}\right) + i\sin\left(\frac{\theta + 2\pi k}{n}\right)\right] \tag{10.14}$$

where k is any integer.

Now, as k takes on successive values from 0 to $n - 1$, Equation (10.14) gives different complex numbers. However, when k reaches n, the numbers given in (10.14) begin to repeat. For example, when $k = n$, we get

$$\begin{aligned} w &= \sqrt[n]{r}\left[\cos\left(\frac{\theta + 2\pi n}{n}\right) + i\sin\left(\frac{\theta + 2\pi n}{n}\right)\right] \\ &= \sqrt[n]{r}\left[\cos\left(\frac{\theta}{n} + 2\pi\right) + i\sin\left(\frac{\theta}{n} + 2\pi\right)\right] \\ &= \sqrt[n]{r}\left[\cos\left(\frac{\theta}{n}\right) + i\sin\left(\frac{\theta}{n}\right)\right] \end{aligned}$$

which is the same number that we get when $k = 0$. Similarly, when $k = n + 1$, we will get the same number as when $k = 1$; when $k = n + 2$, we will get the same number as when $k = 2$, and so on. We will also get the same numbers when we take negative values of n.

Thus, Equation (10.13) produces exactly n different complex numbers, one for each of the values $k = 0, 1, \ldots, n - 1$. In other words, there are exactly n different nth roots of the number z, all of which are given by Equation (10.13), for $k = 1, 2, \ldots, n - 1$. Let us state this important fact in a theorem.

THEOREM

Let $z = r(\cos\theta + i\sin\theta)$ be a nonzero complex number, and let n be a positive integer. Then z has *exactly* n different nth roots, given by

$$\sqrt[n]{r}\left[\cos\left(\frac{\theta + 2\pi k}{n}\right) + i\sin\left(\frac{\theta + 2\pi k}{n}\right)\right]$$

for $k = 0, 1, \ldots, n - 1$.

This theorem is one of the most satisfying in all of mathematics. For it shows quite clearly that the complex number system has a certain "completeness" about it that the real number system does not have. For it is certainly not true that every real number has exactly n different *real* nth roots. (For example, the real number -1 has *no* real square roots.)

Now let us consider some examples of the use of our theorem.

EXAMPLE 2: Find the 3 cube roots of the complex number i.

Solution: The trigonometric form of the number i is

$$i = \cos\frac{\pi}{2} + i\sin\frac{\pi}{2}$$

and so, according to our theorem, the cube roots of i are given by

$$\cos\left(\frac{\pi/2 + 2\pi k}{3}\right) + i\sin\left(\frac{\pi/2 + 2\pi k}{3}\right) = \cos\left(\frac{\pi}{6} + \frac{2\pi}{3}k\right) + i\sin\left(\frac{\pi}{6} + \frac{2\pi}{3}k\right)$$

for $k = 0$, 1, and 2. Writing out each root separately, we get

$$\cos\frac{\pi}{6} + i\sin\frac{\pi}{6} = \frac{\sqrt{3}}{2} + \frac{1}{2}i$$

$$\cos\left(\frac{\pi}{6} + \frac{2\pi}{3}\right) + i\sin\left(\frac{\pi}{6} + \frac{2\pi}{3}\right) = \cos\frac{5\pi}{6} + i\sin\frac{5\pi}{6} = -\frac{\sqrt{3}}{2} + \frac{1}{2}i$$

and

$$\cos\left(\frac{\pi}{6} + \frac{2\pi}{3}\cdot 2\right) + i\sin\left(\frac{\pi}{6} + \frac{2\pi}{3}\cdot 2\right) = \cos\frac{3\pi}{2} + i\sin\frac{3\pi}{2} = -i$$

▶ **Study Suggestion 10.19:** Find the 4 fourth roots of i. ◀

These values can easily be checked by cubing them to get i.

Try Study Suggestion 10.19. □

EXAMPLE 3: Find the 4 fourth roots of the complex number $z = -8(1 + \sqrt{3}i)$.

Solution: The trigonometric form of z is

$$z = 16\left(\cos\frac{4\pi}{3} + i\sin\frac{4\pi}{3}\right)$$

(See Figure 10.7.)

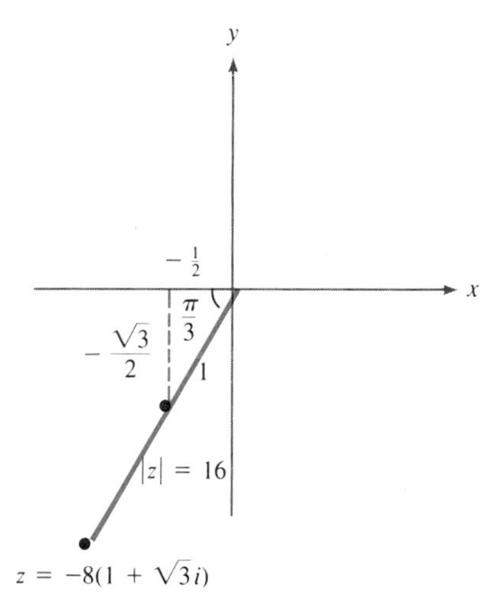

FIGURE 10.7

Therefore, according to our theorem, the 4 fourth roots of z are

$$2\left[\cos\left(\frac{(4/3)\pi + 2\pi k}{4}\right) + i\sin\left(\frac{(4/3)\pi + 2\pi k}{4}\right)\right]$$

$$= 2\left[\cos\left(\frac{\pi}{3} + \frac{\pi}{2}k\right) + i\sin\left(\frac{\pi}{3} + \frac{\pi}{2}k\right)\right]$$

where $k = 0, 1, 2,$ and 3. Writing out each root separately, we get

$$2\left[\cos\frac{\pi}{3} + i\sin\frac{\pi}{3}\right] = 1 + \sqrt{3}i$$

$$2\left[\cos\frac{5\pi}{6} + i\sin\frac{5\pi}{6}\right] = -\sqrt{3} + i$$

$$2\left[\cos\frac{4\pi}{3} + i\sin\frac{4\pi}{3}\right] = -1 - \sqrt{3}i$$

and

$$2\left[\cos\frac{11\pi}{6} + i\sin\frac{11\pi}{6}\right] = \sqrt{3} - i$$

Try Study Suggestion 10.20. □

▶ **Study Suggestion 10.20:** Find the 2 square roots of the number $z = -8(1 + \sqrt{3}i)$. ◀

The nth roots of 1 are especially important in applications, and they are usually referred to as the ***n*th roots of unity.**

EXAMPLE 4: Find the 5 fifth roots of unity.

Solution: The trigonometric form of the complex number 1 is

$$1 = \cos 0 + i\sin 0$$

and so, according to our theorem, the 5 fifth roots of unity are

$$\cos\frac{2\pi}{5}k + i\sin\frac{2\pi}{5}k$$

where $k = 0, 1, 2, 3,$ and 4. Writing these roots separately, we get

$$w_1 = \cos 0 + i\sin 0 = 1$$

$$w_2 = \cos\frac{2\pi}{5} + i\sin\frac{2\pi}{5}$$

$$w_3 = \cos\frac{4\pi}{5} + i\sin\frac{4\pi}{5}$$

$$w_4 = \cos\frac{6\pi}{5} + i\sin\frac{6\pi}{5}$$

and

$$w_5 = \cos\frac{8\pi}{5} + i\sin\frac{8\pi}{5}$$

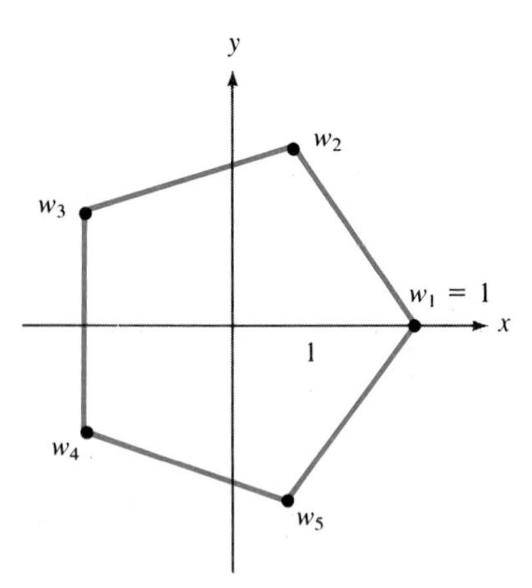

The 5 fifth roots of unity

FIGURE 10.8

We have plotted these five roots of unity in the complex plane in Figure 10.8. As you can see, the first one occurs on the positive real axis, and the others are equally spaced around the unit circle, where the angle separating two adjacent roots is $2\pi/5$ radians. (All of these roots are on the unit circle since they all have modulus equal to 1.) □

It is interesting to observe that, since the 5 fifth roots of unity are equally spaced around the unit circle, they form the vertices of a regular pentagon, one of whose vertices is the number 1. (By *regular*, we mean that all sides of the pentagon have the same length.) Actually, there is nothing special about the fifth roots of unity.

For each integer $n \geq 2$, the n different nth roots of unity are equally spaced around the unit circle, starting with the real number 1. Thus, the angle between adjacent roots is equal to

$$\frac{2\pi}{n}$$

radians, and these roots form the vertices of a regular $n$gon (for $n \geq 3$).

Try Study Suggestions 10.21 and 10.22.

▶ **Study Suggestion 10.21:** Find the 3 cube roots of unity. Then plot them in the same plane, and draw a regular polygon using the roots as vertices. What figure do you get? ◀

▶ **Study Suggestion 10.22:** Find the 4 fourth roots of unity. Then plot them in the same plane, and draw a regular polygon using the roots as vertices. What figure do you get? ◀

Ideas to Remember

- The trigonometric form of a complex number is extremely useful for finding the powers and roots of that number.
- Every nonzero complex number has exactly n different nth roots.

EXERCISES

In Exercises 1–12, use De Moivre's formula to evaluate the given expression.

1. i^{12}

2. $(1 + i)^{20}$

3. $(-7)^3$

4. $(1 - \sqrt{3}i)^3$

5. $(\sqrt{3} + i)^4$

6. $(i - 1)^5$

7. $(-\sqrt{3} - i)^{20}$

8. $\left(\frac{\sqrt{2}}{2} + \frac{\sqrt{2}}{2}i\right)^5$

9. $(1 + \sqrt{3}i)^5$

10. $(\sqrt{5} - \sqrt{15}i)^5$

11. $(\cos 60° + i \sin 60°)^4$

12. $(\cos 15° + i \sin 15°)^8$

13. Find the 2 square roots of i.

14. Find the 4 fourth roots of $-i$.

15. Find the 3 cube roots of $2i$.

16. Find the 4 fourth roots of $1 + i$.

17. Find the 2 square roots of $3 - 3i$.

18. Find the 5 fifth roots of $-3 - \sqrt{3}i$.

19. Find the 3 cube roots of $-2 - 2i$.

20. Find the 6 sixth roots of unity. Then plot them in the same plane, and draw a regular polygon using the roots as vertices. What figure do you get?

21. Find the 7 seventh roots of unity. Then plot them in the same plane, and draw a regular polygon using the roots as vertices. What figure do you get?

22. Find the 8 eighth roots of unity. Then plot them in the same plane, and draw a regular polygon using the roots as vertices. What figure do you get?

In Exercises 23–28, find all complex number solutions to the given equation.

23. $x^6 = 64$

24. $(x - 2)^3 = i.$

25. $ix^3 = -1$

26. $(i + x^2)^2 = -1$

27. $x^4 + 6ix = 0$

28. $x^4 = x$

29. If z is an nth root of unity, show that $1/z$ is also an nth root of unity.

10.6 Review

CONCEPTS FOR REVIEW

Complex number
Complex number system
Conjugate
Absolute value
Modulus
Trigonometric form
Polar form
Argument
Principal value of the argument
De Moivre's formula
Roots of unity

REVIEW EXERCISES

In Exercises 1–10, simplify as much as possible.

1. $(3 + 2i) - (6 - i) + (7 + 3i)$

2. $(3 - 3i)(4 + i) - 3i(5 - i/3)$

3. $i^2(1 - i) + (1 - i)^2$

4. $(\sqrt{2} - \sqrt{-2})^2$

5. $(\sqrt{-3}i - \sqrt{-3})^3$

6. $1 - i - i^2 + i^3 + i^4 - i^5 - i^6 + i^7 + i^8$

7. i^{17398}

8. $\dfrac{i}{3 + i}$

9. $\dfrac{2 + 6i}{3 - 9i}$

10. $\dfrac{1 + i + i^2}{3 - i} + \dfrac{2}{i} - \dfrac{1 - 3i}{1 - i}$

11. Find all real numbers x and y for which $2x^2 - 4y^2i = 6 - 2yi$.

12. Find all real numbers x and y for which $(2x + 3yi)^2 = x + y$.

13. Find all complex numbers z for which $z^2 = -4$.

14. Find all complex numbers z for which $z^2 = 2z$.

15. Find all complex numbers z for which $(z + 2i)^2 = 3$.

16. Find all complex numbers z for which $(z + 2i)(z - 2i) = z^2$.

17. Find the modulus of the complex number

$$\frac{1 + i}{3 - 2i}$$

18. Find the modulus of the complex number

$$\frac{2 + i}{3i} - \frac{1 + i}{i}$$

In Exercises 19–22, find all complex numbers z that satisfy the given equation.

19. $2iz + 1 = 3i - z$

20. $\bar{z} = 2z - 1$

21. $|z - 1| = |z + 1|$

22. $(z - 2i)^2 = (z + 1)(z + i)$

In Exercises 23–26, solve for x.

23. $x^2 + \sqrt{8} = 0$

24. $x^4 = \sqrt{5}$

25. $3\sqrt{3}x^2 - 6x + 28\sqrt{3} = 0$

26. $400x^2 + 160x + 41 = 0$

In Exercises 27–30, find a quadratic equation with real coefficients that has the given complex number as a solution.

27. $1 - \sqrt{2}i$

28. i^6

29. $1 + \dfrac{1}{2i}$

30. $\dfrac{7 - i}{2 + i}$

In Exercises 31–33, find a quadratic equation whose roots are the given complex numbers.

31. $2,\ 3i$

32. $i,\ \dfrac{1}{1+i}$

33. $\dfrac{2+3i}{1-i},\ \dfrac{1+i}{1-i}$

34. Convert the complex number $11[\cos(-\pi/3) + i\sin(-\pi/3)]$ to the form $a + bi$.

35. Convert the complex number $\sqrt{13}(\cos 120° + i\sin 120°)$ to the form $a + bi$.

In Exercises 36–39, find the trigonometric form of the given complex number.

36. $5i$

37. $\pi + \pi i$

38. $-\pi\sqrt{3} + \pi i$

39. $\sqrt{6} + \sqrt{2}i$

In Exercises 40–43, find the product zw and the quotient z/w by using the trigonometric forms of the given complex numbers.

40. $z = \sqrt{2}(\cos 1° + i\sin 1°),\ w = \sqrt{3}(\cos \pi/2 + i\sin \pi/2)$

41. $z = \sqrt{3}/2 + i/2,\ w = 1 - i$

42. $z = i,\ w = 1 + i$

43. $z = 1 + i,\ w = 1 + \sqrt{3}i$

44. Use De Moivre's theorem to compute $(1 + i)^{50}$.

45. Use De Moivre's theorem to compute $\left(-\dfrac{1}{2} - \dfrac{\sqrt{3}}{2}i\right)^8$.

46. Use De Moivre's theorem to compute $\left(-\dfrac{1}{\sqrt{2}} + \dfrac{i}{\sqrt{2}}\right)^{14}$.

47. Find the 5 fifth roots of -10.

48. Find the 4 fourth roots of $-1 - \sqrt{3}i$.

49. Find the 9 ninth roots of unity. Then plot them in the same plane, and draw a regular polygon using the roots as vertices. What figure do you get?

50. Solve the equation $x^5 + i = 0$.

51. Solve the equation $(1 + ix^3)^2 = 1$.

11
CONIC SECTIONS

11.1 Introduction

As we have seen in Chapter 3, a parabola is a very special type of curve. However, it is just one example of a group of curves known as **conic sections,** each of which has some very special properties.

There are four types of conic sections: the circle, the ellipse, the parabola, and the hyperbola. These curves are pictured in Figure 11.1. They are called conic sections because they arise from the intersection of a plane and a cone (with two nappes). (There are also certain types of *degenerate* conic sections,

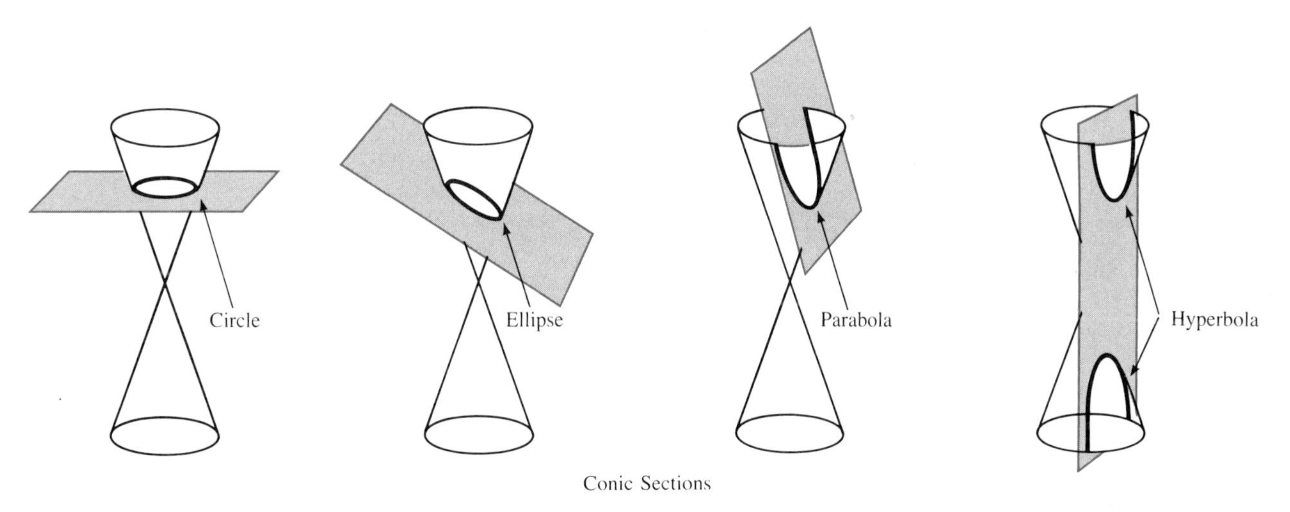

FIGURE 11.1

such as two intersecting straight lines, that we will discuss as they arise in the text.)

Conic sections play an especially important role in the physical world. For example, the planets travel in elliptical orbits around the sun, and most comets seem to travel in parabolic orbits around the sun. (Elliptical and hyperbolic orbits do occur, but parabolic orbits are the most common for comets.)

Each of the conic sections can be described geometrically, as the *locus* (that is, *set*) of all points in the plane that satisfy certain conditions, as well as algebraically, as the graph of a certain equation in two variables. For example, a circle can be described geometrically as the locus of all points in the plane that are a fixed distance (the radius) from a given point (the center). An algebraic description of a circle can be derived from this geometric description by noting that if r is the radius and (h, k) is the center, then according to the distance formula

$$\sqrt{(x-h)^2+(y-k)^2}=r$$

Squaring both sides of this gives

$$(x-h)^2+(y-k)^2=r^2$$

which is the equation of a circle of radius r and center (h, k).

Our plan in this chapter is to discuss the other three conic sections. In each case, we will first give the geometric description of the curve, and then use this description to obtain the algebraic one.

Incidentally, the technique of using algebra to describe geometric concepts is known as *analytic geometry*, and was given its major development by the French mathematicians René Descartes (1596–1650) and Pierre de Fermat (1601–1665). It is interesting to note, however, that the study of conic sections dates back to the mathematician and astronomer Apollonius (ca. 262–190 B.C.), who wrote an eight volume treatise entitled *Conic Sections*, for which he received the title *The Great Geometer*. Apollonius also coined the phrases "ellipse," "parabola," and "hyperbola" for the conic sections. (Actually, he borrowed these terms from the Greeks, who used them for somewhat different purposes.)

Ideas to Remember

Each of the conic sections can be described geometrically, as the locus of all points in the plane satisfying certain conditions, as well as algebraically, as the graph of a certain equation in two variables.

11.2 The Parabola

If F is a point in the plane, and L is a line that does not contain F, then the **distance** $d(F, L)$ from F to L is the length of the line segment from F to L that is perpendicular to L. This is pictured in Figure 11.2. Using this concept, we can give the geometric definition of a parabola.

DEFINITION

> A **parabola** is the locus of all points in the plane that are equidistant from a given point F and a given line L that does not contain F. (See Figure 11.3.)

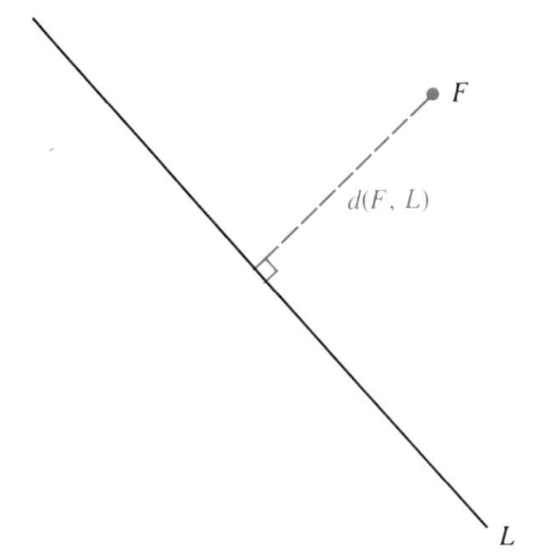

FIGURE 11.2

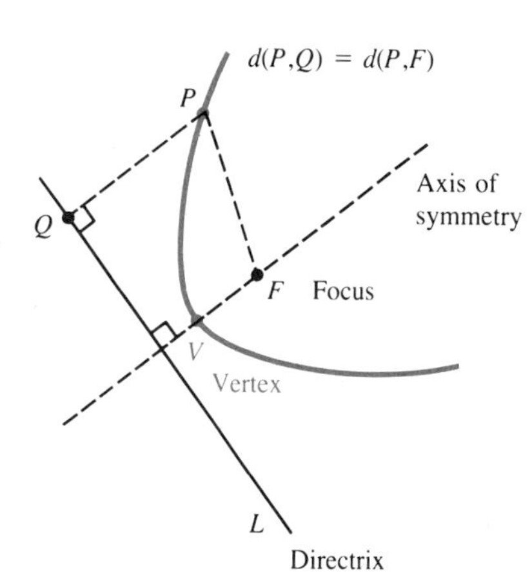

FIGURE 11.3

The point F is called the **focus of the parabola,** and the line L is called the **directrix of the parabola.** The line through the focus F perpendicular to the directrix L is the **axis of symmetry** (or simply **axis**) of the parabola. The point V that lies on both the parabola and the axis of symmetry is called the **vertex** of the parabola. (See Figure 11.3.)

Now let us derive the equation of a parabola whose vertex is at the origin of a Cartesian coordinate system, as illustrated in Figure 11.4a and b. In this case, if we label the focus $F = (0, p)$, then the directrix is the line $y = -p$. (Figure 11.4a shows the case $p > 0$, and Figure 11.4b shows the case $p < 0$.)

In order to derive the equation of this parabola, we pick a "variable" point $P = (x, y)$ on the parabola, and apply the geometric definition of a parabola. The distance from the point $P = (x, y)$ to the focus $F = (0, p)$ is

$$\sqrt{(x - 0)^2 + (y - p)^2}$$

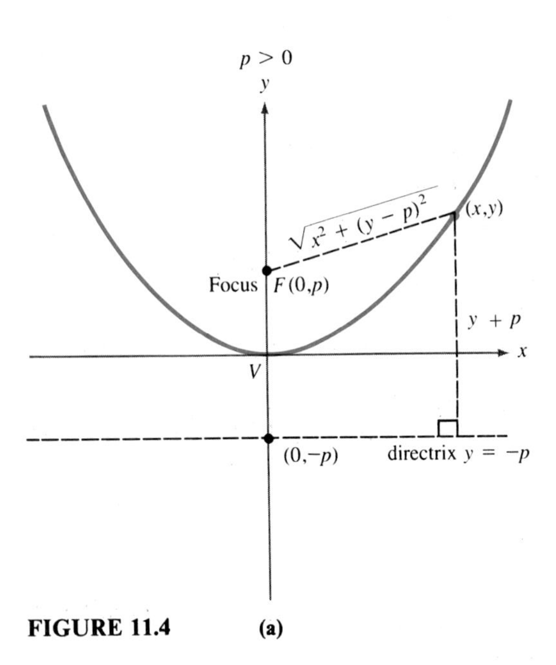

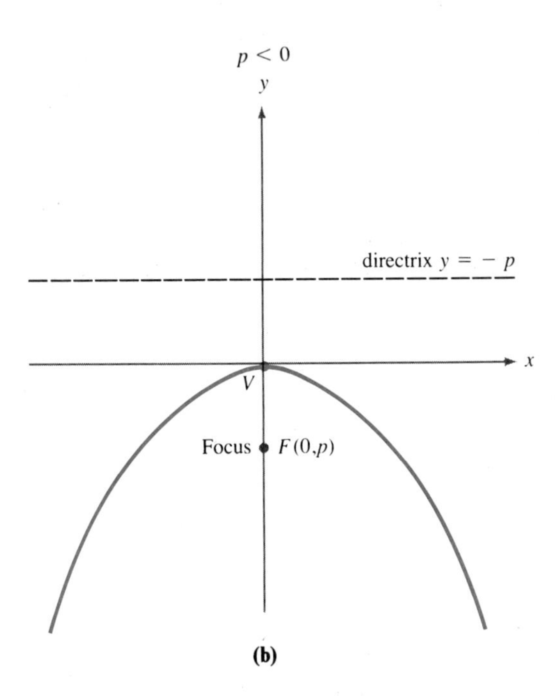

FIGURE 11.4 **(a)** **(b)**

and the distance from the point $P = (x, y)$ to the directrix L is the same as the distance from the point P to the point $(x, -p)$, which is

$$\sqrt{(x - x)^2 + (y + p)^2}$$

Since by definition these two distances are equal, we have

$$\sqrt{(x - 0)^2 + (y - p)^2} = \sqrt{(x - x)^2 + (y + p)^2}$$

Now all we have to do is simplify,

$$\sqrt{x^2 + (y - p)^2} = \sqrt{(y + p)^2}$$

Squaring both sides gives

$$x^2 + (y - p)^2 = (y + p)^2$$
$$x^2 + y^2 - 2py + p^2 = y^2 + 2py + p^2$$
$$x^2 = 4py$$

or

$$y = \frac{1}{4p} x^2 \tag{11.1}$$

A similar calculation can be made in case the vertex is at the origin but the focus is on the x-axis, as shown in Figure 11.5a and b. In this case, we will leave it as an exercise to show that the resulting equation is

$$x = \frac{1}{4p} y^2$$

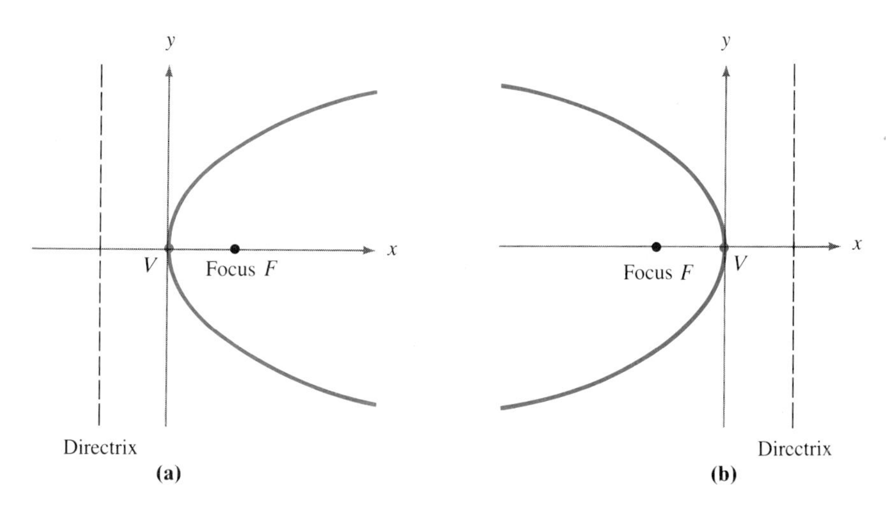

FIGURE 11.5

which is the same as Equation (11.1) except that the roles of x and y are interchanged. Let us summarize our results in a theorem.

THEOREM

1. The equation

$$y = \frac{1}{4p} x^2$$

is the equation of a parabola whose focus is the point $F = (0, p)$, whose directrix is the horizontal line $y = -p$, and whose vertex is at the origin. (See Figure 11.4.)

2. The equation

$$x = \frac{1}{4p} y^2$$

is the equation of a parabola whose focus is the point $F = (p, 0)$, whose directrix is the vertical line $x = -p$, and whose vertex is at the origin. (See Figure 11.5.)

EXAMPLE 1: Find the focus and directrix of the parabola whose equation is $-2x = y^2$. Sketch the graph of this parabola.

Solution: Since $-2 = 4(-1/2)$, the equation of the parabola can be written in the form

$$x = \frac{1}{4(-1/2)} y^2$$

Hence, according to part 2 of the theorem, this is the equation of a parabola whose focus is $F = (-1/2, 0)$ and whose directrix is the line $x = 1/2$, as shown in Figure 11.6. *Try Study Suggestion 11.1.* □

▶ **Study Suggestion 11.1:** Find the focus and directrix of the parabola whose equation is $3y = 2x^2$. Sketch the graph of this parabola. ◀

EXAMPLE 2: Find the equation of the parabola that has its vertex at the origin, opens up, and goes through the point (3, 2).

Solution: Since the parabola has its vertex at the origin and opens up, it has the form shown in Figure 11.4a. Therefore, according to our theorem, its equation has the form

$$y = \frac{1}{4p} x^2$$

In order to determine the value of p, we use the fact that the point (3, 2) is on the parabola, and hence satisfies this equation. In other words, we have

$$2 = \frac{1}{4p} 3^2$$

and so $4p = 9/2$. Hence, the equation of this parabola is

$$y = \frac{2}{9} x^2$$

Try Study Suggestion 11.2. □

▶ **Study Suggestion 11.2:** Find the equation of the parabola that has its vertex at the origin, opens down, and goes through the point $(1, -2)$. ◀

Now let us turn our attention to parabolas whose vertices are not at the origin, as in Figure 11.7. In order to determine the equation of the parabola in this figure, we use the results of Section 2.6 concerning the translation of graphs, to obtain the equation

$$y - k = \frac{1}{4p} (x - h)^2$$

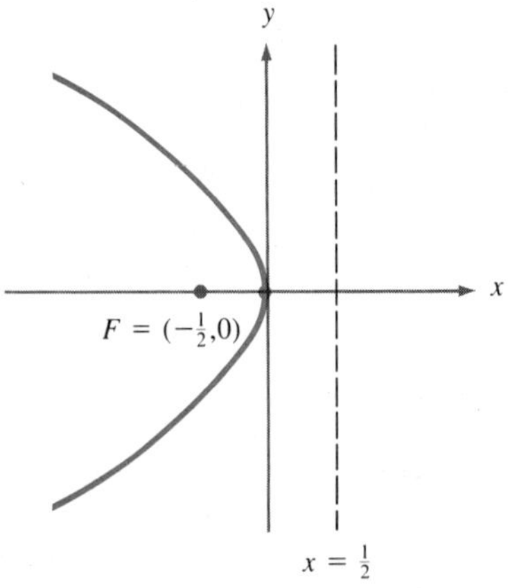

FIGURE 11.6

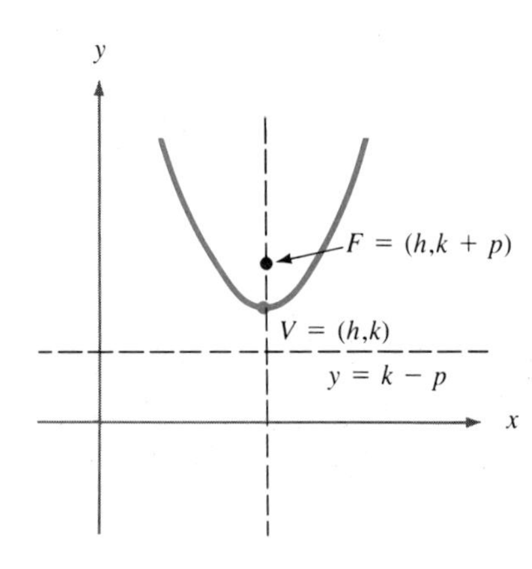

FIGURE 11.7

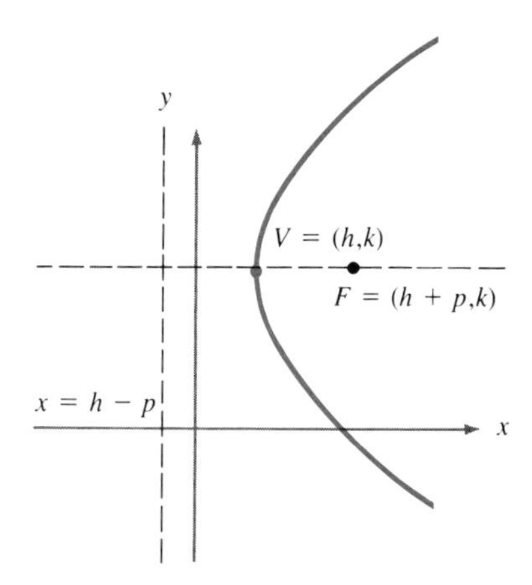

FIGURE 11.8

This is known as the **standard form of the equation of a parabola.** A similar equation holds for parabolas of the form shown in Figure 11.8. Let us summarize our results in another theorem.

THEOREM

1. The equation

$$y - k = \frac{1}{4p}(x - h)^2$$

is the equation of a parabola with vertex $V = (h, k)$, focus $F = (h, k + p)$, and directrix $y = k - p$. This is shown in Figure 11.7 for the case $p > 0$.

2. The equation

$$x - h = \frac{1}{4p}(y - k)^2$$

is the equation of a parabola with vertex $V = (h, k)$, focus $F = (h + p, k)$, and directrix $x = h - p$. This is shown in Figure 11.8 for the case $p > 0$.

EXAMPLE 3: Find the equation of the parabola with vertex $V = (2, -3)$ and directrix $x = 3$. Also, find the focus of this parabola, and sketch the graph.

Solution: In this case, since $V = (2, -3)$, we have $h = 2$ and $k = -3$. Also, since the directrix is of the form $x = h - p = 2 - p$, we have $2 - p = 3$ or $p = -1$. Hence, according to part 2 of the previous theorem, the focus of the parabola is $F = (h + p, k) = (1, -3)$, and the equation of the parabola is

$$x - 2 = -\frac{1}{4}(y + 3)^2$$

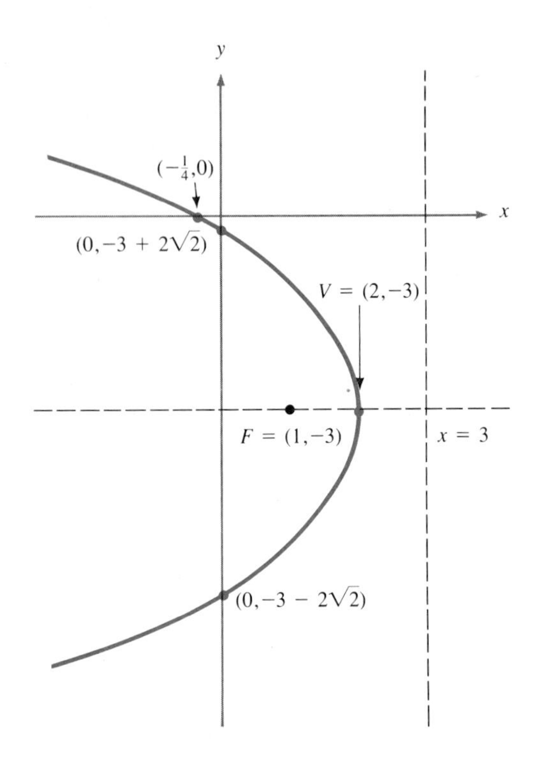

FIGURE 11.9

The graph of this parabola is shown in Figure 11.9. (The y-intercepts of the graph were found by setting $x = 0$ and solving for y.)

Try Study Suggestion 11.3. □

▶ **Study Suggestion 11.3:** Find the equation of the parabola with vertex $V = (-4, -1)$ and directrix $y = -6$. Also, find the focus of this parabola, and sketch the graph. ◀

EXAMPLE 4: Find the vertex, focus, and directrix of the parabola whose equation is

$$y = x^2 - 4x + 3$$

Sketch the graph of this parabola.

Solution: We begin by completing the square in the variable x (see Chapter 1 for a review of completing the square),

$$y - 3 = x^2 - 4x$$
$$y - 3 + 4 = x^2 - 4x + 4$$
$$y + 1 = (x - 2)^2$$

Looking at part 1 of the previous theorem, we see that $(h, k) = (2, -1)$ and $4p = 1$, that is $p = 1/4$. Hence, this is the equation of a parabola whose vertex is $V = (h, k) = (2, -1)$, whose focus is $F = (h, k + p) = (2, -3/4)$, and whose directrix is the line $y = k - p = -1 - (1/4) = -5/4$. The graph is sketched in Figure 11.10. *Try Study Suggestion 11.4.* □

▶ **Study Suggestion 11.4:** Find the vertex, focus, and directrix of the parabola whose equation is $x = 4y^2 - 24y + 34$. Sketch the graph of this parabola. ◀

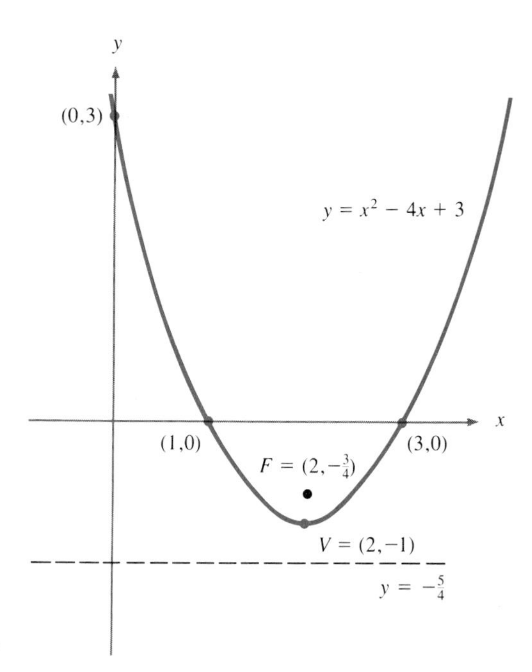

FIGURE 11.10

Ideas to Remember

The parabolic shape is very important in nature. For instance, most comets seem to travel in parabolic orbits around the sun.

EXERCISES

In Exercises 1–10, graph the given equation. Give the vertex, focus, and directrix in each case.

1. $2x = y^2$

2. $(x - 2)^2 = 3y$

3. $y = -4x^2 + 10x + 1$

4. $x = (1/2)(1 - 2y^2)$

5. $-2x + 3y = y^2$

6. $x = y^2 + y + 1$

7. $y = 4x^2 + x + 1$

8. $2y = x^2 + 2x$

9. $y + 10 = x - x^2$

10. $x = 2y^2 + 8y + 8$

In Exercises 11–28, find the equation of the parabola that satisfies the given conditions.

11. (a) Vertex = (0, 0); directrix: $y = -5$

(b) Vertex = (0, 0); directrix: $y = 5$

12. (a) Vertex = (0, 0); directrix: $x = -5$

(b) Vertex = (0, 0); directrix: $x = 5$

13. Vertex = (1, 3); directrix: $y = 0$

14. Vertex = (−2, −2); directrix: $x = -1$

15. Vertex = (0, −1/2); directrix: $x = -2$

16. Vertex = (0, 0); focus = (0, 1)

17. Vertex = (3, −3); focus = (0, −3)

18. Vertex = (1, 1); focus = (1, 5)

19. Vertex = (−2, −2); focus = (−2, 2)

20. (a) Focus = (0, 0); directrix: $y = -3/2$

(b) Focus = (0, 0); directrix: $y = 3/2$

21. (a) Focus = (0, 0); directrix: $x = -3/2$
    (b) Focus = (0, 0); directrix: $x = 3/2$
22. Focus = (−2, 3); directrix: $y = 1$
23. Focus = (−1/2, −1/2); directrix: $x = 1/2$
24. Focus = (π, π); directrix: $y = -\pi$
25. Vertex = (0, 0); opens up; goes through (1, 1)
26. Vertex = (1, 1); opens to right; goes through (3, 6)
27. Vertex = (−2, 0); opens to left; goes through (−4, −1)
28. Vertex = (−4, 4); opens down; goes through the origin
29. *Derive* the equation of a parabola with vertex at the origin and focus on the x-axis.
30. Find the equation of the parabola that goes through the points (0, 0), (1, 2), and (3, 4) and that has a vertical axis of symmetry. (*Hint:* start with the equation $y = ax^2 + bx + c$, and obtain three equations in the three variables a, b, and c.)
31. Find the equation of the parabola that goes through the points (1, 1), (2, −1), and (3, 1) and has a vertical axis of symmetry.
32. Graph the equations $y = 4 - x^2$ and $y = x^2 - 4$ on the same plane. Then shade in the region that lies below the first curve and above the second curve. Label the intersection points of the two curves.
33. Graph the equations $3y = -2x^2 + 9x + 5$ and $3y = 2x^2 - 11x + 21$ on the same plane. Then shade in the region that lies below the first curve and above the second curve. Label the intersection points of the two curves.
34. A certain archway has the shape of a parabola that opens down, as shown in the following figure. Notice that the height of the archway at its highest point is 10 feet and the base of the archway is 16 feet across. Suppose that it is necessary to slide rectangular concrete blocks through the archway. If the blocks have a height of 4 feet, how wide can they be and still fit through the archway?

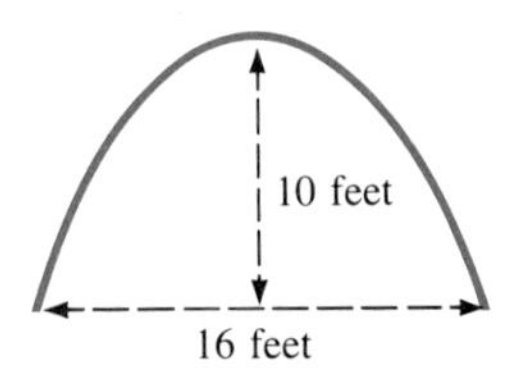

11.3 The Ellipse

Let us begin with the geometric definition of an ellipse.

DEFINITION

An **ellipse** is the locus of all points in the plane the *sum* of whose distances from two fixed points F_1 and F_2 is equal to a fixed positive constant. (See Figure 11.11.) The points F_1 and F_2 are called the **foci** (plural of focus) of the ellipse. The **center** of the ellipse lies at the midpoint of the line segment connecting the two foci. The two solid lines in Figure 11.11 are called the **axes** of the ellipse. The longer of the two axes is called the **major axis,** and the shorter of the two is called the **minor axis.** (When the two axes are the same length, we have a circle!) As we will see, the axis that contains the foci of the ellipse is *always* the major axis.

In order to derive the equation of an ellipse, let us look at Figure 11.12a, where we have placed the coordinate axes so that the foci are at the points

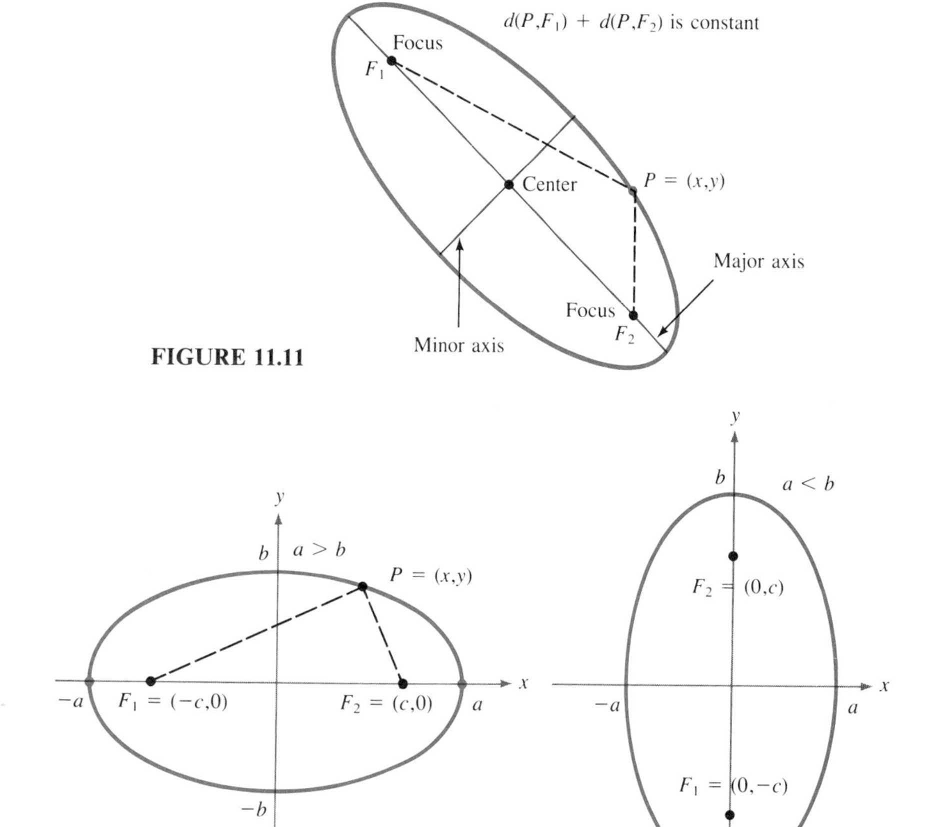

FIGURE 11.11

FIGURE 11.12 **(a)** **(b)**

$F_1 = (-c, 0)$ and $F_2 = (c, 0)$, for $c > 0$. Hence, the center of the ellipse is at the origin, and the major axis is along the x-axis.

Now, according to the definition, we have

$$d(P, F_1) + d(P, F_2) = 2a \tag{11.2}$$

where $2a$ is our fixed positive constant. (The following calculations are made a bit simpler by letting the constant have the form $2a$, rather than a.)

Notice that, since the distance between the two foci is $2c$, there will be no points P satisfying Equation (11.2) unless $2a \geq 2c$; that is, $a \geq c$. Furthermore, if $a = c$, then the *only* point P that will satisfy (11.2) is the origin $P = (0, 0)$. Hence, from now on we will assume that $a > c$.

Using the distance formula, Equation (11.2) becomes

$$\sqrt{(x + c)^2 + y^2} + \sqrt{(x - c)^2 + y^2} = 2a$$

Our job now is to simplify this equation. First, we write it in the form

$$\sqrt{(x + c)^2 + y^2} = 2a - \sqrt{(x - c)^2 + y^2}$$

and then we square both sides

$$(x + c)^2 + y^2 = 4a^2 - 4a\sqrt{(x - c)^2 + y^2} + [(x - c)^2 + y^2]$$

or, after subtracting y^2 from both sides and rearranging

$$4a\sqrt{(x - c)^2 + y^2} = 4a^2 + (x - c)^2 - (x + c)^2$$

which simplifies to

$$4a\sqrt{(x - c)^2 + y^2} = 4a^2 - 4cx$$

or

$$a\sqrt{(x - c)^2 + y^2} = a^2 - cx$$

Again we must square both sides, to get

$$a^2[(x - c)^2 + y^2] = a^4 - 2a^2cx + c^2x^2$$

or

$$a^2[x^2 - 2cx + c^2 + y^2] = a^4 - 2a^2cx + c^2x^2$$

or

$$a^2x^2 - 2a^2cx + a^2c^2 + a^2y^2 = a^4 - 2a^2cx + c^2x^2$$

which simplifies to

$$(a^2 - c^2)x^2 + a^2y^2 = a^4 - a^2c^2$$

or

$$(a^2 - c^2)x^2 + a^2y^2 = a^2(a^2 - c^2)$$

At this point we can divide both sides by $a^2(a^2 - c^2)$ to get

$$\frac{x^2}{a^2} + \frac{y^2}{a^2 - c^2} = 1$$

For convenience, we can now let $b^2 = a^2 - c^2$. We are relying here on the fact that $a > c$, so that $a^2 - c^2$ is positive. Thus, it can be written in the form b^2. It is important also to notice that $a^2 > b^2$, and so $a > b$. This gives the equation

$$\frac{x^2}{a^2} + \frac{y^2}{b^2} = 1 \tag{11.3}$$

In order for this equation to be useful, we must find a simple geometric meaning for the numbers a and b. Fortunately, this is easy to do. For if we set $x = 0$ in Equation (11.3), we get

$$\frac{y^2}{b^2} = 1 \qquad \text{or} \qquad y^2 = b^2 \qquad \text{or} \qquad y = \pm b$$

Thus, we see that the numbers $\pm b$ are the *y-intercepts* of the ellipse. Similarly, by setting $y = 0$ in Equation (11.3), we get $x = \pm a$, and so the numbers $\pm a$ are the *x-intercepts* of the ellipse. These points are labeled in Figure 11.12(a).

A similar result can be obtained when the foci of the ellipse lie on the y-axis, as shown in Figure 11.12b. In this case, we have $a < b$. Let us summarize our results.

THEOREM

The equation

$$\frac{x^2}{a^2} + \frac{y^2}{b^2} = 1$$

where a and b are positive constants is the equation of an ellipse whose center is at the origin, whose x-intercepts are $\pm a$, and whose y-intercepts are $\pm b$. Furthermore,

1. if $a > b$, then the major axis of the ellipse is along the x-axis, and the foci of the ellipse are at the points $(-c, 0)$ and $(c, 0)$, where $c^2 = a^2 - b^2$. (See Figure 11.12a.)
2. if $a < b$, then the major axis of the ellipse is along the y-axis, and the foci of the ellipse are at the points $(0, -c)$ and $(0, c)$, where $c^2 = b^2 - a^2$. (See Figure 11.12b.)

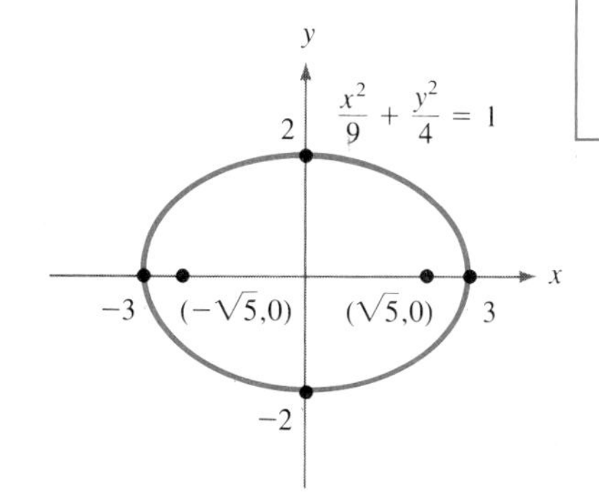

FIGURE 11.13

▶ **Study Suggestion 11.5:** Find the equation of the ellipse whose center is at the origin, whose x-intercepts are $\pm 1/2$, and whose y-intercepts are ± 1. Sketch the graph of this ellipse and label the foci. ◀

Now let us do some examples.

EXAMPLE 1: Find the equation of the ellipse whose center is at the origin, whose x-intercepts are ± 3, and whose y-intercepts are ± 2. Sketch the graph of this ellipse and label the foci.

Solution: In this case, we have $a = 3$ and $b = 2$, and so the equation of the ellipse is

$$\frac{x^2}{9} + \frac{y^2}{4} = 1$$

The graph is shown in Figure 11.13. In order to determine the foci, we recall that $c^2 = a^2 - b^2 = 9 - 4 = 5$, and so $c = \sqrt{5}$. (Remember that $c > 0$.) Thus, the foci are at the points $(-\sqrt{5}, 0)$ and $(\sqrt{5}, 0)$. *Try Study Suggestion 11.5.* □

EXAMPLE 2: Sketch the graph of the equation

$$3x^2 + y^2 = 9$$

Also, label the intercepts and the foci.

Solution: The first step is to write the equation in the form of Equation (11.3). This can be done by dividing both sides by 9 (which will make the right-hand side equal to 1)

$$\frac{x^2}{3} + \frac{y^2}{9} = 1$$

and then writing this in the form

$$\frac{1}{(\sqrt{3})^2} + \frac{y^2}{3^2} = 1$$

Now we see that this is the equation of an ellipse whose center is at the origin, whose x-intercepts are $\pm\sqrt{3}$ and whose y-intercepts are ± 3. The graph is shown in Figure 11.14. The coordinates of the foci are found by using the fact that $c^2 = b^2 - a^2 = 9 - 3 = 6$ and so $c = \sqrt{6}$.

Try Study Suggestion 11.6. □

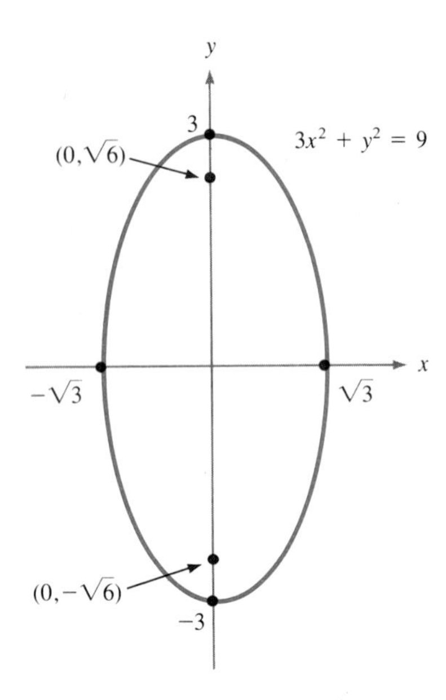

FIGURE 11.14

▶ **Study Suggestion 11.6:** Sketch the graph of the equation $x^2 + 100y^2 = 25$. Also, label the intercepts and the foci. ◀

Using the technique of translating a curve that we used in Section 2.6, we can easily determine the equation of an ellipse whose center is not necessarily at the origin.

THEOREM

The equation

$$\frac{(x-h)^2}{a^2} + \frac{(y-k)^2}{b^2} = 1 \qquad \textbf{(11.4)}$$

where a and b are positive constants is the equation of an ellipse whose center is at the point (h, k). Furthermore,

1. if $a > b$, then the major axis is horizontal, and the foci lie at the points $(h - c, k)$ and $(h + c, k)$, where $c^2 = a^2 - b^2$. (See Figure 11.15a.)

2. if $a < b$, then the major axis is vertical, and the foci lie at the points $(h, k - c)$ and $(h, k + c)$, where $c^2 = b^2 - a^2$. (See Figure 11.15b.)

Equation (11.4) is called the **standard form of the equation of an ellipse.** As you can see from Figure 11.15, the quantity a is *one-half* the length of the axis of the ellipse that is parallel to the x-axis, and the quantity b is *one-half*

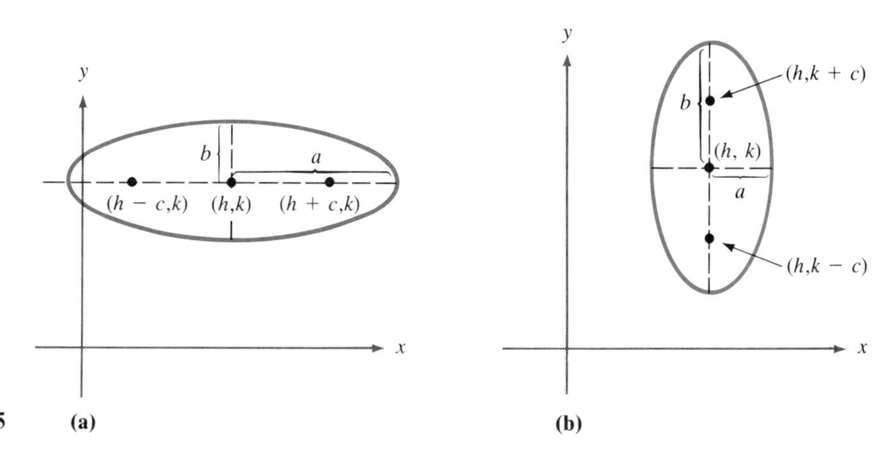

FIGURE 11.15 (a) (b)

the length of the axis of the ellipse that is parallel to the y-axis. (Which of these axes is the major axis and which is the minor axis will depend on whether $a > b$ or $a < b$.)

EXAMPLE 3: Find the equation of the ellipse with center $(3, -4)$ whose major axis is parallel to the x-axis and has length 12, and whose minor axis has length 5. Find the foci of the ellipse.

Solution: Since $a = 6$ and $b = 5/2$, the standard form of the equation of this ellipse is

$$\frac{(x-3)^2}{6^2} + \frac{(y+4)^2}{(5/2)^2} = 1$$

or

$$\frac{(x-3)^2}{36} + \frac{(y+4)^2}{25/4} = 1$$

In order to locate the foci of this ellipse, we first find c,

$$c^2 = a^2 - b^2 = 36 - \left(\frac{25}{4}\right) = \frac{119}{4}$$

and so $c = \sqrt{119}/2$. Thus the foci, lying on the horizontal (major) axis, have coordinates $(3 - \sqrt{119}/2, -4)$ and $(3 + \sqrt{119}/2, -4)$.

Try Study Suggestion 11.7. □

▶ **Study Suggestion 11.7:** Find the equation of the ellipse whose center is at $(0, 2)$, whose major axis is parallel to the y-axis and has length 4, and whose minor axis has length 2. Find the coordinates of the foci. ◀

EXAMPLE 4: Graph the equation

$$9x^2 + 4y^2 - 18x - 24y + 9 = 0$$

Solution: The first step is to put the equation into standard form by completing the square in both x and y. Subtracting 9 from both sides and rearranging terms gives

$$9x^2 - 18x + 4y^2 - 24y = -9$$

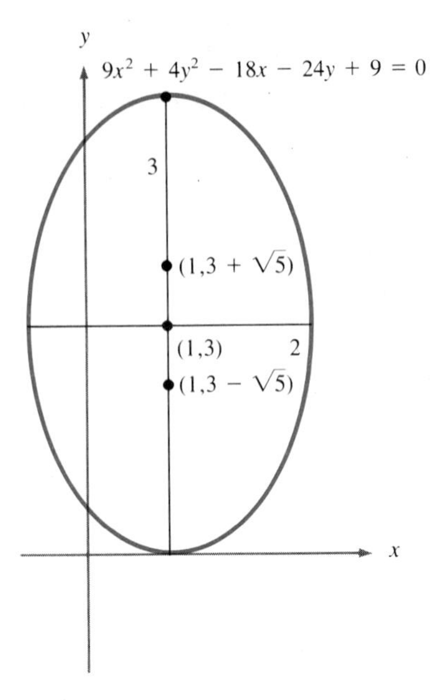

FIGURE 11.16

▶ **Study Suggestion 11.8:** Graph the equation $x^2 + 2y^2 - 2x - 4y - 3 = 0$. ◀

Next, we factor out the coefficients of x^2 and y^2

$$9(x^2 - 2x) + 4(y^2 - 6y) = -9$$

Then we complete the square inside each of the parentheses. As always, we must perform the same operations on both sides of the equation, and since we are actually adding 9 and 36 to the left side, we must also add these numbers to the right side

$$9(x^2 - 2x + 1) + 4(y^2 - 6y + 9) = -9 + 9 + 36$$

This is equivalent to the equation

$$9(x - 1)^2 + 4(y - 3)^3 = 36$$

Now we divide both sides by 36, in order to make the right side equal to 1

$$\frac{(x - 1)^2}{4} + \frac{(y - 3)^2}{9} = 1$$

This is the standard form of the equation of an ellipse, with center $(h, k) = (1, 3)$ and $a = 2$, $b = 3$. (See Figure 11.16.) In order to find the coordinates of the foci, we observe that $c^2 = b^2 - a^2 = 9 - 4 = 5$ and so $c = \sqrt{5}$. Hence the foci, lying on the vertical (major) axis, have coordinates $(1, 3 - \sqrt{5})$ and $(1, 3 + \sqrt{5})$. *Try Study Suggestion 11.8.* □

EXAMPLE 5: Graph the equation

$$2x^2 + y^2 - 4x + 2y + 3 = 0$$

Solution: In this case, completing the square in both variables gives

$$(x - 1)^2 + \frac{(y + 1)^2}{2} = 0$$

But this equation is not in standard form (and can never be put in standard form.) In fact, since both of the terms of the left-hand side are nonnegative, their sum can equal zero if and only if both terms are zero. In other words, the *only* solution to this equation is $(x, y) = (1, -1)$. Hence, the entire graph consists of the single point $(1, -1)$. This is an example of a **degenerate ellipse.** □

Ideas to Remember

The elliptical shape is very important in nature. For instance, the orbit of each of the planets around the sun is an ellipse with the sun at one of the foci.

EXERCISES

In Exercises 1–14, graph the given equation. Also give the center and foci of the ellipse.

1. $x^2 + \dfrac{y^2}{2} = 1$

2. $\dfrac{(x-3)^2}{4} + \dfrac{(y+1)^2}{9} = 1$

3. $4y^2 + 3x^2 = 1$

4. $3(x+1)^2 + 4(y-2)^2 = 12$

5. $x^2 + 3y^2 - 4x - 6y + 7 = 0$

6. $(x+1)^2 = 4 - 2(y-2)^2$

7. $x^2 + 2y^2 - 2x - 4y = 1$

8. $(2x+1)^2 + (y-4)^2 = 2$

9. $7x^2 + 6y^2 - 42 = 0$

10. $-9x^2 = 3y^2 + 18y$

11. $x^2 + 5y^2 + 50y - 125 = 0$

12. $4x^2 - 4x = 7 - 8y^2$

13. $10x^2 + 8y^2 - 10x - 4y - 37 = 0$

14. $2x^2 + 3y^2 - 12x + 18y + 39 = 0$

In Exercises 15–22, find the equation of the ellipse with the given properties. Give the center and foci.

15. Center (0, 0), x-intercepts ± 3, y-intercepts ± 10.

16. Center (0, 0), x-intercepts ± 5, y-intercepts ± 2.

17. **(a)** Center (1, 1), major axis horizontal, axes of length 1 and 2.

(b) Center (1, 1), major axis vertical, axes of length 1 and 2.

18. **(a)** Center $(-1, 2)$, major axis horizontal, axes of length 2 and 1/2.

(b) Center $(-1, 2)$, major axis vertical, axes of length 2 and 1/2.

19. Center (0, 5), minor axis vertical, axes of length $\sqrt{2}$ and 5.

20. Foci at $(-1, -1)$ and $(3, -1)$, axes of length 4 and 6.

21. Foci at (0, 0) and (0, 6), major axis of length 8.

22. Foci at (1, 1) and (5, 1), minor axis of length 6.

23. Find the points of intersection of the ellipses whose equations are

$$\frac{x^2}{4} + \frac{y^2}{16} = 1 \quad \text{and} \quad \frac{x^2}{16} + \frac{y^2}{4} = 1$$

24. Find the points of intersection of the ellipses whose equations are $4x^2 + 9y^2 - 8x + 18y = 23$ and $2x^2 + 10y^2 - 4x + 20y = 9$.

25. Is knowing the center and foci enough to determine the ellipse? Justify your answer.

26. A certain archway has the shape of the upper half of an ellipse, as shown in the following figure. Suppose that the base of the archway is 10 feet across and the height of the archway at its highest point is 4 feet. Suppose also that it is necessary to slide rectangular concrete blocks through the archway. If the blocks are 6 feet wide, how high can they be and still fit through the archway?

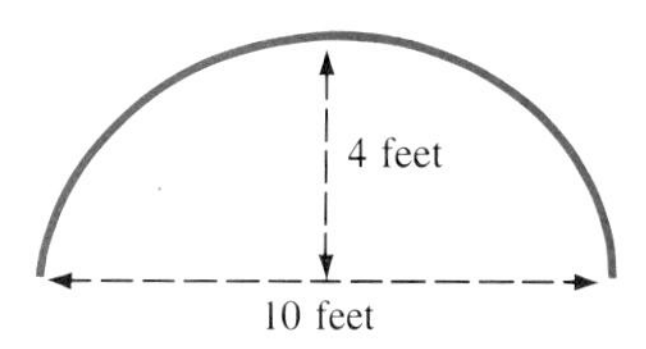

27. A square whose sides are parallel to the coordinate axes is inscribed inside the ellipse whose equation is

$$\frac{x^2}{a^2} + \frac{y^2}{b^2} = 1$$

Express the area of the square in terms of a and b.

28. Find the equations of the inscribed and circumscribed circles of the ellipse whose equation is

$$\frac{x^2}{a^2} + \frac{y^2}{b^2} = 1$$

29. Consider the point $P = (x, y)$ shown in part (a) of the following figure, where we assume that $a \neq b$. As the line segment AB moves so that A and B remain on the coordinate axes, as shown in parts (a) through (d), show that the point P traces out an ellipse. (*Hint:* determine the length of the line segment CB in part (a) by two different methods.) The **eccentricity** of an ellipse whose equation is

$$\frac{x^2}{a^2} + \frac{y^2}{b^2} = 1$$

is defined to be the number $e = \sqrt{|a^2 - b^2|}/a$. (Notice the absolute value bars to make the quantity under the square root sign nonnegative.) As the next exercises show, the eccentricity is a measure of how "flat" the ellipse is.

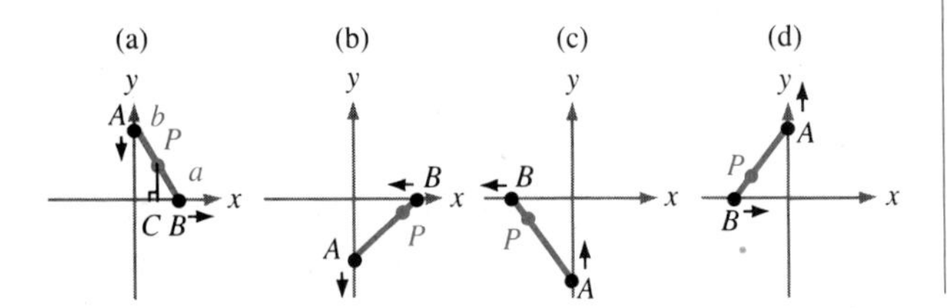

30. Show that the eccentricity satisfies $0 \le e \le 1$.

31. What is the eccentricity of a circle? (Take $a = b = r$.)

32. (a) What is the eccentricity of the nearly "round" ellipse

$$\frac{x^2}{100} + \frac{y^2}{101} = 1$$

Sketch the graph of this ellipse.

(b) What is the eccentricity of the "flat" ellipse

$$\frac{x^2}{100} + y^2 = 1$$

Sketch the graph of this ellipse.

33. Give an argument to show that the "flatter" the ellipse, the closer the eccentricity is to 1, and the "rounder" the ellipse, the closer the eccentricity is to 0.

11.4 The Hyperbola

We now turn to the last of the conic sections.

DEFINITION

A **hyperbola** is the locus of all points in the plane the *difference* of whose distances from two distinct points F_1 and F_2 is equal to a fixed positive constant. (See Figure 11.17.) The points F_1 and F_2 are called the **foci** of the hyperbola. The midpoint of the line segment that connects the two foci is the **center** of the hyperbola, and the points where this line segment meet the curve itself are the **vertices** of the hyperbola. Finally, the line segment between the two vertices is called the **transverse axis** of the hyperbola.

Notice from Figure 11.17 that the hyperbola consist of two *branches*, one of which contains the points P that satisfy the equation

$$d(P, F_1) - d(P, F_2) = 2a$$

and the other contains the points that satisfy the equation

$$d(P, F_2) - d(P, F_1) = 2a$$

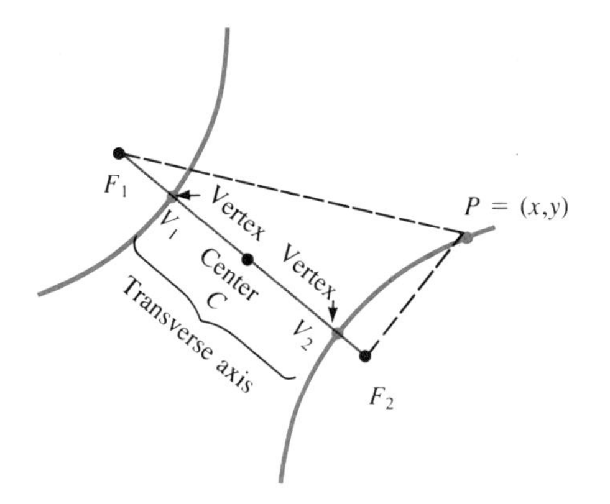

FIGURE 11.17

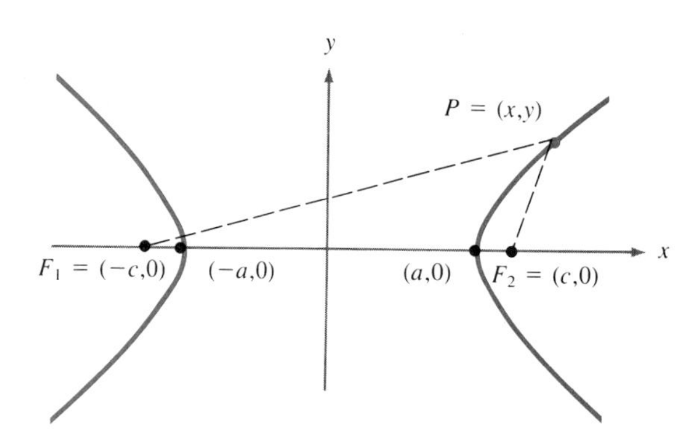

FIGURE 11.18

As with the ellipse, we use $2a$ to denote the positive distance mentioned in the definition. These two equations can be combined into one equation by writing

$$d(P, F_1) - d(P, F_2) = \pm 2a \tag{11.5}$$

In order to obtain the equation of a hyperbola, we place the curve as shown in Figure 11.18, so that the foci are on the x-axis. (Hence, the vertices and the center are also on the x-axis.) Now, a point $P = (x, y)$ will be on the hyperbola if and only if it satisfies Equation (11.5). Using the distance formula, we get

$$\sqrt{(x + c)^2 + y^2} - \sqrt{(x - c)^2 + y^2} = \pm 2a$$

This equation can be solved in a manner similar to that used in the previous section (we will omit the details) to obtain

$$\frac{x^2}{a^2} - \frac{y^2}{c^2 - a^2} = 1 \tag{11.6}$$

At this point, we should determine the relationship between c and a, which we can do by considering the triangle PF_1F_2 in Figure 11.18. Since the sum of the lengths of any two sides of a triangle is always greater than the length of the third side, we have

$$d(P, F_1) < d(P, F_2) + d(F_1, F_2)$$

or

$$d(P, F_1) - d(P, F_2) < d(F_1, F_2)$$

But $d(F_1, F_2) = 2c$ and so we get

$$d(P, F_1) - d(P, F_2) < 2c$$

Similarly, we have $d(P, F_2) - d(P, F_1) < 2c$ and so from (11.5) we can conclude that $2a < 2c$; that is, $a < c$. Since this is the case, we may set

$b^2 = c^2 - a^2$ where $b > 0$, and Equation (11.6) becomes

$$\frac{x^2}{a^2} - \frac{y^2}{b^2} = 1 \tag{11.7}$$

If we set $y = 0$ in Equation (11.7), we get $x^2/a^2 = 1$, and so $x = \pm a$. In other words, the x-intercepts of the hyperbola are $\pm a$. (Because of the position of the hyperbola, the x-intercepts also happen to be the vertices of the hyperbola.) However, the role of the number b is not so obvious. In fact, if we set $x = 0$ in Equation (11.7), we get $-y^2/b^2 = 1$, which has no solutions. Hence, the hyperbola does *not* cross the y-axis (as is clear from Figure 11.18).

In order to determine the role of b in Equation (11.7), we first solve this equation for y,

$$y = \pm\frac{b}{a}\sqrt{x^2 - a^2} = \pm\frac{bx}{a}\sqrt{1 - \frac{a^2}{x^2}}$$

Now, as x gets very large, we see that the expression $1 - a^2/x^2$ under the square root sign approaches $1 - 0 = 1$ and so y approaches $\pm bx/a$. In terms of the graph, this says that as x gets very large (either in the positive or negative sense) the graph approaches the straight lines

$$y = \frac{b}{a}x \qquad \text{and} \qquad y = -\frac{b}{a}x$$

This is shown in Figure 11.19a. Because of this behavior, we say that the lines $y = bx/a$ and $y = -bx/a$ are **diagonal asymptotes** of the graph. (Recall that the graphs of rational functions may have vertical or horizontal asymptotes.)

An easy way to remember the position of the diagonal asymptotes is to remember that the asymptotes go through the points $(\pm a, \pm b)$, as shown in

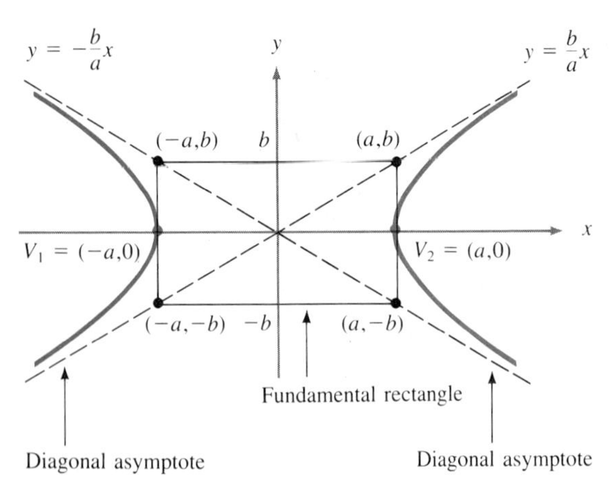

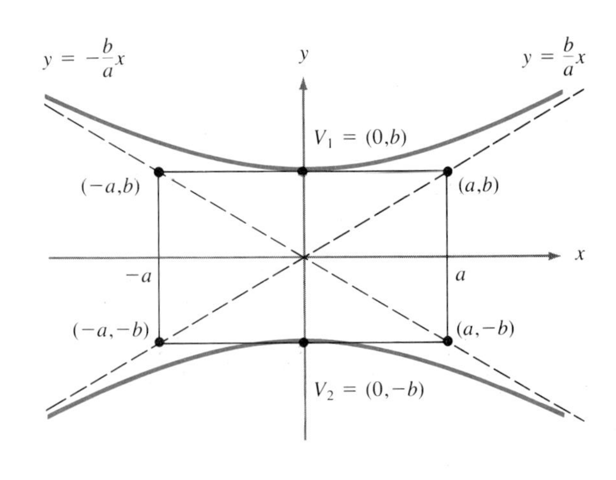

FIGURE 11.19 **(a)** **(b)**

Figure 11.19a. In fact, the asymptotes are simply an extension of the diagonals of the rectangle whose corners are at $(\pm a, \pm b)$. This rectangle is called the **fundamental rectangle** of the hyperbola, and is also shown in Figure 11.19a. Notice that the hyperbola is actually tangent to the fundamental rectangle at each of the x-intercepts. This makes it very easy to sketch the graph of the hyperbola, for we can simply

1. draw the fundamental rectangle;
2. extend its diagonals to obtain the diagonal asymptotes of the hyperbola; and
3. sketch the hyperbola using the asymptotes and the fact that the x-intercepts are $\pm a$.

EXAMPLE 1: Graph the equation

$$\frac{x^2}{9} - \frac{y^2}{16} = 1$$

Solution: First we sketch the fundamental rectangle, whose corners are at the points $(\pm 3, \pm 4)$. (See Figure 11.20.) Then we extend the diagonals of this rectangle to obtain the diagonal asymptotes of the graph. Finally, we can sketch the graph as shown in Figure 11.20. *Try Study Suggestion 11.9.* □

▶ **Study Suggestion 11.9:** Graph the equation $x^2 - 2y^2 = 4$. (*Hint:* first put this equation in the form of Equation (11.7).) ◀

Similar results can be obtained for the hyperbola shown in Figure 11.19b. In this case, the curve has two y-intercepts, but no x-intercepts, and the

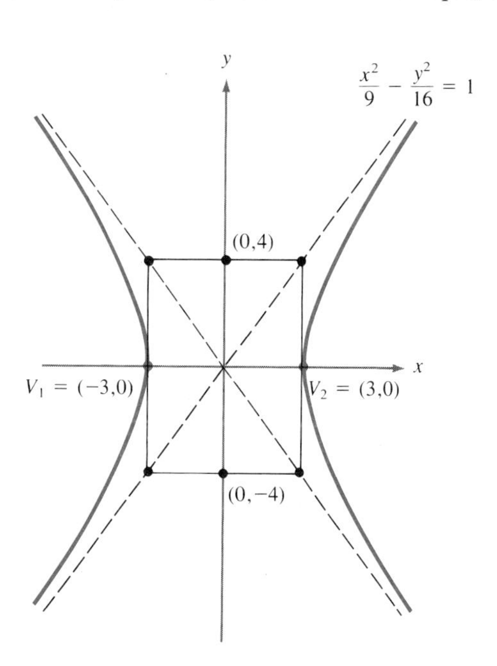

FIGURE 11.20

equation of the curve is

$$\frac{y^2}{b^2} - \frac{x^2}{a^2} = 1$$

(The difference between this equation and Equation (11.7) is that here the minus sign is associated with the term involving x, whereas in Equation (11.7), the minus sign was associated with the term involving y.) Let us summarize in a theorem.

THEOREM

1. The equation

$$\frac{x^2}{a^2} - \frac{y^2}{b^2} = 1$$

is the equation of a hyperbola whose center is at the origin, whose x-intercepts (vertices) are $\pm a$, whose foci are at the points $(-c, 0)$ and $(c, 0)$, where $c^2 = a^2 + b^2$, and whose diagonal asymptotes go through the points $(\pm a, \pm b)$. (See Figure 11.19a.)

2. The equation

$$\frac{y^2}{b^2} - \frac{x^2}{a^2} = 1$$

is the equation of a hyperbola whose center is at the origin, whose y-intercepts (vertices) are $\pm b$, whose foci are at the points $(0, -c)$ and $(0, c)$, where $c^2 = a^2 + b^2$, and whose diagonal asymptotes go through the points $(\pm a, \pm b)$. (See Figure 11.19b.)

EXAMPLE 2: Graph the equation $y^2 - 5x^2 = 25$.

Solution: First we put the equation into the form given in the theorem, by dividing both sides by 25,

$$\frac{y^2}{25} - \frac{x^2}{5} = 1$$

Thus, we see that $a = \sqrt{5}$ and $b = \sqrt{25} = 5$. Now we can draw the fundamental rectangle, and extend its diagonals to get the asymptotes of the hyperbola. Finally, we can draw the hyperbola using the asymptotes and the fact that the y-intercepts are $\pm b = \pm 5$. (See Figure 11.21.)

▶ **Study Suggestion 11.10:** Sketch the graph of the equation $y^2 - 3x^2 = 3$. ◀

Try Study Suggestion 11.10. □

By using the method of translating, we can obtain the equation of a hyperbola whose center is not necessarily at the origin. Let us put these facts in a theorem.

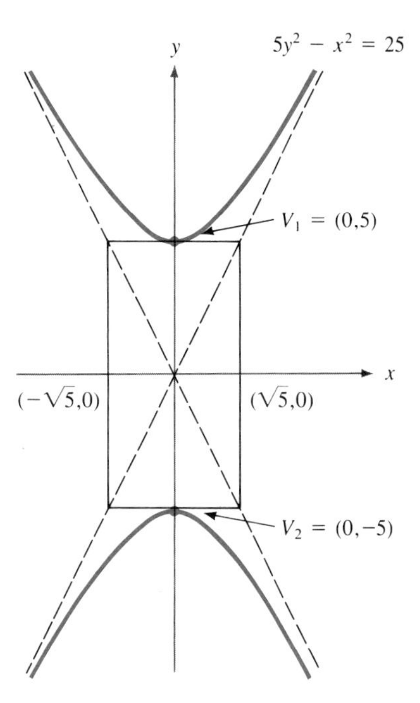

FIGURE 11.21

THEOREM

1. The equation

$$\frac{(x-h)^2}{a^2} - \frac{(y-k)^2}{b^2} = 1 \qquad \textbf{(11.8)}$$

is the equation of a hyperbola whose center is at the point (h, k), whose vertices are at the points $(h \pm a, k)$, whose foci are at the points $(h \pm c, k)$ where $c^2 = a^2 + b^2$, and whose diagonal asymptotes go through the points $(h \pm a, k \pm b)$. (See Figure 11.22a.)

2. The equation

$$\frac{(y-k)^2}{b^2} - \frac{(x-h)^2}{a^2} = 1 \qquad \textbf{(11.9)}$$

is the equation of a hyperbola whose center is at the point (h, k), whose vertices are at the points $(h, k \pm b)$, whose foci are at the points $(h, k \pm c)$ where $c^2 = a^2 + b^2$, and whose diagonal asymptotes go through the points $(h \pm a, k \pm b)$. (See Figure 11.22b.)

Equations (11.8) and (11.9) are called the **standard forms of the equation of a hyperbola.** The following procedure describes how to graph a hyperbola whose center is not necessarily at the origin.

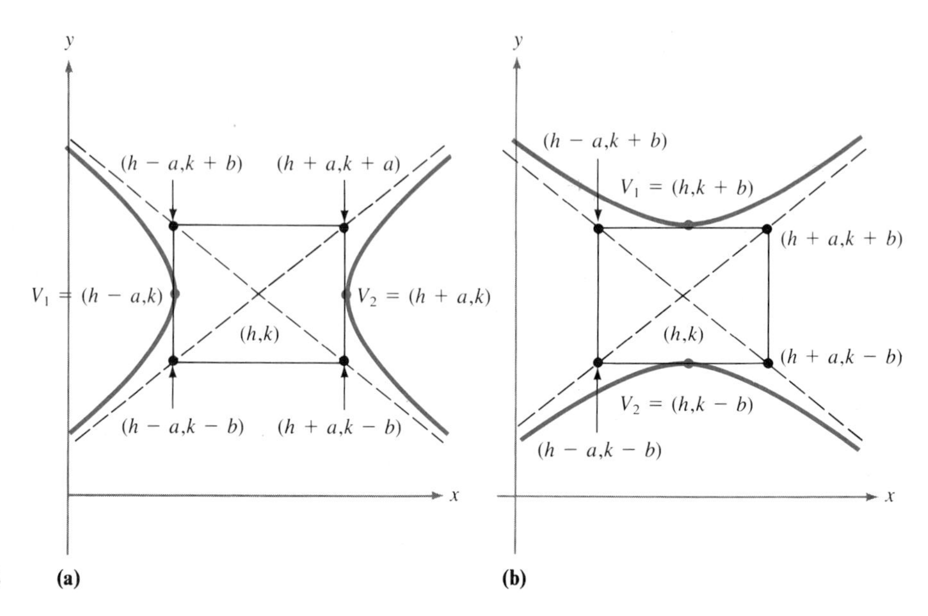

FIGURE 11.22 **(a)** **(b)**

Procedure for Graphing a Hyperbola

STEP 1. Put the equation into standard form. This may involve completing the square in one or both variables.

STEP 2. Locate the center of the hyperbola.

STEP 3. Sketch the fundamental rectangle, and extend the diagonals of this rectangle to get the diagonal asymptotes of the hyperbola.

STEP 4. Sketch the hyperbola, using the asymptotes and the fact that the curve is tangent to the fundamental rectangle at the vertices.

Let us try some examples of this procedure.

EXAMPLE 3: Graph the equation

$$\frac{(x-2)^2}{4} - \frac{(y+1)^2}{16} = 1$$

Solution:

STEP 1. The equation is already in standard form.

STEP 2. The center of the hyperbola is the point $(h, k) = (2, -1)$. (See Figure 11.23.)

STEP 3. The corners of the fundamental rectangle are

$$(h \pm a, k \pm b) = (2 \pm 2, -1 \pm 4)$$

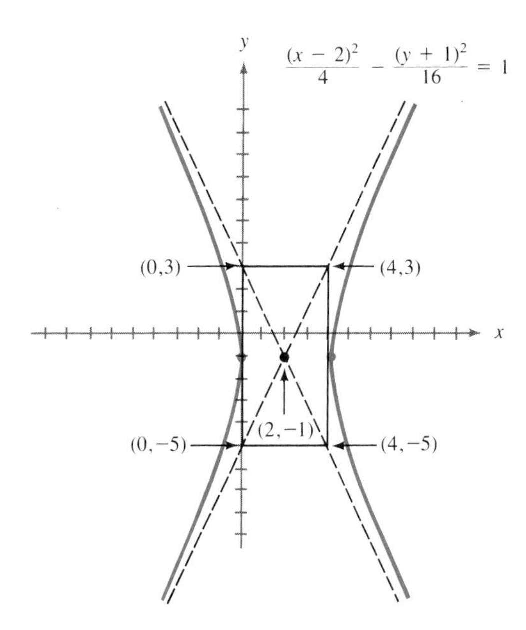

FIGURE 11.23

or

$$(0, -5), (0, 3), (4, -5), \quad \text{and} \quad (4, 3)$$

The fundamental rectangle and the diagonal asymptotes are shown in Figure 11.23.

STEP 4. Now we can draw the graph shown in Figure 11.23. Since $c = \sqrt{a^2 + b^2} = \sqrt{20} = 2\sqrt{5}$ the foci are at $(h \pm c, k) = (2 \pm 2\sqrt{5}, -1)$.

▶ Study Suggestion 11.11: Graph the equation

$$\frac{(y+1)^2}{4} - \frac{(x-2)^2}{9} = 1 \quad ◀$$

Try Study Suggestion 11.11. □

EXAMPLE 4: Graph the equation

$$4y^2 - x^2 - 24y + 6x + 23 = 0$$

Solution:

STEP 1. The first step is to put this equation into standard form. We will leave it to you to fill in the explanations.

$$4y^2 - x^2 - 24y + 6x + 23 = 0$$
$$4y^2 - 24y - x^2 + 6x = -23$$
$$4(y^2 - 6y) - (x^2 - 6x) = -23$$
$$4(y^2 - 6y + 9) - (x^2 - 6x + 9) = -23 + 36 - 9$$
$$4(y-3)^2 - (x-3)^2 = 4$$
$$\frac{(y-3)^2}{1} - \frac{(x-3)^2}{4} = 1$$

STEP 2. The center of the hyperbola is at the point $(h, k) = (3, 3)$. (See Figure 11.24.)

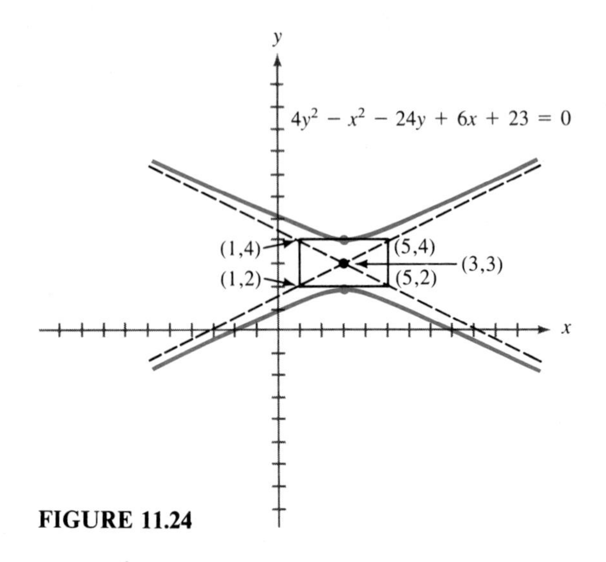

FIGURE 11.24

STEP 3. The corners of the fundamental rectangle are

$$(h \pm a, k \pm b) = (3 \pm 2, 3 \pm 1)$$

or

$$(1, 2), (1, 4), (5, 2), \text{ and } (5, 4)$$

The fundamental rectangle and the diagonal asymptotes are shown in Figure 11.24.

STEP 4. Now we can draw the graph shown in Figure 11.24. Since $c = \sqrt{a^2 + b^2} = \sqrt{5}$ the foci are at $(h, k \pm c) = (3, 3 \pm \sqrt{5})$.

▶ Study Suggestion 11.12: Graph the equation

$$x^2 - 3y^2 - 2x + 6y - 5 = 0.$$ ◀

Try Study Suggestion 11.12. □

EXAMPLE 5: Graph the equation $9x^2 - y^2 - 18x - 4y + 5 = 0$.

Solution:

STEP 1. Completing the square in both variables gives

$$9(x^2 - 2x) - (y^2 + 4y) = -5$$
$$9(x^2 - 2x + 1) - (y^2 + 4y + 4) = -5 + 9 - 4$$
$$9(x - 1)^2 - (y + 2)^2 = 0$$

As you can see, this is not the equation of a hyperbola, since the constant term on the right is equal to 0. (Dividing both sides by 9 will not help!) However, we can still graph this equation. Before doing this however, we should first solve for y,

$$(y + 2)^2 = 9(x - 1)^2$$
$$y + 2 = \pm 3(x - 1)$$
$$y = -2 \pm 3(x - 1)$$
$$y = 3x - 5 \quad \text{or} \quad y = -3x - 1$$

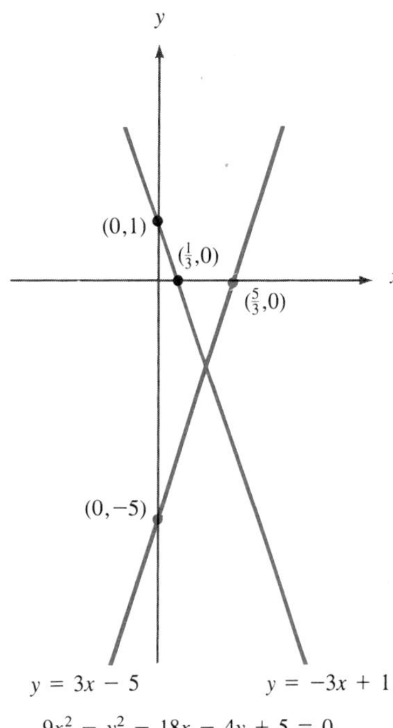

FIGURE 11.25 $9x^2 - y^2 - 18x - 4y + 5 = 0$

▶ **Study Suggestion 11.13:** Graph the equation

$$4y^2 - x^2 - 24y - 2x + 35 = 0.$$

Thus we see that the graph consists of two straight lines, as shown in Figure 11.25. A graph such as this is an example of a **degenerate hyperbola.**

Try Study Suggestion 11.13. □

Ideas to Remember

The hyperbolic shape has many practical uses. As already mentioned, the orbits of certain comets about the sun are hyperbolic. Also, hyperbolas are used in navigation.

EXERCISES

In Exercises 1–15, graph the given equation. Give the center and foci (provided that the hyperbola is not degenerate).

1. $\dfrac{(x-3)^2}{9} - \dfrac{(y-1)^2}{25} = 1$

2. $\dfrac{(y-1)^2}{8} - x^2 = 1$

3. $y^2 - x^2 = 1$

4. $(y+2)^2 = 3 + (x+1)^2$

5. $2(x-1)^2 - 3(y+2)^2 = 1$

6. $x^2 - y^2 - 4x + 4y = 0$

7. $(6y-1)^2 - 2(3x+2)^2 = 2$

8. $2x^2 - y^2 = 2$

9. $3y^2 - 4x^2 - 16x - 8 = 0$

10. $4y^2 - 3x^2 + 2x = 0$

11. $x^2 - y^2 - 10x - 6y + 15 = 0$

12. $x^2 - 3y^2 + 2x - y = 0$

13. $x^2 - y^2 - 2\pi x + 2\pi y = 1$

14. $3x^2 - 2y^2 - 6x - 8y - 11 = 0$

15. $2x^2 - 3y^2 - 4x - 12y - 10 = 0$

16. Graph the equation $x^2 - y^2 = 0$.

17. Graph the equation $xy = 1$. Do you recognize the shape of this curve?

In Exercises 18–23, find the equation of the hyperbola satisfying the given conditions. Also, give the center, vertices, and foci.

18. **(a)** Center (0, 0); Vertex (0, 2); Focus (0, 3)

(b) Center (0, 0); Vertex (2, 0); Focus (3, 0)

19. Center (2, 1); Vertex (3, 1); Focus (−4, 1)

20. Center (−5, 1); Vertex (−5, 2); Focus (−5, 7)

21. Asymptotes $y = \pm 2x$; x-intercept ± 5

22. Asymptotes $y = 2x$ and $y = 4 - 2x$; x-intercept $1 + \sqrt{2}$

23. Center (0, 0); transverse axis horizontal; passing through (1, 1) and (2, 4).

24. Find the equations and sketch the graphs of all hyperbolas with fundamental rectangle having corners (1, 3) and (4, 5).

25. Graph both of the equations $x^2/8 + y^2/4 = 1$ and $y^2 - x^2/2 = 1$ in the same plane. Then label the points of intersection of the two curves.

26. Graph the hyperbola whose equation is

$$\frac{y^2}{4} - \frac{x^2}{16} = 1$$

Find the equation of the parabola whose focus is the same as the upper focus of the hyperbola and whose vertex is the same as the upper vertex of the hyperbola. Then graph the parabola on the same plane as the hyperbola, and carefully compare the parabola with the upper branch of the hyperbola.

27. Find the equation of the locus of all points P in the plane with the property that the difference in the distances from P to (1, 2) and (1, 8) is equal to 4.

28. The graphs of the equations

$$\frac{x^2}{a^2} - \frac{y^2}{b^2} = 1 \quad \text{and} \quad \frac{y^2}{b^2} - \frac{x^2}{a^2} = 1$$

are called **conjugate hyperbolas.** Sketch these two hyperbolas on the same plane for $a = 4$ and $b = 9$. Describe the relationship between these curves.

11.5 The General Quadratic Equation and Rotation of Axes

In this section, we want to discuss the equation

$$Ax^2 + Bxy + Cy^2 + Dx + Ey + F = 0 \qquad (11.10)$$

where at least one of A, B, or C is nonzero. We will assume in our discussion that all of our equations involve *both* of the variables x and y. This will exclude equations such as $Ax^2 + Dx + F = 0$ and $Cy^2 + Ey + F = 0$ from consideration.

Equation (11.10) is the most general form of a quadratic equation in two variables. Our goal is to show that, *if* such an equation has a graph, then the graph must be a conic section. (This includes the possibility of a degenerate ellipse or hyperbola.) We say "if" because not all equations of the form (11.10) have a graph. For example, the equation

$$x^2 + y^2 + 1 = 0$$

does not have a graph since there are no real numbers x and y that satisfy this equation. As we will see, the presence of the **cross term** Bxy in Equation (11.10) indicates that the graph may be *rotated* in such a way that its axis (or axes) are neither vertical nor horizontal.

Before discussing the question of rotation, let us consider the simpler equation

$$Ax^2 + Cy^2 + Dx + Ey + F = 0 \tag{11.11}$$

where at least one of A or C is nonzero. This equation is similar to Equation (11.10) except that it has no cross term, and is the type of equation that we have been graphing in this chapter. In all examples so far, the graph of this equation turned out to be a conic section. Actually, it is not hard to see that, if Equation (11.11) has a graph then this graph must be a conic section.

This can be seen by reasoning as follows. If one of A or C is zero (in other words, if $AC = 0$), then Equation (11.11) is the equation of a parabola. On the other hand, if neither A nor C equals zero, then we may complete the square in both variables as follows

$$A\left(x^2 + \frac{D}{A}x\right) + C\left(y^2 + \frac{E}{C}y\right) = -F$$

$$A\left(x^2 + \frac{D}{A}x + \frac{D^2}{4A^2}\right) + C\left(y^2 + \frac{E}{C}y + \frac{E^2}{4C^2}\right) = -F + \frac{D^2}{4A} + \frac{E^2}{4C}$$

Letting $R = -F + \dfrac{D^2}{4A} + \dfrac{E^2}{4C}$, we can write this in the form

$$A\left(x + \frac{D}{2A}\right)^2 + C\left(y + \frac{E}{2C}\right)^2 = R$$

Now if $R = 0$, then we have a degenerate conic section (either a degenerate ellipse or a degenerate hyperbola). However, if $R \neq 0$, then we may divide both sides of this equation by R, to get

$$\frac{(x + (D/2A))^2}{R/A} + \frac{(y + (E/2C))^2}{R/C} = 1 \tag{11.12}$$

But this is the equation of either a circle, an ellipse, or a hyperbola.

In fact, provided that the graph of Equation (11.11) exists, we can say that

1. **if exactly one of A or C is zero, then Equation (11.11) is the equation of a parabola; otherwise**

2. **if A and C have the same sign, then Equation (11.11) is the equation of an ellipse, or a circle if $A = C$; and**

3. **if A and C have opposite signs, then Equation (11.11) is the equation of a hyperbola.**

EXAMPLE 1: Determine the shape of the graphs of each of the following equations.

(a) $2x^2 + 4y^2 - 3x + 7y - 9 = 0$

(b) $x^2 + 3x - 6y^2 = 12$

(c) $4y^2 - 12x + 9y - 3 = 0$

Solutions:

(a) In this case $A = 2$ and $C = 4$, and so A and C have the same sign. Hence, the graph (if it exists) is an ellipse.

(b) In this case $A = 1$ and $C = -6$, and so A and C have opposite signs. Hence, the graph (if it exists) is a hyperbola.

(c) In this case $A = 0$ and $C = 4$, and so the graph is a parabola.

Try Study Suggestion 11.14. □

▶ **Study Suggestion 11.14:** Determine the shape of the graphs of each of the following equations.

(a) $5x^2 - 4x + 5y^2 - 3y = 10$

(b) $-3x^2 + x - 2y + 9 = 0$

(c) $1 - 4x^2 + 4y^2 = 0$

(d) $2y^2 + 3x^2 - x + 2 = 0$ ◀

As we mentioned earlier, our goal is to show that if the general quadratic equation (11.10) has a graph, then it must be a conic section, where the presence of the cross term Bxy indicates that the graph may be rotated so that its axes are neither vertical nor horizontal. In order to understand the role of the cross term, we must first discuss the rotation of coordinate systems.

Let us begin by looking at Figure 11.26, which shows two coordinate systems that share a common origin (ϕ is the Greek letter "phi"). As you can see, the x', y'-coordinate system has been obtained by rotating the x, y-coordinate system about this origin through an angle θ. Now suppose that a point P has coordinates (x, y) in one coordinate system and (x', y') in the other coordinate system. In order to find a relationship between these two sets of coordinates, we observe from the figure that

$$x' = r\cos\phi, \; y' = r\sin\phi \tag{11.13}$$

and

$$x = r\cos(\theta + \phi), \; y = r\sin(\theta + \phi)$$

Using the addition formulas for the sine and cosine, along with Equations (11.13), we get

$$\begin{aligned} x &= r\cos(\theta + \phi) \\ &= r[\cos\theta\cos\phi - \sin\theta\sin\phi] \\ &= r\cos\phi\cos\theta - r\sin\phi\sin\theta \\ &= x'\cos\theta - y'\sin\theta \end{aligned}$$

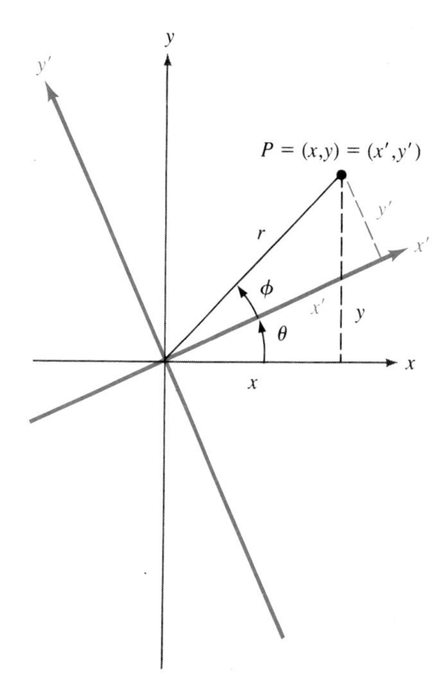

FIGURE 11.26

and

$$\begin{aligned} y &= r\sin(\theta + \phi) \\ &= r[\sin\theta\cos\phi + \cos\theta\sin\phi] \\ &= r\cos\phi\sin\theta + r\sin\phi\cos\theta \\ &= x'\sin\theta + y'\cos\theta \end{aligned}$$

Thus, we have

$$x = x'\cos\theta - y'\sin\theta$$

and

$$y = x'\sin\theta + y'\cos\theta$$

By solving these equations we can express x' and y' in terms of x and y. Let us summarize this in a theorem.

THEOREM

Suppose that the x', y'-coordinate system is obtained by rotating the x, y-coordinate system about the origin through an angle θ. Then the coordinates (x, y) and (x', y') of a point P are related as follows:

1. $x = x'\cos\theta - y'\sin\theta$, $y = x'\sin\theta + y'\cos\theta$
2. $x' = x\cos\theta + y\sin\theta$, $y' = -x\sin\theta + y\cos\theta$

As the next example shows, this theorem can be very useful in helping to graph equations that involve a cross term.

EXAMPLE 2: Consider the graph of the equation $xy = 1$. Suppose that we introduce a new coordinate system obtained from the x, y-coordinate system by rotating through an angle of $\theta = 45°$. Both of these coordinate systems are shown in Figure 11.27. Then the original coordinates (x, y) of a point are related to the new coordinates (x', y') by the equations

$$x = x' \cos 45° - y' \sin 45° = \left(\frac{\sqrt{2}}{2}\right)(x' - y')$$

and

$$y = x' \sin 45° + y' \cos 45° = \left(\frac{\sqrt{2}}{2}\right)(x' + y')$$

Substituting these values of x and y into the equation $xy = 1$ gives

$$\left(\frac{\sqrt{2}}{2}\right)(x' - y')\left(\frac{\sqrt{2}}{2}\right)(x' + y') = 1$$

or after simplifying

$$\frac{(x')^2}{2} - \frac{(y')^2}{2} = 1$$

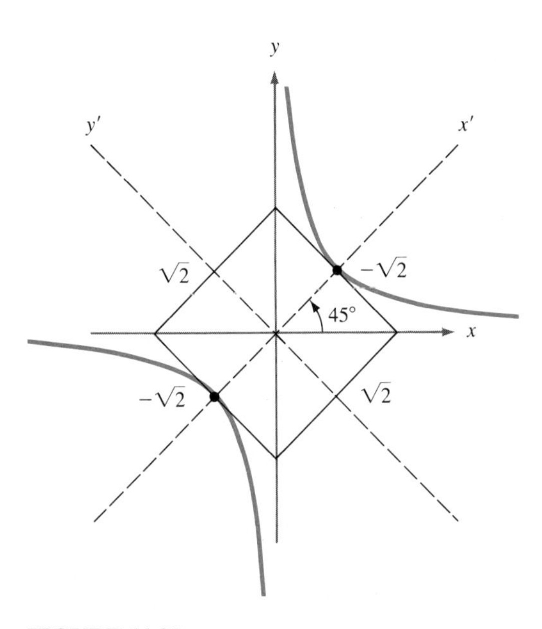

FIGURE 11.27

Now we recognize this as the standard form of the equation of a hyperbola, *when viewed from the point of view of the x', y'-coordinate system*, and we can sketch its graph using the techniques of the previous section, as shown in Figure 11.27. Of course, since both coordinate systems appear in Figure 11.27, we have also obtained the graph of the equation $xy = 1$ in terms of the original x, y-coordinate system. *Try Study Suggestion 11.15.* □

▶ **Study Suggestion 11.15:** Introduce a new coordinate system by rotating the x, y-coordinate system through an angle of 30°. Express the equation $x^2 + 2\sqrt{3}xy - y^2 = 4$ in terms of the new coordinate system. What shape is the graph of this equation? Sketch the graph on a plane with both sets of coordinate axes. ◀

Now let us turn to Equation (11.10). Our plan is to determine an angle θ through which we can rotate the x, y-coordinate system to obtain a new coordinate system in the hopes that when we express Equation (11.10) in the new system, there will be no cross term. If we can accomplish this, then we can graph the resulting equation in the new coordinate system, using the methods of the previous sections. This will also give us the graph of Equation (11.10) in the original coordinate system.

If the x', y'-coordinate system is obtained by rotating the x, y-coordinate axes through an angle θ, then according to the previous theorem, Equation (11.10) becomes

$$\begin{aligned} A(x'\cos\theta - y'\sin\theta)^2 &+ B(x'\cos\theta - y'\sin\theta)(x'\sin\theta + y'\cos\theta) \\ &+ C(x'\sin\theta + y'\cos\theta)^2 + D(x'\cos\theta - y'\sin\theta) \\ &+ E(x'\sin\theta + y'\cos\theta) + F = 0 \end{aligned}$$

After considerable simplification (we will omit the details), this equation can be written in the form

$$A'(x')^2 + B'x'y' + C'(y')^2 + D'x' + E'y' + F = 0$$

where the coefficient of $x'y'$ (which is the only one that we are interested in now) is equal to

$$B' = 2(C - A)\sin\theta\cos\theta + B(\cos^2\theta - \sin^2\theta)$$

Using the double angle formulas

$$2\sin\theta\cos\theta = \sin 2\theta \qquad \text{and} \qquad \cos^2\theta - \sin^2\theta = \cos 2\theta$$

this becomes

$$B' = (C - A)\sin 2\theta + B\cos 2\theta$$

Now we want to choose θ so that $B' = 0$; that is, we want to choose θ so that

$$(C - A)\sin 2\theta + B\cos 2\theta = 0$$

If $A \neq C$, this can be written in the form

$$\tan 2\theta = \frac{B}{A - C}$$

and if $A = C$, then a solution to this equation is $\theta = 45°$. This gives us the following theorem.

THEOREM

Let the x', y'-coordinate system be obtained by rotating the x, y-coordinate system through an angle θ satisfying

$$\tan 2\theta = \frac{B}{A - C} \qquad \text{if } A \neq C$$

or

$$\theta = 45^\circ \qquad \text{if } A = C$$

Then if we express Equation (11.10) in the x', y'-coordinate system, the resulting equation will have no cross term. Hence, if Equation (11.10) has a graph, this graph must be a conic section when viewed in the x', y'-coordinate system, and so also when viewed in the x, y-coordinate system.

Let us see how we can use this theorem to help graph equations of the form (11.10).

EXAMPLE 3: Identify and sketch the graph of the equation

$$x^2 + 2\sqrt{3}xy + 3y^2 + \sqrt{3}x - y + 2 = 0 \tag{11.14}$$

Solution: We seek an angle θ through which to rotate the x, y-coordinate system in order to eliminate the cross term $2\sqrt{3}xy$. According to the previous theorem (since $A \neq C$), such an angle must satisfy the equation

$$\tan 2\theta = \frac{B}{A - C} = \frac{2\sqrt{3}}{-2} = -\sqrt{3}$$

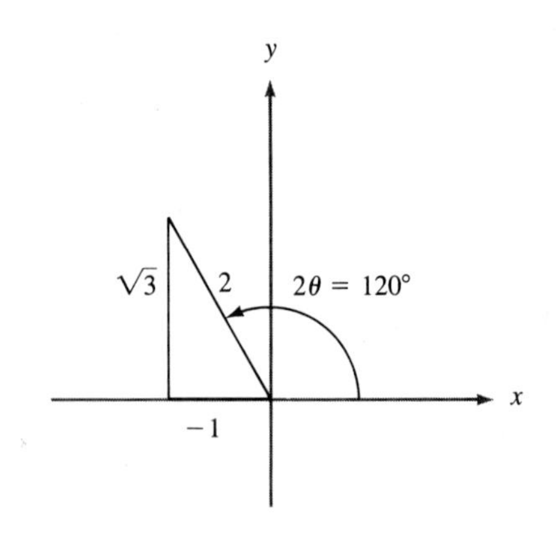

FIGURE 11.28

Figure 11.28 shows one possible choice for the angle 2θ, namely $2\theta = 120^\circ$. (It is always possible to choose 2θ so that $0 < \theta < 90^\circ$.) Hence $\theta = 60^\circ$ and according to the second theorem in this section, if we rotate the x, y-coordinate system through an angle of 60°, the resulting x', y'-coordinate system is related to the original one by the equations

$$x = x' \cos 60^\circ - y' \sin 60^\circ = \left(\frac{1}{2}\right)(x' - \sqrt{3}y')$$

$$y = x' \sin 60^\circ + y' \cos 60^\circ = \left(\frac{1}{2}\right)(\sqrt{3}x' + y')$$

Substituting these values of x and y into Equation (11.14) gives

$$\frac{1}{4}(x' - \sqrt{3}y')^2 + \left(\frac{\sqrt{3}}{2}\right)(x' - \sqrt{3}y')(\sqrt{3}x' + y') + \left(\frac{3}{4}\right)(\sqrt{3}x' + y')^2$$
$$+ \left(\frac{\sqrt{3}}{2}\right)(x' - \sqrt{3}y') - \left(\frac{1}{2}\right)(\sqrt{3}x' + y') + 2 = 0$$

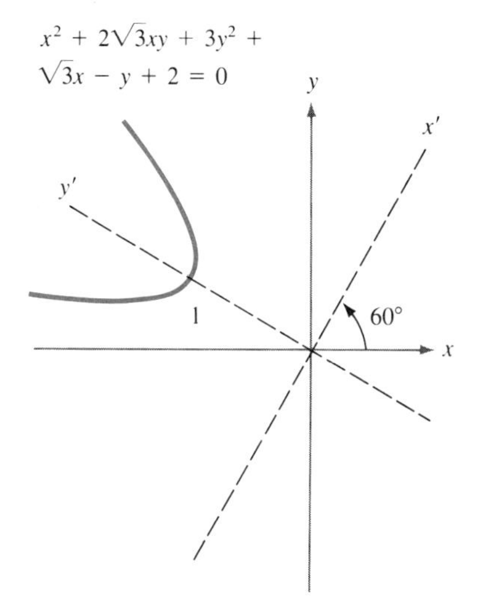

FIGURE 11.29

This equation can be simplified as follows (first we multiply both sides by 4)

$$\begin{aligned}
&(x' - \sqrt{3}y')^2 + 2\sqrt{3}(x' - \sqrt{3}y')(\sqrt{3}x' + y') + 3(\sqrt{3}x' + y')^2 \\
&\qquad + 2\sqrt{3}(x' - \sqrt{3}y') - 2(\sqrt{3}x' + y') + 8 = 0 \\
&((x')^2 - 2\sqrt{3}x'y' + 3(y')^2) + 2\sqrt{3}(\sqrt{3}(x')^2 - 2x'y' - \sqrt{3}(y')^2) \\
&\qquad + 3(3(x')^2 + 2\sqrt{3}x'y' + (y')^2) - 8y' + 8 = 0 \\
&(x')^2 - 2\sqrt{3}x'y' + 3(y')^2 + 6(x')^2 - 4\sqrt{3}x'y' - 6(y')^2 \\
&\qquad + 9(x')^2 + 6\sqrt{3}x'y' + 3(y')^2 - 8y' + 8 = 0 \\
&16(x')^2 - 8y' + 8 = 0 \\
&\quad 2(x')^2 - y' + 1 = 0 \\
&\qquad\qquad y' = 2(x')^2 + 1 \qquad (11.15)
\end{aligned}$$

We recognize this as the equation of a parabola in the x', y'-coordinate system. Now we draw both the x, y- and x', y'-coordinate systems on the same plane, and graph Equation (11.15) *relative to the x', y'-coordinate system.* This is done in Figure 11.29. (You may want to rotate the book so that the x'-axis is horizontal.) Of course, this is also the graph of Equation (11.14) *relative to the x, y-coordinate system.* Notice that when we make the substitution $y = 0$ in Equation (11.14), the result is $x^2 + \sqrt{3x} + 2 = 0$, and since this equation has no solutions, the graph has no x-intercepts.

▶ **Study Suggestion 11.16:** Identify and sketch the graph of the equation $x^2 - 2xy + y^2 - 8\sqrt{2}x - 8\sqrt{2}y = 0$. ◀

Try Study Suggestion 11.16. □

EXAMPLE 4: Identify and sketch the graph of the equation

$$6x^2 + 4xy + 9y^2 - 20 = 0 \qquad (11.16)$$

Solution: Since $A \neq C$, we again seek an angle θ for which

$$\tan 2\theta = \frac{B}{A - C} = \frac{4}{-3}$$

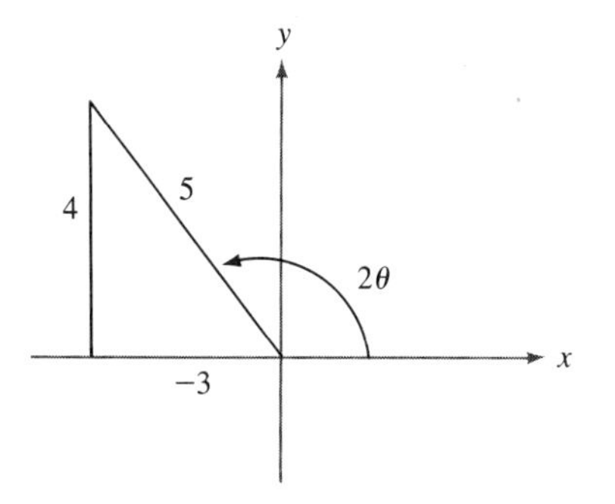

FIGURE 11.30

Figure 11.30 shows one possible choice for the angle 2θ. In this case, it is not so easy to determine the value of θ. Fortunately however, we only need to determine the values of $\sin\theta$ and $\cos\theta$ in order to employ the formula for rotation of coordinate axes. From Figure 11.30 we see that $\cos 2\theta = -3/5$, and so we may use the trigonometric identities

$$\sin^2\theta = \frac{1 - \cos 2\theta}{2} \qquad \text{and} \qquad \cos^2\theta = \frac{1 + \cos 2\theta}{2}$$

to compute $\sin\theta$ and $\cos\theta$. In particular, since $0 < \theta < 90°$, we see that

$$\sin\theta = \sqrt{\frac{1 + 3/5}{2}} = \frac{2}{\sqrt{5}}$$

and

$$\cos\theta = \sqrt{\frac{1 - 3/5}{2}} = \frac{1}{\sqrt{5}}$$

Now, if we rotate the x, y-coordinate system through the angle θ, the resulting x', y'-coordinate system is related to the original one by the equations

$$x = x'\cos\theta - y'\sin\theta = \left(\frac{1}{\sqrt{5}}\right)(x' - 2y')$$

$$y = x'\sin\theta + y'\cos\theta = \left(\frac{1}{\sqrt{5}}\right)(2x' + y')$$

Substituting these values of x and y into Equation (11.16) gives

$$\left(\frac{6}{5}\right)(x' - 2y')^2 + \left(\frac{4}{5}\right)(x' - 2y')(2x' + y') + \frac{9}{5}(2x' + y')^2 - 20 = 0$$

Simplifying this equation gives

$$2(x')^2 + (y')^2 - 4 = 0$$

or in standard form

$$\frac{(x')^2}{2} + \frac{(y')^2}{4} = 1 \qquad \textbf{(11.17)}$$

We recognize this as the equation of an ellipse in the x', y'-coordinate system. Now we draw the x, y- and x', y'-coordinate systems on the same plane. (See Figure 11.31.) Since we do not know the exact value of the angle θ, we content ourselves with using a calculator to obtain the *approximate* value $\theta \approx$

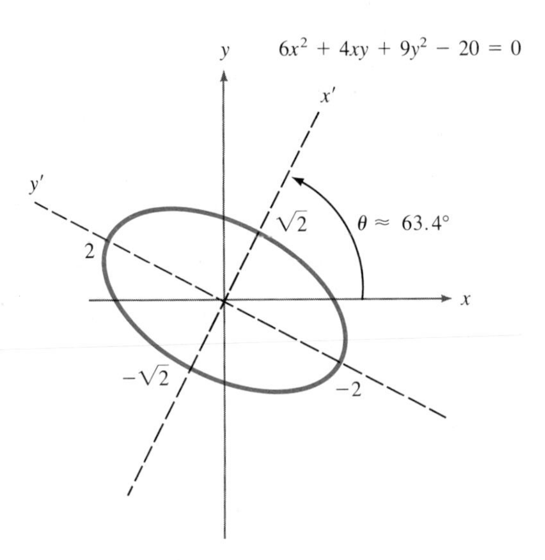

FIGURE 11.31

63.4°. Using the x', y'-coordinate system, we can sketch the graph of Equation (11.17). Of course, this is also the graph of Equation (11.16) with respect to the x, y-coordinate system. *Try Study Suggestion 11.17.* □

▶ **Study Suggestion 11.17:** Identify and sketch the graph of the equation

$$3x^2 - 4xy + 4 = 0.$$ ◀

Let us conclude this section by remarking that it is possible to determine the shape of the graph of the general quadratic (11.10) without having to actually find the graph. The key to making this determination is the expression $B^2 - 4AC$, which is called the **discriminant** of the general quadratic Equation (11.10).

THEOREM

If the equation

$$Ax^2 + Bxy + Cy^2 + Dx + Ey + F = 0$$

has a graph, and if at least one of A and C is not zero, then we have the following possibilities:

1. If $B^2 - 4AC = 0$, the graph is a parabola.
2. If $B^2 - 4AC < 0$, the graph is an ellipse or circle.
3. If $B^2 - 4AC > 0$, the graph is a hyperbola.

In each case, a degenerate graph is possible.

EXAMPLE 5: Determine the shape of the graph of each of the following equations without drawing the graph.

(a) $12x^2 + 12xy + 3y^2 - 15x + 25y - 32 = 0$

(b) $x^2 - 3xy + 2y^2 + 5x - y = 12$

(c) $2x(2x - y) = (1 - 2y)(1 + 2y)$

Solutions:

(a) In this case, the discriminant is $B^2 - 4AC = (12)^2 - 4(12)(3) = 0$, and so the graph (if it exists) is a parabola.

(b) In this case, the discriminant is $B^2 - 4AC = (-3)^2 - 4(1)(2) = 1 > 0$, and so the graph (if it exists) is a hyperbola.

(c) Multiplying out both sides of this equation, and bringing it into the form (11.10) gives $4x^2 - 2xy + 4y^2 = 1$. Since the discriminant in this case is $B^2 - 4AC = (-2)^2 - 4(4)(4) = -60 < 0$, the graph (if it exists) is an ellipse. *Try Study Suggestion 11.18.* □

▶ **Study Suggestion 11.18:** Determine the shape of the graph of each of the following equations without drawing the graph.

(a) $x^2 + xy + y^2 - 3x + y = 10$

(b) $x^2 - 2xy + y^2 - 3x + 5y - 12 = 0$

(c) $x(2x + y) = 1 + 5y$ ◀

Ideas to Remember

If the graph of the quadratic equation

$$Ax^2 + Bxy + Cy^2 + Dx + Ey + F = 0$$

exists (where at least one of A and C is nonzero), then this graph must be a conic section.

EXERCISES

In Exercises 1–6, determine the shape of the graph of each equation without drawing the graph.

1. (a) $5x^2 + 5xy + 2y^2 - 12x - 1 = 0$
   (b) $x^2 + 2xy + y^2 - 3x + 4y - 12 = 0$
   (c) $2x^2 + 5xy + y^2 - x + y - 3 = 0$
2. (a) $2x^2 + xy + y^2 - 3x + 4y = 15$
   (b) $-2x^2 - 2\sqrt{6}xy + 3y^2 = 0$
   (c) $4x^2 - 3xy - y^2 + 12x - 14y = 1$
3. (a) $y(5y + 2) - 3x - 4 = 0$
   (b) $2x^2 + 8xy + 8y^2 + 2x - 3y + 5 = 0$
   (c) $-x^2 + 3xy - 2y^2 + 4x - 3 = 0$
4. (a) $y(2x + y) = 3$
   (b) $-x^2 + 2xy - 2y^2 + 4y - 2 = 0$
   (c) $2x(2x - 1) = -y(3y + 1)$
5. (a) $2x^2 + 12xy + 18y^2 - 36x + 48y - 60 = 0$
   (b) $(x + 3y)^2 - 10y + 10 = 0$
   (c) $5xy = x - 3y + 1$
6. (a) $5x^2 + 10xy + 4y^2 - 5x + 4y - 20 = 0$
   (b) $3x^2 + 3xy + 3y^2 - 5x + 5y - 5 = 0$
   (c) $-32x^2 + 16xy - 2y^2 + 1 = 0$

In Exercises 7–33, identify and sketch the graph of the given equation.

7. $3x^2 + 2xy + y^2 + 5x - 2y + 7 = 0$
8. $34x^2 + 24xy + 41y^2 = 1$
9. $xy = 4$
10. $x^2 + 4\sqrt{3}xy - 5y^2 - 8 = 0$
11. $3x^2 + 6xy + 3y^2 - \sqrt{2}x + \sqrt{2}y = 0$
12. $6x^2 + 4\sqrt{3}xy + 2y^2 - x + \sqrt{3}y - 4 = 0$
13. $xy - 2y - x + 2 = 0$
14. $25x^2 - 14xy + 25y^2 = 288$
15. $3x^2 - 2\sqrt{3}xy + y^2 + 3x + 3\sqrt{3}y - 4 = 0$
16. $4x^2 - 4xy + y^2 - \sqrt{5}x - 2\sqrt{5}y + 10 = 0$
17. $9x^2 + 24xy + 16y^2 + 40x - 30y + 100 = 0$
18. $3x^2 - 10xy + 3y^2 + 10\sqrt{2}x + 6\sqrt{2}y - 5 = 0$
19. $7x^2 - 48xy + 25y^2 = 50$
20. $x^2 + 2xy + y^2 - \sqrt{2}x - \sqrt{2}y = 0$
21. $17x^2 + 6xy + 11y^2 - 40 = 0$
22. $35x^2 - 12xy + 30y^2 = 13$
23. $-17x^2 + 18xy + 7y^2 - 40 = 0$
24. $x^2 + 2xy + y^2 - \sqrt{2}x - 3\sqrt{2}y + 8 = 0$
25. $7x^2 - 6\sqrt{3}xy + 13y^2 - 4\sqrt{3}x - 4y - 12 = 0$
26. $3x^2 + 8xy - 3y^2 - 4\sqrt{5}x - 2\sqrt{5}y = 0$
27. $2x^2 + 5xy + 2y^2 = 0$
28. $2x^2 + 12xy + 18y^2 + 3\sqrt{10}x - \sqrt{10}y - 40 = 0$
29. $9x^2 + 6\sqrt{3}xy + 3y^2 + 2x - 2\sqrt{3}y + 4 = 0$
30. $x^2 + 2xy + y^2 - x + 5y = 0$
31. $5x^2 - 6\sqrt{3}xy - y^2 + 16\sqrt{3}x - 16y + 24 = 0$

32. $6x^2 + 4xy + 9y^2 + 16\sqrt{5}x - 8\sqrt{5}y + 80 = 0$

33. $7x^2 - 2\sqrt{3}xy + 5y^2 + 4x + 4\sqrt{3}y = 0$

34. Consider the equations

$$Ax^2 + Bxy + Cy^2 + Dx + Ey + F = 0$$

and

$$A'(x')^2 + B'x'y' + C'(y')^2 + D'x' + E'y' + F' = 0$$

Assume that the second equation can be derived from the first one by rotating the x, y-coordinate system through an angle θ.

(a) Show that $A + C = A' + C'$

(b) Show that $B^2 - 4AC = (B')^2 - 4A'C'$. This shows that the discriminant is *invariant under a rotation.*

11.6 Review

CONCEPTS FOR REVIEW

Conic section
Parabola
Focus of a parabola
Directrix of a parabola
Axis of a parabola
Vertex of a parabola
Standard form of the equation of a parabola
Ellipse
Foci of an ellipse
Center of an ellipse
Major axis of an ellipse
Minor axis of an ellipse
Standard form of the equation of an ellipse
Hyperbola
Foci of a hyperbola
Center of a hyperbola
Transverse axis of a hyperbola
Asymptotes of a hyperbola
Fundamental rectangle of a hyperbola
Standard form of the equation of a hyperbola
Rotation of axes

REVIEW EXERCISES

In Exercises 1–9, determine the shape of the graph of the given equation without actually graphing the equation.

1. $5x^2 - 2y + 3x + 1 = 0$

2. $6x^2 + 10xy + 2y^2 - 10x + 20y = 50$

3. $3x^2 - 2xy + 5y^2 + 6x - 3y - 9 = 0$

4. $4y^2 - 3y = 2x^2 + 4x - 5$

5. $4x(3x + 12y - 1) = -y(2 + 3y)$

6. $3x^2 + 2y^2 - 4x + 8y - 3 = 0$

7. $x(x + 5y) - 3x + y = 0$

8. $4x^2 + 20xy + 25y^2 - 3x + 2y - 10 = 0$

9. $2x^2 - 6xy + 3y^2 - 8x = 5$

In Exercises 10–13, find the equation of the parabola with the given properties.

10. Vertex (2, 1); directrix: $y = -3$

11. Vertex (3, 4); focus $(-2, 4)$

12. Focus (1, 1); directrix: $x = 0$

13. Vertex $(1, -2)$; opens to the left; goes through the origin

In Exercises 14–16, find the equation of the ellipse with the given properties.

14. Center (1, 2); major axis vertical; axes of length 3 and $\sqrt{5}$

15. Foci (0, 0) and (2, 0); major axis of length 3

16. Foci (0, 0) and (2, 0); minor axis of length 3

In Exercises 17–19, find the equation of the hyperbola with the given properties.

17. Center (1, 1); vertex (1, 2); focus (1, 3)

18. Asymptotes: $y = \pm 3x$, y-intercepts ± 4

19. Center (0, 0); transverse axis vertical; passing through the points (1, 2) and (2, 3)

In Exercises 20–32, sketch the graph of the given equation.

20. $x = 2y^2 + y - 2$

21. $\frac{x^2}{2} + y^2 = 1$

22. $3x^2 + 3y^2 = 1$

23. $\frac{(x + 1)^2}{2} - \frac{(y - 1)^2}{4} = 1$

24. $\frac{(y + 1)^2}{2} - \frac{(x - 1)^2}{4} = 1$

25. $3x^2 + 5y^2 = 1$

26. $x + 3y^2 - y = 0$

27. $y = x^2 + 3x - 1$

28. $y + x^2 + 2x + 2 = 0$

29. $4x^2 - 9y^2 + 16x + 18y + 6 = 0$

30. $xy = 12$

31. $x^2 - 2x + 3 = y^2 + 4y + 6$

32. $4x^2 + 4y^2 - 40x + 8y + 103 = 0$

In Exercises 33–40, graph the given equation.

33. $5x^2 + 12xy - 4 = 0$

34. $34x^2 - 24xy + 41y^2 = 200$

35. $2x^2 - \sqrt{3}xy + y^2 = 20$

36. $x^2 - 2\sqrt{3}xy - y^2 - 2 = 0$

37. $9x^2 + 24xy + 16y^2 + 35x + 5y + 25 = 0$

38. $5x^2 - 10xy + 5y^2 + \sqrt{2}x + \sqrt{2}y = 0$

39. $21x^2 - 16xy + 9y^2 = 5$

40. $2xy + 2y - x + 1 = 0$

41. Is it possible for the fundamental rectangle of a hyperbola to be a square? Explain your answer.

12 SEQUENCES, SERIES AND OTHER TOPICS

12.1 Mathematical Induction

In this section, we will discuss a very special technique for proving certain types of mathematical statements, known as the *principle of mathematical induction.* Let us begin by illustrating the technique with an example.

Suppose that we wish to compare the numbers $2 \cdot n$ and 2^n, for all positive integers n. For the first few values of n, we observe that

$$\begin{array}{llll} 2 \cdot 1 = 2, & 2^1 = 2 & \text{and so} & 2 \cdot 1 = 2^1 \\ 2 \cdot 2 = 4, & 2^2 = 4 & \text{and so} & 2 \cdot 2 = 2^2 \\ 2 \cdot 3 = 6, & 2^3 = 8 & \text{and so} & 2 \cdot 3 < 2^3 \\ 2 \cdot 4 = 8, & 2^4 = 16 & \text{and so} & 2 \cdot 4 < 2^4 \\ 2 \cdot 5 = 10, & 2^5 = 32 & \text{and so} & 2 \cdot 5 < 2^5 \end{array}$$

From these examples, it appears that $2 \cdot n$ might always be less than or equal to 2^n, that is,

$$2n \leq 2^n \tag{12.1}$$

for all positive integers n. However, since Inequality (12.1) is actually an infinite number of different inequalities—one for each value of n, we cannot *prove* it simply by substituting larger and larger values of n.

But suppose we can show that

STATEMENT 1. Inequality (12.1) is true when n is replaced by 1;

STATEMENT 2. *If* (12.1) is true when n is replaced by a positive integer k, then it must also be true when n is replaced by the *next larger* positive integer $k + 1$.

Then, by repeatedly using Statement 2, we can conclude that (12.1) must be true for *all* positive integers n. After all, (12.1) is true for $n = 1$ by Statement 1. Then we may apply Statement 2 to deduce that (12.1) is true for the next larger integer $n = 2$. Using Statement 2 again, we see that (12.1) is true for the next larger integer $n = 3$, and so on. Continuing in this way, we see that (12.1) holds for all positive integers n. This is the idea behind mathematical induction.

So let us show that Statements 1 and 2 are true. Statement 1 says that

$$2 \cdot 1 \leq 2^1$$

which is certainly true. In order to show that Statement 2 is true, we must show that *if* (12.1) is true when n is replaced by a positive integer k, then it is also true when n is replaced by the next larger integer $k + 1$. That is, we must show that if

$$2k \leq 2^k$$

then

$$2(k + 1) \leq 2^{k+1} \tag{12.2}$$

But using $2k \leq 2^k$, we have

$$2(k + 1) = 2k + 2 \leq 2^k + 2 \leq 2^k + 2^k = 2 \cdot 2^k = 2^{k+1}$$

that is,

$$2(k + 1) \leq 2^{k+1}$$

and so (12.2) does hold. Hence, Statement 2 is true.

Thus, by the reasoning above, we conclude that (12.1) holds for all positive integers n. This completes the proof of Equation (12.1) by mathematical induction.

It may help to understand the principle of mathematical induction to think of (12.1) as an entire sequence, or list, of *propositions*, one for each value of n. If we denote the nth proposition by $P(n)$, then (12.1) represents the sequence of propositions

$$\begin{aligned}
&P(1)\colon 2 \cdot 1 \leq 2^1 \\
&P(2)\colon 2 \cdot 2 \leq 2^2 \\
&P(3)\colon 2 \cdot 3 \leq 2^3 \\
&P(4)\colon 2 \cdot 4 \leq 2^4 \\
&\quad\vdots
\end{aligned}$$

Statement 2 provides us with a link between successive propositions in the list, for it tells us that if one of the propositions is true, then the *next* proposition in the list is also true. Notice that this statement, by itself, does not tell us that any of the propositions are necessarily true. However, Statement 1 tells us that the first proposition $P(1)$ is true, and so we can use Statement

2 to go down the entire list of propositions, one at a time, to conclude that each proposition is true.

Using this terminology, we can describe the principle of mathematical induction in the following theorem.

THEOREM

Principle of mathematical induction Let $P(n)$ be a sequence of propositions—one for each positive integer n. If we can show that

STATEMENT 1. Proposition $P(1)$ is true;

STATEMENT 2. *If* proposition $P(k)$ is true for any positive integer k, then the next proposition $P(k + 1)$ is also true;

then we can conclude that $P(n)$ is true for *all* positive integers n.

Let us consider some other examples of mathematical induction.

EXAMPLE 1: Use mathematical induction to show that

$$1 + 2 + 3 + \cdots + n = \frac{n(n + 1)}{2} \tag{12.3}$$

for all positive integers n, where the left side of Equation (12.3) is meant to denote the sum of the positive integers between 1 and n. For example, if $n = 1$, the left side of (12.3) is 1, if $n = 2$, the left side is $1 + 2$, and if $n = 3$, the left side is $1 + 2 + 3$.

Solution: In this case, we let $P(n)$ be the proposition

$$P(n)\colon \quad 1 + 2 + 3 + \cdots + n = \frac{n(n + 1)}{2}$$

Our plan is to show that

STATEMENT 1. $P(1)$ is true;

STATEMENT 2. *If* $P(k)$ is true, then $P(k + 1)$ must also be true.

Then, according to the principle of mathematical induction, $P(n)$ will be true for all positive integers n; that is, Equation (12.3) will be true for all positive integers n.

But $P(1)$ is the proposition that

$$1 = \frac{1(2)}{2}$$

which is certainly true. Now we must prove that *if* $P(k)$ is true, then so is

$P(k+1)$. That is, we must prove that if

$$1 + 2 + 3 + \cdots + k = \frac{k(k+1)}{2} \tag{12.4}$$

then

$$1 + 2 + 3 + \cdots + (k+1) = \frac{(k+1)(k+2)}{2} \tag{12.5}$$

Using Equation (12.4), we have

$$\begin{aligned} 1 + 2 + 3 + \cdots + (k+1) &= (1 + 2 + 3 + \cdots + k) + (k+1) \\ &= \frac{k(k+1)}{2} + (k+1) \\ &= \frac{k(k+1) + 2(k+1)}{2} \\ &= \frac{(k+1)(k+2)}{2} \end{aligned}$$

which shows that Equation (12.5) is true. Thus, according to the principle of mathematical induction, Equation (12.3) is true for all positive integers n. This completes the proof. □

One of the major points of confusion concerning the principle of mathematical induction is the following. It may seem that, in the process of showing that Statement 2 of the theorem is true, we are *assuming* that $P(k)$ is true for all positive integers k. Surely we cannot be allowed to assume what we are trying to prove!

However, we are *not* really assuming that $P(k)$ is true for all positive integers k. Let us explain what we are doing. In order to prove that Statement 2 is true, we must prove that *if* $P(k)$ is true, then so is $P(k+1)$. In order to do this, we *are* allowed to assume that $P(k)$ is true *strictly for the purpose of showing that* $P(k+1)$ *is true.* Once this is done, we can state that *if* $P(k)$ is true, then so is $P(k+1)$. This does *not* show, by itself, that any of the propositions $P(n)$ are necessarily true, all it shows is that *if* one of the propositions is true, then so is the next one.

The assumption that $P(k)$ is true, made for the purpose of showing that $P(k+1)$ is then true, is called the **induction hypothesis.** Let us do another example of mathematical induction.

EXAMPLE 2: Prove by mathematical induction that

$$\frac{1}{1\cdot 2} + \frac{1}{2\cdot 3} + \frac{1}{3\cdot 4} + \cdots + \frac{1}{n(n+1)} = \frac{n}{n+1} \tag{12.6}$$

for all positive integers n.

Solution: In this case, we let $P(n)$ be the proposition that Equation (12.6) is true. Then $P(1)$ is the proposition that

$$\frac{1}{1 \cdot 2} = \frac{1}{1+1}$$

which is true. Now we must show that if $P(k)$ is true (this is the induction hypothesis), then so is $P(k+1)$. That is, we must show that if

$$\frac{1}{1 \cdot 2} + \frac{1}{2 \cdot 3} + \frac{1}{3 \cdot 4} + \cdots + \frac{1}{k(k+1)} = \frac{k}{k+1} \tag{12.7}$$

then

$$\frac{1}{1 \cdot 2} + \frac{1}{2 \cdot 3} + \frac{1}{3 \cdot 4} + \cdots + \frac{1}{(k+1)(k+2)} = \frac{k+1}{k+2} \tag{12.8}$$

But, assuming that (12.7) is true, we have

$$\begin{aligned}
\frac{1}{1 \cdot 2} + \frac{1}{2 \cdot 3} + \frac{1}{3 \cdot 4} + \cdots + \frac{1}{(k+1)(k+2)} \\
&= \left[\frac{1}{1 \cdot 2} + \frac{1}{2 \cdot 3} + \frac{1}{3 \cdot 4} + \cdots + \frac{1}{k(k+1)}\right] + \frac{1}{(k+1)(k+2)} \\
&= \frac{k}{k+1} + \frac{1}{(k+1)(k+2)} \\
&= \frac{k(k+2)+1}{(k+1)(k+2)} \\
&= \frac{k^2+2k+1}{(k+1)(k+2)} \\
&= \frac{(k+1)^2}{(k+1)(k+2)} \\
&= \frac{k+1}{k+2}
\end{aligned}$$

This shows that (12.8) is true, and so, according to the principle of mathematical induction, $P(n)$ is true for all positive integers n. This completes the proof. *Try Study Suggestion 12.1.* □

▶ **Study Suggestion 12.1:** Use mathematical induction to prove that

$$1^2 + 2^2 + 3^2 + \cdots + n^2 = \frac{n(n+1)(2n+1)}{6}$$

for all positive integers n. ◀

Sometimes we want to prove that a sequence of propositions $P(n)$ is true, where n ranges over all integers greater than or equal to a specific integer m, which could be positive, zero, or negative. In other words, we want to show that the propositions

$$\begin{gathered} P(m) \\ P(m+1) \\ P(m+2) \\ \vdots \end{gathered}$$

are true, where m is some integer. In the examples that we have discussed so far, m is equal to 1, but it is quite common for m to be equal to 0.

We can easily modify the principle of mathematical induction to accommodate this situation as follows.

THEOREM

Principle of mathematical induction, second version Suppose that $P(n)$ is a sequence of propositions—one for each integer $n \geq m$. If we can show that

STATEMENT 1. Proposition $P(m)$ is true;

STATEMENT 2. *If* proposition $P(k)$ is true for any integer $k \geq m$, then the next proposition $P(k+1)$ is also true;

then we can conclude that $P(n)$ is true for *all* integers $n \geq m$.

As the next example shows, mathematical induction can be used to prove the properties of exponents that we discussed in Chapter 1.

EXAMPLE 3: Prove by induction that if a and b are nonzero real numbers, then

$$(ab)^n = a^n b^n$$

for all nonnegative integers n.

Solution: Our plan is to apply the second version of the principle of mathematical induction in the case $m = 0$. If we let $P(n)$ be the proposition that $(ab)^n = a^n b^n$, then $P(0)$ is the proposition that

$$(ab)^0 = a^0 b^0$$

But since $(ab)^0 = 1$ and $a^0 b^0 = 1$, this is certainly true. In order to complete the proof, we must show that if

$$(ab)^k = a^k b^k \tag{12.9}$$

for any integer $k \geq 0$, then

$$(ab)^{k+1} = a^{k+1} b^{k+1} \tag{12.10}$$

But, if (12.9) is true (the induction hypothesis), then we have

$$(ab)^{k+1} = (ab)^k(ab) = a^k b^k(ab) = a^k a b^k b = a^{k+1} b^{k+1}$$

and so (12.10) is true. This shows that $P(n)$ is true for all integers $n \geq 0$, and completes the proof by induction. *Try Study Suggestion 12.2.* □

▶ **Study Suggestion 12.2:** Let a and b be nonzero real numbers. Use mathematical induction to prove that $(a/b)^n = a^n/b^n$ for all nonnegative integers. ◀

Ideas to Remember

Even though the principle of mathematical induction applies only to a very specific type of problem, it is still one of the most powerful techniques that we have, for it allows us to prove an infinite number of different propositions at one time.

EXERCISES

In Exercises 1–10, use mathematical induction to prove the given formula.

1. $2 + 4 + 6 + \cdots + 2n = n(n + 1)$

2. $1 + 3 + 5 + \cdots + (2n - 1) = n^2$

3. $1^3 + 2^3 + 3^3 + \cdots + n^3 = \dfrac{n^2(n + 1)^2}{4}$

4. $(-1)^1 + (-1)^2 + (-1)^3 + \cdots + (-1)^n = \dfrac{(-1)^n - 1}{2}$

5. $1 + 2 + 2^2 + \cdots + 2^n = 2^{n+1} - 1$

6. $1 + 2 \cdot 2 + 3 \cdot 2^2 + 4 \cdot 2^3 + \cdots + n \cdot 2^{n-1} = 1 + (n - 1) \cdot 2^n$

7. $1 \cdot 2 + 2 \cdot 3 + 3 \cdot 4 + \cdots + n(n + 1) = \dfrac{n(n + 1)(n + 2)}{3}$

8. $3 + 9 + 15 + \cdots + (6n - 3) = 3n^2$

9. $\dfrac{1}{1 \cdot 3} + \dfrac{1}{3 \cdot 5} + \dfrac{1}{5 \cdot 7} + \cdots + \dfrac{1}{(2n - 1)(2n + 1)} = \dfrac{n}{2n + 1}$

10. $1^3 + 2^3 + 3^3 + \cdots + n^3 = (1 + 2 + 3 + \cdots + n)^2$

11. Prove that if $r \neq 1$, then

$$1 + r + r^2 + \cdots + r^n = \frac{1 - r^n}{1 - r}$$

12. Prove that

$$1 + 2 + 3 + \cdots + n < (1/8)(2n + 1)^2$$

13. Prove that $2^n > n^2$ for all integers $n \geq 5$.

14. Prove that if $a > -1$, then

$$(1 + a)^n > 1 + na$$

for all integers $n \geq 2$.

15. Prove that 2 is a factor of $n^2 + n$ for all positive integers n.

16. Prove that 3 is a factor of $n^3 + 14n + 3$ for all positive integers n.

17. Prove that 4 is a factor of $5^n - 1$ for all positive integers n.

18. Use the fact that $\log ab = \log a + \log b$, and mathematical induction, to prove that $\log a^n = n \log a$ for all positive integers n.

19. Use the fact that $\log ab = \log a + \log b$, and mathematical induction, to prove that

$$\log a_1 a_2 \cdots a_n = \log a_1 + \log a_2 + \cdots + \log a_n$$

for all positive real numbers $a_1, a_2, \ldots, a_n$.

20. Prove that $\sin(\theta + n\pi) = (-1)^n \sin\theta$ for all nonnegative integers n.

21. Prove that $\cos(\theta + n\pi) = (-1)^n \cos\theta$ for all nonnegative integers n.

22. Prove that if z is a complex number, then $|z|^n = |z^n|$ for all nonnegative integers n.

23. Prove that if $z_1, z_2, \ldots, z_n$ are complex numbers, then $|z_1 z_2 \cdots z_n| = |z_1||z_2| \cdots |z_n|$.

24. Prove that if z is a complex number, then $(\bar{z})^n = (\overline{z^n})$ for all nonnegative integers n.

25. Prove that if $z_1, z_2, \ldots, z_n$ are complex numbers, then $\overline{z_1 z_2 \cdots z_n} = \bar{z}_1 \bar{z}_2 \cdots \bar{z}_n$.

26. Use the distributive law and mathematical induction to prove the **generalized distributive law,** which says that

$$a(b_1 + b_2 + \cdots + b_n) = ab_1 + ab_2 + \cdots + ab_n$$

for all numbers $a, b_1, b_2, \ldots, b_n$.

27. Prove De Moivre's Theorem. That is, prove that

$$[r(\cos\theta + i\sin\theta)]^n = r^n(\cos n\theta + i\sin n\theta)$$

for all nonnegative integers n.

28. Using the fact that $|x+y| \leq |x| + |y|$ for all real numbers x and y, prove by induction that

$$|x_1 + x_2 + \cdots + x_n| \leq |x_1| + |x_2| + \cdots + |x_n|$$

for all real numbers $x_1, x_2, \ldots, x_n$.

12.2 The Binomial Formula

Let us begin this section by introducing a bit of new notation. If k is a positive integer, then we denote the product of the first k positive integers by $k!$. This is read "k factorial." Thus, for example,

$$\begin{aligned} 1! &= 1 \\ 2! &= 1 \cdot 2 = 2 \\ 3! &= 1 \cdot 2 \cdot 3 = 6 \\ 4! &= 1 \cdot 2 \cdot 3 \cdot 4 = 24 \\ 5! &= 1 \cdot 2 \cdot 3 \cdot 4 \cdot 5 = 120 \end{aligned}$$

and in general, if k is a positive integer, then

$$k! = 1 \cdot 2 \cdot 3 \cdots k$$

Also, it will be convenient for us to define $0!$ to be equal to 1, that is,

$$0! = 1$$

Since we will use factorials many times in this chapter, let us pause to consider some examples.

EXAMPLE 1: Evaluate the following expressions.

(a) $\dfrac{6!}{4!}$ **(b)** $\dfrac{8!}{5!3!}$

Solutions: Using the definition of the factorial, we have

(a) $\dfrac{6!}{4!} = \dfrac{1 \cdot 2 \cdot 3 \cdot 4 \cdot 5 \cdot 6}{1 \cdot 2 \cdot 3 \cdot 4} = 5 \cdot 6 = 30$

(b) $\dfrac{8!}{5!3!} = \dfrac{1 \cdot 2 \cdot 3 \cdot 4 \cdot 5 \cdot 6 \cdot 7 \cdot 8}{1 \cdot 2 \cdot 3 \cdot 4 \cdot 5 \cdot 1 \cdot 2 \cdot 3} = \dfrac{6 \cdot 7 \cdot 8}{1 \cdot 2 \cdot 3} = 7 \cdot 8 = 56$

▶ **Study Suggestion 12.3:** Evaluate the following expressions.

(a) $\dfrac{8!}{5!}$ **(b)** $\dfrac{9!}{7!2!}$ ◀

Try Study Suggestion 12.3. □

EXAMPLE 2:

(a) Compare the values of $2! + 3!$ and $(2 + 3)!$

(b) Compare the values of $2!4!$ and $(2 \cdot 4)!$

Solutions:

(a) According to the definition,

$$2! + 3! = 1 \cdot 2 + 1 \cdot 2 \cdot 3 = 2 + 6 = 8$$

and

$$(2 + 3)! = 5! = 1 \cdot 2 \cdot 3 \cdot 4 \cdot 5 = 120$$

and so $2! + 3!$ is not equal to $(2 + 3)!$.

(b) In this case, we have

$$2!4! = (1 \cdot 2)(1 \cdot 2 \cdot 3 \cdot 4) = (2)(24) = 48$$

and

$$(2 \cdot 4)! = 8! = 1 \cdot 2 \cdot 3 \cdot 4 \cdot 5 \cdot 6 \cdot 7 \cdot 8 = 40{,}320$$

Hence, $2!4!$ is not equal to $(2 \cdot 4)!$.

Try Study Suggestion 12.4. □

▶ **Study Suggestion 12.4:**

(a) Compare the values of $9! - 6!$ and $(9 - 6)!$

(b) Compare the values of $\frac{8!}{4!}$ and $\left(\frac{8}{4}\right)!$ ◀

Now we can begin discussing the main topic of this section, namely, a formula for $(x + a)^n$, where n is any positive integer. Let us begin by computing a few powers of the binomial $x + a$. We will leave the actual calculations to you.

$$(x + a)^1 = x + a$$
$$(x + a)^2 = x^2 + 2ax + a^2$$
$$(x + a)^3 = x^3 + 3ax^2 + 3a^2x + a^3$$
$$(x + a)^4 = x^4 + 4ax^3 + 6a^2x^2 + 4a^3x + a^4$$
$$(x + a)^5 = x^5 + 5ax^4 + 10a^2x^3 + 10a^3x^2 + 5a^4x + a^5$$

Now, there is a pattern in each of these formulas, and our goal is to explore this pattern, in the hope of finding a general formula for the nth power $(x + a)^n$.

Let us begin by looking at the exponents in the expansion of, say, $(x + a)^4$. Notice that, in each term on the right-hand side of this expansion, the *sum* of the exponent of a and the exponent of x is equal to 4, which happens to be the exponent of $x + a$ on the left side of the expansion. In fact, each term in this expansion has the form

$$ca^kx^{4-k}$$

where c is a constant (the coefficient) and k is an integer between 0 and 4 (inclusive). For instance, the second term in the expansion of $(x + a)^4$ is $4ax^3$, and so in this case, $c = 4$ and $k = 1$.

A similar statement can be made for each of the other expansions, and so we can reasonably guess that a general formula for $(x + a)^n$ might have the form

$$(x + a)^n = \_\_a^0x^n + \_\_a^1x^{n-1} + \_\_a^2x^{n-2} + \cdots + \_\_a^{n-2}x^2 + \_\_a^{n-1}x^1 + \_\_a^nx^0 \qquad \textbf{(12.11)}$$

where the blanks ___ are the as-yet-to-be-determined coefficients. In particular, the general term of this expansion has the form

$$ca^k x^{n-k}$$

where c is the coefficient and k is an integer between 0 and n (inclusive).

Now let us look at the coefficients in the expansion of $(x + a)^4$

$$1 \quad 4 \quad 6 \quad 4 \quad 1$$

These numbers can be written in terms of factorials as follows

$$\frac{4!}{0!4!} \quad \frac{4!}{1!3!} \quad \frac{4!}{2!2!} \quad \frac{4!}{3!1!} \quad \frac{4!}{4!0!}$$

For instance, the third coefficient is 6, and we have

$$\frac{4!}{2!2!} = \frac{1 \cdot 2 \cdot 3 \cdot 4}{(1 \cdot 2)(1 \cdot 2)} = 6$$

We will leave it to you to check the other coefficients.

Similarly, the coefficients in the expansion of $(x + a)^5$

$$1 \quad 5 \quad 10 \quad 10 \quad 5 \quad 1$$

can be written in the form

$$\frac{5!}{0!5!} \quad \frac{5!}{1!4!} \quad \frac{5!}{2!3!} \quad \frac{5!}{3!2!} \quad \frac{5!}{4!1!} \quad \frac{5!}{5!0!}$$

For instance, the third coefficient is 10, and we have

$$\frac{5!}{2!3!} = \frac{1 \cdot 2 \cdot 3 \cdot 4 \cdot 5}{(1 \cdot 2)(1 \cdot 2 \cdot 3)} = 10$$

Thus, we can write the expansions of $(x + a)^4$ and $(x + a)^5$ in the form

$$(x + a)^4 = \frac{4!}{0!4!} a^0 x^4 + \frac{4!}{1!3!} a^1 x^3 + \frac{4!}{2!2!} a^2 x^2 + \frac{4!}{3!1!} a^3 x^1 + \frac{4!}{4!0!} a^4 x^0$$

and

$$(x + a)^5$$

$$= \frac{5!}{0!5!} a^0 x^5 + \frac{5!}{1!4!} a^1 x^4 + \frac{5!}{2!3!} a^2 x^3 + \frac{5!}{3!2!} a^3 x^2 + \frac{5!}{4!1!} a^4 x^1 + \frac{5!}{5!0!} a^5 x^0$$

A similar pattern appears in the other expansions, and now we can make an intelligent guess as to the coefficients in Equation (12.11),

$$(x + a)^n = \frac{n!}{0!n!} a^0 x^n + \frac{n!}{1!(n-1)!} a^1 x^{n-1} + \frac{n!}{2!(n-2)!} a^2 x^{n-2} + \cdots \tag{12.12}$$

$$\cdots + \frac{n!}{(n-2)!2!} a^{n-2} x^2 + \frac{n!}{(n-1)!1!} a^{n-1} x^1 + \frac{n!}{n!0!} a^n x^0$$

where the general term of this expansion has the form

$$\frac{n!}{k!(n-k)!}a^k x^{n-k}$$

for some integer k between 0 and n.

Formula (12.12) is known as the **binomial formula.** We have seen that it holds for $n = 4$ and $n = 5$, since we used these expansions as a guide to finding (12.12). The general formula is indeed true, and can be proved using mathematical induction, but we will not go into the details.

The coefficients in (12.12) occur so frequently that they have a special notation. If k satisfies $0 \le k \le n$, then we write

$$\binom{n}{k} = \frac{n!}{k!(n-k)!}$$

and call $\binom{n}{k}$ a **binomial coefficient.** This is read "n choose k" or "binomial n, k." Using this notation, we can write the binomial formula in the following way.

THEOREM

The binomial theorem For any positive integer n, we have

$$(x+a)^n = \binom{n}{0}a^0x^n + \binom{n}{1}a^1x^{n-1} + \binom{n}{2}a^2x^{n-2} + \cdots$$

$$\cdots + \binom{n}{n-2}a^{n-2}x^2 + \binom{n}{n-1}a^{n-1}x^1 + \binom{n}{n}a^nx^0$$

where the general term of this expansion has the form

$$\binom{n}{k}a^kx^{n-k} = \frac{n!}{k!(n-k)!}a^kx^{n-k}$$

for some integer k between 0 and n.

EXAMPLE 3: Expand the expression $(2x+3)^4$.

Solution: According to the binomial formula, with x replaced by $2x$ and a replaced by 3, we have

$$(2x+3)^4$$

$$= \binom{4}{0}3^0(2x)^4 + \binom{4}{1}3^1(2x)^3 + \binom{4}{2}3^2(2x)^2 + \binom{4}{3}3^3(2x)^1 + \binom{4}{4}3^4(2x)^0$$

$$= \binom{4}{0}16x^4 + \binom{4}{1}24x^3 + \binom{4}{2}36x^2 + \binom{4}{3}54x + \binom{4}{4}81$$

Now, we must evaluate the binomial coefficients,

$$\binom{4}{0} = \frac{4!}{0!4!} = 1, \quad \binom{4}{1} = \frac{4!}{1!3!} = 4, \quad \binom{4}{2} = \frac{4!}{2!2!} = 6,$$

$$\binom{4}{3} = \frac{4!}{3!1!} = 4, \quad \binom{4}{4} = \frac{4!}{4!0!} = 1$$

Putting these values into our expansion, we get

$$(2x + 3)^4 = 16x^4 + 96x^3 + 216x^2 + 216x + 81$$

Try Study Suggestion 12.5. □

▶ **Study Suggestion 12.5:** Expand the expression $(3x - 2)^4$. ◀

The binomial theorem frequently provides the easiest way (by far) to obtain the coefficient of a particular term in an expansion.

EXAMPLE 4: Find the coefficient of x^6 in the expansion of $(5x^2 - 2)^{10}$.

Solution: Referring to the previous theorem, we see that the general term in the expansion of $(5x^2 - 2)^{10}$ is

$$\binom{10}{k} 2^k(5x^2)^{10-k}$$

Choosing $k = 7$ gives the term involving x^6, which is

$$\binom{10}{7}(-2)^7(5x^2)^3 = \binom{10}{7}(-128)(125)x^6$$

Evaluating the binomial coefficient gives

$$\begin{aligned}\binom{10}{7} &= \frac{10!}{7!3!} \\ &= \frac{1 \cdot 2 \cdot 3 \cdot 4 \cdot 5 \cdot 6 \cdot 7 \cdot 8 \cdot 9 \cdot 10}{(1 \cdot 2 \cdot 3 \cdot 4 \cdot 5 \cdot 6 \cdot 7)(1 \cdot 2 \cdot 3)} \\ &= 120\end{aligned}$$

and so the coefficient of x^6 in this expansion is

$$(120)(-128)(125) = -1{,}920{,}000$$

In other words, the term involving x^6 is $-1{,}920{,}000x^6$. This is certainly a lot easier then expanding $(5x^2 - 2)^{10}$ by hand to get the term involving x^6.

Try Study Suggestion 12.6. □

▶ **Study Suggestion 12.6:** Find the coefficient of x^8 in the expansion of $(2x^2 - 3)^{12}$. ◀

EXAMPLE 5: Expand the expression

$$\left(\frac{1}{x} - \sqrt{x}\right)^3$$

Solution: Using the binomial formula, we have

$$\left(\frac{1}{x}-\sqrt{x}\right)^3$$
$$=\binom{3}{0}\left(\frac{1}{x}\right)^3+\binom{3}{1}(-\sqrt{x})\left(\frac{1}{x}\right)^2+\binom{3}{2}(-\sqrt{x})^2\left(\frac{1}{x}\right)+\binom{3}{3}(-\sqrt{x})^3$$
$$=x^{-3}-3x^{-3/2}+3-x^{3/2}$$

Try Study Suggestion 12.7. ☐

▶ **Study Suggestion 12.7:** Expand the expression

$$\left(\frac{1}{x^2}-\sqrt{x}\right)^4.$$ ◀

EXERCISES

In Exercises 1–10, simplify the given expression as much as possible.

1. $\frac{7!}{3!6!}$ **2.** $\frac{2!}{1!0!}$ **3.** $4!-3!$

4. $(0!)(1!)(2!)$ **5.** $\frac{9!}{10!-9!}$ **6.** $7!-5!+6!$

7. $\frac{n!}{(n-1)!}$ **8.** $\frac{(n+2)!}{(n-1)!}$ **9.** $n!-(n-1)!$

10. $n[n!+(n-1)!]$

In Exercises 11–18, evaluate the given binomial coefficient.

11. $\binom{5}{3}$ **12.** $\binom{6}{2}$ **13.** $\binom{5}{0}$

14. $\binom{8}{8}$ **15.** $\binom{9}{4}$ **16.** $\binom{10}{5}$

17. $\binom{10}{1}$ **18.** $\binom{100}{99}$

19. Show that

$$\binom{n}{k}=\binom{n}{n-k}$$

where $0 \le k \le n$.

20. Show that

$$\binom{n}{k}=\frac{n}{k}\binom{n-1}{k-1}$$

In Exercises 21–29, expand and simplify the given expression

21. $(x+1)^6$ **22.** $(2x-3)^5$ **23.** $(5x+5)^7$

24. $(1-a)^{10}$ **25.** $(u-v^2)^5$ **26.** $(\sqrt{x}+x^{-2})^4$

27. $(\sqrt{x}+\sqrt[3]{y})^4$ **28.** $\left(\frac{1}{x}+x\right)^4$ **29.** $\left(\frac{1}{x^2}+x^2\right)^3$

30. Find the coefficient of x^{12} in $(2x+1)^{15}$

31. Find the coefficient of x^3y^2 in $(2x+3y)^5$

32. Find the coefficient of x^{99} in $(17x+12)^{100}$

33. Find the coefficient of c^8d^{18} in $(2c^2-3d^3)^{10}$

34. Find the coefficient of $x^{2/3}$ in $(\sqrt[3]{x}+2)^5$

35. Find the coefficient of x^{23} in $(2x^2+x)^{15}$

36. Find the constant term in the expansion of $\left(x+\frac{1}{x^2}\right)^{15}$

12.3 Infinite Sequences

Sequences of numbers play a very important role in mathematics and its applications. In this section, we will briefly study *infinite* sequences. We begin with some general remarks, and then consider two special types of sequences.

In general, an infinite sequence has the form

$$a_1, a_2, a_3, \ldots \tag{12.13}$$

where the a_i are real numbers, called the **terms** of the sequence. However, (12.13) alone does not completely describe a sequence, since it does not tell us how to get other terms in the sequence. (From now on, whenever we use the term sequence, we will mean *infinite* sequence.)

There are two common methods for describing a sequence. One method is to give a formula for the nth term a_n. We will refer to the nth term of a sequence as the **general term** of the sequence. Given a formula for the general term of the sequence, we can easily find any specific term in the sequence.

EXAMPLE 1: Find the first three terms a_1, a_2, and a_3 and the 10th term a_{10} of the sequence with general term

(a) $a_n = n^2$ **(b)** $a_n = (-1)^n$ **(c)** $a_n = e^{n\pi}$

Solutions: In each case, we simply set n equal to 1, 2, 3, and 10.

(a) $a_1 = 1^2 = 1$, $a_2 = 2^2 = 4$, $a_3 = 3^2 = 9$, $a_{10} = 10^2 = 100$

(b) $a_1 = (-1)^1 = -1$, $a_2 = (-1)^2 = 1$, $a_3 = (-1)^3 = -1$, $a_{10} = (-1)^{10} = 1$

(c) $a_1 = e^{1 \cdot \pi} = e^{\pi}$, $a_2 = e^{2\pi}$, $a_3 = e^{3\pi}$, $a_{10} = e^{10\pi}$

Try Study Suggestion 12.8. □

▶ **Study Suggestion 12.8:** Find the first three terms a_1, a_2, and a_3 and the 10th term a_{10} of the sequence with general term

(a) $a_n = n^n$ **(b)** $a_n = -(-1)^{2n}$ **(c)** $a_n = \log 2^n$ ◀

Another common method for describing the general term of a sequence is to give a formula for the nth term a_n in terms of *previous* terms in the sequence. Such a formula is called a **recurrence relation** for the sequence. The next example illustrates this idea.

EXAMPLE 2: Suppose that a sequence $a_1, a_2, a_3, \ldots$ satisfies the recurrence relation

$$a_n = 2a_{n-1} \tag{12.14}$$

for all $n \geq 2$. If $a_1 = 3$, find the first five terms of the sequence, and a formula for the general term a_n.

Solution: Using Formula (12.14), and the fact that $a_1 = 3$, we have

$$\begin{aligned} a_1 &= 3 \\ a_2 &= 2a_1 = 2 \cdot 3 = 6 \\ a_3 &= 2a_2 = 2 \cdot 2 \cdot 3 = 2^2 \cdot 3 = 12 \\ a_4 &= 2a_3 = 2 \cdot 2^2 \cdot 3 = 2^3 \cdot 3 = 24 \\ a_5 &= 2a_4 = 2 \cdot 2^3 \cdot 3 = 2^4 \cdot 3 = 48 \end{aligned}$$

There is a pattern developing in these terms, and it looks as though the general term of the sequence might be

$$a_n = 2^{n-1} \cdot 3 \tag{12.15}$$

We can actually *prove* that this is the general term by using mathematical induction.

In order to do this, we let $P(n)$ be the proposition that the nth term a_n is given by Formula (12.15). Then $P(1)$ is the proposition that

$$a_1 = 2^{1-1} \cdot 3$$

and since $2^{1-1} \cdot 3 = 3$, this is indeed true. To complete the induction, we must prove that *if*

$$a_k = 2^{k-1} \cdot 3 \tag{12.16}$$

then

$$a_{k+1} = 2^k \cdot 3 \tag{12.17}$$

But, using (12.14) and then (12.16), we get

$$a_{k+1} = 2a_k = 2(2^{k-1} \cdot 3) = 2^k \cdot 3$$

and so (12.17) is true. This shows that Formula (12.15) is true for all n, and so our guess is correct. *Try Study Suggestion 12.9.* □

▶ **Study Suggestion 12.9:** A sequence $a_1, a_2, a_3, \ldots$ has the property that $a_1 = 5$ and $a_n = (1/2)a_{n-1}$ for all $n \geq 2$. Find the first five terms of the sequence, and a formula for the general term a_n. Finally, prove that your formula is true by using mathematical induction. ◀

Now let us discuss two special types of infinite sequences.

DEFINITION

A sequence $a_1, a_2, a_3, \ldots$ is said to be an **arithmetic sequence** if it has the property that successive terms of the sequence differ by the same amount d; that is, if it has the property that

$$a_{k+1} - a_k = d$$

or equivalently

$$a_{k+1} = a_k + d \tag{12.18}$$

for all $k \geq 1$. In this case, the number d is called the **common difference** of the sequence. (Notice that Equation (12.18) is just a recurrence relation for the sequence.)

An arithmetic sequence can be described by giving the first term of the sequence, along with its common difference. Also, an arithmetic sequence can be described by giving just the first two terms of the sequence. (This is not true for sequences in general.)

EXAMPLE 3:

(a) Find the first five terms of the arithmetic sequence whose first term is $a_1 = 3$ and whose common difference is $d = 5$.

(b) Find the first five terms of an arithmetic sequence whose first two terms are $a_1 = 4$ and $a_2 = -6$.

Solutions:

(a) In this case, we use Equation (12.18) with $d = 5$ to get

$$\begin{aligned} a_1 &= 3 \\ a_2 &= a_1 + 5 = 8 \\ a_3 &= a_2 + 5 = 13 \\ a_4 &= a_3 + 5 = 18 \\ a_5 &= a_4 + 5 = 23 \end{aligned}$$

(b) First we use the terms a_1 and a_2 to compute the common difference

$$d = a_2 - a_1 = -6 - 4 = -10$$

Now we can proceed as in Part (a).

$$\begin{aligned} a_1 &= 4 \\ a_2 &= a_1 + (-10) = -6 \\ a_3 &= a_2 + (-10) = -16 \\ a_4 &= a_3 + (-10) = -26 \\ a_5 &= a_4 + (-10) = -36 \end{aligned}$$

Try Study Suggestion 12.10. □

▶ **Study Suggestion 12.10:**
(a) Repeat Part (a) of Example 3 with $a_1 = -2$ and $d = 3$.
(b) Repeat Part (b) of Example 3 with $a_1 = -1/2$ and $a_2 = 0$. ◀

If $a_1, a_2, a_3, \ldots$ is an arithmetic sequence with common difference d, then according to Equation (12.18), we have

$$\begin{aligned} a_1 &= a_1 \\ a_2 &= a_1 + d \\ a_3 &= a_2 + d = a_1 + d + d = a_1 + 2d \\ a_4 &= a_3 + d = a_1 + 2d + d = a_1 + 3d \\ a_5 &= a_4 + d = a_1 + 3d + d = a_1 + 4d \end{aligned}$$

A pattern is developing here, and we are led to the following formula for the general term of an arithmetic sequence, which can be proved by mathematical induction (Exercise 57).

THEOREM

If $a_1, a_2, a_3, \ldots$ is an arithmetic sequence with common difference d, then the general term of the sequence is given by

$$a_n = a_1 + (n - 1)d \qquad \textbf{(12.19)}$$

EXAMPLE 4:

(a) Find the general term of the sequence in Part (a) of Example 3. Then find the 100th term of the sequence.

(b) Find the general term of the sequence in Part (b) of Example 3. Then find the 100th term of the sequence.

Solutions:

(a) Using Equation (12.19), we get

$$a_n = a_1 + (n - 1)d = 3 + 5(n - 1) = 5n - 2$$

Thus, the 100th term of this sequence is

$$a_{100} = 5(100) - 2 = 498$$

(b) In this case, since the common difference is $d = -10$, Equation (12.19) gives

$$a_n = a_1 + (n - 1)d = 4 + (-10)(n - 1) = -10n + 14$$

In particular,

$$a_{100} = -10(100) + 14 = -986$$

▶ **Study Suggestion 12.11:**

(a) Find the general term and the 100th term of the sequence in Part (a) of Study Suggestion 12.10.

(b) Find the general term and the 100th term of the sequence in Part (b) of Study Suggestion 12.10. ◀

Try Study Suggestion 12.11. □

Equation (12.19) can be used to derive a formula for the common difference d of an arithmetic sequence given *any* two terms of the sequence. Suppose that we are given the terms a_m and a_n, where $m < n$. Then (12.19) gives

$$\begin{aligned} a_n - a_m &= [a_1 + (n - 1)d] - [a_1 + (m - 1)d] \\ &= (n - 1)d - (m - 1)d \\ &= (n - m)d \end{aligned}$$

and so

$$d = \frac{a_n - a_m}{n - m} \tag{12.20}$$

EXAMPLE 5: Find the common difference and the general term of the arithmetic sequence for which $a_5 = 7$ and $a_{10} = 32$.

Solution: Using Equation (12.20), we get

$$d = \frac{a_{10} - a_5}{10 - 5} = \frac{32 - 7}{5} = \frac{25}{5} = 5$$

Next we obtain the first term a_1 of the sequence from Equation (12.19), with $n = 5$

$$a_1 = a_n - (n-1)d = a_5 - (5-1) \cdot 5 = 7 - 20 = -13$$

Hence

$$a_n = a_1 + (n-1)d = -13 + (n-1) \cdot 5 = 5n - 18$$

Try Study Suggestion 12.12. □

▶ **Study Suggestion 12.12:** Find the common difference and the general term of the arithmetic sequence for which $a_3 = 8$ and $a_9 = -10$. ◀

Since the knowledge of *any* two terms of an arithmetic sequence is enough to determine the general term of the sequence we can say that

an arithmetic sequence is completely determined by giving *any* two terms of the sequence.

As you can see from the following definition, geometric sequences are the multiplicative analog of arithmetic sequences.

DEFINITION

A sequence $a_1, a_2, a_3, \ldots$ is said to be a **geometric sequence** if it has the property that the ratio of successive terms of the sequence is constant; that is, if it has the property that

$$\frac{a_{k+1}}{a_k} = r$$

or equivalently

$$a_{k+1} = ra_k \tag{12.21}$$

for all $k \geq 1$, where r is a nonzero constant. The number r is called the **common ratio** of the sequence. (Notice that Equation (12.21) is just a recurrence relation for the sequence.)

As with arithmetic sequences, a geometric sequence can be completely described by giving either the first term of the sequence and the common ratio, or simply the first two terms of the sequence.

EXAMPLE 6:

(a) Find the first five terms of the geometric sequence whose first term is $a_1 = -2$ and whose common ratio is $r = 3$.

(b) Find the first five terms of the geometric sequence whose first two terms are $a_1 = 1$ and $a_2 = 1/2$.

Solutions:

(a) Using Equation (12.21), we have

$$\begin{aligned} a_1 &= -2 \\ a_2 &= 3a_1 = 3(-2) = -6 \\ a_3 &= 3a_2 = 3(-6) = -18 \\ a_4 &= 3a_3 = 3(-18) = -54 \\ a_5 &= 3a_4 = 3(-54) = -162 \end{aligned}$$

(b) First we find the common ratio

$$r = \frac{a_2}{a_1} = \frac{1/2}{1} = \frac{1}{2}$$

and then proceed as in part (a)

$$\begin{aligned} a_1 &= 1 \\ a_2 &= \frac{1}{2}a_1 = \frac{1}{2} \cdot 1 = \frac{1}{2} \\ a_3 &= \frac{1}{2}a_2 = \frac{1}{2} \cdot \frac{1}{2} = \frac{1}{4} \\ a_4 &= \frac{1}{2}a_3 = \frac{1}{2} \cdot \frac{1}{4} = \frac{1}{8} \\ a_5 &= \frac{1}{2}a_4 = \frac{1}{2} \cdot \frac{1}{8} = \frac{1}{16} \end{aligned}$$

Try Study Suggestion 12.13. □

▶ **Study Suggestion 12.13:**

(a) Repeat Part (a) of Example 6 with $a_1 = 1/2$ and $r = -2$.

(b) Repeat Part (b) of Example 6 with $a_1 = 1/2$ and $a_2 = 2$. ◀

If $a_1, a_2, a_3, \ldots$ is a geometric sequence, then according to Equation (12.21), we have

$$\begin{aligned} a_1 &= a_1 \\ a_2 &= ra_1 \\ a_3 &= ra_2 = r(ra_1) = r^2a_1 \\ a_4 &= ra_3 = r(r^2a_1) = r^3a_1 \\ a_5 &= ra_4 = r(r^3a_1) = r^4a_1 \end{aligned}$$

The pattern here leads us to the following result, which can be proved by mathematical inducation (Exercise 58).

THEOREM

If $a_1, a_2, a_3, \ldots$ is a geometric sequence with common ratio r, then the general term of the sequence is given by

$$a_n = r^{n-1}a_1 \tag{12.22}$$

Notice that, when the first term of a geometric sequence is $a_1 = 1$, then the general term of the sequence is $a_n = r^{n-1}$. Thus, the first few terms of the sequence are

$$1, r, r^2, r^3, r^4, \ldots$$

and so this type of geometric sequence is nothing more than the sequence of powers of the number r. (Recall that $1 = r^0$ and $r = r^1$.)

EXAMPLE 7:

(a) Find the general term of the sequence in Part (a) of Example 6. Then find the 10th term of the sequence.

(b) Find the general term of the sequence in Part (b) of Example 6. Then find the 10th term of the sequence.

Solutions:

(a) According to Equation (12.22), we have

$$a_n = r^{n-1}a_1 = (-2)3^{n-1}$$

and so

$$a_{10} = (-2)3^9 = -39{,}366$$

(b) Again using Equation (12.22), and the fact that $r = 1/2$, we get

$$a_n = r^{n-1}a_1 = \left(\frac{1}{2}\right)^{n-1} = \frac{1}{2^{n-1}}$$

In particular,

$$a_{10} = \frac{1}{2^9} = \frac{1}{512}$$

Try Study Suggestion 12.14. □

Study Suggestion 12.14:

(a) Find the general term and the 10th term of the sequence in Part (a) of Study Suggestion 12.13.

(b) Find the general term and the 10th term of the sequence in Part (b) of Study Suggestion 12.13.

Try Study Suggestion 12.15.

Study Suggestion 12.15: As you know, an arithmetic sequence is completely determined by knowledge of *any* two terms of the sequence. Is the same statement true for geometric sequences? That is, is it true that a geometric sequence is completely determined by giving *any* two terms of the sequence? Justify your answer. (*Hint:* consider the geometric sequence whose first term is $a_1 = 1$ and whose common ratio is $r = 2$ as well as the geometric sequence whose first term is $b_1 = 1$ and whose common ratio is $r = -2$.)

Ideas to Remember

- There are two common ways to describe an infinite sequence. One is to give a formula for its general term a_n, and the other is to give the first few terms, along with a recurrence relation for the sequence.
- An arithmetic sequence is completely determined by giving *any* two terms of the sequence.

EXERCISES

In Exercises 1–10, find the first five terms of the sequence whose general term a_n is given.

1. $a_n = (-2)^n$

2. $a_n = n^3 + 3n - 1$

3. $a_n = \log(10^n)$

4. $a_n = \cos n\pi$

5. $a_n = 2^n + (-2)^n$

6. $a_n = \begin{cases} 3 & \text{if } n \text{ even} \\ -3 & \text{if } n \text{ odd} \end{cases}$

7. $a_n = \dfrac{2^n}{n+1}$

8. $a_n = n!$

9. $a_n = 8$

10. $a_n = (10)^n + (0.1)^n$

In Exercises 11–20, find the first five terms of the sequence given by the recurrence relation.

11. $a_1 = 2$; $a_n = 3a_{n-1}$

12. $a_1 = -1$; $a_n = a_{n-1} + 3$

13. $a_1 = 0$; $a_n = 7a_{n-1} + 1$

14. $a_1 = 2$; $a_n = 1/a_{n-1}$

15. $a_1 = 1, a_2 = 1$; $a_n = a_{n-1} + a_{n-2}$

16. $a_1 = 2$; $a_n = a_{n-1}{}^n$

17. $a_1 = r$; $a_n = a_{n-1}{}^{1/n}$

18. $a_1 = 2, a_2 = 3$; $a_n = a_{n-1}a_{n-2}$

19. $a_1 = 1$; $a_n = sa_{n-1}$

20. $a_1 = r$; $a_n = \log a_{n-1}$

21. Find the first eight terms of the sequence with the property that $a_1 = 1$, $a_2 = 3$, and $a_n = 2a_{n-2}$ for all $n \geq 3$.

22. If $a_1 = a$, $a_2 = b$ and the nth term of the sequence is the average of the two previous terms, find the first seven terms of the sequence.

23. If $a_1 = 3$, $a_2 = 6$ and the nth term of the sequence is equal to the square root of the sum of the two previous terms, find the first six terms of the sequence.

In Exercises 24–35, find the common difference, the general term, the first three terms, and the 100th term of the arithmetic sequence described by the given information.

24. $a_1 = 3, d = 4$

25. $a_1 = -5, d = -3$

26. $a_2 = 5, d = 3$

27. $a_7 = 12, d = -1/2$

28. $a_1 = 1, a_2 = 4$

29. $a_1 = 7, a_2 = -3$

30. $a_3 = 5, a_4 = 9$

31. $a_3 = 6, a_{10} = 27$

32. $a_4 = 2, a_8 = 3$

33. $a_2 = 4, a_7 = -1$

34. $a_1 = \log 2, a_2 = \log 8$

35. $a_1 = \log 100, a_2 = \log 10$

In Exercises 36–49, find the common ratio, the general term, the first three terms and the 10th term of the geometric sequence described by the given information.

36. $a_1 = 3, r = 2$

37. $a_1 = -2, r = 1/2$

38. $a_6 = 1/2, r = -1$

39. $a_1 = 1, a_2 = 4$

40. $a_1 = 2, a_2 = -5$

41. $a_2 = 2, a_3 = 18$

42. $a_2 = 2, a_3 = 5$

43. $a_3 = 6, a_4 = 12$

44. $a_4 = -10, a_7 = 1250$

45. $a_3 = 2, a_6 = 1/32$

46. $a_1 = x^{-1}, a_2 = -x^2$

47. $a_1 = xy, a_2 = x^2$

48. $a_1 = 2^x, a_2 = 2^{2x+1}$

49. $a_1 = x, a_2 = a_1{}^2 - a_1$

50. Can a sequence that begins $a_1 = 1$, $a_2 = 3$, $a_3 = 12$ be an arithmetic sequence or a geometric sequence? Justify your answer.

51. Can a sequence with the property that $a_1 = 2$ and $a_5 = 6$ be an arithmetic sequence or a geometric sequence? Justify your answer.

52. If $a_1 = 2$ and $a_n = 3a_{n-1}$, prove by induction that $a_n = 2 \cdot 3^{n-1}$.

53. If $a_1 = -1$ and $a_n = a_{n-1} + 5$, prove by induction that $a_n = 5n - 6$.

54. If $a_1 = 1$ and $a_n = 2a_{n-1} + 1$, prove by induction that $a_n = 2^n - 1$.

55. If $a_1 = 1$ and $a_n = na_{n-1}$, prove by induction that $a_n = n!$. (Recall from Section 12.2 that $n! = 1 \cdot 2 \cdot 3 \cdots n$.)

56. If $a_1 = 2$ and $a_n = a_{n-1} + n$, prove by induction that $a_n = (1/2)n^2 + (1/2)n + 1$.

57. Prove the first of the two theorems in this section.

58. Prove the second of the two theorems in this section.

59. Let $a_1, a_2, a_3, \ldots$ be a sequence with the property that $a_1 = 1$ and $a_n = a_{n-1} + 2^{n-1}$. Write out the first four terms of the sequence and guess a formula for the general term a_n. Prove that your guess is correct by using mathematical induction.

60. Let $a_1, a_2, a_3, \ldots$ be a sequence with the property that $a_1 = 4$ and $a_n = a_{n-1} + 2^{n-1} + 3^n$. Write out the first four terms of this sequence and guess a formula for the general term a_n. Prove that your guess is correct by using mathematical induction.

12.4 The Partial Sums of a Sequence

It often happens in applications that we are interested in adding consecutive terms of a given sequence. Let us illustrate this with an example.

An outdoor stadium has seats arranged in rows. The bottom row has 100 seats, and each successive row has two more seats than the row below it. If the stadium has 50 rows, how many seats does it have?

Since the first row has 100 seats, and each row has two more seats than the previous row, the total number of seats is equal to

$$100 + 102 + 104 + \cdots + 198 \qquad \textbf{(12.23)}$$

where the sum has 50 terms. The problem is, "How do we compute this sum?"

As we will see, there is a simple formula for computing sums of this type, and it rests on the fact that (12.23) is the sum of consecutive terms in an *arithmetic sequence*. In fact, this is precisely how we determined that the last term in this sum is 198. That is, we found the 50th term by using the formula $a_{50} = a_1 + (50 - 1) \cdot d = 100 + 49 \cdot 2 = 198$.

Let us begin our discussion by making the following definition.

DEFINITION

If $a_1, a_2, a_3, \ldots$ is a sequence of numbers, then the sum of the first n terms of this sequence is called the nth **partial sum** of the sequence. We denote the partial sums of a sequence by $s_1, s_2, s_3, \ldots$. The first few partial sums are

$$s_1 = a_1$$
$$s_2 = a_1 + a_2$$
$$s_3 = a_1 + a_2 + a_3$$

and the nth partial sum is

$$s_n = a_1 + a_2 + \cdots + a_n$$

Our plan now is to derive a formula for the nth partial sum of any arithmetic sequence. This will enable us to compute sums such as (12.23). Then we will do the same for geometric sequences.

As we know from the previous section, an arithmetic sequence has general term

$$a_n = a_1 + (n - 1)d$$

where d is the common difference. Using this formula for various values of n, we get

$$\begin{aligned} s_n &= a_1 + a_2 + a_3 + \cdots + a_n \\ &= a_1 + [a_1 + d] + [a_1 + 2d] + [a_1 + 3d] + \cdots + [a_1 + (n-1)d] \\ &= na_1 + d + 2d + 3d + \cdots + (n-1)d \\ &= na_1 + d[1 + 2 + 3 + \cdots + (n-1)] \end{aligned}$$

Now we must evaluate the sum

$$1 + 2 + 3 + \cdots + (n-1)$$

But, according to Example 1 of Section 12.1, with n replaced by $n - 1$, we have

$$1 + 2 + 3 + \cdots + (n-1) = \frac{(n-1)n}{2}$$

Substituting this into our expression for s_n gives

$$\begin{aligned} s_n &= na_1 + d\left[\frac{(n-1)n}{2}\right] \\ &= \frac{n}{2}[2a_1 + (n-1)d] \end{aligned}$$

Before expressing this formula in a theorem, we can derive an alternate formula for s_n. Since the general term of an arithmetic sequence is given by $a_n = a_1 + (n-1)d$, we have $(n-1)d = a_n - a_1$, and so

$$\begin{aligned} s_n &= \frac{n}{2}(2a_1 + a_n - a_1) \\ &= \frac{n}{2}(a_1 + a_n) \end{aligned}$$

Now we can state our theorem.

THEOREM

If $a_1, a_2, a_3, \ldots$ is an arithmetic sequence with common difference d, then its nth partial sum is given by

$$s_n = \frac{n}{2}[2a_1 + (n-1)d] \qquad (12.24)$$

or

$$s_n = \frac{n}{2}(a_1 + a_n) \qquad (12.25)$$

Now we can solve the stadium problem mentioned at the beginning of the section.

EXAMPLE 1: How many seats are there in the stadium described at the beginning of this section?

Solution: In order to answer this question, we must find the sum in (12.23). But this is just the 50th partial sum of the arithmetic sequence whose first term is $a_1 = 100$, and whose 50th term is $a_{50} = 198$. Therefore, according to Formula (12.25), we have

$$s_{50} = \frac{50}{2}(100 + 198) = 25(298) = 7450$$

Hence, there are 7450 seats in the stadium. Isn't using this formula easier than having to add all 50 terms in the sum (12.23)?

Try Study Suggestion 12.16. □

▶ **Study Suggestion 12.16:** A bunch of logs are piled up in the shape of a pyramid. There are 80 logs on the bottom row, and each row has 4 fewer logs than the row directly below it. If there are 15 rows in the pyramid, how many logs are there? (*Hint:* use Formula (12.24)). ◀

Now let us turn our attention to finding a formula for the nth partial sum of a geometric sequence. We know from the previous section that the general term of a geometric sequence whose common ratio is r is given by

$$a_n = r^{n-1}a_1$$

Using this formula, for various values of n, we get

$$\begin{aligned} s_n &= a_1 + a_2 + a_3 + \cdots + a_n \\ &= a_1 + ra_1 + r^2a_1 + r^3a_1 + \cdots + r^{n-1}a_1 \\ &= a_1(1 + r + r^2 + r^3 + \cdots + r^{n-1}) \end{aligned} \tag{12.26}$$

Now we must find a formula for the sum

$$x = 1 + r + r^2 + r^3 + \cdots + r^{n-1}$$

which we have denoted by x. In order to do this, we first multiply both sides of this equation by r, to get

$$rx = r + r^2 + r^3 + r^4 + \cdots + r^n$$

Thus we have

$$\begin{aligned} x - rx &= 1 + r + r^2 + r^3 + \cdots + r^{n-1} - (r + r^2 + r^3 + r^4 + \cdots + r^n) \\ &= 1 + r + r^2 + r^3 + \cdots + r^{n-1} - r - r^2 - r^3 - r^4 - \cdots - r^n \\ &= 1 - r^n \end{aligned}$$

and so

$$x - rx = 1 - r^n$$

or

$$x(1 - r) = 1 - r^n$$

Assuming that $r \neq 1$, we can divide both sides of this by $1 - r$, to obtain

$$x = \frac{1 - r^n}{1 - r}$$

Substituting this into (12.26) gives us the formula

$$s_n = a_1\left(\frac{1 - r^n}{1 - r}\right)$$

valid for $r \neq 1$. Let us express this formula as a theorem.

THEOREM

If $a_1, a_2, a_3, \ldots$ is a geometric sequence with common ratio $r \neq 1$, then its nth partial sum is given by

$$s_n = a_1\left(\frac{1 - r^n}{1 - r}\right) \qquad \textbf{(12.27)}$$

EXAMPLE 3: A chessboard has 64 squares on it. Suppose that a rich uncle decides to give you 1 penny for the first square, 2 pennies for the second square, 4 pennies for the third square, and so on. For each square, you are given twice the number of pennies than for the previous square. How much money has your uncle given you?

Solution: The total number of pennies that you are given is equal to the sum

$$1 + 2 + 4 + 8 + \cdots + 2^{63}$$

of the first 64 terms of the geometric sequence whose first term is 1, and whose common ratio is $r = 2$. Thus, according to Equation (12.27), the total number of pennies is

$$s_{64} = 1\left(\frac{1 - 2^{64}}{1 - 2}\right) = 2^{64} - 1 \approx 1.8 \times 10^{19}$$

That is, your uncle has given you approximately 180,000,000,000,000,000 dollars! (This is one hundred and eighty quadrillion dollars.)

Try Study Suggestions 12.17 and 12.18. □

▶ **Study Suggestion 12.17:** Compute the partial sums s_5 and s_{10} of a geometric sequence whose first term is $a_1 = 2$ and whose common ratio is $r = 3$. ◀

▶ **Study Suggestion 12.18:** Find a formula for the nth partial sum s_n of a geometric sequence whose common ratio is 1. ◀

The previous example illustrates the fact that the partial sums of a geometric sequence whose common ratio is greater than 1 grow *very* fast. The next example illustrates the fact that the partial sums of a geometric sequence whose common ratio is less than 1 (but positive) grow *very* slowly.

EXAMPLE 4: Compute the partial sums s_5, s_{10} and s_{25} of a geometric sequence whose first term is $a_1 = 1$ and whose common ratio is $r = 1/2$.

Solution: According to Formula (12.27), we have

$$s_5 = \frac{1-(1/2)^5}{1-1/2} = 2[1-(1/2)^5] \approx 1.938$$

$$s_{10} = \frac{1-(1/2)^{10}}{1-1/2} = 2[1-(1/2)^{10}] \approx 1.998$$

and

$$s_{25} = \frac{1-(1/2)^{25}}{1-1/2} = 2[1-(1/2)^{25}] \approx 1.99999994$$

▶ **Study Suggestion 12.19:** Compute the partial sums s_5 and s_{10} of a geometric sequence whose first term is $a_1 = 3$ and whose common ratio $r = 1/3$. ◀

Try Study Suggestion 12.19. □

EXAMPLE 5: Find a formula for the sum

$$x^2 + x^4 + x^6 + \cdots + x^{2n}$$

Solution: Writing the sum in the form

$$(x^2)^1 + (x^2)^2 + (x^2)^3 + \cdots + (x^2)^n$$

we see that it is just the nth partial sum of a geometric sequence whose first term is $a_1 = (x^2)^1 = x^2$, and whose common ratio is $r = x^2$. Hence, according to Formula (12.27), the sum is equal to

$$s_n = x^2\frac{1-(x^2)^n}{1-x^2} = x^2\frac{1-x^{2n}}{1-x^2}$$

▶ **Study Suggestion 12.20:** Find a formula for the sum $1 + x^3 + x^6 + \cdots + x^{3n}$. ◀

Try Study Suggestion 12.20. □

Ideas to Remember

It is frequently important in applications to be able to compute the partial sums of a sequence. Fortunately, there are relatively simple formulas for these partial sums in the case of arithmetic and geometric sequences. (These are very special cases. In general, there is no formula for the nth partial sum of a sequence.)

EXERCISES

In Exercises 1–8, find the nth partial sum s_n of the arithmetic sequence described by the given information.

1. $a_1 = 2, d = -4$

2. $a_1 = -5, a_2 = 0$

3. $a_1 = 7, a_2 = 12$

4. $a_3 = 5, a_4 = 7$

5. $a_3 = 8, a_6 = -1$

6. $a_1 = x, a_2 = x + y$

7. $a_1 = x, a_2 = y$

8. $a_5 = 5, a_{10} = 10$

In Exercises 9–16, find the nth partial sum s_n of the geometric sequence described by the given information.

9. $a_1 = 2$, $r = 1/4$

10. $a_1 = -1/2$, $a_2 = 1$

11. $a_1 = 5$, $a_2 = 6$

12. $a_2 = 3$, $r = -x$

13. $a_1 = x$, $a_2 = xy$

14. $a_1 = x$, $a_2 = y$

15. $a_1 = 2^x$, $a_2 = 2^{x-2}$

16. $a_1 = 3^x$, $a_4 = 3^{4x-3}$

In Exercises 17–21, compute the indicated sums.

17. $(0.1)^2 + (0.1)^3 + \cdots + (0.1)^8$

18. $1 + (0.01)^2 + (0.01)^4 + (0.01)^6 + \cdots + (0.01)^{12}$

19. $1 + \sin(\pi/6) + \sin^2(\pi/6) + \cdots + \sin^9(\pi/6)$

20. $e + e^2 + e^3 + \cdots + e^{10}$

21. $\log 2 + \log 4 + \log 8 + \cdots + \log 128$

In Exercises 22–26, find a formula for the indicated sum.

22. $1 + x^4 + x^8 + \cdots + x^{4n}$

23. $1 + \dfrac{1}{1+x} + \dfrac{1}{(1+x)^2} + \cdots + \dfrac{1}{(1+x)^n}$

24. $1 + \log x + \log x^2 + \log x^3 + \cdots + \log x^n$

25. $e^{2x} + e^{3x} + e^{4x} + \cdots + e^{nx}$

26. $\sin x + \sin^2 x + \sin^3 x + \cdots + \sin^n x$

27. Find the sum of the first 11 nonnegative even integers.

28. Find the sum of the odd integers between 22 and 188.

29. Find the sum of every third integer, starting with 17 and ending with 92.

30. A stadium has 45 rows of seats. The bottom row has only 12 seats, and each row has 4 more seats than the row below it. How many seats are there in the stadium?

31. A vacuum pump removes 1/2 of the remaining air in a certain bottle with each stroke. What fraction of the original amount of air is left after 10 strokes? (*Hint:* let a_n be the amount of air in the bottle after the nth stroke.)

32. A house appreciates at the rate of 15% per year. If the house originally cost $50,000, how much will it be worth after 5 years?

33. A hole in a certain dam lets 100 gallons of water through in the first minute. If the hole is widening in such a way that each minute it lets three times the amount of water through than it let through the previous minute, how much water will get through after 1 hour?

34. A certain population of microorganisms doubles every 20 minutes. If the initial population is 1000 microorganisms, find a formula for the total number present after t hours. Then find the number of microorganisms present after 10 hours.

12.5 Infinite Series

Let us take a close look at the first few partial sums of the geometric sequence

$$1, \frac{1}{2}, \frac{1}{4}, \frac{1}{8}, \ldots. \tag{12.28}$$

whose first term is $a_1 = 1$ and whose common ratio is $r = 1/2$. These sums can be written in the form

$$s_1 = 1$$

$$s_2 = 1 + \frac{1}{2} = 1.5$$

$$s_3 = 1 + \frac{1}{2} + \frac{1}{4} = 1 + \frac{3}{4} = 1.75$$

$$s_4 = 1 + \frac{1}{2} + \frac{1}{4} + \frac{1}{8} = 1 + \frac{7}{8} = 1.875$$

$$s_5 = 1 + \frac{1}{2} + \frac{1}{4} + \frac{1}{8} + \frac{1}{16} = 1 + \frac{15}{16} = 1.9375$$

The next few partial sums are

$$s_6 = 1 + \frac{31}{32} = 1.96875$$

$$s_7 = 1 + \frac{63}{64} = 1.984375$$

$$s_8 = 1 + \frac{127}{128} = 1.9921875$$

$$s_9 = 1 + \frac{255}{256} = 1.99609375$$

$$s_{10} = 1 + \frac{511}{512} = 1.998046875$$

Now, as you would expect, these sums are getting larger. However, it seems that, *the partial sums* $s_1, s_2, s_3, \ldots$ *never exceed the number 2!* In fact, they seem to be getting closer and closer to 2 as we add more and more terms.

This is a very curious state of affairs. For it would seem that if we continue to add terms in a sequence such as (12.28), the sum should get larger and larger *without bound*. However, this is not the case. No matter how many of the terms of (12.28) we add together, the result will always be a number less than 2.

This statement is not at all obvious, and requires proof. However, we will not go into the proof here. Suffice it to say that the fact that we add new terms to the sum, which certainly does increase it, is counterbalanced by the fact that the new terms are getting smaller and smaller, and so the result is that the partial sums do increase, but by less and less each time, and in such a way that they *approach* (but never exceed) 2. This situation is described by defining a concept called an *infinite series* (or *infinite sum*), and writing

$$1 + \frac{1}{2} + \frac{1}{4} + \frac{1}{8} + \cdots = 2 \qquad \textbf{(12.29)}$$

It is important to keep in mind that the left-hand side of this equation is not actually the sum of an infinite number of terms, but simply a notational device. After all, it is not possible to add up an infinite number of terms.

In a sense, the left-hand side of (12.29) is a notation used to denote the *behavior of the partial sums* of the sequence (12.28). While a mathematician might read (12.29) by saying "the sum of the terms 1, 1/2, 1/4, 1/8, . . . is equal to 2, what he really means is "as n gets larger and larger, the *partial sums* $s_1, s_2, s_3, \ldots$ *approach* the number 2."

Before considering other examples of infinite series, let us make this idea more general in a definition.

DEFINITION

Let $a_1, a_2, a_3, \ldots$ be a sequence of real numbers. If the partial sums

$$s_1, s_2, s_3, \ldots$$

of this sequence approach a number s as n gets larger and larger, then

$$a_1 + a_2 + a_3 + \cdots = s$$

The expression $a_1 + a_2 + a_3 + \cdots$ is called an **infinite series,** and the number s is called the **sum** of the series. If an infinite series has a sum, then we say that the series **converges.** However, if the partial sums do not approach any number s; that is, if the series has no sum, then we say that the series **diverges.**

If $a_1, a_2, a_3, \ldots$ is a geometric sequence with common ratio r, then we call the infinite series

$$a_1 + a_2 + a_3 + \cdots$$

a **geometric series,** with common ratio r. Thus, according to Equation (12.29), the geometric series

$$1 + \frac{1}{2} + \frac{1}{4} + \cdots$$

converges, and has sum equal to 2. Let us consider another example.

EXAMPLE 1: Examine the geometric series

$$1 + \frac{1}{3} + \frac{1}{9} + \frac{1}{27} + \cdots \qquad \textbf{(12.30)}$$

whose first term is $a_1 = 1$ and whose common ratio is $r = 1/3$.

Solution: We begin by computing the first few partial sums, in the hope of seeing a pattern.

$$s_1 = 1$$

$$s_2 = 1 + \frac{1}{3} \approx 1.333$$

$$s_3 = 1 + \frac{1}{3} + \frac{1}{9} = 1 + \frac{4}{9} \approx 1.444$$

$$s_4 = 1 + \frac{1}{3} + \frac{1}{9} + \frac{1}{27} = 1 + \frac{13}{27} \approx 1.48148$$

$$s_5 = 1 + \frac{1}{3} + \frac{1}{9} + \frac{1}{27} + \frac{1}{81} = 1 + \frac{40}{81} \approx 1.493827$$

In this case, it appears that the partial sums are approaching the number $1.5 = 3/2$. This can be further substantiated by computing a few more partial sums. In any case, we will see later in this section that this is indeed the case, and so the series (12.30) does converge, and we can write

$$1 + \frac{1}{3} + \frac{1}{9} + \frac{1}{27} + \cdots = \frac{3}{2}$$

Try Study Suggestion 12.21. □

Study Suggestion 12.21: Use a calculator to examine the geometric series $1 + 1/4 + 1/16 + \cdots$ whose first term is $a_1 = 1$ and whose common ratio is $r = 1/4$. ◄

Now let us consider an example of an infinite series that diverges.

EXAMPLE 2: Examine the geometric series

$$1 + 2 + 4 + 8 + 16 + \cdots \tag{12.31}$$

whose first term is $a_1 = 1$ and whose common ratio is $r = 2$.

Solution: The first few partial sums of this series are

$$s_1 = 1$$
$$s_2 = 1 + 2 = 3$$
$$s_3 = 1 + 2 + 4 = 7$$
$$s_4 = 1 + 2 + 4 + 8 = 15$$
$$s_5 = 1 + 2 + 4 + 8 + 16 = 31$$

Now it appears that these partial sums are increasing *without bound*. In order to substantiate this, we observe that the nth term of the geometric *sequence* whose first term is $a_1 = 1$ and whose common ratio is $r = 2$ is given by $a_n = 2^n$. Furthermore, since each term a_n is positive, we have

$$s_n = a_1 + a_2 + \cdots + a_n > a_n$$

and so

$$s_n > a_n = 2^n > n$$

for $n \geq 1$. (Recall that in Section 12.1 we showed that $2^n \geq 2n$ for $n \geq 1$. Hence, we certainly have $2^n > n$.) Thus, the partial sum s_n satisfies $s_n > n$, which does indeed show that these sums increase without bound. Hence, according to the definition, the series (12.31) does not have a sum; that is, it diverges.

Try Study Suggestion 12.22. □

► **Study Suggestion 12.22:** Examine the series $2/1 + 3/2 + 4/3 + 5/4 + \cdots$. Give a convincing argument to show that this series diverges. ◄

The previous examples show that some geometric series converge, and some diverge. Using the formula for the nth partial sum of a geometric sequence, we can determine which geometric series converge and which diverge. In fact, we can even find a formula for the sum of a convergent geometric series.

Let us take another look at the formula for the nth partial sum of a geometric sequence, whose common ratio is $r \neq 1$,

$$s_n = a_1\left(\frac{1 - r^n}{1 - r}\right) \tag{12.32}$$

If the common ratio r satisfies the inequalities $-1 < r < 1$, then as n gets larger and larger, the expression r^n will approach 0. Therefore, the numerator on the right side of (12.32) will approach 1, and the partial sums s_n will approach the number

$$a_1\left(\frac{1}{1 - r}\right)$$

In other words, the geometric series $a_1 + a_2 + a_3 + \cdots$ will converge, and we will have

$$a_1 + a_2 + a_3 + \cdots = a_1\left(\frac{1}{1 - r}\right)$$

On the other hand, if r does not satisfy these inequalities, then the absolute value of the partial sums will get larger and larger, *without bound*, as n gets larger. Hence, the geometric series will diverge. Let us summarize our results in a theorem.

THEOREM

If the common ratio r of a geometric sequence $a_1, a_2, a_3, \ldots$ satisfies $-1 < r < 1$, then the infinite series $a_1 + a_2 + a_3 + \cdots$ converges, and we have

$$a_1 + a_2 + a_3 + \cdots = a_1\left(\frac{1}{1 - r}\right) \tag{12.33}$$

On the other hand, if r does not satisfy the inequalities $-1 < r < 1$, then the series diverges.

EXAMPLE 4:

(a) Find the sum of the geometric series whose first term is $a_1 = 1$ and whose common ratio is $r = 1/2$.

(b) Find the sum of the geometric series whose first term is $a_1 = 1$ and whose common ratio is $r = 1/3$.

(c) Find the sum of the geometric series whose first term is $a_1 = 3$ and whose common ratio is $r = -1/10$.

Solutions:

(a) According to Formula (12.33), we have

$$a_1 + a_2 + a_3 + \cdots = a_1\left(\frac{1}{1-r}\right) = 1\left(\frac{1}{1-1/2}\right) = 2$$

which confirms our earlier results.

(b) In this case,

$$a_1 + a_2 + a_3 + \cdots = a_1\left(\frac{1}{1-r}\right) = 1\left(\frac{1}{1-1/3}\right) = \frac{3}{2}$$

which also confirms our earlier results.

(c) Here we have

$$a_1 + a_2 + a_3 + \cdots = a_1\left(\frac{1}{1-r}\right) = 3\left(\frac{1}{1-(-1/10)}\right) = \frac{30}{11}$$

This example shows the power of Formula (12.33). For we would have a very hard time guessing, and then proving, that the partial sums of this sequence approach the number 30/11.

Try Study Suggestion 12.23. □

▶ **Study Suggestion 12.23:** Find the sum of the geometric series whose first term is $a_1 = 5$ and whose common ratio is $r = 1/4$. ◀

Ideas to Remember

The concept of an infinite series having a sum is simply a way to describe the fact that the partial sums of a sequence approach a certain number.

EXERCISES

In Exercises 1–12, determine whether or not the given geometric series has a sum, and if so, find that sum.

1. $1 - \frac{1}{2} + \frac{1}{4} - \frac{1}{8} + \cdots$

2. $\sqrt{2} + 2 + 2\sqrt{2} + \cdots$

3. $2^{-1} + 2^{-2} + 2^{-3} + \cdots$

4. $1000 + 100 + 10 + \cdots$

5. $1 + 1 + 1 + \cdots$

6. $147 + 42 + 12 + \cdots$

7. $81 - 54 + 36 - 24 + \cdots$

8. $\frac{1}{4} + \frac{1}{8} + \frac{1}{16} + \cdots$

9. $\frac{1}{2^3} + \frac{1}{2^6} + \frac{1}{2^9} + \cdots$

10. $\frac{1}{1000} + \frac{1}{100} + \frac{1}{10} + \cdots$

11. $3^{1/3} + 3^{2/3} + 3 + \cdots$

12. $3^{1/3} + 1 + 3^{-1/3} + \cdots$

13. A bouncing ball attains a height of 3 feet on its first bounce. On each of its successive bounces, it attains only 1/2 the height as on its previous bounce. Use infinite series to estimate the *total* distance traveled by the ball.

14. A pendulum travels 10 feet on its first swing. Suppose that on each successive swing, it travels only 2/5 as far as on the previous swing. Use infinite series to estimate the total distance traveled by the pendulum.

15. A certain race covers a distance of 1 mile. In order for a runner to complete the race, he must first complete 1/2 of the race; that is, he must first run 1/2 mile. Then, he must complete 1/2 of the remaining distance; that is, he must run $(1/2)(1/2) = 1/4$ mile. By continuing this reasoning, we see that in order for the runner to complete the 1 mile race, he must run distances equal to 1/2, 1/4, 1/8, ... From this point of view, do you think the runner will ever be able to complete the entire race?

12.6 An Introduction to Limits

In this section, we will give a brief introduction to the concept of a *limit*, which is one of the most important concepts in calculus. Let us begin by considering the function

$$f(x) = \frac{\sin x}{x}$$

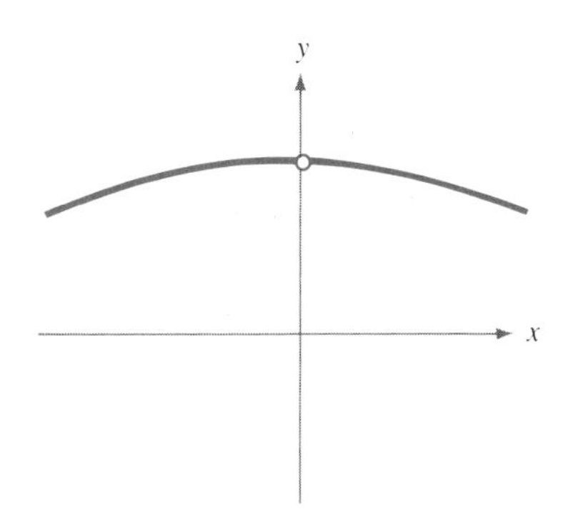

FIGURE 12.1

As you know, the domain of the sine function is the set of all real numbers. Therefore, the domain of the function f is the set of all *nonzero* real numbers. For if we try to set $x = 0$ in $f(x)$, we will get the meaningless expression 0/0.

In terms of the graph, this means that the graph of f will have no y-intercept. In other words, the graph will have a "hole" in it similar to the one pictured in Figure 12.1. (The graph of f may not look exactly like the graph in this figure, but it does have a hole in it just as this graph does.)

How can we determine the coordinates of this hole in the graph of f? Of course, the x-coordinate is 0, but what is the y-coordinate?

Since we can substitute any nonzero value for x in $f(x)$, we may be able to get some idea as to the value of this y-coordinate by choosing a value of x that is very close to 0, and computing $f(x)$. For example, a calculator tells us that

$$f(0.1) = 0.99833$$

which is very close to 1. So perhaps the "missing" y-coordinate is 1.

TABLE 12.1

x	$f(x)$	x
0.5	0.95885	−0.5
0.4	0.97355	−0.4
0.3	0.98507	−0.3
0.2	0.99335	−0.2
0.1	0.99833	−0.1
0.01	0.99998	−0.01
0.001	0.99999	−0.001

We can get an even stronger clue as to the value of the missing y-coordinate by looking at Table 12.1. In this table, we have chosen several values of x, getting closer and closer to 0, with which to compute $f(x)$. (In constructing this table, we have taken advantage of the fact that $f(x) = f(-x)$.)

It is clear from this table that, as we choose values of x closer and closer to 0, either positive or negative, the values of $f(x)$ get closer and closer to 1. In short, *as x approaches* 0, $f(x)$ *approaches* 1. (You may recall that we had a similar discussion in Chapter 4 concerning the behavior of a rational function near its vertical asymptotes.)

Table 12.1 is very convincing evidence that the missing y-coordinate in the graph of f is in fact equal to 1. We should emphasize however, that it is not absolutely conclusive evidence. For we cannot be certain that if we were to expand our table further by choosing values of x even closer to 0, that the values of f might not begin to move away from 1.

Nevertheless, it is possible to show that the evidence given in Table 12.1 is accurate, namely, that as x approaches 0 the values of $f(x)$ do indeed approach 1. (This would be done in a course in calculus.) In view of this behavior, we say that the *limit* of the function f as x approaches 0 is equal to 1, and we write

$$\lim_{x \to 0} \frac{\sin x}{x} = 1$$

Now let us give an informal definition of the term limit. (You will see a more formal definition if you take calculus.)

DEFINITION

Let f be a function, and let a and L be real numbers. If the values of $f(x)$ approach the number L as x approaches the number a, then we say that the **limit** of f as x approaches a is L, and write

$$\lim_{x \to a} f(x) = L$$

(This is read "the limit as x approaches a of $f(x)$ is equal to L.")

Before doing additional examples, let us emphasize a few points.

1. The limit of a function f as x approaches a does not always exist. We will see an example of this later in the section.
2. In order for the limit of a function f as x approaches a to be equal to L, it must be true that the values of $f(x)$ approach L as x approaches a *from either side*, that is, as x approaches a either from the right or from the left (on the real number line).
3. As in the previous example, a table of values can only give us an intelligent guess as to the value of the limit (if it exists). In order to be absolutely certain of this value, more formal methods must be used. However, for this brief introduction, we will content ourselves with guessing the value of the limit from a table of values. (You will see the more formal methods if you take calculus.)

4. The value of the function at a, that is $f(a)$, may or may not be equal to the value of the limit

$$\lim_{x \to a} f(x)$$

In fact, as in our example, $f(a)$ may not even be defined!

Now let us turn to some additional examples of limits.

EXAMPLE 1: Compute the limit

$$\lim_{x \to 1} \frac{\ln x}{1 - x}$$

Solution: We begin by computing the values of the function

$$f(x) = \frac{\ln x}{1 - x}$$

shown in Table 12.2.

TABLE 12.2

x	$f(x)$	x	$f(x)$
1.5	−0.81093	0.5	−1.38629
1.3	−0.87455	0.7	−1.18892
1.1	−0.95310	0.9	−1.05361
1.01	−0.99503	0.99	−1.00503
1.001	−0.99950	0.999	−1.00050
1.0001	−0.99995	0.9999	−1.00005

We can see from this table that as x approaches 1 *from either side*, the values of $f(x)$ approach -1. Again, while this is not absolutely conclusive evidence, it certainly seems reasonable. Therefore, we will accept this evidence and write

$$\lim_{x \to 1} \frac{\ln x}{1 - x} = -1$$

Try Study Suggestion 12.24. □

Study Suggestion 12.24: Compute the limit

$$\lim_{x \to 0} \frac{1 - \cos x}{x}$$ ◀

EXAMPLE 2: Compute the limit

$$\lim_{x \to 2} \frac{x^2 - 4}{x - 2}$$

Solution: As before, we may construct a table of values of the function $f(x) = (x^2 - 4)/(x - 2)$, for values of x that approach 2. However, before doing so we should observe that the numerator of this rational function factors, and we have

$$\frac{x^2 - 4}{x - 2} = x + 2 \quad \text{for} \quad x \neq 2$$

Notice that we have included the condition $x \neq 2$, since we cannot substitute $x = 2$ in the function $f(x)$. However, and this is a very important point, when computing the value of the *limit* of f as x approaches 2, we never actually set x equal to 2, and so the *limit* of $f(x)$ as x approaches 2 is the same as the *limit* of the function $g(x) = x + 2$ as x approaches 2. In short, even though the functions $f(x)$ and $g(x)$ are not the same, since they differ *only* at the value $x = 2$, their *limits* as x *approaches* 2 are the same. In symbols, we have

$$\lim_{x \to 2} \frac{x^2 - 4}{x - 2} = \lim_{x \to 2} (x + 2)$$

Now, it is very easy to evaluate

$$\lim_{x \to 2} (x + 2)$$

For as x gets closer and closer to 2, the quantity $x + 2$ surely gets closer and closer to $2 + 2 = 4$. Hence, we have

$$\lim_{x \to 2} \frac{x^2 - 4}{x - 2} = \lim_{x \to 2} (x + 2) = 4$$

Try Study Suggestion 12.25. ☐

▶ **Study Suggestion 12.25:** Compute the following limits

(a) $\lim_{x \to 1} (x + 2)$

(b) $\lim_{x \to 1} \frac{x^2 + x - 2}{x - 1}$ ◀

EXAMPLE 3: Compute the limit

$$\lim_{x \to 0} \frac{1}{x^2}$$

Solution: Again we prepare a table of values of the function.

x	$f(x)$	x
0.5	4	−0.5
0.3	11.1	−0.3
0.1	100	−0.1
0.01	10,000	−0.01
0.001	1,000,000	−0.001

As we can see from the table, the values of $f(x)$ are not approaching any specific value L, but instead they seem to be getting larger and larger. For this reason, we say that the limit of the function $f(x) = 1/x^2$ as x approaches 0 *does not exist.*

Try Study Suggestion 12.26. ☐

▶ **Study Suggestion 12.26:** Show that the limit

$$\lim_{x \to 1} \frac{x + 1}{(x - 1)^2}$$

does not exist. ◀

Ideas to Remember

The concept of the limit of a function involves the behavior of the function as x *approaches* a specific number.

EXERCISES

In Exercises 1–11, guess the value of the limit.

1. $\lim_{x\to 0}(x+1)$

2. $\lim_{x\to 1} 2x$

3. $\lim_{x\to 1}(-x)$

4. $\lim_{x\to 0} 3x$

5. $\lim_{x\to -2}(1-3x)$

6. $\lim_{x\to 2}\frac{1}{x}$

7. $\lim_{x\to 4}\frac{1-x}{x}$

8. $\lim_{x\to 3} x^2$

9. $\lim_{x\to -2}(x^3+1)$

10. $\lim_{x\to 5}(x^2+3x-1)$

11. **(a)** $\lim_{x\to 2}\frac{1}{2x}$ **(b)** $\lim_{x\to a}\frac{1}{ax}$ $a\neq 0$

In Exercises 12–17, first simplify the function and then guess the limit, as we did in Example 2.

12. $\lim_{x\to 1}\frac{x^2-2x+1}{x-1}$

13. $\lim_{x\to 2}\frac{x^2+x-6}{x-2}$

14. $\lim_{x\to -1}\frac{x^2-1}{x+1}$

15. $\lim_{x\to 2}\frac{x^3-2x^2-x+2}{x-2}$

16. $\lim_{x\to a}\frac{x^2-a^2}{x-a}$ $a\neq 0$

17. $\lim_{h\to 0}\frac{(x+h)^2-x^2}{h}$

In Exercises 18 and 19, construct a table using the values $x=\pm 0.05, \pm 0.03, \pm 0.01, \pm 0.001, \pm 0.0001$ to help guess the value of the limit.

18. $\lim_{x\to 0}\frac{\sin(x^2)}{x^2}$

19. $\lim_{x\to 0}\frac{\ln(1+2\sin x)}{x}$

In Exercises 20–22, construct a table using the values $x=1.5, 1.3, 1.1, 1.01, 1.001, 1.0001, 0.5, 0.7, 0.9, 0.99, 0.999, 0.9999$ to help guess the value of the limit.

20. $\lim_{x\to 1}\frac{x-1}{\ln x}$

21. $\lim_{x\to 1}\frac{x-1}{\ln|\ln x|}$

22. $\lim_{x\to 1}\frac{\ln x}{\sqrt{|x-1|}}$

23. Show that the limit $\lim_{x\to 0}\frac{1}{x^4}$ does not exist.

24. Show that the limit $\lim_{x\to 1}\frac{x-1}{(\ln x)^2}$ does not exist.

12.7 Review

CONCEPTS FOR REVIEW

Principle of mathematical induction
Induction hypothesis
Factorial
Binomial formula
Binomial coefficient
Term of a sequence
Sequence
General term of a sequence
Recurrence relation

Arithmetic sequence
Common difference
Geometric sequence
Common ratio
Infinite series
Infinite sum
Partial sum
Convergent series
Divergent series
Limit of a function

REVIEW EXERCISES

1. Prove by using mathematical induction that

$$2^2 + 4^2 + 6^2 + \cdots + (2n)^2 = \frac{2n(2n+1)(n+1)}{3}$$

2. Prove by using mathematical induction that

$$2 + 5 + 8 + \cdots + (3n-1) = \frac{n(3n+1)}{2}$$

3. Prove by using mathematical induction that

$$1 + 4 + 7 + \cdots + (3n-2) = \frac{n(3n-1)}{2}$$

4. Prove that $1 + 2n \leq 3^n$ for all nonnegative integers n.

5. Prove that $3^n > n^3$ for all integers $n \geq 5$.

6. Prove that 3 is a factor of $n^3 + 8n$ for all integers $n \geq 2$.

7. Prove that $x - y$ is a factor of $x^n - y^n$ for all nonnegative integers n. *Hint*, use the fact that

$$x^{k+1} - y^{k+1} = x^k(x - y) + (x^k - y^k)y$$

8. Prove that $x + y$ is a factor of $x^{2n+1} + y^{2n+1}$.

9. Expand and simplify the expression $(3x^2 - 2y)^6$.

10. Expand and simplify the expression $\left(2x + \frac{1}{x}\right)^5$.

11. Expand and simplify the expression $(1 + \sin x)^4$.

12. Find the coefficient of x^{12} in $(x^3 + 5)^8$.

13. Find the coefficient of x^2 in $\left(2x^3 + \frac{3}{x}\right)^{10}$.

14. Find the constant term in $\left(\frac{x}{2} + \frac{2}{x}\right)^{12}$.

15. Find the coefficient of u^5 in $(3v^2 + 2u - w^4)^8$.

16. Find the first five terms of the sequence whose general term is $a_n = 2^n - 3^n$. (The sequence starts with a_0.)

17. Find the first five terms $a_1, a_2, \ldots, a_5$ of the sequence whose general term is $a_n = \log\log 2^n$. (The sequence starts with a_1.)

18. Find the first five terms of the sequence with the property that $a_1 = a_2 = a$ and $a_n = 2a_{n-1} + 3a_{n-2}$ for $n \geq 3$.

19. Find the first six terms of the sequence with the property that $a_1 = r$, $a_2 = s$, and $a_n = sa_{n-2}$ for all $n > 2$.

20. Find the first seven terms of the sequence with the property that $a_1 = 1$, $a_2 = 2$, $a_3 = 3$ and $a_{n+1} = (n+1)a_n + a_{n-1}{}^2 + a_{n-2}$.

21. If a sequence has the property that $a_1 = 2$ and $a_n = 2na_{n-1}$ for $n \geq 2$, prove by induction that its general term is

$$a_n = 2 \cdot 4 \cdot 6 \cdots (2n)$$

That is, a_n is the product of the even integers from 2 to $2n$.

22. If a sequence has the property that $a_1 = 1$, $a_2 = 2$, and

$$a_n = a_{n-1} + a_{n-2}$$

for $n \geq 3$; that is, if each term is the sum of the previous two terms, prove by induction that its general term is

$$a_n = \frac{1}{\sqrt{5}}\left(\frac{1+\sqrt{5}}{2}\right)^n - \frac{1}{\sqrt{5}}\left(\frac{1-\sqrt{5}}{2}\right)^n$$

In Exercises 23–25, find the general term of the arithmetic sequence described by the given information. Then find the first five terms and the 100th term.

23. $a_1 = 7$, $a_2 = -12$ **24.** $a_1 = 5$, $a_7 = 23$

25. $a_2 = 6$, $a_5 = 8$

In Exercises 26–28, find the general term of the geometric sequence described by the given information. Then find the first five terms and the 10th term.

26. $a_1 = 5$, $a_2 = -5$ **27.** $a_1 = 2$, $a_4 = 54$

28. $a_2 = 8$, $a_5 = -1$

29. Can a sequence that begins $a_1 = x, a_2 = x^2, a_3 = x^2 + x$ be an arithmetic or a geometric sequence? Justify your answer.

In Exercises 30–32, find the nth partial sum s_n in the arithmetic sequence described by the given information.

30. $a_1 = 2,\ a_2 = -1$

31. $a_1 = \frac{1}{3},\ a_2 = \frac{1}{2}$

32. $a_1 = \sin^2 x,\ d = \cos^2 x$

In Exercises 33–35, find the nth partial sum s_n in the geometric sequence described by the given information.

33. $a_1 = \frac{1}{2},\ a_2 = \frac{1}{3}$

34. $a_1 = 5,\ a_2 = -5$

35. $a_1 = x,\ r = \dfrac{1}{(x+1)}$

36. Compute the sum $1 + \cos\frac{\pi}{6} + \cos^2\frac{\pi}{6} + \cdots + \cos^5\frac{\pi}{6}$

37. Compute the sum

$$1 + (-0.1)^3 + (-0.1)^6 + (-0.1)^9 + \cdots + (-0.1)^{15}$$

38. Simplify the sum $e^{3x} + e^{5x} + e^{7x} + \cdots + e^{17x}$

39. Simplify the sum $1 + \log x^k + \log x^{2k} + \cdots + \log x^{nk}$

40. A vacuum pump removes 1/10 of the remaining air from a certain bottle with each stroke. How many strokes are necessary in order to remove 1/2 of the original amount of air from the bottle?

In Exercises 41–43, determine whether the given geometric series has a sum, and if so, find that sum.

41. $1 + \frac{1}{5} + \frac{1}{25} + \cdots$

42. $1 - \frac{1}{7} + \frac{1}{49} - \cdots$

43. $\frac{1}{900} + \frac{1}{810} + \frac{1}{729} + \cdots$

44. For which values of x does the geometric series

$$|x+1| + |x+1|^2 + |x+1|^3 + \cdots$$

have a sum? What is that sum?

45. If an infinite series has terms that increase, can the series possibly have a sum? Justify your answer.

46. If an infinite series has terms that decrease, does that necessarily mean that it has a sum? Justify your answer.

In Exercises 47–49, guess the limit.

47. $\lim_{x \to 5} (2x - 3)$

48. $\lim_{x \to 0} 2x + 1$

49. $\lim_{x \to a} \frac{1}{x} - \frac{1}{a} \qquad a \neq 0$

In Exercises 50 and 51, first simplify the function and then guess the limit.

50. $\lim_{x \to 3} \dfrac{x^2 - 9}{x - 3}$

51. $\lim_{h \to x} \dfrac{\frac{1}{x} - \frac{1}{h}}{x - h}$

52. Construct a table using the values $x = \pm 0.5,\ \pm 0.05,\ \pm 0.03$ and ± 0.01 to guess the value of the following limit

$$\lim_{x \to 0} \frac{\tan x}{x}$$

APPENDIX

TABLE A NATURAL LOGARITHMS

x	0.00	0.01	0.02	0.03	0.04	0.05	0.06	0.07	0.08	0.09
1.0	0.0000	0.0100	0.0198	0.0296	0.0392	0.0488	0.0583	0.0677	0.0770	0.0862
1.1	0.0953	0.1044	0.1133	0.1222	0.1310	0.1398	0.1484	0.1570	0.1655	0.1740
1.2	0.1823	0.1906	0.1989	0.2070	0.2151	0.2231	0.2311	0.2390	0.2469	0.2546
1.3	0.2624	0.2700	0.2776	0.2852	0.2927	0.3001	0.3075	0.3148	0.3221	0.3293
1.4	0.3365	0.3436	0.3507	0.3577	0.3646	0.3716	0.3784	0.3853	0.3920	0.3988
1.5	0.4055	0.4121	0.4187	0.4253	0.4318	0.4383	0.4447	0.4511	0.4574	0.4637
1.6	0.4700	0.4762	0.4824	0.4886	0.4947	0.5008	0.5068	0.5128	0.5188	0.5247
1.7	0.5306	0.5365	0.5423	0.5481	0.5539	0.5596	0.5653	0.5710	0.5766	0.5822
1.8	0.5878	0.5933	0.5988	0.6043	0.6098	0.6152	0.6206	0.6259	0.6313	0.6366
1.9	0.6419	0.6471	0.6523	0.6575	0.6627	0.6678	0.6729	0.6780	0.6831	0.6881
2.0	0.6931	0.6981	0.7031	0.7080	0.7130	0.7178	0.7227	0.7275	0.7324	0.7372
2.1	0.7419	0.7467	0.7514	0.7561	0.7608	0.7655	0.7701	0.7747	0.7793	0.7839
2.2	0.7885	0.7930	0.7975	0.8020	0.8065	0.8109	0.8154	0.8198	0.8242	0.8286
2.3	0.8329	0.8372	0.8416	0.8459	0.8502	0.8544	0.8587	0.8629	0.8671	0.8713
2.4	0.8755	0.8796	0.8838	0.8879	0.8920	0.8961	0.9002	0.9042	0.9083	0.9123
2.5	0.9163	0.9203	0.9243	0.9282	0.9322	0.9361	0.9400	0.9439	0.9478	0.9517
2.6	0.9555	0.9594	0.9632	0.9670	0.9708	0.9746	0.9783	0.9821	0.9858	0.9895
2.7	0.9933	0.9969	1.0006	1.0043	1.0080	1.0116	1.0152	1.0188	1.0225	1.0260
2.8	1.0296	1.0332	1.0367	1.0403	1.0438	1.0473	1.0508	1.0543	1.0578	1.0613
2.9	1.0647	1.0682	1.0716	1.0750	1.0784	1.0818	1.0852	1.0886	1.0919	1.0953
3.0	1.0986	1.1019	1.1053	1.1086	1.1119	1.1151	1.1184	1.1217	1.1249	1.1282
3.1	1.1314	1.1346	1.1378	1.1410	1.1442	1.1474	1.1506	1.1537	1.1569	1.1600
3.2	1.1632	1.1663	1.1694	1.1725	1.1756	1.1787	1.1817	1.1848	1.1878	1.1909
3.3	1.1939	1.1970	1.2000	1.2030	1.2060	1.2090	1.2119	1.2149	1.2179	1.2208
3.4	1.2238	1.2267	1.2296	1.2326	1.2355	1.2384	1.2413	1.2442	1.2470	1.2499
3.5	1.2528	1.2556	1.2585	1.2613	1.2641	1.2669	1.2698	1.2726	1.2754	1.2782
3.6	1.2809	1.2837	1.2865	1.2892	1.2920	1.2947	1.2975	1.3002	1.3029	1.3056
3.7	1.3083	1.3110	1.3137	1.3164	1.3191	1.3218	1.3244	1.3271	1.3297	1.3324
3.8	1.3350	1.3376	1.3403	1.3429	1.3455	1.3481	1.3507	1.3533	1.3558	1.3584
3.9	1.3610	1.3635	1.3661	1.3686	1.3712	1.3737	1.3762	1.3788	1.3813	1.3838
4.0	1.3863	1.3888	1.3913	1.3938	1.3962	1.3987	1.4012	1.4036	1.4061	1.4085
4.1	1.4110	1.4134	1.4159	1.4183	1.4207	1.4231	1.4255	1.4279	1.4303	1.4327
4.2	1.4351	1.4375	1.4398	1.4422	1.4446	1.4469	1.4493	1.4516	1.4540	1.4563
4.3	1.4586	1.4609	1.4633	1.4656	1.4679	1.4702	1.4725	1.4748	1.4770	1.4793
4.4	1.4816	1.4839	1.4861	1.4884	1.4907	1.4929	1.4952	1.4974	1.4996	1.5019
4.5	1.5041	1.5063	1.5085	1.5107	1.5129	1.5151	1.5173	1.5195	1.5217	1.5239
4.6	1.5261	1.5282	1.5304	1.5326	1.5347	1.5369	1.5390	1.5412	1.5433	1.5454
4.7	1.5476	1.5497	1.5518	1.5539	1.5560	1.5581	1.5602	1.5623	1.5644	1.5665
4.8	1.5686	1.5707	1.5728	1.5748	1.5769	1.5790	1.5810	1.5831	1.5851	1.5872
4.9	1.5892	1.5913	1.5933	1.5953	1.5974	1.5994	1.6014	1.6034	1.6054	1.6074
5.0	1.6094	1.6114	1.6134	1.6154	1.6174	1.6194	1.6214	1.6233	1.6253	1.6273
5.1	1.6292	1.6312	1.6332	1.6351	1.6371	1.6390	1.6409	1.6429	1.6448	1.6467
5.2	1.6487	1.6506	1.6525	1.6544	1.6563	1.6582	1.6601	1.6620	1.6639	1.6658
5.3	1.6677	1.6696	1.6715	1.6734	1.6752	1.6771	1.6790	1.6808	1.6827	1.6845
5.4	1.6864	1.6882	1.6901	1.6919	1.6938	1.6956	1.6974	1.6993	1.7011	1.7029

NATURAL LOGARITHMS, *Continued*

x	0.00	0.01	0.02	0.03	0.04	0.05	0.06	0.07	0.08	0.09
5.5	1.7047	1.7066	1.7084	1.7102	1.7120	1.7138	1.7156	1.7174	1.7192	1.7210
5.6	1.7228	1.7246	1.7263	1.7281	1.7299	1.7317	1.7334	1.7352	1.7370	1.7387
5.7	1.7405	1.7422	1.7440	1.7457	1.7475	1.7492	1.7509	1.7527	1.7544	1.7561
5.8	1.7579	1.7596	1.7613	1.7630	1.7647	1.7664	1.7682	1.7699	1.7716	1.7733
5.9	1.7750	1.7766	1.7783	1.7800	1.7817	1.7834	1.7851	1.7867	1.7884	1.7901
6.0	1.7918	1.7934	1.7951	1.7967	1.7984	1.8001	1.8017	1.8034	1.8050	1.8066
6.1	1.8083	1.8099	1.8116	1.8132	1.8148	1.8165	1.8181	1.8197	1.8213	1.8229
6.2	1.8245	1.8262	1.8278	1.8294	1.8310	1.8326	1.8342	1.8358	1.8374	1.8390
6.3	1.8406	1.8421	1.8437	1.8453	1.8469	1.8485	1.8500	1.8516	1.8532	1.8547
6.4	1.8563	1.8579	1.8594	1.8610	1.8625	1.8641	1.8656	1.8672	1.8687	1.8703
6.5	1.8718	1.8733	1.8749	1.8764	1.8779	1.8795	1.8810	1.8825	1.8840	1.8856
6.6	1.8871	1.8886	1.8901	1.8916	1.8931	1.8946	1.8961	1.8976	1.8991	1.9006
6.7	1.9021	1.9036	1.9051	1.9066	1.9081	1.9095	1.9110	1.9125	1.9140	1.9155
6.8	1.9169	1.9184	1.9199	1.9213	1.9228	1.9242	1.9257	1.9272	1.9286	1.9301
6.9	1.9315	1.9330	1.9344	1.9359	1.9373	1.9387	1.9402	1.9416	1.9430	1.9445
7.0	1.9459	1.9473	1.9488	1.9502	1.9516	1.9530	1.9544	1.9559	1.9573	1.9587
7.1	1.9601	1.9615	1.9629	1.9643	1.9657	1.9671	1.9685	1.9699	1.9713	1.9727
7.2	1.9741	1.9755	1.9769	1.9782	1.9796	1.9810	1.9824	1.9838	1.9851	1.9865
7.3	1.9879	1.9892	1.9906	1.9920	1.9933	1.9947	1.9961	1.9974	1.9988	2.0001
7.4	2.0015	2.0028	2.0042	2.0055	2.0069	2.0082	2.0096	2.0109	2.0122	2.0136
7.5	2.0149	2.0162	2.0176	2.0189	2.0202	2.0215	2.0229	2.0242	2.0255	2.0268
7.6	2.0282	2.0295	2.0308	2.0321	2.0334	2.0347	2.0360	2.0373	2.0386	2.0399
7.7	2.0412	2.0425	2.0438	2.0451	2.0464	2.0477	2.0490	2.0503	2.0516	2.0528
7.8	2.0541	2.0554	2.0567	2.0580	2.0592	2.0605	2.0618	2.0631	2.0643	2.0665
7.9	2.0669	2.0681	2.0694	2.0707	2.0719	2.0732	2.0744	2.0757	2.0769	2.0782
8.0	2.0794	2.0807	2.0819	2.0832	2.0844	2.0857	2.0869	2.0882	2.0894	2.0906
8.1	2.0919	2.0931	2.0943	2.0956	2.0968	2.0980	2.0992	2.1005	2.1017	2.1029
8.2	2.1041	2.1054	2.1066	2.1078	2.1090	2.1102	2.1114	2.1126	2.1138	2.1150
8.3	2.1163	2.1175	2.1187	2.1199	2.1211	2.1223	2.1235	2.1247	2.1258	2.1270
8.4	2.1282	2.1294	2.1306	2.1318	2.1330	2.1342	2.1353	2.1365	2.1377	2.1389
8.5	2.1401	2.1412	2.1424	2.1436	2.1448	2.1459	2.1471	2.1483	2.1494	2.1506
8.6	2.1518	2.1529	2.1541	2.1552	2.1564	2.1576	2.1587	2.1599	2.1610	2.1622
8.7	2.1633	2.1645	2.1656	2.1668	2.1679	2.1691	2.1702	2.1713	2.1725	2.1736
8.8	2.1748	2.1759	2.1770	2.1782	2.1793	2.1804	2.1815	2.1827	2.1838	2.1849
8.9	2.1861	2.1872	2.1883	2.1894	2.1905	2.1917	2.1928	2.1939	2.1950	2.1961
9.0	2.1972	2.1983	2.1994	2.2006	2.2017	2.2028	2.2039	2.2050	2.2061	2.2072
9.1	2.2083	2.2094	2.2105	2.2116	2.2127	2.2138	2.2148	2.2159	2.2170	2.2181
9.2	2.2192	2.2203	2.2214	2.2225	2.2235	2.2246	2.2257	2.2268	2.2279	2.2289
9.3	2.2300	2.2311	2.2322	2.2332	2.2343	2.2354	2.2364	2.2375	2.2386	2.2396
9.4	2.2407	2.2418	2.2428	2.2439	2.2450	2.2460	2.2471	2.2481	2.2492	2.2502
9.5	2.2513	2.2523	2.2534	2.2544	2.2555	2.2565	2.2576	2.2586	2.2597	2.2607
9.6	2.2618	2.2628	2.2638	2.2649	2.2659	2.2670	2.2680	2.2690	2.2701	2.2711
9.7	2.2721	2.2732	2.2742	2.2752	2.2762	2.2773	2.2783	2.2793	2.2803	2.2814
9.8	2.2824	2.2834	2.2844	2.2854	2.2865	2.2875	2.2885	2.2895	2.2905	2.2915
9.9	2.2925	2.2935	2.2946	2.2956	2.2966	2.2976	2.2986	2.2996	2.3006	2.3016

TABLE B COMMON LOGARITHMS

x	0.00	0.01	0.02	0.03	0.04	0.05	0.06	0.07	0.08	0.09
1.0	0.0000	0.0043	0.0086	0.0128	0.0170	0.0212	0.0253	0.0294	0.0334	0.0374
1.1	0.0414	0.0453	0.0492	0.0531	0.0569	0.0607	0.0645	0.0682	0.0719	0.0755
1.2	0.0792	0.0828	0.0864	0.0899	0.0934	0.0969	0.1004	0.1038	0.1072	0.1106
1.3	0.1139	0.1173	0.1206	0.1239	0.1271	0.1303	0.1335	0.1367	0.1399	0.1430
1.4	0.1461	0.1492	0.1523	0.1553	0.1584	0.1614	0.1644	0.1673	0.1703	0.1732
1.5	0.1761	0.1790	0.1818	0.1847	0.1875	0.1903	0.1931	0.1959	0.1987	0.2014
1.6	0.2041	0.2068	0.2095	0.2122	0.2148	0.2175	0.2201	0.2227	0.2253	0.2279
1.7	0.2304	0.2330	0.2355	0.2380	0.2405	0.2430	0.2455	0.2480	0.2504	0.2529
1.8	0.2553	0.2577	0.2601	0.2625	0.2648	0.2672	0.2695	0.2718	0.2742	0.2765
1.9	0.2788	0.2810	0.2833	0.2856	0.2878	0.2900	0.2923	0.2945	0.2967	0.2989
2.0	0.3010	0.3032	0.3054	0.3075	0.3096	0.3118	0.3139	0.3160	0.3181	0.3201
2.1	0.3222	0.3243	0.3263	0.3284	0.3304	0.3324	0.3345	0.3365	0.3385	0.3404
2.2	0.3424	0.3444	0.3464	0.3483	0.3502	0.3522	0.3541	0.3560	0.3579	0.3598
2.3	0.3617	0.3636	0.3655	0.3674	0.3692	0.3711	0.3729	0.3747	0.3766	0.3784
2.4	0.3802	0.3820	0.3838	0.3856	0.3874	0.3892	0.3909	0.3927	0.3945	0.3962
2.5	0.3979	0.3997	0.4014	0.4031	0.4048	0.4065	0.4082	0.4099	0.4116	0.4133
2.6	0.4150	0.4166	0.4183	0.4200	0.4216	0.4232	0.4249	0.4265	0.4281	0.4298
2.7	0.4314	0.4330	0.4346	0.4362	0.4378	0.4393	0.4409	0.4425	0.4440	0.4456
2.8	0.4472	0.4487	0.4502	0.4518	0.4533	0.4548	0.4564	0.4579	0.4594	0.4609
2.9	0.4624	0.4639	0.4654	0.4669	0.4683	0.4698	0.4713	0.4728	0.4742	0.4757
3.0	0.4771	0.4786	0.4800	0.4814	0.4829	0.4843	0.4857	0.4871	0.4886	0.4900
3.1	0.4914	0.4928	0.4942	0.4955	0.4969	0.4983	0.4997	0.5011	0.5024	0.5038
3.2	0.5051	0.5065	0.5079	0.5092	0.5105	0.5119	0.5132	0.5145	0.5159	0.5172
3.3	0.5185	0.5198	0.5211	0.5224	0.5237	0.5250	0.5263	0.5276	0.5289	0.5302
3.4	0.5315	0.5328	0.5340	0.5353	0.5366	0.5378	0.5391	0.5403	0.5416	0.5428
3.5	0.5441	0.5453	0.5465	0.5478	0.5490	0.5502	0.5514	0.5527	0.5539	0.5551
3.6	0.5563	0.5575	0.5587	0.5599	0.5611	0.5623	0.5635	0.5647	0.5658	0.5670
3.7	0.5682	0.5694	0.5705	0.5717	0.5729	0.5740	0.5752	0.5763	0.5775	0.5786
3.8	0.5798	0.5809	0.5821	0.5832	0.5843	0.5855	0.5866	0.5877	0.5888	0.5899
3.9	0.5911	0.5922	0.5933	0.5944	0.5955	0.5966	0.5977	0.5988	0.5999	0.6010
4.0	0.6021	0.6031	0.6042	0.6053	0.6064	0.6075	0.6085	0.6096	0.6107	0.6117
4.1	0.6128	0.6138	0.6149	0.6160	0.6170	0.6180	0.6191	0.6201	0.6212	0.6222
4.2	0.6232	0.6243	0.6253	0.6263	0.6274	0.6284	0.6294	0.6304	0.6314	0.6325
4.3	0.6335	0.6345	0.6355	0.6365	0.6375	0.6385	0.6395	0.6405	0.6415	0.6425
4.4	0.6435	0.6444	0.6454	0.6464	0.6474	0.6484	0.6493	0.6503	0.6513	0.6522
4.5	0.6532	0.6542	0.6551	0.6561	0.6571	0.6580	0.6590	0.6599	0.6609	0.6618
4.6	0.6628	0.6637	0.6646	0.6656	0.6665	0.6675	0.6684	0.6693	0.6702	0.6712
4.7	0.6721	0.6730	0.6739	0.6749	0.6758	0.6767	0.6776	0.6785	0.6794	0.6803
4.8	0.6812	0.6821	0.6830	0.6839	0.6848	0.6857	0.6866	0.6875	0.6884	0.6893
4.9	0.6902	0.6911	0.6920	0.6928	0.6937	0.6946	0.6955	0.6964	0.6972	0.6981
5.0	0.6990	0.6998	0.7007	0.7016	0.7024	0.7033	0.7042	0.7050	0.7059	0.7067
5.1	0.7076	0.7084	0.7093	0.7101	0.7110	0.7118	0.7126	0.7135	0.7143	0.7152
5.2	0.7160	0.7168	0.7177	0.7185	0.7193	0.7202	0.7210	0.7218	0.7226	0.7235
5.3	0.7243	0.7251	0.7259	0.7267	0.7275	0.7284	0.7292	0.7300	0.7308	0.7316
5.4	0.7324	0.7332	0.7340	0.7348	0.7356	0.7364	0.7372	0.7380	0.7388	0.7396

COMMON LOGARITHMS, *Continued*

x	0.00	0.01	0.02	0.03	0.04	0.05	0.06	0.07	0.08	0.09
5.5	0.7404	0.7412	0.7419	0.7427	0.7435	0.7443	0.7451	0.7459	0.7466	0.7474
5.6	0.7482	0.7490	0.7497	0.7505	0.7513	0.7520	0.7528	0.7536	0.7543	0.7551
5.7	0.7559	0.7566	0.7574	0.7582	0.7589	0.7597	0.7604	0.7612	0.7619	0.7627
5.8	0.7634	0.7642	0.7649	0.7657	0.7664	0.7672	0.7679	0.7686	0.7694	0.7701
5.9	0.7709	0.7716	0.7723	0.7731	0.7738	0.7745	0.7752	0.7760	0.7767	0.7774
6.0	0.7782	0.7789	0.7796	0.7803	0.7810	0.7818	0.7825	0.7832	0.7839	0.7846
6.1	0.7853	0.7860	0.7868	0.7875	0.7882	0.7889	0.7896	0.7903	0.7910	0.7917
6.2	0.7924	0.7931	0.7938	0.7945	0.7952	0.7959	0.7966	0.7973	0.7980	0.7987
6.3	0.7993	0.8000	0.8007	0.8014	0.8021	0.8028	0.8035	0.8041	0.8048	0.8055
6.4	0.8062	0.8069	0.8075	0.8082	0.8089	0.8096	0.8102	0.8109	0.8116	0.8122
6.5	0.8129	0.8136	0.8142	0.8149	0.8156	0.8162	0.8169	0.8176	0.8182	0.8189
6.6	0.8195	0.8202	0.8209	0.8215	0.8222	0.8228	0.8235	0.8241	0.8248	0.8254
6.7	0.8261	0.8267	0.8274	0.8280	0.8287	0.8293	0.8299	0.8306	0.8312	0.8319
6.8	0.8325	0.8331	0.8338	0.8344	0.8351	0.8357	0.8363	0.8370	0.8376	0.8382
6.9	0.8388	0.8395	0.8401	0.8407	0.8414	0.8420	0.8426	0.8430	0.8439	0.8445
7.0	0.8451	0.8457	0.8463	0.8470	0.8476	0.8482	0.8488	0.8494	0.8500	0.8506
7.1	0.8513	0.8519	0.8525	0.8531	0.8537	0.8543	0.8549	0.8555	0.8561	0.8567
7.2	0.8573	0.8579	0.8585	0.8591	0.8597	0.8603	0.8609	0.8615	0.8621	0.8627
7.3	0.8633	0.8639	0.8645	0.8651	0.8657	0.8663	0.8669	0.8675	0.8681	0.8686
7.4	0.8692	0.8698	0.8704	0.8710	0.8716	0.8722	0.8727	0.8733	0.8739	0.8745
7.5	0.8751	0.8756	0.8762	0.8768	0.8774	0.8779	0.8785	0.8791	0.8797	0.8802
7.6	0.8808	0.8814	0.8820	0.8825	0.8831	0.8837	0.8842	0.8848	0.8854	0.8859
7.7	0.8865	0.8871	0.8876	0.8882	0.8887	0.8893	0.8899	0.8904	0.8910	0.8915
7.8	0.8921	0.8927	0.8932	0.8938	0.8943	0.8949	0.8954	0.8960	0.8965	0.8971
7.9	0.8976	0.8982	0.8987	0.8993	0.8998	0.9004	0.9009	0.9015	0.9020	0.9025
8.0	0.9031	0.9036	0.9042	0.9047	0.9053	0.9058	0.9063	0.9069	0.9074	0.9079
8.1	0.9085	0.9090	0.9096	0.9101	0.9106	0.9112	0.9117	0.9122	0.9128	0.9133
8.2	0.9138	0.9143	0.9149	0.9154	0.9159	0.9165	0.9170	0.9175	0.9180	0.9186
8.3	0.9191	0.9196	0.9201	0.9206	0.9212	0.9217	0.9222	0.9227	0.9232	0.9238
8.4	0.9243	0.9248	0.9253	0.9258	0.9263	0.9269	0.9274	0.9279	0.9284	0.9289
8.5	0.9294	0.9299	0.9304	0.9309	0.9315	0.9320	0.9325	0.9330	0.9335	0.9340
8.6	0.9345	0.9350	0.9355	0.9360	0.9365	0.9370	0.9375	0.9380	0.9385	0.9390
8.7	0.9395	0.9400	0.9405	0.9410	0.9415	0.9420	0.9425	0.9430	0.9435	0.9440
8.8	0.9445	0.9450	0.9455	0.9460	0.9465	0.9469	0.9474	0.9479	0.9484	0.9489
8.9	0.9494	0.9499	0.9504	0.9509	0.9513	0.9518	0.9523	0.9528	0.9533	0.9538
9.0	0.9542	0.9547	0.9552	0.9557	0.9562	0.9566	0.9571	0.9576	0.9581	0.9586
9.1	0.9590	0.9595	0.9600	0.9605	0.9609	0.9614	0.9619	0.9624	0.9628	0.9633
9.2	0.9638	0.9643	0.9647	0.9652	0.9657	0.9661	0.9666	0.9671	0.9675	0.9680
9.3	0.9685	0.9689	0.9694	0.9699	0.9703	0.9708	0.9713	0.9717	0.9722	0.9727
9.4	0.9731	0.9736	0.9741	0.9745	0.9750	0.9754	0.9759	0.9763	0.9768	0.9773
9.5	0.9777	0.9782	0.9786	0.9791	0.9795	0.9800	0.9805	0.9809	0.9814	0.9818
9.6	0.9823	0.9827	0.9832	0.9836	0.9841	0.9845	0.9850	0.9854	0.9859	0.9863
9.7	0.9868	0.9872	0.9877	0.9881	0.9886	0.9890	0.9894	0.9899	0.9903	0.9908
9.8	0.9912	0.9917	0.9921	0.9926	0.9930	0.9934	0.9939	0.9943	0.9948	0.9952
9.9	0.9956	0.9961	0.9965	0.9969	0.9974	0.9978	0.9983	0.9987	0.9991	0.9996

TABLE C EXPONENTIAL FUNCTIONS

x	e^x	e^{-x}	x	e^x	e^{-x}
0.00	1.0000	1.0000	3.0	20.086	0.0498
0.05	1.0513	0.9512	3.1	22.198	0.0450
0.10	1.1052	0.9048	3.2	24.533	0.0408
0.15	1.1618	0.8607	3.3	27.113	0.0369
0.20	1.2214	0.8187	3.4	29.964	0.0334
0.25	1.2840	0.7788	3.5	33.115	0.0302
0.30	1.3499	0.7408	3.6	36.598	0.0273
0.35	1.4191	0.7047	3.7	40.447	0.0247
0.40	1.4918	0.6703	3.8	44.701	0.0224
0.45	1.5683	0.6376	3.9	49.402	0.0202
0.50	1.6487	0.6065	4.0	54.598	0.0183
0.55	1.7333	0.5769	4.1	60.340	0.0166
0.60	1.8221	0.5488	4.2	66.686	0.0150
0.65	1.9155	0.5220	4.3	73.700	0.0136
0.70	2.0138	0.4966	4.4	81.451	0.0123
0.75	2.1170	0.4724	4.5	90.017	0.0111
0.80	2.2255	0.4493	4.6	99.484	0.0101
0.85	2.3396	0.4274	4.7	109.95	0.0091
0.90	2.4596	0.4066	4.8	121.51	0.0082
0.95	2.5857	0.3867	4.9	134.29	0.0074
1.0	2.7183	0.3679	5.0	148.41	0.0067
1.1	3.0042	0.3329	5.1	164.02	0.0061
1.2	3.3201	0.3012	5.2	181.27	0.0055
1.3	3.6693	0.2725	5.3	200.34	0.0050
1.4	4.0552	0.2466	5.4	221.41	0.0045
1.5	4.4817	0.2231	5.5	244.69	0.0041
1.6	4.9530	0.2019	5.6	270.43	0.0037
1.7	5.4739	0.1827	5.7	298.87	0.0033
1.8	6.0496	0.1653	5.8	330.30	0.0030
1.9	6.6859	0.1496	5.9	365.04	0.0027
2.0	7.3891	0.1353	6.0	403.43	0.0025
2.1	8.1662	0.1225	6.5	665.14	0.0015
2.2	9.0250	0.1108	7.0	1096.6	0.0009
2.3	9.9742	0.1003	7.5	1808.0	0.0006
2.4	11.023	0.0907	8.0	2981.0	0.0003
2.5	12.182	0.0821	8.5	4914.8	0.0002
2.6	13.464	0.0743	9.0	8103.1	0.0001
2.7	14.880	0.0672	9.5	13,360	0.00007
2.8	16.445	0.0608	10.0	22,026	0.00004
2.9	18.174	0.0550			

TABLE D TRIGONOMETRIC FUNCTIONS—DEGREE MEASURE

Deg.	sin	tan	cot	cos	
0.0	0.00000	0.00000	∞	1.0000	90.0
.1	.00175	.00175	573.0	1.0000	89.9
.2	.00349	.00349	286.5	1.0000	.8
.3	.00524	.00524	191.0	1.0000	.7
.4	.00698	.00698	143.24	1.0000	.6
.5	.00873	.00873	114.59	1.0000	.5
.6	.01047	.01047	95.49	0.9999	.4
.7	.01222	.01222	81.85	.9999	.3
.8	.01396	.01396	71.62	.9999	.2
.9	.01571	.01571	63.66	.9999	89.1
1.0	0.01745	0.01746	57.29	0.9998	89.0
.1	.01920	.01920	52.08	.9998	88.9
.2	.02094	.02095	47.74	.9998	.8
.3	.02269	.02269	44.07	.9997	.7
.4	.02443	.02444	40.92	.9997	.6
.5	.02618	.02619	38.19	.9997	.5
.6	.02792	.02793	35.80	.9996	.4
.7	.02967	.02968	33.69	.9996	.3
.8	.03141	.03143	31.82	.9995	.2
.9	.03316	.03317	30.14	.9995	88.1
2.0	0.03490	0.03492	28.64	0.9994	88.0
.1	.03664	.03667	27.27	.9993	87.9
.2	.03839	.03842	26.03	.9993	.8
.3	.04013	.04016	24.90	.9992	.7
.4	.04188	.04191	23.86	.9991	.6
.5	.04362	.04366	22.90	.9990	.5
.6	.04536	.04541	22.02	.9990	.4
.7	.04711	.04716	21.20	.9989	.3
.8	.04885	.04891	20.45	.9988	.2
.9	.05059	.05066	19.74	.9987	87.1
3.0	0.05234	0.05241	19.081	0.9986	87.0
.1	.05408	.05416	18.464	.9985	86.9
.2	.05582	.05591	17.886	.9984	.8
.3	.05756	.05766	17.343	.9983	.7
.4	.05931	.05941	16.832	.9982	.6
.5	.06105	.06116	16.350	.9981	.5
.6	.06279	.06291	15.895	.9980	.4
.7	.06453	.06467	15.464	.9979	.3
.8	.06627	.06642	15.056	.9978	.2
.9	.06802	.06817	14.669	.9977	86.1
4.0	0.06976	0.06993	14.301	0.9976	86.0
.1	.07150	.07168	13.951	.9974	85.9
.2	.07324	.07344	13.617	.9973	.8
.3	.07498	.07519	13.300	.9972	.7
.4	.07672	.07695	12.996	.9971	.6
.5	.07846	.07870	12.706	.9969	.5
.6	.08020	.08046	12.429	.9968	.4
.7	.08194	.08221	12.163	.9966	.3
.8	.08368	.08397	11.909	.9965	.2
.9	.08542	.08573	11.664	.9963	85.1
5.0	0.08716	0.08749	11.430	0.9962	85.0
	cos	cot	tan	sin	Deg.

Deg.	sin	tan	cot	cos	
5.0	0.08716	0.08749	11.430	0.9962	85.0
.1	.08889	.08925	11.205	.9960	84.9
.2	.09063	.09101	10.988	.9959	.8
.3	.09237	.09277	10.780	.9957	.7
.4	.09411	.09453	10.579	.9956	.6
.5	.09585	.09629	10.385	.9954	.5
.6	.09758	.09805	10.199	.9952	.4
.7	.09932	.09981	10.019	.9951	.3
.8	.10106	.10158	9.845	.9949	.2
.9	.10279	.10334	9.677	.9947	84.1
6.0	0.10453	0.10510	9.514	0.9945	84.0
.1	.10626	.10687	9.357	.9943	83.9
.2	.10800	.10863	9.205	.9942	.8
.3	.10973	.11040	9.058	.9940	.7
.4	.11147	.11217	8.915	.9938	.6
.5	.11320	.11394	8.777	.9936	.5
.6	.11494	.11570	8.643	.9934	.4
.7	.11667	.11747	8.513	.9932	.3
.8	.11840	.11924	8.386	.9930	.2
.9	.12014	.12101	8.264	.9928	83.1
7.0	0.12187	0.12278	8.144	0.9925	83.0
.1	.12360	.12456	8.028	.9923	82.9
.2	.12533	.12633	7.916	.9921	.8
.3	.12706	.12810	7.806	.9919	.7
.4	.12880	.12988	7.700	.9917	.6
.5	.13053	.13165	7.596	.9914	.5
.6	.13226	.13343	7.495	.9912	.4
.7	.13399	.13521	7.396	.9910	.3
.8	.13572	.13698	7.300	.9907	.2
.9	.13744	.13876	7.207	.9905	82.1
8.0	0.13917	0.14054	7.115	0.9903	82.0
.1	.14090	.14232	7.026	.9900	81.9
.2	.14263	.14410	6.940	.9898	.8
.3	.14436	.14588	6.855	.9895	.7
.4	.14608	.14767	6.772	.9893	.6
.5	.14781	.14945	6.691	.9890	.5
.6	.14954	.15124	6.612	.9888	.4
.7	.15126	.15302	6.535	.9885	.3
.8	.15299	.15481	6.460	.9882	.2
.9	.15471	.15660	6.386	.9880	81.1
9.0	0.15643	0.15838	6.314	0.9877	81.0
.1	.15816	.16017	6.243	.9874	80.9
.2	.15988	.16196	6.174	.9871	.8
.3	.16160	.16376	6.107	.9869	.7
.4	.16333	.16555	6.041	.9866	.6
.5	.16505	.16734	5.976	.9863	.5
.6	.16677	.16914	5.912	.9860	.4
.7	.16849	.17093	5.850	.9857	.3
.8	.17021	.17273	5.789	.9854	.2
.9	.17193	.17453	5.730	.9851	80.1
10.0	0.1736	0.1763	5.671	0.9848	80.0
	cos	cot	tan	sin	Deg.

TRIGONOMETRIC FUNCTIONS—DEGREE MEASURE, *Continued*

Deg.	sin	tan	cot	cos	
10.0	0.1736	0.1763	5.671	0.9848	80.0
.1	.1754	.1781	5.614	.9845	79.9
.2	.1771	.1799	5.558	.9842	.8
.3	.1788	.1817	5.503	.9839	.7
.4	.1805	.1835	5.449	.9836	.6
.5	.1822	.1853	5.396	.9833	.5
.6	.1840	.1871	5.343	.9829	.4
.7	.1857	.1890	5.292	.9826	.3
.8	.1874	.1908	5.242	.9823	.2
.9	.1891	.1926	5.193	.9820	79.1
11.0	0.1908	0.1944	5.145	0.9816	79.0
.1	.1925	.1962	5.079	.9813	78.9
.2	.1942	.1980	5.050	.9810	.8
.3	.1959	.1998	5.005	.9806	.7
.4	.1977	.2016	4.959	.9803	.6
.5	.1994	.2035	4.915	.9799	.5
.6	.2011	.2053	4.872	.9796	.4
.7	.2028	.2071	4.829	.9792	.3
.8	.2045	.2089	4.787	.9789	.2
.9	.2062	.2107	4.745	.9785	78.1
12.0	0.2079	0.2126	4.705	0.9781	78.0
.1	.2096	.2144	4.665	.9778	77.9
.2	.2113	.2162	4.625	.9774	.8
.3	.2130	.2180	4.586	.9770	.7
.4	.2147	.2199	4.548	.9767	.6
.5	.2164	.2217	4.511	.9763	.5
.6	.2181	.2235	4.474	.9759	.4
.7	.2198	.2254	4.437	.9755	.3
.8	.2215	.2272	4.402	.9751	.2
.9	.2233	.2290	4.366	.9748	77.1
13.0	0.2250	0.2309	4.331	0.9744	77.0
.1	.2267	.2327	4.297	.9740	76.9
.2	.2284	.2345	4.264	.9736	.8
.3	.2300	.2364	4.230	.9732	.7
.4	.2317	.2382	4.198	.9728	.6
.5	.2334	.2401	4.165	.9724	.5
.6	.2351	.2419	4.134	.9720	.4
.7	.2368	.2438	4.102	.9715	.3
.8	.2385	.2456	4.071	.9711	.2
.9	.2402	.2475	4.041	.9707	76.1
14.0	0.2419	0.2493	4.011	0.9703	76.0
.1	.2436	.2512	3.981	.9699	75.9
.2	.2453	.2530	3.952	.9694	.8
.3	.2470	.2549	3.923	.9690	.7
.4	.2487	.2568	3.895	.9686	.6
.5	.2504	.2586	3.867	.9681	.5
.6	.2521	.2605	3.839	.9677	.4
.7	.2538	.2623	3.812	.9673	.3
.8	.2554	.2642	3.785	.9668	.2
.9	.2571	.2661	3.758	.9664	75.1
15.0	0.2588	0.2679	3.732	0.9659	75.0
	cos	cot	tan	sin	Deg.

Deg.	sin	tan	cot	cos	
15.0	0.2588	0.2679	3.732	0.9659	75.0
.1	.2605	.2698	3.706	.9655	74.9
.2	.2622	.2717	3.681	.9650	.8
.3	.2639	.2736	3.655	.9646	.7
.4	.2656	.2754	3.630	.9641	.6
.5	.2672	.2773	3.606	.9636	.5
.6	.2689	.2792	3.582	.9632	.4
.7	.2706	.2811	3.558	.9627	.3
.8	.2723	.2830	3.534	.9622	.2
.9	.2740	.2849	3.511	.9617	74.1
16.0	0.2756	0.2867	3.487	0.9613	74.0
.1	.2773	.2886	3.465	.9608	73.9
.2	.2790	.2905	3.442	.9603	.8
.3	.2807	.2924	3.420	.9598	.7
.4	.2823	.2943	3.398	.9593	.6
.5	.2840	.2962	3.376	.9588	.5
.6	.2857	.2981	3.354	.9583	.4
.7	.2874	.3000	3.333	.9578	.3
.8	.2890	.3019	3.312	.9573	.2
.9	.2907	.3038	3.291	.9568	73.1
17.0	0.2924	0.3057	3.271	0.9563	73.0
.1	.2940	.3076	3.251	.9558	72.9
.2	.2957	.3096	3.230	.9553	.8
.3	.2974	.3115	3.211	.9548	.7
.4	.2990	.3134	3.191	.9542	.6
.5	.3007	.3153	3.172	.9537	.5
.6	.3024	.3172	3.152	.9532	.4
.7	.3040	.3191	3.133	.9527	.3
.8	.3057	.3211	3.115	.9521	.2
.9	.3074	.3230	3.096	.9516	72.1
18.0	0.3090	0.3249	3.078	0.9511	72.0
.1	.3107	.3269	3.060	.9505	71.9
.2	.3123	.3288	3.042	.9500	.8
.3	.3140	.3307	3.024	.9494	.7
.4	.3156	.3327	3.006	.9489	.6
.5	.3173	.3346	2.989	.9483	.5
.6	.3190	.3365	2.971	.9478	.4
.7	.3206	.3385	2.954	.9472	.3
.8	.3223	.3404	2.937	.9466	.2
.9	.3239	.3424	2.921	.9461	71.1
19.0	0.3256	0.3443	2.904	0.9455	71.0
.1	.3272	.3463	2.888	.9449	70.9
.2	.3289	.3482	2.872	.9444	.8
.3	.3305	.3502	2.856	.9438	.7
.4	.3322	.3522	2.840	.9432	.6
.5	.3338	.3541	2.824	.9426	.5
.6	.3355	.3561	2.808	.9421	.4
.7	.3371	.3581	2.793	.9415	.3
.8	.3387	.3600	2.778	.9409	.2
.9	.3404	.3620	2.762	.9403	70.1
20.0	0.3420	0.3640	2.747	0.9397	70.0
	cos	cot	tan	sin	Deg.

TRIGONOMETRIC FUNCTIONS—DEGREE MEASURE, *Continued*

Deg.	sin	tan	cot	cos	
20.0	0.3420	0.3640	2.747	0.9397	70.0
.1	.3437	.3659	2.733	.9391	69.9
.2	.3453	.3679	2.718	.9385	.8
.3	.3469	.3699	2.703	.9379	.7
.4	.3486	.3719	2.689	.9373	.6
.5	.3502	.3739	2.675	.9367	.5
.6	.3518	.3759	2.660	.9361	.4
.7	.3535	.3779	2.646	.9354	.3
.8	.3551	.3799	2.633	.9348	.2
.9	.3567	.3819	2.619	.9342	69.1
21.0	0.3584	0.3839	2.605	0.9336	69.0
.1	.3600	.3859	2.592	.9330	68.9
.2	.3616	.3879	2.578	.9323	.8
.3	.3633	.3899	2.565	.9317	.7
.4	.3649	.3919	2.552	.9311	.6
.5	.3665	.3939	2.539	.9304	.5
.6	.3681	.3959	2.526	.9298	.4
.7	.3697	.3979	2.513	.9291	.3
.8	.3714	.4000	2.500	.9285	.2
.9	.3730	.4020	2.488	.9278	68.1
22.0	0.3746	0.4040	2.475	0.9272	68.0
.1	.3762	.4061	2.463	.9265	67.9
.2	.3778	.4081	2.450	.9259	.8
.3	.3795	.4101	2.438	.9252	.7
.4	.3811	.4122	2.426	.9245	.6
.5	.3827	.4142	2.414	.9239	.5
.6	.3843	.4163	2.402	.9232	.4
.7	.3859	.4183	2.391	.9225	.3
.8	.3875	.4204	2.379	.9219	.2
.9	.3891	.4224	2.367	.9212	67.1
23.0	0.3907	0.4245	2.356	0.9205	67.0
.1	.3923	.4265	2.344	.9198	66.9
.2	.3939	.4286	2.333	.9191	.8
.3	.3955	.4307	2.322	.9184	.7
.4	.3971	.4327	2.311	.9178	.6
.5	.3987	.4348	2.300	.9171	.5
.6	.4003	.4369	2.289	.9164	.4
.7	.4019	.4390	2.278	.9157	.3
.8	.4035	.4411	2.267	.9150	.2
.9	.4051	.4431	2.257	.9143	66.1
24.0	0.4067	0.4452	2.246	0.9135	66.0
.1	.4083	.4473	2.236	.9128	65.9
.2	.4099	.4494	2.225	.9121	.8
.3	.4115	.4515	2.215	.9114	.7
.4	.4131	.4536	2.204	.9107	.6
.5	.4147	.4557	2.194	.9100	.5
.6	.4163	.4578	2.184	.9092	.4
.7	.4179	.4599	2.174	.9085	.3
.8	.4195	.4621	2.164	.9078	.2
.9	.4210	.4642	2.154	.9070	65.1
25.0	0.4226	0.4663	2.145	0.9063	65.0
	cos	cot	tan	sin	Deg.

Deg.	sin	tan	cot	cos	
25.0	0.4226	0.4663	2.145	0.9063	65.0
.1	.4242	.4684	2.135	.9056	64.9
.2	.4258	.4706	2.125	.9048	.8
.3	.4274	.4727	2.116	.9041	.7
.4	.4289	.4748	2.106	.9033	.6
.5	.4305	.4770	2.097	.9026	.5
.6	.4321	.4791	2.087	.9018	.4
.7	.4337	.4813	2.078	.9011	.3
.8	.4352	.4834	2.069	.9003	.2
.9	.4368	.4856	2.059	.8996	64.1
26.0	0.4384	0.4887	2.050	0.8988	64.0
.1	.4399	.4899	2.041	.8980	63.9
.2	.4415	.4921	2.032	.8973	.8
.3	.4431	.4942	2.023	.8965	.7
.4	.4446	.4964	2.014	.8957	.6
.5	.4462	.4986	2.006	.8949	.5
.6	.4478	.5008	1.997	.8942	.4
.7	.4493	.5029	1.988	.8934	.3
.8	.4509	.5051	1.980	.8926	.2
.9	.4524	.5073	1.971	.8918	63.1
27.0	0.4540	0.5095	1.963	0.8910	63.0
.1	.4555	.5117	1.954	.8902	62.9
.2	.4571	.5139	1.946	.8894	.8
.3	.4586	.5161	1.937	.8886	.7
.4	.4602	.5184	1.929	.8878	.6
.5	.4617	.5206	1.921	.8870	.5
.6	.4633	.5228	1.913	.8862	.4
.7	.4648	.5250	1.905	.8854	.3
.8	.4664	.5272	1.897	.8846	.2
.9	.4679	.5295	1.889	.8838	62.1
28.0	0.4695	0.5317	1.881	0.8829	62.0
.1	.4710	.5340	1.873	.8821	61.9
.2	.4726	.5362	1.865	.8813	.8
.3	.4741	.5384	1.857	.8805	.7
.4	.4756	.5407	1.849	.8796	.6
.5	.4772	.5430	1.842	.8788	.5
.6	.4787	.5452	1.834	.8780	.4
.7	.4802	.5475	1.827	.8771	.3
.8	.4818	.5498	1.819	.8763	.2
.9	.4833	.5520	1.811	.8755	61.1
29.0	0.4848	0.5543	1.804	0.8746	61.0
.1	.4863	.5566	1.797	.8738	60.9
.2	.4879	.5589	1.789	.8729	.8
.3	.4894	.5612	1.782	.8721	.7
.4	.4909	.5635	1.775	.8712	.6
.5	.4924	.5658	1.767	.8704	.5
.6	.4939	.5681	1.760	.8695	.4
.7	.4955	.5704	1.753	.8686	.3
.8	.4970	.5727	1.746	.8678	.2
.9	.4985	.5750	1.739	.8669	60.1
30.0	0.5000	0.5774	1.732	0.8660	60.0
	cos	cot	tan	sin	Deg.

TRIGONOMETRIC FUNCTIONS—DEGREE MEASURE, *Continued*

Deg.	sin	tan	cot	cos		Deg.	sin	tan	cot	cos	
30.0	0.5000	0.5774	1.7321	0.8660	60.0	35.0	0.5736	0.7002	1.4281	0.8192	55.0
.1	.5015	.5797	1.7251	.8652	59.9	.1	.5750	.7028	1.4229	.8181	54.9
.2	.5030	.5820	1.7182	.8643	.8	.2	.5764	.7054	1.4176	.8171	.8
.3	.5045	.5844	1.7113	.8634	.7	.3	.5779	.7080	1.4124	.8161	.7
.4	.5060	.5867	1.7045	.8625	.6	.4	.5793	.7107	1.4071	.8151	.6
.5	.5075	.5890	1.6977	.8616	.5	.5	.5807	.7133	1.4019	.8141	.5
.6	.5090	.5914	1.6909	.8607	.4	.6	.5821	.7159	1.3968	.8131	.4
.7	.5105	.5938	1.6842	.8599	.3	.7	.5835	.7186	1.3916	.8121	.3
.8	.5120	.5961	1.6775	.8590	.2	.8	.5850	.7212	1.3865	.8111	.2
.9	.5135	.5985	1.6709	.8581	59.1	.9	.5864	.7239	1.3814	.8100	54.1
31.0	0.5150	0.6009	1.6643	0.8572	59.0	36.0	0.5878	0.7265	1.3764	0.8090	54.0
.1	.5165	.6032	1.6577	.8563	58.9	.1	.5892	.7292	1.3713	.8080	53.9
.2	.5180	.6056	1.6512	.8554	.8	.2	.5906	.7319	1.3663	.8070	.8
.3	.5195	.6080	1.6447	.8545	.7	.3	.5920	.7346	1.3613	.8059	.7
.4	.5210	.6104	1.6383	.8536	.6	.4	.5934	.7373	1.3564	.8049	.6
.5	.5225	.6128	1.6319	.8526	.5	.5	.5948	.7400	1.3514	.8039	.5
.6	.5240	.6152	1.6255	.8517	.4	.6	.5962	.7427	1.3465	.8028	.4
.7	.5255	.6176	1.6191	.8508	.3	.7	.5976	.7454	1.3416	.8018	.3
.8	.5270	.6200	1.6128	.8499	.2	.8	.5990	.7481	1.3367	.8007	.2
.9	.5284	.6224	1.6066	.8490	58.1	.9	.6004	.7508	1.3319	.7997	53.1
32.0	0.5299	0.6249	1.6003	0.8480	58.0	37.0	0.6018	0.7536	1.3270	0.7986	53.0
.1	.5314	.6273	1.5941	.8471	57.9	.1	.6032	.7563	1.3222	.7976	52.9
.2	.5329	.6297	1.5880	.8462	.8	.2	.6046	.7590	1.3175	.7965	.8
.3	.5344	.6322	1.5818	.8453	.7	.3	.6060	.7618	1.3127	.7955	.7
.4	.5358	.6346	1.5757	.8443	.6	.4	.6074	.7646	1.3079	.7944	.6
.5	.5373	.6371	1.5697	.8434	.5	.5	.6088	.7673	1.3032	.7934	.5
.6	.5388	.6395	1.5637	.8425	.4	.6	.6101	.7701	1.2985	.7923	.4
.7	.5402	.6420	1.5577	.8415	.3	.7	.6115	.7729	1.2938	.7912	.3
.8	.5417	.6445	1.5517	.8406	.2	.8	.6129	.7757	1.2892	.7902	.2
.9	.5432	.6469	1.5458	.8396	57.1	.9	.6143	.7785	1.2846	.7891	52.1
33.0	0.5446	0.6494	1.5399	0.8387	57.0	38.0	0.6157	0.7813	1.2799	0.7880	52.0
.1	.5461	.6519	1.5340	.8377	56.9	.1	.6170	.7841	1.2753	.7869	51.9
.2	.5476	.6544	1.5282	.8368	.8	.2	.6184	.7869	1.2708	.7859	.8
.3	.5490	.6569	1.5224	.8358	.7	.3	.6198	.7898	1.2662	.7848	.7
.4	.5505	.6594	1.5166	.8348	.6	.4	.6211	.7926	1.2617	.7837	.6
.5	.5519	.6619	1.5108	.8339	.5	.5	.6225	.7954	1.2572	.7826	.5
.6	.5534	.6644	1.5051	.8329	.4	.6	.6239	.7983	1.2527	.7815	.4
.7	.5548	.6669	1.4994	.8320	.3	.7	.6252	.8012	1.2482	.7804	.3
.8	.5563	.6694	1.4938	.8310	.2	.8	.6266	.8040	1.2437	.7793	.2
.9	.5577	.6720	1.4882	.8300	56.1	.9	.6280	.8069	1.2393	.7782	51.1
34.0	0.5592	0.6745	1.4826	0.8290	56.0	39.0	0.6293	0.8098	1.2349	0.7771	51.0
.1	.5606	.6771	1.4770	.8281	55.9	.1	.6307	.8127	1.2305	.7760	50.9
.2	.5621	.6796	1.4715	.8271	.8	.2	.6320	.8156	1.2261	.7749	.8
.3	.5635	.6822	1.4659	.8261	.7	.3	.6334	.8185	1.2218	.7738	.7
.4	.5650	.6847	1.4605	.8251	.6	.4	.6347	.8214	1.2174	.7727	.6
.5	.5664	.6873	1.4550	.8241	.5	.5	.6361	.8243	1.2131	.7716	.5
.6	.5678	.6899	1.4496	.8231	.4	.6	.6374	.8273	1.2088	.7705	.4
.7	.5693	.6924	1.4442	.8221	.3	.7	.6388	.8302	1.2045	.7694	.3
.8	.5707	.6950	1.4388	.8211	.2	.8	.6401	.8332	1.2002	.7683	.2
.9	.5721	.6976	1.4335	.8202	55.1	.9	.6414	.8361	1.1960	.7672	50.1
35.0	0.5736	0.7002	1.4281	0.8192	55.0	40.0	0.6428	0.8391	1.1918	0.7660	50.0
	cos	cot	tan	sin	Deg.		cos	cot	tan	sin	Deg.

TRIGONOMETRIC FUNCTIONS—DEGREE MEASURE, *Continued*

Deg.	sin	tan	cot	cos		Deg.	sin	tan	cot	cos	
40.0	0.6428	0.8391	1.1918	0.7660	50.0	42.5	0.6756	0.9163	1.0913	0.7373	0.5
.1	.6441	.8421	1.1875	.7649	49.9	.6	.6769	.9195	1.0875	.7361	.4
.2	.6455	.8451	1.1833	.7638	.8	.7	.6782	.9228	1.0837	.7349	.3
.3	.6468	.8481	1.1792	.7627	.7	.8	.6794	.9260	1.0799	.7337	.2
.4	.6481	.8511	1.1750	.7615	.6	.9	.6807	.9293	1.0761	.7325	47.1
.5	.6494	.8541	1.1708	.7604	.5	43.0	0.6820	0.9325	1.0724	0.7314	47.0
.6	.6508	.8571	1.1667	.7593	.4	.1	.6833	.9358	1.0686	.7302	46.9
.7	.6521	.8601	1.1626	.7581	.3	.2	.6845	.9391	1.0649	.7290	.8
.8	.6534	.8632	1.1585	.7570	.2	.3	.6858	.9424	1.0612	.7278	.7
.9	.6547	.8662	1.1544	.7559	49.1	.4	.6871	.9457	1.0575	.7266	.6
41.0	0.6561	0.8693	1.1504	0.7547	49.0	.5	.6884	.9490	1.0538	.7254	.5
.1	.6574	.8724	1.1463	.7536	48.9	.6	.6896	.9523	1.0501	.7242	.4
.2	.6587	.8754	1.1423	.7524	.8	.7	.6909	.9556	1.0464	.7230	.3
.3	.6600	.8785	1.1383	.7513	.7	.8	.6921	.9590	1.0428	.7218	.2
.4	.6613	.8816	1.1343	.7501	.6	.9	.6934	.9623	1.0392	.7206	46.1
.5	.6626	.8847	1.1303	.7490	.5	44.0	0.6947	0.9657	1.0355	0.7193	46.0
.6	.6639	.8878	1.1263	.7478	.4	.1	.6959	.9691	1.0319	.7181	45.9
.7	.6652	.8910	1.1224	.7466	.3	.2	.6972	.9725	1.0283	.7169	.8
.8	.6665	.8941	1.1184	.7455	.2	.3	.6984	.9759	1.0247	.7157	.7
.9	.6678	.8972	1.1145	.7443	48.1	.4	.6997	.9793	1.0212	.7145	.6
42.0	0.6691	0.9004	1.1106	0.7431	48.0	.5	.7009	.9827	1.0176	.7133	.5
.1	.6704	.9036	1.1067	.7420	47.9	.6	.7022	.9861	1.0141	.7120	.4
.2	.6717	.9067	1.1028	.7408	.8	.7	.7034	.9896	1.0105	.7108	.3
.3	.6730	.9099	1.0990	.7396	.7	.8	.7046	.9930	1.0070	.7096	.2
.4	.6743	.9131	1.0951	.7385	.6	.9	.7059	.9965	1.0035	.7083	45.1
42.5	0.6756	0.9163	1.0913	0.7373	0.5	45.0	0.7071	1.0000	1.0000	0.7071	45.0
	cos	cot	tan	sin	Deg.		cos	cot	tan	sin	Deg.

TABLE E TRIGONOMETRIC FUNCTIONS—RADIAN MEASURE

x	$\sin x$	$\cos x$	$\tan x$	$\cot x$	$\sec x$	$\csc x$
0.00	0.0000	1.0000	0.0000	—	1.000	—
0.01	0.0100	1.0000	0.0100	99.997	1.000	100.00
0.02	0.0200	0.9998	0.0200	49.993	1.000	50.00
0.03	0.0300	0.9996	0.0300	33.323	1.000	33.34
0.04	0.0400	0.9992	0.0400	24.987	1.001	25.01
0.05	0.0500	0.9988	0.0500	19.983	1.001	20.01
0.06	0.0600	0.9982	0.0601	16.647	1.002	16.68
0.07	0.0699	0.9976	0.0701	14.262	1.002	14.30
0.08	0.0799	0.9968	0.0802	12.473	1.003	12.51
0.09	0.0899	0.9960	0.0902	11.081	1.004	11.13
0.10	0.0998	0.9950	0.1003	9.967	1.005	10.02
0.11	0.1098	0.9940	0.1104	9.054	1.006	9.109
0.12	0.1197	0.9928	0.1206	8.293	1.007	8.353
0.13	0.1296	0.9916	0.1307	7.649	1.009	7.714
0.14	0.1395	0.9902	0.1409	7.096	1.010	7.166
0.15	0.1494	0.9888	0.1511	6.617	1.011	6.692
0.16	0.1593	0.9872	0.1614	6.197	1.013	6.277
0.17	0.1692	0.9856	0.1717	5.826	1.015	5.911
0.18	0.1790	0.9838	0.1820	5.495	1.016	5.586
0.19	0.1889	0.9820	0.1923	5.200	1.018	5.295
0.20	0.1987	0.9801	0.2027	4.933	1.020	5.033
0.21	0.2085	0.9780	0.2131	4.692	1.022	4.797
0.22	0.2182	0.9759	0.2236	4.472	1.025	4.582
0.23	0.2280	0.9737	0.2341	4.271	1.027	4.386
0.24	0.2377	0.9713	0.2447	4.086	1.030	4.207
0.25	0.2474	0.9689	0.2553	3.916	1.032	4.042
0.26	0.2571	0.9664	0.2660	3.759	1.035	3.890
0.27	0.2667	0.9638	0.2768	3.613	1.038	3.749
0.28	0.2764	0.9611	0.2876	3.478	1.041	3.619
0.29	0.2860	0.9582	0.2984	3.351	1.044	3.497
0.30	0.2955	0.9553	0.3093	3.233	1.047	3.384
0.31	0.3051	0.9523	0.3203	3.122	1.050	3.278
0.32	0.3146	0.9492	0.3314	3.018	1.053	3.179
0.33	0.3240	0.9460	0.3425	2.920	1.057	3.086
0.34	0.3335	0.9428	0.3537	2.827	1.061	2.999
0.35	0.3429	0.9394	0.3650	2.740	1.065	2.916
0.36	0.3523	0.9359	0.3764	2.657	1.068	2.839
0.37	0.3616	0.9323	0.3879	2.578	1.073	2.765
0.38	0.3709	0.9287	0.3994	2.504	1.077	2.696
0.39	0.3802	0.9249	0.4111	2.433	1.081	2.630
0.40	0.3894	0.9211	0.4228	2.365	1.086	2.568
0.41	0.3986	0.9171	0.4346	2.301	1.090	2.509
0.42	0.4078	0.9131	0.4466	2.239	1.095	2.452
0.43	0.4169	0.9090	0.4586	2.180	1.100	2.399
0.44	0.4259	0.9048	0.4708	2.124	1.105	2.348
0.45	0.4350	0.9004	0.4831	2.070	1.111	2.299
0.46	0.4439	0.8961	0.4954	2.018	1.116	2.253
0.47	0.4529	0.8916	0.5080	1.969	1.122	2.208
0.48	0.4618	0.8870	0.5206	1.921	1.127	2.166
0.49	0.4706	0.8823	0.5334	1.875	1.133	2.125

TRIGONOMETRIC FUNCTIONS—RADIAN MEASURE, *Continued*

x	$\sin x$	$\cos x$	$\tan x$	$\cot x$	$\sec x$	$\csc x$
0.50	0.4794	0.8776	0.5463	1.830	1.139	2.086
0.51	0.4882	0.8727	0.5594	1.788	1.146	2.048
0.52	0.4969	0.8678	0.5726	1.747	1.152	2.013
$\pi/6$	0.5000	0.8660	0.5774	1.732	1.155	2.000
0.53	0.5055	0.8628	0.5859	1.707	1.159	1.978
0.54	0.5141	0.8577	0.5994	1.668	1.166	1.945
0.55	0.5227	0.8525	0.6131	1.631	1.173	1.913
0.56	0.5312	0.8473	0.6269	1.595	1.180	1.883
0.57	0.5396	0.8419	0.6410	1.560	1.188	1.853
0.58	0.5480	0.8365	0.6552	1.526	1.196	1.825
0.59	0.5564	0.8309	0.6696	1.494	1.203	1.797
0.60	0.5646	0.8253	0.6841	1.462	1.212	1.771
0.61	0.5729	0.8196	0.6989	1.431	1.220	1.746
0.62	0.5810	0.8139	0.7139	1.401	1.229	1.721
0.63	0.5891	0.8080	0.7291	1.372	1.238	1.697
0.64	0.5972	0.8021	0.7445	1.343	1.247	1.674
0.65	0.6052	0.7961	0.7602	1.315	1.256	1.652
0.66	0.6131	0.7900	0.7761	1.288	1.266	1.631
0.67	0.6210	0.7838	0.7923	1.262	1.276	1.610
0.68	0.6288	0.7776	0.8087	1.237	1.286	1.590
0.69	0.6365	0.7712	0.8253	1.212	1.297	1.571
0.70	0.6442	0.7648	0.8423	1.187	1.307	1.552
0.71	0.6518	0.7584	0.8595	1.163	1.319	1.534
0.72	0.6594	0.7518	0.8771	1.140	1.330	1.517
0.73	0.6669	0.7452	0.8949	1.117	1.342	1.500
0.74	0.6743	0.7385	0.9131	1.095	1.354	1.483
0.75	0.6816	0.7317	0.9316	1.073	1.367	1.467
0.76	0.6889	0.7248	0.9505	1.052	1.380	1.452
0.77	0.6961	0.7179	0.9697	1.031	1.393	1.437
0.78	0.7033	0.7109	0.9893	1.011	1.407	1.422
$\pi/4$	0.7071	0.7071	1.000	1.000	1.414	1.414
0.79	0.7104	0.7038	1.009	0.9908	1.421	1.408
0.80	0.7174	0.6967	1.030	0.9712	1.435	1.394
0.81	0.7243	0.6895	1.050	0.9520	1.450	1.381
0.82	0.7311	0.6822	1.072	0.9331	1.466	1.368
0.83	0.7379	0.6749	1.093	0.9146	1.482	1.355
0.84	0.7446	0.6675	1.116	0.8964	1.498	1.343
0.85	0.7513	0.6600	1.138	0.8785	1.515	1.331
0.86	0.7578	0.6524	1.162	0.8609	1.533	1.320
0.87	0.7643	0.6448	1.185	0.8437	1.551	1.308
0.88	0.7707	0.6372	1.210	0.8267	1.569	1.297
0.89	0.7771	0.6294	1.235	0.8100	1.589	1.287
0.90	0.7833	0.6216	1.260	0.7936	1.609	1.277
0.91	0.7895	0.6137	1.286	0.7774	1.629	1.267
0.92	0.7956	0.6058	1.313	0.7615	1.651	1.257
0.93	0.8016	0.5978	1.341	0.7458	1.673	1.247
0.94	0.8076	0.5898	1.369	0.7303	1.696	1.238
0.95	0.8134	0.5817	1.398	0.7151	1.719	1.229
0.96	0.8192	0.5735	1.428	0.7001	1.744	1.221
0.97	0.8249	0.5653	1.459	0.6853	1.769	1.212
0.98	0.8305	0.5570	1.491	0.6707	1.795	1.204
0.99	0.8360	0.5487	1.524	0.6563	1.823	1.196

TRIGONOMETRIC FUNCTIONS—RADIAN MEASURE, *Continued*

x	$\sin x$	$\cos x$	$\tan x$	$\cot x$	$\sec x$	$\csc x$
1.00	0.8415	0.5403	1.557	0.6421	1.851	1.188
1.01	0.8468	0.5319	1.592	0.6281	1.880	1.181
1.02	0.8521	0.5234	1.628	0.6142	1.911	1.174
1.03	0.8573	0.5148	1.665	0.6005	1.942	1.166
1.04	0.8624	0.5062	1.704	0.5870	1.975	1.160
$\pi/3$	0.8660	0.5000	1.732	0.5774	2.000	1.155
1.05	0.8674	0.4976	1.743	0.5736	2.010	1.153
1.06	0.8724	0.4889	1.784	0.5604	2.046	1.146
1.07	0.8772	0.4801	1.827	0.5473	2.083	1.140
1.08	0.8820	0.4713	1.871	0.5344	2.122	1.134
1.09	0.8866	0.4625	1.917	0.5216	2.162	1.128
1.10	0.8912	0.4536	1.965	0.5090	2.205	1.122
1.11	0.8957	0.4447	2.014	0.4964	2.249	1.116
1.12	0.9001	0.4357	2.066	0.4840	2.295	1.111
1.13	0.9044	0.4267	2.120	0.4718	2.344	1.160
1.14	0.9086	0.4176	2.176	0.4596	2.395	1.101
1.15	0.9128	0.4085	2.234	0.4475	2.448	1.096
1.16	0.9168	0.3993	2.296	0.4356	2.504	1.091
1.17	0.9208	0.3902	2.360	0.4237	2.563	1.086
1.18	0.9246	0.3809	2.427	0.4120	2.625	1.082
1.19	0.9284	0.3717	2.498	0.4003	2.691	1.077
1.20	0.9320	0.3624	2.572	0.3888	2.760	1.073
1.21	0.9356	0.3530	2.650	0.3773	2.833	1.069
1.22	0.9391	0.3436	2.733	0.3659	2.910	1.065
1.23	0.9425	0.3342	2.820	0.3546	2.992	1.061
1.24	0.9458	0.3248	2.912	0.3434	3.079	1.057
1.25	0.9490	0.3153	3.010	0.3323	3.171	1.054
1.26	0.9521	0.3058	3.113	0.3212	3.270	1.050
1.27	0.9551	0.2963	3.224	0.3102	3.375	1.047
1.28	0.9580	0.2867	3.341	0.2993	3.488	1.044
1.29	0.9608	0.2771	3.467	0.2884	3.609	1.041

TRIGONOMETRIC FUNCTIONS—RADIAN MEASURE, *Continued*

x	$\sin x$	$\cos x$	$\tan x$	$\cot x$	$\sec x$	$\csc x$
1.30	0.9636	0.2675	3.602	0.2776	3.738	1.038
1.31	0.9662	0.2579	3.747	0.2669	3.878	1.035
1.32	0.9687	0.2482	3.903	0.2562	4.029	1.032
1.33	0.9711	0.2385	4.072	0.2456	4.193	1.030
1.34	0.9735	0.2288	4.256	0.2350	4.372	1.027
1.35	0.9757	0.2190	4.455	0.2245	4.566	1.025
1.36	0.9779	0.2092	4.673	0.2140	4.779	1.023
1.37	0.9799	0.1994	4.913	0.2035	5.014	1.021
1.38	0.9819	0.1896	5.177	0.1931	5.273	1.018
1.39	0.9837	0.1798	5.471	0.1828	5.561	1.017
1.40	0.9854	0.1700	5.798	0.1725	5.883	1.015
1.41	0.9871	0.1601	6.165	0.1622	6.246	1.013
1.42	0.9887	0.1502	6.581	0.1519	6.657	1.011
1.43	0.9901	0.1403	7.055	0.1417	7.126	1.010
1.44	0.9915	0.1304	7.602	0.1315	7.667	1.009
1.45	0.9927	0.1205	8.238	0.1214	8.299	1.007
1.46	0.9939	0.1106	8.989	0.1113	9.044	1.006
1.47	0.9949	0.1006	9.887	0.1011	9.938	1.005
1.48	0.9959	0.0907	10.983	0.0910	11.029	1.004
1.49	0.9967	0.0807	12.350	0.0810	12.390	1.003
1.50	0.9975	0.0707	14.101	0.0709	14.137	1.003
1.51	0.9982	0.0608	16.428	0.0609	16.458	1.002
1.52	0.9987	0.0508	19.670	0.0508	19.695	1.001
1.53	0.9992	0.0408	24.498	0.0408	24.519	1.001
1.54	0.9995	0.0308	32.461	0.0308	32.476	1.000
1.55	0.9998	0.0208	48.078	0.0208	48.089	1.000
1.56	0.9999	0.0108	92.620	0.0108	92.626	1.000
1.57	1.0000	0.0008	1255.8	0.0008	1255.8	1.000
$\pi/2$	1.0000	0.0000	—	0.0000	—	1.000

ANSWERS TO STUDY SUGGESTIONS

Chapter One

1.1.

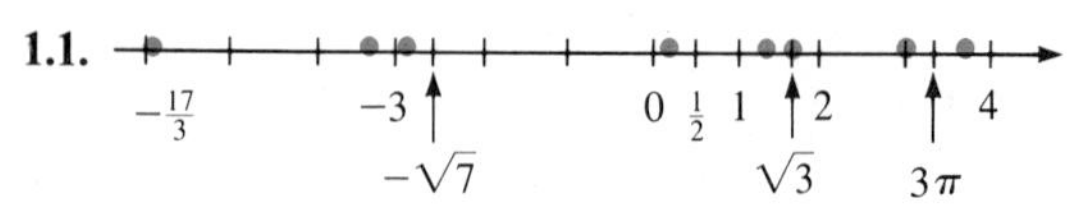

1.2. **(a)** commutative property **(b)** identity property
(c) associative property **(d)** identity property
(e) inverse property **(f)** distributive property
(g) property of 0 **(h)** property of 0

1.3. **(a)** $-1/6$ **(b)** $5/4$ **(c)** $6/5$

1.4. **(a)** true **(b)** false **(c)** false **(d)** true **(e)** false

1.5. $-15 < -10$; $15 > 10$; $0 < 0$; no

1.6. **(a)** All real numbers greater than -4 and less than or equal to 7.

(b) All real numbers greater than or equal to $-2/3$ and less than $-1/2$.

1.7. **(a)** $1/2$ **(b)** 7 **(c)** 1; yes; $|a-b| = |b-a|$ for all real numbers a and b.

1.8. **(a)** 0 **(b)** 5 **(c)** 4 **(d)** $5/6$. Yes, $d(a, b) = |b-a|$ and $d(-a, -b) = |-b-(-a)| = |a-b| = |b-a|$. Hence $d(a, b) = d(-a, -b)$.

1.9. **(a)**

(b)

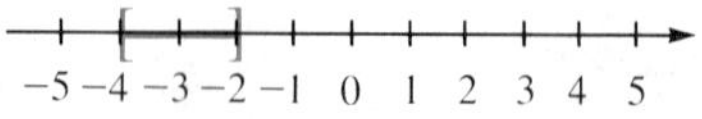

1.10. **(a)**

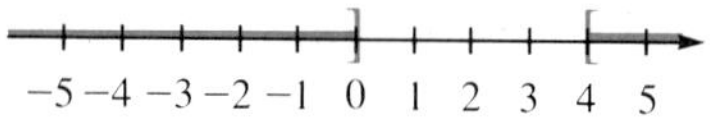

(b)

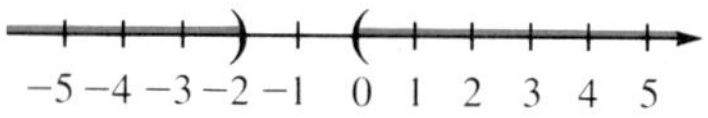

1.11.

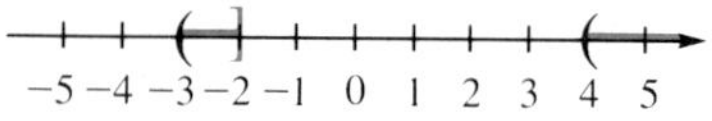

1.12. **(a)** -8 **(b)** $1/27$ **(c)** 1 **(d)** 16

1.13. **(a)** $25/9$ **(b)** $\frac{32}{9}x^3y^5$

1.14. **(a)** 2 **(b)** $1/2$ **(c)** -4 **(d)** $-1/3$

1.15. **(a)** $5\sqrt{2}$ **(b)** $2\sqrt[3]{2}$ **(c)** $\frac{2}{5}\sqrt[4]{375}$ **(d)** 0

1.16. $2(\sqrt{3}-\sqrt{2})$

1.17. **(a)** $\frac{3y^4}{|x|}$ **(b)** $10|x|^3|y|$ **(c)** $\frac{\sqrt{2z}}{4z}$ **(d)** $\frac{x\sqrt[3]{2x^2y^2}}{2y^2}$

1.18. **(a)** 32 **(b)** $1/32$ **(c)** $1/4$

1.19. **(a)** $8^{1/2}x^{3/2}y^2z^3$ **(b)** $\frac{25b^4}{16a^{9/2}}$

1.20. **(a)** $ab^2\sqrt[6]{ab^5}$ **(b)** $yz\sqrt[6]{xz^5}$

1.21. $4 \times 10^{1/2}$

1.22. **(a)** 2.025×10^{11} **(b)** 4.74552×10^{-13}
(c) 3.628411×10^{21} **(d)** 7.32×10^{13}
(e) 7.890481×10^{-22} **(f)** $1.303851604 \times 10^{8}$

1.23. **(a)** 3 **(b)** 2 **(c)** 0 **(d)** no degree

1.24. **(a)** $7x^2 - 4x + 5$
(b) $4x^2 + x - 2$

1.25. $3x^4 - 2x^3 + 6x^2 - 2x + 3$

1.26. $16y^4 - 4x^2$

1.27. **(a)** $x^2y^2 + x^{-2}y^{-2} + 2$
(b) $z + \frac{1}{z} - 2$

1.28. $5x^2y(xy^3 - 3)$

1.29. **(a)** $(2x - 1)(2x + 1)$
(b) $(3x - 4)(3x + 4)$

1.30. **(a)** $(x + 3)(x^2 - 3x + 9)$
(b) $(5z - 1)(25z^2 + 5z + 1)$

1.31. **(a)** $(x - 4)(x + 1)$ **(b)** $\frac{1}{4}(x + 2)(4x - 1)$
(c) $(x + 3a)(x - 2a)$

1.32. $(a - 1)(x - 2y)$

1.33. $(x^3 - 1)(2x - 1)$

1.34. $(x - y + 1)(x - y - 1)$

1.35. **(a)** $x - 3$ for $x \neq -2, 1$ **(b)** x for $x \neq 0$

1.36. $\dfrac{x + y}{x^2}$

1.37. $y(x - 1)$

1.38. $\dfrac{x^2 + 11x + 10}{3(x + 2)(x - 2)^2}$

1.39. $\dfrac{x - 1}{x}$

1.40. $\dfrac{\sqrt{x} + \sqrt{y}}{x - y}$

1.41. $\dfrac{2\sqrt{x}}{1 - x}$

1.42. **(a)** $12x^2 + 8z + 2$ **(b)** $6x^2 - 4x + 3$

1.43. **(a)** $r(s(x)) = x^2$; $s(r(x)) = x^2$
(b) $u(v(x)) = \sqrt{x + 1}$; $v(u(x)) = \sqrt{x} + 1$

1.44. $x = 1/3$; $-1/3$ is *not* a solution

1.45. **(a)** $x = -1, 3$ **(b)** $x = 2$

1.46. **(a)** $x^2 - 3x + 9/4 = (x - 3/2)^2$
(b) $x^2 + (1/3)x + 1/36 = (x + 1/6)^2$

1.47. **(a)** $x = -3, 4$ **(b)** $x = (-5 \pm \sqrt{13})/6$

1.48. **(a)** $x = 1/2, 1$ **(b)** $x = -3/2$

1.49. no real solutions

1.50. **(a)** -20; no real solutions
(b) 0; one solution
(c) -3; no real solutions
(d) 65; two distinct solutions

1.51. **(a)** $\pm\frac{\sqrt{6}}{2}$
(b) 1

1.52. $x = 6$

1.53. $x = -1, 2$

1.54. $x = -2, 12$

1.55. **(a)** $(-\infty, -9/4)$ **(b)** $(-\infty, -3]$

1.56. $[\frac{4}{3}, 2]$

1.57. $(-2, 3)$

1.58. $(-\infty, -4/3] \cup [2, \infty)$

1.59. $(-\infty, -2] \cup [3, \infty)$

1.60. $\{-2\}$

1.61. all real numbers

1.62. **(a)** $(1, 3)$ **(b)** $(1, \infty)$

1.63. $(-\infty, 1/4]$

Chapter Two

2.1. **(a)** not in any quadrant **(b)** not in any quadrant **(c)** not in any quadrant **(d)** IV **(e)** III **(f)** I

2.2. **(a)** $\sqrt{41}$ **(b)** $\sqrt{5}/2$ **(c)** $|b|$

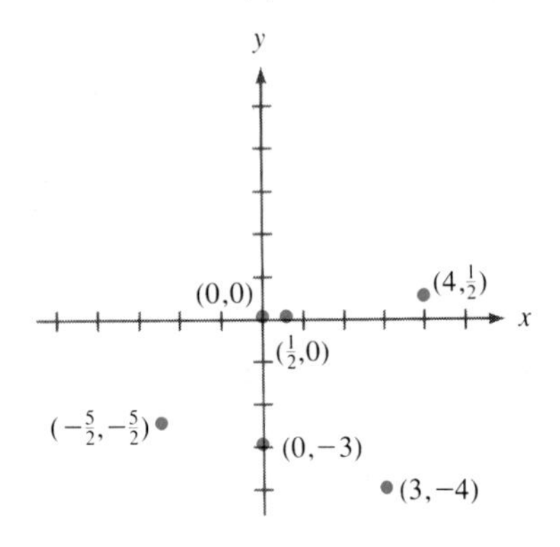

2.3. **(a)** Yes, since $[d(P, Q)]^2 + [d(P, R)]^2 = [d(Q, R)]^2$ **(b)** No

2.4. $(-3/4, 17/8)$

2.5.

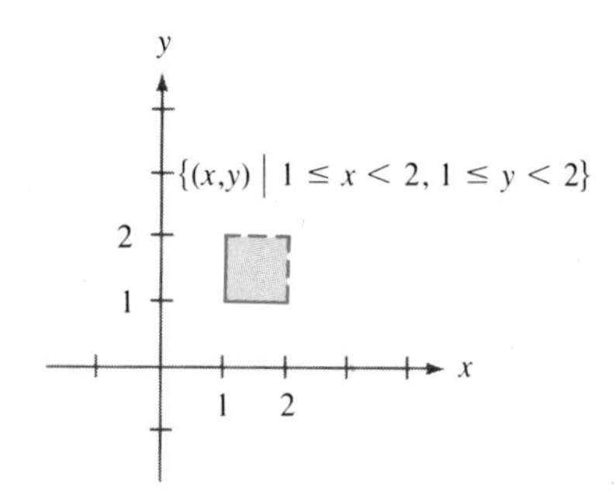

2.6. **(a)**

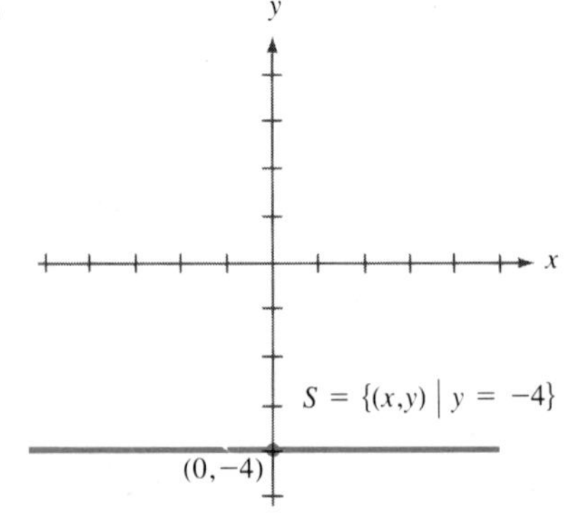

(b)

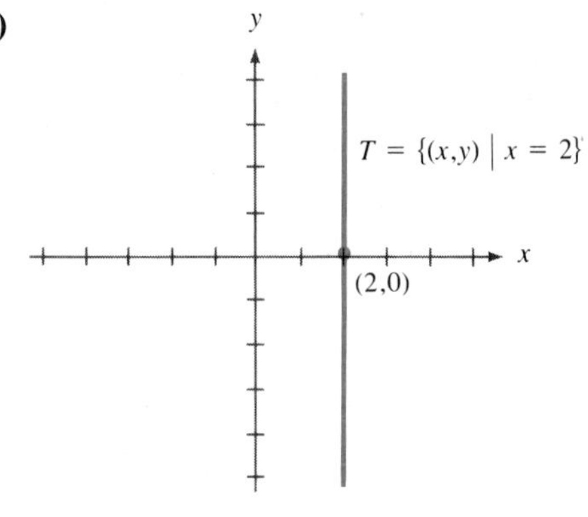

2.7.

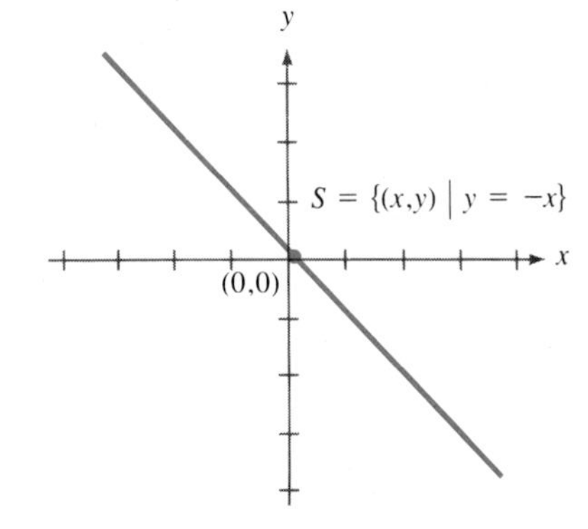

2.8. **(a)**

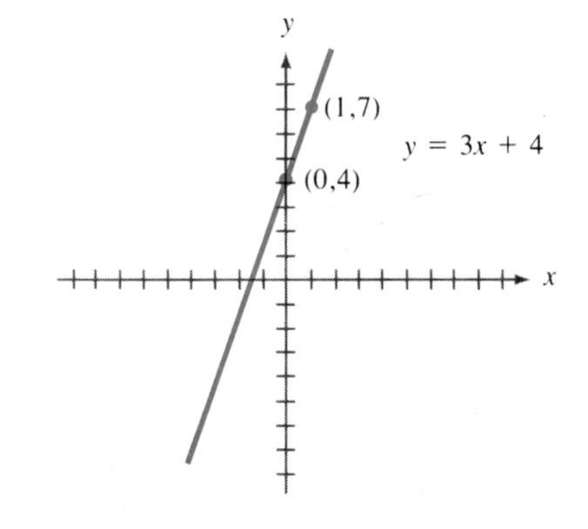

2.8. (b)

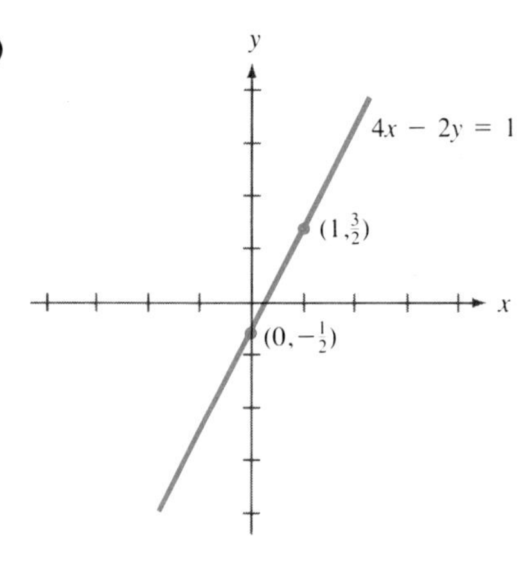

2.9. (a)

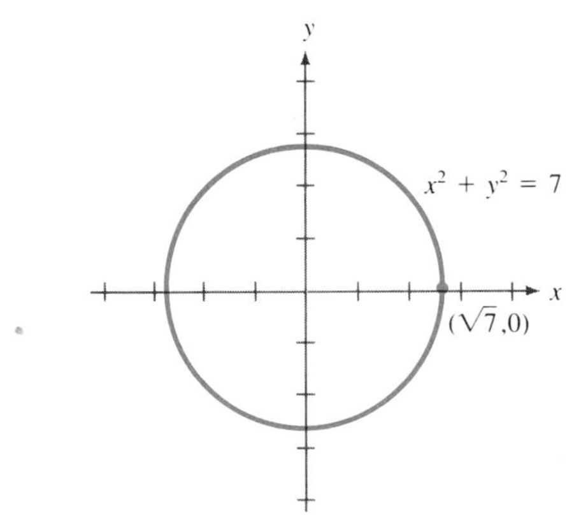

(b)

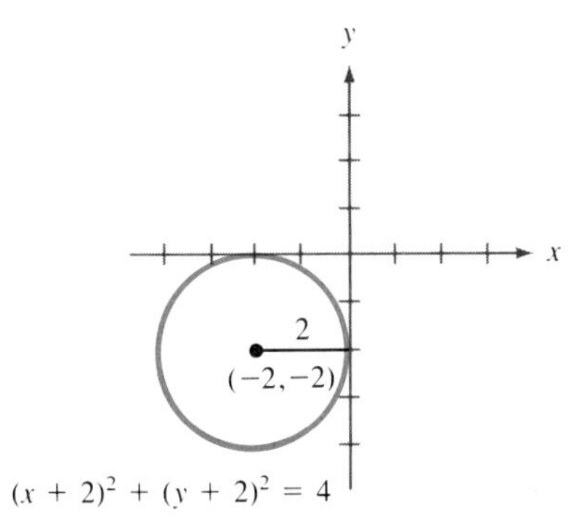

2.10.

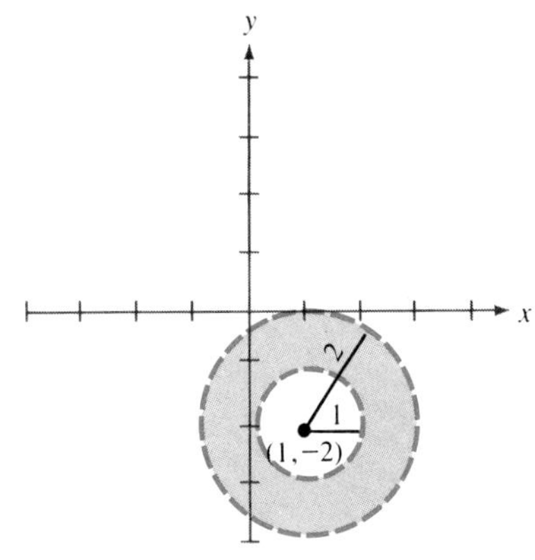

$S = \{(x,y) \mid 1 < (x-1)^2 + (y+2)^2 < 4\}$

2.11.

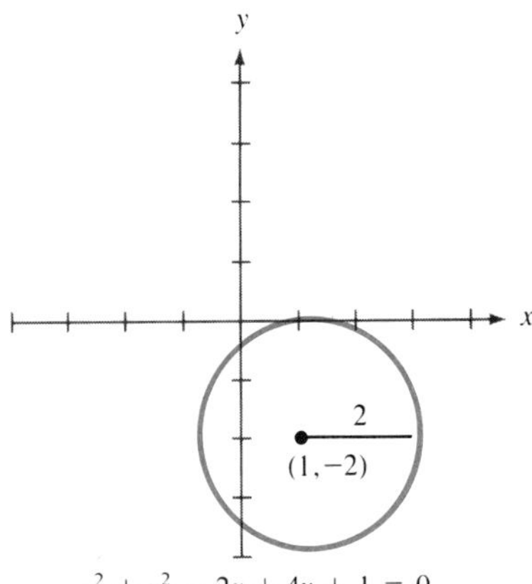

$x^2 + y^2 - 2x + 4y + 1 = 0$

2.12. **(a)** 5 **(b)** 5 **(c)** 1

2.13. **(a)** 2 **(b)** 1/2 **(c)** $\sqrt{2/3}$

2.14. **(a)** 1 **(b)** -2 **(c)** $-4/7$
(d) -1; $1/(1-1) = 1/0$ is not defined

2.15. **(a)** 1/4 **(b)** 0 **(c)** -5 **(d)** 125 **(e)** 4

2.16. $A(x) = 5x$; domain is all positive real numbers

2.17. **(a)** all real numbers
(b) all real numbers except 0
(c) $\{x \mid x \leq -1\} \cup \{x \mid x \geq 1\}$
(d) $\{x \mid -1 < x < 1\}$

2.18. **(a)** $f(2x) = 8x^3$; $2f(x) = 2x^3$; they are not the same
(b) $f(-x) = -x^3$ and $f(x) = x^3$. Since $-x^3 \neq x^3$ for $x \neq 0$, these are not the same.
(c) $3hx^2 + 3h^2x + h^3$

2.19. **(a)** $4x + 2h$ **(b)** $3x^2 + 3hx + h^2$

2.20. $\dfrac{-(2x + h)}{x^2(x + h)^2}$

2.21.

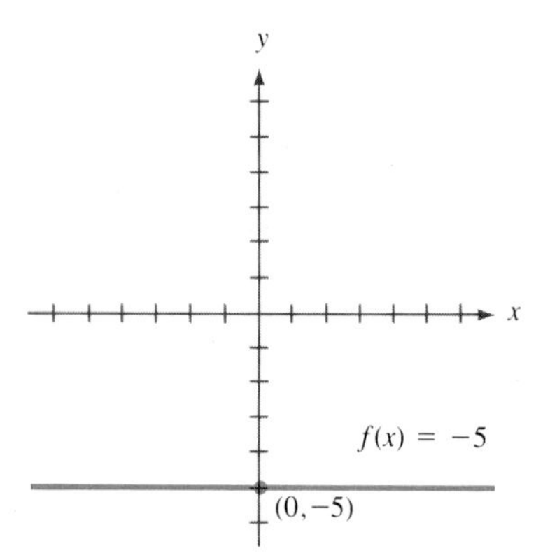

2.22.

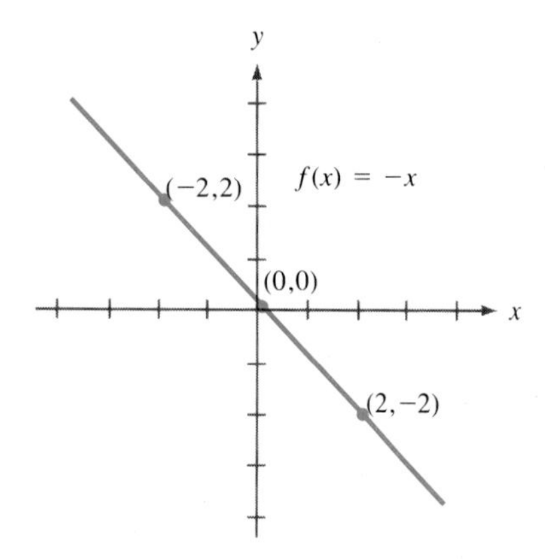

2.23. **(a)**

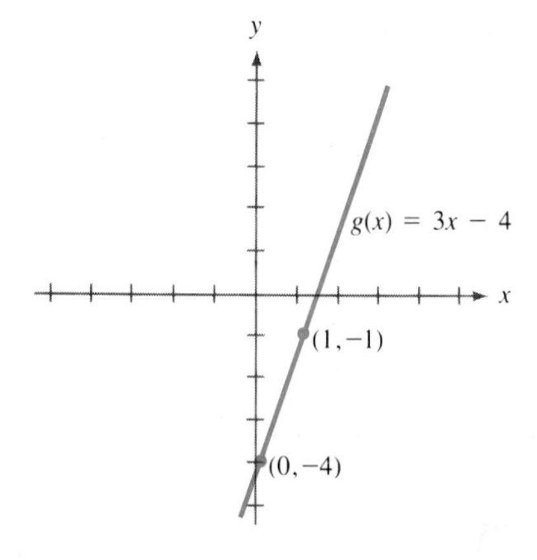

(b)

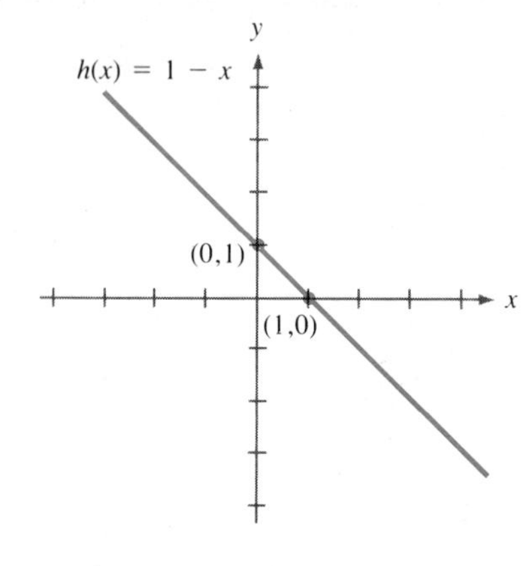

2.24.

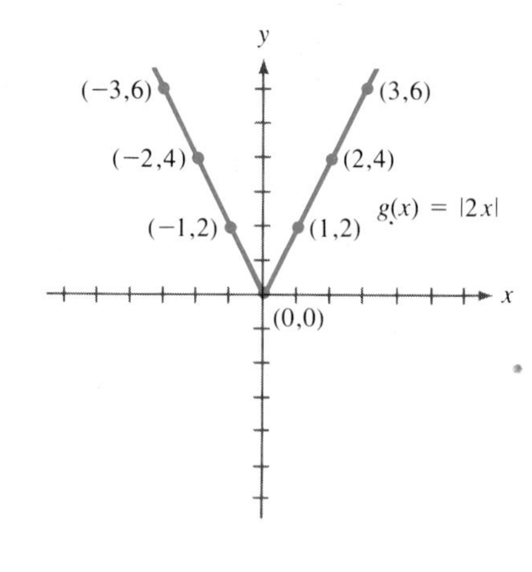

2.25.

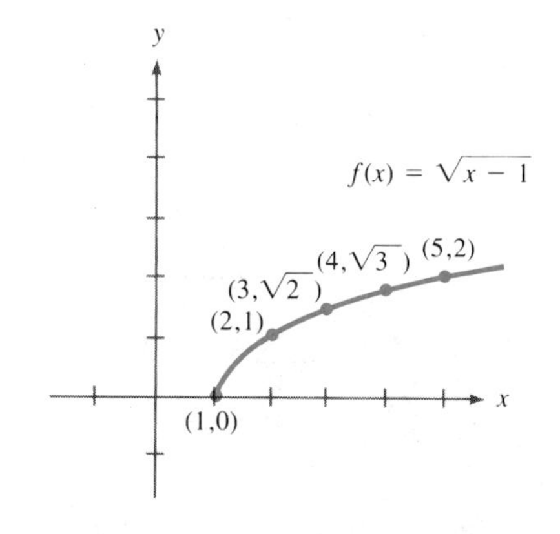

2.26.

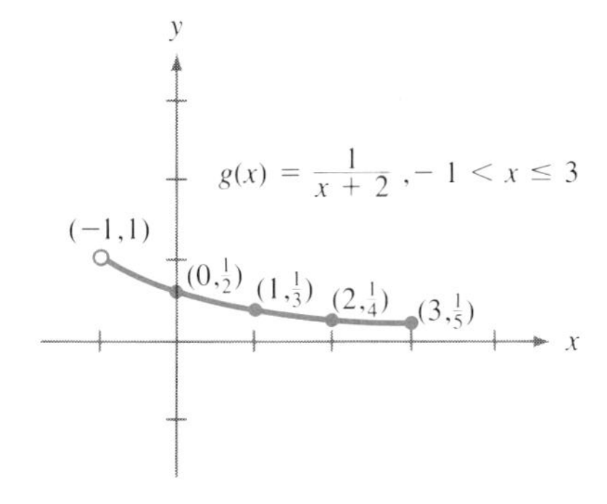

2.27. Domain is all real numbers except -1.

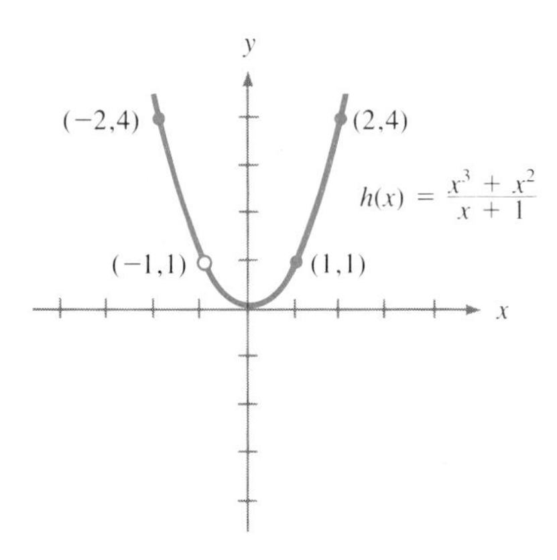

2.28.

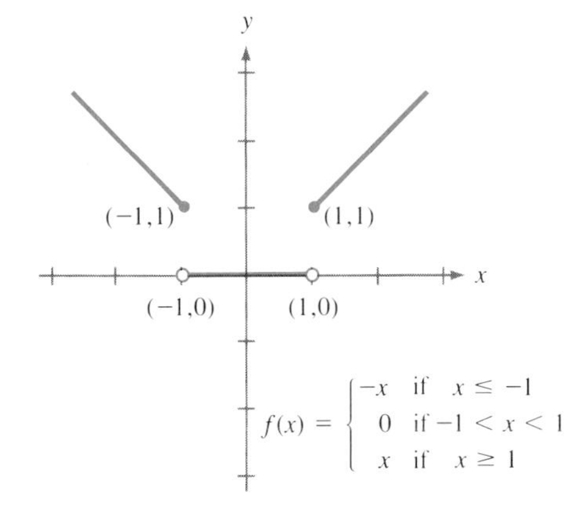

2.29. **(a)** $-13/3, -1, 1, 13/3$

(b) $(-\infty, -13/3] \cup [-1, 1] \cup [13/3, \infty)$

(c) $-4, -2, 2, 4$

(d) $(-4, -2) \cup (2, 4)$

2.30. $g(-x) = (-x)^4 + 1 = x^4 + 1 = g(x)$

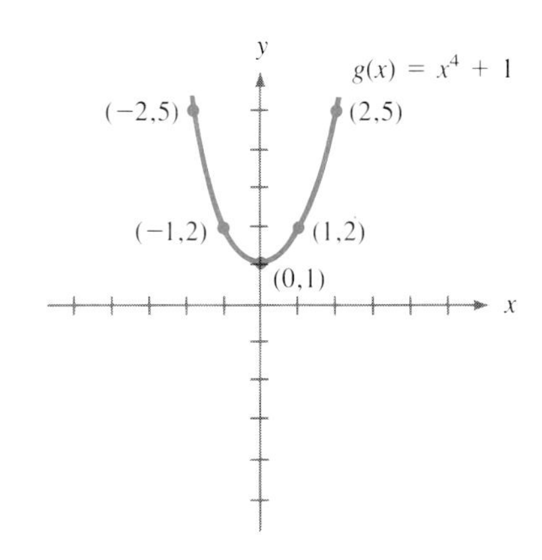

2.31. $g(-x) = -(-x)^3 = -(-x^3) = x^3 = -g(x)$

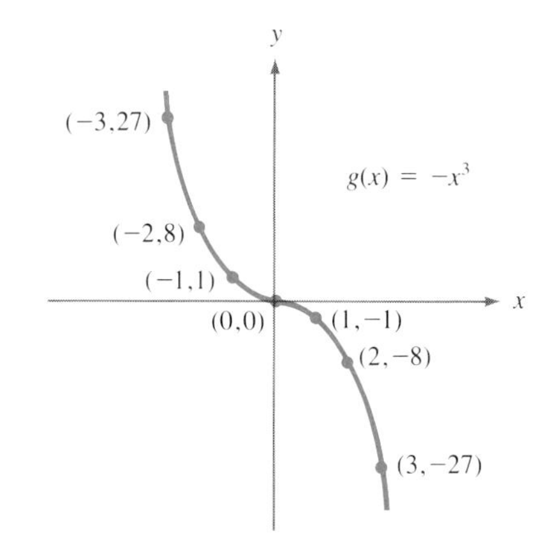

2.32. **(a)** odd **(b)** neither even nor odd **(c)** even **(d)** even

2.33. **(a)**

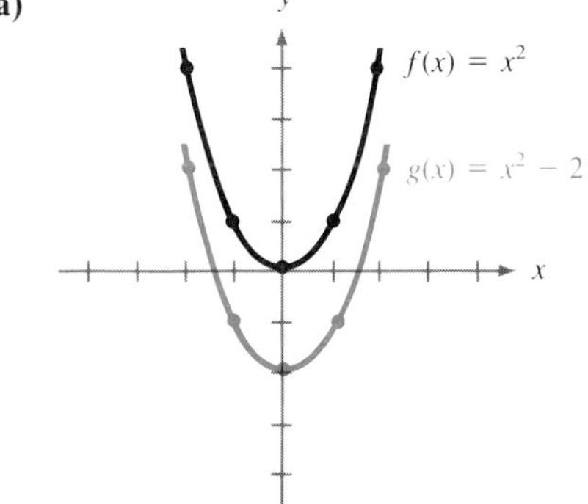

2.33. (b)

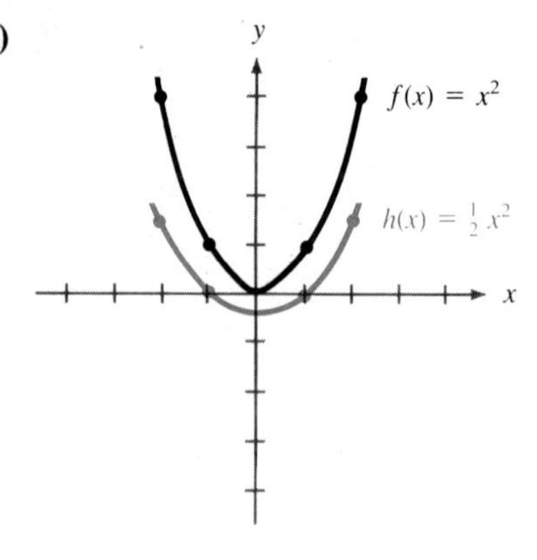

(c)

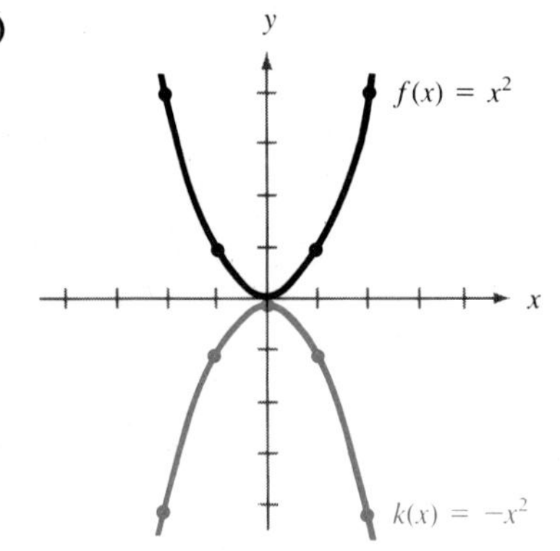

2.34.

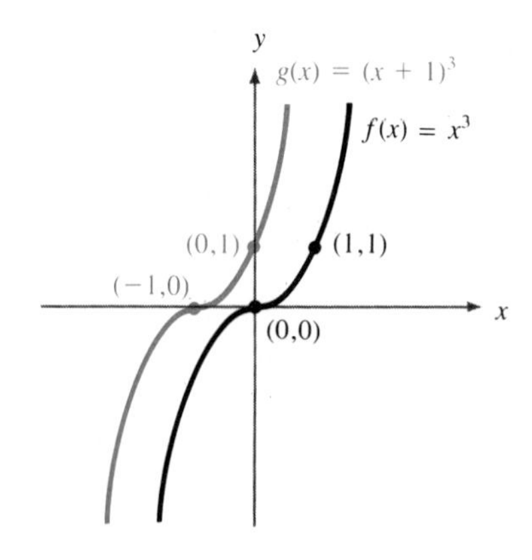

2.35.

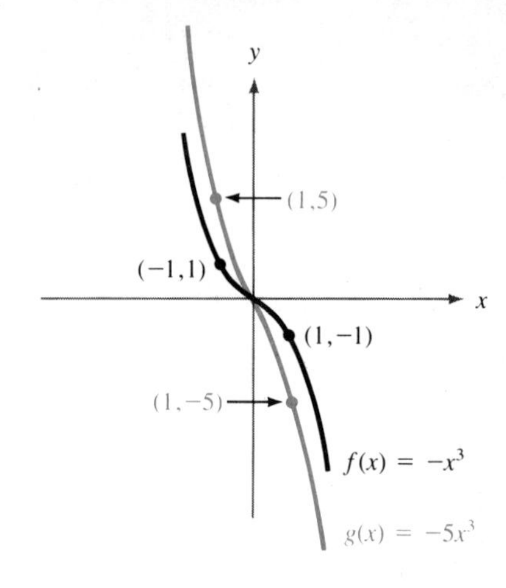

2.36.

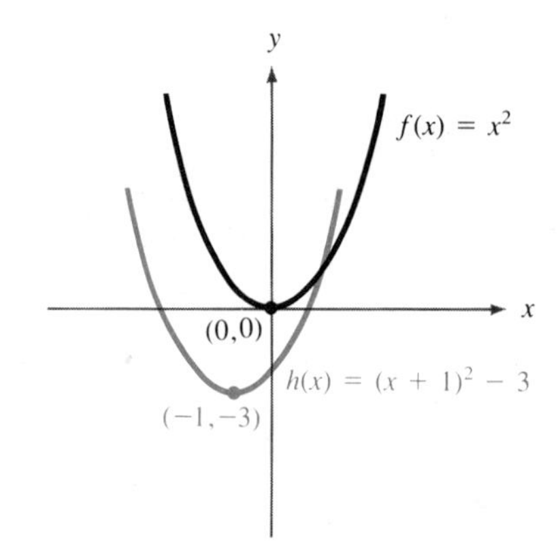

2.37. increasing on $[1, \infty)$; decreasing on $(-\infty, 1]$

2.38. decreasing on $(-\infty, -2]$ and $[0, 2]$; increasing on $[2, 5]$; constant on $[-2, 0]$ and $[5, \infty)$

2.39. **(a)** decreasing **(b)** increasing **(c)** neither

2.40. If $f(a) = f(b)$ then $4a = 4b$ and so $a = b$. Hence f is one-to-one. $g(1) = g(-1)$ but $1 \neq -1$ and so g is not one-to-one.

2.41. f is not one-to-one since $f(-1) = f(1)$ but $1 \neq -1$. g is one-to-one since if $g(a) = g(b)$ then $1/a^3 = 1/b^3$ and so $a^3 = b^3$ and so $a = b$.

2.42. **(a)** $A \cap B = \{1, 5, -3\}$

(b) $A \cap B = \{x \mid -2 < x \leq 1\}$

2.43. **(a)** $(f+g)(x) = 2x^2 - 2$; domain is $\mathbf{R}$

(b) $(f-g)(x) = 2x^2 - 2x$; domain is $\mathbf{R}$

(c) $(fg)(x) = 2x^3 - 3x^2 + 1$; domain is $\mathbf{R}$

(d) $(f/g)(x) = 2x + 1$; domain is $\{x \mid x \neq 1\}$

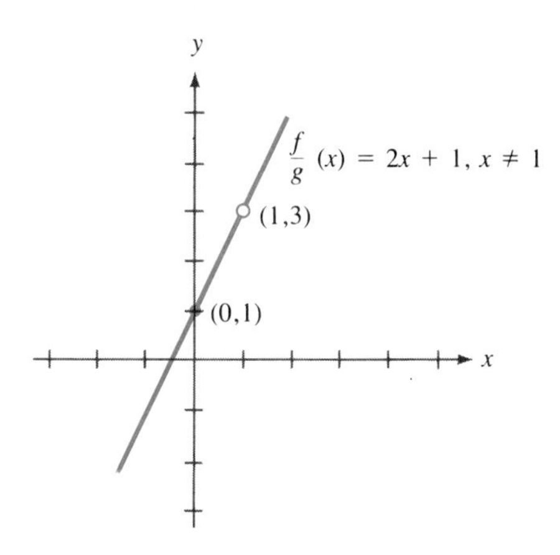

2.44. $(f/g)(x) = x$; domain is $\{x \mid x \neq 0\}$

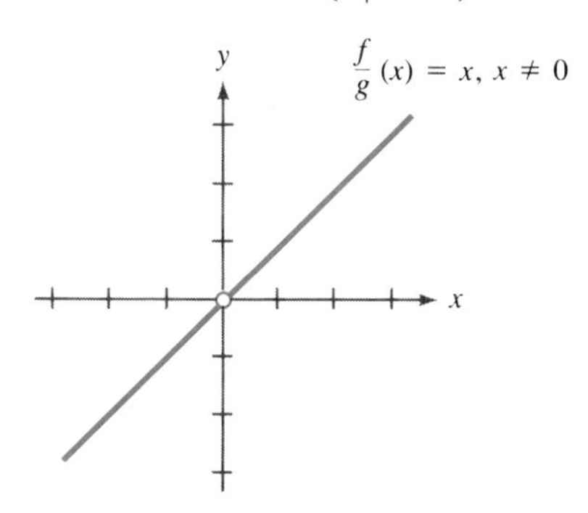

2.45. **(a)** 21 **(b)** 1/21 **(c)** 1

2.46. **(a)** $(f \circ g)(x) = 2x^2 - 11x + 13$

(b) $(g \circ f)(x) = -2x^2 - 3x + 3$

(c) $(g \circ g)(x) = x$

2.47. $(f \circ g)(x) = \dfrac{1}{\sqrt{x} - 1}$; $\{x \mid x \geq 0 \text{ and } x \neq 1\}$

2.48. **(a)** $((f-g) \circ h)(x) = 1 + x - 3/x$

(b) $(g \circ f \circ h)(x) = 7 + 3x$

2.49. **(a)** $g(x) = \sqrt[3]{x}$, $f(x) = 1 - 3x$

(b) $g(x) = x^{-4}$, $f(x) = x^2 - 3x + 1$

(c) $g(x) = \dfrac{\sqrt[4]{x} - 5}{\sqrt{x} - 9}$, $f(x) = x - 2$

2.50. $(f \circ g)(x) = f((1/3)(4-x)) = 4 - 3((1/3)(4-x)) = 4 - (4-x) = x$; $(g \circ f)(x) = g(4-3x) = (1/3)(4-(4-3x)) = (1/3)(3x) = x$. Hence $g = f^{-1}$.

2.51. $(f \circ f)(x) = f\left(\dfrac{x}{x-1}\right)$

$$= \frac{\dfrac{x}{x-1}}{\dfrac{x}{x-1} - 1} = \frac{\dfrac{x}{x-1}}{\dfrac{1}{x-1}} = x$$

Hence $f = f^{-1}$.

2.52. **(a)** $f^{-1}(x) = (x+1)/10$

(b) $f^{-1}(x) = \sqrt[5]{x+1}$

(c) $f^{-1}(x) = \dfrac{x}{x-2}$

2.53. If $f(a) = f(b)$ then $2a^2 - 5 = 2b^2 - 5$ and so $a^2 = b^2$. But since both a and b must be nonnegative, we must have $a = b$. Hence f is one-to-one and so it has an inverse. Using the method of this section, we get $f^{-1}(x) = \sqrt{\dfrac{x+5}{2}}$

2.54. **(a)** If $f(a) = f(b)$ then $2a - 5 = 2b - 5$ and so $a = b$. Hence f is one-to-one and so it has an inverse.

(b) $f^{-1}(x) = \dfrac{x-5}{2}$

(c)

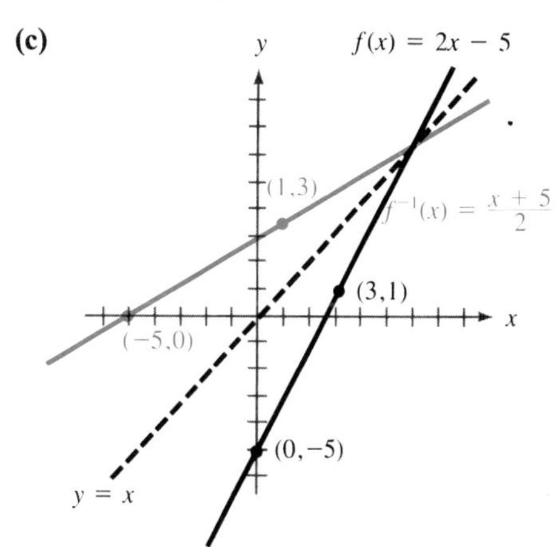

Chapter Three

3.1. **(a)** 7/3 **(b)** 2 **(c)** 0

3.2. **(a)**

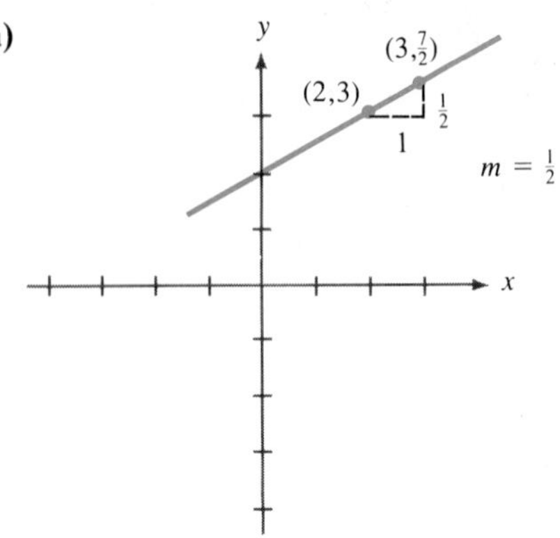

(b)

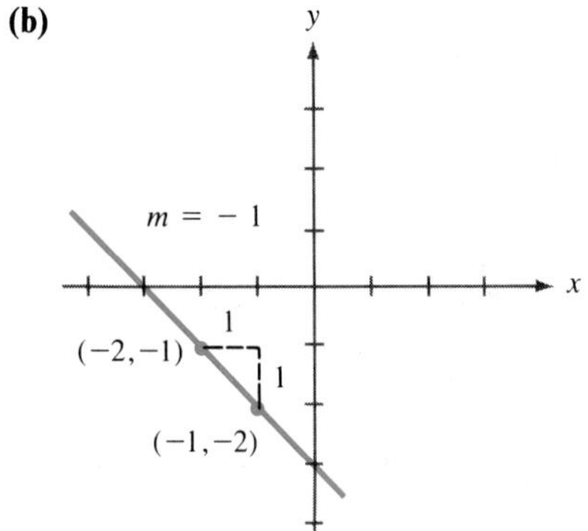

3.3. **(a)** Each line has slope 2. Hence, they are parallel.

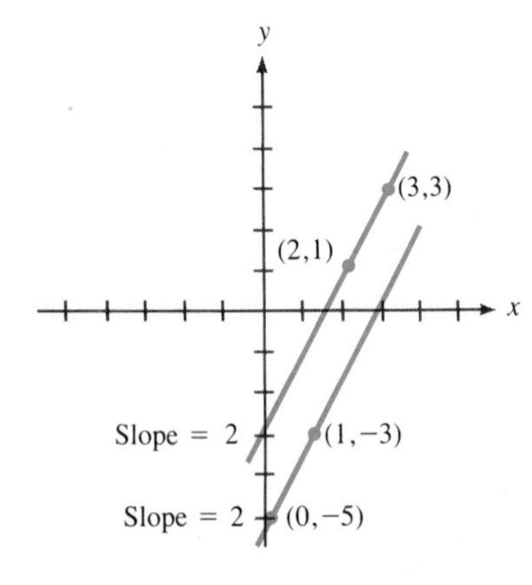

Lines are parallel.

(b) the line containing (5, 2) and (−2, 3) has slope −1/7, and the line containing (−2, 2) and (0, 16) has slope 7. Hence, the lines are perpendicular.

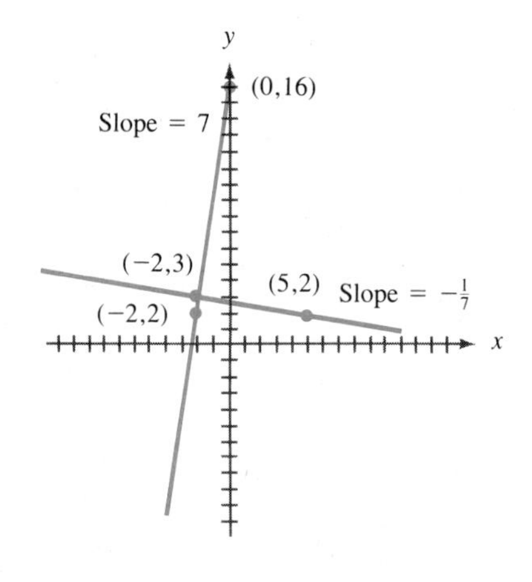

Lines are perpendicular.

(c) The line containing (0, 0) and (1, −3) has slope −3, and the line containing (4, 1/2) and (1/2, 4) has slope −1. Hence, these lines are neither parallel nor perpendicular.

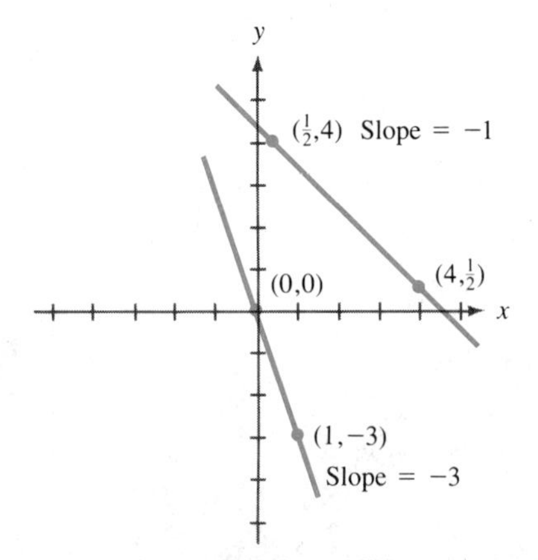

Lines are neither parallel nor perpendicular.

3.4.

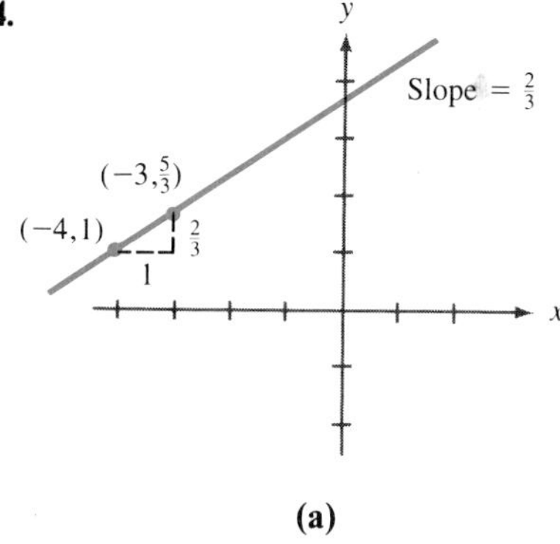

(a)

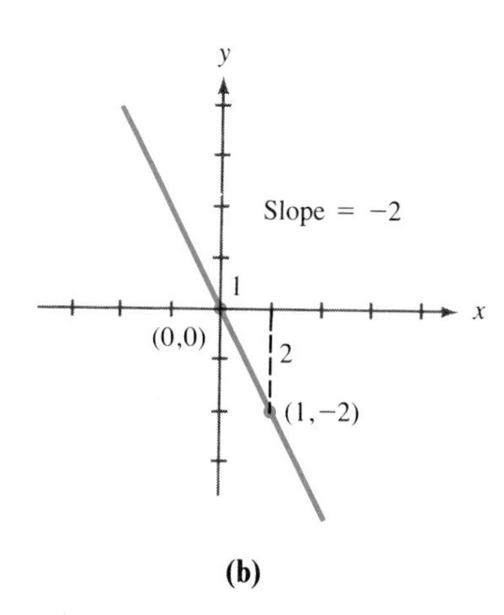

(b)

3.5. $y = -2x + 5$; x-intercept $= 5/2$

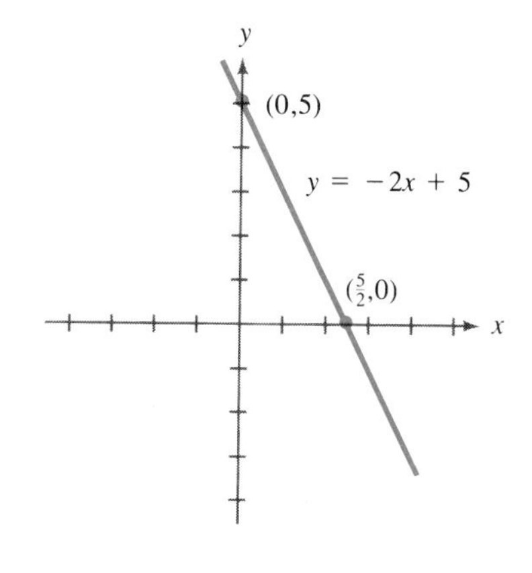

3.6. $y - 1/2 = -2(x + 2)$; y-intercept $= -7/2$

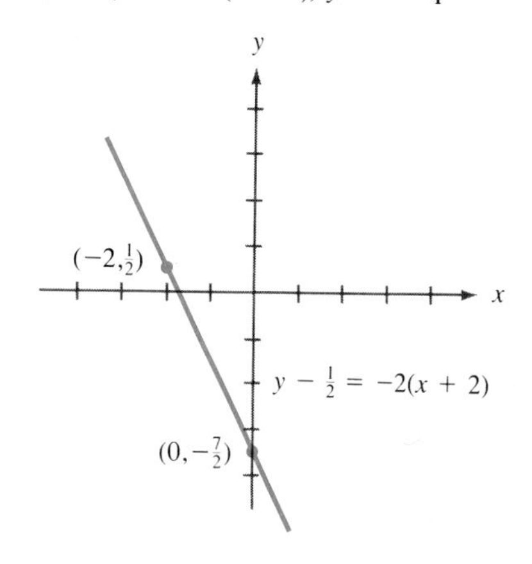

3.7. $y = (3/2)x + 11/4$

3.8. Point-slope form: $y - 3 = x + 1$
Slope-intercept form: $y = x + 4$
Slope = 1, y-intercept = 4

3.9. This equation is equivalent to the equation $y = -(3/2)x + 1/2$ which is the equation of a line with slope $-3/2$ and y-intercept $1/2$.

3.10. **(a)** $y = (x - 3)^2 - 7$

(b) $y = 3(x + 1/3)^2 - 1/3$

3.11. The graph can have at most two x-intercepts since finding these intercepts amounts to solving a quadratic equation, which can have at most two solutions.

3.12. y-intercept $= -4$; x-intercepts $= (1 \pm \sqrt{17})/2$

3.13.

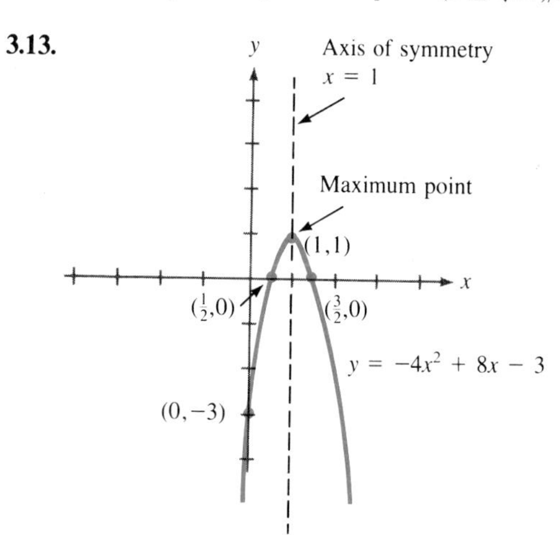

3.14.

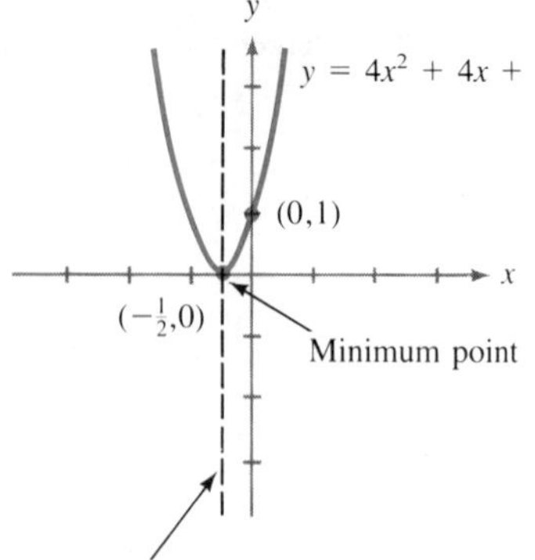

3.15.

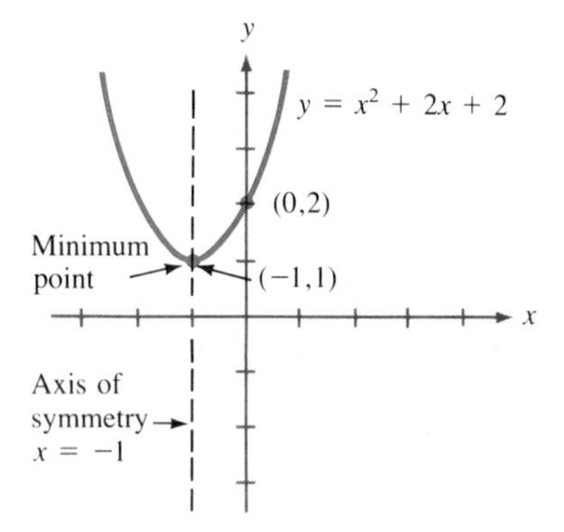

3.16. The closed interval $[-4, 1]$

3.17. $(-\infty, -2) \cup (3, \infty)$

3.18. A square with sides of length 500 feet

3.19. \$4.58; Maximum revenue is approximately \$252.08.

3.20.

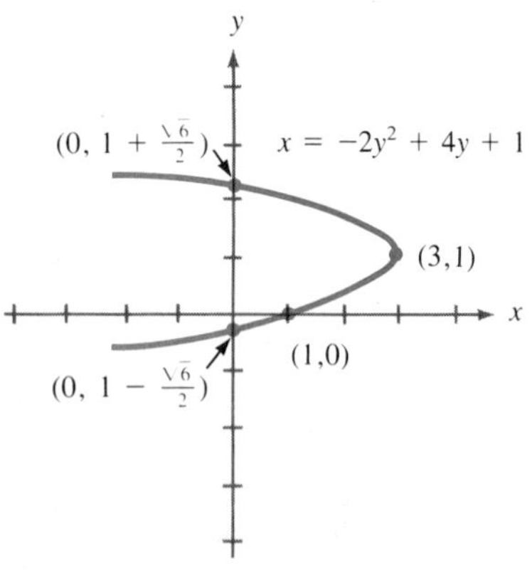

3.21. $x^2 + \dfrac{(y-2)^2}{4} = 1$

3.22.

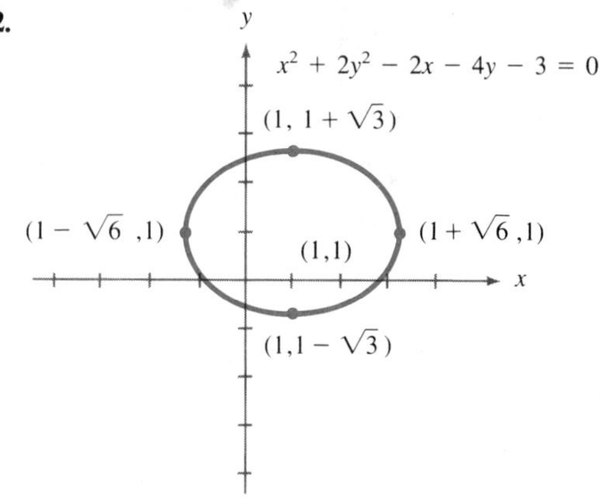

Note: Standard form is $\dfrac{(x-1)^2}{6} + \dfrac{(y-1)^2}{3} = 1$

$\sqrt{6} \approx 2.4$, $\sqrt{3} \approx 1.7$.

3.23.

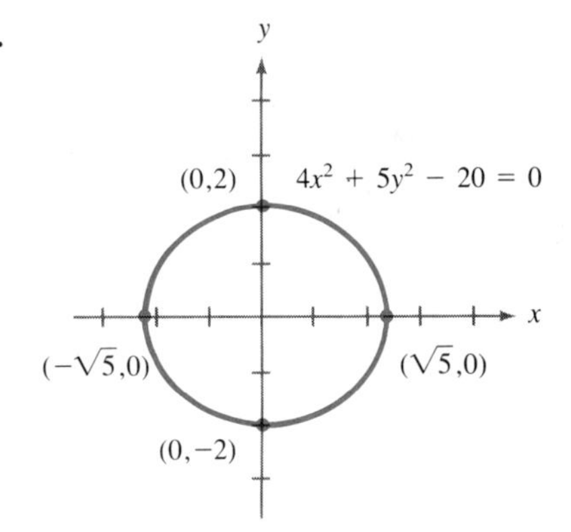

3.24.

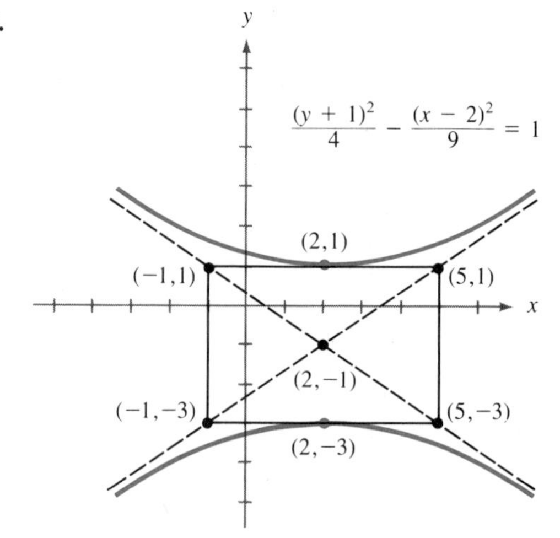

3.25.

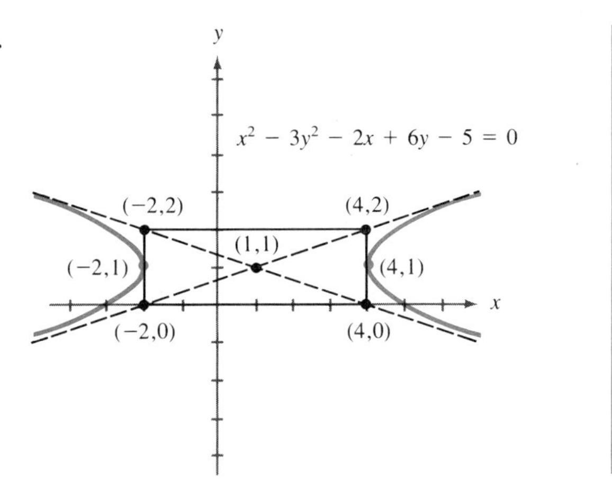

Chapter Four

4.1. $3x^3 - 2x^2 + x - 1 = (3x^2 + 4x + 9)(x - 2) + 17$

4.2. $2x^5 - 3x^2 + x + 1 =$
$(x^3 - (1/2)x - 3/2)(2x^2 + 1) + (3/2)x + 5/2$

4.3.

2	3	−2	1	−1
		6	8	18
	3	4	9	17

4.4.

−3	4	0	0	−2	1	−1
		−12	36	−108	330	−993
	4	−12	36	−110	331	−994

4.5. $p(-3) = -122$

4.6. $-1, \ -2 \pm \sqrt{6}$

4.7. $1, \ -4, \ 3, \ -2$

4.8. **(a)** 2 **(b)** 4

4.9. **(a)** 5 **(b)** no rational solutions

4.10. $5/2$

4.11. 1 (simple root), $-1/3$ (double root)

4.12. $-1/3$

4.13. 0.1875

4.14.

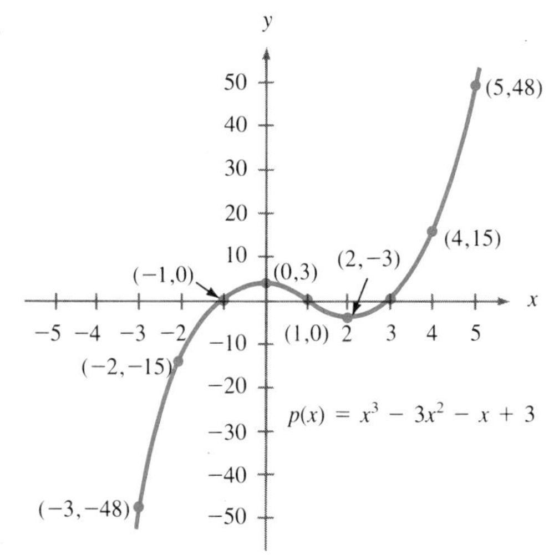

4.15.

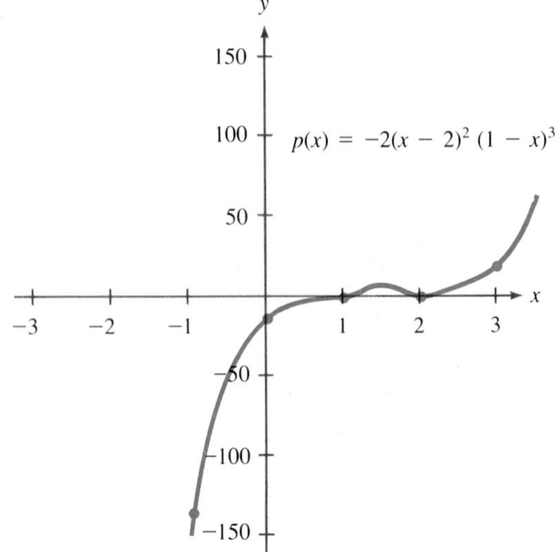

4.16.

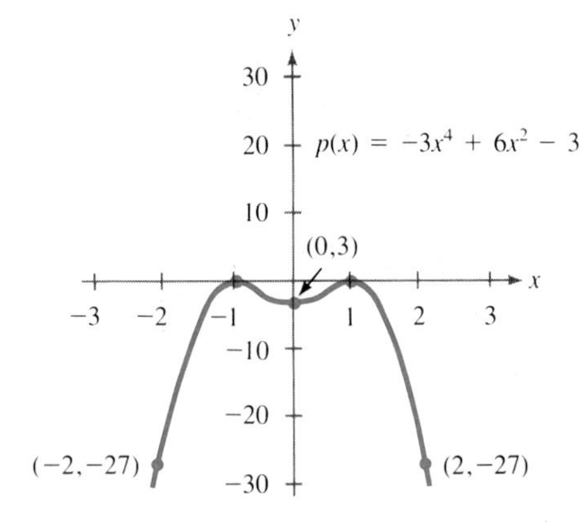

4.17.

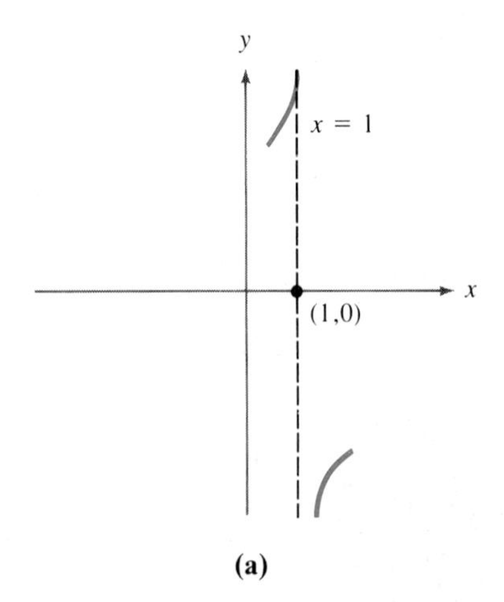

(a)

x = 2

(2,0)

(b)

4.18.

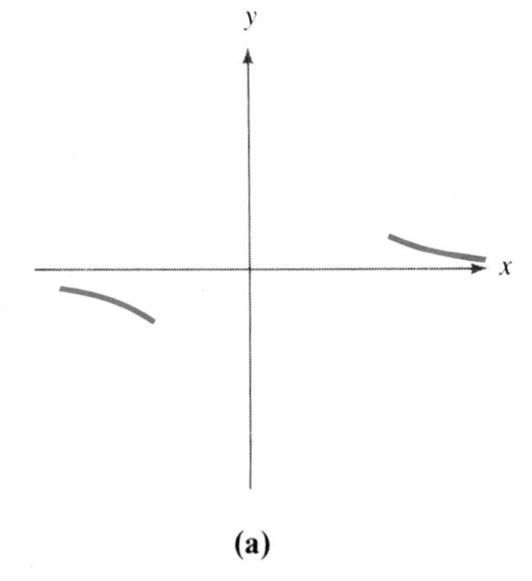

(a)

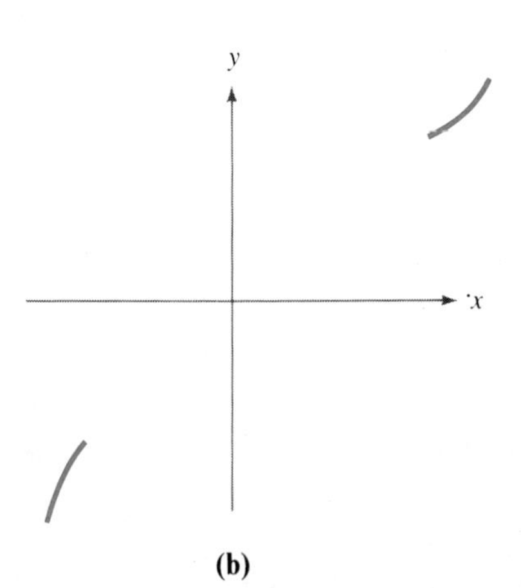

(b)

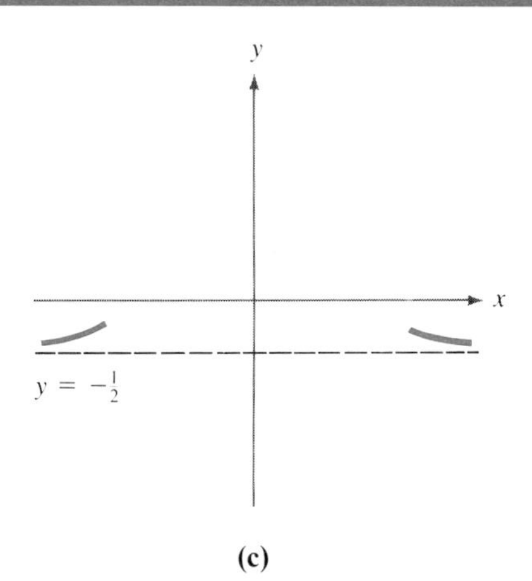

(c)

4.19.

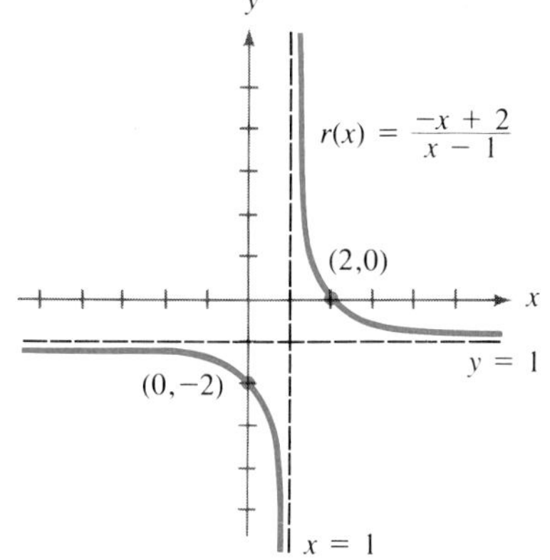

4.20.

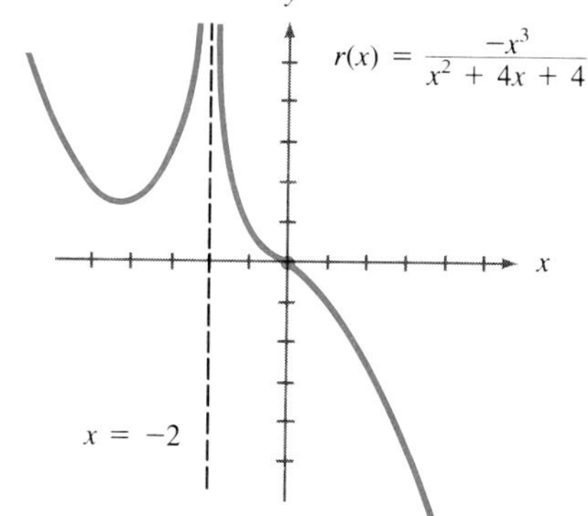

4.21.

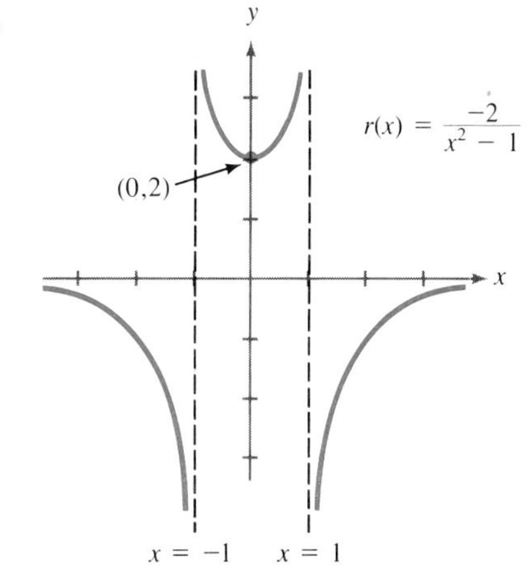

4.22.

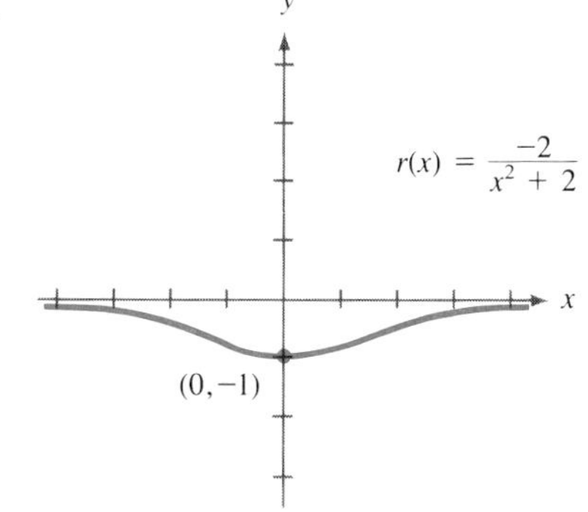

4.23.

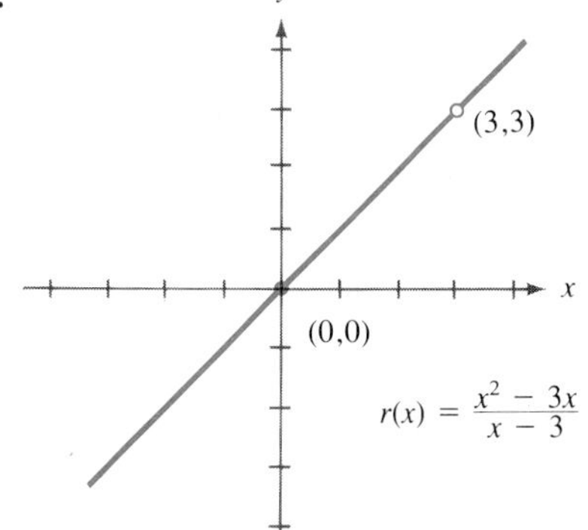

4.24. **(b)** $\dfrac{3}{x^2 - 9} = \dfrac{1}{2(x-3)} - \dfrac{1}{2(x+3)}$

4.25. $\dfrac{-11x^2 + 15x - 6}{x(x-2)(3x-1)} = \dfrac{-3}{x} - \dfrac{2}{x-2} + \dfrac{4}{3x-1}$

4.26. $\dfrac{x^2 - 9x + 3}{x^2(2x-1)} = \dfrac{-5}{2x-1} + \dfrac{3}{x} - \dfrac{3}{x^2}$

4.27. $\dfrac{1}{(x+1)(x^2+1)} = \dfrac{1}{2(x+1)} + \dfrac{-x+1}{2x^2+2}$

Chapter Five

5.1. **(a)** 8.824977827 **(b)** 1.632526919
(c) 5.330288285 **(d)** 0.45985863

5.2.

(a)

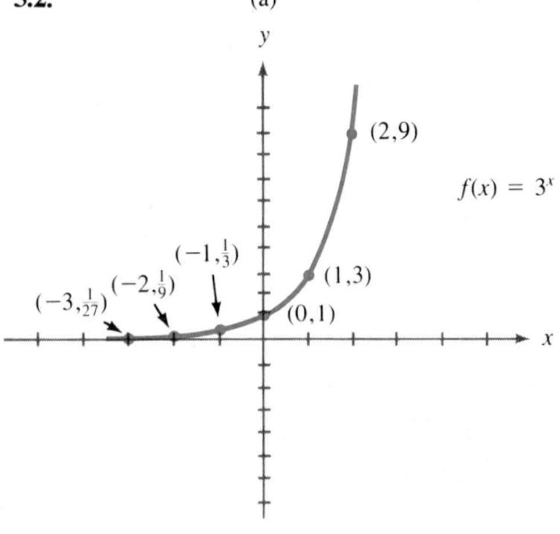

(b)

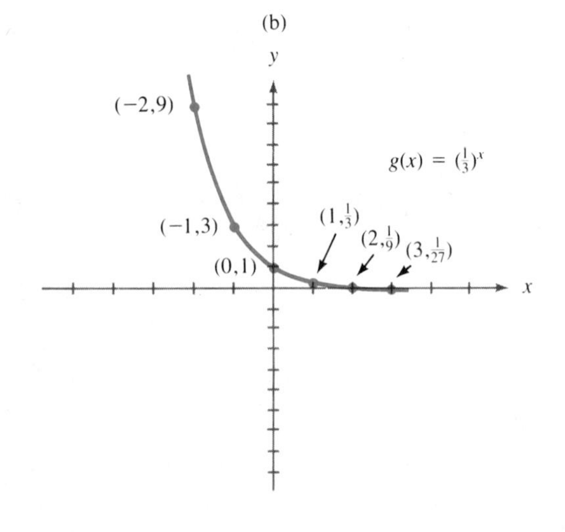

5.3.

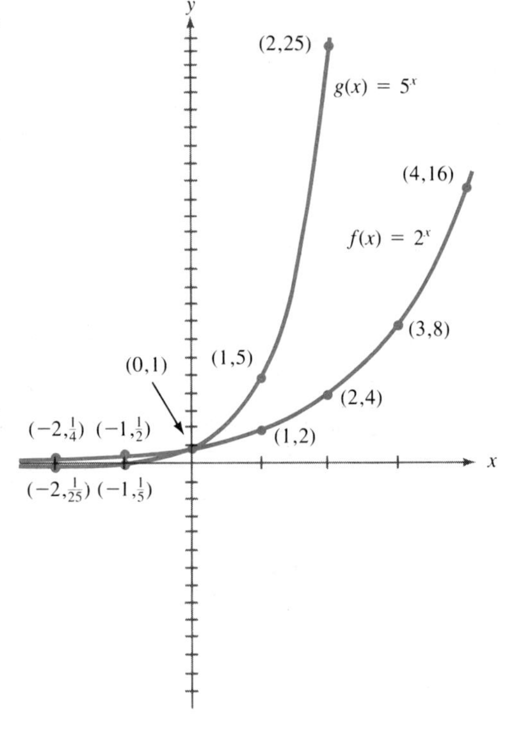

5.4.

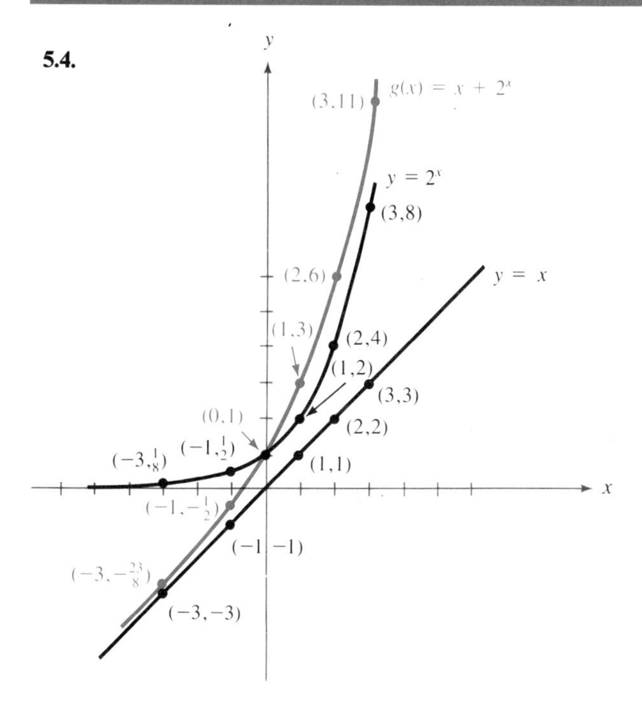

5.5.

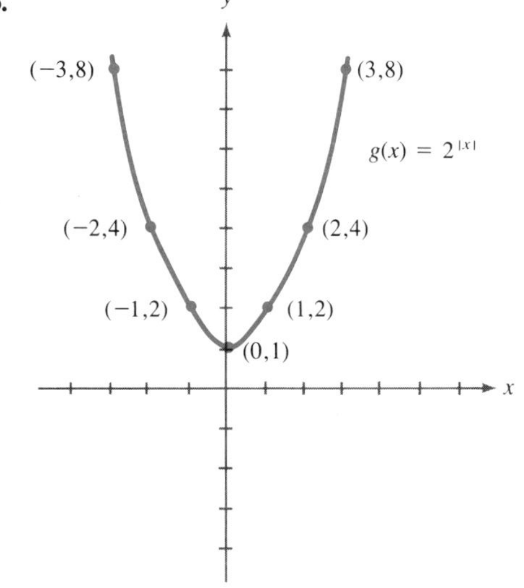

5.6. **(a)** $x = 4$ **(b)** $x = 1$ **(c)** $x = -2$

5.7. **(a)** 2.718281828 **(b)** 7.389056099
(c) 1.648721271 **(d)** 4.113250379
(e) $e^{\pi} > \pi^{e}$

5.8.

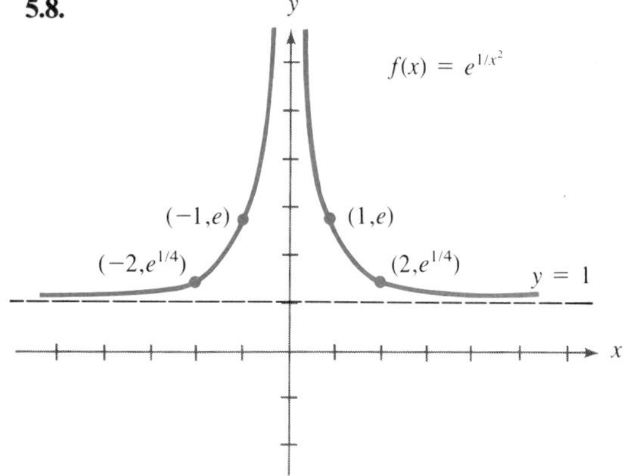

5.9. **(a)** 4 **(b)** −3 **(c)** 0 **(d)** 4 **(e)** 3/4

5.10. **(a)** 1.0000 **(b)** 1.4771 **(c)** 0.3920
(d) 0.2219 **(e)** 0.6990 **(f)** 7.1565

5.11. −1.28

5.12. $\log \dfrac{8x^3}{3y^{4/3}}$

5.13.

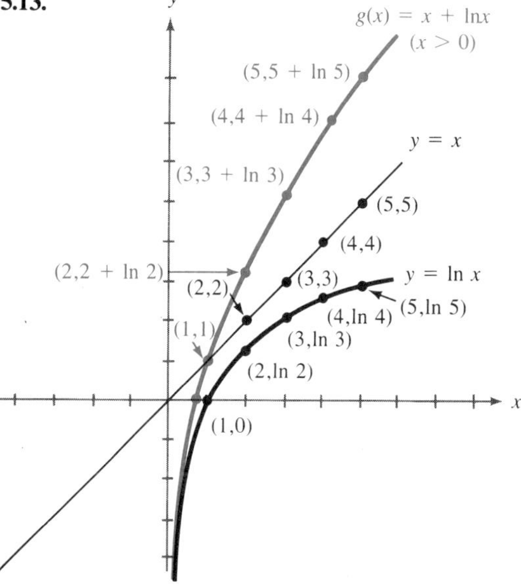

5.14. −2.089267338

5.15. $x = 0.282883312$

5.16. $x = -1.709511291$

5.17. $x = -4$

5.18. The amplitude of the moderately destructive earthquake is $10^4 = 10{,}000$ times greater than the barely perceptible earthquake.

5.19. $10(4 + \log 2) \approx 43$ dbs.

5.20. 10^{10} times greater

5.21. **(a)** \$561.80 **(b)** \$709.26 **(c)** \$595.51

5.22. \$5867.55

5.23. **(a)** 755.87 **(b)** \$763.61 **(c)** \$766.42

5.24. Approximately 7.018 years. This is approximately 0.087 years (or 32 days) longer than at continuous compounding of interest.

5.25. 25 days

5.26. $P(t) = 1000 \cdot 4^{t/100}$; the year 2110

5.27. 7001.52 years

5.28. 1054.85 years

Chapter Six

6.1. **(a)** $-\pi/3$ **(b)** $3\pi/2$ **(c)** $-2\pi/9$

6.2. **(a)** $60°$ **(b)** $2160°$ **(c)** $(360/\pi)°$

6.3. $50\pi/3$ inches (≈ 52.36 inches)

6.4. ≈ 2.29 rotations

6.5. $\approx 11{,}344.64$ yards

6.6. (a)

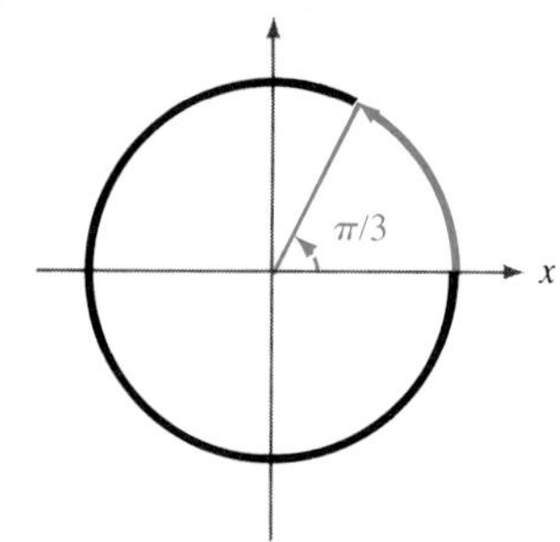

(b)

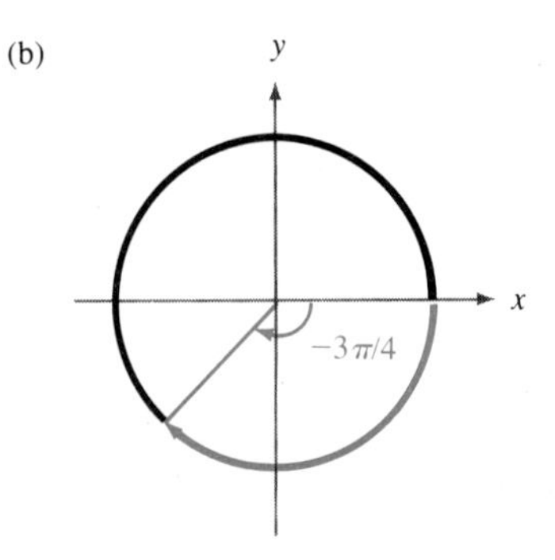

(c)

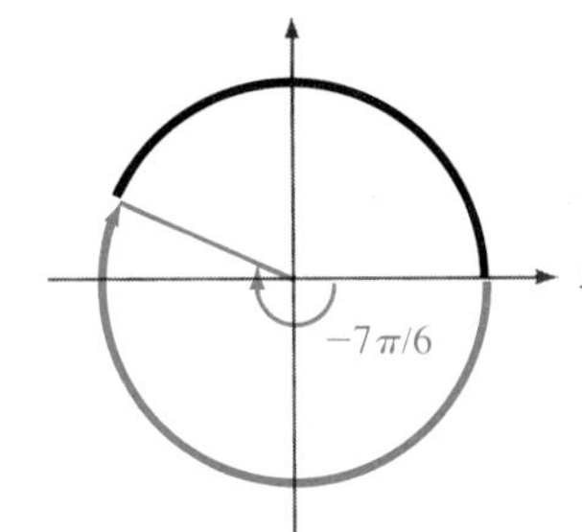

(d)

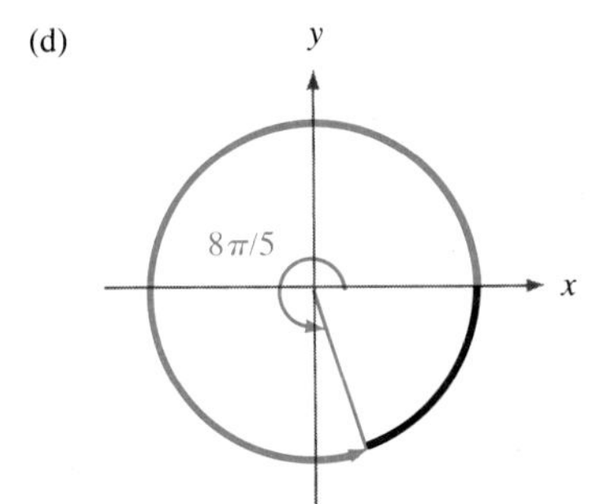

6.7. $P(\pi/3) = (1/2, \sqrt{3}/2)$

6.8. See Figure 6.20

6.9. See Figure 6.20

6.10. **(a)** $(0, 1)$ **(b)** $(-\sqrt{3}/2, 1/2)$ **(c)** $(-1, 0)$

6.11. **(a)** $\sin 2\pi/3 > 0$, $\cos 2\pi/3 < 0$
(b) $\sin 7\pi/4 < 0$, $\cos 7\pi/4 > 0$
(c) $\sin \pi = 0$, $\cos \pi < 0$

6.12. **(a)** $\pi/6$ **(b)** $\pi/6$ **(c)** $\pi/5$ **(d)** $\pi/3$ **(e)** 0

6.13. **(a)** $-\sqrt{2}/2$ **(b)** $-\sqrt{3}/2$ **(c)** $-\sqrt{3}/2$ **(d)** $-\sqrt{2}/2$

6.14. **(a)** 0.587785252 **(b)** 0.540302305 **(c)** -0.97492712
(d) 0.083531936 **(e)** 0.809016994

6.15. **(a)** $-\sqrt{3}/3$ **(b)** $-\sqrt{2}$ **(c)** 1 **(d)** $-\sqrt{3}/3$

6.16. **(a)** 0.726542528 **(b)** -4.584143857
(c) -1.187182563 **(d)** 0.267949192

6.17. **(a)** 30 feet **(b)** 42.5 feet **(c)** 55 feet **(d)** ≈ 10.85 feet

6.18. **(a)** -1 **(b)** $\sqrt{3}/2$ **(c)** $-\sqrt{3}/3$ **(d)** 2

6.19. $\sin^2 2.5 \approx 0.3582$, $\sin(2.5)^2 \approx -0.0332$; they are not equal

6.20. $\cos t = -1/3$, $\sin t = -2\sqrt{2}/3$, $\tan t = 2\sqrt{2}$, $\sec t = -3$, $\csc t = -3\sqrt{2}/4$, $\cot t = \sqrt{2}/4$

6.22. $\sec(t + 2\pi) = 1/\cos(t + 2\pi) = 1/\cos t = \sec t$

6.23. **(a)** $\sqrt{2} + \sqrt{47} \approx 8.27$ inches
(b) $\sqrt{45} \approx 6.71$ inches
(c) $\sqrt{45} \approx 6.71$ inches
(d) $1 + \sqrt{46} \approx 7.78$ inches
The displacement of the piston is $-2\sqrt{3}$ inches.

6.24.

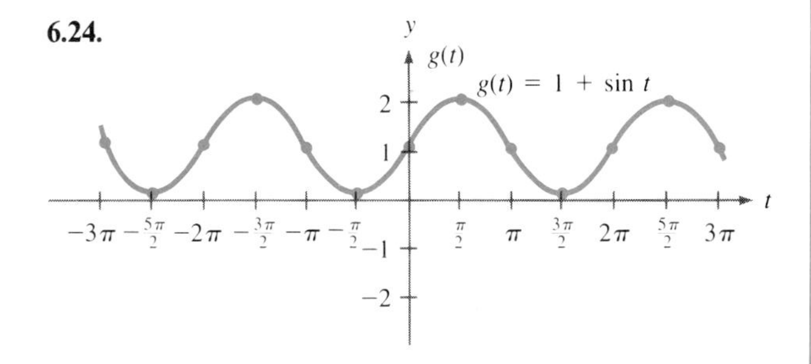

$g(t) = 0$ if and only if t has the form $3\pi/2 + 2\pi n$ for some integer n.

6.25.

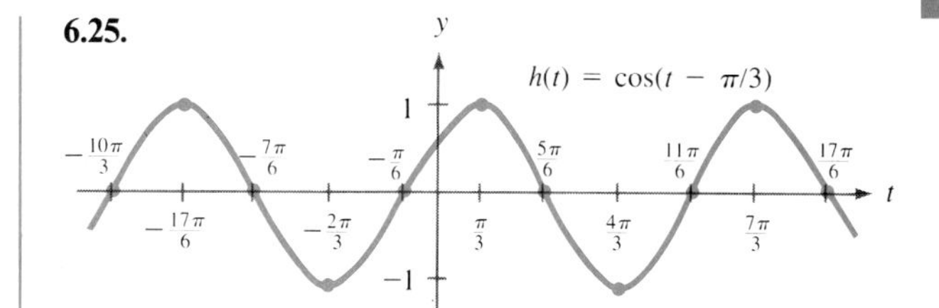

6.26.

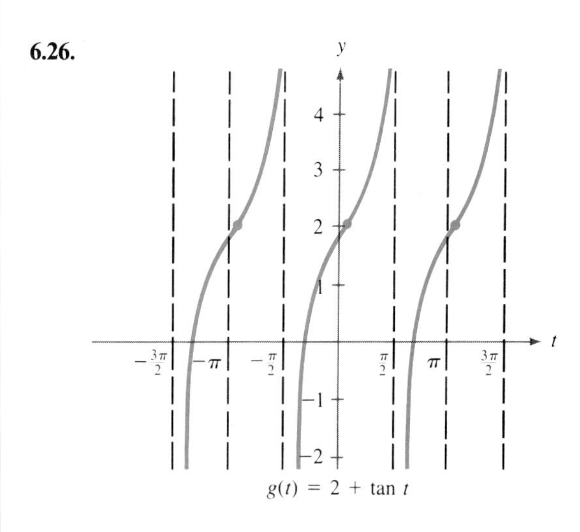

6.27.

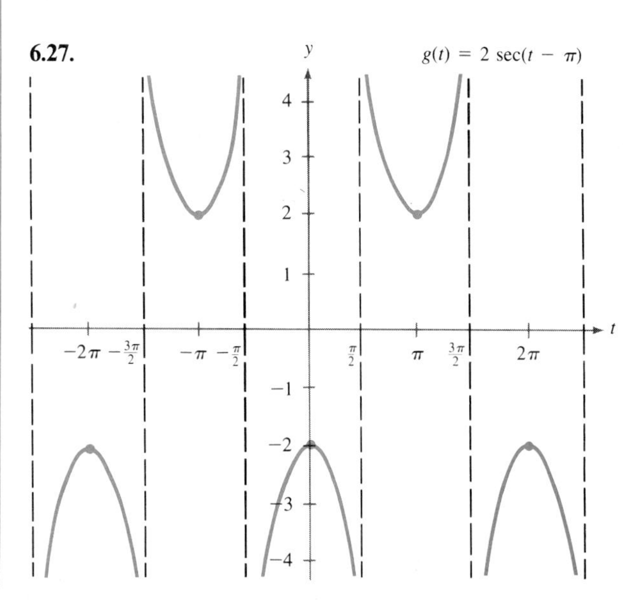

6.29.

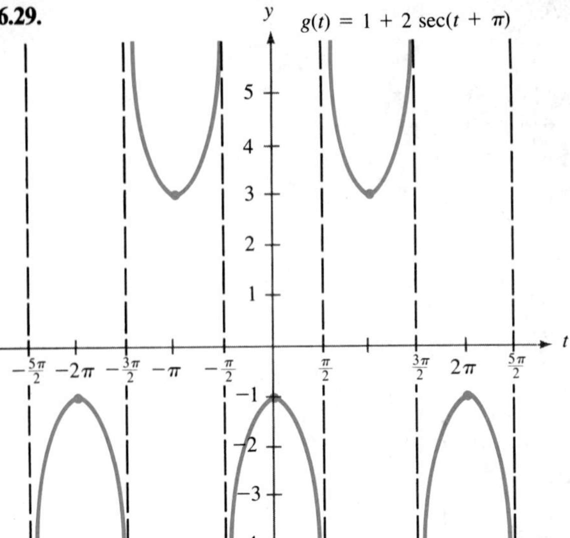

6.30.

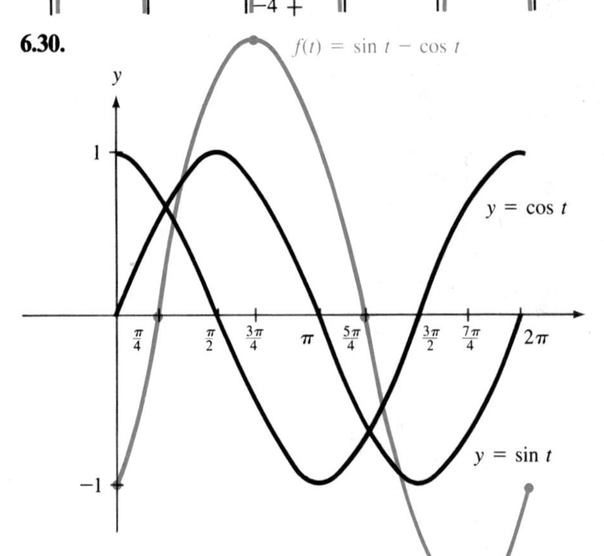

6.31.

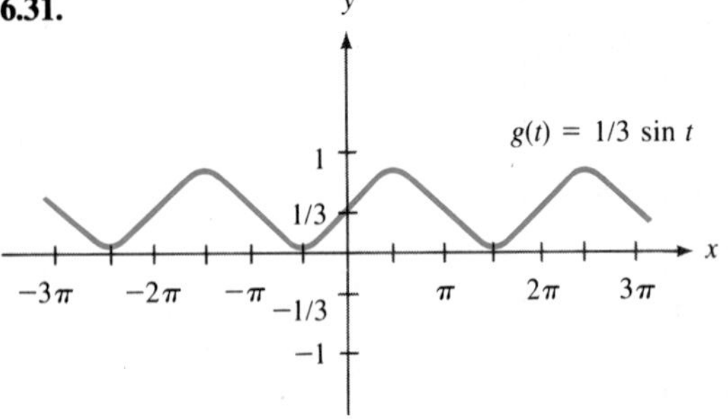

6.32.

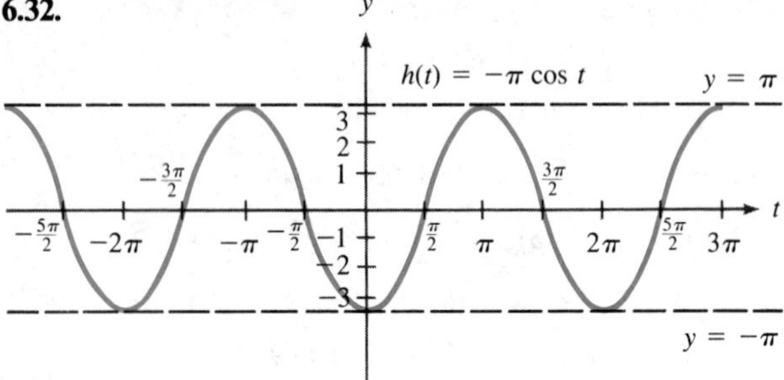

6.33.

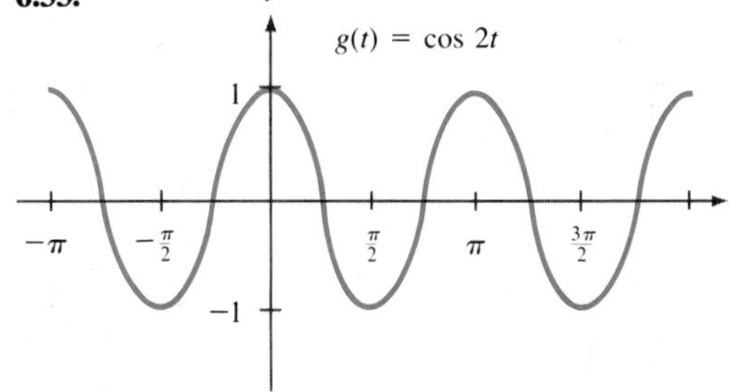

6.34.

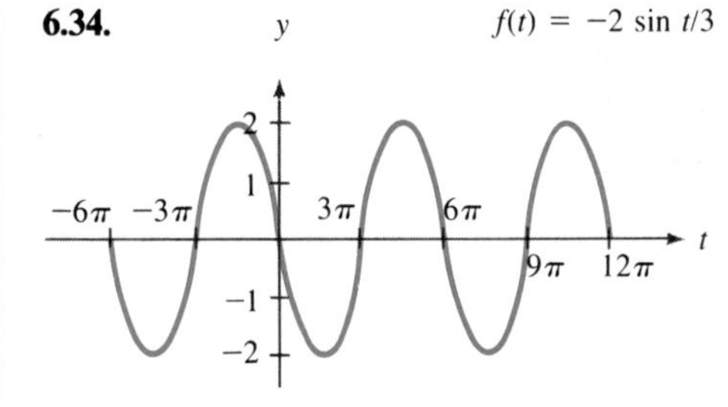

6.35.

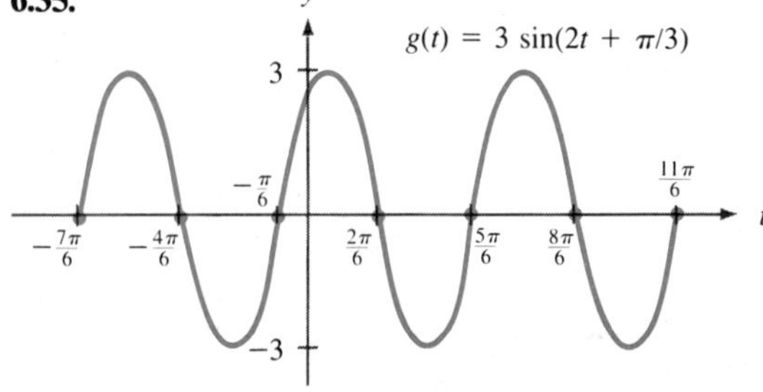

6.36.

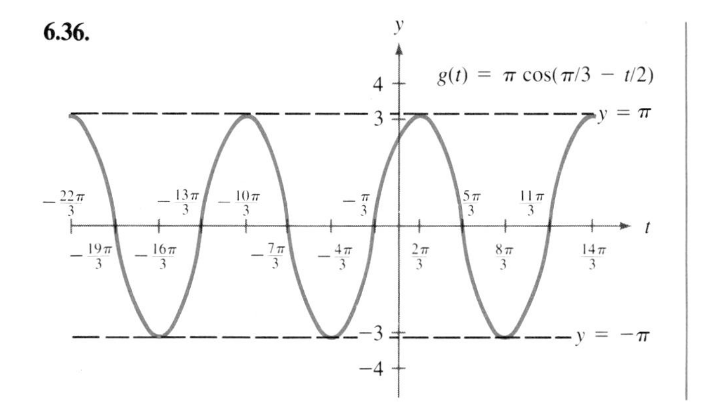

6.37. **(a)** $-\pi/6$ **(b)** $\pi/4$

6.38. **(a)** $-\pi/7$ **(b)** $-\pi/4$ **(c)** $1/4$

6.39. **(a)** $\pi/2$ **(b)** $3\pi/4$

6.40. **(a)** 0.222839704 **(b)** 1.772154248 **(c)** 1.373400767

6.41. $2\sqrt{2}/3$

6.42. $\sqrt{10}$

6.43. 5

Chapter Seven

7.1. **(a)** $\sqrt{2}/2$ **(b)** $\sqrt{3}/2$ **(c)** -1 **(d)** $\sqrt{2}$ **(e)** $-2\sqrt{3}/3$ **(f)** 0

7.2. **(a)** no **(b)** the sine of $\pi/4$ *degrees* ($\approx$ 0.79 degrees) **(c)** the cosine of 45 *radians* ($\approx$ 2578.3 degrees)

7.3. **(a)** 0.5 **(b)** 0.93969262 **(c)** -0.087488663 **(d)** 1.03527618 **(e)** -4.620226315

7.6. Similar triangles have corresponding angles of equal size, but not necessarily corresponding sides of equal length.

7.7. **(a)**

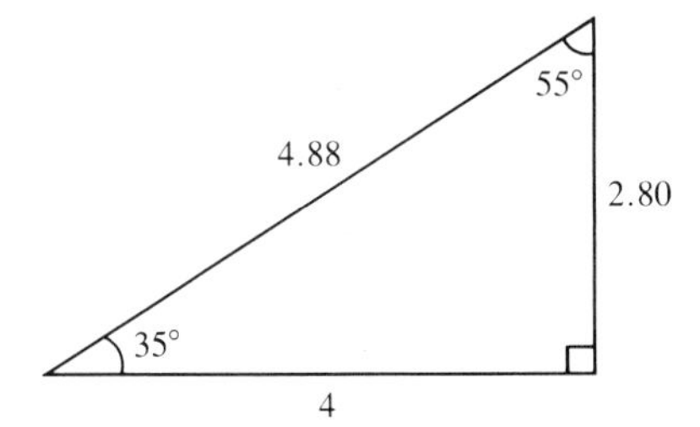

(b)

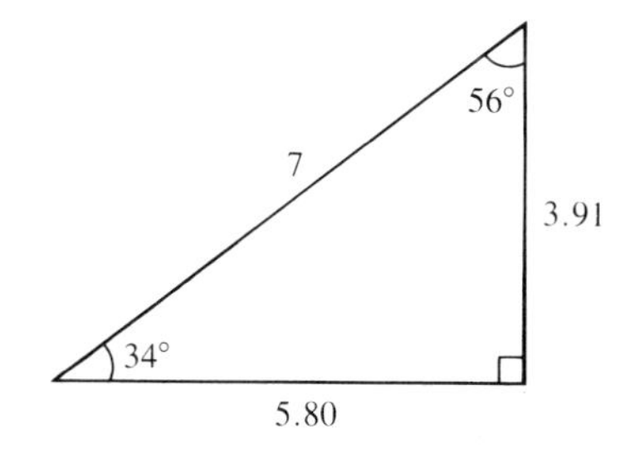

7.8.

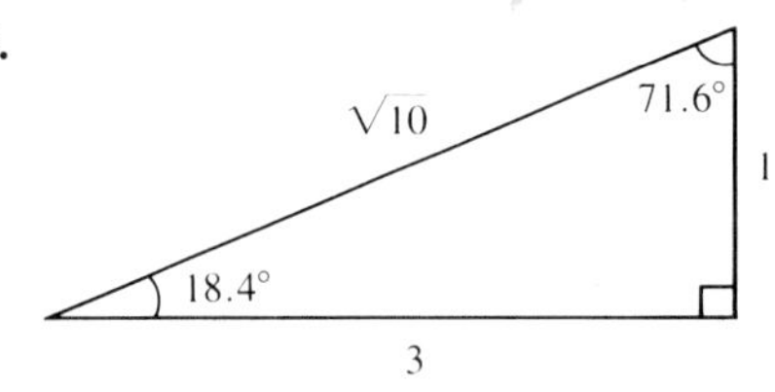

7.9. 10.04 feet

7.10. 54.17 feet

7.11. 14.87 miles from O bearing 109.7°

7.12.

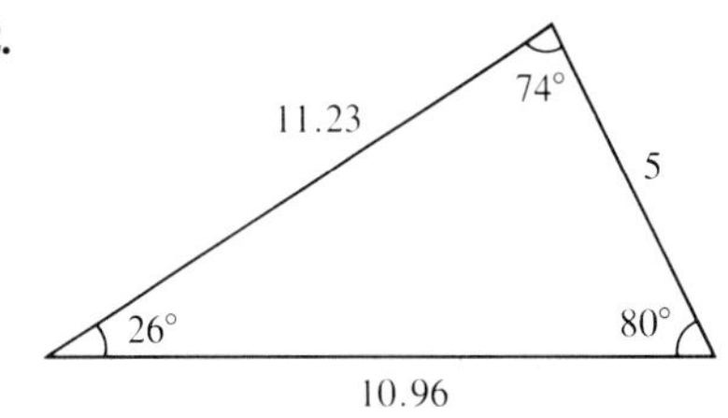

7.13. $\beta_1 \approx 57.7°$, $\gamma_1 \approx 97.3°$, $c_1 \approx 2.347$; $\beta_2 \approx 122.3°$, $\gamma_2 \approx 32.7°$, $c_2 \approx 1.278$

7.14. $\beta \approx 32.6°$, $\gamma \approx 101.4°$, $c \approx 5.451$

7.15. no such triangle

7.16. $c \approx 8.902$, $\alpha \approx 40.1°$, $\beta \approx 104.9°$

7.17. $\alpha \approx 38.9°$, $\beta \approx 109.5°$, $\gamma \approx 31.6°$

7.18. $\approx$ 3.4 miles

7.19.

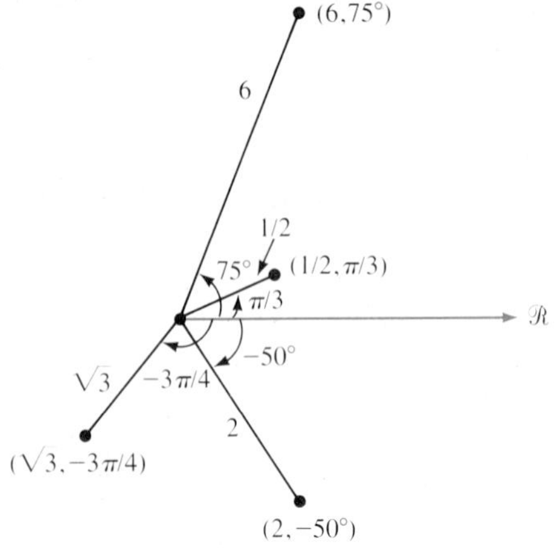

7.20.

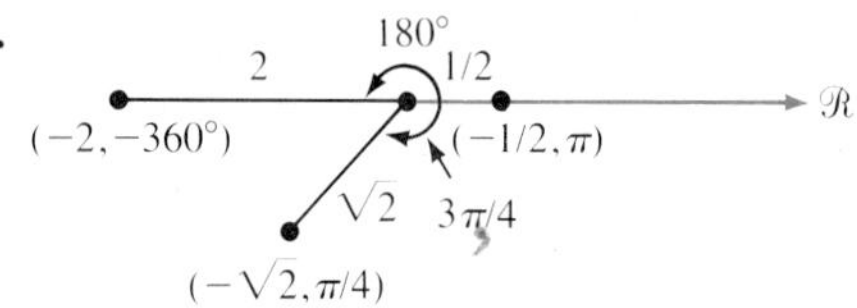

7.21. **(a)** P has polar coordinates $(4, \pi/3 + 2\pi n)$ and $(-4, 4\pi/3 + 2\pi n)$ where n varies over all integers.

(b) Q has polar coordinates $(-2, 110^\circ + 360^\circ n)$ and $(2, 290^\circ + 360^\circ n)$ where n varies over all integers.

7.22. **(a)** $\left(-\dfrac{3\sqrt{3}}{2}, \dfrac{3}{2}\right)$ **(b)** $(0.0868, 0.4924)$

7.23. **(a)** $(2, 150^\circ)$ **(b)** $(4.472, 116.6^\circ)$

7.24. $\mathbf{v} \approx \langle -6.062, 3.5\rangle$

7.25. $|\mathbf{v}| = 5$, $\theta \approx 233.1^\circ$

7.26. $\mathbf{v} + \mathbf{w} = \langle -2, -7\rangle$

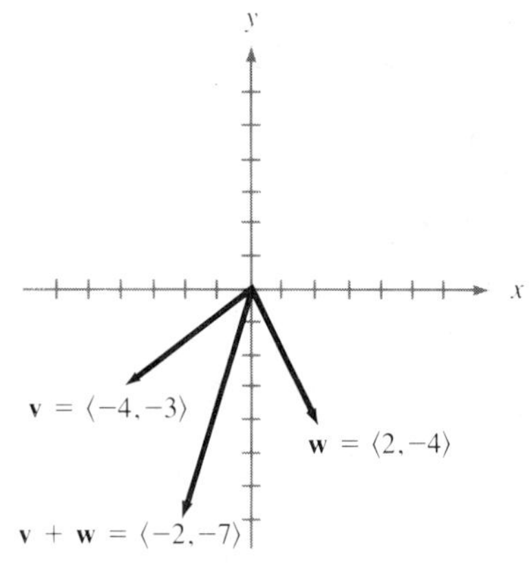

7.27. $\mathbf{v} \approx \langle 2.59, 9.66\rangle$, $\mathbf{w} \approx \langle 11.82, -2.08\rangle$ The total force is $\mathbf{v} + \mathbf{w} \approx \langle 14.41, 7.58\rangle$, which has magnitude $r \approx 16.29$ newtons and direction angle $\theta \approx 27.7^\circ$.

7.28. $\mathbf{v} - \mathbf{w} = \langle 22, -1\rangle$, $\mathbf{w} - \mathbf{v} = \langle -22, 1\rangle$

7.29. $-\mathbf{u} = \langle 1, 4\rangle$, $-2\mathbf{v} = \langle -1, 3/2\rangle$, $-4\mathbf{u} - 8\mathbf{v} = \langle 0, 22\rangle$

7.30. **(a)** $-(1/2)\mathbf{i} - \mathbf{j}$ **(b)** $-5\mathbf{j}$ **(c)** $8\mathbf{i} + \mathbf{j}$

Chapter Eight

8.1. $\cos t$

8.2. **(a)** $\sin t = \pm\sqrt{\dfrac{\sec^2 t - 1}{\sec^2 t}}$ **(b)** $\sin t = 2\sqrt{2}/3$

8.3. $\sin u$

8.10. $t - 5\pi/6, 7\pi/6$

8.11. $t = \pi/12, 3\pi/4, 17\pi/12, 5\pi/12, 13\pi/12, 7\pi/4$

8.12. $t = -\pi/12 + (2/3)\pi n$, $\pi/12 + (2/3)\pi n$ where n is any integer

8.13. $\theta \approx 0.696, 0.696 + \pi/2, 0.696 + \pi, 0.696 + 3\pi/2$

8.14. $t = 0, \pi$

8.15. $x = \pi/2, 3\pi/2, \pi/4, 5\pi/4$

8.16. $t = 7\pi/6, 11\pi/6$

8.17. $t \approx 0.7715 + 2\pi n$, $-0.7715 + \pi + 2\pi n$ where n is any integer

8.18. angle of refraction is $\mu \approx 13.13^\circ$

8.19. **(a)** $\cos 7\pi/12 = (1/4)(\sqrt{2} - \sqrt{6})$

(b) $\cos 75^\circ = (1/4)(\sqrt{6} - \sqrt{2})$

8.20. $\cos(s - t) = (1/30)(\sqrt{24} + \sqrt{35})$. Since this is positive and $-\pi/2 \le s - t \le \pi/2$, we see that $s - t$ lies in the first quadrant.

8.21. $\cos(t + \pi/3) = (1/2)(\cos t - \sqrt{3}\sin t)$

8.23. $\dfrac{-5 + \sqrt{24}}{1 + 5\sqrt{24}}$

8.24. $\sin 2t = -\dfrac{4\sqrt{2}}{9}$, $\cos 2t = \dfrac{7}{9}$, $\tan 2t = -\dfrac{4\sqrt{2}}{7}$

8.25. $\cos 3t = 4\cos^3 t - 3\cos t$

8.26. $\sin^4 t = (1/8)(3 - 4\cos 2t + \cos 4t)$

8.27. **(a)** $\sin 3\pi/8 = \sqrt{\dfrac{2+\sqrt{2}}{4}}$

(b) $\cos 3\pi/8 = \sqrt{\dfrac{2-\sqrt{2}}{4}}$

(c) $\tan 3\pi/8 = 1 + \sqrt{2}$

8.29. $t = \pi, 3\pi, \pi/3, 5\pi/3, 7\pi/3, 11\pi/3$

Chapter Nine

9.1. **(a)** Infinitely many solutions

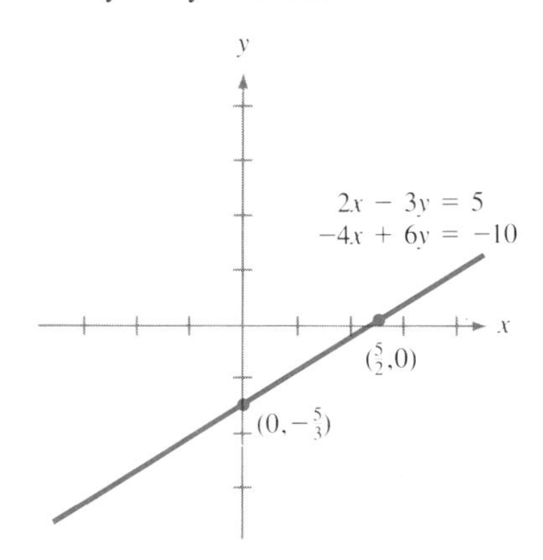

(b) Lines are parallel, no solutions

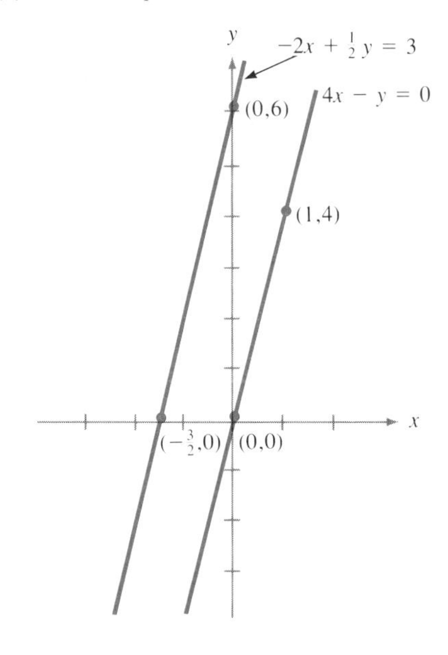

(c) Exactly one solution

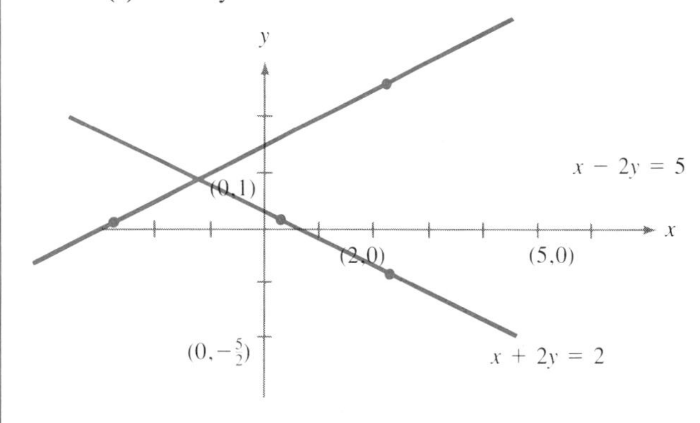

9.2. **(a)** $x = 13/4$, $y = 1/2$ **(b)** $x = 1$, $y = -5$, $z = 3$

9.3. $x = -11/6 + (1/2)u$, $y = 1/3 + 2u$, $z = u$

9.4. $x = 1$, $y = 4$

9.5. $x = 1/2$, $y = -1$, $z = -5/2$

9.6. $x = 1 + (1/2)u$, $y = 2 - 3u$, $z = u$

9.7. no solutions

9.8. $x = 1$, $y = 2$, $z = -3$

9.9. $x = 8$, $y = 4$, $z = -2$, $w = 0$

9.10. $f(x) = -(5/6)x^2 + (19/6)x + 1$

9.11. $x = 1/2$, $y = 2/3$

9.12. **(a)** 7 pounds of Virginia dark, no Indiana one-sucker, 3 pounds of North Carolina bright

(b) same as (a)

9.12. (c) 4 pounds of Virginia dark, 6 pounds of Indiana one-sucker, no North Carolina bright

9.13. No. The solution to the corresponding system of equations is $x = -46/3$, $y = 2$, $z = 55/12$, but x cannot be negative in this problem since it makes no sense to have a negative amount of milk.

9.14. (a) -27 **(b)** -2 **(c)** $a^2 + b^2$

9.15. $x = 1$, $y = -3$

9.16. $M_{1,1} = 8$, $C_{1,1} = 8$; $M_{2,3} = 20$, $C_{2,3} = -20$; $M_{3,2} = -2$, $C_{3,2} = 2$

9.17. -23

9.19. $x = 4$, $y = 5/2$, $z = 7/2$

9.20. $A + B = \begin{pmatrix} 8 & -4 \\ 8 & -\frac{1}{4} \end{pmatrix}$ $\quad A - B = \begin{pmatrix} -4 & -4 \\ 2 & \frac{5}{4} \end{pmatrix}$

9.21. $-4B = \begin{pmatrix} -28 & -4 \\ 0 & -36 \end{pmatrix}$ $\quad -B = \begin{pmatrix} -7 & -1 \\ 0 & -9 \end{pmatrix}$

9.22. $RC = 50$

9.23. (a) $\begin{pmatrix} 0 & -11 \\ 14 & 9 \end{pmatrix}$ **(b)** $\begin{pmatrix} 13 & 2 & 5 \\ 8 & -4 & 2 \\ 24 & 9 & 9 \end{pmatrix}$

9.24. $AB = \begin{pmatrix} 0 & 0 \\ 0 & 0 \end{pmatrix}$ $\quad BA = \begin{pmatrix} 0 & 0 \\ 1 & 0 \end{pmatrix}$

9.25. $\begin{pmatrix} -\frac{1}{10} & \frac{3}{10} & \frac{1}{5} \\ \frac{7}{10} & -\frac{1}{10} & -\frac{2}{5} \\ -\frac{3}{10} & -\frac{1}{10} & \frac{3}{5} \end{pmatrix}$

9.26. (a) $\begin{pmatrix} 3 & -2 \\ 4 & 1 \end{pmatrix}\begin{pmatrix} x \\ y \end{pmatrix} = \begin{pmatrix} 0 \\ 3 \end{pmatrix}$

(b) $\begin{pmatrix} 5 & -4 & 1 \\ 2 & 0 & -3 \end{pmatrix}\begin{pmatrix} x \\ y \\ z \end{pmatrix} = \begin{pmatrix} 5 \\ 8 \end{pmatrix}$

(c) $\begin{pmatrix} 1 & -2 & 3 \\ 1 & 2 & -3 \\ -1 & 0 & 1 \end{pmatrix}\begin{pmatrix} x \\ y \\ z \end{pmatrix} = \begin{pmatrix} 0 \\ 0 \\ 0 \end{pmatrix}$

9.27. $x = -19$, $y = 9$

9.28. $x = 0$, $y = 7$, $z = -4$

9.29. $x = 3\sqrt{3}$, $y = -2\sqrt{3}$; $x = -3\sqrt{3}$, $y = 2\sqrt{3}$;

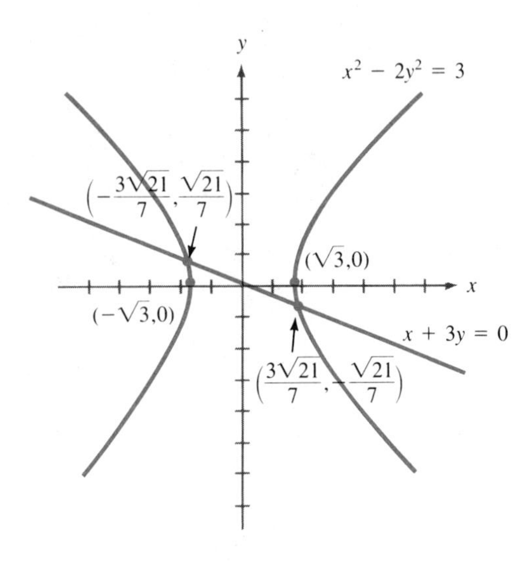

9.30. $x = 1/2$, $y = 1/2$; $x = 1/2$, $y = -1/2$; $x = -1/2$, $y = 1/2$; $x = -1/2$, $y = -1/2$

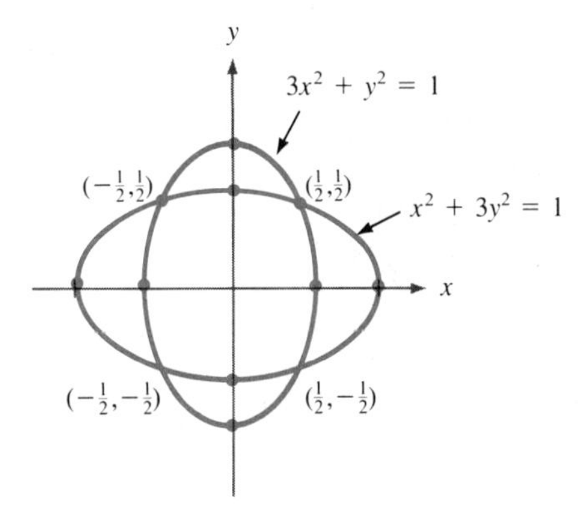

9.31. $x = \pi/2 + 2\pi n$, $y = 0$; $x = 3\pi/2 + 2\pi n$, $y = 0$ where n can be any integer

9.32.

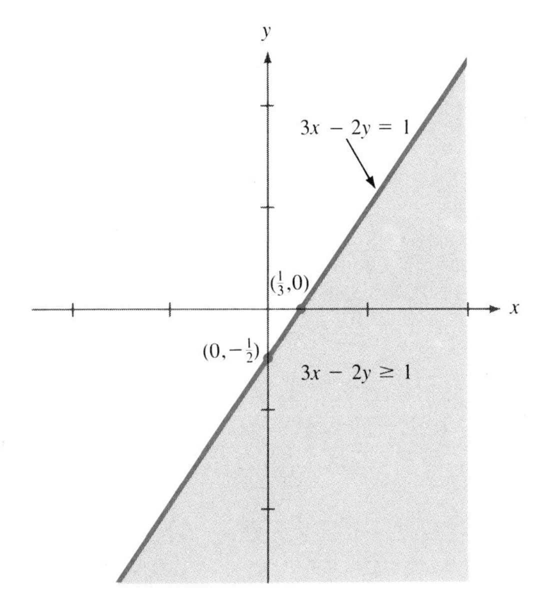

9.34.

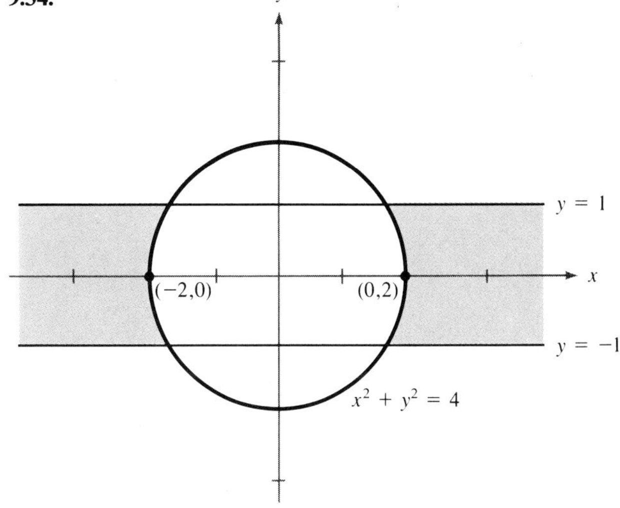

9.33.

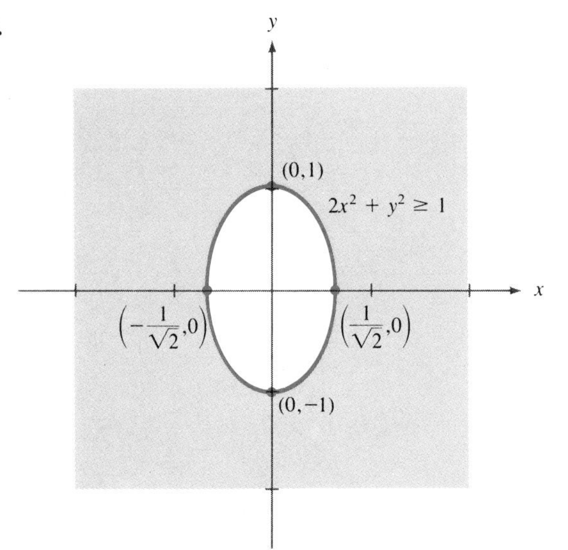

9.35.

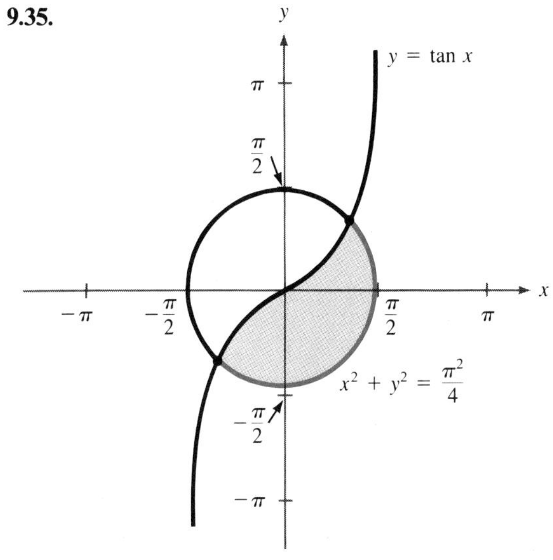

Chapter Ten

10.1. **(a)** $Re(z) = 2$, $Im(z) = -3$

(b) $Re(z) = 1/2$, $Im(z) = \sqrt{2}$

(c) $Re(z) = 0$, $Im(z) = -1/\sqrt{2}$

10.2. **(a)** $4i$ **(b)** $\sqrt{2}i$ **(c)** $5\sqrt{5}i$ **(d)** $(\sqrt{6}/3)i$

10.3. **(a)** $z + w = 3 - (1/2)i$, $z - w = -1 - (3/2)i$

(b) $z + w = 4 + 6i$, $z - w = -4 + 6i$

10.4. **(a)** $10 + 5i$ **(b)** $5/2 - 5i$ **(c)** $2 + 2i$

10.5. **(a)** $2 + i$ **(b)** $\sqrt{2} - \sqrt{3}i$ **(c)** i **(d)** π

10.6. **(a)** 5 **(b)** 34

10.7. **(a)** $2/13 + (3/13)i$ **(b)** $15/17 + (8/17)i$ **(c)** i

10.8. **(a)** $2\sqrt{5}$ **(b)** 1 **(c)** 1024

10.9. $x = -1 \pm 5i$

10.10. $x = -1/2 \pm (\sqrt{3}/2)i$

10.11. $x = \pm i, \pm\sqrt{2}i$

10.12. $x^2 - 4x + 13 = 0$; only one such equation

10.13. $z = \sqrt{2}(\cos 3\pi/4 + i \sin 3\pi/4)$

10.14. $\pi/6 + 2\pi n$ where n varies over all integers. The principal value of the argument is $\pi/6$.

10.15. **(a)** $z = 4(\cos 2\pi/3 + i \sin 2\pi/3)$
(b) $z = \sqrt{5}[\cos(\arctan(-2)) + i \sin(\arctan(-2))]$

10.16. $(1 - i)(-1 + \sqrt{3}i) = 2\sqrt{2}(\cos 15\pi/12 + i \sin 15\pi/12)$

10.17. $zw = 2$, $z/w = i$

10.18. $(2 - 2\sqrt{3}i)^6 = 4096$

10.19. $\cos \pi/8 + i \sin \pi/8$, $\cos 5\pi/8 + i \sin 5\pi/8$, $\cos 9\pi/8 + i \sin 9\pi/8$, $\cos 13\pi/8 + i \sin 13\pi/8$

10.20. $-2 + 2\sqrt{3}i$, $2 - 2\sqrt{3}i$

10.21. $w_1 = 1$, $w_2 = -1/2 + (\sqrt{3}/2)i$, $w_3 = -1/2 - (\sqrt{3}/2)i$. An equilateral triangle.

10.22. $w_1 = 1$, $w_2 = i$, $w_3 = -1$, $w_4 = -i$. A square.

Chapter Eleven

11.1. Focus at (0, 3/8), directrix is $y = -3/8$

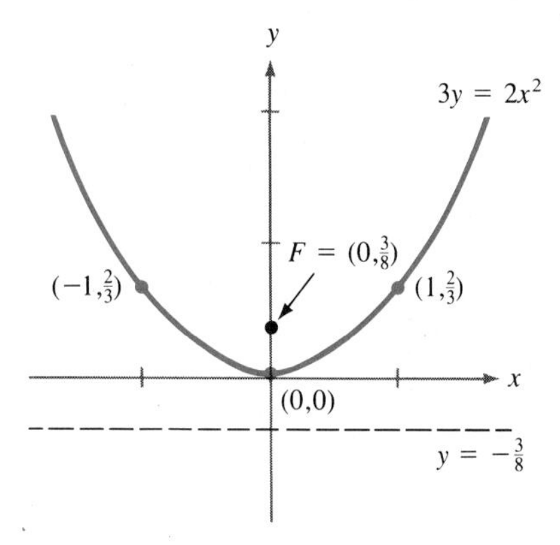

11.2. $y = -2x^2$

11.3. $y + 1 = (1/20)(x + 4)^2$, focus at $(-4, 4)$

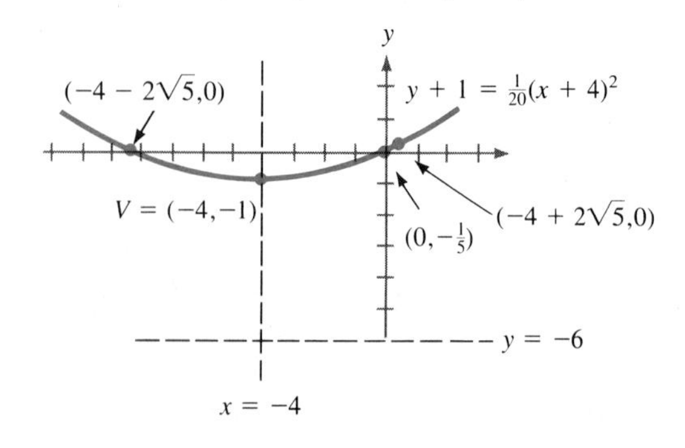

11.4. Vertex at $(-1/4, 3)$, focus at $(1/4, 3)$, directrix is $x = -3/4$

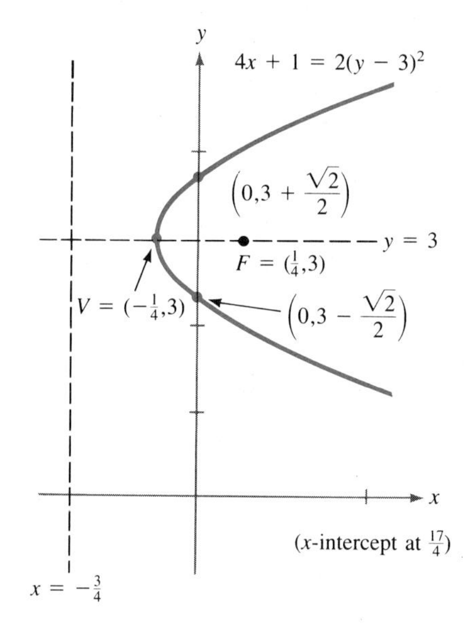

11.5. $4x^2 + y^2 = 1$

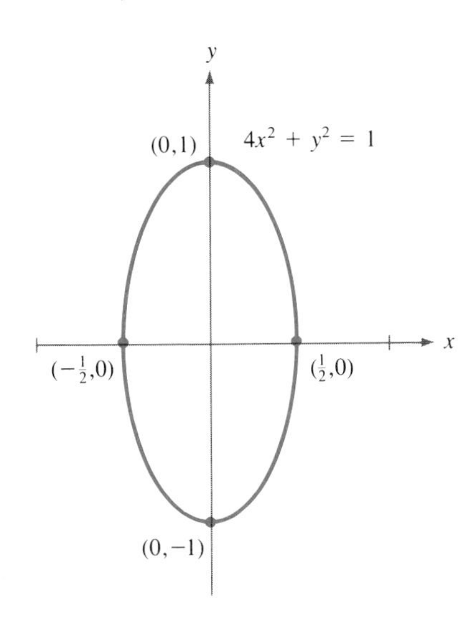

11.6.

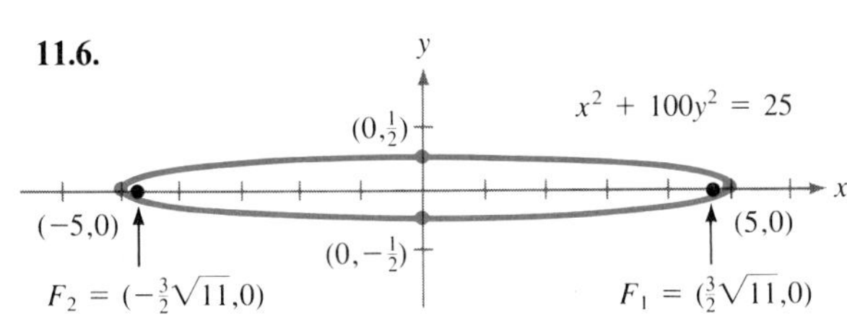

11.7. $x^2 + \frac{(y-2)^2}{4} = 1$, foci at $(0, \sqrt{3})$ and $(0, -\sqrt{3})$

11.8.

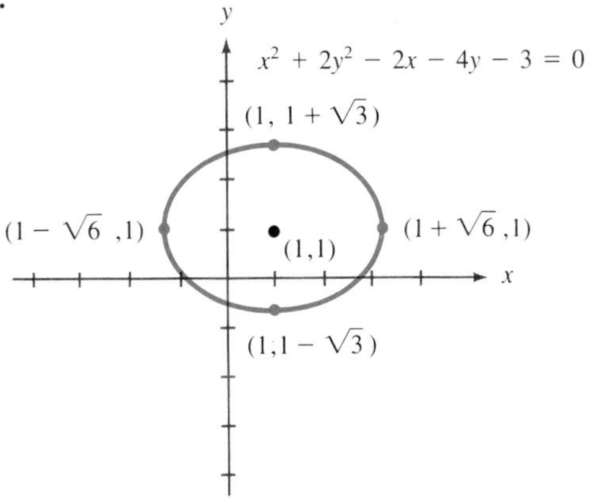

Note: Standard form is $\frac{(x-1)^2}{6} + \frac{(y-1)^2}{3} = 1$

$\sqrt{6} \approx 2.4$, $\sqrt{3} \approx 1.7$.

11.9.

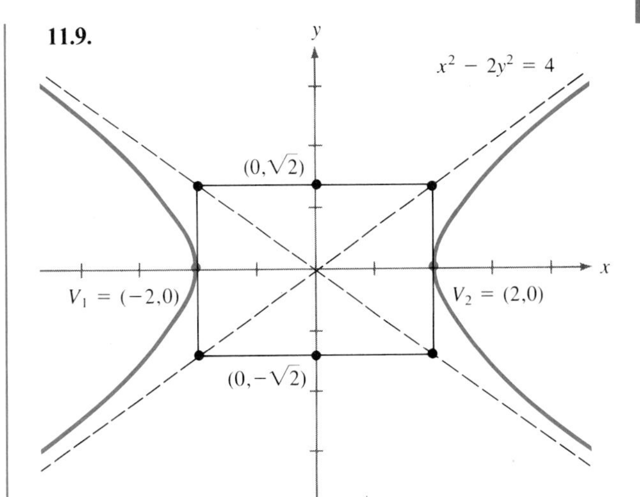

11.10.

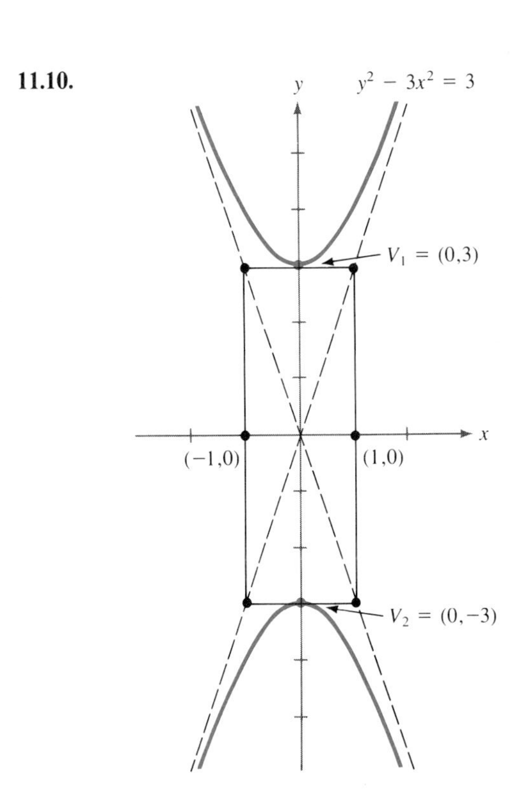

1.11.

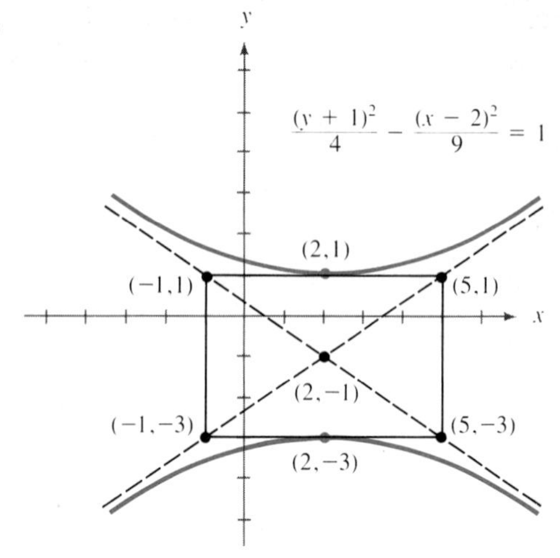

11.12.

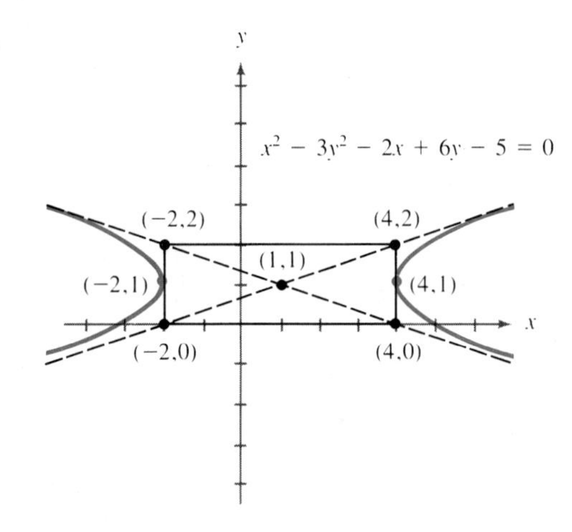

11.13. $4y^2 - x^2 - 24y - 2x + 35 = 0$

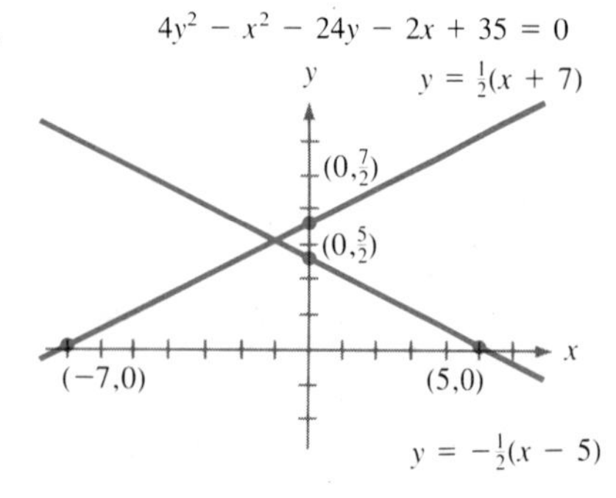

11.14. **(a)** circle **(b)** parabola **(c)** hyperbola **(d)** ellipse

11.15. $(x')^2 - (y')^2 = 2$, graph is a hyperbola

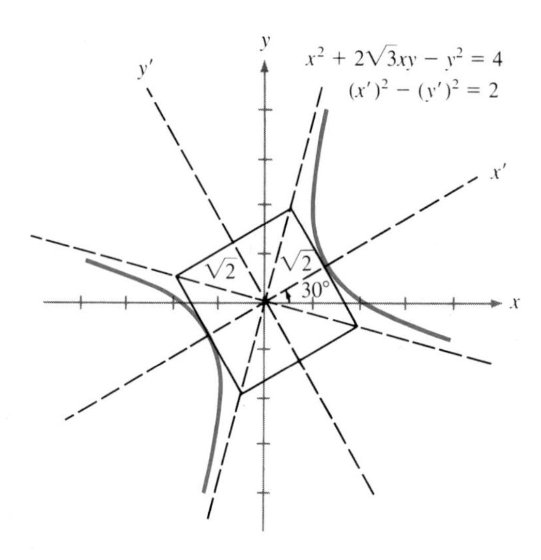

11.16. A parabola

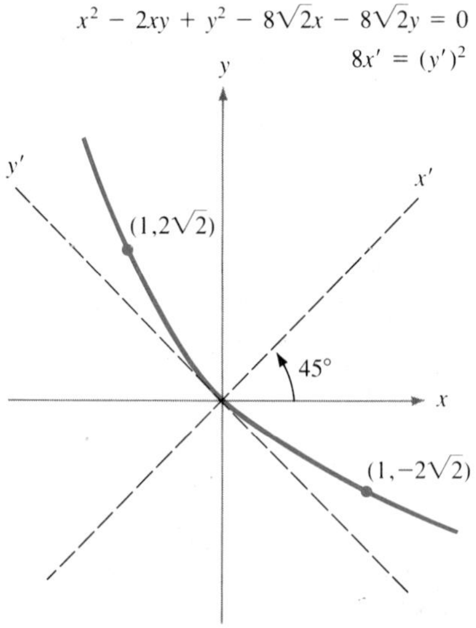

$x^2 - 2xy + y^2 - 8\sqrt{2}x - 8\sqrt{2}y = 0$

$8x' = (y')^2$

y
y′
x′
x
$(1,2\sqrt{2})$
45°
$(1,-2\sqrt{2})$

11.17. A hyperbola

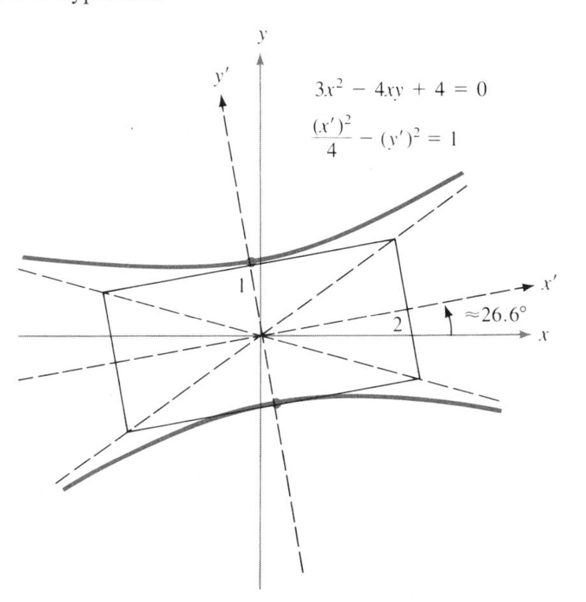

11.18. **(a)** ellipse **(b)** parabola **(c)** hyperbola

Chapter Twelve

12.3. **(a)** 336 **(b)** 36

12.4. **(a)** $9! - 6! = 362{,}160$, $(9 - 6)! = 6$

(b) $8!/4! = 1680$, $(8/4)! = 2$

12.5. $81x^4 - 216x^3 + 216x^2 - 96x + 16$

12.6. $\binom{12}{8}(-3)^8 2^4 = 51{,}963{,}120$

12.7. $x^{-8} - 4x^{-11/2} + 6x^{-3} - 4x^{-1/2} + x^2$

12.8. **(a)** $a_1 = 1$, $a_2 = 4$, $a_3 = 27$, $a_{10} = 10^{10}$

(b) $a_1 = -1$, $a_2 = -1$, $a_3 = -1$, $a_{10} = -1$

(c) $a_1 = \log 2$, $a_2 = 2\log 2$, $a_3 = 3\log 2$, $a_{10} = 10\log 2$

12.9. $a_1 = 5$, $a_2 = 5/2$, $a_3 = 5/4$, $a_4 = 5/8$, $a_5 = 5/16$, $a_n = 5/2^{n-1}$

12.10. **(a)** $a_1 = -2$, $a_2 = 1$, $a_3 = 4$, $a_4 = 7$, $a_5 = 10$

(b) $a_1 = -1/2$, $a_2 = 0$, $a_3 = 1/2$, $a_4 = 1$, $a_5 = 3/2$

12.11. **(a)** $a_n = 3n - 5$, $a_{100} = 295$

(b) $a_n = (1/2)n - 1$, $a_{100} = 49$

12.12. $d = -3$, $a_n = 17 - 3n$

12.13. **(a)** $a_1 = 1/2$, $a_2 = -1$, $a_3 = 2$, $a_4 = -4$, $a_5 = 8$

(b) $a_1 = 1/2$, $a_2 = 2$, $a_3 = 8$, $a_4 = 32$, $a_5 = 128$

12.14. **(a)** $a_n = -(-2)^{n-2}$, $a_{10} = -256$

(b) $a_n = (1/2)(4^{n-1})$, $a_{10} = 131{,}072$

12.15. No

12.16. 780

12.17. $s_5 = 3^5 - 1 = 242$, $s_{10} = 3^9 - 1 = 19{,}682$

12.18. $s_n = na_1$

12.19. $s_5 = \frac{9}{2}\left(1 - \frac{1}{3^5}\right) = \frac{121}{27} \approx 4.481$

$$s_{10} = \frac{9}{2}\left(1 - \frac{1}{3^{10}}\right) \approx 4.500$$

12.20. $1 + x^3\left(\frac{1 - x^{3n}}{1 - x^3}\right)$ or $\frac{1 - x^{3(n+1)}}{1 - x^3}$

12.21. This series converges to 4/3.

12.22. Since $s_n > n$ for all positive integers n, the partial sums s_n increase without bound as n increases. Hence, the series diverges.

12.23. 20/3

12.24. 0

12.25. **(a)** 3 **(b)** 3

12.26. From the table it appears that the values of $(x + 1)/(x - 1)^2$ are getting larger and larger without bound as x approaches 1. Hence the limit does not exist.

x	$\dfrac{x+1}{(x-1)^2}$
1.1	210
1.01	20,100
1.001	2,001,000
1.0001	200,010,000

ANSWERS TO ODD EXERCISES

Chapter One

SECTION 1.1

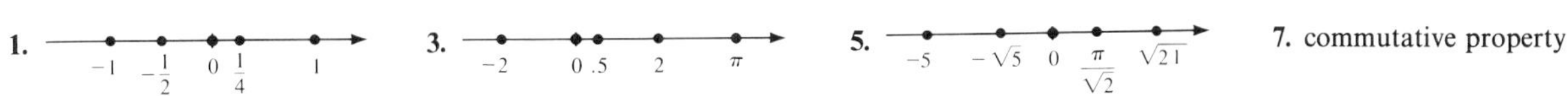

7. commutative property

9. inverse property **11.** identity property **13.** associative property **15.** inverse property **17.** distributive and commutative property **19.** inverse property **21.** distributive and identity properties **23.** $\frac{29}{6}$ **25.** $\frac{7}{10}$ **27.** $\frac{153}{280}$ **29.** $-\frac{31}{78}$ **31.** $\frac{13}{24}$ **33.** $-\frac{1}{15}$ **35.** $-\frac{11}{9}$ **37.** no; in general $a - b \neq b - a$ **39.** $2 < 4$ **41.** $-4 > -5$ **43.** $\frac{1}{2} > -\sqrt{2}$ **45.** $-\pi < -\sqrt{3}$ **47.** $0 \leq x^2$ **49.** 3 **51.** $\sqrt{5} - \sqrt{2}$ **53.** 1 **55.** -1 **57.** -4 **59.** **(a)** $a = 2, b = -1$ **(b)** $a = 0, b = 1$ **61.** **(a)** π **(b)** 2π **63.** **(a)** $\frac{3}{4}$ **(b)** $\frac{2}{9}$ **65.** -1 0 2 **67.** $-\pi$ 0 π **69.** -7 -5 -4 0 3 **71.** -1 0 1

73. **(a)** $\{x \mid -5 \leq x \leq 5\}$ **(b)** $\{x \mid -5 < x < 5\}$ **75.** **(a)** **R** **(b)** $\varnothing$

77. **(a)** $\{x \mid -3 < x < 5\}$ **(b)** $\{x \mid x \leq -3\} \cup \{x \mid x \geq 1\}$ **79.** **(a)** $\{x \mid \frac{1}{6} \leq x \leq \frac{5}{6}\}$ **(b)** $\left\{y \mid y > -\frac{\pi}{2}\right\} \cup \left\{y \mid y < -\frac{3\pi}{2}\right\}$ **81.** no

SECTION 1.2

1. $\frac{1}{27}$ **3.** 1 **5.** 81 **7.** 6 **9.** 1 **11.** $\frac{3}{2}$ **13.** $-\frac{1}{6}$ **15.** $\dfrac{p^2r^4s^2}{2}$ **17.** $-\frac{8}{3}$ **19.** y **21.** $\dfrac{8y^6}{x^9z^3}$ **23.** a^2b^2

25. 1 **27.** 0.000000069 **29.** 28.187 **31.** use x^2 key twice; powers of the form 3^{2^n} where n is any positive integer **33.** 4

35. $-\dfrac{36^{1/3}}{12}$ **37.** 2 **39.** -1 **41.** $2\sqrt[3]{10}$ **43.** not defined **45.** $16\sqrt{3}$ **47.** $2\sqrt[6]{2}$ **49.** $\sqrt{5}$ **51.** $-\sqrt[3]{3}$

53. $\dfrac{-\sqrt[4]{80}}{4}$ **55.** $2 \cdot 5^{1/3}$ **57.** $\dfrac{3}{5}(\sqrt{2} - \sqrt{7})$ **59.** $-5 - 2\sqrt{6}$ **61.** $\dfrac{13}{18}\sqrt[3]{9}$ **63.** $\dfrac{3a^2}{b}$ **65.** $x^3\sqrt[4]{x}$ **67.** $ab\sqrt[6]{a^2b}$

69. $\dfrac{2xz^2}{y}$ **71.** $\dfrac{2y}{x}$ **73.** $\dfrac{-2\sqrt[5]{2y^3}}{y}$ **75.** $\sqrt{x}$ **77.** $\dfrac{b^2}{a^3}$ **79.** $\dfrac{y^2}{46656x^3}$ **81.** $x^{2/3}$ **83.** $2xy^2\sqrt{2}$ **85.** $\dfrac{a^7c^6}{b}$

87. $1 + x^2$ **89.** $2^{3/2}xy$ **91.** $(1 + a)^{5/4}$ **93.** 1 **95.** y **97.** $\frac{1}{p^{2b/3}}$ **99.** $x^{n^3 - 3n}$

101. **(a)** 4.3×10^{-6} **(b)** 1×10^{-11} **103.** **(a)** 2.65001 **(b)** 3.7500009201×10^2 **105.** .00000123 cm **107.** 38.88×10^{-4}

109. $\frac{4}{27} \times 10^7$ **111.** 2.45×10^{-14} **113.** -0.9367 **115.** -1.152 **117.** 0.6542 **119.** 0.2040 **121.** 1.993

123. $(2.78)^{3.14}$ **125.** 7.8155×10^9 **127.** $2.54840104 \times 10^{65}$

SECTION 1.3

1. $x^2 - 3x - 4$; degree 2 **3.** $9u^3 + 3u^2 + 5u$; degree 3 **5.** $x^2 - x - 8$; degree 2 **7.** $-4y^3 + 6y^2 + y$; degree 3
9. $24x^3 - 14x^2 + 9x + 2$; degree 3 **11.** $2x^4 + x^3 - 3x^2 - x + 1$; degree 4 **13.** $28u^{19} - 4u^{11} + 7u^{10} + 4u^9 - u^2 + 1$; degree 19
15. $24y^6 + 58y^4 - 7y^2 - 5$; degree 6 **17.** $2x^2 + 3xz - 2zy - y$ **19.** $u^3 + v^3$ **21.** $xyz + yz + xy + xz + x + y + z + 1$

23. $x + 2\sqrt{x} + 1$ **25.** $y - 2 + \frac{1}{y}$ **27.** $2x + x^{2/3} + x^{1/3} - 1$ **29.** $3x^{-4} + 10x^{-3} + 10x^{-2} + 2x^{-1} - 1$

31. $12u - 20u^{3/4} - 11u^{1/2} + 5u^{1/4} + 2$ **33.** $x^2 + 4x + 6 + \frac{4}{x} + \frac{1}{x^2}$ **35.** $2xy(y + 2)$ **37.** $(x - y)(x + y)$

39. $(v + w)(2u + 3)(u - 1)$ **41.** $(t + 5)(t - 2)$ **43.** $(2y - 1)(y + 5)$ **45.** $2(3a^2 - 3a - 8)$ **47.** $(x - \pi)(x + 2\pi)$
49. $(6a - 7b)(6a + 7b)$ **51.** $(3p^2 - 5q^3)(3p^2 + 5q^3)$ **53.** $2(3a + 2b)(5a - 3b)$ **55.** $(4x + y)(5x + 12y)$ **57.** $(\frac{3}{2} - z)^2$
59. $(6x + 5y^2)^2$ **61.** $(2x + 3w^2)(4x^2 - 6xw^2 + 9w^4)$ **63.** $(2r - s)(r^2 - rs + s^2)$ **65.** $(2x - 3u)(2y + z)$ **67.** $r(\frac{1}{2}s - 1)(t + uv)$
69. $(3x^2 + 2)(2x^2 - 1)$ **71.** $(u - v)(u + v)(u^2 + uv + v^2)(u^2 - uv + v^2)$ **73.** $(2u + 3v - 3)(2u + 3v + 3)$ **75.** $(x + 1)(2x^3 - 1)$

77. $(y + 2)(y - 2)(y^5 + 1)$ **79.** $(2z^2 + 7)(2z^2 - 5)$ **81.** $\left(2s^2 - \frac{9}{2t^2}\right)\left(2s^2 + \frac{9}{2t^2}\right)$

SECTION 1.4

1. $\frac{x + 2}{2x + 3}$ **3.** $\frac{5x + 2}{2(2x + 1)}$ **5.** $\frac{5 + 2y}{3 - 4y}$ **7.** $\frac{z(4z - 1)}{4z + 1}$ **9.** $\frac{8x^2 - 7x - 3}{(2x - 1)(x + 1)}$ **11.** $\frac{6y^2 + y - 23}{(2y + 3)(2y + 4)}$ **13.** $\frac{z^2 + 2z + 3}{(z + 2)(z - 1)(z + 1)}$

15. $\frac{(3x + 1)(x + 4)}{(2x - 1)(3x - 1)}$ **17.** $\frac{(2x + 1)(2x + 7)}{(x + 5)(2x - 1)}$ **19.** $\frac{x(x - 2)}{(x - 1)(x - 3)}$ **21.** $\frac{-30}{(2x + 5)^2}$ **23.** $\frac{2y^3 + 9y^2 + 9y - 3}{3y^4}$

25. $\frac{5w^2 + 3w - 18}{2w(w - 3)(w + 3)}$ **27.** $\frac{-2x^4 + x^3 + 20x^2 + 3x - 36}{2x^2(x + 3)^2}$ **29.** $\frac{-(10t^3 + 7t^2 + 4t + 2)}{2t^3(1 + 2t)}$ **31.** $\frac{-1}{x(x + h)}$

33. $\frac{-4}{(4x + 1)(4x + 4h + 1)}$ **35.** $\frac{-3x^2 - 3xh - h^2}{x^3(x + h)^3}$ **37.** $\frac{x^2 + y^2}{x + y}$ **39.** $\frac{x^2 + xy + y^2}{x^2y^2}$ **41.** $\frac{(x - 3)(x - 1)(3x + 1)}{(5x - 1)(x + 1)(3x - 1)}$

43. $\frac{-t}{s}$ **45.** $\frac{(a - b)(\sqrt{ab})}{ab}$ **47.** $\frac{(\sqrt{a} - \sqrt{z})^2}{a - z}$ **49.** $1 + \sqrt{x}$ **51.** $\frac{x\sqrt{1 - x^2}}{(1 - x^2)^2}$

SECTION 1.5

1. $p(q(x)) = 6x - 7$; $q(p(x)) = 6x - 11$; no **3.** $p(q(x)) = 2x + 3$; $q(p(x)) = 2x + 3$; yes **5.** $p(q(x)) = q(p(x)) = \frac{1}{x^4}$; yes

7. $p(q(x)) = x + 2$; $q(p(x)) = \sqrt{x^2 + 2}$; no **9.** $p(q(x)) = \frac{1}{x^2 - 1}$; $q(p(x)) = \frac{-(2x^2 + 4x + 1)}{(x + 1)^2}$; no **11.** $p(q(x)) = x$; $q(p(x)) = x$; yes

13. $p(q(x)) = -x$; $q(p(x)) = \frac{1}{x}$; no **15.** $4x^2 - 4x + 6 - \frac{2}{x} + \frac{1}{x^2}$ **17.** $3x^2 - 1$ **19.** $x^{40} - 3x^{30} + 5x^{10} - 3$

21. $-2 + 9x^2 - 9x^4 + 3x^6$ **23.** $\dfrac{x}{1 + x + x^2}$ **25.** $\dfrac{x(x^2 - 1)}{-x^4 + 3x^2 - 1}$ **27.** $\dfrac{x^4 + 3x^2 - 1}{4x^4 - x^2 + 1}$ **29.** $\dfrac{1}{x} + x$ **31.** $\dfrac{1}{|x|} + |x|$

33. $8x + 12x^{2/3} + 6x^{1/3} + 1$ **35. (a)** $x^4 + 2x^2 + 2$ **(b)** $x^8 + 4x^6 + 8x^4 + 8x^2 + 5$ **37.** $q(x) = \dfrac{x - 1}{2}$

SECTION 1.6

1. 8; 2 solutions **3.** 25; 2 solutions **5.** 1; 2 solutions **7.** -20; no real solutions **9.** 0; 1 solution **11.** $x = 3, -1$
13. $t = 3$ **15.** $x = -\frac{1}{2}, \frac{1}{5}$ **17.** $x = \pm\sqrt{2}$ **19.** $x = -5, -1$ **21.** no real solutions **23.** $t = 2 \pm \sqrt{3}$ **25.** $x = \frac{5}{2}$

27. $x = \dfrac{-5 \pm \sqrt{35}}{5}$ **29.** $x = -5, -2$ **31.** $t = -3 \pm \sqrt{10}$ **33.** $x = \dfrac{-1 \pm \sqrt{113}}{28}$ **35.** $x = \dfrac{\sqrt{2}}{3}$ **37.** $x = 6, -5$

39. $x = \dfrac{-1 \pm \sqrt{2}}{2}$ **41.** $x = \dfrac{-1 \pm \sqrt{5}}{2}$ **43.** $x = -\frac{1}{3}, -\frac{1}{2}$ **45.** $x = 0.006366, -0.9133$ **47.** $x = 0, 0.03913$ **49.** $x = 7$

51. $s = \frac{333}{130}$ **53.** $t = 3$ **55.** $x = \pm 2$ **57.** $x = \pm 1$ **59.** $z = \pm 1$ **61.** $x = \pm 2$ **63.** $x = 0, \left(\dfrac{2}{3}\right)^9$ **65.** $w = 16$

67. $x\left(\dfrac{-3 \pm \sqrt{17}}{4}\right)^{1/3}$ **69.** $x = -\frac{1}{3}$ **71.** $x = -\frac{43}{10}$ **73.** $t = -\frac{1}{2}$ **75.** $x = \dfrac{3 \pm \sqrt{5}}{4}$ **77.** $x = \dfrac{11 \pm 3\sqrt{13}}{2}$

79. no real solutions **81.** $x = -\frac{5}{4}, \frac{7}{4}$ **83.** $x = -\frac{1}{10}, \frac{3}{4}$ **85.** $x = \pm\sqrt{5}$ **87.** $x = \dfrac{7 \pm \sqrt{33}}{4}, 0, -\dfrac{7}{2}$ **89.** $x = \frac{1}{4}$

SECTION 1.7

1. $[-1, 2]$ **3.** $[-\frac{1}{2}, 0]$ **5.** $(-4, 7)$ **7.** $(-\frac{5}{3}, -\frac{3}{5})$ **9.** $[-12, 4)$ **11.** $(-\infty, -5]$ **13.** $(-\infty, 0)$ **15.** $\{x \mid -3 \le x \le 4\}$
17. $\{x \mid x > 0\}$ **19.** $\{x \mid -\frac{1}{2} < x \le \frac{1}{2}\}$ **21.** $\{x \mid x > \frac{1}{4}\}, (\frac{1}{4}, \infty)$ **23.** $\{x \mid x \ge -\frac{2}{3}\}, [-\frac{2}{3}, \infty)$ **25.** $\{x \mid x \ge \frac{9}{7}\}, [\frac{9}{7}, \infty)$
27. $\{x \mid x \ge -1\}, [-1, \infty)$ **29.** $\{x \mid x \le 1\}, (-\infty, 1]$ **31.** $\{x \mid -1 \le x \le 0\}, [-1, 0]$ **33.** $\{x \mid 1 \le x \le \frac{4}{3}\}, [1, \frac{4}{3}]$

35. $\{x \mid -16 < x < 24\}, (-16, 24)$ **37.** $\left\{x \,\middle|\, \dfrac{1 - \sqrt{2}}{4} \le x \le \dfrac{1 + \sqrt{2}}{4}\right\}, \left[\dfrac{1 - \sqrt{2}}{4}, \dfrac{1 + \sqrt{2}}{4}\right]$ **39.** $\{x \mid 0 < x < 2\sqrt{2}\}, (0, 2\sqrt{2})$

41. $[0, 8]$ **43.** $(0, 1)$ **45.** $(-\infty, \frac{3}{2}] \cup [\frac{5}{2}, \infty)$ **47.** $[\frac{3}{2}, \frac{5}{2}]$ **49.** **R** **51.** $\varnothing$ **53.** $(-\infty, -\frac{7}{2}) \cup (\frac{5}{2}, \infty)$
55. $[1 - 2\sqrt{2}, 1 + 2\sqrt{2}]$

SECTION 1.8

1. $[1, 2]$ **3.** $(-\infty, 4] \cup [-\frac{1}{2}, \infty)$ **5.** $\varnothing$ **7.** $(-\infty, -\frac{1}{2}) \cup (\frac{1}{3}, \infty)$ **9.** $(\frac{3}{2}, 4)$ **11.** **R** **13.** $\varnothing$ **15.** $(-\infty, 1) \cup (5, \infty)$

17. $\{\frac{1}{3}\}$ **19.** $[5, 6]$ **21.** $\left[\dfrac{2 - \sqrt{10}}{6}, \dfrac{2 + \sqrt{10}}{6}\right]$ **23.** $(-\frac{1}{4}, 3)$ **25.** $(-\infty, 2) \cup [\frac{11}{2}, \infty)$ **27.** $(\frac{2}{3}, 1)$ **29.** $(-2, \infty)$

31. $(-\infty, -1) \cup (-1, -\frac{1}{2})$ **33.** $(-\infty, -8) \cup (\frac{4}{5}, 3) \cup (3, \infty)$ **35.** $(0, 1)$ **37.** $\left[-\sqrt{-1 + \sqrt{6}}, \sqrt{-1 + \sqrt{6}}\right]$
39. $(-\infty, -2] \cup [\frac{1}{2}, \frac{3}{2}]$ **41.** $(-\infty, 2)$

SECTION 1.9

1. each step requires inverse, associative, identity **3.** $-\dfrac{32}{63}$ **5.** $\dfrac{x^3 + y^3}{x^2y^2}$ **7.** no **9.** no **11. (a)** $\dfrac{22}{7} - \pi$ **(b)** 1

13. (a) 5 **(b)** $1 + 2\sqrt{2}$ **(c)** 1 **15.** $\{x \mid -2 < x \le -1\} \cup \{x \mid 1 \le x < 2\}$ **17.** $\dfrac{1}{4}$ **19.** $\dfrac{3a^4c^6}{bd^4}$ **21.** $\dfrac{1}{xy}$ **23.** $12\sqrt{2}$

25. 0 **27.** x **29.** $\dfrac{x^4}{x^6+1}$ **31.** 1 provided x and y are not both 0 **33.** $\frac{1}{3} \times 10^{-15/4}$ **35.** $-x^5 + 11x^3 - x^2 + x + 1$

37. $10x^4 - 6x^3 + 11x^2 - 1$ **39.** $2a^2 - 3b^2 + 4c^2 - ab + bc + 6ac$ **41.** $16y^4 - y^{-4}$ **43.** $2ac(2b^2c^2 - 3a + b)$

45. $2(x+1)(6x-5)$ **47.** $(5y-1)(25y^2+5y+1)$ **49.** $(5a+3b)^2$ **51.** $(4x^2+1)(3x^2-2)$ **53.** $(4\sqrt{y}+3)^2$ **55.** $\dfrac{3x-4}{x-1}$

57. $\dfrac{2(2x^2-2x-3)}{(x+1)(2x+1)}$ **59.** $\dfrac{(3y+2)(y-3)}{y+1}$ **61.** $\dfrac{8a^2+2a+3}{(2a+1)^3}$ **63.** $\dfrac{2u-3z}{-u+z}$ **65.** $\dfrac{4xy}{(x-y)^2(x+y)^2}$ **67.** $\dfrac{5+x+2x^2}{x(1+x^2)}$

69. $p(q(x)) = \dfrac{3x-10}{3}$ **71.** $p(q(x)) = |x|$ **73.** $p(q(x)) = \dfrac{(x^2-3)^2}{1-3(x^2-3)^2}$ **75.** $q(x) = \dfrac{x+7}{4}$ **77.** $x = -1, -\frac{3}{4}$ **79.** $x = \frac{5}{7}$

81. $x = \left(\dfrac{a}{(a+1)\sqrt{a}-1}\right)^2$ **83.** $x = -6, 2$ **85.** $t = \dfrac{-\sqrt{3} \pm \sqrt{7}}{2}$ **87.** $x = \dfrac{y}{2}, 2y$ **89.** $x = 0, x = \dfrac{2\sqrt{y}}{y^3}$ for $y > 0$

91. $x = \pm\dfrac{\sqrt{3}}{3}$ **93.** $t = \left(\dfrac{2 \pm \sqrt{2}}{2}\right)^6$ **95.** no real solutions **97.** $x = 0, 4$ **99.** $w = 3 \pm 2\sqrt{2}$

101. If $a = 0$, the solutions are all real numbers. If $a \neq 0$, then the only solution is $x = 0$. **103.** $\{x \mid x > 1\}$
105. $\{x \mid -\frac{5}{2} < x \le 1\} \cup \{x \mid -7 \le x < -\frac{7}{2}\}$ **107.** $\{x \mid -\frac{1}{2} < x \le 3\} \cup \{x \mid -9 \le x < -\frac{11}{2}\}$ **109.** $(-\infty, -1) \cup (\frac{1}{2}, \infty)$

111. $(-\infty, 4 - 3\sqrt{2}) \cup (4 + 3\sqrt{2}, \infty)$ **113.** $(\frac{2}{5}, \frac{1}{2})$ **115.** $\varnothing$ **117.** $\left[-\sqrt{2}, -\dfrac{\sqrt{2}}{2}\right] \cup \left[\dfrac{\sqrt{2}}{2}, \sqrt{2}\right]$

Chapter Two

SECTION 2.1

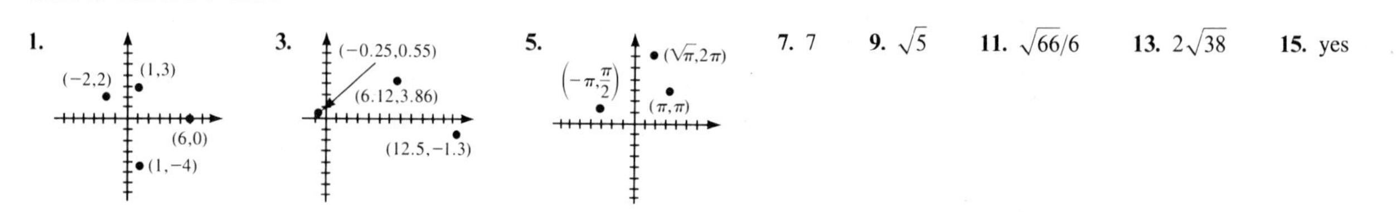

7. 7 **9.** $\sqrt{5}$ **11.** $\sqrt{66}/6$ **13.** $2\sqrt{38}$ **15.** yes

17. no **19.** yes **21.** $(-2, 5/2)$ **23.** $(1/3, 3/16)$ **25.** $(1 + \sqrt{2}, 3/2 - 2\sqrt{3})$ **27.** $(12, 9)$ **29.** $(2\sqrt{3} - \sqrt{2}, 3\pi)$ **31.** yes
33. yes **35.** no **37.** yes **39.** $x = 1/5, x = 1$

SECTION 2.2

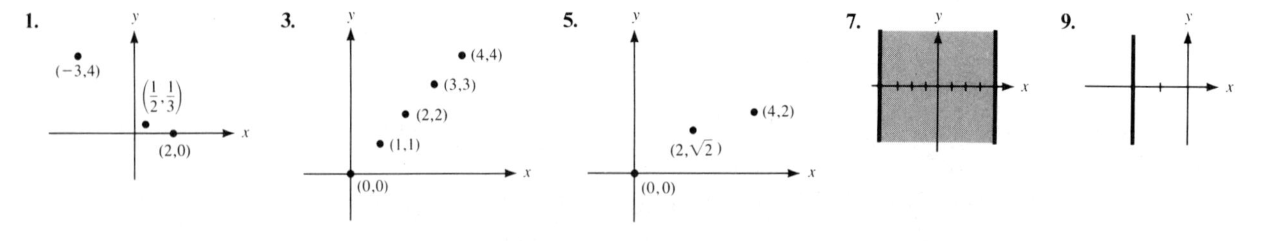

11. 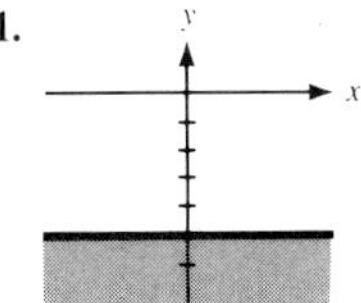

13. 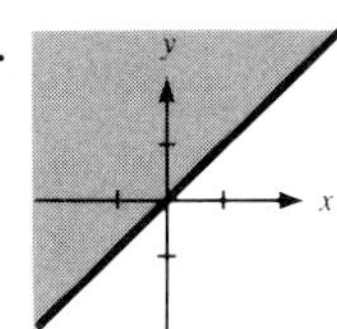

15. 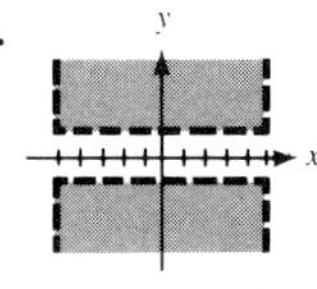

17. 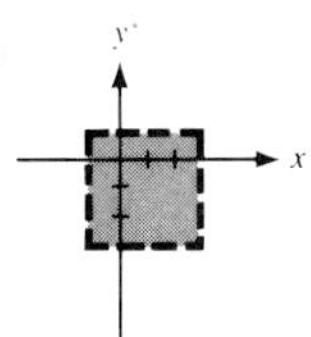

19. 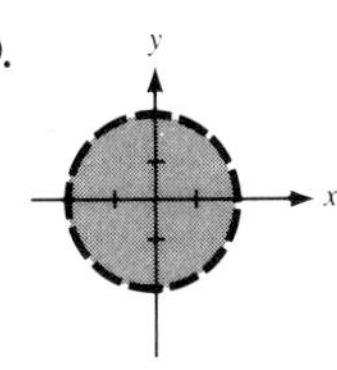

21. $y = x$ 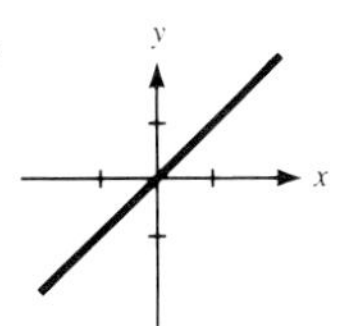

23. 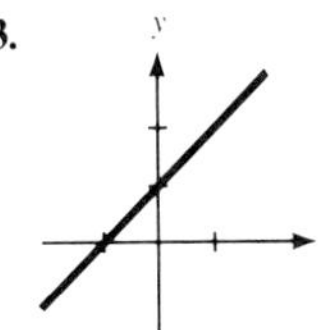

25. 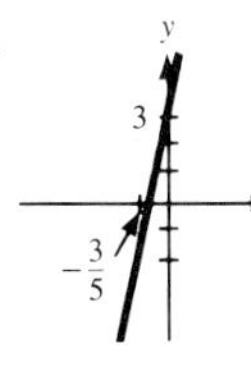

27. 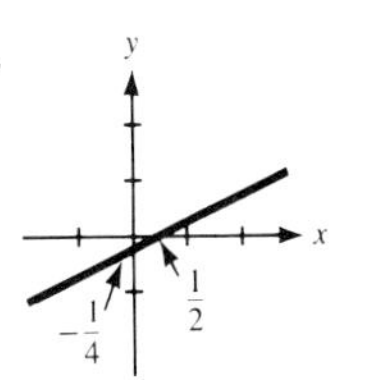

29. 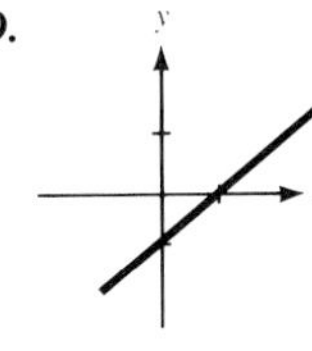

31. 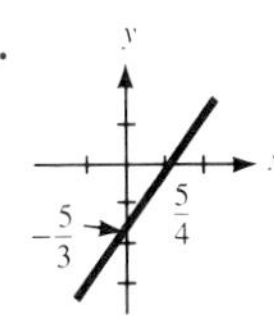

33. 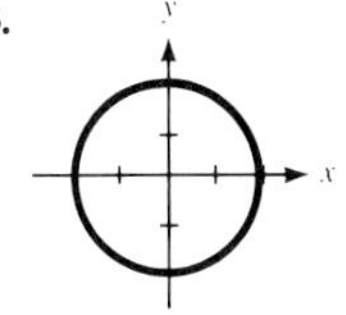

35. 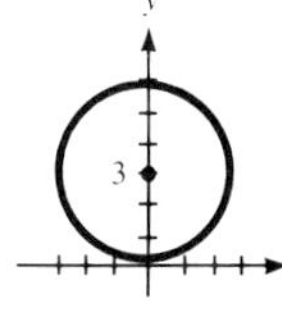

37. 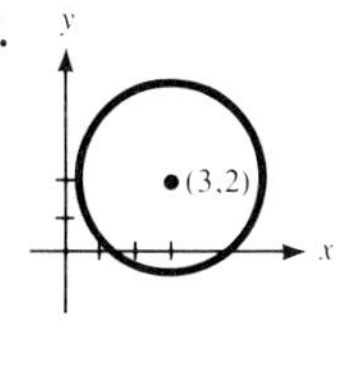

39. 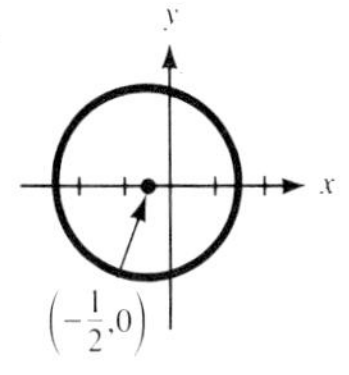

41. $(x-1)^2+(y-4)^2=4$ **43.** $(x-\sqrt{2})^2+(y+\sqrt{2})^2=4$ **45.** $(x+\sqrt{3})^2+(y+\sqrt{3})^2=6$ **47.** $(x+4)^2+(y-3)^2=106$
49. $(x+3)^2+(y+2)^2=9$ **51.** $(x-4)^2+(y-4)^2=16$ **53.** center: $(1,-1)$, $r=\sqrt{3}$ **55.** center: $(0,-3)$, $r=2$
57. center: $(0,0)$, $r=3/2$ **59.** center: $(\frac{1}{6},-\frac{1}{2})$, $r=\sqrt{10}/6$ **61.** center: $(-\frac{1}{2},\frac{1}{3})$, $r=\sqrt{2}$

SECTION 2.3

1. **(a)** -1 **(b)** -7 **(c)** $2\sqrt{2}-1$ **(d)** $2x+1$ **3.** **(a)** $1/2$ **(b)** $1/3$ **(c)** 0 **(d)** $\dfrac{x-1}{x}$

5. **(a)** 0 **(b)** 1 **(c)** 5 **(d)** $2|x|$ **7.** **(a)** 1 **(b)** 1 **(c)** 1 **(d)** $|t|+|t+1|$
9. **(a)** 1 **(b)** 4 **(c)** $-1/27$ **(d)** $\sqrt{2}/2$ **11.** **(a)** $7/2$ **(b)** 5 **(c)** $7/3$
13. **(a)** a^2 **(b)** a^2 **(c)** $-a^2$ **(d)** $a^2+2ab+b^2$ **(e)** a^2+b^2
15. **(a)** $3a^2+2a-1$ **(b)** $3a^2-2a-1$ **(c)** $3a^4+2a^2-1$ **(d)** $(3a^2+2a-1)^2$

17. **(a)** $\dfrac{1}{a(a+1)}$ **(b)** $\dfrac{a+1}{a^2}$ **(c)** $\dfrac{a^2+2a+1}{a+2}$ **(d)** $\dfrac{a^2+a+1}{a+1}$

19. **(a)** $\dfrac{2+x-x^2}{x^2}$ **(b)** $2x^2-11x+14$ **(c)** $2x^4+3x^2+x-1$ **(d)** $2x^4+4x^3+7x^2+5x+2$ **21.** 1 **23.** 3

25. $-4x+1-2h$ **27.** $\dfrac{-2}{(x+1)(x+h+1)}$ **29.** all real numbers **31.** $\{x|x \geq 1/2\}$ **33.** $\{x|x \neq 0\}$ **35.** $\{x|x \neq -3, 2\}$

37. $\{x|x \neq \frac{1}{4}(-1 \pm \sqrt{17})\}$ **39.** $A(s)=\sqrt{3}s^2/4$, $\{s|s>0\}$ **41.** $A(C)=C^2/4\pi$, $\{C|C>0\}$
43. **(a)** $N(p)=-1000p+105000$, $\{p|p \geq 5\}$ **(b)** $I(p)=-1000p^2+105000p$, $\{p|p \geq 5\}$

SECTION 2.4

1. yes; domain: $\{x|0 \leq x \leq 2\}$, range: $\{y|0 \leq y \leq 1\}$ **3.** no **5.** no **7.** no
9. **(a)** no values of x **(b)** $\{x|x \neq 3\}$ **(c)** $\{x|x \leq 1\} \cup \{x|x \geq 11/2\}$ **(d)** $\{x|1 \leq x < 3\} \cup \{x|3 < x \leq 11/2\}$
11. **(a)** $x=3, 0, -3$ **(b)** $x=5$ **(c)** $\{x|-3<x<0\} \cup \{x|x>3\}$ **(d)** $\{x|x \geq 5\}$

13.
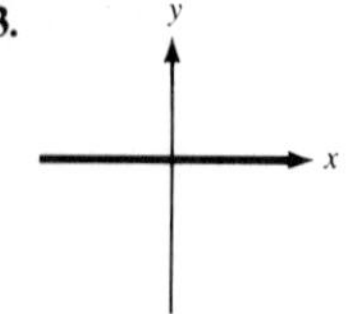
domain: all reals

15.
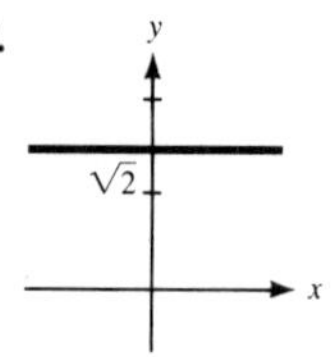

domain: all reals

17.
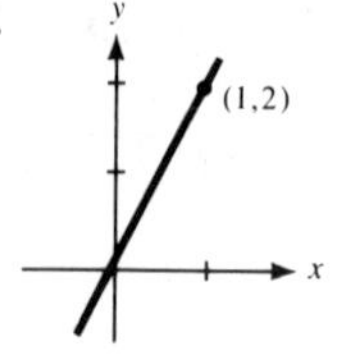

domain: all reals

19.
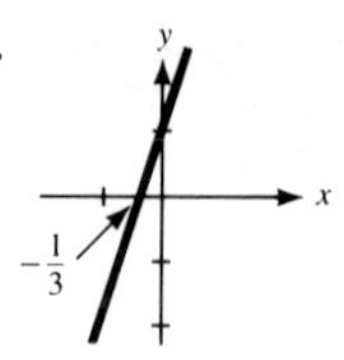

domain: all reals

21.
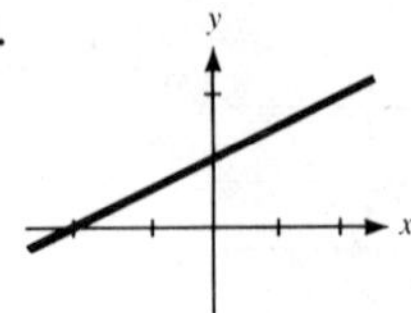
domain: all reals

23.
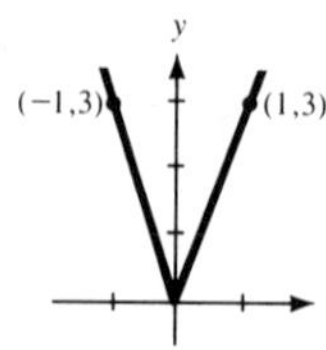

domain: all reals

25.
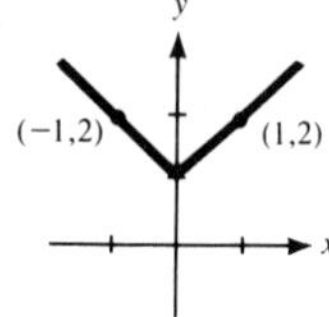

domain: all reals

27.
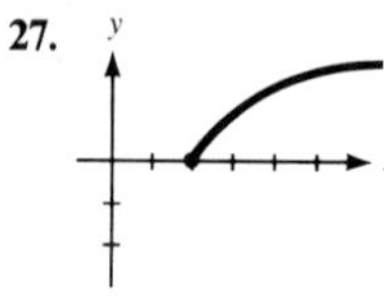
domain: $\{x|x \geq 2\}$

29.
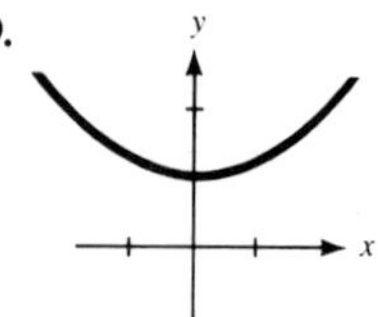
domain: all reals

31.
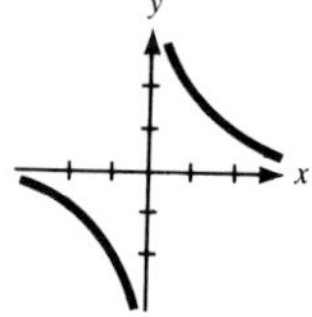
domain: $\{x|x \neq 0\}$

33.
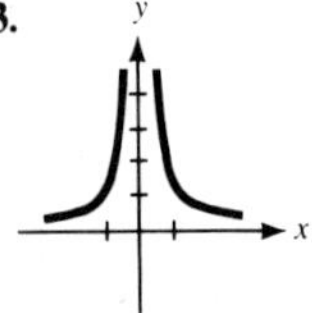
domain: $\{x|x \neq 0\}$

35.
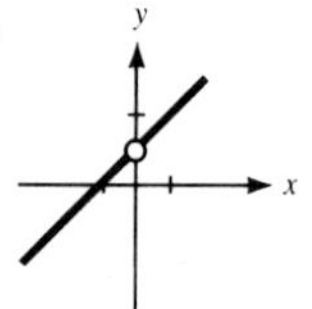
domain: $\{x|x \neq 0\}$

37.
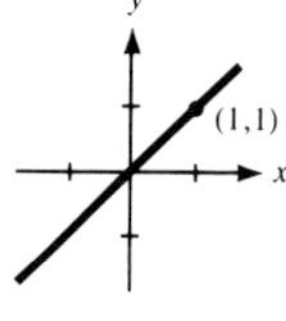

domain: all reals

39.
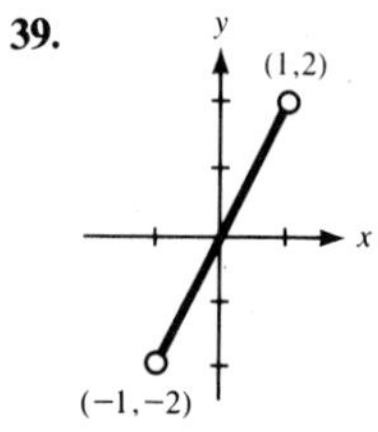

41.
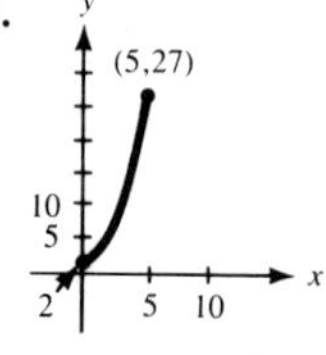

43.
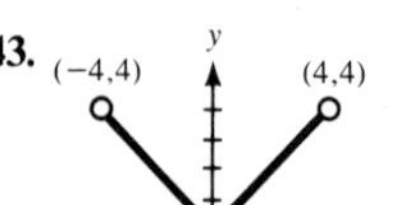

45.
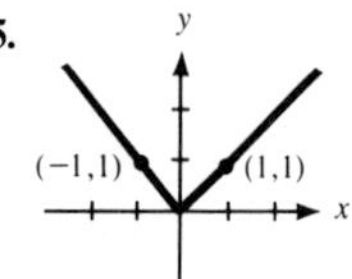

f is the absolute value function

47.
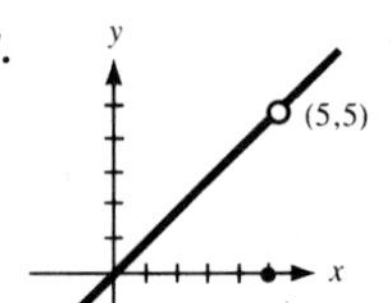

49.
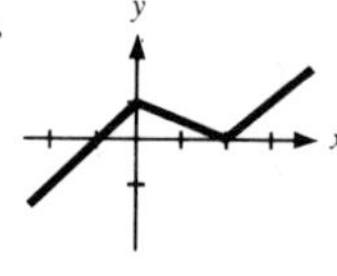

51. **(a)** 1 **(b)** −12 **(c)** 0 **(d)** −1 **(e)** 5 **(f)** −3

SECTION 2.5

1. odd **3.** neither **5.** even **7.** neither **9.** neither **11.** odd **13.** $f(x) = x^3$, $g(x) = x^3 + 2$

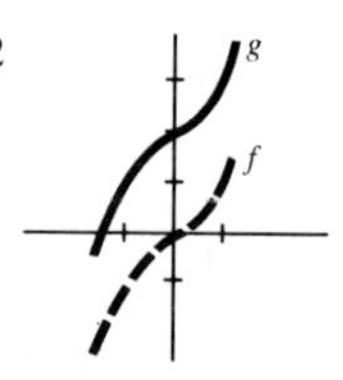

15. $g(x) = (x + 5)^3$, $f(x) = x^3$

g f

17. $G(x) = \frac{1}{2}x^3$, $f(x) = x^3$

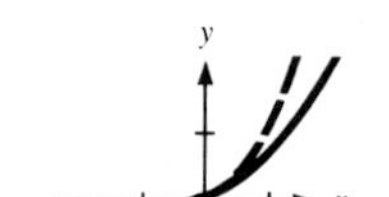

19. $h(x) = (x - 3)^3 + 4$, $f(x) = x^3$

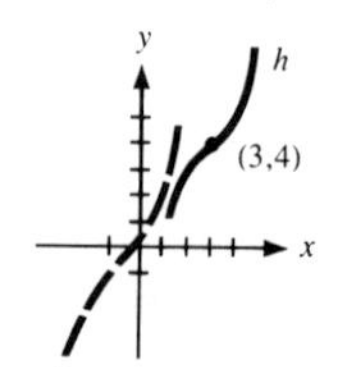

21. $g(x) = 1 - x^3$, $f(x) = x^3$

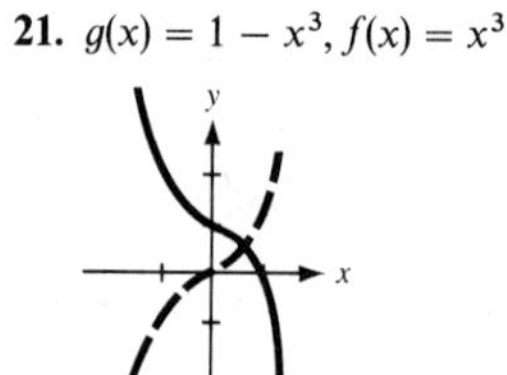

23. $h(x) = |x + \pi|$, $f(x) = |x|$ **25.** $F(x) = |x| + 7$, $f(x) = |x|$

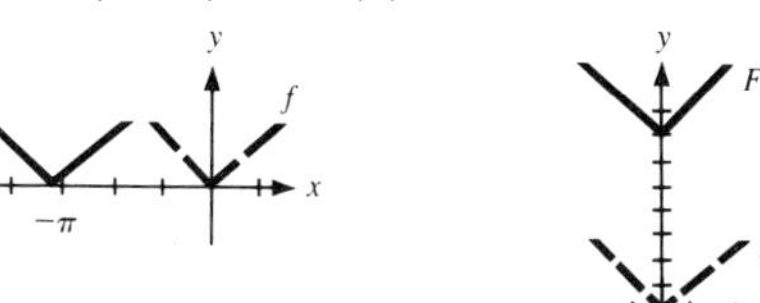

27. $k(x) = -|x|$, $f(x) = |x|$

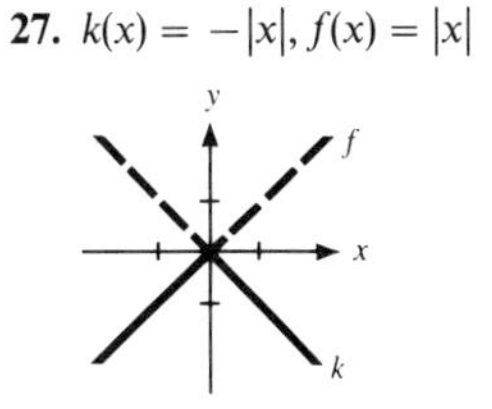

29. $g(x) = |x - 3| + 1$, $f(x) = |x|$

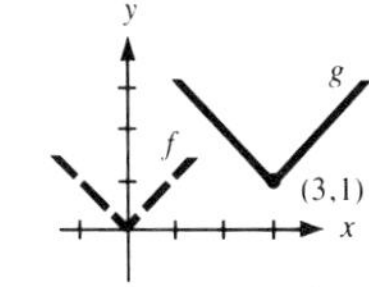
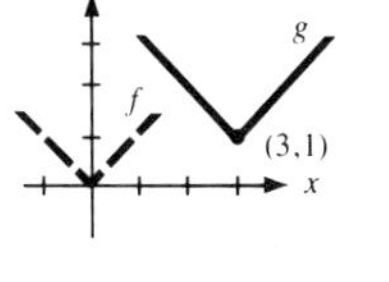

31. $g(x) = \sqrt{x} + 7$, $f(x) = \sqrt{x}$ **33.** $h(x) = \sqrt{x - 5}$, $f(x) = \sqrt{x}$

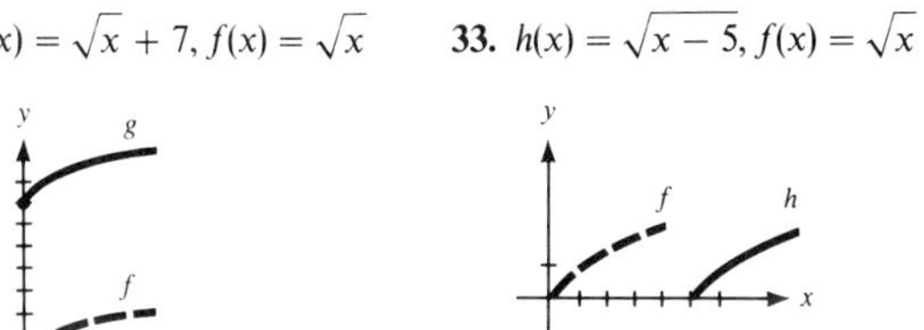

35. $H(x) = -\sqrt{x}$, $f(x) = \sqrt{x}$

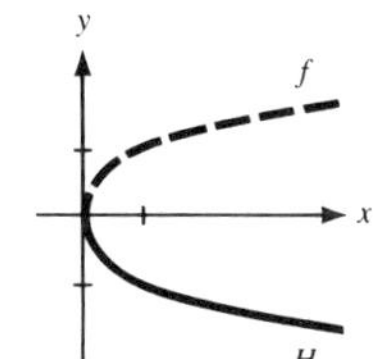

37. $g(x) = -\sqrt{\frac{x}{2} + 1}$, $f(x) = \sqrt{x}$

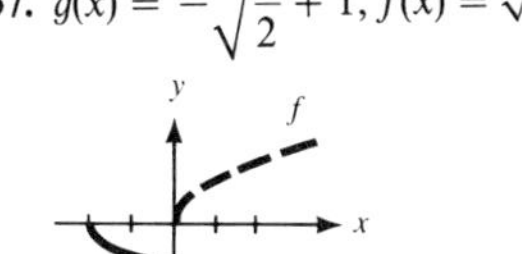

39. $g(x) = \frac{1}{x + 1}$, $f(x) = \frac{1}{x}$

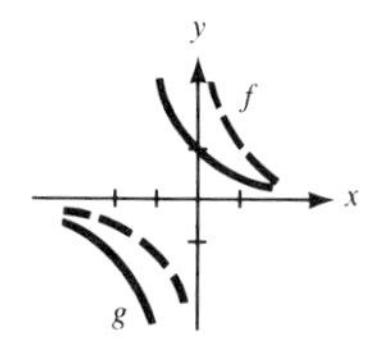

41. $h(x) = \frac{2}{x}$, $f(x) = \frac{1}{x}$ **43.** $A(x) = \frac{1}{x} + 1$, $f(x) = \frac{1}{x}$

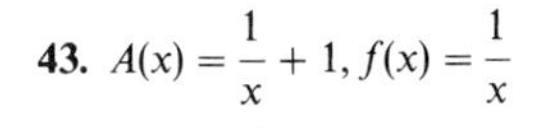

45.

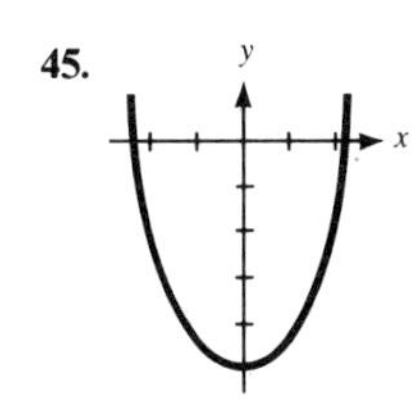

47.

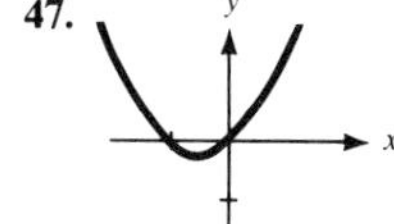

49.

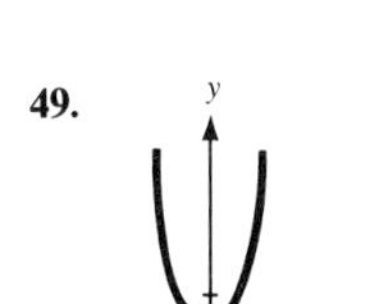

51.

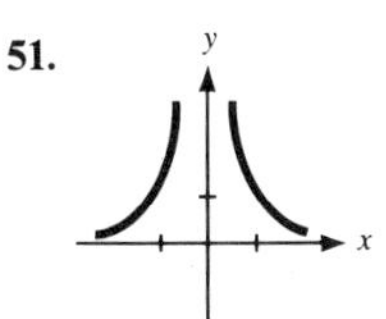

53.

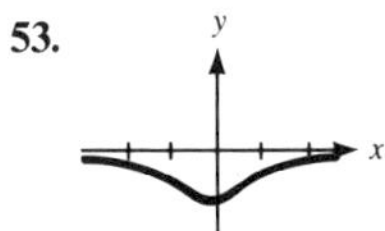

55.

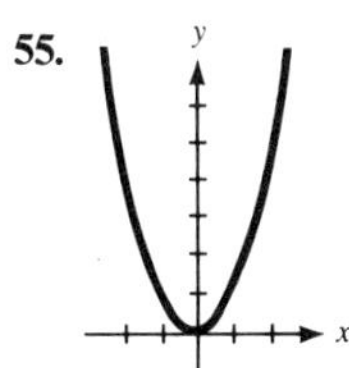

57.

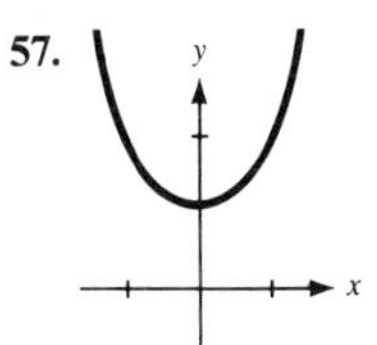

59.

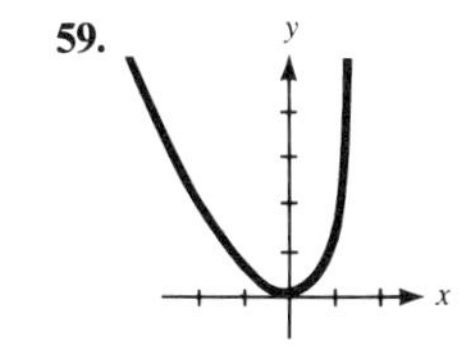

61.

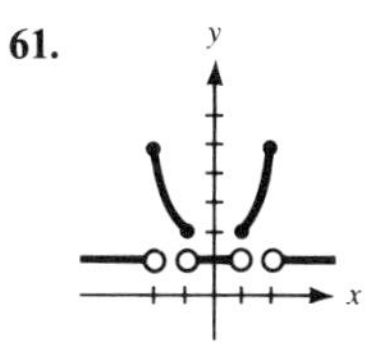

63.

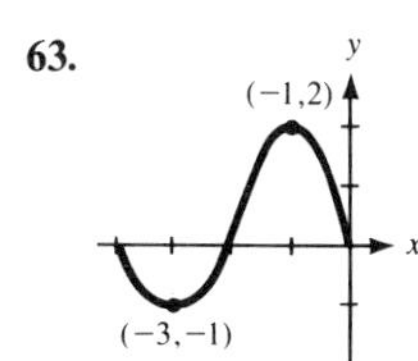

65.

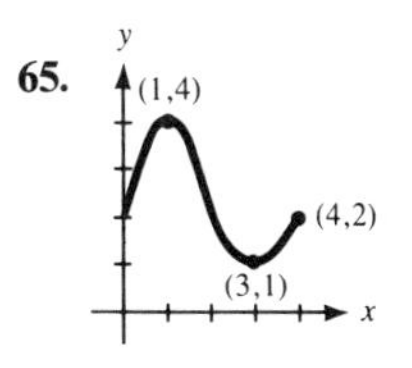

67.

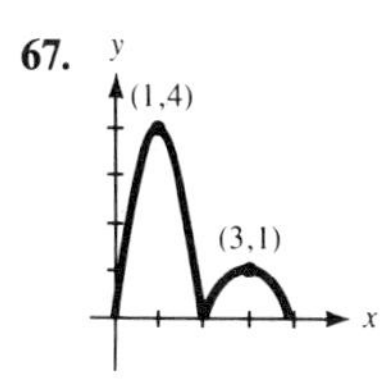

69. Even when n is even, odd when n is odd **71.** Only the zero function $f(x) = 0$

SECTION 2.6

1. increasing: $(-\infty, 0]$, $[1, \infty)$
decreasing: $[0, 1]$
not one-to-one

3. decreasing: $(0, \infty)$
one-to-one

5. increasing: $(-\infty, 0)$
decreasing: $(0, \infty)$
not one-to-one

7. increasing: $[0, \pi/4]$, $[3\pi/4, 5\pi/4]$
decreasing: $[\pi/4, 3\pi/4]$, $[5\pi/4, 3\pi/2]$
not one-to-one

9. increasing: $(-\infty, \infty)$
one-to-one

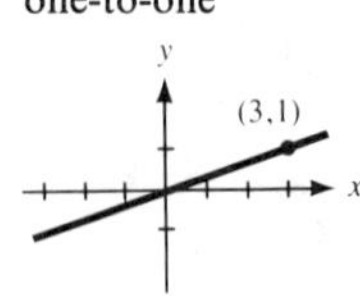

11. increasing: $(-\infty, \infty)$
one-to-one

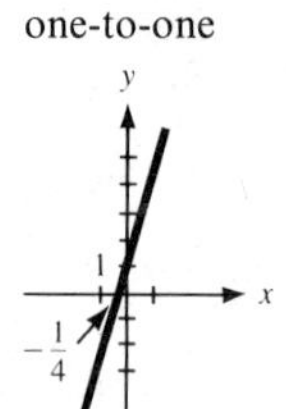

13. increasing: $[0, \infty)$
decreasing: $(-\infty, 0]$
not one-to-one

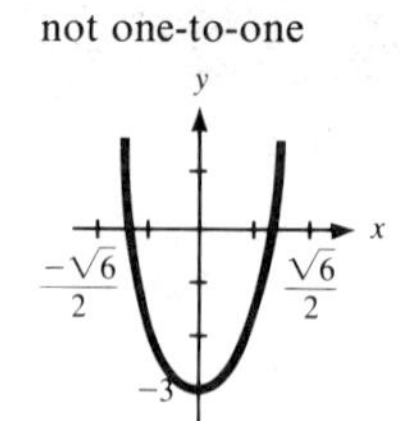

15. increasing: $(-\infty, \infty)$
one-to-one

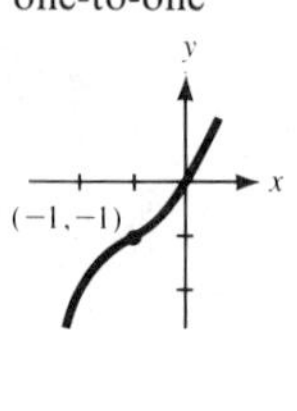

17. decreasing: $(-\infty, 0)$, $(0, \infty)$
one-to-one

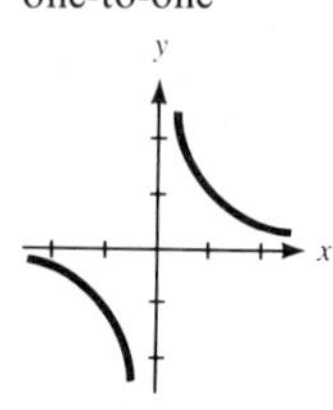

19. decreasing: $[0, \infty)$
one-to-one

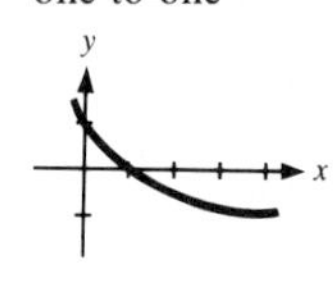

21. increasing: $[3/2, \infty)$
one-to-one

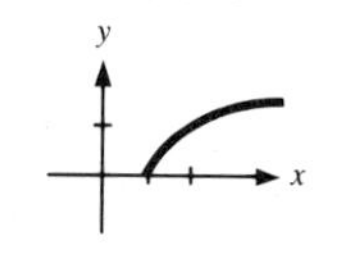

23. increasing: $(-\infty, 0]$
decreasing: $[0, \infty)$
not one-to-one

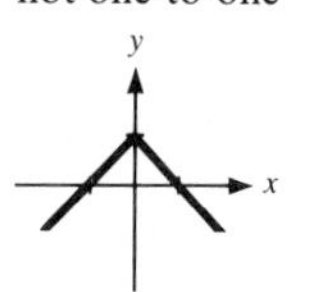

25. increasing: $[3, \infty)$
decreasing: $(-\infty, 3]$
not one-to-one

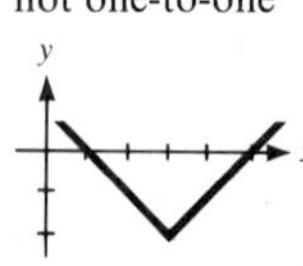

27. constant: $(-\infty, 0)$, $(0, \infty)$
not one-to-one

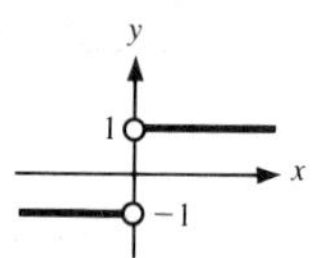

29. increasing: $[3, \infty)$
decreasing: $(-\infty, 2)$
constant: $[2, 3)$
not one-to-one

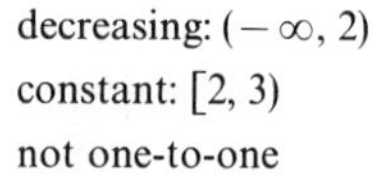

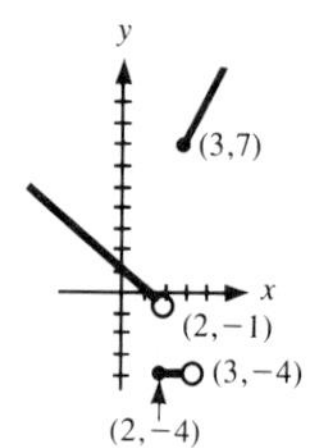

31. increasing: $[0, \infty)$
decreasing: $(-\infty, 0]$
not one-to-one

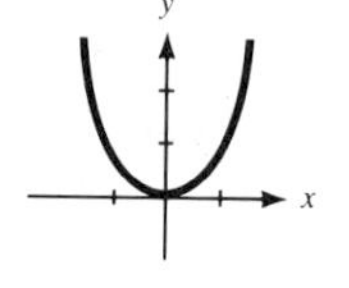

33. constant: $(-\infty, 0)$, $[0, 1)$, $[1, \infty)$
not one-to-one

49. n is odd

51. **(a)** **R** **(b)** no **(c)** yes

SECTION 2.7

1. **(a)** 2; all reals **(b)** $-2x$; all reals **(c)** $1 - x^2$; all reals **(d)** $\dfrac{1-x}{1+x}$; $x \neq -1$

3. **(a)** $3x^2 + 4x - 4$; all reals **(b)** $-x^2 - 2$; all reals **(c)** $2x^4 + 6x^3 - 3x^2 - 8x + 3$; all reals

(d) $\dfrac{x^2 + 2x - 3}{2x^2 + 2x - 1}$; $x \neq (-1 \pm \sqrt{3})/2$

5. **(a)** $\dfrac{1+x^2}{x}$; $x \neq 0$ **(b)** $\dfrac{1-x^2}{x}$; $x \neq 0$ **(c)** 1; $x \neq 0$ **(d)** $1/x^2$; $x \neq 0$

7. **(a)** $\dfrac{1+x^2}{x^2+x}$; $x \neq 0, -1$ **(b)** $\dfrac{1-x^2}{x^2+x}$; $x \neq 0, -1$ **(c)** $\dfrac{1}{(x+1)^2}$; $x \neq 0, -1$ **(d)** $\dfrac{1}{x^2}$; $x \neq 0, -1$

9. **(a)** $x^3 + x^{1/3}$; all reals **(b)** $x^3 - x^{1/3}$; all reals **(c)** $x^{10/3}$; all reals **(d)** $x^{8/3}$; $x \neq 0$

11. **(a)** $2\sqrt{2x+1}$; $x \geq -1/2$ **(b)** 0; $x \geq -1/2$ **(c)** $|2x+1|$; $x \geq -1/2$ **(d)** 1; $x > -1/2$

13. **(a)** 0; $x \geq 0$ **(b)** $2\sqrt{x}$; $x \geq 0$ **(c)** $-|x|$; $x \geq 0$ **(d)** -1; $x > 0$

15. **(a)** $\dfrac{2(x^2+1)}{x^2-1}$; $x \neq 1, -1$ **(b)** $\dfrac{-4x}{x^2-1}$; $x \neq 1, -1$ **(c)** 1; $x \neq 1, -1$ **(d)** $\dfrac{(x-1)^2}{(x+1)^2}$; $x \neq 1, -1$

17. $f^2(x) = x^2(x^2+1)$; $f^3(x) = x^3(x^2+1)\sqrt{x^2+1}$; $f^4(x) = x^4(x^2+1)^2$

19. 101 **21.** 1 **23.** -5 **25.** $\dfrac{20+\sqrt{5}}{395}$ **27.** **(a)** x; all reals **(b)** x; all reals **(c)** $x+2$; all reals

29. **(a)** $4x^2 - 12x + 10$; all reals **(b)** $2x^2 - 1$; all reals **(c)** $2x^4 + 2x^2 + 2$; all reals

31. **(a)** $\dfrac{1}{x^2+1}$; all reals **(b)** $\dfrac{1+x^2}{x^2}$; $x \neq 0$ **(c)** x; $x \neq 0$ **33.** **(a), (b), (c)** x; $x \neq 1$

35. **(a)** $|x|$; all reals **(b)** x; $x \geq 0$ **(c)** $\sqrt[4]{x}$; $x \geq 0$ **37.** **(a)** $|x+1|$; all reals **(b)** $|x| + 1$; all reals **(c)** $|x|$; all reals

39. **(a), (b), (c)** $x^{1/4}$; $x \geq 0$ **41.** **(a), (b), (c)** x; $x \neq 0$ **43.** **(a), (b)** $\dfrac{2-x^2}{x^2}$; yes **45.** $g(x) = |x|$; $f(x) = x - 1$

47. $g(x) = \sqrt{1-x}$; $f(x) = \sqrt{x+5}$ **49.** $g(x) = x^2$; $f(x) = \dfrac{x-1}{x+1}$ **51.** $g(x) = |x| - \sqrt{x}$; $f(x) = x + 1$

53. The sum $(A_1 + A_2)(t) = 1500 + 0.22t$ **55.** The product $(WL)(t) = 20 - 8t + 31t^2 + 2t^3 + 3t^4$

57. **(a)** The size of the barracuda population is the composition $(f \circ g)(x) = 1000(1 + 2\sqrt{2}) + \sqrt{2x}$ where x is the size of the shrimp population. **(b)** $(f \circ g)(1000000) = 1000(1 + 3\sqrt{2})$ barracuda

SECTION 2.8

9. $f^{-1}(x) = \dfrac{2-x}{5}$ **11.** $f^{-1}(x) = \sqrt[3]{x+3}$ **13.** $f^{-1}(x) = \dfrac{2x+1}{7x}$ **15.** $f^{-1}(x) = (x+1)^{-1/5}$ **17.** $f^{-1}(x) = \dfrac{x^2+3}{2}$, $x \geq 0$

19. $f^{-1}(x) = \dfrac{x+1}{x-1}$ **21.** $f^{-1}(x) = (x-1)^2$, $x \geq 1$ **23.** $f^{-1}(x) = \dfrac{x+5}{4}$ **25.** $f^{-1}(x) = \sqrt{x} + 3$ **27.** $f^{-1}(x) = -\dfrac{1}{x}$

29. $f^{-1}(x) = \sqrt[3]{\dfrac{x-1}{2}}$ **31.** $f^{-1}(x) = x^3 - 8$ **35.** $\dfrac{b - dx}{cx - a}$

SECTION 2.9 (REVIEW)

1. $\sqrt{289}$ **3.** $|a|\sqrt{a^2+1}$ **5.** $(1/2, 5/2)$ **7.** $(9, 7)$ **11.** **13.**

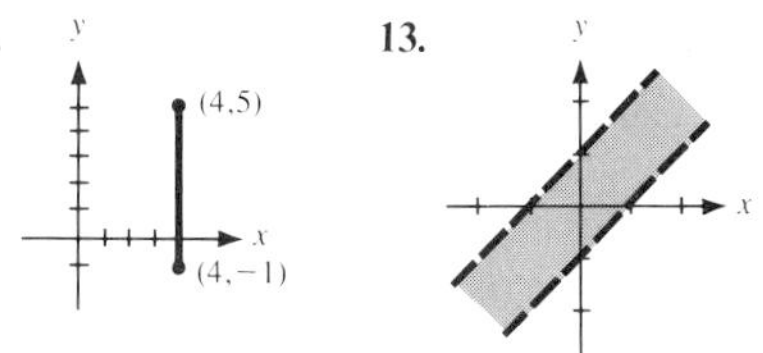

15.

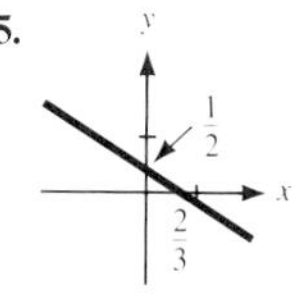

17.

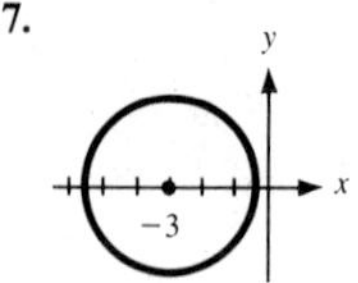

19.

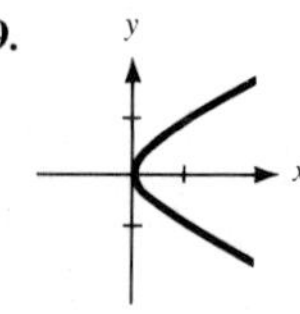

21.

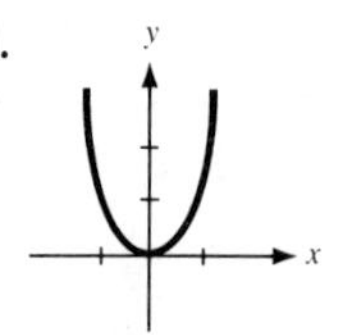

23.

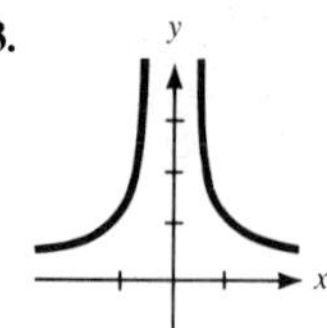

25. $(x + 5)^2 + (y - 5)^2 = 25$

27. **(a)** 1 **(b)** 1/16 **(c)** $2^{-\sqrt{2}}$ **(d)** $x^{-2|x|}$ **31.** $\{x | 0 \le x \le 2\}$ **33.** $\{y | y \ge 1\}$ **35.** even **37.** neither

39. neither **41.** **(a)** $x = -3, 2, 4$ **(b)** $(-\infty, -4] \cup \{3\}$ **(c)** $(-3, 2) \cup (4, \infty)$ **43.**

45.

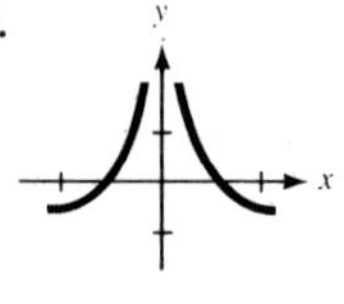

47.

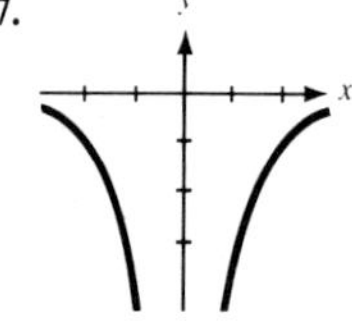

49. increasing: $[0, \infty)$
decreasing: $(-\infty, 0]$

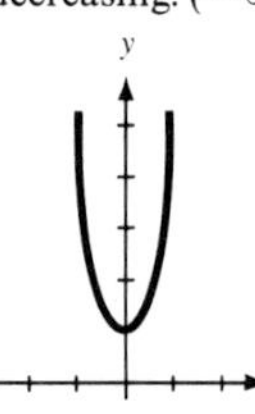

51. increasing: $[-1/2, \infty)$

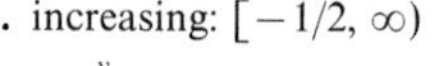

53. decreasing: $(-\infty, \infty)$

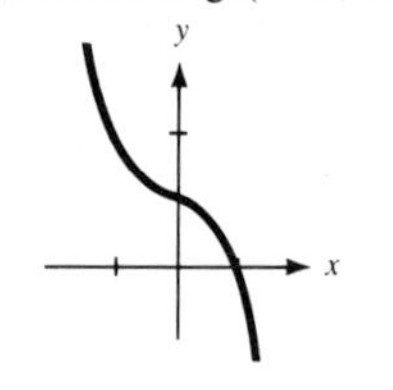

55. decreasing: $(-\infty, 0)$, $[0, \infty)$

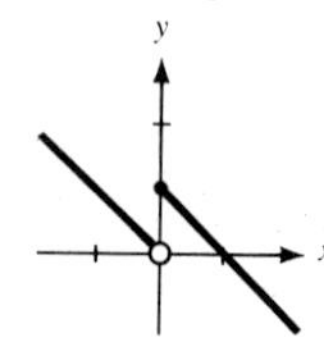

57. increasing on no interval, decreasing on no interval, constant on no interval

59. one-to-one

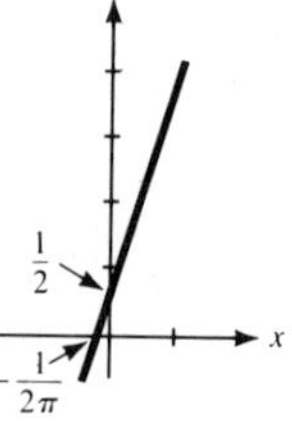

61. one-to-one

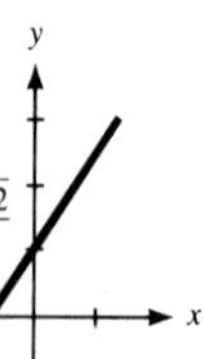

63. not one-to-one

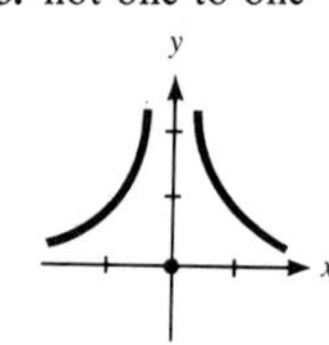

65. one-to-one

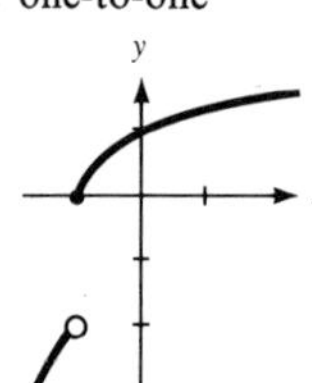

67. not one-to-one **69.** one-to-one

71. not one-to-one **73.** n odd, $a \ne b$ **75.** $(f + g)(x) = x^3 - 2x + 3\sqrt{x}$, $x \ge 0$; $(f - g)(x) = -x^3 + 2x^2 - 4x - \sqrt{x}$, $x \ge 0$

77. $(f + g)(x) = 1$, $x \ne -1$; $(f - g)(x) = \dfrac{x - 1}{x + 1}$, $x \ne -1$ **79.** $(fg)(x) = |(t + 1)^2(t - 1)|$, all reals; $\left(\dfrac{f}{g}\right)(x) = \dfrac{1}{|t - 1|}$, $t \ne 1, -1$

81. $(f \circ g)(x) = 9x^2 + 66x + 110$; all reals **83.** $(f \circ g)(t) = \sqrt{2t}/2$; $t \ge 0$ **85.** $g(x) = \sqrt[3]{x}$; $f(x) = x^2 + x$

87. $g(y) = \sqrt{y^2 - y - 1}$; $f(y) = \sqrt{y + 1}$ **89.** **(a), (b)** $\left(x + \dfrac{1}{x}\right)^2 + \left(x + \dfrac{1}{x}\right) + \dfrac{x}{x^2 + x + 1}$; the distributive property

91. $f^{-1}(x) = x^2/2$, $x \ge 0$ **93.** $f^{-1}(t) = \sqrt[5]{\dfrac{t + 1}{4 - 3t}}$ **95.** $g^{-1}(x) = \sqrt[n]{\dfrac{x + 1}{a - bx}}$

Chapter Three

SECTION 3.1

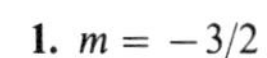

1. $m = -3/2$

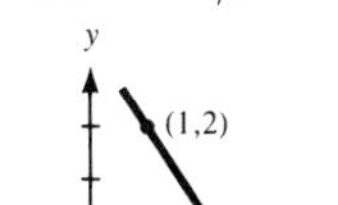

3. slope undefined

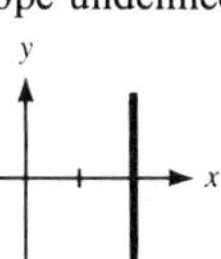

5. $m = 1/2$

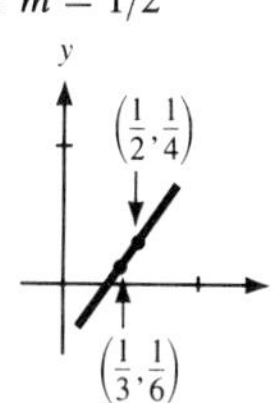

7. $m = -\sqrt{6}$

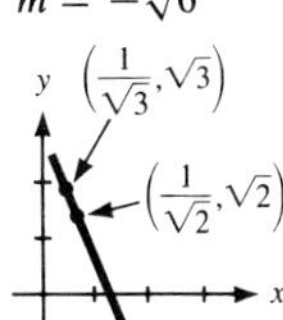
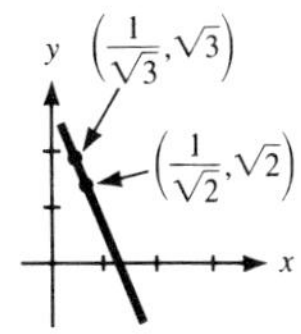

9. $m = b/a,\ a \neq 0,\ m \neq -1$

11. 1, −1, perpendicular

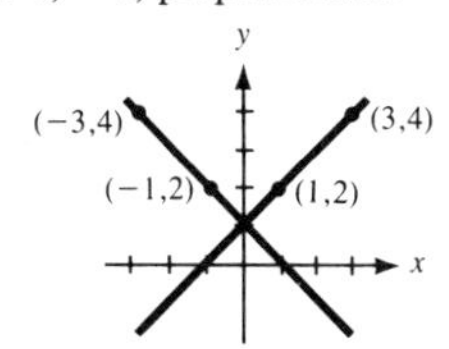

13. 1/3, −1/2, neither

15. 22, −22, neither

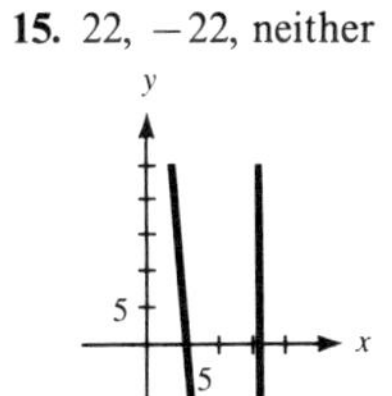

17. −1, −1, parallel

19.

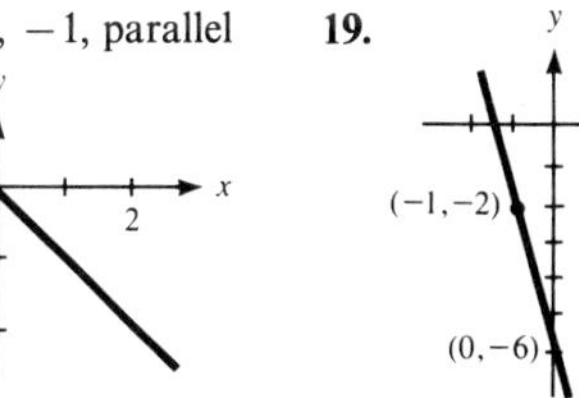

21.

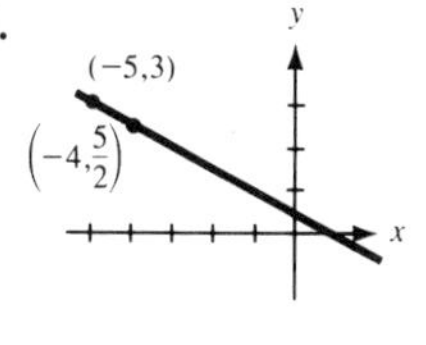

23.

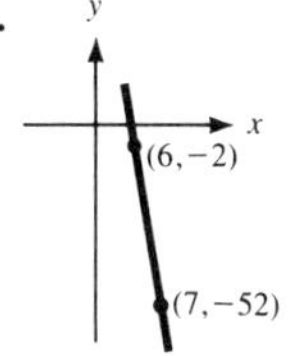

25.

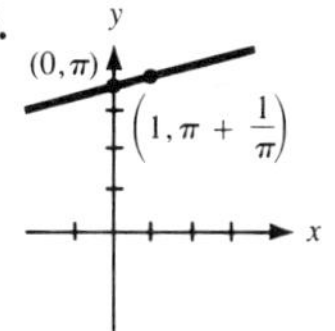

27.

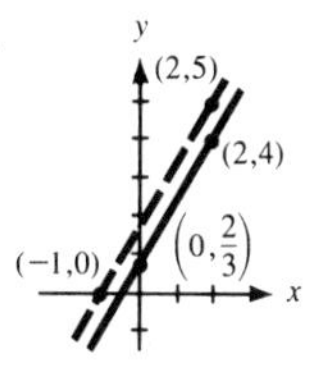

29.

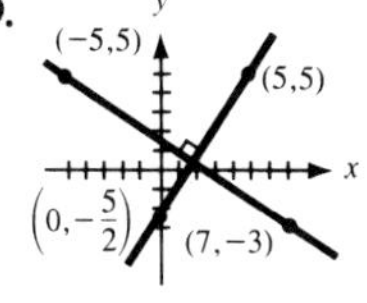

31. No. No two slopes are negative reciprocals of each other. **33.** A square

SECTION 3.2

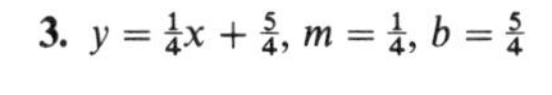

1. $y = 3x - 4,\ m = 3,\ b = -4$ **3.** $y = \frac{1}{4}x + \frac{5}{4},\ m = \frac{1}{4},\ b = \frac{5}{4}$ **5.** $y = \frac{2}{3}x - 2,\ m = \frac{2}{3},\ b = -2$ **7.** $y = x - 1,\ m = 1,\ b = -1$

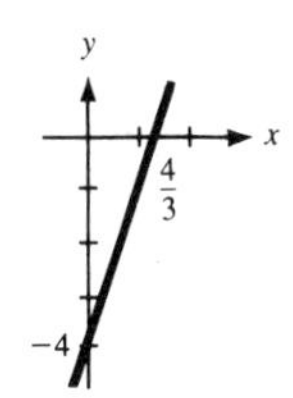

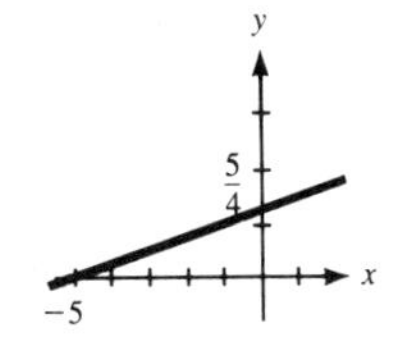

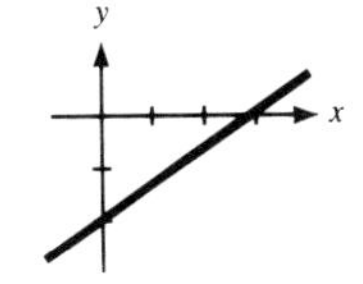

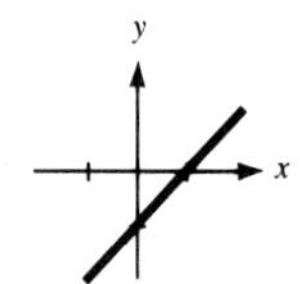

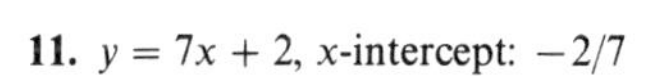

9. $y = \frac{1}{4}x + \frac{1}{4},\ m = \frac{1}{4},\ b = \frac{1}{4}$ **11.** $y = 7x + 2$, x-intercept: $-2/7$ **13.** $y = \sqrt{2}x + 1/2$, x-intercept: $-\sqrt{2}/4$

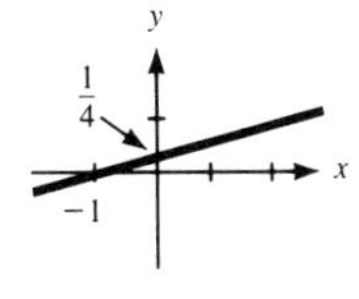

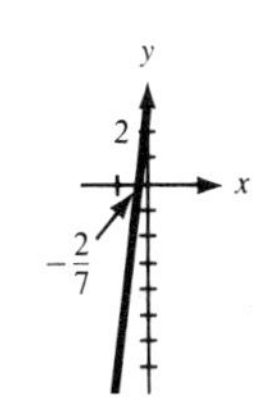

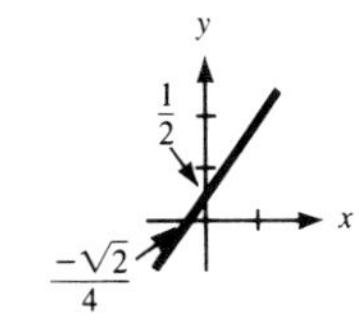

15. $y - 2 = 6(x - 1)$, $b = -4$ **17.** $y - 3 = \frac{1}{2}x$, $b = 3$ **19.** $y = -\frac{2}{3}(x + 4)$, $b = -\frac{8}{3}$ **21.** $y - a = (1/a)(x - a)$, $b = a - 1$

23. $y = -\frac{3}{2}x + \frac{7}{2}$, $m = -\frac{3}{2}$, $b = \frac{7}{2}$ **25.** $y = \frac{4}{3}x - \frac{1}{3}$, $m = \frac{4}{3}$, $b = -\frac{1}{3}$ **27.** $y = -(11/9)x + 109/9$, $m = -11/9$, $b = 109/9$

29. $y = -x + a + b$, $m = -1$, y-intercept $= a + b$ **33.** $\dfrac{x}{-\frac{3}{2}} + \dfrac{y}{3} = 1$ **35.** $\dfrac{x}{-6} + \dfrac{y}{3} = 1$

SECTION 3.3

1. $y = (x - 1)^2 + 2$, $(1, 2)$, minimum, opens up **3.** $y = (x - 3)^2$, $(3, 0)$, minimum, opens up

5. $y = 2(x - 3/4)^2 + 23/8$, $(3/4, 23/8)$, minimum, opens up **7.** $y = 3(x + 1/6)^2 - 1/12$, $(-1/6, -1/12)$, minimum, opens up

9. $y = -(3/2)x^2 + 2$, $(0, 2)$, maximum, opens down **11.** $y = (1/2)(x + 1/2)^2 - 5/8$, $(-1/2, -5/8)$, minimum, opens up

13. $y = \sqrt{2}(x - \frac{1}{2})^2 + \dfrac{4 - \sqrt{2}}{4}$, $\left(1/2, \dfrac{4 - \sqrt{2}}{4}\right)$, minimum, opens up

15. $y = -\sqrt{2}\left(x - \dfrac{\sqrt{2}}{4}\right)^2 + \dfrac{16 + \sqrt{2}}{8}$, $\left(\dfrac{\sqrt{2}}{4}, \dfrac{16 + \sqrt{2}}{8}\right)$, maximum, opens down

17. $y = \dfrac{1}{\pi}\left(x + \dfrac{1}{2}\right)^2 + \dfrac{4\pi - 1}{4\pi}$, $\left(-\dfrac{1}{2}, \dfrac{4\pi - 1}{4\pi}\right)$, minimum, opens up

19. $y = \dfrac{49}{4}\left(x - \dfrac{1}{7}\right)^2$, $\left(\dfrac{1}{7}, 0\right)$, minimum, opens up

21.
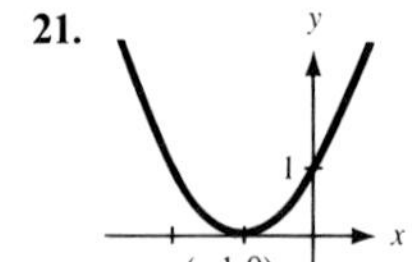

23.
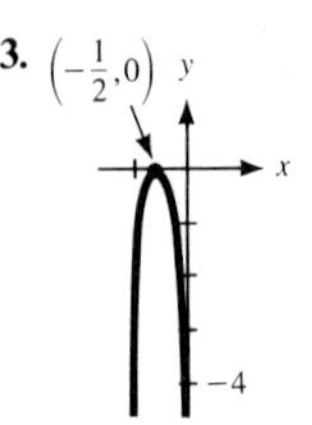

25.
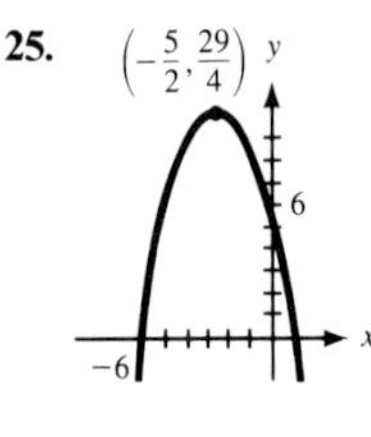

27.
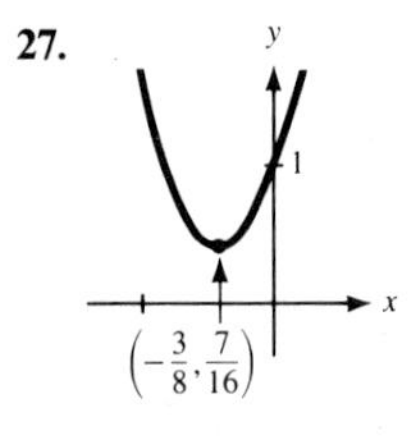

29.
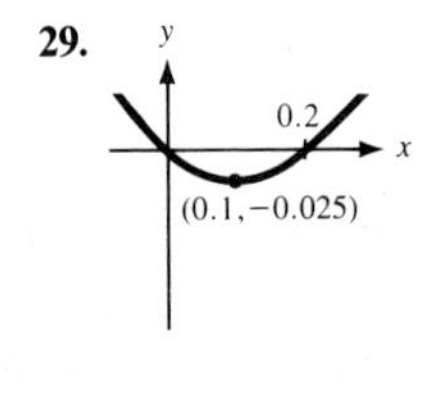

31.
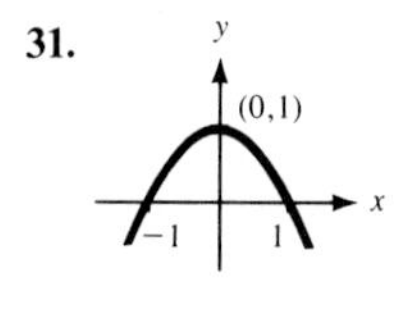

33.
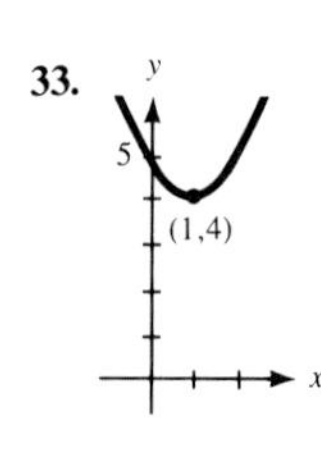

35.
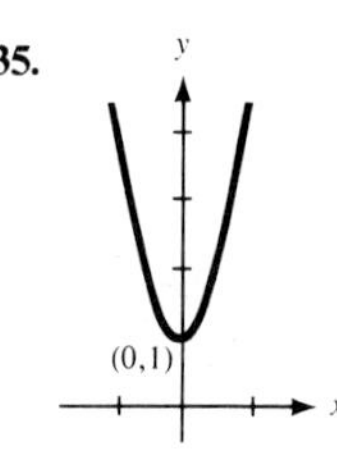

37.
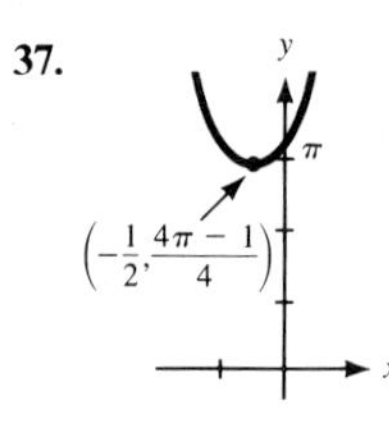

39.
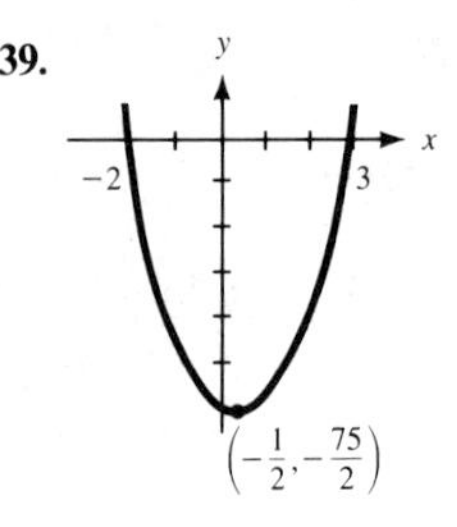

41. $k = 14, -2$ **43.** $\dfrac{3 \pm \sqrt{3}}{2}$

SECTION 3.4

1. $[-2, 2]$ **3.** $\varnothing$ **5.** $(-\infty, -2] \cup [3, \infty)$ **7.** $(-\infty, -1/2] \cup [1/3, \infty)$ **9.** $(-\infty, \infty)$ **11.** $\{3\}$

13. $(-\infty, -1) \cup (0, \infty)$ **15.** $[-3/2, 4]$ **17.** $(-\infty, \infty)$ **19.** $\left(-\infty, \dfrac{5 - \sqrt{17}}{2}\right] \cup \left[\dfrac{5 + \sqrt{17}}{2}, \infty\right)$ **21.** 5/2 ft. by 5 ft.

23. $\dfrac{4}{2 + \pi}$ m. wide, 1 m. high **25.** 200 computers **27.** 16 feet high, return to ground at $t = 2$ sec.

29. \$7.25, \$210.25 **31.** **(a)** $R(x) = 350x - 5x^2$ **(b)**

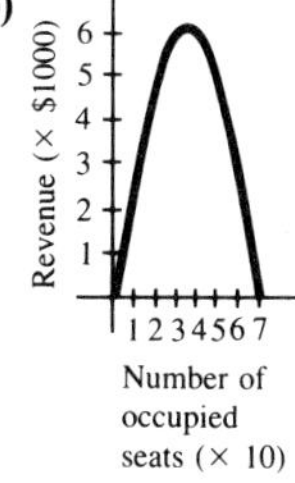

(c) 35 seats **33.** $x = 1/2$

35. Maximum area $= P/4$ when the rectangle is a square. **37.** length = width = 10/3 ft

39. $x = \dfrac{8\pi}{\pi + 4}$, $8 - x = \dfrac{32}{\pi + 4}$, minimum area is $\dfrac{16}{\pi + 4}$ sq. in.

SECTION 3.5

1.

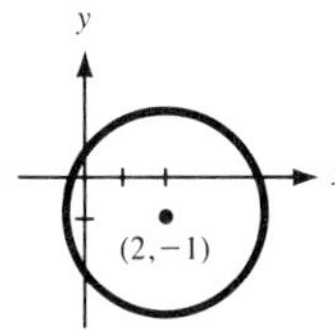

3.

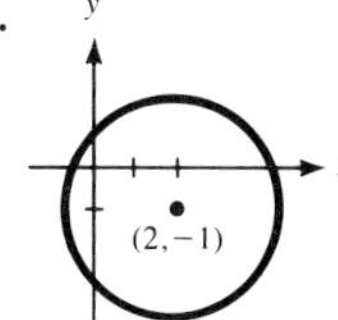

5.

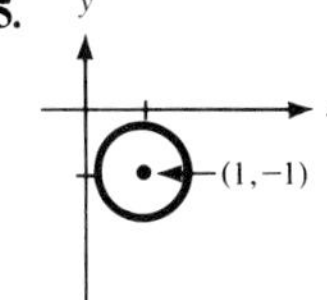

7. There is no graph since the equation is equivalent to $5(x - 1/5)^2 + 5(y - 1/10)^2 = -479/100$, which has no solutions.

9.

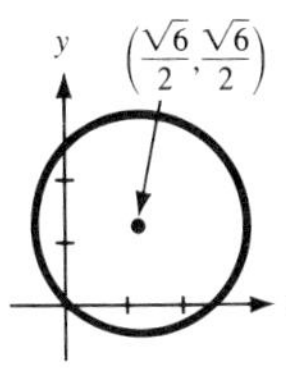

11.

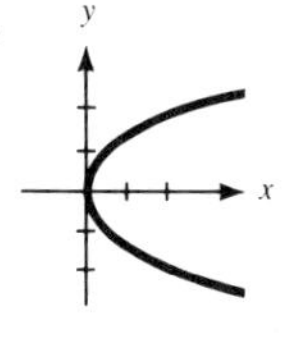

13.

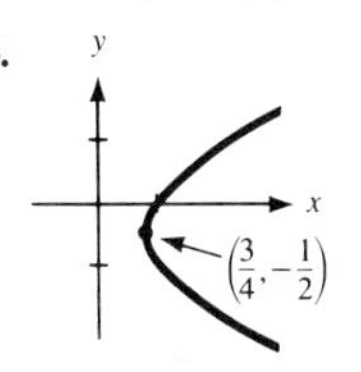

15.

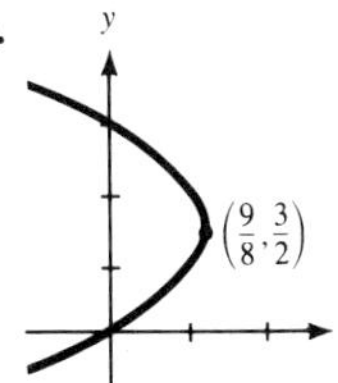

17.

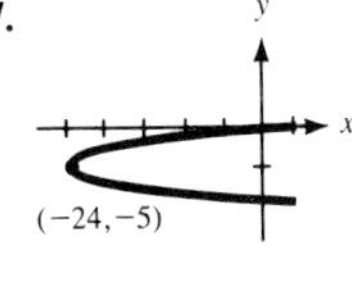

19.

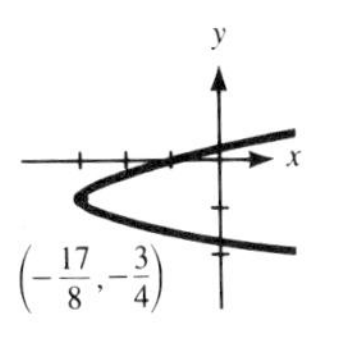

21.

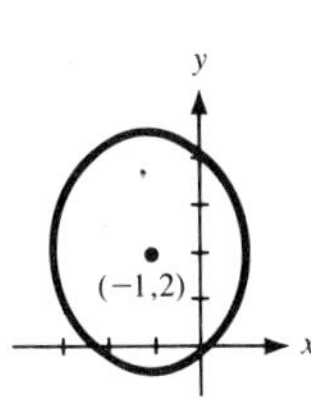

23.

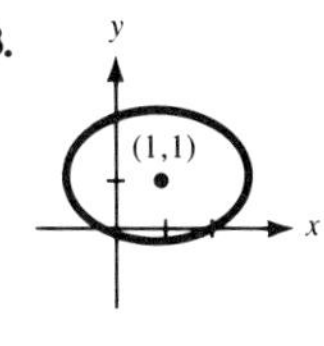

25.

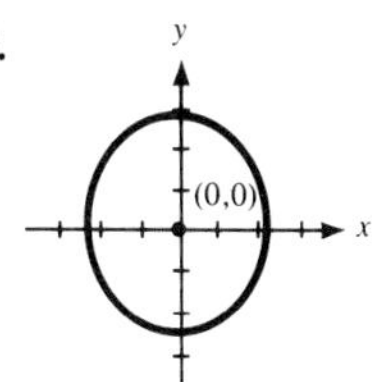

27.

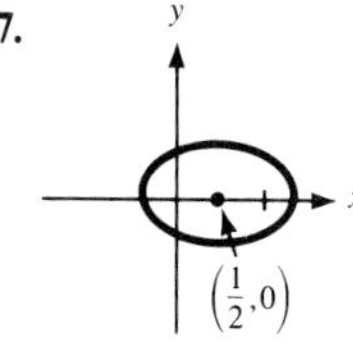

29.

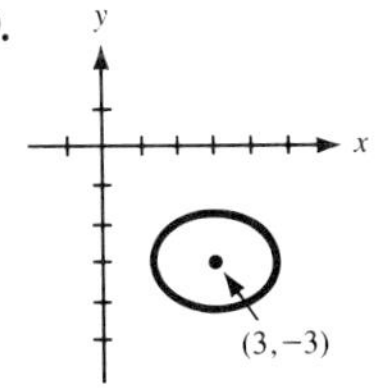

31.

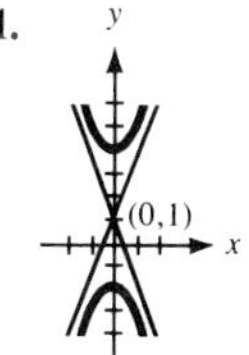

33.

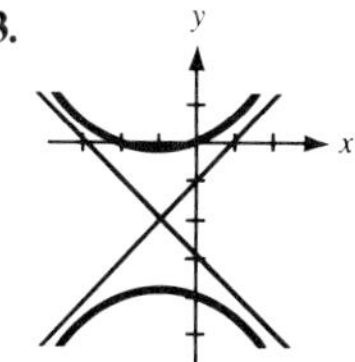

35.

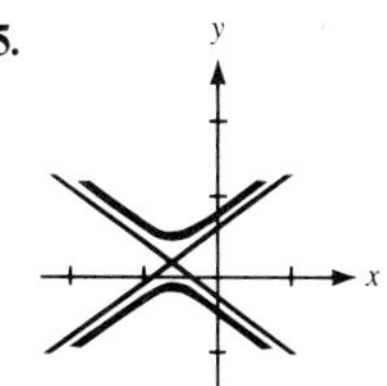

37.

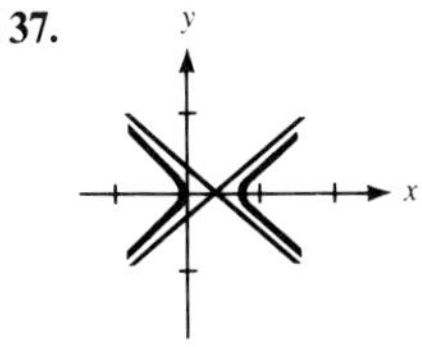

39.

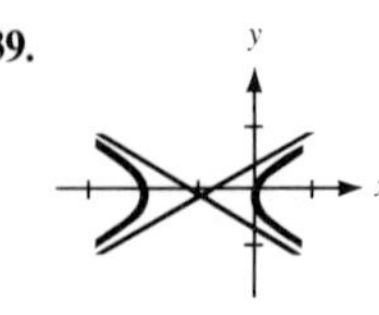

41.

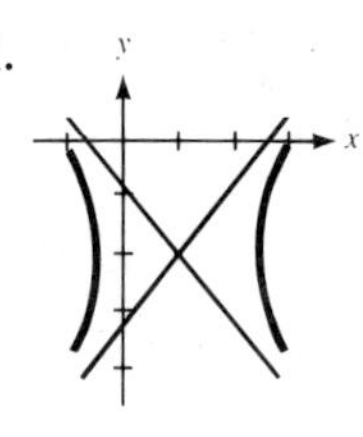

43. a hyperbola

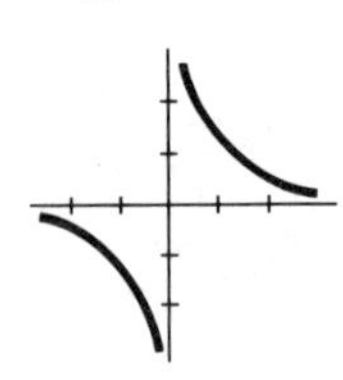

45. $\dfrac{x^2}{25} + \dfrac{y^2}{4} = 1$ **47.** $(x-1)^2 - \dfrac{(y-2)^2}{4} = 1$

SECTION 3.6 (REVIEW)

1. $m = -3/2$

(5, −1) (7, −2)

3. $m = -7/2$

(−5, 4)

5. **(a)** $y = 12x + 7$ **(c)** $12x - y + 7 = 0$

7. **(a)** $y = (2/3)(4 - \sqrt{2})x + (2/3)(2 + \sqrt{2})$ **(b)** $y - 4 = (2/3)(4 - \sqrt{2})(x - 1)$ **(c)** $2(4 - \sqrt{2})x - 3y + 2(2 + \sqrt{2}) = 0$

9. **(a)** $y = -(2/3)x + 8/3$ **(b)** $y - 4 = -(2/3)(x + 2)$ **(c)** $2x + 3y - 8 = 0$ **11.**

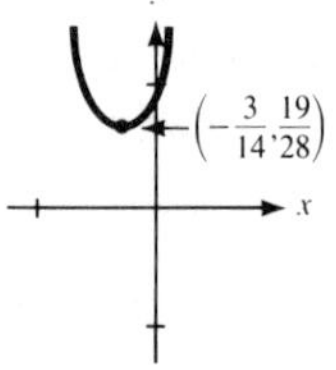

13.

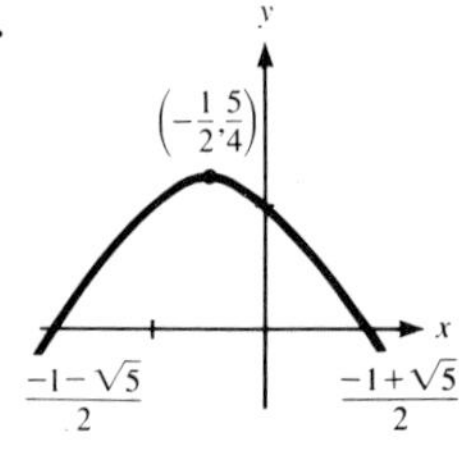

15.

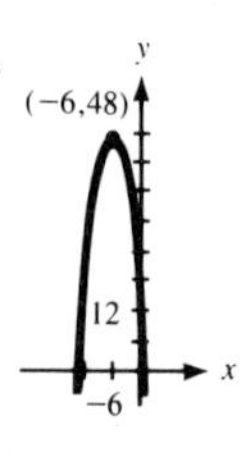

17. all reals **19.** $\left[\dfrac{-3-\sqrt{29}}{2}, -3\right] \cup \left[0, \dfrac{-3+\sqrt{29}}{2}\right]$ **21.** 25/2 ft by 25 ft.

23. Use 18/7 ft for the square and the rest for the rectangle.

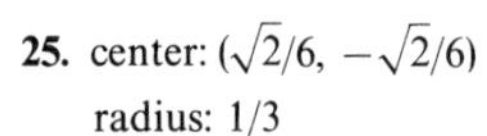

25. center: $(\sqrt{2}/6, -\sqrt{2}/6)$
radius: 1/3

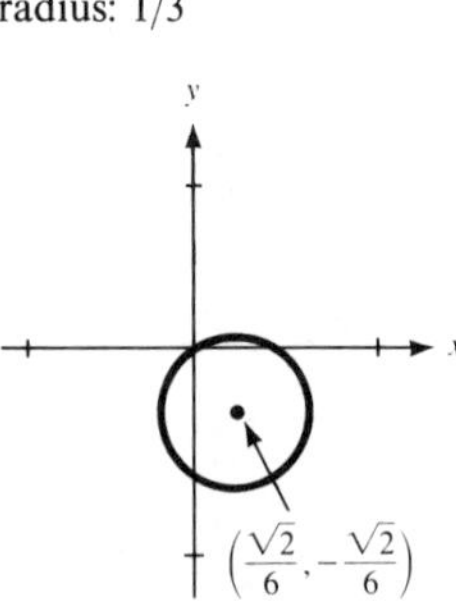

27.

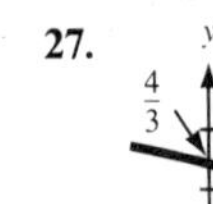

29. $a = \sqrt{3}$, $b = 1$ **31.** $a = \sqrt{17}/2$, $b = \sqrt{24}/4$ **33.** $a = b = \sqrt{6}$ **37.** eccentricity $= 3\sqrt{11}/10$

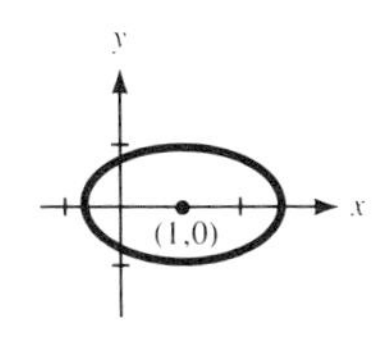

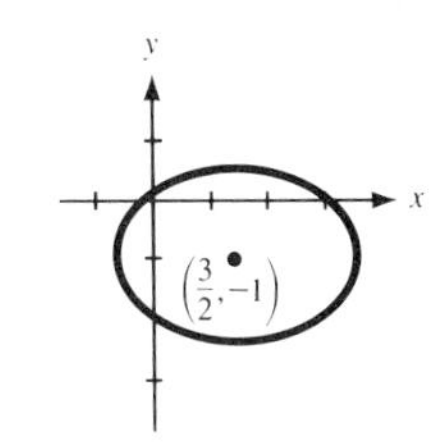

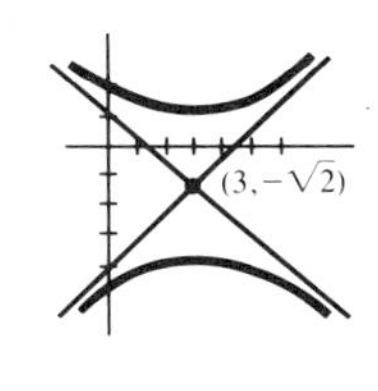

Chapter Four

SECTION 4.1

1. $6x^2 + 5x - 4 = (2x - 1)(3x + 4)$ **3.** $x^3 - 2x^2 + x - 4 = (x^2 - x)(x - 1) - 4$ **5.** $x^3 - 4x^2 + 3x = (x^2 - 3x)(x - 1)$
7. $-12x^3 + 4x^2 - 1 = (-3x + 1)(4x^2 + 1) + 3x - 2$ **9.** $2x^3 + 11x^2 - 7x - 6 = (2x + 1)(x^2 + 5x - 6)$
11. $x - 1 = 0\cdot(2x^2 + 3x - 1) + x - 1$ **13.** $a^2x^2 - 3ax - 10 = (ax + 2)(ax - 5)$
15. $x^3 - a^3 = (x - a)(x^2 + ax + a) + (a^2 - a)x + a^2 - a^3$ **17.** $x^4 - 3x^2 + 2x - 1 = (x)(x^3 - 1) - 3x^2 + 3x - 1$
19. $x^4 + 2x^3 - 3x^2 + x + 1 = (x^3 + 3x^2 + 1)(x - 1) + 2$ **21.** $-x^4 - x^3 + 2x^2 + x + 3 = (-x^2 - 3x - 4)(x^2 - 2x) - 7x + 3$
23. $2x^{12} + 3x^2 - 9 = (2x^8 + 6x^4 + 18)(x^4 - 3) + 3x^2 + 45$ **25.** $x^5 - a = (x^3 + ax)(x^2 - a) + a^2x - a$ **27.** $x^2 = (x + 2)(x - 2) + 4$
29. $x^3 + 2x^2 - x + 1 = (x^2 - 1)(x + 2) + 3$ **31.** $2x^3 + x^2 - x + 1 = (2x^2 - 5x + 14)(x + 3) - 41$
33. $32 - x^5 = (-x^4 - 2x^3 - 4x^2 - 8x - 16)(x - 2)$ **35.** $x^4 - a^4 = (x^3 + ax^2 + a^2x + a^3)(x - a)$
37. $2x^3 - x^2 + 3x - 1 = (2x^2 + 3)(x - \frac{1}{2}) + \frac{1}{2}$ **39.** $2x^3 + x^2 - 3x + 4 = (2x^2 - 3x + 3)(2x + 4) - 4$
41. $4x^2 - x - 1 = (2x + \frac{1}{2})(2x - 1) - \frac{1}{2}$ **43.** $ar^2 + br + c$; $ax^2 + bx + c = (ax + ra + b)(x - r) + ar^2 + br + c$

SECTION 4.2

1. 36 **3.** 5 **5.** -1 **7.** 0 **9.** $\frac{3}{4}$ **25.** n is an odd positive integer **27.** $a = -3, 2$
29. True, since $x^3 - 6x^2 + 12x - 8 = (x - 2)^3$
31. True, since $2x^5 - x^4 + 2x^3 - x^2 + 6x - 3 = (2x^4 + 2x^2 + 6)(x - \frac{1}{2})$ and $2(\frac{1}{2})^4 + 2(\frac{1}{2})^2 + 6 \neq 0$
33. 2 is a double root and -4 is a simple root; $(x - 2)(x - 2)(x + 4)$ **35.** $-4/7$ is a simple root; $7(x + 4/7)(x^2 + x + 2)$
37. 1, -1, 3 and 4 are all simple roots; $(x - 1)(x + 1)(x - 3)(x - 4)$ **39.** -1 and 2 are both double roots; $(x + 1)(x + 1)(x - 2)(x - 2)$

SECTION 4.3

(Note to student: In Exercises 1–31, your answer may differ from the one given.)

1. $\frac{1}{2}$ **3.** $-\frac{1}{3}$ **5.** No rational roots **7.** 1 **9.** -1 **11.** $-\frac{1}{2}$ **13.** $\frac{2}{3}$ **15.** $-\frac{1}{2}$ **17.** No rational roots
19. $\frac{1}{2}$ **21.** $\frac{1}{2}$ **23.** $-\frac{2}{3}$ **25.** No rational roots **27.** No rational roots **29.** No rational roots **31.** 1 **33.** $\frac{5}{2}, -\frac{1}{3}$
35. $-1, \frac{3}{2}, \frac{3}{2}$ **37.** $\frac{1}{2}, -\frac{1}{3}, \frac{5}{2}$ **39.** $3, -\frac{1}{7}, 5$ **41.** $-3, -3$ **43.** No rational roots **45.** $-1, -1, \frac{1}{3}$ **47.** $1, 1, 1, -2, -2$
49. No rational roots **51.** No rational roots.

SECTION 4.4

(Note to student: In Exercises 1–17 your answer may differ from the one given.)

1. -0.6875 **3.** -1.1875 **5.** 1.4375 **7.** -1.5625 **9.** -2.9375 **11.** -1.6875 **13.** 0.78125 **15.** -0.4765625
17. 1.28125

SECTION 4.5

1.
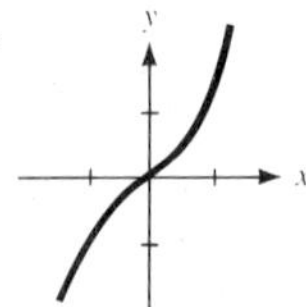

3.
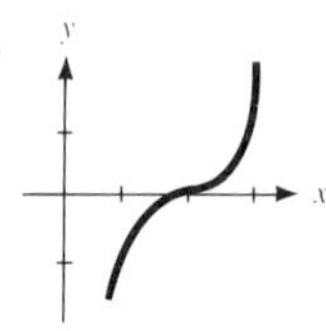

5.
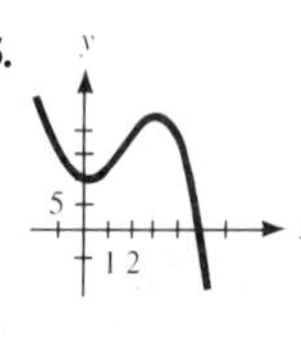

7.
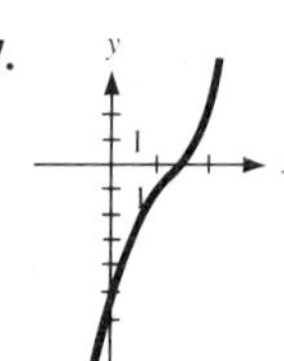

9.
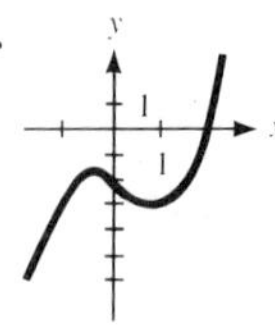

11.
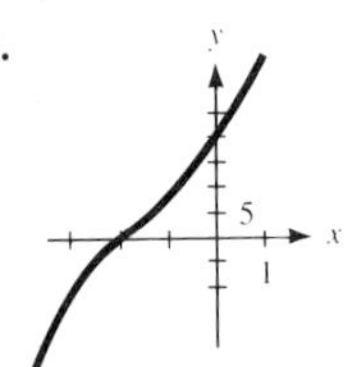

13.
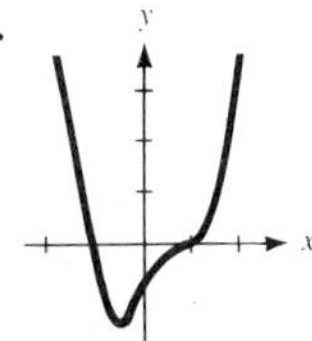

15.
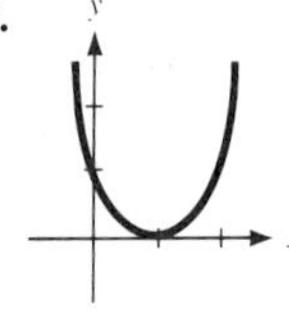

17.
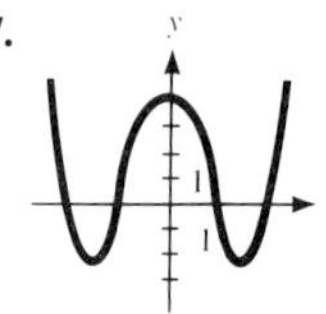

19.
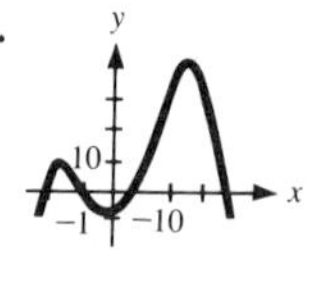

21.
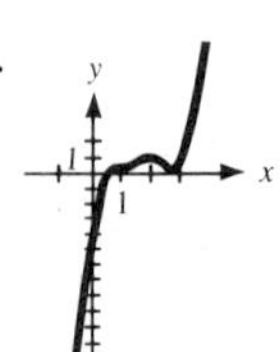

23.
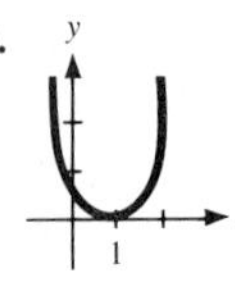

25.
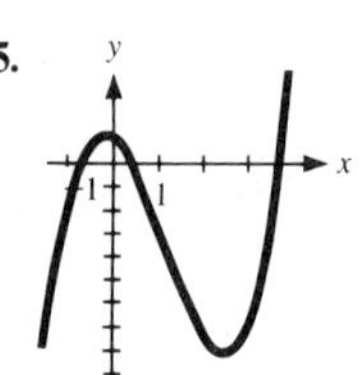

27.
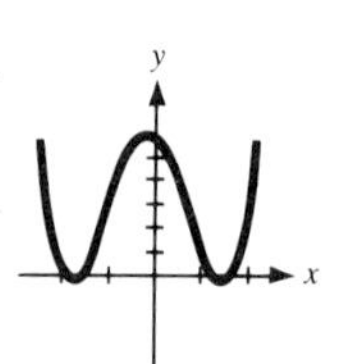

29.
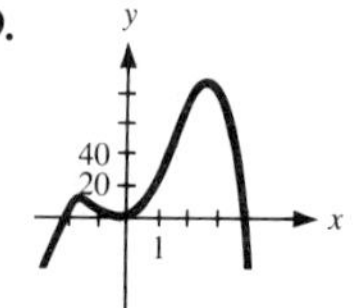

SECTION 4.6

1.
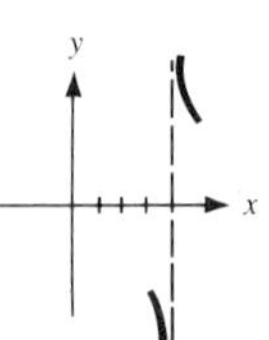

3.
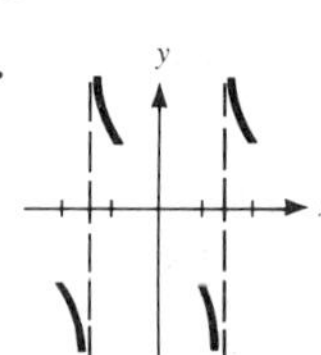

5.
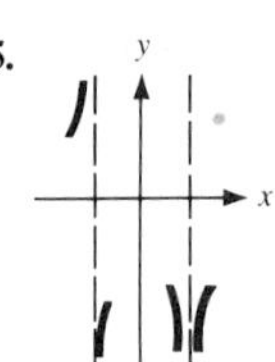

7.
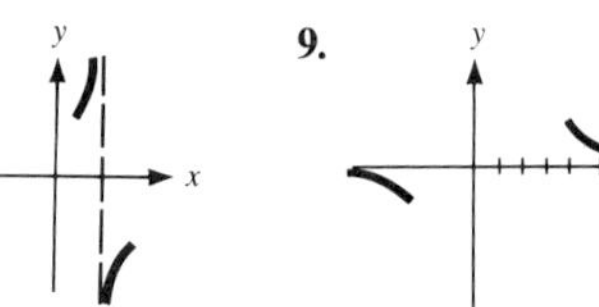

9.
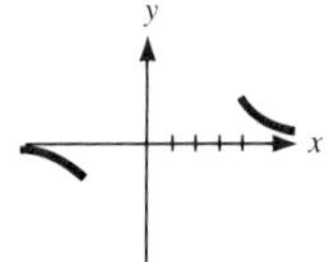

11.
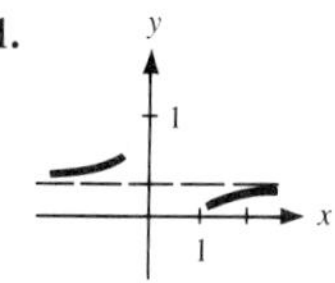

13. No horizontal asymptotes
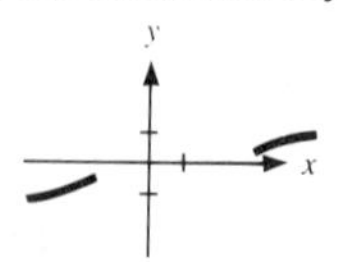

15.
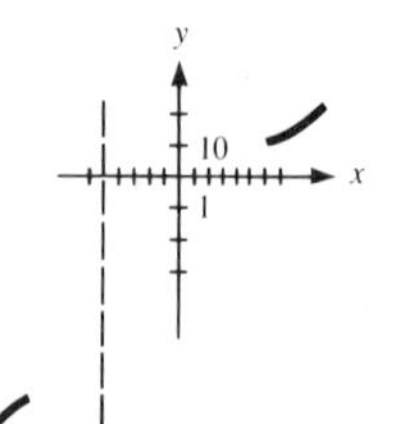

17.
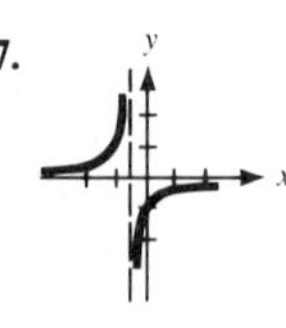

19.
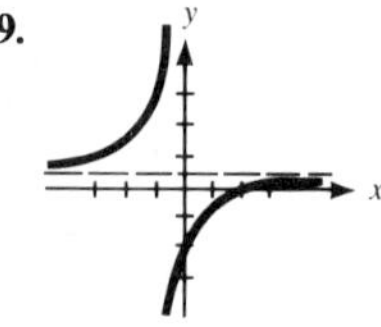

21.
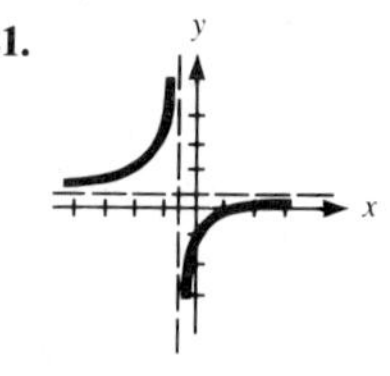

23.
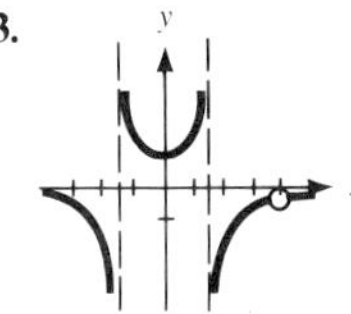

25.
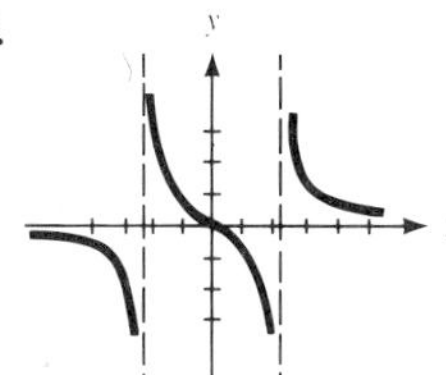

27.
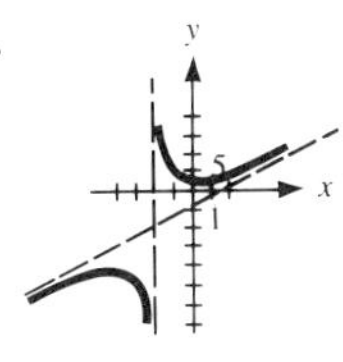

29.
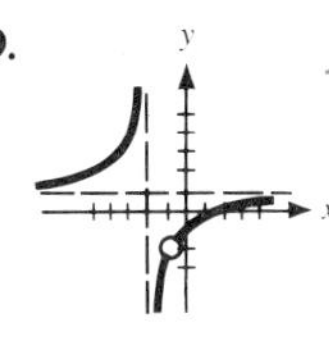

31.
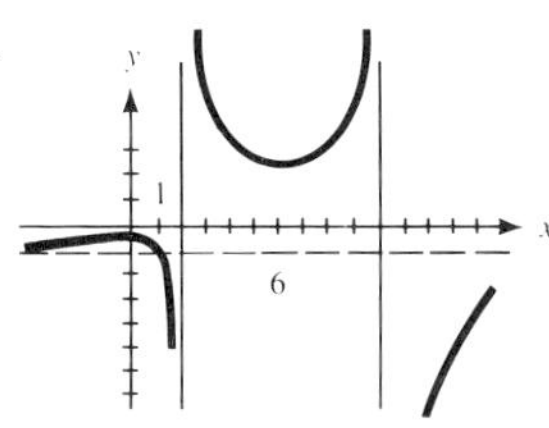

33.
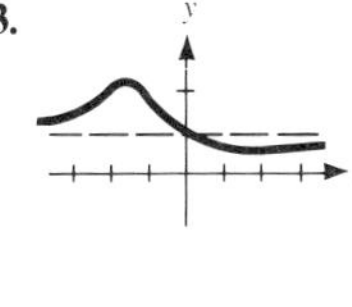

35.
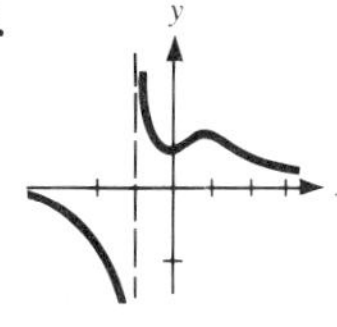

SECTION 4.7

1. $-\frac{1}{2(x-1)}+\frac{1}{2(x-3)}$ **3.** $\frac{\sqrt{3}}{6(x+\sqrt{3})}-\frac{\sqrt{3}}{6(x-\sqrt{3})}$ **5.** $\frac{1}{4(x-3)}+\frac{3}{4(x+1)}$ **7.** $\frac{7}{x+3}-\frac{12}{2x+5}$ **9.** $\frac{2}{x}-\frac{1}{x+1}-\frac{1}{x-1}$

11. $-\frac{1}{6x}+\frac{5}{3(x-3)}-\frac{3}{2(x-2)}$ **13.** $\frac{1}{x}-\frac{1}{x-1}+\frac{1}{(x-1)^2}$ **15.** $\frac{1}{x-1}-\frac{2}{(x-1)^2}+\frac{3}{x-2}$ **17.** $\frac{1}{x}+\frac{x-2}{x^2+x+1}$

19. $\frac{1}{3(x-1)}+\frac{10x+4}{3(2x^2+1)}$ **21.** $-\frac{2}{x}+\frac{1}{x^2}-\frac{x-3}{4x^2+1}$ **23.** $\frac{1}{2(x-1)}-\frac{x+1}{2(x^2+1)}$ **25.** $-\frac{2}{x}+\frac{2x+1}{x^2+1}$

SECTION 4.8 (REVIEW)

1. $2x^2-3x+1=0(x^3-1)+(2x^2-3x+1)$ **3.** $2x^5-12x^4+5x^3-3x^2+2x-1=(2x^3-12x^2+7x-15)(x^2-1)+9x-16$

5. cannot be done by synthetic division **7.** -371 **11.** $\frac{1}{4}, -\frac{2}{3}$ **13.** -3 **15.** $-3, -\frac{2}{3}, \frac{1}{2}$

17. $4, -7, \pm\sqrt{2}$ **19.** -1.8125 **21.** ± 1.34375 **23.**

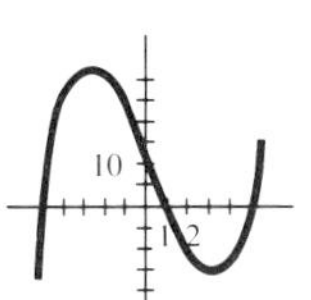

25.
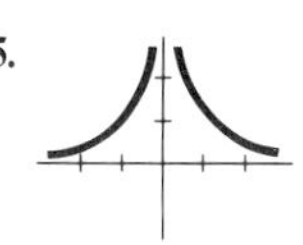

27
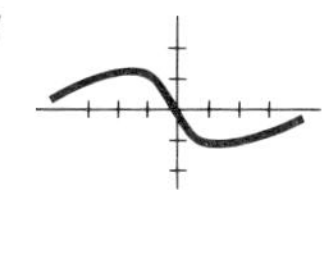

29. $\frac{1}{7(x-4)}-\frac{1}{7(x+3)}$ **31.** $-\frac{1}{x-1}+\frac{4}{2x+3}$ **33.** $-\frac{1}{3x}+\frac{7x}{3(x^2+3x+3)}$

Chapter Five

SECTION 5.1

1. 0.6065 **3.** 0.3626 **5.** -1.0446 **7.** $(2^3)^4=4096$, $2^{(3^4)}\approx 2.418\times 10^{24}$

9. 2, 2.63902, 2.65737, 2.66475, 2.66512, 2.66513 (rounded to 5 decimal places); these numbers seem to be approaching $2^{\sqrt{2}}\approx 2.66514$.

11. 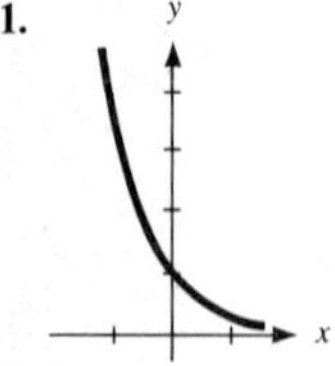

13. 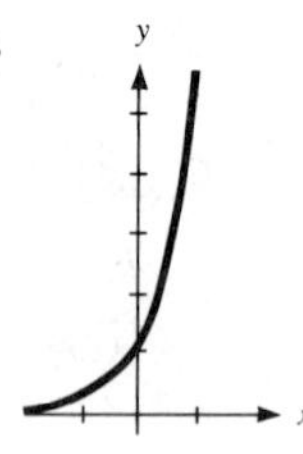

15. 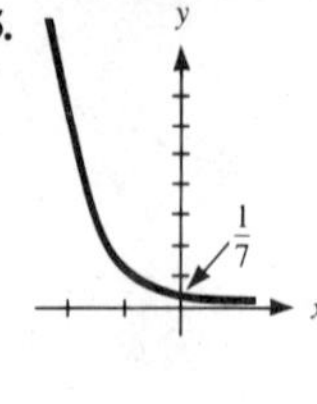

17. 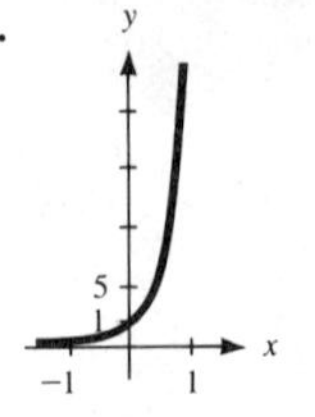

19. 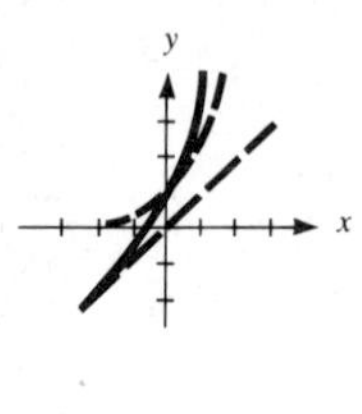

21. 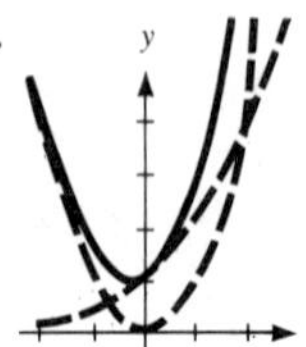

23. 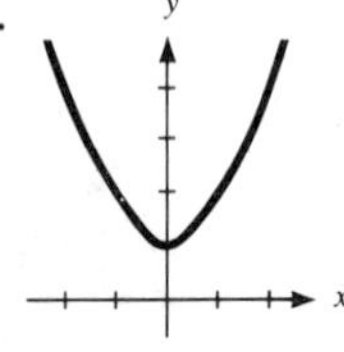

25. 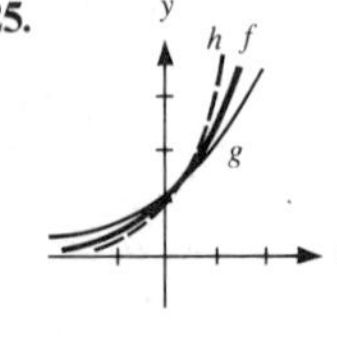

27. 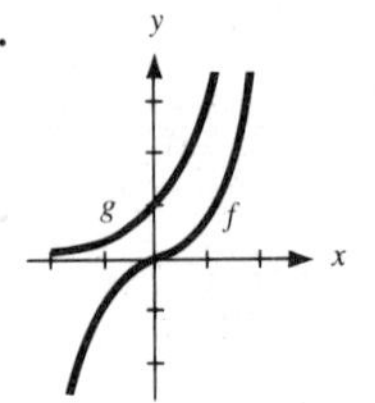

29. 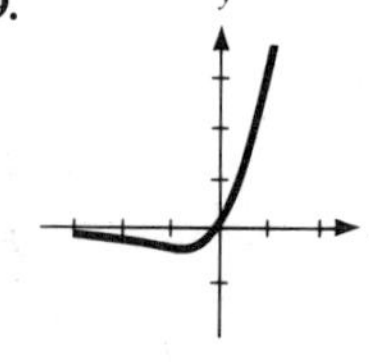

31. 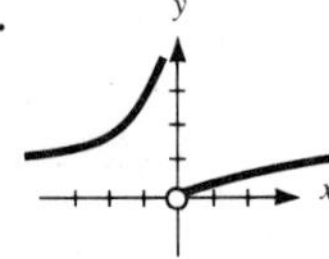

33. $x = 5$ **35.** $x = 2$ **37.** $x = 2$ **39.** $x = 3$ **41.** $x = -1$

43. (a) 5.1615×10^{-2} **(b)** 2.8880×10^{-3} **(c)** 4.2511×10^{-6} **(d)** 0 **(e)** 0 **(f)** 0 **47.** $y = 1$

SECTION 5.2

1. 4 **3.** 1 **5.** undefined **7.** 3 **9.** 6 **11.** 3/2 **13.** 5/2 **15.** $-5/3$ **17.** 25 **19.** 1 **21.** -0.1461
23. 8 **25.** 1.9912 **27.** 0.5229 **29.** -0.03253 **31.** 12.1 **33.** 7.1 **35.** 0.5 **37.** 53.05 **39.** 0 **41.** 0

43. $\log_a \dfrac{x(3x-1)^2}{(3x+1)^2}$ **45.** $\log_a x^{2/3}$ **47.** $x = 5$ **49.** $x = 0$ **51.** $x = 1/2$, $x = 1$ **53.** no solutions **55.** $x = -1$

SECTION 5.3

1. 2.2601 **3.** -1.0792 **5.** -5.3010 **7.** -0.7038 **9.** undefined **11.** 4.0875 **13.** 1.6515 **15.** 0.6309

17. 0.4307 **19.** 0.24 **21.** 1/2 **23.** $\log_b a = \dfrac{1}{\log_a b}$ **25.** domain: $(0, \infty)$ **27.** domain: $(0, \infty)$ **29.** domain: $(-\infty, 0)$

31. 1.0243 **33.** -7.5466 **35.** 3.6102 **37.** -3.3219 **39.** 0.2973 **41.** 2.9534 **43.** approximately 8 million times greater
45. 80 decibels **47.** 10^{11} times as intense **49.** 1.1220 times the standard intensity
51. (a) 3.9×10^{-8} **(b)** 3.98×10^{-13} **(c)** 7.94×10^{-5} **(d)** 1.26×10^{-3} **53.** fewer

SECTION 5.4

1. **(a)** \$1010 **(b)** \$1061.52 **(c)** \$1126.83 **(d)** \$1816.70 **3.** \$28,370.38
5. **(a)** \$3105.58 **(b)** \$3161.68 **(c)** \$3162.59 **(d)** \$3167.96 **(e)** \$3168.15 **7.** \$9070.29 **9.** \$4493.29
11. approximately 6.96 years **13.** approximately 10.98 years **15.** 1.89 years **17.** $r = 11.74\%$, 1.55 years
19. approximately 7% **21.** 10.38%; the second account is better since its effective interest rate is 10.60%
23. $r_{\text{eff}} = (1 + r/4)^4 - 1$

SECTION 5.5

1. $A(t) = 500 \cdot 20^{t/2}$; 80,000,000 bacteria **3.** $A(t) = 1000 \cdot 10^t$; 10^9 viruses
5. approximately 10.87 billion people in 1985, approximately 48.73 billion people in 2000 **7.** 5000 insects
9. For population A we have $A_A(t) = C_A \cdot 2^{t/2}$; for population B we have $A_B(t) = C_B \cdot 3^{t/3}$. Writing $A_A(t) = C_A \cdot (\sqrt{2})^t$ and $A_B(t) = C_B \cdot (\sqrt[3]{3})^t$ and noting that $\sqrt[3]{3} > \sqrt{2}$ we deduce that population B grows faster than population A.

11. 2,471,515 years **13.** 67.01 pounds **15.** 172.74 years **17.** $\dfrac{\log 2}{5600}$ **19.** 24,203 years **21.** 103.28°

SECTION 5.6 (REVIEW)

1. 0.06599 **3.** 2.2804 **5.**

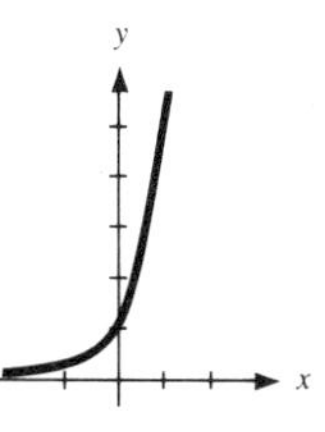

7.

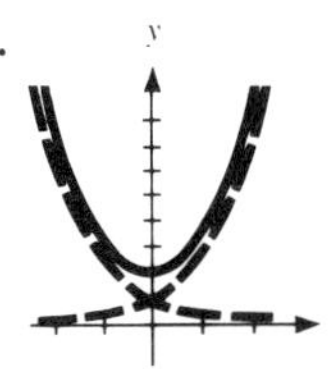

9.

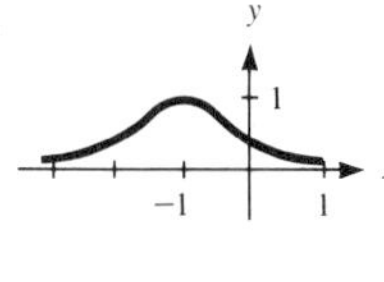

11. $x = -2/3$

13. $x = 2/7$ **15.** 3/2 **17.** 5/2 **19.** $x = 9{,}999{,}999{,}999$ **21.** $x = \sqrt{3}$ **23.** $-31/30$ **25.**

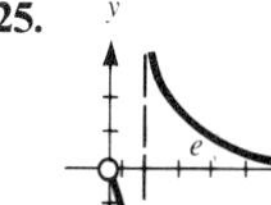

27. 6.0670 **29.** 1.2450 **31.** all real numbers **33.** approximately 15,849 times brighter

35. approximately 39,810,717 times less intense **37.** 11.82% **39.** $t = \dfrac{\log 2}{n \log(1 + r/n)}$

41. 2.25×10^{10} years; this is about 5 times longer than the age of the earth.

Chapter Six

SECTION 6.1

1. $-\dfrac{\pi}{6}$ **3.** $-\dfrac{4\pi}{9}$ **5.** $\dfrac{\pi}{12}$ **7.** $\dfrac{5\pi}{4}$ **9.** 30° **11.** 135° **13.** −60° **15.** 15° **17.** 2.15 **19.** 8.69 **21.** 18°

23. 413.10° **25.** 17,188.73° **27.** $\frac{725\pi}{36} \approx 63.27$ ft **29.** $\frac{3600}{\pi} \approx 1146$ revolutions **31.** $\frac{27}{2\pi} \approx 4.30$ ft **33.** $\frac{3\pi}{10} \approx 0.94$ miles

35. 79.8 miles **37.** $s = \frac{\pi r d}{180}$ **39.** 3,839,724 sq. yds.

SECTION 6.2

1.

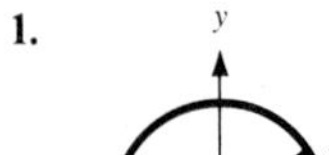

3.

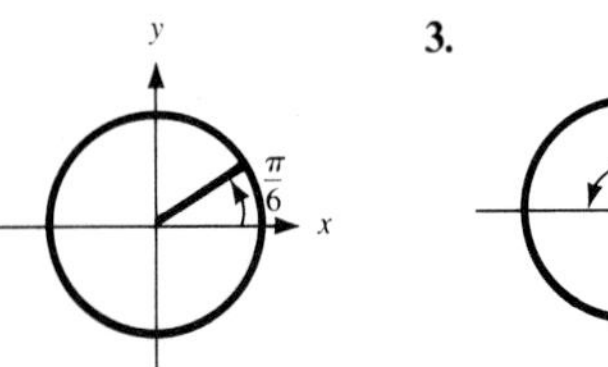

5.

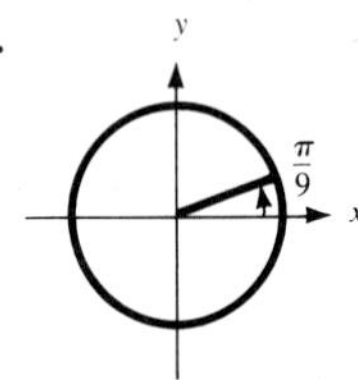

7.

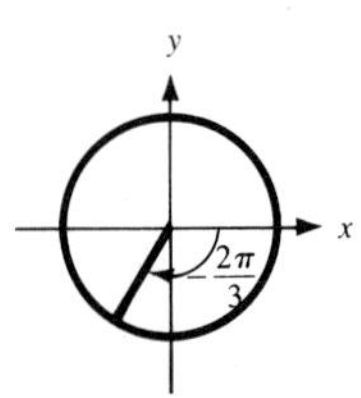

9.

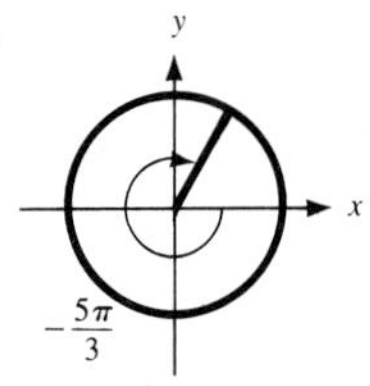

11.

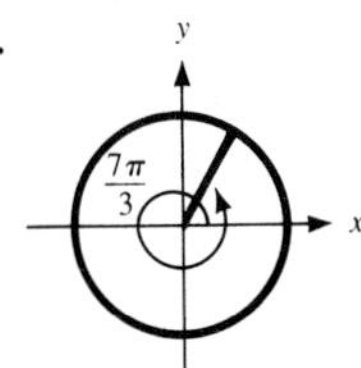

13.

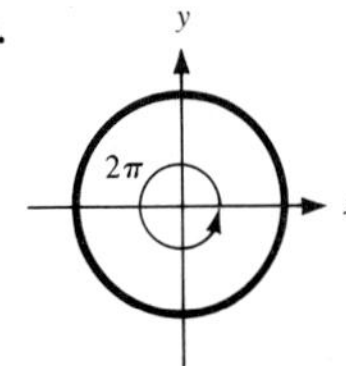

15.

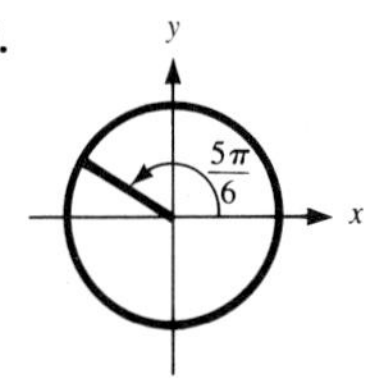

17. $\left(\frac{1}{2}, -\frac{\sqrt{3}}{2}\right)$ **19.** $\left(\frac{\sqrt{3}}{2}, -\frac{1}{2}\right)$

21. $\left(\frac{\sqrt{2}}{2}, \frac{-\sqrt{2}}{2}\right)$ **23.** $\left(\frac{-\sqrt{3}}{2}, \frac{-1}{2}\right)$ **25.** $(0, -1)$ **27.** $(-1, 0)$ **29.** $\left(\frac{\sqrt{3}}{2}, \frac{1}{2}\right)$ **31.** $(1, 0)$ **33.** $\left(\frac{-1}{2}, \frac{\sqrt{3}}{2}\right)$

35.

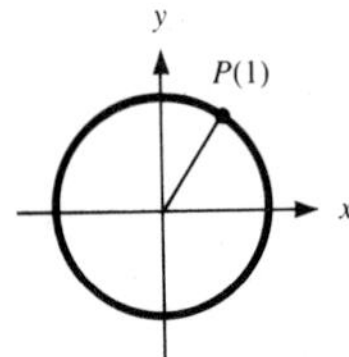

37.

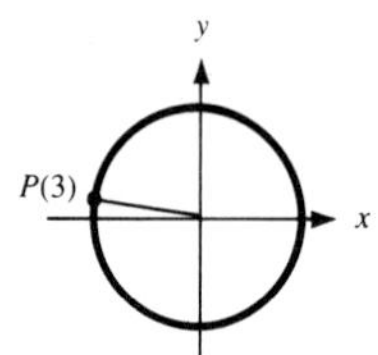

39.

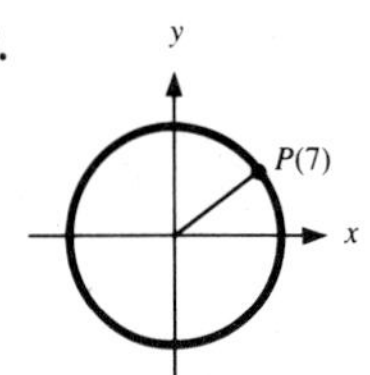

41.

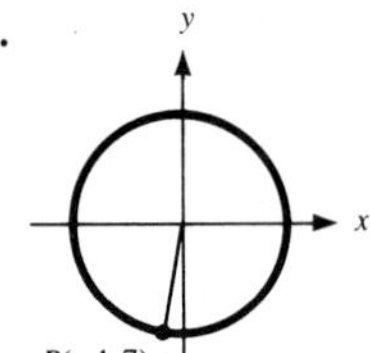

43.

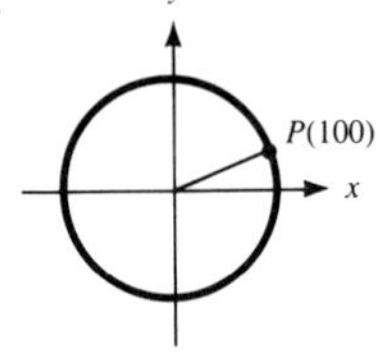

45. $\frac{2\sqrt{2}}{3}$ and $\frac{-2\sqrt{2}}{3}$ **47.** 0 and -1 **49. (a)**

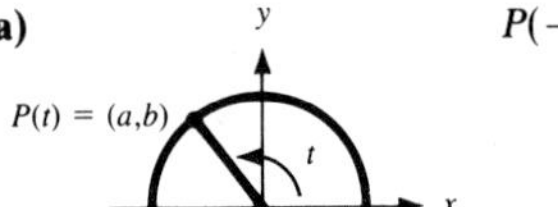

$P(-t) = (a, -b)$ **(b)** $P(-t) = \left(\frac{3}{5}, \frac{4}{5}\right)$

51. (a) $(-b, a)$ **(b)** $\left(\frac{\sqrt{35}}{6}, \frac{1}{6}\right)$

SECTION 6.3

1.

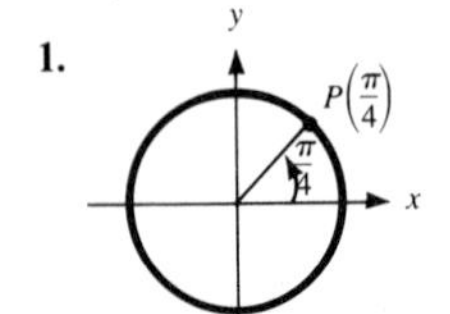

3.

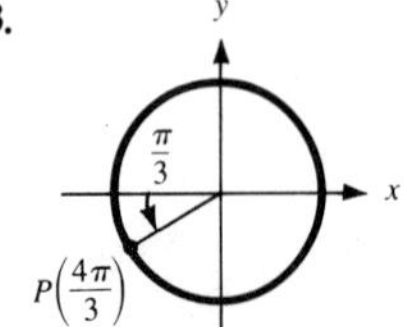

5.

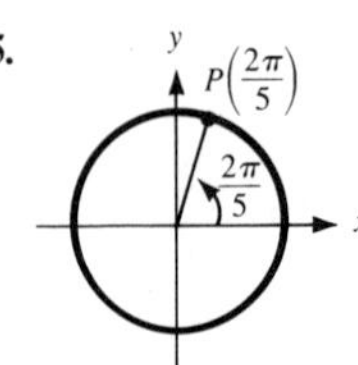

7.

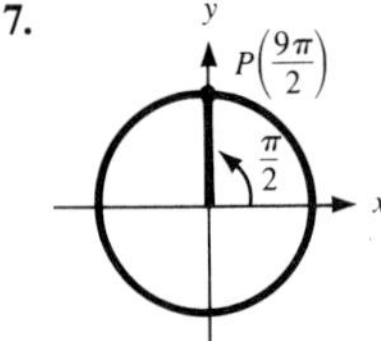

9.

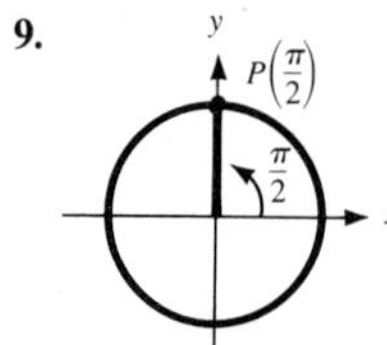

11. 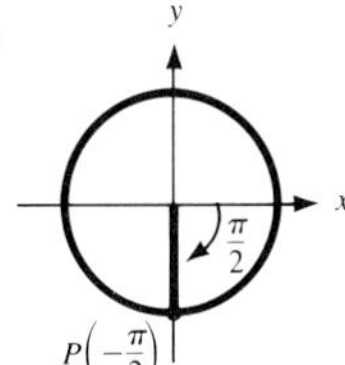

13. All are positive **15.** $\sin t$, $\csc t$ are positive; $\cos t$, $\cot t = 0$; $\tan t$, $\sec t$ are ND (not defined)

17. $\cos t$, $\sec t$ are positive; $\sin t$, $\tan t = 0$; $\text{cst}\, t$, $\cot t$ are ND **19.** $\cos t$, $\sec t$ are positive, $\sin t$, $\tan t$, $\csc t$, $\cot t$ are negative
21. All are positive **23.** $\cos t$, $\sec t$ are positive; $\sin t$, $\tan t$, $\csc t$, $\cot t$ are negative.

25. $\sin t$, $\tan t = 0$; $\cos t$, $\sec t = 1$; $\csc t$, $\cot t$ are ND **27.** $\sin t$, $\cos t = \dfrac{\sqrt{2}}{2}$; $\tan t$, $\cot t = 1$; $\sec t$, $\csc t = \sqrt{2}$

29. $\sin t$, $\csc t = 1$; $\cos t$, $\cot t = 0$; $\tan t$, $\sec t$ are ND **31.** $\sin t = \dfrac{\sqrt{2}}{2}$; $\cos t = \dfrac{-\sqrt{2}}{2}$; $\tan t$, $\cot t = -1$; $\sec t = -\sqrt{2}$; $\csc t = \sqrt{2}$

33. $\sin t$, $\tan t = 0$; $\cos t$, $\sec t = -1$; $\csc t$, $\cot t$ are ND **35.** $\sin t$, $\cos t = \dfrac{-\sqrt{2}}{2}$; $\tan t$, $\cot t = 1$; $\sec t$, $\csc t = -\sqrt{2}$

37. $\sin t$, $\csc t = -1$; $\cos t$, $\cot t = 0$; $\tan t$, $\sec t$ are ND **39.** $\sin t = \dfrac{-\sqrt{2}}{2}$; $\cos t = \dfrac{\sqrt{2}}{2}$; $\tan t = -1$; $\sec t = \sqrt{2}$; $\csc t = -\sqrt{2}$; $\cot t = -1$

41. $\sin t$, $\tan t = 0$; $\cos t$, $\sec t = 1$; $\csc t$, $\cot t$ are ND **43.** 0 **45.** $\dfrac{\sqrt{2}}{2}$ **47.** $\sqrt{3}$ **49.** -2 **51.** $\dfrac{-2\sqrt{3}}{3}$ **53.** -1

55. II **57.** IV **59.** I **61.** I **63.** 0.2588 **65.** 0.9997 **67.** 0.3090 **69.** 1.3764
71. -1.037 **73.** -1.0946 **75.** -1 **77.** 196.46 **79.** -159.15

81. **(a)** $30 + \dfrac{25\sqrt{3}}{2} \approx 51.65$ ft **(b)** same as (a) **(c)** $30 - \dfrac{25\sqrt{3}}{2} \approx 12.32$ ft **(d)** 37.39 ft

SECTION 6.4

1. 0 **3.** $\dfrac{1}{2}$ **5.** $\dfrac{1}{2}$ **7.** $-\dfrac{\sqrt{3}}{2}$ **9.** $-\dfrac{1}{2}$ **11.** $\dfrac{\sqrt{2}}{2}$ **13.** 0 **15.** $\dfrac{\sqrt{2}}{2}$

17. $\cos t = \dfrac{\sqrt{3}}{2}$; $\tan t = \dfrac{\sqrt{3}}{3}$; $\sec t = \dfrac{2\sqrt{3}}{3}$; $\csc t = 2$; $\cot t = \sqrt{3}$

19. $\cos t = -0.9173$; $\tan t = 0.4340$; $\csc t = -2.5119$; $\sec t = -1.0902$; $\cot t = 2.3042$

21. $\sin t = -\dfrac{\sqrt{15}}{4}$; $\tan t = -\sqrt{15}$; $\sec t = 4$; $\csc t = -\dfrac{4\sqrt{15}}{15}$; $\cot t = -\dfrac{\sqrt{15}}{15}$

23. $\cos t = 0.8954$; $\tan t = -0.4972$; $\csc t = -2.2462$; $\sec t = 1.1168$; $\cot t = -2.0112$

25. $\cos t = \dfrac{\sqrt{5}}{3}$; $\tan t = \dfrac{2\sqrt{5}}{5}$; $\csc t = \dfrac{3}{2}$; $\sec t = \dfrac{3\sqrt{5}}{5}$; $\cot t = \dfrac{\sqrt{5}}{2}$

27. $\cos t = -\dfrac{\sqrt{3}}{2}$; $\tan t = -\dfrac{\sqrt{3}}{3}$; $\csc t = 2$; $\sec t = \dfrac{-2\sqrt{3}}{3}$; $\cot t = -\sqrt{3}$

29. $\csc t = -\dfrac{2\sqrt{3}}{3}$; $\cot t = -\dfrac{\sqrt{3}}{3}$; $\cos t = \dfrac{1}{2}$; $\sin t = -\dfrac{\sqrt{3}}{2}$; $\tan t = -\sqrt{3}$ **31.** $-\dfrac{\sqrt{5}}{2}$ **33.** $\dfrac{2\sqrt{5}}{5}$ **35.** $-\dfrac{6\sqrt{37}}{37}$

43. **(a)** $5\sqrt{3} \approx 8.66$ inches **(b)** $-1 + 1\sqrt{46} \approx 5.78$ inches **(c)** 8.41 inches **(d)** 8.94 inches **45.** $-2\sqrt{2}$ inches

SECTION 6.5

1. $\frac{\pi}{2}$ and $-\frac{3\pi}{2}$; max. = 1 **3.** $-\frac{\pi}{2}$ and $\frac{3\pi}{2}$; min. = -1 **5.** $\frac{\pi}{4}$; $\frac{3\pi}{4}$; $-\frac{7\pi}{4}$; $-\frac{5\pi}{4}$ **7.** No **9.** $\frac{\pi}{6}$; $\frac{5\pi}{6}$; $-\frac{11\pi}{6}$; $-\frac{7\pi}{6}$

11. $\frac{\pi}{4}$; $\frac{5\pi}{4}$; $-\frac{3\pi}{4}$; $-\frac{7\pi}{4}$ **13.** 4 **15.** Any odd multiple of $\frac{\pi}{2}$ **17.**

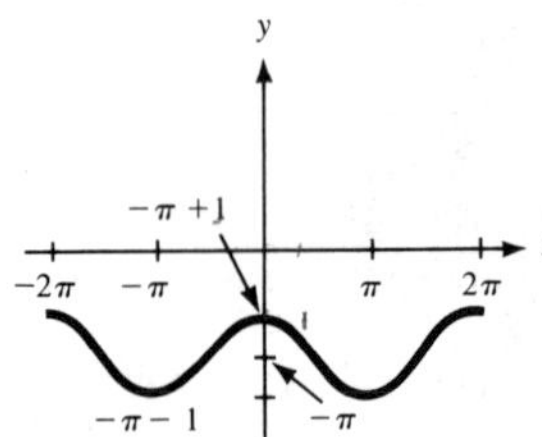

19.

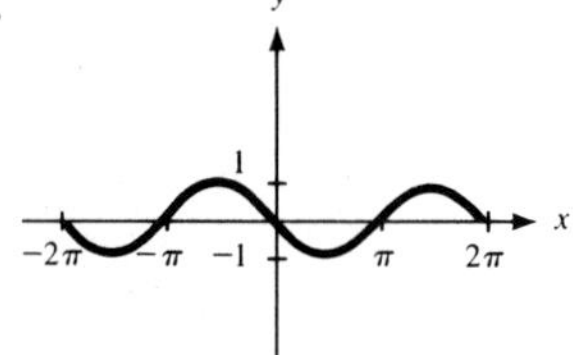

21.

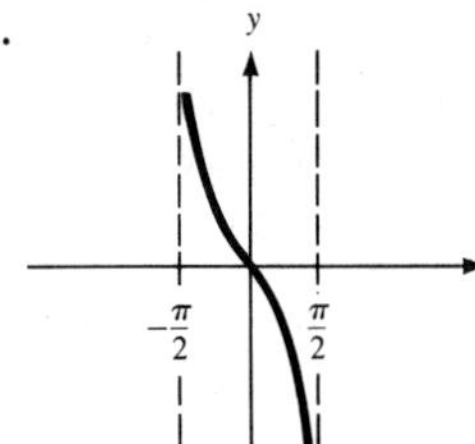

23.

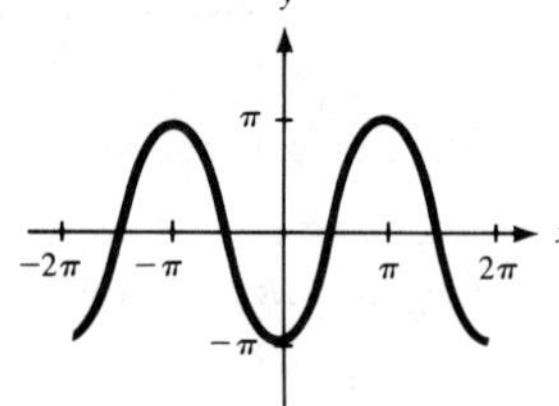

25.

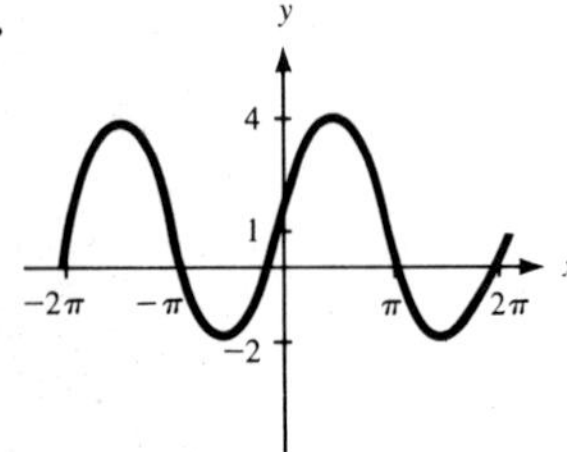

27.

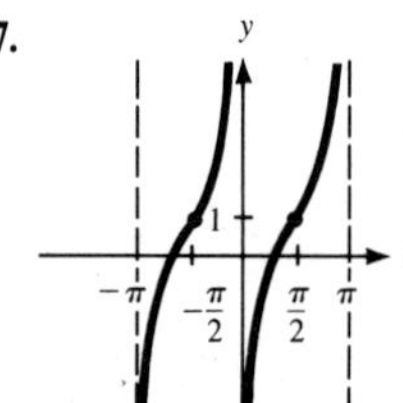

29.

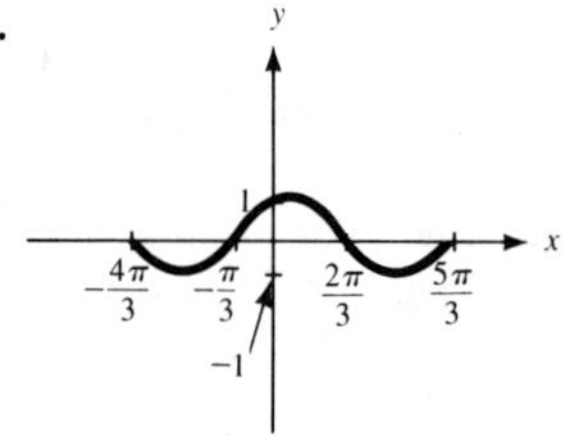

31.

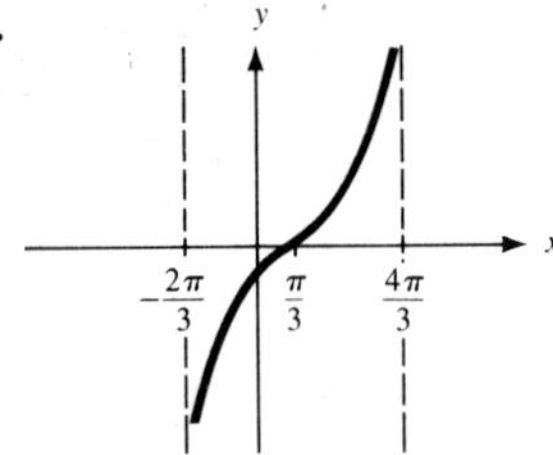

33.

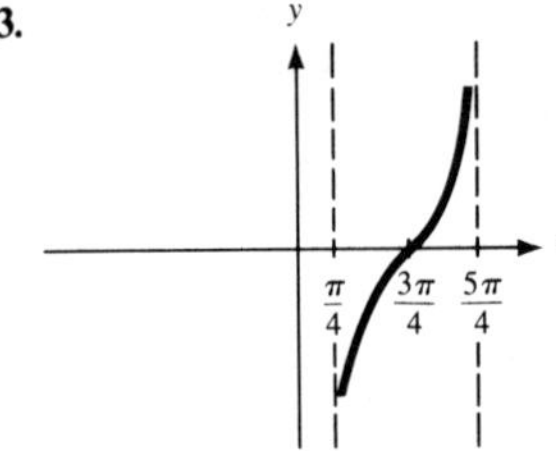

35.

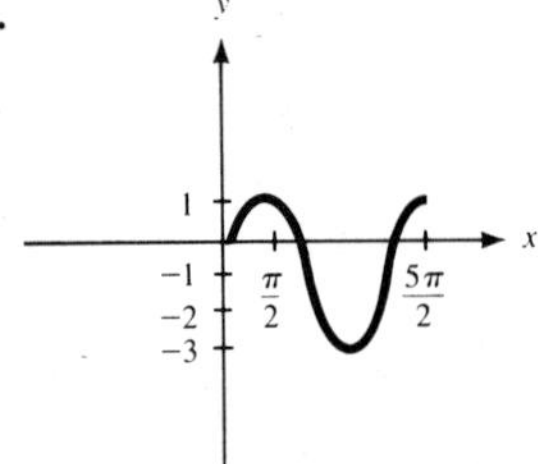

37.

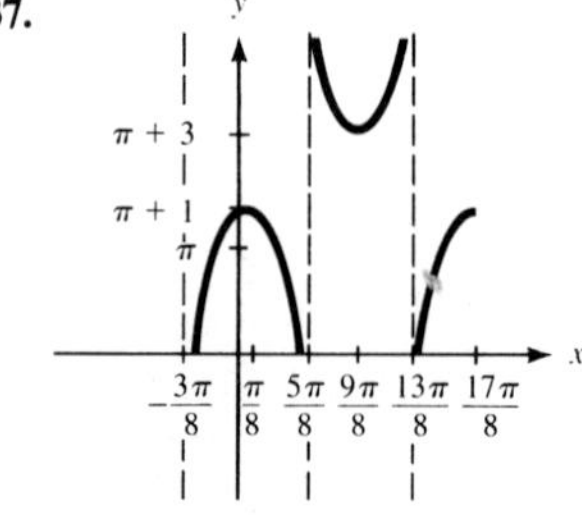

39.

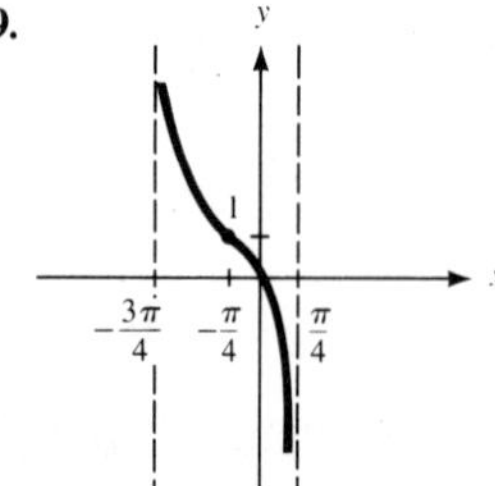

41.

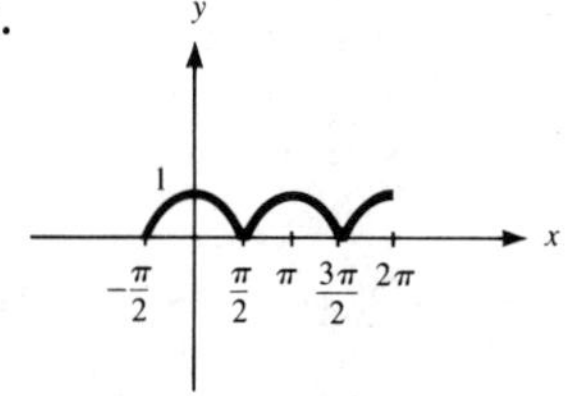

43.

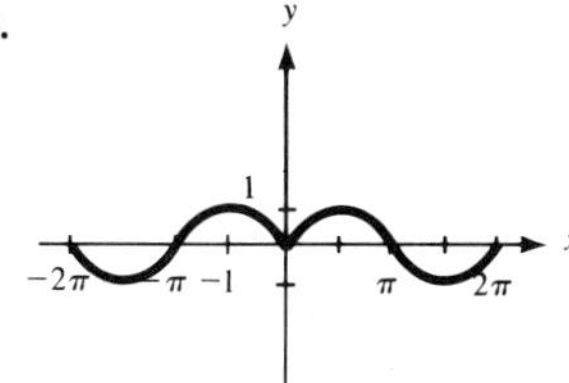

45.

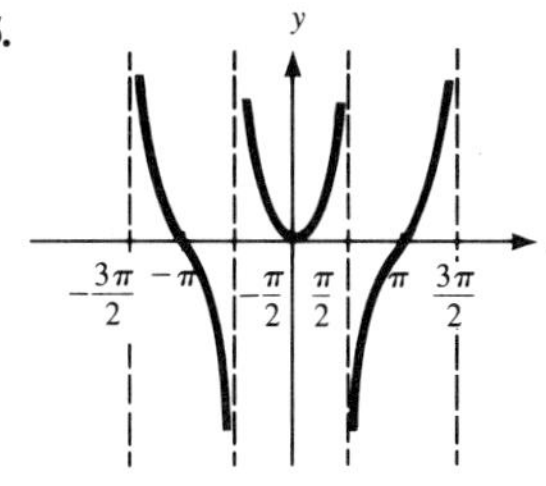

47.

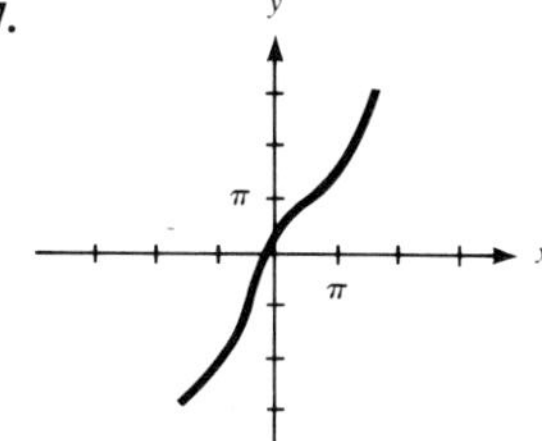

49.

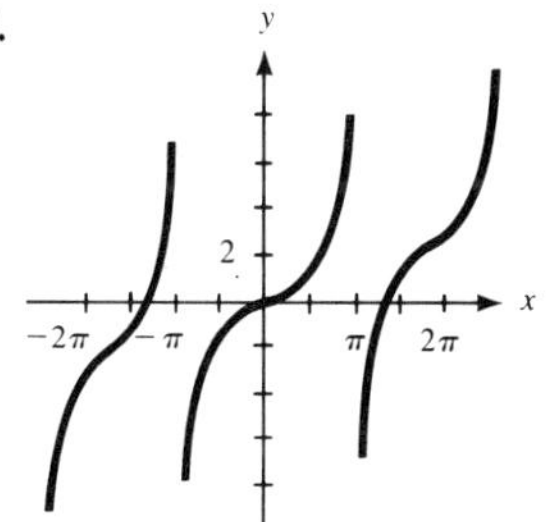

51.

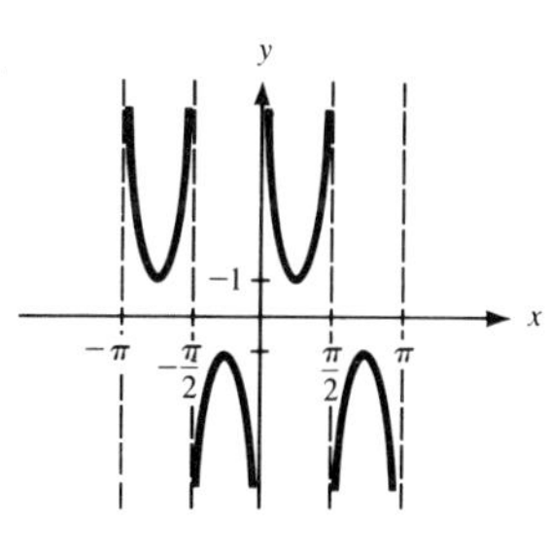

53.

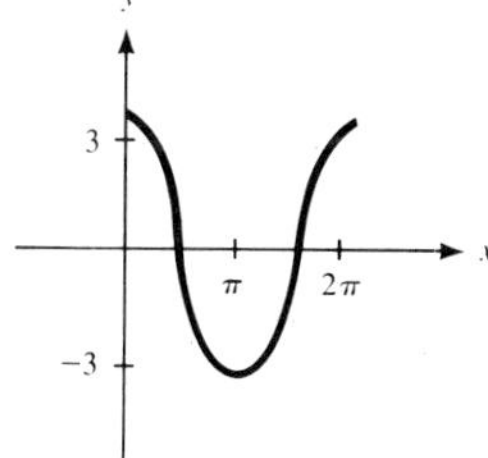

55.

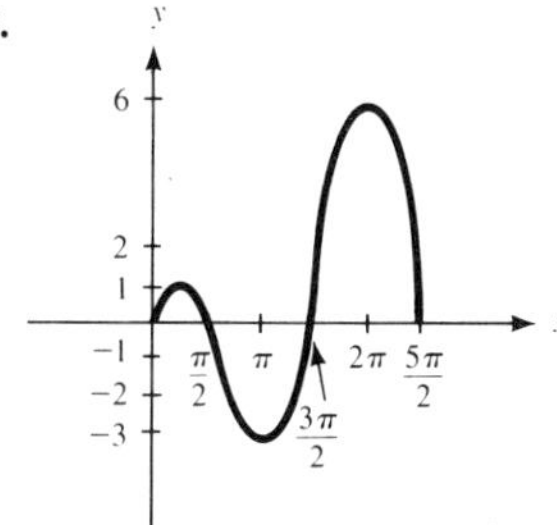

SECTION 6.6

1.

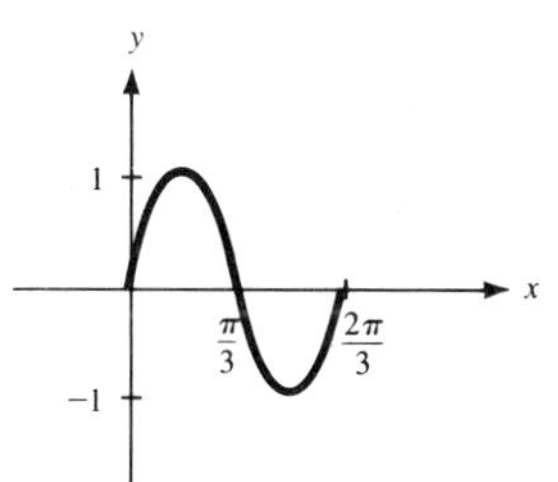

$A = 1$; $p = \dfrac{2\pi}{3}$; $f = \dfrac{3}{2\pi}$

3.

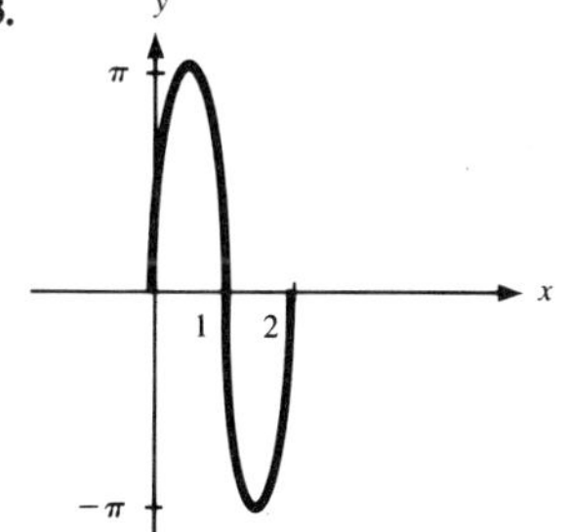

$A = \pi$; $p = 2$; $f = \dfrac{1}{2}$

5.

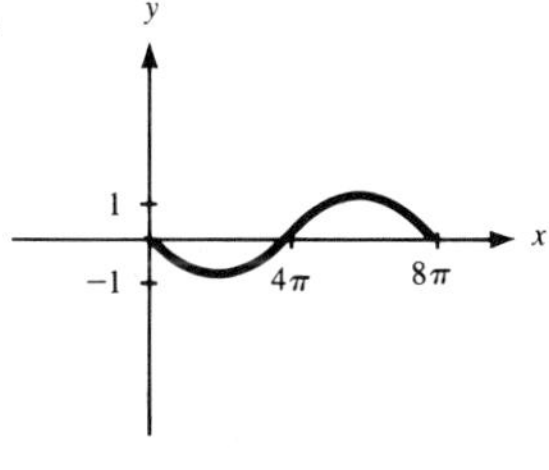

$A = 1$; $p = 8\pi$; $f = \dfrac{1}{8\pi}$

7.

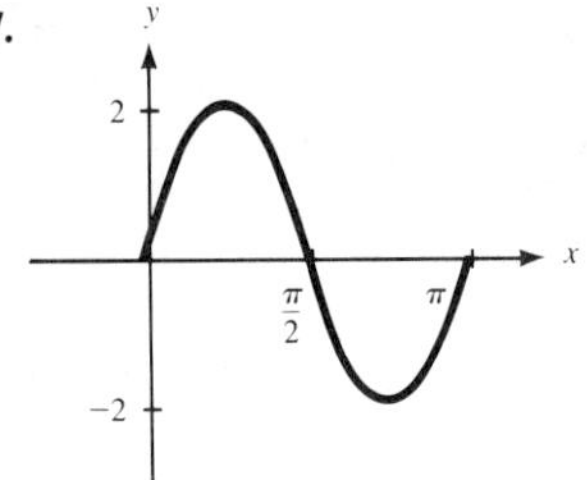

$A = 2;\ p = \pi;\ f = \dfrac{1}{\pi}$

9.

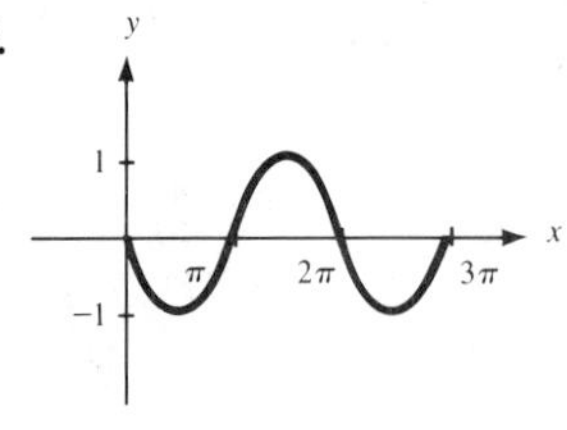

$A = 1;\ p = 2\pi;\ f = \dfrac{1}{2\pi}$

11.

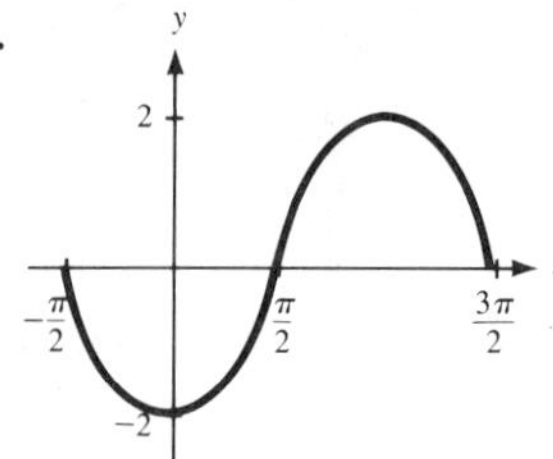

$A = 1;\ p = 2\pi;\ f = \dfrac{1}{2\pi}$

13.

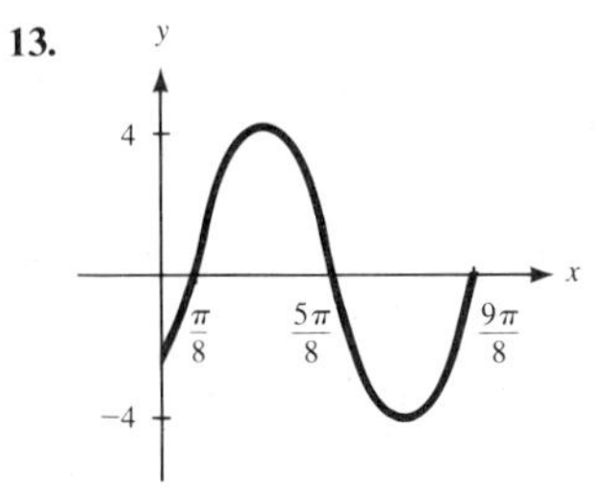

$A = 4;\ p = \pi;\ f = \dfrac{1}{\pi}$

15.

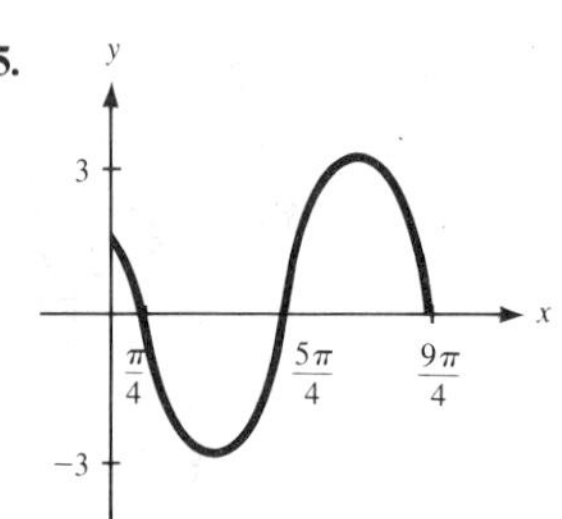

$A = 3;\ p = 2\pi;\ f = \dfrac{1}{2\pi}$

17.

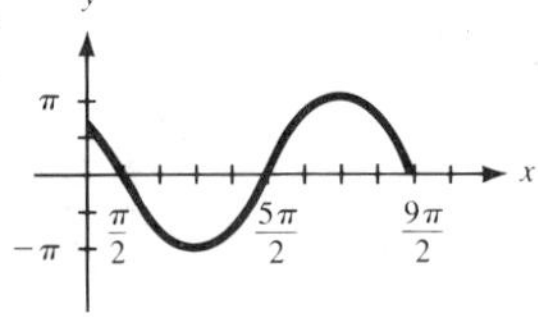

$A = \pi;\ p = 4\pi;\ f = \dfrac{1}{4\pi}$

19.

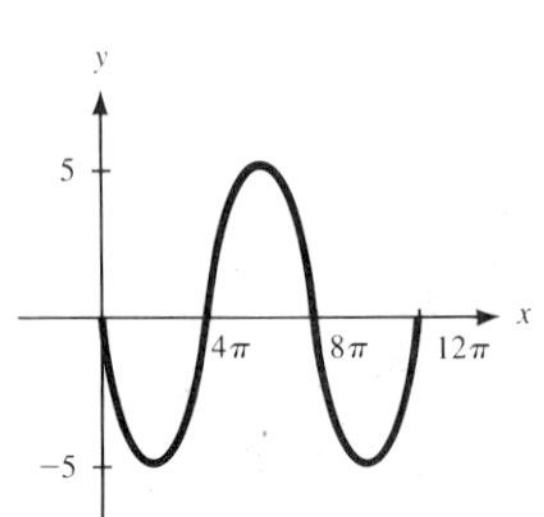

$A = 5;\ p = 8\pi;\ f = \dfrac{1}{8\pi}$

21.

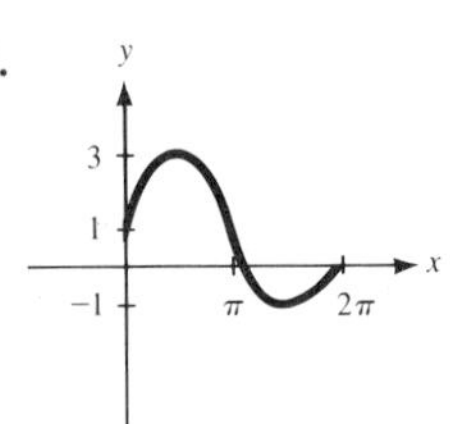

$A = 2;\ p = 2\pi;\ f = \dfrac{1}{2\pi}$

23.

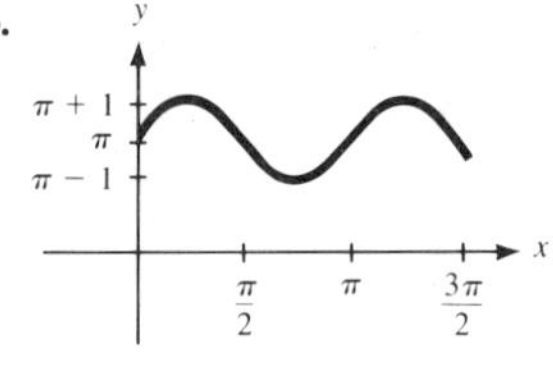

$A = 1;\ p = \pi;\ f = \dfrac{1}{\pi}$

25.

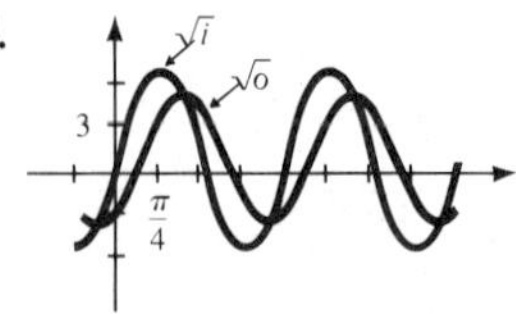

27.

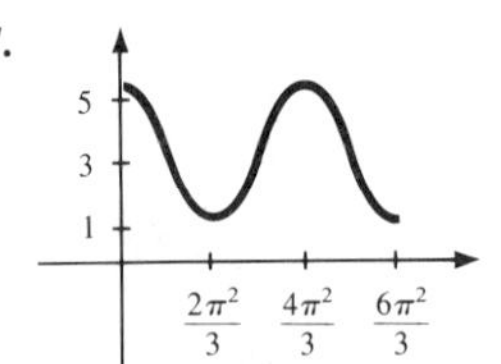

SECTION 6.7

1. $0, \dfrac{\pi}{2}$ **3.** $-\dfrac{\pi}{6}, \dfrac{2\pi}{3}$ **5.** $-\dfrac{\pi}{4}, \dfrac{3\pi}{4}$ **7.** $-\dfrac{\pi}{3}, \dfrac{5\pi}{6}$ **9.** $-\dfrac{\pi}{2}, \pi$ **11.** $\dfrac{3\pi}{4}$ **13.** $\dfrac{\pi}{4}$ **15.** $-\dfrac{\pi}{4}$ **17.** 0.7137

19. $\cos^{-1} 1.1$ is not defined **21.** −0.9553 **23.** 1.1671 **25.** 0.3322 **27.** 1.4924 **29.** 1.5708 **31.** 0.321

33. $\frac{\sqrt{15}}{4}$ **35.** $\frac{-\pi}{10}$ **37.** $\frac{\pi}{5}$ **39.** $\sqrt{3}$ **41.** $\sqrt{5}$ **43.** arcsin 12 is not defined **45.** yes **47.**

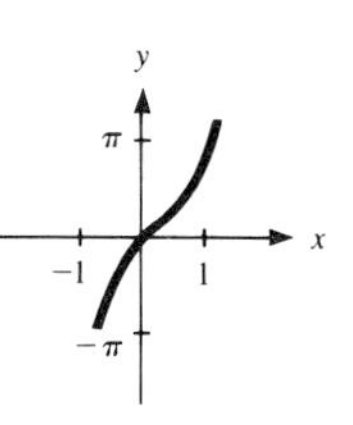

49.

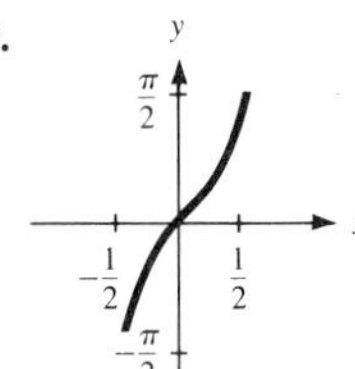

51.

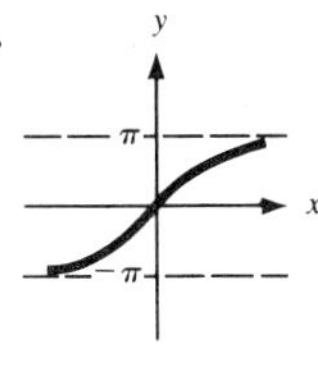

53.

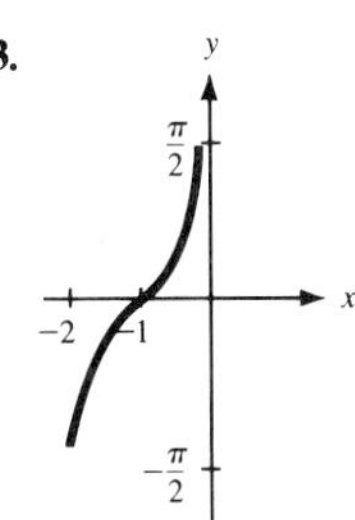

55.

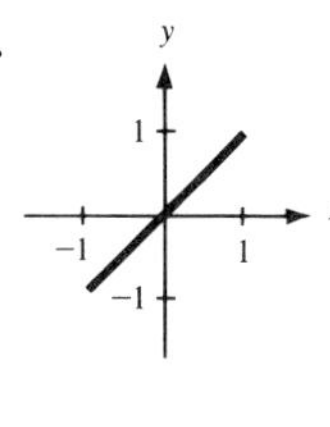

59. Yes.

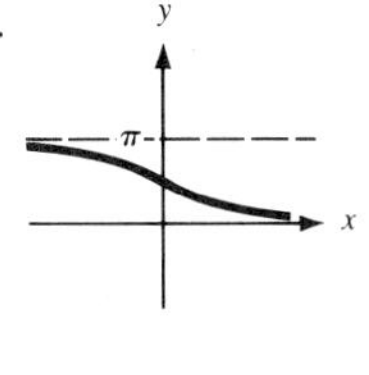

61. Yes

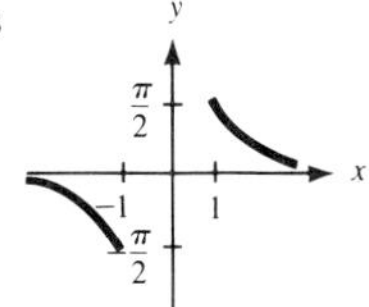

SECTION 6.8 (REVIEW)

1. $\left(\frac{1}{2}, \frac{\sqrt{3}}{2}\right)$ **3.**

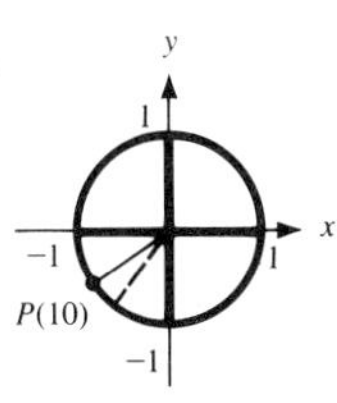

5. (a) $\left(\frac{1}{3}, -\frac{2\sqrt{2}}{3}\right)$ **(b)** $\left(-\frac{1}{3}, -\frac{2\sqrt{2}}{3}\right)$ **(c)** $\left(-\frac{1}{3}, \frac{2\sqrt{2}}{3}\right)$ **7.** $(-b, a)$ **9.** $-\frac{5\pi}{6}$

11. $\frac{4\pi}{5}$ **13.** 7.5° **15.** 227, 520° **17.** $\frac{5}{12}$ radians, $\frac{75}{\pi} \approx 23.87°$ **19.** $\frac{1}{2}$ **21.** $-\frac{\sqrt{2}}{2}$ **23.** $\sqrt{3}$ **25.** -1 **27.** ND

29. $\frac{\sqrt{3}}{3}$ **31.** ND **33.** -0.4310 **35.** 0.7087 **37.** 2 **39.** 0.7236

41. $\cos t = -\frac{3\sqrt{11}}{10}$; $\tan t = -\frac{\sqrt{11}}{33}$; $\csc t = 10$; $\sec t = -\frac{10\sqrt{11}}{33}$; $\cot t = -3\sqrt{11}$

43. $\sin t = -\frac{10\sqrt{101}}{101}$; $\cos t = -\frac{\sqrt{101}}{101}$; $\cot t = \frac{1}{10}$; $\csc t = -\frac{\sqrt{101}}{10}$; $\sec t = -\sqrt{101}$

45. $\sin t \approx -0.645$; $\cos t \approx 0.764$; $\sec t \approx 1.31$; $\tan t \approx -0.844$; $\cot t \approx -1.18$ **47.**

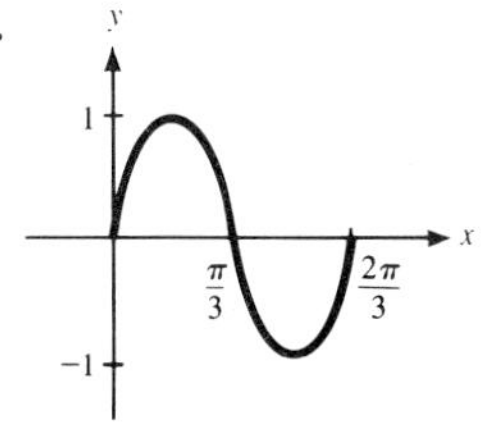

49.

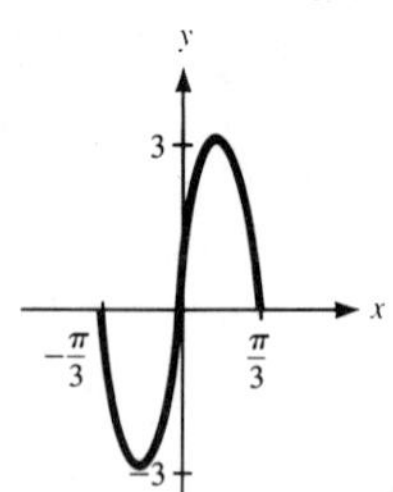

51.

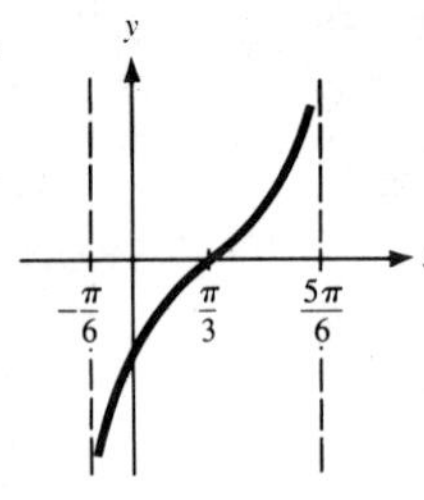

53.

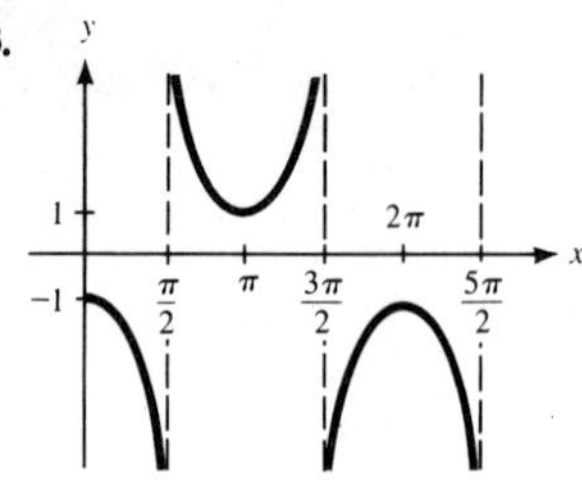

55.

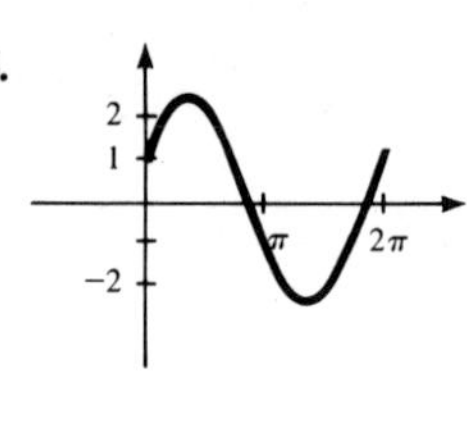

57.

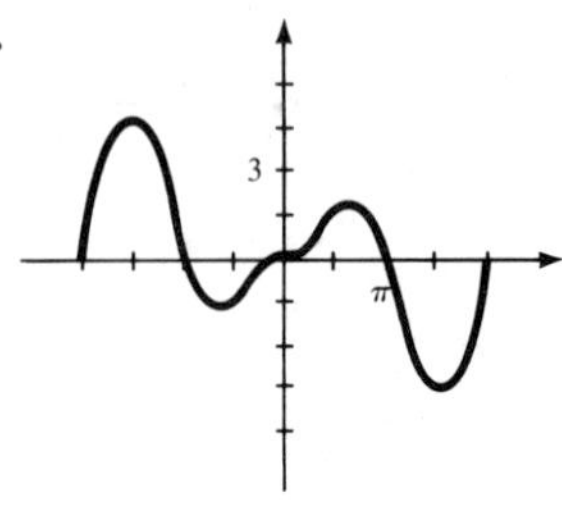

59. **(a)** $f(0.9) = 0.4204$; $f(0.7) = 0.3359$; $f(0.5) = 0.2448$; $f(0.3) = 0.1489$; $f(0.1) = 0.0500$; The function approaches zero.

(b)

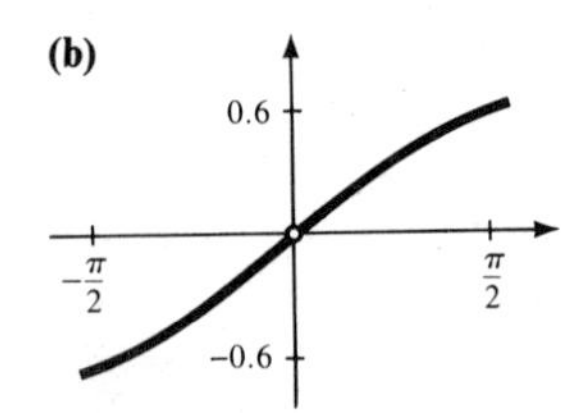

61. $-\frac{\pi}{6}$ **63.** $\frac{\pi}{3}$ **65.** ND **67.** 19 **69.** $\sqrt{5}$ **71.** ND **73.** 1.4940 **75.** 1.533

77. **(a)** -0.5240 **(b)** -0.0003 **(c)** -0.000001545 **(d)** -1.0×10^{-11} **(e)** -1.2×10^{-12}

79. **(a)** 0.03677 **(b)** 1×10^{-10} **(c)** 1×10^{-11}

Chapter Seven

SECTION 7.1

1. 0, 1, 0, 1, ND, ND **3.** $\frac{\sqrt{2}}{2}; \frac{\sqrt{2}}{2}, 1, \sqrt{2}, \sqrt{2}, 1$ **5.** 1, 0, ND, ND, 1, 0 **7.** $\frac{\sqrt{2}}{2}, -\frac{\sqrt{2}}{2}, -1, -\sqrt{2}, \sqrt{2}, -1$

9. 0, -1, 0, -1, ND, ND **11.** $-\frac{\sqrt{2}}{2}, -\frac{\sqrt{2}}{2}, 1, -\sqrt{2}, -\sqrt{2}, 1$ **13.** -1, 0, ND, ND, -1, 0 **15.** $-\frac{\sqrt{2}}{2}, \frac{\sqrt{2}}{2}, -1, \sqrt{2}, -\sqrt{2}, -1$

17. $-\frac{1}{2}$ **19.** $\sqrt{3}$ **21.** 0 **23.** $-\frac{2\sqrt{3}}{3}$ **25.** ND **27.** -2 **29.** $\frac{\sqrt{3}}{2}$ **31.** 1 **33.** 0.9744 **35.** 1.0000

37. -2.9042 **39.** **(a)** yes **(b)** no **41.** $\cos\theta = \frac{3}{5}$; $\sin\theta = \frac{4}{5}$; $\tan\theta = \frac{4}{3}$; $\sec\theta = \frac{5}{3}$; $\csc\theta = \frac{5}{4}$; $\cot\theta = \frac{3}{4}$

43. $\cos\theta = \frac{\sqrt{26}}{26}$; $\sin\theta = \frac{5\sqrt{26}}{26}$; $\tan\theta = 5$; $\sec\theta = \sqrt{26}$; $\csc\theta = \frac{\sqrt{26}}{5}$; $\cot\theta = \frac{1}{5}$

45. $\cos\theta = \frac{1}{2}$; $\sin\theta = \frac{\sqrt{3}}{2}$; $\tan\theta = \sqrt{3}$; $\sec\theta = 2$; $\csc\theta = \frac{2\sqrt{3}}{3}$; $\cot\theta = \frac{\sqrt{3}}{3}$

47. $\cos\theta = \frac{a}{\sqrt{a^2+b^2}}$; $\sin\theta = \frac{b}{\sqrt{a^2+b^2}}$; $\tan\theta = \frac{b}{a}$; $\sec\theta = \frac{\sqrt{a^2+b^2}}{a}$; $\csc\theta = \frac{\sqrt{a^2+b^2}}{b}$; $\cot\theta = \frac{a}{b}$

49. $\cos\theta = \frac{\sqrt{3}}{2}$; $\sin\theta = \frac{1}{2}$; $\tan\theta = \frac{\sqrt{3}}{3}$; $\sec\theta = \frac{2\sqrt{3}}{3}$; $\csc\theta = 2$; $\cot\theta = \sqrt{3}$

51. 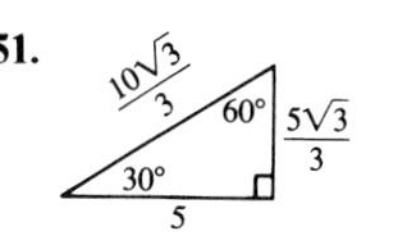

53. 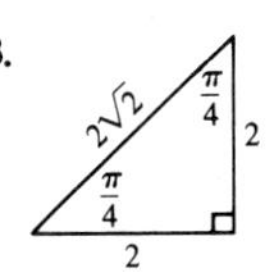

55. 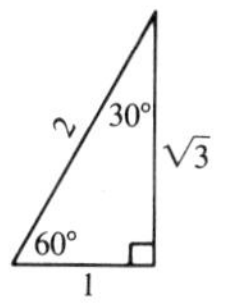

57. 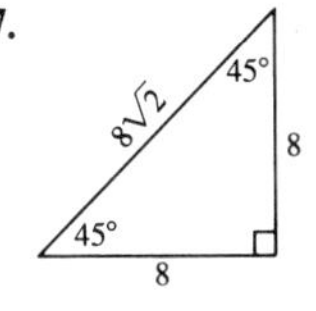

59. 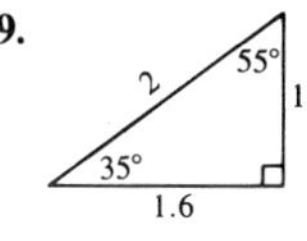

61. 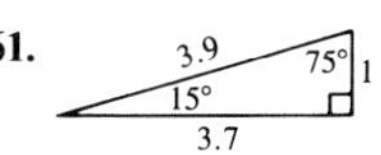

63 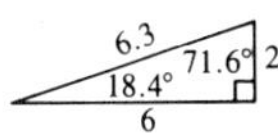

65. 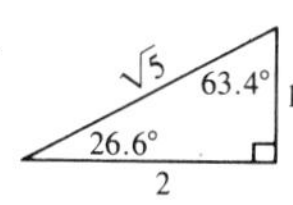

67. 500 yds, 233.1° **69.** 0.52 miles **73.** 1.86 ft **75.** 2930 ft **77.** 295.9 ft, bearing = 38.6°

SECTION 7.2

1. $\gamma = 80°$, $b \approx 13.47$, $c \approx 15.32$ **3.** $\beta = 73°$, $a \approx 1.042$, $c \approx 0.2174$ **5.** $\beta = \frac{2\pi}{7}$, $b \approx 3.604$, $c \approx 4.494$

7. $\beta_1 \approx 74.2°$, $\gamma_1 \approx 45.8°$, $c_1 \approx 3.725$; $\beta_2 \approx 105.8°$, $\gamma_2 \approx 14.2°$, $c_2 \approx 1.275$ **9.** $\alpha \approx 8.3°$, $\beta \approx 161.7°$, $b \approx 54.25$ **11.** there are no triangles
13. $\alpha_1 \approx 84.1°$, $\gamma_1 \approx 15.9°$, $c_1 \approx 2.754$; $\alpha_2 \approx 95.9°$, $\gamma_2 \approx 4.1°$, $c_2 \approx 0.7187$ **15.** $\beta \approx 54.6°$, $\gamma \approx 69.2°$, $c \approx 6.075$
17. $\alpha \approx 24.4°$, $\beta \approx 130.3°$, $b \approx 10.71$ **19.** $\gamma \approx 0.9785$, $\alpha \approx 0.6865$, $a \approx 7.640$ **21.** 1297 ft. **23.** 90.29 ft.
25. 26.6°; $d(A$, hiker$) \approx 316$ meters; $d(B$, hiker$) \approx 141$ meters
27. height ≈ 4.798 ft; lengths of sides ≈ 9.055 ft, 6.457 ft; area ≈ 28.79 sq. ft.
29. speed ≈ 3.29 km/hr; time to shore ≈ 23.2 min.

SECTION 7.3

1. $a \approx 3.050$, $\beta \approx 6.5°$, $\gamma \approx 163.5°$ **3.** $b \approx 5.978$, $\alpha \approx 0.7591$, $\gamma \approx 1.754$ **5.** $c \approx 123.9$, $\alpha \approx 23.8°$, $\beta \approx 126.2°$
7. $a \approx 3.629$, $\beta \approx 1.682$, $\gamma \approx 0.6742$ **9.** $\alpha \approx 46.6°$, $\beta \approx 29°$, $\gamma \approx 104.4°$ **11.** $\alpha \approx 35.3°$, $\beta \approx 100.8°$, $\gamma \approx 43.9°$
13. no such triangle exists **15.** $\alpha \approx 87.1°$, $\beta \approx 87.1°$, $\gamma \approx 5.8°$ **17.** $\alpha \approx 114.7°$, $\beta \approx 27.9°$, $\gamma \approx 37.4°$ **19.** 186 feet
21. 7.613 cm **23.** d(car, balloon) ≈ 14.16 miles; height of balloon = 7.5 miles **25.** 47.8 miles **27.** height ≈ 619.3 feet
29. d(weight, A) ≈ 73.15 feet, d(weight, B) ≈ 107.0 feet

SECTION 7.4

1. 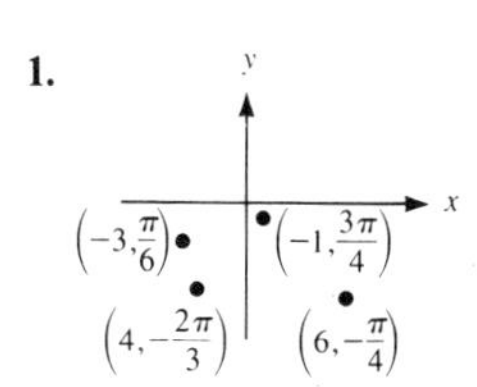

3.

5. $\left(2, \frac{\pi}{4} + 2\pi n\right)$ and $\left(-2, \frac{5\pi}{4} + 2\pi n\right)$ for any integer n

7. $\left(\frac{8}{3}, \frac{-\pi}{3} + 2\pi n\right)$ and $\left(-\frac{8}{3}, \frac{2\pi}{3} + 2\pi n\right)$ for any integer n **9.** $\left(-\frac{1}{2}, \frac{3\pi}{4} + 2\pi n\right)$ and $\left(\frac{1}{2}, \frac{7\pi}{4} + 2\pi n\right)$ for any integer n

11. $(5, 2\pi n)$ and $(-5, \pi + 2\pi n)$ for any integer n **13.** $(-\sqrt{2}, 35° + 360°n)$ and $(\sqrt{2}, 215° + 360°n)$ for any integer n

15. $(-7, -125° + 360°\,n)$ and $(7, 55° + 360°\,n)$ for any integer n **17.** $\left(\frac{1}{2}, \frac{\sqrt{3}}{2}\right)$ **19.** $(0, -4)$ **21.** $\left(\frac{\sqrt{6}}{2}, \frac{\sqrt{2}}{2}\right)$

23. $\left(-\frac{5\sqrt{2}}{2}, \frac{5\sqrt{2}}{2}\right)$ **25.** $\left(-\frac{21\sqrt{3}}{20}, -\frac{21}{20}\right)$ **27.** $\left(\frac{5\sqrt{2}}{2}, \frac{5\sqrt{2}}{2}\right)$ **29.** $(1.809, 1.314)$ **31.** $(-4.669, -11.271)$ **33.** $(1.264, 1.739)$

35. $(6.729, -1.529)$ **37.** $(0.4875, -3.970)$ **39.** $\left(2\sqrt{2}, \frac{3\pi}{4}\right)$ **41.** $\left(\sqrt{2}\pi, \frac{7\pi}{4}\right)$ **43.** $\left(4, \frac{\pi}{6}\right)$ **45.** $\left(\frac{1}{3}, \frac{\pi}{2}\right)$ **47.** $\left(\frac{3}{2}, \frac{3\pi}{2}\right)$

49. $\left(14, \frac{5\pi}{6}\right)$ **51.** $(7.810, 0.8761)$ **53.** $(2.386, 4.501)$ **55.** $(235.8, 1.012)$ **57.** $(6.888, 3.625)$ **59.**

61.

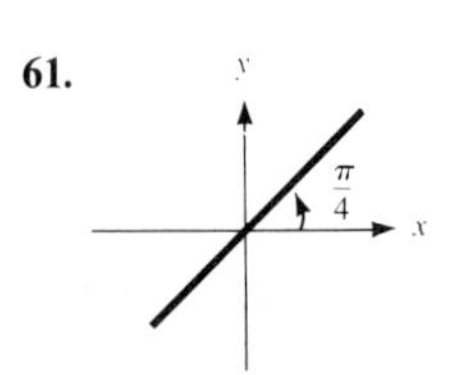

SECTION 7.5

1. $\langle\sqrt{3}, 1\rangle$ **3.** $\left\langle -\frac{\sqrt{2}}{2}, \frac{\sqrt{2}}{2}\right\rangle$ **5.** $\langle 0, -10\rangle$ **7.** $\langle 7.372, -5.162\rangle$ **9.** $\langle 5.273, 2.863\rangle$ **11.** $|\mathbf{v}| = 1, \theta = \frac{\pi}{2}$

13. $|\mathbf{v}| = 2\sqrt{2}, \theta = \frac{\pi}{4}$ **15.** $|\mathbf{v}| = 2\sqrt{3}, \theta = \frac{7\pi}{6}$ **17.** $|\mathbf{v}| = 2, \theta = \frac{2\pi}{3}$ **19.** $|\mathbf{v}| = \sqrt{5}, \theta \approx 2.256$ **21.** $|\mathbf{v}| = 5\sqrt{5}, \theta \approx 3.605$

23. $|\mathbf{v}| = \sqrt{26}\pi, \theta \approx 1.373$ **25.** $\langle 4, -6\rangle$ **27.** $\langle 26, -3\rangle$ **29.** $\langle 3, -2\rangle$ **31.** $2\mathbf{i} - 7\mathbf{j}$ **33.** $6\mathbf{i} - 2\mathbf{j}$ **35.** 0

37. Addition of vectors satisfies the associative property **39.** Magnitude of total force $\approx 9.434n$, direction angle $\approx 2.0°$

41. Magnitude of total force $\approx 101.98n$, direction angle $\approx 258.7°$

SECTION 7.6 (REVIEW)

1. **3.** **5.**

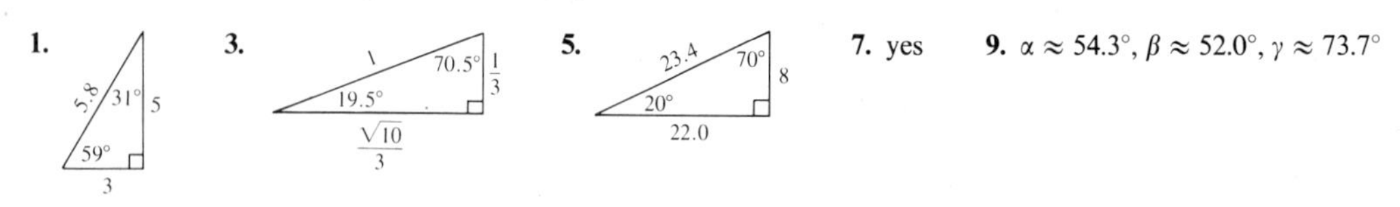

7. yes **9.** $\alpha \approx 54.3°, \beta \approx 52.0°, \gamma \approx 73.7°$

11. $\alpha \approx 20.45°, \beta \approx 97.16°, c \approx 31.80$ **15.** 2429 feet **17.** approximately 81 ft. and 112 ft.

19. Let s be the speed of sound in feet/sec. and let d be the distance from the woman to the first factory. Then the distance between the factories is $2d^2 + 4sd + 4s^2 - 2d(d + 2s)\cos 32°$.

21. $\left(1, \frac{\pi}{2} + 2\pi n\right)$ and $\left(-1, -\frac{\pi}{2} + 2\pi n\right)$ for any integer n **23.** $\left(2, -\frac{4\pi}{3} + 2\pi n\right)$ and $\left(-2, -\frac{\pi}{3} + 2\pi n\right)$ for any integer n

25. $(-20, -20° + 360° n)$ and $(20, 160° + 360° n)$ for any integer n. **27.** $(0, -\sqrt{2})$ **29.** $(0, 0)$ **31.** $(-2.5, 0)$ **33.** $(11.41, 3.71)$

35. $(154.55, 41.41)$ **37.** $\left(\frac{2\sqrt{3}}{3}, \frac{5\pi}{6}\right)$ **39.** $\left(\frac{\sqrt{2}}{3}, \frac{5\pi}{4}\right)$ **41.** $(5.143, 5.820)$ **43.** $(7.001, 6.263)$ **45.** $\left\langle \frac{\sqrt{2}}{4}, \frac{\sqrt{2}}{4} \right\rangle$

47. $\langle 0.3177, 0.0891 \rangle$ **49.** $|\mathbf{v}| = 10;\ \theta = \frac{11\pi}{6}$ **51.** $\langle 12, -8 \rangle$ **53.** $\langle 0, -5 \rangle$

Chapter Eight

SECTION 8.1

1. (a) $\cos t = \pm\sqrt{\frac{\csc^2 t - 1}{\csc^2 t}}$ **(b)** $\frac{2\sqrt{2}}{3}$ **3. (a)** $\cos t = \pm\sqrt{\frac{\cot^2 t}{1 + \cot^2 t}}$ **(b)** $-\frac{5\sqrt{29}}{29}$

5. $\sin t = \pm\sqrt{1 - \cos^2 t}$, $\tan t = \pm\frac{\sqrt{1 - \cos^2 t}}{\cos t}$, $\sec t = \frac{1}{\cos t}$, $\csc t = \pm\frac{1}{\sqrt{1 - \cos^2 t}}$, $\cot t = \pm\frac{\cos t}{\sqrt{1 - \cos^2 t}}$ **7.** $\cos t$ **9.** $\csc t$

11. $\tan\theta$ **13.** $\csc^2\theta$ **15.** $\frac{1 + \cos x}{1 + \sin x}$ **17.** $\sec\theta$ **19.** $\frac{(\sin^3 u)(1 + \cos u)}{\cos u}$ **21.** $\csc u$

SECTION 8.2

Note: In these answers, n varies over all integers.

1. $\frac{7\pi}{6}, \frac{11\pi}{6}$ **3.** π **5.** $\frac{\pi}{3}, \frac{4\pi}{3}$ **7.** $\frac{2\pi}{3}, \frac{5\pi}{3}$ **9.** $\frac{\pi}{9}, \frac{5\pi}{9}, \frac{7\pi}{9}, \frac{11\pi}{9}, \frac{13\pi}{9}, \frac{17\pi}{9}$ **11.** $\frac{\pi}{6}, \frac{5\pi}{6}$ **13.** $\frac{\pi}{2} + 2\pi n$ **15.** $\frac{\pi}{6} + \pi n$

17. $\frac{\pi}{6} + 2\pi n, \frac{11\pi}{6} + 2\pi n$ **19.** 3.6084, 5.8164 **21.** 0.6675, 5.6157 **23.** 1.4711, 4.6127 **25.** 2.3237, 5.4653 **27.** 1.4295, 4.8537

29. 0.3401, 2.8015 **31.** $0, \pi$ **33.** $\pi n, \frac{\pi}{4} + \pi n$ **35.** $\pm\frac{\pi}{2}, \pm\frac{\pi}{3}, \pm\frac{2\pi}{3}$ **37.** $\frac{\pi}{2}$ **39.** $\frac{\pi}{4}, \frac{3\pi}{4}, \frac{5\pi}{4}, \frac{7\pi}{4}$ **41.** $\frac{\pi}{2}$

43. $\frac{5\pi}{6} + 2\pi n, \frac{7\pi}{6} + 2\pi n, 2\pi n$ **45.** $-0.9129, -2.2287, 4.0545, 5.3703$

47. $1.3181, -1.3181, 4.9651, -4.9651, 1.9106, -1.9106, 4.3726, -4.3726$ **49.** $0, \frac{\pi}{2}, \pi, \frac{3\pi}{2}, \frac{\pi}{8}, \frac{3\pi}{8}, \frac{5\pi}{8}, \frac{7\pi}{8}, \frac{9\pi}{8}, \frac{11\pi}{8}, \frac{13\pi}{8}, \frac{15\pi}{8}$

51. 0.3398, 2.8018 **53.** $\pi n, 0.3398 + 2\pi n, 2.8018 + 2\pi n$ **55.** $\frac{\pi}{4} + \pi n$ **57.** 7.9°

59. (a) $\theta_c = \sin^{-1} n$ **(b)** $|n| \leq 1$ **(c)** $\theta_c = 13.5°$

SECTION 8.3

1. $\sin(s + t) = \frac{2 + 6\sqrt{14}}{25}$, $\cos(s + t) = \frac{\sqrt{21} - 4\sqrt{6}}{25}$, quadrant II **3.** $\cos(s - t) = \frac{2\sqrt{2} + \sqrt{15}}{12}$, $\sin(s - t) = \frac{1 - 2\sqrt{30}}{12}$, quadrant III

5. $\frac{\sqrt{6}-\sqrt{2}}{4}$ **7.** $\frac{\sqrt{2}+\sqrt{6}}{4}$ **9.** $\frac{\sqrt{2}-\sqrt{6}}{4}$ **11.** $\frac{-\sqrt{6}-\sqrt{2}}{4}$ **13.** $\frac{\sqrt{2}-\sqrt{6}}{4}$ **15.** $-\frac{\sqrt{2}}{2}$ **17.** $\frac{-\sqrt{6}+\sqrt{2}}{4}$ **19.** $\frac{-\sqrt{2}-\sqrt{6}}{4}$

21. $\cos t$ **23.** $-\cot\theta$ **25.** $\sin x$ **27.** $-\tan t$ **29.** $\sin s$ **31.** $-\cot t$ **33.** $\frac{\sqrt{3}}{2}\cos t + \frac{1}{2}\sin t$

35. $-\frac{\sqrt{2}}{2}\cos t + \frac{\sqrt{2}}{2}\sin t$ **37.** $\frac{1}{2}\cos t + \frac{\sqrt{3}}{2}\sin t$ **47.** $\frac{10+2\sqrt{6}}{5\sqrt{101}}$ **49.** $\frac{\pi-\sqrt{3(16-\pi^2)}}{8}$ **51.** $\frac{-2\sqrt{15}+1}{\sqrt{15}+2}$

53. 0 **55.** $\frac{\sqrt{1-x^2}+x^2}{\sqrt{1+x^2}}$

SECTION 8.4

1. $\sin 2t = \frac{4\sqrt{2}}{9}$; $\cos 2t = \frac{7}{9}$; $\tan 2t = \frac{4\sqrt{2}}{7}$ **3.** $\sin 2t = \frac{-4\sqrt{6}}{25}$; $\cos 2t = \frac{-23}{25}$; $\tan 2t = \frac{4\sqrt{6}}{23}$

5. $\sin 2t = \frac{5}{13}$; $\cos 2t = \frac{-12}{13}$; $\tan 2t = \frac{-5}{12}$ **7.** $\sin 2t \approx -0.4495$; $\cos 2t \approx 0.8933$; $\tan 2t \approx -0.5032$ **9.** $\sqrt{\frac{4-\sqrt{2}-\sqrt{6}}{8}}$

11. $\frac{\sqrt{2}}{2-\sqrt{2}}$ **13.** $\frac{-\sqrt{2-\sqrt{2}}}{2}$ **15.** $\frac{-2}{\sqrt{2+\sqrt{2}}}$ **17.** $-\sqrt{2}-1$ **19.** $\frac{2+\sqrt{2-\sqrt{3}}}{\sqrt{2+\sqrt{3}}}$ **21.** $\frac{-\sqrt{6}-\sqrt{2}}{4}$ **23.** $2+\sqrt{3}$

25. $\sin\frac{t}{2} = \frac{3\sqrt{10}}{10}$, $\cos\frac{t}{2} = \frac{\sqrt{10}}{10}$, $\tan\frac{t}{2} = 3$ **27.** $\sin\frac{t}{2} = \sqrt{\frac{5-\sqrt{5}}{10}}$, $\cos\frac{t}{2} = -\sqrt{\frac{5+\sqrt{5}}{10}}$, $\tan\frac{t}{2} = -\sqrt{\frac{3-\sqrt{5}}{2}}$

29. $\sin\frac{t}{2} \approx 0.7268$, $\cos\frac{t}{2} \approx 0.6869$, $\tan\frac{t}{2} \approx -1.0581$ **31.** **(a)** $\csc 2t = \pm\frac{\csc^2 t}{2\sqrt{\csc^2 t - 1}}$ **(b)** $\pm\frac{25\sqrt{6}}{4}$ **(c)** $\frac{9\sqrt{2}}{8}$

33. $\frac{23}{27}$ **35.** $\frac{9\sqrt{10}}{50}$ **37.** 0.854 **39.** $\cos 3t = 4\cos^3 t - 3\cos t$ **41.** $-\frac{42\sqrt{6}}{125}$ **43.** $\frac{52\sqrt{17}}{289}$ **45.** -0.794

47. $\frac{64}{81}$ **63.** $0, \pi$ **65.** 0, $\arctan 2$, $\pi + \arctan 2$ **67.** $0, \frac{\pi}{3}, \frac{5\pi}{3}$

SECTION 8.6 (REVIEW)

1. $\cos\theta = \frac{1}{2}$, $\csc\theta = -\frac{2\sqrt{3}}{3}$, $\cot\theta = -\frac{\sqrt{3}}{3}$, $\sec\theta = 2$ **3.** $\csc\theta = -\frac{6\sqrt{11}}{11}$, $\cot\theta = \frac{5\sqrt{11}}{11}$, $\cos\theta = -\frac{5}{6}$, $\sin\theta = -\frac{\sqrt{11}}{6}$, $\tan\theta = \frac{\sqrt{11}}{5}$

5. $-\sec t$ **7.** $2\csc\beta$ **9.** valid for all $\theta \neq \pi n$, where n is any integer **21.** $\frac{\pi}{3} + 2\pi n, \frac{5\pi}{3} + 2\pi n$, where n is any integer

23. No solutions **25.** $\sin\theta = \frac{x^2}{\sqrt{x^4-1}}$, $\cos\theta = \frac{1}{\sqrt{x^4-1}}$, $\sin\mu = \frac{\sqrt{x^2-1}}{x}$, $\cos\mu = \frac{1}{x}$ **27.** $\cot\left(t-\frac{\pi}{3}\right) = \frac{1+\sqrt{3}\tan t}{\tan t - \sqrt{3}}$

29. $\frac{30-\sqrt{3}}{3+10\sqrt{3}}$ **31.** $\cot(x+y) = \frac{\cot x \cot y - 1}{\cot x + \cot y}$ **35.** $-\frac{\sqrt{15}}{8}$ **37.** $\frac{36}{323}$ **39.** $\sqrt{\frac{6+\sqrt{36-\pi^2}}{12}}$

41. $\frac{1}{2}\sqrt{2-\sqrt{2+\sqrt{2}}}$ **43.** $\sqrt{\frac{4+\sqrt{6}-\sqrt{2}}{8}}$ **47.** $\tan 3t = \frac{\tan^3 t - 3\tan t}{3\tan^2 t - 1}$ **49.** $-\frac{11\sqrt{15}}{45}$ **51.** $\frac{23\sqrt{2}}{20}$ **53.** 4.224

Chapter Nine

SECTION 9.1

1. yes **3.** no **5.** no **7.** yes **9.** yes **15.** one solution

17. infinite solutions

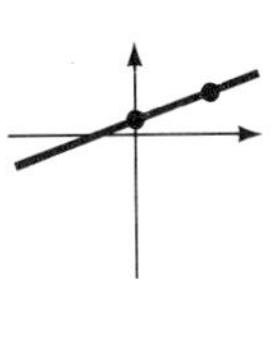

19. no solutions

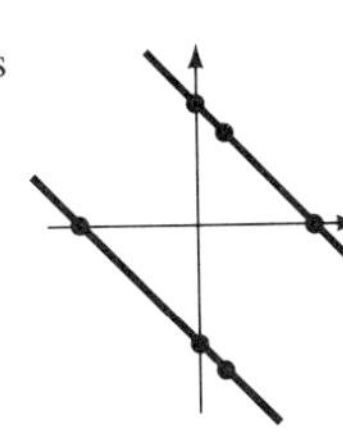

21. $\left(\frac{23}{7}, \frac{5}{7}\right)$ **23.** $x = 0, y = 0, z = 0$ **25.** $x = 12, y = 2, z = 6$

27. $x = 1 + 3u, y = u, z = u$ **29.** $x = -1, y = 2 - u, z = u$ **31.** $x = \frac{1}{2} + \frac{2}{5}u, y = \frac{1}{2}, z = u$

SECTION 9.2

1. $x = 4, y = -4$ **3.** $x = \frac{13}{5}, y = \frac{-9}{5}$ **5.** $x = \frac{2 + u}{3}, y = u$ **7.** $s = \frac{1}{4}, t = \frac{1}{6}$ **9.** $x = \frac{2}{k}, y = \frac{3}{k^2}$ **11.** $x = u, y = u, z = u$

13. no solution **15.** $u = \frac{8k - 5a}{7}, v = \frac{3k + 6a}{7}, w = a$ where a is a parameter **17.** $x = \frac{2u}{3}, y = u, z = \frac{5u}{3}$

19. $x = 1, y = 2, z = -3$ **21.** $x = 0, y = \frac{5}{2}, z = \frac{-3}{2}$ **23.** $x = -1, y = \frac{7}{2}, z = \frac{-1}{2}$ **25.** no solution **27.** no solution

29. no solution **31.** $x = 0, y = 0, z = 0$ **33.** $x = 1, y = 2, z = 3$ **35.** $x = k, y = -k, z = 2k$ **37.** $x = 1, y = 1, z = -1$

39. $x = \frac{2}{7}u, y = -\frac{15}{21}u, z = \frac{2}{7}u, w = u$ **41.** $u = \frac{13 + 7a}{20}, v = \frac{6 - a}{5}, w = a, x = \frac{17 + 23a}{20}$ where a is a parameter

43. $x = 3k - u, y = u - k, z = 4k - u, w = u$ **45.** $x = 2, y = 1, z = -1, w = 4$

47. $x = 1 - c, y = a, z = b, w = c, u = 2 - a - b$ where a, b, c are parameters **49.** $x = 2, y = -3$

SECTION 9.3

1. $\left(\begin{array}{cc|c} 2 & -4 & 16 \\ 2 & \frac{1}{3} & -4 \end{array}\right)$ **3.** $\left(\begin{array}{ccc|c} 1 & 0 & 2 & 0 \\ 0 & 1 & 3 & 5 \\ 0 & 0 & 1 & 6 \end{array}\right)$ **5.** $\left(\begin{array}{cccc|c} 2 & 3 & -2 & 0 & k \\ 0 & 4 & 0 & -\frac{1}{2} & \frac{2}{3} \\ 0 & 0 & -\frac{1}{3} & \frac{1}{12} & 13k \\ 1 & 0 & -k & 0 & -1 \end{array}\right)$

7. $\begin{aligned} 3x + 6y + 7z &= 5 \\ 4x \qquad + 6z &= -2 \end{aligned}$ **9.** $\begin{aligned} x &= 1 \\ 2x &= 3 \\ 4x &= -8 \end{aligned}$

11. $x = 0$ **13.** $\begin{aligned} kx + (k+1)y + (k+2)z &= 0 \\ 2kx + 3ky + 4kz &= 0 \\ 5x + 7y + 4z &= 0 \\ kx + k^2y + k^3z &= k^4 \end{aligned}$ **15.** $\begin{pmatrix} 2 & 3 & 4 \\ 3 & 4 & 5 \\ 4 & 5 & 6 \end{pmatrix}$ **17.** $\begin{pmatrix} 1 & 0 & 0 \\ 0 & 1 & 0 \\ 0 & 0 & 1 \end{pmatrix}$ **19.** $\begin{pmatrix} 0 & 0 & 0 \\ 0 & 0 & 0 \\ 0 & 0 & 0 \end{pmatrix}$

SECTION 9.4

1. $f(x) = 0$ **3.** $f(x) = \frac{11}{4}x - \frac{49}{4}$ **5.** $f(x) = -2x^2 + 7x - 4$ **7.** $f(x) = \frac{23}{12}x^2 - \frac{43}{12}x - \frac{3}{2}$ **9.** $f(x) = 3\sin x + (4 - 3\sqrt{3})\cos x$

11. $f(x) = \frac{2}{2+\sqrt{6}}\sin x + \frac{\sqrt{2}}{2+\sqrt{6}}\cos x$ **13.** $14 + \frac{u}{3}, 7 - \frac{4}{3}u, u$ where u is any real number **15.** 236 **17.** 123 adults, 77 children

19. $33\frac{1}{3}$ ml. of 25% solution, $66\frac{2}{3}$ ml. of 10% solution **21.** \$7,692.32 at 8%, \$3,846.15 at 10% **23.** \$37,500 at 10%, \$62,500 at 18%
25. speed of plane = 550 mph, speed of wind = 50 mph **27.** 50 of type A, 100 of type B

SECTION 9.5

1. 11 **3.** 0 **5.** 1 **7.** $y - x$ **9.** 0 **11.** $x = -8, y = -4$ **13.** $x = \frac{a+b}{6}, y = \frac{4b-2a}{6}$ **15.** $x = \frac{4}{3}, y = \frac{4}{9}$

17. $x = \frac{-3}{k}, y = \frac{5+9k}{2}$ **19.** $x = -\frac{70}{9}, y = -\frac{16}{9}$

SECTION 9.6

1. $M_{2,3} = -6$; $C_{2,3} = 6$; $M_{1,3} = -3$; $C_{1,3} = -3$ **3. (a)** -96 **(b)** -96 **5. (a)** -53 **(b)** -53 **7.** abc **9.** 0

11. -117 **13.** 12 **15.** $\frac{29}{180}$ **17.** $yz^2 - zy^2 - xz^2 + xy^2 + x^2z - x^2y$ **19.** adf **21.** 1 **23.** cannot use

25. $x = \frac{15}{7}, y = -\frac{10}{7}, z = \frac{16}{7}$ **27.** cannot use **29.** $x = \frac{15}{16}, y = \frac{45}{32}, z = \frac{23}{32}$ **31.** $x = \frac{65}{42}, y = \frac{10}{21}, z = -\frac{35}{42}$

SECTION 9.7

1. $\begin{pmatrix} 3 & 2 \\ 5 & 7 \end{pmatrix}$ **3.** $\begin{pmatrix} 3 & 9 \\ 12 & 0 \end{pmatrix}$ **5.** $\begin{pmatrix} 4 & 5 \\ 9 & 7 \end{pmatrix}$ **7.** $\begin{pmatrix} -1 & 10 \\ 8 & -4 \\ 9 & -5 \end{pmatrix}$ **9.** $\begin{pmatrix} -5 & 8 \\ -8 & 8 \\ 1 & -5 \end{pmatrix}$ **11.** $\begin{pmatrix} -18 & 33 \\ -24 & 26 \\ 8 & -20 \end{pmatrix}$ **13.** $\begin{pmatrix} -1 & 0 & 4 \\ 5 & 3 & -1 \\ 2 & 2 & 4 \end{pmatrix}$

15. $\begin{pmatrix} -2\pi & \pi & 3\pi \\ 5\pi & 0 & -\pi \\ \pi & -\pi & 0 \end{pmatrix}$ **17.** $\begin{pmatrix} -10 & 7 & 5 \\ 15 & -12 & -3 \\ -1 & -15 & -16 \end{pmatrix}$ **19.** associative **21.** property of negatives

23. commutative property of multiplication does not hold **25.** $\begin{pmatrix} 14 & 0 \\ 36 & -60 \end{pmatrix}$ **27.** $\begin{pmatrix} 0 & 0 \\ 0 & 0 \end{pmatrix}$ **29.** $\begin{pmatrix} 9 & -12 & 6 \\ -1 & 6 & 4 \end{pmatrix}$

31. $\begin{pmatrix} 7 & -4 \\ 24 & 19 \\ 3 & -4 \end{pmatrix}$ **33.** $\begin{pmatrix} 0 & 0 & 0 \\ 0 & 0 & 0 \\ 0 & 0 & 0 \end{pmatrix}$ **35.** $\begin{pmatrix} k & k & k \\ k & k & k \\ k & k & k \end{pmatrix}$ **37.** $\begin{pmatrix} a & 2a & 3a \\ b & 2b & 3b \\ c & 2c & 3c \end{pmatrix}$ **43.** $\begin{pmatrix} \frac{1}{3} & \frac{1}{3} \\ 0 & -1 \end{pmatrix}$ **45.** $\begin{pmatrix} \frac{1}{2} & 0 \\ -\frac{3}{10} & \frac{1}{5} \end{pmatrix}$ **47.** $\begin{pmatrix} \frac{2}{3} & -\frac{1}{3} \\ -\frac{1}{15} & \frac{2}{15} \end{pmatrix}$

49. $\begin{pmatrix} \frac{2}{3} & \frac{2}{3} & -1 \\ \frac{5}{3} & -\frac{1}{3} & -1 \\ -1 & 0 & 1 \end{pmatrix}$ **51.** $\begin{pmatrix} \frac{1}{2} & 0 & \frac{1}{2} \\ -\frac{9}{4} & 1 & -\frac{5}{4} \\ -2 & 0 & -1 \end{pmatrix}$ **53.** $\begin{pmatrix} -3 & 1 & -5 \\ 4 & -1 & 7 \\ 8 & -2 & 13 \end{pmatrix}$ **55.** $\begin{pmatrix} \frac{3}{32} & \frac{17}{32} & -\frac{3}{8} \\ -\frac{1}{32} & \frac{5}{32} & \frac{1}{8} \\ \frac{1}{4} & -\frac{1}{4} & 0 \end{pmatrix}$

SECTION 9.8

1. $\begin{pmatrix} 2 & 3 \\ 1 & -2 \end{pmatrix}\begin{pmatrix} x \\ y \end{pmatrix} = \begin{pmatrix} 1 \\ 0 \end{pmatrix}$ **3.** $\begin{pmatrix} 10 & -11 \\ 17 & 12 \end{pmatrix}\begin{pmatrix} x \\ y \end{pmatrix} = \begin{pmatrix} 12 \\ \sqrt{2} \end{pmatrix}$ **5.** $\begin{pmatrix} 2 & -3 \\ 1 & 4 \end{pmatrix}\begin{pmatrix} x \\ y \end{pmatrix} = \begin{pmatrix} 7 \\ 2 \end{pmatrix}$ **7.** $\begin{pmatrix} 3 & 2 \\ 5 & 0 \end{pmatrix}\begin{pmatrix} x \\ y \end{pmatrix} = \begin{pmatrix} r \\ r^2 \end{pmatrix}$

9. $\begin{pmatrix} \pi & \sqrt{2} & -1 \\ 4 & -\sqrt{3} & 1 \\ 1 & -1 & 1 \end{pmatrix}\begin{pmatrix} x \\ y \\ z \end{pmatrix} = \begin{pmatrix} 3 \\ \sqrt{5} \\ \pi \end{pmatrix}$ **11.** $\begin{pmatrix} 2 & 3 & -5 \\ 3 & -2 & 9 \\ -8 & 0 & -1 \end{pmatrix}\begin{pmatrix} x \\ y \\ z \end{pmatrix} = \begin{pmatrix} 9 \\ \frac{1}{2} \\ -\frac{1}{2} \end{pmatrix}$ **13.** $\begin{pmatrix} 1 & 0 & 0 \\ 0 & 1 & 0 \\ 0 & 0 & 1 \end{pmatrix}\begin{pmatrix} x \\ y \\ z \end{pmatrix} = \begin{pmatrix} 9 \\ 9 \\ 9 \end{pmatrix}$

15. $x = 0, y = 0$ **17.** $x = 18, y = -11$ **19.** $x = \frac{9}{10}, y = \frac{13}{10}$ **21.** $x = 0, y = 0$ **23.** $x = 0, y = 0, z = 0$

25. $x = 3, y = 0, z = -1$ **27.** $x = 0, y = 0, z = 0$ **29.** $x = 13, y = \frac{-23}{2}, z = -8$ **31.** $x = \frac{-1}{7}, y = \frac{2}{7}, z = \frac{6}{7}$

33. $x = 0, y = 0, z = 0$ **35.** $x = -2, y = 3, z = 7$ **37.** $x = -1 + 2a - 3a^2, y = 2 + a, z = 4 - 2a + 5a^2$

SECTION 9.9

Note: In Exercises 1–23, solutions to systems of equations are given as ordered pairs.

1. $(2, 0); \left(\frac{-10}{13}, \frac{-24}{13}\right)$ **3.** $\left(\frac{7}{5}, \frac{1}{5}\right); \left(-\frac{7}{5}, \frac{1}{5}\right)$ **5.** $(\sqrt{3}, 0); (-\sqrt{3}, 0)$ **7.** $(1, 2); (1, -2)$

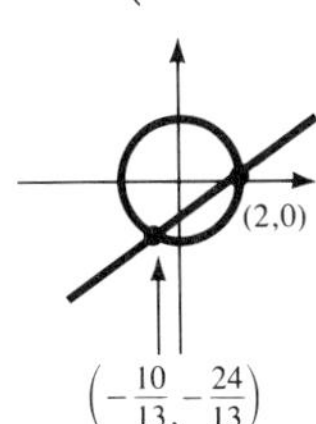

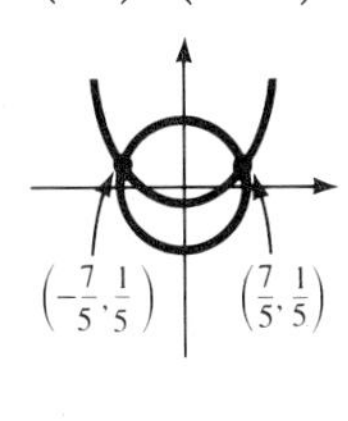

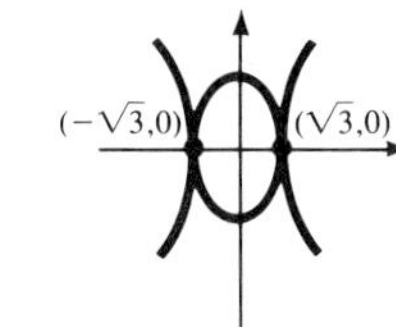

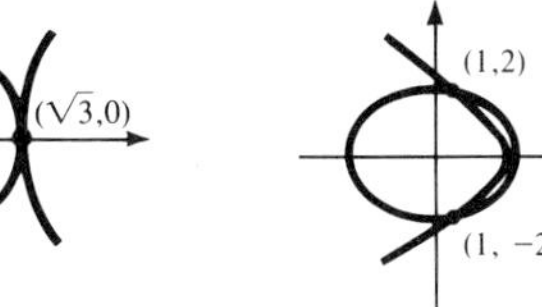

9. $\left(\frac{5 + \sqrt{19}}{3}, \frac{4 - 4\sqrt{19}}{9}\right); \left(\frac{5 - \sqrt{19}}{3}, \frac{4 + \sqrt{19}}{9}\right)$ **11.** no solutions

$\left(\frac{5 - \sqrt{19}}{3}, \frac{4 + \sqrt{19}}{9}\right)$

$\left(\frac{5 + \sqrt{19}}{3}, \frac{4 - 4\sqrt{19}}{9}\right)$

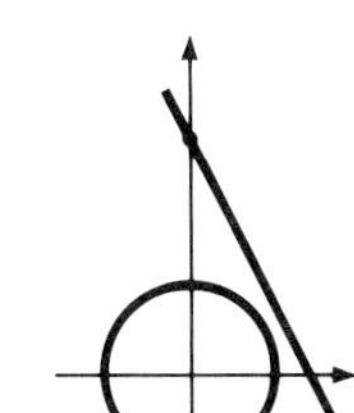

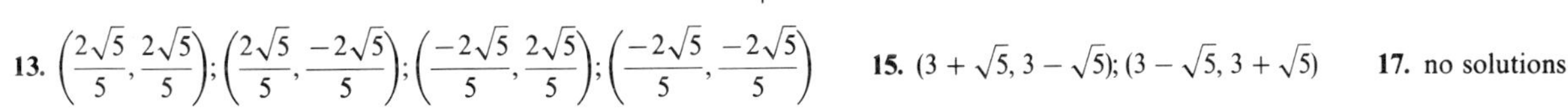

13. $\left(\frac{2\sqrt{5}}{5}, \frac{2\sqrt{5}}{5}\right); \left(\frac{2\sqrt{5}}{5}, \frac{-2\sqrt{5}}{5}\right); \left(\frac{-2\sqrt{5}}{5}, \frac{2\sqrt{5}}{5}\right); \left(\frac{-2\sqrt{5}}{5}, \frac{-2\sqrt{5}}{5}\right)$ **15.** $(3 + \sqrt{5}, 3 - \sqrt{5}); (3 - \sqrt{5}, 3 + \sqrt{5})$ **17.** no solutions

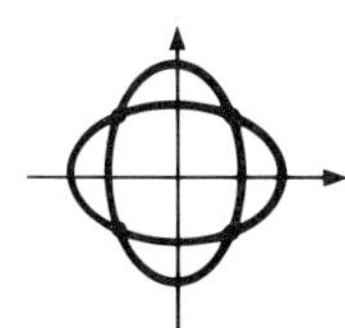

19. $\left(\dfrac{-3+\sqrt{17}}{4}, \dfrac{4}{3-\sqrt{17}}\right); \left(\dfrac{-3-\sqrt{17}}{4}, \dfrac{4}{3+\sqrt{17}}\right)$ **21.** (4, 9) **23.** $\left(\dfrac{\log(-1+\sqrt{11})}{\log 3}, 8-\sqrt{11}\right)$ **25.** no solutions

27. $x = 1, y = \dfrac{1}{\pi}, z = \cos\dfrac{1}{\pi}$ **29.** $\dfrac{1+\sqrt{5}}{2}, \dfrac{-1+\sqrt{5}}{2}$ and $\dfrac{1-\sqrt{5}}{2}, \dfrac{-1-\sqrt{5}}{2}$ **31.** $-2, 3$ **33.** $\frac{1}{2}$ ft. × 2 ft.

35. radius = 10 m., height = 4 m. **37.** $(x-\frac{1}{2})^2 + (y+\frac{1}{2})^2 = \frac{5}{2}$ **39.** width = 2 cm., length = 4 cm.

SECTION 9.10

1.

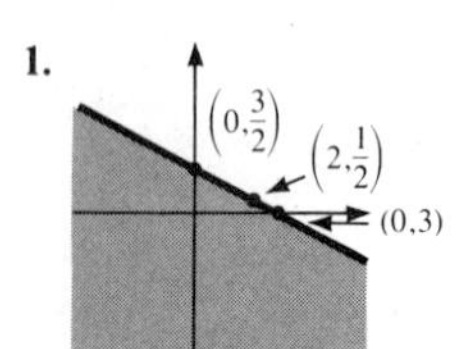

3.

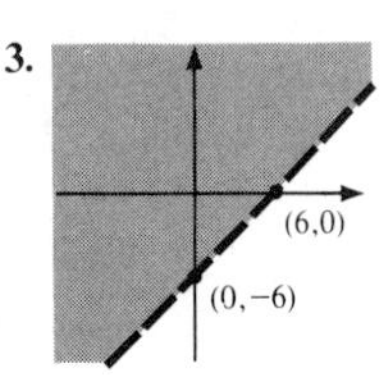

5.

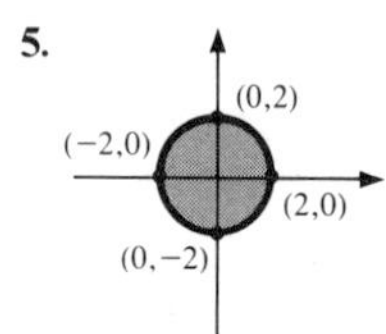

7.

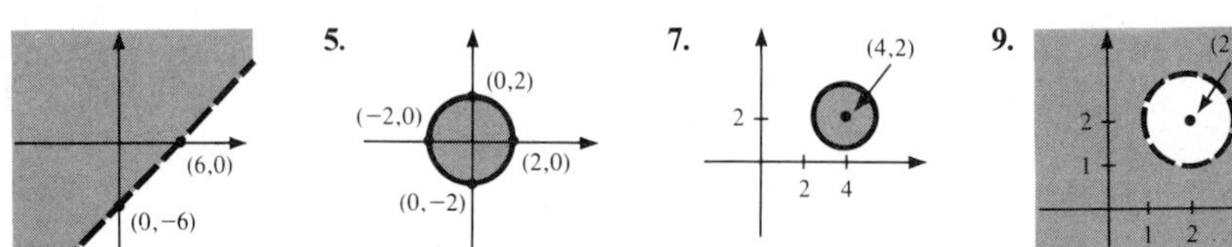

9.

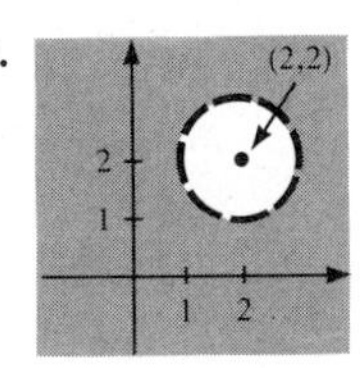

11.

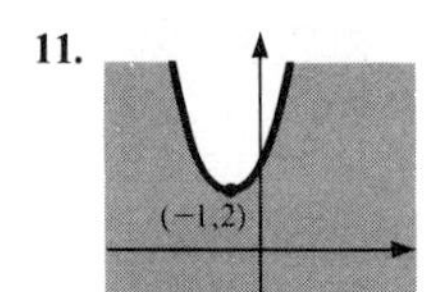

13.

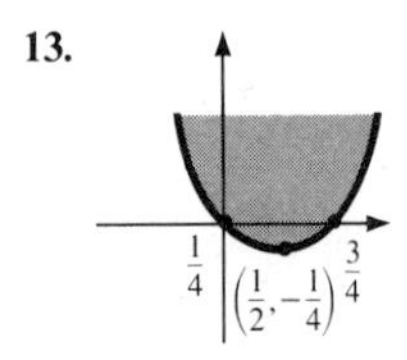

15.

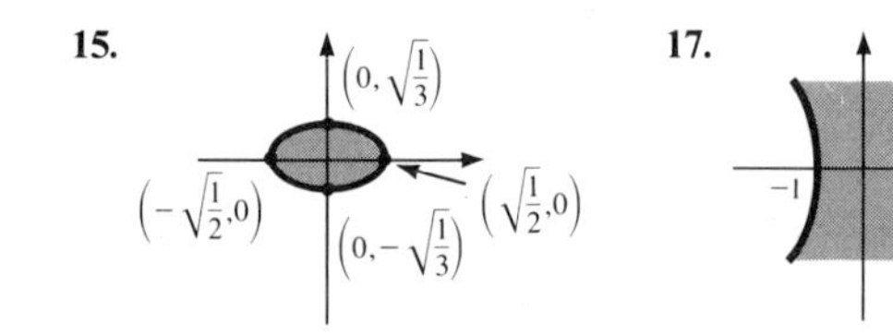

17.

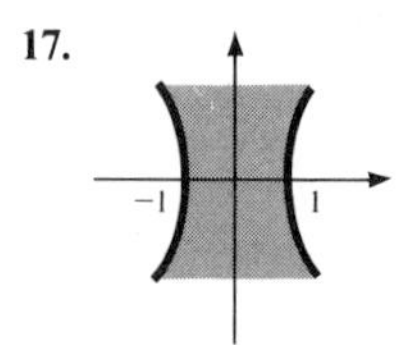

19.

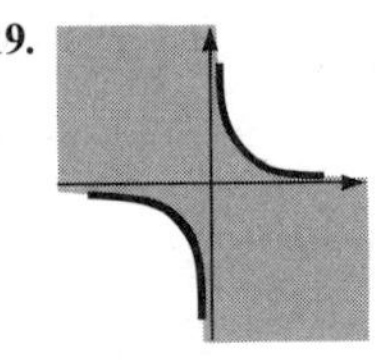

21.

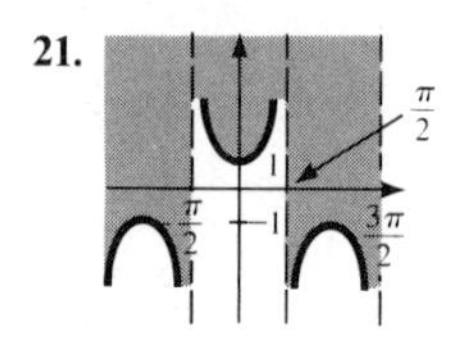

23.

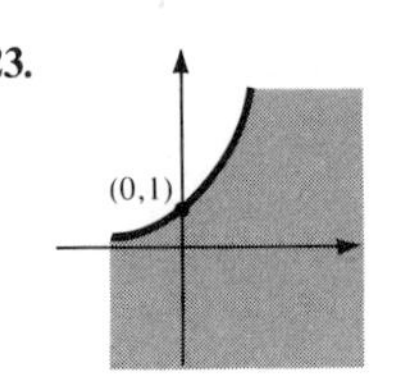

25.

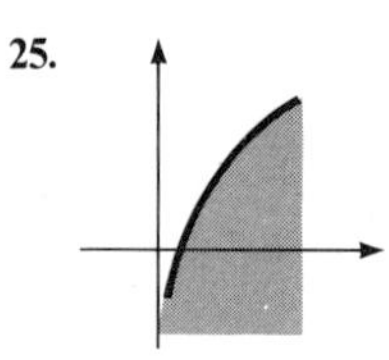

27.

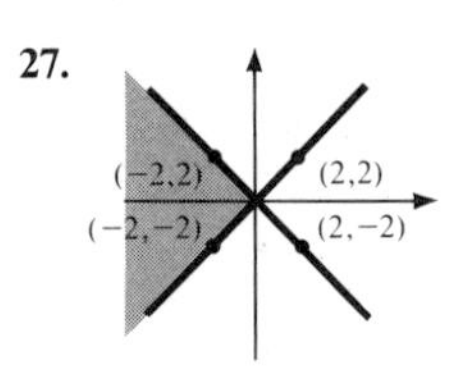

29.

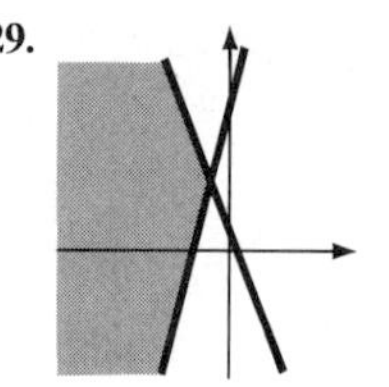

31.

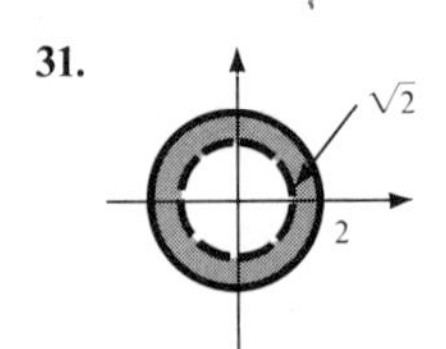

33.

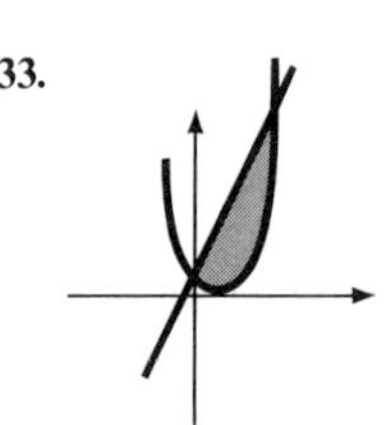

35.

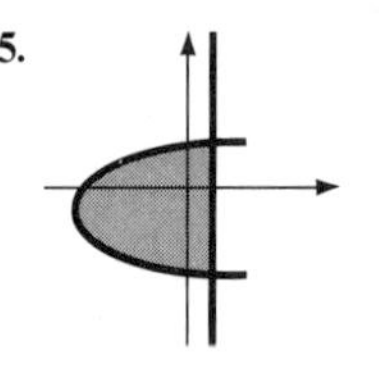

37.

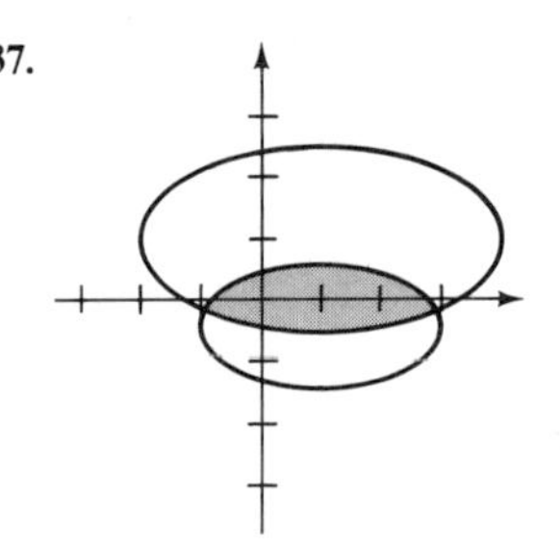

39.

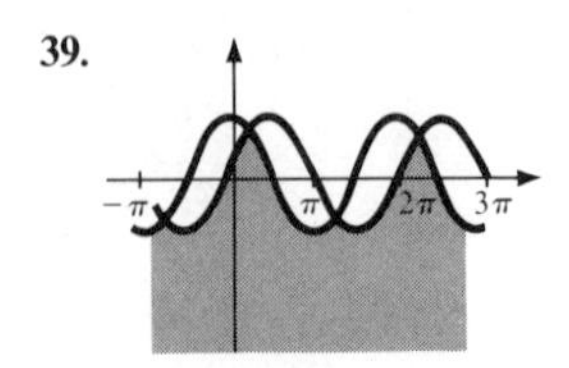

41.

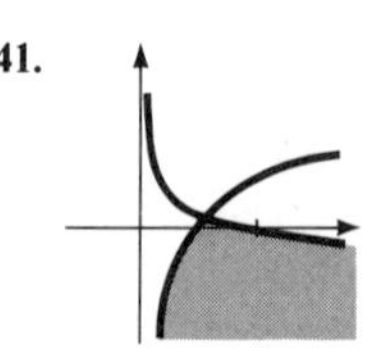

43.

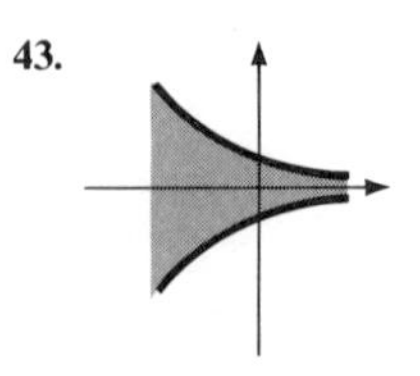

45.

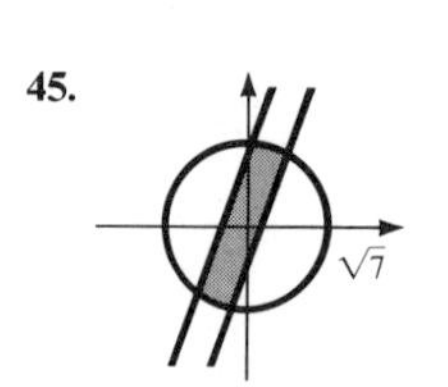

47.

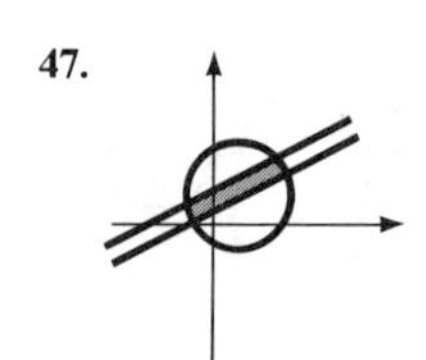

49.

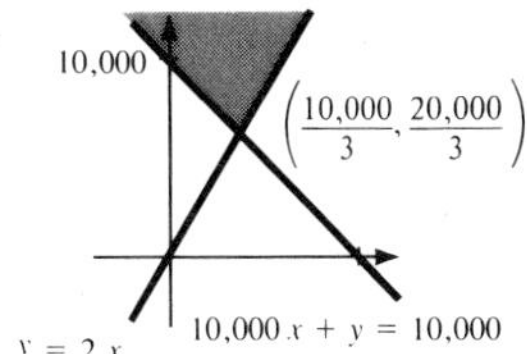

50. no

SECTION 9.11 (REVIEW)

1. no **3.** yes **5.** $x = \dfrac{4a+7}{3}, y = a$ **7.** $x = 1, y = 3, z = -\dfrac{1}{2}$ **9.** $x = 2, y = 2, z = 1$ **11.** $x = 10, y = -9, z = 2, w = 8$

13. if $k = 7$, there are infinitely many solutions; if $k \neq 7$, there are no solutions

15. if $k = 2$, there are infinitely many solution; if $k \neq 2$, there are no solutions **17.** $a = \frac{2}{3}, b = 2; x = 2, y = -2$

19. $a = -2, b = 3$ **21.** 342 **23.** -25 **25.** 1 **27.** $2a^3 - 6a^2 + 6a$ **29.** 1 **31.** 0 **33.** $x = \frac{1}{2}, y = 0$

35. cannot use Cramer's rule **37.** $\begin{pmatrix} -2 & -3 & 3-\sqrt{2} \\ 2 & \pi & 1 \\ 6 & 1 & -7 \end{pmatrix}$ **39.** $\begin{pmatrix} 2 & 4 & 6 \\ 4 & 0 & 0 \\ 8 & -2 & 4 \end{pmatrix}$ **41.** $\begin{pmatrix} -9 & -16 & -2\sqrt{2}-9 \\ -6 & 2\pi & 2 \\ -8 & 7 & -24 \end{pmatrix}$ **43.** $\begin{pmatrix} \frac{7}{4} & \frac{1}{2} \\ \frac{5}{2} & -\frac{1}{4} \end{pmatrix}$

45. $\begin{pmatrix} \begin{pmatrix} 2 & 7 \\ 5 & 2 \end{pmatrix} & \begin{pmatrix} 3 & 8 \\ 5 & 0 \end{pmatrix} \\ \begin{pmatrix} 3 & 6 \\ 2 & 4 \end{pmatrix} & \begin{pmatrix} 0 & 1 \\ 1 & 2 \end{pmatrix} \end{pmatrix}$ **47.** $\begin{pmatrix} \begin{pmatrix} 0 & 6 \\ 12 & 3 \end{pmatrix} & \begin{pmatrix} 0 & 6 \\ 3 & -6 \end{pmatrix} \\ \begin{pmatrix} 6 & 12 \\ 3 & 6 \end{pmatrix} & \begin{pmatrix} 9 & 0 \\ 0 & 9 \end{pmatrix} \end{pmatrix}$ **49.** $\begin{pmatrix} \begin{pmatrix} -2 & -1 \\ 7 & 1 \end{pmatrix} & \begin{pmatrix} -3 & -2 \\ -2 & -6 \end{pmatrix} \\ \begin{pmatrix} 3 & 6 \\ 1 & 2 \end{pmatrix} & \begin{pmatrix} 9 & -1 \\ -1 & 7 \end{pmatrix} \end{pmatrix}$ **51.** $\begin{pmatrix} 0 & 1 \\ 0 & 0 \end{pmatrix}$ **53.** $\begin{pmatrix} 0 \\ 0 \end{pmatrix}$

55. $\begin{pmatrix} 10 & 1 & 5 & 0 \\ -24 & 16 & -24 & 8 \\ -36 & 13 & 29 & 6 \\ 45 & 7 & 40 & 7 \end{pmatrix}$ **57.** $\begin{pmatrix} 3 & 2 \\ 1 & 4 \end{pmatrix}\begin{pmatrix} x \\ y \end{pmatrix} = \begin{pmatrix} 3 \\ 0 \end{pmatrix}$ **59.** $\begin{pmatrix} 9 & 2 & -3 & 9 \\ 1 & -8 & 1 & 0 \\ 1 & 1 & -2 & -1 \\ 2 & 9 & -7 & 0 \end{pmatrix}\begin{pmatrix} x \\ y \\ z \\ w \end{pmatrix} = \begin{pmatrix} 12 \\ 10 \\ 3 \\ 0 \end{pmatrix}$

61. $\begin{pmatrix} -1 & 1 \\ -\frac{1}{5} & \frac{1}{10} \end{pmatrix}$ **63.** $\begin{pmatrix} -5 & 4 & -7 \\ 10 & -7 & 13 \\ 8 & -6 & 11 \end{pmatrix}$ **65.** $x = \dfrac{-39}{10}, y = \dfrac{41}{10}, z = \dfrac{1}{2}$ **67.** $x = -8, y = -16, z = -33, w = 54$

69. $x = \dfrac{1}{3}, y = \dfrac{-16}{3}, z = -1, w = 8$ **71.** $x = 1, y = 100; x = 2, y = 10$ **73.** radius $= \dfrac{\sqrt{6}}{3}$ **75.** $x = 0, y = 0$

Chapter Ten

SECTION 10.1

1. 2 **3.** 0 **5.** 1 **7.** $\sqrt{2}$ **9.** i **11.** 1 **13.** $8 - 2i$ **15.** 3 **17.** $\pi - 2\sqrt{2} + 3i$ **19.** $11 + 10i$ **21.** 11

23. $10 + 26i$ **25.** 10 **27.** $2 - 11i$ **29.** 0 **31.** 0 **33.** $2i$ **35.** 0 **37.** $-2\sqrt{3}$ **39.** $\sqrt{3} + \sqrt{3}i$

41. $7 + (1 - \sqrt{2})i$ **43.** $(2 + \sqrt{2})i$ **45.** $-24\sqrt{3}i$ **47.** no real solutions **49.** $x = \pm 2, y = \pm 1$

51. $x = 3, y = \frac{1}{3}; x = -3, y = -\frac{1}{3}$ **53.** no real solutions **55.** $z = \pm\sqrt{3}i$ **57.** $z = 0, -1$ **59.** $z = 0, -2i$

61. $z = -i$ **63.** $z = 0, -2$ **65.** $z = \dfrac{i}{2}$

SECTION 10.2

1. $2 - 3i$ **3.** i **5.** i **7.** $\frac{i}{2} - 7$ **9.** $\sqrt{2}$ **11.** $\sqrt{11}$ **13.** 10^{10} **15.** $\frac{\sqrt{10}}{2}$ **17.** 1 **19.** $\frac{\sqrt{130}}{10}$ **21.** $\sqrt{2}$

23. 6 **25.** $\frac{5}{61} + \frac{6}{61}i$ **27.** $\frac{21}{25} + \frac{28}{25}i$ **29.** $\frac{5}{13} - \frac{12}{13}i$ **31.** $\frac{\sqrt{2}}{11} + \frac{3}{11}i$ **33.** $\frac{1}{85} + \frac{38}{85}i$ **35.** $-i$ **37.** $\frac{2}{5} - \frac{11}{5}i$

39. $\frac{4}{25} - \frac{3}{25}i$ **41.** $z = a(1 + i)$ where a is any real number **43.** $z = 0$ **45.** $z = r$ where r is any *nonnegative* real number

47. $-\frac{1}{10} + \frac{7}{10}i$

SECTION 10.3

1. $\pm\frac{i}{2}$ **3.** $\pm 2i$ **5.** $1 \pm i$ **7.** $\frac{1}{2}, 3$ **9.** $\frac{1}{2} \pm \frac{1}{3}i$ **11.** $\pm\frac{\sqrt{3}}{3}i$ **13.** $\pm\pi i$ **15.** $2 \pm \sqrt{5}i$ **17.** $1 \pm \sqrt{5}i$

19. $\frac{\sqrt{2}}{2} \pm 2i$ **21.** $\frac{2}{9} \pm \frac{\sqrt{5}}{9}i$ **23.** $\pm 2, \pm\sqrt{2}$ **25.** $\pm\frac{\sqrt{6}}{3}i, \pm\frac{\sqrt{10}}{4}i$ **27.** $\pm\frac{\sqrt{110 + 10\sqrt{161}}}{10}, \pm\frac{\sqrt{10\sqrt{161} - 110}}{10}i$

29. $x = 1, 1 \pm 2i$ **31.** $x^2 + 1 = 0$ **33.** $4x^2 - 24x + 37 = 0$ **35.** $x^2 + 1 = 0$ **37.** $x^2 - 2x + 2 = 0$ **39.** $x^2 - 4x + 4 = 0$

SECTION 10.4

1. $3i$ **3.** $\frac{5\sqrt{2}}{2} + \frac{5\sqrt{2}}{2}i$ **5.** -2 **7.** $-2\sqrt{3} + 2i$ **9.** $\sqrt{2}\left(\cos\left(-\frac{3\pi}{4}\right) + i\sin\left(-\frac{3\pi}{4}\right)\right)$ **11.** $\cos\left(-\frac{\pi}{2}\right) + i\sin\left(-\frac{\pi}{2}\right)$

13. $2\left(\cos\frac{\pi}{3} + i\sin\frac{\pi}{3}\right)$ **15.** $\frac{\sqrt{13}}{2}(\cos(\arctan\frac{2}{3}) + i\sin(\arctan\frac{2}{3}))$ **17.** $8(\cos 0 + i\sin 0)$ **19.** $9\sqrt{2}\left(\cos\frac{3\pi}{4} + i\sin\frac{3\pi}{4}\right)$

21. $0(\cos\theta + i\sin\theta)$ where θ is any real number **23.** $\pi\left(\cos\frac{\pi}{2} + i\sin\frac{\pi}{2}\right)$ **25.** $415(\cos(-\pi + \arctan\frac{4}{3}) + i\sin(-\pi + \arctan\frac{4}{3}))$

27. $\sqrt{5}(\cos(\arctan\frac{1}{2}) + i\sin(\arctan\frac{1}{2}))$ **29.** $\frac{1}{324}\left(\cos\frac{\pi}{2} + i\sin\frac{\pi}{2}\right)$ **31.** $zw = 4(\cos 17° + i\sin 17°), \frac{z}{w} = \frac{1}{4}(\cos 13° + i\sin 13°)$

33. $zw = 9\sqrt{2}\left(\cos\left(-\frac{\pi}{2}\right) + i\sin\left(-\frac{\pi}{2}\right)\right), \frac{z}{w} = \frac{9\sqrt{2}}{2}\left(\cos\frac{\pi}{2} + i\sin\frac{\pi}{2}\right)$ **35.** $zw = -2 + 2i, \frac{z}{w} = 1 + i$ **37.** $zw = 2, \frac{z}{w} = -i$

39. $zw = -4, \frac{z}{w} = -\frac{1}{2} + \frac{\sqrt{3}}{2}i$ **41.** $zw = -i, \frac{z}{w} = -i$ **43.** $z^2 = r^2(\cos 2\theta + i\sin 2\theta), z^3 = r^3(\cos 3\theta + i\sin 3\theta), z^n = r^n(\cos n\theta + i\sin n\theta)$

SECTION 10.5

1. 1 **3.** -343 **5.** $8(-1 + \sqrt{3}i)$ **7.** $-2^{19}(1 + \sqrt{3}i)$ **9.** $16(1 - \sqrt{3}i)$ **11.** $-\frac{1}{2} - \frac{\sqrt{3}}{2}i$ **13.** $\pm\frac{\sqrt{2}}{2}(1 + i)$

15. $-\sqrt[3]{2}i, \frac{\sqrt[3]{2}}{2}(\sqrt{3} + i), \frac{\sqrt[3]{2}}{2}(-\sqrt{3} + i)$ **17.** $\pm\sqrt[4]{18}\left(\cos\left(-\frac{\pi}{8}\right) + i\sin\left(-\frac{\pi}{8}\right)\right)$

19. $\sqrt[6]{8}\left(\frac{\sqrt{2}}{2} - \frac{\sqrt{2}}{2}i\right), \sqrt[6]{8}\left(\cos\frac{5\pi}{12} + i\sin\frac{5\pi}{12}\right), \sqrt[6]{8}\left(\cos\frac{13\pi}{12} + i\sin\frac{13\pi}{12}\right)$

21. $1, \cos\frac{2\pi}{7} + i\sin\frac{2\pi}{7}, \cos\frac{4\pi}{7} + i\sin\frac{4\pi}{7}, \cos\frac{6\pi}{7} + i\sin\frac{6\pi}{7}, \cos\frac{8\pi}{7} + i\sin\frac{8\pi}{7}, \cos\frac{10\pi}{7} + i\sin\frac{10\pi}{7}, \cos\frac{12\pi}{7} + i\sin\frac{12\pi}{7}$; a regular heptagon.

23. $x = \pm 2, \pm(1 + \sqrt{3}i), \pm(1 - \sqrt{3}i)$ **25.** $x = -i, \pm\frac{\sqrt{3}}{2} - \frac{1}{2}i$ **27.** $x = 0, \sqrt[3]{6}i, \sqrt[3]{6}\left(\pm\frac{\sqrt{3}}{2} - \frac{1}{2}i\right)$

SECTION 10.6 (REVIEW)

1. $4 + 6i$ **3.** $-1 - i$ **5.** $6\sqrt{3}(1 - i)$ **7.** -1 **9.** $-\frac{8}{15} + \frac{6}{15}i$ **11.** $x = \pm\sqrt{3}; y = 0, \frac{1}{2}$ **13.** $z = \pm 2i$

15. $z = \pm\sqrt{3} - 2i$ **17.** $\frac{\sqrt{26}}{13}$ **19.** $z = 1 + i$ **21.** $z = bi$ for any real number b **23.** $x = \pm\sqrt[4]{8}i$ **25.** $\frac{\sqrt{3}}{3} \pm 3i$

27. $x^2 - 2x + 3 = 0$ **29.** $4x^2 - 8x + 5 = 0$ **31.** $z^2 - (2 + 3i)z + 6i = 0$ **33.** $-2iz^2 - (7 + i)z + 5i - 1 = 0$

35. $-\frac{\sqrt{13}}{2} + \frac{\sqrt{39}}{2}i$ **37.** $\pi\sqrt{2}\left(\cos\frac{\pi}{4} + i\sin\frac{\pi}{4}\right)$ **39.** $2\sqrt{2}\left(\cos\frac{\pi}{6} + i\sin\frac{\pi}{6}\right)$

41. $zw = \sqrt{2}\left(\cos\left(-\frac{\pi}{12}\right) + i\sin\left(-\frac{\pi}{12}\right)\right), \frac{z}{w} = \frac{\sqrt{2}}{2}\left(\cos\frac{5\pi}{12} + i\sin\frac{5\pi}{12}\right)$

43. $zw = 2\sqrt{2}\left(\cos\frac{7\pi}{12} + i\sin\frac{7\pi}{12}\right), \frac{z}{w} = \frac{\sqrt{2}}{2}\left(\cos\left(-\frac{\pi}{12}\right) + i\sin\left(-\frac{\pi}{12}\right)\right)$ **45.** $-\frac{1}{2} + \frac{\sqrt{3}}{2}i$

47. $\sqrt[5]{10}(\cos\pi + i\sin\pi), \sqrt[5]{10}\left(\cos\frac{7\pi}{5} + i\sin\frac{7\pi}{5}\right), \sqrt[5]{10}\left(\cos\frac{9\pi}{5} + i\sin\frac{9\pi}{5}\right), \sqrt[5]{10}\left(\cos\frac{11\pi}{5} + i\sin\frac{11\pi}{5}\right), \sqrt[5]{10}\left(\cos\frac{13\pi}{5} + i\sin\frac{13\pi}{5}\right)$

49. $1, \cos\frac{2\pi}{9} + i\sin\frac{2\pi}{9}, \cos\frac{4\pi}{9} + i\sin\frac{4\pi}{9}, \cos\frac{2\pi}{3} + i\sin\frac{2\pi}{3}, \cos\frac{8\pi}{9} + i\sin\frac{8\pi}{9}, \cos\frac{10\pi}{9} + i\sin\frac{10\pi}{9}, \cos\frac{4\pi}{3} + i\sin\frac{4\pi}{3},$

$\cos\frac{14\pi}{9} + i\sin\frac{14\pi}{9}, \cos\frac{16\pi}{9} + i\sin\frac{16\pi}{9}$, a regular nonagon

51. $x = 0, -\sqrt[3]{2}i, \sqrt[3]{2}\left(\pm\frac{\sqrt{3}}{2} + \frac{1}{2}i\right)$

Chapter Eleven

SECTION 11.2

1. $2x = y^2$

Vertex: (0, 0), Focus: $(\frac{1}{2}, 0)$

3. $y = -4x^2 + 10x + 1$

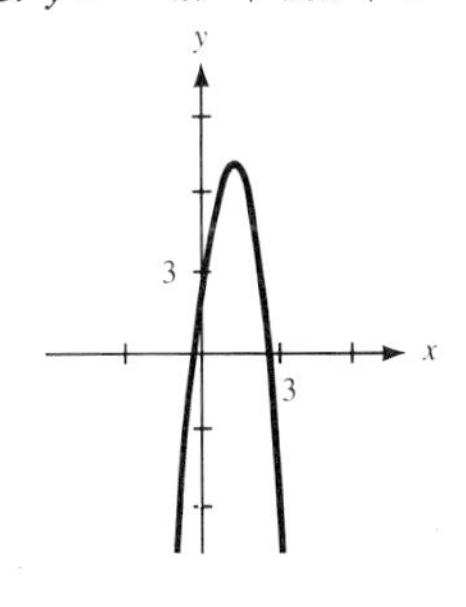

Vertex: $(\frac{5}{4}, \frac{29}{4})$, Focus: $(\frac{5}{4}, \frac{115}{16})$

5. $-2x + 3y = y^2$

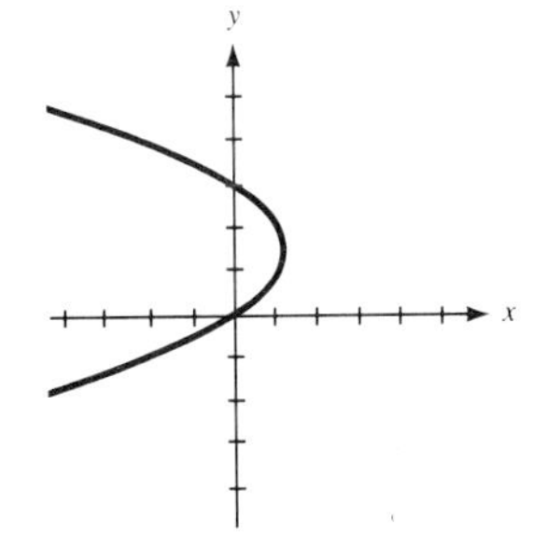

Vertex: $(\frac{9}{8}, \frac{3}{2})$, Focus: $(\frac{5}{8}, \frac{3}{2})$

7. $y = 4x^2 + x + 1$

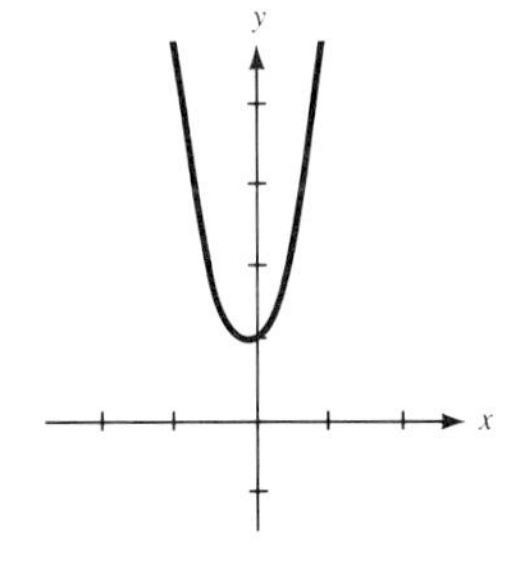

Vertex: $(-\frac{1}{8}, \frac{15}{16})$, Focus: $(-\frac{1}{8}, 1)$

9. $y + 10 = x - x^2$ **11. (a)** $y = \frac{1}{20}x^2$ **(b)** $y = -\frac{1}{20}x^2$ **13.** $y - 3 = \frac{1}{12}(x - 1)^2$ **15.** $x = \frac{1}{8}(y + \frac{1}{2})^2$

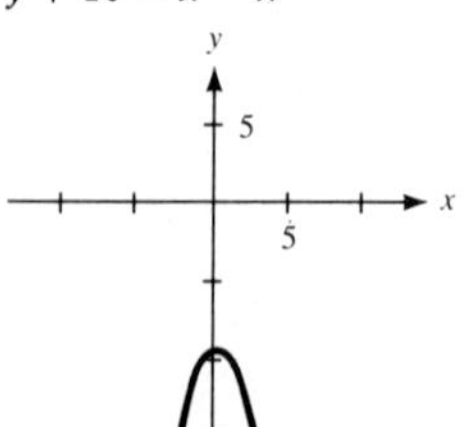

Vertex: $(\frac{1}{2}, -\frac{39}{4})$, Focus: $(\frac{1}{2}, -10)$

17. $x - 3 = -\frac{1}{12}(y + 3)^2$ **19.** $y + 2 = \frac{1}{16}(x + 2)^2$ **21. (a)** $x + \frac{3}{4} = \frac{1}{3}y^2$ **(b)** $x - \frac{3}{4} = -\frac{1}{3}y^2$ **23.** $x = -\frac{1}{2}(y + \frac{1}{2})^2$

25. $y = x^2$ **27.** $x + 2 = -2y^2$ **31.** $y = 2x^2 - 8x + 7$

33. Parabola that opens up has equation $3y = 2x^2 - 11x + 21$
Parabola that opens down has equation $3y = -2x^2 + 9x + 5$

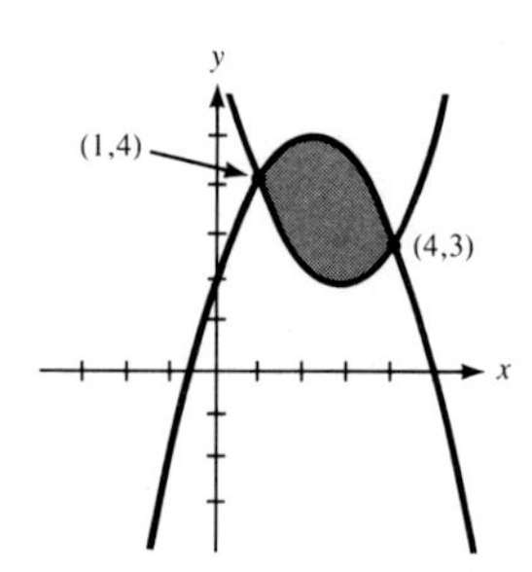

SECTION 11.3

1. $x^2 + \frac{y^2}{2} = 1$ **3.** $4y^2 + 3x^2 = 1$ **5.** $x^2 + 3y^2 - 4x - 6y + 7 = 0$ **7.** $x^2 + 2y^2 - 2x - 4y = 1$

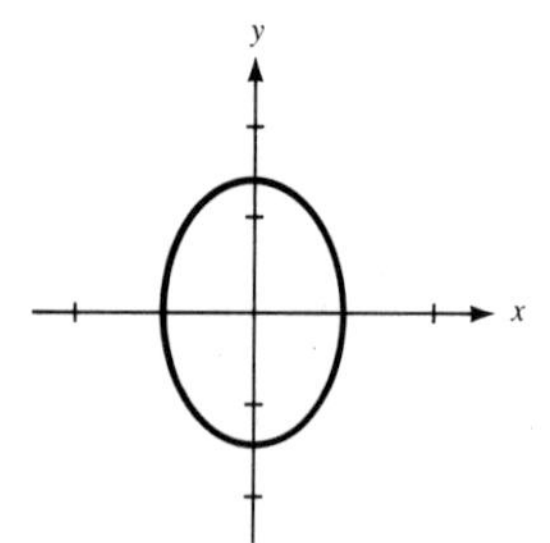

Center: (0, 0)

Foci: (0, 1), (0, −1)

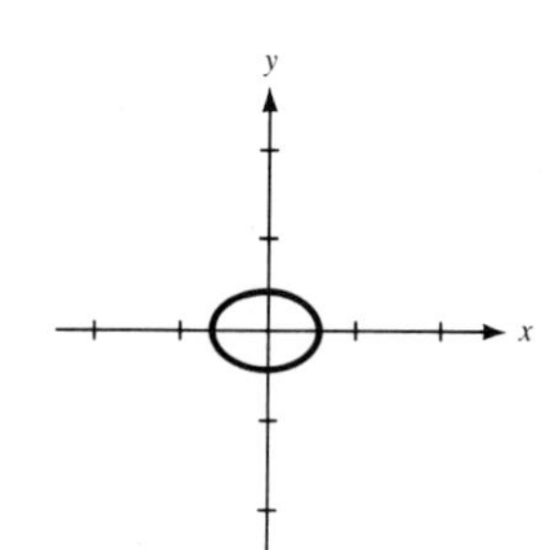

Center: (0, 0)

Foci: $\left(-\frac{\sqrt{3}}{6}, 0\right), \left(\frac{\sqrt{3}}{6}, 0\right)$

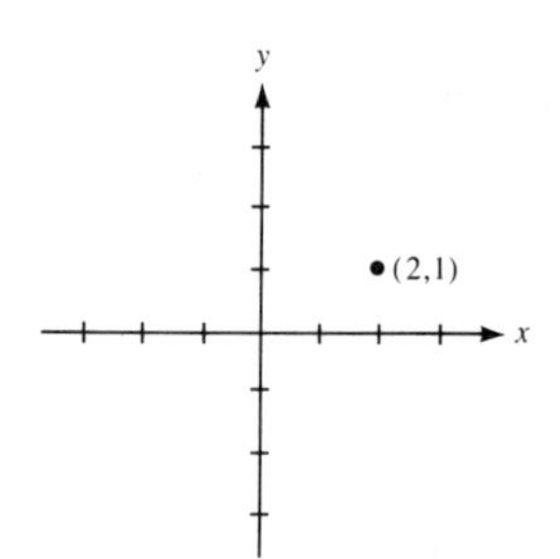

Graph is a point at (2, 1)

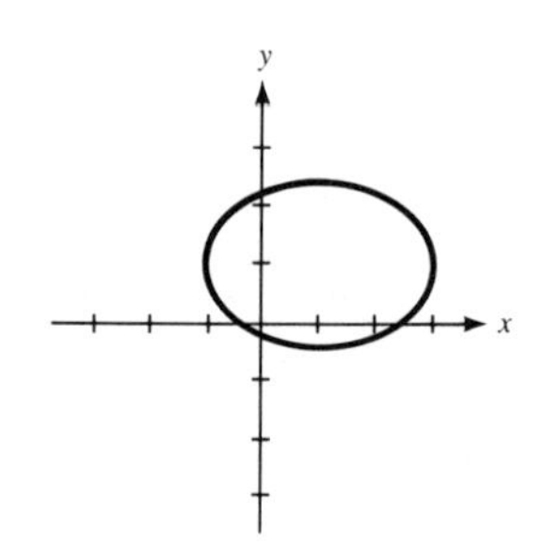

Center: (1, 1)

Foci: $(1 + \sqrt{2}, 1), (1 - \sqrt{2}, 1)$

9. $7x^2 + 6y^2 - 42 = 0$

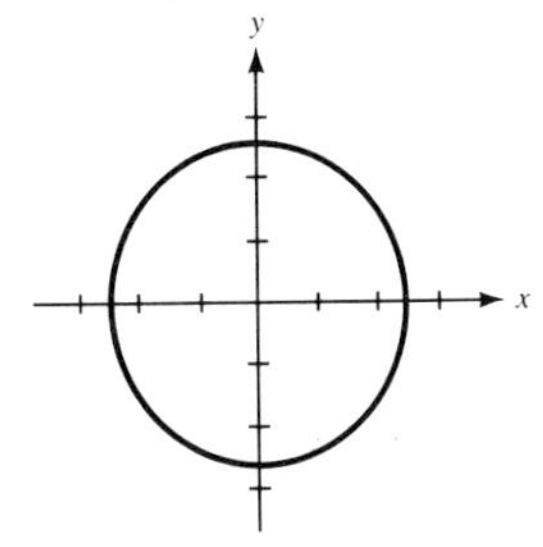

Center: $(0, 0)$

Foci: $(0, 1), (0, -1)$

11. $x^2 + 5y^2 + 50y - 125 = 0$

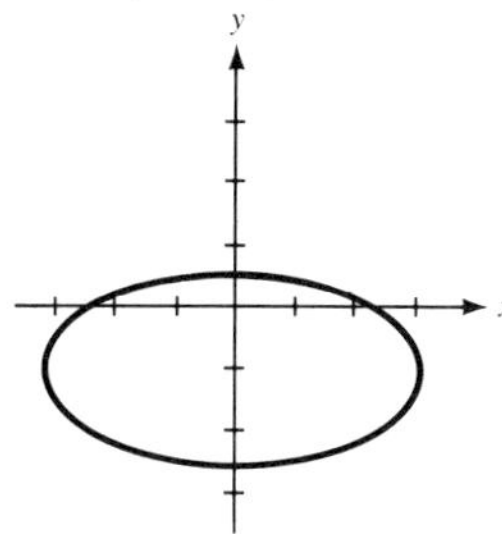

Center: $(0, -5)$

Foci: $(-10\sqrt{2}, 0), (10\sqrt{2}, 0)$

13. $10x^2 + 8y^2 - 10x - 4y - 37 = 0$

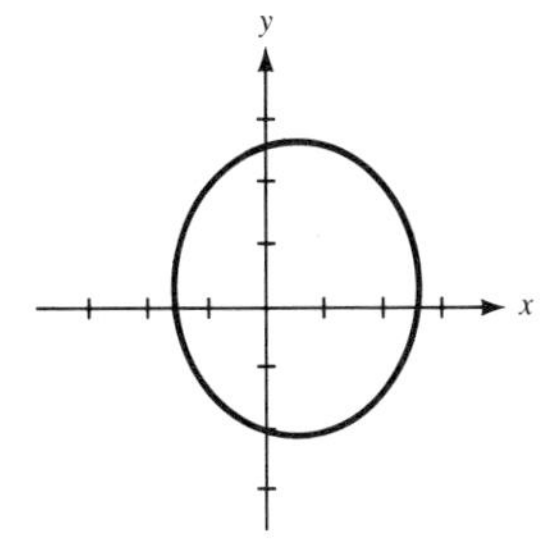

Center: $(\frac{1}{2}, \frac{1}{4})$

Foci: $(\frac{1}{2}, \frac{5}{4}), (\frac{1}{2}, -\frac{3}{4})$

15.  $\dfrac{x^2}{9} + \dfrac{y^2}{100} = 1$

17. **(a)** $(x - 1)^2 + 4(y - 1)^2 = 1$ **(b)** $4(x - 1)^2 + (y - 1)^2 = 1$ **19.** $\dfrac{4x^2}{25} + 2(y - 5)^2 = 1$ **21.** $\dfrac{x^2}{7} + \dfrac{(y - 3)^2}{16} = 1$

23. $\left(\pm\dfrac{4\sqrt{5}}{5}, \pm\dfrac{4\sqrt{5}}{5}\right)$ **25.** no **27.** Area $= \dfrac{4a^2b^2}{a^2 + b^2}$ **31.** 0

SECTION 11.4

1. $\dfrac{(x - 3)^2}{9} - \dfrac{(y - 1)^2}{25} = 1$

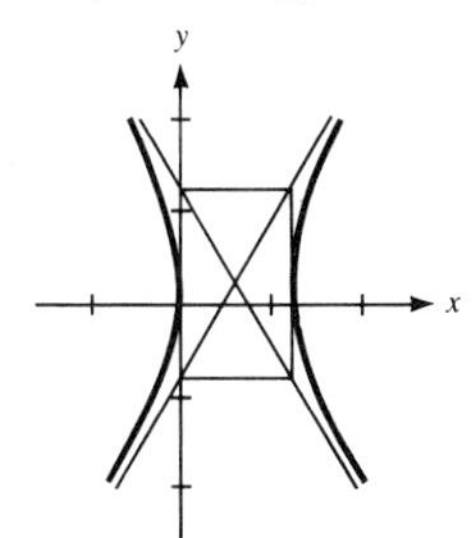

Center: $(3, 1)$

Foci: $(3 - \sqrt{34}, 1), (3 + \sqrt{34}, 1)$

3. $y^2 - x^2 = 1$

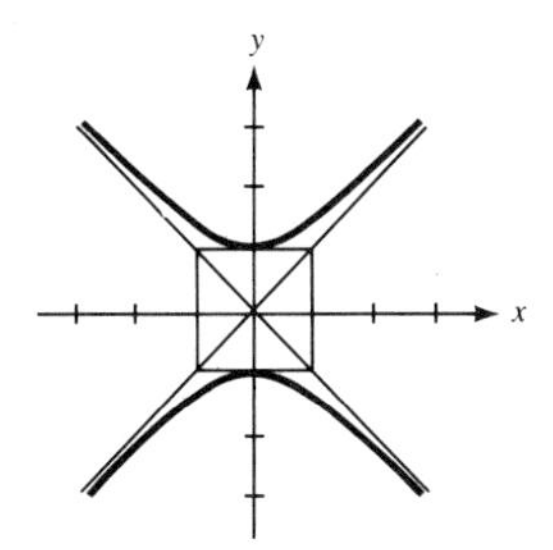

Center: $(0, 0)$

Foci: $(0, -\sqrt{2}), (0, \sqrt{2})$

5. $2(x - 1)^2 - 3(y + 2)^2 = 1$

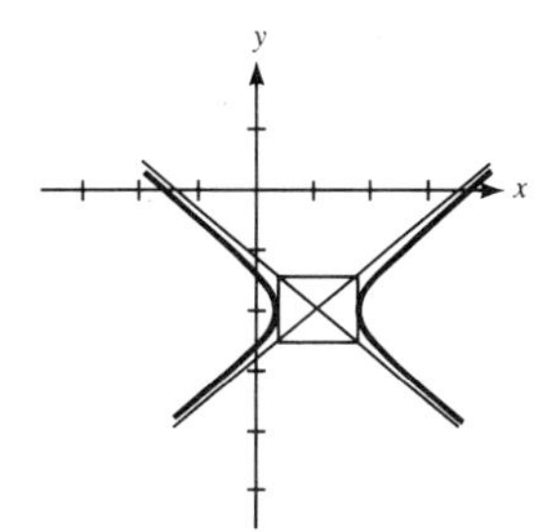

Center: $(1, -2)$

Foci: $\left(1 - \dfrac{\sqrt{30}}{6}, -2\right), \left(1 + \dfrac{\sqrt{30}}{6}, -2\right)$

7. $(6y - 1)^2 - 2(3x + 2)^2 = 2$

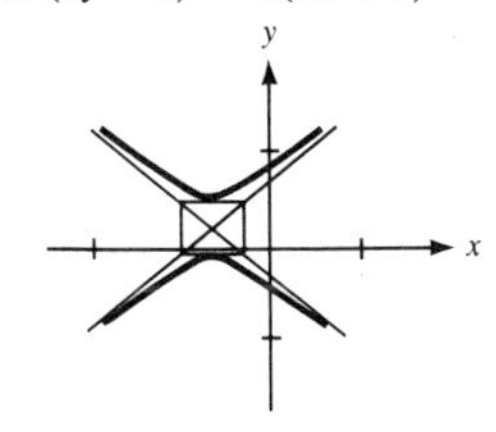

Center: $(-\frac{2}{3}, \frac{1}{6})$

Foci: $\left(-\dfrac{2}{3}, \dfrac{1 - \sqrt{6}}{6}\right), \left(-\dfrac{2}{3}, \dfrac{1 + \sqrt{6}}{6}\right)$

9. $3y^2 - 4x^2 - 16x - 8 = 0$

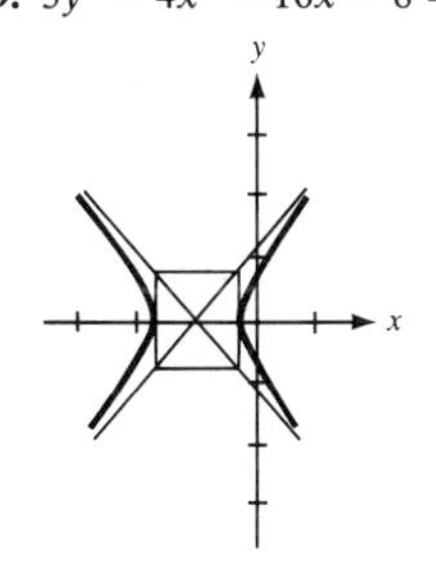

Center: $(-2, 0)$

Foci: $\left(-2 - \dfrac{\sqrt{42}}{3}, 0\right), \left(-2 + \dfrac{\sqrt{42}}{3}, 0\right)$

11. $x^2 - y^2 - 10x - 6y + 15 = 0$

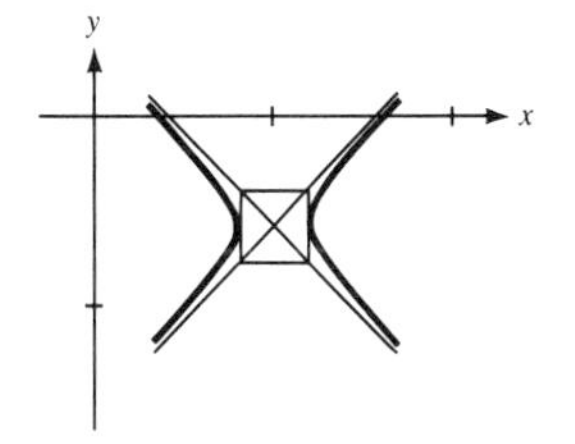

Center: $(5, -3)$

Foci: $(5 - \sqrt{2}, -3), (5 + \sqrt{2}, -3)$

13. $x^2 - y^2 - 2\pi x + 2\pi y = 1$

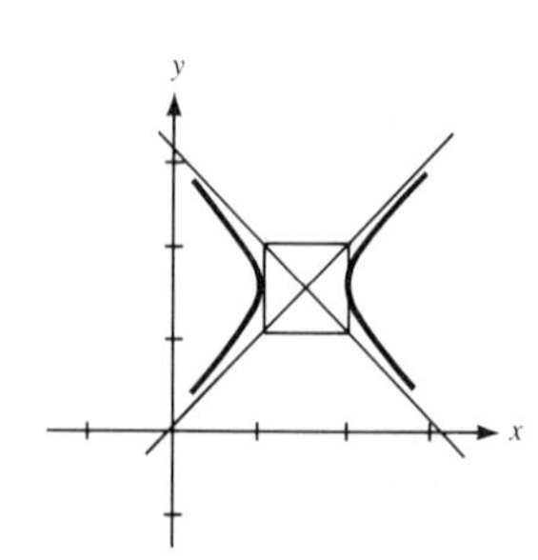

Center: (π, π)
Foci: $(\pi - \sqrt{2}, \pi)$, $(\pi + \sqrt{2}, \pi)$

15. $2x^2 - 3y^2 - 4x - 12y - 10 = 0$

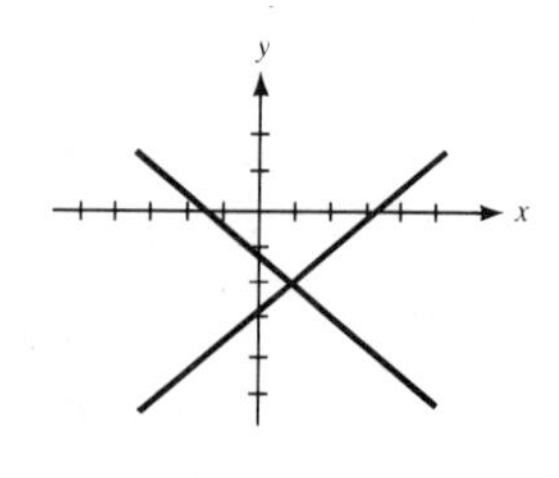

Graph is a pair of straight lines.

17. $xy = 1$

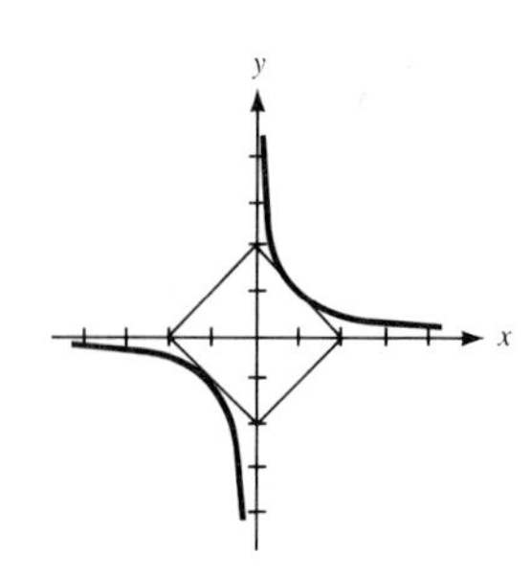

Graph is a hyperbola.

19. $(x - 2)^2 - \dfrac{(y - 1)^2}{35} = 1$ **21.** $\dfrac{x^2}{25} - \dfrac{y^2}{100} = 1$ **23.** $5x^2 - y^2 = 4$

25. Ellipse: $\dfrac{x^2}{8} + \dfrac{y^2}{4} = 1$

Hyperbola: $y^2 - \dfrac{x^2}{2} = 1$

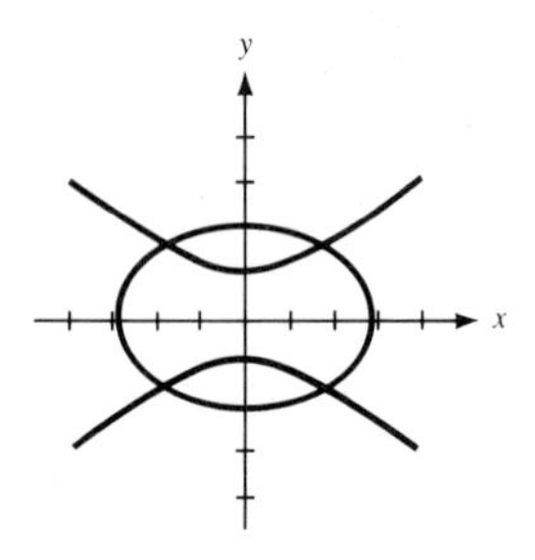

Points of intersection are $\left(\pm\sqrt{3}, \pm\dfrac{\sqrt{10}}{2}\right)$.

27. $\dfrac{(y - 5)^2}{5} - \dfrac{(x - 1)^2}{4} = 1$

SECTION 11.5

1. **(a)** ellipse or circle **(b)** parabola **(c)** hyperbola **3.** **(a)** parabola **(b)** parabola **(c)** hyperbola

5. **(a)** parabola **(b)** parabola **(c)** hyperbola

7.

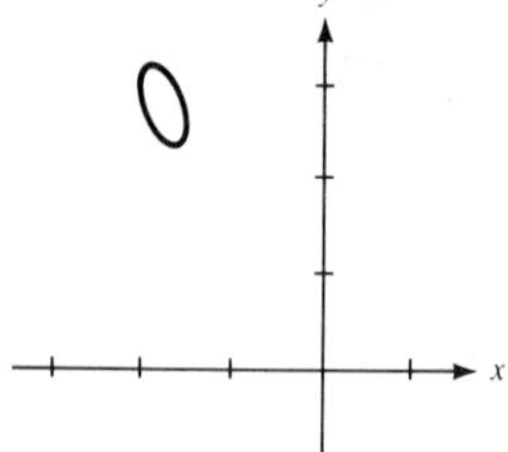

9.

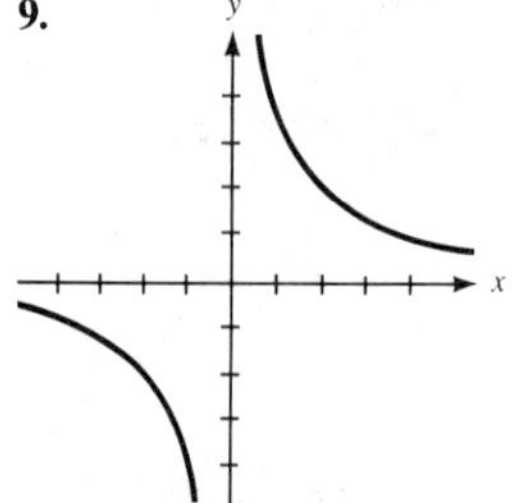

11.

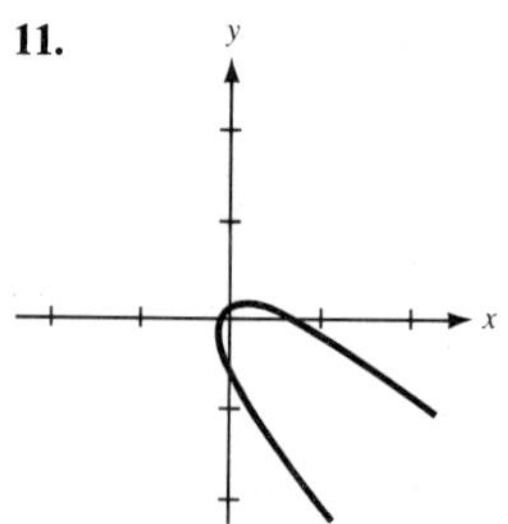

13.

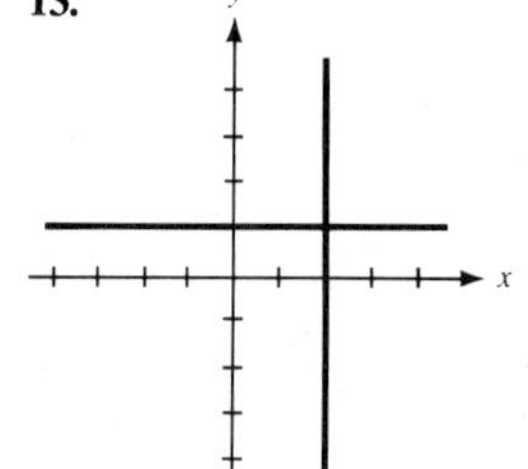

15.

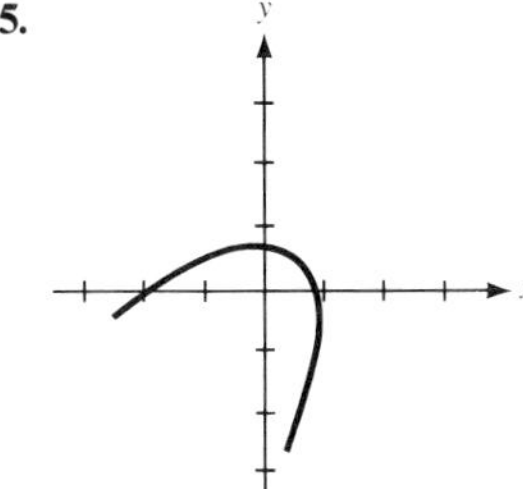

17.

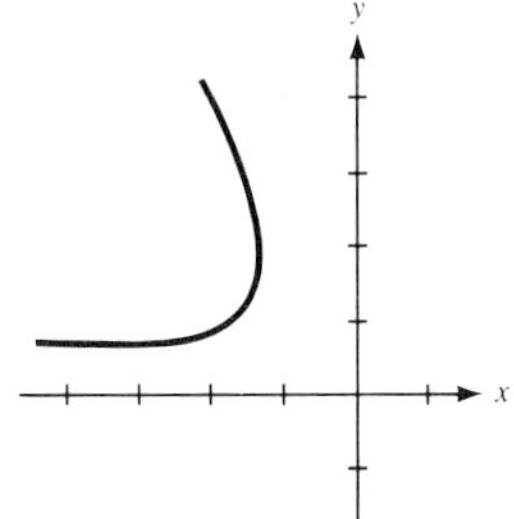

19.

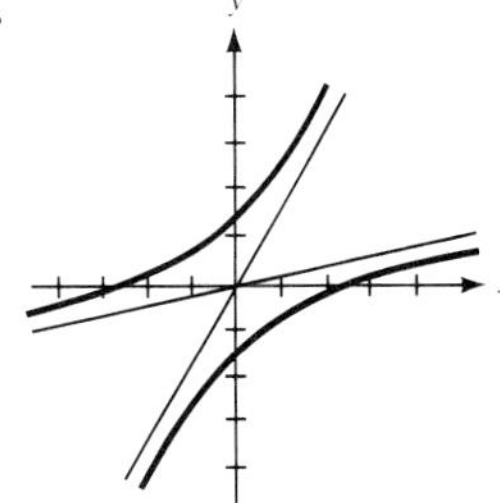

21.

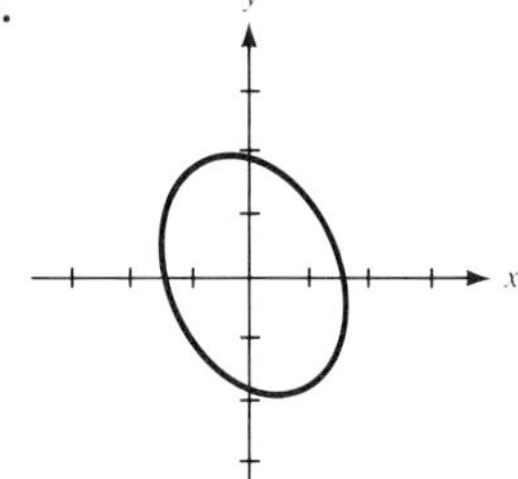

23.

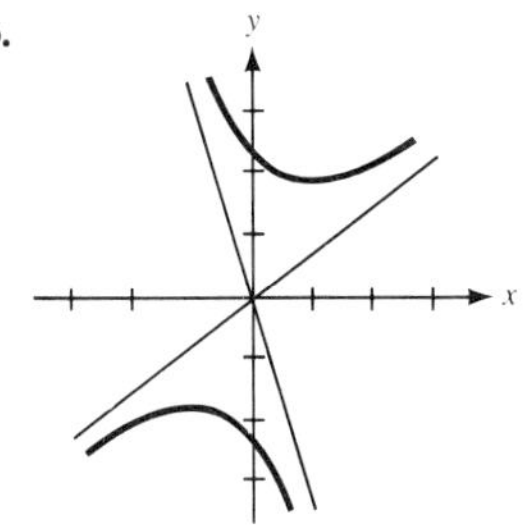

25.

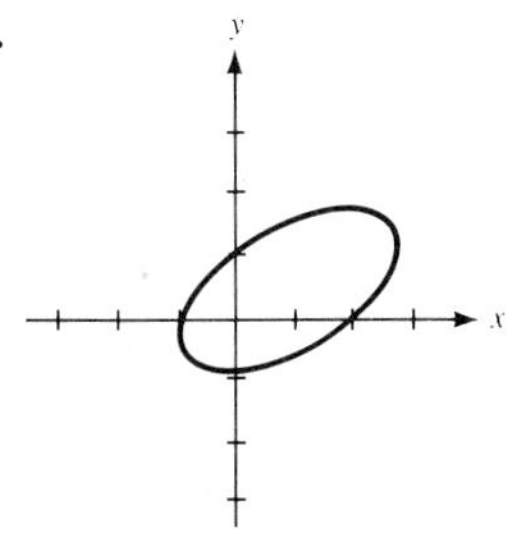

27.

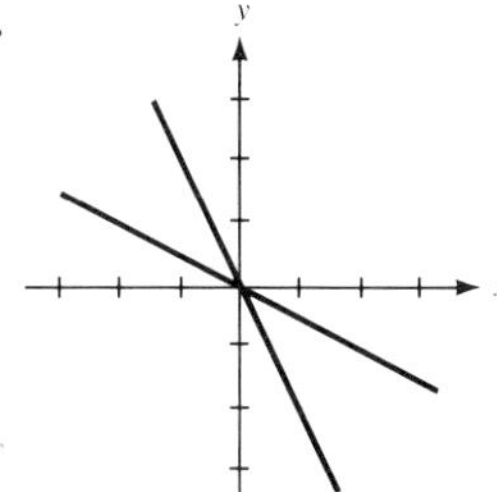

29.

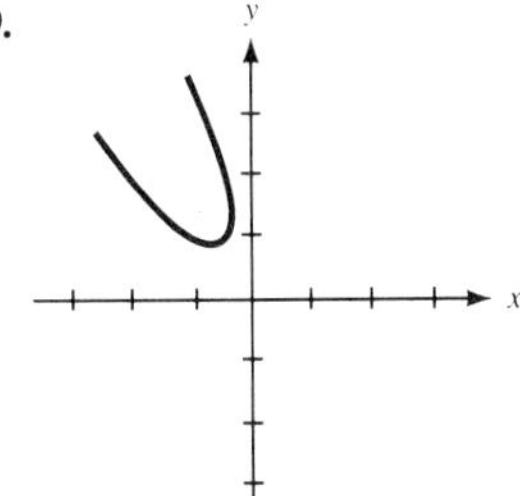

31.

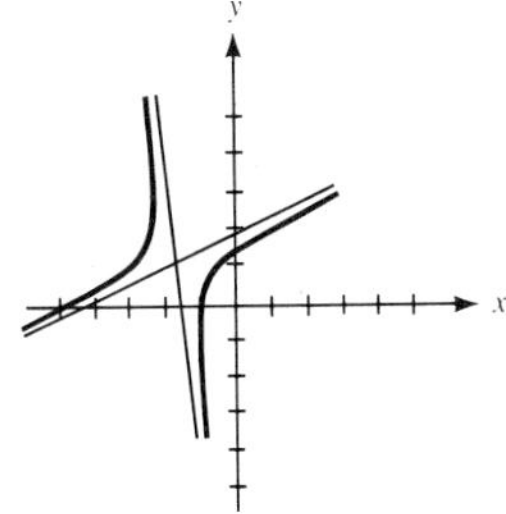

33.

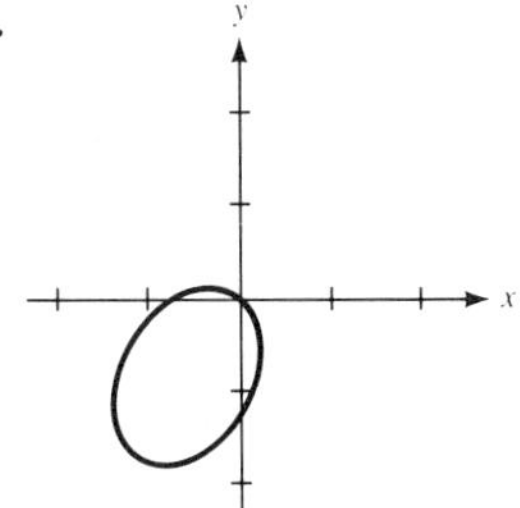

SECTION 11.6

1. parabola **3.** ellipse or circle **5.** hyperbola **7.** hyperbola **9.** hyperbola **11.** $y - 1 = \frac{1}{16}(x - 2)^2$

13. $2x - 1 = (y - 1)^2$ **15.** $\dfrac{4(x-1)^2}{9} + \dfrac{4y^2}{5} = 1$ **17.** $(y - 1)^2 + \dfrac{(x-1)^2}{3} = 1$ **19.** $\dfrac{3y^2}{7} - \dfrac{5x^2}{7} = 1$

21.

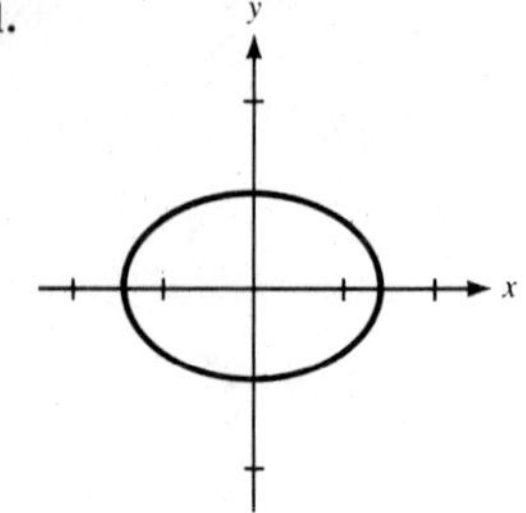

23.

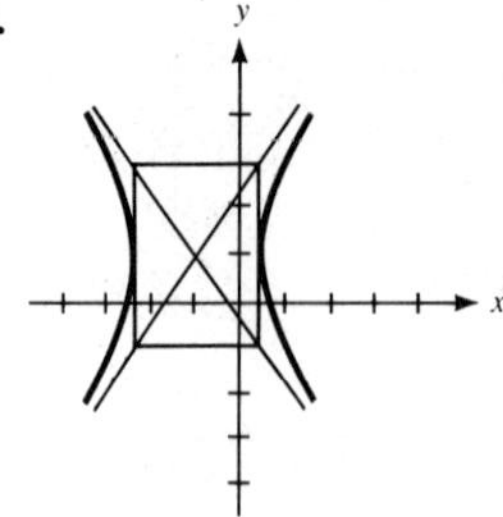

25.

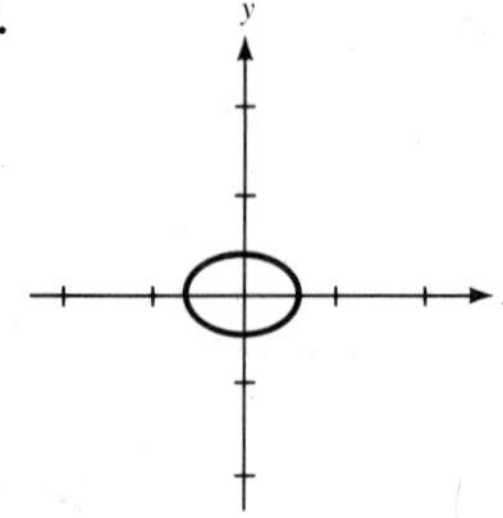

27.

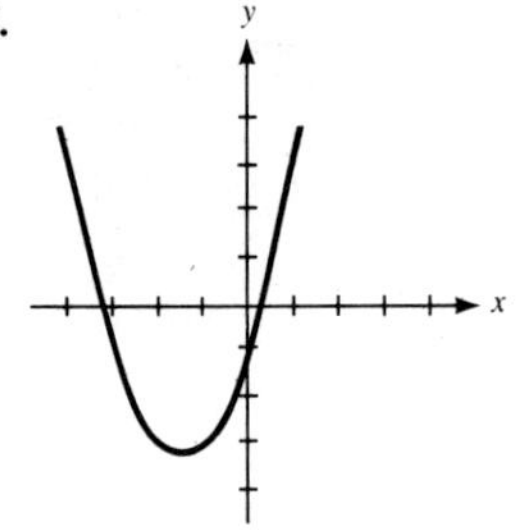

29.

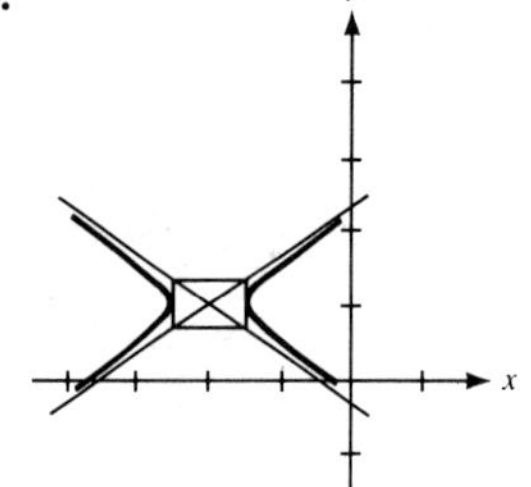

31.

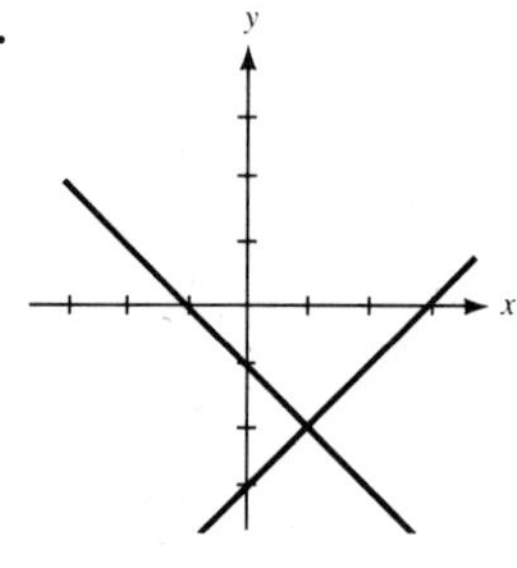

33.

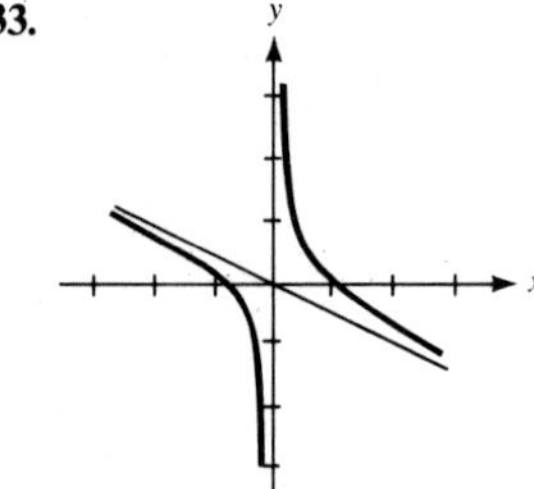

35.

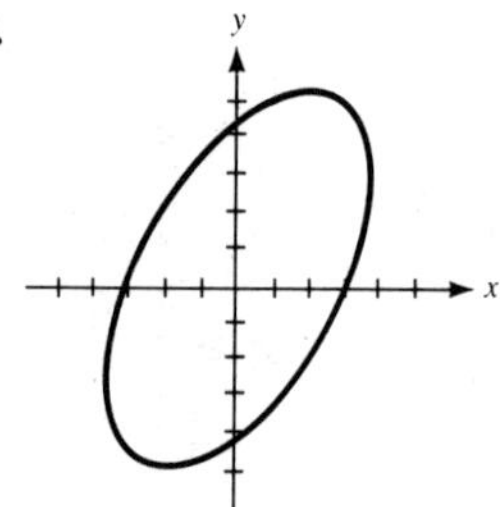

37.

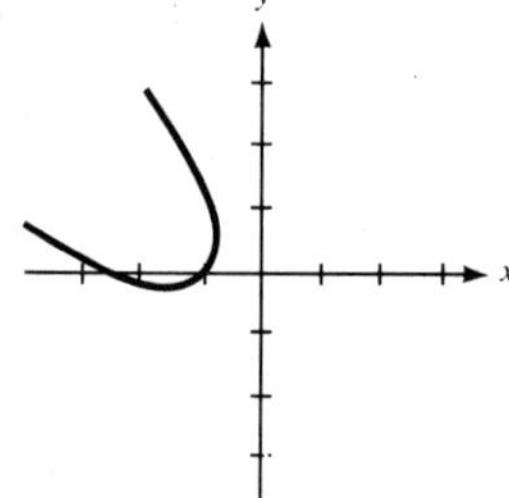

39.

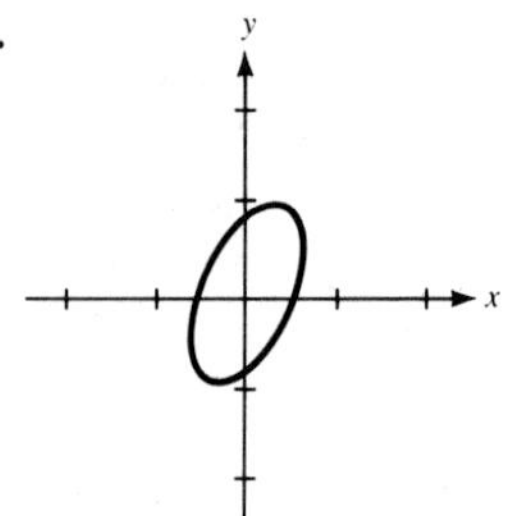

41. yes

Chapter Twelve

SECTION 12.2

1. $\frac{7}{6}$ **3.** 18 **5.** $\frac{1}{9}$ **7.** n **9.** $(n-1)(n-1)!$ **11.** 10 **13.** 1 15. 126 **17.** 10
21. $x^6 + 6x^5 + 15x^4 + 20x^3 + 15x^2 + 6x + 1$ **23.** $5^7[x^7 + 7x^6 + 21x^5 + 35x^4 + 35x^3 + 21x^2 + 7x + 1]$

25. $u^5 - 5u^4v^2 + 10u^3v^4 - 10u^2v^6 + 5uv^8 - v^{10}$ **27.** $x^2 + 4x^{3/2}y^{1/3} + 6xy^{2/3} + 4x^{1/2}y + y^{4/3}$ **29.** $\frac{1}{x^6} + \frac{3}{x^2} + 3x^2 + x^6$

31. 720 **33.** $\binom{10}{4}2^4 3^6 = 2{,}449{,}440$ **35.** $\binom{15}{8}2^8 = 1{,}647{,}360$

SECTION 12.3

1. $-2, 4, -8, 16, -32$ **3.** 1, 2, 3, 4, 5 **5.** 0, 8, 0, 32, 0 **7.** $1, \frac{4}{3}, 2, \frac{16}{5}, \frac{16}{3}$ **9.** 8, 8, 8, 8, 8 **11.** 2, 6, 18, 54, 162
13. 0, 1, 8, 57, 400 **15.** 1, 1, 2, 3, 5 **17.** $r, r^{1/2}, r^{1/6}, r^{1/24}, r^{1/120}$ **19.** $1, s, s^2, s^3, s^4$ **21.** 1, 3, 2, 6, 4, 12, 8, 24
23. $3, 6, 3, 3, \sqrt{6}, \sqrt{3 + \sqrt{6}}$ **25.** $d = -3, a_n = -2 - 3n, a_1 = -5, a_2 = -8, a_3 = -11, a_{100} = -302$
27. $d = -1/2, a_n = \frac{1}{2}(31 - n), a_1 = 15, a_2 = \frac{29}{2}, a_3 = 14, a_{100} = -\frac{69}{2}$
29. $d = -10, a_n = 17 - 10n, a_1 = 7, a_2 = -3, a_3 = -13, a_{100} = -983$ **31.** $d = 3, a_n = 3n - 3, a_1 = 0, a_2 = 3, a_3 = 6, a_{100} = 297$
33. $d = -1, a_n = 6 - n, a_1 = 5, a_2 = 4, a_3 = 3, a_{100} = -94$ **35.** $d = -1, a_n = 3 - n, a_1 = 2, a_2 = 1, a_3 = 0, a_{100} = -97$

37. $r = \frac{1}{2}, a_n = -\frac{1}{2^{n-2}}, a_1 = -2, a_2 = -1, a_3 = -\frac{1}{2}, a_{10} = -\frac{1}{256}$ **39.** $r = 4, a_n = 4^{n-1}, a_1 = 1, a_2 = 4, a_3 = 16, a_{10} = 4^9$

41. $r = 9, a_n = 2 \cdot 9^{n-2}, a_1 = \frac{2}{9}, a_2 = 2, a_3 = 18, a_{10} = 2 \cdot 9^8$ **43.** $r = 2, a_n = 3 \cdot 2^{n-2}, a_1 = \frac{3}{2}, a_2 = 3, a_3 = 6, a_{10} = 768$

45. $r = \frac{1}{4}, a_n = \frac{2}{4^{n-3}}, a_1 = 32, a_2 = 8, a_3 = 2, a_{10} = \frac{2}{4^7}$ **47.** $r = \frac{x}{y}, a_n = \frac{x^n}{y^{n-2}}, a_1 = xy, a_2 = x^2, a_3 = \frac{x^3}{y}, a_{10} = \frac{x^{10}}{y^8}$

49. $r = x - 1, a_n = x(x-1)^{n-1}, a_1 = x, a_2 = x(x-1), a_3 = x(x-1)^2, a_{10} = x(x-1)^9$ **51.** yes; yes
59. $1, 1 + 2, 1 + 2 + 2^2, 1 + 2 + 2^2 + 2^3; a_n = 1 + 2 + 2^2 + \cdots + 2^{n-1} = 2^n - 1$

SECTION 12.4

1. $s_n = n(4 - 2n)$ **3.** $s_n = \frac{n}{2}(9 + 5n)$ **5.** $s_n = \frac{n}{2}(31 - 3n)$ **7.** $s_n = \frac{n}{2}[(3 - n)x + (n - 1)y]$ **9.** $s_n = \frac{8}{3}\left(1 - \frac{1}{4^n}\right)$

11. $s_n = -25\left(1 - \left(\frac{6}{5}\right)^n\right)$ **13.** $s_n = x\left(\frac{1 - y^n}{1 - y}\right)$ **15.** $s_n = \frac{4}{3}2^x\left(1 - \frac{1}{4^n}\right)$ **17.** $\frac{1}{90}\left(1 - \frac{1}{10^8}\right)$ **19.** $2\left(1 - \frac{1}{2^9}\right)$

21. $28 \log 2$ **23.** $\frac{(1 + x)^{n+1} - 1}{x(1 + x)^n}$ **25.** $e^{2x}\left(\frac{1 - e^{(n-1)x}}{1 - e^x}\right)$ **27.** 132 **29.** 1417 **31.** $\frac{1}{1024}$

33. $50(3^{60} - 1) \approx 2 \times 10^{30}$ gallons

SECTION 12.5

1. converges, sum $= \frac{2}{3}$ **3.** converges, sum $= 1$ **5.** diverges **7.** converges, sum $= \frac{243}{5}$ **9.** converges, sum $= \frac{1}{7}$
11. diverges **13.** total distance traveled is 6 feet

SECTION 12.6

1. 1 **3.** -1 **5.** 7 **7.** $-\frac{3}{4}$ **9.** -7 **11.** (a) $\frac{1}{4}$ (b) $\frac{1}{a^2}$ **13.** 5 **15.** 3 **17.** $2x$ **19.** 2 **21.** 0

SECTION 12.7 (REVIEW)

9. $729x^{12} - 2916x^{10}y + 4860x^8y^2 - 4320x^6y^3 + 2160x^4y^4 - 576x^2y^5 + 64y^6$ **11.** $1 + 4\sin x + 6\sin^2 x + 4\sin^3 x + \sin^4 x$

13. $\binom{10}{3} \cdot 8 \cdot 3^7 = 2{,}099{,}520$ **15.** $\binom{8}{3} \cdot 2^5 \cdot (3v^2 - w^4)^3$ **17.** $-0.5214, -0.2204, -0.0443, 0.0807, 0.1776$

19. $a_1 = r, a_2 = s, a_3 = rs, a_4 = s^2, a_5 = rs^2, a_6 = s^3$ **23.** $a_n = 2 + 5n; a_{100} = 502$ **25.** $a_n = \frac{14}{3} + \frac{2}{3}n, a_{100} = \frac{214}{3}$

27. $a_n = 2 \cdot 3^{n-1}, a_{10} = 2 \cdot 3^9$ **29.** yes, yes **31.** $\frac{n(n+3)}{12}$ **33.** $\frac{3}{2}\left(1 - \left(\frac{2}{3}\right)^{n+1}\right)$ **35.** $(x+1)\left(1 - \frac{1}{(x+1)^{n+1}}\right)$

37. $\frac{1000}{999}(1 - (0.1)^{18})$ **39.** $1 + \frac{kn(n+1)}{2}\log x$ **41.** converges, sum $= \frac{5}{4}$ **43.** diverges **45.** No **47.** 7 **49.** 0

51. $-\frac{1}{x^2}$

INDEX

Abscissa, 225
Absolute value
  of a complex number, 487
  equations, 47–48
  function, 85
  properties of, 7
  of real number, 6
Adding ordinates, 225
Addition
  of complex numbers, 479
  of functions, 109
  of matrices, 451
  of vectors, 366
Addition formulas, 394–95, 412
Algebra of functions, 109–11
Algebraic expression, 23
Amplitude modulation (AM), 314
Analytic geometry, 156, 510
Angle, 260
  central, 260
  direction, 364
  full, 260
  of incidence, 392
  initial side, 260
  negative, 260
  positive, 260
  of reflection, 392
  right, 260
  straight, 260
  terminal side, 260
  vertex, 260
Annulus, 74
Arccosine, 320
Arcsine, 317–18
Arctangent, 321
Arc length, 265
Argument, 495
  principal value of, 495
Arithmetic sequence, 563
  common difference of an, 563
Associated equation, 468
Associated quadratic equation, 55
Associative property, 2
Asymptote
  horizontal, 206
  of a hyperbola, 528
  vertical, 204
Augmented matrix, 428
Axis, 65, 486

Back substitution, 420
Base
  of an exponential function, 222
  of a logarithm, 230
Bearing, 339
Binomial coefficient, 559
Binomial formula, 559
Binomial Theorem, 559
Bisection method, 191

Cancellation properties, 3
Cardioid, 362
Carrier wave, 314
Cartesian coordinate system, 65
Center
  of a circle, 73, 518
  of an ellipse, 160
  of a hyperbola, 162
Central angle, 260
Change
  in x, 130
  in y, 130
Circle, 73
Circular function, 259
Coefficient, 23
Cofactor, 443
Cofunction, 397
  identities, 412
  theorem, 398
Common difference, 563
Common factor, 27
Common logarithm, 237
Common ratio, 566
Common products, 25
Commutative property, 2
Completing the square, 43, 45
Complex number, 478
  absolute value of a, 487
  addition of, 479
  argument of, 495
  conjugate of, 482, 485
  division of, 484, 498
  imaginary part of, 478
  modulus of, 487
  multiplication of, 479, 497
  polar form of, 481
  properties of, 481
  purely imaginary, 478
  real part of, 478
  subtraction of, 479
  system, 478
  trigonometric form of, 495
Complex plane, 486
Composition
  of algebraic expressions, 37
  of functions, 111–15
Compound interest, 244–49
Conic sections, 155–66, 509
  degenerate, 509
Conjugate
  of a complex number 482, 485
  hyperbolas, 167
Consistent system, 419
Constant function, 83
Constant polynomial, 24
Constant term, 23, 417
Coordinate(s)
  axes, 65
  Cartesian, 66
  polar, 356
Cosecant, 281
Cosine, 275
Cotangent, 281
Cramer's Rule, 440, 447
Crossterm, 537
Cubic polynomial, 24
Cycle, 296

Damped sine wave, 316
Decibel, 242
Decreasing function, 102–103
Degree, 261
Degree of a polynomial, 23
De Moivre's formula, 500
Dependent variable, 79
Determinant
  2×2, 439
  3×3. 444
Difference quotient, 80

Difference of two cubes, 27
Difference of two squares, 27
Direction angle of a vector, 364
Directrix, 149, 511
Discriminant, 46, 491, 545
Distance, 7
Distance formula, 67
Distributive properties 3
Dividend, 172, 174
Division
  of complex numbers, 484, 498
  of functions, 110
  of polynomials, 171–76
  synthetic, 174
Divisor, 172, 174
Domain, 76
Double angle formulas, 404–405, 412
Double root, 46, 182

e, 227, 247
Eccentricity of an ellipse, 169, 526
Electromagnetic field, 312
Elementary (row) operations, 422, 432
Ellipse, 160, 518–24
  axes, 160, 518
  center, 160, 518
  degenerate, 524
  eccentricity, 169, 526
  foci, 160, 518
  standard form of the equation of, 160, 522
Empty set, 50
Endpoint of a ray, 259
Endpoints of an interval, 51
Equations
  equivalent, 41
  exponential, 238–39
  linear, 417
  polar, 361
  polynomial, 171
  quadratic, 42
  rectangular, 361
  trigonometric, 385–92
Equivalent equations, 41
Even function, 94
Exponent
  integer, 11
  properties of, 11
  rational, 13, 17
Exponential decay, 253
Exponential equation, 226, 238–39
Exponential function, 222
Exponential growth, 251
Extraneous solution, 41
Extreme point of a parabola, 144, 151
Extreme value of a quadratic function, 151

Factor, 26
Factor Theorem, 179
Factorial, 556, 583
Focus
  of an ellipse, 160, 518
  of a hyperbola, 162
  of a parabola, 143, 156, 511
Frequency modulation (FM), 314
Function, 76
  absolute value, 85
  algebra of, 109–11
  circular, 259
  composition of, 111–15
  constant, 83, 102
  decreasing, 102, 103
  domain of a, 76
  even, 94
  exponential, 222
  graph of a, 83
  greatest integer, 93
  identity, 84
  increasing, 102, 103
  inverse of a, 117–23
  invertible, 118
  linear, 84
  logarithmic, 230
  odd, 95
  one-to-one, 104
  periodic, 286
  piecewise defined, 78
  quadratic, 142
  range of a, 76
  rational, 202
  trigonometric. *See* Trigonometric functions
  zero, 101
Fundamental counting principle, 581
Fundamental rectangle, 166, 529

Gaussian elinination, 422
General form of the equation of a line, 141
General term of a sequence, 562
Geometric sequence, 566
  common ratio of a, 566
Geometric series, 577
Graph
  of an equation, 72
  of a function 83
  of a set, 71
  of a system of two equations in two variables, 418
Greater than, 5
Greatest integer function, 93

Half-angle formulas, 407–408, 412
Half-life, 253
Heron's area formula, 415
Hertz, 312
Homogeneous system 422
Horizontal asymptote, 206
Horizontal line test, 105
Hyperbola, 162–64, 526–35
  center, 162, 526
  conjugate, 167, 536
  degenerate, 535
  diagonal asymptotes, 528
  foci, 162, 526
  fundamental rectangle, 166, 529
  procedure for graphing, 164
  standard form of the equation of, 162, 531
  transverse axis, 162, 526
  vertices, 162, 526
Hyperbolic spiral, 362

Identity
  confunction, 398, 412
  double-angle, 404–405, 412
  half-angle, 407–408, 412
  matrix, 454
  product, 413
  property, 2
  sum and difference, 413
  trigonometric, 380
Identity function, 84
Image, 76
Imaginary axis, 486
Imaginary number, 478
Inconsistent system, 419
Increasing function, 102–103
Independent variable, 79
Index of a radical, 14
Index of refraction, 325, 392
Induction hypothesis, 552
Inequalities, 49
  absolute value, 54–56
  linear, 50
  quadratic, 50, 55, 157
  systems of, 467–71
Infinite sequence, 561
  arithmetic, 563
  general term of a, 562
  geometric, 566
  partial sum of a, 570
  recurrence relation for, 562
  term of a, 562
Infinite series, 576–77
  converge, 577
  diverge, 577
  geometric, 577
  sum of an, 577
Infinite sum, 576
Initial point of a vector, 364
Initial side of an angle, 260
Integer, 1
Intercept, 83
Interest, 244–48
  continuous compounding of, 246
Intersection, 109
Interval notation, 51

Inverse function, 117–23
Inverse properties, 2
Inverse trigonometric functions
  inverse cosine, 320
  inverse sine, 317–18
  inverse tangent, 321
  others, 325
Invertible
  function, 118
  matrix, 454
Irrational number, 2
Irreducible polynomial, 26
Iteration, 193

Law of cosines, 351
Law of sines, 342
Leading term, 23
Least common denominator, 33
Lemniscate, 362
Less than, 5
Limit, 582
Line, equation of
  general form, 141
  point-slope form, 139
  slope-intercept form, 138
Linear
  equation, 417
  function, 84
  inequality, 50
  polynomial, 24
Local extreme point, 198
Locus, 510
Logarithm
  common, 237
  natural, 237
  properties of, 233
Logarithmic function, 230
Logarithmic scale, 240–42

Magnitude of a vector, 364
Mathematical induction, 551, 554
Matrices
  addition of, 451
  augmented, 428
  of coefficients, 439
  of constant terms, 439
  entry of a, 428, 450
  equality of, 450
  identity, 454
  inverse of a, 454
  multiplication, 453
  negative of a, 451
  size of a, 428, 450
  square, 450
  subtraction of, 451
  of variables, 439
  zero, 452
Method of plotting points, 83
Midpoint formula, 69
Minor, 443
Modulus, 487
Multiple root, 182
Multiplication
  of complex numbers, 479, 497
  of functions, 109
  of matrices, 453
Multiplicative inverse
  of a matrix, 454
Multiplicity, 182

Natural logarithm, 237
Natural number system, 1
Negative
  angle, 260
  of a matrix, 451
  properties of, 4–5
Nonnegative, 4
Nonpositive, 4
Nonsingular matrix, 454
*n*th root
  of a complex number, 501–505
  of a real number, 14
  of unity, 504–505
Number
  complex, 478
  imaginary, 478
  integer, 1
  irrational, 2
  natural, 1
  purely imaginary, 478
  rational, 2
  real, 1, 2
Number line, 1

Oblique triangle, 342
Odd function, 95
One-to-one function, 104
Ordered pair, 51, 65–66
Ordinate, 225
Origin, 66

Parabola, 142, 156–59, 511–17
  axis of symmetry, 142, 156, 511
  directrix, 511
  extreme point, 144, 151
  focus, 143, 156, 511
  procedure for graphing, 145
  standard form of equation of, 144, 515
  vertex, 157, 511
Parallelogram law, 366
Parameter, 421
Partial fractions, 213
Partial sum of a sequence, 570
Perfect square, 43
Periodic function, 286
pH scale, 244
Phase shift, 308
Piecewise defined function, 78
Point-slope form, 139
Polar coordinates, 356
Polar equation, 361
Polar form of a complex number, 495
Polar ray, 357
Polynomial, 23–24
  common products of, 25
  constant, 24
  cubic, 24
  degree of a, 23
  equation, 171
  irreducible, 26
  linear, 24
  procedure for graphing, 196
  quadratic, 24
  term of a, 23
  zero, 24
Population growth, 250–53
Positive, 4
Positive angle, 260
Principal value of the argument, 495
Principle of Mathematical Induction, 551, 554
Product and difference formulas, 413
Properties of
  exponents, 11
  logarithms, 233
  real numbers, 2–3
Purely imaginary number, 478
Pythagorean identities, 289, 378, 411

Quadrant, 66
Quadratic
  equation, 42, 489, 536
  formula, 45, 491
  function, 142
  inequality, 55
  polynomial, 24
Quotients, properties of, 3

Radian, 261
Radical, 14
  index of a, 14
  properties of, 14
  simplifying, 15
Radicand, 14
Radioactive decay, 253–55
Range of a function, 76
Rational expression, 30
  reduced to lowest terms, 31
Rational function, 202
  procedure for graphing, 207
Rational number, 2
Rational Root Theorem, 184
Rationalizing the denominator, 15
Ray, 259, 357
Real axis, 486
Real number, 1, 2

Real number line, 1
Real number system, 1, 2
Rectangular coordinate system, 65
Rectangular equation, 361
Recurrence relation, 562
Reference number, 277
Reference Number Theorem, 279
Reflection of a graph, 98–99
Remainder, 172, 174
Remainder Theorem, 178
Restricted
  cosine function, 320
  others 325
  sine function, 317–18
  tangent function, 321
Richter scale, 240
Root, 172
Row operations, 432

Scalar, 369
Scalar multiplication, 369
Scientific notation, 19
Secant, 281
Sequence. *See* Infinite sequence
Shrinking or stretching of a graph, 98–100
Simple root, 182
Sine, 275
Sine wave
  amplitude, 304
  crest, 304
  cycle, 304
  frequency, 312
  node, 304
  period, 304
  phase shift, 308
  procedure for graphing, 309
  trough, 304
  wavelength, 311
Slope of a line, 130–36
Slope-intercept form, 138
Snell's law of refraction, 392
Solution
  of an equation, 41, 417
  extraneous, 41
  of an inequality, 50
  of a system of equations, 418
Solution set, 50
Solving a triangle, 335
Sound waves, 312
Square matrix, 450
Standard basis vector, 370
Standard form, equation
  of an ellipse, 160, 522
  of a hyperbola, 162, 531
  of a parabola, 144, 515
Standard position of an angle, 260
Stretching or shrinking of a graph, 98–100
Subtend, 260
Subtraction
  of complex number, 479
  of functions, 109
  of matrices, 451
  of vectors, 368
Subtraction formula, 394–95
Sum and difference formulas, 413
Sum of two cubes, 27
Sweeping out a central angle, 268
Symmetry
  with respect to the origin, 95
  with respect to the *y*-axis, 94
Synthetic division, 174
Systems of inequalities, 467–71
Systems of linear equations
  consistent, 419
  graph, 418
  homogeneous, 422
  inconsistent, 419
  solution, 418
  upper triangular, 419
Systems of nonlinear equations, 463–66

Tangent, 281
Term
  of a polynomial, 23
  of a sequence, 562
Terminal point of a vector, 364
Terminal side of an angle, 260
Test value, 56
Translation of a graph, 98–100
Transverse axis of a hyperbola, 164
Trigonometric equation, 385–92
Trigonometric form of a complex number, 495
Trigonometric functions, definitions
  cosine, 275
  others, 281
  sine, 275
Trigonometric functions, graphs
  cosecant, 300
  cosine, 297
  cotangent, 300
  secant, 299
  sine, 296
  tangent, 298
Trigonometric functions, properties
  cosine, 291
  others, 293
  sine, 291
Trigonometric identity, 380
Triple root, 182

Union, 8
Unit circle, 261
Upper triangular form, 419

Variable
  dependent, 79
  independent, 79
Vector, 364
  addition of, 366
  direction angle, 364
  initial point of, 364
  magnitude of, 364
  scalar multiplication, 369
  standard basis, 370
  subtraction of, 368
  terminal point of, 364
  zero, 365
Vertex of an angle, 260
Vertex of a parabola, 157, 511
Vertical asymptote, 204
Vertical line test, 90

x-axis, 65
x-intercept, 83, 137

y-axis, 65
y-intercept, 83, 137

Zero
  properties of, 3
Zero exponent, 11
Zero function, 101
Zero matrix, 452
Zero of a polynomial, 172
Zero polynomial, 24
Zero vector, 365

A 6
B 7
C 8
D 9
E 0
F 1
G 2
H 3
I 4
J 5

Definitions of the Trigonometric Functions

$$P(t) = (\cos t, \sin t)$$

$$\tan t = \frac{\sin t}{\cos t} \qquad \cot t = \frac{\cos t}{\sin t} \qquad \sec t = \frac{1}{\cos t} \qquad \csc t = \frac{1}{\sin t}$$

Common Values of the Trigonometric Functions

t	$\sin t$	$\cos t$	$\tan t$	$\cot t$	$\sec t$	$\csc t$
0	0	1	0	—	1	—
$\frac{\pi}{6}$	$\frac{1}{2}$	$\frac{\sqrt{3}}{2}$	$\frac{\sqrt{3}}{3}$	$\sqrt{3}$	$\frac{2\sqrt{3}}{3}$	2
$\frac{\pi}{4}$	$\frac{\sqrt{2}}{2}$	$\frac{\sqrt{2}}{2}$	1	1	$\sqrt{2}$	$\sqrt{2}$
$\frac{\pi}{3}$	$\frac{\sqrt{3}}{2}$	$\frac{1}{2}$	$\sqrt{3}$	$\frac{\sqrt{3}}{3}$	2	$\frac{2\sqrt{3}}{3}$
$\frac{\pi}{2}$	1	0	—	0	—	1

Properties of the Sine and Cosine Functions

Range:

$$-1 \le \sin t \le 1 \qquad -\infty < \tan t < \infty \qquad |\sec t| \ge 1$$
$$-1 \le \cos t \le 1 \qquad -\infty < \cot t < \infty \qquad |\csc t| \ge 1$$

Periodicity: For all integers n

$$\sin(t + 2\pi n) = \sin t \qquad \tan(t + n\pi) = \tan t \qquad \sec(t + 2\pi n) = \sec t$$
$$\cos(t + 2\pi n) = \cos t \qquad \cot(t + n\pi) = \cot t \qquad \csc(t + 2\pi n) = \csc t$$

Evenness and oddness:

$$\sin(-t) = -\sin t \qquad \tan(-t) = -\tan t \qquad \sec(-t) = \sec t$$
$$\cos(-t) = \cos t \qquad \cot(-t) = -\cot t \qquad \csc(-t) = -\csc t$$

Relationship between the sine and cosine:

$$\cos t = \sin(t + \pi/2)$$

Miscellaneous properties:

$$\sin(t + \pi) = -\sin t \qquad \cos(t + \pi) = -\cos t$$